T0143257

Distributed Sensor Networks

Second Edition

Sensor Networking and Applications

Distributed Sensor Networks, Second Edition

Distributed Sensor Networks, Second Edition: Image and Sensor Signal Processing

Distributed Sensor Networks, Second Edition: Sensor Networking and Applications

CHAPMAN & HALL/CRC COMPUTER and INFORMATION SCIENCE SERIES

Series Editor: Sartaj Sahni

PUBLISHED TITLES

ADVERSARIAL REASONING: COMPUTATIONAL APPROACHES TO READING THE OPPONENT'S MIND
Alexander Kott and William M. McEneaney

DISTRIBUTED SENSOR NETWORKS, SECOND EDITION
S. Sitharama Iyengar and Richard R. Brooks

DISTRIBUTED SYSTEMS: AN ALGORITHMIC APPROACH
Sukumar Ghosh

ENERGY-AWARE MEMORY MANAGEMENT FOR EMBEDDED MULTIMEDIA SYSTEMS: A COMPUTER-AIDED DESIGN APPROACH
Florin Balasa and Dhiraj K. Pradhan

ENERGY EFFICIENT HARDWARE-SOFTWARE CO-SYNTHESIS USING RECONFIGURABLE HARDWARE
Jingzhao Ou and Viktor K. Prasanna

FUNDAMENTALS OF NATURAL COMPUTING: BASIC CONCEPTS, ALGORITHMS, AND APPLICATIONS
Leandro Nunes de Castro

HANDBOOK OF ALGORITHMS FOR WIRELESS NETWORKING AND MOBILE COMPUTING
Azzedine Boukerche

HANDBOOK OF APPROXIMATION ALGORITHMS AND METAHEURISTICS
Teofilo F. Gonzalez

HANDBOOK OF BIOINSPIRED ALGORITHMS AND APPLICATIONS
Stephan Olariu and Albert Y. Zomaya

HANDBOOK OF COMPUTATIONAL MOLECULAR BIOLOGY
Srinivas Aluru

HANDBOOK OF DATA STRUCTURES AND APPLICATIONS
Dinesh P. Mehta and Sartaj Sahni

HANDBOOK OF DYNAMIC SYSTEM MODELING
Paul A. Fishwick

HANDBOOK OF ENERGY-AWARE AND GREEN COMPUTING
Ishfaq Ahmad and Sanjay Ranka

HANDBOOK OF PARALLEL COMPUTING: MODELS, ALGORITHMS AND APPLICATIONS
Sanguthevar Rajasekaran and John Reif

HANDBOOK OF REAL-TIME AND EMBEDDED SYSTEMS
Insup Lee, Joseph Y-T. Leung, and Sang H. Son

HANDBOOK OF SCHEDULING: ALGORITHMS, MODELS, AND PERFORMANCE ANALYSIS
Joseph Y.-T. Leung

HIGH PERFORMANCE COMPUTING IN REMOTE SENSING
Antonio J. Plaza and Chein-I Chang

INTRODUCTION TO NETWORK SECURITY
Douglas Jacobson

LOCATION-BASED INFORMATION SYSTEMS: DEVELOPING REAL-TIME TRACKING APPLICATIONS
Miguel A. Labrador, Alfredo J. Pérez, and Pedro M. Wightman

METHODS IN ALGORITHMIC ANALYSIS
Vladimir A. Dobrushkin

PERFORMANCE ANALYSIS OF QUEUING AND COMPUTER NETWORKS
G. R. Dattatreya

THE PRACTICAL HANDBOOK OF INTERNET COMPUTING
Munindar P. Singh

SCALABLE AND SECURE INTERNET SERVICES AND ARCHITECTURE
Cheng-Zhong Xu

SOFTWARE APPLICATION DEVELOPMENT: A VISUAL C++®, MFC, AND STL TUTORIAL
Bud Fox, Zhang Wenzu, and Tan May Ling

SPECULATIVE EXECUTION IN HIGH PERFORMANCE COMPUTER ARCHITECTURES
David Kaeli and Pen-Chung Yew

VEHICULAR NETWORKS: FROM THEORY TO PRACTICE
Stephan Olariu and Michele C. Weigle

Distributed Sensor Networks

Second Edition

Sensor Networking and Applications

Edited by

S. Sitharama Iyengar

Richard R. Brooks

CRC Press is an imprint of the
Taylor & Francis Group, an **informa** business
A CHAPMAN & HALL BOOK

MATLAB® is a trademark of The MathWorks, Inc. and is used with permission. The MathWorks does not warrant the accuracy of the text or exercises in this book. This book's use or discussion of MATLAB® software or related products does not constitute endorsement or sponsorship by The MathWorks of a particular pedagogical approach or particular use of the MATLAB® software.

CRC Press
Taylor & Francis Group
6000 Broken Sound Parkway NW, Suite 300
Boca Raton, FL 33487-2742

Printed in the United States of America on acid-free paper
Version Date: 20120809

International Standard Book Number: 978-1-4398-6287-2 (Hardback)

Library of Congress Cataloging-in-Publication Data

Distributed sensor networks. Sensor networking and applications / editors, S. Sitharama Iyengar and Richard R. Brooks. -- 2nd ed.
p. cm. -- (Chapman & Hall/CRC computer and information science series)
Includes bibliographical references and index.
ISBN 978-1-4398-6287-2 (alk. paper)
1. Sensor networks. 2. Multisensor data fusion. 3. Intelligent agents (Computer software) I. Iyengar, S. S. (Sundararaja S.) II. Brooks, R. R. (Richard R.)

TK7872.D48D5734 2013
681'.2--dc23 2012027630

Visit the Taylor & Francis Web site at
http://www.taylorandfrancis.com

and the CRC Press Web site at
http://www.crcpress.com

I would like to dedicate this book to Professor C.N.R. Rao for his outstanding and sustainable contributions to science in India and around the world. Professor Rao has been able to create and establish opportunities in science education in India and the world.

Last but not the least, I am grateful to my graduate students who make everything possible in my career.

S. Sitharama Iyengar

Dedicated to my graduate students, who make the research possible.

Richard R. Brooks

Contents

PART II Adaptive Tasking

Preface to the Second Edition

The six years since the first edition appeared have seen the field of sensor networks ripen. The number of researchers working in the field and technical (journal and conference) publications related to sensor networks has exploded. Those of us working in the field since its infancy have been gratified by the large number of practical sensor network applications that are being used.

Not surprisingly, many lessons have been learned. The current state of affairs could hardly have been predicted when the first edition went to press. Partly because of this, we were extremely gratified when Chapman & Hall contacted us and suggested printing a revised second edition of *Distributed Sensor Networks*.

It was a daunting task to bring together, once again, the distinguished set of researchers whom we relied on for the first edition. We are proud to have been able to expand and revise the original tome of over 1100 pages. The size of the new book forced us to deliver it in two books:

1. *Distributed Sensor Networks, Second Edition: Image and Sensor Signal Processing* comprises the following parts:
   I. Overview
   II. Distributed Sensing and Signal Processing
   III. Information Fusion
   IV. Power Management
2. *Distributed Sensor Networks, Second Edition: Sensor Networking and Applications* comprises the following parts:
   I. Sensor Deployment
   II. Adaptive Tasking
   III. Self-Configuration
   IV. System Control
   V. Engineering Examples

Although half of the chapters remain the same as in the first edition, 13 chapters have been revised, and there are 22 new chapters. Readers of the first edition will recognize many of the same authors. Several of them were veterans of the DARPA ISO SENSIT program. We were privileged to get inputs from a number of distinguished new contributors. We were also saddened by the demise of Dr. Bose from Penn State who wrote the chapter on multispectral sensing.

We believe that the new books have managed to maintain the feel of our earlier text. The chapters serve both as tutorials and archive research material. Our target audience is students, researchers, and engineers who want a book with practical insights providing current information on the field.

Preface to the First Edition

In many ways, this book started ten years ago, when the editors started their collaboration at Louisiana State University in Baton Rouge. At that time, sensor networks were a somewhat arcane topic. Since then, many new technologies have ripened and prototype devices have emerged in the market. We were lucky enough to be able to continue our collaboration under the aegis of the DARPA IXO Sensor Information Technology Program and the Emergent Surveillance Plexus Multi-disciplinary University Research Initiative.

What was clear ten years ago, and has become more obvious, is that the only way to adequately monitor the real world is to use a network of devices. Many reasons for this will be given in this book. These reasons range from financial considerations to statistical inference constraints. Once you start using a network situated in the real world, the need for adaptation and self-configuration also becomes obvious.

What was probably not known ten years ago was the breadth and depth of research needed to adequately design these systems. The book in front of you contains chapters from acknowledged leaders in sensor network design. The contributors work at leading research institutions and have expertise in a broad range of technical fields.

The field of sensor networks has matured greatly within the last few years. The editors are grateful to have participated in this process. We are especially pleased to have been able to interact with the research groups whose work is presented here. This growth has only been possible with support from many government agencies, especially within the Department of Defense. Visionary program managers at DARPA, ONR, AFRL, and ARL have made a significant impact on these technologies.

It is the editors' sincere hope that the field continues to mature. We also hope that the cross fertilization of ideas between technical fields that has enabled these advances deepens.

The editors thank all the authors who contributed to this book. Jamie Sigal and Bob Stern from CRC Press were invaluable. We also recognize the efforts of Rose Sweeney in helping to organize and maintain this enterprise.

Editors

S. Sitharama Iyengar is the director and Ryder Professor of the School of Computing and Information Sciences at Florida International University, Miami, and is also the chaired professor at various institutions around the world. His publications include 6 textbooks, 5 edited books, and over 400 research papers. His research interests include high-performance algorithms, data structures, sensor fusion, data mining, and intelligent systems. Dr. Iyengar is a world class expert in computational aspects of sensor networks, data structures, and algorithms for various distributed applications. His techniques have been used by various federal agencies (Naval Research Laboratory [NRL], ORNL, National Aeronautics and the Space Administration [NASA]) for their projects. His work has been cited very extensively by researchers and scientists around the world.

Dr. Iyengar is an SIAM distinguished lecturer, ACM national lecturer, and IEEE distinguished scientist. He is a fellow of IEEE, ACM, AAAS, and SDPS. Dr. Iyengar is a recipient of IEEE awards, best research paper awards, the Distinguished Alumnus award of the Indian Institute of Science, Bangalore, and other awards. He has served as the editor of several IEEE journals and is the founding editor in chief of the *International Journal of Distributed Sensor Networks*.

Dr. Iyengar's research has been funded by the National Science Foundation (NSF), Defense Advanced Research Projects Agency (DARPA), Multi-University Research Initiative (MURI Program), Office of Naval Research (ONR), Department of Energy/Oak Ridge National Laboratory (DOE/ORNL), NRL, NASA, U.S. Army Research Office (URO), and various state agencies and companies. He has served on U.S. National Science Foundation and National Institutes of Health panels to review proposals in various aspects of computational science and has been involved as an external evaluator (ABET-accreditation) for several computer science and engineering departments.

Dr. Iyengar has had 40 doctoral students under his supervision, and the legacy of these students can be seen in prestigious laboratories (JPL, Oak Ridge National Lab, Los Alamos National Lab, Naval Research Lab) and universities round the world. He has been the program chairman of various international conferences and has given more than 50 keynote talks at these conferences.

Richard R. Brooks is an associate professor of electrical and computer engineering at Clemson University in Clemson, South Carolina. He received his PhD in computer science from Louisiana State University and his BA in mathematical sciences from The Johns Hopkins University. He has also studied operations research at the Conservatoire National des Arts et Metiers in Paris, France.

Dr. Brooks is a senior member of the IEEE. He has written the book *Disruptive Security Technologies with Mobile Code and Peer-to-Peer Networks* and has cowritten *Multi-Sensor Fusion*. He has coedited both versions of *Distributed Sensor Networks* in collaboration with S. S. Iyengar.

Dr. Brooks was principal investigator (PI) of the Reactive Sensor Networks Project sponsored by the Defence Advanced Research Projects Agency (DARPA) ITO Sensor Information Technology initiative, which has explored collaborative signal processing to aggregate information moving through the network and the use of mobile code for coordination among intelligent sensor nodes. He was co-PI of a DARPA IXO JFACC program that has used distributed discrete event controllers for air combat C2 planning. He has coordinated a DARPA MURI program that uses cooperating automata in a cellular space to coordinate sensor network planning and execution. He is PI of an ONR URI on cybersecurity issues relating to mobile code and the construction of secure information infrastructures.

Dr. Brooks' current research concentrates on adaptation in distributed systems. His research interests include network security, sensor networks, and self-organizing systems. His research has been sponsored by ONR, DARPA, ARO, AFOSR, NIST, U.S. Department of State, NSF, and BMW Manufacturing Corporation.

Dr. Brooks' PhD dissertation has received an exemplary achievement certificate from the Louisiana State University graduate school. He is associate managing editor of the *International Journal of Distributed Sensor Networks*. He has had a broad professional background with computer systems and networks, and was head of the Pennsylvania State University Applied Research Laboratory Distributed Systems Department for over six years. Dr. Brooks was technical director of Radio Free Europe's computer network for many years, and is a consultant to the French stock exchange authority and the World Bank.

Contributors

Joon Ahn
IP Systems S&T Ericsson, Inc.
San Francisco, California

Behtash Babadi
School of Engineering and Applied Sciences
Harvard University
Cambridge, Massachusetts

Devang Bagaria
Department of Electrical and Computer Engineering
Clemson University
Clemson, South Carolina

N. Balakrishnan
Carnegie Mellon University
Pittsburgh, Pennsylvania

and

Indian Institute of Science
Bangalore, India

Doina Bein
Applied Research Laboratory
The Pennsylvania State University
University Park, Pennsylvania

Harakrishnan Bhanu
Holcombe Department of Electrical and Computer Engineering
Clemson University
Clemson, South Carolina

Edo Biagioni
Department of Information and Computer Sciences
University of Hawaii at Manoa
Honolulu, Hawaii

Cliff Bowman
Ember Corporation
Boston, Massachusetts

K.W. Bridges
Department of Botany
University of Hawaii at Manoa
Honolulu, Hawaii

Richard R. Brooks
Holcombe Department of Electrical and Computer Engineering
Clemson University
Clemson, South Carolina

David W. Carman
Johns Hopkins Applied Physics Laboratory
Laurel, Maryland

Krishnendu Chakrabarty
Department of Electrical and Computer Engineering
Duke University
Durham, North Carolina

Eungchun Cho
Kentucky State University
Lexington, Kentucky

Romit Roy Choudhury
Department of Electrical Communication Engineering
Duke University
Durham, North Carolina

Mashrur Chowdhury
Department of Civil Engineering
Clemson University
Clemson, South Carolina

Ryan Craven
Holcombe Department of Electrical and Computer Engineering
Clemson University
Clemson, South Carolina

Sajal K. Das
Department of Computer Science and Engineering
The University of Texas at Arlington
Arlington, Texas

Eiman Elnahrawy
Department of Computer Science
Rutgers University
New Brunswick, New Jersey

Deborah Estrin
Department of Computer Science
University of Southern California
Los Angeles, California

N. Gautam
Department of Industrial and Systems Engineering
Texas A&M University
College Station, Texas

Johannes Gehrke
Department of Computer Science
University of California, Berkeley
Berkeley, California

and

Cornell University
Ithaca, New York

Ramesh Govindan
Department of Computer Science
University of Southern California
Los Angeles, California

P.Y. Govindaraju
Clemson University
Clemson, South Carolina

Christopher Griffin
Applied Research Laboratory
The Pennsylvania State University
University Park, Pennsylvania

John Heidemann
Information Sciences Institute
University of Southern California
Marina del Rey, California

Jun-Won Ho
Department of Information Security
Seoul Women's University
Seoul, South Korea

Yong Huang
Department of Mechanical Engineering
Clemson University
Clemson, South Carolina

S. Sitharama Iyengar
School of Computer Science
Florida International University
North Miami, Florida

Vijay S. Iyer
Super Computing Research Center
Indian Institute of Science
Bangalore, India

Mahmut Kandemir
The Pennsylvania State University
University Park, Pennsylvania

Nathaniel Karst
Babson College
Wellesley, Massachusetts

T. Keiser
Applied Research Laboratory
Department of Distributed Systems
The Pennsylvania State University
University Park, Pennsylvania

James Kelly
Naval Undersea Warfare Center
Newport, Rhode Island

Bhaskar Krishnamachari
Ming Hsieh Department of Electrical Engineering
and
Department of Computer Science
Viterbi School of Engineering
University of Southern California
Los Angeles, California

Teja Phani Kuruganti
Department of Electrical and Computer Engineering
University of Tennessee
Knoxville, Tennessee

Jacob Lamb
Applied Research Laboratory
The Pennsylvania State University
University Park, Pennsylvania

Xin Shane Li
Department of Electrical and Computer Engineering
Louisiana State University
Baton Rouge, Louisiana

Alvin S. Lim
Department of Computer Science and Engineering
Auburn University
Auburn, Alabama

Donggang Liu
Department of Computer Science and Engineering
The University of Texas at Arlington
Arlington, Texas

Bharat B. Madan
Applied Research Laboratory
The Pennsylvania State University
University Park, Pennsylvania

Samuel Madden
Department of Forestry
University of California, Berkeley
Berkeley, California

and

Cornell University
Ithaca, New York

Prakash Manghwani
Bolt, Beranek and Newman Technologies
Cambridge, Massachusetts

Jeff Mazurek
Bolt, Beranek and Newman Technologies
Cambridge, Massachusetts

Gail Mitchell
Bolt, Beranek and Newman Technologies
Cambridge, Massachusetts

Vijaykrishnan Narayanan
Department of Computer Science and Engineering
The Pennsylvania State University
University Park, Pennsylvania

Badri Nath
Department of Computer Science
Rutgers University
New Brunswick, New Jersey

Amiya Nayak
School of Information Technology and Engineering
University of Ottawa
Ottawa, Ontario, Canada

Sotiris Nikoletseas
Computer Engineering and Informatics Department
University of Patras and CTI
Patras, Greece

Peng Ning
Department of Computer Science
North Carolina State University
Raleigh, North Carolina

Shashi Phoha
Applied Research Laboratory
The Pennsylvania State University
University Park, Pennsylvania

Brijesh Pillai
Clemson University
Clemson, South Carolina

Matthew Pirretti
Applied Research Laboratory
Department of Distributed Systems
The Pennsylvania State University
University Park, Pennsylvania

Robert Poor
Ember Corporation
Boston, Massachusetts

Hairong Qi
Department of Electrical and Computer Engineering
University of Tennessee
Knoxville, Tennessee

Suresh Rai
Department of Electrical and Computer Engineering
Louisiana State University
Baton Rouge, Louisiana

Parameswaran Ramanathan
Department of Electrical and Computer Engineering
University of Wisconsin-Madison
Madison, Wisconsin

Nageswara S.V. Rao
Computer Science and Mathematics Division
Oak Ridge National Laboratory
Oak Ridge, Tennessee

Asok Ray
Department of Mechanical Engineering
The Pennsylvania State University
University Park, Pennsylvania

Errol G. Rowe
Naval Undersea Warfare Center
Newport, Rhode Island

Sushmita Ruj
School of Information Technology and Engineering
University of Ottawa
Ottawa, Ontario, Canada

Prince Samar
School of Electrical and Computer Engineering
Cornell University
Ithaca, New York

Jason Schwier
Holcombe Department of Electrical and Computer Engineering
Clemson University
Clemson, South Carolina

Fabio Silva
Information Sciences Institute
University of Southern California
Marina del Rey, California

Ivan Stojmenovic
School of Information Technology and Engineering
University of Ottawa
Ottawa, Ontario, Canada

Ankit Tandon
Microsoft Corporation
Seattle, Washington

Lei Tang
Department of Mechanical Engineering
Clemson University
Clemson, South Carolina

Vahid Tarokh
School of Engineering and Applied Sciences
Harvard University
Cambridge, Massachusetts

Ken Theriault
Bolt, Beranek and Newman Technologies
Cambridge, Massachusetts

Vijay K. Vaishnavi
Department of Computer Information Systems
Georgia State University
Atlanta, Georgia

Kuang-Ching Wang
Department of Electrical and Computer Engineering
Clemson University
Clemson, South Carolina

Anthony G. Warrack
Department of Mathematics
North Carolina Agricultural and Technical State University
Greensboro, North Carolina

Yicheng Wen
Department of Mechanical Engineering
The Pennsylvania State University
University Park, Pennsylvania

Stephen B. Wicker
School of Electrical and Computer Engineering
Cornell University
Ithaca, New York

Matthew Wright
Department of Computer Science and Engineering
The University of Texas at Arlington
Arlington, Texas

Qishi Wu
Department of Computer Science
Louisiana State University
Baton Rouge, Louisiana

and

Computer Science and Mathematics Division
Oak Ridge National Laboratory
Oak Ridge, Tennessee

Yingyue Xu
Department of Electrical and Computer Engineering
University of Tennessee
Knoxville, Tennessee

Fan Yang
Department of Electrical and Computer Engineering
Clemson University
Clemson, South Carolina

Qing Yang
Department of Computer Science
Montana State University
Bozeman, Montana

Lu Yu
Holcombe Department of Electrical and Computer Engineering
Clemson University
Clemson, South Carolina

Lianyu Zhao
Holcombe Department of Electrical and Computer Engineering
Clemson University
Clemson, South Carolina

Tong Zhou
Department of Electrical Communication Engineering
Duke University
Durham, North Carolina

Mengxia Zhu
Department of Computer Science
Southern Illinois University
Carbondale, Illinois

Yi Zou
Department of Electrical and Computer Engineering
Duke University
Durham, North Carolina

Marco Zuniga
University of Duisburg-Essen
Essen, Germany

I

Sensor Deployment

Part I considers how to best deploy sensor networks. A hostile environment can occlude sensors and/or make communications impossible. Hostile opponents can subvert the network. In this part, we discuss important issues related to multiple networking and security. Only a few chapters from the first edition have not been revised in this part, while most of the chapters have been thoroughly revised. In addition, many new chapters have been provided that deal with sensor network security issues.

Zou and Chakrabarty consider how best to place sensors in order to monitor events in a region in Chapter 1. They describe a virtual force algorithm that allows nodes to position themselves in a globally desirable pattern using only local information. In doing so, they introduce many self-organization concepts.

Iyengar et al. provide an overview of the computational problems posed by sensor deployment. This section will provide many different views of these problems. Wu et al. consider data routing in sensor networks using mobile agents in Chapter 3. They phrase routing as an optimization problem that is then solved using genetic algorithms.

Dr. Rai provides a computer-networking tutorial in Chapter 4. This tutorial thoroughly illustrates communications concepts that are used throughout this book. The concepts of protocol layering and data transmission are described in detail. An introduction to wireless communications issues is provided as well.

Drs. Wang and Ramanathan provide a revised chapter (Chapter 5) that discusses location-centric networking. In this approach, the network is separated into distinct regions and manager nodes assigned to coordinate work within the geographic region.

We continue the geometric analysis initiated with location-centric routing with a new chapter (Chapter 6) provided by researchers from the Naval Undersea Warfare Center: Dr. Rowe, Kelly, and Dr. Warrack. They use probability, calculus, and geometry to determine many properties regarding network performance. Where many researchers will, incorrectly, model sensor coverage using disks, or similar geometric constructs, this chapter takes a rigorous approach to analyzing the performance of sensor fields.

Silva et al. explain the concepts behind diffusion routing in Chapter 7. Diffusion routing is a technology that has become strongly identified with sensor networking. It is a data-centric communications technology. The implementation described in this chapter uses a publish–subscribe paradigm that changes the way sensor network applications are designed. (The editors can personally attest to this.) This chapter describes both how the approach is used and its internal design.

In a classic chapter from the first edition, Carman discusses data security issues in sensor networks in Chapter 8. The chapter starts by describing possible attacks on sensor networks and the data security requirements of the systems. What makes these networks unique, from a data security perspective, are the numerous operational constraints that must be maintained. A security architecture is then proposed that fulfills the systems needs without violating the strict resource constraints. Carman was possibly the first security researcher in this field, being part of the seminal Defence Advanced Research Projects Agency (DARPA) SensIT program.

Zhao et al. contribute a chapter on security (Chapter 9). This chapter looks in depth at random key predistribution techniques, where symmetric key cryptography maintains system security and each node has a set of keys drawn as a random sample from a large pool of available keys. It provides an in-depth analysis of this approach and extends it to solve a number of difficult security problems: cloning, sleep deprivation, and Sybil and Byzantine attacks.

Gautam then presents a revised tutorial on network quality of service in Chapter 10. That is, a network needs to be able to fulfill its demands with a reasonable certainty within time constraints. This chapter discusses how this can be quantified and measured. This type of analysis is essential for distributed systems designs.

Rao and Wu discuss netlets concept in Chapter 11. This concept uses small agile processes to overcome many potential network problems. Network daemons are distributed processes that form an overlay network. They cooperate to overcome many potential network contention problems and provide a more predictable substrate.

Dr. Nikoletseas contributes a new (Chapter 12), which investigates a fundamental problem for routing in wireless sensor network. First, they examine static sensor nodes and analyze greedy protocols, suggesting new hybrid combinations. Their results indicate an inherent trade-off between performance measures such as energy and latency. They try to optimize performance. This chapter studies the impact of heterogeneous, highly dynamic mobility. These nodes can lower the cost of connectivity and add redundancy to data propagation.

Yu et al. provide a new chapter (Chapter 13) that discusses a security problem, which is common to most networks. Sensor networks are particularly susceptible to traffic analysis attacks. Because sensor networks are deployed in situ, it is easy for attackers to monitor timing side channels, even when communications are encrypted. This chapter shows how hidden Markov models can be inferred and used to violate many security guarantees in these systems.

Drs. Karst and Wicker contribute a new (Chapter 14) that deals with key management issues of sensor networks. They consider the electric power smart grid as an example sensor network with obvious security needs. Cryptography is the obvious tool for securing this network. It is important to refresh keys in order to maintain network security, and they consider the trade-offs in rekeying methods.

Wang et al. provide another new chapter (Chapter 15) that looks at a factory monitoring system. They monitor machines with high-speed rotations. This chapter looks in detail at the measurement and modeling of the performance of 802.15.4-based networks in factory environments.

Zhou et al. contribute another new chapter (Chapter 16) that deals with Sybil attacks, like Zhao et al., but in this case, the nodes are mobile. The use of vehicular and hoc networks (VANETs) changes the problem and forces the system designer to look at the trade-off between privacy and security. In this approach, road side boxes monitor network traffic to discover the excessive use of pseudonyms.

Part I has considered distributed sensor networks (DSNs) as distributed processes. Many networking technologies have been discussed in tutorial fashion. We have discussed how to position nodes in detail. Network security has been explored in depth, and finally a number of innovative networking technologies are presented.

1

Coverage-Oriented Sensor Deployment

Yi Zou
Duke University

Krishnendu Chakrabarty
Duke University

1.1 Introduction

Wireless sensor networks that are capable of observing the environment, processing data, and making decisions based on these observations, have recently attracted considerable attention [2,16,36,39]. These networks are important for a number of applications such as coordinated target detection and localization, surveillance, and environmental monitoring. Breakthroughs in miniaturization, hardware design techniques, and system software have led to cheaper sensors and fueled recent advances in wireless sensor networks [1,2,16].

In this chapter, we are focusing on coverage-driven sensor deployment. The coverage of a sensor network refers to the extent to which events in the monitored region can be detected by the deployed sensors. We present strategies for enhancing the coverage of sensor networks with low computation cost, a small number of sensors, and low energy consumption. We also present a probabilistic framework for uncertainty-aware sensor deployment, with applications to air-dropped sensors and deployment through dispersal.

Sensor node deployment problems have been studied in a variety of contexts. In the area of adaptive beacon placement and spatial localization, a number of techniques have been proposed for both fine-grained and coarse-grained localization [5,18]. Sensor deployment and sensor planning for military applications are described in Ref. [33], where a general sensor model is used to detect elusive targets in the battlefield. The sensor coverage analysis is based on a hypothesis of possible target movements and sensor attributes. However, the proposed wireless sensor networks framework in Ref. [33] requires a considerable amount of a priori knowledge about possible targets. A variant of sensor deployment has been considered for multi-robot exploration [10,23]. Each robot can be viewed as a sensor node in such systems. An incremental deployment algorithm is used in which sensor nodes are deployed one by one in an adaptive fashion. Each new deployment of a sensor is based on the sensed information from sensors deployed earlier. A drawback of this approach is that it is computationally expensive. As the number of sensors increases, each new deployment results in a relatively large amount of computation.

In Ref. [21], the concept of potential force is used in a distributed fashion to perform sensor node deployment in ad hoc wireless sensor networks. The problem of evaluating the coverage provided by a given placement of sensors is discussed in Refs. [31,32]. The major concern here is the self-localization of sensor nodes; sensor nodes are considered to be highly mobile and they move frequently. An optimal polynomial-time algorithm that uses graph theory and computational geometry constructs is used to determine the best-case and the worst-case coverage. Radar and sonar coverage also present several related challenges. Radar and sonar netting optimization are of great importance for detection and tracking in a surveillance area. Based on the measured radar cross-sections and the coverage diagrams for the different radars, a method has been proposed for optimally locating the radars to achieve satisfactory surveillance with limited radar resources.

Sensor placement on two- and three-dimensional grids has been formulated as a combinatorial optimization problem, and solved using integer linear programming in Refs. [7,8]. This approach suffers from two main drawbacks. First, computational complexity makes the approach infeasible for large problem instances. Second, the grid coverage approach relies on "perfect" sensor detection, that is, a sensor is expected to yield a binary yes/no detection outcome in every case. However, because of the inherent uncertainty associated with sensor readings, sensor detection must be modeled probabilistically [12,13]. It is well known that there is inherent uncertainty associated with sensor readings; hence sensor detections must be modeled probabilistically. A probabilistic optimization framework for minimizing the number of sensors for a two-dimensional grid has been proposed recently [12,13]. This algorithm attempts to maximize the average coverage of the grid points.

There also exists a close resemblance between the sensor placement problem and the art gallery problem (AGP) addressed by the art gallery theorem [34]. The AGP problem can be informally stated as that of determining the minimum number of guards required to cover the interior of an art gallery. (The interior of the art gallery is represented by a polygon.) The AGP has been solved optimally in two-dimensional and shown to be NP-hard in the three-dimensional case. Several variants of AGP have been studied in the literature, including mobile guards, exterior visibility, and polygons with holes.

A related problem in wireless sensor networks is that of spatial localization [18]. In wireless sensor networks, nodes need to be able to locate themselves in various environments and on different distance scales. Localization is particularly important when sensors are not deployed deterministically for example, when sensors are thrown from airplanes in a battlefield, and for underwater sensors that might move due to drift. Sensor networks also make use of spatial information for self-organization and configuration. A number of techniques for both fine and coarse-grained localization have been proposed [5,6].

Other related work includes the placement of a given number of sensors to reduce communication cost [25], and optimal sensor placement for a given target distribution [35]. Sensor deployment for collaborative target detection is discussed in Ref. [11], where path exposure is used as a measure of the effectiveness of the sensor deployment. This method uses sequential deployment of sensors, that is, a limited number of sensors are deployed in each step until the desired minimum exposure or probability of detection of a target is achieved. In most practical applications however, we need to deploy the sensors in advance without any prior knowledge of the target and sequential deployment is often infeasible. Moreover, sequential deployment may be undesirable when the number of sensors or the area of the sensor field is large. Thus a single step-deployment scheme is more advantageous in such scenarios. In Ref. [28], the authors propose a dual-space approach to event tracking and sensor resource management.

Chapter outline: We present a virtual force algorithm (VFA) as a sensor deployment strategy to enhance the coverage after an initial random placement of sensors. The VFA algorithm is based on disk packing theory [29] and the virtual force field concept from physics and robotics [10,23]. For a given number of sensors, the VFA algorithm attempts to maximize the sensor field coverage. A judicious combination of attractive and repulsive forces is used to determine the new sensor locations that improve the coverage. Once the effective sensor positions are identified, a one-time movement with energy consideration incorporated is carried out, that is, the sensors are redeployed, to these positions. The sensor field is represented by a two-dimensional grid. The dimensions of the grid provide a measure of the sensor field. The granularity of the grid, that is, distance between grid points can be adjusted to trade off computation time of the VFA algorithm with the effectiveness of the coverage measure. The detection by each sensor is modeled as a circle on the two-dimensional grid. The center of the circle denotes the sensor while the radius denotes the detection range of the sensor. We first consider a binary detection model in which a target is detected (not detected) with complete certainty by the sensor if a target is inside (outside) its circle. The binary model facilitates the understanding of the VFA model. We then investigate realistic probabilistic models in which the probability that the sensor detects a target depends on the relative position of the target within the circle.

We also formulate an uncertainty-aware sensor deployment problem to model scenarios where sensor locations are precomputed but the sensors are airdropped or dispersed. In such scenarios, sensor nodes cannot be expected to fall exactly at predetermined locations; rather there are regions where there is a high probability of sensor being actually located. Such examples include airdropped sensor nodes and underwater sensor nodes that drift due to water currents. Thus a key challenge in sensor deployment is to determine an uncertainty-aware sensor field architecture that reduces cost and provides high coverage, even though the exact location of the sensors may not be completely controllable. In this proposal, we present two algorithms for sensor deployment wherein we assumed that sensor positions are not exactly predetermined. We assume that the sensor locations are calculated before deployment and an attempt is made during the airdrop to place sensors at these locations; however, the sensor placement calculations and coverage optimization are based on a Gaussian model, which assumes that if a sensor is intended for a specific point P in the sensor field, its exact location can be anywhere in a "cloud" surrounding P. Note that the placement algorithms give us the sensor positions prior to actual placement and we assume that sensors are deployed in a single step.

1.2 Sensor Detection Model

The sensor field is represented by a two-dimensional grid. The dimensions of the grid provide a measure of the sensor field. The granularity of the grid, that is, distance between grid points can be adjusted to trade off computation time of the VFA algorithm with the effectiveness of the coverage measure. The detection by each sensor is modeled as a circle on the two-dimensional grid. The center of the circle denotes the sensor while the radius denotes the detection range of the sensor. We first consider a binary detection model in which a target is detected (not detected) with complete certainty by the sensor if a target is inside (outside) its circle. The binary model facilitates the understanding of the VFA model. We then investigate two types of realistic probabilistic models in which the probability that the sensor detects a target depends on the relative position of the target.

Let us consider a sensor field represented by a $m \times n$ grid. Let s be an individual sensor node on the sensor field located at grid point (x,y). Each sensor node has a detection range of r. For any grid point P at (i,j), we denote the Euclidean distance between s at (x,y) and P at (i,j) as $d_{ij}(x,y)$, that is, $d_{ij}(x,y) = \sqrt{(x-i)^2 + (y-j)^2}$. Equation 1.1 shows the binary sensor model [7] that expresses the coverage $c_{ij}(x,y)$ of a grid point at (i,j) by sensor s at (x,y):

$$c_{ij}(x,y) = \begin{cases} 1, & \text{if } d_{ij}(x,y) < r \\ 0, & \text{otherwise} \end{cases} \tag{1.1}$$

The binary sensor model assumes that sensor readings have no associated uncertainty. In reality, sensor detections are imprecise, hence the coverage $c_{ij}(x,y)$ needs to be expressed in probabilistic terms.

A possible way of expressing this uncertainty is to assume the detection probability on a target by a sensor varies exponentially with the distance between the target and the sensor [12,13]. This probabilistic sensor detection model given in Equation 1.2:

$$c_{ij}(x,y) = e^{-\alpha d_{ij}(x,y)} \tag{1.2}$$

This is also the coverage confidence level of this point from sensor s. The parameter α can be used to model the quality of the sensor and the rate at which its detection probability diminishes with distance. Clearly, the detection probability is 1 if the target location and the sensor location coincide. Alternatively, we can also use another probabilistic sensor detection model given in Equation 1.3, which is motivated in part by Ref. [14]:

$$c_{ij}(x,y) = \begin{cases} 0, & \text{if } r + r_e \le d_{ij}(x,y) \\ e^{-\lambda a^{\beta}}, & \text{if } r - r_e < d_{ij}(x,y) < r + r_e \\ 1, & \text{if } r - r_e \ge d_{ij}(x,y) \end{cases} \tag{1.3}$$

where

- r_e ($r_e < r$) is a measure of the uncertainty in sensor detection
- $a = d_{ij}(x,y) - (r - r_e)$
- λ and β are parameters that measure detection probability when a target is at distance greater than r_e but within a distance from the sensor

This model reflects the behavior of range sensing devices such as infrared and ultrasound sensors. The probabilistic sensor detection model is shown in Figure 1.1. Note that distances are measured in units of grid points. Figure 1.1 also illustrates the translation of a distance response from a sensor to the confidence level as a probability value about this sensor response. Different values of the parameters α and β yield different translations reflected by different detection probabilities, which can be viewed as the characteristics of various types of physical sensors.

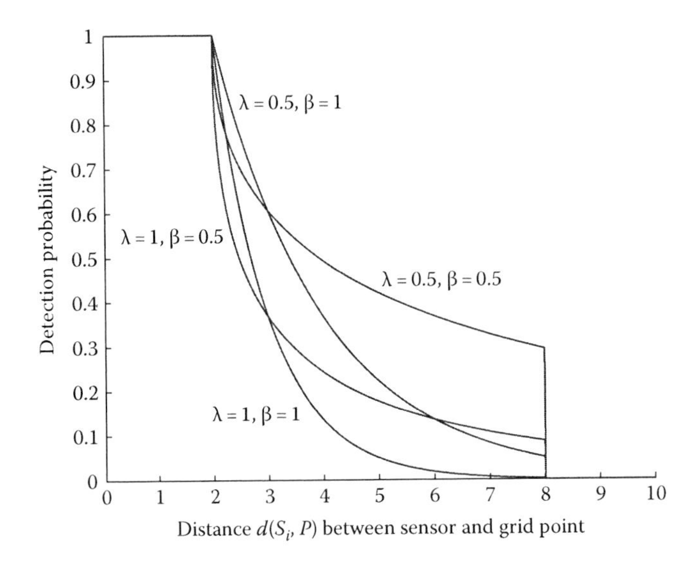

FIGURE 1.1 Probabilistic sensor detection model.

It is often the case that there are obstacles in the sensor field terrain. If we are provided with such a priori knowledge about where obstacles in the sensor field, we can also build the terrain information into our models based on the principle of line of sight. An example is given in Figure 1.2. Some types of sensors are not able to see through any obstacles located in the sensor field; hence models and algorithms must consider the problem of achieving an adequate sensor field coverage in presence of obstacles. Suppose C_{xy} is a $m \times n$ matrix that corresponds to the detection probabilities of each grid point in the sensor field when a sensor node is located at grid point (x,y), that is, $C_{xy} = [c_{ij}(x,y)]_{m \times n}$. To achieve the coverage in presence of obstacles, we need to generate a mask matrix for the corresponding coverage probability matrix C_{xy} to mask out those grid points as the "blocked area," as shown in Figure 1.2. In this way, the sensor node placed at the location (x,y) will not see any grid points beyond the obstacles. We also assume that sensor nodes are not placed on any grid points with obstacles. Figure 1.3 is an example of the mask matrix for a sensor node at (1,1) in a 10 by 10 sensor field grid with obstacles located at (7,3), (7,4), (3,5), (4,5), (5,5).

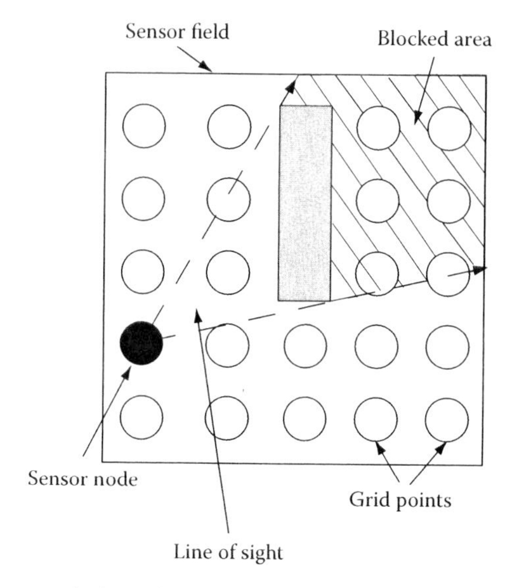

FIGURE 1.2 Example to illustrate the line of sight principle.

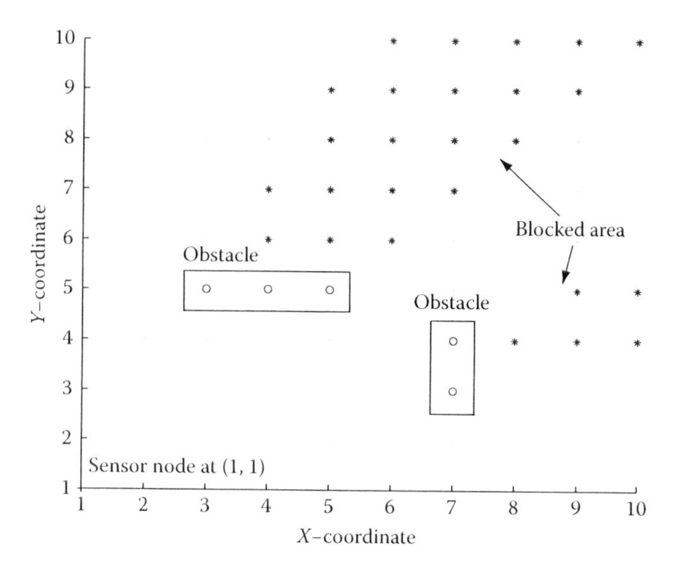

FIGURE 1.3 Obstacle mask matrix example.

1.3 Virtual Force Algorithm for Sensor Node Deployment

As an initial sensor node deployment step, a random placement of sensors in the target area (sensor field) is often desirable, especially if no a priori knowledge of the terrain is available. Random deployment is also practical in military applications, where wireless sensor networks are initially established by dropping or throwing sensors into the sensor field. However, random deployment does not always lead to effective coverage, especially if the sensors are overly clustered and there is a small concentration of sensors in certain parts of the sensor field. However the coverage provided by a random deployment can be improved using a force-directed algorithm. We present the VFA as a sensor deployment strategy to enhance the coverage after an initial random placement of sensors. The VFA algorithm combines the ideas of potential field [10,23] and disk packing [29]. For a given number of sensors, VFA attempts to maximize the sensor field coverage using a combination of attractive and repulsive forces. During the execution of the force-directed VFA algorithm, sensors do not physically move but a sequence of virtual motion paths is determined for the randomly-placed sensors. Once the effective sensor positions are identified, a one-time movement is carried out to redeploy the sensors at these positions. Energy constraints are also included in the sensor repositioning algorithm. In the sensor field, each sensor behaves as a "source of force" for all other sensors. This force can be either positive (attractive) or negative (repulsive). If two sensors are placed too close to each other, the "closeness" being measured by a pre-determined threshold, they exert negative forces on each other. This ensures that the sensors are not overly clustered, leading to poor coverage in other parts of the sensor field. On the other hand, if a pair of sensors is too far apart from each (once again a pre-determined threshold is used here), they exert positive forces on each other. This ensures that a globally uniform sensor placement is achieved. Figure 1.4 illustrates how the VFA algorithm is used for sensor deployment.

1.3.1 Virtual Forces

We now describe the virtual forces and virtual force calculation in the VFA algorithm. In the following discussion, we use the notation introduced in the previous subsection. Let S denote the set of deployed sensors node, that is, $S = \{s_1, \ldots, s_k\}$ and $|S| = k$. Let the total virtual force action on a sensor node $s_p(p = 1, \ldots, k)$ be denoted by $\vec{F}_p$. Note that $\vec{F}_p$ is a vector whose orientation is determined by the vector sum of all the forces acting on s_p. Let the force exerted on s_p by another sensor $s_q(q = 1, \ldots, k, q \neq p)$ be denoted by $\vec{F}_{pq}$. In addition to the positive and negative forces due to other sensors, a sensor s_p is also subjected to forces exerted by obstacles and areas of preferential coverage in the grid. This provides us with a convenient method to model obstacles and the need for preferential coverage. Sensor deployment must take into account the

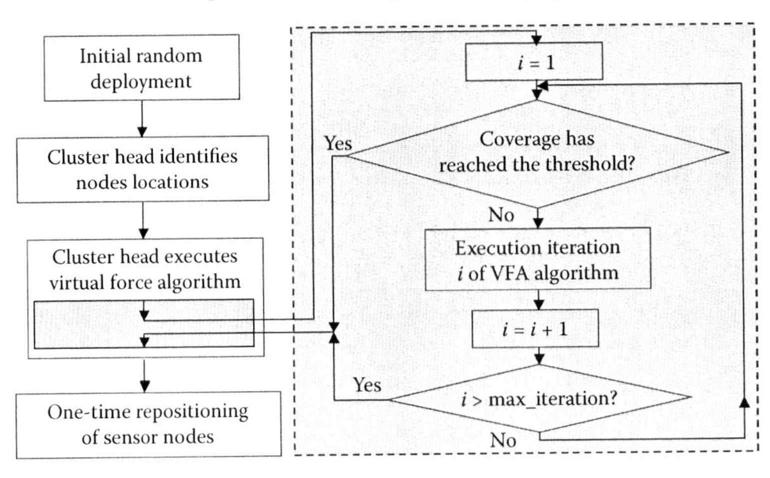

FIGURE 1.4 Sensor deployment with VFA algorithm.

nature of the terrain, for example, obstacles such as building and trees in the line of sight for infrared sensors, uneven surface and elevations for hilly terrain, etc. In addition, based on relative measures of security needs and tactical importance, certain areas of the grid need to be covered with greater certainty.

The knowledge of obstacles and preferential areas implies a certain degree of a priori knowledge of the terrain. In practice, the knowledge of obstacles and preferential areas can be used to direct the initial random deployment of sensors, which in turn can potentially increase the efficiency of the VFA algorithm. In our virtual force model, we assume that obstacles exert repulsive (negative) forces on a sensor. Likewise, areas of preferential coverage exert attractive (positive) forces on a sensor. If more detailed information about the obstacles and preferential coverage areas is available, the parameters governing the magnitude and direction (i.e., attractive or repulsive) of these forces can be chosen appropriately. In this work, we let $\vec{F}_{pA}$ be the total attractive force on s_p due to preferential coverage areas, and let $\vec{F}_{pR}$ be the total repulsive force on s_p due to obstacles. The total force $\vec{F}_p$ on s_p can now be expressed as

$$\vec{F}_p = \sum_{q=1,q\neq p}^{k} \vec{F}_{pq} + \vec{F}_{pR} + \vec{F}_{pA} \tag{1.4}$$

We next express the force $\vec{F}_{pq}$ between s_p and s_q in polar coordinate notation. Note that $\vec{f} = (r, \theta)$ implies a magnitude of r and orientation θ for vector $\vec{f}$

$$\vec{F}_{pq} = \begin{cases} (w_A(d_{pq} - d_{th}), \theta_{pq}) & \text{if } d_{pq} > d_{th} \\ 0, & \text{if } d_{pq} = d_{th} \\ \left(w_R \dfrac{1}{d_{pq}}, \theta_{pq} + \pi \right), & \text{if otherwise} \end{cases} \tag{1.5}$$

where

$d_{pq} = \sqrt{(x_p - x_q)^2 + (y_p - y_q)^2}$ is the Euclidean distance between sensor s_p and s_q
d_{th} is the threshold on the distance between s_p and s_q
θ_{pq} is the orientation (angle) of a line segment from s_p to s_q
w_A (w_R) is a measure of the attractive (repulsive) force

The threshold distance d_{th} controls how close sensors get to each other. As an example, consider the four sensors s_1, s_2, s_3, and s_4 in Figure 1.5. The force $\vec{F}_1$ on s_1 is given by $\vec{F}_1 = \vec{F}_{12} + \vec{F}_{13} + \vec{F}_{14}$. If we assume that

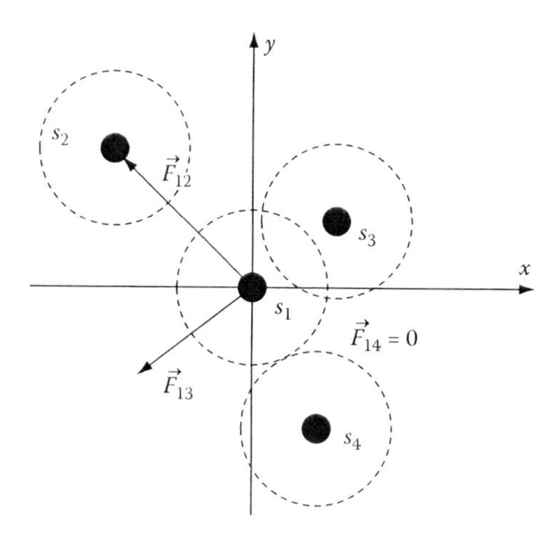

FIGURE 1.5 An example of virtual forces with four sensors.

$d_{12} > d_{th}$, $d_{13} < d_{th}$, and $d_{14} = d_{th}$, s_2 exerts an attractive force on s_1, s_3 exerts a repulsive force on s_1 and s_4 exerts no force on s_1. This is shown in Figure 1.5. Note that d_{th} is a pre-determined parameter that is supplied by the user, who can choose an appropriate value of d_{th} to achieve a desired coverage level over the sensor field.

1.3.2 Overlapped Sensor Detection Areas

If $r_e \approx 0$ and we use the binary sensor detection model given by Equation 1.1, we attempt to make d_{pq} as close to $2r$ as possible. This ensures that the detection regions of two sensors do not overlap, thereby minimizing "wasted overlap" and allowing us to cover a large grid with a small number of sensors. This is illustrated in Figure 1.6a. An obvious drawback here is that a few grid points are not covered by any sensor. Note that an alternative strategy is to allow overlap, as shown in Figure 1.6b. While this approach ensures that all grid points are covered, it needs more sensors for grid coverage. Therefore, we adopt the first strategy. Note that in both cases, the coverage is effective only if the total area $k\pi r^2$ that can be covered with the k sensors exceeds the area of the grid.

If $r_e > 0$, r_e is not negligible and the probabilistic sensor model given by Equation 1.2 or Equation 1.3 is used. Note that due to the uncertainty in sensor detection responses, grid points are not uniformly covered with the same probability. Some grid points will have low coverage if they are covered only by only one sensor and they are far from the sensor. In this case, it is necessary to overlap sensor detection areas in order to compensate for the low detection probability of grid points that are far from a sensor. Consider a grid point with coordinate (i,j) lying in the overlap region of sensors s_p and s_q located at (x_p,y_p) and (x_q,y_q) respectively. Let $c_{ij}(s_p,s_q)$ be the probability that a target at this grid point is reported as being detected by observing the outputs of these two sensors. We assume that sensors within a cluster operate independently in their sensing activities. Thus

$$c_{ij}(s_p,s_q) = 1-(1-c_{ij}(s_p))(1-c_{ij}(s_q)) \tag{1.6}$$

where $c_{ij}(s_p) = c_{ij}(x_p,y_p)$ and $c_{ij}(s_q) = c_{ij}(x_q,y_q)$ are coverage probabilities from the probabilistic sensor detection models. Since the term $1 - (1 - c_{ij}(s_p))(1 - c_{ij}(s_q))$ expresses the probability that neither s_p nor s_q covers grid point at (i,j), the probability that the grid point (i,j) is covered is given by Equation 1.6. Let c_{th} be the desired coverage threshold for all grid points. This implies that

$$\min_{i,j}\{c_{ij}(s_p,s_q)\} \geq c_{th} \tag{1.7}$$

Note that Equation 1.6 can also be extended to a region which is overlapped by a set of k_{ov} sensors, denoted as S_{ov}, $k_{ov} = |S_{ov}|$, $S_{ov} \subseteq \{s_1,s_2, \ldots, s_k\}$. The coverage of the grid point at (i,j) due to a set of sensor nodes S_{ov} in this case is given by

$$c_{ij}(S_{ov}) = 1 - \prod_{s_p \in S_{ov}} (1 - c_{ij}(s_p)) \tag{1.8}$$

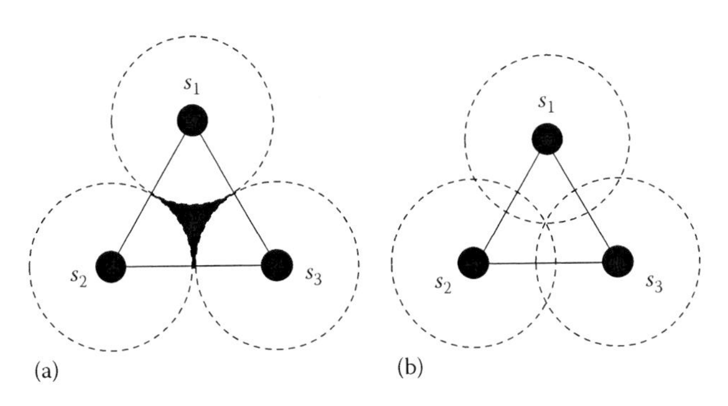

FIGURE 1.6 (a) Non-overlapped and (b) overlapped sensor coverage areas.

As shown in Equation 1.5, the threshold distance d_{th} is used to control how close sensors get to each other. When sensor detection areas overlap, the closer the sensors are to each other, the higher is the coverage probability for grid points in the overlapped areas. Note however that there is no increase in the point coverage once one of the sensors gets close enough to provide detection with a probability of one. Therefore, we need to determine d_{th} that maximizes the number of grid points in the overlapped area that satisfies $c_{ij}(s_p) > c_{th}$.

1.3.3 Energy Constraint on the VFA Algorithm

In order to prolong the battery life, the distances between the initial and final position of the sensors are limited in the repositioning phase to conserve energy. We use $d_{max}(s_p)$ to denote the maximum distance that sensor s_p can move in the repositioning phase. To simplify the discussion without loss of generality, we assume $d_{max}(s_p) = d_{max}(s_q) = d_{max}$, for $p,q = 1,2, \ldots, k$. During the execution of the VFA algorithm, for each sensor node, whenever the distance from the current virtual position to the initial position reaches the distance limit d_{max}, any virtual forces on this sensor are disabled. For sensor s_p, let $(x_p, y_p)_{random}$ be the initial location obtained from the random deployment, and $(x_p, y_p)_{virtual}$ be the location generated by the VFA algorithm. The energy constraint can be described as

$$\vec{F}_p = \begin{cases} 0, & \text{if } d((x_p, y_p)_{rand}, (x_p, y_p)_{virtual}) \geq d_{max} \\ \vec{F}_p, & \text{otherwise (i.e., the force is unchanged)} \end{cases} \quad (1.9)$$

Therefore the virtual force $\vec{F}_p$ given by Equation 1.4 on sensor s_p is ignored whenever the move violates the energy constraint expressed by d_{max}. Note that due to the energy constraint on the one-time repositioning given by Equation 1.9, it might be necessary to trade off the coverage with the energy consumed in repositioning if d_{max} is not large enough.

Note that the VFA algorithm is designed to be executed on the cluster head, which is expected to have more computational capabilities than sensor nodes. The cluster head uses the VFA algorithm to find appropriate sensor node locations based on the coverage requirements. The new locations are then sent to the sensor nodes, which perform a one-time movement to the designated positions. No movements are performed during the execution of the VFA algorithm.

1.3.4 Procedural Description of the VFA Algorithm

We next describe the VFA algorithm in pseudo-code. Figure 1.7 shows the data structure of the VFA algorithm and Figure 1.8 shows the implementation details in pseudo code form. For a n by m grid with a total of k sensors deployed, the computational complexity of the VFA algorithm is $O(nmk)$. Due to the granularity of the grid and the fact that the actual coverage is evaluated by the number of grid points that

VFA Data Structures: Grid, $\{s_1, s_2, \ldots, s_k\}$

/* n_P is the number of preferential area blocks (attractive forces) an n_O is the number of obstacle blocks (repulsive forces). $(x,y)_{VFA}$ is the final position found by the VFA algorithm. d_{max} is the energy constraint on the sensor repositioning phase in the VFA algorithm. */

1 *Grid* structure:

2 Properties: $width, height, k, c_{th}, d_{th}, c(loops), \bar{c}, \Delta c$;

3 Preferential areas: $PA_i(x, y, wx, wy)$, $i = 1,2, \ldots, n_P$;

4 Obstacles areas: $OA_i(x, y, wx, wy)$, $i = 1,2, \ldots, n_O$;

5 Grid points, P_{ij}: $c_{ij}(\{s_1, s_2, \ldots, s_k\})$;

6 Sensor s_p structure: $(x_p, y_p)_{random}$, $(x_p, y_p)_{virtual}$, $(x, y)_{VFA}$, $p, r, r_e, \alpha, \beta, d_{max}$;

FIGURE 1.7 Data structures used in the VFA algorithm.

```
Procedure Virtual_Force_Algorithm (Grid, {s_1, s_2, ..., s_k})
1    Set loops = 0;
2    Set MaxLoops = MAX_LOOPS;
3    While (loops < MaxLoops)
4        /* coverage evaluation */
5        For grid point P at (i, j) in Grid, i ∈ [1, width], j ∈ [1, height]
6            For s_p ∈ {s_1, s_2, ..., s_k}
7                Calculate c_ij (x_p, y_p) from the sensor model using (d_ij (x_p, y_p), c_th, d_th, α, β);
8            End
9        End
10       If coverage requirements are met: |c(loops) − c̄| ≤ Δc
11           Break from While loop;
12       End
13       /* virtual forces among sensors */
14       For s_p ∈ {s_1, s_2, ..., s_k}
15           Calculate F⃗_pq using d(s_p, s_q), d_th, w_A, w_R;
16           Calculate F⃗_pA using d(s_p, PA_1, ..., PA_np), d_th;
17           Calculate F⃗_pR using d(s_p, OA_1, ..., OA_no), d_th;
18           F⃗_p = ΣF⃗_pR + F⃗_pA, q = 1,...,k, q ≠ p;
19       End
20       /*move sensors virtually */
21       For s_p ∈ {s_1, s_2, ..., s_k}
22           /* energy constraint on the sensor movement */
23           If d((x_p, y_p)_random, (x, y)_virtual) ≥ d_max
24               Set F⃗_p = 0;
25           End
26           F⃗_p virtually moves s_p to its next position;
27       End
28       /* continue to next iteration */
29       Set loops = loops + 1;
30   End
```

FIGURE 1.8 Pseudo code of the VFA algorithm.

have been adequately covered, the convergence of the VFA algorithm is controlled by a threshold value, denoted by Δc. Let us use *c*(*loops*) to denote the current grid coverage of the number *loops* iteration in the VFA algorithm.

For the binary sensor detection model without the energy constraint, the upper bound value denoted as $\bar{c}$ is $k\pi r^2$; for the probabilistic sensor detection model or binary sensor detection model with the energy constraint, *c*(*loops*) is checked for saturation by defining $\bar{c}$ as the average of the coverage ratios of the near 5 (or 10) iterations. Therefore, the VFA algorithm continues to iterate until $|c(loops) - \bar{c}| \leq \Delta c$. In our experiments, Δc is set to 0.001.

Note that there exists the possibility of certain pathological scenarios in which the VFA algorithm is rendered ineffective, for example, if the sensors are initially placed along the circumference of a circle such that all virtual forces are balanced. The efficiency of the VFA algorithm depends on the values of the force parameters w_A and w_R. We found that the algorithm converged more rapidly for our case studies if $w_R \gg w_A$. This need not always be true, so we are examining ways to choose appropriate values for w_R and w_A based on the initial configuration.

1.3.5 VFA Simulation Results

In this section, we present simulation results obtained using the VFA algorithm. The deployment requirements include the maximum improvement of coverage over random deployment, the coverage for preferential areas and the avoidance of obstacles. For all simulation results presented in this section, distances are measured in units of grid points. A total of 20 sensors are placed in the sensor field in the random placement stage. Each sensor has a detection radius of 5 units ($r = 5$), and range detection error of 3 units ($r_e = 3$) for the probabilistic detection model. The sensor field is 50 by 50 in dimension. The simulation is done on a Pentium III 1.0 GHz PC using MATLAB®.

1.3.6 Case Study 1

Figures 1.9 through 1.11 present simulation results for the probabilistic sensor model given by Equation 1.3. The probabilistic sensor detection model parameters are set as $\lambda = 0.5$, $\beta = 0.5$, and $c_{th} = 0.7$. The initial sensor placements are shown in Figure 1.9. Figure 1.10 shows the final sensor positions determined

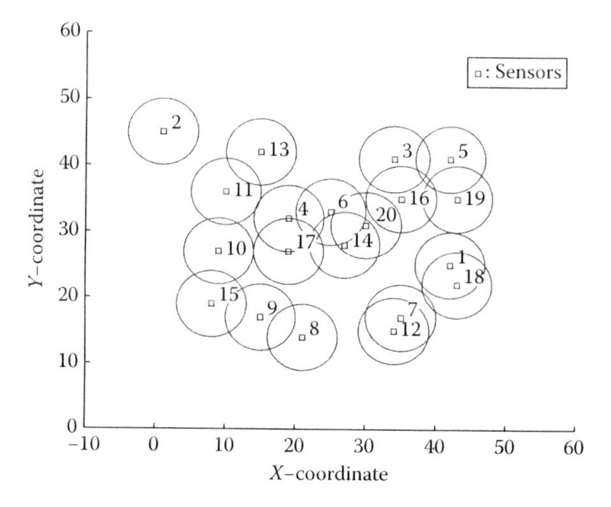

FIGURE 1.9 Initial sensor positions after random placement (probabilistic sensor detection model).

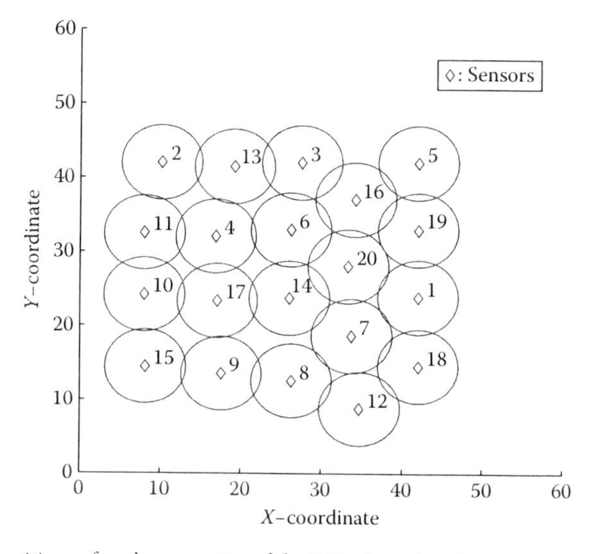

FIGURE 1.10 Sensor positions after the execution of the VFA algorithm (probabilistic sensor detection model).

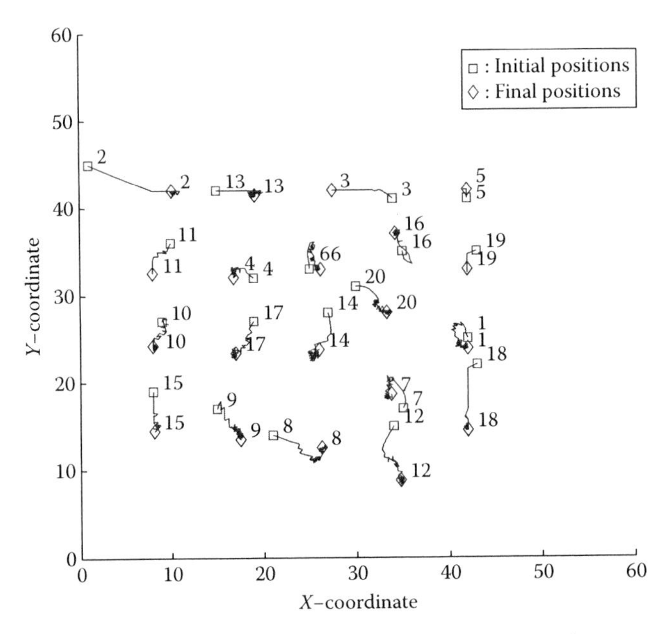

FIGURE 1.11 A trace of virtual moves made by the sensors (probabilistic sensor detection model).

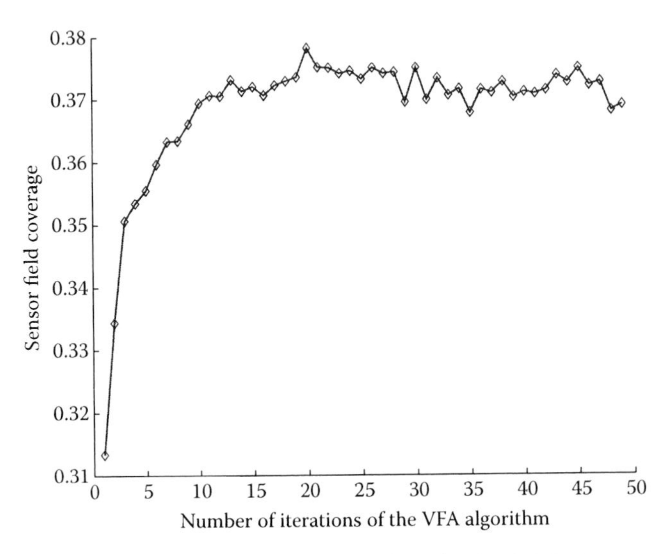

FIGURE 1.12 Sensor field coverage achieved using the VFA algorithm (probabilistic sensor detection model).

by the VFA algorithm. Figure 1.11 shows the virtual movement traces of all sensors during the execution of the VFA algorithm. We can see that overlap areas are used to increase the number of grid points whose coverage exceeds the required threshold c_{th}. Figure 1.12 shows the improvement of coverage during the execution of the VFA algorithm.

1.3.7 Case Study 2

As discussed in Section 1.3, VFA is also applicable to a sensor field containing obstacles and preferential areas. If obstacles are to be avoided, they can be modeled as repulsive force sources in the VFA algorithm. Preferential areas should be covered first, therefore they are modeled as attractive force sources in the VFA algorithm. Figures 1.13 through 1.16 present simulation results for a 50 by 50 sensor field that contains an obstacle and a preferential area. The binary sensor detection model given by Equation 1.1 is used for this simulation. The initial sensor placements are shown in Figure 1.13. Figure 1.14 shows the final sensor positions determined by the VFA algorithm. Figure 1.15 shows the virtual movement traces of all sensors during the execution of the VFA algorithm. Figure 1.16 shows the improvement of coverage during the execution of the VFA algorithm.

The VFA algorithm does not require much computation time. For Case study 1, the VFA algorithm took only 25 s for 30 iterations. For Case study 1, the VFA algorithm took only 3 min to complete 50 iterations. Finally for Case study 2, the VFA algorithm took only 48 s to complete 50 iterations.

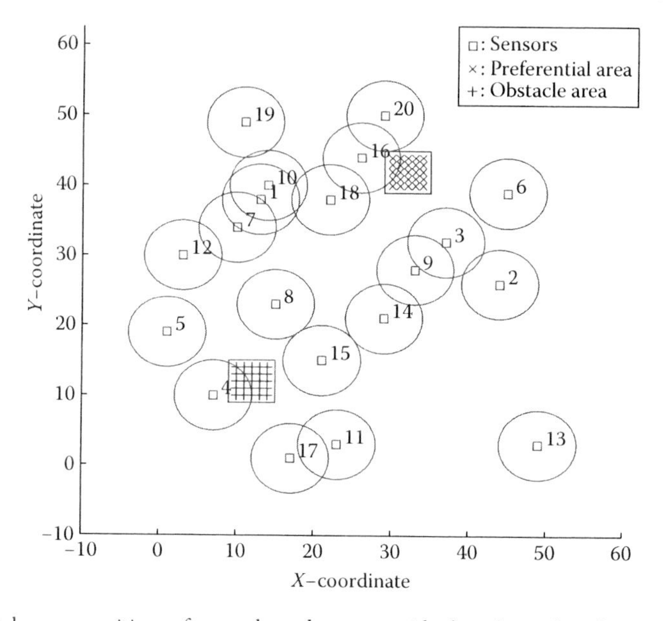

FIGURE 1.13 Initial sensor positions after random placement with obstacles and preferred areas.

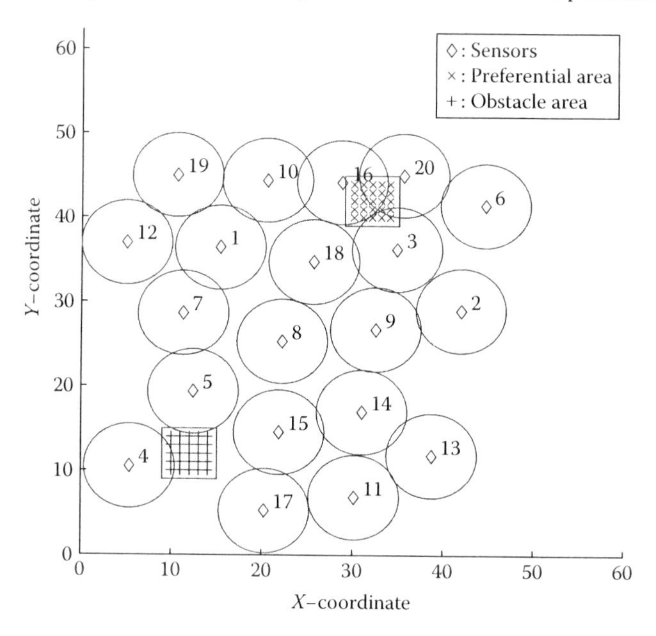

FIGURE 1.14 Sensor positions after the execution of the VFA algorithm with obstacles and preferred areas.

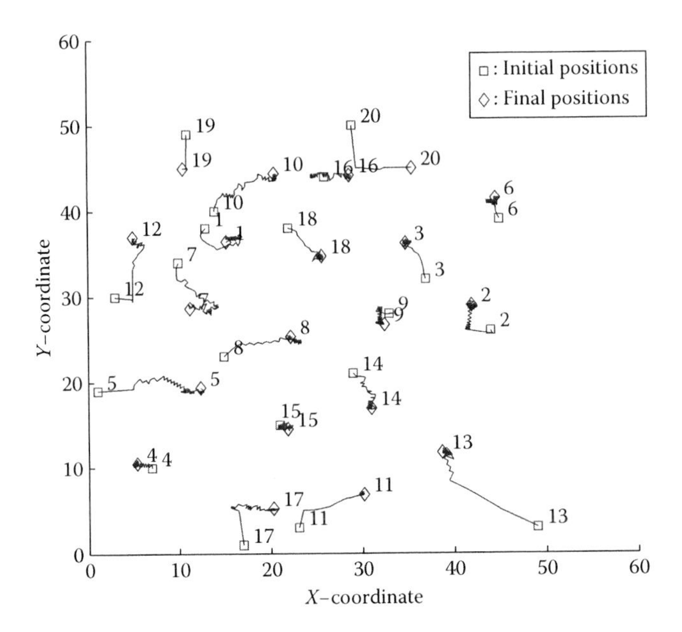

FIGURE 1.15 A trace of virtual moves made by the sensors with obstacles and preferred areas.

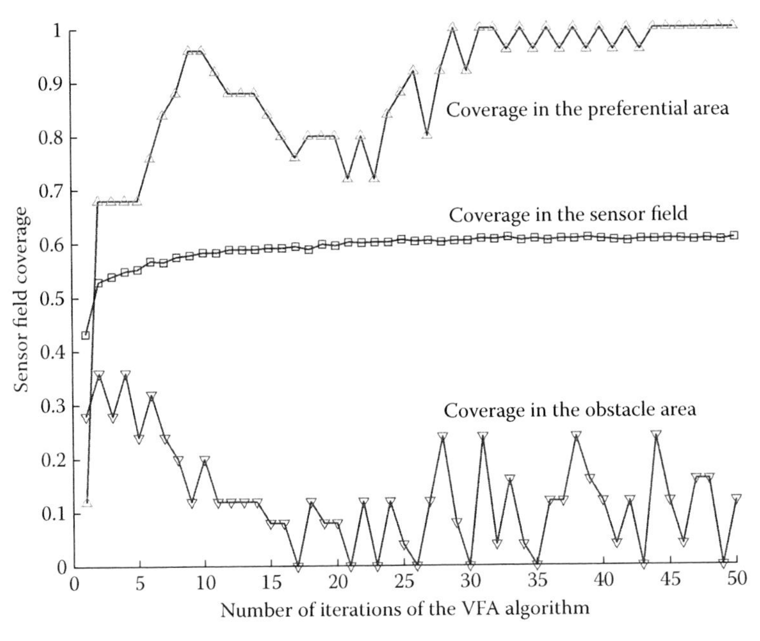

FIGURE 1.16 Sensor field coverage achieved using the VFA algorithm with obstacles and preferred areas.

Note that these computation times include the time needed for displaying the simulation results on the screen. CPU time is important because sensor redeployment should not take excessive time.

In order to examine how the VFA algorithm scales for larger problem instances, we considered up to 90 sensor nodes in a cluster for a 50 by 50 grid, with $r = 3$, $r_e = 2$, $\lambda = 0.5$, and $\beta = 0.5$ for all cases. For a given number of sensor nodes, we run the VFA algorithm over 10 sets of random deployment results and take the average of the computation time. The results, listed in Table 1.1, show that the CPU time grows slowly with the number of sensors k. For a total of 90 sensors, the CPU time is only 4 min on a

TABLE 1.1 The Computation Time for the VFA Algorithm for Larger Problem Instances

k	Binary Model (s)	Probabilistic Model (min)	k	Binary Model (s)	Probabilistic Model (min)
40	21	1.8	70	46	3.6
50	32	2.2	80	59	3.7
60	38	3.1	90	64	4.0

Pentium III PC. In practice, a cluster head usually has less computational power than a Pentium III PC; however, our results indicate that even if the cluster head has less memory and an on-board processor that runs 10 times slower, the CPU time for the VFA algorithm is reasonable.

1.4 Uncertainty Modeling in Sensor Node Deployment

The topology of the sensor field, that is, the locations of the sensors, determines to a large extent the quality and the extent of the coverage provided by the sensor network. However, even if the sensor locations are precomputed for optimal coverage and resource utilization, there are inherent uncertainties in the sensor locations when the sensors are dispersed, scattered, or airdropped. Thus a key challenge in sensor deployment is to determine an uncertainty-aware sensor field architecture that reduces cost and provides high coverage, even though the exact location of the sensors may not be controllable. We consider the sensor deployment problem in the context of uncertainty in sensor locations subsequent to airdropping. Sensor deployment in such scenarios is inherently non-deterministic and there is a certain degree of randomness associated with the location of a sensor in the sensor field. We present two algorithms for the efficient placement of sensors in a sensor field when the exact locations of the sensors are not known.

In applications such as battlefield surveillance and environmental monitoring, sensors may be dropped from airplanes. Such sensors cannot be expected to fall exactly at predetermined locations; rather there are regions where there is a high probability of sensor being actually located (Figure 1.17). In underwater deployment, sensors may move due to drift or water currents. Furthermore in most real-life situations, it is difficult to pinpoint the exact location of each sensor since only a few of the sensors may be aware of their locations. Thus the position of sensors may not exactly known and for every point in the sensor field, there is only a certain probability of a sensor being located at that point.

In this section, we present two algorithms for sensor deployment wherein we assumed that sensor positions are not exactly predetermined. We assume that the sensor locations are calculated before deployment and an attempt is made during the airdrop to place sensors at these locations; however,

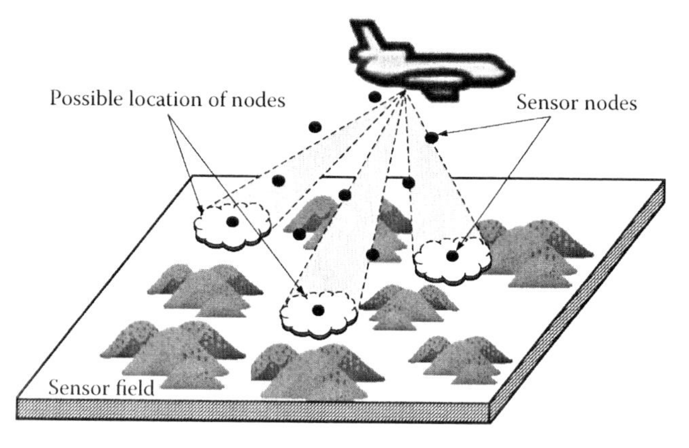

FIGURE 1.17 Sensors dropped from airplanes. The clouded region gives the possible region of a sensor location. The black dots within the clouds show the mean (intended) position of a sensor.

the sensor placement calculations and coverage optimization are based on a Gaussian model, which assumes that if a sensor is intended for a specific point P in the sensor field, its exact location can be anywhere in a "cloud" surrounding P.

1.4.1 Modeling of Non-Deterministic Sensor Node Placement

During sensor deployment, an attempt is made to place sensors at appropriate predetermined locations by air-dropping or other means. This does not guarantee however that sensors are actually placed at the designated positions, due to unanticipated conditions such as wind, the slope of the terrain, etc. In this case, there is a certain probability of a sensor being located at a particular grid point as a function of the designated location. The deviation about the designated sensor locations may be modeled using a Gaussian probability distribution, where the intended coordinates (x, y) serve as the mean values with standard deviation σ_x and σ_y in the x and y dimensions, respectively. Assuming that the deviations in the x and y dimensions are independent, the joint probability density function with mean (x, y) is given by

$$p_{xy}(x', y') = \frac{e^{-\left((x'-x)^2/2\sigma_x^2\right)/\left((y'-y)^2/2\sigma_y^2\right)}}{2\pi\sigma_x\sigma_y} \tag{1.10}$$

Let us use the notation introduced in the previous section. We still consider a sensor field represented by a $m \times n$ grid, denoted as *Grid*, with S denoting the set of sensor nodes. Let L_S be the set that contains corresponding sensor node locations, that is, $L_S = \{(x_p, y_p) | s_p \text{ at } (x_p, y_p), s_p \in S\}$. Let A be the total area encompassing all possible sensor locations. To model the uncertainty in sensor locations, the conditional probability $c_{ij}^{*}(x, y)$ for a grid point (i, j) to be detected by a sensor that is supposed to be deployed at (x, y) is then given by

$$c_{ij}^{*}(x, y) = \frac{\sum_{(x', y') \in A} c_{ij}(x', y') p_{xy}(x', y')}{\sum_{(x', y') \in A} p_{xy}(x', y')} \tag{1.11}$$

Based on Equations 1.10 and 1.11, we define the matrices $C_{xy}^{*} = [c_{ij}^{*}(x, y)]_{m \times n}$ and $P = [p_{xy}(x', y')]_A$.

1.4.2 Uncertainty-Aware Sensor Node Placement Algorithms

In this section, we introduce the sensor placement algorithm with consideration of uncertainties in sensor locations. The goal of sensor placement algorithms is to determine the minimum number of sensors and their locations such that every grid point is covered with a minimum confidence level. The sensor placement algorithms do not give us the actual location of the sensor but only the mean position of the sensor. It is straightforward to define the miss probability in our sensor deployment scenario. The *miss probability* of a grid point (i, j) due to a sensor at (x, y), denoted as $m_{ij}(x, y)$, is given by

$$m_{ij}(x, y) = 1 - c_{ij}^{*}(x, y) \tag{1.12}$$

Therefore the miss probability matrix due to a sensor placed at (x, y) is $M_{xy} = [m_{ij}(x, y)]_{m \times n}$. M_{xy} is associated with each grid point and can be pre-determined based on Equations 1.10 through 1.12. Since a number of sensors are placed for coverage, we would like to know the miss probability of each grid point due to a set of sensors, namely the collective miss probability. We denote the term *collective miss probability* as m_{ij} and define it in the form of maximum likelihood function as

$$m_{ij} = \prod_{(x, y) \in L_S} m_{ij}(x, y) = \prod_{(x, y) \in L_S} \left[1 - c_{ij}^{*}(x, y)\right] \tag{1.13}$$

Accordingly we have $M = [m_{ij}]_{m \times n}$ as the collective miss probability matrix over the grid points in the sensor field.

We determine the location of the sensors one at a time. In each step, we find all possible locations that are available on the grid for a sensor, and calculate the overall miss probability associated due to this sensor and those already deployed. We denote the *overall miss probability* due to the newly introduced sensor at grid point (x, y) as $\tilde{m}(x, y)$, which is defined as

$$\tilde{m}(x,y) = \sum_{(i,j)\in Grid} m_{ij}(x,y)m_{ij} \tag{1.14}$$

Based on the $\tilde{m}(x, y)$ values, where $(x, y) \in Grid$ and $(x, y) \notin L_S$, we can place sensors either at the grid point with the maximum miss probability (the worst coverage case) or the minimum miss probability (the best coverage case). We refer to the two strategies as MAX_MISS and MIN_MISS, respectively. Therefore, the sensor location can be found based on the following rule. For $(x, y) \in Grid$ and $(x,y) \notin L_S$,

$$\tilde{m}(x,y) = \begin{cases} \min\{\tilde{m}(x',y')\}, & \text{if MIN\_MISS is used} \\ \max\{\tilde{m}(x',y')\}, & \text{if MAX\_MISS is used} \end{cases} \tag{1.15}$$

When the best location is found for the current sensor, the collective miss probability matrix M is updated with the newly introduced sensor at location (x, y). This is carried out using Equation 1.16:

$$M = M \cdot M_{xy} = [m_{ij} \cdot m_{ij}(x,y)]_{m\times n} \tag{1.16}$$

There are two parameters that serve as the termination criterion for the two algorithm. The first is k_{max}, which is the maximum number of sensors that we can afford to deploy. The second is the threshold on the miss probability of each grid point, m_{th}. Our objective is to ensure that every grid point is covered with probability at least $c_{th} = 1 - m_{th}$. Therefore, the rule to stop the further execution of the algorithm is

$$m_{ij} < m_{th} \quad \text{for all } (i,j) \in Grid \text{ or } k > k_{max} \tag{1.17}$$

where k is the number of deployed sensors. The performance of the proposed algorithm is evaluated using the *average coverage probability* of the grid defined as

$$c_{avg} = \frac{\sum_{(x,y)\in Grid} c_{ij}}{m \times n} \tag{1.18}$$

where c_{ij} is the *collective coverage probability* of a grid point due to all sensors on the grid, defined as

$$\begin{aligned} c_{ij} &= 1 - \prod_{(x,y)\in L_S} m_{ij}(x,y) \\ &= 1 - \left\{ \prod_{(x,y)\in L_S} \left[1 - c_{ij}^{*}(x,y)\right] \right\} \end{aligned} \tag{1.19}$$

We have thus far considered the coverage of only the grid points in the sensor field. In order to provide robust coverage of the sensor field, we also need to ensure that the region that lies between the grid points is adequately covered, that is, every non-grid point has a miss probability less than the threshold m_{th}. Consider the four grid points in Figure 1.16 that lie on the four corners of a square. Let the distance between these grid points be d^*. The point of intersection of the diagonals of the square is at distance $d^*/\sqrt{2}$ from the four grid points. The following theorem provides a sufficient condition under which the non-grid points are adequately covered by the MIN_MISS and MAX_MISS algorithms.

Theorem 1.1

Let the distance between the grid point P_1 and a potential sensor location P_2 be d. Let the distance between adjacent grid points be d^*. If a value of $d+(d^*/\sqrt{2})$ is used to calculate the coverage of grid point P_1 due to a sensor at P_2, and the number of available sensors is adequate, the miss probability of all the non-grid points is less than a given threshold m_{th} when the algorithms MAX_MISS and MIN_MISS terminate.

Proof

Consider the four grid points in Figure 1.18. The center of square, that is, the point of intersection of diagonals, is at a distance of $d^*/\sqrt{2}$ from each of the four grid points. Every other non-grid point is at a shorter distance (less than $d^*/\sqrt{2}$) from at least one of the four grid points. Thus if a value of $d+(d^*/\sqrt{2})$ is used to determine coverage in the MAX_MISS and MIN_MISS algorithms, we can guarantee that every non-grid point is covered with a probability that exceeds $1 - m_{th}$.

In order to illustrate Theorem 1.1, we consider a 5 by 5 grid with $\alpha = 0.5$, $\lambda = 0.5$, $\beta = 0.5$, and $m_{th} = 0.4$. We use Theorem 1.1 and the MAX_MISS algorithm to determine sensor placement and to calculate the miss probabilities for all the centers of the squares. The results are shown in Figures 1.19 and 1.20 for both sensor detection models. They indicate that the miss probabilities are always less than the threshold m_{th}, thereby ensuring adequate coverage of the non-grid points.

1.4.3 Procedural Description

Note that matrices C_{xy}, M_{xy}, and P_A can all be calculated before the actual execution of the placement algorithms. This is illustrated in Figure 1.21 as the pseudocode for the initialization procedure. The initialization procedure is the algorithm overhead which has a complexity of $O((mn)^2)$, where the dimension of the grid is $m \times n$. Once the initialization is done, we may apply either MIN_MISS or MIN_MISS uncertainty-aware sensor placement algorithm using different values for m_{th} and k_{max}

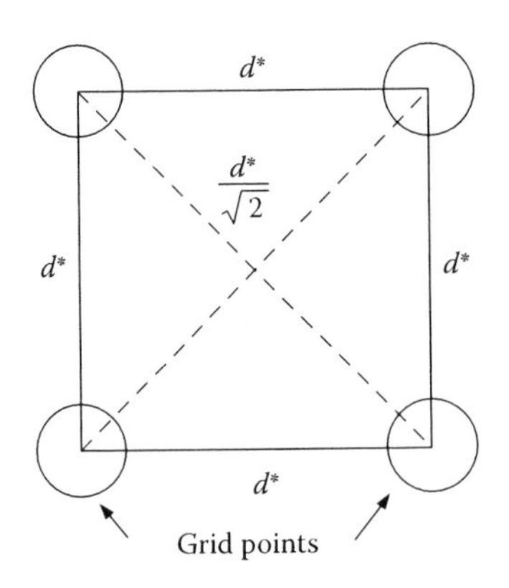

FIGURE 1.18 Coverage of non-grid points.

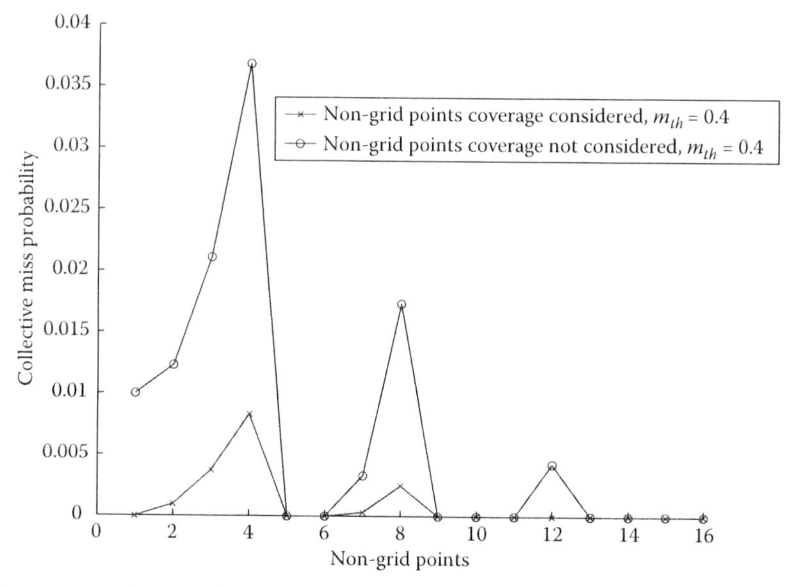

FIGURE 1.19 Coverage of non-grid points for the sensor model given by Equation 1.2.

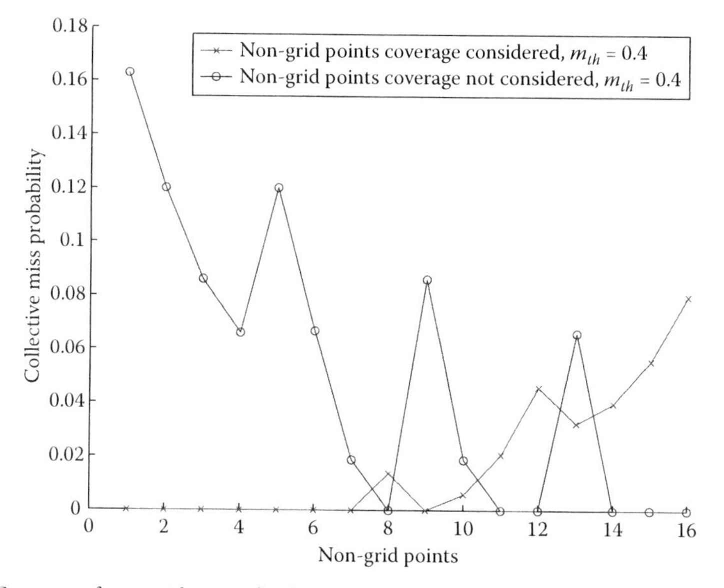

FIGURE 1.20 Coverage of non-grid points for the sensor model given by Equation 1.3.

with the same C_{xy}, M_{xy}, and P_A. Figure 1.22 outlines the main part in pseudocode for the uncertainty-aware sensor placement algorithms. The computational complexity for both MIN_MISS and MAX_MISS is $O(mn)$.

1.4.4 Simulation Results on Uncertainty-Aware Sensor Deployment

Next we present simulation results for the proposed uncertainty-aware sensor placement algorithms MIN_MISS and MAX_MISS using the same testing platform. Note that for practical reasons, we use a truncated Gaussian model because the sensor deviations in a sensor location are

Procedure NDSP_Proc_Init (Grid, σ_x, σ_y, σ, λ, β)

01 /* Build the uncertainty area matrix $P = [p_{xy}(x',y')]A$ */

02 **For** $(x',y') \in A$

03 $$p_{xy}(x',y') = \frac{e^{-\left((x'-x)^2/2\sigma_x^2\right)/\left((y'-y)^2/2\sigma_y^2\right)}}{2\pi\sigma_x\sigma_y}$$

04 **End**

05 /* Build the miss probability matrix for all grid points. */

06 **For** grid point $(x, y) \in$ *Grid*

07 /* Build C_{xy}, C^*_{xy}, and M_{xy} for sensor node at (x, y). */

08 **For** grid point $(i, j) \in$ *Grid*

09 /* Non-grid points coverage are considered based on Theorem 1. */

09 $$d_{ij}(x,y) = \sqrt{(x-i)^2 + (y-j)^2} + \frac{d^*}{\sqrt{2}};$$

10 /* Calculate the grid point coverage probability based on the sensor detection model.*

11 Calculate $C_{ij}(x, y)$:

12 /* Sensor detection model 1. */

13 Model 1: $C_{ij}(x,y) = e^{-\alpha d_{ij}(x,y)}$

14 /* Sensor detection model 2. */

15 $$\text{Model 2: } C_{ij}(x,y) = \begin{cases} 0, & \text{if } r + r_e \le d_{ij}(x,y) \\ e^{-\lambda\alpha^\beta}, & \text{if } |r - d_{ij}(x,y)| < r_e \\ 1, & \text{if } r - r_e \ge d_{ij}(x,y) \end{cases}$$

16 /* Modeling of uncertainty in sensor node locations. */

17 $$c^*_{ij}(x,y) = \frac{\sum_{(x',y')\in A} c_{ij}(x',y')p_{xy}(x',y')}{\sum_{(x',y')\in A} p_{xy}(x',y')}$$

18 /* The miss probability matrix */

19 $m_{ij}(x,y) = 1 - c^*_{ij}(x,y);$

20 **End**

21 /* Use the obstacle mask matrix based on the a priori knowledge about the terrain. */

22 **If** Obstacles exist

23 $C_{xy} = C_{xy}$. *ObstacleMaskMatrix*

24 Revise M_{xy}.

25 **End**

26 **End**

27 /* Initially the overall miss probability matrix is set to *I*. */

28 $M = [m_{ij}]_{m\times n} = [1]_{m\times n}$;

FIGURE 1.21 Initialization pseudocode.

unlikely to span the complete sensor field. Therefore $x' - x$ and $y' - y$ in Equation 1.10 are limited to a certain range, which reflects how large the variation is in the sensor locations during the deployment. The maximum error in *x* direction is denoted as $e^x_{max} = \max(x' - x)$, and the maximum error in *y* direction is denoted as $e^y_{max} = \max(y' - y)$. We then present our simulation results for different sets of parameters in units of grid point where $m = n = 10$, $\sigma_x = \sigma_y = 0.1, 0.32, 1, 2$, and $e^x_{max} = e^y_{max} = 2,3,5$.

Procedure *NDSP_Proc_Main* (type, k_{max}, m_{th}, *Grid*, C_{xy}, M_{xy}, P_A, M)
01 /* Initially no sensors have been placed yet. */
02 Set $S = \phi$; $L_S = \{\phi\}$; $k = |S|$;
03 /* Repeatedly placing sensors until requirement is satisfied.*/
04 **Repeat**
05 /* Evaluate the miss probability due to a sensor at (x, y). */
06 **For** grid point $(x, y) \in Grid$ **And** $(x, y) \notin L_S$
07 Retrieve $M_{xy} = \left[m_{ij}(x,y)\right]_{m\times n}$

$$= \left[1 - c_{ij}^{*}(x,y)\right]_{m\times n}$$

$$= \left[1 - \frac{\sum_{(x',y')\in A} c_{ij}(x',y')p_{xy}(x',y')}{\sum_{(x',y')\in A} p_{xy}(x',y')}\right]_{m\times n}$$

08 /* Miss probability if sensor node is placed at (x, y) */
09 $\tilde{m}(x,y) = \sum_{(i,j)\in Grid} m_{ij}(x,y)m_{ij}$;
10 **End**
11 /* Place sensor node using selected algorithm. */
12 **If** *type* =MIN_MISS
13 Find $(x, y) \in Grid$ and $(x, y) \notin L_s$ such that $\tilde{m}(x,y) = \min\{\tilde{m}(x',y')\}$, $(x',y') \in Grid$;
14 **Else** /* MAX_MISS */
15 Find $(x, y) \in Grid$ and $(x, y) \notin L_s$ such that $\tilde{m}(x,y) = \max\{\tilde{m}(x',y')\}$, $(x',y') \in Grid$;
16 **End**
17 /* Save the information of sensor node just placed. */
18 Set $k = k + 1$;
19 Set $L_S = L_S \cup \{(x, y)\}$
20 Set $S = S \cup \{s_k\}$;
20 /* Update current overall miss probability matrix. */
21 **For** grid point $(i, j) \in Grid$
22 $m_{ij} = m_{ij} . m_{ij}(x, y)$;
23 **End**
24 /* Check if the placement requirement is satisfied. */
25 **Until** $m_{ij} < m_{th}$ for all $(i, j) \in Grid$ **Or** $k > k_{max}$;

FIGURE 1.22 Pseudocode for sensor placement algorithm.

1.4.4.1 Case Study 1

We first consider the probabilistic sensor detection model given by Equation 1.2 with $\alpha = 0.5$. Figure 1.23 presents the result for the two sensor placement algorithms described by Equation 1.15. Figure 1.24 compares the proposed MIN_MISS and MAX_MISS algorithms with the base case where no location errors are considered, that is, an uncertainty-oblivious (UO) strategy is followed by setting $\sigma_x = \sigma_y = 0$. We also consider a random deployment of sensors. The results show that MIN_MISS is nearly as efficient as the base uncertainty-oblivious algorithm, yet it is much more robust. Figure 1.25 presents results for the truncated Gaussian models with different maximum errors. Compared to random deployment, MIN_MISS requires more sensors here but we expect random deployment to perform worse in the presence of obstacles. Figure 1.26 for MIN_MISS and MAX_MISS with coverage obtained without location uncertainty. The results show that the MAX_MISS algorithm, which place more sensors for a given coverage threshold, provides higher overall coverage.

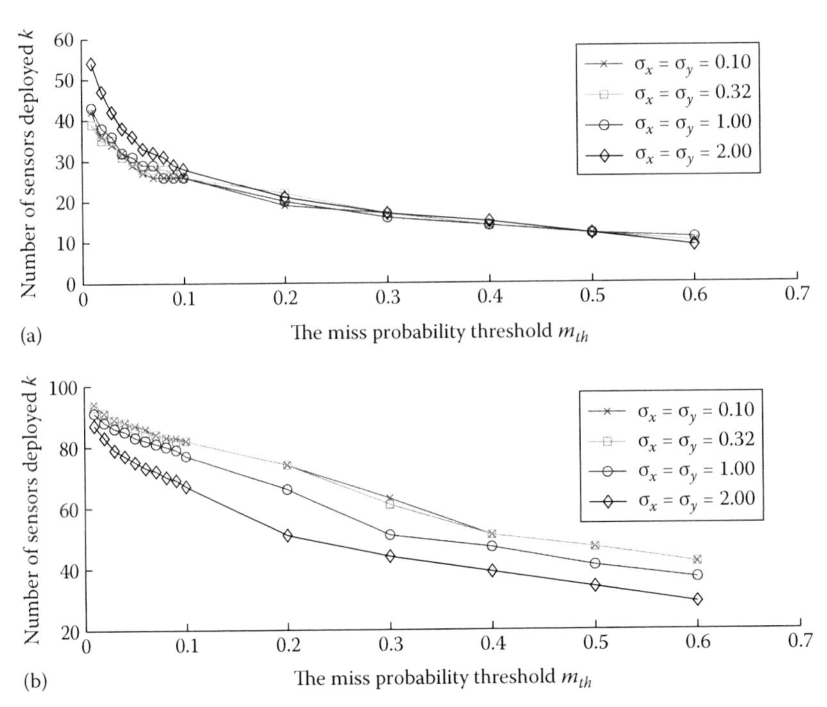

FIGURE 1.23 Number of sensors required as a function of the miss probability threshold with $\alpha = 0.5$, $e^x_{max} = e^y_{max} = 5$, for (a) MIN_MISS (b) MAX_MISS.

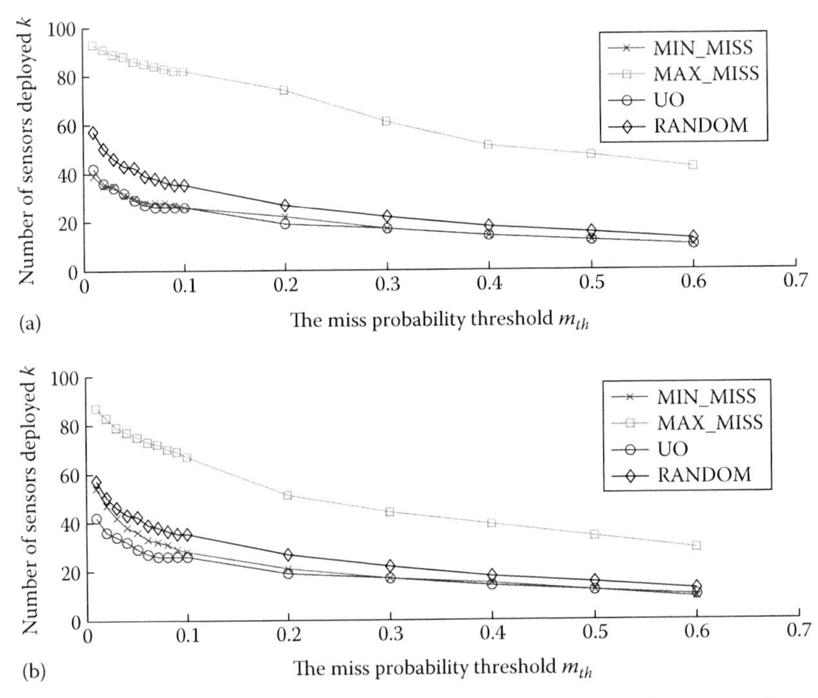

FIGURE 1.24 Number of sensors required for various placement schemes with $\alpha = 0.5$, $e^x_{max} = e^y_{max} = 5$, and (a) $\sigma_x = \sigma_y = 0.32$ (b) $\sigma_x = \sigma_y = 2$.

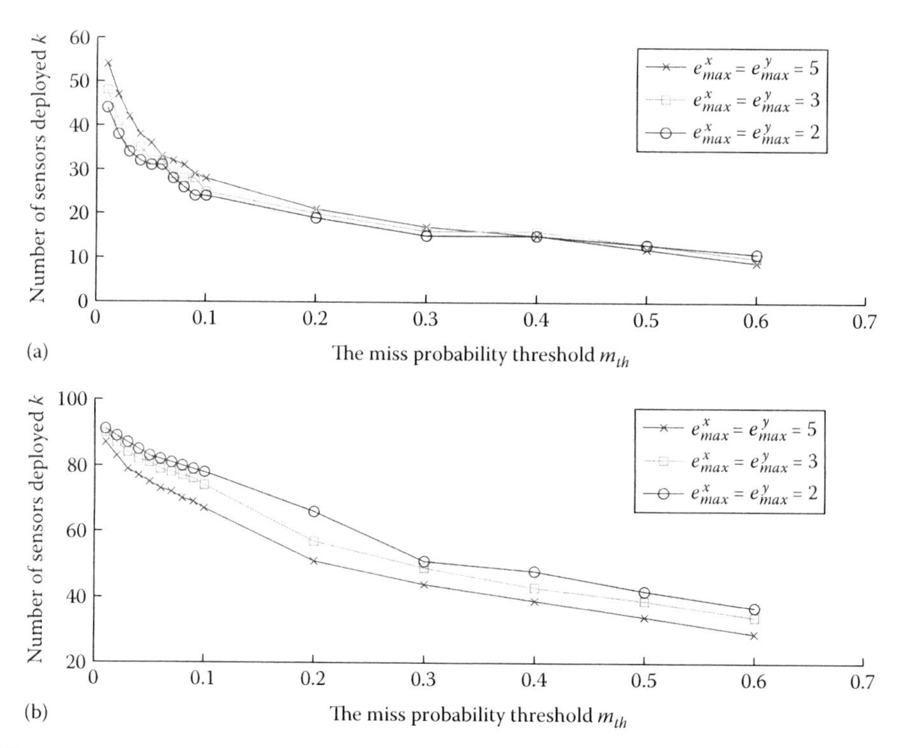

FIGURE 1.25 Comparisons in different truncated Gaussian models with $\alpha = 0.5$, $\sigma_x = \sigma_y = 2$ for (a) MIN_MISS (b) MAX_MISS.

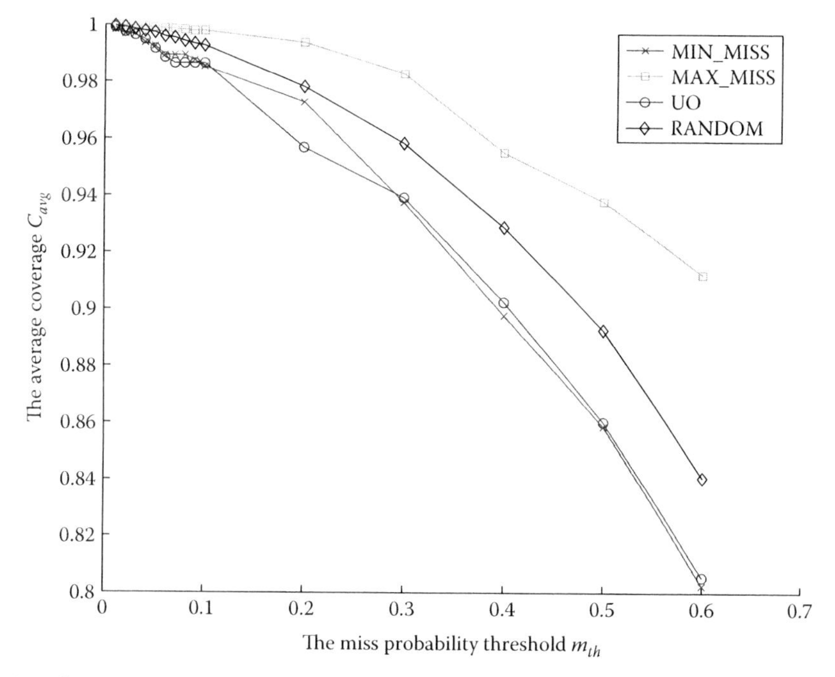

FIGURE 1.26 Comparison in average coverage for various placement schemes with $\alpha = 0.5$, $e^x_{max} = e^y_{max} = 5$, $\sigma_x = \sigma_y = 0.32$.

1.4.4.2 Case Study 2

Next, we consider the probabilistic sensor detection model given by Equation 1.3 with $r = 5$, $r_e = 4$, $\lambda = 0.5$, and $\beta = 0.5$. Figure 1.27 presents the result for the two sensor placement algorithms described by Equation 1.15. Figure 1.28 compares the proposed MIN_MISS and MAX_MISS algorithms with the base case where no location errors are considered. Figure 1.29 presents results for the truncated Gaussian models with different maximum errors. Figure 1.30 compares the coverage based on Equation 1.18 for MIN_MISS and MAX_MISS with coverage obtained without location uncertainty. We notice that due to the different probability values as a reflection of the confidence level in sensor responses from these two different models, the results in sensor placement are also different. Compared with Case study 1, this sensor detection model with the selected model parameters as $\lambda = 0.5$ and $\beta = 0.5$ requires less number of sensor nodes for the same miss probability threshold. Part of the reason is due to the fact that in Equation 1.3, we have full confidence in sensor responses for grid points that are very close to the sensor node, that is, $c_{ij}(x, y) = 1$ if $r - r_e \geq d_{ij}(x, y)$. However, this case study shows that the proposed sensor deployment algorithms do not depend on any specific type of sensor models. The sensor detection model can be viewed as a plug-in module when different types of sensors are encountered in applying the deployment algorithms.

1.4.4.3 Case Study 3

Next we consider a terrain model with the existence of obstacles. We have manually placed one obstacle that occupies grid points (7,3), (7,4), and another obstacle that occupies grid points (3,5), (4,5), (5,5). They are marked as "Obstacle" in Figure 1.3, which gives the layout of the setup for this case study. We have evaluated the proposed algorithms on the sensor detection model in Case study 2, which is given by Equation 1.3 with the same model parameters as $r = 5$, $r_e = 4$, $\lambda = 0.5$, and $\beta = 0.5$. Figure 1.31 presents results for the truncated Gaussian models with different maximum errors. Figure 1.32 compares the coverage based on Equation 1.18 for MIN_MISS and MAX_MISS with coverage obtained without location uncertainty. It is obvious that because of the existence of obstacles, the actual range of sensor detection due to the line-of-sight principle. Therefore, the reduction in sensor detection range causes an increase in the number of sensors required for the same miss probability threshold, as shown in Figures 1.31 and 1.32.

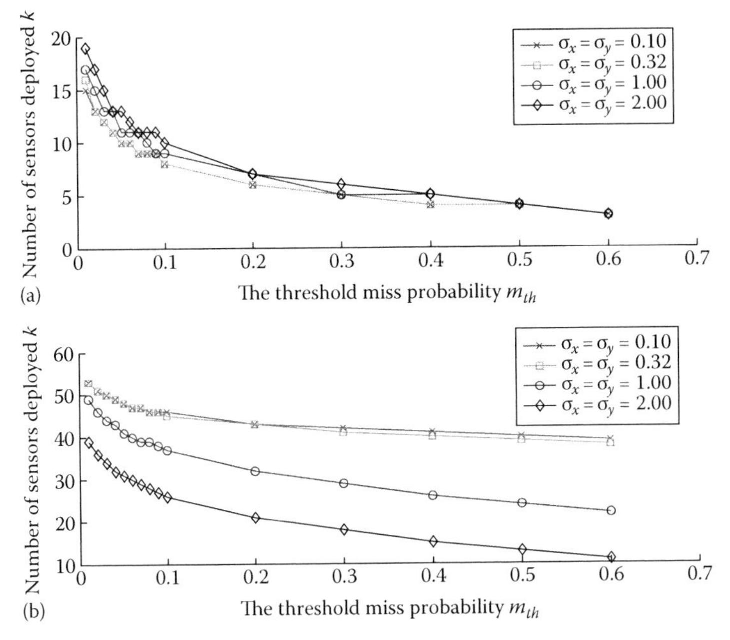

FIGURE 1.27 Number of sensors required as a function of the miss probability threshold with $\alpha = 0.5$, $e^x_{max} = e^y_{max} = 5$, for (a) MIN_MISS (b) MAX_MISS.

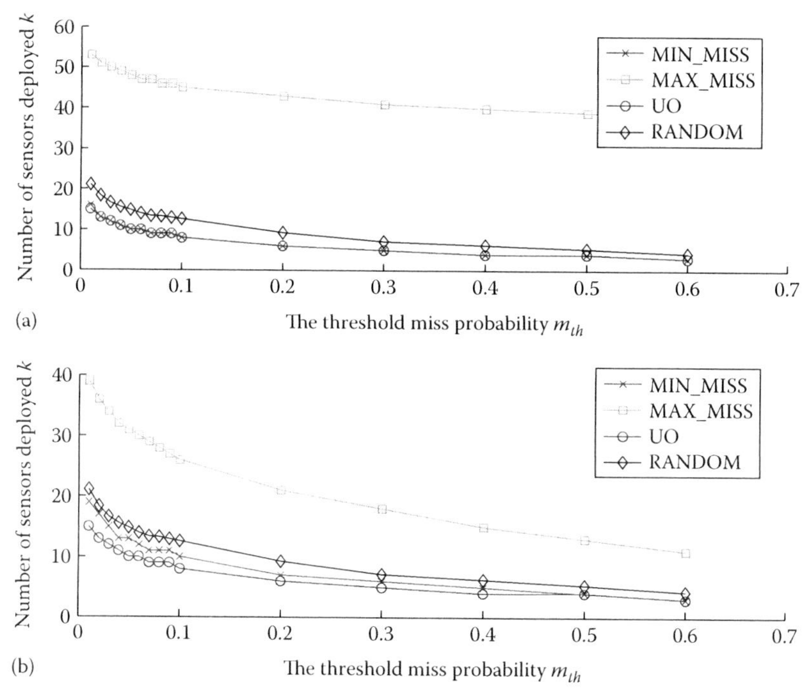

FIGURE 1.28 Number of sensors required for various placement schemes with α = 0.5, $e^x_{max} = e^y_{max} = 5$, and (a) $\sigma_x = \sigma_y = 0.32$ (b) $\sigma_x = \sigma_y = 2$.

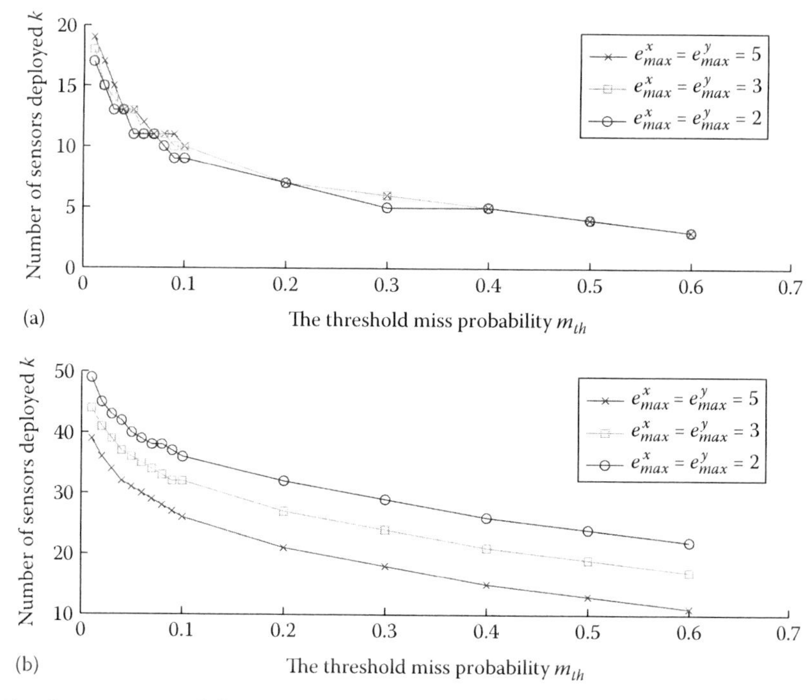

FIGURE 1.29 Comparisons in different truncated Gaussian models with α = 0.5, $\sigma_x = \sigma_y = 2$ for (a) MIN_MISS (b) MAX_MISS.

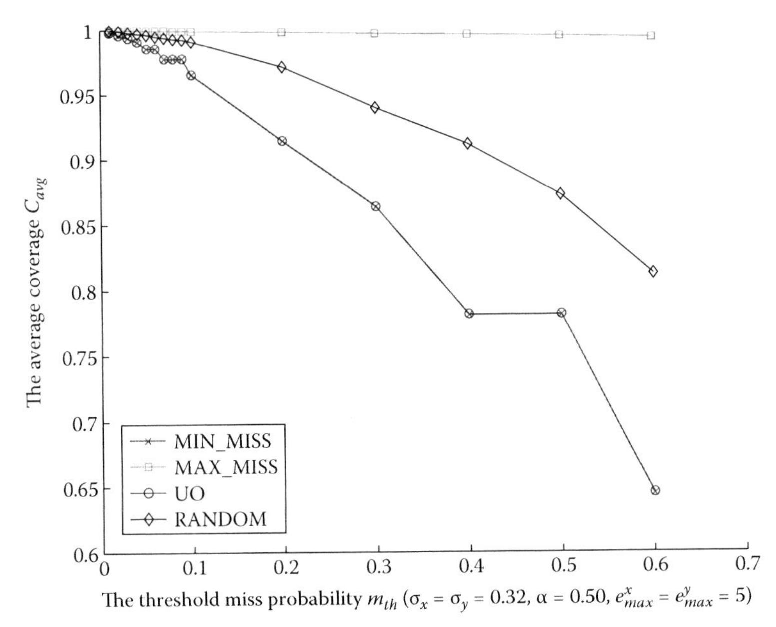

FIGURE 1.30 Comparison in average coverage for various placement schemes with $\alpha = 0.5$, $e^x_{max} = e^y_{max} = 5$, $\sigma_x = \sigma_y = 0.32$.

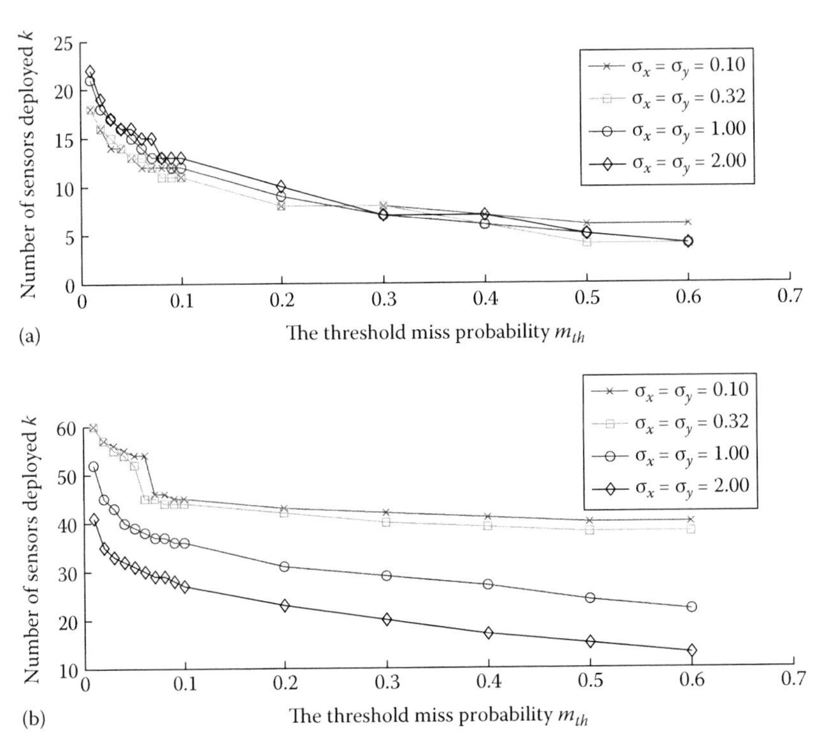

FIGURE 1.31 Number of sensors required as a function of the miss probability threshold in presence of obstacles with $\alpha = 0.5$, $e^x_{max} = e^y_{max} = 5$ for (a) MIN_MISS (b) MAX_MISS.

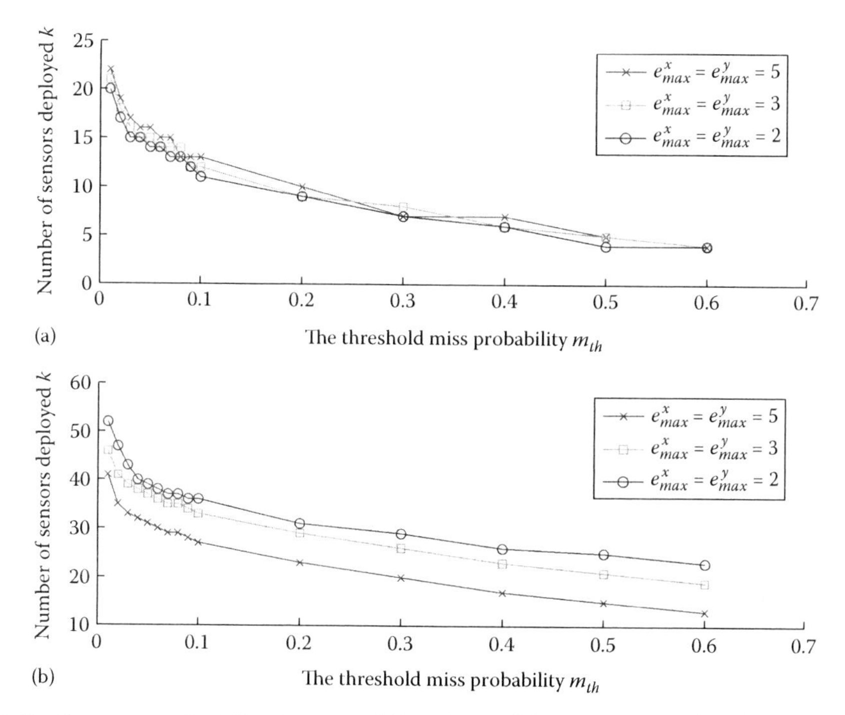

FIGURE 1.32 Comparisons in different truncated Gaussian models in presence of obstacles with $\alpha = 0.5$, $\sigma_x = \sigma_y = 2$ for (a) MIN_MISS (b) MAX_MISS.

1.5 Conclusions

In this chapter, we have discussed two important aspects in sensor node deployment for wireless sensor networks. The proposed VFA algorithm introduced in Section 1.3 improves the sensor field coverage considerably compared to random sensor placement. The sensor placement strategy is centralized at the cluster level since every cluster head makes redeployment decisions for the nodes in its cluster. Nevertheless, the clusters make deployment decisions independently, hence there is a considerable degree of decentralization in the overall sensor deployment. The virtual force in the VFA algorithm is calculated with a grid point being the location indicator and the distance between two grid points being a measure of distance. Furthermore, in our simulations, the preferential areas and the obstacles are both modeled as rectangles. The VFA algorithm however is also applicable for alternative location indicators, distance measures, and models of preferential areas and obstacles. Hence the VFA algorithm can be easily extended to heterogeneous sensors, where sensors may differ from each other in their detection modalities and parameters.

In Section 1.4, we have formulated an optimization problem on uncertainty-aware sensor placement. A minimum number of sensors are deployed to provide sufficient grid coverage of the sensor field though the exact sensor locations are not known. The sensor location has been modeled as a random variable with a Gaussian probability distribution. We have presented two polynomial-time algorithms to optimize the number of sensors and determine their placement in an uncertainty-aware manner. The proposed algorithms address coverage optimization under constraints of imprecise detections and terrain properties.

References

1. J. Agre and L. Clare, An integrated architecture for cooperative sensing networks, *IEEE Computer*, 33, 106–108, 2000.
2. I. F. Akyildiz, W. Su, Y. Sankarasubramaniam, and E. Cayirci, A survey on sensor networks, *IEEE Communications Magazine*, 102–114, August 2002.

3. M. Bhardwaj, T. Garnett, and A. P. Chandrakasan, Upper bounds on the lifetime of sensor networks, *Proceedings of IEEE International Conference on Communications*, Helsinki, Finland, Vol. 3, pp. 785–790, June 2001.
4. M. Bhardwaj and A. P. Chandrakasan, Bounding the lifetime of sensor networks via optimal role assignments, *Proceedings of IEEE Infocom Conference*, New York, pp. 1587–1596, 2002.
5. N. Bulusu, J. Heidemann, and D. Estrin, GPS-less low-cost outdoor localization for very small devices, *IEEE Personal Communication Magazine*, 7(5), 28–34, 2000.
6. N. Bulusu, J. Heidemann, and D. Estrin, Adaptive beacon placement, *Proceedings of International Conference on Distributed Computing Systems*, Phoenix, AZ, pp. 489–498, April 2001.
7. K. Chakrabarty, S. S. Iyengar, H. Qi, and E. Cho, Grid coverage for surveillance and target location in distributed sensor networks, *IEEE Transactions on Computers*, 51, 1448–1453, 2002.
8. K. Chakrabarty, S. S. Iyengar, H. Qi, and E. Cho, Coding theory framework for target location in distributed sensor networks, *Proceedings of International Symposium on Information Technology: Coding and Computing*, Las Vegas, NV, pp. 130–134, 2001.
9. G. Chartrand, *Introductory Graph Theory*, New York: Dover, 1985.
10. M. R. Clark, G. T. Anderson, and R. D. Skinner, Coupled oscillator control of autonomous mobile robots, *Autonomous Robots*, 9(2), 189–198, 2000.
11. T. Clouqueur, V. Phipatanasuphorn, P. Ramanathan, and K. K. Saluja, Sensor deployment strategy for target detection, *Proceedings of First ACM International Workshop on Wireless Sensor Networks and Applications*, New York, pp. 42–48, September 2002.
12. S. S. Dhillon, K. Chakrabarty, and S. S. Iyengar, Sensor placement for grid coverage under imprecise detections, *Proceedings of International Conference on Information Fusion*, Annapolis, MD, pp. 1581–1587, 2002.
13. S. S. Dhillon and K. Chakrabarty, Sensor placement for effective coverage and surveillance in distributed sensor networks, *Proceedings of IEEE Wireless Communications and Networking Conference*, New Orleans, LA, paper ID: TS49-2, 2003.
14. A. Elfes, Occupancy grids: A stochastic spatial representation for active robot perception, *Proceedings of Sixth Conference on Uncertainty in AI*, Cambridge, MA, pp. 60–70, 1990.
15. J. Elson and D. Estrin, Time synchronization for wireless sensor networks, *Proceedings of International Parallel and Distributed Processing Symposium*, San Francisco, CA, pp. 2033–2036, 2001.
16. D. Estrin, R. Govindan, J. Heidemann, and S. Kumar, Next century challenges: Scalable coordination in sensor networks, *Proceedings of IEEE/ACM MobiCom Conference*, Seattle, WA, pp. 263–270, 1999.
17. M. R. Garey and D. S. Johnson, *Computers and Intractability*, San Francisco, CA: W. H. Freeman & Company, 1979.
18. J. Heidemann and N. Bulusu, Using geospatial information in sensor networks, *Proceedings of CSTB Workshop on Intersection of Geospatial Information and Information Technology*, 2001.
19. W. B. Heinzelman, J. Kulik, and H. Balakrishnan, Adaptive protocols for information dissemination in wireless sensor networks, *Proceedings of IEEE/ACM MobiCom Conference*, Seattle, WA, pp. 174–185, 1999.
20. W. B. Heinzelman, A. Chandrakasan, and H. Balakrishnan, Energy-efficient communication protocol for wireless micro sensor networks, *Proceedings of 33rd Annual Hawaii International Conference on System Sciences*, Maui, HI, pp. 3005–3014, 2000.
21. N. Heo and P. K. Varshney, A distributed self spreading algorithm for mobile wireless sensor networks, *Proceedings of IEEE Wireless Communications and Networking Conference*, New Orleans, LA, paper ID: TS48-4, 2003.
22. J. Hill, R. Szewczyk, A. Woo, S. Hollar, D. Culler, and K. Piste, System architecture directions for network sensors, *Proceedings of International Conference on Architectural Support for Programming Languages and Operating Systems*, Cambridge, MA, 2000.

23. A. Howard, M. J. Matarić, and G. S. Sukhatme, Mobile sensor network deployment using potential field: A distributed scalable solution to the area coverage problem, *Proceedings of Sixth International Conference on Distributed Autonomous Robotic Systems*, Fukuoka, Japan, pp. 299–308, 2002.
24. C. Intanagonwiwat, R. Govindan, and D. Estrin, Directed diffusion: A scalable and robust communication paradigm for sensor networks, *Proceedings of IEEE/ACM MobiCom Conference*, Boston MA, 2000.
25. T. Kasetkasem and P. K. Varshney, Communication structure planning for multisensor detection systems, *Proceedings of IEE Conference on Radar, Sonar and Navigation*, Vol. 148, pp. 2–8, 2001.
26. E. Lawler, J. Lenstr, A. R. Kan, and D. Shmoys, *The Traveling Salesman Problem: A Guided Tour of Combinatorial Optimization*, New York: Wiley, 1985.
27. S. Lindsey and C. S. Raghavendra, PEGASIS: Power-efficient gathering in sensor information systems, *Proceedings of IEEE Aerospace Conference*, Big Sky, MT, Vol. 3, pp. 1125–1130, 2002.
28. J. Liu, P. Cheung, L. Guibas, and F. Zhao, A dual-space approach to tracking and sensor management in wireless sensor networks, *Proceedings of First ACM International Workshop on Wireless Sensor Networks and Applications*, Atlanta, GA, pp. 131–139, September 2002.
29. M. Locateli and U. Raber, Packing equal circles in a square: A deterministic global optimization approach, *Discrete Applied Mathematics*, 122, 139–166, 2002.
30. A. Manjeshwar and D. P. Agrawal, TEEN: A routing protocol for enhanced efficiency in wireless sensor networks, *Proceedings of 15th International Parallel and Distributed Processing Symposium*, San Francisco, CA, pp. 2009–2015, 2001.
31. S. Meguerdichian, F. Koushanfar, M. Potkonjak, and M. B. Srivastava, Coverage problems in wireless ad-hoc sensor networks, *Proceedings of IEEE Infocom Conference*, Anchorage, AK, Vol. 3, pp. 1380–1387, 2001.
32. S. Meguerdichian, F. Koushanfar, G. Qu, and M. Potkonjak, Exposure in wireless ad-hoc sensor networks, *Proceedings of Mobicom Conference*, Rome, Italy, pp. 139–150, July 2001.
33. S. A. Musman, P. E. Lehner, and C. Elsaesser, Sensor planning for elusive targets, *Journal of Computer and Mathematical Modeling*, 25, 103–115, 1997.
34. J. O'Rourke, *Art Gallery Theorems and Algorithms*, New York: Oxford University Press, 1987.
35. D. E. Penny, The automatic management of multi-sensor systems, *Proceedings of International Conference on Information Fusion*, 1998.
36. G. J. Pottie and W. J. Kaiser, Wireless sensor networks, *Communications of the ACM*, 43, 51–58, 2000.
37. S. Slijepcevic and M. Potkonjak, Power efficient organization of wireless sensor networks, *Proceedings of IEEE International Conference on Communications*, Helsinki, Finland, pp. 472–476, June 11–14, 2001.
38. C. Srisathapornphat, C. Jaikaeo, and C.-C. Shen, Sensor information networking architecture, *International Workshops on Parallel Processing*, Palo Alto, CA, pp. 23–30, 2000.
39. S. Tilak, N. B. Abu-Ghazaleh, and W. B. Heinzelman, A taxonomy of wireless micro-sensor network models, *ACM Mobile Computing and Communications Review*, 6(2), 2002.

2

Deployment of Sensors: An Overview

S. Sitharama Iyengar
Florida International University

Ankit Tandon
Microsoft Research

Qishi Wu
Louisiana State University and Oak Ridge National Laboratory

Eungchun Cho
Kentucky State University

Nageswara S.V. Rao
Oak Ridge National Laboratory

Vijay K. Vaishnavi
Georgia State University

2.1 Introduction

2.1.1 What Is a Sensor Network?

A distributed sensor network (DSN) is a collection of a large number of heterogeneous intelligent sensors distributed logically, spatially, or geographically over an environment and connected through a high-speed network. The sensors may be cameras as vision sensors, microphones as audio sensors, ultrasonic sensors, infrared sensors, humidity sensors, light sensors, temperature sensors, pressure/force sensors, vibration sensors, radioactivity sensors, seismic sensors, etc. The sensors continuously monitor and collect measurements of respective data from their environment. The collected data are processed by an associated processing element that then transmits it through an interconnected communication network. The information that is gathered from all parts of the sensor network is then integrated using some data-fusion strategy. This integrated information is then useful to derive appropriate inferences about the environment where the sensors are deployed.

2.1.2 Example

With the emergence of high-speed networks and with their increased computational capability, DSNs have a wide range of real-time applications in aerospace, automation, defense, medical imaging, robotics, weather prediction, etc. To elucidate, let us consider sensors spread in a large geographical territory collecting data on various parameters, like temperature, atmospheric pressure, wind velocity, etc. The data from these sensors are not as useful when studied individually; but, when integrated, they give the picture of a large area. Changes in the data across time for the entire region can be used in predicting the weather at a particular location.

DSNs are a key part of the surveillance and reconnaissance infrastructure in modern battle spaces. DSNs offer several important benefits, such as ease of deployment, responsiveness to battlefield situations, survivability, agility, and easy sustainability. These benefits make DSNs a lethal weapon for any army, providing it with the high-quality surveillance and reconnaissance data necessary for any combat operation [1].

2.1.3 Computational Issues

Coordinated target detection, surveillance, and localization require efficient and optimal solutions to sensor deployment problems (SDPs), and have attracted a great deal of attention from several researchers. Sensors must be suitably deployed to achieve the maximum detection probability in a given region while keeping the cost within a specified budget. Recently, SDPs have been studied in a variety of contexts. In the adaptive beacon placement, the strategy is to place a large number of sensors and then shut some of them down based on their localization information. Most of the approaches are based on sensor devices with deterministic coverage capability. In reality, the sensor coverage is not only dependent on the geometric distance from the sensor, but also on other factors, such as environmental conditions and device noise. The deterministic models do not adequately capture the tradeoffs between sensor network reliability and cost. Thus, next-generation sensor networks must go beyond the deterministic coverage techniques to perform the assigned tasks, such as online tracking/monitoring in unstructured environments.

In reality, the probability of successful detection decreases in some way as the target moves further away from the sensor, because of less received power, more noise, and environmental interference. Therefore, the sensor detection is "probabilistic." The sensor deployment is a complex task in DSNs because of factors such as different sensor types and detection ranges, sensor deployment and operational costs, and local and global coverage probabilities. Essentially, the sensor deployment is an optimization problem, which often belongs to the category of multi-dimensional and nonlinear problems with complicated constraints. When the deployment locations are restricted to (discrete) grid points, this problem becomes a combinatorial optimization problem but is still computationally very difficult. In particular, this problem contains a considerable number of local maxima, and it is very difficult for the conventional optimization methods to obtain its global maximum.

Distributed, real-time sensor networks are essential for effective surveillance in a digitized battlefield and environmental monitoring. There are several underlying challenges in the design of a sensor network. A key issue is the layout or distribution of sensors in the environment. The number, type, location, and density of sensors determine the layout of a sensor network. An intelligent placement of sensors can enhance the performance of the system significantly. Some redundancy is also needed for error detection and correction caused by faulty sensors and an unreliable communication network. At the same time, a large number of sensors correspond to higher deployment costs, the need of higher bandwidth, increased collisions in relaying messages, higher energy consumption, and more time-consuming algorithms for data fusion.

Usually, sensors are deployed in widespread hazardous, unreliable or possibly even adversarial environments, and it is essential that they do not require human attention very often. It is necessary that sensors are self-aware, self-configurable, autonomous, and self-powered. They must have enough energy reserves to work for a long period of time or they should be able to recharge themselves. Power in each sensor is finite and precious, and it is extremely essential to conserve it.

Sensors typically communicate through wireless networks, where bandwidth is significantly lower than the wired channels. Wireless networks are more unreliable and data-faulty; therefore, there is a need for robust, fault-tolerant routing and data-fusion algorithms. It is of the utmost importance to use techniques that increase the efficiency of data communication, thus reducing the number of overall bits transmitted and also reducing the number of unnecessary collisions. It has been found that, typically, it requires 100–1000 times more energy to transmit a bit than to execute an instruction, which means that it is beneficial to compress the data before transmitting it. Hence, it is essential to minimize data transfer in the sensor network to make it more energy efficient.

In real-time medical and military applications, it is sometimes essential to have an estimate of the message delay between two nodes of a sensor network. The current algorithms to compute sensor message delay are computationally very expensive and pose a challenge for further study.

2.2 Importance of Sensor Deployment

Sensor placement directly influences resource management and the type of back-end processing and exploitation that must be carried out with the sensed data in a DSN. A key challenge in sensor resource management is to determine a sensor field architecture that optimizes cost, and provides high sensor coverage, resilience to sensor failures, and appropriate computation/communication tradeoffs. Intelligent sensor placement facilitates unified design and operation of sensor/exploitation systems, and decreases the need for excessive network communication for surveillance, target location, and tracking. Therefore, sensor placement forms the essential "glue" between front-end sensing and back-end exploitation.

In a resource-bounded framework of a sensor network, it is essential to optimize the deployment of sensors and their transmission. Given a surveillance area, the most important challenge is to come up with the architecture of a "minimalistic sensor network" that requires the least number of sensors (with the lowest deployment costs) and has maximum coverage. It is also important that the sensors are deployed in such a manner that they transmit/report the minimum amount of sensed data. The ensemble of this data must contain sufficient information for the data-processing center to subsequently derive appropriate inferences and query a small set of sensors for detailed information.

In addition to the above, sensor networks must take into account the nature of the terrain of the environment where they would be deployed. In practical applications, sensors may be placed in a terrain that has obstacles, such as buildings and trees that block the line of vision of infrared sensors. Uneven surfaces and elevations of a hilly terrain may make communication impossible. In battlefields, radio jamming may make communication among sensors difficult and unreliable. Thus, while deploying the sensors, it is necessary to take the above factors into account and to estimate the need for redundancy of sensors due to the likelihood of sensor failures, and the extra power needed to transmit between deployed sensors and cluster heads.

In the case of mobile sensors, the sensor fields are constructed such that each sensor is repelled by both obstacles and by other sensors, thereby forcing the network to spread itself through the environment. However, most of the practical applications, like environmental monitoring, require static sensors, and the above scenario of self-deployment does not provide a solution.

Some applications of sensor networks require target detection and localization. In such cases, deployment of sensors is the key aspect. In order to achieve target localization in a given area, the sensors have to be placed in such a manner that each point is sensed by a unique set of sensors. Using the set of sensors that sense the target, an algorithm can predict or pinpoint the location of the target.

The above issues clearly prove that sensor placement is one of the most key aspects of any DSN architecture, and efficient algorithms for computing the best layout of sensors in a given area need to be researched. Using the concept of Eisenstein integers, one such algorithm that computes efficient placement of sensors of a DSN covering a bounded region on the plane is presented next. The number of sensors required in the DSN based on Eisenstein integers is about $4/3\sqrt{3} \approx 0.77$ of the number of the sensors required by the traditional rectangular grid-point-based networks covering the same amount of area.

2.3 Placement of Sensors in a DSN Using Eisenstein Integers

2.3.1 Introduction

A DSN covering a region in the plane $\mathbb{R}^2$ such that each lattice point (grid point) can be detected by a unique set of responding sensors is convenient for locating stationary or mobile targets in the region.

In such sensor networks, each set of responding sensors uniquely identifies a grid point corresponding to the location of the target [2]. Moreover, the location of the target is easily computed from the set of responding sensors, the locations of which are fixed and known. For simplicity, we assume that both the sensors and the targets are located only on lattice points. More realistically, we may require only sensors to be placed at lattice points and targets are located by finding the nearest lattice points. We consider the ring of Eisenstein integers, which have direct applications to the design of a DSN.

2.3.2 Eisenstein Integers

Gaussian integers are complex numbers of the form $a + bi$, where a and b are integers. Gaussian integers form a standard regular rectangular grid on a plane, which we will call a Gaussian grid. Let G be the set of Gaussian integers

$$G = \{a + bi : a, b \in \mathbb{Z}\}$$

G is closed under addition, subtraction, and multiplication:

$$(a + bi) \pm (c + di) = (a \pm c) + (b \pm d)i$$

$$(a + bi)(c + di) = (ac \pm bd) + (ad \pm bc)i$$

In other words, G is invariant under the addition (translation) and multiplication (dilation) by any Gaussian integer and G is a subring of $\mathbb{C}$, the field of complex numbers. We may consider any point (x, y) in the two-dimensional real plane $\mathbb{R}^2$ as a complex number $x + iy$, that is, as a point in the complex plane $\mathbb{C}$. G is the set of all integer lattice points of $\mathbb{R}^2$.

Recall that i is the primary fourth root of 1, that is, $i^4 = 1$ and any complex number z with $z^4 = 1$ is i^k for some $k \in \mathbb{Z}$. In fact, z is either 1, −1, i or $-i$. If i is replaced by ω, the primary third root of 1, then we get Eisenstein integers. The primary root of ω is of the form

$$e^{2\pi i/3} = \frac{\cos 2\pi}{3} + \frac{i \sin 2\pi}{3} = -\frac{1}{2} + \frac{\sqrt{3}}{2i}$$

and satisfies $\omega^2 + \omega + 1 = 0$. This means that any integer power of ω can be represented as a linear combination of 1 and ω.

Let E be the set of Eisenstein integers

$$E = \{a + b\omega : a, b \in \mathbb{Z}\}$$

E is also invariant under the translation and dilation by any Eisenstein integer and E forms a subring of $\mathbb{C}$, since

$$(a + b\omega) \pm (c + d\omega) = (a \pm c) + (b \pm d)\omega$$

$$(a + b\omega)(c + d\omega) = (a - bd) + (ad + bc - bd)\omega$$

The three solutions of $z^3 = 1$, given by 1, ω, and $\omega^2 f = -1 - \omega$ form an equilateral triangle. The Eisenstein integers ± 1, $\pm\omega$, $\pm(1 + \omega)$ are called the Eisenstein units. Eisenstein units form a regular hexagon centered at the origin [3]. As G yields a tessellation of $\mathbb{R}^2$ by squares, E forms a tessellation of $\mathbb{R}^2$ by equilateral triangles and its dual forms a tessellation of $\mathbb{R}^2$ by regular hexagons. The main theorem of this paper is as follows. A DSN whose sensors (with unit range) are placed at Eisenstein integers of the form $m + n\omega$ with $m + n \equiv 0 \bmod 3$ detects the target on Eisenstein integers uniquely. Each location at an Eisenstein integer $a + b\omega$ is detected by one sensor located at itself, by the set of three sensors placed at $\{(a + 1) + b\omega, a + (b + 1)\omega, (a - 1) + (b - 1)\omega\}$, or by the set of three sensors placed at $\{(a-1)+b\omega, (a+1)+(b+1)\omega, a+(b-1)\omega\}$. In practical applications, the location of the target is easily approximated either by the location of the sensor itself (if there is only one responding sensor) or simply the average $(a_1+a_2+a_3)/3+(b_1+b_2+b_3)\omega/3$ of the three Eisenstein integers $a_i+b_i\omega$. The proof of the theorem will be given after more mathematical background on Eisenstein integers and tessellation is given.

Six equilateral triangles sharing a common vertex form a regular hexagon, which generates a hexagonal tessellation of $\mathbb{R}^2$. E is the subring of $\mathbb{C}$, which means E an additive subgroup of $\mathbb{C}$, is closed under complex multiplication satisfying the usual associative, commutative, and distributive properties. ω generates a multiplicative subgroup $\{1, \omega, \omega^2\}$ of the circle, called a cyclotomic (circle cutting) subgroup of order 3. Eisenstein units $1, -\omega^2, \omega, -1, \omega^2, -\omega$ form a cyclotomic subgroup of order 6 (and a regular hexagon centered at the origin).

Each closed unit disk centered at a Gaussian integer $m+ni$ contains four other Gaussian integers $(m\pm 1)+n\omega$, $m+(n\pm 1)\omega$, and $(m-1)+(n\pm 1)\omega$. Any point in $\mathbb{R}^2$ is within a $1/\sqrt{2}$ radius of a Gaussian integer and within a $1/\sqrt{3}$ radius of an Eisenstein integer. Let $N(e)$ be the neighborhood of $e\in E$ in $\mathbb{R}^2$, defined as the set of all points for which the closest point in E is e, that is, the set of all points which are not farther from e than from any other points in E

$$N(e)=\{x\in\mathbb{R}^2 : \|x-e\|\,\Omega\,\|x-f\|\ \forall f\in E\}$$

$N(e)$ is the regular hexagon centered at e with the edge length $1/\sqrt{3}$ whose vertices are the centers of equilateral triangles of Eisenstein tessellation, and the area of $N(e)$ is $1/\sqrt{3}/2$. The regular hexagons $N(e)$ for $e\in E$ form a tessellation of $\mathbb{R}^2$. Each $N(e)$ contains exactly one Eisenstein integer (namely, e). In this sense, the density of E in $\mathbb{R}^2$ is $2/\sqrt{3}$, the inverse of the area of $N(e)$. A similar argument shows $N(g)$, the set of all points in $\mathbb{R}^2$ for which the closest point in G is g, is a square centered at g with unit side whose vertices are centers of the Gaussian square tessellation. The density of Gaussian integers G is unity, which is lower than the density of E.

2.3.3 Main Theorem

Now we consider a DSN covering the complex plane such that each Eisenstein integer (grid point) can be detected by a unique set of responding sensors. That is, a DSN with the property that for each set of responding sensors there is a unique Eisenstein integer corresponding to the location of the target. Moreover, the location of the target is easily computed from the set of responding sensors that are fixed and known points at Eisenstein integers.

A DSN whose sensors (with unit range) are placed at Eisenstein integers of the form $m+n\omega$ with $m+n\equiv 0 \bmod 3$ detects each Eisenstein integer uniquely. Each Eisenstein integer $a+b\omega$ is detected by one sensor located at itself, by a set of three sensors placed at $\{(a+1)+b\omega, a+(b+1)\omega, (a-1)+(b-1)\omega\}$, or by the set of three sensors placed at $\{(a-1)+b\omega, (a+1)+(b+1)\omega, a+(b-1)\omega\}$.

Proof. The minimum distance between distinct points in E is unity and a sensor placed at a point $e=a+b\omega\in E$ detects six neighbor points in E in addition to itself. The six neighbors are $(a\pm 1)+b\omega$, $a+(b\pm 1)\omega$, and $(a-1)+(b\pm 1)\omega$, which form a regular hexagon centered at e. Consider the hexagonal tessellation of $\mathbb{R}^2$ generated by the regular unit hexagon with vertices at ± 1, $\pm\omega$, and $-1\pm\omega$ with center at $e=0$, the origin of the complex plane. Let V be the set of all vertices of the tessellation and C be the set of all centers of the hexagons of the tessellation. We note $E=V\cup C$ and $V\cap C=\emptyset$, that is, every Eisenstein integer is either a vertex of the tessellation or the center of the hexagons. The minimum distance between distinct points in C is $1/\sqrt{3}$ and every point in C is of the form $e=a+b\omega$ with $a+b\equiv 0 \bmod 3$. For example, $0, 1+2\omega, 2+\omega, 1-\omega, \ldots$. For each v in V, there exist exactly three points c_1, c_2, and c_3 in C such that $\text{dist}(v, c_i)=1$ and $(c_1+c_2+c_3)/3$, with $\text{dist}(v, c_i)=1$. This means that if the sensors are placed at the points in C, the centers of the hexagons tessellating the plane, then every point e in E is detected either by a single sensor (when e belongs to C) or by a set of three sensors (when e belongs to V).

Remark. Hexagonal tessellation is the most efficient tessellation (there are only two more tessellations of a plane by regular polygons: square tessellation and triangular tessellation) in the sense that the vertices belong to exactly three neighboring hexagons (square tessellation requires four and triangular tessellation six) and each set of three neighboring hexagons has only one vertex in common.

2.3.4 Conclusion

In practical applications, the location of a target is easily approximated with such sensor networks. Assuming the targets are located on grid points only, the target location is either the position of the sensor itself (if there is only one responding sensor), or simply the average $(a_1 + a_2 + a_3)/3 + (b_1 + b_2 + b_3)\omega/3$ of the three Eisenstein integers $a_i + b_i\omega$. More generally, the target location is approximated either by the position of the sensor itself (if there is only one responding sensor), or by the average $(a_1 + a_2)/2 + (b_1 + b_2)\omega/2$ of the two Eisenstein integers (if there are two responding sensors) or the average $(a_1 + a_2 + a_3)/3 + (b_1 + b_2 + b_3)\omega/3$ of the three Eisenstein integers $a_i + b_i\omega$ (if there are three responding sensors).

A similar result follows for the sensor network based on a Gaussian lattice whose sensors are placed at Gaussian integers $a + bi$, where $a + b = 0 \bmod 2$. The minimum distance between sensors in this network is $\sqrt{2}$. A target at a Gaussian integer $a + bi$ with $a + b \equiv 0 \bmod 2$ is detected by the sensor placed on it. Otherwise, that is $a + b = 1 \bmod 2$, the target is detected by four sensors placed at the four neighboring Gaussian integers $(a \pm 1) + bi$ and $a + (b \pm 1)i$. The average density of the sensors in the Gaussian-integer-based network is about 1/2, whereas the average density for the network based on the Eisenstein integer is about $2/3\sqrt{3} \approx 0.38$. In other words, the Eisenstein network requires less sensors (about $4/3\sqrt{3} \approx 0.77$) that the former.

2.4 Complexity Analysis of Efficient Placement of Sensors on Planar Grid

One of the essential tasks in the design of distributed sensor systems is the deployment of sensors for an optimal surveillance of a target region while ensuring robustness and reliability. Those sensors with probabilistic detection capabilities with different costs are considered here. An SDP for a planar grid region is formulated as a combinatorial optimization problem to maximize the overall detection probability within a given deployment cost. This sensor placement problem is shown to be NP-complete, and an approximate solution is proposed based on the genetic algorithm method. The solution is obtained by the specific choices of genetic encoding, fitness function, and genetic operators (such as crossover, mutation, translocation, etc.) for this problem. Simulation results are presented to show the benefits of this method, as well as its comparative performance with a greedy sensor placement method.

2.4.1 Introduction

Sensor deployment is important for many strategic applications, such as coordinated target detection, surveillance, and localization. There is a need for efficient and optimal solutions to these problems. Two different, but related, aspects of sensor deployment are the target detection and localization. For optimal detection, sensors must be suitably deployed to achieve the maximum detection probability in a given region while keeping the cost within a specified budget. To localize a target inside the surveillance region, the sensors must be strategically placed such that every point in the surveillance region is covered by a unique subset of sensors [4,5]. The research work presented in this paper is focused on the first aspect.

Optimal SDP have been studied in a variety of contexts. Recently, in adaptive beacon placement, the strategy is to place a large number of sensors and then shut some of them down based on their localization information. In this context, Bulusu et al. [6,7] consider the evaluations for spatial localization based on radio-frequency proximity, and present an adaptive algorithm based on measurements.

In a related area, Guibas et al. [8] present a unique solution to the visibility-based pursuit evasion problem in robotics applications. In this context, Meguerdichian et al. [9] describe coverage problems in wireless ad hoc sensor networks given the global knowledge of node positions, using a Voronoi diagram for maximal breach path for worst-case coverage and Delaunay triangulation for maximal support paths for best-case coverage. These approaches are based on sensor devices with deterministic coverage capability. In practice, the sensor coverage is not only dependent on the geometrical distance from the sensor [9], but also on factors such as environmental conditions and device noise. As such, the deterministic models do not adequately capture the tradeoffs between sensor network reliability and cost. Thus, the next-generation sensor networks must go beyond the deterministic coverage techniques to perform the assigned tasks, such as online tracking/monitoring in unstructured environments.

In practical sensors, the probability of successful detection decreases as the target moves further away from the sensor, because of less received power, more noise, and environmental interference. Therefore, the sensor detection is "probabilistic," which is the focus of this paper. The sensor deployment is a complex task in DSNs because of factors such as different sensor types and detection ranges, sensor deployment and operational costs, and local and global coverage probabilities [10,11]. Essentially, the sensor deployment is an optimization problem, which often belongs to the category of multidimensional and nonlinear problems with complicated constraints. If the deployment locations are restricted to discrete grid points, then this problem becomes a combinatorial optimization problem, but it is still computationally very difficult. In particular, this problem contains a considerable number of local maxima, and it is very difficult for the conventional optimization methods to obtain its global maximum [12].

A generic SDP over the planar grid to capture a subclass of sensor network problems can now be formulated. Consider sensors of different types, wherein each type is characterized by a detection region and an associated detection probability distribution. Thus, each deployed sensor detects a target located in its region with certain probability and incurs certain cost. Also, consider an SDP that deals with placing the sensors at various grid points to maximize the probability of detection while keeping the cost within a specified limit.

In this section, it is shown that this SDP is NP-complete, and hence it is unlikely that one will find a polynomial-time algorithm for solving it exactly. Next, an approximate solution to this problem using the genetic algorithm [13] for the case where the sensor detection distributions are statistically independent is presented. The solution presented is based on specifying the components of the genetic algorithm to suit the SDP. In particular, the genetic encoding and fitness function is specified to match the optimization criterion, and also specify the crossover, mutation, and translocation operators to facilitate the search for the near-optimal solutions. In practice, near-optimality is often good enough for this class of problems. Simulation results are then presented for 50×50 or larger grids with five or more sensor types when the a priori distribution of target is uniform. The solution proposed is quite effective in yielding solutions with good detection probability and low cost. A comparison of the proposed method with a greedy approach of uniformly placing the sensors over the grid follows next. From the comparison, it is found that this method achieved significantly better target detection probability within the budget.

The rest of this text is organized as follows. In Section 2.4.2, a formulation of the SDP is given and it is shown to be NP-complete. In Section 2.4.3, an approximate solution using a genetic algorithm is presented. Section 2.4.4 discusses the experimental results.

2.4.2 The SDP

In this section, the SDP is formulated, and then it is shown to be NP-complete.

2.4.2.1 Surveillance Region

A planar surveillance region R is to be monitored by a set of sensors to detect a target T if located somewhere in the region (our overall method is applicable to three dimensions). The planar surveillance region is divided into a number of uniform contiguous rectangular cells with identical dimensions,

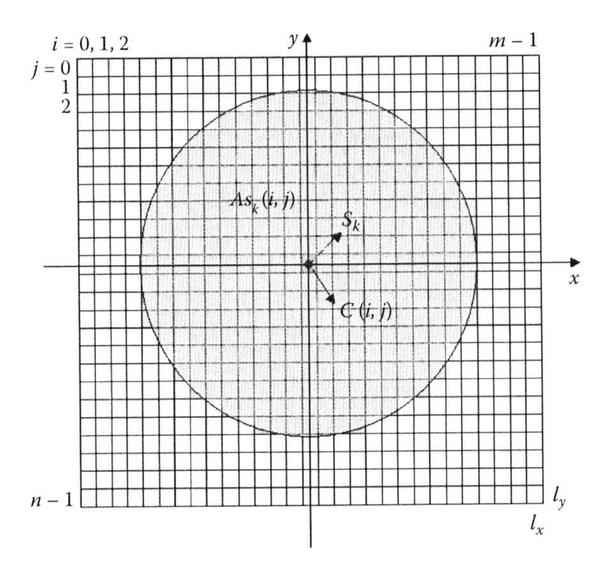

FIGURE 2.1 Surveillance region R is divided into mn rectangular cells, with sensor S_k deployed in cell $C(i, j)$ covering area $As_k(i, j)$ in a probabilistic sense.

as shown in Figure 2.1. Each cell of R is indexed by a pair (i, j), and $C(i, j)$ denotes the corresponding cell. Let l_x and l_y denote the dimensions of a cell along the x and y coordinates respectively. As Figure 2.1 shows, a circular coverage area is approximated by a set of cells within a certain maximum detection distance of sensor S_k.* When the ratio of sensor detection range to cell dimension is very large, the sensor coverage area made up of many tiny rectangular cells will approach the circle.

There are q types of sensor and a sensor of the kth type is denoted by Sk for $k \in \{1, 2, \ldots, q\}$. There are N_k sensors of type k. A sensor S can be deployed in the middle of $C(i, j)$ to cover the discretized circular area $AS(i, j)$ consisting of cells as shown in Figure 2.1. A sensor Sk deployed at cell (i, j) detects the target $T \in As_k(i, j)$ according to the probability distribution $P\{Sk|T \in As_k(i, j)\}$ while incurring the cost $w(k)$.

A *sensor deployment* is a function $\mathfrak{R}$ from the cells of R to $\{\varepsilon, 1, 2, \ldots, q\}$ such that $\mathfrak{R}(i, j)$ is the type of sensor deployed at the cell (i, j); $\mathfrak{R}(i, j) = \varepsilon$ indicates no sensor is deployed, that is, $w(\varepsilon) = 0$. The *cost of a sensor deployment* $\mathfrak{R}$ is the sum of cost of all sensors deployed in region R, given by

$$\text{Cost}(\mathfrak{R}) = \sum_{C(i,j) \in R} w(\mathfrak{R}(i, j))$$

The *detection probability* of deployment $\mathfrak{R}$, given by $P(\mathfrak{R}|T \in R\}$, is the probability that a target T located somewhere in region R will be detected by at least one deployed sensor. The SDP considered in this paper can now be formally stated:

Given a surveillance region R, cost budget Q, q types of sensor, and N_k sensors of type k, find a sensor deployment $\mathfrak{R}$ to maximize detection probability $P\{\mathfrak{R}|T \in R\}$ under the constraint $\text{Cost}(\mathfrak{R}) \leq Q$.

Informally, it is required to locate the sensors of various types on the grid points to achieve the maximum detection probability while keeping the deployment cost within a specified budget. The *decision version* of SDP asks for a deployment with detection probability at least A under the same cost condition, that is, $P\{\mathfrak{R}|T \in R\} \geq A$ and $\text{Cost(Eft)} \leq Q$. The traditional polygon or rectangle coverage problems, studied in VLSI and related areas, focus on covering regions with a minimum number of circles or rectangles and do not incorporate the probabilistic aspects of the sensors [14].

* In Figure 2.1, a cell is shaded if and only if its center is located within the sensor's maximum detection range.

2.4.2.2 Sensor Detection Distributions

Now, one can briefly describe some detection probability distributions used in SDP. The exact form of the distributions is not critical to the discussion in this text, only that they be in computable form.

Consider that each sensor type is specified by its local detection probability of detecting a target at a point within its detection region. With regard to a sensor, detection is more likely as a target approaches the sensor. The cumulative detection probability of a sensor for a region is computed by integrating its local detection probability for detecting a target as the target gets close to the sensor, passes near the sensor, and then leaves it behind. In general, there are two ways of modeling sensor detection performance, based on how the integrated detection probability is approximated [15]:

- *Definite range law approximation* (*Cookie Cutter*): In this model, only one parameter, that is, maximum detection range, is used. A target is always detected if it lies within a certain distance of the sensor, or it is never detected if it lies beyond the sensor's maximum detection range, as Figure 2.2 shows.
- *Imperfect sensor approximation*: Besides the maximum detection range, a second parameter, mean detection probability (less than 1), is specified for such a sensor model, as Figure 2.3 shows.

Comparing the above two approximations, we suggest that the latter models the real situations more reasonably. Based on the imperfect sensor approximation, a more accurate sensor performance model may be specified by a Gaussian cumulative detection probability instead of mean detection probability to approximate a real sensor.

Given the detection probability density function $ps_k(x)$ for a sensor of type k, the detection probability $P\{S_k|T \in C(i,j)\}$ for cell $C(i,j)$ is given by

$$P\{S_k \mid T \in C(i,j)\} = \int_{x \in C(i,j)} ps_k(x)\,dx$$

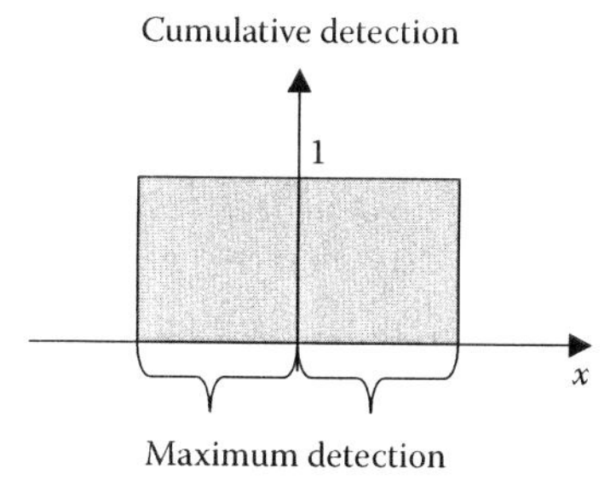

FIGURE 2.2 Definite range law.

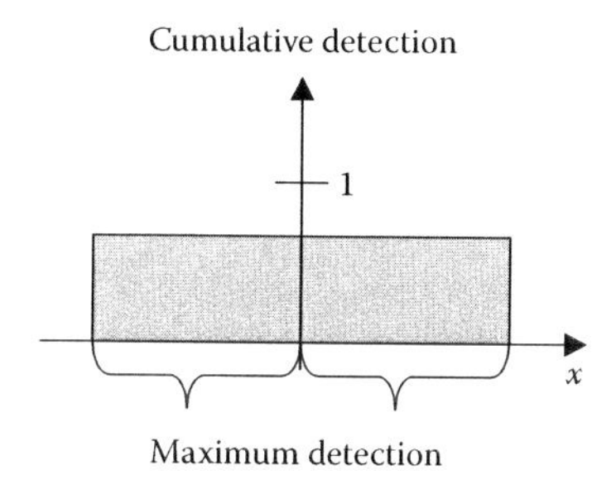

FIGURE 2.3 Imperfect sensor.

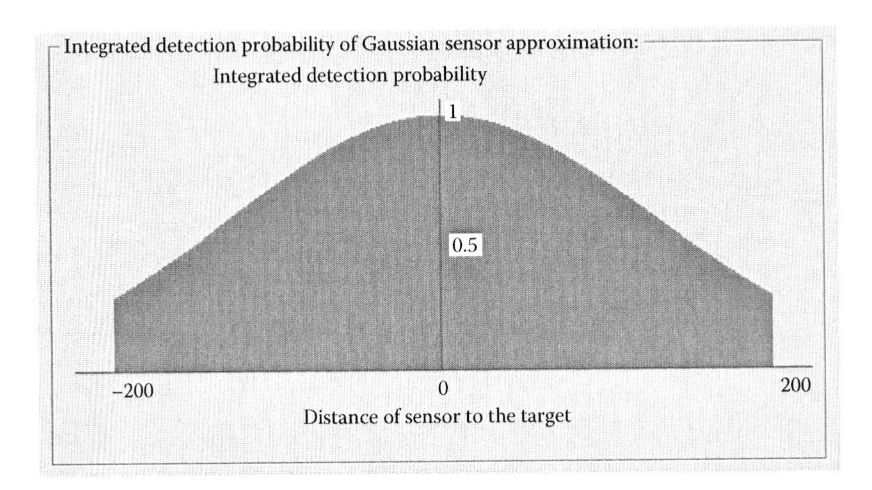

FIGURE 2.4 Integrated detection probability of Gaussian sensor.

After obtaining all individual detection probabilities for the cells covered by sensor S_k, Gaussian function to compute the cumulative detection probability may be used. In this paper, imperfect sensor approximation with a Gaussian cumulative detection probability to abstract a real sensor is used. Consider

$$P(S_k\tau,\alpha_{sk}) = P\{S_k,\tau,\alpha_{sk} \mid T \in A_{sk,\tau}\} = e^{\tau^2/2\alpha_{sK}^2}\tau \in [0,d_{sk}]$$

which is the detection probability for a target located at distance τ from the sensor. The sensor detection quality coefficient α_{sk} determines the shape of the detection probability curve. Distance τ is in the range between zero and the maximum detection distance d_{sk}. A typical integrated detection probability of Gaussian sensor approximation is shown in Figure 2.4, where the measure of detection probability is assumed to reach unity when the target is very close to the sensor. We utilize this distribution in our computations for detection probability, but the genetic algorithm method presented here is applicable to suitably computable sensor distributions.

2.4.2.3 NP-Completeness of Sensor Deployment Problem

The SDP can now be shown to be NP-complete by reducing the Knapsack problem (KP) to a special case of SDP, wherein each sensor monitors a single cell with a specified probability. Consider q-KP. Given a set U of n items such that, for each $u \in U$, we have size $s(u) \in \mathbb{Z}^+$ and the value $v(u) \in \mathbb{Z}^+$, does there exist a subset $V \in U$ of exactly q items such that $\sum_{u\in V} s(u) \le B$ and $\sum_{u\in V} v(u) \ge K$ for given B and K? Note that exactly q items are required, as opposed to an unrestricted value in the usual KP; note that KP and q-KP are polynomially equivalent [16], since $q \le n$ and the input for either problem instance has at least n items.

Consider the decision version of the SDP that asks for exactly q sensors to be deployed. Reduce the q-KP to a particular restriction of the SDP, denoted by q-SDP, such that only one sensor of each type is given, and each sensor S monitors a single cell and when two sensors are located in the same cell only one of them detects the target (that is, suitable conditional probabilities are zero). For this special case, to maximize the detection probability, without the loss of generality, each cell is assumed to be occupied by no more than one sensor. Furthermore, under the uniform prior distribution of target T in cells combined with the nonoverlapping sensor regions, the probability of detection is simply the average of the probability of detection of the sensors deployed. Considering the sensor deployment $\Re$ deploys q sensors, the equation now becomes

$$P\{\Re \mid T \in R\} = \frac{1}{q} \sum_{\Re(i,j)=\kappa;\Re(i,j)\ne\varepsilon} P\{S_k \mid T \in A_{sk}\}$$

Given an instance of q-KP, each $u \in U$ can be mapped to a sensor S_u such that its cost and value are given by $w(u) = s(u)$ and

$$P\{S_u \mid T \in A_{su}\} = \frac{v(u)}{\sum_{a \in U} v(a)}$$

Then specify the sensor cost bound as $Q = B$ and the detection probability as

$$A = \frac{K}{q \sum_{a \in U} v(a)}$$

Given a solution to q-KP, a solution to q-SDP exists by just placing the sensors corresponding to the members of V on nonoverlapping grid points. Let (i_u, j_u) be the cell receiving a sensor due to $u \in V$. Then

$$\sum_{\Re(i,j) \neq \varepsilon; \Re(i,j) = \kappa} w(k) = \sum_{\Re(i_u, j_u): u \in V} w(k) = \sum_{u \in V} s(u) \leq Q$$

which satisfies the first condition for q-SDP. For the second condition

$$P\{\Re \mid T \in R\} = \frac{1}{q} \sum_{\Re(i,j) = \kappa; \Re(i,j) \neq \varepsilon} P\{S_k \mid T \in A_{sk}\} = \frac{1}{q} \sum_{\Re(i_u, j_u) = \kappa; \Re(i_u, j_u) \neq \varepsilon} \frac{v(u)}{\sum_{a \in U} v(a)} \geq A$$

Given a solution to the SDP, the solutions to q-KP can be obtained by choosing the items corresponding to the sensors deployed. Let $u(i, j)$ denote the chosen item in corresponding to sensor located at $\Re(i, j)$. The first condition for q-KP follows from

$$\sum_{u \in V} s(u) = \sum_{u(i,j)} s(u_{(i,j)}) = \sum_{\Re(i,j) = k} w(k) \leq B$$

The second condition for q-KP follows from

$$\sum_{u \in V} v(u) = \sum_{u(i,j)} v(u_{(i,j)}) = \sum_{a \in U} v(a) \sum_{\Re(i,j) = k; \Re(i,j) \neq \varepsilon} P\{S_k \mid T \in A_{sk})$$

$$= q \sum_{a \in U} v(a) P\{\Re \mid T \in R\} \geq Aq \sum_{a \in U} v(a) = K$$

It has been shown that SDP is NP-complete even when severe restrictions are imposed on the joint distributions, which is an indication of the computational complexity of this problem. Thus, it is unlikely that polynomial-time solutions that optimally solve SDP exist, which is a motivation to consider approximate solutions.

2.4.2.4 Sensor Detection Probability under Independence Condition

In this section, a restricted version of the SDP is considered, such that the sensors satisfy a certain statistical independence condition, which enables the joint detection probabilities to be computed efficiently. To guarantee high probability of detection, sensor detection range should overlap to insure that the critical areas of the surveillance region are covered by at least one sensor [17]. The local detection probability $P\{\Re \mid T \in C(i, j)\}$ must be suitably accumulated for each cell $C(i, j)$ covered by two or more sensors.

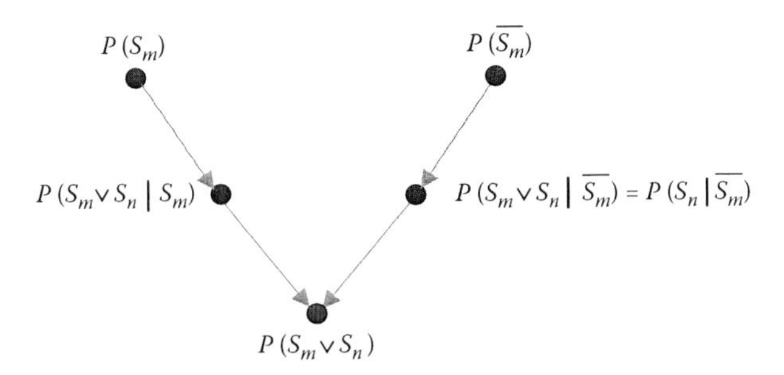

FIGURE 2.5 Multiple cases to successful detection in the simplest case.

To determine the sensor detection probabilities for such cells, we consider a simplest case first with two detection probabilities: $P\{S_m|T \in C(i, j)\}$ and $P\{S_n|T \in C(i, j)\}$, corresponding to sensors S_m and S_n which overlap in $C(i, j)$. The detection probability $P\{S_m \vee S_n|T \in C(i, j)\}$ is the probability of detecting the target successfully by at least one of the two sensors. Let $P(S_l)$ denote $P\{S_l|T \in C(i, j)\}$, for $l = m, n$, and let $P(S_m \vee S_n)$ denote $P\{S_m \vee S_n|T \in C(i, j)\}$. There are two mutually exclusive and collectively exhaustive cases for the successful detection, as Figure 2.5 shows.

Assuming that sensors S_m and S_n are *statistically independent*, such that

$$P(S_m \wedge S_n) = P(S_m)P(S_n)$$

we have

$$P(S_m \vee S_n) = P(S_m) + P(S_n) - P(S_m)P(S_n)$$

For a general case of n sensors covering a cell, by the inclusion-exclusion principle [18] we have

$$P(S_1 \vee S_2 \vee \cdots \vee S_n) = P(S_1 \vee S_2 \vee \cdots \vee S_{n-1}) + P(S_n) - P(S_1 \vee S_2 \vee \cdots \vee S_{n-1})P(S_n)$$

$$\cdots$$

$$= \sum_{i=1}^{n} P(S_i) - \sum_{1 \le i < j \le n} P(S_i)P(S_j) + \sum_{1 \le i < j < k \le n} P(S_i)P(S_j)P(S_k) + \cdots + (-1)^{n-1} P(S_1)P(S_2)\cdots P(S_n) \tag{2.1}$$

The overlap of local detection probabilities for n sensors is computed by applying the simple formula in Equation 2.1 repeatedly for each additional sensor as follows for computing $P\{\Re|T \in C(i,j)\}$ for each cell $C(i, j)$:

Step 1: Initialize local coverage probabilities and total cost.
Step 2: Locate a cell in which a sensor is deployed.
Step 3: Determine the sensor type.
Step 4: Update total cost.
Step 5: Compute the detection area of this sensor using Equation 2.1.
Step 6: For each of the cells within the discretized circular detection area, compute the overlapping detection probability.
Step 7: Update local detection probability for each cell covered by this sensor.
Step 8: Go back to Step 2 until all cells in the whole surveillance region are examined.

The details of the algorithm to compute the local coverage probabilities and total cost outlined above are presented as follows:

Input: sensor deployment scheme $\Re$
Output: local coverage probability $P\{\Re|T \in C(i, j)\}$ for each cell $C(i, j)$ and total cost, where $i = 0, 1, 2, \ldots, m - 1, j = 0, 1, 2, \ldots, n - 1$

```
Begin
Initialize P{ℜ|T∈C(i,j)} to 0;
Initialize Cost(ℜ) to 0;
for i=0 to m−1
{for j=0 to n−1
    {let k=ℜ(i,j);
     if (k==ε)
       continue
     else
       Update(k)
    }
}
End
  Auxiliary function Update (Sensor k)
Update(sensor k)
Begin
  let Cost(ℜ)=Cost(ℜ)+w(S_k);
  let a=⌊d_sk / l_x⌋;
  let b=⌊d_sk / l_y⌋;
  for r=−a to a
      for s=−b to b
      {let τ=√((r l_x)^2+(s l_y)^2);
       if (τ≤d_sk)
       {
         let overlap=P{ℜ|T∈C(i+s)}*P(S_k,τ,α_sk);
         let P{ℜ|T∈C(i+r,j+s)}=P{ℜ|T∈C(i+r,j+s)}+P(S_k,τ,α_sk))−overlap;
       }
      }
End
```

Then compute $P\{\Re|T \in R\}$ by adding all local detection probabilities in the surveillance region given by

$$P\{\Re \mid T \in R\} \sum_{i=0}^{m-1} \sum_{j=0}^{n-1} P\{\Re \mid T \in C(i,j)\} * P\{T \in C(i,j)\} \tag{2.2}$$

which is the objective function to be maximized under the cost condition $\text{Cost}(\Re) \leq Q$. Given the a priori distribution $P\{T \in C(i,j)\}$ of target T in a computable form and the sensor distributions, the objective function can be computed.

This version of the SDP, namely under the statistical independence condition, can be shown to be NP-complete by a simple extension of the results of the last section. Under the statistical independence, within each cell the probability of joint detection is the product of individual probabilities, and hence is smaller than either. Thus, any overlapping sensors within a cell can be separated to increase the probability of detection, and it suffices to consider no more than one sensor per cell. The rest of the proof follows the last section: under the restriction that each sensor detects a target in the cell it is currently located, this problem reduces to q-SDP in the last section, which shows the current problem to be NP-complete by restriction.

2.4.3 Genetic Algorithm Solution

A genetic algorithm is a computational model that simulates the process of genetic selection and natural elimination in biological evolution [19]. It has been frequently used to solve combinatorial and nonlinear optimization problems with complicated constraints or nondifferentiable objective functions [20,21]. The computation of a genetic algorithm is an iterative process toward achieving global optimality. During the iterations, candidate solutions are retained and ranked according to their quality. A fitness value is used to screen out unqualified solutions. Genetic operations of crossover, mutation, translocation, inversion, addition, and deletion are then performed on those qualified solutions to create new candidate solutions of the next generation. The above process is carried out repeatedly until a certain stopping or convergence condition is met. For simplicity, a maximum number of iterations can be chosen to be the stopping condition. The variation difference of the fitness values between two adjacent generations may also serve as a good indication for convergence. To utilize the genetic algorithm method, various parts of the SDP must be mapped to the components of the genetic algorithm, as will be shown in this section.

2.4.3.1 Genetic Encoding for Sensor Deployment

Since a candidate solution to the SDP requires a two-dimensional sensor ID matrix, a two-dimensional numeric encoding scheme needs to be adopted to make up the chromosomes instead of the conventional linear sequence in order. As Figure 2.6 shows, a sensor ID matrix for a possible sensor deployment scheme is constructed. Each element in the matrix on the right-hand side corresponds to a cell within a surveillance region on the left-hand side. As mentioned above, an empty value s in the matrix indicates that its corresponding cell has no sensor deployed and should be covered by the sensors deployed in its neighborhood area. Furthermore, q types of available sensor are arranged in the following order:

$$\frac{ds_1}{w(s_1)} \geq \frac{ds_2}{w(s_2)} \geq \cdots \geq \frac{ds_k}{w(s_k)} \geq \cdots \geq \frac{ds_q}{w(s_q)}$$

Recall that ds_k and $w(s_k)$ are the maximum detection distance and cost of sensor of type k respectively. The rank of ratio is used to decide the probability of type of sensor selected during the population initialization, as the well as in the addition operation.

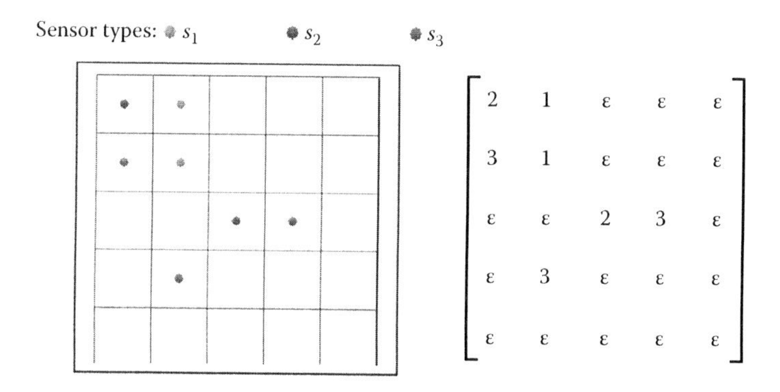

FIGURE 2.6 A candidate deployment solution and its corresponding matrix.

2.4.3.2 Fitness Function

The fitness function from the objective function can be constructed as

$$f(\mathfrak{R}) = P\{\mathfrak{R} \mid T \in R\} + g$$

where g, the penalty function for overrunning the constraint, is defined by

$$g = \begin{cases} 0 & \text{Cost}(\mathfrak{R}) \leq Q \\ \delta E_m(Q - \text{Cost}(\mathfrak{R}))/Q & \text{Cost}(\mathfrak{R}) > Q \end{cases}$$

such that δ is a proper penalty coefficient and is set to 100, and $E_m = \max\{d_{sk}/w(sk)\}$.

2.4.3.3 Selection of Candidates

The selection operation retains good candidates and eliminates others from the population based on the individual fitness values. It is also called a reproduction operation. It aims to inherit good individuals, either directly from the last generation or indirectly from the new individuals produced by mating the old individuals. The frequently used selection mechanisms include the fitness proportional model, the rank-based model, the expected value model, and the elitist model [22].

In this implementation, the survival probability B_i for each individual (solution) $\mathfrak{R}_I$ is computed based on the following fitness proportional model:

$$B_i = \frac{f(\mathfrak{R}_i)}{\sum_{j=1}^{M} f(\mathfrak{R}_i)}$$

where M is the population size. The hybridization individuals are produced according to the selection rule such that the individual with bigger B_i has a higher probability to survive.

2.4.3.4 Implementation of Genetic Operators

The solution set of each new generation after the initial population is generated as follows. Randomly select two hybridization individuals $\mathfrak{R}_u$ and $\mathfrak{R}_v$ and combine them to get two other individuals $\mathfrak{R}'_u$ and $\mathfrak{R}'_v$ of the new generation by using combinatorial rules of crossover, mutation, inversion, translocation, addition, and deletion [23]. Some of these genetic operators are carried out on a two-dimensional basis.

Except for crossover, all the other operators operate on only one parent solution. This process continues until all M individual solutions of the new generation are created:

1. *Crossover*: This is an operation of segment exchange for two solutions. Given two parent (hybridization individuals) solutions on the left side in Figure 2.7, a two-dimensional two-point crossover operation produces two child solutions on the right side, as Figure 2.7 illustrates. Both window size and location for crossover are selected randomly.
2. *Translocation*: The objective of translocation is to exchange information between different segments within a single solution, as Figure 2.8 illustrates. As in the crossover operation, the translocation window size is picked up at random, as well as its source and destination position.
3. *Mutation*: The mutation operator chooses one or more cells randomly in the surveillance region and changes their values by the preset mutation probability. Actually, it is a combination of addition and deletion operators. As Figure 2.9 shows, a sensor of type A is moved to a new place, a sensor of type B is replaced by a sensor of type C, a sensor of type C is deleted, and a sensor of type D is replaced by a sensor of type B. The selection probability of a certain type of sensor for the addition operator depends on the ratio of its detection range to unit price, so does for the population initialization.

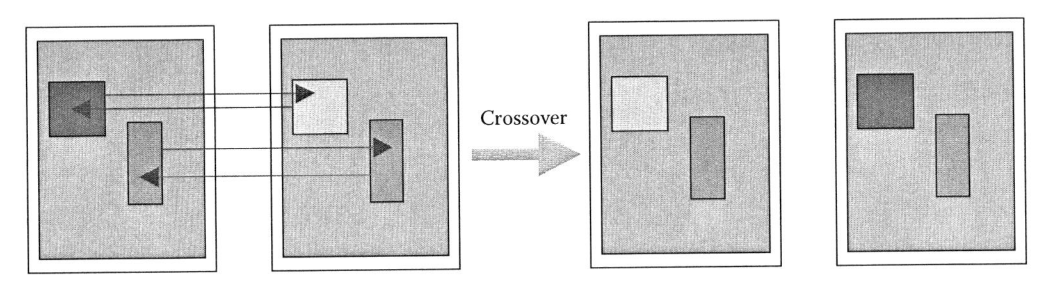

FIGURE 2.7 Two-dimensional two-point crossover.

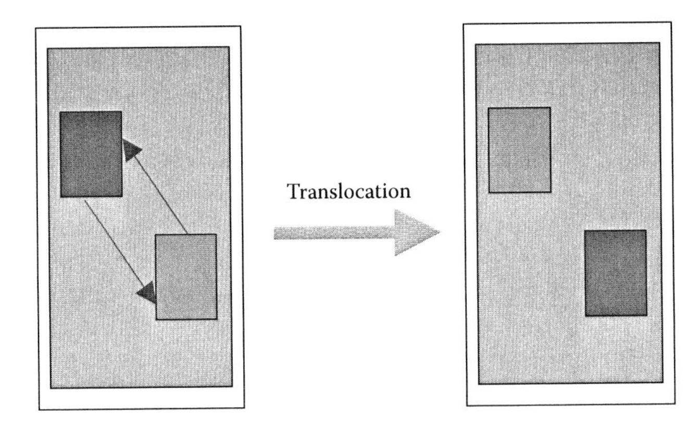

FIGURE 2.8 Two-dimensional translocation.

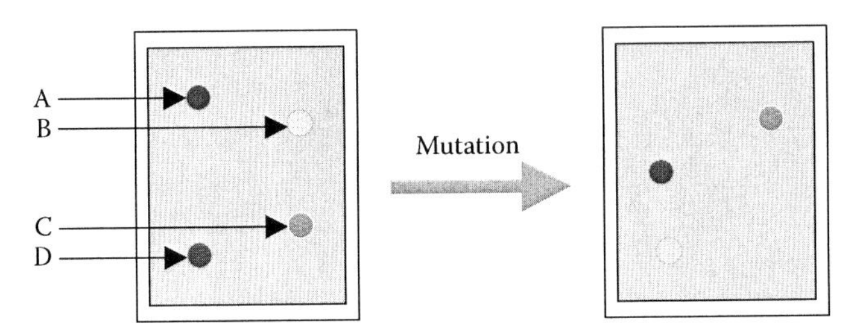

FIGURE 2.9 Mutation operator.

2.4.4 Computational Results

In this section, experimental test results comparing our genetic algorithm described in the last section with the greedy method based on uniform placement (UP) of sensors are presented. Both algorithms are implemented in C++. Consider that the target has a uniform a priori distribution in the surveillance region such that the probability of target T appearing in a cell $C(i, j)$ is $P\{T \in C(i, j)\} = 1/(mn)$. Then, the detection probability, which is the objective function, has the following formula by Equation 2.2:

$$P\{\Re \mid T \in R\} = \sum_{i=0}^{m-1}\sum_{j=0}^{n-1} \frac{P\{\Re \mid T \in C(i,j)\}}{(mn)}$$

with the constraint $\mathrm{Cost}(\Re) \leq Q$. This formula is utilized for the computations in this section, and the case in which the target has other prior distributions can be handled using Equation 2.2 in place of the above expression.

In Case 1, consider a surveillance region of 50 × 50 cells with five types of sensors as listed in Table 2.1. All parameters used by the genetic algorithm are specified in Table 2.2. The investment limit is set to be 1800 units expense and the maximum generation number is set to be 200, also as shown in the upper data part of Figure 2.10. After 200 generations of optimization, an acceptable deployment scheme is achieved with detection probability of 94.52% for the surveillance region within the investment budget. The graph on the right side of Figure 2.10 shows the optimization process curve with the generation number represented on the x-axis and the corresponding fitness value represented on the y-axis. The computational result is illustrated on the left side of Figure 2.10

TABLE 2.1 Parameter Specifications for Five Types of Sensors Used in Case 1

Sensor Type	Sensor ID	Unit Price	Detection Range	Detection Coefficient
Sen1	1	86	124	80
Sen2	2	111	159	78
Sen3	3	113	163	68
Sen4	4	135	195	68
Sen5	5	139	200	84

TABLE 2.2 Parameters Used by the Genetic Algorithm in Case 1

Genetic Algorithm Parameters	Values
Maximum generation number	200
Maximum investment limit	1800
Population size	30
Probability of crossover	0.99
Probability of mutation	0.24
Probability of deletion	0.10
Probability of translocation	0.99
Probability of inversion	0.82
Probability of addition	0.10

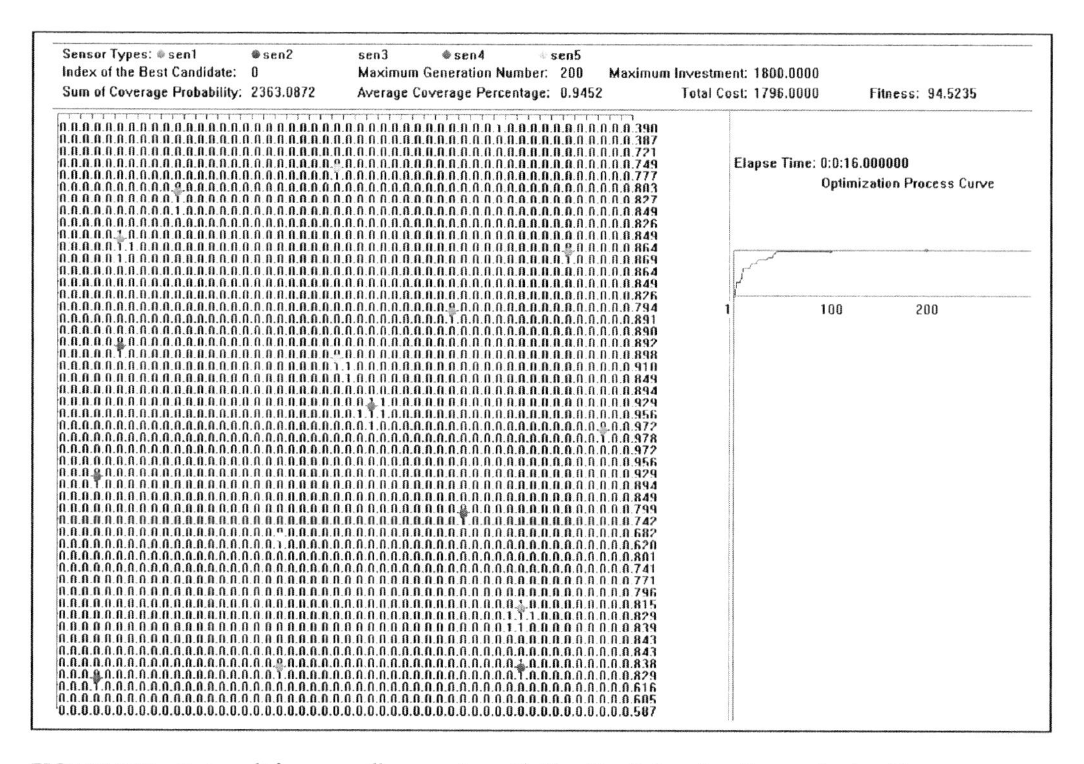

FIGURE 2.10 Test result for a surveillance region with 50 × 50 cells based on the genetic algorithm.

for the sensor deployment based on the genetic algorithm. In each cell, a local detection probability is given for evaluation.* Its corresponding three-dimensional display of the local coverage probabilities of 50 × 50 cells is shown in Figure 2.11.

Figure 2.12 shows the computational result of the same surveillance region based on UP using the sensors of type with the maximum ratio of detection range to unit price. The UP achieves an average detection probability 88.83% within the investment budget. Its corresponding three-dimensional display of the local coverage probabilities of 50 × 50 cells is shown in Figure 2.13.

Figure 2.14 is obtained by carrying out a series of runs for this surveillance region with increasing investment limits. From the plot of detection probability versus investment limit shown in Figure 2.14, it is observed that increasing the investment beyond 1800 units does not pay off any more, since the incremental gain of the detection probability is very small. Not only could this graph help in determining a sensor deployment scheme within any given cost for a given surveillance region, but it could also be used to choose a proper initial investment limit for the given surveillance region.

Now let us consider larger surveillance regions and more sensor types with different parameters. The UP always uses only the sensors of type with the maximum ratio of detection range to unit price. The comparisons of computational results between the genetic algorithm and the UP are shown in Table 2.3. In summary, the genetic algorithm achieved higher probability of detection while satisfying the cost bound.

* The values of local detection probabilities are overwritten by those values of neighbor cells except for the rightmost column due to the relatively small display screen, as is the case in Figure 2.11.

Local coverage probabilities
1
0.9
0.8
0.7
0.6
0.5
0.4
0.3
0.2
0.1
0
50
40
30
20
10
0
Row
0
5
10
15
20
25
30
35
40
45
50
Column

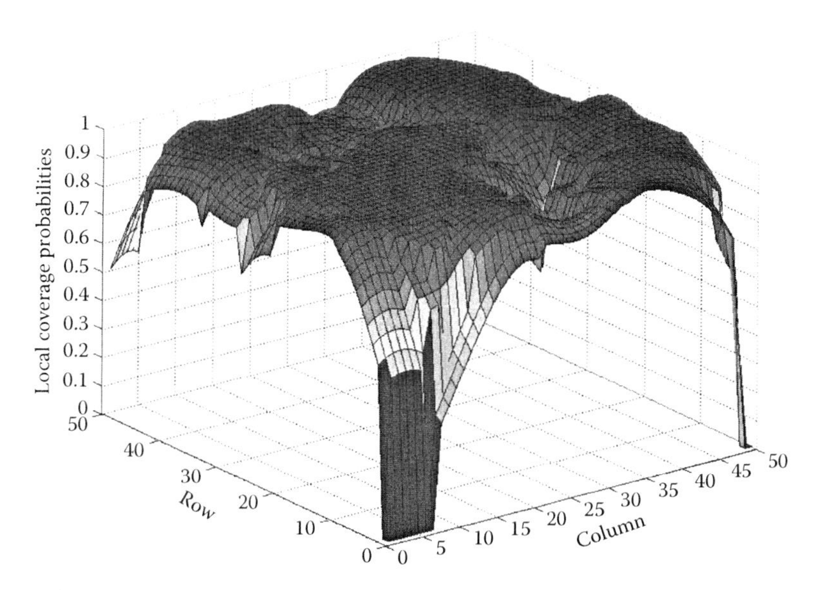

FIGURE 2.11 Three-dimensional display of the local coverage probabilities of 50 × 50 cells computed by the genetic algorithm.

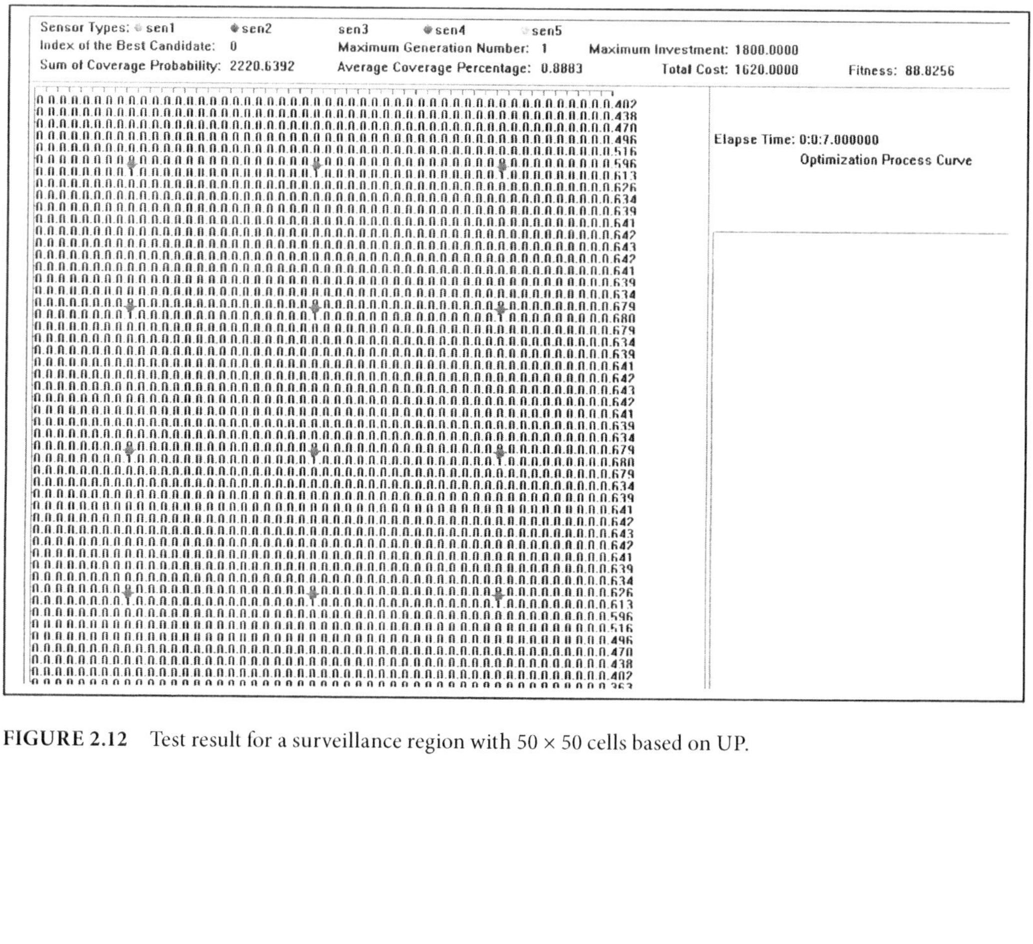

Sensor Types: sen1 sen2 sen3 sen4 sen5
Index of the Best Candidate: 0
Maximum Generation Number: 1
Maximum Investment: 1800.0000
Sum of Coverage Probability: 2220.6392
Average Coverage Percentage: 0.8883
Total Cost: 1620.0000
Fitness: 88.8256
Elapse Time: 0:0:7.000000
Optimization Process Curve

FIGURE 2.12 Test result for a surveillance region with 50 × 50 cells based on UP.

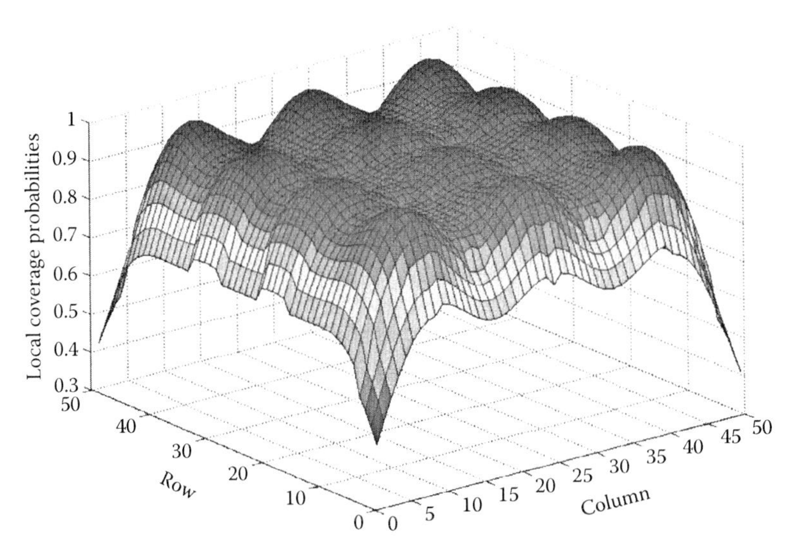

FIGURE 2.13 Three-dimensional display of the local coverage probabilities of 50 × 50 cells computed by UP.

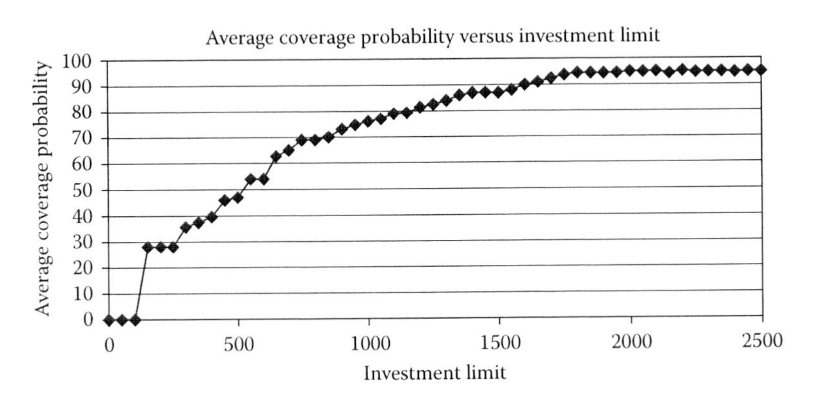

FIGURE 2.14 Detection probability versus investment limit for a region with 50 × 50 cells.

TABLE 2.3 Comparison of Performance of Genetic Algorithm and UP

			Genetic Algorithm		Uniform Placement	
Surveillance Region Size	Number of Sensor Types	Maximum Investment Limit	Total Cost	Ave. Detection Probability (%)	Total Cost	Ave. Detection Probability (%)
50 × 50	5	1800	1796	94.52	1620	88.83
100 × 100	5	2100	2081	93.03	1920	84.41
120 × 120	7	2250	2226	94.18	2160	87.44
150 × 150	7	2350	2340	93.12	2187	85.96
200 × 200	8	2600	2587	93.97	2430	88.61
300 × 300	5	3900	3861	93.64	3630	88.75
600 × 600	6	4600	4598	96.84	4400	87.69
750 × 750	5	6000	5995	93.81	5670	87.40
900 × 900	8	9000	8949	95.16	8993	86.61
1000 × 1000	9	9700	9698	93.70	9680	88.58

2.4.5 Conclusions

Optimal detection and target localization are two critical but difficult tasks of sensor deployment, particularly if the sensors are of different types and incur different costs. A general SDP for a planar grid region is formulated with the objective of maximizing the detection probability within a given deployment cost. It is shown that this problem is NP-complete, and then an approximate solution is presented using a genetic algorithm for the case where the sensor distributions are statistically independent. Computational results are presented when the target has uniform prior distribution and Gaussian approximation for sensor distributions, which shows that this solution performs favorably in solving the SDP. This solution is applicable to more general cases in which the target's a priori distribution is not uniform and the sensor distributions are more complicated but easily computable. In general, however, the computational cost of such extensions would be correspondingly higher.

There are a number of avenues for further research. First, it would be interesting to see whether analytical performance bounds can be placed on the solution computed by our method. Also, extensions of the proposed method when the statistical independence is not satisfied would be applicable to wider classes of SDPs. The challenge in this case is to ensure low computational complexity by utilizing the domain-specific knowledge of the sensors. In particular, the simple incremental formula in Equation 2.1 is no longer valid, and in the worst case this computation may have an exponential complexity for arbitrary distributions. From an algorithmic perspective, polynomial-time (deterministic) approximations to the SDP that are guaranteed to be provably close-to-optimal will be of future interest.

Acknowledgment

This research is sponsored by the Material Science and Engineering Division, Office of Basic Energy Sciences, U.S. Department of Energy, under contract No. DE-ACC5-00OR22725 with UT-Battelle, LLC, the Defence Advanced Projects Research Agency under MIPR No. K153, and by National Science Foundation under Grants No. ANI-0229969 and No. ANI-335185.

References

1. Dhillon, S.S. and Chakrabarty, K., A fault-tolerant approach to sensor deployment in distributed sensor networks, in *The 23rd Army Science Conference*, Orlando, FL, December 2–5, 2002.
2. Chakrabarty, K. et al., Coding theory framework for target location in distributed sensor networks, in *Proceedings of International Symposium on Information Technology: Coding and Computing*, Las Vegas, NV, 2001, p. 130.
3. Convay, J.H. and Guy, R.K., *The Book of Numbers*, Copernicus Books, New York, 1996, p. 220.
4. Chakrabarty, K. et al., Grid coverage for surveillance and target location in distributed sensor networks, *IEEE Transactions on Computers*, 51(12), 1448, 2002.
5. Chakrabarty, K. et al., Coding theory framework for target location in distributed sensor networks, in *Proceedings of the IEEE International Conference on Information Technology: Coding and Computing*, Las Vegas, NV, April 2001, p. 157.
6. Bulusu, N. et al., Adaptive beacon placement, in *Proceedings of ICDCS-21*, Phoenix, AZ, April 2001.
7. Bulusu, N. and Estrin, D., Scalable *ad hoc* deployable RF-based localization, in *Proceedings of the Grace Hopper Celebration of Women in Computing Conference 2002*, Vancouver, British Columbia, Canada, October 2002.
8. Guibas, L. et al., Visibility-based pursuit evasion in a polygonal environment, *International Journal of Computational Geometry Applications*, 9(4/5), 471, 1999.
9. Meguerdichian, S. et al., Coverage problems in wireless *ad hoc* sensor networks, in *Proceedings of IEEE Infocom 2001*, Anchorage, AK, April 22–26, 2001.

10. Iyengar, S.S. et al., *Advances in Distributed Sensor Integration: Application and Theory*, Prentice Hall, Upper Saddle River, NJ, 1995, p. 130.
11. Brooks, R.R. and Iyengar, S.S., *Multi-Sensor Fusion: Fundamentals and Applications with Software*, Prentice Hall PTR, Upper Saddle River, NJ, 1998.
12. Liu, Y. et al., *Non-Numerical Parallel Algorithms (II)—Genetic Algorithms*, Science Press, Beijing, China, 1995.
13. Han, Z.X. and Wen, F.S., Optimization method simulating evolution and its application, *Computer Science*, 22(2), 47–56, 1995.
14. A compendium of NP optimization problems, http://www.nada.kth.se/~viggo/problemlist/compendium.html
15. http://www.nosc.mil/robots/research/manyrobo/detsensors.html
16. Garey, M.R. and Johnson, D.S., *Computers and Intractability: A Guide to the Theory of NP-Completeness*, W.H. Freeman, New York, 1979.
17. http://www.nlectc.org/perimetr/full2.htm
18. Liu, C.L., *Introduction to Combinatorial Mathematics*, McGraw-Hill, New York, 1968.
19. Holland, J.H., *Adaptation in Nature and Artificial Systems*, University of Michigan Press, Ann Arbor, MI, 1975 (Reprinted by MIT Press, 1992).
20. Coley, D.A., *An Introduction to Genetic Algorithms for Scientists and Engineers*, World Scientific, River Edge, NJ, 1999.
21. Winston, P.H., *Artificial Intelligence*, 3rd edn., Addison-Wesley, Reading, MA, 1993.
22. Chen, G.L., *Genetic Algorithm and Its Application*, People's Post Publishing House, Beijing, China, 1996.
23. Goldberg, D.E., *Genetic Algorithms in Search, Optimization and Machine Learning*, Addison-Wesley, Reading, MA, 1989.

3

Genetic Algorithm for Mobile Agent Routing in Distributed Sensor Networks*

Qishi Wu
Louisiana State University

S. Sitharama Iyengar
Florida International University

Nageswara S.V. Rao
Oak Ridge National Laboratory

3.1 Introduction

In the past decade, sensor networks have become an active area of research for computer scientists and network engineers due to their wide usage in both military and civilian applications [1]. The increasing sophistication of multi-sensor systems for state estimation, region surveillance as well as target detection and tracking has generated a great deal of interest in the development of new computational structures and networking paradigms [2].

A distributed sensor network (DSN) consists of intelligent sensors that are geographically dispersed in a region of interest and interconnected via a communication network. The sensed data of different types (such as acoustic, seismic, and infrared, etc.) is preprocessed by sensor nodes and then transmitted over the network to participate in the data integration at processing elements (PE), based on which appropriate inferences can be derived about the environment for certain purposes.

* Part of this text was taken from "On Computing the Route of a Mobile Agent for Data Fusion in a Distributed Sensor Network" by Q. Wu, S.S. Iyengar, N.S.V. Rao, J. Barhen, V.K. Vaishnavi, H. Qi, K. Chakrabarty, which is under consideration for publication in the *IEEE Transactions on Knowledge and Data Engineering*.

The study of information fusion methods have been the research focus since the early stage of DSN development [3–5]. Recent advances in sensor technologies make it possible to deploy a large number of inexpensive and small sensors to "achieve quality through quantity" in very complex scenarios, which necessitates applying new computing techniques such as genetic algorithm (GA) to meet some theoretical and methodological challenges. This chapter provides a general introduction to GA and its application to mobile agent routing problem (MARP) in DSN with a special networking paradigm.

DSNs are typically deployed for remote operations in large unstructured geographical areas, where wireless networks with low bandwidth are usually the only means of communication among the sensors. The communication consumes the limited power available at sensor nodes, and thus power consumption is to be restricted. Furthermore, the massively deployed sensors usually supply huge amount of data of various modalities, which makes it critical to collect only the information that is most desired and to collect it efficiently. Despite the abundance of deployed sensors, not all the information from these sensors need to be collected to ensure the quality of the fused information such as adequate detection energy for target detection or tracking.

Instead of sending all sensor data to the processing element, which performs a one-time data fusion as in a conventional server/client system, the mobile agent-based distributed sensor network (MADSN) proposed in Ref. [6] enables the computation to be spread out onto the participating leaf nodes with the aim of decreasing the consumption of scarce network resources (mostly the bandwidth) and the risk of being spied with hostile intent. In such a network scheme, a mobile agent carrying the executable instructions of data integration is dispatched from the processing element and selectively visits the leaf sensors along a certain path to fuse the data incrementally on a sequential basis.

The path quality of a mobile agent significantly affects the overall performance of MADSN implementation because the communication cost and detection accuracy depend on the order and the number of nodes to be visited. We formulate the MARP as a combinatorial optimization problem with certain constraints and construct an appropriate objective function that reflects the routing requirements. We show the MARP's NP-hardness by reducing to it a variation of the 3D traveling salesman problem, which rules out any polynomial solutions. Therefore, we propose an approximate solution based on a two-level GA and compare the simulation results with those computed by two other deterministic heuristics, namely Local Closest First (LCF) and Global Closest First (GCF).

The rest of this chapter is organized as follows: In Section 3.2, we introduce a general computing technique based on GA. In Section 3.3, we describe the models for sensor nodes and wireless communication links and then formulate the MARP. The details of the solution using GA are given in Section 3.4, including the design of a two-level encoding scheme, derivation of the objective function, and implementations of genetic operators. Simulation results are presented and discussed in Section 3.5. Concluding remarks are provided in Section 3.6.

3.2 Computational Technique Based on Genetic Algorithm

3.2.1 Introduction to Genetic Algorithm

GA is a computational model simulating the process of genetic selection and natural elimination in biologic evolution. Pioneering work in this field was conducted by Holland in 1960s [7,8]. GA was proposed to find global or local optima in the large search space. Comparing to traditional search algorithms in artificial intelligence, GA is able to automatically acquire and accumulate the necessary knowledge about the search space during its search process, and self-adaptively control the entire search process through random optimization techniques. A computational technique based on GA is particularly useful to avoid combinatorial explosion, which is always caused by disregarding the inherent knowledge within the enormous search space. In addition, the GA is characterized by its simplicity, flexibility, robustness, and adaptability to parallel process. It has found many successful applications in various areas solving combinatorial optimization problems and non-linear problems with complicated constraints or non-differentiable objective functions.

3.2.2 A General Method Using Genetic Algorithm

The computation of GA is an iterative process that simulates the process of genetic selection and natural elimination in biologic evolution. In each iteration cycle, good candidate solutions are retained and any unqualified solutions are screened out according to their corresponding fitness values. Genetic operators, such as crossover, mutation, translocation, and inversion, are then performed on those surviving solutions to produce a next generation of new candidate solutions. The above process is carried out repeatedly until certain convergence condition is met. To illustrate the principle of the algorithm, we take the classical knapsack problem as an example [9], which is formulated as follows:

$$\text{Maximize: } \sum_{i=1}^{n} B_i X_i \tag{3.1}$$

$$\text{Constraint: } \sum_{i=1}^{n} S_i X_i \leq C \quad X_i \in \{0,1\}, \quad 1 \leq i \leq n \tag{3.2}$$

where

S_i represents the resource consumption for the i-th activity
C represents the total available resources
B_i represents the gained profit from the i-th activity
X_i holds binary values: if i-th activity is carried out, $X_i = 1$; otherwise, $X_i = 0$

The essence of the knapsack problem is to pursue the maximum profit with the constraint of limited total available resources. We now describe the standard steps taken by GA to find a solution to the knapsack problem:

1. Initialization: a set of M random solutions T_k $(1 \leq k \leq M)$ are generated, where M is an appropriately selected initial population size
2. Genetic encoding: a string T of n binary bits is used to represent one possible solution. If the i-th activity is carried out, $T(i)$ $(1 \leq i \leq n) = 1$; otherwise $T(i)$ $(1 \leq i \leq n) = 0$
3. Fitness value calculation: the objective function of the knapsack problem can be defined as

$$J(T_k) = \sum_{i=1}^{n} T_k(i) B_i \tag{3.3}$$

$$\text{Subject to: } \sum_{i=1}^{n} T_k(i) S_i \leq C \quad 1 \leq k \leq M \tag{3.4}$$

where C is a bounding constant. We construct a fitness function for the knapsack problem as follows to compute the fitness value for each individual solution:

$$f(T_k) = J(T_k) + g(T_k), \quad 1 \leq k \leq M \tag{3.5}$$

where $g(T_k)$ is the penalty function when T_k violates the constraints, which may take the following form:

$$g(T_k) = \begin{cases} 0 & C \geq \sum_{i=1}^{n} T_k(i) S_i \\ \beta E_m \left(C - \sum_{i=1}^{n} T_k(i) S_i \right) & C < \sum_{i=1}^{n} T_k(i) S_i \end{cases} \tag{3.6}$$

where

E_m is the maximum value of $B_i/S_i (1 \leq i \leq n)$
β is a proper penalty coefficient

4. Survival probability calculation: the survival probability P_k for each individual (solution) T_k can be calculated based on the following fitness proportional model:

$$P_k = \frac{f(T_k)}{\sum_{j=1}^{M} f(T_j)} \tag{3.7}$$

A random selector is then designed to produce the hybridization individuals according to each P_k.

5. New generation production: two hybridization individuals T_u and T_v are combined to create two individuals T_u' and T_v' of new generation by applying combinatorial rules of selection, crossover, mutation and inversion. This process continues until all M individual solutions of new generation are produced. There are several genetic operators involved in this procedure.

   Crossover is an operation of segment exchange for two solutions. Given two parents (hybridization individuals) solutions with their crossover points represented by "/":

$$T_u = 01010/1011100$$

$$T_v = 10100/1101010$$

   a one-point crossover operator produces two children solutions:

$$T_u' = 01010/1101010$$

$$T_v' = 10100/1011100$$

   Inversion is to reverse the order of data in a solution segment. Given one parent solution with the inversion segment enclosed by a pair of "/":

$$T_u = 010/01101/1100$$

   an inversion operator produces the following child:

$$T_u' = 010/10110/1100$$

   The mutation operator chooses one or more gene loci randomly in the individual string and changes their values (e.g., 0–1 reverses) with the preset mutation probability. Given one parent solution as follows:

$$T_v = 01\underset{\uparrow}{0}10101\underset{\uparrow}{1}100$$

two gene loci before mutation

   a two-point mutation operator produces the following child:

$$T_v' = 01\underset{\uparrow}{1}10101\underset{\uparrow}{0}100$$

two gene loci after mutation

6. Repeat Step (2) to Step (5) until the predefined convergence condition is met, for example, the maximum generation number is reached or the solution quality is satisfied.

A general description of the above iterative process is given in C language as follows:

```
main()
  {
    int gen_no;
    initialize();
    generate(oldpop);
    for(gen_no=0;gen_no<maxgen;gen_no++)
    {
      evaluate(oldpop);
      newpop=select(oldpop);
      crossover(newpop);
      mutation(newpop);
      inversion(newpop);
      oldpop=newpop;
    }
  }
```

The above pseudo code only depicts major steps of a GA. Some auxiliary functions are needed to implement a complete GA for a certain application. During the search process, GA does not require any outside knowledge except fitness values to select qualified solutions. Therefore, the design of fitness function has a significant impact on the overall algorithmic performance.

3.2.3 Parameters and Operators in Genetic Algorithm

3.2.3.1 Population Size

Population size is a key parameter in GA. The Schema Theorem [10] establishes that given the population size M, the genetic operators are able to produce M^3 schemas, which ensures that the number of building blocks is dramatically increased when the search for the optimal solution progresses. Obviously, a GA with a larger population size is more likely to obtain the global optimum because a larger population size produces a wider variety of individuals and therefore the search process has higher probability to avoid being trapped into local optima. On the contrary, a small population size limits the search space and hence the premature convergence may occur under this circumstance, which may greatly impair the performance.

However, a large population size also brings some disadvantages. For instance, the computation complexity increases as a result and some good individuals with high fitness values may be eliminated during the selection operation.

3.2.3.2 Crossover, Mutation and Inversion

Similar to the gene recombination, which plays an essential role during the natural biologic evolution process, the crossover in GA is the most critical operator in the genetic search strategy, which guides the main behavior of the optimization process. There are several commonly used crossover schemes such as one-point crossover, two-point crossover, and multi-point crossover. A good design of any crossover operator must ensure that the desirable gene segments of old individuals be properly inherited by the new individuals of new generation. A high crossover probability may improve GA's capability to explore new solution space, while increase the likelihood of disordering the combination of good gene segments. An inappropriately low crossover probability may cause the search process to be trapped in a dull status and be prone to ceasing.

The main purpose of a mutation operator is to maintain the variety of the population by preventing a single important gene segment from being corrupted. In practice a relatively small mutation probability, such as 0.001, is favorable because GA may tend to be a random search if too frequent mutation operations are conducted.

The inversion is actually a special form of mutation. It is designed to carry out reordering operation and improve the local search ability. Either crossover or mutation is not adequate for search in the local solution space: the search activities of the crossover operator span in the whole feasible solution space while the local search ability of the mutation operator is always suppressed by the genetic selection and natural elimination.

3.2.3.3 Encoding and Fitness Function

Since GA is unable to manipulate the parameters in the problem space directly, it is necessary to convert them to individuals made up of genes in the GA domain. This mapping from problem space to algorithm domain is called encoding. Actually, the robustness of GA reduces the reliance of performance on encoding schemes, as long as minimum three encoding criteria, such as completeness, soundness and non-redundancy, are satisfied [11].

In general, the control of the search process in GA does not need any information from outside but fitness values (or objective values). The objective function of a complex system usually has discontinuous or non-differentiable constraints. For a general optimization problem with complicated constraints, the penalty method is often used in the design of a fitness function. For example, an original minimization problem with constraints can be described as follows [10]:

$$\text{Minimize: } F(x) \tag{3.8}$$

$$\text{With constraints: } b_l(x) \geq 0 \quad l = 1, 2, \ldots, p \tag{3.9}$$

where
$F(x)$ is the objective function
$b_l(x)$ are a group of constraint functions

By applying the penalty method, we are able to convert the above problem to a non-constraint problem:

$$\text{Minimize: } F(x) + \lambda \cdot \sum_{l=1}^{p} \Phi[b_l(x)] \tag{3.10}$$

where
λ is the penalty coefficient
Φ is the penalty function, which may take the form of Equation 3.6

3.2.3.4 Selection Mechanism

The selection operation, also referred to as reproduction operation, is to select good individuals and eliminate bad individuals from the population according to individual fitness values. A good selection mechanism is able to inherit good individuals directly from last generation or indirectly from the new individuals produced by mating the old individuals. The commonly used selection mechanisms include fitness proportional model, rank-based model, expected value model, and elitist model, etc.

3.3 Mobile Agent Routing Problem

We now briefly describe the architecture of a MADSN to motivate the later formulation of the optimization problem. A MADSN typically consists of three types of components: *processing elements*, *sensor nodes*, and *communication network* [12]. The various PE and sensors are usually interconnected via a wireless communication network. A group of neighboring sensor nodes that are commanded by the same PE forms a *cluster*.

3.3.1 Sensor Nodes

A sensor node, also referred to as a leaf node, is the basic functional unit for data collection in a MADSN. A sensor node may have several channels with different sensors connected to each of them. Sensor nodes are always geographically distributed to collect different types of measurements such as acoustic, seismic, and infrared from the environment. The data acquisition is controlled by a sampling subsystem, which provides the acquired data to the main system processor [13]. The signal energy from each channel can be detected individually and processed in the analog front end. A mobile agent migrates among the sensor nodes via the network, integrates local data with a desired resolution sequentially, and carries the final result to the originating PE. The fused data may be used to derive appropriate inferences about the environment for a certain civilian or military application.

We now provide object-oriented descriptions of sensor and PE nodes, which are used in our implementation. The sensor label is a unique ID of a sensor node, which corresponds to its static *Internet Protocol* (IP) address in the sensor network. We assume that a PE with label "0" remains active during the period of operation of MADSN. Some sensor nodes may be shut down or go to sleep due to intermittent faults or power considerations, and may be brought back up later if necessary. The sensor location is determined by its longitude and latitude obtained from the embedded *Global Position System* (GPS) module. The abstract sensor class, defined in C++ language, is listed in Appendix. Both the leaf node and processing element are derived from the abstract sensor class.

The signal energy, which is detected in real time at each local node and broadcast over the whole *cluster*, is an indicator of how close the node is to a potential target. In target detection and tracking applications, a leaf node with higher signal energy carries more information and should have higher priority of being visited. To simplify computation, we use a quantitative value to represent the level of signal energy detected by a local sensor node. The *setup* time spent by a PE accounts for loading the mobile agent code and performing other initialization tasks.

3.3.2 Communication Links

Wireless communication links need to be established between neighboring nodes as the mobile agent migrates along a route. The embedded RF modems of a sensor node provide such a networking capability with low power requirement. On the WINS NG 2.0 platform, each node is equipped with two RF modems, both of which support 2.4 GHz *frequency-hopped spread spectrum* (FHSS) communication [13]. The different *clusters* select different "network numbers," which correspond to separate hopping pseudo-noise sequences to avoid interferences. The detailed radio configuration and wireless link establishment is beyond the scope of this chapter. We define an abstract link class with only the parameters we are interested in (see Appendix for details).

It is worthwhile to note that the message transmission time between two sensor nodes depends not only on the physical distance between them, but on the channel bandwidth and the data packet loss rate as well as the size of messages to be transmitted, which includes partially integrated data and mobile agent code itself. In general, the electromagnetic propagation time is almost negligible in short-range wireless communication. Hence the physical distance is not explicitly considered in our model but incorporated as a part of the path loss (PL) representing the signal attenuation. The received signal strength below a certain level due to PL may not be acceptable. The system loss factor is a parameter of the free space propagation model, which is not necessarily related to the physical propagation [14].

3.3.3 Mobile Agent Routing

A mobile agent is dispatched from the *processing element* and expected to visit a subset of sensors within the *cluster* to fuse data collected in the coverage area. Generally speaking, the more sensors visited, the higher will be the accuracy achieved using any reasonable data fusion algorithm [15]. It is important to

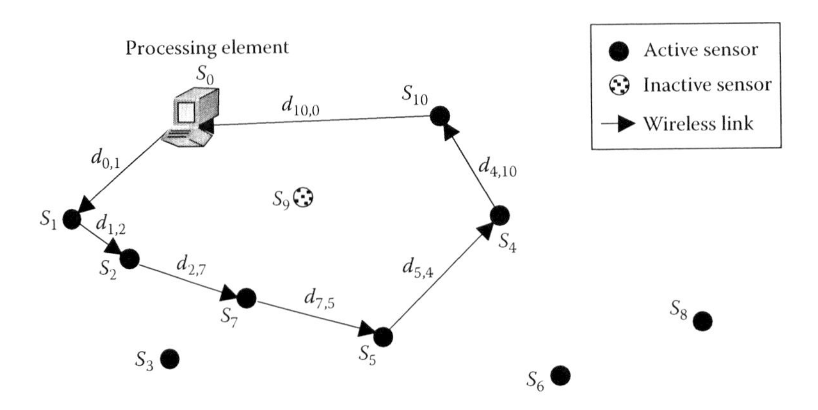

FIGURE 3.1 A simple MADSN with 1 PE and 10 leaf nodes.

select an appropriate route so that the required signal energy level can be achieved with a low cost in terms of total energy consumption and PL.

A MADSN with a simple configuration is shown in Figure 3.1 for illustrative purposes. The sensor network contains one PE, labeled as S_0, and $N = 10$ leaf nodes, labeled as S_i, $i = 1, 2, \ldots, N$, one of which is down. The sensor nodes are spatially distributed in a surveillance region of interest, each of which is responsible for collecting measurements in the environment. The signal energy detected by sensor node S_i is denoted by e_i, $i = 1, 2, \ldots, N$. Sensor node S_i takes time t_i,acq for data acquisition and time $t_{i,proc}$ for data processing. The wireless communication link with physical distance $d_{i,j}$ between sensor node S_i and S_j has channel width W bits and operates at frequency B Hz. Some sensor nodes may be down temporarily due to intermittent failures, as sensor S_9 in Figure 3.1.

The routing objective is to find a path for a mobile agent that satisfies the desired detection accuracy, while minimizing the energy consumption and PL. The energy consumption depends on the processor operational power and computation time, and the PL is directly related to the physical length of the selected path. We define these quantities in the next section.

3.3.4 Objective Function

The objective function for the MARP is based on three aspects of a routing path: energy consumption, PL, and detected signal energy.

1. *Energy consumption*: The energy consumption at a sensor node is determined by the processing speed and the computation time. If an energy-driven *Real-Time Operating System* (RTOS) is installed on the sensor node, the processor speed can be dynamically scaled depending on workload and task deadlines [16]. For wireless message transmissions, the energy consumption depends on the sensor's transmission power and message transmission time. We assume that the message includes the mobile agent code of size M bits and measured data of size D bits. For a given desired resolution, a fixed data size D is used to store the partially integrated data at each sensor. The time for the message to be transmitted over a wireless channel of bandwidth BW bps is calculated as

$$t_{msg} = \frac{1}{B} \cdot \left\lceil \frac{M+D}{W} \right\rceil \tag{3.11}$$

The energy consumption EC of path P, consisting of nodes $P[0], P[1], \ldots, P[H-1]$, is defined as

$$EC(P) = a \cdot (t_{0,setup} + t_{0,proc}) \cdot F_0^2 + P_{0,t} \cdot t_{msg} + \sum_{k=1}^{H-1} \left\{ b \cdot \left\{ (t_{P[k],acq} + t_{P[k],proc}) \cdot F_{P[k]}{}^2 \right\} + P_{P[k],t} \cdot t_{msg} \right\} \tag{3.12}$$

where k-th leaf node $S_{P[k]}$ on path P has data acquisition time $t_{P[k],acq}$, data processing time $t_{P[k],proc}$, operational level $F_{P[k]}$, transmitting power $P_{P[k],t}$, $i = 1, 2, \ldots, N$, and node $P[0] = 0$ corresponds to the PE. Coefficients a and b are chosen to "normalize" the processor speed to its power level.

2. *Path loss*: The power received by sensor S_j has the following relation with the power transmitted by sensor S_i according to the Friis free space propagation model [14]:

$$P_{j,r}(d_{i,j}) = P_{i,t} \cdot \frac{G_{i,t}G_{j,r}\lambda^2}{(4\pi)^2 {d_{i,j}}^2 \beta}, \tag{3.13}$$

where

$G_{i,t}$ is the gain of sensor S_i as a transmitter
$G_{j,r}$ is the gain of sensor S_j as a receiver

Wavelength λ is the ratio of speed of light c and carrier frequency f, and β is the system loss factor. The physical distance $d_{i,j}$ between S_i and S_j is computed from their spatial locations. PL represents the signal attenuation as a positive quantity measured in dB, and is defined as the difference (in dB) between the effective transmitted power and the received power:

$$PL(d_{i,j}) = 10\log\frac{P_{i,t}}{P_{j,r}} = 10\log\left[\frac{(4\pi)^2\beta}{G_{i,t}G_{j,r}\lambda^2} \cdot {d_{i,j}}^2\right] \tag{3.14}$$

Therefore, the total PL along path P can be calculated as

$$PL(P) = \sum_{k=0}^{H-1}\left[10\log\left(\frac{(4\pi)^2\beta}{G_{P[k],t}G_{P[(k+1)\bmod H],r}\lambda^2} \cdot d^2_{P[k],P[(k+1)\bmod H]}\right)\right] \tag{3.15}$$

3. *Signal energy*: An active sensor detects a certain amount of energy emitted by the potential target, which may or may not be used by a mobile agent for data integration. A mobile agent always tries to accumulate as much signal energy as possible for accurate decision in target classification or tracking application. The sum of the detected signal energy SE along path P is defined as

$$SE(P) = \sum_{k=1}^{H-1} s_{P[k]} \tag{3.16}$$

where $s_{P[k]}$ is the signal energy detected by the k-th sensor node on path P.

By combining the above three aspects of a routing path, we consider an objective function as follows:

$$O(P) = SE(P)\left(\frac{1}{EC(P)} + \frac{1}{PL(P)}\right) \tag{3.17}$$

wherein three terms $SE(P)$, $EC(P)$, and $PL(P)$ are first normalized to appropriately reflect the contribution by various loss terms. This objective function prefers paths with higher signal energies by penalizing those with high path losses and energy consumption. A path providing high signal energy at the expense of a considerable amount of energy consumption and PL may not be preferable. Alternative objective functions may be used as long as they correctly reflect the tradeoff between detected signal energy, energy consumption, and PL.

To facilitate the GA algorithm in Section 3.4, we define a fitness function based on the objective function as follows:

$$f(P) = O(P) + g \tag{3.18}$$

where g is the penalty function for overrunning the constraint defined by

$$g = \begin{cases} 0, & SE(P) \geq E \\ \delta \cdot (SE(P) - E)/E & SE(P) < E \end{cases} \tag{3.19}$$

where

E is the desired detection accuracy or signal energy level

δ is a properly selected penalty coefficient

3.3.5 NP-Hardness of Mobile Agent Routing Problem

The MARP is to compute a path P in a MADSN such that $O(P) > \tau$ and *k-hop mobile agent routing problem* (k-MARP) additionally requires that the path P have exactly k edges. We now show the latter to be NP-hard by reducing the *3D Maximum Traveling Salesman Problem* (MTSP) to it, which is an indication of the intractability of MARP. We first present the definition of MTSP. Given a completely connected graph $G = (V, E)$, and a nonnegative real number α, does there exist a closed-loop path P, with nodes $P[0], P[1], \ldots, P[n-1], P[n] = P[0]$, such that $\sum_{i=0}^{n-1} l_3(P[i], P[(i+1) \bmod n]) \geq \alpha$? Here, each vertex corresponds to a point in three-dimensional Euclidean space R^3. The starting point $v_{P[0]}$ and ending point $v_{P[n]}$ in the space refer to the same vertex in the graph.

The quantity $l_3(P[i], P[i+1]) = \sqrt{(x_{P[i]} - x_{P[i+1]})^2 + (y_{P[i]} - y_{P[i+1]})^2 + (z_{P[i]} - z_{P[i+1]})^2}$ is the Euclidean distance between two adjacent vertices $P[i]$ and $P[i+1]$ on path P. The MTSP under Euclidean distances in R^d for any fixed $d \geq 3$ is proved to be NP-hard in Ref. [17]. The conventional traveling salesman problem requires that path length be minimized and the cities are defined for dimension $d = 2$. On the contrary, MTSP requires maximization of path length and is known to be intractable in three or higher dimensions. Note that k-MARP requires maximization of $O(P)$ but is defined for $d = 2$, which makes a direct reduction from MTSP non-trivial.

Given an instance of MTSP, we generate an instance of k-MARP as follows. We create a graph for the k-MARP with $k = n$ using only x and y coordinates of vertices of MTSP (without loss of generality we assume that all coordinate values are distinct). We consider the objective $O(P) = (SE(P)/PL(P)) = \left(\sum_{i=0}^{k-1} s_{P[i]} \Big/ PL(P)\right)$ by ignoring the energy consumption component. Recall that the PL is given by

$$PL(P) = \sum_{i=0}^{k-1} 10 \cdot \log\left(A \cdot d^2_{P[i], P[(i+1) \bmod k]}\right) \tag{3.20}$$

Let $e_{P[i]}$ represent the edge between vertices $P[i]$ and $P[i+1]$ on path P. We define $s_{P[i]} = s(e_{P[i]}) = l_3(P[i], P[i+1]) + \tau\left(10 \cdot \log(A \cdot d^2_{P[i], P[(i+1) \bmod k]})\right) - \alpha/k$. A solution to the k-MARP is a path P with k hops such that $O(P) = \left(\sum_{i=0}^{k-1} s_{P[i]} \Big/ \sum_{i=0}^{k-1} d(v_{P[i]}, v_{P[i+1]})\right) \geq \tau$, τ is a given nonnegative real number. After reorganizing, the objective function can be rewritten as

$$\sum_{i=0}^{k-1} (s(e_{P[i]} - \tau \cdot d(e_{P[i]})) = \sum_{i=0}^{k-1} \left(l_3(P[i], P[(i+1) \bmod k]) - \frac{\alpha}{k} \right) > 0 \tag{3.21}$$

which guarantees the condition necessary for a solution to the corresponding MTSP.

On the other hand, if there exists a solution to the MTSP, that is, a closed-loop path P consisting of n edges such that $\sum_{i=0}^{n-1} l_3(P[i], P[(i+1) \bmod k]) \geq \alpha$, this path can be used to solve the corresponding n-MARP such that $O(P) \geq \tau$. Note that above reduction from MTSP to n-MARP is polynomial-time computable, and hence NP-hardness of the latter follows from that of the former.

The restriction of n-MARP is studied in Ref. [6], where two heuristics LCF and GCF are proposed. In the next section we propose a GA based method for MARP and show that it outperforms LCF and GCF.

3.4 Genetic Algorithm for Mobile Agent Routing Problem

3.4.1 Two-Level Genetic Encoding

We design a two-level encoding scheme to adapt the GA to the MARP in MADSN. The first level is a numerical encoding of the sensor (ID) label sequence L in the order of sensor nodes being visited by mobile agent. For the MADSN shown in Figure 3.1, the sensor label sequence L has the following contents:

0	1	2	3	7	5	6	8	4	10	9

The first element is always set to be "0" for reason that a mobile agent starts from the PE S_0. The mobile agent returns to S_0 from the last visited sensor node, which is not necessarily the last element of the label sequence if there are any inactive sensor nodes in the network. This sequence consists of a complete set of sensor labels because it takes part in the production of a new generation of solutions through the genetic operations. It is desired to inherit as much information as possible in the new generation from the old one. For example in Figure 3.1, although node 3, 6, 8, and 9 are not visited in the given solution (the second level sequence is designed to do so), they or some of them may likely make up a segment of a better solution than the current one in the new generation.

The second level is a binary encoding of the visit status sequence V in the same visiting order. For the MADSN in Figure 3.1, the visit status sequence V contains the following binary codes:

1	1	1	0	1	1	0	0	1	1	0

where

"1" indicates "visited"

"0" indicates "unvisited"

The first bit corresponds to the PE and is always set to be "1" because the PE is the starting point of the itinerary. If a sensor is inactive, its corresponding bit remains "0" until it is reactivated and visited.

Masking the first level of numerical sensor label sequence L with the second level of binary visit status sequence V yields a candidate path P for mobile agent. In the above example, the path P is obtained as

0	1	2	3	7	5	6	8	4	10	9

These two levels of sequences are arranged in the same visiting order for the purpose of convenient manipulations of visited/unvisited and active/inactive statuses in the implementation of GA. The number of hops H in a path P can be easily calculated from the second level of binary sequence as follows:

$$H = \sum_{i=0}^{N} V[i], \quad \begin{cases} V[i] = 1, & \text{sensor } S_i \text{ is active and visited} \\ V[i] = 0, & \text{sensor } S_i \text{ is inactive or unvisited} \end{cases} \tag{3.22}$$

3.4.2 Implementations of Genetic Operators

We now describe the genetic operators. These operators are similar to those used in the conventional solution to *Traveling Salesman Problem*. However, we adapt the details to the current routing problem.

3.4.2.1 Selection Operator

As discussed above, the purpose of the selection operation is to select good individuals and at the same time eliminate bad individuals from the population based on the evaluation of individual fitness. In our implementation, each pair of individuals is selected randomly from the old generation to perform the crossover, mutation, and inversion operations. The fitness is computed for every newly generated child for evaluation. To maintain the same population sizes for each generation, the fitness of every newly generated child is compared with the minimum fitness of the whole population. If it is bigger than the minimum fitness value, then this child is added to the population and the individual with the minimum fitness is removed; otherwise, the new child is discarded.

3.4.2.2 Crossover Operator

We design a two-point crossover operator in our implementation for both levels of sequences. These two crossover points are selected randomly. Given two parents as follows:

Parent 1: First level sequence: 0-2-7-3-/-5-1-6-/-4-9-8 Second level sequence: 1-0-1-1-/-1-0-0-/-1-1-1
Parent 2: First level sequence: 0-3-5-2-/-9-6-4-/-1-7-8 Second level sequence: 1-0-0-0-/-1-0-1-/-1-0-1

where "/" represents the crossover points, the crossover operator produces two children:

Child 1: First level sequence: 0-9-6-4-2-7-3-5-1-8 Second level sequence: 1-0-1-1-/-1-0-1-/-1-1-1
Child 2: First level sequence: 0-5-1-6-3-2-9-4-7-8 Second level sequence: 1-0-0-0-/-1-0-0-/-1-0-1

For the first level of label sequence, the crossover portion (between the two crossover points) of one individual is copied and inserted at the front of the other individual (immediately after label 0). All the duplicate genes in the resulting individual are knocked out to guarantee that each node appears exactly once in that individual. For the second level of visit status binary sequence, the crossover portions are simply exchanged between two individuals.

3.4.2.3 Mutation Operator

We implement a two-point mutation operator that randomly selects two points and exchanges the values of these two points in both strings. As an example, consider the following parent individual:

First level sequence: 0-2-9-3-7-1-4-5-8-6
Second level sequence: 1-0-1-1-0-1-0-0-1-1
↑ ↑
two selected gene loci

The mutation operator produces the following child:

First level sequence: 0-2-9-3-8-1-4-5-7-6
Second level sequence: 1-0-1-1-1-1-0-0-0-1

3.4.2.4 Inversion Operator

We implement the inversion operator as follows. At a time, two inversion points are selected randomly to determine the inversion portion of the individual. The inversion operation is executed by reversing the order of the inversion portion of the original individual. Given one parent as follows:

First level sequence: 0-5-7-/-1-2-8-9-/-6-3-4
Second level sequence: 1-0-1-/-1-0-1-1-/-0-0-1

where the inversion portions are enclosed by two "/" signs, the inversion operator produces the following child:

First level sequence: 0-5-7-/-9-8-2-1-/-6-3-4
Second level sequence: 1-0-1-/-1-1-0-1-/-0-0-1

3.4.3 Parameter Selection for Genetic Algorithm

We usually select a high probability value above 0.9 for a genetic operator like crossover, which controls the main direction of evolution process. A low probability value below 0.1 is appropriate for genetic operators like mutation or inversion to reduce the risk of destroying the good gene segments in later generations. Observed from experimental data, small variation of these probabilities does not have significant impact on the performance of GA. With respect to the maximum generation number, we select different values for different test examples in order to ensure that the optimization process approaches a steady state eventually. The difference of the best fitness values between two adjacent generations may be used as an alternative convergent indicator. In this case, the GA does not have to wait for a long time to reach the pre-specified maximum generation number if the optimization process converges quickly. Its disadvantage is that the program may prematurely terminate if the optimization process does not converge quickly.

3.5 Simulation Results and Algorithm Analysis

3.5.1 Simulation Results

We compare the search results of the GA with those computed by LCF and GCF. In most of cases, LCF is able to deliver satisfying route for mobile agent, hence it is a comparable algorithm with GA. GCF may find a path with less number of hops than LCF, but it usually has significantly longer path length resulting in unacceptable PL. A series of experimental networks of different sensor node sizes and distribution patterns are created to conduct the optimal routing. The spatial locations of all the nodes are randomly selected. The LCF and GCF algorithm pick up the center node as the starting point in each network. About 1%–10% of the sensors are shut down uniformly over the surveillance region. All sensor parameters of data acquisition and wireless channel in the MADSN use the real-life data of the field demo listed in Table 3.1.

In order to make a visual comparison, the search results computed by GA, LCF, and GCF for the first relatively small sensor network are shown in Figures 3.2 through 3.4, respectively. This sensor network consists of 200 nodes, 8 of which are in the sleep state. A quantified amount of signal energy associated with each active sensor ranging from 0 to 64 is displayed under the corresponding sensor node. The minimum acceptable amount of signal energy detected by an individual sensor node is 5, and inactive nodes do not detect any signal energy. There is one potential target located in the region. The sensor nodes in the vicinity of the targets detect higher signal energy than other nodes. The total detected signal energy is 1467 units, and the acceptable signal level for correct inference is set to be 1200 units.

TABLE 3.1 Parameters of the MADSN

Sensor Node Processor Type	Hitachi SuperH Processor SH-4 Architecture
Sensor node processor speed	200 MHz
Mobile agent sizes	400 bytes
Ave. data sizes	100 bytes
Carrier frequency band	2.4 GHz
Transmitting power	100 mW
Transmitter gain	2
Receiving power	80 mW
Receiver gain	2
Channel operation frequency	20 kHz
Channel width	16 bits
Data sampling rate	20 kHz
Sample data format	16-bit

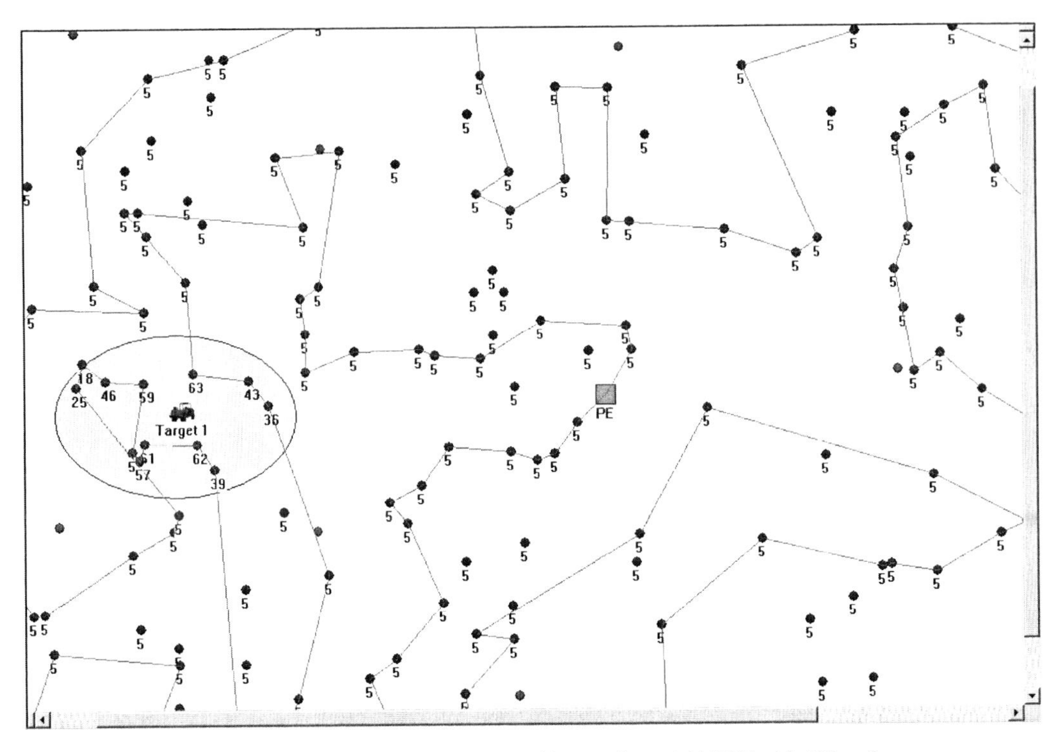

FIGURE 3.2 Visualization of the search result computed by GA for an MADSN with 200 nodes.

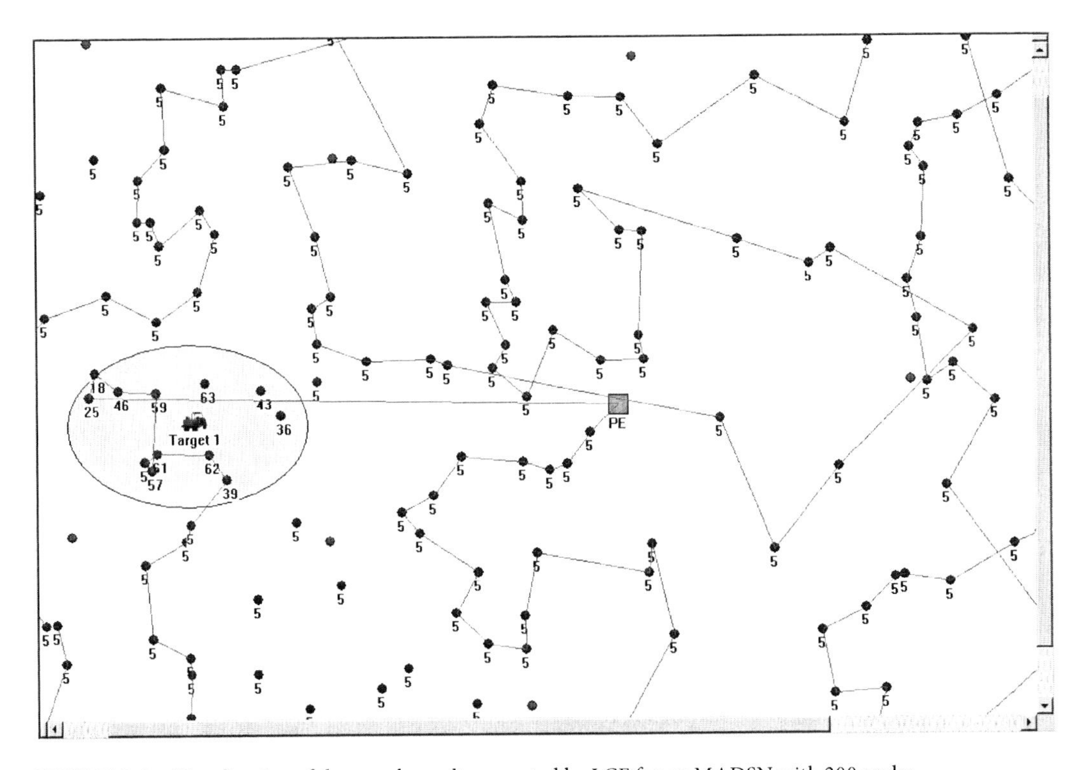

FIGURE 3.3 Visualization of the search result computed by LCF for an MADSN with 200 nodes.

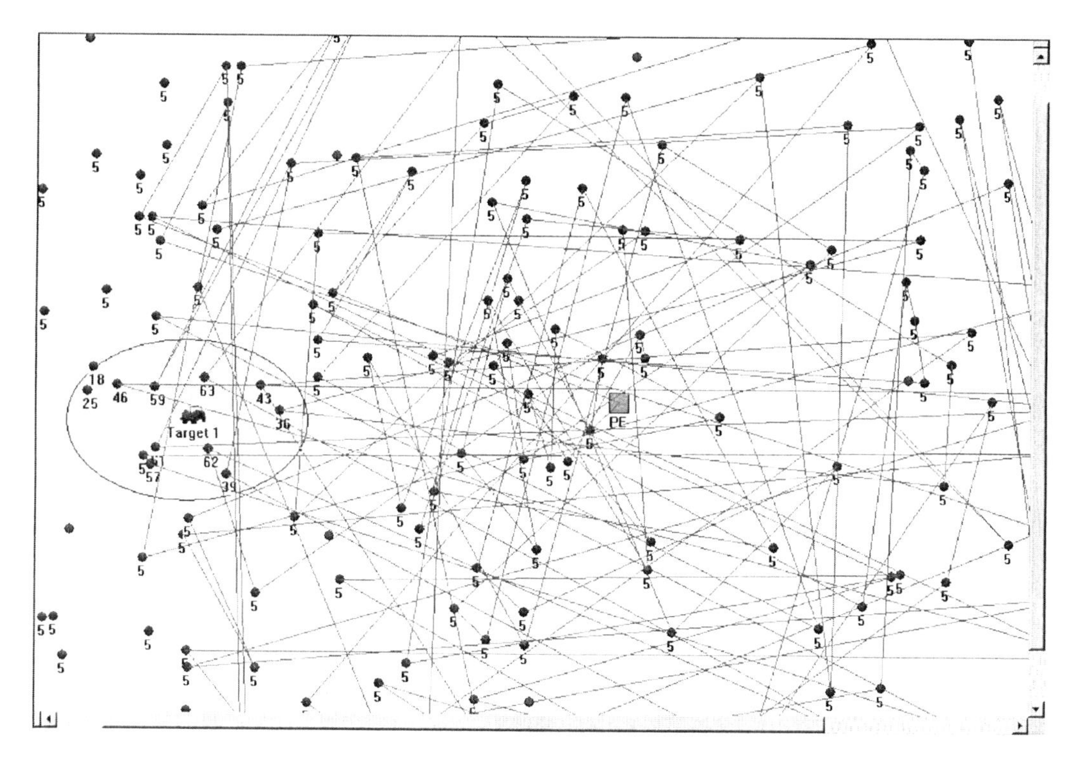

FIGURE 3.4 Visualization of the search result computed by GCF for an MADSN with 200 nodes.

The maximum generation number 200 is specified for GA as the convergent indicator, which informs the program when to stop searching process.

It has been observed that the optimization process of GA moves forward rapidly in the beginning, and becomes slow and stable in the later stages of computation, especially after the generation number reaches 100.

Table 3.2 shows that GA uses 140 hops to achieve the acceptable signal level, while LCF uses 168 hops and GCF uses 147 hops, respectively. The path losses of GA, LCF and GCF are 22,919 units, 26,864 units, and 28,748 units, respectively. The migration of mobile agent along the path computed by GA consumes 407 units of energy, while LCF consumes 489 units of energy, and GCF consumes 428 units of energy. The simulation results of other sensor networks with larger node sizes and randomized distribution patterns are also tabulated in Table 3.2.

In Table 3.2, the total detected signal energy represents the maximum energy detected by all active sensors deployed in the region under surveillance. The acceptable signal energy level is a given value desired for a specific application.

Figures 3.5 through 3.8 illustrate the corresponding curves of node sizes vs. number of hops, node sizes vs. PL, node sizes vs. energy consumption, and node sizes vs. objective values, respectively. From Table 3.2 and Figures 3.5 through 3.7, it has been seen that in most of cases GA is able to find a satisfying path with less number of hops, less energy consumption, and less PL than LCF and GCF algorithm. Figure 3.8 clearly shows that the GA has a superior overall performance over two other heuristics in terms of the objective function defined in this implementation.

The current GA program was implemented in C++ using MFC with GUI and per-generation result display. The code takes a few seconds to run the first 100 generations for a network of hundreds of nodes. GA runs much faster and its executable code size decreases significantly when GUI is not implemented. In such cases, the GA run time may not be a serious problem for semi-dynamic routing, which will be discussed in the next subsection.

TABLE 3.2 Comparisons of Search Results of GA, LCF, and GCF for Networks of Different Node Sizes and Distribution Patterns

Case #	Node Sizes	# of Dead Sensors	# of Potential Targets	Total Detected Signal Energy	Acceptable Signal Energy Level
1	200	8	1	1,467	1,200
2	300	4	2	2,985	2,750
3	400	5	2	4,000	3,680
4	500	8	3	4,080	3,710
5	600	9	4	4,980	4,800
6	700	10	5	5,340	5,190
7	800	11	4	6,200	5,950
8	900	13	3	7,000	6,380
9	1000	14	5	7,525	7,000
10	1100	13	4	8,515	7,990
11	1200	18	5	10,050	9,425
12	1300	16	4	11,410	9,600
13	1400	15	4	11,500	9,800
14	1500	19	5	12,380	11,000
15	1600	25	6	13,505	12,210

	GA				LCF				GCF			
Case #	No. Hops	Path Loss	Energy Cnsmpt	Achieved Signal Energy	No. Hops	Path Loss	Energy Cnsumpt	Achieved Signal Energy	No. Hops	Path Loss	Energy Cnsumpt	Achieved Signal Energy
1	140	22,919	407	1,202	168	26,864	489	1,215	147	28,748	428	1,204
2	216	34,154	670	2,785	273	42,027	796	2,757	228	43,191	664	2,763
3	300	47,100	875	3,691	379	57,842	1106	3,684	291	55,897	849	3,688
4	424	64,302	1237	3,716	460	68,645	1342	3,721	424	82,817	1237	3,714
5	483	76,205	1389	4,815	561	82,627	1637	4,807	490	95,417	1430	4,819
6	544	88,344	1552	5,195	663	96,370	1935	5,197	566	110,972	1652	5,194

7	613	98,722	1803	5,963	757	109,877	2186	5,952	708	139,179	2093	5,958
8	703	113,790	2079	6,400	847	121,082	2473	6,390	774	151,392	2260	6,385
9	838	122,087	2481	7,001	948	134,053	2768	7,003	929	183,379	2713	7,004
10	931	137,906	2704	8,002	1059	150,332	3123	7,990	988	207,397	2845	8,005
11	1067	152,397	3116	9,425	1146	161,248	3347	9,460	1117	221,073	3262	9,450
12	1009	145,899	2946	9,650	1208	171,032	3528	9,600	1138	231,898	3301	9,610
13	1105	158,471	3227	9,840	1287	175,492	3693	9,800	1174	239,843	3459	9,820
14	1270	170,029	3634	11,005	1381	192,300	4099	11,019	1315	258,942	3832	11,004
15	1369	182,382	3823	12,210	1534	211,317	4480	12,210	1495	295,711	4366	12,240

Case #	GA Objective Value $O(P) = SE(P)(1/EC(P) + 1/PL(P))$	LCF Objective Value $O(P) = SE(P)(1/EC(P) + 1/PL(P))$	GCF Objective Value $O(P) = SE(P)(1/EC(P) + 1/PL(P))$
1	3.005763	2.52989	2.854965
2	4.238259	3.529169	4.225116
3	4.296651	3.394613	4.409913
4	3.061832	2.826934	3.047271
5	3.529708	2.994646	3.420435
6	3.406098	2.739716	3.190872
7	3.367668	2.776951	2.88944
8	3.134647	2.63668	2.867397
9	2.87919	2.582226	2.619838
10	3.017345	2.611586	2.852306
11	3.086556	2.885079	2.939742
12	3.34177	2.777218	2.95268
13	3.111365	2.709512	2.879914
14	3.093068	2.745518	2.914104
15	3.260774	2.783227	2.844873

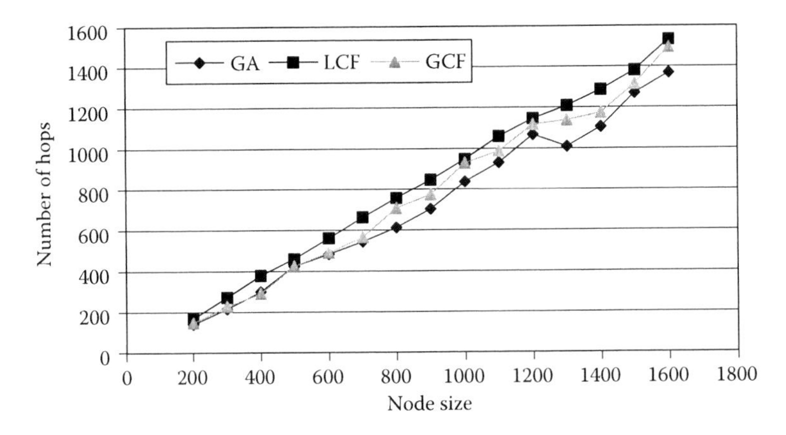

FIGURE 3.5 Node sizes vs. number of hops for the three algorithms.

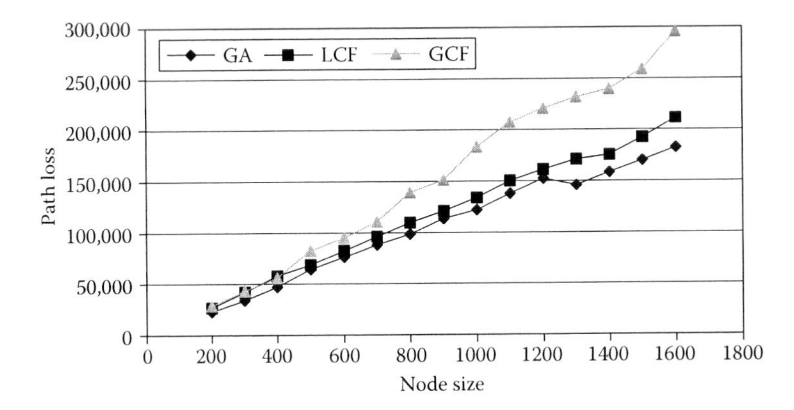

FIGURE 3.6 Node sizes vs. path loss for the three algorithms.

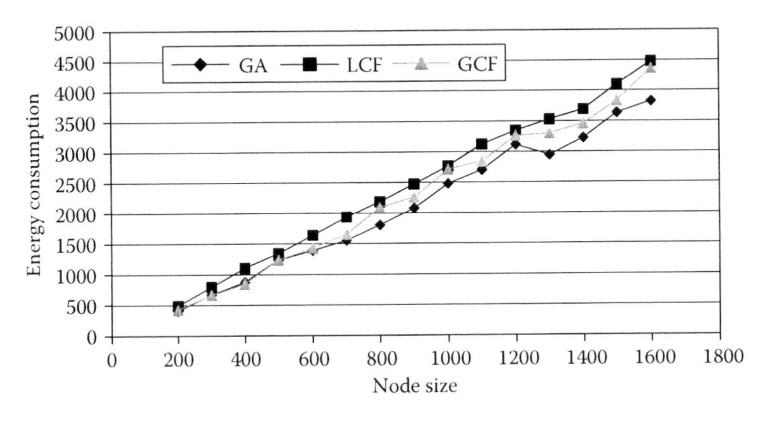

FIGURE 3.7 Node sizes vs. energy consumption for the three algorithms.

3.5.2 Algorithm Comparison and Discussion

GCF algorithm is relatively simple and fast but suffers from poor performance in terms of PL. GCF algorithm essentially utilizes sorting to compute the path. Its computational complexity is $O(N \log N)$ if using a comparison based sorting algorithm. LCF algorithm has the computation complexity of $O(N^2)$ if the closest neighbor node is obtained by simple comparison in each step. The analysis for

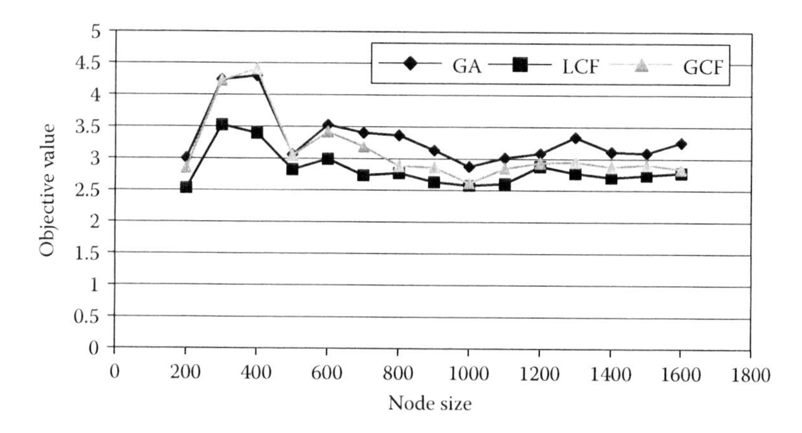

FIGURE 3.8 Node sizes vs. objective values for the three algorithms.

the computation complexity of the GA is more complicated. After making some simplifications on its implementation, the computation complexity for the GA is $O(NMG)$, where N is the number of nodes in the network, M is the initial population size, and G is the maximum generation number used to indicate the end of the computation.

As is the case for GCF algorithm, the performance of LCF algorithm also depends significantly on the network structure [6]. In some bad cases, it may end up with unacceptable results. Comparatively, the network structure has much less influence on the performance of the GA due to its random search technique.

Unlike the LCF and GCF algorithms, it is not necessary to specify a starting node for the implementation of GA. Actually, any active sensor node can be designated as a starting node in the GA, while the performance of LCF and GCF algorithms are crucially dependent on the location of its starting node. In addition, no matter in what order the nodes are visited, GA always comes up with a closed route for the mobile agent to come back to its starting node.

Mobile agent routing algorithm can be classified as dynamic and static routing according to the place where routing decisions are taken. A dynamic method determines the route locally on the fly at each hop of the migration of a mobile agent among sensor nodes. A static method uses centralized routing, which computes the route at PE node in advance of mobile agent migration. For different sensor network applications, either dynamic or static method can be applied. For example, it might be sufficient to use a static routing for target classification, but target tracking may require a dynamic routing due to its real-time constraint.

LCF algorithm is suitable for carrying out dynamic routing because each step of its computation depends only on the location of the current node, while the GCF is in favor of static routing, whose computation can be carried out offline based on the global network structure. Both LCF and GCF are deterministic routing, which always supply the same path between a source/destination pair in a given network. The GA collects information about the network status from all sensor nodes so that it is able to conduct adaptive routing. However, broadcasting the detected signal energy produces extra communication overhead.

Since it is desired to keep the mobile agent code as compact as possible, the GA may be used to implement a semi-dynamic routing. In this routing scheme, the routing code does not go with the mobile. If the network system is notified of some events (e.g., some nodes are shut down or activated, or do not have enough remaining energy to transmit the signal along the previously designated link), which causes the previously computed route to be invalid, the routing code is rerun based on the updated system information, and the new route is transmitted to the mobile agent for its further migration.

This semi-dynamic routing scheme is supported by the robustness of sensor network. According to the experience from the field demo, sensor nodes usually function well once brought up, and the network may remain stable continuously for 1–2 h of sessions.

3.6 Conclusions

We presented a mobile agent-based paradigm for data fusion in DSNs. By utilizing a simplified analytical model for the DSN, we formulated a route computation problem for the mobile agent in terms of maximizing the received signal strength while keeping PL and energy consumption low. This route computation problem turned out to be NP-hard thereby making it highly unlikely to develop a polynomial-time algorithm to compute an optimal route. Hence we proposed a GA to solve this problem by employing a two-level genetic encoding and suitable genetic operators. Simulation results are presented for comparison between our GA, and existing LCF, and GCF heuristics. Various aspects of the proposed algorithm such as computational complexity, impact of network structure and starting node, dynamic and static routing, are discussed.

Future research work is to be focused on exploring more complex routing models with more general objective function. For example, in the current model we assume that the sensor locations are fixed once they are manually deployed, which is the case in the field demo. However, in a real world sensor network, sensors could be airborne or installed on vehicles or robots. The mobility of sensors brings new challenges to the design of dynamic routing algorithm for the mobile agent. Besides, instead of using the simple free space propagation model to compute PL, more complex empirical propagation models may be studied and applied in the construction of objective function.

3.A Appendix

Class definitions of abstract sensor, leaf node, processing element, and wireless link used in the algorithm implementations are listed as follows:

```
class CSensor:
{
      unsigned int   m_sensorLabel;        // a unique sensor ID: 0 for PE,
                                              else for leaf nodes
      bool           m_sensorStatus;       // TRUE: active; FALSE: inactive
      double         m_processorSpeed;
      double         m_locationLongitude
      double         m_locationLatitude;
      double         m_dataProcessingTime;
      double         m_transmittedPower;
      double         m_transmitterGain;
      double         m_receivedPower;
      double         m_receiverGain;
}
```

3.A.1 Definition of Sensor Class

```
class CLeafNode: public CSensor
{
          bool      m_visited;
          double    m_dataAcquisitionTime;
          double    m_detectedSignalEnergy;
          double    m_dataSamplingRate;
          double    m_sampleDataFormat;
}

class CProcessingElement: public CSensor
{
       double m_setupTime;
}
```

3.A.2 Leaf Node Class Derived from CSensor and Processing Element Class Derived from CSensor

```
class CLink:
{
        CSensor*        m_pSensorTransmitter;
        CSensor*        m_pSensorReceiver;
        double          m_linkDistance;
        double          m_bandWidth;
        double          m_channelWidth;
        double          m_operateFrequency;
        double          m_carrierFrequency;
        double          m_linkPropagationTime;
        double          m_msgTransmissionTime;
        double          m_linkPowerLoss;
        double          m_systemLossFactor;
}
```

3.A.3 Definition of Wireless Link Class

Most of the attributes defined in these classes are self-explanatory. Leaf node and processing element classes are in turn derived from the abstract sensor class CSensor. In the definition of leaf node class, the attribute *m_dataSamplingRate* represents the frequency at which the signal data is sampled. The amount of memory space used to store the sampled data is determined by the attribute *m_sampleDataFormat*. In the definition of wireless link class, *m_systemLossFactor* is a parameter of the free space propagation model, which is not necessarily related to the physical propagation.

References

1. A.K. Hyder, E. Shahbazian, E. Waltz, *Multisensor Fusion*, Kluwer Academic Publishers, New York, 2002.
2. S.S. Iyengar, Q. Wu, Computational aspects of distributed sensor networks, *Proceedings of International Symposium on Parallel Architectures, Algorithms and Networks*, May 22–24, Manila/Makati, Philippines, IEEE Computer Society Press (I-SPAN 2002), Washington, DC, 2002.
3. D.N. Jayasimha, S.S. Iyengar, Information integration and synchronization in distributed sensor networks, *IEEE Transactions on Systems, Man, and Cybernetics*, 21(5), 1991, 1032–1043.
4. Y.F. Zheng, Integration of multiple sensors into a robotics system and its performance evaluation, *IEEE Transactions on Robotic Automation*, 5, 1989, 658–669.
5. R.C. Luo, M.G. Kay, Multisensor integration and fusion in intelligent systems, *IEEE Transactions on System, Man, and Cybernetics*, 19, 1989, 901–931.
6. H. Qi, S.S. Iyengar, K. Chakrabarty, Multi-resolution data integration using mobile agents in distributed sensor networks, *IEEE Transactions on Systems, Man, and Cybernetics Part C: Applications and Reviews*, 31(3), 2001, 383–391.
7. J.H. Holland, *Adaptation in Nature and Artificial Systems*, The University of Michigan Press, Ann Arbor, MI, 1975, reprinted by MIT Press, Cambridge, MA, 1992.
8. D.A. Coley, *An Introduction to Genetic Algorithms for Scientists and Engineers*, World Scientific, Singapore, 1999.
9. P.H. Winston, *Artificial Intelligence*, 3rd edn., Addison-Wesley Publishing Company, Reading, MA, 1993.
10. G. Chen, *Genetic Algorithm and Its Applications*, People's Post Publishing House, Beijing, China, 1996.
11. D.E. Goldberg, *Genetic Algorithms in Search, Optimization and Machine Learning*, Addison-Wesley Publishing Company, Reading, MA, 1989.

12. S.S. Iyengar, D.N. Jayasimha, D. Nadig, A versatile architecture for the distributed sensor integration problem, *IEEE Transactions of Computers*, 43(2), 1994, 175–185.
13. Rev. A, *WINS NG 2.0 User's Manual and API Specification*, Sensoria Corporation, San Diego, CA, May 30, 2002.
14. T.S. Rappaport, *Wireless Communications Principles and Practice*, 2nd edn., Prentice Hall PTR, Upper Saddle River, NJ, 2002.
15. N.S.V. Rao, Multisensor fusion under unknown distributions: Finite sample performance guarantees, in *Multisensor Fusion*, A.K. Hyder, E. Shahbazian, E. Waltz (eds.), Kluwer Academic Publisher, Boston, MA, 2002.
16. V. Swaminathan, K. Chakrabarty, Real-time task scheduling for energy-aware embedded systems, *IEEE Real-Time Systems Symposium (Work-in-Progress Sessions)*, Orlando, FL, November 2000.
17. A. Barvinok, S.P. Fekete, D.S. Johnson et al., The geometric maximum traveling salesman problem, combined journal version of previous paper and Fekete's 1999 SODA paper on the maximum TSP, including a faster algorithm for arbitrary polyhedral metrics. Submitted to *JACM*, 2002.

4

Computer Network: Basic Principles

Suresh Rai
Louisiana State University

Today, the word computer network is synonym to Internet or information super-highway and is a household name. Its mass appeal among novices, nerds, and pundits is primarily due to the applications like electronic mail (e-mail), world wide web (*www*), remote terminal access (*telnet*), and different protocols such as file transfer protocol (*ftp*), network file system (*NFS*), network news transfer protocol (*NNTP*), etc. These applications have made information dissemination easy, timely, and cool. Not long ago, when the Mars Path Finder Lander (nicknamed Sagan Memorial Station) landed on the Mars surface on July 4, 1997 (Independence day for the United States), almost everybody got surprised from the interest of a vast number of people who wanted to know the results and look at the first-ever high-resolution color images of the Martian surface themselves. This "get-self-involved" urge took them not in front of TV sets to watch a reporter narrate the story, but to the Internet where they felt satisfied by watching the story revealing itself through images that were posted on the Internet almost instantaneously by NASA scientists. More than one million people visited the Jet Propulsion Laboratory web page and its various mirror sites created for this purpose using their web browsers and service providers*; this was a record in itself. Since 1997, the popularity of Internet has grown exponentially. It is a worldwide collection of more than 250,000 networks, public and private, that have agreed to use common protocols and exchange traffic. An internet (lowercase "i") or internetwork refers to a set of networks connected by routers and appears to its users as a single network. A network can be taken as a set of machines or hosts linked by repeaters or bridges.

* Some typical service providers include America On Line (AOL), Juno On Line, CompuServe, AT&T, MCI, etc.

4.1 Layered Architecture and Network Components

The basic objective of a computer network is for an application on one node to communicate with another application or device on another node. An application could be a file transfer, terminal access, electronic mail, resource sharing, etc. While this may sound simple, some complexities are involved with the possibility of different implementations of WAN (wide area network), MAN (metropolitan area network), and LAN (local area network) systems. This section, first, discusses a reference model from International Standards Organization (ISO) and, then, introduces the Transmission Control Protocol/Internet Protocol (TCP/IP) architecture (commonly known as the Internet architecture). We also describe some typical internetworking components such as repeater, bridge, and router.

4.1.1 Layering and OSI Model

The basic purpose of layering is to separate network specific functions to help make the implementation transparent to other components. Thus, elements of layered specifications allows standard interfaces for "plug-and-play" compatibility and multi-vendor integration. Other reasons for having layered systems include enhancing and information-hiding features, ease of modification and testing, and portability. As an example of adding features or facilities, consider an unreliable physical layer that is made reliable through the use of a data link layer supporting an automatic repeat request (ARQ) scheme. Finally, layering allows us to reuse functionality. This is typically evident with an operating system design where a lower layer implements functionality once, which can then be shared by many upper layer application programs.

Layering is a form of "information hiding," which offers advantages but also leads to some problems. A lower layer provides a service interface to an upper layer, hiding the details of "how is it done?" Further, it is sometimes difficult to allow efficient performance out of a layer without violating the layer boundary. As an example, consider the protocol stack with error and flow control functions. The error control feature deals with the ability to retransmit packets corrupted on a link; the flow control, on the other hand, relates to the rate at which packets are placed by the source on the link. A flow control that relies on network congestion and offers a better job from performance view point poses a threat to layer boundary violation. It is because the flow control layer needs to know the details of data transfer over the local links in the network. To obtain an efficient performance, it becomes imperative to "leak" enough information between layers. Choosing a balance between the "leak" and "information hiding" feature of the layer model provides a hallmark of good protocol stack design.

Most communication environments use layering to separate the communication functions and application processing. For example, the OSI* reference model divides the communication between any two networked computing devices into seven layers or categories. Commonly used architectures such as AppleTalk and TCP/IP, on the other hand, has only five layers. It is important to remember that a layered model does not constrain an implementation; it provides a framework. Implementations, thus, do not conform to the OSI reference model, but they do conform to the standards developed from the OSI reference.

In the following, we consider a top-down approach to describe seven OSI layers and their functionality. Layer 7 (1) stands for the top (bottom) layer.

Layer 7 (application layer): This layer contains the programs that perform the tasks desired by users. Such tasks may include file transfer, electronic mail, printer service, remote job entry, and resource allocation. Special purpose applications such as Gopher, Fetch, and Wide Area Information Server (WAIS)

* The ISO Open System Interconnection (OSI) reference model is termed *open* because, unlike proprietary architectures such as IBM's System Network Architecture (SNA) and Digital Equipment Corporation's DECnet, its details are publicly available to anyone at little or no charge.

help navigate the way to resources on the Internet. Similarly, the World Wide Web links thousands of servers using a variety of formats including text, graphics, video, and sound. Browsers such as Netscape and Internet Explorer are used for the purpose.

Layer 6 (presentation layer): The presentation layer accepts a message from the application layer; provides formatting functions such as encryption, code conversion, and data compression; and then passes the message to the session layer. Encryption is needed for security reasons. Note that data can be represented by any of several different codes, the two most common being American Standard Code for Information Interchange (ASCII) and Extended Binary Coded Decimal Interchange Code (EBCDIC). Also, different computers store data differently in the memory. The big (little) endian strategy considers high-order (low-order) byte first in the memory. Code conversion is essential to take care of all these variations. Other Layer 6 standards guide graphic and visual presentation. PICT is a picture format used to transfer QuickDraw graphics between Power PC or Macintosh programs. Tagged Image File Format (TIFF) is a standard graphics format for high resolution, bit-mapped images. JPEG standards come from Joint Photographic Experts Group. For sound and movies, presentation layer standards include MIDI (Musical Instrument Digital Interface) for digitized music, and MPEG (Motion Pictures Expert Group) standard for compression and coding of motion video for CDs and digital storage.

Example 1.1

To illustrate the need for MPEG data compression, consider a frame for network video having 352×240 pixels (a pixel refers to one dot on the display). If we use 24 bits/pixel (to help maintain the color), each frame will necessitate 247.5 kB. To make it video quality, one requires a rate of 25–30 frames/s, which produces 247.5×30 kB/s or approximately 60 Mbps. Obviously, some method such as the MPEG technique is needed to compress the bit stream to reduce its rate up to 1.5 Mbps. Equipping multimedia clients and servers with compression/decompression capability will, thus, lower the demand on the medium.

Layer 5 (session layer): This layer enables two applications to establish, manage, and terminate a session (or logical connection) to facilitate communication across the network on an end-to-end basis (refer to Figure 4.1). Essentially, the session layer co-ordinates service requests that occur when applications communicate between different hosts. For example, a user may "logon" to a remote system and may communicate by alternately sending and receiving messages. The session layer helps coordinate the process by informing each end when it can send or must listen. This is a form of synchronization. The session layer also controls data flow (which can be either full or half duplex) and provides recovery if a failure occurs. As an example, let a user be sending the contents of a large file over a network that suddenly fails. Instead of retransmitting the file from the beginning, the session layer allows the user to insert checkpoints in a long stream. If a network crashes, then only the data transmitted since the last checkpoint is lost.

Following are typical examples of session-layer protocols and interfaces:

- NFS, developed by Sun Microsystems, allows transparent access to remote network-based resources. It is used with TCP/IP and UNIX.
- RPC (Remote Procedure Call) provides a general redirection mechanism for distributed service environments. RPC procedures are built on clients, and then executed on servers.
- ASP (AppleTalk Session Protocol) is used to establish and maintain sessions between AppleTalk client and a server.
- X-Window system permits intelligent terminals to communicate with remote UNIX computers as if they were directly attached monitors.

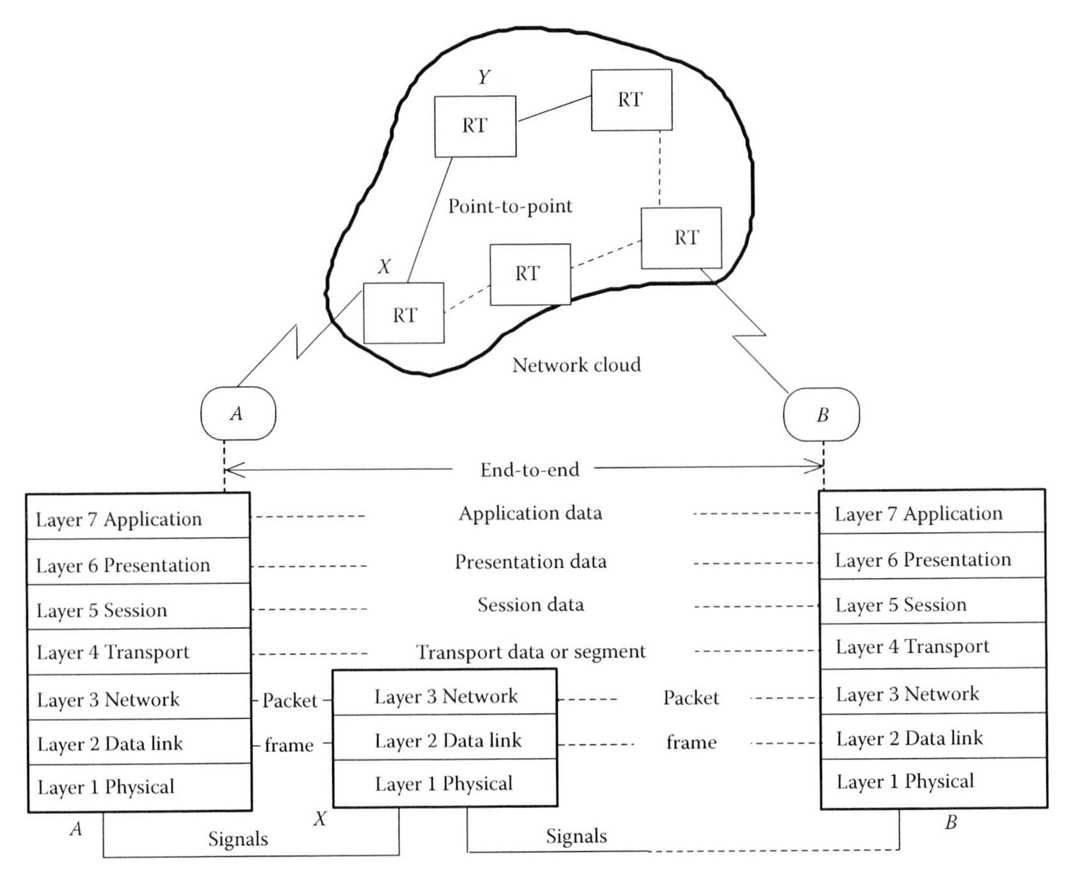

<table>
<tr><td></td><td>Wide area network (WAN)</td><td colspan="4">Local area network (LAN)</td></tr>
<tr><td>Data link layer</td><td>SDLC, HDLC, X.25, Frame relay, ISDN, PPP</td><td colspan="3">LLC (802.2)</td><td rowspan="2">Ethernet</td></tr>
<tr><td>Physical layer</td><td>V.24, EIA-232</td><td>802.3</td><td>802.5</td><td>FDDI</td></tr>
</table>

(b)

FIGURE 4.1 (a) A network cloud, scope of end-to-end and point-to-point communication, and OSI layers (RT means a router). (b) Physical and data link standards.

Layer 4 (*transport layer*): The fourth layer arranges for end-to-end delivery of messages between *transport service access points* (TSAPs) or *ports*. For this, a transport layer often fragments messages into segments and reassembles them into messages, possibly after resequencing them. (A layer 4 packet is called "segment.") It also oversees data integrity. In other words, layer 4 ensures that no messages are lost or duplicated and that messages are free of errors. As an example, a typical protocol in this layer may use TSAPs of the source and destination session entities, together with a checksum to detect errors, and message sequence numbers to ensure that messages are successfully received and are in order. Further, the transport layer delivery of messages is either *connection-oriented* or *connectionless*.

A connection-oriented service provides a reliable transport and is modeled after the telephone system. Its two typical examples include remote login and digitized voice. In such cases, the communicating end systems accomplish the following:

- Ensure that segments delivered will be acknowledged back to the sender
- Provide the retransmission of any segments that are not acknowledged
- Put segments back into their correct sequence at the destination
- Provide congestion avoidance and control

Contrary to this, a service such as electronic (junk) mail does not require connections. All that is needed is a best effort strategy, that is, a high probability of arrival, but no guarantee. It is achieved through a connectionless service that is modeled after the postal system.

Layer 3 (network layer): The network layer supports naming, addressing, accounting, and routing functions. Naming provides a way to identify a host in the network and is mainly for human usage. The addressing at Layer 3 provides an alternative means to locate computer systems on an internetwork and is primarily used by the machine or computer. A name server then resolves between the host name and its (logical) address. There are several addressing schemes, depending on the protocol family being used. In other words, AppleTalk addressing is different from TCP/IP addressing, which in turn is different from OSI addressing, and so on. Regardless of the protocol type, a network layer address is generally called virtual or logical address and is often hierarchical. This greatly enhances with which addresses could be assigned and also helps enable scaleable aggregation of addressing and routing information. Upon receiving a packet from the transport layer, the network layer logs the event to the accounting system and then prepares the packet for transmission to the next node on the path to the destination. It does this by looking up the destination address in its network routing table to find the next address along the path. (This approach is called destination-based routing.) The function—finding the path the packets must follow—is called *routing*.* A proper routing strategy such as RIP (routing information protocol), OSPF (open shortest path first), BGP (border gateway protocol), and PIM (protocol independent multicasting) helps build the routing table that provides a fastest, cheapest, or safest path.

Layer 2 (data link layer): This layer is responsible for the transmission of a frame on point-to-point links or between two adjacent nodes (see Figure 4.1). Typically, a frame encapsulates a packet by appending additional fields such as a unique hardware† (physical or link layer) address of the network interface card (NIC) that identifies the machine attached to a shared link (technically, a physical address is not needed for a point-to-point link), cyclic redundancy check (CRC) bits to accomplish low-level transmission error detection, and control fields to help perform error recovery and flow control. In this layer, we also consider methods such as bit stuffing, byte stuffing, or code violation for achieving data transparency, which refers to the ability of the data link to transmit any bit combination (arising from binary or text file). Figure 4.1 shows standards for data link layer.

---

* In fact, network layer routing has two components: forwarding and control. The forwarding component deals with actual forwarding of packets and works on destination basis; that is, the decision about where to forward a packet is made based only on its destination address. The control component is responsible for the construction and maintenance of the forwarding (routing) table and consists of one or more routing protocols.

† Contrary to logical address, a hardware address is flat. One good example of a flat address space is the U.S. social security numbering system, where each person has a single, unique social security number. The hardware address is unique for each network connection. The NIC is the hardware in the computer that enables one to connect it to a network. On most LAN-interface cards, the physical address is burned into ROM; when the NIC is initiated, this address is copied into RAM. A host or computer has only one NIC because most computer systems have one physical network connection, they have only a single link-layer address. Routers and other systems connected to multiple physical networks can have multiple link-layer addresses. Such machine, having more than one network interface, is called a multi-homed host (MH). However, an MH does not work as a router (which always has more than one NIC). Some typical examples of the MH include servers such as NFS, database, and firewall gateways because all these are configured as multi-homed hosts.

Layer 1 (*physical layer*): The physical layer pertains to the transmission of bits of data over a physical medium such as twisted pair wire, cable, optical fiber, or satellite link and, hence, characterizes the transmission medium, the nature of the signals, the data rate, and related matters. It also defines the mechanical, electrical, functional, and procedural aspects of the interface between the data terminal equipment (DTE), such as a computer or workstation, and data circuit-terminating equipment (DCE), such as a modem. For example, a commonly used physical layer interface in the United States is EIA-232-D. Its mechanical specifications include the physical dimensions, latching and mounting arrangements, and so forth of its *D*-shaped 25 pin interface connector. The electrical characteristics provide voltage levels and timing (such as pulse rise times and duration). Functional specifications assign meaning to circuits or pins which are divided into four subgroups: data, control, timing (bit and byte), and grounds. Finally, procedural sequences provide handshaking rules to set up, maintain, and deactivate physical level interconnections that help accomplish data transfer.

4.1.2 TCP/IP Layering

The TCP/IP protocol suite was developed as part of the research done by the DARPA. Later, TCP/IP was included with the Berkeley Software Distribution of UNIX. The Internet protocol suite includes not only Layer 3 and 4 specifications (such as IP and TCP), but also specifications for such common applications as e-mail, remote login, terminal emulation, and file transfer. Loosely, the TCP/IP refers to the Internet architecture and upholds only five layers of the OSI reference model (Figure 4.2).

The application layer in TCP/IP combines the features of Layers 5 through 7 of the OSI model and supports protocols such as file transfer, e-mail, remote login, etc. Network management is also an important ingredient of this layer. As is obvious from Figure 4.2, the TCP/IP protocol stack maps closely to the OSI reference model in the layers 4 and 3. TCP and user datagram protocol (UDP) represent two protocol examples working at Level 4. The transport layer of TCP performs two functions, namely, flow control and reliability. The flow control is provided by sliding windows and the reliability is achieved through sequence numbers and acknowledgments. As mentioned earlier, the UDP is a connectionless protocol and uses no windowing or acknowledgments. In this case, an application layer protocol such as Trivial FTP, Simple Network Management Protocol, Network File Server, and Domain Name System is responsible for providing the reliability. The network layer of TCP/IP contains following protocols:

- IP (Internet Protocol) provides connectionless, best effort delivery routing of datagrams. Note that a packet at Layer 3 is called datagram. Further, the IP layer is not concerned with the content of the datagrams. It looks for a way to move the datagrams to their destination.
- ICMP (Internet Message Control Protocol) provides control and messaging capabilities and is implemented by all TCP/IP hosts. An ICMP message is carried in IP datagram.

		Internet or TCP/IP Suite		
Layer	OSI	TCP/IP	Data Format	Protocols
7	Application	Application	Messages or streams	Telnet, FTP, TFTP, SMTP, SNMP, etc
6	Presentation			
5	Session			
4	Transport	Transport	Segment	TCP, UDP
3	Network	Network	Datagram	IP
2	Data link	Data link	Frame	PPP*
1	Physical	Physical	Bits	

FIGURE 4.2 Internet protocol suite vs. OSI layers. (*Besides PPP, the data link layer also supports other layers depending on the type of networking hardware being used: Ethernet, token ring, FDDI, etc.)

- ARP (Address Resolution Protocol) determines the data link layer address for a known IP addresses. RARP (Reverse ARP), on the other hand, obtains network addresses when data link layer addresses are known. ARP and RARP are implemented directly on top of the data link layer, especially over a multi-access medium such as Ethernet.

The network interface at layers 2 and 1 are technically not defined as part of the TCP/IP stack. Hence all the variations given in Figure 4.1b can be applied for this interface, which could be a LAN or PPP (Point-to-Point Protocol) connection. For example, Figure 4.3a shows two hosts A and B connected to an Ethernet segment. Here, a model illustrating layers at each node explains the basic concepts including the peer (or similar) protocols. Thus, the TCP at A has a peer relationship with the TCP at B and so does the IPs at both hosts. Note that peers communicate to each other through entities supported by their lower levels. To explain this, we have shown the encapsulation or how application data is packaged for transmission. As is obvious from Figure 4.3b, an application data is handled by TCP/UDP as a segment, in which the corresponding transport layer header is attached to the data. The network layer, in turn, takes the segment and passes as a datagram to the data link layer after appending it with an IP header. The data link layer encapsulates the datagram and creates a frame to be handled by the physical layer. The encapsulation requires applying both header and trailer, depending on the type of data link used. At the receiving end, the message flows similarly; but this time the specific layer header is removed and the user data is eventually passed on to the peer application.

The generic term for information combined with an appropriate layer header is Protocol Data Unit (PDU). For example, a TCP segment is a transport layer PDU, and an IP datagram is a network layer PDU.

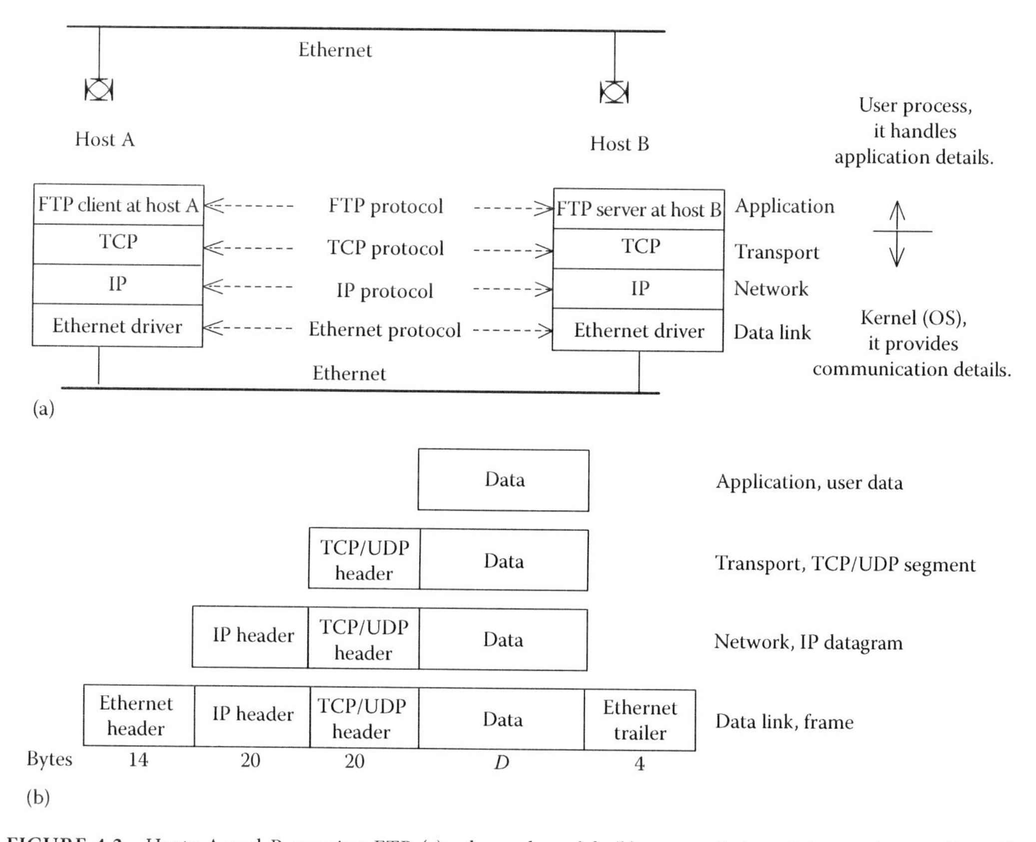

FIGURE 4.3 Hosts A and B running FTP (a) a layered model, (b) encapsulation of data as it goes down the protocol stack. (Note $20 + 20 + D = 46–1500$ bytes, supported by the Ethernet.)

4.1.3 Internetworking Components

In the following, we consider internetworking components (also called *protocol converters*) such as repeater, bridge, and router and explain their functionality vis-à-vis the OSI model. These components are used to provide an establishment-wide interconnected system. Alternatively, technologies based on FDDI and ATM can also be used for this purpose. Note that two networks may differ at any of the OSI layers. But the design of the protocol converter that establishes a connection becomes more involved as the layer where they differ increases. We also lose forwarding speed as we progress from repeater through gateways. However, we gain better functionality with this change.

Repeater
A repeater that realizes only the physical layer of the OSI model duplicates, amplifies, or regenerates the transmission signal from one network segment onto another network segment. It only interconnects two homogeneous (identical) networks making a single, larger network (Figure 4.4a). For instance, a repeater helps overcome the segment-length limitation of 500 m on an Ethernet. Since the repeater forwards each bit it receives on the connected segment, the loading or traffic on the entire LAN increases with an increase in demand in the form of number of nodes on the network. This, obviously, causes deterioration in the overall network response time. Also a LAN has constraints concerning the use of repeaters. Because the repeater takes a finite time to sample a pulse rise and to regenerate the received pulse, it introduces a slight pulse delay—known as *jitter*. As jitter accumulates, it adversely affects the ability of network hosts (stations) to receive data.

Bridge
A bridge is a device or layer of software that allows the interconnection of two or more LANs at the media-access (MAC) sublayer of the data link layer of the OSI model. As shown in Figure 4.4b, different types of LANs are interconnected through a bridge; a LAN segment joins the bridge through

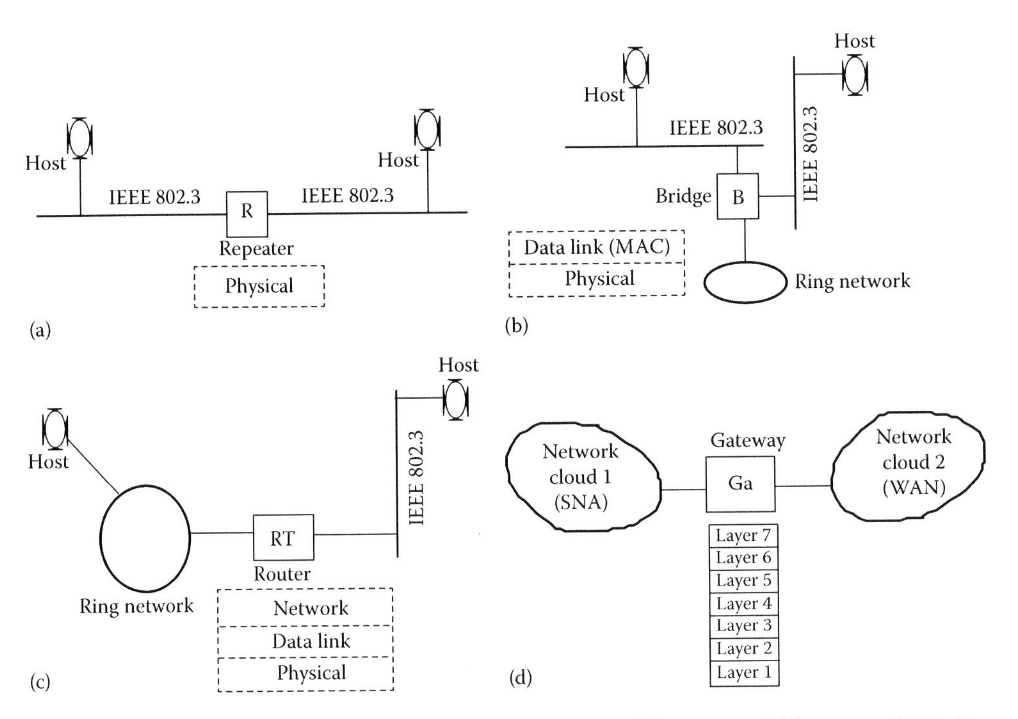

FIGURE 4.4 Internetworking components (a) repeater, (b) bridge, (c) router, and (d) gateway (SNA: System Network Architecture, WAN: Wide Area Network).

a port; and as many segments can be connected as the number of ports in the bridge. A bridge acts in the promiscuous mode which means it receives and buffers all frames in their entirety on each of its ports. If the frame is error-free and the node to which it is addressed is on a different segment, then the bridge forwards the frame onto that segment. Thus, bridges are superior to repeaters because they do not replicate noise or malformed frames; a completely valid frame must be received before it will be relayed. It is important to note that the bridge routing uses the hardware address (of the NIC).

Router

As shown in Figure 4.4c, a router operates at the network layer (or OSI layer 3) and is not sensitive to the details of the data link and physical layers. It contains two or more NICs just like a bridge. (Any system with multiple interfaces is called *multihomed*.) Unlike bridges, an IP address, described in Chapter 4, identifies the interface. Thus it can be used to connect different types of networks, such as a token ring LAN to an IEEE 802.3 LAN or a LAN to a WAN (see Figure 4.5). A router maintains routing information in routing tables. These tables contain the IP addresses of the hosts and routers on the networks to which the router is connected. The tables also contain pointers to these networks. When a router gets a packet, it consults its routing table to see if it lists the destination address in the header. If the table does not contain the destination address, then the router forwards the packet to a default router listed in its routing table.

Figure 4.4d, which incorporates all the seven layers, is known as gateway. Common examples of gateways include web proxy gateways and transcoders, which convert multimedia data from one format to another. All these example gateways operate on entire messages.

Routers offer an important capability that bridges do not have. It relates to frame segmentation for transmission between LANs. Note that the token ring (Ethernet) supports a data field of up to 4500 (1500) bytes. When a bridge is used to connect these two networks, software must be set at each token ring host to limit all frames to a maximum of 1500 bytes, to accommodate their flow onto a bridged Ethernet network. (Here the software can be set to limit only those frames destined for an Ethernet host, while still allowing the token ring capability on local hosts.) In contrast, a router can divide a token ring frame into two or more Ethernet frames eliminating the need of resetting software on each token ring host.

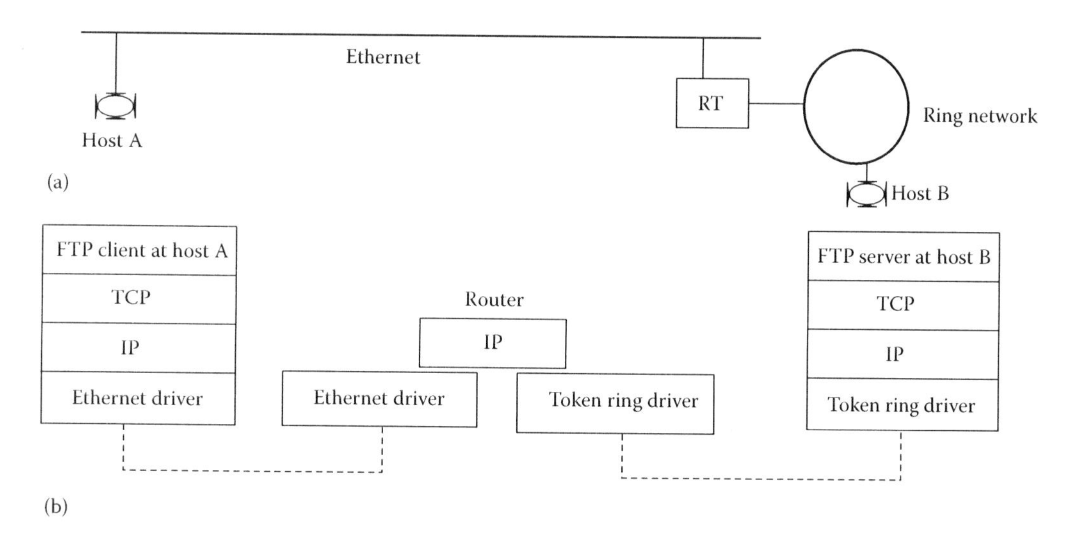

FIGURE 4.5 Client/server model (a) a network of Ethernet and token ring, (b) layered representation.

4.2 Link Sharing: Multiplexing and Switching

A high bandwidth* (or capacity) line is a costly resource. The cost can be spread among many users by allowing them to share the line by *multiplexing* and *switching*. Multiplexing is, generally, described in the frequency or time domain where signals of many users are combined into one large bandwidth or long duration, respectively. Note that users are geographically distributed and are not confined to a few locations. Switching allows us to bring together the signals of these scattered users by assigning the line on a demand basis for the duration of time needed, then returning the line back to the pool of available resources after communications have been completed.

On the basis of a frequency or time domain view, a high bandwidth link is considered as a group of frequency or time slots, respectively. Let us call it a "channel." In frequency division multiplexing (FDM), each user gets a channel on the high bandwidth link, thus allowing a link to be shared among several users. With time division multiplexing (TDM), a time slot allows the user to utilize the entire bandwidth of the link, but for a very small amount of time (usually of the order of a few ms or less). Contrary to FDM, a signal in TDM is in digital form. Switching, on the other hand, uses switches to allow the link sharing. Different switching techniques are described in this section.

4.2.1 Multiplexing Techniques

Multiplexing contains two Latin words *multi* and *plex* which mean *many mixing*. It, thus, allows two or more low-speed channels† (signals) to share a single high-speed transmission medium such as wire, cable, optical fiber, etc., of equivalent bandwidth.‡ A multiplexer denotes a device that performs multiplexing according to a frequency or time domain view of a signal. Thus there are three basic kinds of multiplexing methods, *frequency-division multiplexing*, *time-division multiplexing*, and *code division multiplexing*. They differ from each other in the following ways (refer to Figure 4.6):

1. FDM is an analog methodology allowing users to share frequency slots from the channel's bandwidth. TDM is a digital technique; it means bits and bytes share the time slots on a high speed line.
2. FDM uses broadband transmission (which means the line bandwidth is divided into multiple channels). It is achieved by translating in frequency or modulating the band-limited signals by separate carriers employing amplitude, frequency, or phase modulation methods, etc. TDM, on the other hand, uses baseband transmission in which signals are applied to the transmission medium without being translated in frequency.
3. Most FDM units are used to combine very low-speed circuits onto single voice-grade lines for transmission to a central site, while TDMs are used for both low- and high-speed lines for the same purpose.
4. Code-division multiplexing is completely different from FDM and TDM. It allows a station to use all the bandwidth all the time. The basic concept uses a technique called spread-spectrum communication, which was developed to send military signals having antijam capability and covert operation or low-probability of intercept capability. It was also useful for multiple-access, where many users share a band of frequencies. Spread spectrum communication provides the simultaneous use of a wide frequency band via code-division–multiple-access (CDMA) technique. In code-division multiplexing, a bit duration is subdivided into n very small intervals; $n = 64$ or 128. The intervals

* We shall use the terms bandwidth (a measure of a link's frequency band in Hz) and capacity (or speed or bit rate in bps) interchangeably. Strictly speaking, these terms differ and one can refer to any data communication book for more details.

† A channel is a communication link of fixed bandwidth or capacity.

‡ The term equivalent bandwidth characterizes the source and provides a conservative estimate of its capacity.

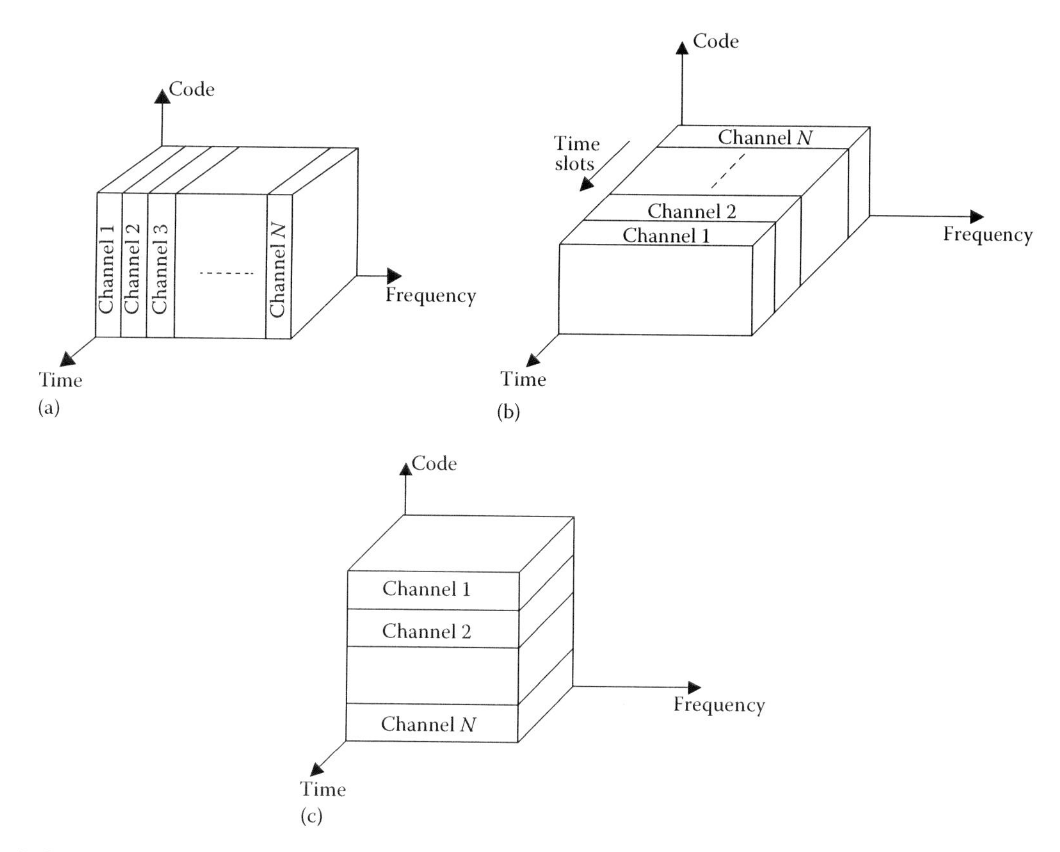

FIGURE 4.6 (a) Frequency-division, (b) time-division and (c) code-division multiplexing concepts.

are called *chips*. Each station is assigned an *n*-bit code or chip sequence. All chip sequences are pairwise orthogonal, such that their normalized inner product is zero. This property is useful as it helps decoding process or recovering the correct bit from among the received bits.

4.2.1.1 Frequency Division Multiplexing

As stated above, FDM involves dividing the total bandwidth of a channel into a number of frequency slots, with each slot assigned to a pair of communicating entities. Let the bandwidth of an analog signal $s_i(t)$ be b_i. If the available bandwidth of the high-speed transmission medium is B, then it is divided into n channels such that

$$B > \sum_{i=1}^{n} b_i$$

The difference in bandwidths, $B - \sum b_i$, is used for guard bands that prevent a channel bleeding into adjoining channels and, thus, help maintain the integrity of the input signals. Figure 4.7 shows some typical aspects of FDM such as multiplexing (MUX) and demultiplexing, (deMUX) frequency translation using modulation, and frequency-time view. Examples of FDM include broadcast radio (both AM and FM) and TV transmissions. For instance, the bandwidth for broadcast TV (54–806 MHz) is partitioned into 68 channels, each having 6 MHz. The very high frequency (VHF) (ultra high frequency [UHF]) channels 2–13 (14–69) span between 54–215 MHz (470–806 MHz). FDM offers a specific advantage as it does not need any addressing. The filter and carrier assigned to each frequency slot suffice to

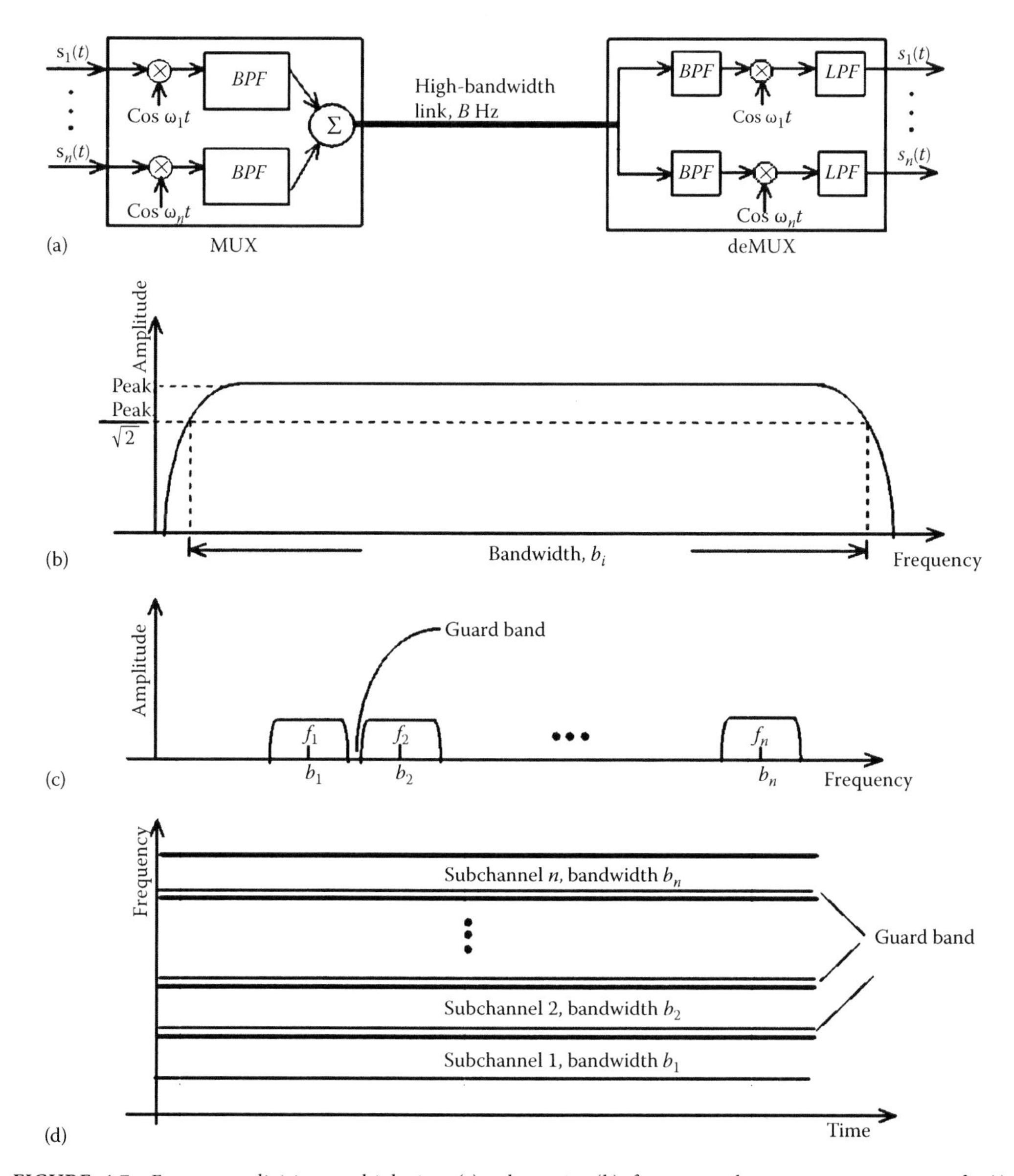

FIGURE 4.7 Frequency division multiplexing (a) schematic, (b) frequency domain representation of $s_i(t)$, (c) multiplexed channel, and (d) frequency-time view (BPF: band pass filter, LPF: low pass filter, and f_i: ith carrier frequency).

separate the signals and direct them to their destination nodes. Note that FDM is effective for analog signals, and is not useful for data communication where we have digital signals. Furthermore, a variation of FDM called *wavelength division multiplexing* (WDM) is used for fiber optic channels.

4.2.1.2 Time Division Multiplexing

TDM assigns each signal, preferably in a round-robin fashion, the entire high-speed bandwidth (or capacity, to be very specific) for a *time slot* (TS), which is a very small time duration of the order of millionths of a second. TDM uses digital signals. Depending on the width of a slot, which could be a bit, byte, or group of bytes, TDMs have three basic modes of operation: bit multiplexing, byte multiplexing, and block multiplexing. A T-1 carrier scheme represents a typical example of

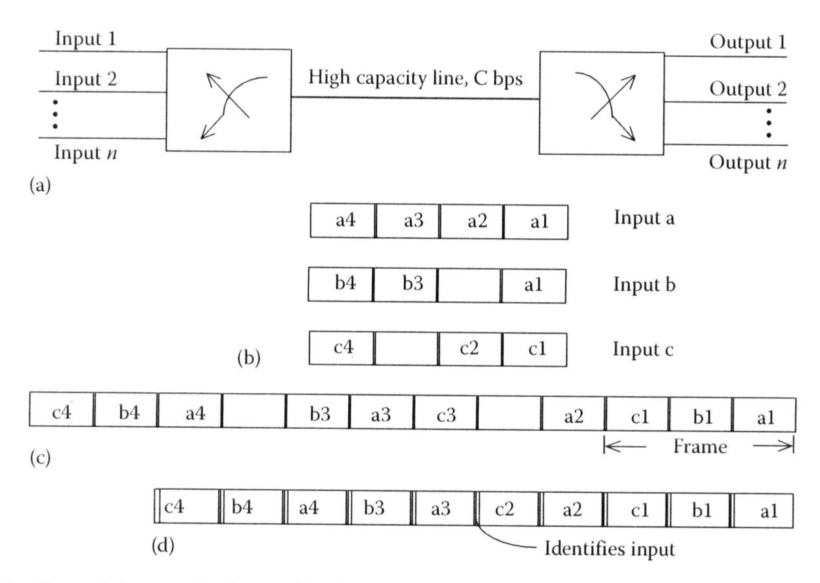

FIGURE 4.8 Time division multiplexing (a) schematic, (b) slots from three inputs a, b, and c, (c) synchronous TDM signal, and (d) asynchronous TDM signal on high capacity line.

byte multiplexing. Others are functionally implemented in exactly the same way. Further, the way a TS is allocated to a channel produces two TDM techniques: *synchronous* and *asynchronous*. Refer to Figure 4.8 for details.

In *synchronous* TDM, or just TDM, each input has its own pre-assigned time slot in a frame* of size $n \times \tau$, where n is the number of time multiplexed input signals and τ is the size of a TS. Since the timing of the slot is fixed, it is named synchronous. If the input has no data to send at that moment when its turn comes, then the TS remains empty because no one else can use it. This is also the case with FDM. In both cases, the capacity of a costly resource is wasted. Figure 4.8 shows a synchronous TDM scheme. It is obvious from this illustration that if c_i is the input rate of the ith source and C is the capacity of the aggregated link, then $C \geq \sum c_i$. Synchronous TDM is good for delay-sensitive traffic such as voice and broadcast TV signals. If used with data communication that is generally bursty, the link utilization† will be less because there exist longer quiet periods (when there is no data on the line) with burst traffic.

In an *asynchronous* (or *statistical*) TDM, a time slot can be used by any input as long as it is not-occupied, and hence the name asynchronous. This is achieved by assigning an identifier for the particular input in each time slot. Figure 4.8d shows an asynchronous TDM scheme. It is obvious from this illustration that if c_i is the input rate of the ith source and C is the capacity of the aggregated link, then $C < \sum c_i$. Thus the ratio of $\sum c_i/C$, called *multiplexing gain,* is greater than one (note that it is at most 1 for synchronous TDM). Alternatively, it states that more than n inputs can be accommodated over the channel having a capacity of C bps. This, though, imposes a performance restriction in terms of *bit error rate* (BER) tolerance. Also, buffering is needed to avoid overflows, especially when several incoming traffic bursts coincide. Asynchronous TDM is useful for delay-insensitive data traffic because it helps improve link utilization by dynamically assigning time slot duration to active users.

* A frame consists of a sequence of slots: slot 1, slot 2, ..., slot n. A *logical channel* employs every nth slot. Thus, a logical channel i occupies slots i, $i + n$, $i + 2n$, ..., and so on.

† The link or channel *utilization* refers to the average data rate divided by channel capacity or the fraction of time a channel is busy.

4.2.2 Switching Techniques

Recall switching helps apportion a line/channel* among users on a demand basis. À la carte, if available, a channel is assigned to a user for the duration of time needed. Once the duration is over, the channel is returned back into a common resource pool where it is accessible to other users. A switch† or a router performs the switching function which includes: (a) access control (that is, whether or not to accept the call from the requesting user), (b) determining the appropriate route, and (c) resource allocation (means assigning a bandwidth or capacity) along the route to be used to transfer the data stream. Switching techniques, described below, can be distinguished based on these three features. Another point of differentiation comes from the concept of channel utilization, defined as the ratio of busy time to the total duration or fraction of time the channel is busy. Using this, the transmission of a voice signal has a utilization of around 50%. A data communication, generally termed *bursty*, can have utilization that ranges from 0% to 100%. It is very low in the case of interactive data (telnet and rlogin, for example) and high in the case of bulk data (*ftp*, e-mail, Usenet news). Besides utilization, other parameters that describe a call include origin and destination, bandwidth (peak, average, or effective), delay constraint, and admission error rate. Considering the scope of the book, we shall exclude these other parameters from our discussion below. We now describe the three most important types of switching techniques: *circuit switching*, *message switching*, and *packet switching*.

4.2.2.1 Circuit Switching

Circuit switching, a method popularized by telephony, requires three steps. Step 1 is a *call* or *connection* set up phase. Here a source attempts to establish an end-to-end physical connection by generating a call request signal to the nearest switch of the network to which it is connected. As the request hops from node to node toward its destination, a circuit from each link along the route, if available, is reserved by the network. The circuit may contain a frequency band or a slot depending on FDM or TDM being used. The route decision utilizes a simple strategy such as the first path available from a fixed list or a complex measure like taking a path with the largest residual capacity. If the circuit assignment on links en route to the destination is successful, the network is said to *admit the call*. However, if it fails on any one of the links, the call is said to be blocked or rejected (depending on the fact that it is made to wait until the resource becomes available, or is cleared). The network is then termed *busy*. Step 2 is called the *call holding* phase and performs data transfer once the connection is established. Thus an admitted call has a fixed route and bandwidth during the data transfer phase. A particular frequency band or a time slot assigned in step 1 is not available to anyone else as long as the data is transmitted. This makes flow control within the call unnecessary. Further, during this phase, the data stream does not undergo any buffering at intermediate nodes. This eliminates queuing and processing delays at these switches. There exists propagation delay, however, which is about 0.5 ms for every 100 km line segment. Circuit switching is ideally suited for audio and video transmissions that represent delay sensitive signals. Step 3 is known as the *call clearing* phase and is performed when the data transfer is complete. In this step, the connection is deallocated, which means the circuit is returned back to the bank of available resources for future assignments. Figure 4.9 provides an example of a circuit switched network where (s,t) refers to a source-destination node pair.

Circuit switching is not efficient for data traffic. As mentioned earlier, data stream is bursty. It means transmissions are short and occur irregularly. In other words, periods of peak transmission are followed by intervals when there is no data on the line. Note that a circuit switching scheme reserves the line/channel for the entire duration of the transfer; thus a user is billed for the data transfer periods in which

* We assume that the capacity of a channel is partitioned into a number of fixed-rate *circuits* or *logical channels*. The apportionment is usually achieved using TDM. For example, $T-1$ contains 24 voice or data circuits.

† The term *switch* is used in telephony, while a *router* is employed in computer communication. We shall use switch and router in the same sense.

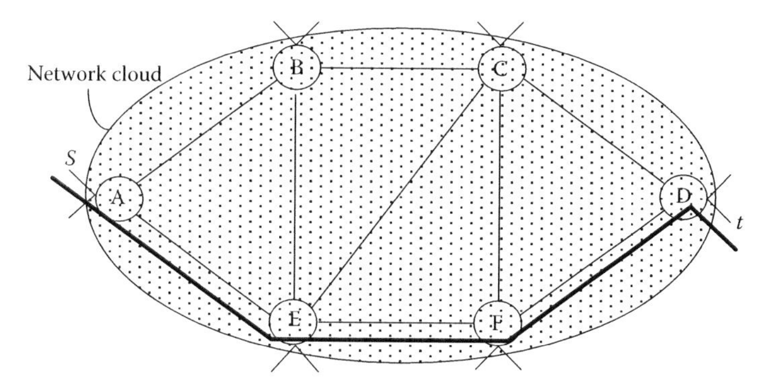

FIGURE 4.9 Illustrating circuit switching.

the line remains used. The problem becomes acute with interactive data transfer. Another reason that makes circuit switching unattractive for data traffic is the varying rate of data transmission. Note, data rates differ widely–from hundreds of bps (bits/s) between a terminal and the computer to millions of bps between two computers. A system that should provide circuit switched connectivity to varying rates needs to be designed keeping maximum rate in view. It is because with peak demand the system will fail to work as circuit switched data cannot be delayed. Designing the entire system to take care of this situation, evidently, presents a waste on resources.

4.2.2.2 Message Switching

CYBERNET, implemented by Control Data Corporation (CDC) using switched, leased, and private lines and satellite links, provides an example of a message switched network. A message is a variable length bit string and characterizes a logical unit of information for the purpose of communicating to one or more destinations. Programs and data files present two typical examples of messages. Similar to processing telegrams in the telegraph industry, message switching is based on a store and forward concept. This means that message switching does not require establishing an a priori physical connection between a source-destination pair. Here, the sender prefixes the destination address to a block of data and sends it to its neighbor. The neighbor stores the data and later forwards it to its neighbor en route to the destination. The message is thus transmitted in its entirety and hop by hop through the network until it finally reaches the destination. This scheme provides efficient channel utilization. Since messages wait in a queue before being transmitted and they also arrive at a node randomly depending on the congestion in the network, a message in this switching technique encounters queuing and transmission delays in addition to propagation delay. Further, no restriction is imposed on message lengths, which can differ vastly. This implies an intermediate node storing the information before forwarding to its neighbor should have enough disk space for the varying message sizes. To get around this problem, the concept of packet switching is used.

4.2.2.3 Packet Switching

Similar to message switching, a packet is stored and then forwarded by successive nodes en route to the destination node in the packet switched network. A *packet* is a small fixed-size block of the message. Each packet contains a sequence number and a label specifying the destination address, and is individually transmitted. Such *fragmentation* of a message offers the following advantages. Because the size of a packet is known a priori, its storage at intermediate nodes during transmission between source and destination is manageable. Also, a packet i that has been received and processed at some node j can be forwarded before packet $(i + 1)$ fully arrives at j. It obviously reduces delays and improves the throughput. Figure 4.10 illustrates the working of a packet-switched network. We assume that a message containing five packets, labeled 1 through 5, traverses between switches A and D. These packets are generated at source s and are destined for destination t. Node A transmits alternately to nodes B and E,

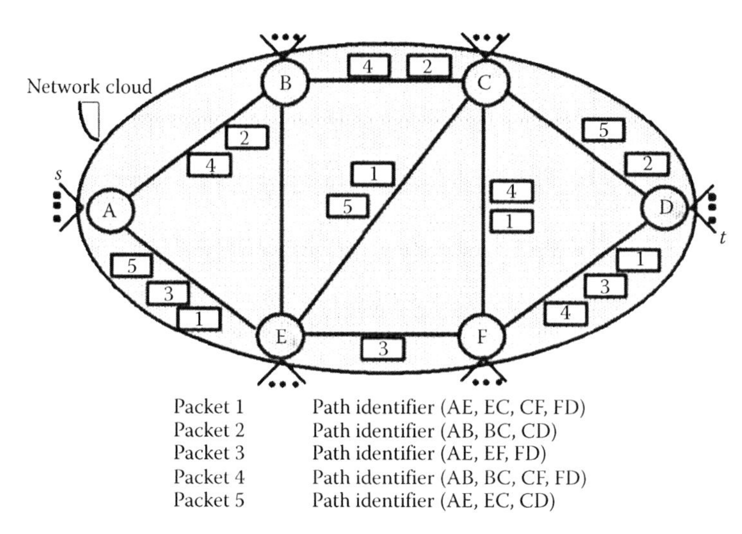

FIGURE 4.10 Packet switching concept and path identifiers.

while intermediate switches decide the next hop depending on a routing strategy in accordance with traffic congestion, the shortest end-to-end path, and other criteria.

Let us now consider functions such as access control, routing, and resource allocation to help compare packet and circuit switching methods. The process of call acceptance and routing for packet switched calls is similar to the corresponding processes for circuit switched calls. For example, a switch in packet switching may generally accept all arriving packets or may use some discretion based on parameters such as cost, delay, throughput, reliability, etc. A routing protocol, then, employs these measures to determine a source to destination path. Typically, the call is connected on the paths with the minimum expected end-to-end delay* within a given list of available paths. The capacity allocation problem provides an important point of difference between packet and circuit switching techniques. We make a network-wide resource allocation in circuit switching and that, too, for the complete duration of the call. Thus there exists a real or deterministic commitment of resources. In packet switching, an allocation is done for individual packets for the duration of the transmission from one switch to the next as they progress within the network. The commitment is therefore virtual or stochastic. To illustrate this point, if the capacity allocation is made on an average bandwidth basis and bursts occur at peak rates, then the network depending on the policy that is currently being enforced may drop all or part of the data packet stream. Also, this necessitates the use of a flow control scheme, mostly on a link basis using on-off or continuous (selective or go-back to *n* type) ARQ scheme.

We have already mentioned that a switch decides a bandwidth or capacity and an appropriate path to the requesting user. In packet switched networks, packets from different sources that share a common link are statistically multiplexed. It helps optimize the bandwidth allocation by providing a multiplexing gain as discussed earlier in Section 4.2.1.2.

The selection of a route that packets should follow is determined in one of two ways: *virtual circuit* and *datagram*. Figure 4.11 illustrates these services. The *virtual circuit* (VC) or connection-oriented transport service is similar to circuit switching. In this case, a path called a virtual circuit is set up through the network for each session (Figure 4.11a). Switches along the route then contain a mapping between incoming and outgoing VCs. Different packets that are part of the same data transfer use this information for their routing decision. Thus they follow the same path and packets arrive in order at the destination, eliminating the need of a packet assembler/disassembler (PAD). At the end of the session,

* Typical acceptable delay depends on the type of data stream. For real time applications, it is about 200 ms; while for interactive traffic (non-interactive services such as e-mail), it is a few (many) seconds.

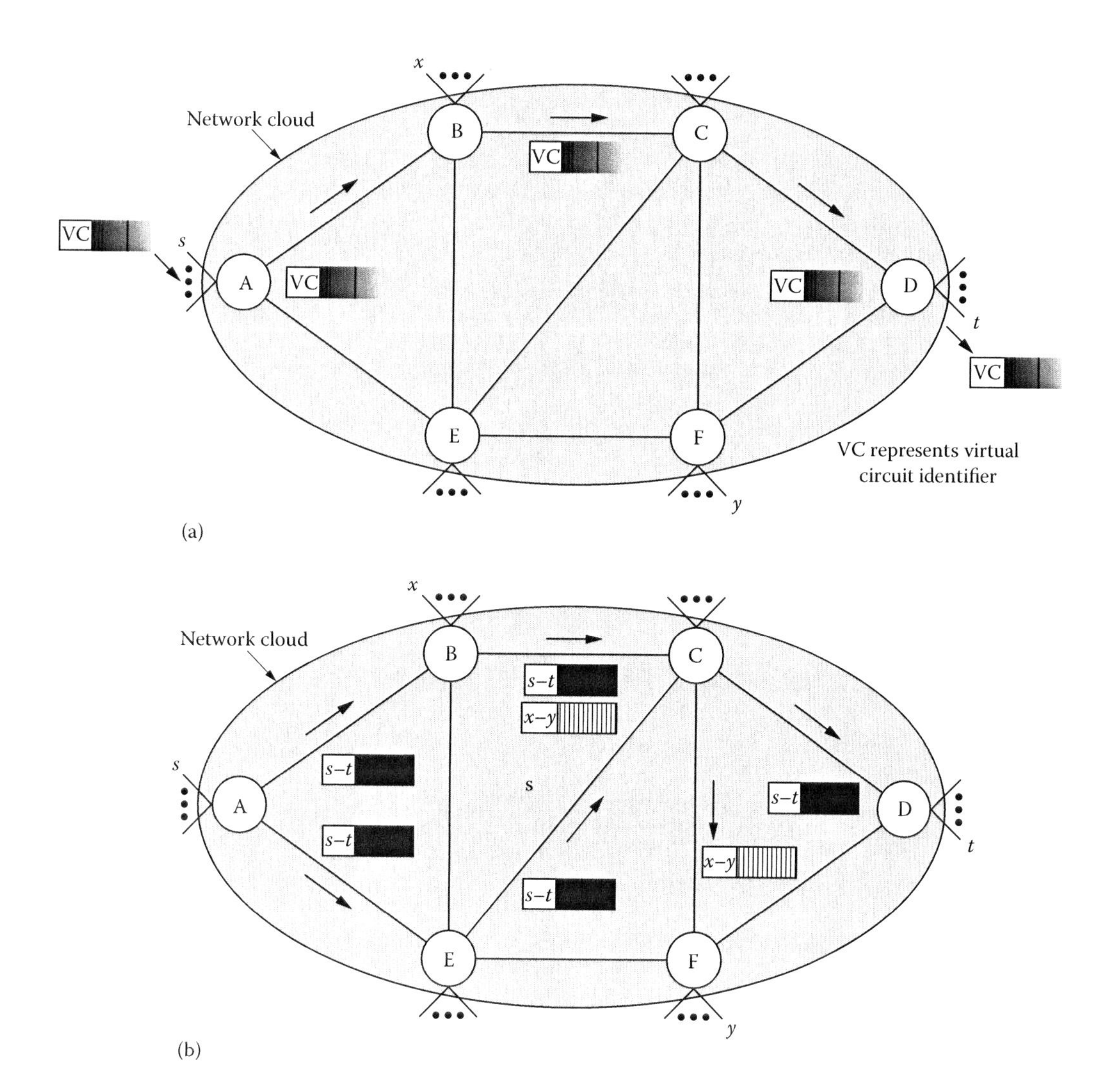

FIGURE 4.11 Two packet switching techniques—(a) virtual circuit and (b) datagram.

like circuit switching, the VC is cleared. A VC service is further classified as switched or permanent. The *switched* VC (SVC) is similar to a dial-up or public connection, while the *permanent* VC (PVC) is analogous to a leased or private line. Thus, PVC does not need a call set up phase. There is, though, a subtle difference between a private line and a PVC connection. In PVC, the bandwidth of the line is shared among multiple users, while the entire bandwidth of the line is dedicated when it is leased. TCP (transport control protocol) provides a typical example of VC service.

Datagram (or *connectionless*) transport service, on the other hand, is very similar to the way letters are handled by the post office. Each letter, marked with its destination address, goes from one post office to other post office, and so on until it reaches its destination. In datagram switching, a packet is labeled with its source and destination addresses and is transmitted individually (see Figure 4.11b for *s–t* and *x–y* traffic). The inclusion of address bits (which are often quite long) with each packet provides an overhead. The routing decision for each packet is made at every intermediate switch using a static or dynamic (meaning periodically updated) routing table. Note: the table is created based on some optimality criterion such as cost, delay, distance, throughput, etc. It is likely that a routing table may change because of changing conditions when a node or link fails or is unavailable because of congestion.

Datagram packets react quickly to the changes, meaning thereby that packets from the same source may follow different paths to arrive at the destination and their arrival could be out of order, too. Hence, a destination node should be capable of re-sequencing the packets; an expensive PAD handles this job. The datagram transport service does not need any prior path set up, hence it is quite suitable for short transmissions of a few packets. User datagram protocol (UDP) provides an illustrative example.

Example 1.2 [10]

To illustrate the computation of delays in virtual circuit and datagram services, consider a typical route (AB, BC, CD) in Figure 4.11 taken by a packet having 500 bits of data in either of the services. In addition to data bits, we assume in the virtual circuit case a header of 5 bytes for VC number and a trailer of 2 bytes to support error correction. Further, let the virtual circuit set up time be 250 ms. Thus, the transmission time in VC is $(500 + 8 \times 5 + 8 \times 2)/56{,}000$ which is 9.93 ms. Here, we assume that the capacity of each link is 56 kbps. To transmit N packets, we require a total time $T(\text{VC}) = 250 + (N + 3) \times 9.93$ ms because three links AB, BC, and CD are involved. To compute these parameters for datagram service, consider the header part as 10 bytes instead of 5 bytes as it indicates both source and destination addresses. The trailer part, however, remains unchanged as 2 bytes. Thus, the transmission time in datagram is $(500 + 8 \times 10 + 8 \times 2)/56{,}000$ which is 10.64 ms and $T(\text{datagram}) = (N + 3) \times 10.64$ ms.

Example 1.3

Obtain N for which $T(\text{VC}) < T(\text{datagram})$ in Example 1.2.

4.3 Data Transmission Basics

Several parameters such as medium, bit and character encoding schemes, half- or full-duplex, serial or parallel transfer, and asynchronous or synchronous type help characterize a data communication arrangement such as point-to-point protocol (PPP). A twisted wire, cable, and optical fiber form some typical examples of the communication medium. Unshielded twisted wire* pair (UTP) is mainly utilized in telephone networks. A LAN based on 10BaseT† also uses UTP. A shielded twisted wire pair, cable, or optical fiber provides better bandwidth. For example, 10Base2 employs a thin coaxial cable while 10Base5 utilizes a thick coaxial cable. The distance limit for thin (thick) cable is 185 (500) m/segment. Both 10Base2 and 10Base5 are used in LAN. Line codes such as non-return to zero (NRZ), Manchester, alternate mark inversion (AMI), B8ZS, 4B/5B etc., characterize typical bit encoding schemes. ASCII and EBCDIC codes represent the two most popular character encoding schemes in data communication. They provide unique 7 or 8 bit patterns in order to identify all the characters on the keyboard. A simplex mode refers to data communication in only one direction. On the contrary, the capability of data transmission alternatively or simultaneously in both directions is designated as half- or full-duplex mode, respectively. In the following, we describe fundamental concepts associated with the bit-serial transmission because of its ubiquitous usage in computer networks.

* A twisted wire pair is better than a two-wire medium because any noise interference affects both wires (not just one) in the pair and hence gets canceled. In addition, the effects of crosstalk can be minimized by enclosing a number of twisted pairs within the same cable.

† Each IEEE 802.3 physical layer has a name that summarizes three characteristics of a LAN. For example, the first parameter 10 in 10Base5 refers to the speed of the LAN in Mbps. The word "Base" identifies a baseband technique; "Broad" denotes broadband method. Finally, the 5 provides the LAN segment length, in multiples of 100 m. Here "*T*" means a twisted wire pair.

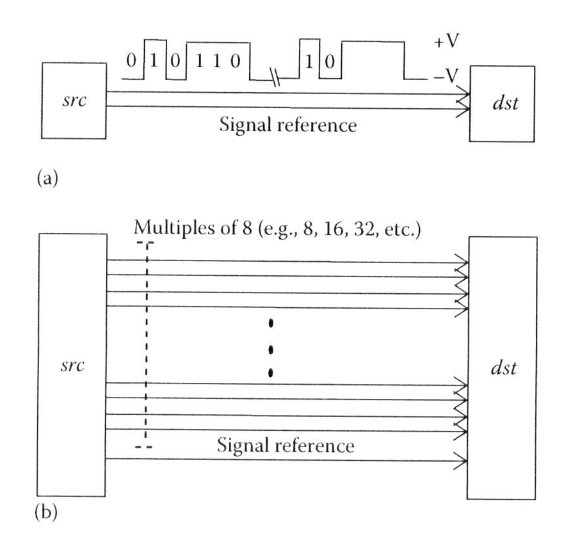

FIGURE 4.12 Illustrating (a) bit-serial and (b) bit-parallel transmission modes. (*src* denotes source, *dst* means destination.)

4.3.1 Serial and Parallel Modes

There are two ways to transmit the bits that make up a data character: *serial* or *parallel*. Serial transmission sends bits sequentially, one after the other, over a single wire (or channel). It is further classified as *asynchronous* or *synchronous* type. In parallel transmission, bits of a data character are transmitted simultaneously over a number of channels, which are generally organized as a multiple of 8. Figure 4.12 provides an illustration for bit-serial and bit-parallel transmission modes. Although serial transmission is slower than parallel, it is predominantly used between host and computer today. A typical example of parallel transmission is found with address, data, and control buses between a microprocessor and external devices such as memory and input/output modules. Considering the scope of the chapter, we will not further elaborate on parallel transmission mode such as the High Performance Parallel Interface (HIPPI), which is an ANSI (American National Standard Institute) standard and is adopted by supercomputer, workstation, and peripheral manufacturers as the high-performance interface of choice.

4.3.2 Transmission Type

As we mentioned at the beginning of this chapter, a frame is the logical group of information sent as a data link layer unit over a transmission medium. The frame thus contains bytes, and bytes embody bits (Figure 4.13). For the receiver to interpret the bit pattern correctly, the bit (clock) period, start and end of byte (character), and start and end of frame must be uniquely resolved. These tasks are called *bit (clock) synchronization*, *byte (character) synchronization*, and *frame synchronization*, respectively. They are accomplished in one of two ways, depending on bit-serial transmission being *asynchronous* or *synchronous* type.

The asynchronous mode treats each character (byte) independently and hence embeds them with additional control bits (refer to Figure 4.13a). This makes the approach useful when the data to be transmitted is of *low-volume* or generated randomly. The asynchronous mode of communication is normally used for low data rates, up to 19.2 kbps.* A typical example of data being generated randomly is a user sending data from the keyboard to the computer.

* Note that EIA-232-D interface specifications limit the data rate to 19.2 kbps with a 50 ft cable. But, due to improved techniques, higher data rates such as 28.8, 33.6, and 56 kbps are available these days.

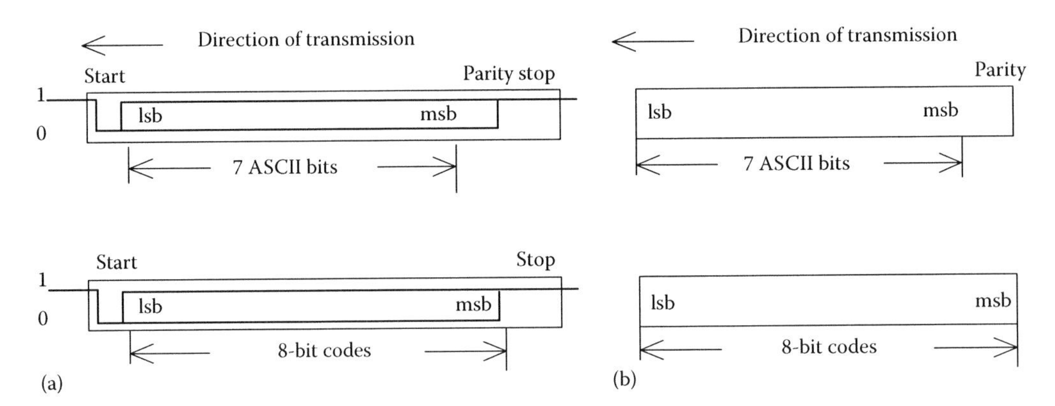

FIGURE 4.13 Byte structure in a bit-serial (a) asynchronous and (b) synchronous transmission.

Synchronous transmission, on the contrary, permits the transmission of a complete frame as a contiguous string of bits, and hence it does not use the control bits on per-character basis (see Figure 4.13b). It is primarily used in a large-volume data transfer, as could happen if a file is being transferred between two computers. Synchronous transmission is also an alternative to transmit data at higher bit rates.

4.3.2.1 Asynchronous Transmission

We have just indicated that each character in an asynchronous mode is embedded with control bits called start, stop, and, if applicable, parity bits (see Figure 4.13a). One start and 1, 1.5, or 2 stop bits designate the beginning and ending of bytes. The presence of an even, odd, MARK, or SPACE parity* bit helps detect the transmission errors at the byte level. The accepted mode of asynchronous transmission is to hold the serial line at a logic 1 level, until data is to be transmitted. The transmission begins with a logic zero for one bit time (start bit). The principle of bit and character synchronization is as follows: The receiver contains a circuit having a serial-in parallel-out shift register, a local clock that is N times faster than the transmitted bit rate (usually $N = 16$), and a divide by N counter. The local clock is chosen to be faster because the receiver clock runs asynchronously with respect to the transmitter clock. After detecting a $1 \rightarrow 0$ transition associated with the start bit of a byte, the first bit is sampled at $N/2$ clock cycles and then subsequently after N cycles for each bit in the character. In this way, we ensure that each bit is sampled approximately at the center of the bit cell. This procedure is repeated at the start of each character received. The sampled bits, retained with the serial-in parallel-out shift register, define a character byte. The frame is encapsulated using reserved (transmission control) characters such as the END byte in SLIP (serial line IP) and mark in Kermit. Since the frame is transmitted character by character, its synchronization is automatically achieved when bits and bytes are synchronized. The need of additional bits at both character and frame levels makes the asynchronous mode inefficient in its use of the line capacity. For example if we assume that 100 data characters are to be transmitted using SLIP, then the minimum transmission overheads at the byte level add up to 200 bits (with 1 start bit and 1 stop bit per 8-bit character) and at frame level to 16 bits (considering that no byte stuffing is required). It means the *transmission efficiency*, defined as the ratio of data bits to the total number of bits actually transmitted, is 800/1016 or 78.7%. This will further decrease if the inclusion of parity bits and character stuffing is assumed.

* The parity is termed *even* if the modulo-2 sum of all the bits, including the parity bit, is even. For an *odd* parity, the sum is odd. MARK parity, regardless of data bits, is always 1. The SPACE parity is similarly always 0.

Example 1.4

Let R bps be the data rate in an asynchronous transmission. If CK denotes the receiver clock frequency, then bit synchronization requires $CK = N \times R$. It means the bit period is N/CK. The mid-point sampling of the subsequent bit cells depends on the detection of $1 \rightarrow 0$ transition associated with the start bit in each byte. Thus, a detection error will cause sampling to deviate from the center of the bit cell. Halshall [13] has shown that the worst-case deviation, D, from the nominal bit cell center is approximately one cycle of the receiver clock ($1/CK$). Alternatively, $D = (1/N) \times$ bit period. Consider $CK = 19.2$ kHz and transmission data rate R in bps is (i) 2,400, (ii) 9,800, and (iii) 19,200. The parameter D, in terms of corresponding bit period, is (i) 1/8, (ii) 1/2, and (iii) 1, respectively. The chance of missing a bit is 50% in case (ii) and 100% in case (iii). The corresponding data rates are unacceptable with CK being 19.2 kHz.

4.3.2.2 Synchronous Transmission

From Example 1.4, it is evident that the bit synchronization in an asynchronous mode becomes unreliable with an increase in data rate. Synchronous transmission is used to overcome this problem. As its name suggests, synchronous mode is the operation of a network system wherein events at both transmitter and receiver occur with precise clocking. There are two types of synchronous transmission, namely, byte-oriented and bit-oriented. Both use the same bit synchronization methods that could be based on a circuit having a digital phase-lock loop which exploits $1 \rightarrow 0$ and $0 \rightarrow 1$ transitions present in the received bit stream. The byte and frame synchronization are, however, handled differently in two systems. In byte-oriented transmission, the transmitter sends two or more ASCII SYN characters preceding each frame. Its bipolar coding, on one hand, helps maintain synchronization of transmitter and receiver clocks. On the other hand, the SYN character itself provides the correct byte boundaries. With bit-oriented transmission, the transmitter continuously sends idle bytes (0111 1111) on the line during the period between the transmission of successive frames. Here also a bipolar encoding of the idle bytes either using the Manchester scheme or its variants provides clock alignment. Note, in either type of synchronous transmission, the characters are not embedded with control bits because the frame is transmitted as a bit stream (Figure 4.13b). Nonetheless, similar to asynchronous mode, a reserved transmission control byte or character is used for frame encapsulation in both types of synchronous transmissions. With byte-oriented, it is generally ASCII start-of-text (STX) and end-of-text (ETX) codes, while it is a flag pattern 0111 1110 in case of bit-oriented type. IBM's Binary Synchronous protocol (BiSync) provides an example of byte-oriented protocol. Considering the scope of the book, we have excluded its description. A typical example of bit-oriented transmission is PPP or HDLC (high level data link control). Both byte-oriented and bit-oriented types of synchronous protocol contain additional fields that make them flexible and help support a variety of link types and topologies. For instance, HDLC can be used with point-to-point and multipoint links, half- and full-duplex transmission facilities, and circuit-switched and packet-switched networks. Further, with the help of CRC and ARQ schemes, the synchronous transmission scheme achieves link error recovery.

4.4 Wireless Networks

Flexibility and mobility make wireless networks become one of the fastest growing areas in the telecommunication industry. Wireless networks link users and information services through a wireless communication path or channel. They offer both LAN and WAN connectivity for business and home users. Different types of wireless networks have been used for the different types of services. Satellite systems provide high-speed broadcast services, and low-speed long distance (even international) digital voice services. Cellular system provides radio coverage to a wide area, such as a city. Sensor network is formed when a set of small sensor devices that are deployed in an ad hoc fashion cooperate on sensing a physical phenomenon. Sensor networks greatly extend the ability to monitor and control the physical environment from remote locations, for example, it can be used to analyze the motion of a tornado, detect fires in a forest etc. Wireless LAN (WLAN) is designed to operate in a small area such as a building or office, and allows computers and workstations to communicate with each other using radio signals to transfer high-speed digital information. This section emphasize some principles and technologies of wireless networks, based on cellular systems.

4.4.1 Terminology and Model

Various wireless communication systems, such as remote controllers, cordless telephones, pagers, walkie-talkies, cellular telephones, and so on are currently being used. Their mobility allow users move during operation. Unlike wired links, which usually provide one-to-one communication without interference, wireless links use one-to-many communication that exists some problems, such as noise, interference, bandwidth limitations and security etc. Mobile radio transmission systems may be classified as *simplex*, *half-duplex* and *full-duplex*. Simplex systems provide only one-way communication, such as paging systems. Half-duplex systems allow two-way communication, but use the same radio channel for both transmission and reception, such as "push-to-talk" and "release-to-listen" systems. At any time, users can only either transmit or receive information. Full-duplex systems, on the other hand, allow simultaneous radio transmission and reception by using two separate channels. All wireless networks that we consider in this section are full-duplex systems.

The mobile stations communicate to fixed base stations (BS), which are located at the center or on the edge of a coverage region and consist of radio channels and transmitter and receiver antennas on a tower. Mobile Switching Center (MSC) coordinates the activities of all of BS in a large service area. Service area is the total region over which the BS are distributed. Coverage area is the actual region over which communications can be provided. The ratio of coverage area and service area is called *area availability*. Radio channels used for transmission of information from the base station to the mobile station are called Forward Channel (Downlink); radio channels used for transmission of information from the mobile station to the base station are called Reverse Channel (Uplink). Some specific radio channels are assigned as control channels, which are used for transmission of call setup, call request and other beacon or control purpose. Similarly, control channels include forward control channels and reverse control channels. System capacity is be defined as the largest number of users that could be handled by a system. Figure 4.14 shows a most common wireless network model.

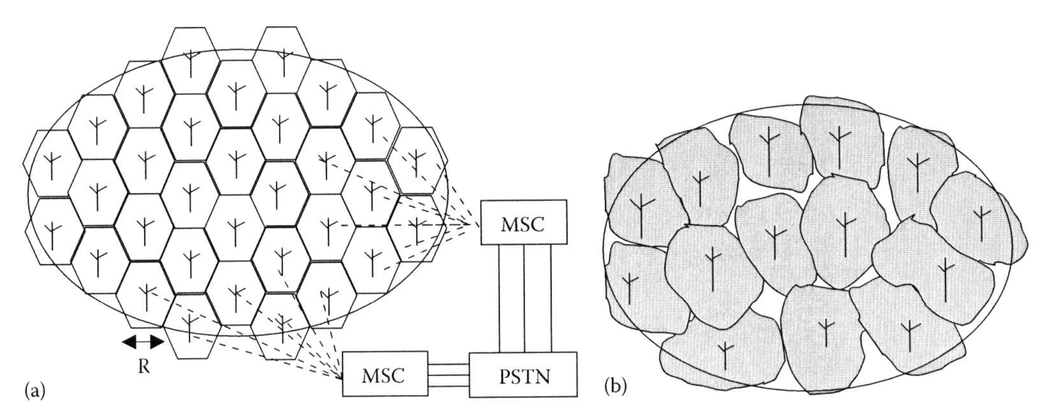

FIGURE 4.14 (a) A typical wireless network model (cellular system) with uniform hexagonal shape areas called cells, MSC. (b) An illustration of coverage areas in a real-world wireless network system.

4.4.2 Frequency Reuse and Channel Assignment

A frequency spectrum allocated to wireless communication is limited. By replacing a single, high power transmitter (covering a large service area) with many low power transmitters (a large service area being divided into many small cells), the cellular networks reuse the frequency at smaller distance such that system capacity is increased.

Frequency reuse is possible due to the propagation properties of radio waves. The transmitted power is chosen to be just large enough to help communication with mobile units located near the edges of the cell. Each cell is assigned a certain number of channels. To minimize the interference between BS, the channels assigned to one cell must be completely different from the channels assigned to its neighboring cells. By carefully spacing BS, properly setting the distance D and assigning their channel groups, the same frequency can be used in the two (or more) cells simultaneously as long as the interference between co-channel stations is kept below acceptable levels. Interference coming from a reused frequency is called *co-channel interference*. The distance D is called the reuse distance. Figure 4.15 illustrates the concept of frequency reuse. Cells with the same number use the same group of channels (frequencies). A cell cluster is outlined in bold and replicated over the coverage area.

For example, a cellular system has a total of S duplex channels available for use. S channels are divided among N cells into unique and disjoin channel groups, each of which have the same number of channels and the number is k ($k < S$). N is called the cell cluster size (refer to Figure 4.15, $N = 7$). The reciprocal $1/N$ is known as the frequency reuse factor. If a cluster is replicated M times within the system, the total number of duplex channel (a measure of system capacity) is $C = M \times K \times N = M \times S$.

Maximizing the number of times each channel may be reused and minimizing interference at the same time in a given geographic area is the key to an efficient cellular system design. Co-channel interference ratio at the desired mobile receiver depends on cell radius R and the reuse distance D, and is

$$\frac{S}{I} = \frac{S}{\sum_{k=1}^{N_I}(I_k)} = \frac{1}{\sum_{k=1}^{N_I}(D_k/R)^{-\gamma}}$$; $S(I)$ is signal (interference) power and I_k and D_k refer to kth cell

where

γ is the propagation path-loss slope which depends on the terrain environment

N_I is the number of co-channel interfering cells, for example, in a fully equipped hexagonal-shaped cellular system, $N_I = 6$ for the first tier

Channel assignment strategies are classified into two groups: fixed or dynamic. In a fixed channel assignment (FCA), each cell is assigned a predetermined set of channels. FCA is simple and works well when the traffic follows a uniform distribution. But, it behaves very poorly in the worst case. In a dynamic channel assignment (DCA), channels are not assigned to cells permanently. All channels are kept in a central pool and are assigned dynamically to cells as new calls arrive in the system. A channel is eligible for use in any cell as long as interference constraints are met.

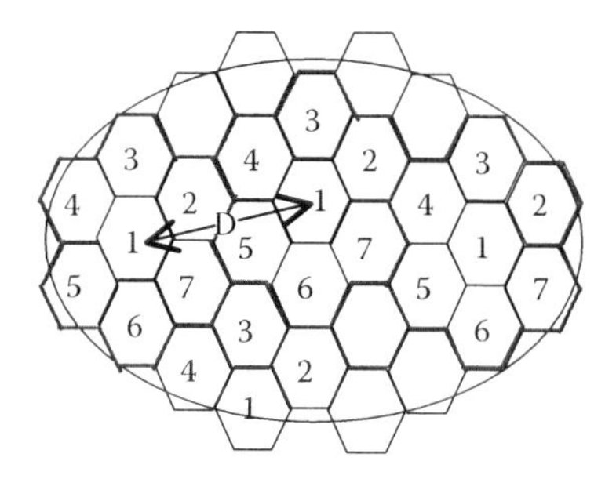

FIGURE 4.15 Illustrating frequency reuse.

4.4.3 Handoff

The handoff process mainly consists of four steps: initiation, resource reservation, execution and completion. It is used to automatically switch channels to maintain a conversation in progress as the mobile terminal moves into a different cell. Handoff is very important in any cellular radio system and must be performed successfully as infrequently as possible, and imperceptible to the users. The operation not only involves identifying a new base station, but also requires that the voice and control signals be allocated to channels associated with the new base station.

Handoffs can be classified as soft handoff or hard handoff: In a soft handoff, the mobile terminal can communicate with two radio ports simultaneously such that the commutation won't be interrupted when the mobile terminal moves into a different cell. A hard handoff occurs when the old connection is broken before a new connection is activated, due to disjoint radio systems, different frequency assignments, or different air interface features. It is a "break-before-make" process at air interface.

4.4.4 Multiple Access Technologies

Section 4.2.1 has discussed FDM and TDM systems. Applying these principles, we have FDMA (frequency division multiple access) and TDMA (time division multiple access) technologies to handle multiple access where many users share a finite amount of radio spectrum simultaneously.

FDMA systems divide bandwidth of the air interface between the mobile station and the base station into multiple equal analog channels, each of which occupies one part of a larger frequency spectrum. Then, FDMA assigns individual channels to individual users that is, each user is allocated a unique frequency band. These channels are assigned on demand to users who request service. For example, in 1983, the first U.S. cellular telephone system—Advanced Mobile Phone System (AMPS), which is based on FDMA, divided 40 MHz of spectrum into 666 duplex channels with a one-way bandwidth of 30 kHz ($666 \times 2 \times 30\,\text{kHz} \approx 40\,\text{MHz}$).

TDMA systems share a single RF (Radio Frequency) channel with several users, but divide the radio spectrum into non-overlapping time slots. In each digital time slot only one user is allowed to either transmit or receive, and the slots are rotated among the users during a periodic time. Consequently, packet transmission in a TDMA system occurs in a serial fashion, with each user taking turns accessing the channel. For example, in the beginning of 1990s, the first U.S. Digital Cellular (USDC) system made use of TDMA, digital modulation (a type of DQPSK), and speech coding implementing three times capacity than that of AMPS. It replaced single-user analog channels with digital channels which can support three users in the same 30 kHz bandwidth.

Besides FDMA and CDMA, a *code division multiple access* (CDMA) system places all users on the same frequency spectrum at the same time, and use different pseudorandom codes to distinguish between the different users (see also Section 4.2.1 and Figure 4.6). Each code is approximately orthogonal to all other codes. The receiver performs a time correlation to detect only the desired code. The advantage of using CDMA over FDMA and TDMA is that it provides more capacity. There is no absolute limit on the number of users in CDMA, but the channel performance limits the practical capacity of CDMA, which gradually degrades for all users as their number increases. For example, in 1993, a cellular system based on CDMA was developed by Qualcomm, Inc. and standardized by the TIA as an Interim Standard (IS-95). The system supports a variable number of users in 1.25 MHz wide channels. IS-95, discussed below, has been successfully installed in many areas of the world and it is possible to be migrated to 3G wireless networks with CDMA2000 to give the higher data rates that are needed for video and data transfer, and retain compatibility with the existing networks at the same time.

Note that CDMA access is a form of spread spectrum communication, which allows multiple users to share the same frequency band by spreading the information signal for each user over a wide frequency bandwidth (several order of magnitude greater than the minimum required RF bandwidth). A pseudonoise (PN) or pseudorandom sequence is used as a spreading code to covert a narrowband signal to a wideband noise-like signal before transmission. A PN code is a binary sequence that appears random but can be reproduced in a deterministic manner by intended receivers. At the receiver, crosscorrelation with the correct PN sequence de-spreads the spread spectrum signal and restores the modulated message in the same narrowband as the original data. By using a group of approximately orthogonal spreading codes, it is possible that the receiver picks out a signal based on its code from many other signals with different spreading codes. That is, up to a certain number of users, interference between spread spectrum signals using the same frequency is negligible by using different orthogonal codes. The system has been likened to hearing many people in a room speaking different languages. Despite a very high noise level it is possible to pick out the person speaking your own language, for example, English. Since many users can share the same spread spectrum bandwidth without interfering with one another, spread spectrum systems become bandwidth efficient in a multiple user environment.

There are two main types of spread spectrum techniques: Frequency hopped spread spectrum (FHSS) and direct sequence spread spectrum (DSSS). In a FHSS system, the carrier frequencies of the individual users are varied in a pseudorandom manner within a wideband channel. The digital data is broken into uniform sized bursts which are transmitted on different carrier frequencies. A DSSS system spreads the baseband data by directly multiplying the baseband data pulses with a PN sequence. CDMA uses DSSS.

One of the advantages in CDMA is that handoff can be made easier and more reliable. Unlike the wireless systems that assign different radio channel during a (hard) handoff, it is possible that spread spectrum mobiles share the same channel (code) in every cell since adjacent cells use the same RF band in IS-95. There is a degree of risk with hard handoff approach because occasionally a handoff does not proceed smoothly. IS-95 system provides soft handoff capacity. Soft handoff is a make-before-break system. When a call is in a soft handoff condition, the two BS transmit the same signal enabling the mobile to receive the signal via two routes at the same time. The system takes advantage of the moment-by-moment changes in signal strength at each of the two cells to pick out the best signal.

In addition to the soft handoff, IS-95 also supports a hard handoff, which is typically done when (1) moving between BSs served by different switches; (2) encountering BSs of different vendors; (3) encountering BSs that support different air interface such as CDMA-to-AMPS.

4.4.5 New Generation Wireless Networks

First generation (1G) wireless system uses analog signaling (analog FDMA*) for the user traffic on the air interface, like AMPS. Second generation (2G) wireless system uses digital voice coding and digital modulation (such as TDMA, CDMA) for the user traffic. Compared to 1G system, all 2G systems have improved voice privacy and authentication capability as well as system capacity and signal quality.

Besides voice and short messages, the growing demands for wireless networks providing high-rate data service and better spectrum efficiency are driving the deployment of new wireless technologies.

The aim of third generation (3G) wireless system is to provide a single set of standards that can meet a wide range of wireless applications and provide universal access throughout the world. One standard is called universal mobile telecommunications system (UMTS) given by the European Telecommunication Standard Institute (ETSI). Three main objectives [16] for UMTS systems are

1. Full coverage and high data rate
2. Use of different sized cells (Macrocell: suburban area; microcell: urban area; and picocell: in building) for indoor and outdoor application, with seamless handover between them
3. High spectrum efficiency

Three main components constitute the UMTS system: Core network (CN), UMTS radio access network, and Mobile equipment (ME). The general architecture of 3G wireless networks is illustrated in Figure 4.16.

* Typically, analog FDMA systems allow a single mobile terminal to occupy on a radio channel at a specific time, while the digital FDMA systems allow multiple users to share a single radio channel. The latter one is used in 2G wireless systems.

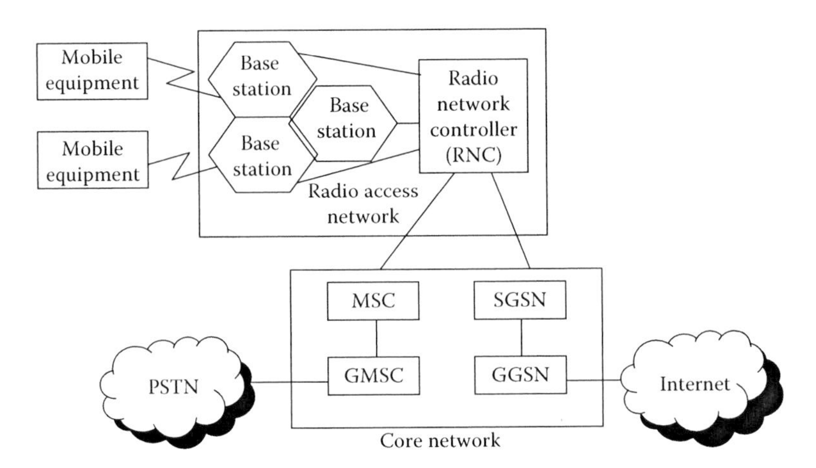

FIGURE 4.16 The general architecture of 3G wireless networks.

RNC is the radio network controller for a group of adjacent BS. It coordinates the overall operation of each base station within its radio subsystem, such as the procedure of handoff decisions. SGSN (the serving general packet radio service node) coordinates the active wireless data devices in its area. GGSN (Gateway general packet radio service support node) is a packet switch that routes packets between UMTS CN and external data network. GMSC (gateway mobile switch center) connect UMTS network to the public circuit switch networks, such as PSTN. Until 2001, there were three different system specifications for multiple access schemes in 3G wireless systems: wideband CDMA (WCDMA), time division CDMA (TD/CDMA), and CDMA 2000.

4.5 Wireless Local Area Networks

Wireless technologies are spanning from wide area technologies (for example, satellite-based networks, cellular networks) to local and personal area networks. WLAN offer high flexibility and ease of network installation with respect to wired LAN infrastructures. Two main standards for WLAN are IEEE 802.11 and HiperLAN (high performance radio LAN, promoted by ETSI). In this section, we present a brief introduction to IEEE 802.11.

The IEEE 802.11 technology operates in the 2.4 GHz industrial, scientific, and medicine band and provides wireless connectivity for fixed, portable, and mobile station in a local area. IEEE 802.11 protocol defines the MAC and physical layer. At the physical layer, two RF transmission methods and one infrared method are defined. The RF transmission standards are Frequency Hopping Spread Spectrum (FHSS) and DSSS. The most popular method is DSSS. Some similar to the 802.3 Ethernet wired line standard, the MAC layer specification for 802.11 uses a protocol scheme known as carrier-sense, multiple access, and collision avoidance (CSMA/CA). 802.11 avoids collisions instead of detecting a collision used in 802.3. The MAC layer defines two access methods: the distributed coordination function and the point coordination function. Figure 4.17 illustrates the layer architecture of IEEE 802.11.

The IEEE 802.11 standard defines the protocol for two types of networks: infrastructure-based WLAN and Ad-hoc WLAN. In an infrastructure-based network, there is a centralized controller for each cell, referred to an access point. Mobile nodes communicate with fixed network access points

Date link layer	802.2		
	802.11 MAC		
Physical layer	FHSS	DSSS	IR

FIGURE 4.17 IEEE 802.11 layer architecture.

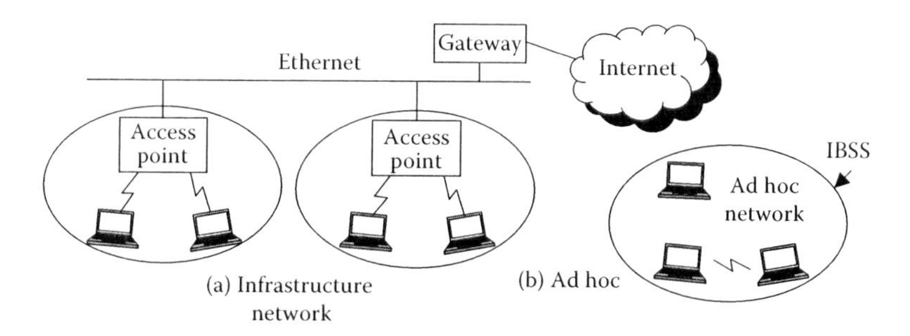

FIGURE 4.18 (a) Infrastructure and (b) ad hoc networks. (From Stojmennovic, I., *Handbook of Wireless Networks and Mobile Computing*, John Wiley Inc., New York, 2002.)

which are usually connected to the wired network to provide Internet access for mobile devices. In an Ad-hoc network, every 802.11 device in the same cell, or independent basic service set (IBSS) can directly communicate with every other 802.11 device within the cell, without the use of an access point or server. There is no structure to the network; there are no fixed points (see Figure 4.18). In addition, the Bluetooth technology can also be used for building a mobile ad hoc network.

Currently, a huge amount of new industry standards, new techniques, new devices are in the development. In the near future, more applications and services in wireless networks will effect on various aspects of our life.

Acknowledgment

The author acknowledges the support from NSF (grant CCR 0073429) and help from Y. Zhao, a graduate student, during the preparation of this chapter.

References

1. S. Feit, *TCP/IP*, 2nd edn., McGraw Hill, New York, 1997.
2. B. Halabi, *Internet Routing Architectures*, New Riders Publishing, Indianapolis, IN, 1997.
3. S. Keshav, *An Engineering Approach to Computer Networking: ATM Networks, the Internet, and the Telephone Network*, Addison-Wesley, Reading, MA, 1997.
4. S. A. Thomas, *IPng and the TCP/IP Protocols*, John Wiley & Sons, New York, 1996.
5. W. R. Stevens, *TCP/IP Illustrated*, Addison-Wesley Publishing Company, Reading, MA, 1994.
6. P. Miller, *TCP/IP Explained*, Digital Press, Boston, MA, 1997.
7. U. Black and S. Waters, *SONET and T1: Architectures for Digital Transport Networks*, Prentice Hall PTR, Upper Saddle River, NJ, 1977.
8. S. W. Seetharam, G. J. Minden, and J. B. Evans, A parallel SONET scrambler/descrambler architecture, *IEEE International Symposium in Circuits and Systems*, Chicago, IL, 1993, pp. 2011–2014.
9. *Synchronous Optical Network Transport Systems: Common Generic Criterion*, TR-TSY-000253, Bellcore, Morristown, NJ, 1993.
10. J. Walrand, *Communication Networks: A First Course*, McGraw Hill, Boston, MA, 1998.
11. J. Walrand and P. Varaiya, *High Performance Communication Networks*, Morgan Kaufmann Publishers Inc., San Francisco, CA, 1996.
12. A. S. Tannenbaum, *Computer Networks*, Prentice Hall PTR, Upper Saddle River, NJ, 2003.
13. F. Halshall, *Data Communications, Computer Networks and Open Systems*, Addison-Wesley Publishing Company, Reading, MA, 1996.
14. W. J. Goralski, *SONET: A Guide to Synchronous Optical Network*, McGraw Hill, New York, 1997.
15. J. F. Kurose and K. W. Ross, *Computer Networking: A Top-Down Approach Featuring the Internet*, Addition Wesley, Boston, MA, 2003.

16. I. Stojmennovic, *Handbook of Wireless Networks and Mobile Computing*, John Wiley Inc., New York, 2002.
17. T. S. Rappaport, *Wireless Communications: Principles and Practice*, Prentice Hall PTR, Upper Saddle River, NJ, 1996.
18. U. D. Black, *Second Generation Mobile and Wireless Networks*, Prentice Hall PTR, Upper Saddle River, NJ, 1999.
19. L. Harte, *3G Wireless Demystified*, McGraw-Hill, New York, 2002.
20. V. K. Garg and J. E. Wilkes, *Wireless and Personal Communications Systems*, Prentice Hall PTR, Upper Saddle River, NJ, 1996.
21. P. Brenner, A technical tutorial on the IEEE 802.11 protocol, Breeze COM, 1997.
22. L. Harte, *CDMA IS-95 for Cellular and PCS*, McGraw-Hill, New York, 1999.
23. C.-Y. Lin, *IS-95 North American Standard—A CDMA Based Digital Cellular System*, Columbia University, New York, 1996.

5

Location-Centric Networking in Distributed Sensor Networks

Kuang-Ching Wang
Clemson University

Parameswaran Ramanathan
University of Wisconsin-Madison

5.1 Introduction

Sensor networks are an emerging technology that interconnects spatially distributed sensing devices to monitor and interact with the surrounding physical world [1,5,8]. Each device (node) in a sensor network senses a limited neighborhood around itself with its sensing modalities. A single node, however, is often unable to sufficiently characterize an area of interest due to its limited sensing capability and reliability. Therefore, collaboration among a set of nodes is necessary in order to produce a complete picture of the area. Most of the time, applications are interested in retrieving information about a particular geographic region. The scope of collaboration is thus determined based on geographic locations of the nodes. Conventionally, network designers assign unique addresses to individual nodes and application programmers design actions of each node based on its address. We consider such a networking approach to be *node centric*, where every node has its unique identity and must be individually addressed in a network application. This is, however, inefficient in designing sensor network applications. Sensor network applications are *location centric* as they pose interest in *region-based* information that requires collaboration among a set of nodes based on their geographic locations rather than their addresses. This chapter introduces a *location-centric networking* model useful in designing collaborative applications in distributed sensor networks.

The location-centric networking approach aims to address a number of challenges in designing collaborative applications in a large-scale distributed sensor network:

1. The need of a simple and flexible programming model
   Considering a huge number of nodes distributed in a large geographic area. It is a formidable challenge for application designers to develop efficient algorithms if they have to manage individual node addresses, locations, and operational status. Moreover, nodes in a dynamic network may be relocated or become nonoperational at any time. Therefore, a simple and flexible programming model is needed for application programmers to easily address nodes in regions of interest and to facilitate collaboration among them even in a dynamic network.
2. The need of energy and bandwidth efficiency
   Sensor nodes have limited energy (batteries) and limited communication bandwidth. Moreover, wireless devices consume most of their energy in communication. While collaborative applications require intensive communication among sensor nodes, it is challenging for application designers to ensure their communication efficiency. As a matter of fact, most application communication requirements are similar. We thus contend that it is essential to provide application designers with a common set of energy and bandwidth efficient communication services in a distributed sensor network.
3. The need of application robustness
   Sensor networks are potentially deployed in harsh environments. At any given time, a large fraction of nodes may be nonoperational or malfunctioning. Collaborative applications must be robust to node failures and ensure correctness. We contend that application robustness is necessary in all collaborative applications in a distributed sensor network, and fault tolerance capabilities must be incorporated in its communication services.

At the University of Wisconsin, Madison, researchers pursue a location-centric networking approach with an application programmers' interface, UW-API, and a location-centric ad hoc routing protocol, UW-Routing [2,12,13]. Together, UW-API and UW-Routing address the challenges and provide the platform for location-centric application designs.

UW-API is motivated by the well-known message-passing interface standard MPI-1.1 [14] proposed by the MPI forum in 1995. In MPI-1.1, message passing primitives are defined to facilitate distributed computations in parallel computer systems and networks of workstations. In a distributed sensor network, applications require collaborative computations among a set of nodes similar to those in a parallel computer system. However, MPI-1.1 is node centric and is used to facilitate communication between a fixed set of nodes. UW-API, on the other hand, is location centric. Communication primitives in UW-API facilitate communication among *regions* instead of nodes. In UW-API, nodes are not individually addressable while regions are the only addressable entities. Application designers specify regions to communicate with rather than particular nodes. Once the regions are specified, communication takes place among all nodes in the regions. In a dynamic network, the set of nodes participating in such primitives naturally adapts to changes in node locations and operational status.

To facilitate region-based communication in UW-API, the underlying routing scheme has to be *location aware*. Several location-aware routing schemes have been studied for ad hoc wireless networks [6,7]. In these schemes, a node obtains its geographic location with the global positioning system (GPS) [11]. In Ref. [6], the on-demand location-aided routing (LAR) utilizes geographic positions of a sender and its receiver to define a constrained zone for route discovery. The flooding route request messages are forwarded and processed only by nodes within this zone for energy and bandwidth conservation. Similarly, in Ref. [7], the GeoTORA routing protocol constrains flooding operations in TORA [10] with location information and demonstrates enhanced efficiencies. Nevertheless, they are node-centric and route messages between a pair of nodes. UW-Routing is developed in conjunction with UW-API. Quite differently, UW-Routing routes messages among regions rather than nodes. The scheme facilitates region-based communication in the following sense:

- From any node, a message is delivered to a region addressed with its geographic coordinates.
- A message for a region is delivered to every node in the region.

The protocol is on demand combining unicast and constrained broadcast. In general, a message is delivered from a node to a region in two steps. First, the message is unicast to any node in the destination region. Second, the message is flooded within the region such that the same message is delivered to all nodes in the destination region.

Together, UW-API and UW-Routing provide a location-centric networking approach that addresses the aforementioned challenges. With UW-API, collaborations among nodes in a region or among multiple regions are readily programmable with the provided primitives. Each primitive is accomplished with a series of message exchanges using UW-Routing. Communication efficiency is addressed in two aspects. UW-API optimizes the sequence of message exchanges needed to accomplish each of its primitives; on the other hand, UW-Routing optimizes the number of physical transmissions needed for each message exchange. Application robustness is addressed in two aspects as well. First, every primitive considers a region rather than any particular node. Individual node failures are not expected to pose significant impacts on the collaboration computation results. Second, each primitive is inherently prone to failures. Time-out constraints are explicitly specified for all primitives such that application performance and robustness are easily adapted and failed primitives are handled without halting the application. To demonstrate the usage of UW-API and UW-Routing, a target-tracking application was developed [12]. The application defined multiple regions and utilized the collaborative primitives in UW-API to detect, localize, classify, and track ground vehicle targets along their trajectories. The application was implemented and evaluated on various sensor network test beds and also software simulators on workstations.

The remainder of this chapter is organized as follows. Section 5.2 describes the location-centric computing model in a distributed sensor network. The network model is defined in Section 5.3. In Section 5.4, we propose the location-centric networking mechanisms UW-API and UW-Routing. The exemplary target-tracking application is introduced in Section 5.5. Evaluation studies on sensor network test beds are summarized in Section 5.6. Section 5.7 concludes the chapter.

5.2 Location-Centric Computing

The location-centric computing model is based on the premise that sensor network applications are typically interested in acquiring information about a region rather than a particular node. For example, there are queries such as the following:

- What is the concentration profile of a certain biochemical agent in a given area?
- What is the temperature or pressure variation in an area?
- Have there been any unauthorized entries into an area and how many are there?

Oftentimes, such queries cannot be answered by any single node but require the collaboration among *nodes in a certain geographic area* to establish a complete picture. This requirement of *location-based* collaboration among nodes does not exist in conventional network applications.

Traditional parallel computing models consider applications designed with a fixed set of nodes. Designers partition the workload among the nodes and define the communication patterns for each node. In such an application, each node serves an indispensable role and the set of nodes is not expected to change throughout the application. The location-centric computing model, however, considers otherwise. The model is not concerned about any particular nodes; instead, it considers regions to be the entities of interest to most applications. As an application specifies a region of interest, all nodes in the region start to participate in this application. Since the model does not assume the existence of any particular node, it naturally accommodates programming in a dynamic network where nodes may be relocated or become nonoperational. At any given time, a node participates in collaborative computations only if it resides in an active region of an application. Similarly, an application only works with nodes in regions it specifies.

At the University of Wisconsin, Madison, we recently proposed the network application programmers' interface called UW-API [13] that is particularly well-suited for location-centric computing. In UW-API, a geographic region plays the role of a node in traditional network interfaces. A region represents a set of nodes residing in a specific geographic area, and it is explicitly *created* and *deleted* with primitives in the API. Currently, we consider a region to occupy a rectangular geographic area, which can be represented with its corner coordinates. A region does not necessarily contain all nodes in the area. Constraints such as sensor types can be used to define regions to be a subset of all nodes in the geographic area. UW-API considers a region to be the only addressable entity in any applications. The API not only supports interregion communication among regions, it also supports communication within a region. To coordinate communication within a region, it further defines the *manager subregion* of a region. The manager subregion is a subset of nodes in the corresponding region. Nodes in the manager subregion coordinate all collaborative communications in its region.

To better understand the location-centric computing model, consider the following example. Let there be a *controller node* in a distributed sensor network. A temperature monitoring application in this sensor network allows users to retrieve average temperature information from a particular area covered by the network. At the controller node, a user issues the query that specifies an area of interest and expects the answer to be returned. This application can be designed as follows. The controller process (at the controller node) first *creates* a region covering the area of interest. Then, the controller must *send* the query to the created region. As all nodes in the region *receive* the query, a collaborative algorithm is used to compute the average temperature in the region. This is done by all nodes reporting their temperature readings to nodes in the manager subregion, which collect all readings and compute the average. The operation is known as a *reduce* operation in parallel applications. Finally, the computed result is sent back to the controller node.* The example demonstrates how a sensor network application can be implemented without the burden of exploiting the network topology or managing individual nodes. As are shown in this example, several communication and collaborative operations are useful in designing sensor network applications. The proposed UW-API perfectly addresses these requirements with its primitives.

5.3 Network Model

We consider a sensor network with a large number of nodes distributed in a large geographic area. Every node is aware of its geographic location and is capable of sensing, information processing, and wireless processing. The radio range of a node is limited, and nodes may communicate in different channels. Thus, it is not necessarily possible for a node to directly communicate with all other nodes in the network. For a node to communicate with nodes other than its immediate neighbors, multihop forwarding is needed. The network is, as a result, a wireless multihop ad hoc network.

5.4 Location-Centric Networking

UW-API provides application programmers with a set of primitives to design their collaborative applications in a distributed sensor network. Section 5.4.1 introduces six primitives defined in the current API. These primitives are supported by the underlying location-aware routing scheme called UW-Routing. The routing scheme efficiently delivers a message from a node to a region. Section 5.4.2 presents the UW-Routing protocol in detail.

5.4.1 UW-API

As of today, six primitives are defined in UW-API. For region management, two primitives are defined as *SN_CreateRegion* and *SN_DeleteRegion*. For message passing among regions, *SN_Send* and *SN_Recv* are defined. Two primitives for collaborative computing are defined. *SN_Reduce* allows nodes

* Since regions are the only addressable entities in the location-centric computing model, the controller node is not addressable, either. We create a controller region, to which the controller node belongs. All messages for the controller node are sent toward the controller region

in a region to aggregate individual information into defined forms such as average, minimum, and maximum values. *SN_Barrier* provides the mechanism to synchronize program execution of all nodes in the same region.

1. *SN_CreateRegion*: The primitive prototype is as follows:*
   RegionID = SN_CreateRegion(SourceRegionID,Range,ManagerRange,SensorType,Timeout)

   SN_CreateRegion creates a region in *Range* with its manager subregion in *ManagerRange*, where *Range* and *ManagerRange* are corner geographical coordinates of the square areas. A region can be constrained to have only nodes with the specified *SensorType*. The caller of the primitive must specify the region it belongs to with its region handle *SourceRegionID* and once the region is successfully created, an acknowledgment will be sent by the manager subregion just created toward the source region. *Timeout* specifies the amount of time the caller is willing to wait for the acknowledgment before the call returns as failed. Once created, a region is assigned an integer handle *RegionID*.
2. *SN_DeleteRegion*: The primitive prototype is as follows:
   Status = SN_DeleteRegion(SourceRegionID,RegionID,Timeout)

   SN_DeleteRegion deletes a region with its handle *RegionID*. The caller of the primitive also specifies its own region handle *SourceRegionID*. Once the region is successfully deleted, each manager node in the deleted region sends an acknowledgment to the source region and terminates the call. *Timeout* specifies the maximum amount of time the call waits for the acknowledgment before it returns with *Status* as failure.
3. *SN_Send*: The primitive prototype is as follows:
   Status = SN_Send(SourceRegionID,DestinationRegionID,Data,DataType,Size,Tag)

   SN_Send delivers a message named *Data* from the caller node to a specified region. The regions are referred to with their handles *SourceRegionID* and *DestinationRegionID*. *DataType* and *Size* specify the type and size of the message. The *Tag* provides auxiliary information of the message such as data encoding options or application types. SN_Send returns with *Status*, successful or not, once the message is handed to the routing agent. SN_Send does not guarantee end-to-end delivery, however, to the remote region.
4. *SN_Recv*: The primitive prototype is as follows:
   Status = SN_Recv(SourceRegionID,Data,DataType,Size,Tag,Timeout)

   SN_Recv is the matching primitive that receives a message sent with SN_Send. SN_Recv receives a message named *Data* with *DataType* and *Size* from a specified region with its handle *SourceRegionID*. If *Tag* is specified, it further constrains the scope of messages to be received. Both *SourceRegionID* and *Tag* can be set to accept any incoming messages. *Timeout* specifies the amount of time the caller is willing to wait for a matching message to be returned. If no matching message is received before it times out, the call returns null and the *Status* indicates failure.
5. *SN_Reduce*: The primitive prototype is as follows:
   Status = SN_Reduce(RegionID,Data,DataType,Count,Operation,NumNodes,Timeout,Tag)

   In parallel computing terminologies, *reduce* is a collective communication primitive that collects data from all nodes in a specified group and performs an operation such as sum, min, or max on all collected data. The same collective communication is supported in UW-API with SN_Reduce. When an application needs to perform a reduce operation, it programs all nodes in a specific region to call this primitive. Each participating node delivers its *Data*, which is an array in *DataType* with *Count* elements, to nodes in the manager subregion, and the specified *operation* is performed on all collected data. There is potentially more than one node in the manager subregion, and all such nodes perform the same data collection and computation. Results may differ among these nodes and some may be faulty in a harsh environment. To enhance robustness and

* Types of variables are omitted in the following presentation. Interested readers may refer to Ref. [14].

fault tolerance of the computation, there exist sophisticated algorithms for information fusion [3]. The call termination depends on *NumNodes* and *Timeout*. Nodes in the manager subregion are blocked in the call until at least *NumNodes* reports are collected or a *Timeout* period has elapsed. All other nodes return after sending their data. Again, *Tag* provides auxiliary information for the operation if necessary.

6. *SN_Barrier*: The primitive prototype is as follows:
   Status = SN_Barrier(RegionID,NumNodes,Timeout,Tag)

   In parallel computing terminologies, *barrier* is a collective communication primitive that synchronizes a group of processes at a certain point in an application. A process becomes blocked when it calls the primitive, and the block is released once all processes in the group have called the primitive or *arrived at the same point of an application*. For processes to know how many peer processes have arrived, each process broadcasts a message in the group once it calls the primitive. In conventional parallel applications, the number of processes *N* is statically known. Thus, all processes complete the call at hearing the *N*th broadcast.

In UW-API, SN_Barrier synchronizes all nodes in a specified region. Similarly, each node gets blocked and broadcasts a message as it calls SN_Barrier. Each node also counts the number of broadcast messages to determine the call termination. However, in a distributed sensor network, the number of nodes in a region cannot be known a priori. Furthermore, the number may change as nodes relocate, enter a sleep mode, or even expire. Similar to SN_Reduce, *NumNodes* and *Timeout* are specified for call termination. Unlike SN_Reduce, SN_Barrier blocks all nodes until at least *NumNodes* broadcasts are heard or a *Timeout* period has elapsed. *Tag* provides auxiliary information for the operation if necessary.

5.4.2 UW-Routing

The UW-Routing protocol provides region-based communication that facilitates UW-API. As an on-demand protocol, UW-Routing delivers a message to a specific region in two steps. First, it *unicasts* the message to an arbitrary node in the specified region. Once the message arrives at the region, the second step is to broadcast the message to all nodes in the region.

UW-Routing uses message flooding in various places. For energy and bandwidth efficiency, all flooding operations are geographically constrained. Similar to Ref. [6], each flooded message has a square *forwarding zone* specified in geographic coordinates. The forwarding zone must cover the source and destination regions. Figure 5.1 shows our strategy in deciding a message's forwarding zone. The controllable margin accommodates situations where feasible routes do not exist in a minimum forwarding zone. A node receiving a flooded message continues the forwarding if and only if it resides within the message's forwarding zone. The following describes the message formats, route discovery for a remote region, and routing within a region.

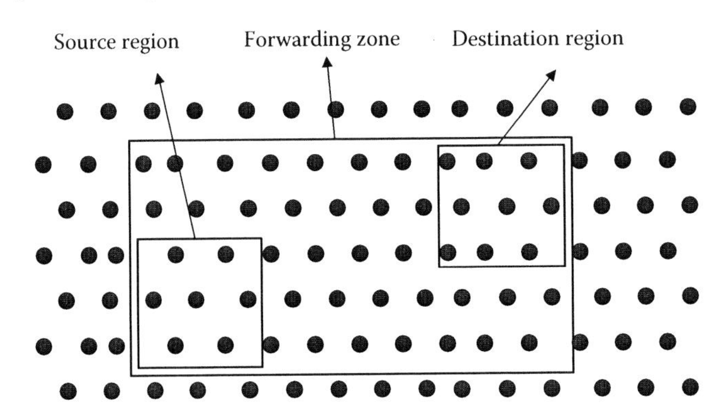

FIGURE 5.1 Message forwarding zone covering source and destination regions.

5.4.2.1 Message Formats and Address Resolution

Figure 5.2 summarizes the message formats. Each message serviced by UW-Routing carries a common header containing its message identification number, message type, *network address* of its previous forwarder, its forwarding zone, and a time stamp. For each message type, a specific extended header follows. Four message types are defined: route request (RREQ), route reply (RREP), manager route construction (MRC), and payload.

RREQ and RREP messages are used in route discovery. They have the same extended header containing the geographic coordinates as well as the region handles for the source and destination regions. Geographic region coordinates are explicitly specified only in RREQ, RREP, and the payload for SN_CreateRegion. Once the route discovery and creation complete for a region, further routing for the region is based on the region handle only instead of geographic coordinates. An MRC message is used in constructing routes in a region to facilitate communication between manager and nonmanager

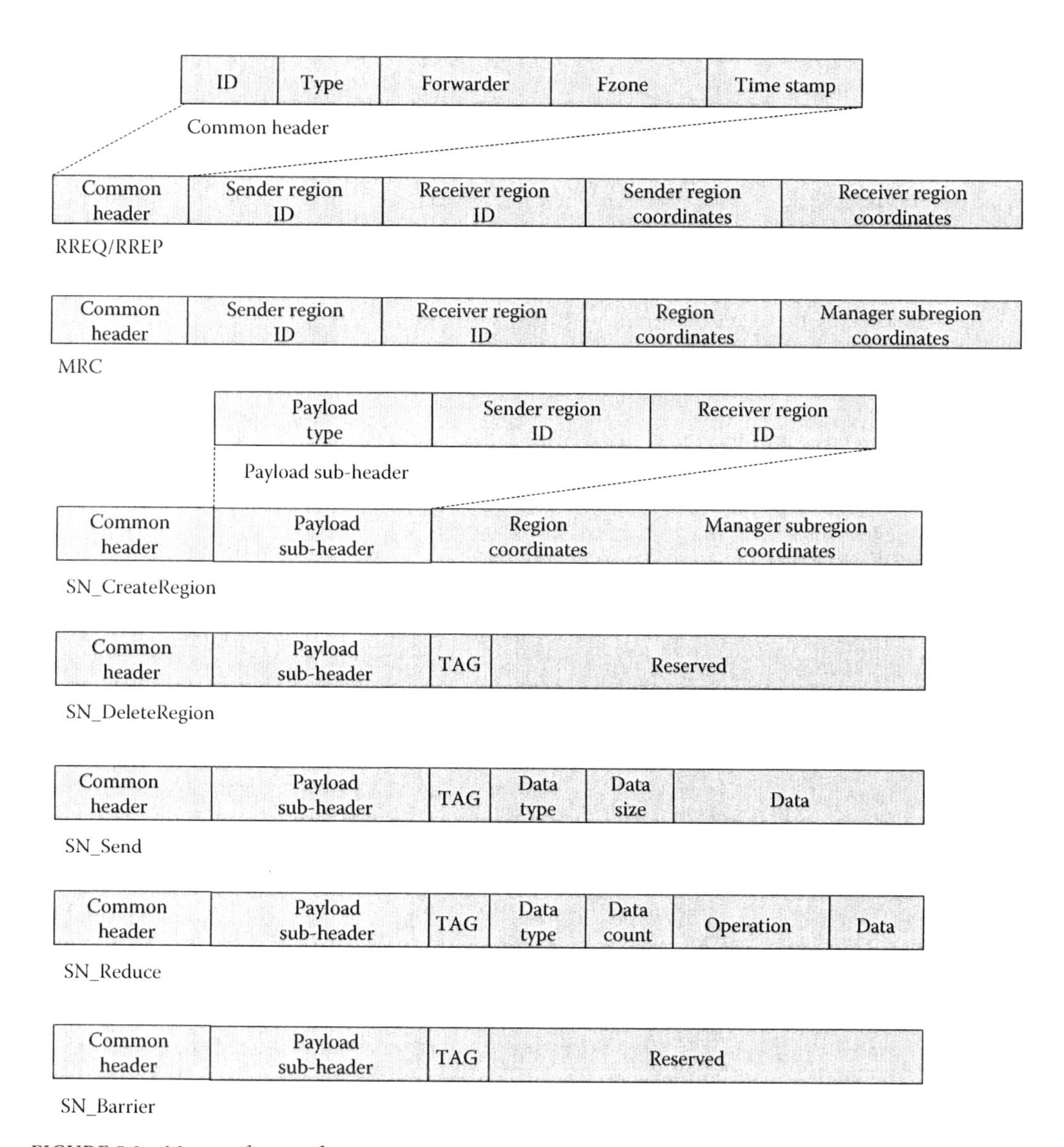

FIGURE 5.2 Message formats for UW-Routing.

Entry Number	Node Network Address	Link Layer Address	Interface
1	1	10.0.1	A
2	7	10.0.23	B
3	5	10.0.29	B
4	12	10.0.42	A
5	15	10.0.37	B
...	...	...	...

FIGURE 5.3 Address resolution table for UW-Routing.

nodes. Its extended header carries the handle of the respective region. Payload messages associated with respective UW-API primitives are also shown in Figure 5.2 and self-explanatory according to their functionalities.

Network addresses must be associated with corresponding *link layer addresses* before messages can be physically transmitted through a node's wireless interface. The address resolution is done in conjunction with route discovery with no additional overhead. Note that each message carries the forwarding node's network address, and its link layer header carries link layer addresses of both sending and receiving nodes. During route discovery, RREQ and RREP are flooded. By associating the network and link layer addresses of the sender of each message, address resolution for all neighboring nodes is done after one flooding operation. As a node potentially has more than one wireless interfaces, an interface is associated to each neighbor as well. Figure 5.3 shows one example of an address resolution table.

5.4.2.2 On-Demand Route Discovery

Each router maintains a routing table as illustrated in Figure 5.4.

For each region with a known route, the table has an entry recording the network address of the next-hop node of the route. The *Timeout* for each entry indicates the amount of time it remains valid and an expired route is removed from the table. When the router receives a message whose destination region does not exist in the table, the message is withheld in queue and the route discovery procedure is initiated. The route discovery is based on RREQ and RREP messages.

Entry Number	Region ID	Coordinates	Next Hop Node	TTL
1	1	[0, 0, 10, 10]	7	70
2	1379	[10, 20, 70, 80]	7	25
3	6517	[25, 75, 75, 125]	5	82
4	7482	[250, 10, 350, 110]	12	13
5	1010	[100, 150, 200, 250]	15	2
...	...	...	...	...

FIGURE 5.4 Routing table for UW-Routing.

The process starts with the router creating and flooding an RREQ message in the network. As mentioned earlier, the flooding is constrained within its forwarding zone. If a node receives an RREQ and it currently has a valid route for the requested region, the node generates an RREP and floods it back to the RREQ source region. If a node receives an RREQ but does not already know of the region, it simply continues the flooding. If a node that belongs to the requested region receives the RREQ, it generates an RREP and floods it back to the RREQ source region.

The routing table is updated at each node by monitoring RREQ and RREP messages. When an RREQ is received and the source region is not known, the source region is added to the table with the RREQ forwarding node as its next hop. When an RREP is received and its source region (that sends the RREP) is not known, the source region is also added to the table with the RREP forwarding node as its next hop. Whenever a new entry is added, the router inspects all messages in queue and services those messages whose destination region is now known to the node.

5.4.2.3 Routing within a Region

Within a region, two communication patterns frequently arise. First, it is necessary to disseminate messages from any node to all other nodes in the region. This occurs when a message arrives at a node in the destination region and the node must disseminate the message to all other nodes in the region. This essentially requires flooding a message to all nodes in a region. UW-Routing floods a message within a region if it is destined for all nodes in the region.

The second communication pattern is to collect messages from each node to nodes in the manager subregion. This occurs as SN_Reduce is called and all nodes report their data messages to the manager nodes. In this case, flooding certainly is not the most efficient solution. Instead, UW-Routing constructs a tree that roots at one of the manager nodes and spans all other nodes in the region. Thus, all SN_Reduce messages are delivered to the manager node following the tree structure. The manager node then floods all messages it receive within the manager subregion such that all manager nodes receive the same set of messages.

To construct a tree, a manager node floods an MRC message in the region after the region is created and a random time period has elapsed. The random wait is to reduce the probability of more than one node initiating the tree construction. Thus, a manager node receiving an MRC message will no longer initiate another tree construction itself. A node receiving an MRC message records the forwarding node as its parent in the tree and continues flooding the message. A tree is constructed once all nodes determine their parents. Still, there can potentially be more than one manager node constructing multiple trees in the region. This is equally acceptable as long as each node belongs to one tree. Therefore, each node accepts the first MRC message it receives and joins the respective tree. Figure 5.5 illustrates two such tree structures.

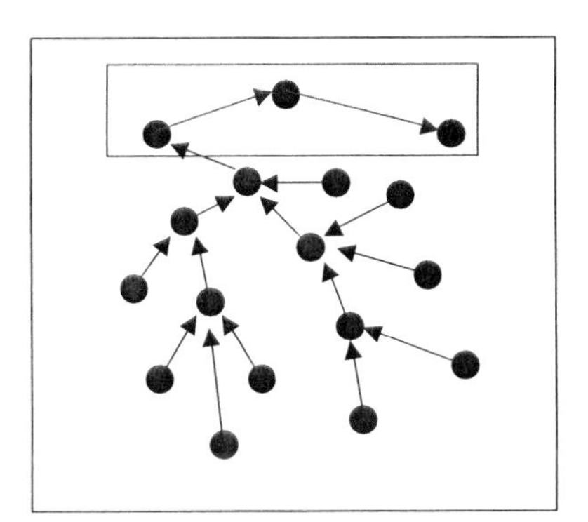

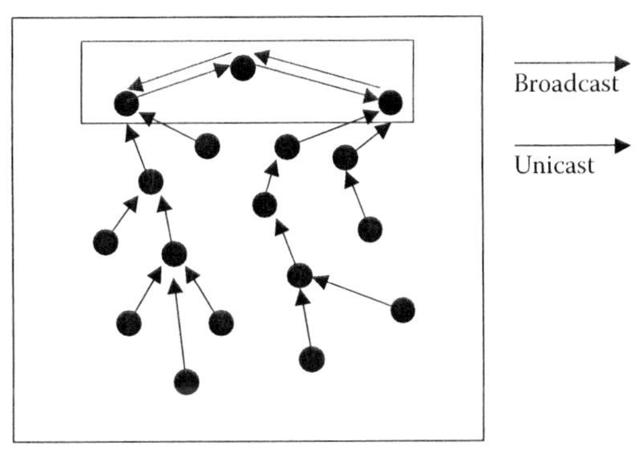

FIGURE 5.5 Message delivery trees rooted at nodes in a manager subregion.

5.5 Target-Tracking Application

To demonstrate the location-centric computing approach for collaborative application design in a distributed sensor network, a target-tracking application was developed at the University of Wisconsin with UW-API and UW-Routing. Figure 5.6 shows a typical sensor network where we detect and track potential target activities. In this application, the sensor network is tasked to detect the presence of a certain type of ground vehicle and to track its movement through the sensor field. With UW-API, regions are first created around potential target entry areas. Only sensor devices in these initial regions are active before any target enters the area. All other devices passively wait to be activated by these initial regions when needed. In an active region, the following operations are performed:

- *Per-node detection*: Each device constantly attempts to detect the presence of a target with its own sensing modalities. Detection decisions of each node are periodically reported to its manager subregion.
- *Region detection*: Each device in the manager subregion collects detection decisions from all nodes in the region. Individual decisions are combined with robust information fusion algorithms to arrive at a consensus decision for the region as a whole.
- *Per-node classification*: When a device decides that a target is present, it further attempts to determine the target vehicle type with its classifier algorithm. Classification results of each node are periodically reported to its manager subregion.
- *Region classification*: Each device in the manager subregion collects classification results from all nodes in the region. Individual results are combined to arrive at a consensus of the classification result for the region and whether the vehicle is of the desired type.
- *Target localization*: Once region detection and region classification confirm the presence of a desired type of target, the target must be localized. This is done collaboratively with all nodes in the region using an energy-based localization algorithm [9]. In this algorithm, all nodes report to nodes in the manager subregion with their energy measurements, based on which the target location is estimated.
- *Target tracking*: With a series of estimated target locations at consecutive time instances, nodes in the manager subregion are able to estimate the target speed and direction. Based on this information, near-term target locations can be estimated. If the predicted target location lies beyond the current region, additional regions are to be created along potential future target trajectories and the regions are to be tasked with this same tracking application.

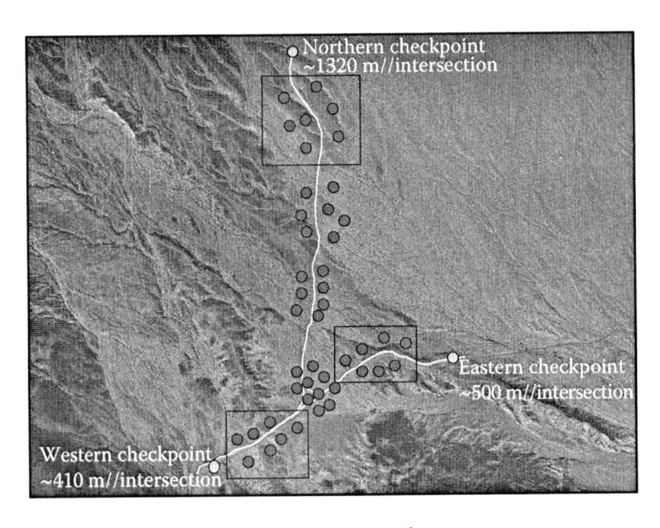

FIGURE 5.6 Typical sensor network for the target-tracking application.

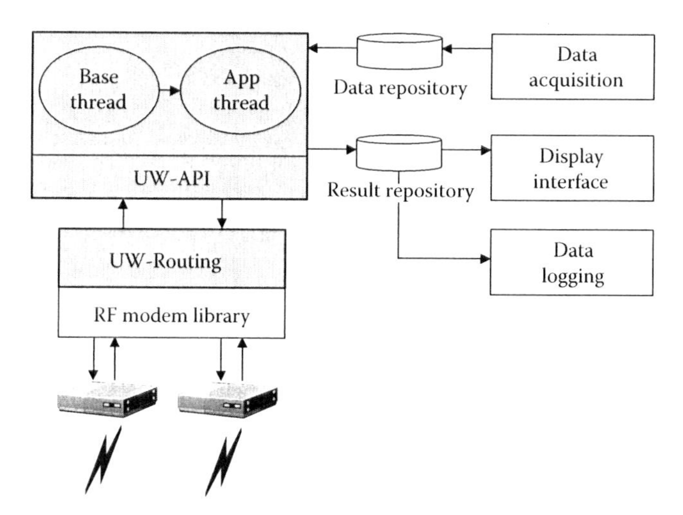

FIGURE 5.7 UW software architecture for location centric computing.

Figure 5.7 shows the software architecture. The shaded portions are location-centric components that interact with wireless communication interfaces and data peripherals on each device through UW-API and UW-Routing. Applications are managed with threads. On each device, the base thread is responsible for application creation, termination, and accepting incoming messages from the router. Each application is designed as a thread. When a device receives a message commanding it to initiate an application, the base thread spawns the corresponding thread. Through UW-API, an application can carry out collaboration within each region as well as among multiple regions. In our tracking application, various UW-API primitives were used to facilitate both intraregion and interregion collaborations. Figure 5.8 shows the pseudocode of the tracking application consisting of the following six software modules corresponding to the aforementioned operations:

- *NodeDet*: Each device runs a constant false alarm rate (CFAR) energy detector based on its own sensor measurements in NodeDet. Periodically, all nodes call SN_Reduce with its summing operation on their binary detector decisions. The sum of all decisions are thus computed at nodes in the manager subregion for later use in the DetFus module. At the same time, each node uses SN_Send to report the average energy detected during the period to nodes in the manager subregion for later use in the TarLoc module.
- *DetFus*: At nodes in the manager subregion, the reduced sums of detector decisions are used in a robust fusion algorithm to arrive at a consensus of the region detection decisions [3].
- *NodeClass*: If NodeDet decides a target is present, NodeClass is invoked to classify the type of target with its detector energy series. Each node uses SN_Send to periodically report the classifier result to nodes in the manager subregion.
- *ClassFus*: Periodic reports of classifier results are collected with SN_Recv at nodes in the manager subregion. A robust fusion algorithm [4] is used to aggregate these reports into a consensus whether the target is of the desired type.
- *TarLoc*: Periodic energy reports are collected with SN_Recv at nodes in the manager subregion. If DetFus decides there is a target and ClassFus confirms the desired type, the energy reports are used to estimate the target location [9].
- *TarTrak*: TarTrak tracks a confirmed target by recording its consecutive location estimates and predicting its near-term future positions. This is done at nodes in the manager subregion. If the predicted location lies beyond the current region, SN_CreateRegion is used to create a new region around the predicted location. Finally, a command message is sent with SN_Send to the new region to initiate the same application in the new region.

Nodes in manager sub-region:

For every interval of T,

```
Manager_track(vehicle_type){
   //DetFus
       SN_Reduce(my_region, detection_result);
   //ClassFus
       SN_Reduce(my_region, classification_result);
   //Collect energy reports from nodes
       report_deadline=set_report_deadline();
       while current_time<=report_deadline,
              SN_Recv(node_energy);
              energy_reports=energy_reports ∪node_energy;
   //TarLoc and TarTrak if desired type of target exists
   if detection_result==True and classification_result==vehicle_type,
       Location_estimate=TarLoc(energy_reports);
       Predicted_location=TarTrak(Location_estimate);
       if Predicted_location is beyond this region,
              New_region=Predict_new_region(Predicted_location);
              SN_CreateRegion(New_region);
}
```

Nodes not in manager sub-region:

For every interval of T,

```
Nonmanager_track(){
   //NodeDet
       detection_result=NodeDet(sensor_reading);
   //NodeClass
       classification_result=NodeClass(sensor_reading);
   //DetFus
       SN_Reduce(my_region, detection_result);
   //ClassFus
       SN_Reduce(my_region, classification_result);
   //Send energy report to manager sub-region
       SN_Send(my_manager_region, node_energy);
}
```

FIGURE 5.8 Pseudocode of the target-tracking application.

5.6 Testbed Evaluation

UW-API, UW-Routing, and the target-tracking application have been implemented on dedicated sensor devices developed for SensIT. These devices are equipped with acoustic, seismic, and infrared sensors. They are also equipped with GPS and wireless communication interfaces. To evaluate the location-centric approach, we deployed these devices in two test beds and measured the network performances. The two test beds are

- A test bed of 20 nodes deployed in Waltham, Massachusetts
- A test bed of 70 nodes deployed in Twentynine Palms, California

The testbed topologies are shown in Figure 5.9. In these test beds, the tracking application was launched in three scenarios and the network performance was evaluated. The three scenarios were as follows:

1. *Cross-run*: In this scenario, the network tracked two vehicles (assault amphibious vehicle [AAV] and dragon wagon [DW]) at the same time as they crossed each other in the 20-node test bed.
2. *Turn-back*: In this scenario, the network tracked two vehicles (assault amphibious vehicle [AAV] and dragon wagon [DW]) at the same time as they met and turned around in the 20-node test bed.
3. *Single-AAV*: In this scenario, a single AAV ran from east to west in the 70-node test bed.

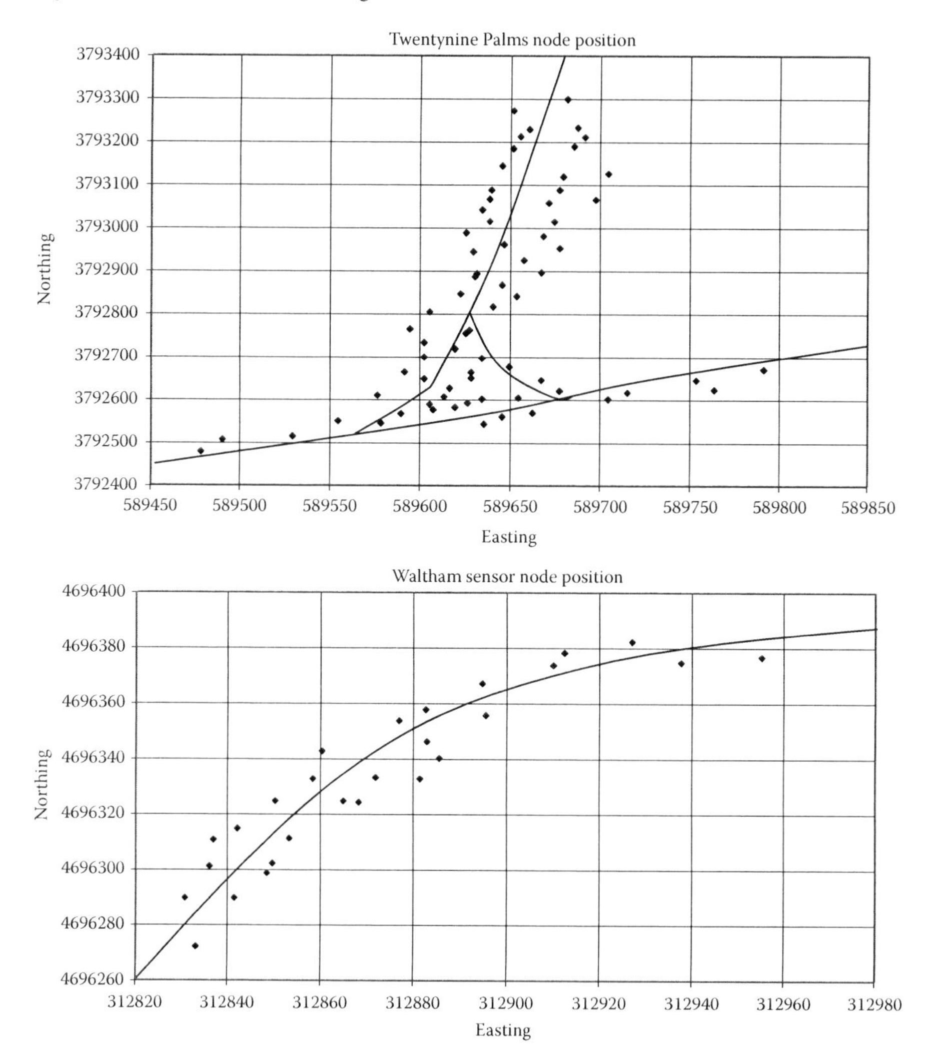

FIGURE 5.9 Sensor network testbed topologies.

The network performance was evaluated in the following aspects:

- Control messages versus payload messages forwarded
- In-region messages versus out-of-region messages forwarded
- Overall per-node bandwidth consumption

5.6.1 Control Messages versus Payload Messages

Location-centric applications require control message exchanges only during region creation and initial setup stages. Figure 5.10 summarizes the number of control and payload messages ever transmitted by

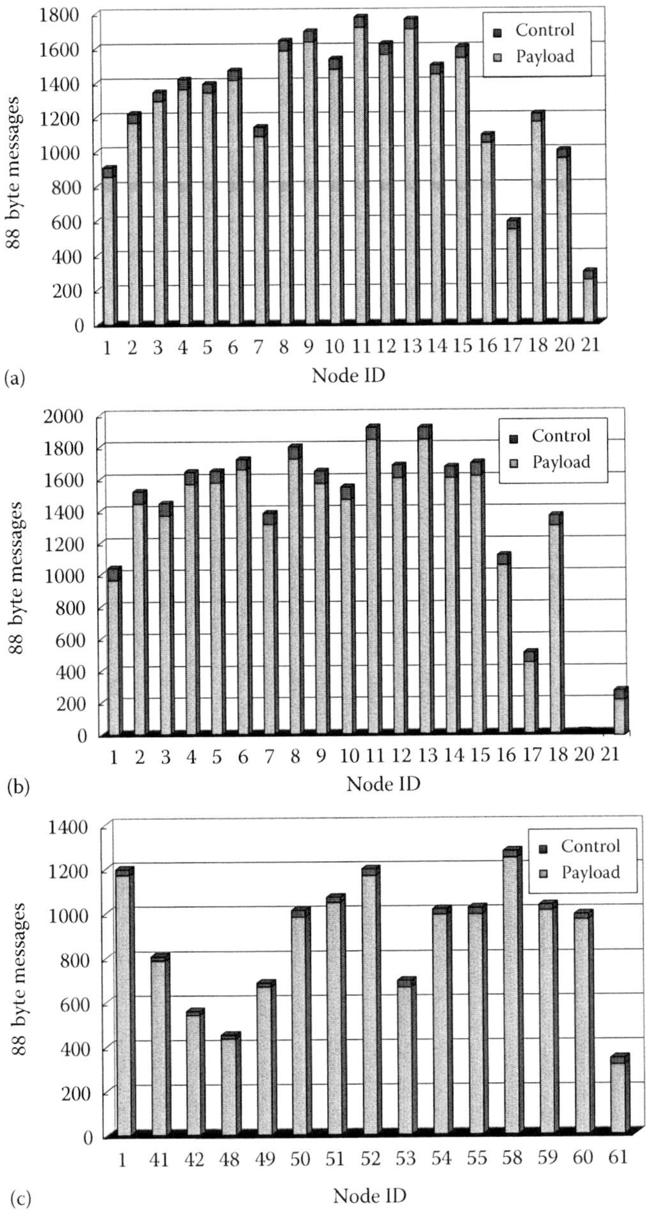

FIGURE 5.10 The number of control and payload messages transmitted by each node in the three scenarios: (a) cross-run, (b) turn-back, and (c) single-AAV.

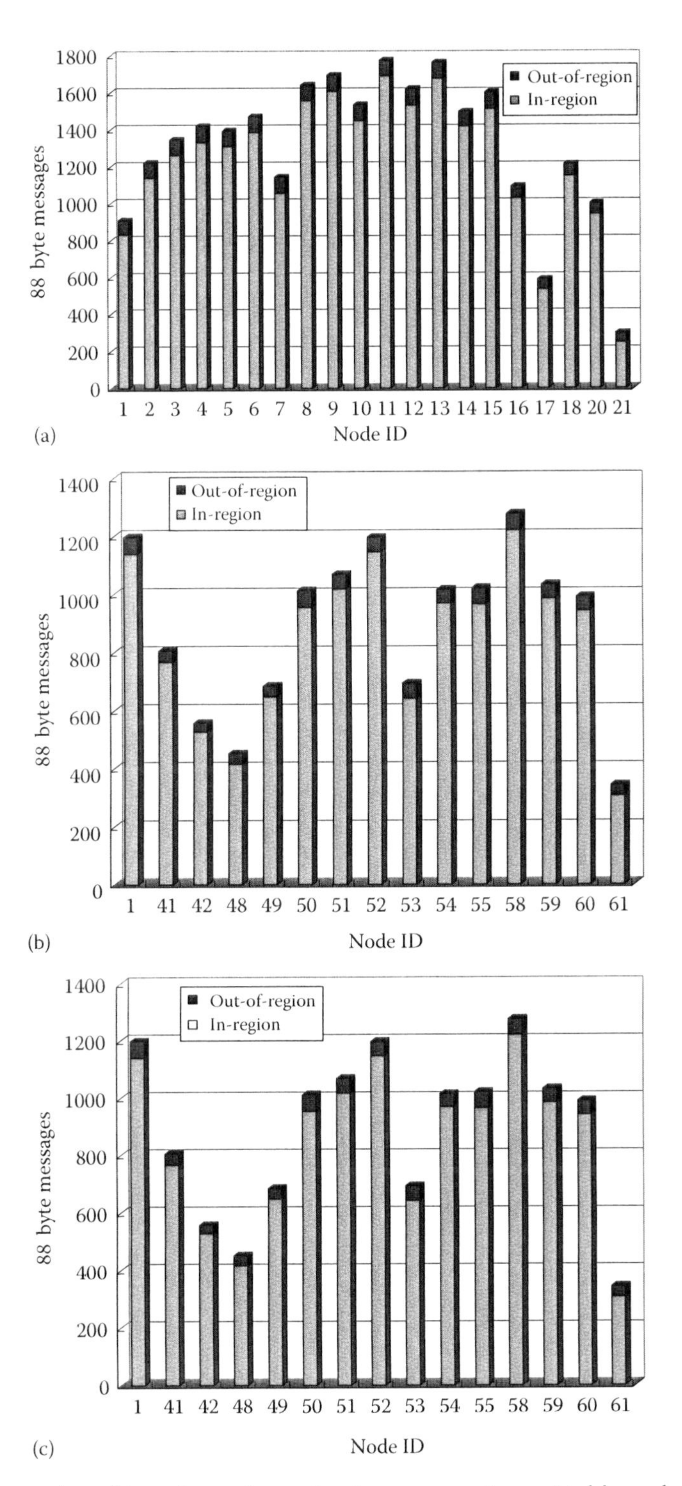

FIGURE 5.11 The number of in-region and out-of-region messages transmitted by each node in the three scenarios: (a) cross-run, (b) turn-back, and (c) single-AAV.

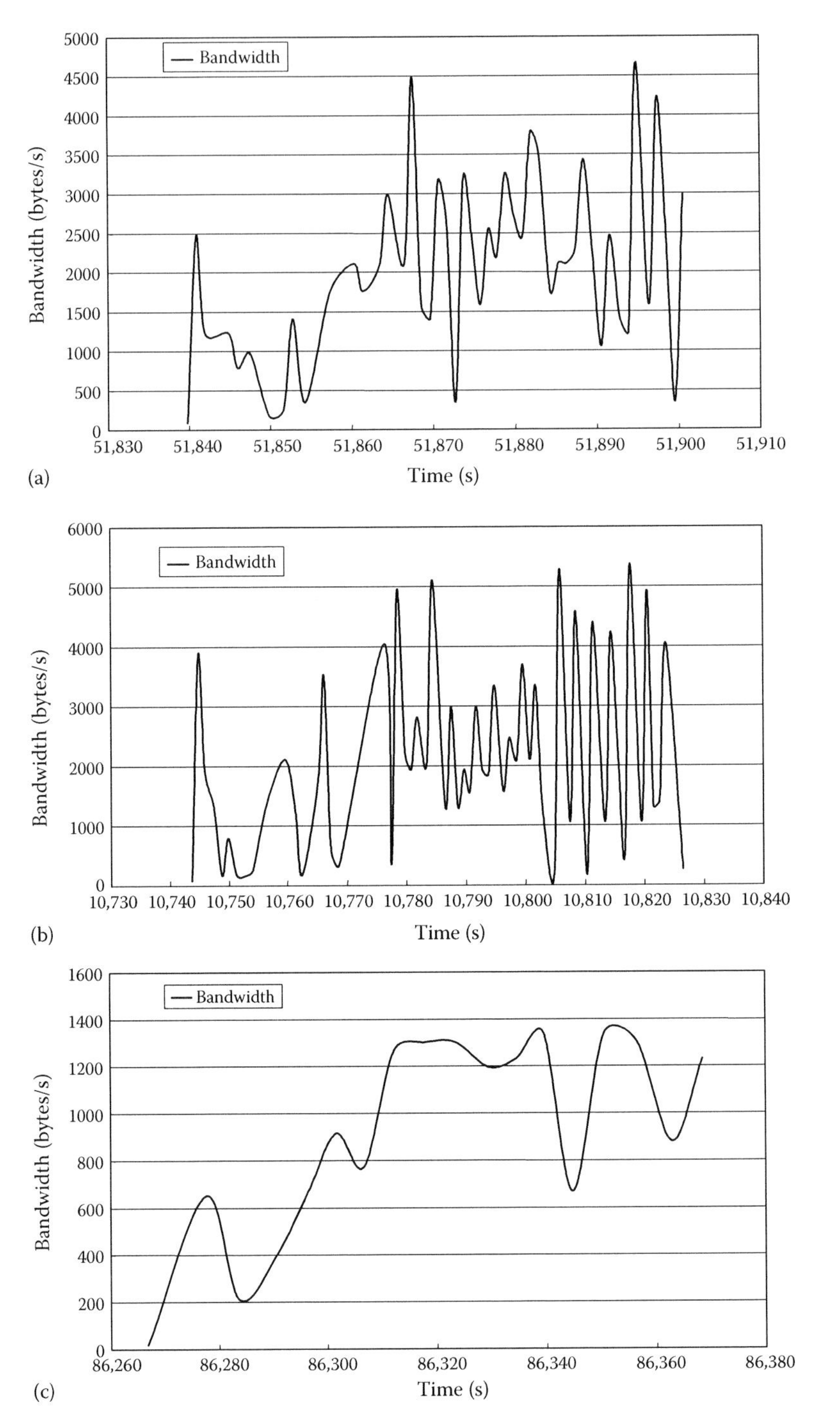

FIGURE 5.12 Bandwidth consumption of a node in the manager subregion in the three scenarios: (a) cross-run, (b) turn-back, and (c) single-AAV.

each node in all scenarios. All messages had the same size of 88 bytes. The number of control messages was far smaller than that of payload messages, indicating the relatively low control overhead required in a location-centric application.

5.6.2 In-Region Messages versus Out-of-Region Messages

Location-centric applications are expected to have primarily localized computation and communication. We analyzed for each node the number of transmitted messages destined for its own region (in-region) versus those destined for a foreign region (out-of-region). Figure 5.11 summarizes the comparison results for three scenarios.

As expected, the fraction of out-of-region messages serviced by each node was low.

5.6.3 Overall Per-Node Bandwidth Consumption

Figure 5.12 shows the typical bandwidth consumption of a manager node in each scenario. The bandwidth profiles were actually similar among the scenarios. All of them started with a short surge. After the surge, they stayed low for a certain amount of time before the final phase began where bandwidth consumption rose and fell in a periodic pattern. The profile closely reflected the tracking application's communication pattern. The initial surge indicates control message exchanges involved in region creation. Once a region is created, nodes remain inactive if no tasks are issued. As soon as a region is tasked with the tracking application, the collaborative algorithm requires nodes in a region periodically collect detection and classification decisions in the manager subregion. The periodic bandwidth profile is thus explained.

5.7 Conclusion

This chapter presents a location-centric approach for programming collaborative applications in a large-scale distributed sensor network. The location-centric approach relieves the burden for programmers to keep track of individual nodes' status and operations; instead, it allows programmers to define regions based on geographic areas of substance to the distributed algorithms. UW-API and UW-Routing serve as an example implementation of this new programming paradigm. The live tests of a distributed sensing, detection, classification, and tracking of military vehicles reveal the ease of use of this approach and various typical performance metrics of interest to collaborative distributed applications.

Acknowledgments

The work reported here was supported in part by the Defense Advanced Research Projects Agency (DARPA) and Air Force Research Laboratory, Air Force Material Command, USAF, under agreement number F30602-00-2-0555, and by U.S. Army Research grant DAAD19-01-1-0504 under a subrecipient agreement S01-24 from the Pennsylvania State University. The views and conclusions contained herein are those of the authors and should not be interpreted as necessarily representing the views of funding agencies.

References

1. J. Agre and L. Clare. An integrated architecture for cooperative sensing networks. *Computer*, 33:106–108, May 2000.
2. P. Brooks, P. Ramanathan, and A. Sayeed. Distributed target classification and tracking in sensor networks. *Proceedings of IEEE*, 81(8):1163–1171, 2003.

3. T. Clouqueur, P. Ramanathan, K. Saluja, and K.-C. Wang. Value-fusion versus decision-fusion for fault-tolerance in collaborative target detection in sensor networks. In *Proceedings of International Conference on Information Fusion*, Montreal, Quebec, Canada, August 2001.
4. A. D'Costa and A. Sayeed. Collaborative signal processing for distributed classification in sensor networks. In *Proceedings of International Workshop on Information Processing in Sensor Networks*, Palo Alto, CA, pp. 1–16, 2003.
5. D. Estrin, L. Girod, G. Pottie, and M. Srivastava. Instrumenting the world with wireless sensor network. In *Proceedings of ICASSP 2001*, Salt Lake City, UT, pp. 2675–2678, 2001.
6. Y.-B. Ko and N.H. Vaidya. Location-aided routing (LAR) in mobile ad hoc networks. In *Proceedings of ACM MOBICOM 1998*, Dallas, TX, pp. 66–75, October 1998.
7. Y.-B. Ko and N.H. Vaidya. GeoTORA: A protocol for geocasting in mobile ad hoc networks. In *Proceedings of ICNP*, Riverside, CA, pp. 240–250, 2000.
8. S. Kumar, F. Zhao, and D. Shepherd (Eds.). Special issue on collaborative signal and information processing in microsensor networks. *IEEE Signal Processing Magazine*, 19(2), pp. 13–14, March 2002.
9. D. Li, K. Wong, Y.-H. Hu, and A. Sayeed. Detection, classification, tracking of targets in microsensor networks. *IEEE Signal Processing Magazine*, 19(2), 17–29, March 2002.
10. V.D. Park and M.S. Corson. A highly adaptive distributed routing algorithm for mobile wireless networks. In *Proceedings of INFOCOM*, Kobe, Japan, pp. 1405–1413, April 1997.
11. N. Parkinson and S. Gilbert. NAVSTAR: Global positioning system—Ten years later. In *Proceedings of IEEE*, Atlanta, GA, pp. 1177–1186, 1983.
12. P. Ramanathan. Location-centric approach for collaborative target detection, classification, and tracking. In *Proceedings of IEEE CAS Workshop on Wireless Communication and Networking*, Pasadena, CA, September 2002.
13. P. Ramanathan, K. Saluja, K.-C. Wang, and T. Clouqueur. UW-API: A network routing application programmer's interface. Technical Documentation for DARPA SensIT Program, http://www.ece.wisc.edu/~sensit, October 2001.
14. M. Snir, S. Otto, S. Huss-Lederman, D. Walker, and J. Dongarra. *MPI—The Complete Reference*. MIT Press, Cambridge, MA, 1998.

6

Application of Geometric Probability and Integral Geometry to Sensor Field Analysis

Errol G. Rowe
Office of Naval Research Global-Asia

James Kelly
Naval Undersea Warfare Center

Anthony G. Warrack
North Carolina Agricultural and Technical State University

6.1 Geometric Probability

6.1.1 Introduction

In seeking to maximize the probability of target detection by sensors randomly distributed over some region, one might consider factors such as the numbers of sensors, the detective power of the sensor, and the way in which the sensors are spatially distributed throughout the region. Many of these factors have a strong probabilistic and geometric nature. In this chapter, we use techniques from geometric probability and integral geometry to investigate coverage and detection problem for sensor fields. Methods of integral geometry have been used in computing coverage probabilities in heterogeneous sensor networks by Lazos et al. [15] using many of the results in the classic work of Santaló [23]. Similarly, Rowe and Wettergren [22] used geometric probability methods to examine the reliability over time of randomly distributed heterogeneous sensor networks.

This section outlines some techniques from geometric probability that can be used to examine the coverage and performance of randomly distributed heterogeneous fields consisting of a large number of sensors. The remainder of this chapter uses techniques from integral and stochastic geometry to examine fields consisting (usually) of a small number of sensors. Section 6.2 considers various measures that can be used in detection problems. In particular, it demonstrates why it is essential to use an invariant measure when considering these types of problems. Section 6.3 considers the detection of stationary and moving targets. This section is the most technically challenging; however, it is the most rewarding as it demonstrates how one can use results from integral geometry (e.g., Steiner's formula and Blaschke's kinematic formula) to obtain results that are extremely difficult to obtain using other methods.

6.1.2 Spatial Poisson Processes

This section commences with a review of the more familiar temporal Poisson point process, followed by a description of its spatial equivalent. Recall that the usual (i.e., temporal) Poisson point process N_t, with rate λ, is a stochastic process in which N_t is the number of occurrences of an event during t units of time. Moreover, N_t follows a Poisson distribution with mean λt. Finally, the event occurrences in disjoint time intervals are independent. More formally, the counting process $N = \{N_t, t \geq 0\}$ is a Poisson process with rate $\lambda > 0$, if it possesses the following properties*:

1. $N_0 = 0$.
2. $P\{N_h = 1\} = \lambda h + o(h)$.
3. $P\{N_h \geq 2\} = o(h)$.
4. N_t satisfies the stationary and independent increment property.

Property 2 says that, over a very short time period, the probability of an event occurrence is approximately linear with time. Property 3 says that the probability of two or more occurrences over a very short time period is essentially zero. And according to property 4, the process N_t is independent of the process $N_{t+h} - N_t$, that is, the number of occurrences up to time t is independent of the number of occurrences after time t.

For a spatial Poisson process, instead of counting the number of occurrences over a time interval, one is interested in the number of points associated with an event over a region of space. Let S be an n-dimensional set and suppose $A \subset S$, that is, A is a subset of S. Consider points scattered randomly throughout S and let $N(A)$ denote the number of points from the scattered set that are contained in A. The stochastic process $N(A)$ is called a *point process* in S. Depending on the dimension of S, let $\|A\|$ denote the length, area, volume, etc., of A. The stochastic process $\mathrm{N} = \{N(A), A \subset S\}$ is an n-dimensional Poisson counting process with parameter $\lambda > 0$ if

a. $N(A)$ follows a Poisson distribution with mean $\lambda\|A\|$
b. The number of points occurring in disjoint subsets of S is mutually independent

Thus, for the spatial Poisson process, the statements equivalent to properties 1–4 mentioned earlier are (here $\emptyset$ denotes the empty set) as follows:

1. $N(\emptyset) = 0$.
2. $P\{N(A) = 1\} = \lambda \|A\| + o(\|A\|)$.
3. $P\{N(A) \geq 2\} = o(\|A\|)$.
4. $N(A)$ satisfies the stationary and independent increment property.

In particular, property (2) says that if $\|A\|$ (the volume of the set A) is small, then the probability of one occurrence is approximately linear with respect to volume; property (3) says that for small volumes, the probability of two or more occurrences is approximately zero. Finally, property (4) says that if $A, C \subset S$, then $N(A)$ is independent of $N(C \cap A^c)$.

---

* The function $g(h)$ is said to be $o(h)$, written $g(h) = o(h)$ if $\lim_{h \to 0} g(h)/h = 0$. Thus, according to property 3, as h goes to zero, the probability of two or more occurrences goes to zero faster than any linear function.

The definitions are summarized as follows: If $N = \{N(A), A \subset S\}$ is an n-dimensional Poisson counting process with parameter $\lambda > 0$, then the probability of k occurrences in the subset A is given by*

$$P\{N(A) = k\} = \frac{(\lambda \| A \|)^k e^{-\lambda \|A\|}}{k!}$$

See, for example, Resnick [21] or Kingman [14] for further information on spatial Poisson processes.

6.1.3 Sensor Coverage

This section contains the main result of this report, with the remainder of the report addressing the consequences of this result. Most of the discussion concerns coverage in three-dimensional space; however, all arguments apply to two dimensions. A two-dimensional equivalent of the main result is also stated. The proof is technical and is postponed until the Appendix 6.A. A restatement of the problem can be found in Chapter 16 of Karlin and Taylor [12], where instead of sensors and range sensitivities, spherical centers and corresponding radii are considered.

Theorem 6.1

Consider a set of omnidirectional sensors in three-dimensional space randomly scattered throughout some region $S \subset R^3$. Suppose that the sensors are spatially distributed according to a Poisson point process with parameter λ. Suppose also that the range sensitivity of all sensors is distributed according to the cumulative distribution function $F(r)$ with density $F'(r) = f(r)$ having finite third moment. Finally, assume that range sensitivity is independent of the sensor location. Then the number of sensors that will detect a target located at some point $\mathbf{x} \in S$ is a Poisson random variable with mean

$$\Lambda = \frac{4}{3}\lambda\pi \int_0^\infty r^3 f(r)\,dr. \tag{6.1}$$

Remark 1
Given an arbitrary point $\mathbf{x} \in S \subset R^n$,

$$P\{\text{target at position } \mathbf{x} \text{ detected by } k \text{ sensors}\} = \frac{\Lambda^k e^{-\Lambda}}{k!}$$

The probability that the point is not detected is $\exp(-\Lambda)$. Hence, the probability that at least one sensor covers the point is $\rho = 1 - \exp(-\Lambda)$.

[*The case of fixed range sensitivity*]: The reference to range sensitivity distribution alludes to the fact that the sensors are not perfect. Thus, instead of a few very expensive sensors with fixed range sensitivity r_o, the network may contain several less expensive sensors with variable range sensitivities. Nevertheless, through sampling, one can get an estimate of the sensitivity distribution. In the idealized case in which all sensors have identical fixed range sensitivity r_o, one obtains

$$\Lambda = \frac{4}{3}\lambda\pi \int_0^\infty r^3 \delta(r - r_o)\,dr$$

$$= \lambda \frac{4}{3}\pi r_o^3$$

* Recall that if X is a Poisson random variable with mean λ, then $P\{X = k\} = \lambda^k e^{-\lambda}/k!$. According to property (a), therefore, one can substitute $N(A)$ for X and $\lambda\|A\|$ for λ to get the equivalent probabilities for the spatial random variable.

Remark 2

If an arbitrary point in the volume is to be detectable by at least some fixed number of sensors, then the result shows us what intensity level (i.e., Λ) is required to achieve this. Since the mean number of sensors that detect a target at some point in S is $\frac{4}{3}\lambda\pi\int_0^{\infty} r^3 f(r)dr$, one can adjust λ (by increasing the number of sensors scattered throughout S) to achieve the desired coverage. The only quantity that is required is some estimate of $f(r)$, the range sensitivity density—sampling.

6.1.3.1 Problem Statement for R^2

Proposition 6.1

Consider disks in two-dimensional space with centers distributed according to a Poisson distribution with mean $\lambda\|A\|$, where $\|A\|$ represents the area of the set A. Suppose that the radii of all disks are independent of the location of the center of the disk and distributed according to $F(r)$ with density $f(r)$ and finite second moment. Then the number of disks that cover a point **x** is a Poisson random variable with parameter

$$\Lambda = \lambda\pi\int_0^{\infty} r^2 f(r)dr. \tag{6.2}$$

Definition 6.1

Throughout the remainder of this chapter, the statistic $\rho = 1 - \exp(-\Lambda)$ will be referred to as the *predicted coverage* for the sensor network. The coverage statistic ρ is a measure of the probability that an arbitrary point is detected by at least one sensor.

6.1.4 Coverage Comparison Study

This section provides a comparison of the coverage provided by two systems of randomly distributed sensors. Throughout this section, it is assumed that the sensors are distributed over a planar region, that is, $S \subset R^2$. Case one considers a system in which all sensors have identical range sensitivity: $r_o = 3$ units. Thus, the range sensitivity density for this system is the Dirac delta function $\delta(r - r_o)$. For case two, the gamma density is used to model the range sensitivity over all sensors, that is,

$$f(r) = \frac{1}{\Gamma(\alpha)\beta^{\alpha}} r^{\alpha-1} e^{-r/\beta}, \quad r \geq 0$$

In the example given, $\alpha = 2.5$ and $\beta = 1.2$; thus, the mean range sensitivity is also $\alpha\beta = 3.0$. The gamma density is used to model range sensitivity to convey the idea that some sensors will be better than others. Upon deployment, some sensors will be defective or give very poor performance, whereas others will perform very well. And, of course, several will provide average performance (Figure 6.1).

In general, for f equal to the gamma density, one has*

$$\Lambda = \frac{C}{\Gamma(\alpha)\beta^{\alpha}} \int_0^{\infty} r^{\alpha+\gamma-1} e^{-r/\beta} dr$$

$$= \frac{C}{\Gamma(\alpha)\beta^{\alpha}} \Gamma(\alpha+\gamma)\beta^{\alpha+\gamma} \tag{6.3}$$

* Here, the parameters C and γ depend on the dimension: if the search region is in R^2, then $\gamma = 2$ and $C = \lambda\pi$; if the search region is a subset of R^3, then $\gamma = 3$ and $C = \frac{4}{3}\lambda\pi$.

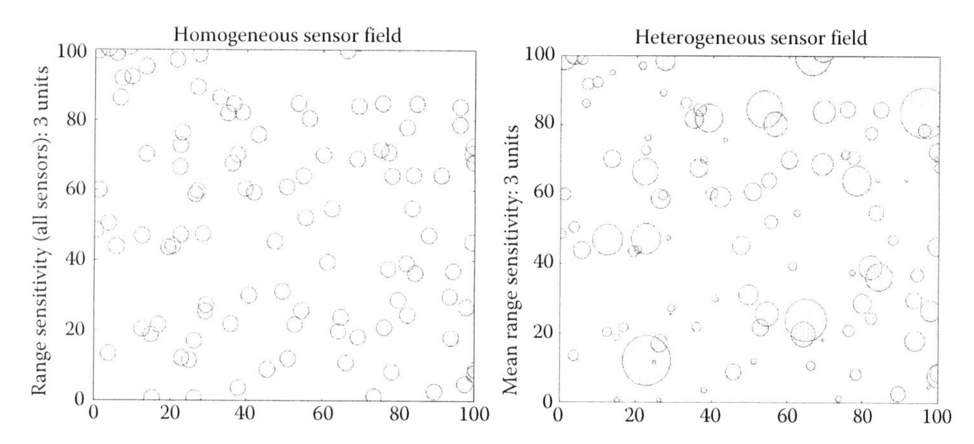

FIGURE 6.1 On the left, 100 nodes randomly distributed over a 100 by 100 square unit area. The same nodes are depicted on the right, each with omnidirectional communication range randomly selected from a gamma distribution with density $f(r) = \frac{1}{\Gamma(\alpha)\beta^{\alpha}} r^{\alpha-1} \exp\{-r/\beta\}$, $r \geq 0$, $\alpha = 2.5$, $\beta = 0.75$.

From Section 6.3.2, it follows that the mean parameter for the homogeneous case is $\Lambda = \lambda\pi \int_0^\infty r^2 \delta(r - r_0)\, dr = \lambda\pi r_0^2$.

In this simulation, 100 sensors are uniformly distributed over a 100 by 100 square unit area. For the homogeneous system, with all sensors having range up to r_0 = 3 units, the predicted coverage* is $\rho = 1 - \exp(-\lambda\pi r_0^2) = 0.246$. On the other hand, for the heterogeneous system with range sensitivity following a gamma distribution with parameters $\alpha = 2.5$, $\beta = 1.2$, the predicted coverage is $\rho = 1.0 - \exp\left\{-\frac{\lambda\pi}{\Gamma(\alpha)\beta^{\alpha}} \Gamma(\alpha+2)\beta^{\alpha+2}\right\} = 0.326$. Note that the mean sensitivity range for the heterogeneous sensors is $\alpha\beta = 3$. Figure 6.1 compares the coverage provided by just one simulation of a randomly distributed system of homogeneous and heterogeneous sensors. The plot on the left depicts 100 randomly placed sensors. The actual coverage provided by this system is 0.256. The plot on the right contains 100 sensors with 0.339 coverage. This is, of course, just one instantiation of the two systems. In the next section, data from several instantiations are used to test the long-term accuracy of the simulations.

Figure 6.2 compares the number of sensors deployed to the predicted coverage. From the formula for ρ, it is clear that the critical factor in the coverage statistic (for the case of a planar field of sensors†) is the mean coverage area of the sensors: $\int_0^\infty r^2 f(r)\, dr$.

6.1.5 Monte Carlo Study of the Coverage Statistic

Suppose that, for each simulation, one randomly distributes sensors over a search region and then randomly chooses a point in this region. The randomly chosen point should be covered by at least one sensor approximately $\rho = 1 - \exp(-\Lambda)$ percent of the time.‡ For the ith simulation, let $X_i = 1$ if the randomly chosen point is within range of at least one sensor; otherwise, set $X_i = 0$. The random variables X_i are binomial with success probability ρ. The sample average $\hat{\rho} = \frac{1}{N}\sum_{i=1}^{N} X_i$ of the simulations can be

* The term *predicted coverage* denotes the probability that an arbitrary point in the space is within range of at lease one sensor.

† For a network of sensors over a volume, the mean volume $\int_0^\infty r^3 f(r)\, dr$ is the critical statistic.

‡ Recall from Proposition 1 that the number of sensors that detect a randomly chosen point in space is Poisson with mean Λ. Hence, the probability that the point is not detected is $\exp(-\Lambda)$. The probability that at least one sensor covers the point is $1 - \exp(-\Lambda)$.

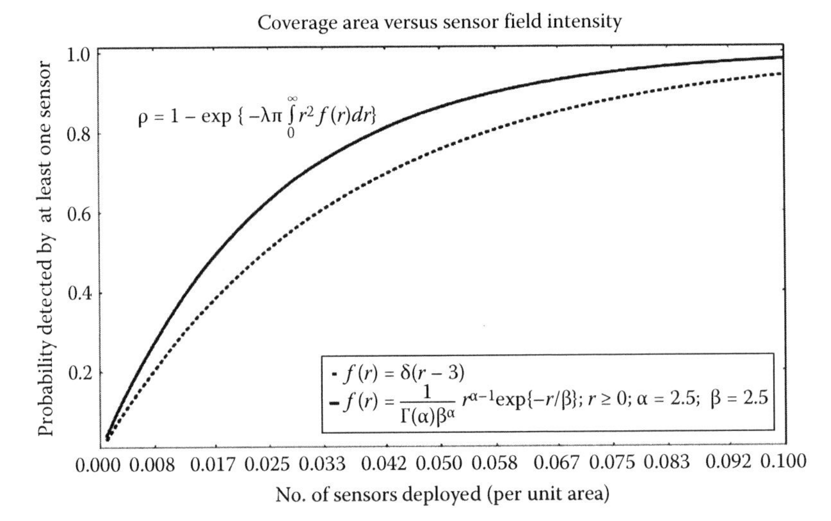

FIGURE 6.2 Sensors are randomly distributed over the surveillance region. In the homogeneous field, all nodes have communication range of 3 units. Communication range in the heterogeneous field varies according to a gamma distribution with parameters $\alpha = 2.5$ and $\beta = 1.2$. In this case, the heterogeneous system provides better coverage, even though its mean communications range is $\alpha\beta = 3$ units also. This follows from the observation that coverage is related to the second moment of the communications range and not the mean value.

used to determine the number of simulations required so that $\hat{\rho}$ is within 5% of ρ with 90% confidence. Specifically, one wants to estimate N such that

$$P\left(|\hat{\rho}-\rho| \le 0.05\rho\right) = 0.90$$

Note that

$$\mu_{\hat{\rho}} = \rho$$

$$\sigma_{\hat{\rho}}^2 = \frac{\rho(1-\rho)}{N}$$

By the central limit theorem, for N large, $\hat{\rho}$ is approximately normally distributed with mean ρ and variance $\rho(1-\rho)/N$. Hence,

$$\begin{aligned} 0.90 &= P\left(|\hat{\rho}-\rho| \le 0.05\rho\right) \\ &= P\left(\left|\frac{\hat{\rho}-\rho}{\sqrt{\rho(1-\rho)/N}}\right| \le 0.05\rho\frac{\sqrt{N}}{\sqrt{\rho(1-\rho)}}\right) \\ &= P\left(|Z| \le 0.05\rho\frac{\sqrt{N}}{\sqrt{\rho(1-\rho)}}\right) \end{aligned}$$

where Z is the standard normal random variable. In order for $\hat{\rho}$ to be within 5% of ρ (90% of the time), it is required that

$$0.05\rho\frac{\sqrt{N}}{\sqrt{\rho(1-\rho)}} \approx 1.645$$

that is, $N \approx (1.645/0.05)^2(1-\rho)/\rho$.

TABLE 6.1 Monte Carlo Simulation of Coverage Probability

System	No. Simulations	ρ	$\hat{\rho}$	Percentage Difference
Homogeneous	348	0.757	0.744	1.72
Heterogeneous	174	0.862	0.845	1.97

The analysis was performed for $\lambda = 0.05$ over a 100 by 100 square unit area. Thus, each simulation has roughly $10{,}000\lambda = 500$ sensors. As in the previous section, each sensor in the homogeneous field has a range sensitivity of $r_0 = 3$ units, and the sensor range density for the heterogeneous system is gamma with parameters $\alpha = 2.5$, $\beta = 1.2$. The predicted coverage provided by the two systems is 0.757 and 0.862, respectively. Note that, for this example, detection in the homogeneous case is a more rare event, hence more simulations are required to approximate the true value: $N = (1.645/0.05)^2(1 - 0.757)/0.757 = 348$ versus $N = (1.645/0.05)^2(1 - 0.862)/0.862 = 174$ for the heterogeneous case (Table 6.1).

6.1.6 Simple Network Communications Application

We start by deriving the density function that describes the distance from an arbitrary node to its k nearest neighboring node. For an arbitrary node, p, let $D_p(k)$ denote the distance to its kth nearest node. This distance is a random variable, dependent on the particular distribution of all nodes. Moreover, the probability that this distance is less than or equal to the range r is

$$\begin{aligned} P\{D_p(k) \le r\} &= 1 - P\{D_p(k) > r\} \\ &= 1 - P\,\{\text{at most } (k-1) \text{ within distance } r\} \\ &= 1 - \sum_{n=0}^{k-1} \frac{(\lambda\pi r^2)^n}{n!} \exp\{-\lambda\pi r^2\} \end{aligned} \tag{6.4}$$

The derivative of the cumulative distribution function (6.4) gives the density for the kth nearest neighbor:

$$\begin{aligned} g_k(r) &= -\sum_{n=1}^{k-1} \frac{(\lambda\pi r^2)^{(n-1)}}{(n-1)!}(2\lambda\pi r)\exp\{-\lambda\pi r^2\} + \sum_{n=0}^{k-1} \frac{(\lambda\pi r^2)^n}{n!}(2\lambda\pi r)\exp\{-\lambda\pi r^2\} \\ &= \frac{2(\lambda\pi)^k r^{2k-1}}{(k-1)!} \exp\{-\lambda\pi r^2\} \end{aligned} \tag{6.5}$$

For example, the density function for the distance from an arbitrary node to its nearest neighbor is

$$g_1(r) = 2\pi\lambda r \exp\{-\lambda\pi r^2\}, \quad r \ge 0 \tag{6.6}$$

Remark

Similar results can be given for the three-dimensional case. For example, the kth nearest neighbor density in the three dimensional case is given by

$$g_k(r) = \frac{3\left(\frac{4}{3}\lambda\pi\right)^k r^{3k-1}}{(k-1)!} \exp\left\{-\frac{4}{3}\lambda\pi r^3\right\}$$

We now extend the previous results to consider problems related to communications between nodes. In particular, we are interested in the percentage of nodes that can communicate directly with at least one

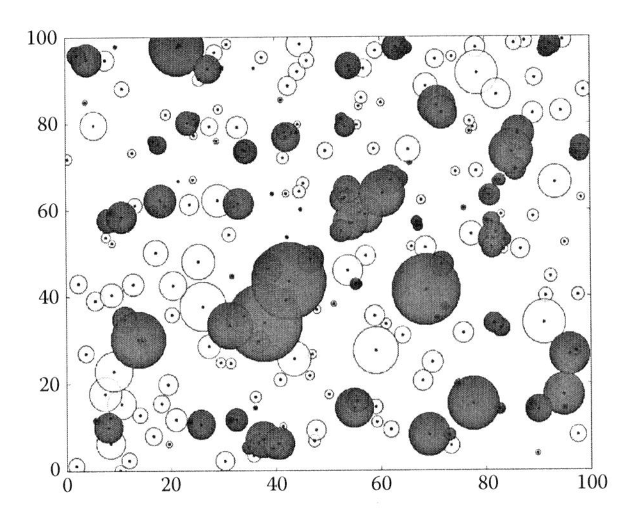

FIGURE 6.3 Two hundred and fifty nodes are randomly distributed over a 100 by 100 square unit area. Each node has omnidirectional communication range randomly selected from a gamma distribution with density $f(r) = \frac{1}{\Gamma(\alpha)\beta^{\alpha}} r^{\alpha-1} \exp\{-r/\beta\}, r \geq 0, \alpha = 2.5, \beta = 75$. Nodes that can communicate with at least one other node are shaded communications range.

other node in the network. For example, in Figure 6.3, 250 nodes are randomly distributed over a 10,000 square unit area. Each node has a random communication range selected from a gamma distribution. The communication ranges of connected nodes have been shaded. Thus, in this simple example, two nodes are said to communicate if the communications range of one extends as far as the other node.

Suppose now that s is an arbitrary node in the field and that s_k is the kth nearest node to s. Then s detects s_k if the random variable $D_s(k)$ is within the detection range, R_s, of s. The probability of this occurring is

$$P\{R_s > D_s(k)\} = P\{D_s(k) - R_s < 0\} \tag{6.7}$$

Recall that the detection ranges, R_s, vary from node to node according to some density $f(r)$ and, more importantly, are independent of node locations, that is, they are independent and identically distributed (iid); henceforth, we use the symbol R to denote the detection range random variable. Let us temporarily use the substitution $H_k = -D_s(k)$ and denote the density of H_k by h_k. Also, let $\Upsilon_{R-D_s(k)}$ be the density associated with the random difference $R - D_s(k)$. Then

$$\begin{aligned}
\Upsilon_{R-D_s(k)} &= \Upsilon_{R+H_k} \\
&= \int_{-\infty}^{\infty} h_k(r-u) f(u)\, du \\
&= \int_{-\infty}^{\infty} h_k(-\nu) f(r+\nu)\, d\nu, \quad -\nu = r - u \\
&= \int_{-\infty}^{\infty} g_k(\nu) f(r+\nu)\, d\nu \\
&= \int_{0}^{\infty} g_k(\nu) f(r+\nu)\, d\nu
\end{aligned} \tag{6.8}$$

TABLE 6.2 Direct Communications Probabilities, Calculated with Formula 6.9

Direct Communications Probabilities: 1 Node per 20 Square Units			
*k*th Nearest Sensor	Communications Range of *s* Extends to *k*th Nearest Sensor	Range of *k*th Nearest Sensor Extends to *s*	Sensor *s* Communicates with *k*th Nearest Sensor
1	0.354	0.125	0.582
2	0.156	0.024	0.287
3	0.078	0.006	0.149
4	0.043	0.002	0.084
5	0.026	0.001	0.051

Five hundred nodes were randomly distributed over a 100 by 100 square node area. The values 0.555, 0.287, 0.154, 0.085, and 0.047, respectively, were obtained through simulations for the direct communications probabilities.

Thus, the probability that the range of an arbitrary node extends beyond the *k*th nearest neighboring node is

$$P\{R > D_s(k)\} = \int_0^\infty \Upsilon_{R-D_s(k)}(r)\,dr$$

$$= \int_0^\infty \int_0^\infty g_k(\nu) f(r+\nu)\,d\nu\,dr \tag{6.9}$$

Remark

Formula 6.8 is the cross-correlation function. It is used to determine densities for the differences of independent random variables.

Table 6.2 shows direct connection probabilities calculated with Formula 6.9. Column 2 is the probability that an arbitrary node's communications range extends as far as its *k*th nearest neighboring node. Numbers in column 3 are the square of the corresponding column 2 numbers. This is the probability that the range of sensor *s* extends to its *k*th nearest sensor and the *k*th nearest sensor also extends as far as *s*. The final column is the probability that an arbitrary node's communications range extends out to its *k*th nearest neighboring node, *or vice versa*. This is the probability that the two nodes directly communicate. For this example, 500 nodes were randomly distributed over a 100 by 100 square node area, that is, an average of 1 per 20 square units.

6.2 Integral Geometry I

In this section, we consider the application of basic ideas of geometric probability and integral geometry in computing sensor detection probabilities. We assume that we have a convex area (e.g., a circle, a square, or an ellipse), where sensors are to be distributed, possibly at random. Each sensor has some radius of detection r_d, which may either be fixed or may vary according to some known probability distribution. We assume that intruders pass through the area in straight line trajectories. We assume initially that we have no a priori knowledge of where intruders may appear, and we wish to compute the probability that a random trajectory will pass through the detection area of a sensor or sensors. We must, however, be careful in using phrases like *random trajectory*. As Kendall and Moran [13] observe

> In the theory of geometrical probabilities the random elements are not quantitative but geometrical objects such as points, lines, and rotations... The ascription of a measure to such elements is not quite an obvious procedure. A number of paradoxes can be produced by failure to distinguish the reference set.

Of these paradoxes, the following is perhaps the best known.

6.2.1 Bertrand Paradox

The problem is generally stated as follows: if a chord is chosen at random on the unit circle, what is the probability its length will be at least $\sqrt{3}$?

This of course begs the question of how one defines "at random." There are three ways to do this, and each gives a different probability:

1. Given a point on the circle, draw the equilateral triangle one of whose vertices is that point. The sides will have length $\sqrt{3}$. Imagine that a second point on the circle is chosen to generate the given chord. Only one-third of the circle will give a chord that has length greater than $\sqrt{3}$. Thus, the answer is 1/3.
2. Imagine a radius is given and then a point on the radius is chosen at random, and a chord perpendicular to the radius is drawn. Only chords that are at a distance of less than 1/2 from the center will have length greater than $\sqrt{3}$. Thus, the answer is 1/2.
3. Suppose a point within the circle is chosen at random. A radius is drawn through the point, and the chord passing through the point and perpendicular to the radius is drawn. The chord will have length greater than $\sqrt{3}$ if and only if the distance from the point to the center is less than 1/2. The probability of this happening is $\pi(1/2)^2/\pi = 1/4$.

6.2.2 Geometric Probability and Integral Geometry

As seen earlier, it is important to clarify what we mean when we refer to some geometric object, for example, a line, being chosen *at random*. For example, a line may be parameterized in several different ways. In two dimensions, we may parameterize it by means of the slope and the intercept, $y = mx + b$, or by a point and the slope, $y - y_0 = m(x - x_0)$, or it may be written in "standard form," $ux + vy + 1 = 0$. A line may also be uniquely determined by two parameters p and θ where p is the length of a perpendicular drawn from the line to the origin, and θ is the angle between this perpendicular line and the positive part of the x-axis. Using this parameterization, it may be shown that the probability distribution associated with the three solutions to the Bertrand paradox is as follows. For the first $dP_1(p, \theta) = dpd\theta/2\pi\sqrt{1-p^2}$, for the second $dP_2(p,\theta) = (2\pi)^{-1}dpd\theta$, and for the third $dP_3(p,\theta) = (\pi)^{-1}pdpd\theta$.

Of these three probability distributions, only the second has the property of being invariant under translations, rotations, and reflections. Thus, we will use the measure proportional to $dpd\theta$, $p \geq 0$, $0 \leq \theta \leq 2\pi$ for the measure of a set of lines contained in a set A. So, for example, if A is the set of all lines passing through a circle of radius r centered at the origin we would obtain

$$m\{A\} = \int_0^r \int_0^{2\pi} d\theta\, dp = 2\pi r$$

In other words, the measure of the set of lines passing through the circle is equal to the perimeter of the circle. In fact, more general result may be shown (see Santaló [23], p. 30, Solomon [27], p. 30, or Kendall and Moran [13], p. 58), namely, that the measure of the set of lines intercepting any bounded convex set is equal to the length of the perimeter of the set.

It should be noted that although we have discussed measures, they have not necessarily been probability measures, and indeed the measure defined by $dpd\theta$ is unbounded. However, we may define a conditional probability measure by the following statement: If a convex set K_1 is contained in the bounded convex set K, the probability that a random line G intersects K_1 given that it intersects K is $p = L_1/L$, where L_1 and L are the perimeter lengths of K_1 and K, respectively.

As an application of this result, consider a case where we have a circular area with radius R in which a sensor with detection radius r is contained. Then given that an intruder enters on a straight line trajectory at some point on the outer radius, the probability, p, that he would be detected is $p = 2\pi r/2\pi R = r/R$, just the ratio of the two radii.

The study of geometrical probability when combined with that of invariance properties under certain transformations is generally referred to as *integral geometry*. In the next section, we summarize some useful results pertaining to the invariant measure of straight lines, most of which may be found in Santaló [23].

6.2.3 Some Useful Theorems in Integral Geometry

Theorem 6.2

The measure of the set of lines intercepting any bounded convex set is equal to the length of the perimeter of the set.

Theorem 6.3

Let K_1 be a convex set contained in the bounded convex set K. The probability that a random line G intersects K_1 given that it intersects K is $p = L_1/L$, where L_1 and L are the perimeter lengths of K_1 and K, respectively [23].

If a straight line of length l is contained in a convex set K, it may be surrounded by a convex set whose perimeter is arbitrary close to $2l$. Taking limits we obtain the following theorem [13], p. 58.

Theorem 6.4

If a line of length l is contained in a bounded convex set K, the probability that a random line G intersects the line given that it intersects K is $p = 2l/L$, where L is the perimeter length of K.

The following enables us to compute probabilities of lines intersecting both or at least one of a pair of nonoverlapping convex sets. Let K_1 and K_2 be nonintersecting convex sets with perimeter lengths L_1 and L_2. Let the length of the perimeter of the convex hull (the smallest convex set containing K_1 and K_2) of K_1 and K_2 be L_e. Now define the *internal cover* of K_1 and K_2 by C_i which may be described as follows: take a piece of elastic string drawn about K_1 and K_2, and crossing over at a point O between K_1 and K_2. Let L_i be the length of this internal cover. Then the following may be shown [23], p. 30:

Theorem 6.5

The measure of all lines that meet K_1 and K_2 is $L_i - L_e$.

Theorem 6.6

The measure of the set of lines that meet K_1 without meeting K_2 is $L_1 - (L_i - L_e)$ and the measure of the set of lines that meet K_2 without meeting K_1 is $L_2 - (L_i - Le)$.

Theorem 6.7

The measure of the set of lines that separate K_1 and K_2 is $L_i - (L_1 + L_2)$.

Using Theorems 6.2 and 6.7 we have

Theorem 6.8

The measure of the set of lines that meet at least one of the sets K_1 or K_2 is $L_1 + L_2 + L_e - L_i$.

We have assumed that $K_1 \cap K_2 = \varnothing$. However if they intercept it is easy to see that Theorems 6.5 through 6.7 still hold with $L_1 + L_2$ substituted for L_i.

In Ref. [17], Nelson et al. consider the case of a single sensor with detection radius R centered in the middle of a square with sides equal to $2U$. They consider the case of an intruder entering the sea base on a straight line trajectory at a point randomly and uniformly distributed along the base and at an angle randomly and uniformly distributed over the region $(0, \pi)$. They analytically derive the probability that the intruder is detected (i.e., crosses the circle generated by the sensor's radius of detection). It is interesting to compare the results they obtained with those that would be obtained from Theorem 6.3, which would give the ratio of the perimeters, $2\pi R/8U = \pi R/4U$. For example, for $R = 1$ and $U = 3$, this would give a detection probability of $p = 0.2618$. For the same values, they would obtain a probability of $p = 0.3022$. It is important to realize that the discrepancy is not due to any mistake in computation but that the "random" intruder trajectories are being generated by different probability measures.

6.2.4 Sensor Deployment and Detection Probabilities: Some Examples

6.2.4.1 Two Sensors with Equal Detection Radius

We start with the relatively simple case of two sensors with detection radius r deployed in a circular area of radius R, where $R > 2r$, and compute the probability that a random straight line trajectory is detected by at least one of the sensors, that is to say it passes through the circle represented by the sensor's radius of detection.

In this situation, we can use Theorems 6.5 and 6.8. Suppose we have two sensors with centers c units apart (we assume that $c > 2r$) and wish to compute the probability that a random straight line trajectory is detected by at least one of them. We have $L_1 = L_2 = 2\pi r$, $L_e = \pi r + \pi r + c + c = 2\pi r + 2c$, and after some trigonometry, we have $L_i = 2\pi r + 4\sqrt{c^2/4 - r^2} + 4r\sin^{-1}(2r/c)$. Thus, using Theorem 6.5, we obtain the following expression for the measure of the set of paths crossing both of the two sensor radii

$$m(S_1 \cap S_2) = L_i - L_e = 4\sqrt{\left(\frac{c}{2}\right)^2 - r^2} + 4r\sin^{-1}\left(\frac{2r}{c}\right) - 2c \tag{6.10}$$

And using Theorem 6.8 for the measure the set of trajectories detected by at least one of the sensors we have

$$m(S_1 \cup S_2) = L_1 + L_2 - (L_i - L_e) = 4\pi r + 2c - 4\sqrt{\frac{c^2}{4} - r^2} - 4r\sin^{-1}\frac{2r}{c} \tag{6.11}$$

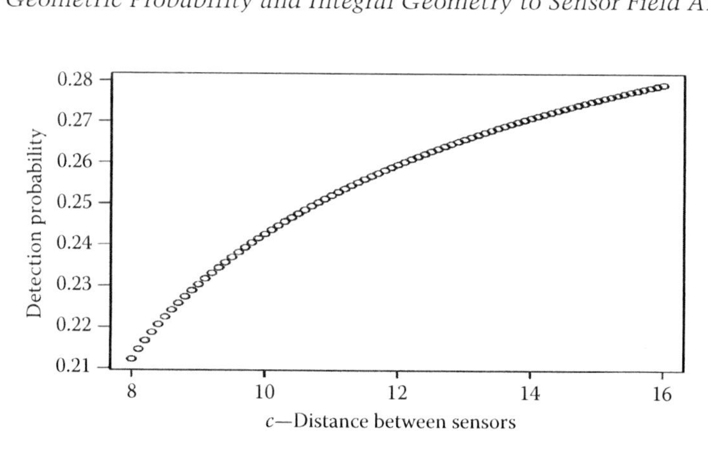

FIGURE 6.4 Detection probability by at least one sensor.

So using 6.11 and Theorem 6.3, the probability that a trajectory would be detected by at least one sensor given that it entered the area would be given by $p = m(S_1 \cup S_2)/2\pi R$. This turns out to be an increasing function of c, as can be seen from the following picture, plotted for $r = 4$ (Figure 6.4).

This result should not be too surprising. If we let S_i be the event that a track is detected by sensor i, we have $P(S_1 \cup S_2) = P(S_1) + P(S_2) - P(S_1 \cap S_2)$ and the intersection probability obviously decreases as the sensors become further apart. Also since $P(S_1 \cup S_2) \leq P(S_1) + P(S_2) = r/R + r/R = 2r/R$, we see that two sensors with detection radius r will always have a lower probability of detection than one sensor with detection radius $2r$.

The following table gives the probability of a track being detected by at least one sensor of two with detection radius r units, with centers c units apart in a circular region of radius $R = 20$. Note that for $r = 4$ and $c = 16$, the sensors are as far apart as they could possibly be. The detection probability of $p = 0.367$ is 91.8% of the detection probability of one sensor with detection radius $2r = 8$ for which we would have $p = 8/20 = 0.40$.

$R = 20$	$r = 2$	$r = 3$	$r = 4$
$c = 8$	0.184	0.262	0.327
$c = 12$	0.189	0.276	0.356
$c = 16$	0.192	0.282	0.367

6.2.4.2 Two Sensors with Unequal Detection Radius

Now suppose the two sensors have radii of detection r_1 and r_2. Computations proceed along similar lines to the previous section, but the expression 6.11 becomes more complicated. We find the measure $m(S_1 \cup S_2)$ of all tracks passing through at least one of the sensor radii. This can be shown to be

$$m(S_1 \cup S_2) = 2\pi(r_1 + r_2) + 2(r_1 - r_2)\sin^{-1}\left(\frac{r_1 + r_2}{c}\right) + 2\sqrt{c^2 - (r_1 - r_2)^2} - 2(r_1 + r_2)\sqrt{\left(\frac{c}{r_1 + r_2}\right)^2 - 1} \quad (6.12)$$

and the probability that a trajectory would be detected by at least one sensor given that it entered the detection region would be given by $p = m(r_1, r_2, c)/2\pi R$.

The following table is similar to the previous one, except that instead of using detection radii of 2, 3, and 4, we used pairs of radii, whose averages were 2, 3, and 4. Although every entry in the following table is higher than the corresponding entry in the table given earlier, the differences are negligible, ranging from 0.002 to 0.004.

$R = 20$	$r_1 = 1, r_2 = 3$	$r_1 = 2, r_2 = 4$	$r_1 = 3, r_2 = 5$
$c = 8$	0.187	0.266	0.331
$c = 12$	0.192	0.278	0.358
$c = 16$	0.194	0.284	0.369

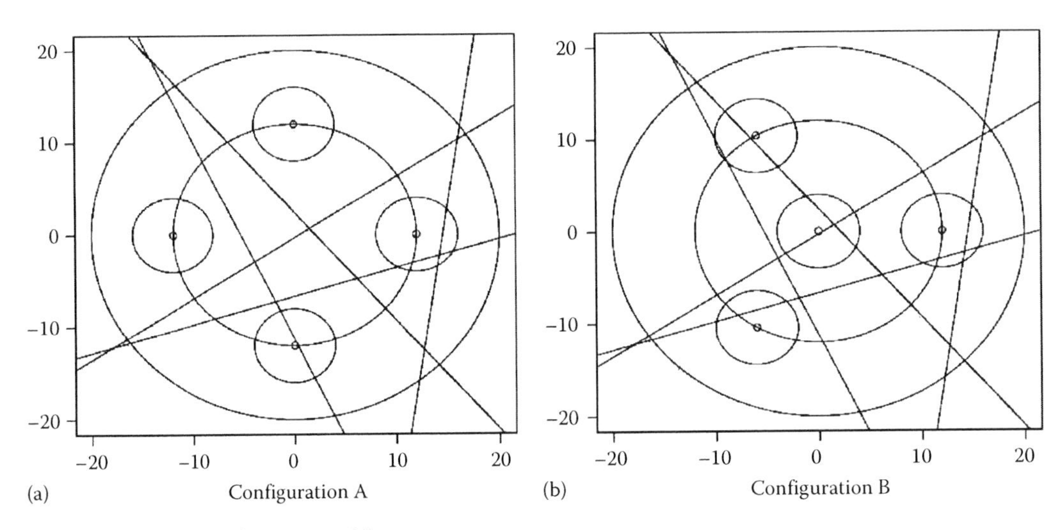

FIGURE 6.5 Two configurations of four sensors.

6.2.4.3 Four Sensors: Comparison of Two Configurations

The following table compares two configurations: *configuration A* and *configuration B*. In *configuration A*, the sensors are deployed round a circle, and in *configuration B*, three of the sensors are deployed around the circle, and the fourth is placed in the middle (Figure 6.5). Probabilities were computed for a circular region of radius $R = 20$ with sensors located at the following distances d from the center of the region $d = 8$, $d = 12$, and $d = 16$. All of the sensors had radius of detection $r = 4$. For each value configuration A gives a higher detection probability.

$R = 20$	$d = 8$	$d = 12$	$d = 16$
Configuration A	0.545	0.635	0.677
Configuration B	0.530	0.603	0.648

6.2.5 Computing the Probability of at Least One Detection

In general, it is not possible to find a closed form solution. But the following result is useful.

Let a set of n sensors, each with radius of detection r_c, be distributed at equal intervals around the perimeter of a circle with radius r. Denote the probability of detection by sensor number i by $P(S_i)$. Suppose that $P(S_i \cap S_j \cap S_k) = 0$ for all i, j, k, and suppose the sensors are situated in a circular region of radius R. Then the probability of detection by at least one sensor is given by

$$P(S_1 \cup S_2 \cup \cdots S_n) = \frac{1}{2\pi R}\left(2n\pi r_c - \sum_{k=1}^{n-1}\left\{4\sqrt{\left(\frac{2r\sin(k\pi/n)}{2}\right)^2 - r_c^2} + 4r_c\sin^{-1}\left(\frac{2r_c}{2r\sin(k\pi/n)}\right) - 4r\sin\left(\frac{k\pi}{n}\right)\right\}\right) \tag{6.13}$$

Given the condition $P(S_i \cap S_j \cap S_k) = 0$ we have

$$P(S_1 \cup S_2 \cup \cdots S_n) = \sum_{i=1}^{n} P(S_i) - \sum_{i<j} P(S_i \cap S_j) \tag{6.14}$$

Note that if the n sensors are distributed at equal distances around the circle, the distance between any two consecutive sensor centers is $c = 2r\sin(\pi/n)$. Pick one sensor and label it S_1, then going clockwise name the rest $S_2, \ldots, S_n$. Let $p_k = P(S_i \cup S_{i+k})$ for any i, and $k \leq n - i$. Using Equation 6.11, we have

$$p_k = \frac{\left(4\sqrt{(c_k/2)^2 - r_c^2} + 4r_c \sin^{-1}(2r_c/c_k) - 2c_k\right)}{2\pi R} \tag{6.15}$$

where c_k is the distance between two sensors which have $k - 1$ sensors between them and simple trigonometry gives $c_k = 2r\sin(k\pi/n)$. When summing over all the $n(n - 1)/2$ intersection probabilities, we obtain

$$\sum_{i<j} P(S_i \cap S_j) = (n-1)p_1 + (n-2)p_2 + \cdots + 2p_{n-2} + p_{n-1} \tag{6.16}$$

Combining these last three equations we obtain 6.13.

The following table gives probabilities, p, and Monte Carlo estimates, $\hat{p}$ (based on $N = 100{,}000$ iterations for the probability of detection by at least one sensor in a circle of radius $R = 20$ for $n = 5$ sensors equally spaced around the center. Distances from the center of $d = 12$ and $d = 16$ are compared. In each case, the sensor radius of detection was $r = 3$. The analytic values and the Monte Carlo estimates agree well.

$d = 12$	$p = 0.58372$	$\hat{p} = 0.58349$
$d = 16$	$p = 0.62596$	$\hat{p} = 0.62527$

The following subroutine in the R programming language computes the probability:

```
#R-function computes the probability of detection by at least 1 of n sensors
#Sensors are located at equal distances from each other, detection radius=rc
#Sea Bed is a circular region of radius R. n is number of sensors, r is
#radius of circle on which sensors located.
#
#NOTE: This subroutine is only valid for the case where P(Si and Sj and
Sk)=0.
###########################################################################
##
probsense <- function(R,r,rc,n){
   c <- 2*r*sin( (1:(n-1) )*pi/n)
   p2 <- 4*sqrt(c^2/4-rc^2)+ 4*rc*asin(2*rc/c)-2*c ##intersection probability
   prob <- n*2*pi*rc - sum( (n-1):1*p2)
   prob <- prob/(2*pi*R)
   return(prob)
}
###########################################################################
##
```

6.3 Integral Geometry II

6.3.1 Problem Formulation: Detection of a Stationary Target by Sensors Randomly Located

Here we consider the detection and localization of a stationary or very slow-moving target that may be present in some convex region. An arbitrary, but known, number of sensors are placed at random throughout the region. We seek to rigorously characterize this kind of operation, which applies to various kinds of naval missions and oceanographic scientific surveys. There is a wide body of literature that

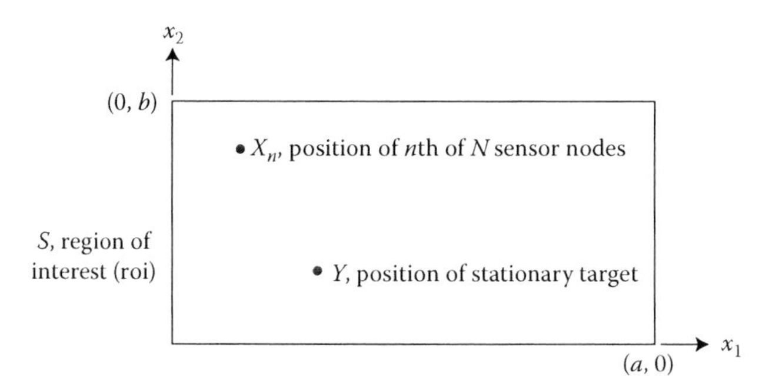

FIGURE 6.6 roi and nomenclature for a stationary target.

addresses this problem; representative work can be found among [1,4,5,7,11,16,26,30,33,34]. Additional work is specifically cited in what follows.

Let a region of interest (roi) be the compact, convex region $S \subset \mathbb{R}^2$. In Figure 6.6, the roi is shown, for example, as a rectangle. Let $\boldsymbol{X}_n \in S$ and $\boldsymbol{Y} \in S$ denote, respectively, the fixed position of the nth of N identical sensor nodes and the position of a fixed (or slowly moving) target, which is a source of radiation. Each of the N nodes has a sensor with associated electronics, a communication device, and an embedded detector of the received energy radiating from the target at $\boldsymbol{Y}$. Moreover, we assume that each node is equipped with a GPS receiver or other devices for accurately determining the location of the given node.

The local detector at each node tests the alternative hypothesis $\mathcal{H}_1$ (*target present* in S) versus the null hypothesis $\mathcal{H}_0$ (*target absent* from S) and makes a decision represented by $D_n \in \{0, 1\}$. For $n = 1, 2, \ldots, N$, $D_n = 1$ denotes a *target* decision, that is, a detection (true or false), and $D_n = 0$ denotes a *no-target* decision, that is, a rest (true or false). $D \in \{0, 1\}$ is similarly defined as the output of a fusion center's detector that processes the $\{D_n\}_{n=1}^N$, which have been transmitted to it in parallel, without error. The local detection decisions are assumed to be mutually and conditionally independent random variables under $\mathcal{H}_0$ and $\mathcal{H}_1$.

Regarding the probability of a local false alarm, D_n has the *zero–one* distribution:

$$\Pr\{D_n = d_n \mid \mathcal{H}_0\} = p_f^{d_n}(1 - p_f)^{1-d_n}, \quad d_n \in \{0, 1\} \tag{6.17}$$

where, for all $n = 1, 2, \ldots, N$, p_f is the probability of a local false alarm, arising only from random noise (or interference) sensed at the nodes. We assume for now that p_f is constant for all n.

For $n = 1, 2, \ldots, N$, let $p_d(\boldsymbol{x}_n, \boldsymbol{y}, n)$ denote the probability (conditioned on the locations in S of the node and the target) of a local detection: $p_d(\boldsymbol{x}_n, \boldsymbol{y}, n) := \Pr\{D_n = 1 | \mathcal{H}_1; \boldsymbol{X}_n = \boldsymbol{x}_n, \boldsymbol{Y} = \boldsymbol{y}\}$. The nodal index n indicates the possible dependence of $p_d(\boldsymbol{x}_n, \boldsymbol{y}, n)$ on n other than through only $\boldsymbol{x}_n$. The sensors, the modes of operation, or certain parameters, such as the detection thresholds, may not be identical among the nodes. In the last case, the nodal probability of a false alarm would vary with n; thus, we would rescind the assumption underlying (6.17). For the detection of a signal in additive noise, $p_d(\boldsymbol{x}_n, \boldsymbol{y}, n) \geq p_f$, for all $\boldsymbol{x}_n, \boldsymbol{y} \in S$ and $n = 1, 2, \ldots, N$.

The conditional (zero–one) distribution of each of the conditionally, mutually independent $\{D_n\}_{n=1}^N$, for $n = 1, 2, \ldots, N$, is given by

$$\Pr\{D_n = d_n \mid \mathcal{H}_1; \boldsymbol{X}_n = \boldsymbol{x}_n, \boldsymbol{Y} = \boldsymbol{y}) = p_d(\boldsymbol{x}_n, \boldsymbol{y}, n)^{d_n}(1 - p_d(\boldsymbol{x}_n, \boldsymbol{y}, n))^{1-d_n}, \quad d_n \in \{0, 1\} \tag{6.18}$$

From (6.18), the likelihood, under $\mathcal{H}_1$, $\ell(\{D_n\}_{n=1}^N \mid \mathcal{H}_1; \{\boldsymbol{X}_n = \boldsymbol{x}_n\}_{n=1}^N, \boldsymbol{Y} = \boldsymbol{y})$, is given by

$$\ell(\{D_n\}_{n=1}^N \mid \mathcal{H}_1; \{\boldsymbol{X}_n = \boldsymbol{x}_n\}_{n=1}^N, \boldsymbol{Y} = \boldsymbol{y}) = \prod_{n=1}^{N} p_d(\boldsymbol{x}_n, \boldsymbol{y}, n)^{D_n}(1 - p_d(\boldsymbol{x}_n, \boldsymbol{y}, n))^{1-D_n} \tag{6.19}$$

The corresponding log-likelihood $L(\{D_n\}_{n=1}^N \mid \mathcal{H}_1; \{X_n = x_n\}_{n=1}^N, Y = y)$ is given by

$$L(\{D_n\}_{n=1}^N \mid \mathcal{H}_1; \{X_n = x_n\}_{n=1}^N, Y = y) = \sum_{n=1}^N D_n \log p_d(x_n, y, n) + \sum_{n=1}^N (1 - D_n)\log(1 - p_d(x_n, y, n)) \quad (6.20)$$

As observed by Artés-Rodríguez [2], the first term on the right side of (6.20) corresponds to the nodes that locally decide *target present* and the second term to those that decide *target absent*; thus, all nodes figure into the likelihood.

Now consider that the $\{X_n = x_n\}_{n=1}^N$ are initially iid, each with pdf $f_X(x)$. However, when deployed and subsequently activated, each of the N nodes transmits its location, which, we assume, is received without error by the fusion center. Therefore, the locations $\{X_n = x_n\}_{n=1}^N$ become precisely known to the fusion center.

From (6.17) and (6.20), the likelihood ratio, denoted by $\lambda(\{D_n\}_{n=1}^N \mid \{X_n = x_n\}_{n=1}^N, Y = y)$, for the test of $\mathcal{H}_1$ versus $\mathcal{H}_0$ is given by

$$\lambda(\{D_n\}_{n=1}^N \mid \{X_n = x_n\}_{n=1}^N, Y = y) = \prod_{n=1}^N \left[\frac{p_d(x_n, y, n)}{p_f}\right]^{D_n} \left[\frac{1 - p_d(x_n, y, n)}{1 - p_f}\right]^{1-D_n} \quad (6.21)$$

The corresponding log-likelihood ratio $\Lambda(\{D_n\}_{n=1}^N \mid \{X_n = x_n\}_{n=1}^N, Y = y)$ is given by

$$\Lambda(\{D_n\}_{n=1}^N \mid \{X_n = x_n\}_{n=1}^N, Y = y) = \sum_{n=1}^N D_n \log\left[\frac{p_d(x_n, y, n)}{p_f}\right] + (1 - D_n)\log\left[\frac{1 - p_d(x_n, y, n)}{1 - p_f}\right] \quad (6.22)$$

This is the Chair–Varshney statistic for the fusion of distributed, binary detectors under the aforementioned assumptions [28,29]. The sufficient statistic for the test of $\mathcal{H}_1$ (*target present* in S) versus $\mathcal{H}_0$ (*target absent* from S) can also be written as

$$T(\{D_n\}_{n=1}^N \mid \{X_n = x_n\}_{n=1}^N, Y = y) := \sum_{n=1}^N \beta_1(x_n, y, n) D_n + \beta_0(x_n, y, n) \quad (6.23)$$

where

$$\beta_1(x, y, n) := \log\left[\frac{p_d(x, y, n) / p_f}{(1 - p_d(x, y, n)) / (1 - p_f)}\right]$$

$$\beta_0(x, y, n) := \log\left[\frac{1 - p_d(x, y, n)}{1 - p_f}\right]$$

Because $\beta_1(x, y, n)$ is an increasing function of $p_d(x, y, n)$, for all $x, y \in S$, it weighs the output D_n according to the nth (of N) node's probability of detecting a target at any position $y \in S$.

6.3.1.1 Non-Bayesian Case

We assume either that $Y \in S$ is a random vector, independent of each of $\{X_n = x_n\}_{n=1}^N$, with prior pdf $f_Y(y)$ (Bayesian case), or that $Y = y$, a nonrandom vector that is a priori unknown (non-Bayesian case). If $Y = y$ is known, the problem is one of detection, conditioned on the target being located at $\boldsymbol{y}$. First, we briefly consider the non-Bayesian case. The maximum likelihood (non-Bayesian) estimator of $\boldsymbol{y}$ using (6.20) is given by

$$\widehat{\mathbf{Y}}_{ml} := \arg\max_{\mathbf{y}} L(\{D_n\}_{n=1}^{N} \mid \mathcal{H}_1; \{\mathbf{X}_n = \mathbf{x}_n\}_{n=1}^{N}, \mathbf{Y} = \mathbf{y})$$

$$= \arg\max_{\mathbf{y}} \left\{ \sum_{n=1}^{N} D_n \log p_d(\mathbf{x}_n, \mathbf{y}, n) + (1 - D_n)\log(1 - p_d(\mathbf{x}_n, \mathbf{y}, n)) \right\} \tag{6.24}$$

By substituting $\widehat{\boldsymbol{Y}}_{ml}$ for $\boldsymbol{y}$ in (6.23), we obtain the test statistic for the *generalized* likelihood ratio test (glrt). This glrt is evidently a *localize-before-detect* procedure.

Artés-Rodríguez [2] addressed a special case (the "Spanish hat model") of the isotropic and homogeneous form: $p_d(\boldsymbol{x}, \boldsymbol{y}, n) = p_d(|\boldsymbol{x} - \boldsymbol{y}|)$, for all $\boldsymbol{x}, \boldsymbol{y} \in S$ and $n = 1, 2, \ldots, N$. For known node locations $\{\boldsymbol{X}_n = \boldsymbol{x}_n\}_{n=1}^{N}$, he computed the (asymptotically Gaussian) performance of a threshold detector based on (6.22) when $\boldsymbol{y}$ is an a priori known parameter vice estimated by $\widehat{\boldsymbol{Y}}_{ml}$ as in (6.24). In the event $\boldsymbol{y}$ is an unknown, nonrandom variable, Artés-Rodríguez et al. [3] compute $\widehat{\boldsymbol{Y}}_{ml}$ for the case of what they call an "exponential l_2 squared-norm model" of the special form $p_d(|\boldsymbol{x} - \boldsymbol{y}|) = p_f + (p_d - p_f)e^{-c\|\boldsymbol{x}-\boldsymbol{y}\|}$, where $p_f, p_d := p_d(0)$, and c are specified parameters.

6.3.1.2 Bayesian Case

Although we briefly addressed the non-Bayesian case, we will continue for the remainder of this chapter only with the Bayesian method. Let $\boldsymbol{Y} \in S$ be a random vector, independent of each $\{\boldsymbol{X}_n = \boldsymbol{x}_n\}_{n=1}^{N}$, with a known prior pdf $f_Y(\boldsymbol{y})$. A Bayesian estimator of $\boldsymbol{Y}$ is determined from its posterior pdf (under $\mathcal{H}_1$), the log of which is given by

$$\log f_Y(\boldsymbol{y} \mid \{\boldsymbol{X}_n = \boldsymbol{x}_n\}_{n=1}^{N}, \{D_n = d_n\}_{n=1}^{N}) = L(\{D_n\}_{n=1}^{N} \mid \mathcal{H}_1; \{\boldsymbol{X}_n = \boldsymbol{x}_n\}_{n=1}^{N}, \boldsymbol{Y} = \boldsymbol{y}) + \log f_Y(\boldsymbol{y}) - \log C \tag{6.25}$$

The normalizing constant C is the total probability.

Therefore, the maximum a posteriori (map) estimator of $\boldsymbol{Y}$, again using (6.20), is given by

$$\widehat{\boldsymbol{Y}}_{map} := \arg\max_{\boldsymbol{y}} \left\{ \log f_Y(\boldsymbol{y}) + \sum_{n=1}^{N} D_n \log p_d(\boldsymbol{x}_n, \boldsymbol{y}, n) + (1 - D_n)\log(1 - p_d(\boldsymbol{x}_n, \boldsymbol{y}, n)) \right\} \tag{6.26}$$

The log-likelihood ratio is given by, wherein $E_Y[\cdot]$ denotes expectation over $\boldsymbol{Y}$,

$$\Lambda(\{D_n\}_{n=1}^{N} \mid \{\boldsymbol{X}_n = \boldsymbol{x}_n\}_{n=1}^{N}) := E_Y[\Lambda(\{D_n\}_{n=1}^{N} \mid \{\boldsymbol{X}_n = \boldsymbol{x}_n\}_{n=1}^{N}, \boldsymbol{Y} = \boldsymbol{y})] \tag{6.27}$$

From (6.22), (6.27) becomes

$$\Lambda(\{D_n\}_{n=1}^{N} \mid \{\boldsymbol{X}_n = \boldsymbol{x}_n\}_{n=1}^{N}) = \sum_{n=1}^{N} \alpha_1(\boldsymbol{x}_n, n) D_n + \alpha_0(\boldsymbol{x}_n, n) \tag{6.28}$$

where $\alpha_i(\boldsymbol{x}, n) := \int_S \beta_i(\boldsymbol{x}, \boldsymbol{y}, n) f_Y(\boldsymbol{y}) d\boldsymbol{y}$, for $i = 0, 1$. The sufficient statistic for the test is

$$T(\{D_n\}_{n=1}^{N} \mid \{\boldsymbol{X}_n = \boldsymbol{x}_n\}_{n=1}^{N}) := \sum_{n=1}^{N} \alpha_1(\boldsymbol{x}_n, n) D_n \tag{6.29}$$

Given the condition $\{X_n = x_n\}_{n=1}^N$ the term $\sum_{n=1}^{N} \alpha_0(x_n, n)$, is, in theory, known; however, it is independent of the data $\{D_n\}_{n=1}^N$ and can be ignored in the test.

The problem of determining the weights $\{\alpha_1(\boldsymbol{x}_n, n)\}_{n=1}^N$ remains. Theoretically, these weights can be calculated, but the distribution of the statistic $T(\{D_n\}_{n=1}^N \,|\, \{\boldsymbol{X}_n = \boldsymbol{x}_n\}_{n=1}^N)$ becomes unwieldy. We address this issue in what follows. Note that the test statistic in the non-Bayesian case given by (6.23) has the data-independent term $\sum_{n=1}^{N} \beta_0(\boldsymbol{x}_n, \boldsymbol{y}, n)$; it, however, cannot be ignored because it depends on the unknown $\boldsymbol{y}$.

6.3.1.3 Some Approximations

Suppose that, as it depends on $\boldsymbol{y} \in S$, $\beta_1(\boldsymbol{x}, \boldsymbol{y}, n)$ has significant value only within a small ε-neighborhood of $\boldsymbol{x}$, $s(\boldsymbol{x}, \epsilon) \subset S$, where $s(\boldsymbol{x}, \epsilon) := \{\boldsymbol{y} : |\boldsymbol{y} - \boldsymbol{x}| < \epsilon > 0\}$. Furthermore, suppose that $f_Y(\boldsymbol{y})$ is nearly constant over $s(\boldsymbol{x}, \epsilon)$, for all $\boldsymbol{x}, \boldsymbol{y} \in S$. Then, from the definition of $\alpha_1(\boldsymbol{x}, n)$ in (6.28),

$$\alpha_1(\boldsymbol{x}, n) \approx f_Y(\boldsymbol{x}) \int_S \beta_1(\boldsymbol{x}, \boldsymbol{y}, n)\, d\boldsymbol{y} \tag{6.30}$$

for all $\boldsymbol{x} \in S$ and $n = 1, 2, \ldots, N$. Equation 6.30 reveals that the $\boldsymbol{x}$-dependency of $\alpha_1(\boldsymbol{x}, n)$ follows, in part, the pdf $f_Y(\boldsymbol{x})$. Also, the integral expression in (6.30) represents the $\boldsymbol{x}$-dependent edge effect induced as a node's location nears the boundary of S.

If $\boldsymbol{Y}$ is uniformly distributed over S, then, for all $\boldsymbol{x} \in S$ and $n = 1, 2, \ldots, N$,

$$\alpha_1(\boldsymbol{x}, n) = \|S\|^{-1} \int_S \beta_1(\boldsymbol{x}, \boldsymbol{y}, n)\, d\boldsymbol{y} \tag{6.31}$$

where $\|S\|$ is the area of S. The $\boldsymbol{x}$-dependency, in this case, is induced by only the edge effect.

6.3.1.3.1 Cookie-Cutter Detection Model

Consider the case of the so-called cookie-cutter detection model. Let $\{S_n \subset \mathbb{R}^2\}_{n=1}^N$ be a set of compact regions, the *shapes* of the *cookie cutters*. For $\boldsymbol{x}_n$ and $n = 1, 2, \ldots, N$, define $C_{\boldsymbol{x}}^n$ by the *Minkowski sum**:

$$\begin{aligned} C_{\boldsymbol{x}}^n &:= \boldsymbol{x}_n \oplus S_n \\ &= \{\boldsymbol{x}_n + \boldsymbol{z} : \boldsymbol{z} \in S_n\} \end{aligned}$$

In general, the Minkowski sum of two sets A and B is $A \oplus B := \{\boldsymbol{a} + \boldsymbol{b} : \boldsymbol{a} \in A, \boldsymbol{b} \in B\}$. Also $A \oplus B = \bigcup_{\boldsymbol{a} \in A} (\boldsymbol{a} \oplus B\}$, the union of copies of B, one for each $\boldsymbol{a} \in A$, as shown, for example, in O'Rourke ([19], Section 8.3). Santaló ([23], Sections I.1.4 and III.13.3) uses the later form to define an exterior parallel set if A is convex and B is a disk, in the plane case. $C_{\boldsymbol{x}}^n$ is the Minkowski sum of S_n and the singleton set $\{\boldsymbol{x}_n\}$; thus, it is, in this case, the translation of S_n by $\boldsymbol{x}_n$.

The collection of sets $\{C_{\boldsymbol{x}}^n\}_{n=1}^N$ is called a coverage process with centers $\{\boldsymbol{X}_n = \boldsymbol{x}_n\}_{n=1}^N$ per Hall [10]. Observe that $p_d(\boldsymbol{x}_n, \boldsymbol{y}, n)$ depends on both $\boldsymbol{x}_n$ and S_n as well as $\boldsymbol{y}$. Furthermore, for $\boldsymbol{x}_n, \boldsymbol{y} \in S$ and $n = 1, 2, \ldots, N$, let

$$p_d(\boldsymbol{x}_n, \boldsymbol{y}, n) = (p_d - p_f)\mathcal{X}_{C_{\boldsymbol{x}}^n}(\boldsymbol{y}) + p_f, \quad \text{where } p_d \geq p_f \tag{6.32}$$

* With this definition of $C_{\boldsymbol{x}}^n$, we abuse notation to keep it concise. Strictly, we should write $C_{\boldsymbol{x}_n}^n$, where the center, $\boldsymbol{x}_n$ is explicitly denoted, and the superscript n refers to the dependence on S_n.

The indicator function $\mathcal{X}_C(\boldsymbol{y})$ of any closed region $C \subset \mathbb{R}^2$ and any vector $\boldsymbol{y} \in \mathbb{R}^2$ is defined by

$$\mathcal{X}_C(\boldsymbol{y}) := \begin{cases} 1 & \text{if } \boldsymbol{y} \in C \\ 0 & \text{if } \boldsymbol{y} \notin C \end{cases} \tag{6.33}$$

From (6.32) and the definition in (6.23), $\beta_i(\boldsymbol{x}_n, \boldsymbol{y}, n) = \beta_i \mathcal{X}_{C_{\boldsymbol{x}}^n}(\boldsymbol{y})$, for $i = 0, 1$ and $n = 1, 2, \ldots, N$, where $\beta_1 := \log\left[\dfrac{p_d/p_f}{(1-p_d)/(1-p_f)}\right]$ and $\beta_0 := \log\left[\dfrac{1-p_d}{1-p_f}\right]$. Then, from the definition in (6.28),

$$\begin{aligned} \alpha_i(\boldsymbol{x}_n, n) &:= \beta_i \int_S \mathcal{X}_{C_{\boldsymbol{x}}^n}(\boldsymbol{y}) f_Y(\boldsymbol{y}) d\boldsymbol{y} \\ &= \beta_i \int_{C_{\boldsymbol{x}}^n \cap S} f_Y(\boldsymbol{y}) d\boldsymbol{y}, \quad \text{for } i = 0, 1 \end{aligned} \tag{6.34}$$

If $\boldsymbol{Y}$ is uniformly distributed over S, then, for $n = 1, 2, \ldots, N$ and for $\boldsymbol{x}_n \in S$,

$$\alpha_i(\boldsymbol{x}_n, n) = \beta_i \, \|S\|^{-1} \|C_{\boldsymbol{x}}^n \cap S\|, \quad \text{for } i = 0, 1 \tag{6.35}$$

Figure 6.7 is an example of shapes $\{S_n\}_{n=1}^N$ that are congruent but have varying orientation. Each of the $\{C_{\boldsymbol{x}}^n\}_{n=1}^N$ depends on n via both $\boldsymbol{x}_n$ and S_n. If the $\{S_n\}_{n=1}^N$ are, indeed, congruent, then $\|S_n\| = \|S_1\|$, for all $n = 1, 2, \ldots, N$.

Not all of the $\{C_{\boldsymbol{x}}^n\}_{n=1}^N$ are contained within S. For example, in Figure 6.7, $C_{\boldsymbol{x}}^2$ intersects the boundary of S, illustrating the edge effect. For any $n \in \{1, 2, \ldots, N\}$ that $C_{\boldsymbol{x}}^n \subset S$, then $\|C_{\boldsymbol{x}}^n \cap S\| = \|C_{\boldsymbol{x}}^n\|$ Since $\|C_{\boldsymbol{x}}^n\| = \|S_n\|$, from (6.35), $\alpha_i(\boldsymbol{x}_n, n) = \beta_i\|S_n\|/\|S\|$, for $i = 0, 1$. In this case, for $i = 0, 1$, $\alpha_i(\boldsymbol{x}_n, n)$ is proportional to the relative area $\|S_n\|/\|S\|$; it is not dependent on $\boldsymbol{x}_n$, but still depends on n. If the shapes have identical areas (but not necessarily congruent), for any $n \in \{1, 2, \ldots, N\}$ that $C_{\boldsymbol{x}}^n \subset S$, then $\alpha_i(\boldsymbol{x}_n, n) = \alpha_i(\boldsymbol{x}_1, 1) = \beta_i\|S_1\|/\|S\|$, a constant for $i = 1, 0$. Otherwise, near the boundary of S, for any $n \in \{1, 2, \ldots, N\}$ that $\|C_{\boldsymbol{x}}^n \cap S\| < \|C_{\boldsymbol{x}}^n\|$, then $\alpha_i(\boldsymbol{x}_n, n) < \alpha_i(\boldsymbol{x}_1, 1) = \beta_i\|S_1\|/\|S\|$, a constant for $i = 1, 0$.

Equation 6.35 includes that case when the detection regions of the individual nodes are not identically oriented or congruent. Even though the computation of the test statistic in (6.29) is straightforward for this ostensibly simple case, computation of the field performance is difficult. It is less difficult if we obviate the edge effect as we do later.

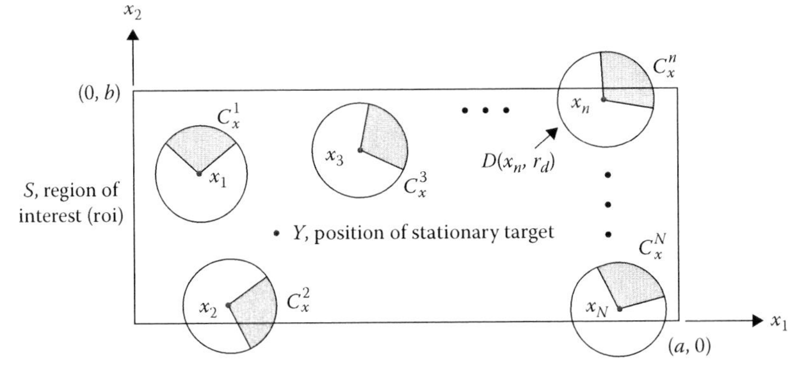

FIGURE 6.7 Example of a coverage $\{C_{\boldsymbol{x}}^n\}_{n=1}^N$ with shapes $\{S_n\}_{n=1}^N$ (shaded) and corresponding centers at $\{\boldsymbol{x}_n\}_{n=1}^N$. In this example, the shapes are congruent sectors of a disk $D(\boldsymbol{0}, r_d)$ but are oriented differently. The circles are the boundaries of the disks $\{D(\boldsymbol{x}_n), r_d)\}_{n=1}^N$ of constant detection radius r_d.

6.3.1.3.2 *Target-Centered Detection Model*

Equation 6.32, for $\boldsymbol{x}_n, \boldsymbol{y} \in S$ and $n = 1, 2, \ldots, N$, can also be expressed as

$$p_d(\boldsymbol{x}_n, \boldsymbol{y}, n) = (p_d - p_f)\mathcal{X}_{R_y^n}(\boldsymbol{x}_n) + p_f \tag{6.36}$$

where $R_y^n := \boldsymbol{y} \oplus \breve{S}_n = \{\boldsymbol{y} + \boldsymbol{z} : \boldsymbol{z} \in \breve{S}_n\}$, and $\breve{S}_n$ is the *reflection* of the set $S_n \in \mathbb{R}^2$: $\breve{S}_n = \{\boldsymbol{x}: -\boldsymbol{x} \in S_n\}$. (For a symmetric shape, $\breve{S}_n = S_n$.)

Equations 6.32 and 6.36 are equivalent because of the following lemma.

Lemma 6.1

$$\{\boldsymbol{x}, \boldsymbol{y} \in S : \boldsymbol{x} = \boldsymbol{x}_n, \boldsymbol{y} \in C_x^n\} = \{\boldsymbol{x}, \boldsymbol{y} \in S : \boldsymbol{x} = \boldsymbol{x}_n \in R_y^n\}$$

Proof

For $\boldsymbol{x}_n \in S$ and $n = 1, 2, \ldots, N$, $C_x^n = \boldsymbol{x}_n \oplus S_n = \{\boldsymbol{x}_n + \boldsymbol{u} : \boldsymbol{u} \in S_n\}$; therefore, $\{\boldsymbol{x}, \boldsymbol{y} \in S : \boldsymbol{x} = \boldsymbol{x}_n, \boldsymbol{y} \in C_x^n\} = \{\boldsymbol{x}, \boldsymbol{y} \in S : \boldsymbol{x} = \boldsymbol{x}_n, \boldsymbol{y} - \boldsymbol{x}_n \in S_n\} = \{\boldsymbol{x}, \boldsymbol{y} \in S : \boldsymbol{x} = \boldsymbol{x}_n, \boldsymbol{x}_n - \boldsymbol{y} \in \breve{S}_n\} = \{\boldsymbol{x}, \boldsymbol{y} \in S : \boldsymbol{x} = \boldsymbol{x}_n \in R_y^n\}$.

Lemma 6.1 holds for any compact, not necessarily convex, shape. Moreover, it shows the equivalence of a node-centered detection region C_x^n to a target-centered detection region R_y^n. We will return later to this useful relation.

6.3.1.3.3 *Illustration of the Edge Effect*

We illustrate the edge effect on $\alpha_1(\boldsymbol{x}, n)$ with a simple numerical example. First, define the disk in $\mathbb{R}^2$ of radius r: $D(\boldsymbol{x}, r) = \{\boldsymbol{u} \in \mathbb{R}^2 : |\boldsymbol{u} - \boldsymbol{x}| \leq r\}$ for $\boldsymbol{x} \in S$. Let the shapes be identical and isotropic (i.e., identical disks *vice* the sectors shown in Figure 6.7). For $n = 1, 2, \ldots, N$, let $S_n = D(\boldsymbol{0}, r_d)$ where $r_d > 0$ is some constant detection range. Thus, $C_x^n = D(\boldsymbol{x}_n, r_d)$ and $\alpha_1(\boldsymbol{x}, n) = \alpha_1(\boldsymbol{x})$ for $\boldsymbol{x} \in S$ and $n = 1, 2, \ldots, N$, where

$$\alpha_1(\boldsymbol{x}) := \beta_1 \int_S \mathcal{X}_{D(\boldsymbol{0}, r_d)}(\boldsymbol{y} - \boldsymbol{x}) f_Y(\boldsymbol{y}) d\boldsymbol{y} \tag{6.37}$$

For $\boldsymbol{Y}$ uniformly distributed over S,

$$\begin{aligned} \alpha_1(\boldsymbol{x}) &= \beta_1 \, \| S \|^{-1} \int_S \mathcal{X}_{D(\boldsymbol{0}, r_d)}(\boldsymbol{y} - \boldsymbol{x}) d\boldsymbol{y} \\ &= \beta_1 \, \| S \|^{-1} \| D(\boldsymbol{x}, r_d) \cap S \| \end{aligned} \tag{6.38}$$

Figure 6.8 is a plot of $\alpha_1(\boldsymbol{x})/\alpha_1(\boldsymbol{x}_m)$ for all $\boldsymbol{x} \in S$ computed with (6.38) for $r_d = 0.75$ nm and S a square ($a = b = 12$ nm); $\boldsymbol{x}_m := [a/2, b/2]^T$. Evidently, in this example, $\alpha_1(\boldsymbol{x})/\alpha_1(\boldsymbol{x}_m)$ varies only within a distance r_d of the boundary.

If for $n = 1, 2, \ldots, N$, $\alpha_1(\boldsymbol{x}_n, n)$ depends only weakly on both $\boldsymbol{x}_n$ and n, the test statistic of (6.29) is approximated by

$$T := \sum_{n=1}^{N} D_n \tag{6.39}$$

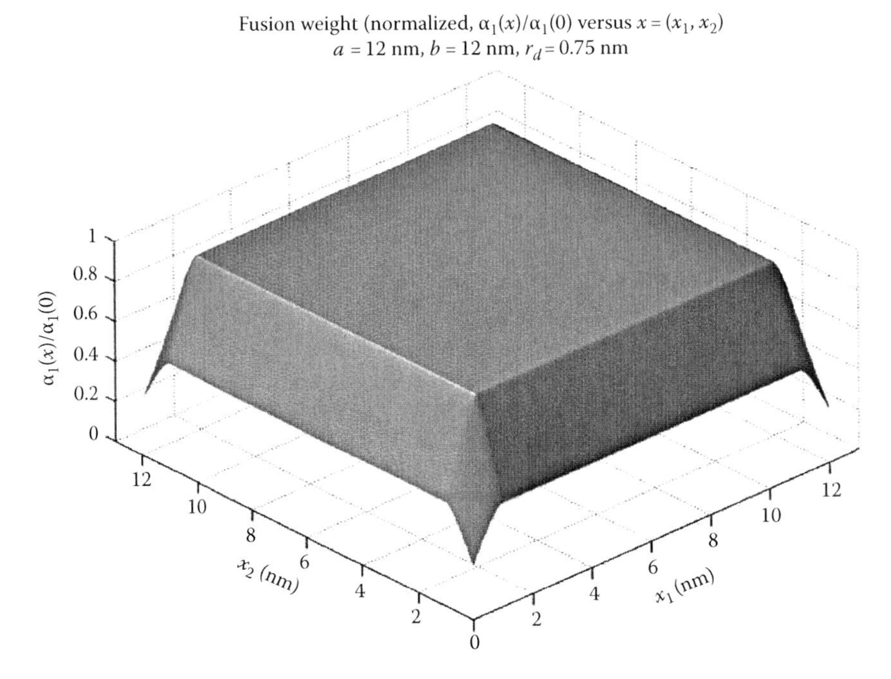

FIGURE 6.8 $\alpha_1(\boldsymbol{x})/\alpha(\boldsymbol{x}_m)$ versus $\boldsymbol{x}$ for a square S roi, with midpoint at $\boldsymbol{x}_m := [a/2, b/2]^T$. The detection region is a disk of radius r_d.

where $T \in \{1, 2, \ldots, N\}$. This fusion rule is adopted exclusively in the sequel. We do this for several reasons: (1) For identical areas of the sensors' shapes, the values of $\alpha_1(\boldsymbol{x}_n, n)$ are constant over most of S, varying (decreasing) only near the edges of S. (2) The test statistic in (6.39) is suboptimal and thus leads to conservative predictions of performance. (3) The $\{D_n\}_{n=1}^N$ are conditionally independent (i.e., conditioned on the target's position $\boldsymbol{Y} = \boldsymbol{y}$) and in some special cases distributed identically. Thus, the performance calculations are less difficult than they would otherwise be. (4) The statistic T is simple; it does not depend on prior knowledge of the environment or estimates of detection performance. Niu and Varshney [18] make similar arguments and, in effect, postulate (6.39).

6.3.1.4 Performance Evaluation

From Equation 6.39, the binary test of $\mathcal{H}_1$ versus $\mathcal{H}_0$ for the field of N sensors is as follows: decide $\mathcal{H}_1$ if

$$T = \sum_{n=1}^{N} D_n \geq k \tag{6.40}$$

otherwise, choose $\mathcal{H}_0$, where $k \in \{1, 2, \ldots, N\}$ is the detection threshold. The test is a k-of-N fusion rule. For $k = 1$, the rule is an *or* rule; for $k = N$, it is an *and* rule.

If under $\mathcal{H}_0$ the $\{D_n\}_{n=1}^N$ are iid, then T is distributed as a binomial with parameters p_f and N. In this case, the probability of false alarm $P_f(k, N)$: = $\Pr\{T \geq k | \mathcal{H}_0; N\}$ for the field of N nodes is

$$P_f(k,N) = \sum_{n=k}^{N} \binom{N}{n} p_f^n (1 - p_f)^{N-n} \tag{6.41}$$

6.3.1.4.1 *Generalized Binomial*

The probability of detection by the field of N nodes is $P_d(k, N) := Pr\{T \geq k | \mathcal{H}_1; N\}$, which we determine as in Niu and Varshney [18]. Under $\mathcal{H}_1$, the characteristic function (cf) of T, conditioned on both $\{\boldsymbol{X}_n = \boldsymbol{x}_n\}_{n=1}^{N}$ and $\boldsymbol{Y} = \boldsymbol{y}$, is defined by the conditional expectation:

$$M_T\left(\omega \middle| \mathcal{H}_1; \{\boldsymbol{X}_n = \boldsymbol{x}_n\}_{n=1}^{N}, \boldsymbol{Y} = \boldsymbol{y}\right) := E_T\left[\exp\left(i\omega \sum_{n=1}^{N} D_n\right) \middle| \mathcal{H}_1; \{\boldsymbol{X}_n = \boldsymbol{x}_n\}_{n=1}^{N}, \boldsymbol{Y} = \boldsymbol{y}\right]. \tag{6.42}$$

Therefore,

$$\begin{aligned} M_T\left(\omega \middle| \mathcal{H}_1; \{\boldsymbol{X}_n = \boldsymbol{x}_n\}_{n=1}^{N}, \boldsymbol{Y} = \boldsymbol{y}\right) &= \prod_{n=1}^{N} E_{D_n}\left[\exp(i\omega D_n) \mid \mathcal{H}_1; \{\boldsymbol{X}_n = \boldsymbol{x}_n\}_{n=1}^{N}, \boldsymbol{Y} = \boldsymbol{y}\right] \\ &= \prod_{n=1}^{N} [p_d(\boldsymbol{x}_n, \boldsymbol{y}_n, n)\exp(i\omega) + (1 - p_d(\boldsymbol{x}_n, \boldsymbol{y}_n, n)] \end{aligned} \tag{6.43}$$

The cf $M_T(\omega \mid \mathcal{H}_1; \{\boldsymbol{X}_n = \boldsymbol{x}_n\}_{n=1}^{N}, \boldsymbol{Y} = \boldsymbol{y})$ in (6.43) has no simple inverse. As described in Ref. [8], it is the cf of the so-called *generalized* binomial distribution (variously called the *inhomogeneous* or *nonstationary* binomial). It is the distribution of the number of successes in N Bernoulli trials with variable probabilities of success, $\{p_d(\boldsymbol{x}_n, \boldsymbol{y}_n, n)\}_{n=1}^{N}$. In this case, methods such as the fast Fourier transform or series expansion using *Maxima*, say, must be used to invert the cf. This becomes important if we consider feedback from the fusion center, which could, for example, alter the threshold of the local detectors at certain nodes forming subsets of the total of N. In such a case, the receiver operating characteristic could vary among the subsets.

6.3.1.4.2 *Poisson Limit*

Let $\eta := \max\{p_d(\boldsymbol{x}_n, \boldsymbol{y}_n, n)\}_{n=1}^{N}$ and $\mu := \sum_{n=1}^{N} p_d(\boldsymbol{x}_n, \boldsymbol{y}_n, n)$. If, when $N \to \infty$ and $\eta \to 0$, μ remains bounded, then, using the cf of (6.43) as shown in Ref. [8], the conditional distribution of T converges to the Poisson with intensity μ:

$$\Pr\left\{T = j \middle| \mathcal{H}_1; \{\boldsymbol{X}_n = \boldsymbol{x}\}_{n=1}^{N}, \boldsymbol{Y} = \boldsymbol{y}\right\} = \frac{\mu^j e^{-\mu}}{j!}, \quad j = 0, 1, 2, \ldots \tag{6.44}$$

The result provides the rationale for the Poisson detection model of networks of sensors distributed within a bounded region, especially in the case of so-called large N.

6.3.1.4.3 *Extended roi*

For $n = 1, 2, \ldots, N$, we assume that the boundary of each shape S_n is a sensor's directivity pattern, which is the sensitivity to received radiation as a function of the angle of incidence to the sensor. For example, let S_n be a sector of the disk $D(\boldsymbol{0}, r_d)$ with constant angle $\Theta \in (0, 2\pi]$ and variable (with n) orientation $\Phi_n \in (0, 2\pi]$, as in Figure 6.9. (The most sensitive region of a realistic directivity pattern will be usually bounded by a sector.) Thus, the sector $S_n \subseteq D(\boldsymbol{0}, r_d)$ and $C_x^n \subseteq D(\boldsymbol{x}_n, r_d)$, with identity in both cases for $\Theta = 2\pi$.

We now define the region $S^+ \supset S$, as the *exterior parallel* set (in the distance r_d) of the convex set S, that is, $S^+ := S \oplus D(\boldsymbol{0}, r_d)$, the Minkowski sum of S and the disk $D(\boldsymbol{0}, r_d)$. Thus, as shown in Figure 6.9, $S^+ = \bigcup_{x \in S}(\boldsymbol{x} \oplus D(\boldsymbol{0}, r_d)) = \bigcup_{x \in S} D(\boldsymbol{x}, r_d)$. We continue to assume that $\boldsymbol{y} \in S$, that is, the support of $f_Y(\boldsymbol{y})$ remains as S. However, we now assume that $\boldsymbol{x}_n \in S^+$, for $n = 1, 2, \ldots, N$. The region S^+ is the *extended* roi and the support of $f_X(\boldsymbol{x})$.

By inspecting Figure 6.9, we see that for S rectangular,

$$||S^+|| = ab + 2(a + b)r_d + \pi r_d^2 \tag{6.45}$$

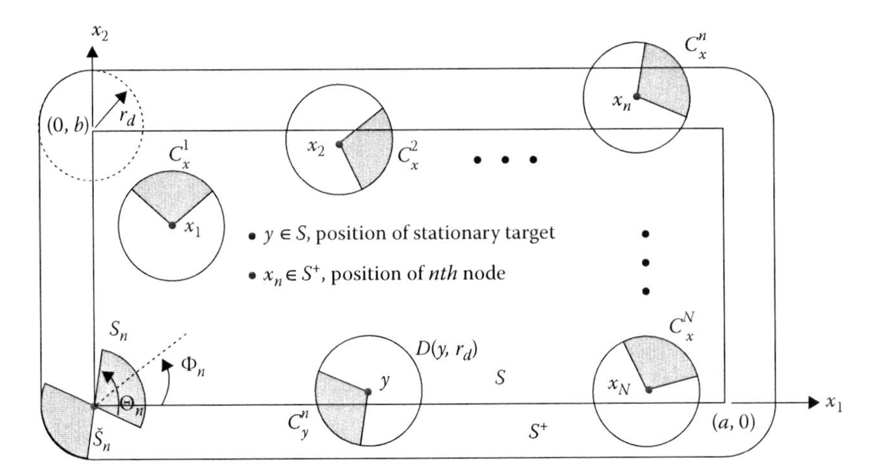

FIGURE 6.9 Example of a coverage $\{C_x^n\}_{n=1}^N$ with shapes $\{S_n\}_{n=1}^N$ (shaded) and corresponding centers at $\{\boldsymbol{x}_n\}_{n=1}^N$. In this example, the shapes are congruent sectors of a disk $D(\mathbf{0}, r_d)$ but are oriented differently. The circles are the boundaries of the disks $\{D(\boldsymbol{x}_n), r_d\}_{n=1}^N$ of constant detection radius r_d.

More generally, if A and $B(\Phi)$ are any compact, convex sets in $\mathbb{R}^2$, with $B(\Phi)$ oriented by an angle Φ that is random and uniformly distributed over $(0, 2\pi]$, then

$$E_\Phi\left[\|A \oplus B(\Phi)\|\right] = \|A\| + \frac{1}{2\pi}\|\partial A\|\|\partial B\| + \|B\| \tag{6.46}$$

where ∂A and $\|\partial A\|$ denote, respectively, the boundary of a set A and its perimeter. Equation 6.46 is *Steiner's formula* for convex domains in the plane (see, e.g., [23], Section III.13.3, or [25], Section IV, C2).

When $A = S$, and B is a disk, (6.46) reduces to $|S^+| = |S| + r_d|\partial A| + \pi r_d^2$. An S^+ can be determined for arbitrary shapes by letting $r_d = \min\{r: D(\mathbf{0}, r) \supseteq S_n, n = 1, 2, \ldots, N\}$; that is, the boundary of $D(\mathbf{0}, r_d)$ is the circle of minimum radius that contains all of the $\{S_n\}_{n=1}^N$.

Within this framework, we accordingly alter the domain of $\beta_i(\boldsymbol{x}, \boldsymbol{y}, n)$ and $\alpha_i(\boldsymbol{x}, n)$ defined in (6.23) and (6.28): $\boldsymbol{x} \in S^+$, and, as before, $\boldsymbol{y} \in S$, for $n = 1, 2, \ldots, N$.

For the *cookie-cutter* detection model of (6.32), $p_d(\boldsymbol{x}_n, \boldsymbol{y}, n) = (p_d - p_f)\mathcal{X}_{C_x^n}(\boldsymbol{y}) + p_f$,

$$\beta_i(\boldsymbol{x}, \boldsymbol{y}, n) = \beta_i \mathcal{X}_{C_x^n}(\boldsymbol{y}) \tag{6.47}$$

$$\alpha_i(\boldsymbol{x}, n) = \beta_i \int_{S \cap C_x^n} f_Y(\boldsymbol{y}) d\boldsymbol{y} \tag{6.48}$$

for $x \in S^+$, $\mathbf{y} \in S$, and $i = 0, 1$. If $\boldsymbol{Y}$ is uniformly distributed over S, then (6.48) becomes

$$\alpha_i(\boldsymbol{x}, n) = \beta_i \|S \cap C_x^n\| \|S\|^{-1} \tag{6.49}$$

where now $\boldsymbol{x}_n \in S^+$, vice $\boldsymbol{x} \in S \subset S^+$, as before; and $i = 0, 1$.

6.3.1.4.4 *Conditional Binomial*

If the $\{X_n\}_{n=1}^N$ are iid with common pdf $f_X(\boldsymbol{x})$ for $\boldsymbol{x} \in S^+$, then (6.43) yields

$$M_T\left(\omega \middle| \mathcal{H}_1; Y = \boldsymbol{y}\right) = \prod_{n=1}^{N} [p_d(\boldsymbol{y}, n)\exp(i\omega) + (1 - p_d(\boldsymbol{y}, n))] \tag{6.50}$$

where $p(\boldsymbol{y},n) := \int_{S^+} p_d(\boldsymbol{x},\boldsymbol{y},n) f_X(\boldsymbol{x}) d\boldsymbol{x}$. If $p(\boldsymbol{y},n)$ is independent of $n = 1, 2, \ldots, N$, then (6.50) reduces to the binomial (conditioned on $\boldsymbol{Y} = \boldsymbol{y}$) with parameters $p(\boldsymbol{y}) := p(\boldsymbol{y},n)$ and N:

$$M_T\left(\omega \middle| \mathcal{H}_1; \boldsymbol{Y} = \boldsymbol{y}\right) = [p_d(\boldsymbol{y})\exp(i\omega) + (1 - p_d(\boldsymbol{y}))]^N \tag{6.51}$$

The distribution of T under $\mathcal{H}_1$ is then given by $\Pr\{T = j | \mathcal{H}_1; N\} = p_d(j, N)$, where

$$p_d(j,N) := \int_S \binom{N}{j} p(\boldsymbol{y})^j (1 - p(\boldsymbol{y}))^{N-j} f_Y(\boldsymbol{y}) d\boldsymbol{y} \tag{6.52}$$

for $j = 0, 1, 2, \ldots, N$. Furthermore, $P_d(k,N) := \sum_{j=k}^{N} p_d(j,N)$, for $k \in \{1, 2, \ldots, N\}$, is given by

$$P_d(k,N) := \sum_{j=k}^{N} \binom{N}{j} \int_S p(\boldsymbol{y})^j (1 - p(\boldsymbol{y}))^{N-j} f_Y(\boldsymbol{y}) d\boldsymbol{y} \tag{6.53}$$

For $\boldsymbol{Y}$ uniformly distributed over S,

$$P_d(k,N) := \| S \|^{-1} \sum_{j=k}^{N} \binom{N}{j} \int_S p(\boldsymbol{y})^j (1 - p(\boldsymbol{y}))^{N-j} d\boldsymbol{y} \tag{6.54}$$

6.3.1.4.5 Cookie-Cutter Detection Model

Again, let $p_d(\boldsymbol{x}_n, \boldsymbol{y}, n) = (p_d - p_f)\mathcal{X}_{C_{\boldsymbol{x}}^n}(\boldsymbol{y}) + p_f$, for $n = 1, 2, \ldots, N$. Assume that $f_X(\boldsymbol{x})$ is the uniform pdf over S^+. Then, for $\boldsymbol{y} \in S$, from the definition of $p(\boldsymbol{y}, n)$ in (6.50), (6.36), from $R_{\boldsymbol{y}}^n \cap S^+ = R_{\boldsymbol{y}}^n$, and from $| R_{\boldsymbol{y}}^n | = | S_n |$,

$$\begin{aligned} p(\boldsymbol{y}, n) &= \|S^+\|^{-1} \int_{S^+} \left[(p_d - p_f)\mathcal{X}_{C_{\boldsymbol{x}}^n}(\boldsymbol{y}) + p_f\right] d\boldsymbol{x}_n \\ &= \|S^+\|^{-1} \int_{S^+} \left[(p_d - p_f)\mathcal{X}_{R_{\boldsymbol{y}}^n}(\boldsymbol{x}_n) + p_f\right] d\boldsymbol{x}_n \\ &= p_n \end{aligned} \tag{6.55}$$

where

$$\begin{aligned} p_n &:= p_d \frac{\| S_n \|}{\|S^+\|} + p_f \left(1 - \frac{\| S_n \|}{\|S^+\|}\right) \\ &= (p_d - p_f) \| S_n \| \| S^+ \|^{-1} + p_f, \quad n = 1, 2, \ldots, N \end{aligned}$$

Therefore, in this case, $p(\boldsymbol{y}, n)$ no longer depends on $\boldsymbol{y}$; it depends only on the areas of the shapes. This simple form, devoid of an edge effect, results from choosing S^+ and S as the supports of $f_X(\boldsymbol{x})$ and $f_Y(\boldsymbol{y})$, respectively.

In the Poisson limit of (6.44), the intensity μ becomes $\mu = \rho^+[p_d\bar{A} + (A^+ - \bar{A})p_f]$, where $\bar{A} = \frac{1}{N}\sum_{n=1}^{N} \| S_n \|$, $A^+ = \|S^+\|$, and $\rho^+ = N/A^+$, the density of nodes over S^+.

6.3.1.4.6 *Homogeneous Binomial*

If the shapes have equal areas, that is, if $||S_n|| = ||S_1||$, then $p_n = p_1$, a constant for $n = 1, 2, \ldots, N$, where

$$p_1 = (p_d - p_f)\,||S_1||\,||S||^{-1} + p_f \tag{6.56}$$

Equation 6.54 then reduces to the case of the homogeneous binomial, for $k \in \{1, 2, \ldots, N\}$:

$$P_d(k, N) = \sum_{j=k}^{N} \binom{N}{j} p_1^j (1 - p_1)^{N-j} \tag{6.57}$$

The homogeneous binomial obtains because the shapes of the detection region have the same area, although they are not necessarily congruent.

6.3.1.4.7 *Shapes Dependent on Random Parameters*

Let the shapes depend on parameters that are random variables; that is, we write the set of shapes as $\{S_n(\boldsymbol{U}_n)\}_{n=1}^N$ where the $\{\boldsymbol{U}_n\}_{n=1}^N$ are iid random variables that are independent of the $\{\boldsymbol{X}_n\}_{n=1}^N$. As before, we assume the aforementioned *cookie-cutter* detection model and that $f_X(\boldsymbol{x})$ is the uniform pdf over S. Then we can rewrite (6.55) as

$$p_n = (p_d - p_f)\,||S_c^+||^{-1}\, E_{U_n}\left[||S_n(\boldsymbol{U}_n)||\right] + p_f \tag{6.58}$$

where S_c^+ is a modified extension of the roi S and is the exterior parallel set of S in some distance r_c; that is, S_c^+ is the dilation of S by the disk $D(\mathbf{0}, r_c)$. The constant r_c may be defined, for example, as the radius such that, for all $n = 1, 2, \ldots, N$, $\Pr\{S_n \subset D(\mathbf{0}, r_c)\} \geq c < 1$. We thus avoid randomizing the extended roi.

Consider the case where the shapes are sectors as shown in Figure 6.9. Let $\boldsymbol{U} := [\Theta_n, R_n]^T$, where Θ_n and R_n are, respectively, the angular width and radius (vice the constant radius r_d) of the nth of the N sectors; assume that they are random variables, independent for a given n. Then the area $||S_n(\cdot)|| = \frac{1}{2}\Theta_n R_n^2$ for each orientation, $n = 1, 2, \ldots, N$. Since the $\{\boldsymbol{U}_n\}_{n=1}^N$ are iid, it follows from (6.58) that p_n is constant for all $n = 1, 2, \ldots, N$:

$$p_n = \frac{1}{2}(p_d - p_f)\,||S_c^+||^{-1}\,\mu_{R^2}\mu_\Theta + p_f \tag{6.59}$$

where μ_{R^2} and μ_Θ denote, respectively, the expected values of R_n^2 and Θ_n.

6.3.2 Problem Formulation: Detection of a Moving Target by Sensors Randomly Located

6.3.2.1 Target Motion

We turn to the case of target motion in the $x_1 - x_2$ plane:

$$\boldsymbol{Y}(t) = \int_{T_0}^{t} \boldsymbol{V}(\tau)d\tau + \boldsymbol{Y}(T_0), \quad t \in [T_0, T_f] \tag{6.60}$$

where

$\boldsymbol{Y}(t) \subset S \subset \mathbb{R}^2$ is the target's position at time t

T_0 and T_f are the epochs of entrance to and exit from S of the target

The target's instantaneous velocity is $\boldsymbol{V}(t) = [V_1(t), V_2(t)]^T$. In general, $\boldsymbol{V}(t)$ is a stochastic process. Although T_0 and T_f are random variables, the following results are conditioned on $\{T_0 = t_0\}$ and $\{T_f = t_f\}$. The initial position $\boldsymbol{Y}(t_0)$ is unknown, but with a presumably known pdf $f_{\boldsymbol{Y}}(\boldsymbol{y})$ over S.

Let the set $\Gamma(t, t_0) := \{\boldsymbol{y} \in S : \boldsymbol{y} \in \{\boldsymbol{Y}(\tau)\}_{t_0}^{t}\}$, and let $\gamma(t, t_0) := \{\boldsymbol{y} \in S : \boldsymbol{y} \in \{\boldsymbol{y}(\tau)\}_{t_0}^{t}\}$, a realization of $\Gamma(t, t_0)$. We assume that $\gamma(t, t_0)$ is piecewise smooth and simple in that it does not cross itself during $[t_0, t_f]$; furthermore, we assume that $V_2(t) > 0$. Thus, we are addressing a target that is transiting, vice patrolling the region S. (Implicitly, constraints imposed by Equation 6.60 and the accompanying remarks hold with probability 1.)

6.3.2.2 Local Detection of the Track

The conditional probability of the local detection by a node at $\boldsymbol{x}_n \in S^+$ given a realized track is

$$p_d(\boldsymbol{x}_n, \gamma, n) = \Pr\{D_n = 1 \mid \mathcal{H}_1; \boldsymbol{X}_n = \boldsymbol{x}_n, \Gamma = \gamma\} \tag{6.61}$$

where the t-dependence notation of $\Gamma(t, t_0)$ and $\gamma(t, t_0)$ is suppressed. Under the alternative hypothesis $\mathcal{H}_1$, $\Gamma \cap S \neq \emptyset$. Conversely, under the null hypothesis $\mathcal{H}_0$, $\Gamma \cap S = \emptyset$.

Assume that the nth of N nodes has a *detection opportunity (do)*, a fortuitous geometric relationship between the target's track and the node necessary for a significant probability of a true detection (as opposed to a false alarm) by that node. For $n = 1, 2, \ldots, N$ and $t \in t_0, t_f$, given S_n, and given the condition $\Gamma = \gamma$, define the region S_n^γ:

$$S_\gamma^n := \breve{S}_n \oplus \gamma \tag{6.62}$$

the Minkowski sum of the sets $\breve{S}_n$ and γ, where $\breve{S}_n := \{\boldsymbol{x} : -\boldsymbol{x} \in S_n\}$ is the *reflection* of S_n; therefore, $S_\gamma^n = \{\boldsymbol{x} = \boldsymbol{y} + \boldsymbol{z} : \boldsymbol{y} \in \gamma, \boldsymbol{z} \in \breve{S}_n\}$.* Assuming again that $S_n \subseteq D(\boldsymbol{0}, r_d)$, then $\breve{S}_n \subseteq D(\boldsymbol{0}, r_d)$. Since $\gamma \subset S$ and $S^+ = S \oplus D(\boldsymbol{0}, r_d)$, $S_n^\gamma \subset S^+$, for $n = 1, 2, \ldots, N$.

In mathematical morphology, S_γ^n is called the *dilation* of γ by the structuring element S_n (see, e.g., [25]). If the shape S_n depends on random parameters, the appropriate expectation is applied to the right of (6.62).

A detection opportunity occurs iff the target's track Γ ($\Gamma \subset S$ with probability 1) intersects (or *hits*) the region C_x^n. For each $n = 1, 2, \ldots, N$, define the variable Z_n: $Z_n = 1$ iff $\Gamma \cap C_x^n \neq \emptyset$; otherwise, $Z_n = 0$. The probability of a detection opportunity is defined by

$$p_{do}(\boldsymbol{x}_n, \gamma, n) = \Pr\{Z_n = 1 \mid \mathcal{H}_1; \boldsymbol{X}_n = \boldsymbol{x}_n, \Gamma = \gamma\} \tag{6.63}$$

Thus,

$$p_{do}(\boldsymbol{x}_n, \gamma, n) = \Pr\{\Gamma \cap C_x^n \neq \emptyset \mid \mathcal{H}_1; \boldsymbol{X}_n = \boldsymbol{x}_n, \Gamma = \gamma\} \tag{6.64}$$

It is easy to show that given that $\boldsymbol{X}_n = \boldsymbol{x}_n$ and $\Gamma = \gamma$, then, for $n = 1, 2, \ldots, N$,

$$\{\boldsymbol{x}_n \in S^+, \gamma \subset S : \gamma \cap C_x^n \neq \emptyset\} = \{\boldsymbol{x}_n \in S^+, \gamma \subset S : \boldsymbol{x}_n \in S_\gamma^n\} \tag{6.65}$$

Therefore, for $n = 1, 2, \ldots, N$,

$$\begin{aligned} p_{do}(\boldsymbol{x}_n, \gamma, n) &= \Pr\{\boldsymbol{x}_n \in S_\gamma^n \mid \mathcal{H}_1; \boldsymbol{X}_n = \boldsymbol{x}_n, \Gamma = \gamma\} \\ &= \mathcal{X}_{S_\gamma^n}(\boldsymbol{x}_n) \end{aligned} \tag{6.66}$$

where $\mathcal{X}_{S_\gamma^n}(\cdot)$ is the indicator (set) function of S_γ^n.

* If the target is stationary, γ is a singleton $\boldsymbol{y}$, and S_γ^n reverts to R_y^n, as previously defined for the case of a stationary target.

Let $p_{do}(\gamma, n) := \Pr\{\mathbf{X}_n \in S_\gamma^n \mid \mathcal{H}_1; \Gamma = \gamma\}$. If the $\{\mathbf{X}_n\}_{n=1}^N$ are identically distributed with pdf $f_X(\mathbf{x})$, for $\mathbf{x} \in S^+$, then, for $n = 1, 2, \ldots, N$ and $\gamma \in S$,

$$p_{do}(\gamma, n) = \int_{S^+} p_{do}(\mathbf{x}, \gamma, n) f_X(\mathbf{x}) d\mathbf{x} \tag{6.67}$$

and, from (6.66),

$$p_{do}(\gamma, n) = \int_{S_\gamma^n} f_X(\mathbf{x}) d\mathbf{x} \tag{6.68}$$

Moreover, if the $\{\mathbf{X}_n\}_{n=1}^N$ are uniformly distributed over S^+, then (6.68) becomes, for $\gamma \subset S$ and $n = 1, 2, \ldots, N$,

$$p_{do}(\gamma, n) = |S_\gamma^n| |S^+|^{-1} \tag{6.69}$$

(Recall that we have suppressed the t-dependence in these expressions. Explicitly, we can write $S_\gamma^n(t, t_0) = \breve{S}_n \oplus \gamma(t, t_0)$, when necessary for clarity.)

6.3.2.3 Special Case: The Shapes Are Randomly Oriented, Convex Domains, and $\gamma = \ell$, a Segment of a Line

Suppose that the shapes, now denoted by $\{S_n(\Phi_n)\}_{n=1}^N$, are known except for their individual random orientations $\{\Phi_n\}_{n=1}^N$, which are iid random variables uniform over $(0, 2\pi]$. Consider the special case when the target's track γ is a line segment ℓ of length s; that is, it is a convex domain with perimeter $\|\partial\gamma\| = 2s$. We invoke, once again, Steiner's formula for convex domains in the plane: for $n = 1, 2, \ldots, N$, and from (6.62),

$$\begin{aligned} E_{\Phi_n}(\| S_\gamma^n \|) &= E_{\Phi_n}(\| \ell \oplus \breve{S}_n(\Phi_n) \|) \\ &= \| S_n \| + \frac{1}{\pi} s \| \partial S_n \| \end{aligned} \tag{6.70}$$

given that the area $\|\breve{S}_n\| = \|S_n\|$, for all $\Phi_n \in (0, 2\pi]$. Again observe that the expressions depend implicitly on t and t_0.

We further specialize this case to where the $\{S_n(\Phi_n)\}_{n=1}^N$ are sectors of a disk of radius r_d and known angles $\{\theta_n\}_{n=1}^N$, as in Figure 6.10. Then, for $n = 1, 2, \ldots, N$, each $S_n(\Phi_n)$ is convex iff $\theta_n \leq \pi$ or $\theta_n = 2\pi$. Either $\|\partial S_n\| = r_d(\theta_n + 2)$ or $\|\partial S_n\| = 2\pi r_d$, and $\| S_n \| = \frac{1}{2} r_d^2 \theta_n$. Finally, Equation 6.70 becomes, for $n = 1, 2, \ldots, N$,

$$E_{\Phi_n}(\| S_\gamma^n \|) = \frac{1}{2} r_d^2 \theta_n + \frac{1}{\pi} r_d(\theta_n + 2)s, \quad \text{for } \theta_n \leq \pi \tag{6.71}$$

for $\theta_n = 2\pi$, $\| S_\gamma^n \| = \pi r_d^2 + 2 r_d s$, the area of the "pill-shaped region," as in Wettergen [32].

6.3.2.4 Blaschke's Kinematic Formula

We invoke the kinematic formula of W. Blaschke, a fundamental theorem of integral geometry, to determine the $\{E_{\Phi_n}(\| S_\gamma^n \|)\}_{n=1}^N$ for cases more general than when γ is the segment of a line. Once again, we use as our principal source Santaló [23], in which Blaschke's original work is extensively cited. The monographs by Ren Delin [20] and Chern [6] are also very useful. For a more recent mathematical treatment, see Schneider and Weil [24].

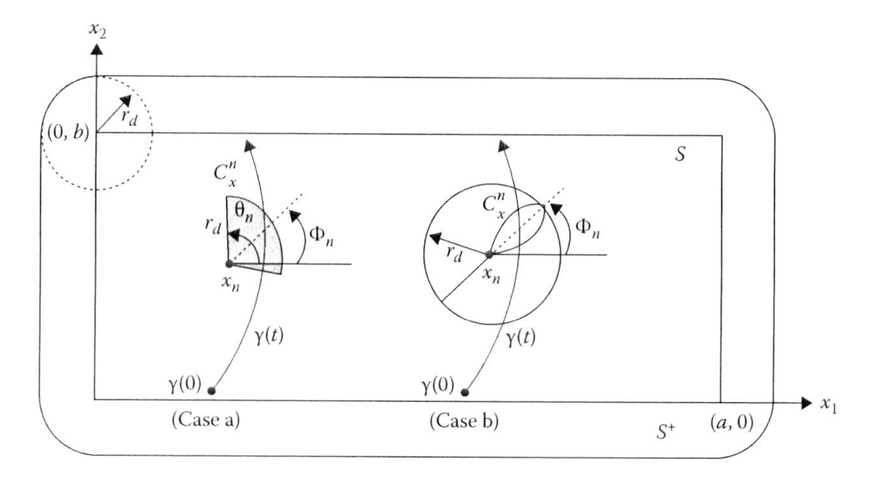

FIGURE 6.10 Examples of special cases of $\gamma \cap C_x^n \neq \varnothing$, when $\nu = 1$: case (a) C_x^n is a sector; case (b) C_x^n is a smooth directivity lobe.

First, we summarize the definition of *total curvature*, denoted by c, of a simple (i.e., without double points), closed, piecewise smooth curve denoted by $A \subset \mathbb{R}^2$. Herein, we represent A as the union of a finite number m of arcs $\{a_i\}_{i=1}^m$ (each of class C^2; i.e., $A = \bigcup_{i=1}^n a_i$ with corners at the points $a_i \cap a_{i-1}$ for $i = 1, 2, \ldots, m$; and $a_0 = a_m$) because A is closed. The total curvature c is defined by

$$c = \sum_{i=1}^{m} \int_{a_i} d\tau + \sum_{i=1}^{m} \alpha_i \tag{6.72}$$

where

τ is the angle (made with the x_1-axis) of the oriented tangent to a given arc
$\{\alpha_i\}_{i=1}^m$ are the exterior angles at the corresponding corners $\{a_i \cap a_{i-1}\}_{i=1}^m$

As in Ref. [6] (p. 19), by the *theorem of turning tangents*, the total curvature of a simple, closed, piecewise smooth curve is $\pm 2\pi$, the sign depending upon the orientation of the curve.

Consider now a domain $D \in \mathbb{R}^2$ bounded by a finite number of simple, closed, piecewise smooth curves each like the curve A. The total curvature of D, denoted by $c(D)$, is given by $c(D) = 2\pi\chi(D)$, where $\chi(D)$ is an integer, the *Euler constant* of D. Note that D is not necessarily convex; moreover, it may be multiply connected (in our application, we need consider only simply connected domains). We extract (with suitable editing) a statement of Blaschke's theorem from Santaló ([23], Section I.7.4), which has its proof.

Theorem 6.9 (Blaschke's kinematic formula)

Let $D, D' \subset \mathbb{R}^2$ denote the domains of the kind described earlier. Let c be the total curvature, $\|D\|$ the area, and $\|\partial D\|$ the perimeter of D. Similarly define c', $\|D'\|$, and $\|\partial D'\|$ for D'. Assume that D' moves in the plane (i.e., rigid body motion with translation $\mathbf{x}$ and rotation ϕ) with kinematic density $dK' = d\mathbf{x}d\phi = dx_1dx_2d\phi$. Letting c'' denote the total curvature of the domain $D \cap D'$, then

$$\frac{1}{2\pi} \int_{D \cap D' \neq \varnothing} c'' dK' = \|D\| c' + \|D'\| c + \|\partial D\| \|\partial D'\|$$

The following particular cases of this theorem obtain, again as per Santaló ([23], Section I7.4.1):

1. Let both D and D' be convex. Then $D \cap D'$ is also convex, and $c = c' = c'' = 2\pi$, in which case the theorem reduces to

$$\frac{1}{2\pi} \int_{D \cap D' \neq \varnothing} dK' = \| D \| + \| D' \| + \frac{1}{2\pi} \| \partial D \| \| \partial D' \| \tag{6.73}$$

2. Let both the domains D and D' each be bounded by a single closed curve that is simple, and piecewise smooth. Then $c = c' = 2\pi$, and $D \cap D'$ is composed of a certain number ν of separate domains, each bounded by a single closed curve of total curvature 2π. Then the theorem reduces to

$$\frac{1}{2\pi} \int_{D \cap D' \neq \varnothing} \nu dK' = \| D \| + \| D' \| + \frac{1}{2\pi} \| \partial D \| \| \partial D' \| \tag{6.74}$$

3. Let each domain D and D' degenerate to simple curves A and A' of lengths L and L', respectively. Then $\|D\| = \|D'\| = 0$; furthermore, the degenerate domains have perimeters $\|\partial D\| = 2L$ and $\|\partial D'\| = 2L'$, respectively. The total curvature of $D \cap D'$ becomes $c'' = 2\pi n$, where n is the number of points in the intersection $A \cap A'$. Therefore, in this case, the theorem reduces to Poincaré's formula:

$$\int_{A \cap A'} n dK' = 4LL' \tag{6.75}$$

6.3.2.5 Application of Blaschke's Formula

We now apply these results to determine $E_{\Phi_n}(\| S_\gamma^n \|)$, using, in particular, the case addressed by (6.74), that is, when the domains are each simply connected. Let the (degenerate) domain $D = \gamma$, a curve segment of length l_γ. Thus, $\|D\| = \|\gamma\| = 0$ and the perimeter $\|\partial D\| = \|\partial\gamma\| = 2l_\gamma$. Furthermore, let the domain $D' = C_x := \boldsymbol{x} \oplus S(\phi)$, where $S(\phi)$ is any plane, compact shape with orientation ϕ. Then (6.74) becomes

$$\int_{\gamma \cap C_x \neq \varnothing} \nu d\boldsymbol{x} d\phi = 2\pi \| S \| + 2 l_\gamma \| \partial S \| \tag{6.76}$$

We additionally assume that S is convex and that $\nu = 1$. The later assumption is when there exists only one segment of the curve γ in $\gamma \cap C_x \neq \varnothing$. Then (6.76) becomes

$$\int_{\gamma \cap C_x \neq \varnothing} d\boldsymbol{x} d\phi = 2\pi \| S \| + 2 l_\gamma \| \partial S \| \tag{6.77}$$

Before applying the aforementioned results, we first prove the following simple lemma.

Lemma 6.2

Let $C_x^n = \boldsymbol{x}_n \oplus S_n(\phi_n)$, where ϕ_n denotes the orientation of $S_n(\phi_n)$, for $n = 1, 2, \ldots, N$. In accord with (6.62), *$S_\gamma^n := \breve{S}_n \oplus \gamma$. Let ϕ_n be the realization of the random angle Φ_n that is uniformly distributed over* $(0, 2\pi]$. *If $\nu_n = 1$, for all $\gamma \cap C_x^n \neq \varnothing$,* then by Blaschke's kinematic formula as reduced to (6.77), the following holds, for $n = 1, 2, \ldots, N$

$$E_{\Phi_n}\left(\| S_\gamma^n \| \right) = \| S_n \| + \frac{1}{\pi} l_\gamma \| \partial S_n \| \tag{6.78}$$

Proof

From (6.65) $\{x_n \in S^+, \gamma \subset S : \gamma \cap C_x^n \neq \varnothing\} = \{x_n \in S^+, \gamma \subset S : x_n \in S_\gamma^n\}$. Therefore,

$$\int_{\gamma \cap C_x^n \neq \varnothing} dx d\phi_n = \int_{x_n \in S_\gamma^n} dx_n d\phi_n = \int_0^{2\pi} S_\gamma^n d\phi_n = 2\pi E_{\Phi_n}\left(\| S_\gamma^n \|\right) \tag{6.79}$$

The result follows from (6.77) and (6.79).

The assumption that $\nu = 1$ over the (non-null) intersection restricts γ and $S_n(\phi_n)$, for all $\phi \in (0, 2\pi]$ and $n = 1, 2, \ldots, N$. For example, in Figure 6.10, we show a special case where $\nu = 1$. Observe that the result depends upon *only* the (realized) length of the target's path, for rather general motion.

6.3.2.5.1 *Simple Example: Randomly Oriented, Directional Sensors*

Consider the case when the orientation Φ_n of the shape S_n (located at x_n) is a random variable uniformly distributed over $(0, 2\pi]$. We, thus, can write Equation 6.69 conditioned on $\Phi_n = \phi_n$ as

$$p_{do}(\gamma, n \mid \Phi_n = \phi_n) = \| S_\gamma^n(\phi_n) \| \| S^+ \|^{-1} \tag{6.80}$$

Then

$$\begin{aligned} p_{do}(\gamma, n) &= E_{\Phi_n}\left[p_{do}(\gamma, n \mid \Phi_n = \phi_n) \right] \\ &= E_{\Phi_n}\left(\| S_\gamma^n(\Phi_n) \|\right) \| S^+ \|^{-1} \end{aligned} \tag{6.81}$$

which from (6.78) becomes

$$p_{do}(\gamma, n) = \| S^+ \|^{-1} \left(\| S_n \| + \frac{1}{\pi} l_\gamma \| \partial S_n \| \right) \tag{6.82}$$

As in Figure 6.10, we let the shape S_n be the sector with known respective angle $\theta_n \leq 2\pi$ and detection radius r_d, for all $n = 1, 2, \ldots, N$. Then, in this special case, (6.82) becomes

$$p_{do}(\gamma, n) = \| S^+ \|^{-1} \left[\frac{1}{2} r_d^2 \theta_n + \frac{1}{\pi} r_d (\theta_n + 2) l_\gamma \right], \quad \text{for } \theta_n \leq \pi \tag{6.83}$$

for an arbitrary (smooth) path of a transiting target. For an extended roi as in Figure 6.10, $\| S^+ \| = \| S \| + \| \partial A \| r_d + \pi r_d^2$. Note that S is any compact, convex region, not necessarily a rectangle.

Finally, $p_{do}(\gamma, n)$ is given by

$$p_{do}(\gamma, n) = \left[\| S \| + \| \partial A \| r_d + \pi r_d^2 \right]^{-1} \left[\frac{1}{2} r_d^2 \theta_n + \frac{1}{\pi} r_d (\theta_n + 2) l_\gamma \right], \quad \text{for } \theta_n \leq \pi \tag{6.84}$$

To summarize this example, (6.84) applies to any convex, compact roi S, dilated by the disk $D(\mathbf{0}, r_d)$, with N randomly located and randomly oriented, nonisotropic sensors. Moreover, the target motion is limited to a smooth, transiting (realized) path γ of length l_γ, but otherwise arbitrary.

If the sensors have identical directivity, that is, if $\theta_n = \theta \le 2\pi$, then $p_{do}(\gamma, n) = p_{do}(\gamma)$, for all $n = 1, 2, \dots, N$; and the distribution of the test statistic $T = \sum_{n=1}^{N} D_n$, given by (6.83), is the homogeneous binomial.

6.A Appendix

6.A Proof of Theorem 6.1

We give a proof for Theorem 6.1. But first, we need the following lemma:

Lemma 6.A.1

Let N be a spatial Poisson process with parameter λ. Then, under condition $N(A) = k$ for $||A|| > 0$, the k points are independent and uniformly distributed.

Proof

Let $A_1, A_2, \dots, A_n$ be n disjoint regions with $\bigcup_1^n A_i = A$. Suppose also that the integers $k_1, k_2, \dots, k_n$ satisfy $k_1 + k_2 + \cdots + k_n = k$. Then

$$
\begin{aligned}
&P\{N(A_1) = k_1, N(A_2) = k_2, \dots, N(A_n) = k_n \hat{u} N(A) = k\} \\
&= \frac{P\{N(A_1) = k_1, N(A_2) = k_2, \dots, N(A_n) = k_n\}}{P\{N(A) = k\}} \\
&= \frac{\left[\frac{[\lambda||A_1||]^{k_1}}{k_1!} e^{-\lambda||A_1||}\right]\left[\frac{[\lambda||A_2||]^{k_2}}{k_2!} e^{-\lambda||A_2||}\right]\cdots\left[\frac{[\lambda||A_n||]^{k_n}}{k_n!} e^{-\lambda||A_n||}\right]}{\frac{[\lambda||A||]^{k}}{k!} e^{-\lambda||A||}} \\
&= \frac{k!}{k_1!k_2!\dots k_n!}\left[\frac{||A_1||}{||A||}\right]^{k_1}\left[\frac{||A_2||}{||A||}\right]^{k_2}\cdots\left[\frac{||A_n||}{||A||}\right]^{k_n}
\end{aligned} \tag{6.A.1}
$$

In particular,

$$
\begin{aligned}
P\{N(A_1) = 1 \hat{u} N(A) = 1\} &= P\{N(A_1) = 1, N(A_1^c) = 0 \hat{u} N(A) = 1\} \\
&= \frac{P\{N(A_1) = 1, N(A_1^c) = 0\}}{P\{N(A) = 1\}} \\
&= \frac{1!}{1!0!}\left[\frac{||A_1||}{||A||}\right]^{1}\left[\frac{||A_1^c||}{||A||}\right]^{0} \\
&= \frac{||A_1||}{||A||}
\end{aligned} \tag{6.A.2}
$$

According to (6.A.2), *given that an event has occurred in volume A, it is equally likely to be found anywhere in A*. Also, note that

$$P\{N(A_1) = k_1 \hat{u} N(A) = k_1\} = \left[\frac{\| A_1 \|}{\| A \|}\right]^{k_1}$$

Thus, Equation 6.A.1 implies independence. Note that the leading coefficient in (6.A.1) is the sum over all the configurations in which the k points can be divided into n subgroups with k_1 points in the first group, k_2 in the second, etc. Each of these configurations has probability $\| A_1 \|^{k_1} \| A_2 \|^{k_2} \ldots \| A_n \|^{k_n} / \| A \|^k$.

Theorem 6.A.1

Consider a set of omnidirectional sensors in three-dimensional space scattered throughout some region $S \subset R^3$. Suppose that the sensors are spatially distributed according to a Poisson point process with parameter λ. Suppose also that the range sensitivity of all sensors is distributed according to the cumulative distribution function $F(r)$ with density $F'(r) = f(r)$ having finite third moment. Finally, assume that range sensitivity is independent of the sensor location. Then the number of sensors that will detect a target located at some point $\boldsymbol{x} \in S$ is a Poisson random variable with mean

$$\Lambda = \frac{4}{3}\lambda\pi \int_0^{\infty} r^3 f(r)\,dr$$

Independence Assumptions: (1) Sensors are distributed according to a Poisson random variable. *This means that the number of sensors in nonintersecting regions of space (e.g., nonintersecting spherical shells centered about any point) are independent.* It is useful to think of each sensor and its sensitivity extent as a sphere, with the center of the sphere at the sensor location, and radius equal to the sensitivity range of the sensor. (2) According to the theorem, the radius of each sphere is assumed to be a random variable independent of the location of the sphere.

Proof

Fix the origin* at any point in R^3. Let $S(r)$ be the sphere of radius r with center at the origin. Also, let $S(r, r + \Delta r)$ denote the volume of the shell, or region, between two concentric spheres centered at the origin and having radii r and $r + \Delta r$, respectively. Then

$$\begin{aligned} S(r, r+\Delta r) &= \| S(r+\Delta r) \| - \| S(r) \| \\ &= \frac{4}{3}\pi(r+\Delta r)^3 - \frac{4}{3}\pi r^3 \\ &= \frac{4}{3}\pi\left[r^3 + 3r^2\Delta r + 3r(\Delta r)^2 + (\Delta r)^3 - r^3\right] \\ &= \frac{4}{3}\pi\Delta r\left[3r^2 + 3r\Delta r + (\Delta r)^2\right] \\ &= 4\pi r^2 \Delta r + o(\Delta r) \end{aligned}$$

* For argument's sake, the origin is used, although any fixed point would suffice. It might be useful for the reader to consider this point a target, as it will be shown that the number of sensors that detect this point follows a Poisson distribution.

Therefore, the probability of a sphere occurring in the shell (i.e., having center in the shell $S(r, r + \Delta r)$) with radius extending out to the origin is the product of (1) the probability that a spherical center occurs within the shell (i.e., $\lambda(4\pi r^2\Delta r + o(\Delta r))$) and (2) the probability that the sphere's radius extends at least out to r (i.e., $\int_r^\infty f(\rho)d\rho$):

$$\lambda\left(4\pi r^2\Delta r + o(\Delta r)\right)\int_r^\infty f(\rho)d\rho = \left(\lambda 4\pi r^2\int_r^\infty f(\rho)d\rho\right)\Delta r + o(\Delta r)$$

That is,

$$P\{N(S(r, r+\Delta r)) = 1\} = \Lambda(r)\Delta r + o(\Delta r) \tag{6.A.3}$$

where $N(S(r, r+\Delta r))$ is the number of sensors (i.e., sphere centers) to occur in the shell $S(r, r+\Delta r)$ with radius extending out to the origin

$$\Lambda(r) = 4\pi\lambda r^2\int_r^\infty f(\rho)d\rho$$

It remains to be shown that the probability of the occurrence of two or more spheres with centers in the shell $S(r, r+\Delta r)$ and radii extending out to the origin is $o(\Delta r)$:

$$P\{N(S(r, r+\Delta r)) \geq 2\} = \exp\left(-4\pi\lambda r^2\Delta r\right)\sum_{k=2}^{\infty}\frac{\left[4\pi\lambda r^2\Delta r\right]^k}{k!}\left(\int_r^\infty f(\rho)d\rho\right)^k$$

$$= \exp\left(-4\pi\lambda r^2\Delta r\right)\sum_{k=2}^{\infty}\frac{\left[4\pi\lambda r^2\Delta r\int_r^\infty f(\rho)d\rho\right]^k}{k!}$$

$$\leq \exp\left(-4\pi\lambda r^2\Delta r\right)\frac{\left[4\pi\lambda r^2\Delta r\int_r^\infty f(\rho)d\rho\right]^2}{2!}\times\sum_{k\geq 0}\frac{\left[4\pi\lambda r^2\Delta r\int_r^\infty f(\rho)d\rho\right]^k}{k!}$$

$$= \exp\left(-4\pi\lambda r^2\Delta r\right)\exp\left(-4\pi\lambda r^2\Delta r\int_r^\infty f(\rho)d\rho\right) \tag{6.A.4}$$

$$= o(\Delta r) \tag{6.A.5}$$

Note that for small Δr, the exponential factors in (6.A.4) are approximately 1. The coefficient of $(\Delta r)^2$ in the third factor is a constant.

From the two independence assumptions, it follows that $S(r)$ has independent increments. Now, (6.A.3), (6.A.5), and the independent increment property imply that the number of spheres with centers approximately a distance r from the origin and having radius out to the origin is a Poisson random variable with mean $\Lambda(r)$. That is, the number of sensors with range out to the origin is a nonhomogeneous Poisson random variable with local intensity function $\Lambda(r)$. Finally, one integrates $\Lambda(r)$ over all ranges to get the cumulative intensity function:

$$\int_0^\infty \Lambda(r)dr = \int_0^\infty 4\pi\lambda r^2 \int_r^\infty f(\rho)d\rho dr$$
$$= \int_0^\infty f(\rho)\int_0^\rho 4\pi\lambda r^2 dr d\rho$$
$$= \int_0^\infty \frac{4}{3}\pi\lambda\rho^3 f(\rho)d\rho$$

Since the point chosen as origin was arbitrary, the result follows for any point in R^3.

References

1. Aldosari, S. and Moura, J., Detection in sensor networks: The saddlepoint approximation, *IEEE Trans. Signal Process.*, 55(1), 327–340, January 2007.
2. Artés-Rodríguez, A., Decentralized detection in sensor networks using range information, *ICASSP 2004*, Vol. 2, pp. 265–268, Montreal, Quebec, Canada, 2004.
3. Artés-Rodríguez, A., Lázaro, M., and Tong, L., Target location estimation in sensor networks using range information, *IEEE Sensor Array and Multichannel Signal Processing Workshop*, pp. 608–612, Montreal, Quebec, Canada, 2004.
4. Barndorff-Nielson, O.E. et al. (Eds), *Stochastic Geometry-Likelihood and Computation*, Chapman & Hall, New York, 1999.
5. Blum, R., Kassam, S., and Poor, H.V., Distributed detection with multiple sensors: Part II, *Proc. IEEE*, 85(1), 64–79, January 1997.
6. Chern, S.S., Curves and surfaces in Euclidian space, in *Studies in Global Geometry and Analysis*, S.S. Chen (ed.), Vol. 4, pp. 16–56, Mathematical Association of America, Washington, DC, 1967.
7. Djurić, P.M. et al., Target tracking by particle filtering in binary sensor networks, *IEEE Trans. Signal Process.*, 56(6), 2229–2238, June 2008.
8. Feller, W., *An Introduction to Probability Theory and Its Applications*, Vol. I, 3rd edn., Wiley, New York, 1957.
9. Grimmett, G. and Stirzaker, D., *Probability and Random Processes*, 3rd edn., Oxford University Press, Oxford, U.K., 2005.
10. Hall, P., *Introduction to the Theory of Coverage Processes*, Wiley, New York, 1988.
11. Hero, A., III. and Blatt, D., Sensor network source localization via projection onto convex sets, *ICASSP*, pp. 689–692, Philadelphia, PA, 2005.
12. Karlin, S. and Taylor, H.M., *A Second Course in Stochastic Processes*, Academic Press, San Diego, CA, 1981.
13. Kendall, M.G. and Moran, P.A.P., *Geometric Probability*, Griffin's Statistical Monographs and Courses, London, U.K., 1963.
14. Kingman, J.F.C., *Poisson Processes*, Oxford, U.K., 1993.
15. Lazos, L., Poovendran, R., and Ritcey, J., Probabilistic detection of mobile targets in heterogeneous sensor networks, *IPSN07 (ACM)*, pp. 519–528, Cambridge, MA, April 2007.

16. Miguez, J. and Artés-Rodríguez, A., Particle filtering algorithms for tracking a maneuvering target using a network of wireless dynamic sensors, *EURASIP J. Appl. Signal Process.*, 2006, article ID 83042, 1–16, 2006.
17. Nelson, J.K., Rowe, E.G., and Carter, G.C., Detection capabilities of randomly-deployed sensor fields, *Int. J. Distrib. Sensor Netw.*, 5(6), 708–728, 2009.
18. Niu, R. and Varshney, P., Performance analysis of distributed detection in a random sensor field, *IEEE Trans. Signal Process.*, 56(1), 339–349, January 2008.
19. O'Rourke, J., *Computational Geometry*, 2nd edn., Cambridge University Press, Cambridge, U.K., 1998.
20. Ren, D., *Topics in Integral Geometry*, World Scientific, Singapore, 1999.
21. Resnick, S.I., *Adventures in Stochastic Processes*, Birkhauser, Boston, MA, 1992.
22. Rowe, E.G. and Wettergren, T.A., Coverage and reliability of randomly distributed sensor systems with heterogeneous detection range, *Int. J. Distrib. Sensor Netw.*, 5(4), 303–320, July 2009.
23. Santaló, L., *Integral Geometry and Geometric Probability*, 2nd edn., Cambridge University Press, Cambridge, U.K., 2004.
24. Schneider, R. and Wolfgang, W., *Stochastic and Integral Geometry*, Springer-Verlag, Berlin, Germany, 2008.
25. Serra, J., *Image Analysis and Mathematical Morphology*, Vol. 1, Academic Press, San Diego, CA, 1982.
26. Sheng, X. and Hu, Y.H., Maximum likelihood multiple source localization using acoustic energy measurements with wireless sensor networks, *IEEE Trans. Signal Process.*, 53(1), 44–53, January 2005.
27. Solomon, H., *Geometric Probability*, SIAM CBMS-NSF Regional Conference Series in Applied Mathematics, Philadelphia, PA, 1978.
28. Varshney, P. and Chair, Z., Optimal data fusion in multiple sensor detection systems, *IEEE Trans. AES*, AES-22(1), 98–101, January 1986.
29. Varshney, P., *Distributed Detection and Data Fusion*, Springer-Verlag, New York, 1997.
30. Viswanathan, R. and Varshney, P., Distributed detection with multiple sensors—Part I, *Proc. IEEE*, 85(1), 54–63, January 1997.
31. Wettergren, T., Performance of search via track-before-detect for distributed sensor networks, *IEEE Trans. AES*, 44(1), 314–324, 2008.
32. Wettergren, T. and Walsh, M., Localization accuracy of track-before-detect search strategies for distributed sensor networks, *EURASIP J. Adv. Signal Proc.*, 2008, Article ID 264638, pp. 1–15, 2008.
33. Yan, Q. and Blum, R., Distributed signal detection under the Neyman-Pearson Criterion, *IEEE Trans. Inf. Theory*, 47(4), May 2001.
34. Zhou, S. and Willett, P., Submarine location estimation via a network of detection-only sensors, *IEEE Trans. Signal Process.*, 55(6), pp. 3104–3115, June 2007.

7

Directed Diffusion*

Fabio Silva
University of Southern California/Information Sciences Institute

John Heidemann
University of Southern California/Information Sciences Institute

Ramesh Govindan
University of Southern California

Deborah Estrin
University of Southern California

7.1 Introduction

Traditional sensing models assume one or a few powerful sensors and centralized computation. Today, technological trends enable the creation of inexpensive, small, intelligent devices for sensing and actuation. If many small sensors can work together as a *sensor network*, they provide several advantages over traditional centralized sensing. By placing the sensor close to the object being sensed, signal processing and target discrimination problems in sensing can be greatly simplified. By communicating over several short hops rather than one long hop, energy consumed in communication can be reduced [29]. Moreover, by processing data *in* the network, often the amount of data transferred can be reduced, further saving energy [22].

Motivated by robustness, scaling, and energy efficiency requirements, this paper examines a new data dissemination paradigm for such sensor networks. This paradigm, which we call *directed diffusion*,† is data-centric. Data generated by sensor nodes is named by attribute-value pairs. A node requests data by sending *interests* for named data. Data matching the interest is then "drawn" down toward that node. Intermediate nodes can cache, or transform data, and may direct interests based on previously cached data (Section 7.3).

Directed diffusion is significantly different from IP-style communication where nodes are identified by their end-points, and inter-node communication is layered on an end-to-end delivery service

* This work was supported by DARPA under grant DABT63-99-1-0011 as part of the SCAADS project.

† Van Jacobson suggested the concept of "diffusing" attribute named data for this class of applications that later led to the design of directed diffusion.

provided within the network. In directed diffusion, nodes in the network are application-aware as we allow application-specific code to run in the network and assist diffusion in processing messages. This allows directed diffusion to cache and process data in the network (aggregation), decreasing the amount of end-to-end traffic, and resulting in higher energy savings. We show that using directed diffusion one can realize robust multi-path delivery, empirically adapt to a small subset of network paths, and achieve significant energy savings when intermediate nodes aggregate responses to queries (Section 7.5).

This chapter describes diffusion, starting from the point of view of an application (Section 7.2) and naming (Section 7.2.2). We realize these abstractions with lower-level primitives and several different data dissemination algorithms described in Section 7.3, and show how applications can influence routing (Section 7.4). We summarize simulation and experimentation results in Section 7.5.

7.2 Programming a Sensor Network

The innovations of diffusion are approaches to allow applications to process data as it moves through the network, and dissemination algorithms that select efficient paths through the network. Although these topics are important and we explore them in the following chapter, applications require abstractions over these details. This section presents an *application-level* view of diffusion, looking at our publish/subscribe-based API and how applications name data in the network.

7.2.1 The Publish/Subscribe API

We have adopted a publish/subscribe-based API for diffusion, shown in Figure 7.1.* To receive data, users or programs *subscribe* to a particular set of attributes, becoming data *sinks*. A callback function is then invoked whenever relevant data arrives at the node. Sensors *publish* data that they have, becoming data *sources*. In both cases, what data is provided or received is described by an attribute-based naming scheme described in Section 7.2.2. It is the job of the diffusion dissemination algorithms (Section 7.3) to ensure that data is communicated efficiently from sources to sinks across a multi-hop network. In general, publishing and subscribing sends messages across the network. The exact cost of these operations depends on which diffusion algorithm is used.

Publish/Subscribe APIs:

```
handle NR::subscribe(NRAttrVec *subscribe_attrs, const NR::Callback * cb);
int NR::unsubscribe(handle subscription_handle);
handle NR::publish(NRAttrVec *publish_attrs);
int NR::unpublish(handle publication_handle);
int NR::send(handle publication_handle,
          NRAttrVec *send_attrs);
```

Filter-specific APIs:

```
handle NR::addFilter(NRAttrVec *filter_attrs, u_int16_t priority, FilterCallback *cb);
int NR::removeFilter(handle filter_handle);
void NR::sendMessage(Message *msg, handle h, u_int16_t priority = 0);
```

FIGURE 7.1 Basic diffusion APIs for sending and receiving data, and for adding filters.

* This API was originally designed in collaboration with Coffin and Van Hook [14]; we have since extended it to support filters.

To allow applications to influence data as it moves through the network, users can create *filters* at each sensor node with the filter APIs in the bottom of Figure 7.1. Filters indicate what messages they are interested in by attributes; each time a matching message arrives at that node the filter is allowed to inspect and alter its progress in any way. Filters can suppress messages, change where they are sent next, or even send other messages in response to one (perhaps triggering further sensors to satisfy a query).

A more complete reference to directed diffusion APIs and example code is available in the diffusion manual [30].

7.2.2 Naming Concepts

Diffusion uses an attribute-based naming scheme to associate sources and sinks and to trigger filters. This flexible approach to naming is important in several ways. First, attribute-based naming is consistent with the publish/subscribe application-level interface (Section 7.2) and many-to-many communication. Diffusion's naming scheme is *data-centric*, allowing applications to focus on what data is desired rather than on individual sensor nodes. The approach also supports multiple sources and sinks, rather than simple point-to-point communication. Thus applications may subscribe to "seismic sensors in the southeast region" rather seismic sensors #15 and #35, or hosts 10.1.2.40 and 10.2.1.88.

Second, diffusion attributes provide some structure to a message. By identifying separate fields, data dissemination algorithms can use application data to influence routing. For example, application-specific, geographic information can limit where diffusion must look for sensors. In addition, treating messages as sets of attributes simplifies application and protocol extensions (a need also suggested for future Internet-based protocols [7]).

Finally, attributes serve to associate messages with sources, sinks, and filters via *matching*. If the attributes in a sink's subscription match those of source's publication, diffusion must send any published data to the sink.

7.2.3 Matching in Naming

Each set of attributes in diffusion is a set of (key, type, operator, value) tuples. The most important parts of an attribute are the *key* and *value*, which together specify the meaning of the data (longitude, temperature, detection confidence, etc.) and its actual contents (118.40817°, 98.6°, 80%, etc.). The *type* defines how the value field is interpreted: as a string, integer or floating point type, or as uninterpreted binary data (blobs).

The operator field allows attributes to not only contain data, but to express simple constraints. There are two classes of operators: first, IS, the *actual* operator, is used to indicate a specific value. The second group includes binary comparisons (EQ, NE, LE, GT, LE, GE, corresponding to equality, inequality, less than, etc.) and "EQ_ANY" (which matches anything); these are collectively called *formal* operators.

Actuals are statements about data. So "latitude IS 33.9425, longitude IS 118.40817" might indicate a location, or "sensor IS seismic, value IS 7.0, confidence IS 80" might indicate a specific sensor reading.

Formals allow one to select sets of sensors, thus indicating which publish and subscribe operations should be connected. Thus, a subscription might indicate "latitude GT 33.5, latitude LT 34.0, sensor EQ seismic" to indicate seismic sensors in some area.

Formals and actuals can be mixed and used in publications, subscriptions, or filters.

The exact process of determining which publications and subscriptions are related is called matching. A *one-way match* compares all formal parameters of one attribute set against the actuals of the others (Figure 7.2). Any formal parameter that is missing a matching actual in the other attribute set causes the one-way match to fail (e.g., "confidence GT 0.5" must have an actual such as "confidence IS 0.7" and would not match "confidence IS 0.3," "confidence LT 0.7," or "confidence GT 0.7"). Two sets of attributes have a *complete match* if one-way matches succeed in both directions. In other words, attribute sets A and B match if the one-way match algorithm succeeds from both A to B and B to A.

```
one-way match:
given two attribute sets A and B
for each attribute a in A where a.op is a formal {
    matched = false
    for each attribute b in B where a.key = b.key and b.op is an actual
        if a.val compares with b.val using a.op, then matched = true
    if not matched then return false (no match)
}
return true (successful one-way match)
```

FIGURE 7.2 Our one-way matching algorithm.

Matching is used to associate publications and subscriptions and to activate filters as messages flow through the network.

Although matching is reasonably powerful, it does not perfectly cover all scenarios or tasks. Matching strikes a balance between ease of implementation and flexibility. For example, while attributes can easily define a square, they cannot directly operate on arbitrarily complex sensor detection regions. We expect applications to use attributes for rough matching and refine matching with application-specific code (such as with filters, Section 7.4).

For detailed examples of naming in diffusion, please see the diffusion manual [30] or [18].

7.3 Directed Diffusion Protocol Family

Publish/subscribe provides an application's view to a sensor network, and attribute-based naming a detailed way to specify which sources and sinks communicate. The "glue" that binds the two are the directed diffusion algorithms for data dissemination. In a traditional network, communication is effected by routing, usually based on global addresses and routing metrics. Instead, we use the term data dissemination to emphasize the lack of global addresses, reliance on local rules, and, as described in the Section 7.4, the use of application-specific in-network processing.

The original, two-phase directed diffusion uses several control messages to realize our publish/subscribe API: sinks send *interest* messages to find sources, sources use *exploratory data* messages to find sources, and positive and negative *reinforcement* messages select or prune parts of the path. Early work [22] identified these primitives, described the concept of diffusion, and evaluated a specific algorithm that we now call *two-phase pull* diffusion. We found this algorithm ideal for some applications but as our experience with sensor networks applications grew, we found two-phase pull a poor match for other classes of applications.

We see diffusion not as a single algorithm, but as a *family* of algorithms built from these primitives. Other algorithms provide better performance for some applications. We have recently made two additions to the diffusion protocol family: *one-phase push* and *one-phase pull* [17].

Another way to optimize diffusion performance is to use physical or application-specific information. The physical nature of a sensor network's deployment makes geographically scoped queries natural, prompting the development of geographically-aided routing protocols such as *GEAR* [34], *GPSR* [26], and *rumor routing* [8]. Application-specific information can also be exploited using filters (described in Section 7.4).

We expect application designers to match an appropriate algorithm with their application's requirements. Table 7.1 compares the interactions of the algorithms; we describe them below in more detail and review their performance in Section 7.5.4. More detail is the subject of current [17,22,24] and ongoing research.

TABLE 7.1 Comparison of Interactions in Diffusion Algorithms

Protocol	Sink	Source
Two-phase pull	Interest* (every interest interval)	
		Exploratory data* (every exploratory interval)
	Positive reinforcement (response to exp. data)	
		Data (rate defined by app.)
One-phase pull	Interest* (every interest interval)	
		Data
Push		Exploratory data* (every exploratory interval)
	Positive reinforcement (response to exp. data)	
		Data

Asterisks (*) indicate messages that are sent to all nodes (flooded or geographically scoped). All algorithms also have negative reinforcement messages.

7.3.1 Two-Phase Pull Diffusion

The purpose of directed diffusion is to establish efficient n-way communication between one or more sources and sinks. Directed diffusion is a data-centric communication paradigm that is quite different from host-based communication in traditional networks. To describe the elements of diffusion, we take the simple example of a sensor network designed for tracking animals in a wilderness refuge.

Suppose that a user in this network would like to track the movement of animals in some remote sub-region of the park. The user would subscribe to "animal-track" information, specified by a set of attributes. Sensors across the network publish animal-track information.

The user's application subscribes to data using a list of attribute-value pairs that describe a task using some task-specific naming scheme. Intuitively, attributes describe the data that is desired by specifying sensor types and possibly some geographic region. The user's node becomes a *sink*, creating an *interest* of attributes specifying a particular kind of data.

The interest is propagated from neighbor-to-neighbor towards sensor nodes in the specified region. A key feature of directed diffusion is *that every sensor node can be task-aware*—by this we mean that nodes store and interpret interests, rather than simply forwarding them along. In our example, each sensor node that receives an interest remembers which neighbor or neighbors sent it that interest. To each such neighbor, it sets up a *gradient*. A gradient represents both the direction towards which data matching an interest flows, and the status of that demand (whether it is active or inactive and possibly the desired update rate). After setting up a gradient, the sensor node redistributes the interest to its neighbors. When the node can infer where potential sources might be (e.g., from geographic information or existing similar gradients), the interest can be forwarded to a subset of neighbors. Otherwise, it will simply broadcast the interest to all of its neighbors.

Sensors indicate what data they may generate by publishing with an appropriate set of attributes. They thus become potential *sources*. As interests travel across the network, sensors with matching publications are triggered and the application activates its local sensors to begin collecting data. (Prior to activation we expect the node's sensors would be in a low-power mode.) The sensor node then generates *data* messages matching the interest. In directed diffusion, data is also represented using an attribute-based naming scheme.

Data is cached at intermediate nodes as it propagates toward sinks. Cached data is used for several purposes at different levels of diffusion. The core diffusion mechanism uses the cache to suppress duplicate messages and prevent loops, and it can be used to preferentially forward interests. (Since the filter core is primarily interested in an exact match, as an optimization, hashes of attributes can be computed and compared rather than complete data.) Cached data is also used for application-specific, in-network processing. For example, data from detections of a single object by different sensors may be merged to a single response based on sensor-specific criteria.

The initial data message from the source is marked as *exploratory* and is sent to all neighbors for which it has matching gradients. The initial flooding of the interest, together with the flooding of the exploratory data, constitutes the first phase of two-phase pull diffusion. If the sink has multiple neighbors, it chooses to receive subsequent data messages for the same interest from a preferred neighbor (e.g., the one which delivered the first copy of the data message). To do this, the sink *reinforces* the preferred neighbor, which, in turn reinforces its preferred upstream neighbor, and so on. The sink may also *negatively reinforce* its current preferred neighbor if another neighbor delivers better (lower latency) sensor data. This negative reinforcement propagates neighbor-to-neighbor, removing gradients and tearing down and existing path if it is no longer needed [22]. Negative reinforcements suppress loops or duplicate paths that may arise due to changes in network topology.

After the initial exploratory data message, subsequent messages are sent only on reinforced paths. (The path reinforcement, and the subsequent transmission of data along reinforced paths, constitutes the second phase of two-phase pull diffusion). Periodically the source sends additional exploratory data messages to adjust gradients in the case of network changes (due to node failure, energy depletion, or mobility), temporary network partitions, or to recover from lost exploratory messages. Recovery from data loss is currently left to the application. While simple applications with transient data (such as sensors that report their state periodically) need no additional recovery mechanism, we are also developing retransmission scheme for applications that transfer large, persistent data objects [31].

This simplified description points out several key features of diffusion, and how it differs from traditional networking. First, diffusion is data-centric; all communication in a diffusion-based sensor network uses interests to specify named data. Second, all communication in diffusion is neighbor-to-neighbor or hop-by-hop, unlike traditional data networks with end-to-end communication. Every node is an "end" in a sensor network. A corollary to this previous observation is that there are no "routers" in a sensor network. Each sensor node can interpret data and interest messages. This design choice is justified by the task-specificity of sensor networks. Sensor networks are not general-purpose communication networks. Third, nodes do not need to have globally unique identifiers or globally unique addresses for regular operation. Nodes, however, do need to distinguish between neighbors. Fourth, because individual nodes can cache, aggregate, and more generally, process messages, it is possible to perform coordinated sensing close to the sensed phenomena. It is also possible to perform in-network data reduction, thereby resulting in significant energy savings. Finally, although our example describes a particular usage of the directed diffusion paradigm (a query-response type usage, see Figure 7.3), the paradigm itself is more general than that; we discuss several other usages next.

7.3.2 Push Diffusion

Two-phase pull diffusion works well for applications where a small number of sinks collects data from the sensor net, for example, a user querying a network for detections of some tracked object. Another class of applications involves sensor-to-sensor communication within the sensornet. A simple example of this class of application might have sensors operating at a low duty cycle most of the time, but when one sensor detects something it triggers nearby sensors to become more active and vigilant. Push diffusion was motivated by applications such as these being developed at Sensoria, University of Wisconsin, and PARC. A characteristic of this class of application is that there are many sensors interested in data

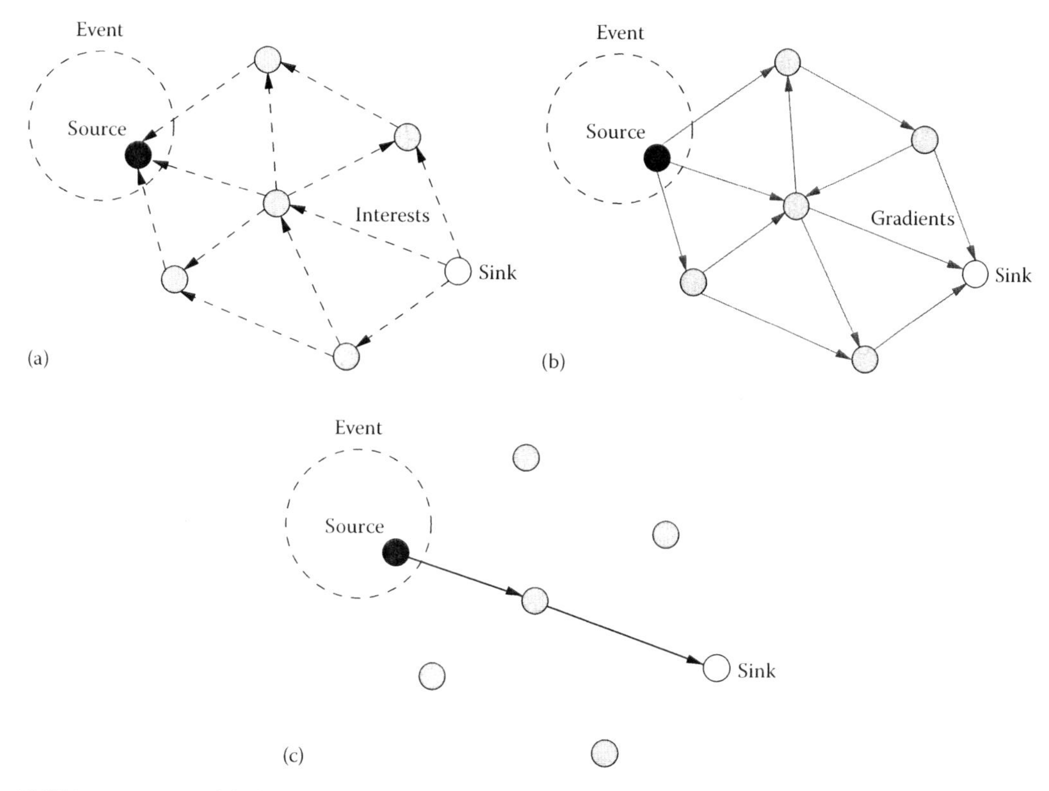

FIGURE 7.3 A simplified schematic for directed diffusion.

(activation triggers), and many that can publish such data, but the frequency of triggers actually being sent is fairly rare. Two-phase pull diffusion behaves poorly for this application, because all sensors actively send interests and maintain gradients to all other sensors even though nothing is detected.

One-phase push diffusion (or just push diffusion) was designed for this application. Although the API is the same as two-phase pull diffusion (except for a flag to indicate "push"), in the implementation, the roles of the source and sink are reversed. Sinks become passive, with interest information kept local to the node subscribing to data. Sources become active; exploratory data is sent throughout the network without interest-created gradients. As with two-phase pull, when exploratory data arrives at a sink a reinforcement message is generated and it recursively passes back to the source creating a reinforced gradient, and non-exploratory data follows only these reinforced gradients. Push can also take advantage of GEAR-style geographic optimizations.

Push is thus optimized for a different class of applications from two-phase pull: applications with many sources and sinks, but where sources produce data only occasionally. Push is not a good match for applications with many sources continuously generating data since such data would be sent throughout the network even when not needed. Section 7.5.4.1 presents a performance comparison of push and two-phase pull diffusion for such an application.

7.3.3 One-Phase Pull Diffusion

A benefit of push diffusion compared to two-phase pull is that it has only one case where information is sent throughout the network (exploratory data) rather than two (interests and exploratory data). In large networks without geographically scoped queries, minimizing flooding can be a significant benefit. Inspired by efficiency of pull for some applications, we revisited two-phase pull to eliminate one of its phases of flooding.

One-phase pull is a subscriber-based system that avoids one of the two phases of flooding present in two-phase pull. As with two-phase pull, subscribers send interest messages that disseminate through the network, establishing gradients. Unlike two-phase pull, when an interest arrives at a source it does not mark its first data message as exploratory, but instead sends data only on the preferred gradient. The preferred gradient is determined by the neighbor who was the first to send the matching interest, thus suggesting the lowest latency path. Thus one-phase pull does not require reinforcement messages, and the lowest latency path is implicitly reinforced.

One-phase pull has two disadvantages compared to two-phase pull. First, it assumes symmetric communication between nodes since the data path (source-to-sink) is determined by lowest latency in the interest path (sink-to-source). Two-phase pull reduces the penalty of asymmetric communication since choice of data path is determined by lowest-latency exploratory messages, both in the source-to-sink direction. However, two-phase pull still requires some level of symmetry since reinforcement messages travel reverse links. Although link asymmetry is a serious problem in wireless networks, many other protocols require link symmetry, including 802.11 and protocols that use link-level acknowledgments. As such, it is reasonable to assume that detecting and filtering such links will be done at the MAC layer, allowing one-phase diffusion to work.

Second, one-phase pull requires interest messages to carry a flow-id. Although flow-id generation is relatively easy (uniqueness can be provided by MAC-level addresses or probabilistically with random assignment and periodic reassignment), this requirement makes interest size grow with number of sinks. By comparison, though, with two-phase pull the number of interest messages grows with proportion to the number of sinks, so the cost here is lower. Second, the use of end-to-end flow-ids means that one-phase pull does not use only local information to make data dissemination decisions.

7.3.4 Using Geographic Cues to Limit Flooding

The physical nature of a sensor network's deployment makes geographically scoped queries natural. If nodes know their locations, then geographic queries can influence data dissemination, limiting the need for flooding to the relevant region.

GEAR (Geographic and Energy-Aware Routing) extends diffusion when node locations and geographic queries are present [34]. GEAR is an extension to existing diffusion algorithms that replaces network-wide communication with geographically constrained communication. When added to one-phase or two-phase pull diffusion, GEAR's subscribers actively send interests into the network. However, queries expressing interest in a region are sent *toward* that region using greedy geographic routing (with support for routing around holes); flooding occurs only when interests reach the region rather than sent throughout the whole network. Exploratory data is sent only on gradients set up by interests, so the limited dissemination of interests also reduces the cost of exploratory data.

For one-phase push diffusion, GEAR uses the same mechanism to send exploratory data messages containing a destination region towards that region. This avoids flooding by allowing data senders to push their information only to subscribers within the desired region, which in turn will send reinforcements resulting in future data messages following a single path to the subscriber. In Section 7.5.4.2, we present a field experiment showing a performance comparison of push diffusion with and without GEAR using the PARC IDSQ application.

We have also implemented GPSR [26] in the filter framework as an alternative to GEAR.

7.4 Facilitating In-Network Processing

Filters are our mechanism for allowing application-specific code to run in the network and assist diffusion and processing. Applications provide filters before deployment of a sensor network, or in principle filters could be distributed as mobile code packages at run-time. Filters register what kinds of data they handle through matching; they are then triggered each time that kind of data enters the node. When invoked, a filter can arbitrarily manipulate the message, caching data, influencing how or where it is sent onward, or generating new messages in response. Uses of filters include routing, in-network aggregation, collaborative signal processing, caching, and similar tasks that benefit from control over data movement, as well as debugging and monitoring.

Filters use only one-way matching. A message entering a node triggers a filter if the attributes specified by the filter match the attributes in the message, but it does not require matching in the other direction. This approach allows filters to process data more generally with the publish-subscribe API.

The filter core is the system component responsible for interconnecting all hardware devices, applications, and filters. Even though logically messages pass from filter to filter, in practice, all messages pass through the filter core, which shepherds messages from filter to filter, according to filter priorities.

Priorities, defined at filter configuration, give a total ordering of all filters in a system. While message attributes select which filters can process a message, priorities specify the *order* in which those filters act.

Priorities are needed because the attributes of an incoming message may match multiple filters. In this case, filter priorities indicate which filter is invoked first. As described later in Section 7.2, once a filter receives a message, it has total control over where the message will go next. A filter can pass the message to the next filter, modify the message and then send it, suppress it, generate messages in response of it, etc. Filters can also use the filter API to override the order of message processing by changing the priority field and/or messages attributes. Thus a knowledgeable filter can direct a message anywhere in the diffusion stack. Since the contents or priority can change any time a message leaves a filter, all messages are always sent to the filter core, not immediately to the next filter.

7.4.1 Implemented Filters

In this section we describe the set of filters that we have implemented and designed. As shown in Figure 7.4, the filter core interacts with all filters (rectangles), applications (circles at the top right), and radio hardware (the lozenge at the bottom). Solid and dashed rectangles represent existing and planned filters, respectively. The core is responsible for dispatching all messages as they pass through the system and for suppressing duplicate messages.

Basic diffusion is implemented in the *two-phase pull* filter. This filter maintains gradients representing the state of any existing flows to all neighbors and is responsible for forwarding data messages using reinforced paths, in addition to periodically send out reinforcement messages and interests.

GEAR is a pair of filters that can optionally surround the two-phase pull filter to implement GEAR [34]. Lacking prior information (such as geographic information or prior saved state), basic diffusion floods interests to all nodes in the network. GEAR overrides this behavior to forward messages with geographic assistance (interests are sent basically toward their geographic destination, but around any holes in the topology). GEAR consists of two filters, a *pre-processing* filter that sits above the two-phase pull filter to handle GEAR-specific beacon messages and to remove transient geographic information on arrival, and a *geographic routing* filter that acts after the two-phase pull filter to forward interests in a good direction.

Ben Greenstein and Xi Wang have each implemented versions of *GPSR* [26] as filters. GPSR, like GEAR, uses geographic information to make informed neighbor selection when forwarding packets. One implementation was done as an extension of diffusion (as described above), another as a standalone routing module (independent of diffusion).

Reliable Multi-Segment Transport (RMST) is a module that allows reliable transfers of large (multi-packet), uninterpreted data across unreliable links [31]. RMST is being used to investigate the trade-offs among MAC, transport, and application reliability. As a filter, it has two interesting characteristics. First, it caches data locally to support loss recovery, similar to approaches taken in reliable multicast [15] and SNOOP TCP [3], but at all hops rather than at the end-points or at

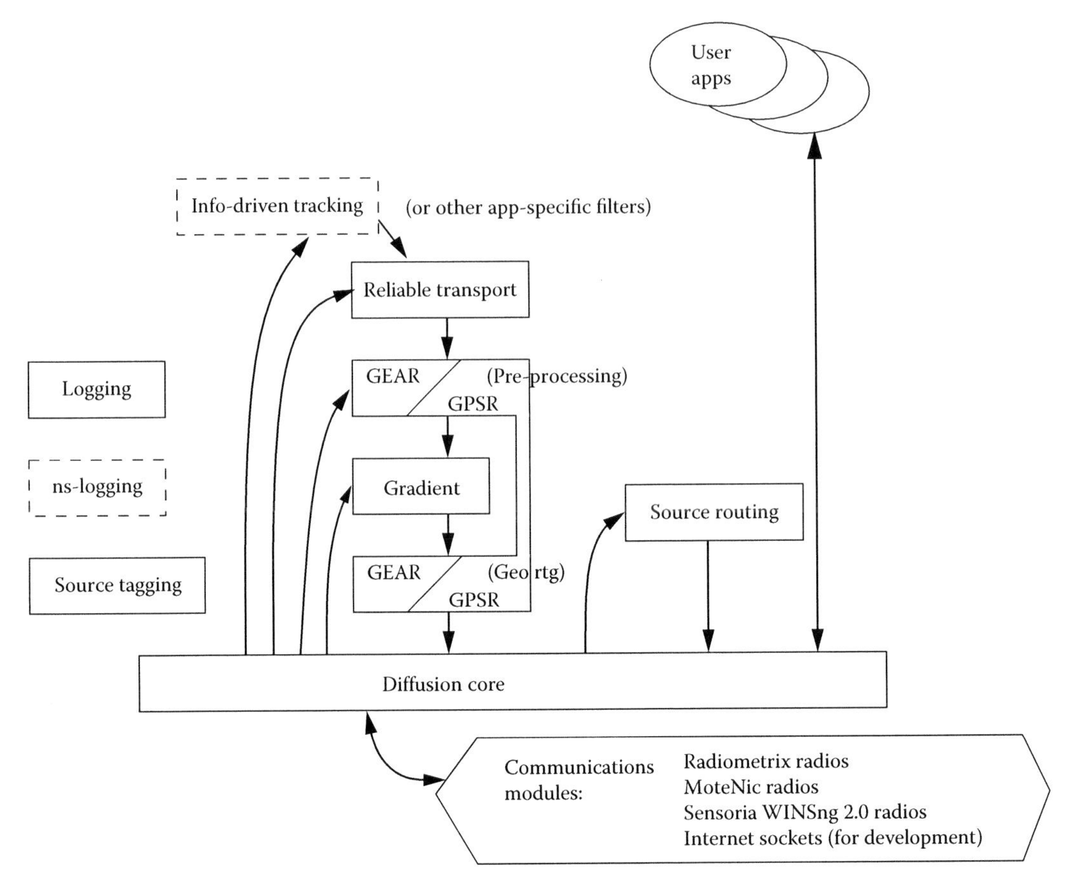

FIGURE 7.4 Current and planned filters in diffusion and how they interact.

base-stations only. Second, it implements a *back channel*, the reverse of the reinforced path created by the gradient filter. This back channel is used to propagate negative acknowledgment messages from the receiver to the sender.

The *information-driven tracking filter* is an example of how application-specific information can assist routing, proposed by researchers at Xerox PARC [35]. An important application of sensor networks is object tracking—multiple sensors may collaborate to identify one or more vehicles, estimating their position and velocity. Which sensors collaborate in this case is dependent on the direction of vehicle movement. They have proposed using current vehicle estimates (or "belief state") to involve the relevant sensors in this collaboration while allowing other sensors to remain inactive (conserving network bandwidth and battery power). While GEAR uses generic (geographic) information to reduce unnecessary communication, the information-driven tracking filter uses application-specific information to further reduce communication. As other applications are explored we expect to develop other application-specific filters similar to the information-driving tracking filter.

One use of filters is logging information for debugging. We have implemented a *logging* filter for this purpose, and we are considering implementing an *ns-logging* filter for simulator-specific logging. These filters are shown to the left of diffusion stack because they can be placed between any two modules.

Although this architecture was built to explore diffusion-style routing, for debugging purposes we also developed support for source routing. Source routing is provided as two filters: the *source tagging* filter functions similar to the logging filters in that it can be configured anywhere in the diffusion stack. This filter adds a record of each node that the message passes through, much like the traceroute command used on the Internet. The *source routing* filter provides the opposite function. It takes a message that includes an attribute listing the path of nodes the message should take through the network and dispatches it along that path. A design principle of directed diffusion is local operation—nodes should not need to know information about neighbors multiple hops away. While source routing is directly opposite to this goal, it can be provided within our software framework, and is still sometimes a useful debugging tool.

7.5 Evaluation

In this section we evaluate diffusion using simulations, and real-life experiments. We start by presenting simulation results showing the performance impact of diffusion. Then, we describe nested queries, a new query model that reduces end-to-end traffic by doing application-level aggregation using diffusion's in-network processing capabilities. Later, we show examples where application performance is highly affected by choosing the best diffusion algorithm.

7.5.1 Implementation Experience

Several implementations of diffusion have existed in simulation and on several hardware platforms. Diffusion was first implemented by Chalermek Intanagonwiwat in simulation with ns-2 [22]. The first implementations for native hardware were for Linux/x86 (desktop computers) and WINSng 1.0 sensor nodes running Windows CE (Figure 7.5c). More recent implementations added support for filters, PC/104 hardware with several kinds of radios (Figure 7.5c), and WINSng 2.0 nodes (Figure 7.5a) based on the SH-4 processor. The most recent release (3.2 as of this writing) includes nearly source-compatible support for the ns-2 simulator.

Researchers at UCLA have implemented *Tiny Diffusion*. Tiny-diffusion is a simplified version of diffusion, which runs on the resource-constrained Mica motes (Figure 7.5e) running TinyOS [20] with a limited amount of memory and processing power. Although it does not include support for filters, this simplified version does support attributes as well as a simplified version of the publish/subscribe API. Different versions of tiny diffusion have implemented both two-phase and one-phase pull diffusion.

7.5.2 Evaluation of Diffusion Design Choices

In this section, we use packet-level simulation to explore, in some detail, the implications of some of our design choices. Such an examination complements and extends our description for the two-phase pull diffusion from Section 7.3.1. This section describes our methodology, compares the performance of diffusion against some idealized schemes, then considers impact of network dynamics on simulation. Please refer to [22] for a more detailed description of the simulations.

7.5.2.1 Goals, Metrics, and Methodology

We implemented a vehicle tracking instance of directed diffusion in the ns-2 [2] simulator (the current ns release with diffusion support can be downloaded from http://www.isi.edu/nsnam/ns). Our goals in conducting this evaluation study were to: verify and complement our analytic evaluation, understand the impact of dynamics—such as node failures—on diffusion, and study the sensitivity of directed diffusion performance to the choice of parameters.

We choose two metrics to analyze the performance of directed diffusion and to compare it to other schemes: mean dissipated energy and mean delay. *Mean dissipated energy* measures the ratio of total dissipated energy *per node* in the network to the number of *distinct* events seen by sinks. This metric computes the mean work done by a node in delivering useful tracking information to the sinks. The metric also indicates the overall lifetime of sensor nodes. *Mean delay* measures the mean

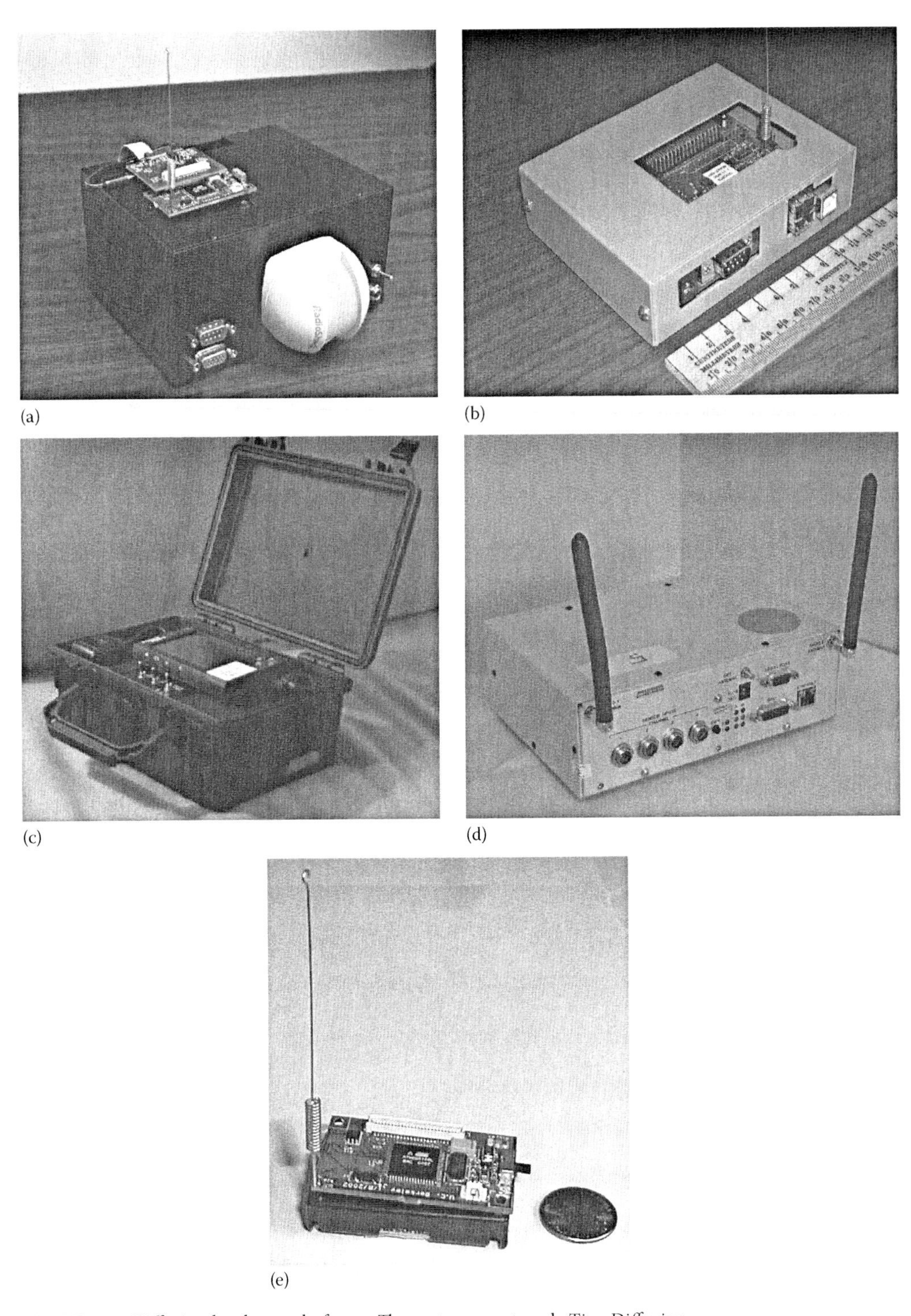

FIGURE 7.5 Diffusion hardware platforms. The mote supports only Tiny Diffusion.

one-way latency observed between transmitting an event and receiving it at each sink. This metric defines the temporal accuracy of the location estimates delivered by the sensor network. We study these metrics as a function of sensor network size.

In order to study the performance of diffusion as a function of network size, we generate a variety of sensor fields of different sizes. In each of our experiments, we study five different sensor fields, ranging from 50 to 250 nodes in increments of 50 nodes. Our 50 node sensor field generated by randomly placing the nodes in a 160 m by 160 m^2. Each node has a radio range of 40 m. Other sizes are generated by scaling the square and keeping the radio range constant in order to approximately *keep the average density of sensor nodes constant*. We do this because the macroscopic connectivity of a sensor field is a function of the average density. If we had kept the sensor field area constant but increased network size, we might have observed performance effects not only due to the larger number of nodes but also due to increased connectivity. Our methodology factors out the latter, allowing us to study the impact of network size alone on some of our mechanisms.

The ns-2 simulator implements a 1.6 Mbps 802.11 MAC layer. Our simulations use a modified 802.11 MAC layer. To more closely mimic realistic sensor network radios [25], we altered the ns-2 radio energy model such that the idle time power dissipation was about 35 mW, or nearly 10% of its receive power dissipation (395 mW), and about 5% of its transmit power dissipation (660 mW). This MAC layer is not completely satisfactory, since energy efficiency provides a compelling reasons for selecting a TDMA-style MAC for sensor networks rather than one using contention-based protocols [29]. Briefly, these reasons have to do with energy consumed by the radio during idle intervals; with a TDMA-style MAC, it is possible to put the radio in standby mode during such intervals. By contrast, an 802.11 radio consumes as much power when it is idle as when it receives transmissions. In Section 7.5.2.3, we analyze the impact of a MAC energy model in which listening for transmissions dissipates as much energy as receiving them.

Finally, data points in each graph represent the mean of 10 scenarios with 95% confidence intervals. Please refer to Ref. [22] for a more detailed description of the methodology used.

7.5.2.2 Comparing Diffusion with Alternatives

Our first experiment compares diffusion to omniscient multicast and flooding scheme for data dissemination in networks. Figure 7.6a shows the average dissipated energy per packet as a function of network size. Omniscient multicast dissipates a little less than a half as much energy per packet per node than flooding. It achieves such energy efficiency by delivering events along a single path from each source to every sink. Directed diffusion has noticeably better energy efficiency than omniscient multicast. For some sensor fields, its dissipated energy is only 60% that of omniscient multicast. As with omniscient multicast, it also achieves significant energy savings by reducing the number of paths over which redundant data is delivered. In addition, diffusion benefits significantly from *in-network aggregation*. In our experiments, the sources deliver identical location estimates, and intermediate nodes *suppress* duplicate location estimates. This corresponds to the situation where there is, for example, a single vehicle in the specified region.

Figure 7.6b plots the average delay observed as a function of network size. Directed diffusion has a delay comparable to omniscient multicast. This is encouraging. To a first approximation, in an uncongested sensor network and in the absence of obstructions, the shortest path is also the lowest delay path. Thus, our reinforcement rules seem to be finding the low delay paths. However, the delay experienced by flooding is almost an order of magnitude higher than other schemes. This is an artifact of the MAC layer: to avoid broadcast collisions, a randomly chosen delay is imposed on all MAC broadcasts. Flooding uses MAC broadcasts exclusively. Diffusion only uses such broadcasts to propagate the initial interests. On a sensor radio that employs a TDMA MAC-layer, we might expect flooding to exhibit a delay comparable to the other schemes.

In summary, directed diffusion exhibits better energy dissipation than omniscient multicast and has good latency properties.

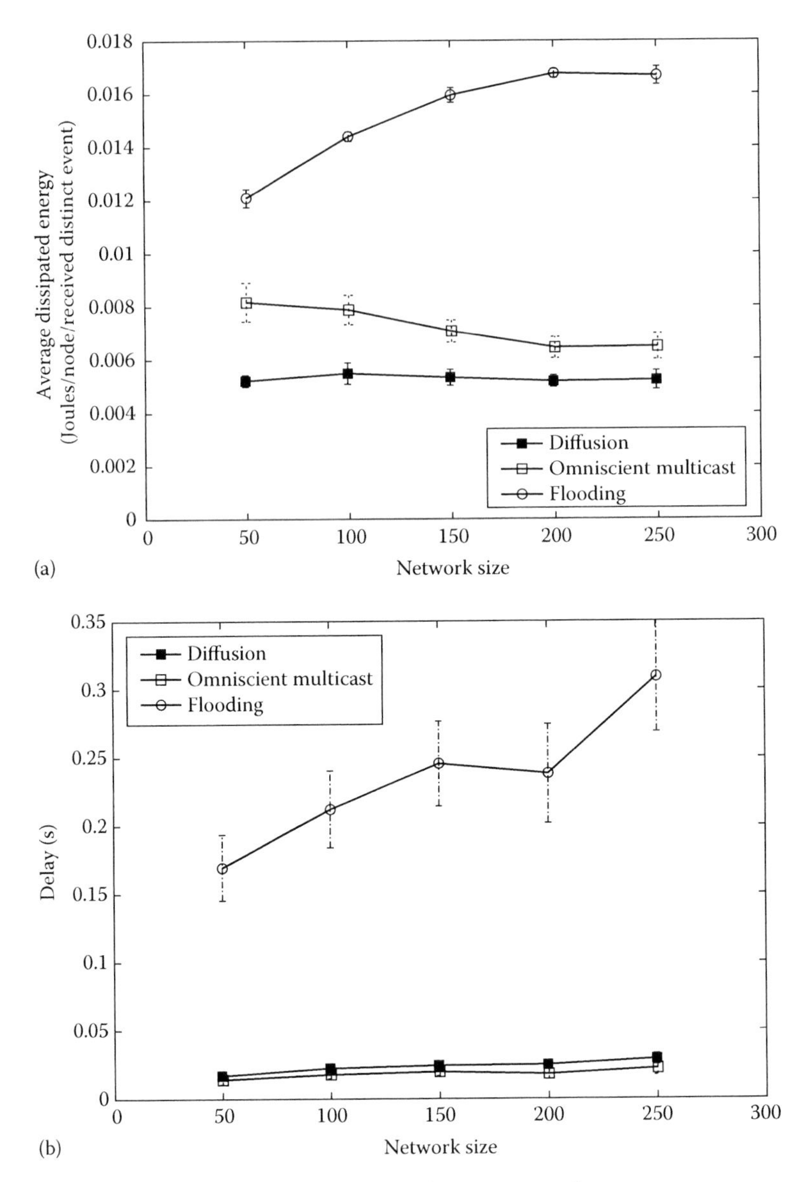

FIGURE 7.6 Directed diffusion compared to flooding and omniscient multicast.

7.5.2.3 Effects of Data Aggregation

To explain what contributes to directed diffusion's energy efficiency, we now describe two separate experiments. In both of these experiments, we do not simulate node failures. First, we compute the energy efficiency of diffusion with and without aggregation. Recall from Section 7.5.2.2 that in our simulations, we implement a simple aggregation strategy, in which a node suppresses identical data sent by different sources. As Figure 7.7a shows, diffusion expends nearly five times as much energy, in smaller sensor fields, as when it can suppress duplicates. In larger sensor fields, the ratio is 3. Our conservative negative reinforcement rule accounts for the difference in the performance of diffusion without suppression as a function of network size. With the same number of sources and sinks, the larger network has longer alternate paths. These alternate paths are truncated by negative reinforcement because they consistently deliver events with higher latency. As a result, the larger

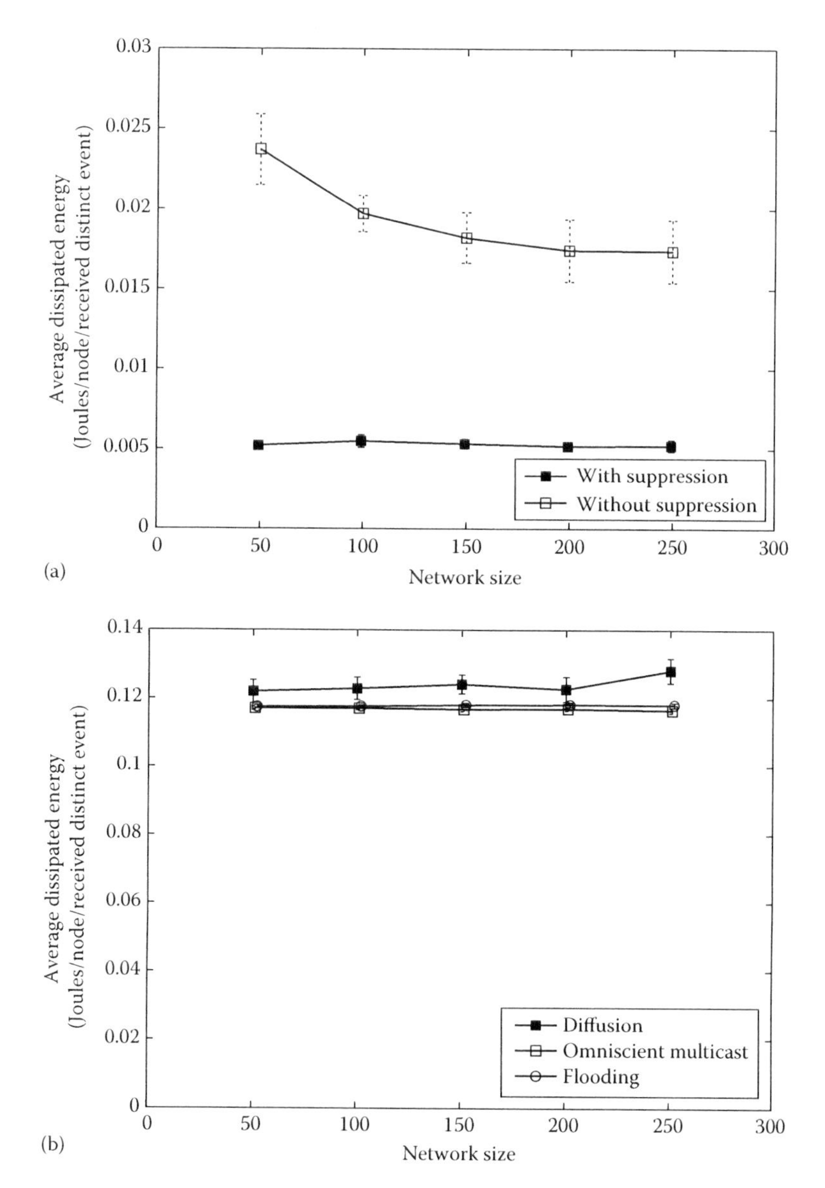

FIGURE 7.7 Impact of various factors on directed diffusion.

network expends less energy without suppression. We believe that suppression also exhibits the same behavior, but the energy difference is relatively small.

7.5.2.4 Effects of Radio Energy Model

Finally, we evaluate the sensitivity of our comparisons (Section 7.5.2.2) to our choice of energy model.

Sensitivity of diffusion to other factors (numbers of sinks, size of source region) is discussed in greater detail in Ref. [23].

In our comparisons, we selected radio power dissipation parameters to more closely mimic realistic sensor radios [25]. We re-ran the comparisons of Section 7.5.2.2, but with power dissipation comparable to the AT&T Wavelan: 1.6 W transmission, 1.2 W reception and 1.15 W idle [32]. In this case, as Figure 7.7b shows, the distinction between the schemes disappears. In this regime, we are better off flooding all events. This is because idle time energy utilization completely dominates the performance of all schemes. This is the reason why sensor radios try very hard to minimize listening for transmissions.

7.5.3 Evaluation of In-Network Processing

Real-world events often occur in response to some environmental change. For example, a person entering a room is often correlated with changes in light or motion, or a flower's opening with the presence or absence of sunlight. Multi-modal sensor networks can use these correlations by triggering a secondary sensor based on the status of another, in effect nesting one query inside another. Reducing the duty cycle of some sensors can reduce overall energy consumption (if the secondary sensor consumes more energy than the initial sensor, e.g., as an accelerometer triggering a GPS receiver) and network traffic (e.g., a triggered imager generates much less traffic than a constant video stream). Alternatively, in-network processing might choose the best application of a sparse resource (e.g., a motion sensor triggering a steerable camera).

Figure 7.8 shows two approaches for a user to cause one sensor to trigger another in a network. In both cases we assume sensors know their locations and not all nodes can communicate directly. Part (a) shows a direct way to implement this: the user queries the initial sensors (small squares), when a sensor is triggered, the user queries the triggered sensor (the small gray circle). The alternative shown in part (b) is a nested, two-level approach where the user queries the triggered sensor which then sub-tasks the initial sensors. This nested query approach grew out of discussions with Philippe Bonnet and embedded database query optimization in his COUGAR database [6].

The advantage of a nested query is that data from the initial sensors can be interpreted directly by the triggered sensor, rather than passing through the user. In monitoring applications the initial and triggered sensors would often be quite close to each other (to cover the same physical area), while the user would be relatively distant. A nested query localizes data traffic near the triggering event rather than sending it to the distant user, thus reducing network traffic and latency. Since energy-conserving networks are typically low-bandwidth and may be higher-latency, reduction in latency can be substantial, and reductions in aggregate bandwidth to the user can mean the difference between an overloaded and operational network. The challenges for nested queries are how to robustly match the initial and triggered sensors and how to select a good triggered sensor if only one is desired.

Implementation of direct queries is straightforward with attribute-addressed sensors. The user subscribes to data for initial sensors and when something is detected he requests the status of the triggered sensor (either by subscribing or asking for recent data). Direct queries illustrate the utility of predefined attributes identifying sensor types. Diffusion may also make use of geography to optimize routing.

Nested queries can be implemented by enabling code at each triggered sensor that watches for a nested query. This code then sub-tasks the relevant initial sensors and activates its local triggered sensor on demand. If multiple triggered sensors are acceptable but there is a reasonable definition of which one is best (perhaps, the most central one), it can be selected through an election algorithm. One such

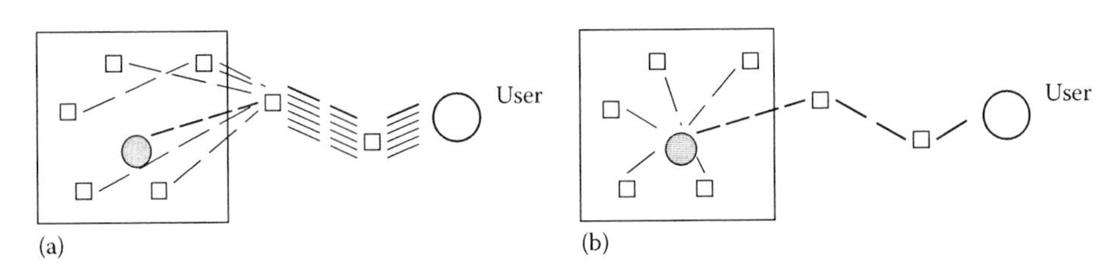

FIGURE 7.8 Two approaches to implementing nested queries. Squares are initial sensors, gray circles are triggered sensors, and the large circle is the user. Thin dashed lines represent communication to initial sensors; bold lines are communication to the triggered sensor.

algorithm would have triggered sensors nominate themselves after a random delay as the "best," informing their peers of their location and election (this approach is inspired by SRM repair timers [16]). Better peers can then dispute the claim. Use of location as an external frame of reference defines a best node and allows timers to be weighted by distance to minimize the number of disputed claims.

In the next section we evaluate nested queries with experiments in our testbed.

7.5.3.1 Goals and Methodology

To validate our claim about the potential performance benefits of this implementation we measure the performance of an application that uses nested queries against one that does not.

The application is similar to that described in Figure 7.8: a user requests acoustic data correlated with (triggered by) light sensors. For this experiment, we used our testbed of 14 PC/104 sensor nodes distributed on two floors of ISI (Figure 7.9). These sensors are connected by Radiometrix RPC modems (off-the-shelf, 418 MHz, packet-based radios that provide about 13 kbps throughput) with 10 dB attenuators on the antennas to allow multi-hop communications in our relatively confined space. The exact topology varies depending on the level of RF activity, and the network is typically five hops across. We placed the user "U" at node 39, the audio sensor "A" at node 20, and light sensors "L" at nodes 16, 25, 22, and 13. It is one hop from the light sensors to the audio sensor, and two hops from there to the user node. To provide a reproducible experiment we simulate light data to change automatically every minute on the minute. Light sensors report their state every 2 s (no special attempt is made to synchronize or desynchronize sensors). Audio sensors generate simulated audio data each time any light sensor changes state. Light and audio data messages are about 100 bytes long.

7.5.3.2 Nested Queries Benefits

Figure 7.10 shows the percentage of light change events that successfully result in audio data delivered to the user. (Data points represent the mean of three 20-min experiments and show 95% confidence intervals.) The total number of possible events are the number of times all light sources change state and a successful event is audio data delivered to the user. These delivery rates do not reflect per-hop message delivery rates (which are much higher), but rather the cumulative effect of sending best-effort data across three or five hops for nested or flat queries, respectively.

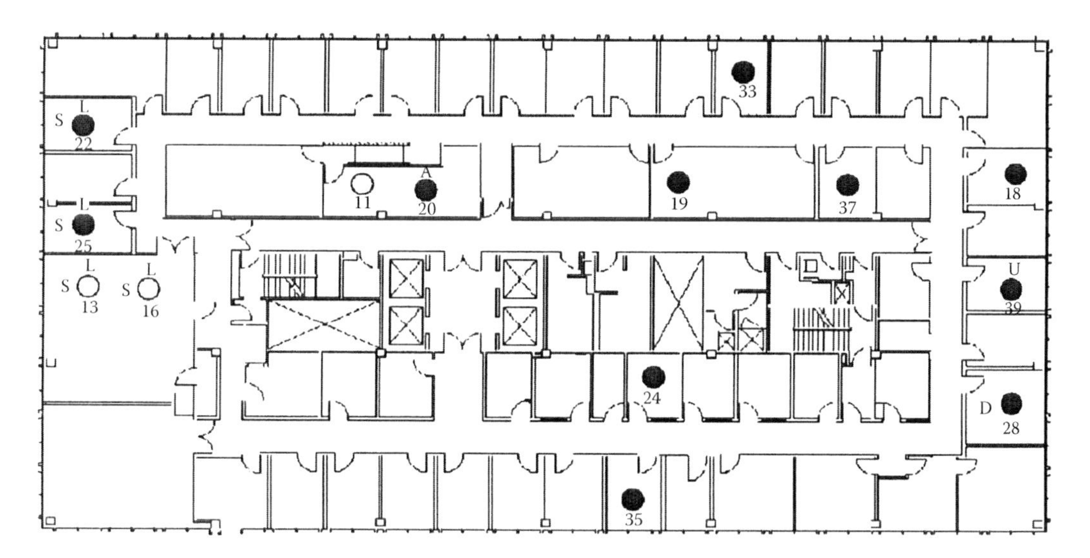

FIGURE 7.9 Node positions in our sensor testbed. Light nodes (11, 13, 16) are on the 10th floor; the remaining dark nodes are on the 11th floor. Radio range varies greatly depending on node position, but the longest stable link was between nodes 20 and 25.

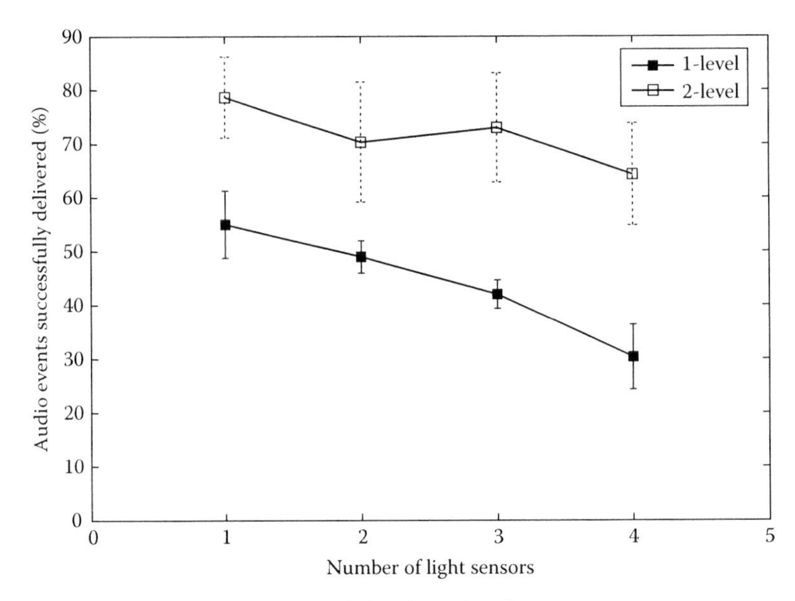

FIGURE 7.10 Percentage of audio events successfully delivered to the user.

This system is very congested and the network exhibits very high loss rates. Our current MAC is quite unsophisticated, performing only simple carrier detection and lacking RTS/CTS or ARQ. Since all messages are broken into several 27-byte fragments, loss of a single fragment results in loss of the whole message, and hidden terminals are endemic to our multi-hop topology, this MAC performs particularly poorly at high load. Missing events translate into increased detection latency. Although a sensor network could afford to miss a few events (since they would be retransmitted in the next time the sensor is measured), these loss rates are unacceptably high for an operational system.

However, this experiment sharply contrasts the bandwidth requirements of nested and flat queries. Even with one sensor the flat query shows significantly greater loss than the nested query because both light and audio data must travel to the user. Both flat and nested queries suffer greater loss when more sensors are present, but the one-level query falls off further. Comparing the delivery rates of nested queries with one-level queries shows that localizing the data to the sensors is very important to parsimonious use of bandwidth. In an uncongested network we expect that nested queries would allow operation with a lower level of data traffic than one-level queries and so would allow a lower radio duty cycle and a longer network lifetime.

7.5.4 Application Performance with Different Diffusion Algorithms

In Section 7.3 we described a series of diffusion algorithms that were designed in response to application needs. This section describes two applications developed or inspired by other researchers that benefit from push and GEAR, and it quantifies the performance gains in switching diffusion algorithms.

7.5.4.1 One-Phase Push vs. Two-Phase Pull Diffusion

Our first application considers trade-offs in push against two-phase pull versions of diffusion. In two-phase pull, data sinks are active, sending out interests, while sources are passive until interests arrive. By contrast, with push, data sources are active, sending out data when it arrives. Push is designed for the case when there are many active sinks (listening for data), but relatively few nodes actually generating data. A common case of this kind of application is where many nodes are *cross-subscribed* to each other but mostly quiescent, all waiting for a triggering event to happen.

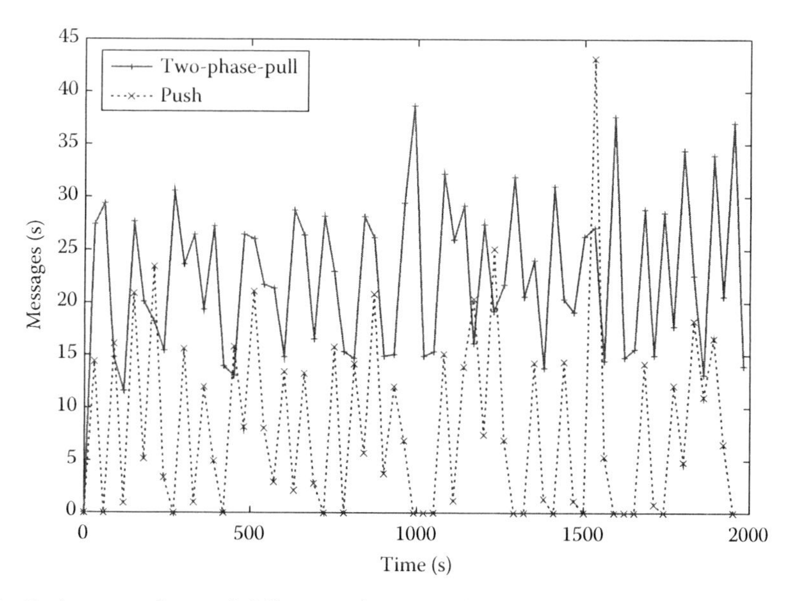

FIGURE 7.11 Push vs. two-phase pull diffusion with a cross-subscription application.

We explored this kind of application in the BAE sensor network testbed composed of 15 Sensoria WINSng 2.0 nodes. (These are 32-bit embedded computers with megabytes of memory and two independent, frequency-hopping radios that send data at about 20 kbps.) The application was inspired by applications at University of Wisconsin and PARC that employ cross-subscription. However, because those applications were not available to us at the time, we implemented a comparable application with a field of seven sensor nodes, all cross-subscribed to each other. When any one sensor changes state, all sensors send their readings to a triggered node that aggregates these readings and sends the aggregated result to the user. To control traffic, sensors were set to generate readings every 5 s and to change state every minute.

Figure 7.11 shows a trace of communication rates across this experiment, where each point represents the number of packets sent over the last 30 s. Two things stand out about this graph. First, the application's traffic is quite bursty. Second, push (the dotted line) is able to consistently out-perform two-phase pull (the solid line), transferring the same data with about 60% fewer messages.

Part of the saving in this experiment is because push is better suited to this application than two-phase pull. With many nodes cross-subscribed to each other, each will be frequently sending out interested messages to the network. With push, these interests are not sent; the only flooded messages are exploratory data.

If the sensors pushed relatively few detection events, the benefits of push would be greater still. In this case, data is sent every 5 s from each sensor to the others and so sensors are not quiescent.

7.5.4.2 Geographic Constraints

Researchers at Xerox PARC have suggested Information Driven Sensor Querying [13], an information-theoretic approach to sensornet tracking. With their approach, one node (the leader) keeps track of the current target estimate. It periodically computes which other sensor can add the most information about the target location and then transfers leadership to that node through a process called *state transfer*. To keep system state consistent, leader election includes a suppression process where a leader informs other nodes not to become active, duplicate leaders themselves. Suppression messages are sent when the target is first detected and as it moves through the network. State transfer messages occur twice each second.

This application should benefit from push in the same way as the previous application (Section 7.5.4.1). In addition, suppression and state transfer are both geographically-scoped actions. To investigate the

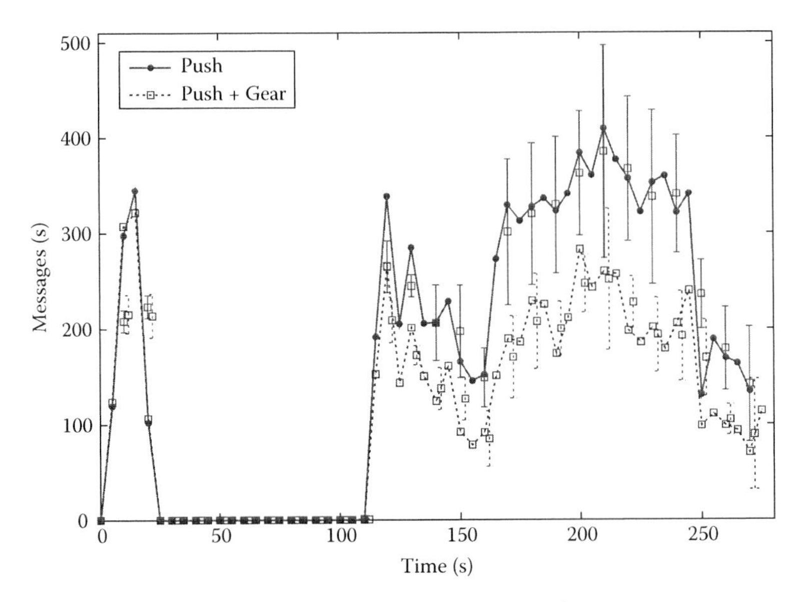

FIGURE 7.12 Push diffusion with and without GEAR over the IDSQ application.

benefits of geographically scoped communications jointly with them we evaluated this application both with and without GEAR [34]. This application runs over 18 WINSng 2.0 nodes in the PARC sensor network testbed. Sensor data in this case is generated by one or two human pulling a cart with pre-recorded acoustic data mimicking a large vehicle. The first simulated vehicle starts at 120 s, the second at about 170 s.

Figure 7.12 shows the message rates for this application. As can be seen, geographic scoping reduces message counts by 40%. This reduction is due to scoping of suppression messages. State transfer messages in this application are sent to a single point and so are also geographically directed, however this early implementation of push with GEAR did not support constraint of messages to a single point, only to regions, and so state transfer messages were flooded. We would expect a larger reduction in control overhead now that push with GEAR constrains control traffic directed to a point.

7.5.4.3 Discussion

These case studies illustrate the importance of matching the application to an appropriate data dissemination algorithm.

They also illustrate the complexity of selecting the best algorithm for a given application. Application designers are experts in their field, not networking, and so do not always have the best perspective to chose between several similar algorithms. The effects of selecting a diffusion algorithm can easily be masked by application errors. Our comparison of algorithms below is a first step to provide guidance to application designers, but an important area of future work is tools to help visualize and debug communication patterns in distributed, sensor-network applications.

To some extent it is a misstatement to suggest that there is a best algorithm for a single application. A sophisticated application like IDSQ has different patterns of communication in different parts of the application, and so requires *different* diffusion algorithms for different parts of the application. This supports our claim that a range of general and application-specific communication protocols are required for efficient data dissemination in sensor networks, both for different applications, and even in a single application.

A more specific result of these field studies concerns the appropriate means to select between algorithms. We had originally assumed that diffusion could infer the correct algorithm from the user's commands. For example, if geographic information was present, GEAR optimizations would be used. This approach proved too fragile for several reasons. First, it is prone to error. A misconfigured set of attributes can be syntactically correct but will not select the intended algorithm. The application will still run, but at greatly reduced performance. This problem is quite difficult to identify and correct, because performance of a distributed system can be difficult to measure, poor performance can be due to many causes, and the difference between correct code and incorrect is subtle. Second, as the number of alternative algorithms grow, it is no longer possible to distinguish between them automatically. Often the choice between algorithms depends on characteristics of the application known only to the programmer such as the communications patterns. A self-tuning system would be ideal, but collecting information for tuning requires communication itself and so will add its own overhead. For these reasons we now select algorithms explicitly as an attribute to publish and subscribe calls. We view the algorithm attribute as a programmer-provided assertion, much as annotations are used in distributed-shared-memory systems (e.g., Munin [11]).

7.6 Related Work

Space constraints preclude a detailed summary of related work; for a more detailed study, see the related work sections of prior papers [17,18,22,24].

Our publish/subscribe API was designed with researchers from MIT's Lincoln Labs [14]. The approach was inspired by prior, Internet-based publish/subscribe systems (e.g., include [5,12,27,28]). The concept of formals and actuals in matching is derived from Linda [10], and our application of attribute based naming to sensor networks was inspired by SRM [16].

The diffusion approach to data dissemination can be compared to ad hoc routing protocols (Broch et al. survey several protocols [9] including DSR and AODV). Unlike end-to-end Internet protocols, diffusion encourages in-network processing. In-network processing in diffusion is similar to active networking [33], although the domain is quite different. DataSpace provided geographic routing [21], but not tied with attributes.

In sensor networks, Piconet provided a fairly static system with devices, concentrators, and hosts [4]. SPIN evaluates several variants of flooding for wireless sensor networks [19], and INS designed a naming system for Internet-based hosts [1]. Neither exploited in-network processing. COUGAR adopted database-like approach to sensor networks [6] and inspired our approach to nested queries.

7.7 Conclusion

We have described directed diffusion, a data-centric approach to information dissemination for sensor networks. Building on a publish/subscribe API and attribute-based naming, the diffusion primitives support a family of routing algorithms optimized for different applications. Filters support in-network processing to allow applications to manipulate data as it flows through the network.

Acknowledgments

Directed diffusion builds on the work of many people. Van Jacobson suggested the concept of diffusion as a communication strategy. Chalermek Intanagonwiwat designed and evaluated the basic algorithm. Dan Coffin collaborated in definition of the publish/subscribe API. Philippe Bonnet inspired our approach to nested queries. Yan Yu, Fred Stann, Ben Greenstein, Xi Wang developed filters. Philippe Bonnet and Joe Reynolds were early users of diffusion, Jim Reich and Julia Liu early users of push, and Eric Osterweil of one-phase pull diffusion.

References

1. W. Adjie-Winoto, E. Schwartz, H. Balakrishnan, and J. Lilley. The design and implementation of an intentional naming system. In *Proceedings of the 17th Symposium on Operating Systems Principles*, pp. 186–201, Kiawah Island, SC, December 1999. ACM Press, New York.

2. S. Bajaj, L. Breslau, D. Estrin, K. Fall, S. Floyd, P. Haldar, M. Handley et al. Improving simulation for network research. Technical Report 99-702b, University of Southern California, Los Angeles, CA, March 1999. Revised September 1999, to appear in *IEEE Computer*.
3. H. Balakrishnan, S. Seshan, E. Amir, and R. H. Katz. Improving TCP/IP performance over wireless networks. In *Proceedings of the First ACM Conference on Mobile Computing and Networking*, pp. 2–11, Berkeley, CA, November 1995. ACM Press, New York.
4. F. Bennett, D. Clarke, J. B. Evans, A. Hopper, A. Jones, and D. Leask. Piconet: Embedded mobile networking. *IEEE Personal Communications Magazine*, 4(5):8–15, October 1997.
5. K. P. Birman. The process group approach to reliable distributed computing. *Communications of the ACM*, 36(12):36–53, December 1993.
6. P. Bonnet, J. Gehrke, T. Mayr, and P. Seshadri. Query processing in a device database system. Technical Report TR99-1775, Cornell University, Ithaca, NY, October 1999.
7. R. Braden, T. Faber, and M. Handley. From protocol stack to protocol heap–role-based architecture. In *Proceedings of the ACM Workshop on Hot Topics in Networks I*, Princeton, NJ, October 2002. ACM Press, New York.
8. D. Braginsky and D. Estrin. Rumor routing algorithm for sensor networks. In *Proceedings of the First ACM Workshop on Sensor Networks and Applications*, pp. 22–31, Atlanta, GA, October 2002. ACM Press, New York.
9. J. Broch, D. A. Maltz, D. B. Johnson, Y.-C. Hu, and J. Jetcheva. A performance comparison of multi-hop wireless ad hoc network routing protocols. In *Proceedings of the ACM/IEEE International Conference on Mobile Computing and Networking*, pp. 85–97, Dallas, TX, October 1998. ACM Press, New York.
10. N. Carriero and D. Gelernter. The S/Net's Linda kernel. In *Proceedings of the 10th Symposium on Operating Systems Principles*, pp. 110–129, Orcas Island, WA, December 1985. ACM Press, New York.
11. J. B. Carter, J. K. Bennett, and W. Zwaenepoel. Implementation and performance of Munin. In *Proceedings of the 13th Symposium on Operating Systems Principles*, pp. 152–164, October 1991. ACM Press, New York.
12. A. Carzaniga, D. S. Rosenblum, and A. L. Wolf. Design and evaluation of a wide-area event notification service. *ACM Transactions on Computer Systems*, 19(3):332–383, August 2001.
13. M. Chu, H. Haussecker, and F. Zhao. Scalable information-driven sensor querying and routing for ad hoc heterogeneous sensor networks. Technical Report P2001-10113, XEROX Palo Alto Research Center, Palo Alto, CA, May 2001.
14. D. A. Coffin, D. J. Van Hook, S. M. McGarry, and S. R. Kolek. Declarative ad-hoc sensor networking. In *Proceedings of the SPIE Integrated Command Environments Conference*, San Diego, CA, July 2000. SPIE, Bellingham, WA. (Part of *SPIE International Symposium on Optical Science and Technology*.)
15. S. Floyd, V. Jacobson, C. Liu, S. McCanne, and L. Zhang. A reliable multicast framework for light-weight sessions and application level framing. In *Proceedings of the ACM SIGCOMM Conference*, pp. 342–356, Cambridge, MA, August 1995. ACM, New York.
16. S. Floyd and V. Jacobson. Link-sharing and resource management models for packet networks. *ACM/IEEE Transactions on Networking*, 3(4):365–386, August 1995.
17. J. Heidemann, F. Silva, and D. Estrin. Matching data dissemination algorithms to application requirements. In *Proceedings of the ACM SenSys Conference*, Los Angeles, CA, November 2003. ACM, New York.
18. J. Heidemann, F. Silva, C. Intanagonwiwat, R. Govindan, D. Estrin, and D. Ganesan. Building efficient wireless sensor networks with low-level naming. In *Proceedings of the Symposium on Operating Systems Principles*, pp. 146–159, Chateau Lake Louise, Banff, Alberta, Canada, October 2001. ACM, New York.
19. W. R. Heinzelman, J. Kulik, and H. Balakrishnan. Adaptive protocols for information dissemination in wireless sensor networks. In *Proceedings of the ACM/IEEE International Conference on Mobile Computing and Networking*, pp. 174–185, Seattle, WA, August 1999. ACM, New York.
20. J. Hill, R. Szewczyk, A. Woo, S. Hollar, D. Culler, and K. Pister. System architecture directions for network sensors. In *Proceedings of the Ninth International Conference on Architectural Support for Programming Languages and Operating Systems*, pp. 93–104, Cambridge, MA, November 2000. ACM, New York.

21. T. Imielinski and S. Goel. DataSpace: Querying and monitoring deeply networked collections in physical space. *IEEE Personal Communications*. Special Issue on Smart Spaces and Environments, 7(5):4–9, October 2000.
22. C. Intanagonwiwat, R. Govindan, and D. Estrin. Directed diffusion: A scalable and robust communication paradigm for sensor networks. In *Proceedings of the ACM/IEEE International Conference on Mobile Computing and Networking*, pp. 56–67, Boston, MA, August 2000. ACM, New York.
23. C. Intanagonwiwat, R. Govindan, and D. Estrin. Directed diffusion: A scalable and robust communication paradigm for sensor networks. Technical Report 00-732, University of Southern California, Los Angeles, CA, March 2000.
24. C. Intanagonwiwat, R. Govindan, D. Estrin, J. Heidemann, and F. Silva. Directed diffusion for wireless sensor networking. *ACM/IEEE Transactions on Networking*, 11(1):2–16, February 2003.
25. W. J. Kaiser. WINS NG 1.0 transceiver power dissipation specifications. Sensoria Corporation, San Diego, CA.
26. B. Karp and H. T. Kung. GPSR: Greedy perimeter stateless routing for wireless networks. In *Proceedings of the ACM/IEEE International Conference on Mobile Computing and Networking*, pp. 243–254, Boston, MA, August 2000. ACM, New York.
27. B. Oki, M. Pfluegl, A. Siegel, and D. Skeen. The information bus—An architecture for extensible distributed systems. In *Proceedings of the 14th Symposium on Operating Systems Principles*, pp. 58–68, Asheville, NC, December 1993. ACM, New York.
28. L. L. Peterson. A yellow-pages service for a local-area network. In *Proceedings of the ACM SIGCOMM Conference '87*, pp. 235–242, August 1987.
29. G. J. Pottie and W. J. Kaiser. Embedding the internet: Wireless integrated network sensors. *Communications of the ACM*, 43(5):51–58, May 2000.
30. F. Silva, J. Heidemann, and R. Govindan. *Network Routing Application Programmer's Interface (API) and Walk Through 9.0.1*. USC/Information Sciences Institute, Marina Del Rey, CA, December 2002.
31. F. Stann and J. Heidemann. Rmst: Reliable data transport in sensor networks. In *Proceedings of the First International Workshop on Sensor Net Protocols and Applications*, pp. 102–112, Anchorage, AK, April 2003. USC/Information Sciences Institute, IEEE, Fort Worth, TX.
32. M. Stemm and R. H. Katz. Measuring and reducing energy consumption of network interfaces in hand-held devices. *IEICE Transactions on Communications*, E80-B(8):1125–1131, August 1997.
33. D. L. Tennenhouse, J. M. Smith, W. D. Sincoskie, D. J. Wetherall, and G. J. Minden. A survey of active network research. *IEEE Communications Magazine*, 35(1):80–86, January 1997.
34. Y. Yu, R. Govindan, and D. Estrin. Geographical and energy aware routing: A recursive data dissemination protocol for wireless sensor networks. Technical Report TR-01-0023, University of California, Los Angeles, CA, 2001.
35. F. Zhao, J. Shin, and J. Reich. Information-driven dynamic sensor collaboration for tracking applications. *IEEE Signal Processing Magazine*, March 2002.

8

Data Security Perspectives

David W. Carman
Johns Hopkins Applied Physics Laboratory

8.1 Introduction

Why do we need data security? Data generated and communicated by a sensor network can be thought of as property. Sensor network owners own this property. Like a house or car, an owner wishes to protect property from theft, damage, or destruction. Sensor network owners usually seek at least three security goals: (1) *preventing data disclosure to non-owners*, (2) *preserving data authenticity*, and (3) *preserving data availability*. Sensor network owners may also wish to prevent the unauthorized use of the sensor network by others (i.e., *theft of service*).

The consequences of not securing data in a government or military environment are severe. Losing information security battles can cause loss of military initiative, territory, resources, or even life [Denning]. Failure to ensure data security for sensor intelligence systems can cause grave damage to national security, sometimes for decades to come.

What makes sensor network data security unique and challenging is the unattended nature of the nodes, and the severe energy and communications constraints under which they must operate. This chapter examines common threats to sensor network data, corresponding security requirements, and the constraints of sensor environments that affect data security. We examine approaches to address these requirements and identify security mechanisms applicable to this setting.

8.2 Threats

Threats are the potential for an adversary or the environment to prevent the sensor network owners from achieving their data security goals. Passive attacks are threats that do not reveal the attacker (e.g., eavesdropping), whereas active attacks, such as transmitting electronic information or noise, alter or prevent data reception at the receiver.

The primary threat to privacy is eavesdropping using a compatible communications device. This threat can be significant for networks using common protocols such as IEEE 802.11 where millions of users have compatible wireless LAN cards. Conversely, a system interconnected using a proprietary laser communications system will only be intercepted by very capable, well-financed, and determined adversaries.

Adversaries might use eavesdropping to thwart various sensor network goals, including avoiding detection by "counter-detecting" where the sensor network is located, learning when and where the sensor network has detected him, or learning the sensor network's capabilities (e.g., detection range). The adversary may obtain message contents from unencrypted communications, by thwarting any encryption of the data through cryptanalysis or other means, or by performing traffic analysis of the encrypted communications. Traffic analysis uses message size, frequency, source and other data to garner information without knowledge of the unencrypted data traffic.

Data authenticity can be threatened by both malicious adversaries and unintentional environmental noise in the communications medium. Adversaries may attempt to forge sensor network messages by fabricating or altering the data ultimately received by the receiver. Environmental noise will cause random data errors that may prevent reconstruction of the original message at the receiver.

Adversaries may deny communications service through active jamming or subversion of network routing protocols [Wood]. Jamming refers to the act of generating electronic noise to prevent successful reception of communications by one or more receivers. However, jamming is inherently dangerous to the adversary, since it exposes the transmitter's whereabouts. During Operation Iraqi Freedom, the perils of an adversary attempting active jamming attacks were reinforced when U.S. Air Force Major General Victor Renuart reported that six Global Positioning System (GPS) jammers were located and destroyed via coalition air strike [Wilkison].

Vulnerabilities in conventional ad hoc routing protocols provide the opportunity for adversaries to delay or prevent sensor network traffic from reaching its intended destination [Law, Papadimitratos, Murphy]. Adversaries can forge bogus routing information to cause all data to be forwarded to an eavesdropping node. Bogus routing information can make sensor nodes believe routing paths do not exist, when in fact they do. Compromise of the ad hoc routing protocol may even allow an adversary to usurp the sensor network to pass its own traffic.

8.3 Security Requirements

To counter the threats to the sensor network, we describe security requirements in terms of the security services they must provide. Although data privacy and message authentication are the two main goals, we additionally examine other objectives in achieving sensor network security. A detailed treatment of general security requirements and services is provided by Stallings [Stallings].

8.3.1 Confidentiality

Also called privacy, confidentiality is a commonly desired security service that thwarts eavesdropping by an adversary. Confidentiality is often provided by encrypting data using a cipher algorithm such as the Advanced Encryption Standard (AES) [NIST1]. Encryption obscures the data, preventing an eavesdropper from discovering the original values without the appropriate decryption key. However, simple encryption does not prevent an eavesdropper from learning the size of messages, the frequency of the messages, and sometimes even the source and destination.

If a data-owner wishes to prevent an eavesdropper from even knowing that communications is occurring, *low probability of detection* (LPD) characteristics may be desirable. LPD refers to a communications system's ability to thwart an adversary's detection of transmitted signals as information bearing. Similarly, *low probability of intercept* (LPI) refers to the ability to prevent an eavesdropper from determining the modulation, modulated data, or origin of detected communications. These properties are particularly desirable in military or covert intelligence environments where the inability to detect or locate the sensor network is necessary to ensure its survival.

8.3.2 Message Authentication

Also called data origin authentication, message authentication ensures that data has not been altered from its original source. The properties of data integrity and authentication of the source together comprise the message authentication security service.

Traditionally authentication binds an assertion (e.g., "Alice owes Bob $10") to a person (e.g., "Alice"). In sensor networks, authentication binds a sensor data assertion (e.g., "Sensor #123456 sees a tank at Lat X Lon Y") with either a general identity (e.g., "I'm a U.S. Army, 3rd Division sensor with the network-wide key") or a specific identity (e.g., "I'm sensor #123456 and only I have this particular key").

A further desirable property, that an authentic message is original and has not been replayed, is called *transaction authentication*. This property is often provided by additionally binding some time-varying parameter to the data and identity assertions (e.g., "Sensor #123456 sees a tank at Lat X Lon Y at date/time Z"). Anti-replay protection requires the sender to apply the time-varying parameter to the message, and for the receiver to correspondingly verify the parameter using its own counter state or onboard clock.

8.3.3 Availability

The ability of sensor networks to reliably provide their data to the sensor network owners is termed *availability*. Sensor networks must provide the means to communicate between nodes to perform networking, and to route data through the network to the intended destination. The ad hoc routing protocols that enable sensor data to transit multiple hops to their destination must be protected from subversion. Since adversaries can disrupt communications at multiple layers, security mechanisms to provide availability must be allocated at each vulnerable layer.

8.3.4 Nonrepudiation

When sensor data is used for financial or legal matters, sensor data owners may need to prove its authenticity to third parties. *Nonrepudiation* ensures that the originator of the sensor data message cannot later deny having signed the message. Digital signatures combined with public key infrastructures provide a way of satisfying such requirements.

8.4 Constraints

Security solutions for the sensor network environment are constrained by limited battery energy, limited computational and communications capability, and the unattended nature of sensor node devices.

8.4.1 Limited Battery Energy

The computations and communications required to perform security mechanisms consume energy, reducing the lifetime of sensors with non-rechargeable batteries. Security-related computations, particularly modular exponentiations associated with public key operations, can cause the microprocessor to delay returning to an energy-conserving sleep mode. The potentially larger impact of security is the additional transmit and receive energy consumed due to communications from key management, message authentication tags, and other communicated data.

8.4.2 Limited Computational Capability

Cost, size, and energy considerations encourage sensor device designers to choose microprocessors with the minimum requirements to perform sensor processing functions. Unfortunately, these limited processors are often ill-suited to performing the computationally intensive modular exponentiation operations needed for key management and authentication. As a result, security solutions need to take into account the latency impact of cryptographic computations.

8.4.3 Limited Communications Capability

Several factors limit the communications capability of sensor nodes, including

- *Near-earth communications paths* between ground-based sensors cause transmit signal energy to attenuate more rapidly then free space propagation [Asada]
- *Small omni-directional antennas* cause antenna gain reduction [Asada]
- *Limited battery energy* causes budgeting of transmit signal amplification
- *Multi-user interference*, which causes similar effects as jamming, limits the effective channel rate of each communicant
- *Hostile jamming*, or even just the threat of it, results in waveform, receiver processing, and signal-to-jammer-ratio designs that limit throughput

The result of these limitations is that the energy consumed per communicated bit is much greater than many conventional communication environments. When cryptographic protocols add more traffic to this communications-constrained system, the impact on energy consumption and latency can be dramatic. Latency in distributing key management information impacts the ability of the network to establish and re-establish keys in a timely fashion [Carman1].

8.4.4 Unattended Nature of Sensor Devices

Sensor devices will almost always be located somewhere beyond the immediate control of sensor owners. This unattended nature of sensor devices makes them vulnerable to physical compromise by adversaries. Compromise of the device may include extraction of the sensor data or cryptographic keys, and may even allow the adversary to operate the device in a manner of its choosing. Compromise of cryptographic keys must be appropriately factored in to the sensor data security architecture.

8.5 Architecting a Solution

Like most disciplines, architecting a data security solution is best done by breaking a large problem into many smaller solvable problems. For a sensor network, this is typically done by allocating requisite security services to various layers of the International Standards Organization's [ISO7498] Open Systems Interconnect (OSI) model [ISO]. How to establish and manage the keys for all of these security services must also be planned.

Architecting an efficient security solution is highly dependent on the unique attributes of each network's sensor devices, applications, and performance requirements. Key considerations for this design are whether sensor data fusion will be required and whether sensor devices can efficiently perform public key processing.

8.5.1 Physical Layer

Allocating security services to the physical layer is generally difficult since communications systems are routinely optimized for parameters such as throughput, error rate, bandwidth consumption, etc. Confidentiality is the sole security service routinely provided at the physical layer, and is seldom used outside the military. Sensor networks requiring LPD/LPI protection use transmission security (TRANSEC) techniques such as spreading the signal over a large bandwidth, reducing the transmit power, and/or "hopping" the frequency. When these techniques use a pseudorandom code or cryptographic key, the key management function must establish the secret value with all potential communicants prior to use.

Recently, techniques has been developed to provide authentication at the physical layer, including (1) RF watermarking, where the legitimacy of the sender can be verified, and (2) Message Authentication Streams [Carman2], where the sender and message data can be verified as authentically bound.

8.5.2 Link Layer

Sensor network security designs often allocate confidentiality and message authentication security services to the link layer. Link-layer protection protects the message from the sender to the receiver over each "hop" of a multi-hop communication. In sensor networks that fuse data at intermediate points, link-layer encryption provides protection of over-the-air traffic while allowing intermediate access for sensor data fusion. In contrast, network-layer encryption provided between two multi-hop endpoints generally prevents intermediate sensor network nodes from performing data fusion.

Message authentication at the link layer provides the important capability of verifying the legitimacy of a multi-hop message at each intermediate hop. If only network-layer message authentication is provided, errors or forgeries will generally not be detected until the message arrives at its end destination. If verification fails at the end destination, the resulting negative acknowledgement (NACK) and/or retransmission messages incur significant latency and consume considerable communications energy. Hop-by-hop verification helps maintain a reliable link layer that reduces end-to-end communications failures.

Both unicast and broadcast traffic can be protected at the link layer provided corresponding key management support is implemented. Unicast traffic is protected with a pairwise key shared by the sender and receiver. Broadcast traffic is protected by a group key shared by the sender and all receivers within reception range.

8.5.3 Network Layer

The network layer is a popular layer to allocate sensor network security services including confidentiality, message authentication, and availability. Confidentiality and message authentication can be provided for unicast traffic by protocols such as the popular IETF standard IPsec [RFC2401]. End-to-end confidentiality at the network layer prevents intermediate nodes from accessing the sensor data. However, in cases when sensor fusion is beneficial, end-to-end encryption must be avoided. The Internet Key Exchange (IKE) [RFC2409] protocol establishes IPsec keys for pairs of communicating nodes.

Multicast traffic can be similarly protected at the network layer, although corresponding protocol standardization is less mature than for unicast traffic. No single group key management protocol is optimal for all sensor networks. For large multicast groups, hierarchical key management protocols are usually preferable since they provide efficient additions and deletions of group members.

Availability is provided at the network layer through routing protocol security techniques. Routing protocol security is either an inherent element of the networking protocol, or a separately added protocol-specific security design. In either case, confidentiality and message authentication mechanisms are uniquely applied to each routing protocol to thwart eavesdropping and forgery by malicious network nodes. We do not examine each of these solutions here, and instead refer the reader to the many research efforts identified by Zhu [Zhu].

8.5.4 Transport Layer and Above

Most sensor networks will not have elaborate network applications operating above the network layer due to the device resource constraints. However, sensor network services above the network layer may require additional confidentiality, authentication, or access control protection.

The Transport Layer Security (TLS) Protocol standard [RFC2246], formerly called the Secure Sockets Layer (SSL) protocol, specifies protocols and mechanisms for key management, confidentiality, and message authentication for unicast traffic. As with IKE, TLS key management incurs a significant latency and communications energy penalty that may not be suitable for lower data rate sensor networks. The Wireless Application Protocol (WAP) Forum created a version of TLS [WTLS] that reduces transmitted data, providing a security protocol more suitable to the wireless environment.

8.5.5 Key Management

The primary security support service needed for sensor data security is *key management*. As the backbone for all security services, key management provides cryptographic keys needed for encryption, message authentication, non-repudiation, and other security services. Key management functionality occurs, and provides keys for security services, at multiple network layers.

Key management establishes keys for pairs of communicating nodes, small groups, large groups, or even the entire sensor network. To maximize sensor data security in an environment where sensor devices are often left unattended and strong physical security is uneconomical, encryption and message authentication keys should be shared amongst the fewest nodes possible. Thus, for a unicast communication, only the sender and receiver should possess the corresponding encryption and message authentication keys. For multicast communications, the group leader should carefully manage which nodes possess the group key, including changing the key when group members join and leave. By managing keys with this fine "granularity" and changing keys regularly, the effects of an enemy compromise can be mitigated to impact only a small portion of the sensor network. Depending on the key management mechanism, however, this enhanced security may incur considerable communications and computational complexity.

Key management can be provided by a variety of mechanisms including preloading, key transport, and key agreement. *Preloading* refers to the practice of inserting encryption and authentication keys into a device prior to it being deployed. *Key transport* involves one node securely communicating an encryption or message authentication key to one or more other nodes. *Key agreement* describes a process whereby two or more nodes engage in a protocol that establishes a common secret key. Whereas key management methods such as key transport and key agreement require some type of communications interchange to establish encryption and authentication keys between sensor nodes, traditional preloading schemes do not. Many practical key management solutions employ a mix of preloading and either key transport or key agreement.

If a key transport or key agreement scheme is used, some type of trust management mechanism must be employed. *Trust management* comprises the set of protocols and actions necessary to ensure parties are legitimate. Trust management steps of verifying exchanged certificates or other credentials are achieved in concert with key management as an integrated solution to establishing keys with legitimate communicants.

Physical layer TRANSEC keys are usually established via preloading, since key transport and key agreement cannot occur until communications are established. Link layer keys may be established in a variety of ways, dependent on the requirements and constraints of the given sensor network. Network and higher-layer keys are usually established via key transport or key agreement protocols such as IKE and TLS. These protocols often leverage public key infrastructures that contain certifying authorities or other trusted third parties. However, public-key-based key agreement protocols generally require several communications exchanges that incur a sizable latency and consume considerable communications energy.

8.6 Security Mechanisms

Once the required security services have been determined and their application within the communications architecture designed, the actual security mechanisms that perform the security services can be selected. Fortunately for sensor network security engineering, the maturing of the cryptography field has provided an array of algorithm choices for various security services. In this section, we discuss various security mechanisms and their suitability to sensor networks.

Despite the maturity of the cryptographic field, designing strong security mechanisms is difficult. A solid cryptography background might be sufficient to design an algorithm or protocol that has no obvious weaknesses, but almost all efforts fail to withstand the cryptanalysis of expert cryptologists. The prudent engineering approach is to use standardized algorithms and protocols for which quantitative security claims can be made. Beware the new protocol that has not been subjected to real scrutiny, or the home-grown algorithm that is patented or must remain a trade secret—such mechanisms almost always fail to provide security as strong and free as standardized mechanisms.

8.6.1 Low Probability of Detection/Interception

The communications capability of sensor devices will generally determine whether confidentiality can be provided by LPD/interception techniques. Spread spectrum modulation techniques, which are becoming common in even low-cost devices such as sensor nodes, are a practical method of providing LPD and LPI in the sensor network domain. For instance, direct sequence spread spectrum communications provides confidentiality when the data signal is modulated or "spread" by a high-rate non-linear pseudorandom noise (PN) code known only to the sender and receiver. The PN code is commonly loaded prior to deployment, but active key management techniques could be used as well. Security for broadcast messages can be provided by sharing the PN code with all potential friendly receivers.

Frequency hopping, where the carrier frequency of the modulated signal is periodically changed, provides a deterrent to enemy interception. For this technique to work, the sender and receiver must agree on a frequency hopping pattern and maintain accurate time synchronization. The adversary must listen on a wide range of frequencies and quickly synchronize to the signal to effectively eavesdrop. Security is achieved by hopping between frequencies in a manner unknown to the adversary, over a large bandwidth, and at sufficiently fast rates that an adversary can not gather any useful information.

In addition to modulation techniques, there are two general methods of reducing the amount of signal power received by the adversary. First, the common technique of reducing transmit power to conserve precious battery energy has the side benefit of reducing an adversary's ability to detect and intercept communications as well. Second, in the rare case where the sensor device employs a directional antenna, an adversary not in the main lobe of the transmit antenna gain will encounter a further reduction in received signal power.

8.6.2 Encryption

At the link, network, and higher layers, confidentiality is provided by transforming data using encryption algorithms. The sender employs an encryption algorithm and a key to convert the original data or *plaintext* into an unintelligible jibberish called *ciphertext*. This ciphertext is transmitted to the receiver without concern that an eavesdropper can derive any beneficial information. Upon receipt of the ciphertext, the receiver employs the corresponding decryption algorithm and key to transform the ciphertext back to plaintext. For multicast operations, the key may be established between the senders and multiple receivers.

Modern encryption algorithms come in two basic types: block ciphers and stream ciphers. Both types employ a secret key known only to the sender and receiver to transform data between plaintext and ciphertext. Secret keys that are 128 bits or more provide sufficient security for sensor data security.

Block ciphers encrypt one block of data at a time, usually 64 or 128 bits. *Stream ciphers* create a key stream that is exclusive-ORed with one bit or byte of data at a time to transform data from plaintext to ciphertext.

For most sensor encryption applications, employing a block cipher is better than a stream cipher. In general, block and stream ciphers encrypt and decrypt at roughly the same speeds. From a security perspective however, block ciphers such as Triple DES [FIPS46-3] and AES have been thoroughly studied for cryptographic weaknesses and are highly regarded within the research community. No stream ciphers have successfully withstood such scrutiny. This is not to say that stream ciphers such as RC4 are not secure, it is simply to point out that with all other considerations being equal, using well-analyzed block ciphers is the prudent approach. Furthermore, one of the unfortunate properties of stream ciphers is that repeated use of the same key stream allows an eavesdropper to compromise the encryption process. Although proper key management can prevent key stream reuse, there is generally no reason to choose stream ciphers over block ciphers for sensor network security.

8.6.2.1 Block Cipher Encryption Modes

Another important aspect of block cipher encryption is the choice of mode. Modes describe the method in which the block cipher algorithm and key are used to transform multiple blocks of plaintext to and from multiple blocks of ciphertext. Modes have different properties regarding error propagation, performance, and security. NIST has published a standard on the various block cipher encryption modes [NIST1].

The most basic block cipher mode is Electronic Code Book (ECB), where encryption is accomplished by transforming each block of plaintext into ciphertext and vice versa. If the amount of data to be encrypted does not equal an exact number of blocks, a simple padding scheme is usually used to complete the remainder of the plaintext block. Upon decryption, this padding is discarded.

An undesirable security property of ECB is that two identical plaintext blocks encrypted using the same key generate the same two identical ciphertext blocks. An eavesdropper that detects the repeated ciphertext blocks at least learns that the plaintext blocks were repeated. If the eavesdropper had learned the value of one of the plaintext blocks through other means, it could use the repetition of ciphertext blocks to learn the value of the other.

Cipher Block Chaining (CBC) is a block cipher encryption mode designed to overcome the repeated block vulnerability of ECB. CBC encrypts by combining the plaintext block with the previous ciphertext block in a "chain." To start the chain, an initialization vector (IV) is established by the sender and made available to the receiver—either implicitly or by explicitly transmitting it. With CBC, the chance that two identical plaintext blocks will encrypt to the same ciphertext block is astronomically small. Although CBC is probably the most popular block encryption mode in use today, it has two notable drawbacks: (1) encryption operations can not be performed in parallel since each block encryption requires the result of the previous block in the chain; and (2) messages that do not end on an even block boundary require padding as described for ECB above.

Counter Mode (CTR) is a relatively new block encryption mode that overcomes the two notable drawbacks of CBC. CTR uses the block cipher, a key, and a numerical counter to generate a key stream that is exclusive-ORed with the plaintext to generate the ciphertext. The receiver performs the same operation and transforms the ciphertext to plaintext. First, we note that the block encryption operations, other than the exclusive-OR, can be performed in parallel by the sender, since the input to the block cipher is known even before the plaintext. Similarly, if the receiver knows the counter and key beforehand, it could precompute the key stream before message receipt. The second benefit of CTR is that messages that do not end on an even block boundary need not be padded. This second benefit can be especially important in resource-constrained sensor networks where every additionally communicated bit consumes precious bandwidth and battery energy. The one major drawback of CTR mode is similar to a stream cipher; the key stream generated from a given key and counter can not be reused. This drawback can usually be easily mitigated through proper management of the counter values.

8.6.2.2 Block Encryption Algorithms

Choosing the best block encryption algorithm has become much easier with the National Institute of Standards and Technology's (NIST) adoption of the AES in 2001. The AES block encryption algorithm provides strong confidentiality with exceptional performance on a wide range of microprocessors. AES is faster and generally regarded as more secure than Triple DES, which in turn succeeded the venerable Data Encryption Standard (DES) some years prior [FIPS46-3].

The security of encryption algorithms is determined by key size and the strength of the underlying algorithm to resist cryptanalytic attack. Key sizes of 128 bits and greater are sufficient to prevent brute force searching for the correct key. Determining the quantitative security of the underlying encryption algorithm remains elusive, however. The best the current cryptographic community can currently achieve with most encryption algorithms is to demonstrate how a given algorithm is resistant to known cryptanalytic attacks such as linear [Matsui] and differential cryptanalysis [Biham]. Only after years of analysis by the research community are encryption algorithms such as AES regarded as secure.

Rare exceptions to choosing AES for encryption include some military applications and some extremely memory-limited or computationally-limited processors. In U.S. military applications, Type I classified algorithms might be used, but due to their unattended nature, use of classified algorithms in sensor devices will be rare. Although not a memory "hog," there may be extremely resource constrained devices for which AES is too big or slow, and a small footprint algorithm must be used instead. Great care must be taken when considering alternate encryption algorithms, since without the proper amount of cryptanalytic investigation, no guarantee regarding the security of the algorithm can be made.

8.6.3 Message Authentication

There are two main types of algorithms that provide message authentication: digital signatures and message authentication codes (MACs). *Digital signatures* are created by cryptographically hashing the message data and then signing the hash value using a public key algorithm. A cryptographic hash function takes a variable-sized data input and produces a small fixed-length output such as 160 bits. MD5 [RFC1321] was the preferred hash for many years, but security concerns [Dobbertin] have caused most cryptographic engineers to choose the slower but more secure SHA-1 [FIPS180-2] for use with digital signatures. One of the main benefits of traditional digital signatures over MACs is that the signer uniquely holds the private signing key, thus enabling nonrepudiation in addition to individual source authentication of the signed message. Digital signatures are performed over sensor data or sensor network routing information just before transmission, and are appended (or prepended) to the transmitted message.

8.6.3.1 Digital Signature Algorithms

The two most popular digital signature algorithms (DSAs) today are the DSA [FIPS186-2] and RSA [PKCS1]. Both are public key algorithms where the basis of security is the difficulty of computing discrete logarithms (DSA) and factoring (RSA). Like AES, DSA is a NIST standard that is widely used and well studied by the cryptanalytic community. The NIST Digital Signature Standard (DSS) explicitly specifies SHA-1 hash as the accompanying hashing algorithm and generates a 320-bit signature tag regardless of the size of the corresponding public key. A major drawback for using DSA for sensor data security is that the signature generation operation performed by the sender and signature verification operation performed by the receiver are computationally intensive. Even using modern high-speed processors, millions of multiply instructions must be executed for each DSA signature generation or verification operation. For the computationally-limited microcontrollers and embedded microprocessors of most sensor devices, DSA is untenable save for infrequent operations.

RSA signature generation also is computationally intensive and unattractive for routine message authentication. Unlike DSA however, RSA signature verification is over an order of magnitude easier to compute than signature generation, thus making RSA the DSA of choice when only verifications, such as checking certificates, will be performed by sensor devices. One additional drawback of RSA is its signature size, which is the same number of bits as the modulus, and is thus at least 1024 bits for strong security. Appending such a long signature to each message is unattractive in most sensor networks.

A variant of DSA based on computations performed over elliptic curves has been standardized by NIST and called ECDSA [FIPS186-2]. This public key DSA offers reduced computations over DSA without sacrificing security. Some methods of implementing elliptic curve operations are patented and the software to perform elliptic curve cryptography (ECC) is generally more complex than DSA and RSA, but ECDSA's speed and small key size makes it an attractive alternative for digital signature operations in sensor devices. Although computationally superior to DSA and RSA, ECDSA is still a computationally expensive operation for frequent sensor data security use.

8.6.3.2 Message Authentication Codes

MACs provide computational and bandwidth efficiency and required authentication for routine sensor data messages. Unlike public-key-based DSAs, MACs use a secret key shared by the sender and one or more receivers. There are two popular types of MACs: block-cipher-based MACs and hash-based-MACs.

Block-cipher-based MACs use a special block cipher mode to compute an authentication tag instead of ciphertext blocks. Cipher-block-chaining MACs (CBC-MACs) use the encryption algorithm, key, plaintext and the preceding ciphertext block to compute a ciphertext block to be fed into the next link of the chain. A portion of the final ciphertext block constitutes the authentication tag. The sender generates the authentication tag and appends it to the sensor data message. The receiver computes the authentication tag using the received data, and compares the computed and received authentication tags, determining the sensor data message to be authentic if the tags match. Although block-cipher-based MACs have existed for some time, their development remains an active field of cryptographic research as increased performance and provable security claims are pursued. NIST has proposed [NIST2] One-Key CBC MAC (OMAC) [Iwata] as a standard block-cipher-based MAC.

Hash-based MACs (HMACs) use cryptographic hash functions to compute the authentication tag. The sensor data to be authenticated is first hashed with a initial key to create an intermediate result. The intermediate result is then hashed with a second key to create an output that is the same size as the hash function output. The authentication tag is a truncation of the output generated by the second hash invocation.

HMACs are generally faster than block-cipher-based MACs since hash functions usually process data faster than block ciphers. Since most sensor processors are computationally limited, HMACs are the preferred choice over block-cipher-based MACs when only message authentication is being performed on the sensor data. However, if both encryption and message authentication are to be performed on a sensor data message, at least three block-cipher-based MACs [Jutla, Gligor, Rogaway] have been designed that perform both functions in a single pass through the data. Although these *authenticated-encryption* schemes are patent pending, they are attractive for sensor security use since they require fewer computations compared to separate invocations of a block cipher for encryption and an HMAC or other block-cipher-based MAC for message authentication. The memory constraints of sensor devices may also favor the use of a single cryptographic primitive, the block cipher, over using two cryptographic primitives, a block cipher for encryption and a hash function for HMAC-based message authentication.

Transaction authentication is provided by including a monotonically increasing counter or timestamp in the MAC computation. A 32- or 64-bit counter maintained and sent by the sender is usually sufficient to uniquely mark each outgoing sensor data message. The receiver keeps track of the received counters, making sure the counter value always increases. Received messages that use "old" counter values are assumed to be replay attacks and are discarded. Accommodations for out-of-order packets that occur due to disparate network routing paths can be made using windowing methods suggested by Kent and Atkinson [RFC2401]. The sender may also include a timestamp in each sensor data message that is verified by the receiver. Some level of clock synchronization is needed in this case to avoid discarding legitimate messages due to the sender or receiver having an inaccurate clock.

8.6.4 Key Management

A critical component of encryption and message authentication security is how the mechanism's key is established and maintained. A small key that is fixed for a long period of time and shared with a large number of sensor nodes provides less security than a large key that is changed often and shared among only the few sensor nodes that need it. Larger keys are more secure than smaller keys (to a point), since larger keys require an attacker to attempt more keys to guess the key by brute force. Most military systems use cryptoperiods, where a given cryptographic key is used for only a limited period of time. Active key management, where the key is established by the sensor nodes themselves while deployed, is an effective way of limiting access to the key.

8.6.4.1 Preloading

The simplest method of providing needed keys to sensor devices is *preloading* the entire sensor network with the same encryption and MAC keys. Preloading refers to the practice of loading keys into the sensor nodes prior to them being deployed. A major advantage of preloading is that little or no key management messages need be exchanged during deployment. However, there are several disadvantages of preloading, including the susceptibility to compromise when using a network-wide key and the inability to change keys in the advent of such a compromise. Two preloading schemes [Eschenauer1, Perrig] avoid use of a network-wide key by randomly predistributing many keys to subsets of nodes in a manner that probabilistically assures that any two communicants will share a common key, while dramatically reducing the potential exposure of each key. A disadvantage of the random preloading schemes is that discovery of which keys the other communicant possesses requires energy and time-consuming communications to occur.

8.6.4.2 Public Key Cryptography

To achieve a greater resistance to compromise, security architectures can forgo the use of a widely-deployed key and instead use key management protocols based on public key cryptography. A public key cryptography system uses two keys: a widely known public key and a private key known only by the possessor. The public and private keys are mathematically related, but in such a way that prevents an adversary from determining the private key from the public key. Most public key cryptography systems are based on one of three hard mathematical problems: (1) *integer factorization*, which is the security basis of RSA; (2) *discrete logarithm*, which is the security basis of Diffie–Hellman key agreement; or (3) *ECC*, which is the security basis of Elliptic Curve-based Diffie–Hellman key agreement.

RSA and Diffie–Hellman are the most popular public key methods and are the basis for most modern key management protocols such as IKE [RFC2409] and TLS [RFC2246]. These protocols interactively exchange public keys and other key management information and can be used to establish a key *after* sensor devices are deployed. Established keys are unique to the communicants that hold the private keys and participate in the protocol, and are thus less susceptible to disclosure. Key management protocols based on public key algorithms may be invoked more than once, changing encryption and authentication keys regularly to limit the amount of data protected, and thus limiting the amount of sensor data that could be compromised.

For public keys to be useful, nodes must trust the association between a public key and an entity. Modern key management protocols leverage the use of public key infrastructures that allow nodes to trust the binding between the identity of a node and its public key. This cryptographic binding of information such as identity, role, or other attributes is instantiated in a public key *certificate*. A unique certificate is generated for each sensor device by the sensor data owner, who acts as a *certifying authority* for the sensor network. Sensor devices usually obtain the public key certificate via an online exchange with the node to which the certificate refers. Eschenauer, Gligor, and Baras suggest alternate methods of generating, distributing, and discovering trust information [Eschenauer2]. Stajano and Anderson offer a radical approach that may be practical for sensor networks where nodes spend the majority of their time in an energy-conserving sleep state [Stajano].

Once trust is established, key agreement or key transport protocols may be used to establish keys. Key agreement allows both nodes to contribute key management information that is used to create the key. Mutual contributions prevent one node from causing a weak key to be used, whether intentionally or unintentionally. Key transport occurs when one node generates a secret key and sends it to another node. RSA is most often used to protect key transport by encrypting the secret key using the public key of the recipient. A version of Diffie–Hellman called ElGamal [ElGamal] may also be used for this purpose. Although key transport is not contributory as with key agreement, it can reduce the number of interactive key management messages that must be exchanged.

RSA and Diffie–Hellman require significant computations, making them unattractive for most sensor network applications. Both protocols perform modular exponentiations that require millions of multiply instructions on the computationally-limited microprocessors present in most sensor devices. As a result, RSA and Diffie–Hellman operations sometimes take seconds (or longer) to be performed, thus significantly contributing to the latency in establishing keys. Since most sensors return their processors to an energy-conserving sleep state when not being used, the additional computation time also consumes more energy.

A potentially larger issue than computations is the energy consumed and latency incurred due to the communications required by the RSA and Diffie–Hellman protocols. The public key certificates and other public key management information communicated require several thousands of bits to be exchanged. These communications consume both transmit energy by the sender and receive energy by the receiver. Moreover, a considerable latency is introduced due to the time required to send the information between nodes. This latency is actually much greater than the number of bits multiplied by the data rate since the effects of the media access control layer and retransmissions must also be considered. The sum of computational and communications effects on energy consumed and latency incurred usually render these protocols unsuitable for use in sensor networks.

ECC [Menezes1] is an attractive alternative to RSA and Diffie–Hellman for sensor key management. Diffie–Hellman key agreement based on elliptic curve operations requires fewer computations and communications for the same level of security. For instance, a 161-bit ECC key is roughly equivalent in security strength to a 1024-bit RSA key [Johnson]. Minimal certificates for ECC systems are about 62 bytes, whereas RSA systems are 256 bytes [Johnson]. ECC-based schemes have been added to IKE and other standard protocols as these techniques have matured and gained wider acceptance.

8.6.4.3 Identity-Based Cryptography

[Shamir] is especially attractive for use in sensor networks since it significantly reduces key management messaging. A trusted authority generates a private key for each sensor node based on its identity. Only the trusted authority knows how to perform this function. Sensor nodes derive public keys from sensor device identities, thus eliminating the exchange of public key certificates when executing a key management protocol. Maurer and Yacobi described a scheme [Maurer] that eliminates the need to exchange anything other than the identities. Although their scheme requires significant offline computations for private key generation, Matt [Matt] provides an overview of recently developed identity-based elliptic curve schemes that do not suffer from this problem.

8.6.4.4 Group Key Management

When encryption or message authentication is required to secure broadcast or multicast communications, a group key management protocol is needed. Few group key management protocols have been standardized, let alone targeted for sensor network environments. Group key agreement protocols such as Group Diffie–Hellman [Steiner] and Burmester–Desmedt [Burmester] require considerable communications, even when ECC keys are used. A group keying scheme that combines identity-based cryptography and group key transport was invented by Matt and examined for use in an Army sensor network [Carman3].

8.7 Other Sources

An excellent scientific treatment of the theory and application of cryptography can be found in *Handbook of Applied Cryptography* [Menezes2]. *Applied Cryptography* [Schneier] provides a very good introduction to cryptography with an encyclopedia of references to cryptographic algorithms, protocols, and applications.

8.8 Summary

How best to provide data security for a given sensor network is highly dependent on the unique requirements and constraints of the target environment. However, we have identified approaches and standard mechanisms that provide effective data security against a variety of threats to data security. Encryption and message authentication are best provided at the link and network layers, depending on whether sensor data fusion is being performed. When sensor data fusion is needed, hop-by-hop encryption and message authentication should be implemented. When data fusion is not required, hop-by-hop authentication and end-to-end encryption and message authentication are an effective multi-layer protection.

When selecting security algorithms, a compelling reason is needed *not* to use AES for encryption. HMAC-SHA-1, HMAC-MD5, or AES-based OMAC provides strong authentication of routine sensor data messages when keys of at least 128 bits are used. ECDSA should be used to sign data during infrequent but important operations such as key management, routing, and special sensor data operations.

Key management and routing protocol security are challenging due to the limited bandwidth, energy and unattended nature of nodes in the sensor network environment. Strong key management requires occasional exchange of key management messages and public key computations, which consume energy and incur latency in resource-constrained sensor devices. ECC, whether used in conjunction with a public key infrastructure or using an identity-based scheme, provides an efficient alternative to the popular RSA and Diffie–Hellman algorithms. Group key management, needed to secure link-layer broadcast or network-layer multicast, is even more challenging. Conventional ad hoc routing protocols are subject to subversion and need protocol-specific protections to successfully provide sensor data availability.

References

[Asada] Asada, G. et al., Wireless integrated network sensors: Low power systems on a chip, *European Solid State Circuits Conference*, The Hague, the Netherlands, October 1998.

[Biham] Biham, E. and Shamir, A., *Differential Cryptanalysis of the Data Encryption Standard*, Springer-Verlag, Berlin, Germany, 1993.

[Burmester] Burmester, M. and Desmedt, Y., A secure and efficient conference key distribution system, *Proceedings of Eurocrypt '94*, Lecture Notes in Computer Science 950, pp. 275–286, 1995.

[Carman1] Carman, D. and Cirincione, G., Energy and latency costs of communicating certificates during secure network initialization of sensor networks, *Proceedings of Collaborative Technology Alliance Conference 2003*, College Park, MD, 2003.

[Carman2] Carman, D. and Boncelet, C., A new message authentication approach with less overhead and greater reliability, *Proceedings of Collaborative Technology Alliances (CTA) Communications & Networks (C&N) Alliance 2003 Annual Symposium*, College Park, MD, April 2003.

[Carman3] Carman, D., Cirincione, G., and Matt, B., Energy-efficient and low-latency key management for sensor networks, *Proceedings of the 23rd Army Science Conference*, Orlando, FL, December 2–5, 2002.

[Denning] Denning, D., *Information Warfare and Security*, Addison-Wesley, Reading, MA, 1999.

[Dobbertin] Dobbertin, H., Cryptanalysis of MD5 compress, German Information Security Agency, May 1996.

[Dworkin] Dworkin, M., NIST Special Publication 800-38A—Recommendation for block cipher modes of operation—Methods and techniques, National Institute of Standards and Technology (NIST), Gaithersburg, MD, December 2001.

[ElGamal] ElGamal, T., A public key cryptosystem and a signature scheme based on the discrete logarithm, *IEEE Transactions on Information Theory*, 31(4):469–472, 1985.

[Eschenauer1] Eschenauer, L. and Gligor, V., A key-management scheme for distributed sensor networks, *Proceedings of the Ninth ACM Conference on Computer and Communications Security 2002*, Washington, DC, 2002.

[Eschenauer2] Eschenauer, L., Gligor, V., and Baras, J., On trust establishment in mobile ad-hoc networks, *Proceedings of the Security Protocols Workshop*, Cambridge, U.K., April 2002.

[FIPS46-3] U.S. Department of Commerce, National Institute of Standards and Technology (NIST), Data Encryption Standard (Federal Information Processing Standards Publication 46-3), October 1999.

[FIPS186-2] U.S. Department of Commerce, National Institute of Standards and Technology (NIST), Digital Signature Standard (DSS) (Federal Information Processing Standards Publication 186-2), January 2000.

[FIPS197] U.S. Department of Commerce, National Institute of Standards and Technology (NIST), Advanced Encryption Standard (AES) (Federal Information Processing Standards Publication 197), November 26, 2001.

[FIPS180-2] U.S. Department of Commerce, National Institute of Standards and Technology (NIST), Secure Hash Standard (Federal Information Processing Standards Publication 180-2), August 2002.

[Gligor] Gligor, V. and Donescu, P., Fast encryption and authentication: XCBC encryption and XECB authentication modes, Presented at the *Second NIST Workshop on AES Modes of Operation*, Santa Barbara, CA, August 24, 2001.

[ISO7498] International Organisation for Standardization (ISO), Information processing systems open systems interconnection basic reference model, ISO 7498, 1984.

[Iwata] Iwata, T. and Kurosawa, K. OMAC: One-key CBC MAC, *Proceedings of Fast Software Encryption 2003*, Lecture Notes in Computer Science, Lund, Sweden, Springer-Verlag, Berlin, Germany, February 24–26, 2003.

[Johnson] Johnson, D., *ECC, Future Resiliency and High Security Systems*, Certicom Corporation, Mississauga, Ontario, Canada, March 1999.

[Jutla] Jutla, C., Encryption modes with almost free message integrity, *Advances in Cryptology—EUROCRYPT '01*, Lecture Notes in Computer Science, Vol. 2045, Springer-Verlag, Berlin, Germany 2001.

[Law] Law, Y. et al., Assessing security-critical energy-efficient sensor networks, TR-CTIT-02-18, June 2002.

[Matt] Matt, B., Efficient pairwise key establishment for battlefield networks, *Proceedings of Collaborative Technology Alliance Conference 2003*, College Park, MD, 2003.

[Matsui] Matsui, M., Linear cryptoanalysis method for DES cipher, *Proceedings of EUROCRYPT 1993*, Lofthus, Norway, Springer-Verlag, Berlin, Germany, pp. 386–397, 1993.

[Maurer] Maurer, U. and Yacobi, Y., A non-interactive public-key distribution system, *Designs, Codes and Cryptography*, 9(3):305–316, 1996.

[Menezes1] Menezes, A., *Elliptic Curve Public Key Cryptosystems*, Kluwer Academic Publishers, Norwell, MA, 1993.

[Menezes2] Menezes, A., van Oorschot, P., and Vanstone, S., *Handbook of Applied Cryptography*, CRC Press, Boca Raton, FL, 1996.

[Murphy] Murphy, S. and Weiler, S., Overview of potential compromises and security paradigms in wireless routing protocols, *Proceedings of Collaborative Technology Alliance Conference 2003*, College Park, MD, 2003.

[NIST2] National Institute of Standards and Technology (NIST), Rationale for the selection of the OMAC variation of XCBC, 2003.

[Papadimitratos] Papadimitratos, P. and Haas, Z., Secure routing for mobile ad hoc networks, *SCS Communication Networks and Distributed Systems Modeling and Simulation Conference (CNDS 2002)*, San Antonio, TX, January 27–31, 2002.

[Perrig] Perrig, A., Chan, H., and Song, D., Random key predistribution schemes for sensor networks, *IEEE Symposium on Security and Privacy 2003*, 2003.

[PKCS1] RSA Laboratories, PKCS #1 v2.1: RSA Cryptography Standard, June 2002.

[RFC2401] Kent, S. and Atkinson, R., Security architecture for the internet protocol, RFC 2401, November 1998.

[RFC2409] Harkins, D. and Carrel, D., The internet key exchange (IKE), RFC 2409, November 1998.

[RFC2246] Dierks, T. and Allen, C., The TLS protocol version 1.0, RFC 2246, January 1999.

[RFC1321] Rivest, R., The MD5 message-digest algorithm, RFC 1321, April 1992.

[Rogaway] Rogaway, P. et al., OCB: A block-cipher mode of operation for efficient authenticated encryption, *Eighth ACM Conference on Computer and Communications Security (CCS-8)*, Philadelphia, PA, ACM Press, New York, pp. 196–205, 2001.

[Schneier] Schneier, B., *Applied Cryptography*, 2nd edn., John Wiley & Sons, New York, 1996.

[Shamir] Shamir, A., Identity-based cryptosystems and signature schemes, *Proceedings of Crypto'84*, New York, pp. 47–53, 1985.

[Stallings] Stallings, W., *Cryptography and Network Security: Principles and Practice*, 3rd edn., Prentice Hall, Upper Saddle River, NJ, 2003.

[Stajano] Stajano F. and Anderson, R., The resurrecting duckling: Security issues for ad-hoc wireless networks, *Security Protocols, Seventh International Workshop*, 1999.

[Steiner] Steiner, M., Tsudik, G., and Waidner, M., Key agreement in dynamic peer groups, *IEEE Transactions on Parallel and Distributed Systems*, 11(8):769–780, 2000.

[Wilkison] Wilkison, R., U.S. says Iraq GPS jamming sites destroyed, Iraq Crisis Bulletin/(INEWS) International News E-Wire Service, Doha, Qatar, March 26, 2003.

[Wood] Wood, A. and Stankovic, J., Denial of service in sensor networks, *IEEE Computer*, 35(10):54–62, 2002.

[WTLS] Wireless Application Protocol Forum, Ltd., WAP transport layer end-to-end security, 2001.

[Zhu] Zhu, F., Security for ad hoc networks (including sensor networks) web page, http://www.ccs.neu.edu/home/zhufeng/security_manet.html

9

Sensor Network Security with Random Key Predistribution

Lianyu Zhao
Clemson University

Richard R. Brooks
Clemson University

Brijesh Pillai
Clemson University

P.Y. Govindaraju
Clemson University

Matthew Pirretti
The Pennsylvania State University

Vijaykrishnan Narayanan
The Pennsylvania State University

Mahmut Kandemir
The Pennsylvania State University

9.1 Introduction

Although sensor networks are supposed to be highly independent and require little supervision, sensor networks deployed in hostile environment are subjected to all kinds of attacks. For example, in ColTraNe [1], sensors are deployed in hostile terrain to track the movements of enemy tanks. Attacks could be initialized by the enemy to maliciously capture, modify the compromised sensor nodes, or even inject information to them.

Public key approaches consume too much energy [2]; improvements such as cluster-based key establishment are based on identities, which pose constraints on data-centric schemes [1,3]. On the other hand, symmetric keys require less energy but the use of a single key makes the network vulnerable to key leakage by compromised nodes. To conquer these problems, *random key distribution* is proposed by Eschenauer and Gligor [4], in which a large number of keys are generated and each node is assigned a small portion of all keys in the key pool. Adversaries can only obtain a fraction of keys by compromising a node.

Cloning attacks physically introduce cloned nodes into the network. If the sensors are running commodity hardware and operating system, it is easy for the attacker to clone the captured nodes by copying data and reprogram them to subvert the network. In Ref. [5], Chan and Perrig explain how clones can falsify sensor data, extract data from the network, and/or stage denial of service attacks. Cloned nodes can provide false target information and spurious network traffic; thus, nodes can lose tracks or be kept from entering sleep modes. In this chapter, we focus on the detection of clones and countermeasures using random key predistribution.

Another attack prevents sensor nodes from sleeping. Sensor nodes enter sleep mode to save battery power and increase network longevity. However, the limited tamper resistance of low-cost sensor nodes leaves them vulnerable to compromise [5,6]. In this chapter, we discuss the *barrage attack* and the *sleep deprivation* attack. We show the latter one is harder to be detected and requires more effort.

We use multicast with random key predistribution to build a key management infrastructure, which partitions the network into multicast regions. Nodes in each region are managed by a randomly selected keyserver node. We show how to organize and maintain these regions; quantify energy savings; show that this design is resistant to many attacks including Byzantine, Sybil, and Cloning attacks; and show that this design has no single point of failure. More analysis and discussions on the power consumption issues that reveal the trade-off for these parameters are presented.

9.2 Sensor Networks and Random Key Predistribution

9.2.1 Sensor Network Security

Security is a major issue in sensor networks, since adversaries can potentially manipulate sensors to disseminate incorrect information. However, sensors are usually inexpensive and thus have several significant constraints. Lightweight processors and cheap hardware components are used. To support extended network lifetimes, nodes must be frugal with their battery power. To prevent detection of a sensor network by an adversary, military applications will use low probability of detection/interception communications with short communication ranges. Because of these constraints, general-purpose security protocols are rarely appropriate. For example, the cost required to incorporate tamper-resistant hardware into sensor nodes is likely to be prohibitive. Public key cryptography approaches attempted to date are based on the RSA approach. Their energy consumption and computational latency makes them inappropriate for sensor network applications. Symmetric key encryption requires less energy and computation, but using a single key for the network makes the system vulnerable to key disclosure when nodes are physically compromised.

9.2.2 Random Key Predistribution

In order to avoid the aforementioned disadvantages in traditional security algorithms, Eschenauer and Gligor [4] proposed a simple key predistribution scheme that provides a relatively high level of security. In traditional key preloading, each node is given a systemwide key, which is used to communicate with any other node. This supports network self-organization, which is a necessity when network deployments are not rigidly structured. However, traditional preloading is not very secure. If an adversary captures a single node and finds the key, he/she will be able to disrupt the entire network.

In their proposed scheme, a large pool of P cryptographic keys and key identifiers is generated offline. Each of the n sensor nodes in the system randomly selects a subset of k keys (key ring) from the key pool without replacement (i.e., each node has a set of k distinct keys from P). The keys in the key ring serve as authentication tokens that are used to attest that nodes can be trusted. Each node is also given a systemwide symmetric key that is used to communicate with the base station. After sensor nodes are deployed, there is a key discovery phase where each node attempts to communicate with all other nodes in radio range. Links are established between any two nodes that find a common key in their set of k keys. All further communications use this shared secret key. Key discovery is followed by a path key establishment phase, where nodes within communication range without keys in common establish direct links by exchanging a key. This is possible as long as there is a path of nodes sharing common keys between them. Eschenauer and Gligor also provide a mechanism for key revocation when a sensor node is compromised. When a sensor node is compromised, the base station transmits a revocation message with the identifiers of the k keys on the compromised node. Each node checks to see if any key in its key ring is in the revocation list. Offending keys are removed and connections using compromised keys terminated. They assume that since only a small subset k of pool P is removed from the system, only a small number of nodes are affected.

9.3 Cloning Attack Detection

9.3.1 Clone Detection Using Random Key Predistribution

Clone detection approach in this section is summarized in Figure 9.1. This approach is based on analyzing node authentication statistics. Since each node randomly selects k keys from the pool, we can predict the distribution of the number of nodes processing a given key and the distribution of the number of times a key is used is a direct consequence of the number of nodes possessing the key. When clones are inserted into the network, the key usage distribution is skewed. Cloned keys are present on a greater number of nodes than normal and are, therefore, used more frequently than keys that have not been cloned. By collecting key usage statistics, we can determine which keys have been cloned.

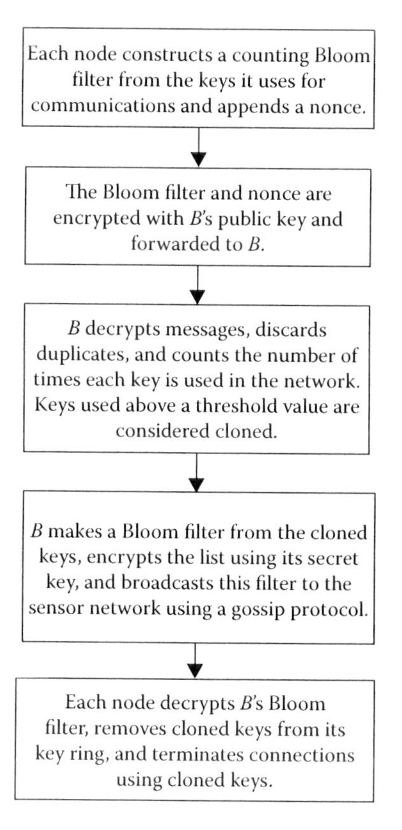

FIGURE 9.1 Clone detection protocol.

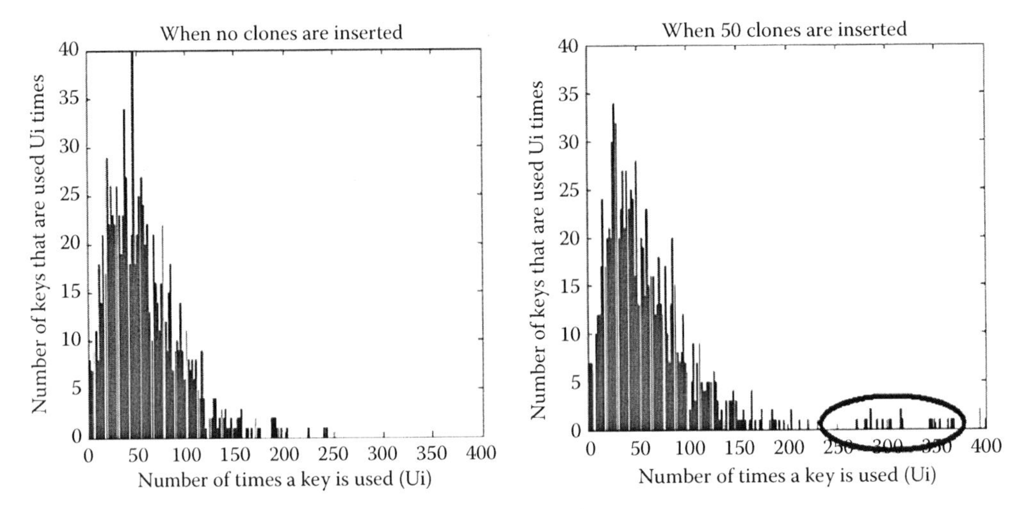

FIGURE 9.2 (Left) Normal key usage distribution and (right) key usage in a network containing 50 clones.

Let us consider a sensor network has a pool of P keys for a set of N authorized nodes. Each of the N nodes has a key ring of k keys taken randomly from the pool. C cloned nodes are added to the network. The total number of nodes is $T = N + C$. A base station B is outside the network with a copy of the entire key pool. Every node has a public key that can be used to communicate securely with the base station. B collects key usage statistics from the network and detects the keys that are used by cloned nodes. However, an extension of this approach [7] exists, which requires neither public key encryption nor a key server. In that approach, keyservers are chosen at random from the network using secure cluster head protection scheme, which guarantees that all nodes have an equal chance of becoming keyservers.

Although cloned nodes could collude to use their keys differently from legitimate nodes, for example, initially using only a small subset of the keys that process to establish connectivity with the network and switch to other keys if these keys are detected, this approach would make clones easier to detect, since the keys that they collude to use would be used even more often than normally. Therefore, whenever a node authenticates itself using a key, such information should be analyzed to guard against the insertion of a clone. Consider a network with $P = 1000$, $N = 250$, and $k = 50$, the left part of Figure 9.2 shows that key usage distribution with no clones, and the right part shows the same network with 50 clones. The circled region contains the cloned keys, which are used more often.

9.3.1.1 Bloom Filters

Counting Bloom filters can be used to collect key usage statistics. A Bloom filter is an approximate representation of a set that supports membership queries. It is a vector of m bits. Each member of the set is hashed using h hash functions, each with range $[1, \ldots, m]$. The bit correspondingly to each hash value is set to 1. Membership queries hash an element with the h hash functions. If the corresponding bit positions in the filter vector are all 1, the element is said to be a member of that set.

Each node makes a counting Bloom filter of the keys it uses to communicate with neighboring nodes. It appends a random number (nonce) to the Bloom filter and encrypts the result using B's public key. This encrypted data structure is forwarded to B. B decrypts the Bloom filters it receives. The nonces help foil replay attacks, since messages with duplicate nonces are discarded. B performs membership queries on the counting Bloom filters it receives and calculates how many times each key is used in the network. If a Bloom filter is found to be invalid (i.e., some bits do not belong to valid membership queries), it is discarded and not used in calculating key usage statistics.

9.3.1.2 Threshold Determination

For both consistency with the existing literature and ease of presentation, we assume the network has an Erdos–Renyi topology, the key pool contains P keys, and each node has a key ring of k keys chosen at random from the pool. Considering in the clone detection protocol, the legitimate connections are reported twice (once by each node using the connection). Brooks et al. [8] prove that the expected number of times a key used for communications in the network is

$$\mu_0 = 2 * \mu_k = 2 * \frac{\sum_{j=1}^{n} M_j * (N_j/M_j)}{P} \tag{9.1}$$

with variance

$$s^2 = 4 * V_k = 4 * \frac{\sum_{j=1}^{n} M_j * ((N_j/M_j) - \mu_k)^2}{P} \tag{9.2}$$

where

M_j is the number of keys on exactly j nodes
N_j is the expected number of times a key on exactly j nodes is used

Therefore, given the central limit theorem, one can create a normal curve to represent the usage distribution of uncloned keys. Given the number of times key i is used to establish connections gathered using Bloom filters. By using the Neyman–Pearson test, one can determine if a given key has been cloned, by determining if the probability that the key fits a normal curve with a fixed false detection rate α. Figure 9.3 shows the number of cloned keys detected for different α settings and different numbers of clones in the network.

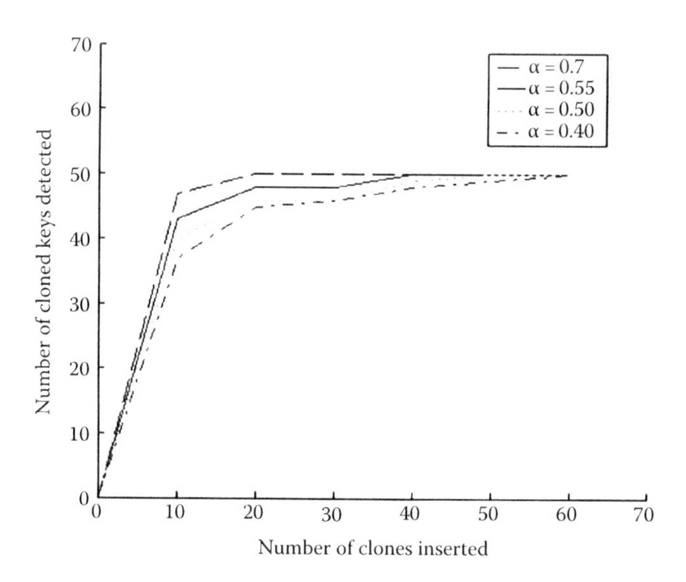

FIGURE 9.3 Analytical prediction of the number of keys detected as cloned as the number of clones inserted into the network varies.

To analyze the security effectiveness, consider the situation from cloned node n_cs view point. If n_c reports its key usage to *B*, the presence of cloned keys will become more apparent. Since n_c cannot decrypt the packets, tampering is futile. An adversary can attempt to disrupt the protocol by randomly generating a string of bits and passing it off as a valid Bloom filter. This works only if there is a reasonable probability that a randomly generated bit string is a valid Bloom filter for some combination of keys. In Ref. [8], it is proved that such probability is less than

$$P_b = \sum_{i=1}^{k} \binom{P}{i} \Big/ \binom{m}{L_i} \tag{9.3}$$

if L_i is the expected number of bits set in the Bloom filter.

9.3.2 Clone Detection in Grid and Ad Hoc Networks

In the detection of clones introduced in the previous section, the network is assumed to follow Erdos–Renyi topology. However, this is not realistic for wireless sensor networks. It posits that a connection between any two nodes is equally likely, which does not match wireless communication with limited communication ranges. Grid networks are highly structured; each node can communicate only with a fixed number of immediate neighbors. When users control node placement, nodes are often distributed in a regular pattern.

Ad hoc networks are less structured; nodes are placed at random. They can only communicate with the other nodes within communication range *r*. Each node's neighborhood size is defined by a random process. There is no control over node placement. An ad hoc model introduced in Ref. [9] assumes the sensor network has *N* nodes that are uniformly, randomly distributed in a square region, and they can communicate only with nodes within communication range. If we assume nodes within range of each other can communicate with probability p_c in a network with key pool size *P* and *k* keys in the key ring, then

$$p_c = 1 - \frac{(1-(k/P))^{2(P-k+1/2)}}{(1-(2k/p))^{(P-2k+1/2)}} \tag{9.4}$$

If $P_d(j)$ is the probability, a key is on exactly *j* nodes and C_j denotes the number of times that key, k_j, on *j* nodes is used for making connections in the network, then

$$N_j = P * P_d(j) * C_j * \sum_{i=0}^{k-1} \frac{P_c(i)}{i+1} \tag{9.5}$$

In Ref. [8], the value of C_j is given by $j(j-1)/2$. However, in a two-dimensional grid network where the neighborhood size is 4, it is proved in Ref. [10] that

$$C_j = j * \sum_{i=1}^{4} i * \binom{4}{i} * \left(\frac{j-1}{N-1}\right)^i * \left(\frac{N-j}{N-1}\right)^{4-j} \tag{9.6}$$

Figure 9.4 shows simulation results and predicted values for mean key usage and variance as network size increases for grid networks.

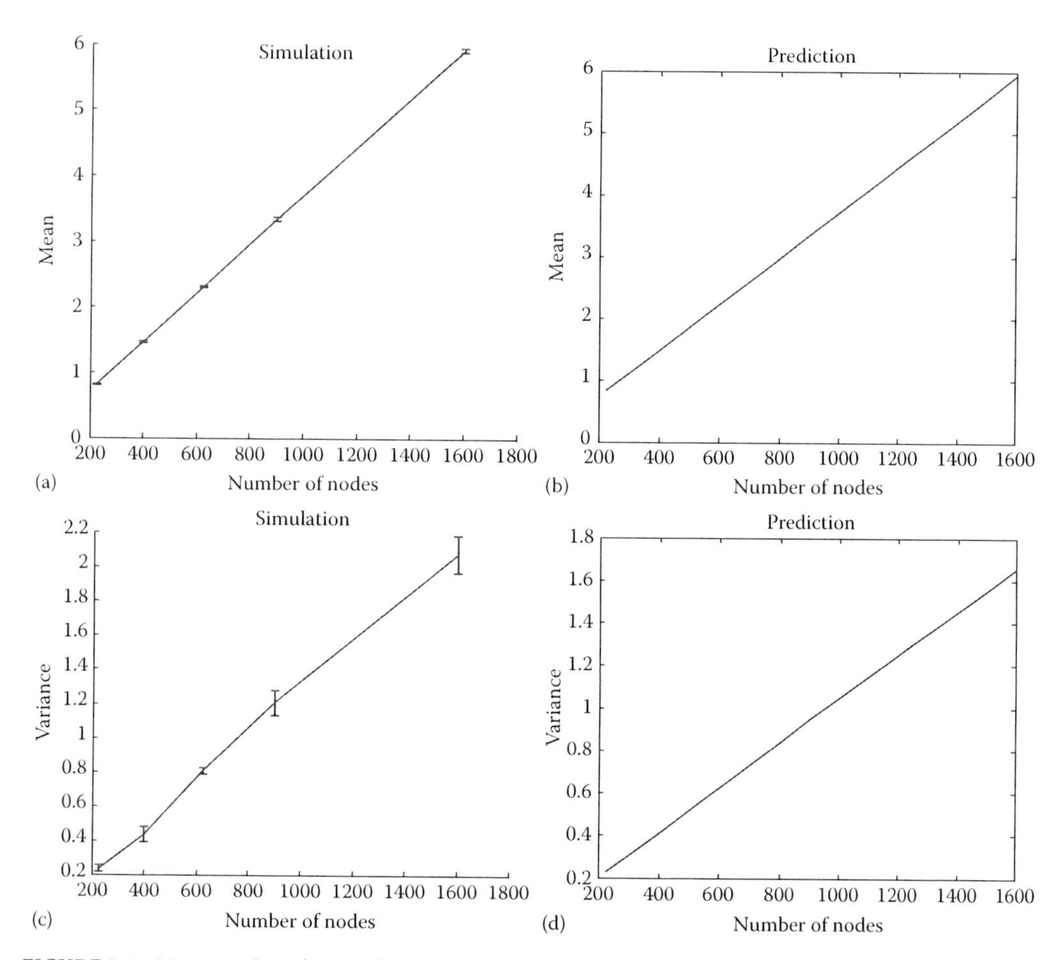

FIGURE 9.4 Mean number of times a key is used to create communication links in a grid sensor network: simulation (a) and analytical prediction (b); the associated variance: simulation (c) and analytical prediction (d). For all the graphs, $P = 1000$, $k = 50$, 35 repetitions, and 95% confidence interval.

For an ad hoc network,

$$C_j = j * \sum_{i=1}^{4} i * \binom{n}{i} * \left(\frac{j-1}{N-1}\right)^i * \left(\frac{N-j}{N-1}\right)^{n-j} \tag{9.7}$$

where $n = \sum_{i=0}^{N-1} i * p_i * p_c$ is the probability i nodes are in range of the node and p_c is given by Equation 9.4.

Figure 9.5 shows simulation results and predicted values for mean key usage and variance as network size increases for grid networks.

Figure 9.6 shows the number of cloned keys detected for different α settings and different numbers of clones for both ad hoc and grid networks. The keys that are detected are the keys that are most highly used in the network.

9.4 Sleep Deprivation Attacks

The victim of this attack is a battery-powered computing device, such as a sensor node, which attempts to remain in a low-power sleep mode for as long as would be possible without adversely affecting the node's applications. The attacker launches a sleep deprivation attack by interacting with the victim in a

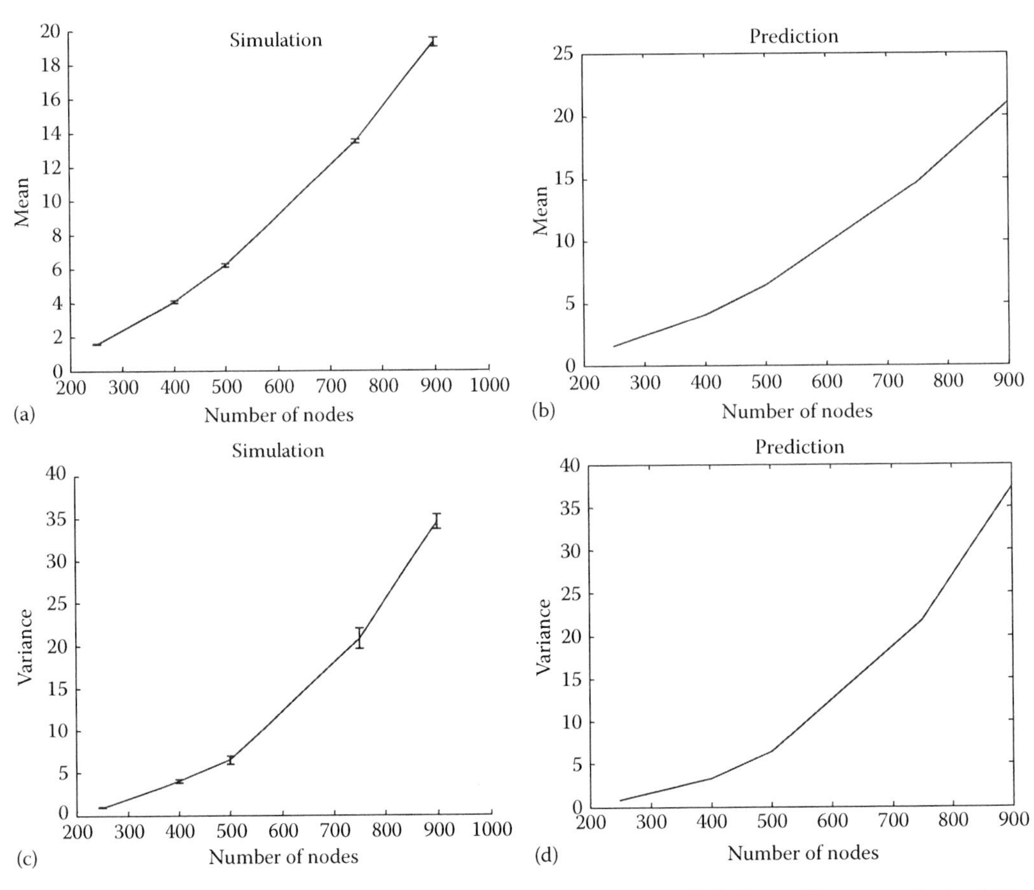

FIGURE 9.5 Mean number of times a key is used to create communication links in an ad hoc network: simulation (a) and analytical prediction (b); the associated variance: simulation (c) and analytical prediction (d). For all the graphs, $P = 1000$, $k = 50$, 35 repetitions, and 95% confidence interval.

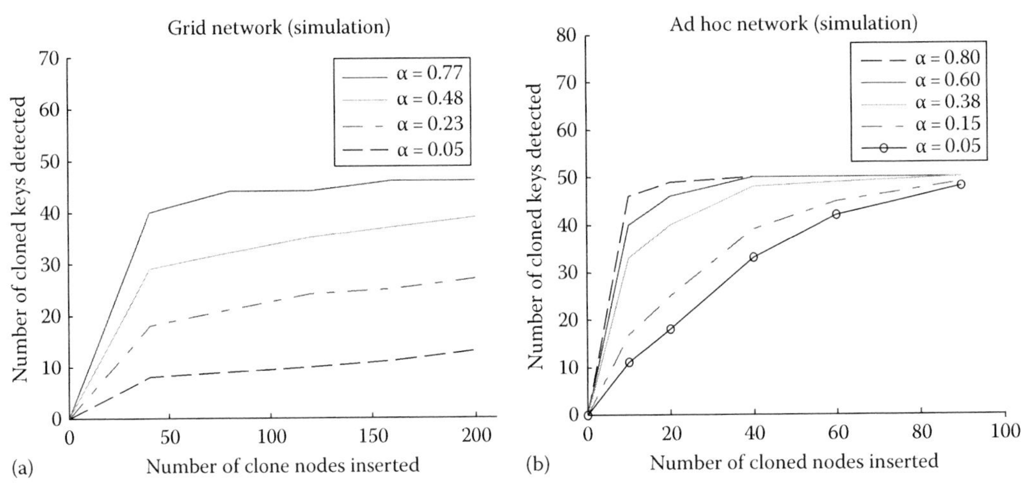

FIGURE 9.6 Number of cloned keys detected versus α. The key ring size k is 50. (a) Grid networks and (b) ad hoc networks.

manner that appears to be legitimate; however, the purpose of the interactions is to keep the victim node out of its power-conserving sleep mode. Thus, this attack can be used to dramatically reduce the lifetime of the victim. Further, this attack is difficult to detect given that it is carried out solely through the use of seemingly innocent interactions.

9.4.1 System Modeling

9.4.1.1 Assumptions

Let us consider the placement of sensor nodes to be spread uniformly at random throughout a square region. The sensor nodes operate on nonrenewable batteries; once a node exhausts its battery, it is considered to be dead. To preserve their battery power, sensor nodes will cycle in and out of a low-power sleep state. Assuming the presence of publish/subscribe routing protocol [11], sensor nodes subscribe to all track records within *z* meters of their position. When a local group of nodes detects a target, they will share sensor readings. One of these nodes is then elected as cluster head. This node aggregates the sensor data on behalf of the cluster and forms a track record. The track record is sent to all subscribed nodes by repeatedly invoking the publish method. The publish method is invoked one time for each subscribed node. We assume a homogeneous network, where a cluster head is identical to any other node in the cluster in terms of capacity and resources. For simplicity, we shall assume that nodes within the same cluster can communicate directly.

On security and attack model, we assume an adversary is capable of compromising sensor nodes. Compromised nodes (including cluster heads) may launch various attacks; each sensor node is assigned a unique ID prior to deployment. Further, sensor nodes cannot impersonate uncompromised nodes so that a compromised node can only submit a single *vote* in a cluster head selection algorithm.

9.4.1.2 Sensor Node States

We consider the behavior of sensor nodes to cycle between the sleep state and the idle state, staying in each state for fixed time quanta. A sensor node in the idle state will enter the receive state upon receipt of a track record. This will only occur when the sensor node is subscribed to an adversary node. Once a sensor node has received a track record, it will stay in idle mode for an extended period of time. The adversary nodes are also partitioned into two states: sleep and transmit. In Figure 9.7, we illustrate what we consider normal sensor node behavior when no tracking is occurring. A sensor node alternates between sleep mode for t_{slp} seconds and idle mode for t_{idle} seconds.

In Figure 9.8, we illustrate the behavior that a sensor node would exhibit upon receiving a single track record (represented pictorially by TR in the figure).

Figure 9.9 illustrates a sensor node being victimized by a sleep deprivation attack. The attacker sends the victim a track record every $t_{ar} + t_{rs}$ seconds. Each time the victim is just about to decide that it can go to sleep mode, it receives another track record. Thus, the victim is kept perpetually awake.

FIGURE 9.7 Timing diagram of normal node behavior when no tracking is occurring.

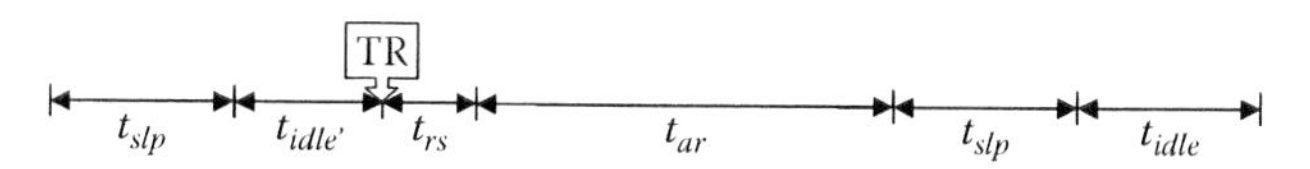

FIGURE 9.8 Timing diagram of normal node behavior after receiving track record.

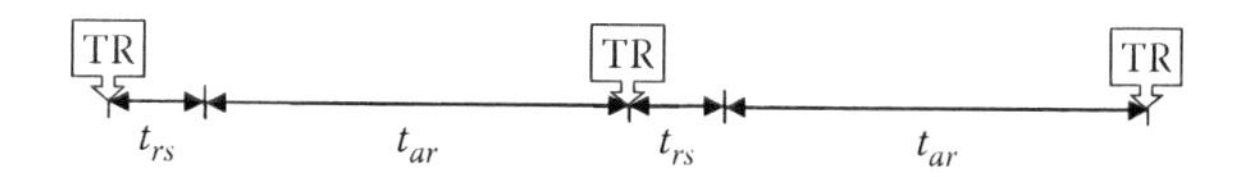

FIGURE 9.9 Timing diagram of node under sleep deprivation attack.

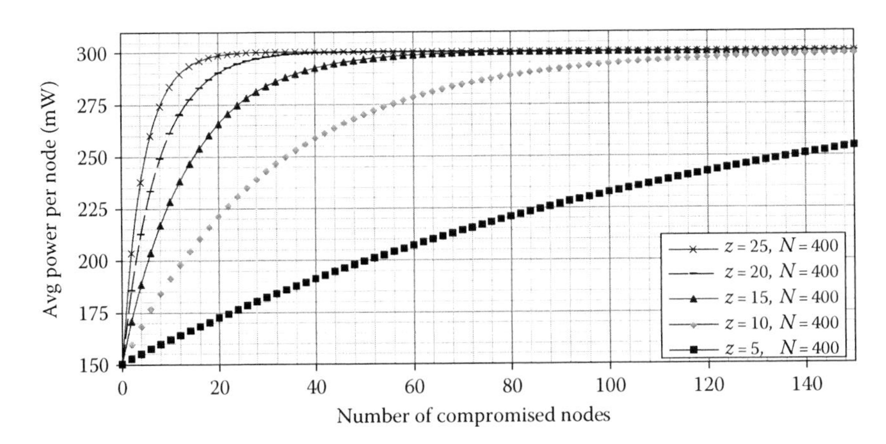

FIGURE 9.10 Effect of subscription radius in sleep deprivation attack.

9.4.1.3 Derivation of Model

The probability that a sensor node will be subscribed to an adversary cluster head, given that each node is deployed in a square region of length L and each node subscribes to all track records within a radius of size $z < L$, is $(\pi z^2)/L^2$. Given C adversary nodes with uniformly random placement, the probability that a given node will be subscribed to at least on malicious cluster head is

$$p = 1 - \left(1 - \frac{(\pi z^2)}{L^2}\right)^C \tag{9.8}$$

From the deduction in Ref. [12], Figure 9.10 illustrates the effect of malicious nodes on a network containing 400 legitimate nodes for different subscription radii (i.e., z).

It is also shown in Ref. [12] that an adversary node could simultaneously attack up to 150 nodes before its power consumption became higher than that of its victims, and thus from a power perspective, this attack does not require the attacker to expend a great deal of energy. Therefore, we conclude that the sleep deprivation attack enables the adversary to cause a lot of damage without expending a lot of effort.

9.4.2 Sleep Deprivation Detection with Random Vote

We have observed that clustering algorithms rely on the honesty of all participating nodes, allowing a malicious node to generate false information to ensure its selection as cluster head. The random vote scheme counteracts this characteristic by randomizing cluster head selection.

In this scheme, a group of local nodes using the random vote scheme form a cluster by performing the following steps:

1. Each node locally broadcasts its unique ID.
2. Each node uses a pseudorandom number generator to pick the ID of the local node it desires to become the next cluster head.
3. Nodes locally broadcast the ID of their desired cluster head.
4. Each node repeats step 2 until a single node attains a majority of votes. This node will become the next cluster head.

Each sensor node is only allowed to cast one vote at a time. However, in the case of a tie, the nodes will restart the algorithm at step 2. We refer to the execution of steps 2 through 4 as a round or iteration. Thus, the random vote algorithm is potentially composed of several rounds.

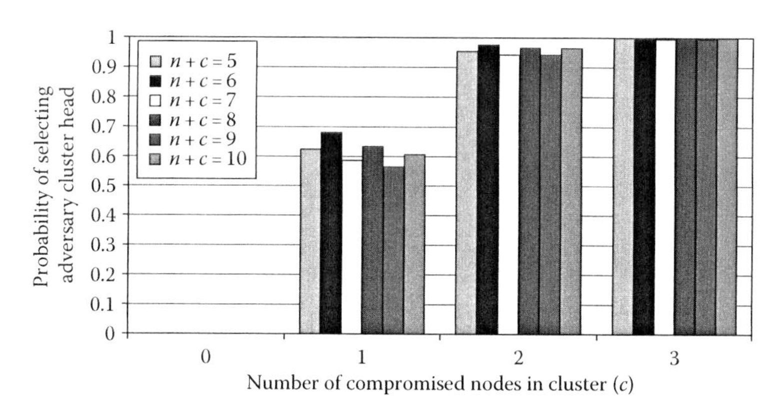

FIGURE 9.11 Probability that the random vote scheme selects an adversary cluster head.

In an argument similar to what was used to formulate p, the probability of a legitimate node being subscribed to at least one cluster containing adversary nodes is

$$p' = 1 - \left(1 - \frac{(\pi z^2)}{L^2}\right)^{C/2} \tag{9.9}$$

Without any defense mechanism in place, a malicious node is selected as cluster head with a probability of 1. Figure 9.11 shows that given a single adversary node in a cluster, the random vote scheme nearly halves the probability that the adversary node will be selected as cluster head. However, this probability jumps to nearly 1 when multiple adversary nodes are located in the same cluster. It may startle the reader to see that when $n + c$ is even, the adversary appears to have a better chance of becoming a cluster head. This is a direct result of the difference between attaining a majority for odd and even numbers.

9.4.3 Sleep Deprivation Detection with Round Robin

The lack of scalability in the random vote clustering algorithm caused us to consider another approach to secure cluster formation. The *round robin* scheme is based on the observation that if each node maintained more state, that is, if clusters were maintained for long periods of time, a more scalable solution would be possible. With such a scheme, cluster heads could be elected in a round robin fashion. The round robin scheme operates in two phases. The first phase is a bootstrapping phase where the initial clusters are formed. The second phase is a maintenance phase, during which the precise membership of each cluster is updated due to node mobility, addition of new nodes to the network, and removal of nodes from the network.

Assuming an average of c adversary nodes and n legitimate nodes in each cluster, the proportion of time that the c adversary nodes can launch a sleep deprivation attack from a particular cluster is

$$p_{c,rr} = \frac{c}{c+n} \tag{9.10}$$

In Figure 9.12, we have used the equation for $p_{c,rr}$ to see how likely the round robin scheme is to elect an adversary cluster head. Comparison of Figures 9.11 and 9.12 illustrates that the round robin scheme makes it much more difficult for the adversary to become cluster head. In fact, the adversary nodes are just as likely to become cluster head as is any other node.

A nice property of the round robin scheme is that it only requires a single iteration to select a cluster head. Each node must keep track of exactly which nodes are in the cluster and which node is the current

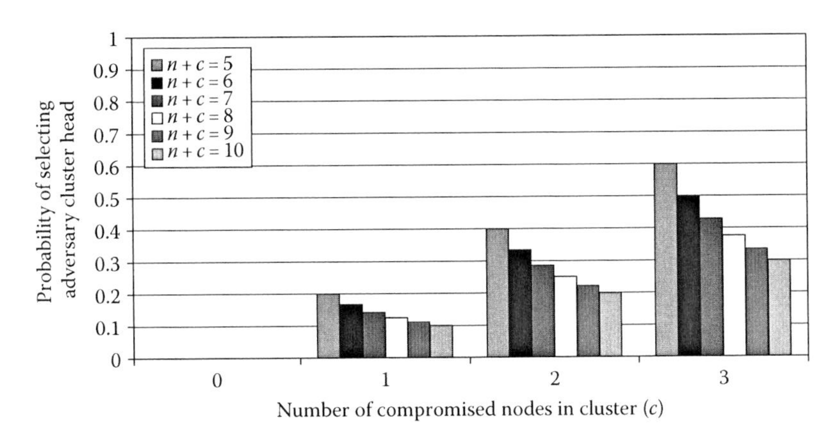

FIGURE 9.12 Probability that the round robin scheme selects an adversary cluster head.

cluster head. Thus, when a new cluster head is needed, each node can independently determine who the new cluster head should be. However, this scheme requires that each sensor node must maintain a list indicating which nodes are in its cluster at all times. Such a list would require an unrealistic amount of per-node storage for larger clusters. This shortcoming motivated us to come up with the hash-based cluster head selection scheme, which we shall call the *hash-based* scheme for brevity.

9.4.4 Sleep Deprivation Detection with Hash

Hash-based scheme performs dynamic clustering in an attack and fault-tolerant manner without excessive overhead. In this scheme, each node generates a random number and then broadcasts their number's hash. This allows each node to commit to their particular number without revealing it until all participating nodes have likewise committed.

The hash algorithm operates by having each participating node execute the following steps:

1. Generate an integer, r_i, using a pseudorandom number generator and locally broadcast a commit message of the form $\langle ID,H(ID,r_i)\rangle$, where ID denotes the node's identifier and $H(\cdot)$ denotes a fixed length collision-resistant hash function.
2. Wait for enough time to pass, T_o, that it is sufficiently unlikely that any more commit messages will be received. Then locally broadcast a list message $\langle(ID_1, \ldots, ID_Y)\rangle$, where $(ID_1, \ldots, ID_Y)$ is a list of all IDs extracted from step 1, including the node's own ID as well.
3. For each list message received, verify that the local list of IDs is the same as the received list of IDs. Send a request commit message to any node whose ID is listed in another node's list message but is not listed in the node's own list of IDs: $\langle ID_{dest}\rangle$.
4. For each request commit message received whose ID_{dest} field matches the node's own ID, reply by repeating the original commit message.
5. Wait for T_o seconds to pass. Then locally broadcast a reveal message to disclose r_i: $\langle ID,r_i\rangle$.
6. Verify each r_i with its associated commit message.
7. Wait for T_o seconds to pass. If any reveal messages have not been received, transmit a request reveal message: $\langle H(ID,r_i)\rangle$.
8. Any node that has a verified r_i can respond to a request reveal message.
9. Given Y participating nodes, the cluster head will be the node whose ID matches the following result:

$$\text{Mod}\left(\left(\sum_{i=1}^{n} r_i\right), Y\right) + 1 \tag{9.11}$$

The key observation regarding the security of this scheme is that every participating node must send out its commit message prior to receiving any reveal messages. This can be ensured with a high likelihood by selecting a sufficiently high value for T_0. The probability that an adversary node is selected to become a cluster head is the same as any node in the cluster, which is the same as in the round robin scheme. Thus, Figure 9.12 also applies to the hash-based scheme.

9.5 Multicast Encryption with Random Key Predistribution

In this section, we show how to use random key predistribution to authenticate nodes and initialize a multicast key management infrastructure in a sensor network [13]. Figure 9.13 shows a summary.

9.5.1 Multicast Tree Management

The approach we present uses two separate tree structures to organize communications among the nodes in the multicast region as shown in the following (Figure 9.14):

1. A *communication tree* that defines the connections used to transmit packets between nodes
2. A *binary key tree* that is used to group nodes so that they share common keys, which allows the number of encryptions to be kept to a minimum

The first step in our approach is determining the membership of each multicast region by choosing the keyserver. Approaches in Refs [7,14] can be used to determine the number of keyservers k and

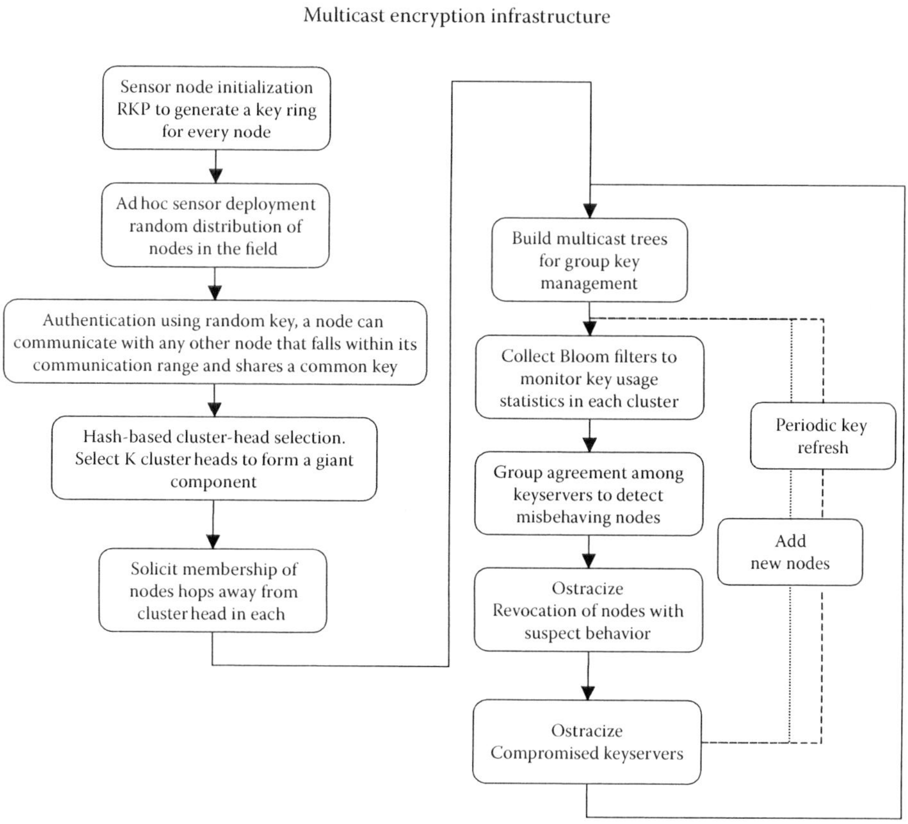

FIGURE 9.13 Flowchart of sensor network security approach presented in this paper.

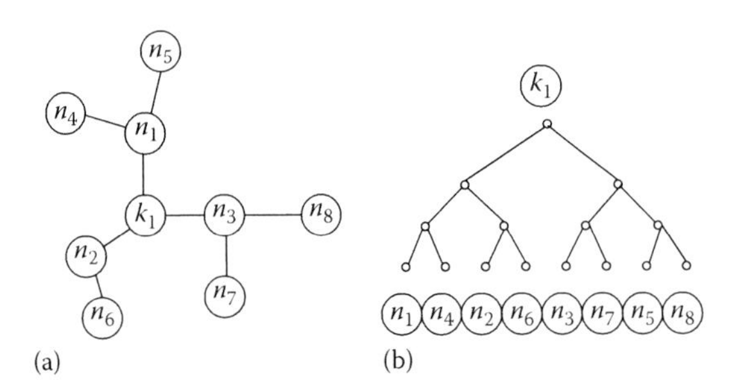

FIGURE 9.14 Tree structures: (a) communication tree and (b) binary key tree.

number of hops h to use for a given network. Keyservers maintain the security of the multicast region by controlling group formation and membership. When selecting the keyservers, each node

1. Generates a random number n
2. Calculates a hash value from the number $h(n)$—any hashing algorithm ranging from modulo arithmetic to SHA1 could be used
3. Broadcasts the hash value to the participating nodes—this commits the node to its number without revealing its value
4. Waits on an agreed-upon time-out period for every participating node to broadcast its hash value
5. Broadcasts a list of all nodes that have transmitted hash values
6. Matches the lists it receives to its local list and requests a hash value from any missing node
7. Waits on an agreed-upon time-out period, and then broadcasts its random number
8. Verifies the random numbers against precommitted hash values to ensure integrity

The keyserver is then chosen using agreed-upon criteria based on these random numbers.

The second step of the process is creating a binary key tree structure rooted at the keyserver. It can be created using a two-pass protocol.

Phase (i) constructs the initial tree T_{bd} that is used to construct the binary key tree T_b in Phase (ii):

1. Initialize T_{bd} by making the keyserver the root node. It is both root and leaf at this point.
2. Each node not in the tree T_{bd}, that is, one hop from a leaf node, becomes its child.
3. Repeat step 2, $(h - 1)$ times where h is the maximum number of hops from the keyserver to any sensor node in the multicast group. Techniques for computing the parameter h are given in Ref. [7].

Phase (ii) constructs the binary key tree T_b from the initial tree T_{bd}:

1. Assign any leaf node of the tree T_{bd} as n_t. (Let n_t be node 5 as shown in Figure 9.15b.)
2. Count the number of siblings of n_t (child nodes of its parent node). Node 5 has 2 siblings: nodes 6 and 7.
   a. If (even) group parent with n_t (node 1 gets grouped with node 5), group all siblings into pairs (nodes 6 and 7 are grouped together).
   b. If (odd) do not group parent node, group all siblings and n_t into pairs.
3. Remove nodes that were grouped in step 2 from consideration and repeat steps 1 and 2 until all nodes are grouped. (The iterations group nodes 8 and 9, 10 and 3, 11 and 12, 2 and 4 as shown in Figure 9.15c.)
4. Fuse grouped nodes into single nodes, reconstruct a balanced tree of the fused nodes, and repeat steps 1 through 3. (Figure 9.15d shows how nodes 5, 1, 6, 7 get grouped.)
5. Repeat step 4 until only one node exists in the initial tree. Construct the key tree according to the grouping of nodes at every iteration in step 4. (Further iterations group nodes 8, 9, 10, 3 into one group and 11, 12, 2, 4 into another to finally form the binary tree as shown in Figure 9.15e.)

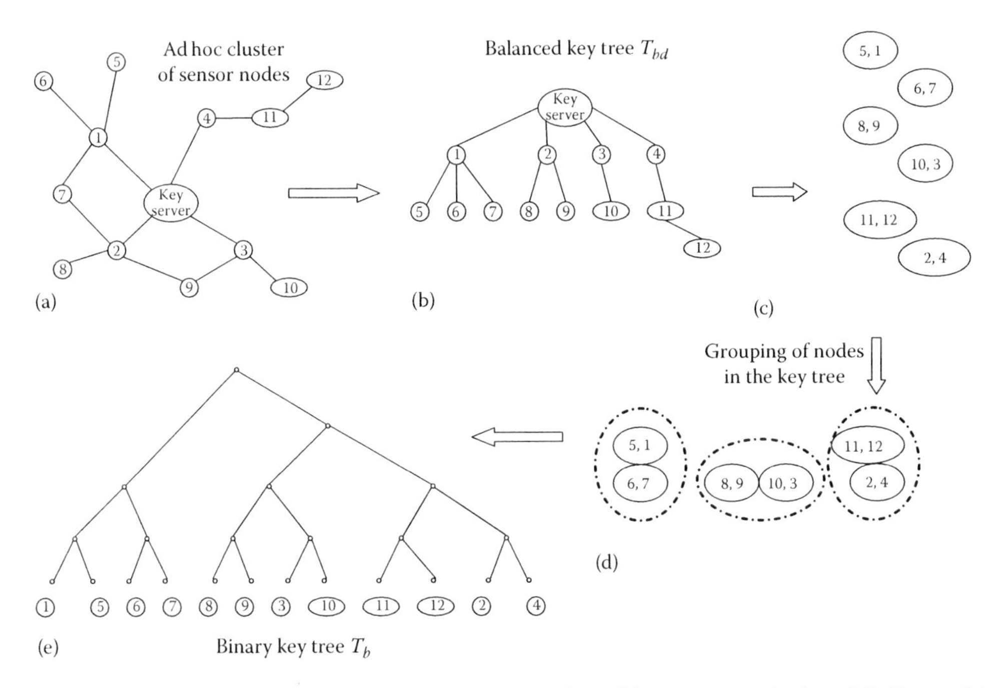

FIGURE 9.15 From the ad hoc network (a), we create an initial tree (b). We group nodes (c and d). The result is a binary tree (e) that groups nodes by the keys they will have in common.

9.5.2 Group Key Agreement Protocol

To reach consensus on the set of cloned keys in the network

1. Each node transmits the counting Bloom filter to its keyserver for the keys used by the node.
2. The keyserver transmits the counting Bloom filters from all the nodes within its multicast region to every other keyserver using an authenticated channel.
3. Every keyserver now has counting Bloom filters for every multicast region. Each keyserver computes key usage statistics for keys in its key ring to identify cloned nodes.
4. The keyservers use a Byzantine agreement protocol [15] to reach consensus on key usage statistics:
   a. The keyserver computes a vector v of usage statistics from the Bloom filters provided by the keyservers.
   b. The vector v is sorted, and the lowest and the highest τ values are discarded to give rise to a new vector v' containing $(k - 2 * \tau)$ entries, where τ is the number of adversaries the network can tolerate as keyservers.
   c. The key usage value is the mean of the vector v'.
   d. This protocol is guaranteed to be correct when $k \geq 3\tau + 1$. Performance degrades slowly until $k = 2\tau + 1$, at which point the majority of the information is false and the opponents control the network.

Suppose that the system is designed to tolerate up to $\tau = 10$ subverted keyservers. The key usage reported by the subverted keyservers is questionable and should not be taken into account in the decision process. We discard 20 values (the first 10 and last 10 from the sorted vector v of key usage statistics) so that values with highest deviation from the mean do not affect the clone detection protocol. Any adversary that launches a Byzantine attack will have to report values near to those reported by the legitimate nodes if their values need to be a part of the decision-making process. The protocol is thus robust against Byzantine attacks. The average number of nodes in a single multicast group is n_c. Hence, $(n_c - 1)$

messages have to be transmitted for the keyserver to collect the counting Bloom filters from every node in its multicast group. This amounts to $k * (n_c - 1)$ for the entire network of k keyservers.

A secure private key has to be established between every pair of keyservers to exchange the compressed counting Bloom filters. Using Shamir's three-pass protocol, $3k(k - 1)/2$ messages have to be transmitted to set up a secure connection between every pair of keyservers. We require additional $k(k - 1)$ messages to exchange counting Bloom filters.

Total number of messages for group agreement is $k(n_c - 1) + 5(k^2 - k)/2$ (derived in Ref. [7]).

9.5.3 Security Analysis

9.5.3.1 Byzantine Attack

In Byzantine attacks, malicious nodes try to force disagreement among legitimate nodes. Consider a network of N nodes scattered randomly on unit area. Let c of the n nodes be the adversaries. The probability that an illegitimate node gets selected as the key server is the same as any other node in the network, that is, c/n. If we pick k nodes at random as keyservers, the expected number of cloned keyservers is $(c * k)/n$. According to the group agreement protocol in Section 9.5.2, we have to introduce redundancy in the network. To tolerate c clones in the entire network, we have to select $\max\{((3c * k)/n) + 1, k\}$ keyservers.

9.5.3.2 Sybil Attack

In Sybil attacks, compromised nodes try to maintain multiple identities. Multiple Sybil identities can drain resources from legitimate nodes by sending spurious traffic. However, a Sybil attack from any node picked as a keyserver on the sensor network makes it easier to detect that adversary [13]. The reason is that a compromised keyserver reports the existence of multiple keyservers. This leads to an increase in the density of reported keyservers. We have a prior threshold on the density of keyservers in the sensor network. Since the keyservers are selected at random, their density should remain within a predictable variance from the mean. A Sybil attack will increase the keyserver density when the multicast region is contained entirely in the field.

9.5.3.3 Cloning Attacks

In a cloning attack, the adversary fills the network with copies of compromised nodes. However, the keyservers come to a consensus on which keys are cloned on the basis of predefined thresholds. Any key that is used more than a given threshold times is a candidate to be a suspected cloned node. If the mean usage and variance of a key is above the specified threshold, the keyserver calculates a confidence on the suspected cloned key. The cloned nodes are removed from the network by terminating all connections that use the set of cloned keys.

9.5.4 Optimal Layout of Multicast Groups

It is important to find the number of keyservers (k) the network needs and the size in hops (h) of the multicast region they should manage. A larger overview of this multicast sensor network security scheme is available in Ref. [7].

In the use of multicast communications to secure sensor networks, a node transmits messages securely within a local multicast group by encrypting the message using a shared symmetric key. Each member of the multicast group reads the message by decrypting it locally. A packet is reencrypted only when moving between different multicast regions. When data are shared within regions, approaches using multicast communication require fewer encryptions for secure message exchange, resulting in a net power savings. Each multicast region has a single keyserver that manages key distribution within the region. The trade-off between the number of multicast regions and the size of each region becomes vital to maintaining minimum message transmission overhead and reducing power consumption due to computation, while at the same time ensuring security.

9.5.4.1 Phase Change Analysis

Consider a surveillance network charged with reporting when a member of a class of objects (targets) traverses a given surveillance domain (terrain). Reports are sent to a user community that we assume, for the sake of discussion, is external to the terrain. The network will be *viable* as long as it assures that (1) an object traversing the terrain is detected (with acceptable error rates) and (2) the user community is alerted.

Ad hoc networks with range-limited communications exhibit phase change phenomena like those found in random graph and percolation theories. Random graph theory is a branch of graph theory that assigns probability distributions to the existence of edges between vertices. Percolation theory, a branch of physics, studies fluid flows in random media. In these models, network behavior has two phases. In the first phase, the probability of connection between nodes is small and the network has a large number of isolated components. As the connection probability grows, the expected size of the largest component grows logarithmically. In the second phase, the network is dominated by a unique giant component that contains most of the system nodes. There are still isolated holes in the network. The size of the largest hole shrinks logarithmically as the connection probability increases. The transition between these two phases, called the phase change, is extremely steep.

It is thus vital for any application to predict this phase transition since it defines the viability of the network. In this chapter, we use the random graph models shown in Figure 9.16 to model sensor networks.

Formally, we model the sensor network as a random graph $G = (V, E)$. The set of vertices V corresponds to the set of sensor nodes, and the elements of the set of edges E are communication links between the sensor nodes.

To analyze the graph, we use a probabilistic connectivity matrix M_h where each element (i,j) is the probability of a connection between nodes i and j in h hops. We assume the graph is undirected. The communication radius is denoted by r. A node can establish a connection with another node in a single hop only if both nodes fall within this communication radius. We normalize the value of r from [0 ... 1]. For range-limited graphs, element (i,j) of the probabilistic connectivity matrix (probability of an edge connecting node j to node i) has the following value (see Ref. [14] for derivation):

$$(2c - c^2) \tag{9.12}$$

where c is a constant defined as

$$c = \begin{cases} r^2 - \left(\dfrac{i}{n+1} - \dfrac{j}{n+1}\right)^2 & \text{if } r^2 \geq \left(\dfrac{i}{n+1} - \dfrac{j}{n+1}\right)^2 \\ 0 & \text{otherwise} \end{cases} \tag{9.13}$$

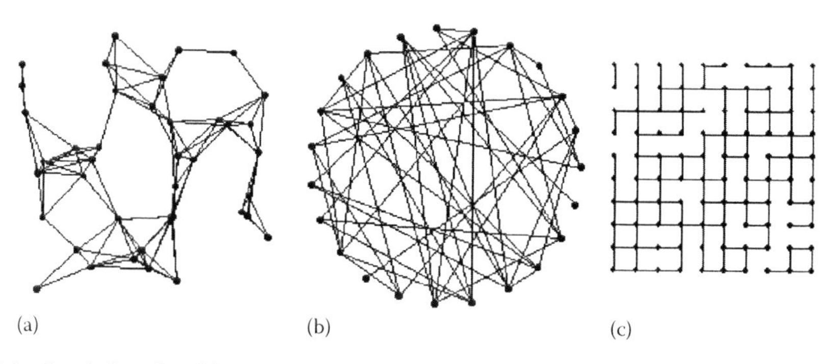

FIGURE 9.16 Graphs based on (a) range-limited model, (b) Erdos–Renyi model, and (c) percolation theory (regular tessellation of nodes).

Equation 9.14 looks for paths from node i to node j by considering paths passing through all possible intermediate nodes. Constraining diagonals in the connectivity matrix to zero removes consideration of a node as its own intermediate node:

$$p_{ij}^{(2)} = 1 - \Pi_{l=1,l\neq i,l\neq j}^{n}(1 - p_{il}^{(1)} * p_{lj}^{(1)}) \tag{9.14}$$

where

$p_{ij}^{(2)}$ is the probability a walk of two hops exists edge between nodes i and j
$p_{ij}^{(1)}$ is the probability an edge exists between nodes i and l
$p_{lj}^{(1)}$ is the probability an edge exists between nodes l and j

To avoid edge effects, we consider nodes $i = [n/2]$ and $j = [n/2] + 1$ to find the phase change. Erdos and Renyi [16] defined a graph topology where there is an equal probability an edge exists between any two vertices. Reference [7] provides theorems that map the connectivity graph for multicast regions to an Erdos–Renyi topology. The likelihood a path of $2h - 1$ hops exists between any two nodes chosen at random on the range-limited graph will be the same. We can therefore consider the keyserver connectivity graph as an Erdos–Renyi graph of k nodes, where k is the number of keyservers. This network of keyservers is modeled as an Erdos–Renyi graph overlaid on the ad hoc network. The phase change for the secure communication network occurs when

$$k = 2 + \frac{\log(1 - p_{ij}^{(2h-1)})}{\log(1 - (p_{ij}^{(2h-2)})^2)} \tag{9.15}$$

where k is the number of keyservers, the keyserver serves all nodes within h hops, and p_h is the probability of a walk of h or fewer hops existing between nodes with the labels $i = [n/2]$ and $j = [n/2] + 1$ for the ad hoc network model.

Figure 9.17 shows phase change for the ad hoc network with communication range $r = 0.07$. The filled circle is the predicted inflection point. Error bars for 95% confidence intervals are shown. The

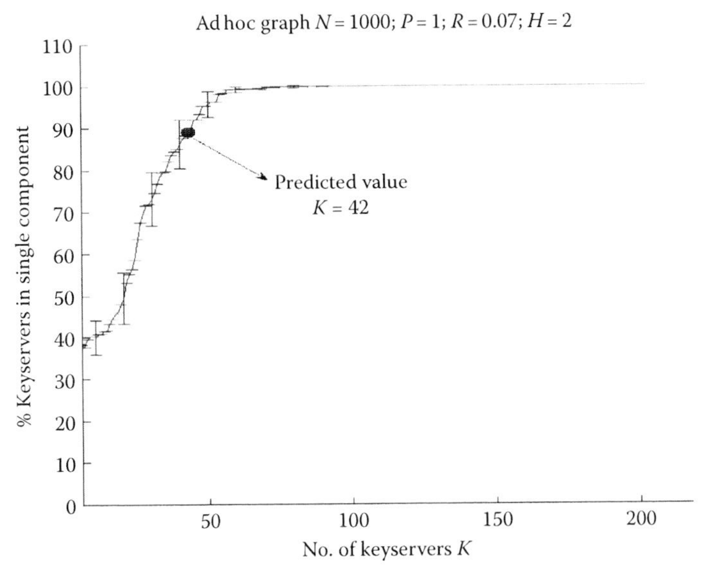

FIGURE 9.17 Percent of keyservers included in the giant component versus the number of keyservers in a network above the percolation threshold.

graphs show the mean of 35 repetitions. The approach predicts that the point of phase change is at 42 keyservers. At this point, 88% of the keyservers are in the same component.

9.5.4.2 Determination of Multicast Parameters

Security is maintained in the multicast network by a group agreement protocol that requires message overhead proportional to k^2 and n_c; k being the number of keyservers and n_c the number of sensor nodes in a multicast region. These security features favor a network with minimum keyservers to reduce the message overhead for group agreement. The number of sensor nodes in a multicast region is proportional to the number of hops h. The multicast architecture and group agreement protocol thus set up a trade-off between the number and size multicast regions, viz., k and h to minimize the message overhead.

We compute n_c as a function of h. The network of keyservers is modeled as an Erdos–Renyi random graph overlaid on a range-limited graph of sensor nodes. Every keyserver has an equal probability of communicating with another keyserver in the network within $2h$ hops. Each node can communicate with all other sensor nodes physically located within its communication range in a single hop. The area covered by this range is $\pi * r^2$. The mean field approximation p, the probability that any node is within range of a given node, is $\pi * r^2/A$, where A is the size of the field or region being surveyed. Each keyserver serves all nodes within h hops. The likelihood that a node is within h hops can therefore be estimated as $\pi * (h * r)^2/A$. The communication range of a node is a circle of radius r around it. The network of numerous sensor nodes is laid within a bounded region called the field. The sensing area does not completely overlap with the field for nodes scattered at the edges. If the size of a multicast region is h hops, then the region outside the field would be within a radius of at most $(h * r)$. We compensate for edge effects that are an artifact of our model by inflating the area considered by a factor of $(h * r)$. Node placement follows a binomial distribution. The probability density function of the number of nodes within h hops becomes phk, where

$$phk = \binom{n}{k} P^k (1-p)^{1-k} \quad \text{and} \quad P = \frac{\pi * (h * r)^2}{A * (1 + h * r)^2} \tag{9.16}$$

The expected number of nodes within h hops of the keyserver becomes

$$n_c = \sum_{k=0}^{n-1} k * phk \tag{9.17}$$

With counting Bloom filters in Section 9.3.1.1, a round of key agreement needed to detect compromised nodes requires an exchange of $k(n_c - 1) + 5(k^2 - k)/2$ messages [7]. The message overhead for the distributed key agreement protocol increases with the number of keyservers. Byzantine agreement protocol ensures that cloned keyservers do not falsify information when exchanging the counting Bloom filters. This can be done by introducing redundancy and allowing nodes to be served by multiple keyservers. This redundancy also improves the accuracy of the key usage statistics reported by the Bloom filters. Thus, both improve the accuracy of the key usage statistics reported by the Bloom filters. Thus, both improve the accuracy of the key usage statistics reported by the bloom filters. To minimize traffic for the agreement protocol, we need to minimize the number of keyservers. Thus, a trade-off exists between the number of keyservers and the number of hops. Since the total messages to set up k multicasts is $k(5(n_c - 1))$, the total number of messages for both is $Ms = k(6(n_c - 1) + 5(k - 1)/2)$. Thus, the optimization problem of minimizing Ms subject to k and h can be solved using gradient descent or any numerical optimization algorithm [17].

TABLE 9.1 Number of Messages Necessary to Establish a Network for Different Network Definition Parameters

Hops in Region	Number of Keyservers	Expected Number of Nodes in Region	Keyservers Needed to Tolerate 25 Clones	Messages
1	8	10.3	16	1492.8
2	4	25.8	8	1330.4
3	3	45.6	7	1978.2
4	3	56.3	7	2427.6

Nodes = 100; range = 0.2.

Table 9.1 shows the results for a network of 100 nodes with a communication range of 0.2. It is clear from the data that, for this instance, the network can be established with a minimum number of messages with four legitimate keyservers and two hops from each keyserver. Cluster size estimates from Ref. [7] give a cluster size of 26 nodes for a network with 100 nodes with a communication radius of 0.2. Hence, the optimal parameters for this network are to have at least four keyservers each with a cluster of all nodes within two hops of the keyserver.

However, assume that c nodes in the network are compromised. Since every node is equally likely to be elected as the keyserver (proved in Ref. [7]), the expected number of compromised keyservers is $[(c/n) * k']$. We need k legitimate keyservers to maintain the giant component so that the network is viable.

We pick k' keyservers to introduce redundancy required by the Byzantine agreement protocol [18]. The protocol discards 2τ (the τ largest and τ smallest) values, where τ indicates the number of adversaries the network can tolerate as keyservers. Hence, extra $2 * [(c/n) * k']$ keyservers are introduced. To tolerate c clones in the network of n nodes, we pick k' keyservers such that

$$k' = 2 * \left\lceil \frac{c}{n} * k' \right\rceil + k \tag{9.18}$$

Figure 9.18 plots the total messages required to initialize the network versus the number of hops in a multicast region. In this graph, when $h = 5$, the number of nodes in a single multicast region is on the order of total nodes in the network. The number of multicast regions reduces with increasing region size. However, the number of messages neither decreases nor increases uniformly with region size. In this

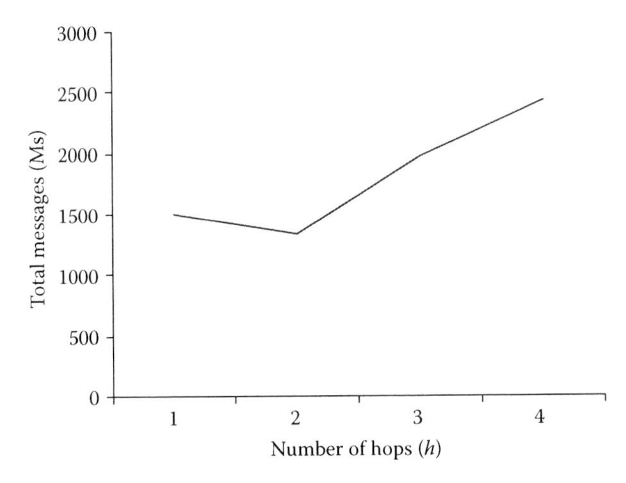

FIGURE 9.18 Plot of number of messages for multicast (communication range = 0.20) initialization for varying multicast size (number of hops).

instance, use of $h = 1$ results in 16 multicast regions. The number of regions drops to 8 when h increases to 2. Larger values of h do not significantly reduce the number of regions but result in additional overhead for key maintenance. This highlights the importance of calculating the minima to reduce overhead and power consumption.

Acknowledgments

This material is based upon the work supported in part by the Air Force Office of Scientific Research contract/grant number FA9550-09-1-0173, NSF grant EAGER-GENI Experiments on Network Security and Traffic Analysis contract/grant number CNS-1049765 and NSF-OCI 1064230 EAGER: Collaborative Research: A Peer-to-Peer based Storage System for High-End Computing. Opinions expressed are those of the author and neither the National Science Foundation nor the U.S. Department of Defense.

References

1. R. Brooks, D. Friedlander, J. Koch, and S. Phoha. Tracking multiple targets with self-organizing distributed ground sensors* 1. *Journal of Parallel and Distributed Computing*, 64(7):874–884, 2004.
2. D.W. Carman, P.S. Kruus, and B.J. Matt. Constraints and approaches for distributed sensor network security (final). DARPA Project Report (Cryptographic Technologies Group, Trusted Information System, NAI Labs), 2000.
3. F. Zhao and L.J. Guibas. *Wireless Sensor Networks: An Information Processing Approach*. Morgan Kaufmann Publishers, Waltham, MA, 2004.
4. L. Eschenauer and V.D. Gligor. A key-management scheme for distributed sensor networks. In *Proceedings of the 9th ACM Conference on Computer and Communications Security*, Washington, DC, November 18–22, 2002, pp. 41–47, ACM, New York.
5. H. Chan and A. Perrig. Security and privacy in sensor networks. *IEEE Computer Magazine*, 36(10):103–105, 2003.
6. A.D. Wood and J.A. Stankovic. Denial of service in sensor networks. *Computer*, 35(10):54–62, 2002.
7. B. Pillai. Network embedded support for sensor network security. PhD thesis, Clemson University, Clemson, SC, 2006.
8. R. Brooks, P.Y. Govindaraju, M. Pirretti, N. Vijaykrishnan, and M.T. Kandemir. On the detection of clones in sensor networks using random key predistribution. *IEEE Transactions on Systems, Man, and Cybernetics, Part C: Applications and Reviews*, 37(6):1246–1258, 2007.
9. B. Krishnamachari, S.B. Wicker, and R. Bejar. Phase transition phenomena in wireless ad hoc networks. In *Global Telecommunications Conference, 2001, GLOBECOM'01*. Vol. 5, pp. 2921–2925, IEEE, New York, 2001.
10. R.R. Brooks, P.Y. Govindaraju, M. Pirretti, N. Vijaykrishnan, and M. Kandemir. Clone detection in sensor networks with ad hoc and grid topologies. *International Journal of Distributed Sensor Networks*, 5(3):209–223, 2009.
11. J. Heidemann, F. Silva, C. Intanagonwiwat, R. Govindan, D. Estrin, and D. Ganesan. Building efficient wireless sensor networks with low-level naming. In *Proceedings of the Eighteenth ACM Symposium on Operating Systems Principles, SOSP'01*, Banff, Alberta, Canada, pp. 146–159, ACM, New York, 2001.
12. M. Pirretti, S. Zhu, N. Vijaykrishnan, P. McDaniel, M. Kandemir, and R. Brooks. The sleep deprivation attack in sensor networks: Analysis and methods of defense. *International Journal of Distributed Sensor Networks*, 2(3):267, 2006.
13. R.R. Brooks, P. Brijesh, P. Matthew, C. Michele et al. Multicast encryption infrastructure for security in sensor networks. *International Journal of Distributed Sensor Networks*, 5(2):139–157, 1900.
14. R.R. Brooks, B. Pillai, S. Racunas, and S. Rai. Mobile network analysis using probabilistic connectivity matrices. *IEEE Transactions on Systems, Man, and Cybernetics, Part C: Applications and Reviews*, 37(4):694–702, 2007.

15. M. Barborak, A. Dahbura, and M. Malek. The consensus problem in fault-tolerant computing. *ACM Computing Surveys (CSUR)*, 25(2):171–220, 1993.
16. P. Erdős and A. Rényi. On the evolution of random graphs. *Bulletin de L' Institut International de Statistique*, 38(4):343–347, 1960.
17. P. Wolfram, et al. *Mathematica Documentation*. Wolfram Research Inc., 2012. http://reference.wolfram.com/legacy/V5_2/book/section-3.9.8.
18. R.R. Brooks and S.S. Iyengar. *Multi-Sensor Fusion: Fundamentals and Applications with Software*. Prentice Hall, Inc., Upper Saddle River, NJ, 1998.

10

Quality of Service Metrics with Applications to Sensor Networks

N. Gautam
Texas A&M University

10.1 Service Systems

The phrase "quality of service" (QoS) has been popular for over 20 years; however, there has been little or no consensus in terms of what QoS actually is, what various QoS metrics are, and what QoS specifications are. Yet, QoS has spread far and wide, beyond the realm of networking (where the term QoS was first used), into areas such as transportation, health care, hospitality, manufacturing, etc. In fact, the author believes that it may be better to introduce QoS using examples from the service industry to provide appropriate analogies in the hope of giving the study of QoS more structure as well as discover newer ways of providing QoS in computer networks and then finally get more specific to discuss QoS in distributed sensor networks. To define a service industry, one must first differentiate between goods that are usually tangible and services that are typically intangible. In fact, several organizations that have been traditionally concentrating on their goods (such as cars at vehicle manufacturers, food at restaurants, books at bookstores, etc.) are now paying a lot of attention to service (such as on-time delivery, availability, warranties, return policies, overall experience, etc.). These typically fall under the realm of providing QoS.

10.1.1 Elements of a Service System

Examples of service systems range from complex interconnected networks such as computer communication networks, transportation systems, theme parks, etc., to simpler individual units such as a barber shop, repair shops, theaters, restaurants, hospitals, hotels, etc. In all these examples, two key players emerge, namely, the service provider and users. As the names suggest, the users receive service provided by the service provider. Users (also called customers if there is money involved) do not necessarily have to be humans; they could be other living or nonliving entities. Further, users do not have to be single individuals, they could be part of a group (such as in a multicast session, in a restaurant, at a play, etc.). On the same token, for a given system, there could be zero, one, or many service providers. Although most services are such that they are owned by a single entity (the one to blame if things go wrong), there are some (including the Internet) that are owned by several groups.

QoS can be defined as a set of measures that the users "want" from the system (or sometimes what the service provider wants to give the users). What the users eventually "get" is called performance. From a physical goods standpoint, QoS is equivalent to specifications (or specs as they are usually abbreviated). Some QoS measures are qualitative (such as taste, ambiance, etc.) and these are hard to provide since different users perceive them differently. Other QoS measures that are quantitative also have some fuzziness attached. For example, on one day a user might find a 90 ms latency intolerable, and on another day the user may find 100 ms latency tolerable. There could be several reasons for that including the mood of the user, the expectations of the user, etc. Capturing such cognitive aspects is beyond the scope of this chapter. We focus on systems where user requirements (i.e., QoS) are known precisely and users are satisfied or unsatisfied if the requirements are met or not met, respectively. That means if 100 ms is the tolerance for latency, then QoS is met (not met) if latency is lesser (greater) than 100 ms.

In some service systems, the users and the service providers negotiate to come up with what is known as a *service level agreement* (SLA). For example, years ago, a pizza company promised to deliver pizzas within 45 min, or the pizzas are free. That is an example of an SLA, which is also called a QoS guarantee. In many service systems, there is not an explicit guarantee, but a QoS indication such as "your call will be answered in about 3 minutes," "the chances of a successful surgery is 99.9%," "the number of defective parts is in the order of one in a million," etc. In many systems, it is not possible to provide absolute QoS guarantees due to the dynamic nature of the system, but it may be feasible to deliver relative QoS. This is typically known as *level of service* (LoS) where, for example, if there are three LoSs, gold, silver, and bronze, then at a given time instant, gold level will get better QoS than silver level which would get a better QoS than bronze level.

10.1.2 Customer Satisfaction

Although many consider QoS and customer satisfaction as one and the same, here QoS is thought of as only a part of customer satisfaction. However, it is not assumed here that providing QoS is the objective of a service provider, but providing customer satisfaction is. With that understanding, the three components of customer satisfaction are as follows: (1) QoS, (2) availability, and (3) cost. The service system (with its limited resources) can be considered either physically or logically as one where customers arrive, enter the system, if resources are available, obtain service for which they incur a cost, and then leave the system (see Figure 10.1). One definition of availability is the fraction of time arriving customers enter the system. Thereby, QoS is provided only for customers that "entered" the system. From an individual customer's standpoint, the customer (i.e., user or application) is satisfied if the customer's requirements over time on QoS, availability, and cost are satisfied. Some service providers' objective is to provide satisfaction aggregated over all customers (as opposed to providing absolute service to an individual customer). Services such as restaurants provide both: they cater to their frequent customers on one hand, and on the other hand they provide overall satisfaction to all their customers.

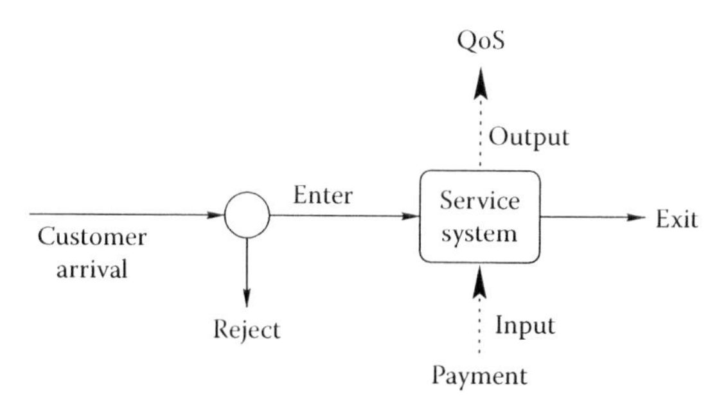

FIGURE 10.1 Customer satisfaction in a service system.

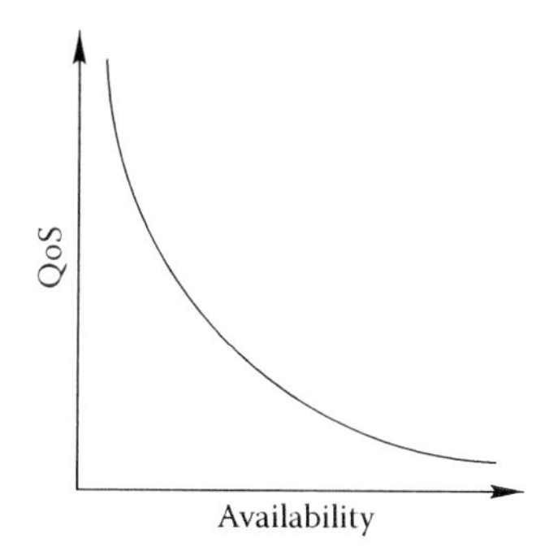

FIGURE 10.2 Relationship between QoS and availability, given cost.

The issue of QoS (sometimes called conditional QoS as the QoS is conditioned upon the ability to enter the system) versus availability needs further discussion. Consider the analogy of visiting a medical doctor. The ability to get an appointment translates to availability; however, once an appointment is obtained, QoS pertains to the service rendered at the clinic such as waiting time, experience, healing time, etc. Another analogy is airline travel. Getting a ticket on an airline at a desired time from desired source to desired destination is availability. QoS measures include delay, smoothness of flight, in-flight service, etc. One of the most critical business decisions is to find the right balance between availability and QoS. The two are inversely related as illustrated in Figure 10.2. A service provider can increase availability by decreasing QoS and vice versa. A major factor that could affect QoS and availability is cost. Usually with cost (somewhat related to pricing), there is a need to segregate the customers into multiple classes. It is not necessary that classes are based on cost, they could also depend on customer type (i.e., applications) and QoS requirements. The ability to provide appropriate customer satisfaction based on class (and relative to other classes) is a challenging problem especially under conditions of stress, congestion, unexpected events, etc. For example, if an airplane encounters turbulence, all customers experience the bumpy ride, irrespective of the class of service.

10.1.3 Effect of Resources and Demand

Customer satisfaction is closely related to both resources available at the service provider as well as demand from the customers. It is of grave importance to understand the relationship and predict or estimate it. First, consider the relationship between resource and performance. The graph of resources available at a service provider versus the performance the service provider can offer is usually as described in Figure 10.3. From Figure 10.3, the following are evident: (1) it is practically impossible to get extremely high performance, and (2) to get a small increase in performance, it would sometimes even require twice the amount of resources, especially when the available performance is fairly high in the first place.

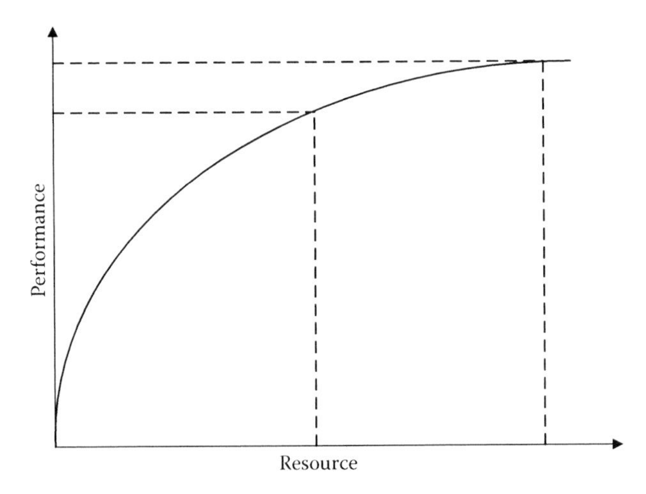

FIGURE 10.3 Resource versus performance.

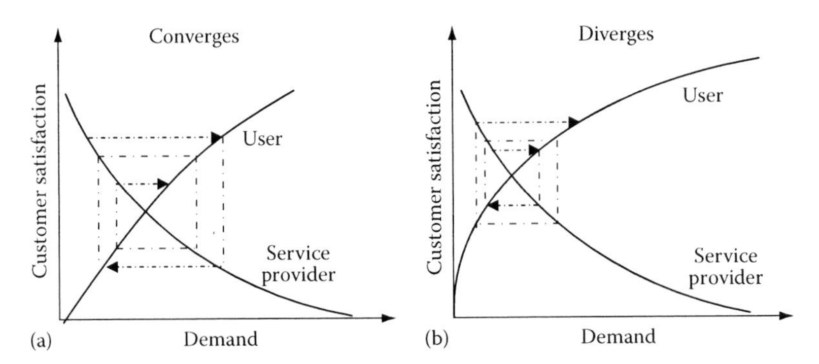

FIGURE 10.4 Demand versus customer satisfaction under scenarios of convergence (a) and divergence (b).

Now consider the relationship between customer satisfaction and demand. If a service offers excellent customer satisfaction, very soon its demand would increase. However, if the demand increases, the service provider would no longer be able to provide the high customer satisfaction that eventually deteriorates. Thereby, demand decreases. This cycle continues until one of two things happens, either the system reaches an equilibrium or the service provider goes bankrupt. The situations are depicted in Figure 10.4. Notice the relation between customer satisfaction and demand: they are inversely related from the service provider's standpoint and directly related from a customer's standpoint.

After studying some general aspects of providing QoS in the service industry, we now turn our attention to QoS provisioning in networking in the next section.

10.2 QoS in Networking

Although the Internet started as a free and best-effort service, the next generation Internet is showing signs of becoming a network based on both pricing issues as well as QoS. The concept, need, and application of QoS have also quickly spread to other high-speed networks such as telephony, peer-to-peer networks, sensor networks, ad hoc networks, private networks, etc. For the remainder of this chapter on QoS, we use networking rather loosely in the context of any of the aforementioned networks. We will not be restricting ourselves to any particular network or protocols running on the network. It may be best to think of an abstract network with nodes (or vertices) and arcs (or links). In fact, the nodes and arcs can be either physical or logical in reality. At each node, there are one or more queues and processors that forward information. With this setting in mind, we proceed to investigate the broad topic of QoS in high-speed networks.

10.2.1 Introduction

The plain old wired telephone network, one of the oldest computer communication networks, is the only high-speed network that can provide practically perfect QoS, at least in the United States. One of the positive aspects of this is that they give other networks (such as the Internet) time and would eventually become as effective as the telephone networks in terms of providing QoS. Having said that, it is important to notice that under some cases the telephone network does struggle to provide QoS; one example is during the terrorist attacks on September 11, 2001, it was impossible to dial into several areas; another example is international phone calls. There is a long way to perfecting QoS for that. With the advent of cell phones, the telephone networks are now facing a new challenge of providing QoS to wireless customers.

From a networking standpoint (for Internet-type networks), one of the difficulties for providing QoS is the presence of multiple classes of traffic such as voice, video, multimedia, data, web, etc. Unlike wired telephone networks that offer a bandwidth of about 60kbps for whatever type of call (regular phone calls, fax, modem call, etc.), in networking, various types of traffic have varying requirements. In fact, real-time traffic can tolerate some loss but very little delay; however, non-real-time traffic cannot tolerate loss but can take a reasonable amount of delay. The number of network applications is increasing steadily; however, it is not practical to support more than a handful of classes (two to four classes). Therefore, clever traffic aggregation schemes need to be developed.

In order to provide QoS in a multiclass network, some of the important aspects to consider and optimize are scheduling, switching, and routing. For differentiating the various classes of traffic as well as providing different QoS, information needs to be processed in a manner other than first-come, first-served (FCFS). The essence of scheduling is to determine what information to serve next. The telephone network that essentially has only one class of traffic does not do any special scheduling. It just does switching and routing. Switching is done using circuit switching policies where, upon dialing a number, a virtual path is created from the source to the destinations through which information is transmitted. An appropriate routing algorithm is used to efficiently send information from the source to the destination. From a networking standpoint, doing appropriate scheduling, switching, and routing, it would be possible to provide QoS to the users. How to do that is being pursued actively by the research community.

10.2.2 Characteristics of Network QoS Metrics

Before looking at how to provision QoS in networking applications, it is important to understand what the QoS metrics are in the first place and how to characterize them. There are four main QoS metrics in networking: delay, jitter, loss, and bandwidth. There are other derived metrics used in sensor networks, but they are all a function of some or all of the main QoS metrics.

- *Delay*: It is defined as the time elapsed between when a node leaves a source and reaches a destination. Though the term delay implies there is a target time and the information comes after the target time elapses, that really is not the case. It is just a measure of travel time from source to destination, which is also called latency or response time.
- *Jitter*: The variation in the delay is termed as jitter. If a stream of packets are sent from a source to a destination, typically all packets do not face the same delay. Some packets experience high delays and others experience low delays. Applications such as video-on-demand can tolerate delays but not jitter. A simple way of eliminating or reducing jitter is to employ a jitter buffer at the destination. All packets are collected and then transmitted. This does increase the delay though. For that reason, it is not common to see jitter, instead most articles focus on delay.

- *Loss*: When a piece of information arrives at a node at a time when the queue at the node is full (i.e., full buffer) or the node is not available for other reasons, then the information is dropped (or lost). This is known as loss. There are several measures of loss including loss probability (the probability that a piece of information can be lost along its way from its source to its destination) and loss rate (the average amount of information lost per unit time in a network or node).
- *Bandwidth*: Several real-time applications such as voice over IP, video-on-demand, etc., require a certain bandwidth (in terms of bytes per second) to be available for successful transmission. In fact, the only QoS guarantee a telephone network provides is bandwidth (of about 60 kbps).

Note: As alluded to in Section 10.1.2, while studying QoS, the concept of availability is skipped; however, it is very important from a customer satisfaction standpoint to consider availability.

Performance metrics can be typically classified into three parts: additive, multiplicative, and minimal. In order to explain them, consider a traffic stream that originates at a particular node and traverses through N nodes before reaching its destination. The objective is to obtain end-to-end performance metrics given metrics across nodes. For example, consider node i (for $i \in [1,N]$), let d_i, ℓ_i, and b_i, respectively, be the delay, loss probability, and bandwidth across node i. Assume that the performance metrics across node i are independent of all other nodes. To compute end-to-end performance measures, the following are used:

- *Additive*: The end-to-end performance measure is the sum of the performance measures over the individual nodes along the path or route. The end-to-end delay (D) for our aforementioned example is obtained as

$$D = d_1 + d_2 + \cdots + d_N = \sum_{j=1}^{N} d_i$$

- *Multiplicative*: The end-to-end performance measure is the product of the performance measures over the individual nodes along the path or route. The end-to-end loss (L) for our aforementioned example is obtained as

$$L = 1 - (1 - \ell_1)(1 - \ell_2) \cdots (1 - \ell_N)$$

  Note that the multiplicative metric can be treated as an additive metric by taking the logarithm of the performance measure.
- *Minimal*: The end-to-end performance measure is the minimum of the performance measures over the individual nodes along the path or route. The end-to-end bandwidth (B) for our aforementioned example is obtained as the minimum bandwidth available across all the nodes in its path. In particular,

$$B = \min\{b_1, b_2, \ldots, b_N\}$$

Although all performance metrics are inherently stochastic and time varying, in order to keep analysis tractable, the following are typically done: replace a metric by its long run or steady state or stationary equivalent; use an appropriate deterministic value such as maximum or minimum or mean or median or mode; use a range of meaningful values. Now, in order to guarantee QoS, depending on whether a deterministic or stochastic performance metric is used, the guarantees are going to be either absolute or probabilistic, respectively. For example, you could give an absolute guarantee that the mean delay is going to be less than 100 ms. Or you could say that the probability that the delay is greater than 200 ms is less than 5%. It is also possible to get bounds on the performance, and it is important to note that giving deterministic bounds could mean underutilization of resources and thereby very poor availability. Once again, it is crucial to realize that for a given infrastructure, the better QoS guarantees one can provide, the worse off will be availability (see Figure 10.2).

10.3 Systems Approach to QoS Provisioning

In this section, we focus on obtaining performance metrics that form the backbone of QoS analysis. There are several methodologies to evaluate the performance of a system and they can be broadly classified into experimental, simulation-based, and analytical methods. Experimental methods tend to be expensive and time consuming, whereas they require very little approximations. The analytical models are just the opposite. Simulation-based techniques fall in the middle of the spectrum. This chapter focuses on obtaining analytical results that would be mainly used in making optimal design and admission control decisions. These can be appropriately used for strategic, tactical, and operational decisions depending on the time granularity. In this section, we present two main performance analysis tools based on queueing theory and large deviations theory.

10.3.1 Performance Analysis Using Queueing Models

We begin this section by considering a single-station queue and then extend the theory to a network of queues. From an analysis standpoint, the most fundamental queueing system is the *M/M/*1 queue. Input to the queue is according to a Poisson process with average rate λ per unit time (i.e., interarrival times exponentially distributed). Service times are exponentially distributed with mean $1/\mu$. It is a single-server queue with infinite waiting room and FCFS service. The following performance measures can be derived (under the assumption $\lambda < \mu$): average number in the system is $\lambda/(\mu - \lambda)$ and average waiting time in the system is $1/(\mu-\lambda)$. For the *M/M/*1 queue, distribution of the waiting times is given by

$$P\{\text{waiting time} \le x\} = 1 - e^{-(\mu-\lambda)x}$$

In this way, other generalizations to this model such as a different arrival process, a different service time distribution, or more number of servers; finite waiting room; different order of service, etc., can be studied. The reader is referred to one of several standard texts on queues (such as Refs [1,10]).

Now we turn to a network of queues, specifically what is known as a Jackson Network. The network consists of N service stations (or nodes). There are s_i servers at node i. Service times at node i are exponentially distributed and independent of those at other nodes. Each node has infinite waiting room. Externally, customers arrive at node i according to a Poisson process with mean rate θ_i. Upon completion of service at node i, a customer departs the system with probability r_i or joins the queue at node j with probability p_{ij}. Assume that at least one node has arrivals externally and at least one node has customers departing the system. The vector of effective arrival rates $\lambda = (\lambda_1\lambda_2 \ldots \lambda_N)$ can be obtained using

$$\lambda = \theta(I - P)^{-1}$$

where

$\theta = (\theta_1\theta_2\ldots\theta_N)$
P is the routing probability matrix composed of various $[p_{ij}]$ values
I is an $N \times N$ identity matrix

Then each queue i can be modeled as independent single station queues. Jackson's theorem states that the steady-state probability of the network can be expressed as the product of the state probabilities of the individual nodes. An application of Jackson networks is illustrated in Section 10.4.1.

10.3.2 Performance Analysis Using Large Deviations Theory

In this section, we focus on using the principle of large deviations for performance analysis of networks based on fluid models. Although large deviations do not require fluid traffic, the reason we pay attention to it is that fluid models represent correlated and long-range-dependent traffic very well. In fact, the simple on–off source could be thought of as one that generates a set of packets back to back when it is on and nothing flows when it is off.

Let $A(t)$ be the total amount of traffic (fluid or discrete) generated by a source (or flowing through a pipe) over time $(0,t]$. For the following analysis, consider a fluid model. Note that it is straightforward to perform similar analysis for discrete models as well. However, the results will be identical. Consider a stochastic process $\{Z(t), t \geq 0\}$ that models the traffic flow. Also let $r(Z(t))$ be the rate at which the traffic flows at time t. Then

$$A(t) = \int_0^t r(Z(u))du$$

The asymptotic log-moment generating function (ALMGF) of the traffic is defined as

$$h(v) = \lim_{t\to\infty} \frac{1}{t} \log E\{\exp(vA(t))\}$$

Using the previous equation, it is possible to show that $h(v)$ is an increasing, convex function of v and for all $v > 0$,

$$r^{mean} \leq h'(v) \leq r^{peak}$$

where

$r^{mean} = E(r(Z(\infty)))$ is the mean traffic flow rate
$r^{peak} = \sup_z\{r(z)\}$ is the peak traffic flow rate
$h'(v)$ denotes the derivative of $h(v)$ with respect to v

The effective bandwidth of the traffic is defined as

$$eb(v) = \lim_{t\to\infty} \frac{1}{vt} \log E\{\exp(vA(t))\} = \frac{h(v)}{v}$$

It can be shown that $eb(v)$ is an increasing function of v and

$$r^{mean} \leq eb(v) \leq r^{peak}$$

Also,

$$\lim_{v\to 0} eb(v) = r^{mean} \quad \text{and} \quad \lim_{v\to\infty} eb(v) = r^{peak}$$

It is not easy to calculate effective bandwidths using the formula provided earlier. However, when $\{Z(t), t \geq 0\}$ is a continuous time Markov chain [6,12] or a semi-Markov process [15], one can compute the effective bandwidths more easily. Also, see Ref. [14] for the calculation of effective bandwidths for traffic modeled by a fractional Brownian motion.

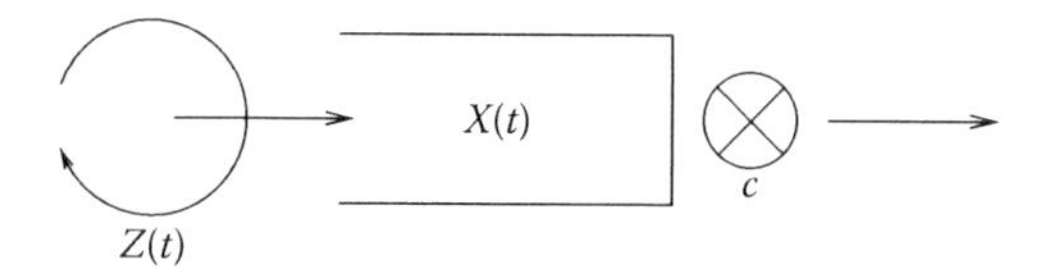

FIGURE 10.5 Single buffer fluid model.

Consider a single buffer fluid model as depicted in Figure 10.5. Input to a buffer of size B is driven by a random environment process $\{Z(t), t \geq 0\}$. When the environment is in state $Z(t)$, fluid enters the buffer at rate $r(Z(t))$. The output capacity is c. Let $X(t)$ be the amount of fluid in the buffer at time t. We are interested in the limiting distribution of $X(t)$, that is,

$$\lim_{t \to \infty} P\{X(t) > x\} = P\{X > x\}$$

Assume that the buffer size is infinite. In reality, the buffer overflows (hence packets/cells are lost) whenever $X(t) = B$ and $r(Z(t)) > c$. Note that the buffer content process $\{X(t), t \geq 0\}$ (when $B = \infty$) is stable if the mean traffic arrival rate is less than c, that is,

$$E\{r(Z(\infty))\} < c$$

Then $X(t)$ has a limiting distribution. From the limiting distribution, use $P\{X > B\}$ as an upper bound for loss probability (remember that B is the actual buffer size). This can also be used for delay QoS. Fluid arriving at time t waits in the buffer (hence faces a delay) for $X(t)/c$ amount of time. Therefore, the long-run probability that the delay across the buffer is greater than δ is

$$P\{\text{delay} > \delta\} = P\{X > c\delta\}$$

Using results from large deviations, it is possible to show that for large values of x (specifically as $x \to \infty$),

$$P\{X > x\} \approx e^{-\eta x}$$

where η is the solution to

$$eb(\eta) = c$$

Note that the previous expression is an approximation, and, in fact, researchers have developed better approximations (see Ref. [7]) and bounds (see Ref. [9]). In fact, Elwalid and Mitra [6] derive exact expressions for $P\{X > x\}$ for any continuous time Morkov chain (CTMC) environment $\{Z(t), t \geq 0\}$ process. We use these results and extensions in an example in Section 10.4.2.

To extend the single node results to a network of nodes, we need two important results. They are summarized as follows:

- *Effective bandwidth of output*: Refer to Figure 10.5. Let $D(t)$ be the total output from the buffer over $(0,t]$. The ALMGF of the output is

$$h_D(v) = \lim_{t \to \infty} \frac{1}{t} \log E\{\exp(vD(t))\}$$

The effective bandwidth of the output traffic from the buffer is

$$eb_D(v) = \lim_{t \to \infty} \frac{1}{vt} \log E\{\exp(vD(t))\}$$

Let the effective bandwidth of the input traffic be $eb_A(v)$. Then the effective bandwidth $eb_D(v)$ of the output can be written as

$$eb_D(v) = \begin{cases} eb_A(v) & \text{if } 0 \le v \le v^* \\ c - \frac{v^*}{v}\{c - eb_A(v^*)\} & \text{if } v > v^* \end{cases}$$

where v^* is obtained by solving for v in the equation

$$\frac{d}{dv}\left[h_A(v)\right] = c$$

For more details refer to Refs. [2–4].

- *Multiplexing independent sources*: Consider a single buffer that admits a single-class traffic from K independent sources. Each source k ($k = 1, \ldots, K$) is driven by a random environment process $\{Z^k(t), t \ge 0\}$. When source k is in state $Z^k(t)$, it generates fluid at rate $r^k(Z^k(t))$ into the buffer. Let $eb_k(v)$ be the effective bandwidths of source k such that

$$eb_k(v) = \lim_{t\to\infty} \frac{1}{vt} \log E\{\exp(vA_k(t))\}$$

where

$$A_k(t) = \int_0^t r^k(Z^k(u))du$$

Let η be the solution to

$$\sum_{k=1}^{K} eb_k(\eta) = c$$

Notice that the effective bandwidth of independent sources multiplexed together is the sum of the effective bandwidths of the individual sources. The effective bandwidth approximation for large values of x yields

$$P\{X > x\} \approx e^{-\eta x}$$

10.4 Case Studies

In this section, case studies are presented to illustrate various performance analysis methodologies as well as to obtain various QoS metrics. The examples are kept simple purely for the purpose of illustration.

10.4.1 Case 1: Delay and Jitter QoS Metrics Using Queueing Networks

PROBLEM 10.1

Consider a system of servers arranged as shown in Figure 10.6. Assume that requests arrive according to a Poisson process and enter node 1 at an average rate of 360/min. These requests can exit the system from nodes 2, 4, 5, 6, or 7. The processing time for each request in node j (for $j = 1, \ldots, 7$) is exponentially

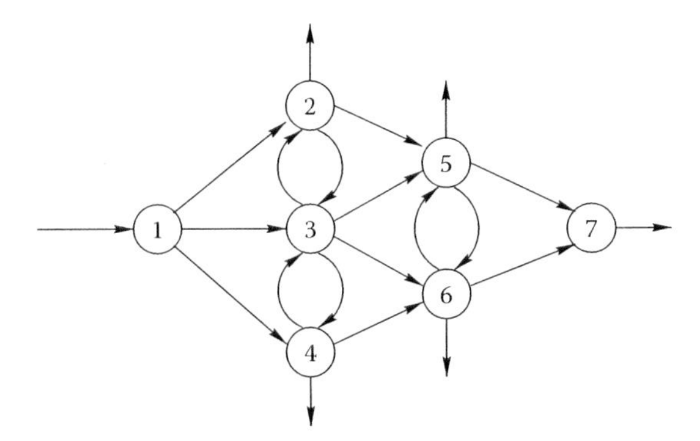

FIGURE 10.6 Server configuration.

distributed with mean $1/\mu_j$ min. The vector of processing rates assume the following numerical values $[\mu_1\ \mu_2\ \mu_3\ \mu_4\ \mu_5\ \mu_6\ \mu_7] = [400\ 200\ 300\ 200\ 200\ 200\ 150]$. There is a single server at each node. When processing is complete at a node, the request leaves through one of the out-going arcs (assume that the arcs are chosen with equal probability). The objective is to obtain the average delay and jitter experienced by all requests. In addition, the average delay and jitter experienced by a particular arriving request that goes through nodes 1–2–5–7 and then exits the network is to be determined. Note that in this example, jitter is defined as the standard deviation of delay.

Solution

The system can be modeled as a Jackson network with $N = 7$ nodes or stations. The external arrival rate vector $\theta = [\theta_1\ \theta_2\ \theta_3\ \theta_4\ \theta_5\ \theta_6\ \theta_7]$ is

$$\theta = [360\ 0\ 0\ 0\ 0\ 0\ 0]$$

The routing probabilities are

$$P = \begin{bmatrix} 0 & 1/3 & 1/3 & 1/3 & 0 & 0 & 0 \\ 0 & 0 & 1/3 & 0 & 1/3 & 0 & 0 \\ 0 & 1/4 & 0 & 1/4 & 1/4 & 1/4 & 0 \\ 0 & 0 & 1/3 & 0 & 0 & 1/3 & 0 \\ 0 & 0 & 0 & 0 & 0 & 1/3 & 1/3 \\ 0 & 0 & 0 & 0 & 1/3 & 0 & 1/3 \\ 0 & 0 & 0 & 0 & 0 & 0 & 0 \end{bmatrix}$$

The effective arrival rate into the seven nodes can be calculated using $\theta(I - P)^{-1}$ as

$$[\lambda_1\ \lambda_2\ \lambda_3\ \lambda_4\ \lambda_5\ \lambda_6\ \lambda_7] = [360\ 180\ 240\ 180\ 180\ 180\ 120]$$

Now, each of the seven nodes can be considered as independent $M/M/1$ queues due to Jackson's theorem. Let L_i be the number of requests in node i in the long run. For $i = 1,2, \ldots, 7$, the mean and variance of the number of requests in node i are

$$E[L_i] = \frac{\lambda_i}{\mu_i - \lambda_i}$$

and

$$\text{Var}[L_i] = \frac{\lambda_i \mu_i}{(\mu_i - \lambda_i)^2}$$

respectively. Plugging in the numerical values, the average number of pending requests in the seven nodes can be computed as $[E[L_1]\ E[L_2]\ E[L_3]\ E[L_4]\ E[L_5]\ E[L_6]\ E[L_7]] = [9\ 9\ 4\ 9\ 9\ 9\ 4]$. Likewise the variance of the number of pending requests in the seven nodes can be computed as $[\text{Var}[L_1]\ \text{Var}[L_2]\ \text{Var}[L_3]\ \text{Var}[L_4]\ \text{Var}[L_5]\ \text{Var}[L_6]\ \text{Var}[L_7]] = [90\ 90\ 20\ 90\ 90\ 90\ 20]$. Let L be the total number of requests in the system of seven nodes in the long run. Due to the fact that $L = L_1 + \cdots + L_7$ and independence between nodes, we have

$$E[L] = \sum_{i=1}^{7} E[L_i] = 53$$

and

$$\text{Var}[L] = \sum_{i=1}^{7} \text{Var}[L_i] = 490$$

Let W be the time spent in the network by a request. The performance metrics of interest, namely, delay and jitter are $E[W]$ and $\sqrt{\text{Var}[W]}$, respectively. Using Little's formula and its extensions (see Ref. [10]) we have

$$E[W] = \frac{E[L]}{\sum_i \theta_i}$$

and

$$\text{Var}[W] = \frac{\text{Var}[L] + \{E[L]\}^2 - E[L]}{\left(\sum_i \theta_i\right)^2} - \{E[W]\}^2$$

Therefore, the average delay and jitter experienced by all requests are 0.1472 and 0.0581 min, respectively.

Now, in order to determine the average delay and jitter experienced by a particular arriving request that goes through nodes 1–2–5–7, we use the fact that the time spent by a request in node i is exponentially distributed with parameter $(\mu_i - \lambda_i)$. Therefore, the mean and variance, respectively, of the time spent in nodes 1, 2, 5, and 7 are [0.025 00.0500 0.0500 0.0333] and [0.0006 0.0025 0.0025 0.0011]. Since the nodes are independent, the mean and variance of the total times are the sum of those spent at the individual nodes. Therefore, the average delay and jitter experienced by a particular arriving request that goes through nodes 1–2–5–7 are 0.1583 and 0.0821 min, respectively.

10.4.2 Case 2: Loss QoS Metrics Using Fluid Models

PROBLEM 10.2

Consider a centralized sensor network as shown in Figure 10.7. The six source nodes send sensor data to a sink that processes the data. The intermediary nodes are responsible for multiplexing and forwarding information. The top three sources generate low-priority traffic and the bottom three sources generate

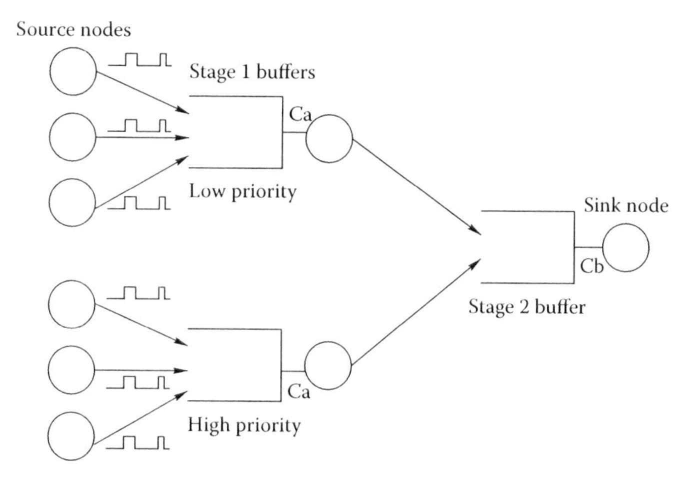

FIGURE 10.7 Sensor network configuration.

high priority. Subscript 1 is used for high priority and subscript 0 for low priority. The sensor data traffic can be modeled using on–off sources such that when the source is on, data are being generated at rate r_i for priority i and when it is off, no data is generated. Let the on and off times be exponentially distributed with parameters α_i and β_i, respectively, for priority i. Let c_a and c_b be the channel capacity of the buffers as shown in Figure 10.7. All buffers are of size B. However, assume that the second stage buffer is partitioned such that a maximum of B_i amount of priority i traffic can be stored in the buffer of size $B = B_0 + B_1$. In addition, at the second-stage buffer, the high-priority traffic is given all available/required processing capacity, anything remaining is given to the low-priority traffic. For this system, it is to be determined what loss probability QoS requirements can be met for both priorities. We use the following numerical values: $\alpha_0 = 1$, $\alpha_1 = 2$, $\beta_0 = 0.2$, $\beta_1 = 0.1$, $r_0 = 1$, $r_1 = 2.8$, $c_a = 1.5$, $c_b = 1.35$, $B_0 = 9$, $B_1 = 3$, and $B = B_0 + B_1 = 12$.

Solution

First, consider the stage 1 buffers. The two buffers are identical in all respects except that the two buffers serve different classes of traffic. The main difference is in the subscript. So, we perform the analysis using a subscript i to denote the priority. Let $eb_{i1}(v)$ be the effective bandwidth of the ith priority traffic into the corresponding stage 1 buffer. The sources are exponential on–off sources, that is, they can be modeled as a CTMC. Therefore, we can use Ref. [6] to obtain the effective bandwidth. Since the effective bandwidth of three independent sources multiplexed together is the sum of the effective bandwidths, we have

$$eb_{i1}(v) = \frac{3}{2v}\left(r_i v - \alpha_i - \beta_i + \sqrt{(r_i v - \alpha_i - \beta_i)^2 + 4\beta_i r_i v}\right)$$

Further, since the source is a CTMC, we can use exact analysis (from Ref. [5]) as opposed to the large deviations' result. The loss probability for priority i traffic at stage 1 buffer (ℓ_{i1}) is given by

$$\ell_{i1} = \frac{3\beta_i r_i}{c_a(\alpha_i + \beta_i)} e^{-\eta_{i1} B}$$

where η_{i1} is the solution to $eb_{i1}(\eta_{i1}) = c_a$ that yields

$$\eta_{i1} = \frac{3\alpha_i}{(3r_i - c_a)} - \frac{3\beta_i}{c_a}$$

Notice that if we had used large deviations, ℓ_{i1} would have been just $e^{-\eta_{i1}B}$ without the constant in front. Plugging in the numerical values (for the exact result, not large deviations), we get $\ell_{01} = 4.59 \times 10^{-9}$ and $\ell_{11} = 3.24 \times 10^{-4}$.

Now consider the stage 2 buffers. In order to differentiate the traffic, we will continue to use subscript i to denote the priority. Let $eb_{i2}(v)$ be the effective bandwidth of the ith priority traffic into the stage 2 buffer. Using the result for effective bandwidth for the output from buffer of stage 1, we have

$$eb_{i2}(v) = \begin{cases} eb_{i1}(v) & \text{if } 0 \le v \le v_i^\star \\ c_a - \dfrac{v_i^\star}{v}\{c_a - eb_{i1}(v_i^\star)\} & \text{if } v > v_i^\star \end{cases}$$

with

$$v_i^\star = \frac{\beta_i}{r_i}\left(\sqrt{\frac{c_a\alpha_i}{\beta_i(3r_i - c_a)}} - 1\right) + \frac{\alpha_i}{r_i}\left(1 - \sqrt{\frac{\beta_i(3r_i - c_a)}{c_a\alpha_i}}\right)$$

For the numerical values mentioned earlier, we get $v_0^\star = 0.8$ and $v_1^\star = 0.4105$. The loss probability for priority i traffic at stage 2 buffer (ℓ_{i2}) is given by

$$\ell_{i2} = e^{-\eta_{i2}B}$$

where

η_{12} is the solution to $eb_{12}(\eta_{12}) = c_b$

η_{02} is the solution to $eb_{02}(\eta_{02}) + eb_{12}(\eta_{02}) = c_b$

This is based on the results in Refs. [7,16]. Plugging in the numerical values, we get $\ell_{02} = 0.0423$ and $\ell_{12} = 0.003$.

Assuming that the loss probability are independent across the two stages, the loss QoS requirements that can be satisfied are $1 - (1 - \ell_{01})(1 - \ell_{02})$ and $1 - (1 - \ell_{11})(1 - \ell_{12})$ for priorities 0 and 1, respectively. Therefore, the QoS guarantee that can be provided is that priority 0 and priority 1 traffic will face a loss probability of not more than 4.23% and 0.33%, respectively.

10.5 Balancing QoS and Power in Multi-Hop Wireless Sensor Networks

All the QoS considerations thus far have only implicitly used power or energy. Power issues become crucial, especially, in sensor networks that deploy battery power for sensing and transmission. In terms of the terminology used thus far, power translates to resource that is directly responsible for node availability in a sensor network. With that understanding, we consider a wireless network that uses battery-powered sensors and transmits using a multi-hop peer-to-peer technology. As an illustration, we show four independent examples of trading off performance and power in various multi-hop wireless sensor domains.

10.5.1 Gossip-Based Information Dissemination

Consider a multi-hop wireless sensor network where the objective of each node is to transmit all the information it has sensed and received to other nodes in a distributed fashion. These transmissions are usually based on what is known as gossip protocols. A gossip protocol typically starts with one node that has sensed information that gets passed along to another node. Now two nodes have that information and they pass it along to two other nodes, and so on. We assume that the communication is line of sight or point to point as opposed to a broadcast where a node can gossip to all the nodes in its range (this is an immediate extension to the scenario here). We assume that nodes only have local information due to the distributed nature and also that nodes could go down due to several reasons (due to attacks, malfunction, or battery running down).

It is crucial to realize that all nodes are continuously sensing and spreading gossip with just the local information. Thus, it becomes important to develop a stopping rule to decide when to stop spreading the gossip. If the stopping time is too short, then only a few nodes would get the gossip. This is equivalent to having a poor loss QoS (because not having a gossip is equivalent to loss of information). However, a long stopping time would not only imply wastage of battery power, but it also increases congestion in the network and hence delays. The stopping criterion typically is based on a time-to-live parameter that could be (1) the number of times a gossip has been transmitted by a node, (2) a threshold number of seconds, or (3) the number of times gossip has been transmitted to nodes that already posses the gossip (based on acknowledgment packets).

Therefore, given a stopping criterion, analytical models can be built to obtain metrics such as the distribution of the number of nodes that get a gossip, distribution of the time for a certain fraction of the nodes to get the gossip, etc. For example, in Ref. [13], we consider a highly mobile network wherein the nodes have an equal probability of meeting any of the other nodes during a transmission opportunity. Further, we consider stopping criteria where a node stops spreading gossip when it encounters another node that already has the gossip. For such a system we show that for any practically sized network (i.e., more than five nodes), then irrespective of the number of nodes in the network, the average fraction of nodes that get a gossip is 0.82, a constant. We also compute the average time a gossip lasts (i.e., there is at least one active node spreading it). There are several opportunities to extend the results to other assumptions regarding mobility, transmission, and stopping rules.

10.5.2 Routing in Underwater Sensor Networks

Underwater wireless sensor networks are usually based on hydrophones and geophones that perform point-to-point communication (and not broadcast). The energy dissipation is based on the distance between the transmitter and the receiver. Although there are several articles on power-aware or energy-aware routing in multi-hop wireless networks, the underwater networks (especially on the ocean bottom) have their own unique characteristics that do not enable the use of existing algorithms. First, the network lifetimes vastly exceed the battery life, so batteries have to be replaced from time to time. However, since the batteries are in the bottom of the ocean and not easily accessible, usually a large number of batteries are simultaneously replaced. Thus, synchronizing battery failures becomes an important criterion.

In Ref. [17], we consider a seismic monitoring application of underwater sensor networks. The monitoring is periodic and fairly deterministic. The nodes are laid out on a grid structure and the sink node is in the center of the grid. Since all the information finally reach the sink node, nodes closer to the sink tend to lose battery life faster than others. An interesting observation is that if all nodes send their traffic through the shortest path (which would also result in the minimum consumption of energy), then nodes do not fail at the same time. Instead, if nodes with lesser traffic take a longer path, then the timing of failures can be more or less synchronized. Thus, the total cost of operation per unit time is minimized. It is also worthwhile to comment that even in the stochastic sensing and transmission case, since the battery consumption is tiny for each transmission, the battery life due to a sum of a large number of transmissions is fairly predictable. In summary, this is a rich problem with a radically different set of conditions than its terrestrial counterpart.

10.5.3 QoS Considerations in Network Coding

The concept of reverse-carpooling using network coding is an efficient way to reduce the number of transmissions in systems where transmissions are the dominant consumer of energy. We assume that the transmission is based on broadcast. As an example, consider three nodes in series. We call them node 1, node R (for relay), and node 2. Nodes 1 and 2 cannot reach each other and hence must transmit only through the relay R. Say node 1 has a binary string x_1 to transmit to node 2 and node 2 has a binary string x_2 to transmit to node 1. First, nodes 1 and 2 transmit x_1 and x_2 to the relay. Then the relay, instead of transmitting x_1 to node 2 and using another transmission to send x_2 to node 1, codes the strings as $x_1 \oplus x_2$ and transmits once. Since nodes 1 and 2 have $x_1 \oplus x_2$ and the strings they transmitted, they can immediately retrieve x_2 and x_1, respectively.

This is certainly an effective way of reducing the number of transmissions and thereby power consumption. However, in a sensor network, information arrives randomly and the relay node may not always have packets from opposite sides to code. Thus, a trade-off needs to be made whether it is worth waiting for an opportunity to code or whether it is better to send off a packet uncoded. We are essentially balancing latency against power consumption. In Ref. [11], we address this problem and show that there exists a threshold policy so that if the number of packets of one type is fewer than a threshold, we must wait and otherwise we must transmit. With a disclaimer that the problem has certain restrictions in terms of the traffic arrival process and transmissions, we find an interesting result, that is, it is not necessary to know how long a packet has been waiting to make that decision. Instead all we need is how many packets are waiting. We are exploring several extensions to this problem of distributed decision making in wireless sensor networks.

10.5.4 Dynamic Voltage Scaling and QoS

In multi-hop wireless sensor networks with significant processor capabilities, the operating system can be controlled by adjusting the voltage dynamically, which results in significant energy savings. This technique is called dynamic voltage scaling (DVS), and simplistically, if the voltage setting is high, then power consumption is high and vice versa. However, if the voltage setting is high, then the processing is also fast, thus the latency is low. For over a decade, energy management has been used in mobile and resource-constrained environments that are limited by battery capacities. By appropriately building software interfaces for DVS, one can obtain power savings without compromising on performance (see Ref. [8]).

To determine the voltage setting to be used at each node, we can formulate an optimization problem that would minimize the long-run average energy cost per unit time subject to satisfying QoS requirements such as average latency. Usually, the optimal policy is of threshold type on the workload. In other words, if there are voltage settings $v_1, v_2, \ldots, v_m$, such that $v_1 < v_2 < \cdots < v_m$, then there exist thresholds on the workload $\theta_1, \theta_2, \ldots, \theta_m$, such that $0 \leq \theta_1 \leq \theta_2 \leq \cdots \leq \theta_m$ so that if the workload at any time is between θ_i and θ_{i+1}, then use voltage setting v_i. Thus, by using only local information, each of the nodes can appropriately tune their processors and efficiently run the system. Of course, opportunities exist for improving these mechanisms by monitoring state information of other nodes, etc.

10.6 Concluding Remarks

In this chapter, abstract models of networks were studied and how to guarantee QoS for them was analyzed. The work focused on methodologies rather than applications. The objective was to develop a set of common tools that are applicable to various types of networks. In particular, the study of QoS is important and extends to sensor networks. Although mentioned in the earlier part of the chapter that issues such as pricing and availability are very crucial for customer satisfaction, we have not paid much attention to them in the case studies on determining QoS metrics. However, while extending the performance analysis to solve design and control problems, it is very important to take pricing and availability into account as well. Another critical aspect that has been left out of this chapter is degradation and failure of resources. A very active research topic that combines issues of availability, QoS, and resource degradation/failure, which, in total, is called survivability or robustness, has been given a lot of attention recently by both government and industry. Since degradation and failure cannot be quantitatively modeled very well, survivable or robust system designs end up being extremely redundant. Therefore, building cost-effective systems that can be robust or survivable is of utmost importance. Interestingly, sensor networks can be used to address several of the issues.

Acknowledgment

The author was partially supported by NSF grant CMMI-0946935.

References

1. G. Bolch, S. Greiner, H. de Meer, and K.S. Trivedi. *Queueing Networks and Markov Chains, Modeling and Performance Evaluation with Computer Science Applications*. John Wiley & Sons, New York, 1998.
2. C.S. Chang and J.A. Thomas. Effective bandwidth in high-speed digital networks. *IEEE Journal on Selected Areas in Communications*, 13(6), 1091–1100, 1995.
3. C.S. Chang and T. Zajic. Effective bandwidths of departure processes from queues with time varying capacities. In *INFOCOM'95*, Boston, MA, pp. 1001–1009, 1995.
4. G. de Veciana, C. Courcoubetis, and J. Walrand. Decoupling bandwidths for networks: A decomposition approach to resource management. In *INFOCOM'94*, Toronto, ON, Canada, pp. 466–473, 1994.
5. A.I. Elwalid and D. Mitra. Analysis and design of rate-based congestion control of high speed networks, Part I: Stochastic fluid models, access regulation. *Queueing Systems, Theory and Applications*, 9, 29–64, 1991.
6. A.I. Elwalid and D. Mitra. Effective bandwidth of general Markovian traffic sources and admission control of high-speed networks. *IEEE/ACM Transactions on Networking*, 1(3), 329–343, June 1993.
7. A.I. Elwalid and D. Mitra. Analysis, approximations and admission control of a multi-service multiplexing system with priorities. In *INFOCOM'95*, Boston, MA, pp. 463–472, 1995.
8. K. Flautner, S. Reinhardt, and T. Mudge. Automatic performance setting for dynamic voltage scaling. In *Proceedings of the 7th Annual International Conference on Mobile Computing and Networking*, Rome, Italy, pp. 260–271, 2001.
9. N. Gautam, V.G. Kulkarni, Z. Palmowski, and T. Rolski. Bounds for fluid models driven by semi-Markov inputs. *Probability in the Engineering and Informational Sciences*, 13, 429–475, 1999.
10. D. Gross and C. M. Harris. *Fundamentals of Queueing Theory*, 3rd edn. John Wiley & Sons Inc., New York, 1998.
11. Y. Hsu, S. Ramasamy, N. Abedini, N. Gautam, A. Sprintson, and S. Shakkottai. Opportunities for network coding: To wait or not to wait. In *IEEE ISIT*, St. Petersburg, Russia, pp. 791–795, 2011.
12. G. Kesidis, J. Walrand, and C.S. Chang. Effective bandwidths for multiclass Markov fluids and other ATM sources. *IEEE/ACM Transactions on Networking*, 1(4), 424–428, 1993.
13. Y.M. Ko and N. Gautam. Epidemic-based information dissemination in wireless mobile sensor networks. *IEEE/ACM Transactions on Networking*, 18(6), 1738–1751, 2010.
14. K.R. Krishnan, A.L. Neidhardt, and A. Erramilli. Scaling analysis in traffic management of self-similar processes. In *Proceedings of 15th International Teletraffic Congress*, Washington, DC, pp. 1087–1096, 1997.
15. V.G. Kulkarni. Effective bandwidths for Markov regenerative sources. *Queueing Systems, Theory and Applications*, 24, 137–153, 1997.
16. V.G. Kulkarni and N. Gautam. Admission control of multi-class traffic with service priorities in high-speed networks. *Queueing Systems, Theory and Applications*, 27, 79–97, 1997.
17. A. Mohapatra, N. Gautam, and R. Gibson. Combined routing and node replacement in energy-efficient underwater sensor networks for seismic monitoring. *IEEE Journal of Ocean Engineering*, 2011 (To appear).

11

Network Daemons for Distributed Sensor Networks

Nageswara S.V. Rao
Oak Ridge National Laboratory

Qishi Wu
Louisiana State University

11.1 Introduction

There is a wide spectrum of scenarios in which *distributed sensor networks* (DSNs) are deployed, ranging from radar sites located across the country that track aircrafts to small robot teams that explore urban areas [3]. Consequently, the networks that underlie DSNs are just as varied, ranging from the long-haul wireline networks to small-area wireless networks. In the former networks, the sustained bandwidth for data transfers and stable channels for control and high priority traffic are required, while in the latter, it is of interest to sustain message delivery under dynamic node movements with very limited or no network infrastructure. Despite the operational diversity, the current DSNs are often deployed by utilizing the present Internet technologies, in part due to their wide availability. For example, the wide-area DSNs are often connected over the Internet (or networks that are characteristically similar) using the hosts equipped with the conventional protocol stacks, and the wireless DSNs deployed in unstructured areas are often connected using IEEE 802.11 technologies. Consequently, the resultant networks often do not exactly match the DSN requirements, since the Internet technologies are geared toward the best effort services with the end hosts having very limited control at the network core [12].

The reliance on the Internet technologies by DSNs often manifests in severe performance limitations. For example, in wide-area DSNs, the nodes cannot control routes to avoid the congested regions or accumulate bandwidths over multiple paths (without drastically changing the infrastructure) because the routing at the network core is solely determined by the underlying routers, which are exclusively controlled by the different service providers. Furthermore, there is no support for realizing stable channels for control purpose over such wide-area networks. The throughput achieved for a control channel that employs the most widely used Transmission Control Protocol (TCP) typically underflows under high traffic and overflows under low traffic. Moreover, such a TCP-based control channel may experience very complicated end-to-end transport dynamics in time-varying network conditions. In IEEE 802.11 networks for a team of mobile robots, which, for example, are deployed to assess the radiation levels of a remote area, an infrastructure of access points must be setup prior to the operation [5]. Such requirement is obviously meaningless if the very goal of the robot team is to assess the suitability of the region for human operation. More generally, in DSNs of mobile nodes, the challenges are to form an ad hoc wireless network without the infrastructure [4] and to cope with the dynamic changes in network connectivity. Note that the node movements are treated as aberrations in the current Internet environments, where as such movements are shown to improve the message delivery in mobile ad hoc networks [6].

The above limitations of wireline and wireless networks are not inherent to DSNs, and in fact, the opposite is true: DSNs offer conducive environments for the end hosts and core nodes to cooperate in overcoming several of them. Note that in principle, networks can be designed from scratch to suit each DSN scenario, but such an approach is too expensive at least in the short term since it involves the development of special-purpose hardware and software. We adopt a more pragmatic approach here by utilizing a framework of *network daemons* to enhance the network functionalities to address DSN requirements. These daemons contain modules for measurement, path computation, routing and transport adaptation, which are all implemented at the application level. These application-level solutions are easily deployable over the current infrastructures and can also be tailored to meet the specific needs of various classes of DSNs.

In this chapter, we consider two specific classes of networks for DSNs: (a) wide-area networks with the requirements of sustained bandwidth and stable control channels, and (b) small-area ad hoc wireless networks of mobile nodes deployed in unstructured areas. To address the first class, we employ the regression-based path computation [15] and transport stabilization based on stochastic approximation method [18]. These daemons collect link delay measurements to compute the best paths and use each other to route around high traffic areas as well as to accumulate bandwidth using multiple paths. For implementing stable control channels, we adopt a source controller that stabilizes a flow using User Datagram Protocol (UDP). We analytically justify both techniques under fairly general conditions. For the second class of networks, the daemons dynamically track the connectivity changes and exploit the node movements to improve the message delivery [19]; this is a departure from the Internet-based approaches that treat connectivity changes as undesirable.

We present a general network daemon framework in Section 11.2 [14,16], which encompasses link measurement, path computation, transport control, and data routing modules. The analytical and experimental results for the wide-area wireline networks are discussed in Section 11.3. Small-area wireless networks are discussed in Section 11.4. The presentation here is tutorial in nature and the details of various parts can be found in Refs [14–16,18] for wireline networks and in Ref. [19] for wireless networks.

11.2 Network Daemons

The network daemons are deployed at DSN nodes, for example, distributed over either wide-area networks across thousands of miles or ad hoc mobile networks confined to small regions. Each daemon consists of four main components as shown in Figure 11.1:

1. *Link measurement module* collects delay or connectivity measurements using test messages that are actively or passively transmitted along virtual links for estimating available link bandwidths and minimum link delays based on linear regression method or determining the network connectivity.
2. *Path computation module* maintains a routing table with link information (either bandwidth or connectivity) from the measured network topology and computes single/multiple paths for the data to be transmitted.

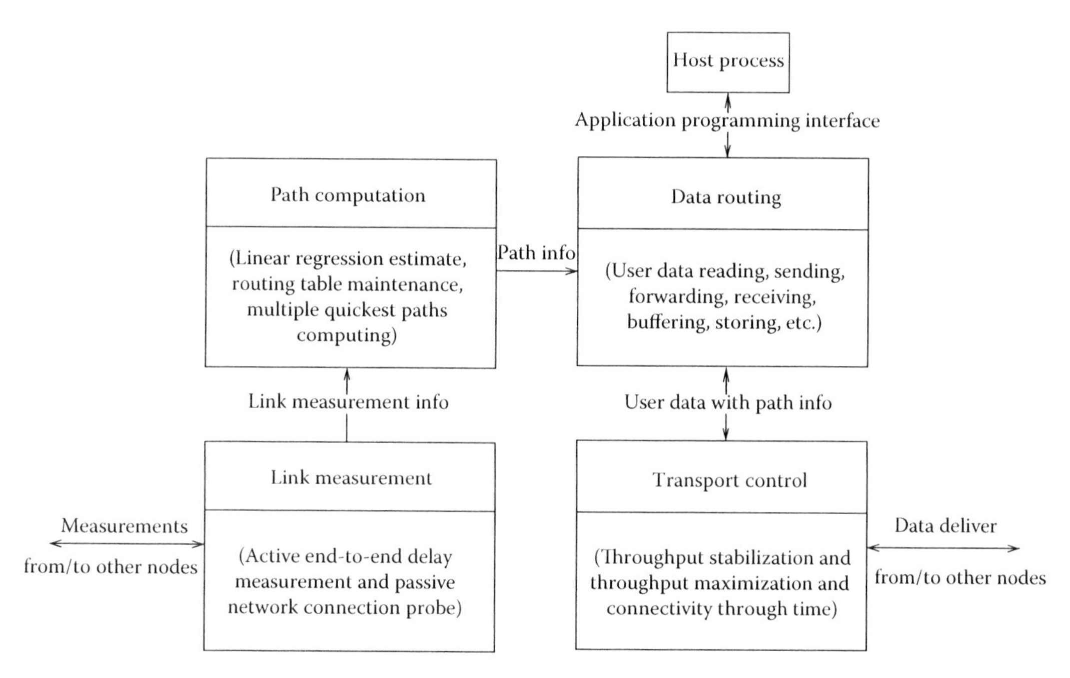

FIGURE 11.1 Functional block diagram of a daemon.

3. *Data routing module* routes the data to the destination via the other daemons using path information specified under this framework. The data may be buffered at the intermediate nodes in ad hoc mobile networks due to the highly dynamic connectivity.
4. *Transport control module* provides the customized data transfer services to meet performance requirements such as throughput stabilization and connectivity through time, which are specific to a particular DSN.

To illustrate the concept of network daemons, consider a computation distributed at several nodes over the Internet, where the messages are communicated as per the task structure shown in Figure 11.2a. The communication between processes, for example between P_1 and P_2, could be handled by a process-to-process TCP stream. The paths traversed by the data packets are decided by the network routers based on a best-effort basis, hence are usually not optimized for the host performance. For example, when the background traffic increases on a shared link, a TCP stream experiences longer delays and higher packet losses, thereby curtailing its throughput. The current Internet typically does not allow the hosts to enforce rerouting to avoid highly congested network segments. Although some level of rerouting might be performed by routers, optimizing the end-to-end performance of any particular host process is not their primary goal.

Now we consider that network daemons are deployed to assist data transfer as shown in Figure 11.2b. In one scenario, if link $R_1 - R_2$ is congested but not the other links, the messages can be sent from R_1 to the routing daemon R_3 then to R_2 via the link $R_3 - R_2$. In another scenario, if the available bandwidth of link $R_1 - R_2$ is not sufficient, a multiple path consisting of $R_1 - R_2$, $R_1 - R_3 - R_2$, $R_1 - R_4 - R_2$ and $R_1 - R_5 - R_2$, can be utilized. The paths here are implemented using the application-level daemon routers, and each virtual link in such paths may correspond to a number of physical paths via the underlying Internet routers. The decisions about how to choose the appropriate paths may be made based on the measurements collected by the daemons as will be illustrated in the next section. Since a DSN consists of a number of nodes under a single control, the required daemons can be easily executed on various nodes.

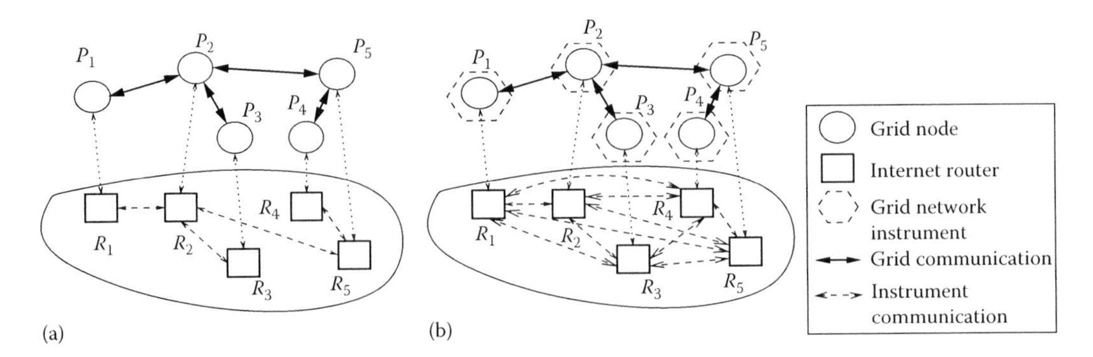

FIGURE 11.2 Typical scenario of wide-area sensor network over wireline network.

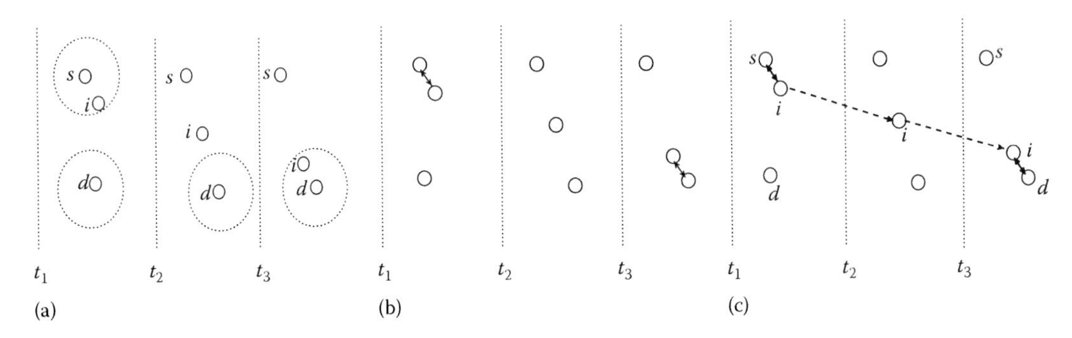

FIGURE 11.3 Typical scenario of dynamic sensor nodes over wireless network.

We now consider an ad hoc wireless network consisting of three nodes (mobile robots) as shown in Figure 11.3 to illustrate how daemons can be utilized to overcome the reachability limitations of the current methods. Here there is no backbone network of access points and the nodes communicate with others within a certain range, that is, only with directly connected neighbors. Consider that a message is to be delivered from robot s to robot d within the time interval $[t_1, t_3]$. In this scenario, the robots s and d are never within the transmission range of each other, which makes the existing protocols inapplicable. TCP cannot deliver the message since it requires that destination be reachable, and the same is true for more special protocols such as TORA [11] that are designed for mobile networks. We now illustrate that using suitable daemons, it is indeed possible to deliver the message. Consider Figure 11.3, which shows the robot movements. Initially, robots s and i are directly connected. Then i moves away from s and all three robots are disconnected during this period. Finally i moves within the range of d and hence is connected to it. A message sent from s at time t_1 can be sent to i initially where it can be buffered at the routing daemon and then delivered to d at time t_3. Essentially, the movements of i are utilized to deliver the message to d, where the daemons are used to buffer the message in time. More details of this class of applications are discussed in Section 11.4.

11.3 Daemons for Wide-Area Networks

We consider a subclass of DSN networks that are connected over wide-area networks with two types of requirements. Firstly, messages of various sizes must be transported quickly between various nodes, for example, sensor measurements from end nodes to fusion centers. Secondly, some of the sensors must be interactively controlled from remote nodes.

11.3.1 Path Computation

For the first task, the variability of message sizes must be explicitly accounted for due to the non-monotonicity of end-to-end delays: a path with high bandwidth (suited for bulk transfers) is not necessarily suitable for transmitting small messages. Indeed, small messages may be delivered more quickly via paths with smaller bandwidth and latency. In the current methods the messages are typically sent as single or parallel TCP streams, and the protocol stack can be optimized to account for various host parameters [22]. Such host-based methods are very effective for the Internet but do not exploit the physical diversity of paths in the network. Since the network paths are solely determined by the routers on a best effort basis, it is quite possible that traffic is routed via congested paths while there are other underutilized paths.

The end-to-end delays are also subject to limitations imposed by the queuing policies and traffic loads at the routers, in addition to the bandwidth limits of the links. While the latter delays are somewhat measurable and predictable, those at the routers cannot be very easily modeled. Consequently, the end-to-end delays of messages contain significant random components, whose distributions can be highly complicated [23]. In such cases, the usual formulation of path computing with the least expected end-to-end delay is not viable since the required distributions are very difficult to estimate. In this section, we adopt a purely measurement-based method wherein the required paths are computed using in-situ measurements.

An *overlay network of daemons* is represented by a graph $G = (V, E)$ with n nodes and m virtual links. Here each node represents a daemon while each link represents a communication channel such as a TCP connection. A message of size r must be transmitted from a source node s to a destination node d, which incurs three types of delays:

1. *Link delay*: For each link $e = (v_1, v_2)$, there is a *link-delay* $d(e) \geq 0$ such that the leading edge of a message sent via e from node v_1 at time t will arrive at node v_2 at time $t + d(e)$.
2. *Bandwidth constrained delay*: Each link $e \in E$ has a deterministic "effective" *bandwidth* $b(e) \geq 0$. Once initiated, a message of r (constant) units can be sent along link e in $r/b(e) + d(e)$ time.
3. A message of size R (random variable) arrives at the source s according to an *unknown* distribution P_R. At any node v, Q_v and R_v are the random variables denoting the queuing delay and message size distributed according to *unknown* distributions $\mathbf{P}_{Q_v}$ and $\mathbf{P}_{R_v}$, respectively. No information about the distributions of R_v and Q_v, $v \in V$, is available. Instead, the *measurements* $(Q_{v;1}, R_{v;1}), (Q_{v;2}, R_{v;2}), \ldots, (Q_{v;l}, R_{v;l})$ that are independently and identically distributed (iid) according to the distribution $\mathbf{P}_{Q_v, R_v}$, are known at each node $v \in V$.

According to the above link model, the bandwidth and minimum link delay of a virtual link in the overlay network can be estimated through active measurements using the following steps:

Step 1. The source node generates a set of test messages of various sizes.

Step 2. The source node divides each message into a number of components of a certain read/send buffer size and transmits them to the destination node through a TCP channel. Note that internally all message components are chunked into segments of MSS at the TCP layer, each of which is probably further fragmented into data packets at the IP layer, depending on the underlying link MTU.

Step 3. The destination node receives message components and acknowledges to the source node the completion of transmission.

Step 4. Upon the receipt of acknowledgments, the source node calculates the end-to-end message delays and apply a linear regression to fit the measured points of message size and end-to-end delay pair. The first order approximate of the available bandwidth and the minimum link delay are then estimated by the slope and intercept of the regression line, respectively.

Such link measurement examples are shown in Figure 11.4, where we consider messages with widely ranging sizes transmitted between ORNL and a number of universities. In the left figure, each cluster

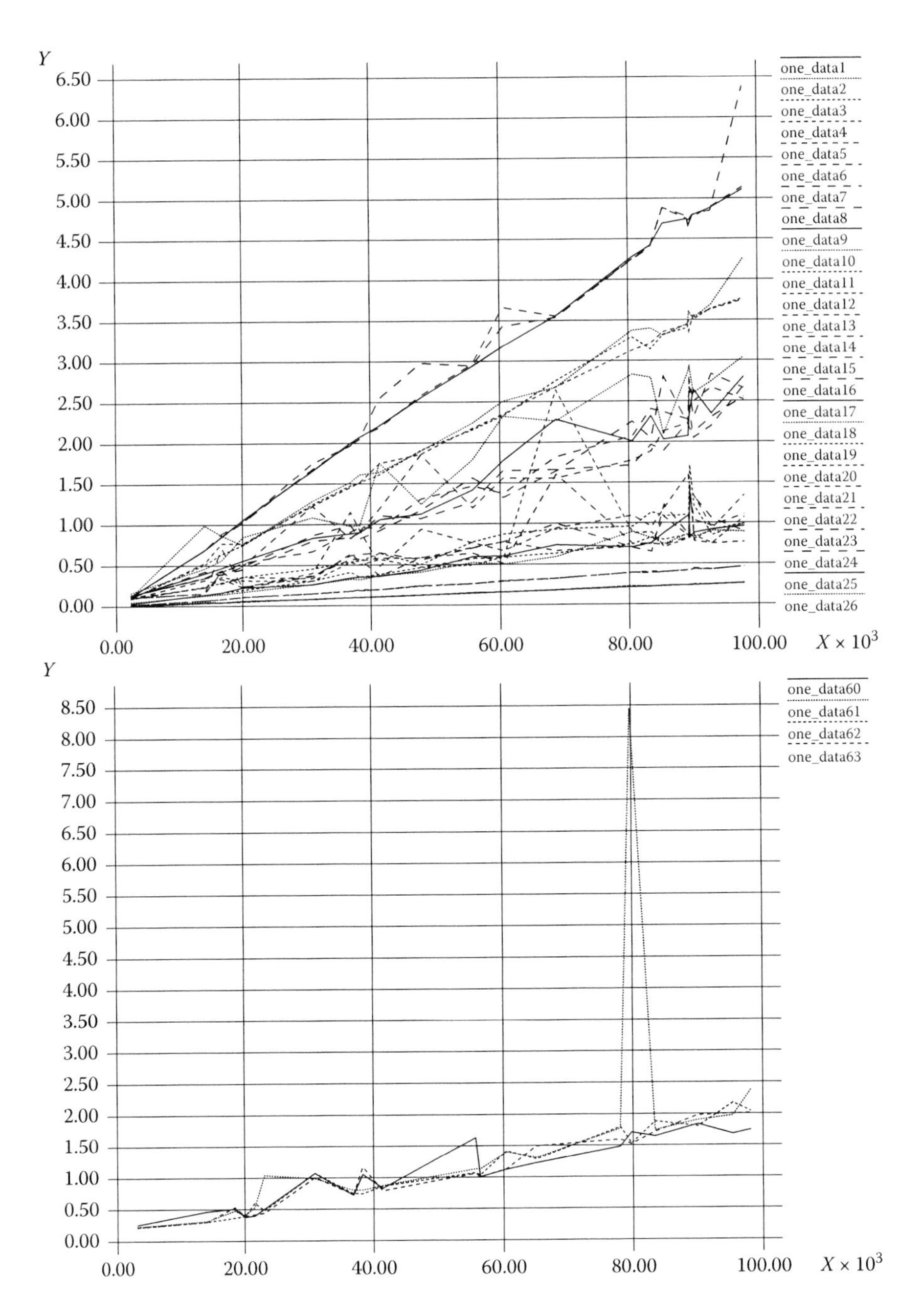

FIGURE 11.4 ORNL-OU: End-to-end delays for large messages. *X*-axis: message size in bytes; *Y*-axis: end-to-end delay in seconds.

corresponds to a single destination. For each destination, each plot corresponds to the measurements collected within the span of a few minutes by randomly picking the message sizes and sending them to the destination. In the right, we show the delays between ORNL and University of Oklahoma (OU). As portrayed by these measurements, our model captures the essence of end-to-end delays: in each plot, the "slope" corresponds to the effective link bandwidth and the additional variation corresponds to the random queuing delay Q_v.

11.3.1.1 Probabilistic Delay Guarantees

Consider a path P, from source $s = v_0$ to destination $d = v_k$, given by $(v_0, v_1), (v_1, v_2), \ldots, (v_{k-1}, v_k)$, where $(v_j, v_{j+1}) \in E$, for $j = 0, 1, \ldots, (k-1)$. The *bandwidth* of this path is $b(P) = \min_{j=0}^{k-1} b(e_j)$, and the *delay due to bandwidth* is $r/b(P)$. The *link delay* of this path is $d(P) = \sum_{j=0}^{k-1} D(e_j)$. The *end-to-end delay* of path P in transmitting a message of size R is the sum of these three delay components

$$T(P,R) = \frac{R}{b(P)} + d(P) + \sum_{j=0}^{k-1} Q_{v_j|R} \tag{11.1}$$

where $Q_{vj|R}$ is the conditional queuing delay at node v_j given that a message of size R arrived at the node. The *expected end-to-end delay* of path P for the given message size R is given by

$$\bar{T}(P,R) = \frac{R}{b(P)} + d(P) + \sum_{j=0}^{k-1} \int Q_{v_j} dP_{Q_{v_j}|R} \tag{11.2}$$

which is a random variable (of R) for a fixed path P. Let $\mathcal{P}$ denote the set of all paths from s to d. Let $P_R^\star$ denote a path with the minimum expected end-to-end delay for the given message size R such that $\bar{T}(P_R^\star, R) = \min_{P \in \mathcal{P}} \bar{T}(P,R)$. If the error distributions are known, $P_R^\star$ can be computed using deterministic optimization methods. Such an approach is infeasible here since in practice the error distributions of queuing delays are so complicated that they are essentially unknown. We compute an estimator $\hat{P}_R$ of $P_R^\star$ using a regression estimator such that

$$\boldsymbol{P}\left\{ \boldsymbol{E}_R[\bar{T}(\hat{P}_R, R) - \bar{T}(P_R^\star, R)] \geq \epsilon \right\} \leq \delta \tag{11.3}$$

for a sufficiently large sample size, which depends on ϵ, δ, n, and a suitably chosen function family $\mathcal{Q}_v$ that contains the regression function. Informally, this condition guarantees that: *the expected delay of* $\hat{P}_R$ *is within* ϵ *of that of* $P_R^\star$ *with probability* $1 - \delta$, *irrespective of the delay distributions.*

This is the best guarantee possible using a measurement-based approach, whose derivation details can be found in Ref. [15]. Often in networks, measurements are collected to estimate the distributions, which are then used to compute best paths. In the present problem, such approach can only result in guarantees, which are strictly weaker than that defined in Equation 11.3 mainly because Q_v can have an arbitrary distribution. Informally speaking, the estimation of distributions involves an infinite dimensional quantity whereas the computation of $\hat{P}_R$ involves minimization over the finite set $\mathcal{P}$.

This guarantee is possible because the measurements are the actual delays collected by the daemons. Traditionally, ICMP-based mechanisms such as ping and traceroute are used to collect measurements. However, the end-to-end guarantees in Equation 11.3 cannot be provided based on such data because some firewalls disable responses to ping and traceroute and sometimes even deliberately send incorrect responses. Also, some firewalls enforce rate controls on ICMP traffic but not on TCP, in which case the delay measurements collected through ping and traceroute could be highly misleading. In certain sense, our approach not only provides analytical guarantees but also provides guidance for the appropriate measurements.

An algorithm was presented in Ref. [15] to compute the best empirical path $\hat{P}_R$ based on a regression estimator, The complexity of this algorithm is $O(m^2 + mn\log n + nf(l))$, where $f(l)$ is the complexity of computing the regression at a given value r. Thus, a polynomial-time (in l) regression estimator results in a polynomial-time (both in n and l) path computation method.

11.3.1.2 Multiple Path Computation

A *multiple path* from s to d, denoted by MP, consists of a set of simple bandwidth-disjoint paths from s to d. For simplicity, consider a network $G = (V, E)$ with zero queuing delays $Q_v = 0$ for all $v \in V$ such that a message of r units can be sent along the edge P in $r/B(P) + D(P)$ time. The *end-to-end delay*, denoted by $T(MP, r)$, of a multiple path MP from s to d is defined as the time required to send a massage of size r from s to d, wherein the message is subdivided and transmitted via the constituent paths. The multiple paths often provide more bandwidth compared to a single path if the message can be suitably divided into parts.

Consider a network consisting of two paths P_1 and P_2 such that $B_1 = 10$ units/s, $B_2 = 20$ units/s, $D_1 = 2$ s, and $D_2 = 12$ s. For a message of size $r = 100$ units, $T(P_1, 100) = 12$ and $T(P_2, 100) = 17$. If a single path is used, P_1 will be chosen for this message size. If 99 units are sent on P_1 and 1 unit is sent on P_2, the corresponding delays are given by $99/10 + 2 = 11.9$ s and $1/20 + 12 = 12.05$ s respectively, resulting in an end-to-end delay of 12.05 s. Hence, two-path $\{P_1, P_2\}$ is not a good choice for this message size. For message of size $r = 1000$ units, the end-to-end delays of single path P_1 and P_2 are calculated as 102 and 62 s, respectively. Thus if a single path is used, then P_2 will be chosen. For a two-path $\{P_1, P_2\}$ for $r = 1000$ such that 400 and 600 units are sent via P_1 and P_2 respectively, it results in individual delays of 42 s for each path. Hence the resultant end-to-end delay is 42 s, which is smaller than that of P_1 or P_2, which are 102 and 62 s, respectively.

In general, for two paths P_1 and P_2, we have $T(\{P_1, P_2\}, r) \leq \min\{T(P_1, r), T(P_2, r)\}$ if and only if the condition $C(P_1, P_2)$ given by

$$D(P_1) + \frac{r}{B(P_1)} \geq D(P_2) \quad \text{and} \quad D(P_2) + \frac{r}{B(P_2)} \geq D(P_1)$$

is satisfied. Under this condition, the minimum end-to-end delay of $\{P_1, P_2\}$ is achieved by dividing the message into two parts of sizes r_1 and r_2, $r_1 + r_2 = r$, which are sent via P_1 and P_2 respectively. The sizes of the two parts are given by

$$r_1 = \frac{B_1 r}{B_1 + B_2} + \frac{B_1 B_2 (D_2 - D_1)}{B_1 + B_2} \quad \text{and} \quad r_2 = \frac{B_2 r}{B_1 + B_2} - \frac{B_1 B_2 (D_2 - D_1)}{B_1 + B_2}$$

The general conditions for dividing the message among p paths are presented in Ref. [15].

The path computation daemon computes the constituent paths of a multiple path for a given message size by repeatedly computing the quickest path and removing it from the graph by reducing the appropriate bandwidths of the links. Then the extracted quickest paths are combined as per the above conditions. Note that the resultant multiple path is not always guaranteed to be optimal in a strict sense because of the additional unaccounted randomness in the delays. But such paths yielded very good results in actual implementations as shown in the next section.

11.3.1.3 Internet Implementation

A distributed computing environment of four sites is shown in Figure 11.5, which is used in our implementation. The server is located at OU and client is located at ORNL. The daemons are implemented using socket programming in C++ under linux/unix operating system. The delay regression estimation is based on potential function method as described in Ref. [15]. Daemons are executed at the sever and client, and at two additional locations at Louisiana State University (LSU) and Old Dominion University (ODU). Typical experimental results are shown in Figure 11.5 in the right, where the upper curve represents the delays in a single TCP stream ORNL-OU plotted as a function of randomly chosen message sizes. The lower curve corresponds to the multiple path consisting of TCP streams ORNL-ODU-OU, ORNL-LSU-OU and two direct parallel TCP streams ORNL-OU. All the overheads of the daemons, namely the path computation and routing times, are included in the measured end-to-end delays. The messages are divided into four parts as per the delay curves of the paths, and are sent along the respective paths. The overall end-to-end delay when daemons are employed is much lower than the delay with a single TCP stream, except for some smaller sizes. The multiple path resulted in an average

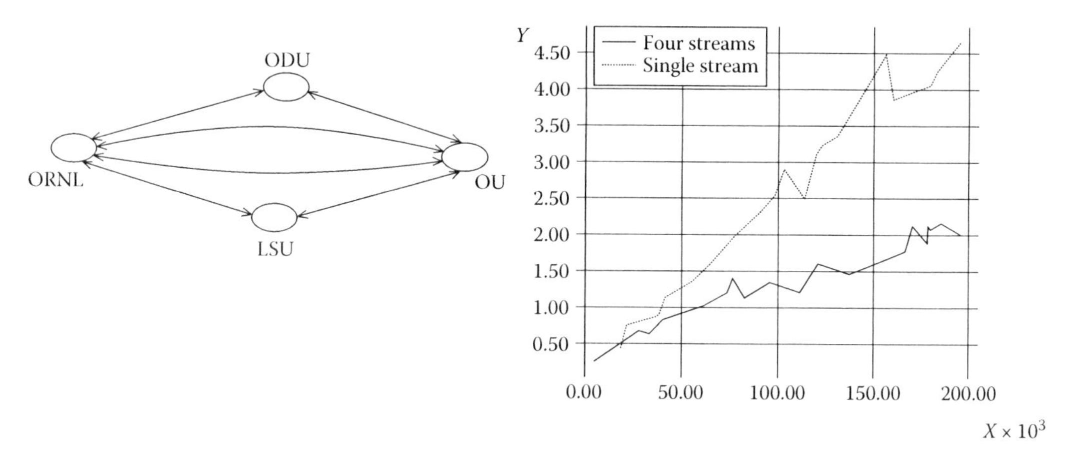

FIGURE 11.5 ORNL-OU: End-to-end delays for large messages. *X*-axis: message size in bytes; *Y*-axis: end-to-end delay in seconds.

improvement of about 35% in the end-to-end delays in the cases we studied. The details of the experimentation and more extensive measurements are provided in Ref. [15].

11.3.2 Transport Control for Throughput Stabilization

We now address the second task of implementing a control channel over wide-area networks. The bandwidth of the control channel is typically only a small fraction of the available bandwidth, but it is extremely important that the throughput rate at remote site(s) be stable in presence of dynamic traffic conditions. Large amount of jitter in throughput can destabilize the control loops needed for remote robots, possibly causing severe damages to them. TCP is not designed for providing such stable throughput. It always continues to increase its throughput until losses are encountered, which often results in much higher throughput than needed. Furthermore, it drastically reduces throughput in response to bursty losses, which might result in long delays in control messages. In general the non-linear TCP dynamics make throughput stabilization very challenging [17].

We describe in this section transport daemons based on the stochastic approximation method which achieve provably stable throughput under very general conditions. Consider stabilizing a transport stream from a source node S to a destination node D over a wide-area network, typically the Internet. The objective is to achieve a *target* throughput rate τ at D by dynamically adjusting the sending rate $r_S(t)$ at S in response to network conditions. Packets are sent from transport daemon at S and are acknowledged by the daemon at D. Both the original packets and their acknowledgments can be delayed or lost altogether during the transmission due to a variety of reasons, including buffer occupancy levels at routers and hosts, and link level losses. Let $r_S(t)$ and $t_D(t)$ denote the *sending rate* at S and *throughput* or *goodput* at D, respectively. The *response plot* corresponds to values of $t_D(.)$ plotted against $r_S(.)$.

In practice, one only has access to various measurements at the source (including the ones that are sent by destination) that need to be utilized to adjust $r_S(t)$. Consider the measurements collected over the Internet shown in Figure 11.6 between ORNL and LSU. In the horizontal plane each point corresponds to window-size and waiting-time (or idle-time) pair, the ratio of which specifies $r_S(t)$; the top and bottom plots represent $t_D(t)$ and *loss rate*, respectively. For illustration purpose, let us fix the waiting-time and increase the window-size which corresponds to taking vertical slices of the plots parallel to the window-size axis. There are three important features: (a) There is an overall trend of increase followed by decrease in t_D as r_S is increased; this overall behavior is quite stable although the transition points vary over time. (b) The plot is quite non-smooth mostly because of the randomness involved in packet delays and losses; derivation of smooth utility functions from the response plots is inherently approximate and

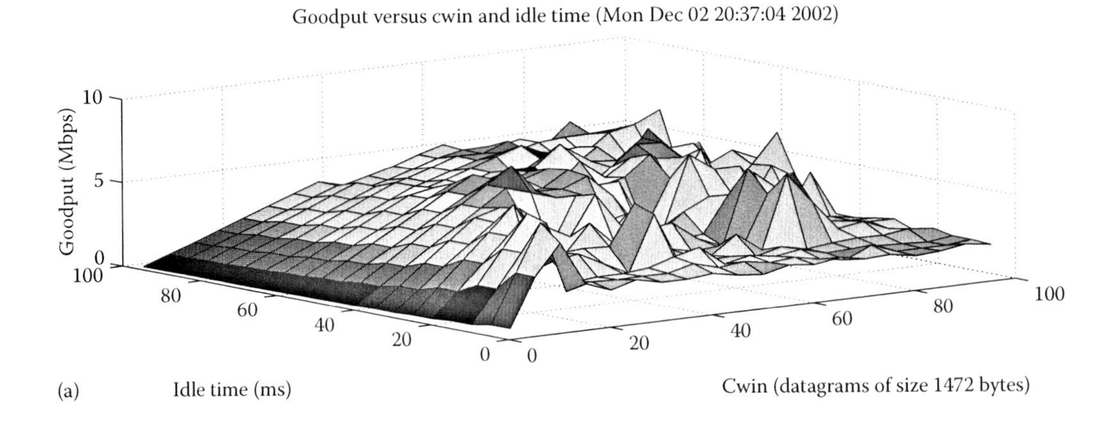

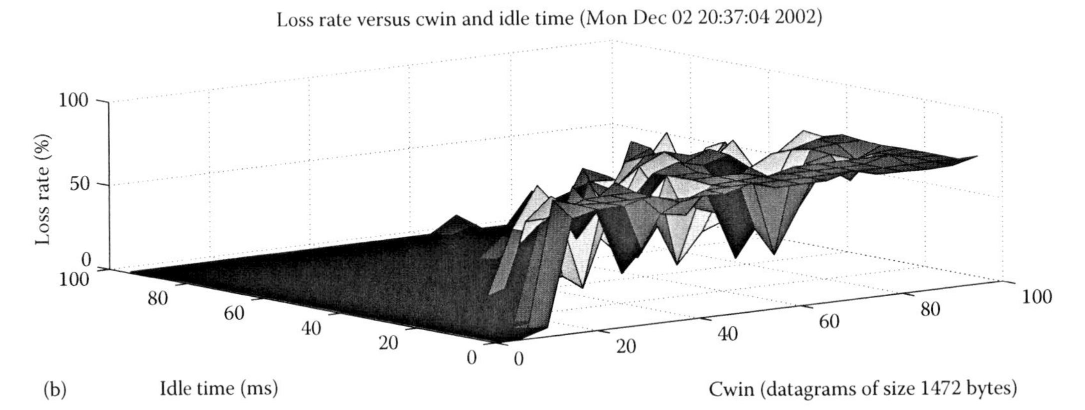

FIGURE 11.6 Internet measurements with sending rate along horizontal plane: (a) top plot: throughput along vertical axis (b) bottom plot: loss rate along vertical axis.

requires a large number of observations in a preprocessing stage. (c) In practice, S has only an approximation $\hat{t}_s(t)$ to $t_D(t)$ typically computed based on acknowledgments.

We assume that the target throughput τ is specified much below the peak of the response plot. Let $\mathcal{T}(r)$ be the *response regression* given by the expected value of t_D corresponding to fixed $r_S(t) = r$, that is,

$$E[t_D(t) \mid r_S(t) = r] = \mathcal{T}(r)$$

Let the *stabilization rate* r_τ be given by $\mathcal{T}(r_\tau) = \tau$. Here τ is chosen such that r_τ is within the initial increasing part of $\mathcal{T}(.)$. We assume that $\mathcal{T}(r)$ is locally monotonic in the neighborhood of τ such that: $\mathcal{T}(r) > \tau$ for $r > r_\tau$ and $\mathcal{T}(r) < \tau$ for $r < r_\tau$. This assumption is consistent with the measurements in Figure 11.6 and, also with the concavity assumptions in Refs [8,10]. Note that $t_D(t)$ can be highly non-smooth and $\mathcal{T}(.)$ is not known.

11.3.2.1 Throughput Stabilization

Our method is based on a simple flow control mechanism at S, where $r_S(t)$ is adjusted in response to a dynamically computed estimate $\hat{t}_s(t)$ of $t_D(t)$ based on acknowledgments received from D. At time t_i, $W(t_i)$ denotes the number of packets to be sent followed by a waiting-time $T(t_i)$ such that

$$r_S(t) = \frac{W(t)}{t_a + T(t)}$$

for $t_i \leq t < t_{i+1}$ where t_a is time needed for transmitting the packets at S and t_{i+1} is the time when $W(.)$ is updated next. We fix $T(t) = T_0$ and update the window size as follows:

$$W(t_{i+1}) = W(t_i) - \gamma_i[\hat{t}_S(t_i) - \tau] \quad (11.4)$$

where $\gamma_i = K(t_a + T_0)/i^{\alpha}$ for $0.5 < \alpha < 1$ and $K > 0$ a suitably chosen constant. This method is a specific form of the well-known Stochastic Approximation (SA) algorithm [9]. Intuitively, this algorithm increases $W(.)$ if the estimate of the throughput is below τ and decreases it otherwise. Initially at $t = t_0$, T_0 and $W(t_0)$ are chosen based on our initial measurements so that $r_S(t_0)$ is within the vicinity of τ.

Let $W_\tau = r_\tau(t_a + T_0)$ correspond to the ideal window size that achieves the stabilization rate τ in an expected sense. We assume that $\mathrm{Var}[\hat{W}_S(t_{i+1}) \mid W(t_1), W(t_2), \ldots, W(t_i)] \leq \sigma^2$ for some σ, and there exist K_0 and K_1 such that $K_0 \,|r - r_\tau| \leq |T(r) - \tau| \leq K_1 \,|r - r_\tau|$. Under these conditions, we have the stability result [18]: $E[(W(t_i) - W_\tau)^2] = O(i^{-\alpha})$. By taking into account the scale factor, we have $E[(r_S(t_i) - r_\tau)^2] = O(i^{-\alpha})$. This result is valid even when r_τ varies over time but somewhat slowly. Since $\hat{t}_S(t)$ is a noisy estimate of a random quantity, it is very critical that the step size γ_i in Equation 11.1 be chosen to satisfy the classical Robbins-Monro conditions [21]: (a) $\gamma_i \to 0$ as $i \to \infty$, (b) $\sum_{i=1}^{\infty} \gamma_i = \infty$. The above algorithm and its stability analysis can be repeated for the case $W(t) = W_0$ is fixed and $T(t)$ is changed in a manner similar to Equation 11.1. Our experimental results are qualitatively identical in both cases.

11.3.2.2 Experimental Results

Our method is tested extensively between ORNL and LSU. During the testing, ORNL is connected to ESnet, which peers with Abilene network in New York. Abilene runs from New York via Washington DC and Atlanta to Houston, where it connects to LSU via a regional network. In terms of network distance, these two sites are separated by more than two thousand miles, and both ESnet and Abilene have significant traffic. Figure 11.7 shows typical results for target throughput at 2.5 Mbps below the peak bandwidth but above throughput 1.09 Mbps achieved by default TCP. In each plot, top and bottom curves correspond to $r_S(t)$ and $\hat{t}_S(t)$, respectively, which often overlap indicating low loss conditions. The stabilization typically occurred under very low albeit non-zero packet loss. The throughput was remarkably robust and was virtually unchanged when transfers of various file sizes using FTP were made at local and other LAN hosts together with various web browsing operations.

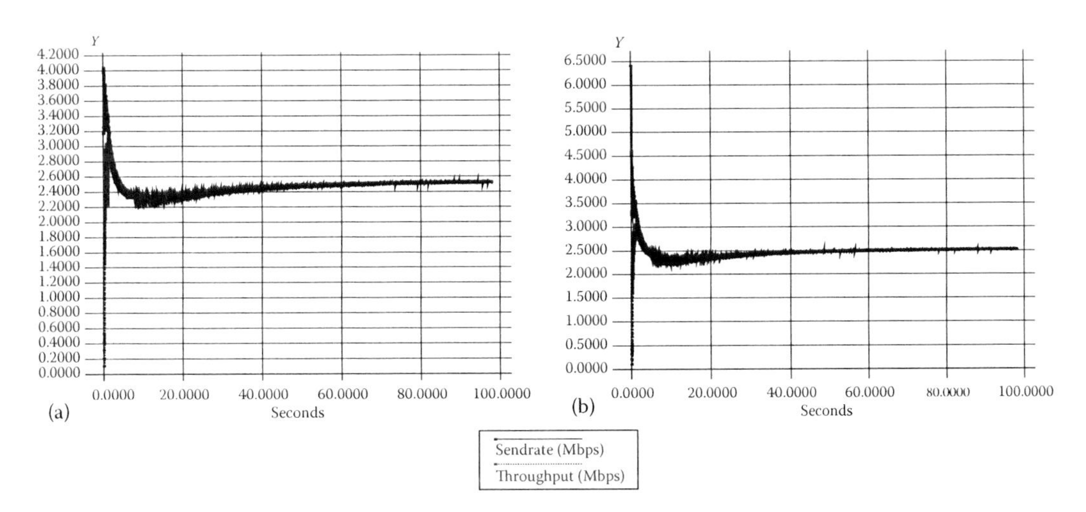

FIGURE 11.7 Stabilization at 2.5 Mbps throughput under various background traffic: (a) large file ftp at host with web browsing, and (b) large file ftp from different LAN nodes with web browsing.

11.4 Daemons for Ad Hoc Mobile Networks

The DSNs of mobile nodes can be applied in a wide variety of scenarios. For example, a robot team can be deployed (perhaps air-dropped) to build a radiation map of an urban area suspected of nuclear or chemical contamination before human operators are allowed into the area [1]. Typically, in these applications there is a need for the robots to effectively communicate to coordinate their activities as well as to combine the gathered information. The networking needs for this class of applications are quite specific and are not adequately addressed by the existing wireless ad hoc networking technologies.

In general, various types of scenarios call for different types of DSN wireless networks [4]. In ad hoc wireless networks the challenge is to form and operate a network without the infrastructure. In dynamic networks the additional challenge is to cope with the changes in network connectivity. Several network protocols have been developed for various sensor network scenarios (see [2,20] and references therein). The specific class of wireless ad hoc networks discussed above lead to the following considerations:

- *Small number of nodes:* We consider networks of tens of nodes which cooperatively perform a task. Its primary focus is to execute a cooperative mission, and the node movements are not tasked exclusively for communication purposes.
- *No infrastructure:* The sensor nodes operate over a wireless network in areas that are typical indoor or urban environments. The radio connectivity is highly dynamic and unpredictable due to the unstructured nature of the terrain and node movements.
- *No special hardware:* We consider that the sensor nodes are equipped with IEEE 802.11 wireless cards, and no special communication hardware is available.

In existing wireless sensor networks it is common to employ Internet wireless network technologies, typically IEEE 802.11 wireless cards and default TCP/IP stack. In the default infrastructure mode nodes communicate exclusively through the access points, which requires a backbone of the access points to connect various nodes. The 802.11 cards can be operated in the ad hoc mode in which case the nodes that are within the radio range can communicate with each another but their connectivity is restricted to pairs that are within the radio range. In the Internet-based technologies, the connectivity changes are treated as aberrations and are handled as exceptions. On the other hand, in the above scenarios the connectivity changes are integral parts of the operation rather than exceptions. More importantly, if suitable protocols are employed, the connectivity changes can actually improve the network throughput as analytically shown in Ref. [6]. In this section, we show that the connectivity-through-time concept provides a way to conceptualize such phenomenon and to design protocols to exploit the node movements [19].

11.4.1 Connectivity-through-Time Concept

The graph $G(t) = (V, E(t))$ represents the connectivity of the network at time t with node $v \in V$ representing a robot and edge $(u, v) \in E(t)$ representing a direct wireless communication link between nodes u and v. At time t a path from nodes s to d in $G(t)$ represents a multi-hop network connection since a message can be routed along nodes of the path. If this path persists for a time interval $[T_1, T_2]$ a message with the end-to-end delay of $T_2 - T_1$ can be successfully delivered from s to d. On the other hand, if there is no path from s to d in $G(t)$ for any t, it does not necessarily mean that a message cannot be delivered.

To discuss the performance of a protocol that achieves such delivery we need to identify a reasonable performance criterion. Since the topology is dynamic, it is too weak to expect that a datagram be delivered from s to d only if they are connected at some time (as is done in TORA [11], for example). On the other hand, it is unreasonable to expect messages to wait indefinitely long in the network; if d is not reachable from s at all, flooding the messages could lead to inordinate amounts of datagrams being generated, thereby causing the denial of service between the nodes that are connected. To address the issue, the concept of *connectivity-through-time* was proposed in Ref. [13].

Let the topology changes occur at unique times, denoted in increasing order by $t_1, t_2, \ldots, t_k$ for $t_i \in [0, T]$. Note that $G(t)$ remains constant for all $t \in [t_i, t_{i+1})$ and is given by $G(t_i)$. We define that s and d are *0-connected-through-time* for interval $[T_L, T_H]$ if they are connected in $G(t)$ for some $t \in [T_L, T_H]$. Consider $[T_L, T_H] \in (t_{i-1}, t_{i+1})$ containing t_i. We define that s and d are 1-*connected-through-time* for interval $[T_L, T_H]$ if

1. They are 0-connected through-time for $[T_L, T_H]$
2. There exists a node v such that: (a) s and v are connected in $G(t_i)$, and (b) v and d are connected in $G(t_{i+1})$

The *time-path* in $[T_L, T_H]$ is represented by the composition of path from s to v in $G(t_i)$, followed by *time-edge* $(v;t_i, v;t_{i+1})$, and followed by path from v to d in $G(t_{i+1})$. The time interval corresponding to $T_H - T_L$ is called the *hold-time* of the path. This definition is recursively applied to an interval containing more than one t_i's as follows. We define that s and d are k-*connected-through-time* for interval $[T_L, T_H]$ containing $t_1, t_2, \ldots, t_k$ if they are

1. 1-connected-through-time for $[T_L, t_1)$
2. $(k - 1)$-connected-through-time for $[t_1, T_H]$

Then s and d are *connected-through-time* for interval $[T_L, T_H]$ if they are k-connected-through-time. We consider that each node v is connected to itself at all times through time-edges denoted by $(v;t_i, v;t_{i+1})$.

One can visualize a time-expanded graph $EG([0, T]) = (EV, EE)$ of $G(t)$ as follows. For each interval $[t_i, t_{i+1})$: (a) each $v \in V$ of $G(t_i)$ corresponds to node $v;t_i$ in EV; and (b) each edge $(u, v) \in E(t_i)$ of $G(t_i)$ is represented by the edge $(u;t_i, v;t_{i+1})$ in $EG([0, T])$. Additionally, for a node v of V, we place the time-edge $(v;t_i, v;t_{i+1})$ for each interval $[t_i, t_{i+1}]$. We define a *time-path* from s to d in the expanded graph as a path with the condition that time intervals of all time-edges be (a) disjoint, and (b) their beginning times be strictly increasing as we move along the path from s to d. Thus a time-path typically consists of the usual graph paths in $G(t_i)$'s interconnected by time-edges. In Figure 11.3c, the time path is denoted by $(s;t_1, i;t_1), (i;t_1, i;t_2), (i;t_2, i;t_3), (i;t_3, d;t_3)$. The hold-time of a time-path is the sum of the hold-times of all its time-edges. Intuitively speaking, if s and d are connected-through-time in $[0, T]$, a datagram from s can be delivered to d by transmitting along graph paths and buffering along the time-edges for a time period given by the hold-time.

There are two practical considerations in implementing the above approach. First, the nodes have finite buffers and packets cannot be indefinitely stored. Second, the transmission time is non-zero and could be significant for newly made connections. As a result, not all messages in the buffers may be delivered during the time a connection is available. We parameterize the packets delivery along the connectivity-through-time with two parameters, *time-to-live* and *minimum-connection time*. The first parameter specifies the time during which the current message is useful. For example, the location information of a moving robot is obsolete after certain time. So we delete the messages from the buffers after the expiry of their time-to-live values. Then packets with appropriate time-to-live value can be delivered along a time-path with sufficient minimum-connection time. Note that the minimum-connection time depends on the robot movements and time-to-live is a protocol parameter.

11.4.2 CTIME-Protocol

The overall idea of this protocol is to track the connectivity and route the packets by suitably buffering them if there is no path to the destination. Each network node acts as a router in delivering the messages. The source nodes decompose the messages as UDP datagrams and send them over the network.

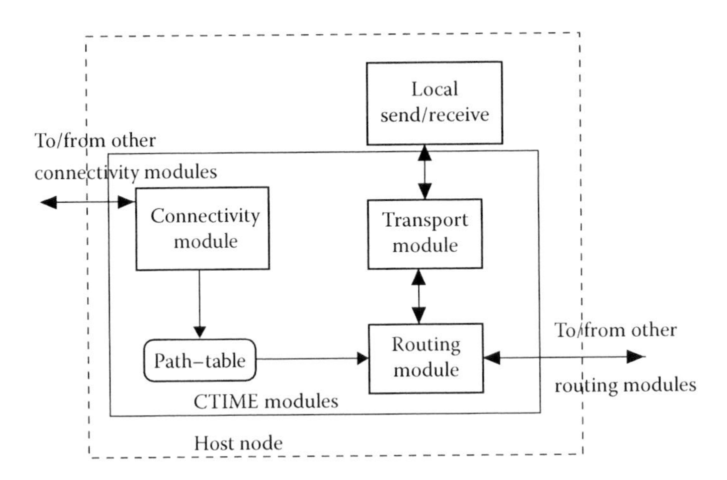

FIGURE 11.8 Daemons for wireless ad hoc networks.

The received datagrams are reassembled at the destination. This protocol is specified by two variables *time-to-live* and *minimum-connection time* both determined empirically. Each packet is given the same time-to-live value. The minimum-connection time is assumed to be sufficient to clear the buffered packets.

The CTIME protocol is implemented using daemons shown in Figure 11.8.

1. *Connectivity computation*: The direct and multiple-hop connectivity at each node is continually updated in response to the I-am-here datagrams broadcast periodically by each node.
2. *Message routing*: The messages are decomposed into UDP datagrams, which are then routed by suitably buffering when needed.
3. *Message transport*: Packet losses, throughput rates and duplicates are handled by using a window-based mechanism akin to TCP but adapted to the current environment.

11.4.2.1 Connectivity Computation

Each node v maintains a list of *direct neighbors* $DN(v)$ that it is in direct contact with and list of *multiple hop neighbors* $MN(v)$ that it is connected with via other nodes. Each node periodically broadcasts an *I-am-here* UDP broadcast packet, which is heard by all nodes within the direct range. This packet includes the list of all direct neighbors as well as multiple hop neighbors. Using these messages from the neighbors, each node computes its direct and multi-hop neighbors using the distributed transitive closure algorithm. This information is periodically updated as the I-am-here messages from other nodes are received. The counting to infinity problem is avoided by not sending the connectivity information to its original source as is usually done.

11.4.2.2 Routing

The message at a source node is decomposed into fixed sized datagrams, which are routed along the nodes. The datagrams are written to the local routing daemon. The routing module also receives packets from other nodes to be routed or buffered. The packets targeted to the local node are simply sent to the send/receive module. For other datagrams if a route to destination exits, then it is sent along the known path by writing it to the next node on the path to destination. That is, if the destination is directly connected, it is sent to it, and if not, it is written to the next node on the path to destination. If no path to the destination exists, then it is buffered locally if not already buffered and then is broadcast to all its immediate neighbors. When new connections are made, this module examines the list of buffered packets and routes them as above. Also, the buffered packets are periodically examined and those that outlived their time-to-live values are simply deleted.

11.4.2.3 Transport Method

The transport module at the source generates the UDP datagrams from the message and keeps track of packets that have not been acknowledged. It maintains a buffer of unacknowledged packets and sends them to the router module at the appropriate rates as described below. It also resends the unacknowledged packets after a time out period. A simple window-based flow control strategy is used at the source and intermediate nodes. Each node maintains a window-size w and window-time T_w to compute its throughput in terms of the number of packets sent during T_w. The preference of this method over TCP for transport is dictated by the following considerations:

1. *High physical-layer losses*: The usual implementation of congestion control is not suited for this environment due to high packet loss and low probability of simultaneous transmissions at the physical layer. TCP interprets the physical layer losses as congestion signals and reduces its throughput. In the current scenario, however, the opposite is needed: the throughput must be increased to account for the packet loss.
2. *Graceful disconnection*: Due to high rate of disconnection, TCP based method will wait for connection time out.
3. *Application-level tuning*: Most TCP parameters are not available to be tuned to suit the application without the kernel modification. For example, it is not easy to select the congestion window size based on the current connection parameters.

At each source, we specify the throughput rate depending on the connectivity to the destination. We collected the throughput rates at the source and destination which showed a unimodal behavior (akin to those used in TCP models [7,10]). In the left plot of Figure 11.9, we show the receiver throughput as a function of the sending rate when both nodes are stationary. While one or more robots are in motion, there are somewhat higher losses and lower throughput as shown in the right plot of Figure 11.9. We choose an appropriate sending rate for direct connections based on whether the robots are moving or not.

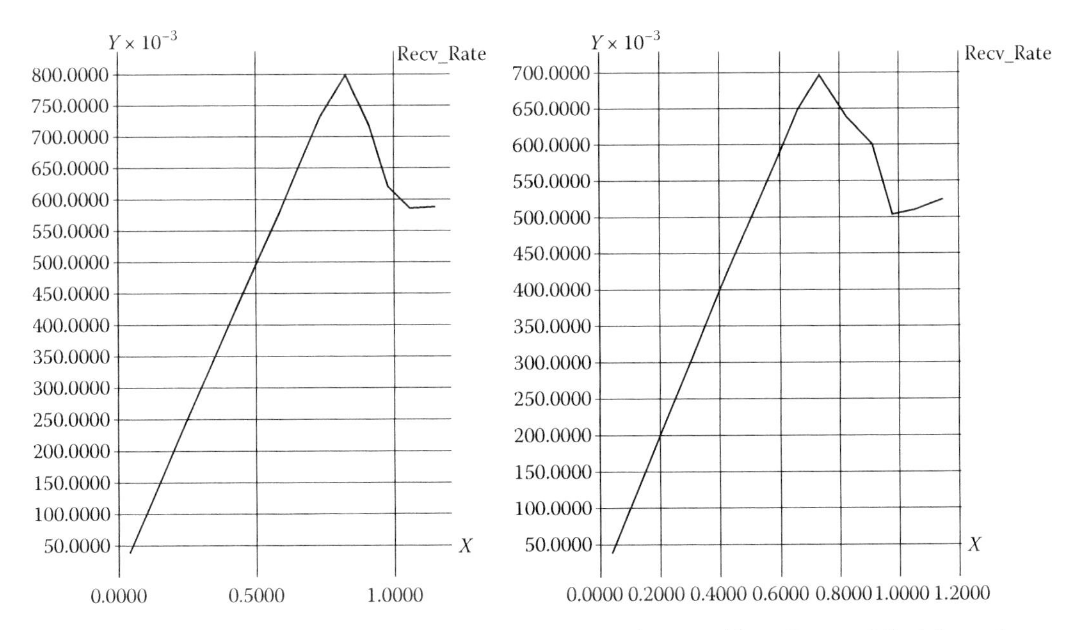

FIGURE 11.9 Destination throughput (Mbps) versus source sending rate (along X-axis in Mbps) for stationary and moving nodes, respectively.

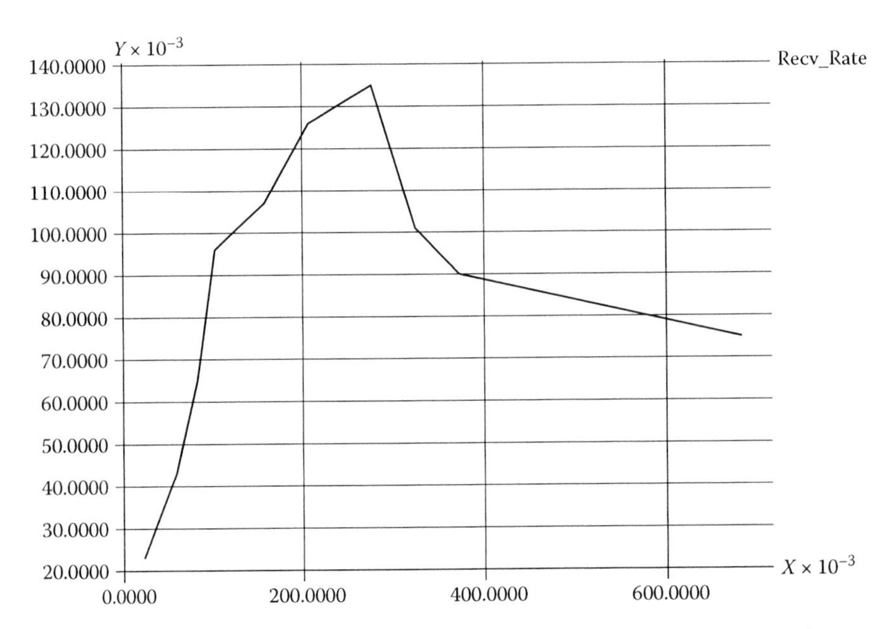

FIGURE 11.10 Destination throughput (Mbps) versus source sending rate (Mbps) for multiple hop connections.

There is a significant reduction in the overall throughput when packets are routed via other robots, because (a) same physical channel is used for two connections at the intermediate node thereby reducing the available raw bandwidth to at most half the peak; and (b) the overheads of routing introduce delays thereby further reducing the bandwidth. Note that TCP is not capable of sending packets using other nodes as routers. Generally TCP only needs to take care of the transport at the two ends, while the intermediate nodes in CTIME protocol have much more complicated transport controls to ensure reliable delivery. The throughput at the destination as a function of source sending rate is shown in Figure 11.10 when the packets are routed via an intermediate node. This plot also shows unimodal behavior but at significantly lower source rate compared to a direct connection. It is important to ensure good throughput to distinguish between the connection from source to destination and employ suitable sending rate. In CTIME, we apply the direct sending rate (a) in transmissions to the destination when directly reachable, or (b) in broadcasting if the destination is not reachable. If the destination is reachable via multiple hops, the lower sending rate is employed as per the observations shown in Figure 11.10.

11.4.3 Experimental Results

We present experimental results based on the implementation on a team of mobile robots. The protocol is implemented in C++ under Linux OS using socket-level programming. The testing is carried out on a team of four Mini ATRV mobile robots equipped with 802.11 wireless cards.

In scenario 1, we demonstrate that the robots are used as routers to deliver messages when there is no direct path from source to destination nodes, but the intermediate node falls within the intersection of the radio ranges of source and destination. As shown in Figure 11.11, the intermediate node receives datagrams from the source and forwards them to the destination. Since the connections of two hops exist at all time, the datagrams are continuously transmitted through the path until they arrive at the destination. The corresponding average throughput is calculated and plotted in

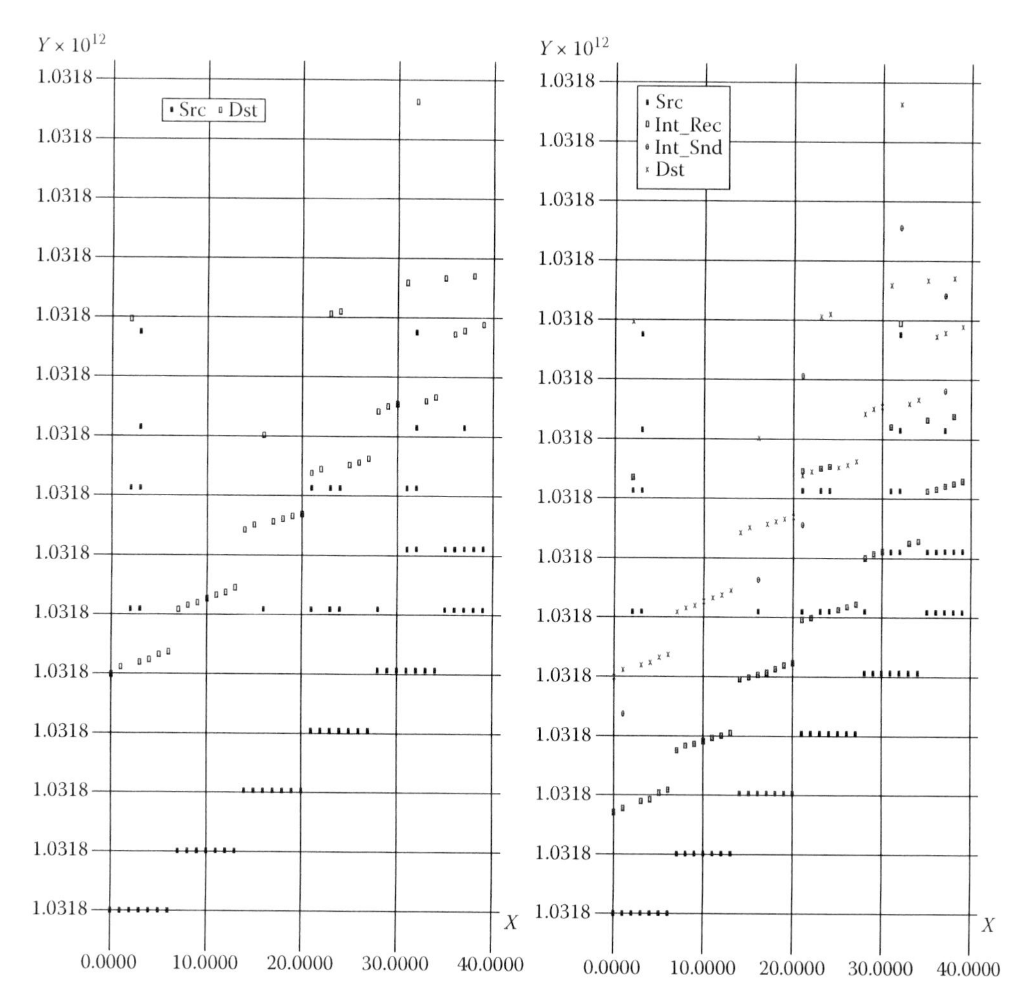

FIGURE 11.11 Scenario 1: Robot serves as a router. Packet number (X-axis) versus send/receive time.

Figure 11.12. Since the incoming and outgoing connections at the intermediate node contend with each other for the physical channel bandwidth, the intermediate and destination nodes have lower receiving rate than the source and intermediate sending rate.

In scenario 2, we illustrate that messages will be buffered where the path to the destination breaks. As shown in Figure 11.13, the data transmission has three stages. The first stage is similar to scenario one: the source and destination are connected to the intermediate node but there is no direct connection between them. The intermediate node serves as a router receiving and forwarding the first set of datagrams. In the second stage, the connection between the intermediate node and destination breaks, so the intermediate node starts buffering incoming datagrams until the connection is brought back up. The second set of datagrams are delivered in the third stage where the connection of the second hop resumes. Figure 11.14 shows the average throughput calculated for scenario two. During the second stage, the throughput of the destination decreases because it does not receive any data from the intermediate node, while the intermediate node has a higher sending rate because the preset timeout incurs broadcasting.

In scenario three, we show that the messages are delivered between source and destination, which are never connected to each other even via multiple hops. As shown in Figure 11.15, this scenario can

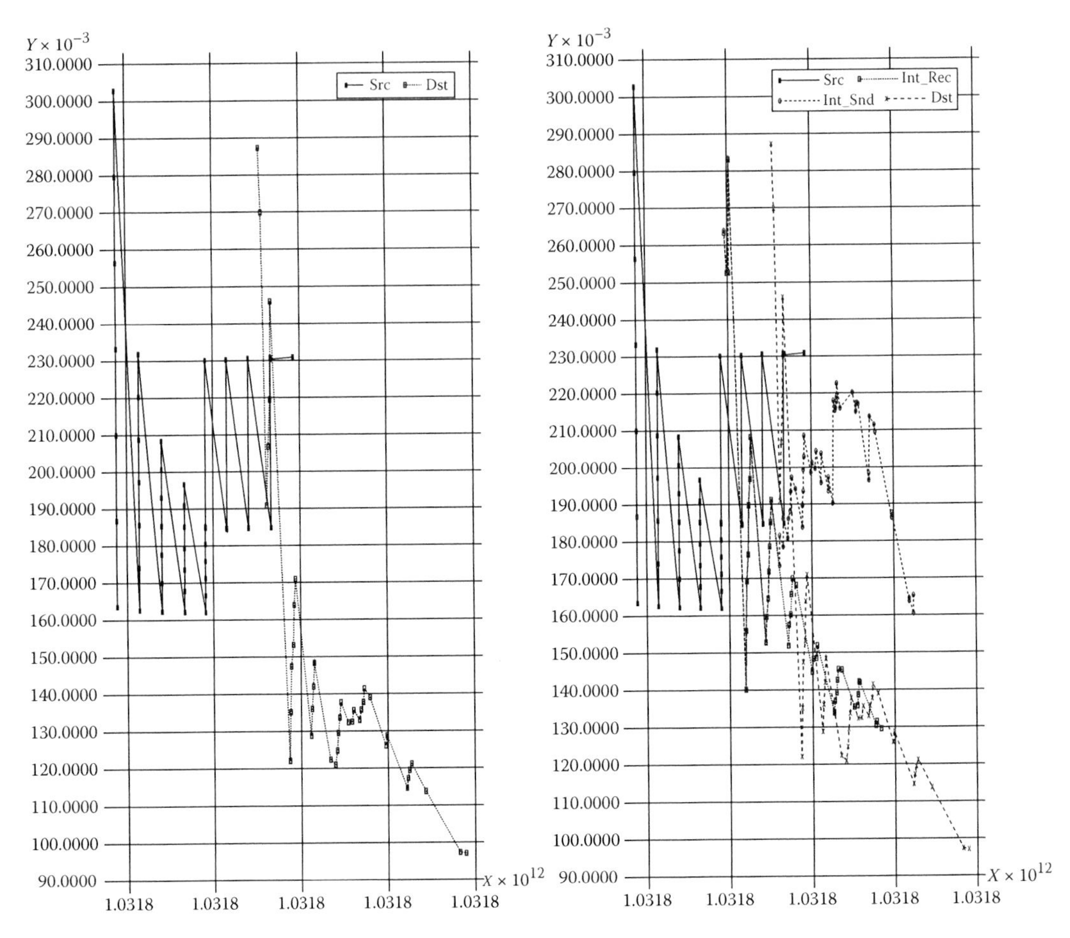

FIGURE 11.12 Average throughput versus time in scenario 1.

also be divided into three stages. In the first stage, only the connection between the source and intermediate node exists, and the datagrams are broadcast after a preset timer expires. This connection breaks in the second stage where the intermediate node is the only active node performing broadcasts. In the third stage, the connection between the intermediate node and destination comes up so that a new path is found to deliver the datagrams from intermediate node to the destination. As a matter of fact, some of the datagrams are received by the destination through broadcast right after the second hop connection is created and before the new path is computed. The corresponding average throughput is shown in Figure 11.16. Observe from the throughput curves that the destination has almost the same throughput as the intermediate node. The explanation for this observation is that the two hop connections never exist at the same time so that each of them has the exclusive bandwidth utilization at different times.

11.5 Conclusions

Many DSNs deployed in the fields utilize technologies developed for the Internet and hence inherit their limitations. But, they offer operational environments that are fundamentally different from the Internet, wherein application-level daemons can be deployed and customized for various classes of sensor networks. We described a generic framework of network daemons that perform the tasks of link measurement, transport control, data routing, and path computation to overcome several throughput

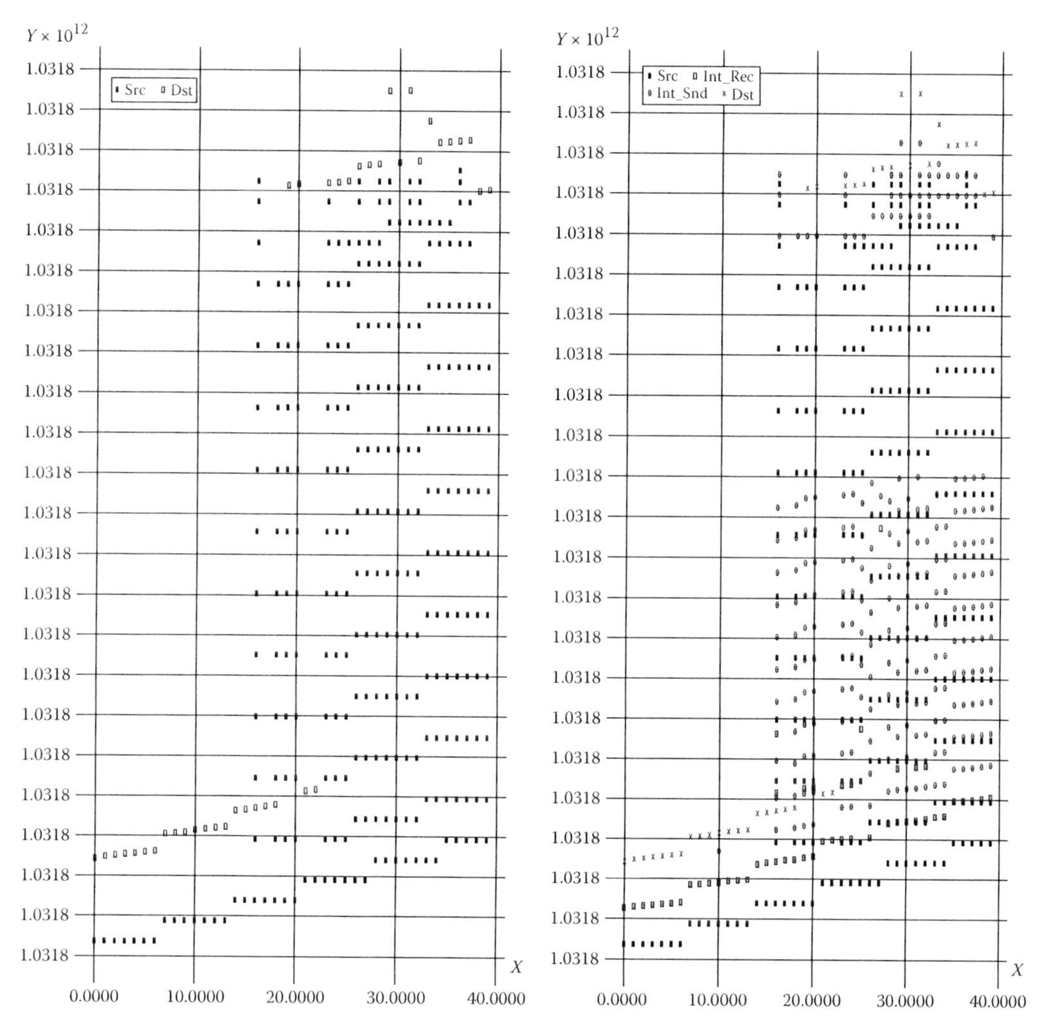

FIGURE 11.13 Scenario 2: path to the destination breaks. Packet number (*X*-axis) versus send/receive time.

and connectivity limitations. We then described two special classes of daemon for wide-area wireline networks and small-area wireless networks. The first class provided bandwidth aggregation using multiple paths and throughput stabilization using stochastic approximation method. The second class provided enhanced multi-hop connectivity by exploiting the node movements.

There are several topics to be further investigated. We discussed two very dissimilar classes of DSNs supported by different networks. It would be interesting to see other classes of DSNs to which the generic daemons can be naturally tailored. Another area of interest is to develop daemons that can automatically adapted to the DSN particularly in cases where the underlying network consists of wireless and wireline networks.

Acknowledgments

This research is sponsored by the Defense Advanced Projects Research Agency under MIPR No. K153, National Science Foundation, and the Engineering Research Program and High-Performance Networking Program of the Office of Science, U.S. Department of Energy, under Contract No. DE-AC05-00OR22725 with UT-Battelle, LLC.

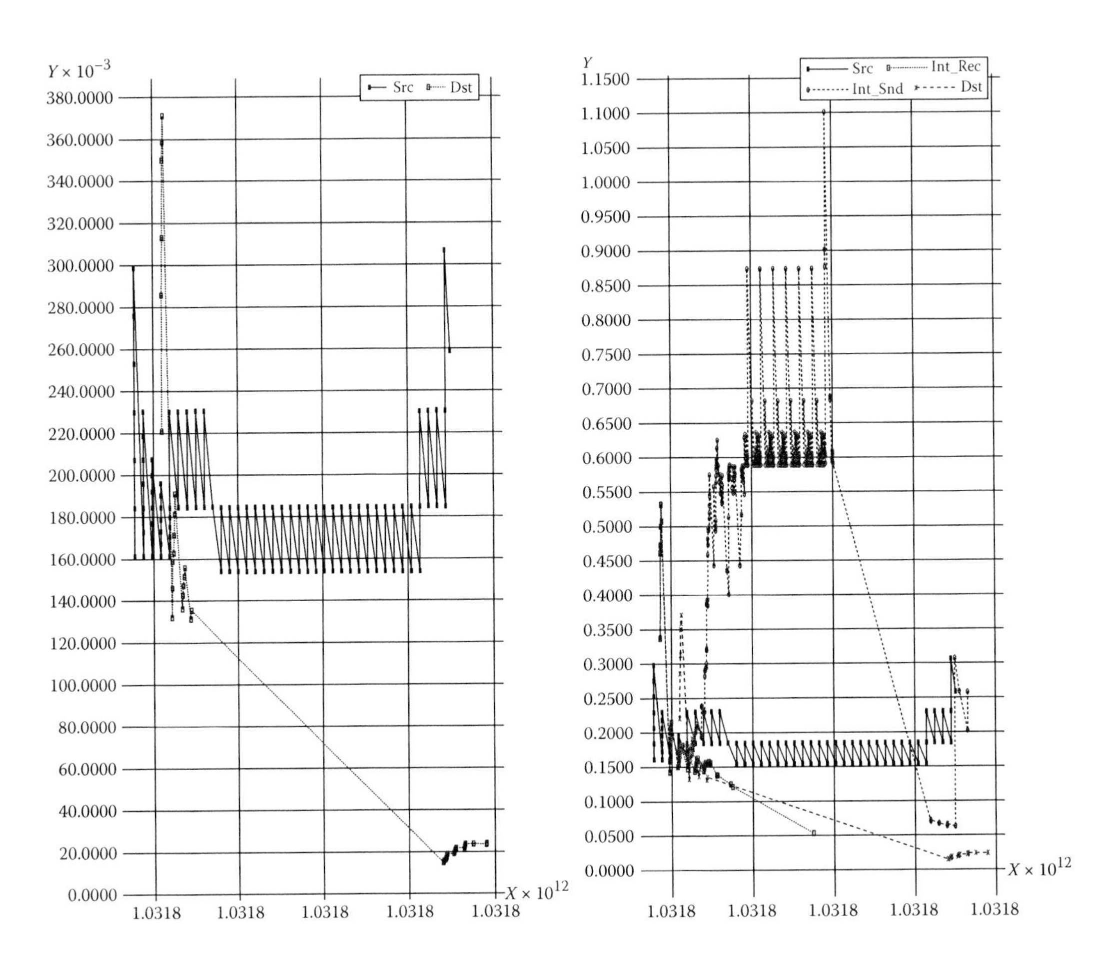

FIGURE 11.14 Average throughput versus time in scenario 2.

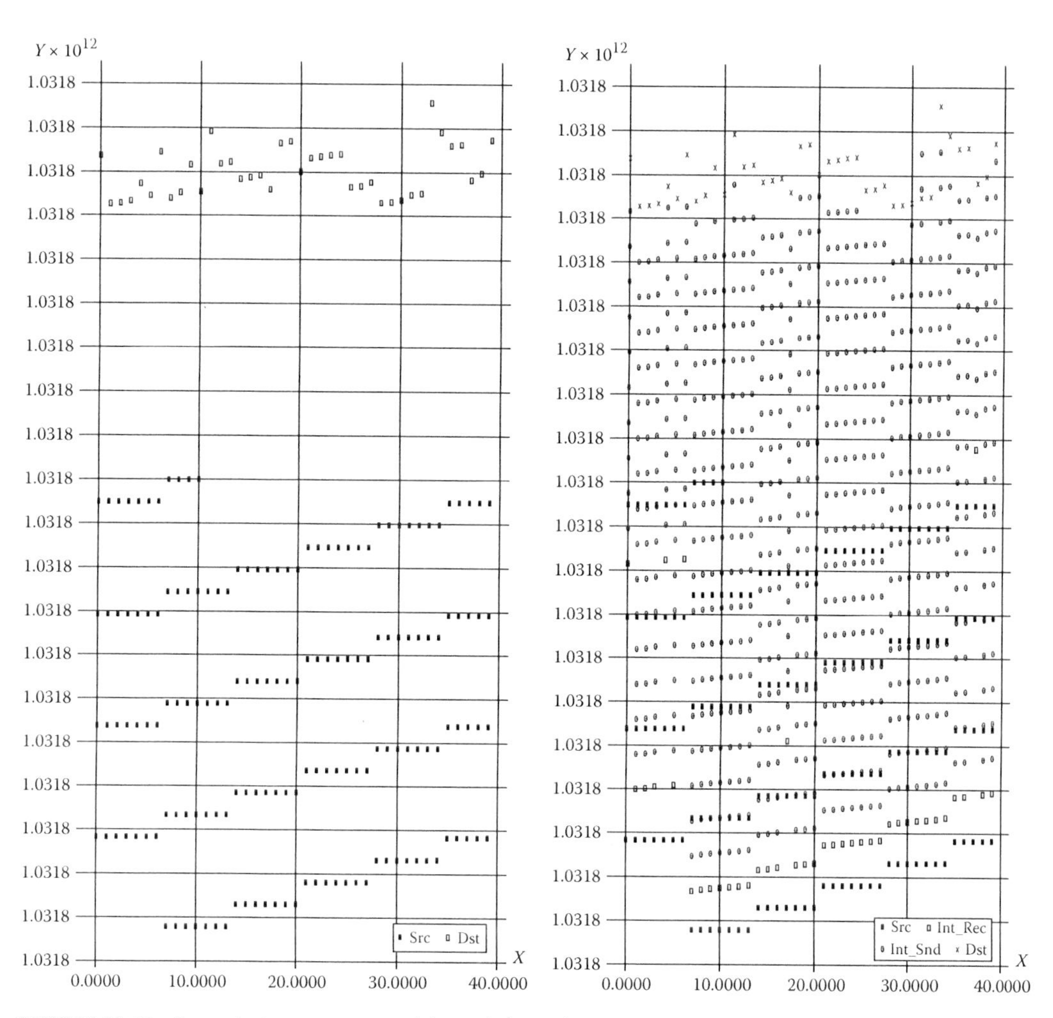

FIGURE 11.15 Scenario 3: messages are delivered through time-connectivity. Packet number (X-axis) versus send/receive time.

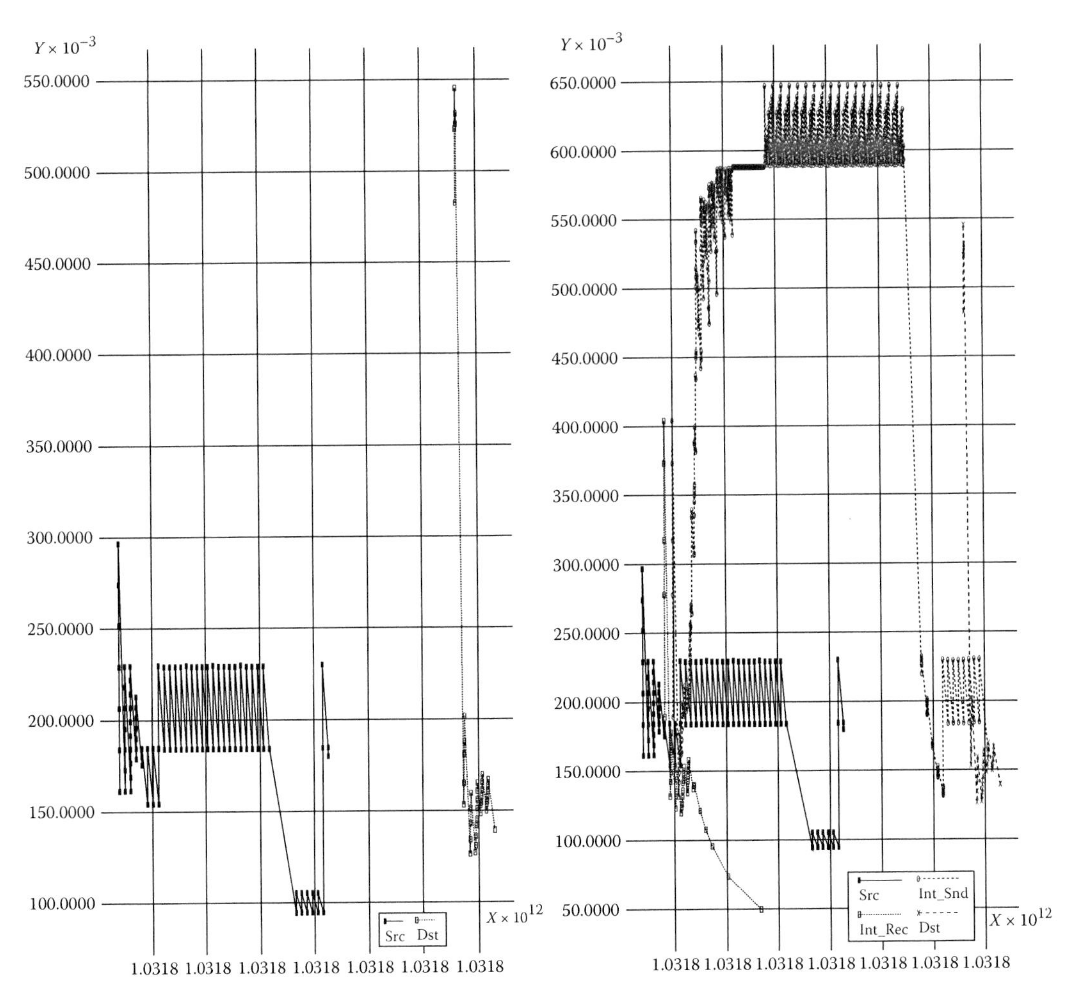

FIGURE 11.16 Average throughput versus time in scenario 3.

References

1. T. Balch and L. E. Parker, eds. *Robot Teams: From Diversity to Polymorphism*. AK Peter Publications, Natick, MA, 2002.
2. D. Braginsky and D. Estrin. Rumor routing algorithm for sensor networks. In *First Workshop on Sensor Networks and Applications*, Atlanta, GA, October 2002.
3. R. R. Brooks and S. S. Iyengar. *Multi-Sensor Fusion: Fundamentals and Applications with Software*. Prentice-Hall, Upper Saddle River, NJ, 1998.
4. D. Estrin, L. Girod, G. Pottie, and M. Srivastava. Instrumenting the world with wireless sensor networks. In *Proceedings of the International Conference on Acoustics, Speech and Signal Processing*, Salt Lake City, UT, May 2001.
5. M. S. Gast. *802.11 Wireless Networks*. O'Reilly, Sebastopol, CA, 2002.
6. M. Grossglauser and D. Tse. Mobility increases the capacity of ad-hoc networks. *IEEE Transactions on Networking*, 10(4), 2002.
7. F. P. Kelley. Mathematical modelling of the Internet. In *Proceedings of Fourth International Congress of Industrial Applied Mathematics*, Edinburgh, U.K., 1999.

8. F. P. Kelly. Mathematical modelling of the Internet. In *Proceedings of International Congress on Industrial and Applied Mathematics*, 1999. http://www.statslab.cam.ac.uk/ frank/mmi.html
9. H. J. Kushner and C. G. Yin. *Stochastic Approximation Algorithms and Applications*. Springer-Verlag, New York, 1997.
10. S. H. Low, F. Paganini, and J. C. Doyle. Internet congestion control. *IEEE Control Systems Magazine*, 2002.
11. V. C. Park and M. S. Corson. A performance comparison of the temporally-ordered routing algorithm and ideal link-state routing. In *Proceedings of INFOCOM'98*, San Francisco, CA, 1998.
12. L. L. Peterson and B. S. Davie. *Computer Networks*, 2nd edn. Morgan Kaufman Publications, San Francisco, CA, 2000.
13. S. Radhakrishnan, G. Racherla, N. Sekharan, N. S. V. Rao, and S. G.Batsell. DST—A routing protocol for ah-hoc networks using distributed spanning trees. In *Proceedings of 1999 IEEE Wireless Communications and Networking Conference*, New Orleans, LA, September 22–27, 1999.
14. N. S. V. Rao. NetLets for end-to-end delay minimization in distributed computing over Internet using two-paths. *International Journal of High Performance Computing Applications*, 16(3), 2002.
15. N. S. V. Rao. Overlay networks of in-situ instruments for probabilistic guarantees on message delays in wide-area networks. *IEEE Journal on Selected Areas in Communications*, 22, 503–523, 2003.
16. N. S. V. Rao, Y. C. Bang, S. Radhakrishnan, Q. Wu, S. S. Iyengar, and H. Cho. Netlets: Measurement-based routing daemons for low end-to-end delays over networks. *Computer Communications*, 26(8), 834–844, 2003.
17. N. S. V. Rao and L. O. Chua. On dynamics of network transport protocols. In *Proceedings of Workshop on Signal Processing, Communications, Chaos and Systems*, pp. i29–i43, 2002.
18. N. S. V. Rao, Q. Wu, and S. S. Iyengar. On throughput stabilization of network transport, ORNL manuscript, 2003.
19. N. S. V. Rao, Q. Wu, S. S. Iyengar, and A. Manickam. On throughput stabilization of network transport. In *Proceedings of International Conference on Robotics and Automation*, 2003.
20. C. K. Toh. *Ad Hoc Mobile Wireless Networks*. Prentice-Hall, Upper Saddle River, NJ, 2002.
21. M. T. Wasan. *Stochastic Approximation*. Cambridge University Press, Cambridge, U.K., 1969.
22. Web100 concept paper, 1999. http://www.web100.org
23. W. Willinger and V. Paxson. Where mathematics meets the Internet. *Notices of the American Mathematical Society*, 45(5), 961–970, September 1998.

12

Energy-Latency Trade-Off for Routing in Sensor Networks and the Positive Impact of Mobility

Sotiris Nikoletseas
University of Patras and CTI

12.1 Routing in Static Networks

12.1.1 Introduction and Contribution Summary

A wireless sensor network (WSN) is an ad hoc collection of large numbers of geographically distributed autonomous nodes that communicate over a wireless link. Each node can directly communicate with other nodes lying within its transmission range. In greedy data propagation, for a packet to reach the destination (e.g., sink or base station), a node forwards the packet to a suitably chosen neighbor, which, in turn, forwards data to one of its neighbors, and so on, until the data reach the final destination.

We study the problem of routing to a static sink by performing greedy, local next-node choices, when all the sensors in the network are also static. Our goal is, first, to mathematically analyze for the first time, experimentally evaluate, and compare the strengths and weaknesses of characteristic greedy protocols such as the location-aware routing (LAR), the direction-aware routing (DAR), and the nearest with forward progress (NFP) protocols. Motivated by this study, we propose two hybrid protocols (hybridization of location aware and direction aware) toward a satisfactory performance trade-off: (1) an energy-aware hybrid protocol and (2) a threshold-based hybrid protocol. We note that greedy protocols must be localized and light weight in order to fit to the resource-constrained nature of sensor networks. We do not consider protocols with guaranteed delivery.

12.1.1.1 Related Work

In greedy algorithms [22,25,28,33,39], a node possessing data under propagation can make a locally optimal, greedy choice (with respect to a specific local criterion) according to the information it has about its one-hop neighbors in order to forward data along a path toward the sink.

A widely used approach of greedy routing is maximum horizontal progress (MHP). In MHP, each node search for the neighbor who corresponds to the MHP, on the line between the current node and the sink. Our analysis and experiments suggest that MHP and LAR are similar and behave almost identically for medium or large densities of sensor deployment.

In the local target protocol (LTP) [11], the model is slightly different than our model and that used in LAR and MHP. The sink is not a single point but a receiving "wall" W (a line segment), each node is aware of the direction toward W and no geolocation abilities are assumed. Just as normal gradient routing, the search phase of LTP finds out the direct neighbor closer to the destination (in this case W), and the direct transmission phase sends the data out. LTP performs well for dense networks, but its performance drops in sparse or faulty networks.

When enhancing the communication model with varying transmission range capabilities, local probabilistic algorithmic design choices can simultaneously satisfy fast forwarding and desired global network properties, such as energy balance toward prolonging the network lifetime [17,35].

12.1.1.2 Our Approach

We mathematically analyze and experimentally evaluate the energy efficiency of three characteristic protocols: (1) the NFP in which the node with the minimum projection is chosen as the next hop node (node s selects node a in Figure 12.1), (2) the LAR where the node with the minimum distance to the destination is chosen as the next hop node (node s selects node c in Figure 12.1), and (3) the DAR where the next hop node is the neighboring node with the minimum angle (node s selects node b in Figure 12.1).

In Figure 12.2 we present an example of the routing paths that are generated using NFP, LAR, and DAR in a square (200×200) network area, where the nodes are placed randomly and uniformly. The source of the data is located at (0, 200) (top left corner) and the sink is placed at (200, 0) (bottom right corner). We observe that the path generated using NFP consists of a large number of small hops (low energy consumption and very high data latency), the path of LAR consists of a small number of large hops (high energy consumption and low data latency), and the path of DAR contains less but larger hops than the NFP's path and more but shorter hops than the LAR's path.

The main advantage of LAR is that it minimizes data delivery latency, but it increases a lot the energy consumption, while DAR reduces energy consumption over LARs; however, it increases data delivery latency.

Our analytic and simulation findings suggest that any single criterion does not simultaneously satisfy both energy efficiency and low latency. Toward a satisfactory energy-latency trade-off, we propose two hybrid protocols (that assume different model strength) in order to use the advantages of both LAR

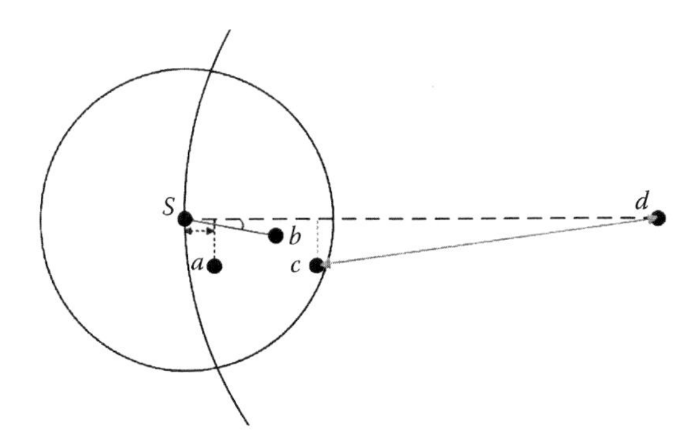

FIGURE 12.1 Illustration of the three greedy routing protocols.

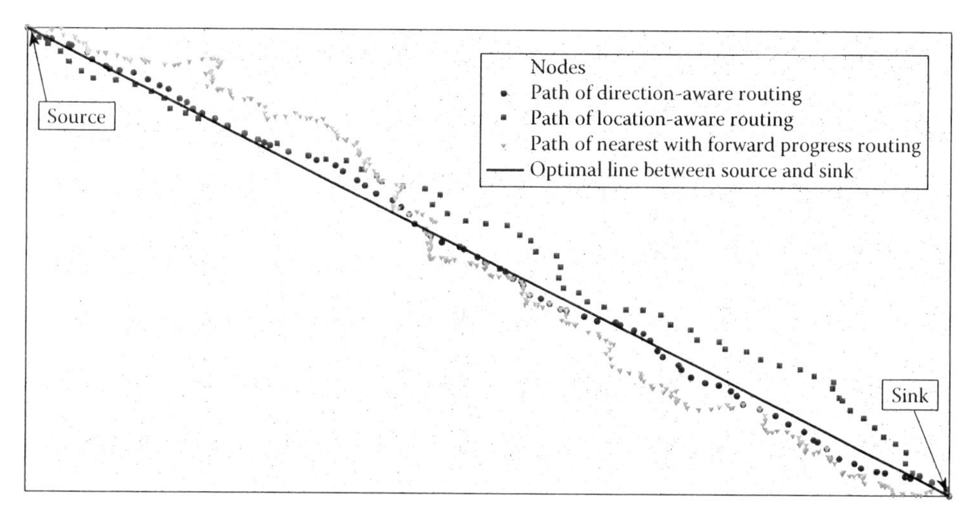

FIGURE 12.2 Comparing the routing paths of DAR, LAR, and NFP with the optimum path.

and DAR. In the first protocol before each transmission, we perform a probabilistic choice according to probability p_i (see Section 12.1.2.6). More concretely, we use location greedy routing with probability p_i and directional greedy routing with probability $1 - p_i$, where p_i depends on the residual energy and the distance to the sink; for large distance and high energy, it tends to perform location-aware choices to accelerate data propagation, while for small distance and low energy, it rather employs direction awareness to save energy. In the second hybrid protocol, we decide to transmit data using LAR or DAR by comparing the latency ratio between the current and the optimal data delivery latency with a predetermined threshold value t_{value}. If the current ratio is lower than the t_{value}, we use location greedy routing, otherwise we use directional greedy routing, that is, the rationale is to guarantee a certain level for the one metric while optimizing the other. Our findings demonstrate in detail the performance characteristics of each protocol and show that significant performance gains are achieved by the hybrid design. We note that the energy-aware heuristic needs stronger modeling assumptions (e.g., distance evaluation) and has better energy balance properties. Still, the threshold-based heuristic (which does not use distance knowledge) achieves similar performance.

A preliminary version of this significantly extended work has appeared in Ref. [16].

12.1.2 Routing Protocols

12.1.2.1 Model

We consider a two-dimensional (plane) sensor network, in which the sensors and the single sink node are static. We abstract the network by a graph $G(V,E)$, where V denotes the set of nodes (sensors), while $E \subseteq V^2$ represents the set of edges (wireless links). The deployment of the sensors is random uniform, and various densities (low, medium, and high) are considered. An edge between two nodes in the graph exists iff the distance between the corresponding sensors in the network is below a certain limit, capturing the wireless transmission range R. The distance between nodes is the euclidean distance, and the path length is the sum of the distances of the intermediate pairs of subsequent nodes (hops).

Nodes are localized; localization can be achieved by either GPS technology (based on localization of a few reference nodes only) or by a system of virtual coordinates. Nodes are aware only of their one-hop away (immediate) neighbors, as well as their locations. Localization allows some direction sensing capabilities, for example, nodes know the direction toward the sink, and can estimate angles around a certain direction. Finally, we assume a setup phase initiated by the sink during which some (limited) global network information is diffused.

Our protocols operate at the network layer, so we are assuming appropriate underlying data link, MAC (medium access control), and physical layers. The nodes' memory is assumed limited, for example, we allow messages to piggyback a constant number of bits of information only, encoding the position of the last node visited and the position of the sink.

Sensor networks are characterized by high network dynamics such as frequent, dense failures. So we investigate the detailed impact of failures on protocol performance. In particular, for each unit of time, failures occur at randomly chosen nodes, instantly, and no further computation and/or communication can be performed by these failed nodes. We examine a broad set of failure probabilities, including low, medium, and high ones.

12.1.2.2 Location-Aware Routing Protocol

A commonly used method of location-aided (or position-based) greedy routing is LAR [43] and its variations. In LAR, each node basically forwards a packet to the neighbor that is closest to the destination. LAR provides a way to deliver a packet to a destination location (e.g., to a static sink) based only on local information without the need for any extra infrastructure or information. For this reason, the LAR protocol is very suitable and efficient in resource-constrained wireless networks, such as WSNs. In this single-path multi-hop routing protocol, each node needs to know only the location information of its direct neighbors in order to forward data packets. This approach basically attempts to find the shortest path to the destination, in terms of either distance or number of wireless hops toward minimizing data propagation latency.

As LAR tries to find the next node that is the closest to the sink in order to transmit the message to, a node considers only those neighbors that are closer to the destination than itself. Sensor nodes can calculate their position in some common coordinate system (e.g., by using navigational equipment or running a virtual coordinates algorithm) and are aware of the sink's position. The aforementioned assumptions can be relaxed in many ways. We note that LAR has not been analyzed mathematically as far as energy efficiency in sensor context is concerned.

12.1.2.3 Direction-Aware Routing Protocol

The main goal of DAR is to minimize energy consumption following an efficient path, which is as close as possible to the optimal (with regard to both energy and latency) direct line that connects the current node to the sink. The basic idea in the protocol is doing shorter transmissions of "nice" direction to neighbors that are closer to the current node in order to keep the energy consumption at low levels (the energy consumed to transmit a message is assumed proportional to some power of the distance between sender and receiver). As mentioned earlier, the main idea is to transmit the data to the node that has the minimum divergence (in terms of angle of direction) from the optimal transmission line that is the line that connects the current node to the sink. More precisely, the neighboring node that forms the minimum angle a_i between the line that connects itself with the current node and the line that connects the current node with the sink is chosen, see Figure 12.3. We note that although the selection criterion is not based on progress to the sink, the fact that data propagation is kept within the optimal transmission zone leads indirectly to not increasing the path length and number of hops too much. DAR is, in fact,

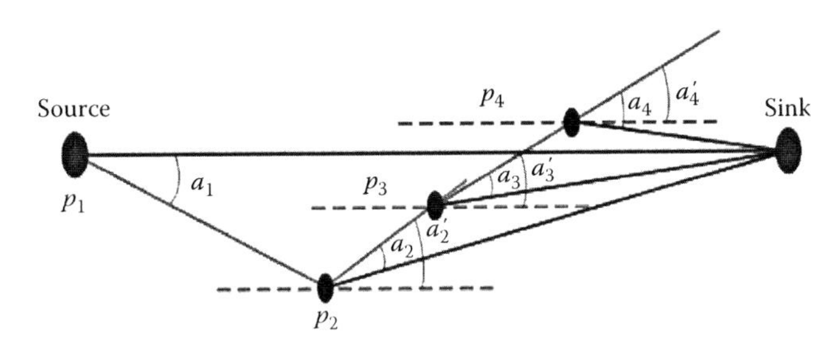

FIGURE 12.3 Example of DAR.

very similar to the compass routing protocol [33]; however Ref. [33] analyzes the correctness and time efficiency of the protocol while the energy efficiency is not studied.

12.1.2.4 Nearest with Forward Progress Protocol

NFP [24] is an energy-aware protocol that tries to minimize the energy consumption by sending the message to the closest node in the direction of the sink. The main advantage of NFP is that it makes collisions less likely as a node will adjust its transmission power to be just strong enough to reach the nearest neighbor that will result in forward progress. This leads to a succession of a large number of small hops (thus, high data delivery latency) and less energy consumption than long hops. We note that NFP has not been analyzed mathematically.

12.1.2.5 Hybrid Routing Protocols

To handle the arising performance trade-off between energy and latency, we propose two hybrid heuristic routing protocols. In the first protocol, each node decides whether to forward data using LAR or DAR by taking into account the distance of the node from the sink and the residual energy of the node. The second protocol takes its decision based on a threshold choice criterion, that is, it tries to optimize one metric as long as the performance of the other one is kept above a certain threshold that can be set by the network implementor.

12.1.2.6 Energy-Aware Hybrid Routing Protocol

Each node forwards data with a probability p_i to that neighbor that is closest to the sink (using LAR) and with probability $1 - p_i$ to that neighbor that forms the minimum angle between the line that connects itself with the current node and the line that connects the current node with the sink (using DAR).

$$\text{Forwarding decision} = \begin{cases} \text{LAR}, & \text{with probability } p_i \\ \text{DAR}, & \text{with probability } 1 - p_i \end{cases}$$

We take

$$p_i = \sqrt{\frac{D_i}{D_{max}} \cdot \frac{E_{current}}{E_{initial}}}$$

where

- D_i is the distance of the node from the sink
- D is the "dimension" of the network (e.g., the "width" of the rectangular network region)
- $D_{max} = \sqrt{2} \cdot D$ is a node's maximum distance from the sink for the rectangular network region $D \times D$
- $E_{current}$ is the residual energy of the node
- $E_{initial}$ is the initial energy of the node

The rationale of *hybrid forwarding* is to forward data using LAR in order to accelerate data transmission when data are far away from the sink and has much residual energy. The opposite happens when data are close to the sink or have little residual energy; henceforth, data tend to be forwarded using DAR so as to reduce energy consumption in the critical, bottleneck region close to the sink. This bottleneck is created because the sensors close to the sink always relay the data for other sensors, resulting in an unequal distribution of network residual energy.

12.1.2.7 Threshold-Based Hybrid Routing Protocol

In this protocol, we wish to guarantee the minimum energy consumption for a predetermined latency ratio between the current and the optimal data delivery latency:

$$\text{Forwarding decision} = \begin{cases} \text{LAR}, & \text{if } Latency_{ratio} < t_{value} \\ \text{DAR}, & \text{if } Latency_{ratio} \geq t_{value} \end{cases}$$

We define $Latency_{ratio}$ as follows:

$$Latency_{ratio} = \frac{Latency_{current}}{Latency_{optimum}}$$

where

$Latency_{current}$ is the total packet travel time from source to the current node
$Latency_{optimum}$ is the optimum packet travel time from source to the current node
t_{value} is the predefined threshold value that can take values between 0 and 1, for example, by choosing $t_{value} = 0.75$ we declare that $Latency_{current}$ at every routing step must be at least 75% of the $Latency_{optimum}$

The rationale of this threshold-based protocol is to forward data using cheap but slow transmissions (DAR) when the $Latency_{ratio}$ is higher than the t_{value}. This means that we are able to conserve energy by transmitting to a short-hop neighbor without increasing data latency. On the other hand, we forward data using fast but expensive transmissions (LAR) when the $Latency_{ratio}$ is lower than the t_{value}, in order to improve the latency and approach the desirable threshold value t_{value}.

12.1.3 Rigorous Performance Analysis

We analyze some important performance properties (energy dissipation and data propagation latency) of each of the three basic routing methods in the following.

12.1.3.1 Analysis of the Direction-Aware Routing Protocol

Let S_1 be the current node possessing data and S_2 the next node.

Definition 12.1

Let a_i or a_{min} (as depicted in Figure 12.3) be the angle between the line that connects the current node to the sink and the line that connects the current node with the next node of the propagation path. It is equal to the minimum angle of all angles that are formed between the line that connects each neighbor with the current node i and the line that connects the current node with the sink.

This angle corresponds to the direction of the node that is chosen for the data propagation. a_i is a random variable since the neighboring nodes are positioned at random, so the angles are also random.

Angle a_i', as depicted in Figure 12.3, is the angle between the line that connects the current node with the next node of the propagation path, with the line that passes through the current node and is parallel to the source–destination line.

To simplify our analysis, we assume that the line x_2 (see Figure 12.4) is parallel to the line x_1 as

$$\sin(\theta) = \frac{p}{x_1} \leq \frac{R}{k \cdot R} = \frac{1}{k} \rightarrow_{k\rightarrow\infty} 0$$

where p is the vertical distance from the line x_1 to the line x_2, p's length is $O(R)$ and x_1's length is $O(k \cdot R)$, where k indicates how many times longer is the source–destination line than radius R, where R is the maximum transmission range. So it is $a_j' \simeq a_i$.

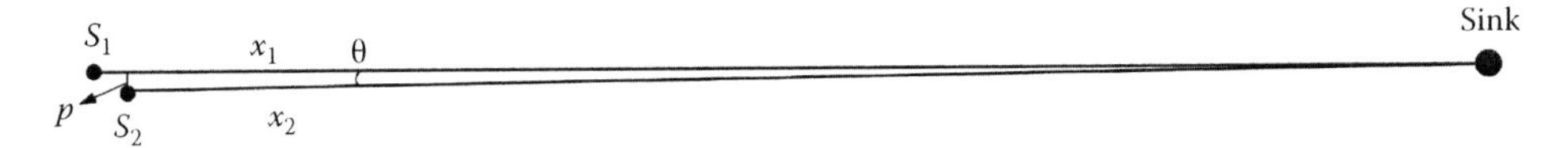

FIGURE 12.4 Example of x_2 being parallel to the line x_1.

Definition 12.2

Let n be the number of neighbors a node has, that is, the number of nodes are positioned on the interior of the circular disk with the node as its center and R as its radius. Let *den* be the density of sensors in the random uniform placement. So

$$n = \pi \cdot R^2 \cdot den$$

Definition 12.3

Let a_{ij} be the random variable that is equal to the angle corresponding to the neighbor j of a node i.

In order to evaluate the performance of the direction-aware protocol, we first must calculate the expected value of this minimum angle.

$$F_{a_i}(x) = P(a_i \le x) = 1 - P(a_i \ge x) = 1 - P(a_{i_1} \ge x,\ a_{i_2} \ge x,\ \ldots,\ a_{i_n} \ge x)$$

The random variables a_{ij} are independent, since the nodes are positioned independently of each other and uniformly at random. As we assume that we investigate nodes only in the positive half-circle of the neighborhood region, the random variables are uniformly distributed in $\left[0, \frac{\pi}{2}\right]$. So

$$F_{a_i}(x) = 1 - P(a_{i_1} \ge x) \cdot P(a_{i_2} \ge x) \cdots P(a_{i_n} \ge x) = 1 - \left(1 - \frac{x}{\pi/2}\right)^n = 1 - \left(1 - \frac{2 \cdot x}{\pi}\right)^n$$

The probability density function of random variable a_i is

$$f_{a_i}(x) = \frac{dF_{a_i}(x)}{dx} = \frac{2 \cdot n}{\pi} \cdot \left(1 - \frac{2 \cdot x}{\pi}\right)^{n-1}$$

We now calculate the expected value of random variable a_i:

$$E(a_i) = \int_0^{\pi/2} x \cdot \frac{2 \cdot n}{\pi}\left(1 - \frac{2 \cdot x}{\pi}\right)^{n-1} \cdot dx$$

$$= \left[-x \cdot \left(1 - \frac{2 \cdot x}{\pi}\right)^n\right]_0^{\pi/2} - \int_0^{\pi/2} \left(1 - \frac{2 \cdot x}{\pi}\right)^n \cdot dx$$

$$= 0 - \left[\frac{\left(1 - \frac{2 \cdot x}{\pi}\right)^{n+1}}{2 \cdot (n+1)}\right]_0^{\pi/2} = \frac{\pi}{2 \cdot (n+1)}$$

Thus, we get the following lemma.

Lemma 12.1

The expectation of angle a_i is

$$E(a_i) = \frac{\pi}{2 \cdot (n+1)}$$

Definition 12.4

Let d_i be the random variable that is equal to the euclidean distance between the current node i and the node (from the set of neighbors of the current node) the protocol chooses to be the one the data are propagated to.

We calculate the distribution function of random variable d_i:

$$F_{d_i}(x) = P(d_i \leq x) = \frac{\pi \cdot x^2}{\pi \cdot R^2} = \frac{x^2}{R^2}$$

The probability density function of random variable d_i is

$$f_{d_i}(x) = \frac{dF_{d_i}(x)}{dx} = \frac{2 \cdot x}{R^2}$$

We now calculate the expected value of random variable d_i:

$$E(d_i) = \int_0^R \frac{2 \cdot x^2}{R^2} dx = \left[\frac{2 \cdot x^3}{3 \cdot R^2} \right]_0^R = \frac{2}{3} \cdot R \tag{12.1}$$

Also, we have

$$E(d_i^2) = \int_0^R \frac{2 \cdot x^3}{R^2} dx = \left[\frac{2 \cdot x^4}{3 \cdot R^2} \right]_0^R = \frac{1}{2} \cdot R^2$$

Thus, we get the following:

Lemma 12.2

The expectation of distance d_i is

$$E(d_i) = \frac{2}{3} \cdot R$$

while the expectation of d_i^2 is

$$E(d_i^2) = \frac{1}{2} \cdot R^2$$

Theorem 12.1

Let the source–destination distance be $dis = k \cdot R$. The mean number of hops required for DAR is upper bounded by the quantity

$$\frac{3 \cdot k}{2 \cdot \left(1 - 2 \cdot \sin \frac{\pi}{4 \cdot (n+1)} \right)}$$

where n is the expected number of sensors within each transmission range of radius R.

Proof

First, we must calculate the expected value of random variable xi, where xi is the projection of the line segment (p_i, p_{i+1}) on the line that is parallel to the line from the source S to the sink. We use the expectation of a_i and the expectation of d_i, as derived in Lemma 12.1 and 12.2. Note that d_i and a_i are independent since the distance d_i of a neighbor to the current node does not give us any information about the angle. So, we have

$$E(x_i) = E(d_i \cdot \cos a_i) = E(d_i) \cdot E(\cos a_i) \tag{12.2}$$

$$E(\cos a_i) = E\left(1 - 2 \cdot \sin^2 \frac{a_i}{2}\right) = 1 + E\left(-2 \cdot \sin^2 \frac{a_i}{2}\right) \geq 1 + E\left(-2 \cdot \sin \frac{a_i}{2}\right)$$

Because $-2 \cdot \sin^2 \frac{a_i}{2}$ is convex as $f''\left(-\sin \frac{a_i}{2}\right) = \frac{\sin \frac{a_i}{2}}{4} > 0 \text{ f } a_i \in \left[0, \frac{\pi}{2}\right]$ we get from Jensen's inequality:

$$1 + E\left(-2 \cdot \sin \frac{a}{2}\right) \geq 1 - 2 \cdot \sin \frac{E(\alpha)}{2} = 1 - 2 \cdot \sin \frac{\pi}{4 \cdot (n+1)} \tag{12.3}$$

So (12.1) through (12.2) and (12.3) $E(x_i) = \frac{2}{3} \cdot R \cdot \left(1 - 2 \cdot \sin \frac{\pi}{4 \cdot (n+1)}\right)$

Let h be the number of hops needed to get to the destination. Clearly,

$$\sum_{i=1}^{h} x_i \leq k \cdot R < \sum_{i=1}^{h+1} x_i$$

so h is a stopping time for the sequence of random variables x_i. So from Wald's equation on the expectation of a random number of random variables [41] we get

$$E(h)E(x_i) \leq k \cdot R < E(h+1)E(x_i)$$

so

$$E(h) \leq \frac{k \cdot R}{E(x_i)}$$

and this concludes the proof.

Now we calculate the expected value of energy consumption of the protocol. Let En be the r.v., that is, equal to the energy consumed at a data propagation. Since we assumed that the energy consumed for a hop is proportional to the second power of the distance traversed, we have

$$E[En] = O\left(E\left[\sum_{i=1}^{h} d_i^2\right]\right)$$

We also get (via Wald's equation, similarly to Theorem 12.1) that

$$E[En] = O(E[h]E[d^2]) \leq O\left(\frac{3 \cdot k}{2 \cdot \left(1 - 2 \cdot \sin\frac{\pi}{4 \cdot (n+1)}\right)} \frac{R^2}{2}\right)$$

12.1.3.2 Analysis of the Location-Aware Routing Protocol

Let us now analyze the location-aware algorithm in terms of hop count and energy consumption. Let Xp be the random variable that is equal to the length of the projection of the distance traversed by a hop to the line that connects the source to the sink. We show the following.

Lemma 12.3

If the source–destination distance is $dis = k \cdot R$, the mean number of hops h required to reach the destination is equal to the quantity

$$\frac{\pi k R}{\int_0^R 2n \left(1 - \frac{2\arccos\left(\frac{x}{R}\right) - \frac{2x\sqrt{1-\frac{x^2}{R^2}}}{R}}{\pi}\right)^{n-1} \left(\frac{1}{R\sqrt{1-\frac{x^2}{R^2}}} + \frac{\sqrt{1-\frac{x^2}{R^2}}}{R} - \frac{x^2}{R^3\sqrt{1-\frac{x^2}{R^2}}}\right) dx}$$

where n is the number of sensors in each transmission range of radius R.

Proof

First, we must compute the distribution function of r.v. Xp.

Let *Area A* be the area that is seen as dark gray at the Figure 12.5 and *Area B* the light gray area. The distribution function of Xp is the following:

$$F(x) = P(Xp \leq x) = \frac{Area\,A}{Area\,A + Area\,B}$$

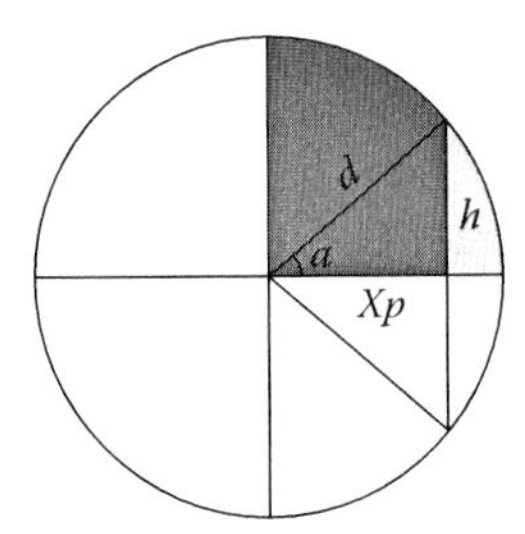

FIGURE 12.5 Distribution of Xp in LAR.

Obviously, dark gray and light gray area form a quadrant. So

$$Area\,A + Area\,B = \frac{\pi R^2}{4}$$

$$\begin{aligned}
Area\,B &= \frac{R^2}{4}(2a - \sin 2a) \\
&= \frac{R^2}{4}\left(2\arccos\frac{Xp}{R} - 2\sin a\cos a\right) \\
&= \frac{R^2}{4}\left(2\arccos\frac{Xp}{R} - 2\sin\left(\arccos\frac{Xp}{R}\right)\cos\left(\arccos\frac{Xp}{R}\right)\right) \\
&= \frac{R^2}{4}\left(2\arccos\frac{Xp}{R} - 2\sin\left(\arcsin\sqrt{1-\left(\frac{Xp}{R}\right)^2}\right)\frac{Xp}{R}\right) \\
&= \frac{R^2}{4}\left(2\arccos\frac{Xp}{R} - 2\left(\sqrt{1-\left(\frac{Xp}{R}\right)^2}\right)\frac{Xp}{R}\right)
\end{aligned}$$

$$\begin{aligned}
F(x) = P(Xp \le x) = 1 - P(Xp \ge x) &= 1 - \frac{Area\,B}{Area\,A + Area\,B} \\
&= 1 - \frac{2}{\pi}\left(\arccos\frac{Xp}{R} - \left(\sqrt{1-\left(\frac{Xp}{R}\right)^2}\right)\frac{Xp}{R}\right) \qquad (12.4)
\end{aligned}$$

Then, we compute the distribution function of the maximum projection from all the projections that are formed with every neighbor of a current node, in other words, the length of the hop in a location-aware transmission:

$$\begin{aligned}
F_{Xpmax}(x) &= P(Xp_{max} \le x) = (P(Xp \le x))^n \\
&= \left(1 - \frac{2}{\pi}\left(\arccos\frac{Xp}{R} - \left(\sqrt{1-\left(\frac{Xp}{R}\right)^2}\right)\right)\frac{Xp}{R}\right)^n
\end{aligned}$$

So,

$$f_{Xpmax}(x) = F'_{Xpmax}(x) = \frac{2n\left(\dfrac{2\arccos\left(\frac{x}{R}\right) - \dfrac{2x\sqrt{1-\frac{x^2}{R^2}}}{R}}{\pi}\right)^{n-1}\left(\dfrac{1}{R\sqrt{1-\frac{x^2}{R^2}}} + \dfrac{\sqrt{1-\frac{x^2}{R^2}}}{R} - \dfrac{x^2}{R^3\sqrt{1-\frac{x^2}{R^2}}}\right)}{\pi}$$

The expected value is

$$E[Xp_{max}] = \int_0^R x f_{Xpmax}(x)dx$$

and

$$E[Xp_{max}^2] = \int_0^R x^2 f_{Xpmax}(x)dx$$

If now we take into account that $E[h] = \frac{dis}{E[Xp]}$, we have completed our proof.

At the end of this section, we numerically estimate the integral. Finally, the energy consumed by the location-aware protocol can be computed as follows:

$$E[En] = O(E[h]E[d^2]) \geq O(E[h]E[Xp_{max}^2])$$

12.1.3.3 Analysis of the Nearest with Forward Progress Routing

Let us now focus at the analysis of the NFP routing. Let again Xp be the random variable that is equal to the length of the projection of the distance traversed by a hop to the line that connects the source to the sink. In the case of computing the expected hop count, we will conclude to a similar Lemma to Lemma 12.3 in the previous section.

Lemma 12.4

If the source–destination distance is $dis = k \cdot R$, the mean number of hops h required to reach the destination in the case of the NFP routing is equal to the quantity

$$\frac{k \cdot R}{\int_0^R xn\left(\frac{2\arccos\left(\frac{x}{R}\right)}{\pi} - \frac{2x\sqrt{1-\frac{x^2}{R^2}}}{\pi R}\right)^{n-1}\left(\frac{2}{\pi R\sqrt{1-\frac{x^2}{R^2}}} + \frac{2\sqrt{1-\frac{x^2}{R^2}}}{\pi R} - \frac{2x^2}{\pi R^3\sqrt{1-\frac{x^2}{R^2}}}\right)dx}$$

where n is the number of sensors in each transmission range of radius R.

Proof

We have shown in the previous section that the distribution function of Xp is the following:

$$F(x) = 1 - \frac{2}{\pi}\left(\arccos\frac{Xp}{R} - \left(\sqrt{1-\left(\frac{Xp}{R}\right)^2}\right)\frac{Xp}{R}\right) \tag{12.5}$$

We now compute the distribution function of the minimum projection from all the projections that are formed with every neighbor of a current node, in other words, the length of the hop in transmission of the NFP routing protocol:

$$F_{Xp_{min}}(x) = P(Xp_{min} \le x) = 1 - P(Xp_{min} \ge x)$$
$$= 1 - (P(Xp \ge x))^n = 1 - (1 - P(Xp \le x))^n$$
$$= 1 - \left(\frac{2}{\pi} \left(\arccos \frac{Xp}{R} - \left(\sqrt{1 - \frac{Xp^2}{R}} \right) \right) \frac{Xp}{R} \right)^n$$

It follows that the probability distribution function of Xp is

$$f_{Xpmin}(x) = F'_{Xpmin}(x) = n \left(\frac{2\arccos\left(\frac{x}{R}\right)}{\pi} - \frac{2x\sqrt{1 - \frac{x^2}{R^2}}}{\pi R} \right)^{n-1} \left(\frac{2}{\pi R \sqrt{1 - \frac{x^2}{R^2}}} + \frac{2\sqrt{1 - \frac{x^2}{R^2}}}{\pi R} - \frac{2x^2}{\pi R^3 \sqrt{1 - \frac{x^2}{R^2}}} \right)$$

The expected value is

$$E[Xp_{min}] = \int_0^R x f_{Xpmax}(x) dx \qquad (12.6)$$

The expected number of hops is $E[h] = \frac{dis}{E[Xp_{min}]}$ and so the proof is completed.

Let us now focus at the expected energy consumption:

$$E[En] = O(E[h]E[d^2])$$

To compute the energy, we first define the random variable $Z = d^2 = Xp_{min}^2 + H^2$ (where H and Xp_{min} are both random variables, and H is equal to the height of the node as seen in Figure 12.5). We now compute $E[Z]$.

From probability theory, we know that if we have a random variable Z, that is, a function of two other random variables Xp_{min} and H, so that $Z = g(Xp_{min}, H)$ the expectation of Z is given by the formula

$$E[Z] = \int_{-\infty}^{\infty} \int_{-\infty}^{\infty} g(xp_{min}, h) f_{Xp_{min}}, H(xp_{min}, h) dxp_{min} dh \qquad (12.7)$$

So, we have

$$E[Z] = \int_{-\infty}^{\infty} \int_{-\infty}^{\infty} g(xp_{min}, h) f_H(h|xp_{min}) f_{Xp_{min}}(xp_{min}) dxp_{min} dh \qquad (12.8)$$

We already have computed $f_{Xpmin}(x)$. From Figure 12.5, we can see that h reaches its maximum value when $d = 1$.

When $d = 1$, we have $1 = Xp_{min}^2 + h^2$, it follows that $h = \sqrt{1 - Xp_{min}^2}$. So h is uniformly distributed in $\left[0, \sqrt{1 - Xp_{min}^2}\right]$ and

$$f_H(h|xp_{min}) = \frac{1}{\sqrt{1 - Xp_{min}^2}}$$

12.1.3.4 Numerical Evaluation

Now we have all the information needed to compute the double integral and get the expected value. In fact, the double integral, as well as the integral in the Lemma and the integrals in the previous section, which are used to compute the expected values, are precisely estimated with an numerical integration technique, in particular the adaptive Simpson's method via MATLAB®, having an error less than 10^{-6}. Without loss of generality, we set $R = 1$ and we get the upper bound on the average number of hops in the case of the direction-aware protocol and the estimation on the other values discussed earlier, when the distance is equal to 100. We do this for different values of n, as seen in Figures 12.6 and 12.7.

In particular, in Figures 12.6 and 12.7, we can see two upper bounds (on the average energy consumption and the average hop count) of the direction-aware protocol, a lower bound on the average energy consumption and an accurate estimation of the average hop count of the location-aware protocol.

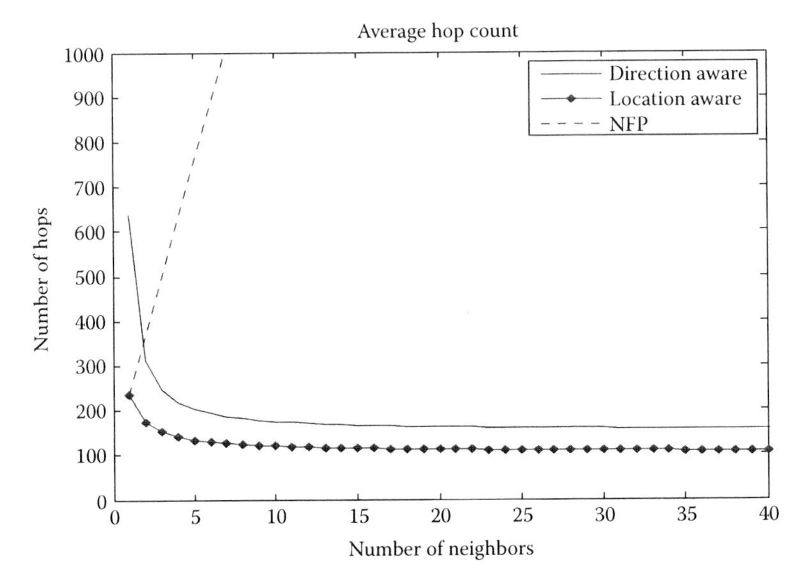

FIGURE 12.6 Average hop count analysis.

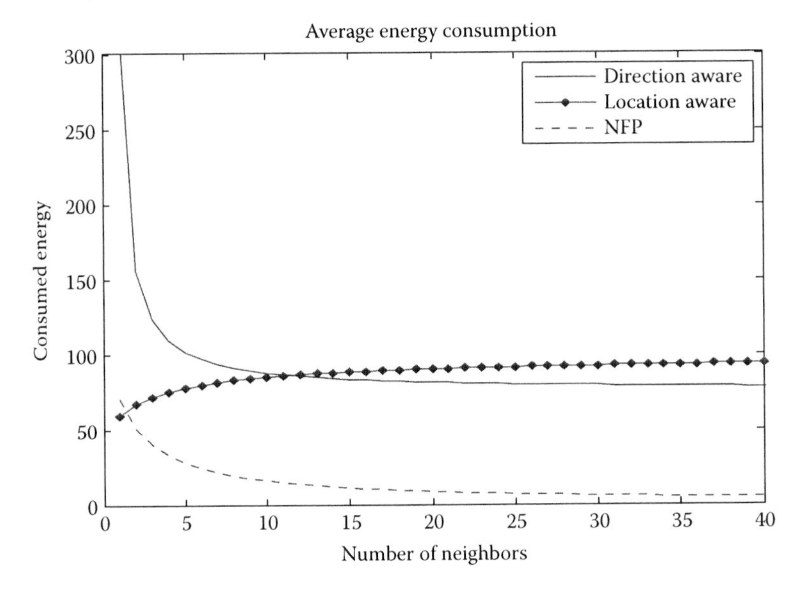

FIGURE 12.7 Average energy consumption analysis.

We can see also an estimation of the average hop count and average energy consumption of the NFP. We see that the upper bound of the average number of hops in the direction aware protocol is in the [160, 180] regime, that is, 1.6–1.8 times the network dimension (k = 100). We note that in routing with fixed transmission range a number of hops 100 to get to the sink represents a minimum needed. This optimal value can only be achieved if we can always get a next hop sensor on the direct source–destination line at pairwise distances R = 1. However, by examining the energy consumption, we observe that the energy consumption upper bound in the direction-aware protocol is low, especially as the network density increases. In contrast to that, the location-aware protocol energy consumption approaches a high value, as the density gets higher. Overall, the analysis validates the properties of our protocol, namely, that it saves a lot of energy while increasing latency compared to the location-aware protocol.

The results also show us that the NFP is an extreme case. As the number of neighbors increases, the hop count increases extremely as the energy consumption falls at an extremely low point. One can explain this by considering the NFP is always choosing the neighbor with the smallest horizontal projection, so when there are many neighbors, there will be extremely many small hops made, which would also result to an extremely low energy consumption.

12.1.4 Experimental Evaluation

12.1.4.1 Simulation Environment

Our simulation environment for making the experiments is the environment of MATLAB 7.9.0. We deploy uniformly at random nodes in the network area. We choose as a communication model the unit disk graph. This means that each node is able to send a message to another iff the distance between them is at most a given threshold (in particular, the wireless transmission range R is taken 5). Using the unit disk graph means that the expected number of neighbors per node is close to $d\pi$, where d is the global density of sensors in the network.

In detail, the network area is rectangular, with length and width equal to 40 units. We apply several times the deployment of nodes in the network and repeat the experiments, for statistical smoothness. We consider different numbers of nodes (with a range from 5,000 up to 30,000 nodes), forming each time a network of different density d (diversely ranging from 10 to 60). For statistical reasons, we take 50 random uniform deployments and for each deployment 1000 data propagations are simulated and the average value is taken. The statistical analysis of the findings (the median, lower, and upper quartiles, outliers of the samples) demonstrates very high concentration around the mean, so in Figures 12.8 and 12.9 we only depict average values.

For each deployment, the source is chosen from the nodes of the network randomly uniformly and the sink is placed at the center (100, 100) of the 200 × 200 deployment area. We measure the average number of hops needed to reach the sink, the average energy consumed in the network, and the average success rate of each algorithm. The success rate is taken as the percentage of generated events that are reported to the sink:

We assume an energy model in which the energy consumed by a message transmission between two nodes is considered the square of the distance between the nodes. For our three metrics, the average is taken over all sensor deployments and algorithm's repetitions. Also, the initial energy available at the sensor devices was set to 1000 energy units at the start of the simulation.

We investigate the performance of the protocols in the presence of permanent node failures during protocol evolution. We study the characteristic cases where 10%, 25%, and 50% of the network nodes fail during the simulation time. For each unit of time, failures occur at randomly chosen nodes, instantly, and no further computation and/or communication can be performed by these failed nodes.

12.1.4.2 Findings

We compare our hybrid protocols with the NFP, direction-aware and baseline location-aware protocol.

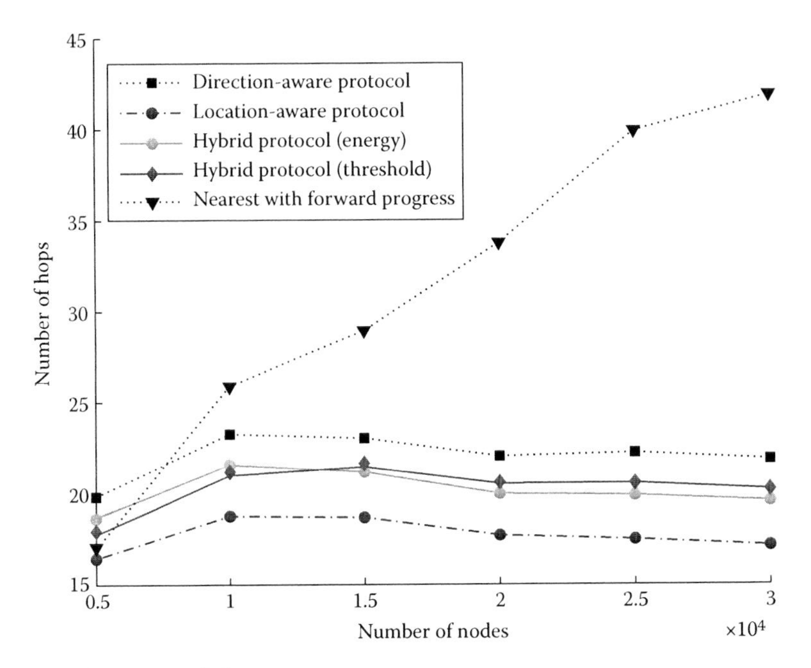

FIGURE 12.8 Data latency—no failures.

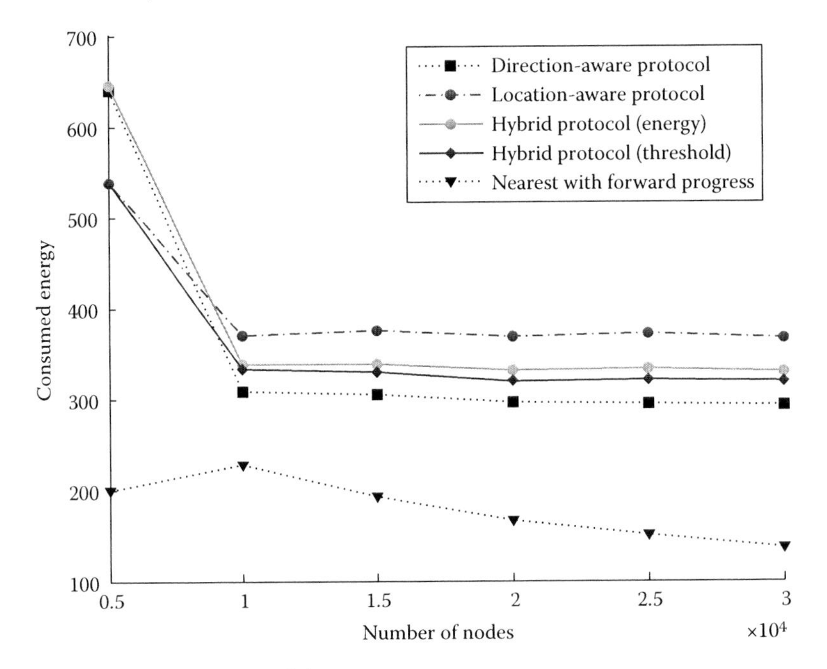

FIGURE 12.9 Energy consumption—no failures.

12.1.4.3 No Failures

We examine the mean number of hops the algorithms need to reach the sink (only the successful trials count here) and Figure 12.8 depicts this metric. We note that, in terms of data packet latency NFP has the highest latency, DAR has the second higher latency, and LAR is the fastest way to deliver the data to the sink. Our hybrid protocols are in between the latency values of the DAR and the LAR. That is an expected result, since in the hybrid protocols, a fraction of the hops are made through the LAR and the remaining hops through the DAR. More specific, the threshold hybrid protocol is a little bit slower, in terms of latency, than the energy hybrid protocol.

We now compare the protocols' performance regarding the energy consumption, Figure 12.9. In this figure we can see that NFP protocol is the most energy conservative in contrast to the LAR which is the most energy consuming. This result is also expected, since the NFP, as mentioned before, chooses as next-hop node the node with the minimum projection. Short hops means less energy, since the energy consumed in a hop is considered to be proportional to the second power of the wireless transmission distance. DAR is the second energy conservative protocol as it makes short hops with a path close to the optimal line that connects the current node with the sink. Regarding the hybrid protocols, the values of the findings are again in-between that of the DAR and the LAR for the same reasons mentioned in the previous paragraph. However, the threshold hybrid protocol is cheaper than the Energy Hybrid protocol and consumes almost as much energy as DAR.

In general, the NFP protocol increases dramatically the data latency while it is very cheap in terms of energy consumption due to the short-range transmissions of data messages. The DAR protocol increases the latency, but it has better results regarding the energy consumption. More specifically, we can see from the figures that the gain we have (in percentage) in energy is the same as the loss we have in latency and this is the basic trade-off. So we can conclude that DAR could be used as a routing strategy in applications where we can afford a small increase in latency, but the energy consumption is the most important factor. NFP could be used in applications where latency is not critical factor and our main goal is to conserve energy.

In Figure 12.10, we observe that the success delivery ratio is almost identical for DAR, LAR, and hybrid protocols. For sparse networks, NFP has the worse behavior. In contrast with the DAR protocol, where the next-hop node is the node with the minimal directional divergence from the optimal source–destination line, in NFP there is no guarantee that the chosen next-hop node will be close to the optimal source–destination line, so the followed path can deviate a lot from reaching the sink and the data can be trapped in a routing hole. For high densities (over 20,000 nodes), the success ratio is, in fact, 1. This is because the network is dense enough, so there is a path from every possible source toward the sink.

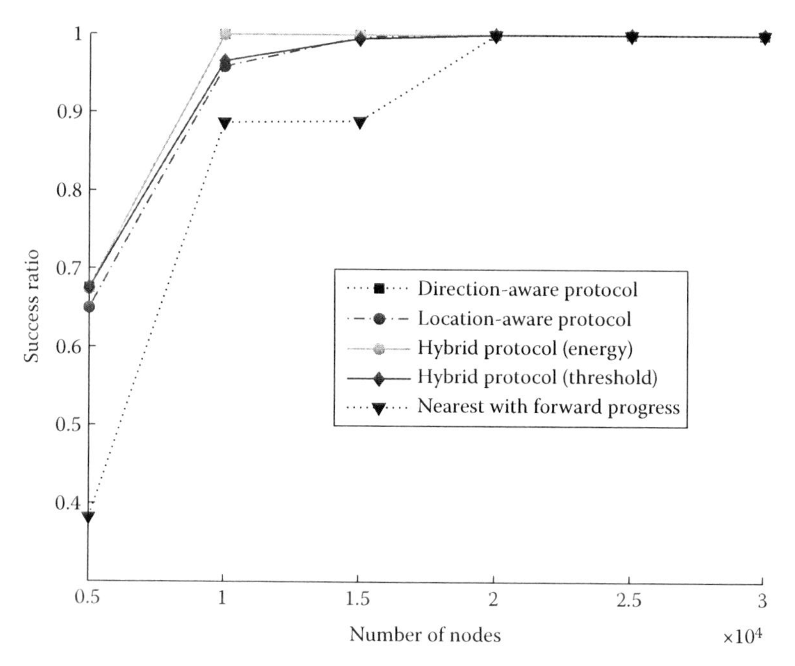

FIGURE 12.10 Success ratio—no failures.

12.1.4.4 Impact of Failures

We examine the cases where the 10%, 25%, and 50% of the network nodes fail during the simulation time. In Figures 12.11 through 12.19, we depict the average number of hops needed to reach the sink, the average energy consumed in the network and the average success rate of each algorithm. We observe that the failure rate mainly affects the success rate of the protocols, as expected. While the absolute values of the energy consumption and the delivery delay are increased, we notice a consistent behavior of all five protocols when the failure rate increases.

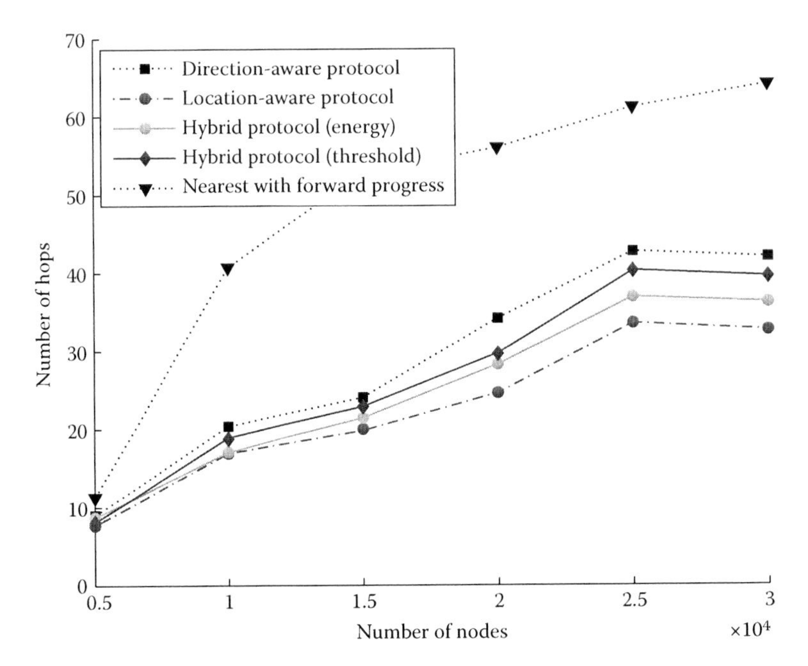

FIGURE 12.11 Data latency—10% failed particles.

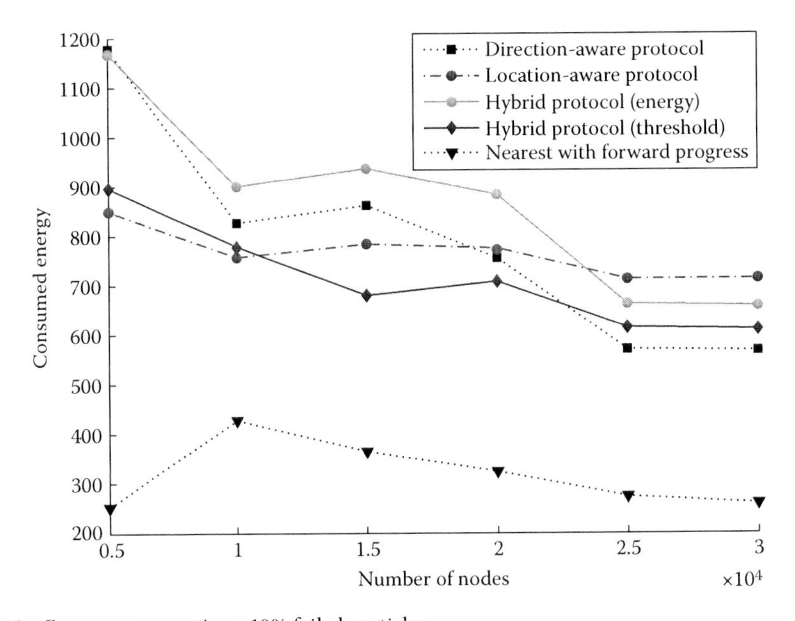

FIGURE 12.12 Energy consumption—10% failed particles.

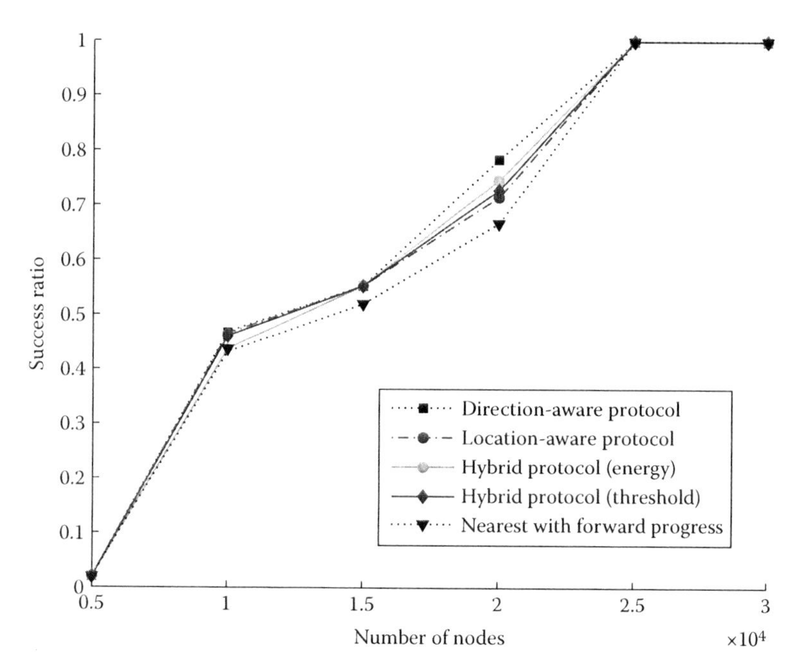

FIGURE 12.13 Success ratio—10% failed particles.

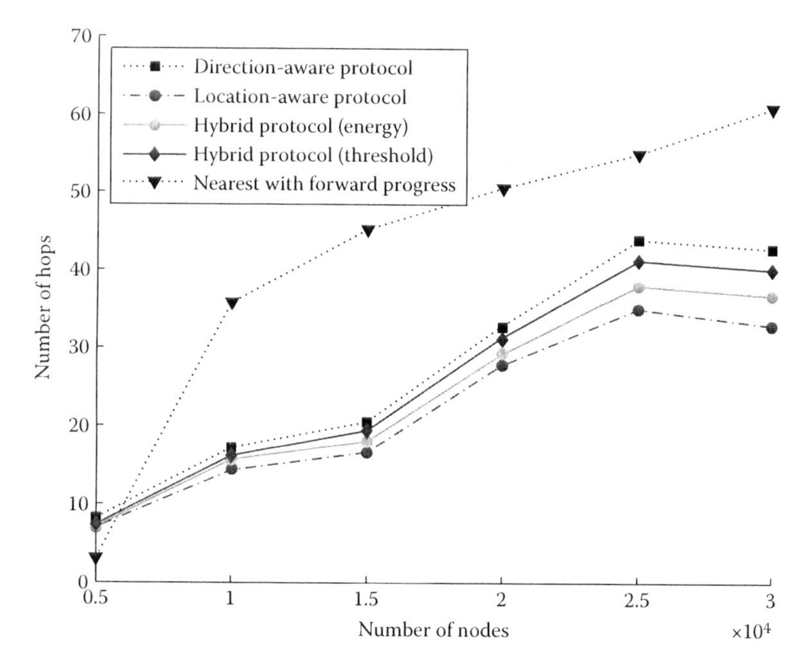

FIGURE 12.14 Data latency—25% failed particles.

For all failure rates, NFP is the most energy-consuming protocol, but on the other hand it has the highest delivery delays. The most energy-consuming protocol is our energy-based hybrid. This can be explained by the fact that the protocol favors the node with the maximum residual energy as the next-hop node, so this protocol does not use the same nodes but diverse nodes. In addition, we observe that our threshold-based hybrid protocol performs well and for some densities it is more energy conservative from DAR (see Figures 12.12, 12.15, and 12.18).

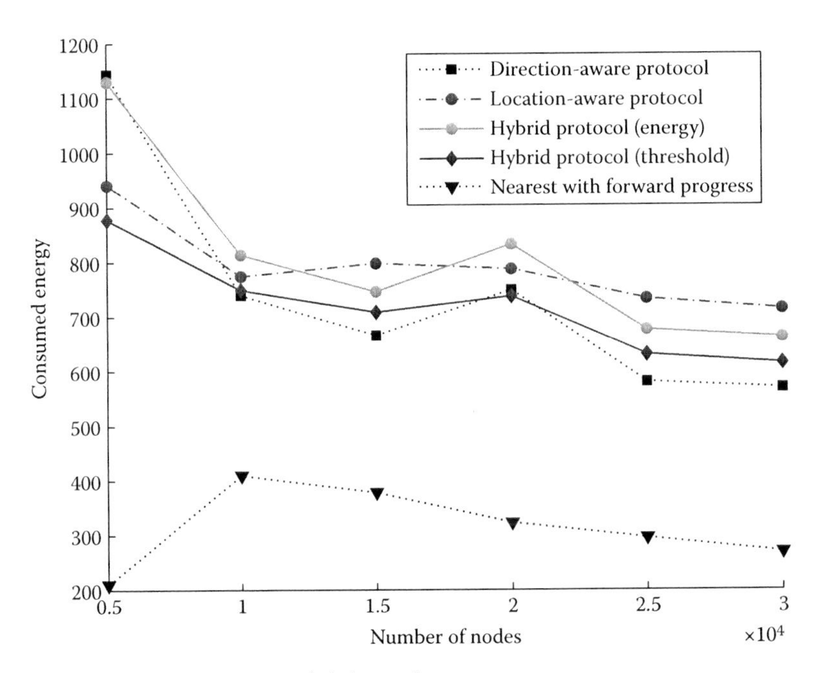

FIGURE 12.15 Energy consumption—25% failed particles.

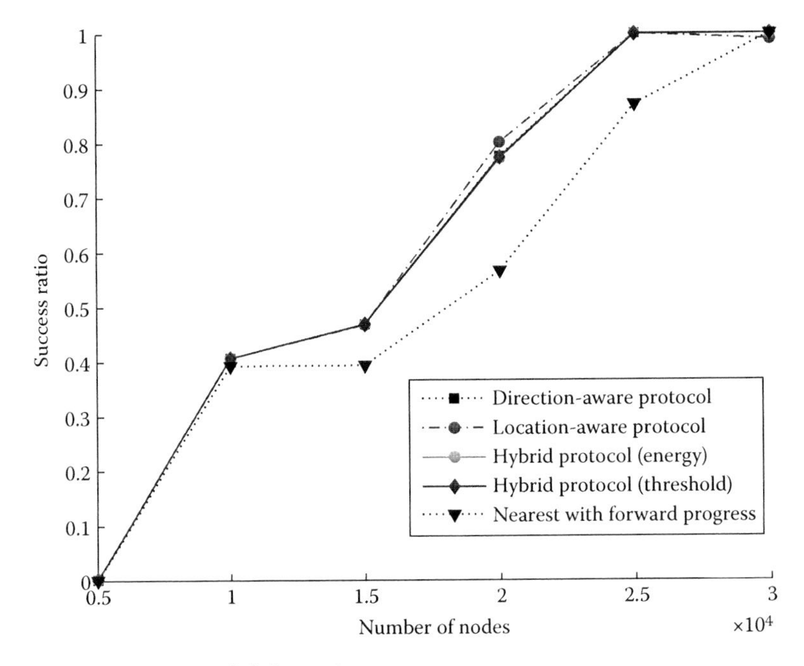

FIGURE 12.16 Success ratio—25% failed particles.

As the number of nodes that fail during the simulation period increases, we observe that the success rate gets worse and we need higher number of nodes (network density) to converge to the optimum success rate (100%). In addition, more node failures results in increased latency and energy consumption. As the network becomes more sparse, this leads to longer paths, which incur higher energy dissipation and thus higher latency.

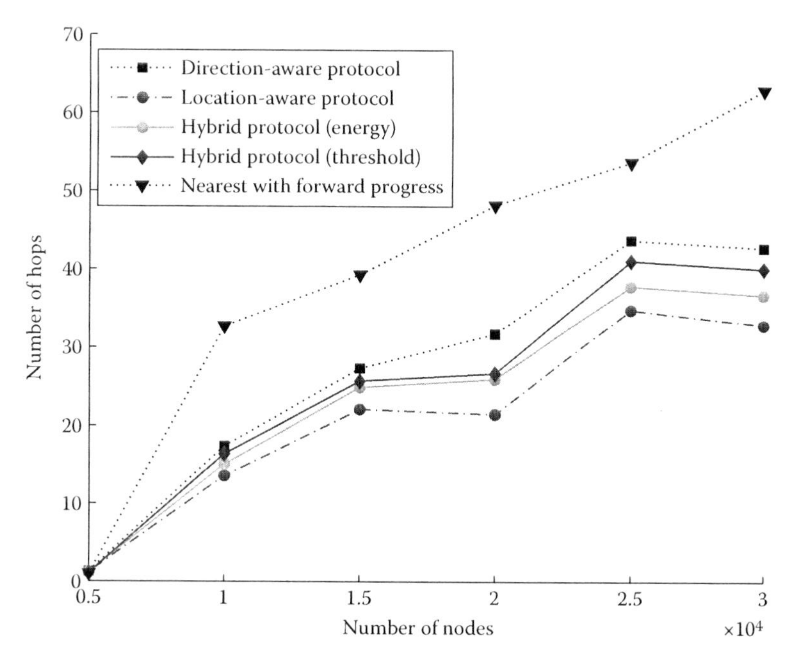

FIGURE 12.17 Data latency—50% failed particles.

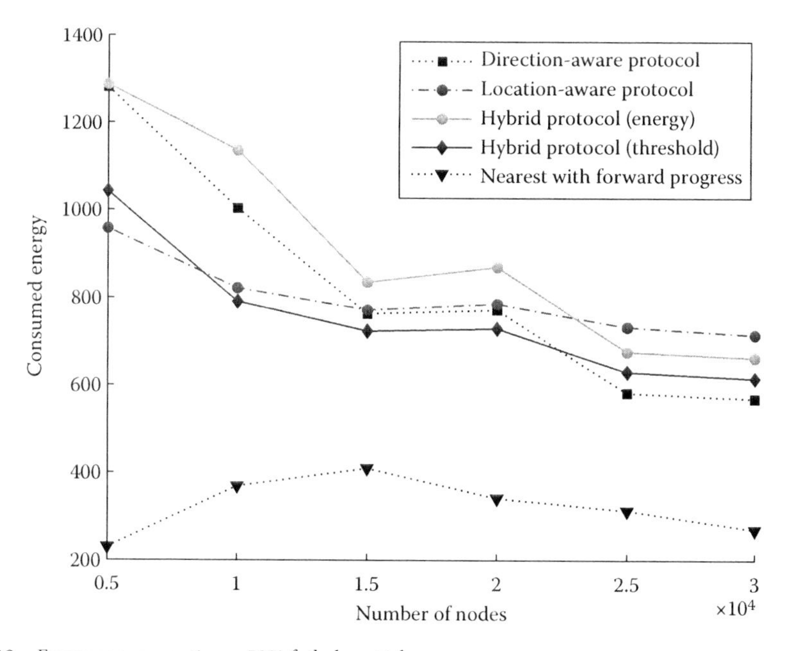

FIGURE 12.18 Energy consumption—50% failed particles.

12.1.4.5 Energy Balance Aspects

In Figures 12.20 and 12.21, we present the spatial evolution of energy dissipation in a network of 30,000 nodes after 4,000 data propagations. The initial energy available to the sensors was set to 1000 energy units at the start of the simulation. Nodes with high-energy dissipation are depicted with dark colors. In contrast, nodes with high residual energy are depicted with bright colors.

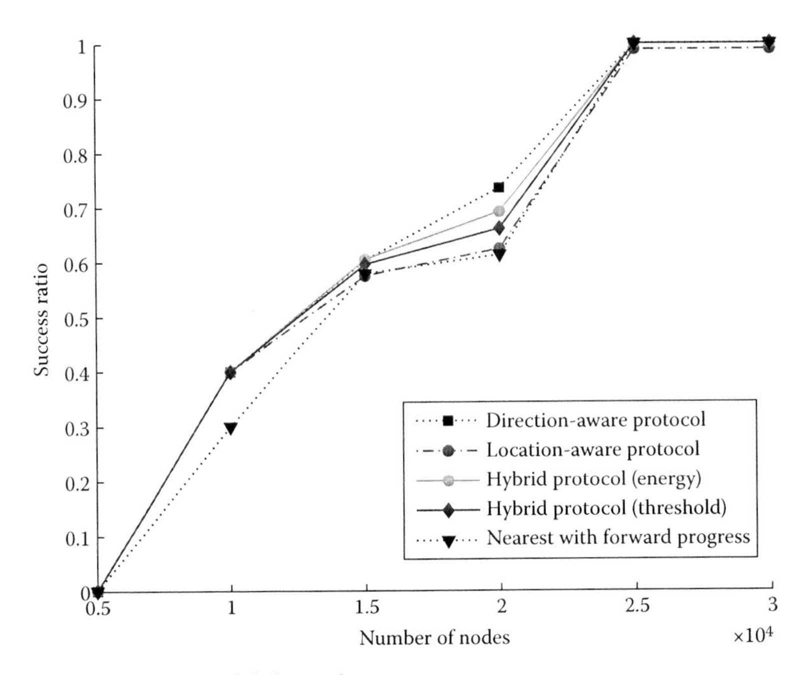

FIGURE 12.19 Success ratio—50% failed particles.

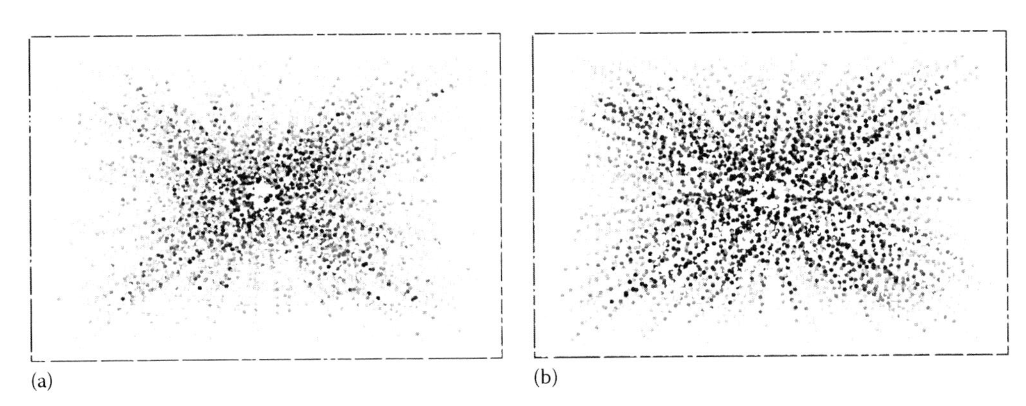

FIGURE 12.20 Energy dissipation in nodes using energy-based hybrid protocol (a) and location-aware protocol (b).

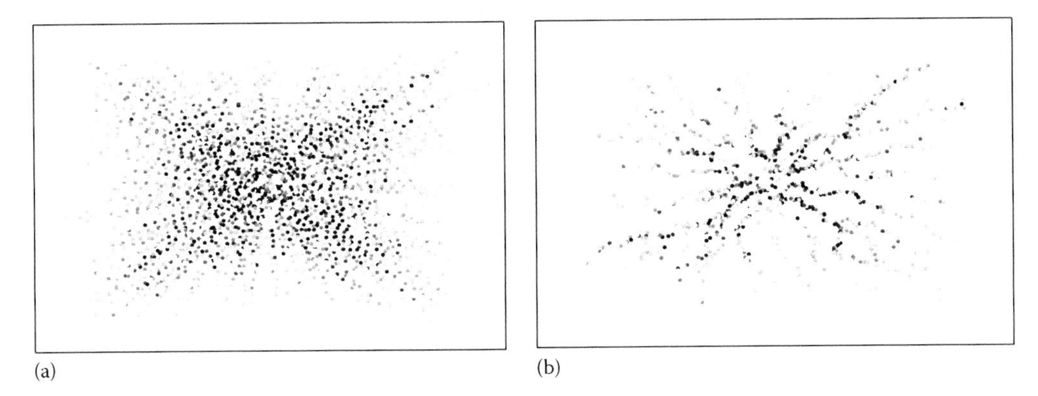

FIGURE 12.21 Energy dissipation in nodes using direction-aware protocol (a) and NFP (b).

TABLE 12.1 Distribution of Node's Energy Dissipation Using the Energy-Based Hybrid Protocol

Energy	0–40	40–80	80–120	120–160	160–200	200–240	240–280	280–1000
Nodes (#)	21,442	1,716	1,755	892	536	316	239	648

TABLE 12.2 Distribution of Node's Energy Dissipation Using the LAR Protocol

Energy	0–40	40–80	80–120	120–160	160–200	200–240	240–280	280–1000
Nodes (#)	23,733	1,236	886	543	390	252	249	1,182

TABLE 12.3 Distribution of Node's Energy Dissipation Using the DAR Protocol

Energy	0–40	40–80	80–120	120–160	160–200	200–240	240–280	280–1000
Nodes (#)	24,728	897	889	545	369	277	186	777

TABLE 12.4 Distribution of Node's Energy Dissipation Using the NFP Protocol

Energy	0–40	40–80	80–120	120–160	160–200	200–240	240–280	280–1000
Nodes (#)	27,477	518	439	280	174	115	83	215

We observe in Figure 12.20a that our energy-based hybrid protocol achieves better energy balancing than the LAR protocol (Figure 12.20b) and DAR protocol (Figure 12.21a); however, in Figure 12.21b, it is obvious that the NFP protocol is the most energy conservative. We obtained two interesting observations from the Tables 12.1 through 12.4: (1) the number of the nodes that have used little or no energy is smaller in the hybrid (21,442 nodes) than the LAR (23,733 nodes), the DAR (24,728 nodes), and the NFP protocol (27,477 nodes); and (2) the number of the nodes that have used most of their initial energy is smaller in the hybrid (234 nodes) than the LAR (554 nodes) and the DAR protocol (388 nodes) but is larger than the slow NFP protocol (94 nodes).

By using the energy-based hybrid protocol instead of the LAR or the DAR protocol, we can achieve more balanced energy dissipation per node, as in the operation of the protocol participate more nodes than in the LAR or the DAR protocol in order to prolong the network lifetime by avoiding early network disconnection. Moreover, LAR and DAR tend to overuse the nodes in the critical region around the sink. Finally, NFP is the most energy-conservative protocol, but it is the slowest protocol among all.

12.2 Impact of Mobility

12.2.1 Introduction

Until the recent past, wireless sensor networks were usually viewed as static collections of very large numbers of smart sensor nodes. In the prevailing scenario, sensors provide fine grained monitoring of a region of interest by being largely deployed in the region and remaining static. The collected data are disseminated to a static data sink in the network area, using redundant multi-hop data propagation [12]. Most research in the relevant state of the art that involves mobility restricts mobility to the data sink(s), that is, the sink(s) may move but the sensors themselves are assumed static.

However, recent technology advances will make wireless sensors ubiquitously present in the ambient environment, enabling new types of important applications, where motion becomes a fundamental, inherent characteristic. In such applications, sensor devices will be attached to moving entities (vehicles, animals, rivers, and people moving around large regions), while robotic elements may also move in the region. Sensory data exchange between sensors and control nodes will drive important applications such as traffic and ecosystem monitoring, smart homes, and pollution control.

Another application scenario that node mobility is crucial for is vehicular ad hoc networks (VANETs). A VANET [34] is a distributed, self-organizing communication network that is formed by moving vehicles and is characterized by high node mobility and limited degrees of freedom in mobility patterns.

These networks exhibit dynamic topology changes and potentially frequent disconnections. In particular, the mobile nodes (vehicles) follow a certain mobility pattern that is a function of the underlying roads, the traffic lights, the speed limit, traffic condition, and drivers' driving behaviors. Examples of VANETs' scenarios are safety applications and applications for intelligent transportation systems (ITS).

In such scenarios, mobility is a crucial characteristic of the system. Usually, such applications aim at gathering and processing of large amounts of sensed data; however, immediate delivery of recorded data is difficult and has high energy cost. Instead, measurements can be cached at the sensors (and also ferried to their neighbors) so as to be delivered to the sink whenever the nodes move within its range. Also, the data acquisition process must not interfere with the network, for example, people and other moving entities should not be asked explicitly to move close to the sink to deliver the collected data. Thus, the mobility is uncontrollable and should be left as such. Furthermore, the mobile sensors may follow several heterogeneous mobility patterns that also change a lot with time. The aforementioned remarks demonstrate the opportunities created by mobility, which basically stem from its ability to serve as a low-cost replacement for connectivity (due to sensor movement to sparse, disconnected areas) and data propagation redundancy (data ferrying by sensors). Still, the mobility management must judiciously cope with complications arising, such as increased data delivery times (high latency), the lack of permanent reference points and the difficulty of maintaining system integrity.

12.2.1.1 Related Work and Comparison

Under highly mobile conditions, the protocols and research recommendations for static wireless sensors networks cannot be directly applied (if at all). Data propagation protocols based on multi-hop data forwarding paths or clustering algorithms do not apply, since under mobility the network topology changes frequently. Also, coverage and localization problems become more difficult under mobility. Even very efficient and robust algorithms need to be redesigned, such as in Ref. [5] where the authors provide a special leader election algorithm suitable for mobile networks in particular. Furthermore, new problems arise due to the high network dynamics: preserving system integrity [18] and security [7] becomes more complicated. Probabilistic propagation protocols may be suitable in such settings since they are local and require a minimum amount of energy dissipation knowledge only. Also, randomness better balances the load in the network [40].

In Ref. [44] the authors introduce epidemic routing where random pairwise exchanges of messages among mobile hosts ensure eventual message delivery and develop techniques to deliver messages in the case where there is never a connected path from source to destination or when a network partition exists at the time a message is originated.

Mobility in sensor networks has been studied mainly, assuming that the (one or more) sinks are mobile and collect data traversing the network region and getting within the range of individual sensors. Different data collection and movement strategies have been proposed, as well as routing methods based on the sink stopping at certain positions to collect data [10]. Also, adaptive scheduling algorithms for the motion of mobile sinks to visit nodes according to the data traffic they produce have been suggested in Ref. [21]. Optimal trajectories of a mobile sink that minimize energy consumption have been shown in Ref. [37]. Repositioning of multiple mobile sinks has been proposed to maximize the network lifetime, as well as randomized distributed coordination mechanisms between mobile sinks [30].

Considering *sensor mobility*, Ref. [26] presents a case study of mobile sensor networks designed for wildlife position tracking. The authors assume varying mobility and propagate data to the node most likely to meet the sink; this likelihood is, however, based on previous history and not the current dynamics. In the shared wireless infostation model (SWIM) [42], a network of intermittently connected nodes is used to gather oceanographic data from sensors attached to whales through an infrastructure of buoys to which the whales can upload their data. The authors in Ref. [46] propose a data dissemination mechanism that also chooses the fittest nodes. Furthermore, they assume limited queues on each node and propose a mechanism to drop messages from the queues based on the likelihood of delivery of each message. We also try to select the best candidates for delivering messages, but we assume that node behavior changes, thus instead of using history we choose the best nodes based on their, dynamically calculated, mobility level. Also, we propose and examine more elaborate variations of mobility patterns, while we adaptively select the amount of redundancy (i.e., the number of message ferrying nodes), in terms of the mobility levels in the network (to benefit from high mobility by reducing redundancy).

In Ref. [36], the authors propose a geometric data dissemination mechanism for delivering data to a mobile sink. Assuming bounded motion of the sink in a specific but arbitrary area, they characterize the motion using geometric criteria. In our work, we propose more elaborate methods for characterizing mobility that capture more subtle variations both in speed and trajectory. Also, in Ref. [45], the authors investigate the trade-off between mobility of nodes and coverage of the network area. Our approach, in fact, exploits such trade-offs in the sense that we handle high mobility as a "replacement" for connectivity and coverage.

In Ref. [29], a related but different mobility-level notion as well as adaptive dissemination schemes have been proposed. Our mobility level here does not take into account (in contrast to Ref. [29]) dislocation from the origin but is instead direction aware, in the sense that it captures how close to the sink a mobile sensor tends to become. Also, although our protocols are also adaptive on the mobility level, we here introduce long-distance transmissions to accelerate data propagation in the case of low mobility. Our approach has been inspired by the research in Ref. [14] that, although referring to a different network type, emphasizes the importance and impact of high network dynamics, also by Ref. [15] that formally shows that high mobility can strongly accelerate communication while saving energy.

Other studies assume the mobile agent approach where one or more mobile agents may visit a number of sensors and progressively aggregate retrieved sensory data prior to returning back to the processing element (e.g., sink) to deliver the data. In Refs [31,32], the authors presented an algorithm for finding near-optimal routes for mobile agents that incrementally aggregate data as they visit sensor nodes in a wireless sensor network. This algorithm gradually builds a number of trees whose traversal will eventually determine the itineraries followed by the mobile agents. In Ref. [20], the authors introduced the design decisions and implementation aspects of a complete mobile agent platform (MAP) research prototype. Reference [38] presents a protocol that proposes the use of urban buses to carry mobile agents that retrieve information from isolated parts of wireless sensor networks. This protocol aims at minimizing the overall network overhead and energy expenditure associated with the multi-hop data retrieval process while also ensuring balanced energy consumption among network nodes and prolonged network lifetime.

Reference [13] presents a model-based routing that takes advantage of the predictable node moments along a highway. The authors have verified the hypothesis that the motion of vehicles on a highway can contribute to successful message delivery, provided that messages can be relayed and stored temporarily at moving nodes while waiting for opportunities to be forwarded further. Our method does not assume predictions, while the network type (sensors) is different.

A first discussion on how to incorporate controllable mobile relays into the network infrastructure has been presented in Ref. [27]. The authors describe an implementation of a sensor network with an autonomous mobile relay (a robot) that visits the static sensor network in order to collect their data and deliver the collected data to the sink. The robot traverses networks with different densities following a straight trail, collects the data, and ferries them to the sink. In contrast, our method does not control the nodes' motion but adaptively exploits this motion.

In Ref. [47], the authors proposed the message ferrying approach, which is a mobility-assisted approach that utilizes a set of special mobile nodes called message ferries to provide communication service for nodes in the deployment area. The main idea behind this approach is to introduce nonrandomness in the movement of nodes and exploit such nonrandomness to help deliver data. In our method, we do not assign ferrying roles to a specific of nodes, but all nodes can ferry messages.

Some works based on creation of dynamic clustering of nodes to speedup and support data dissemination in the vehicular sensor networks' scenarios can be found in Refs. [1–3,19].

Our research is actually related to similar work in the context of delay-tolerant networks (see Refs [23,46]). However, we note that in our case the node mobility is uncontrollable, more random and unpredictable than in the case of a classic delay-tolerant network.

12.2.1.2 Our Contribution

Until recently, most papers on sensor networks mobility have studied the case where only the sinks are mobile. However, the research investigating networks where the sensors themselves move receives growing attention. Motivated by important relevant applications where sensory mobility itself is a dominating aspect, we focus on heterogeneous, highly changing mobility profiles. In particular, (1) we propose a new (locally computable) network parameter, the *direction-aware mobility level*, which quite accurately captures how fast and how close a mobile sensor is expected to get to the data destination (sink), taking into account not only its speed but also the direction of its motion. (2) We exploit sensory mobility as a low-energy replacement for connectivity and data propagation redundancy: we propose adaptive protocols that propagate less data in the presence of high mobility with "good" direction toward the sink and favor relay sensors with "nice" mobility levels. To cope with low-mobility neighborhoods, we also introduce (either deterministically or probabilistically) the alternative of long-distance transmissions (jumps) that, although expensive, propagate data very fast toward the sink. (3) We also propose a progress-sensitive message flooding inhibition scheme that further reduces communication cost by purging obsolete messages. (4) We implement our protocols and relevant state of the art methods under heterogeneous, dynamic mobility scenarios and perform extensive simulations examining realistic cases where sensors have limited queues for buffering messages. (5) The simulation findings demonstrate the efficiency of our mobility-sensitive adaptation schemes, since they manage to reduce latency a lot (even by four times) while also reducing energy dissipation, compared to nonadaptive protocols and even adaptive ones that, however, are not direction aware. One major goal in the design of routing protocols for the constrained sensor networks is energy efficiency in order to reduce power consumption to levels suitable for devices powered by small batteries or energy harvesting supplies.

A preliminary version of this significantly extended work has appeared in Ref. [6].

12.2.2 Model

We assume that the network area $\mathcal{A}$ is a flat square region of size $D \times D$; this assumption can be easily relaxed to include general network areas of arbitrary shapes. The initial positions of sensor nodes within the network area are random and in the general case follow a uniform distribution. *Let n be the number of sensors* spread in the network area and *let d be the density of sensors in that area* (usually measured in numbers of *sensors/m*^{2}). There is a special node within the network area, which we call the sink $\mathcal{S}$, that represents a control center where data should be collected. $\mathcal{S}$ is immobile and passively awaits nodes to pass by it and transmit their data. In order to be detected by the nodes, the sink transmits beacon messages at a rate of $.\lambda_{Beacon}$ messages per second.

Each sensor device has a *broadcast* (digital radio) *beacon mode* of fixed wireless transmission range R and is powered by a battery. Also a sensor is equipped with *a general purpose storage memory (e.g., FLASH) of size* C. This storage is used to cache messages that need delivery or forwarding.

Let ε_i *be the available energy supplies of sensor i at a given time instance.* At any given time, each sensor can be in one of three different modes, regarding the energy consumption: (1) transmission of a message, (2) reception of a message, and (3) sensing of events. For transmitting and receiving a message, we assume that the radio module dissipates an amount of energy proportional to the message's size. To transmit a k-bit message, the radio expends $E_T(k) = (\epsilon_{trans} \cdot k)$ and to receive a k-bit message, the radio expends $E_R(k) = (\epsilon_{recv} \cdot k)$, where ϵ_{trans} and ϵ_{recv} are constants that depend on the radio module hardware, while ϵ_{trans} is also dependent on the square of the transmission range R of the sensors. When the radio is idle, the energy consumed is constant and equals E_{elec}. Overall, there are three different types of energy dissipation: (1) E_T, the energy dissipation for transmission, (2) E_R, the energy dissipation for receiving, and (3) E_{idle}, the energy dissipation for idle state. We note that in our simulations we *explicitly measure the above energy costs.*

We differentiate from most standard models by assuming *mobility* of the sensors. Sensor nodes can calculate their position in some common coordinate system (e.g., by using navigational equipment or running a virtual coordinates algorithms) and are aware of the dimensions of the network area. Sensors are attached to mobile objects; we model their movement through a high-level mobility function, which we symbolize by $\mathcal{M}$. Note that nodes generally follow different mobility functions and in fact a single node may follow different mobility functions from one time to another. We consider several types of random motion with respect to speed (low, medium, and high) and "locality" (local motion within a limited area or global motion covering a large part of the network). We discuss several aspects of mobility modeling in Section 12.3.5. The movement of each sensor node i at time t is characterized by a mobility level $M\ell_i(t)$, which is dependent on the current time.

The mobility function returns a position p_t that the node should move to and the speed of the node to reach p_t. Note that the mobility function can be invoked at anytime even before reaching the designated point. The actual mechanism that moves the mobile entity from position p_{t-1} to position p_t is beyond the scope of this chapter. However, in order to simplify our model we assume that all changes in speed and direction can be done instantly. More information about the calculation of $M\ell_i(t)$ and the movement of nodes is given in the following sections.

Finally, we assume that a specific, high-level application is executed by the sensors that form the network. We model the application by the message generation rate per second λ_i in each sensor i.

12.2.3 Computing the Direction-Aware Mobility Levels

The mobility parameters studied, in previous approaches such as Ref. [29], like distance traveled, current speed, or area covered are not enough to fully capture the ability of a node to arrive close to the sink quite fast. Nodes with the same speed will travel the same distance independently of the trajectory followed, thus overall distance traveled is not characteristic. Furthermore, depending on the type of the mobility pattern, the area covered may vary significantly. Thus, the speed or the area covered by a node gives us only partial information about the time needed to approach the sink. For example, compare a node with high speed that covers a large area that moves in opposite direction from the sink against a node that moves with lower speed and covers a smaller area than the first node, however, with direction toward to the sink. Clearly, based on the earlier-provided information, the latter's progress toward the sink is higher despite the fact that it moves at slower speed and covers a smaller area. We define a new parameter for characterizing mobility that captures the differences between the speed and the direction in the trajectory followed.

The computation of the mobility level can be done easily with information locally obtainable by each node. Each sensor node i is responsible for computing its own mobility level. Consider a time interval t_l; every t_l seconds node i records its speed and the angle $d_i(t)$ between its direction of movement and the line connecting the current position and the sink (see Figure 12.22).

Consider an integer $K \geq 1$. Let $v_i(t)$ be the exponential weighted moving average speed of the last K samples, that is, a sample recorded i time intervals previously, so $t_l = i$ will be multiplied by a weight $w_i = e^{-i}/\sum_{n=1}^{K} e^{-n}$. Let $d_i(t)$ be the exponential weighted moving average direction of the last K samples. Measurements of speed and direction are smoothed by applying an exponential weighted moving

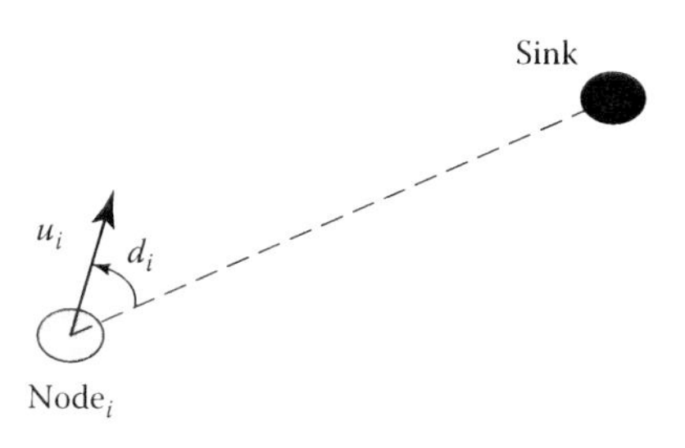

FIGURE 12.22 Example of estimating direction d_i.

average to avoid violent fluctuations of speed or/and direction that can be misleading about the real mobility level of a node and exploit recent history information (e.g., a node that moves toward sink for the latest K samples, is more possible to keep its direction). The mobility level of node i at time t is calculated as follows:

$$Ml_i(t) = v_i(t) * \left(1 - \frac{d_i(t)}{\pi}\right)$$

which is actually equivalent to the following:

$$Ml_i(t) = v_i(t) * \cos(d_i)$$

12.2.4 Adaptive Data Propagation

To accelerate data propagation while keeping energy dissipation and delivery success at satisfactory levels, each node decides based on some criteria, between three options: ferrying the message, transmitting the message to several direct neighbors, or transmitting to one distant neighbor closer to the sink by doing a jump. Ferrying spends no energy at all and is fast when the nodes have high mobility level, while transmission to many neighbors incurs redundancy and long transmissions accelerate data propagation toward the sink at a high energy cost, however; so, the trade-off between these options must be judiciously handled.

Each node can ferry or disseminate the information it records to a number β of its neighbors or to a distant neighbor at the given time. Thus, a node can carry copies of data recorded from itself or from other nodes and deliver them along with their own as soon as they encounter the sink. When a node receives a beacon from the sink, it immediately starts to unicast to the sink and the contents of its data cache. After a data message is successfully delivered, it is removed from the cache and no further attempts to disseminate this message are done from the node that delivered this message to the sink. Other nodes that have a copy of this message continue to disseminate the message until they reach the sink or they need to discard it because of cache overflow.

However, in the case of transmitting to direct neighbors, the selection of β and which particular β neighbors get a copy of a datum requires careful design. By setting $\beta = \infty$, the protocol degenerates to flooding, thus expending a lot of energy of the nodes because of the many redundant packets; however, delay should be minimal since the data will follow all possible paths to the sink. On the other hand, setting β to a small number will decrease communication cost a little but at the same time the data will take a long time to reach the sink. Moreover, in the case of transmitting to a distant neighbor, the selection of which distant neighbor gets a copy and the length TR_i of the wireless jump done by the node is very important in terms of energy consumption. By setting $TR_i = \infty$, every node in the network has direct access to the sink, thus expending a lot of energy of the nodes because of the long transmissions; however, delay should be minimal since there is only one-hop transmissions. On the other hand, setting TR_i to a small number will decrease communication cost a little but at the same time the data will take a long time to reach the sink.

Intuitively, for redundant transmission, slow nodes that make no or small progress toward the sink, nodes that go away from the sink with high speed or nodes that are far away from the sink (i.e., nodes i with small mobility level Ml_i) should choose a larger β to speed up the deliver of data. For jump transmission, nodes that are far away from the sink, have a very low mobility level, and are in a "bad" neighborhood should choose a larger TR_i to overcome the "trap" consisting of nodes with low mobility level. This is because such nodes will take a long time to move close enough to the sink to deliver the data on their own. By spreading out the information to other nodes the probability of meeting the sink increases and consequently the delay of delivery decreases. On the other hand, faster nodes that move toward the sink can benefit of their high mobility and decide to ferry data or to disseminate data while using a smaller value for β. Such nodes are more capable at getting close to the sink; thus, their messages have a high probability of being delivered rapidly. Similarly, when selecting to which nodes to disseminate a message we face again a dilemma. Intuitively, the faster nodes that move toward the sink are more appropriate for message ferrying because they can traverse the network faster and thus may approach the sink more frequently.

12.2.4.1 Dissemination Scheme

The general dissemination algorithm followed by the nodes in our approach is the following:

1. New messages describing events that need reporting or messages that were forwarded by other nodes are stored in a FIFO queue, called the forward queue. The queue has limited size depending on the cache memory of the node. Each node prioritizes its own data over collected from others. In case the queue becomes full, older messages belonging to another node are discarded to make room for new messages. If messages belonging to another node are not available, it will delete its own oldest data.
2. The node pops the next message from the front of the forward queue and decides to act suitably according to the following scenarios:
   a. *Data ferrying*: If the node has high mobility level, it decides to ferry/carry the data instead of transmitting to other nodes (to save energy while propagating data fast).
   b. *Data transmission*: If the node is not ideal to ferry/carry the data, it transmits data using one of the following choices:
      i. *Redundancy:* If at least one direct neighbor of the node has a high mobility level ("good" neighborhood), the node disseminates the information to a number β of its neighbors at the given time (toward a higher success ratio).
      ii. *Jump:* If all of the node's direct neighbors have low mobility level ("bad" neighborhood), the node transmits data not to one-hop neighbors, but to TR_i-hop closer to sink neighbor in order to avoid the "trap" ("bad" neighborhood) (trade-off between latency and energy dissipation).
3. Forwarded messages and messages that for some reason have not been sent successfully (due to collisions, absence of neighbors, etc.) are then stored in a delivery queue that has the same characteristics (size, FIFO, etc.) as the forward queue. The detection of an unsuccessful transmission and the retransmission of the message is assured by the under lying lightweight MAC protocol.
4. For the redundancy transmission scenario, if a beacon from the sink is received, then the node switches to a *connected* operational state and begins the delivery of the messages. For the jump transmission scenario, if the sink is in a node's new transmission range, then the node switches to a *connected* operational state and begins the delivery of the messages. First, it selects messages from the delivery queue to transmit directly to the sink. If the delivery queue is empty, it selects messages from the forward queue. Note that in this case messages are not sent to β neighbors. After successful delivery, messages are erased from the memory of the nodes, thus freeing resources.
5. If the delivery of a message to the sink fails, the node reverts to the *disconnected* state and operates according to steps 1, 2, 3.

Note that messages received from other nodes are discarded, in case they already exist in the forward queue.

12.2.4.2 Decision Criterion

First, node i decides whether to ferry data or to forward data based on a hard threshold choice criterion. The two different scenarios are explained in the following.

We define γ for node i and for its neighborhood, respectively, in terms of the current and maximum mobility levels as follows:

$$\gamma_i = \frac{Ml_i(t)}{Ml_{max}(t)}$$

$$\gamma_{neigh\,i} = \frac{Ml_{max\,neighbor}(t)}{Ml_{max}(t)}$$

where

$Ml_{max}(t)$ is the maximum possible mobility level of a node, which is moving with a maximum velocity $v_{max}(t)$ and with direction $d_i(t) = 0$ between its direction of movement and the line connecting the current position and the sink

$Ml_{max\,neighbor}(t)$ is the maximum mobility node of a neighboring node

1. *Ferrying*: Node i decides to ferry data based on probability γ_i.

$$\gamma_i \geq 1 - \alpha$$

where $\alpha \in [0, 0.7]$ is a constant value. Clearly, ferrying is suitable when γ_i is relatively high. After many simulations tries for various α in this domain ([0, 0.7]), it is demonstrated that the particular value of $\alpha = \frac{1}{2}\left(\gamma = \frac{1}{2}\right)$ optimizes performance. In fact, larger γ_i values lead to many redundant transmissions (see the following), while smaller γ_i values result in ferrying by low mobility nodes, thus increasing latency.

2. *Transmission*: If $\gamma_i < 1 - \alpha$, then no ferrying is done. Interestingly, the decision on whether to do ferrying or transmission exhibits a certain threshold behavior around a constant probability (which is optimized for $\gamma_i = 0.5$). We propose two methods for deciding whether the node will redundantly propagate data or forward to TR_i-hop remote neighbors in the following.
   a. *Hard threshold*: Node i selects to forward data to β_i neighbors or to TR_i-hop neighbors based on hard thresholds γ_i and $\gamma_{neigh\,i}$. Assuming that each node has a maximum mobility capacity, that is bounded by its maximum speed v_{max} when its speed's direction is $d_i = 0$, so it is moving exactly on the line between i and the sink, and then the maximum mobility level a node i can reach is

$$ML_{max} = v_{max} \cdot 1 = v_{max}$$

The same assumption can be applied for each neighbor's maximum mobility capacity, and then the maximum mobility level a neighbor of node i can reach is

$$ML_{max\,i\,neighbor} = max_{j \in neigh\,i} ML_j$$

According to γ_i and $\gamma_{neigh\,i}$ values, there are two scenarios:

   i. *Redundancy*: The condition that has to be satisfied for transmitting to β_i neighbors is the following:

$$\gamma_{neigh\,i} \geq \frac{1}{k}$$

where $k \geq 5$ is a large constant. Intuitively, node i decides to transmit to β_i neighbors, when node i and at least one of its neighbors have a relative high mobility level, in order to provide limited redundancy only when it can be useful. To be more specific, "bad" neighbors make no or very small progress toward the sink, so it is a waste of resources to transmit data if the node i has higher mobility level than them.

The particular $\gamma_{neigh\ i}$ value of 0.1 ($k = 10$) has been selected after extensive simulations, which suggested that this value leads to best possible results. In fact, larger values result to more long-distance transmissions (thus increasing energy dissipation too much), while smaller values lead to too few jumps, thus latency is high.

ii. *Jump*: The condition that has to be satisfied for transmitting to TR_i- hop neighbors is the following:

$$\gamma_{neigh\ i} < \frac{1}{k}$$

where $k \geq 5$ is a large constant. Intuitively, node i decides to transmit to a TR_i-hop neighbor, when node i and all of its neighbors have a relatively low mobility level, in order to avoid a "bad" neighborhood trap and make the maximum possible progress to the sink (trading off with energy cost).

b. *Probabilistic*: Node i examines the neighborhood information and performs a probabilistic choice using p_i in order to forward data to β_i neighbors or to TR_i-hop neighbors. We define p_i for node i in terms of the current and maximum mobility levels of node i and its neighborhood as follows:

$$p_i = \left(1 - \frac{Ml_i(t)}{Ml_{max}}\right) \cdot \left(1 - \frac{Ml_{neigh\,i}(t)}{Ml_{max}(t)}\right)$$

Let p_i be the probability of transmitting to TR_i-hop neighbors and $1 - p_i$ be the probability of transmitting to β_i direct neighbors; the following choice is performed in node i:

$$Transmission = \begin{cases} \text{Redundancy}, & 1 - p_i \\ \text{Jump}, & p_i \end{cases}$$

The hard threshold choice is deterministic and each time selects the best possible option; but this can be myopic in contrast to the randomized decision that has "balancing" properties and can perform better in the long run avoiding anomalies.

12.2.4.3 Calculation of Data Redundancy β

We propose two methods for selecting the number of neighbors β to disseminate a message in the following. The first one is completely local and low cost, while the second collects additional information to improve the decision. The cornerstone of our methods is the use of the mobility level and the distance from the sink of the nodes involved in the process to estimate the requirement for redundancy of message transmissions.

12.2.4.3.1 Completely Local Protocol

A node that moves at maximum mobility level is considered capable of delivering messages, practically without disseminating them to the rest of the nodes. Furthermore, it is more crucial to have larger redundance β to regions far away from the sink, rather in regions close to the sink. We define β for node i in terms of the current and maximum mobility levels and current and maximum distance from the sink:

$$\beta_i = \left\lceil \left(1 - \frac{Ml_i(t)}{Ml_{max}}\right) \cdot \left(\frac{D_i}{D}\right) \cdot \delta_i \right\rceil$$

where

D is the dimension of the $D \times D$ network area
D_i is the current distance from sensor i to the sink
δ_1 represents the maximum possible redundancy as given by

$$\delta_1 = \left\lfloor \frac{dist_{sink}(i)}{R} \right\rfloor$$

where

R is the transmission range of a mobile node
$dist_{sink}(i)$ is the euclidean distance of node i from the sink

For example, assuming that the transmission range of both nodes and sink is set to $R = 70$ m, for a flat square region of size $1000 \times 1000\,m^2$ and that the node i is placed in the borders of the network $\left(dist_{sink}(i) = \frac{D}{2}\right)$, the maximum possible redundancy δ_1 can be calculated by setting $dist_{sink}(i) = \frac{D}{2}$, then $\delta_1 = \left(\frac{D}{2}\right)/2 = 7$. The first term $1 - \frac{Ml_i(t)}{Ml_{max}}$ estimates how close the node's mobility level is to the maximum mobility level. The fraction $\frac{D_i}{D}$ estimates how close node i is to the data sink. Finally, the product of the two previously mentioned terms takes values between 0 and 1. This product is multiplied by $\delta_1 = 7$, which represents the maximum possible redundancy, respectively. The rationale of this function is to calculate large values of β for "slow," moving in "bad" direction and distant from the sink nodes. The opposite happens for "fast," moving in "good" direction and close to the sink nodes: as $Ml_i(t)$ approaches Ml_{max} and/or D_i is small relatively to D, the value of β_i approaches zero, meaning that the node will not redundantly disseminate the message to other nodes but instead transmit directly to the sink as soon as it is within range. β_i is dependent on Ml_i and D_i, and its value also changes over time to reflect the changes in these two metrics; thus, the behavior of the node is adapting to its mobility, direction of movement, and distance from the sink.

12.2.4.3.2 Neighbor Discovery Protocol

Node i transmits a beacon message announcing its mobility level and its id. Nodes that receive the beacon of i respond with a message containing their id and mobility level.

Node i then calculates the average mobility level in the neighborhood; assuming $neigh_i(t)$ is the set of all neighbors of node i (at circular disk of radius R, where R is the transmission range) at time t we have

$$Ml_i^{avg}(t) = \frac{\sum_{j \in neigh_i(t)} Ml_j(t)}{|neigh_i(t)|}$$

In essence, $Ml_i^{avg}(t)$ captures the available mobility at the neighborhood of i at time t. Using $Ml_i^{avg}(t)$ node i can calculate its β as follows:

$$\beta_i = \left\lceil \left(1 - \frac{Ml_i^{avg}(t)}{Ml_{max}}\right) \cdot \left(\frac{D_i}{D}\right) \cdot \delta_1 \right\rceil$$

Note that $Ml_i^{avg}(t)$ encapsulates only the mobility level of the neighbors but not the mobility level of the node *i* itself. There is no need to include the average distance D_i^{avg} from the sink, because nodes in a neighbor have approximately the same distance from the sink. As before, the product of the two first terms takes values between 0 and 1 and approaches 0 when the average mobility approaches Ml_{max}.

12.2.4.4 Calculation of Length of Jump TR_i

We propose two methods for selecting the length of the jump to transmit a message in the following. As in the calculation of redundancy β, the cornerstone of the first method is the use of the mobility level and the distance from the sink of the nodes involved in the process and the core idea of the second method is an expanding ring search (ERS). The first method is local and low cost, while the second is more detailed but can become energy expensive. Note that a node after transmitting a message to a long neighbor, it discards the message from its cache.

12.2.4.4.1 Neighbor Discovery Protocol

Node *i* transmits a beacon message announcing its mobility level and its id. Nodes that receive the beacon of *i* respond with a message containing their id and mobility level. Using $Ml_i(t)$ and $Ml_i^{avg}(t)$ node *i* calculates its TR_i as follows:

$$TR_i = \left\lceil \left(1 - \frac{Ml_i(t) + Ml_i^{avg}(t)}{Ml_{max}}\right) \cdot \left(\frac{D_i}{D}\right) \cdot \frac{\delta_1}{2} \right\rceil$$

As in β_i calculation, the product of the first two terms is multiplied by $\frac{\delta_1}{2} = 3$, where 3 represents the maximum possible jump range in the particular network setting. The rationale of this function is to calculate large values of TR_i for "slow," moving in "bad" direction, distant from the sink nodes that are in relatively "bad" neighborhood. In this case, node *i* makes a jump of TR_i length toward the sink by transmitting directly to TR_i-hop neighbors. The opposite happens for "fast," moving in "good" direction, close to the sink nodes that are in relatively "good" neighborhood: as the sum of $Ml_i(t)$ and $Ml_i^{avg}(t)$ approaches Ml_{max} and/or D_i is small relatively to D, the value of TR_i approaches zero, meaning that the node will not transmit in long range the message to other nodes but instead transmit directly to the sink as soon as it is within range. TR_i is dependent on Ml_i and D_i, and its value also changes over time to reflect the changes in these two metrics; thus, the behavior of the node is adapting to its mobility, direction of movement and distance from the sink.

12.2.4.4.2 Expanding Ring Search

The problem we want to solve is to find the TR_i-hop neighbor that has relatively high mobility level and will be a "good" candidate in order to transmit data, so as to avoid the bad neighborhood. ERS successively searches larger areas until a node with mobility level higher than a hard threshold is located. The complexity of this algorithm can be easily bounded by putting an upper threshold on the number of search iterations; but it can be high anyway, and this is the weakness of this method.

Node *i*, which is in a "bad" neighborhood, begins the search with a time to live (TTL) taken as 2 for the query. Each time the query is forwarded by a node, the TTL value *i* is decremented by 1. When TTL reaches zero, the query packet is not forwarded any further. Thus, by setting the appropriate TTL value in the query packet, the source node can control the search radius. After sending a query with a given TTL, the source node waits for a time-out period to receive a reply. If there is no reply within the time-out period, the source increments the TTL by 1 and reinitiates the query. The repetitive search process continues until the TTL reaches a threshold value *L*. If no reply is received after *L* successive searches, the query is broadcasted through the *L*-hop neighborhood.

In order to decrease communication complexity, the algorithm given earlier can search for "good" neighbors only in a sector of the circular disk of radius $k \cdot R$, where k is the current round of the algorithm. To be more specific, node i will pay the transmission cost of broadcasting the query, but only the nodes in the sector will pay the cost of receiving it. Furthermore, only the nodes in the sector will have to answer back to node i and node i will receive answer from a small partition of nodes on the circular disk of radius $k \cdot R$.

12.2.4.5 Neighbor Selection

Our protocol has to do neighbor selection in two cases, when selecting direct neighbors in order to do redundancy and when jumping to a long neighbor so as to avoid bad neighborhood.

12.2.4.6 Direct Neighbor Selection

After calculating β_i, node i needs to select the particular β_i neighbors to deliver the message to. As mentioned earlier, this selection can influence the overall performance of the protocol; intuitively "fast" moving in "good" direction nodes at high mobility level should be preferred. However, always selecting the same "fast" nodes will result in uneven workload and strain their resources. We present three different strategies in the following for selecting the nodes to disseminate a message to.

Completely random selection: Node i selects β_i of its neighbors randomly. In order to do so, the node uses the neighborhood information gathered by the neighbor discovery protocol. This simple method probabilistically guarantees that the load distribution will be equally shared by the nodes. It is also particularly relevant in cases of limited network knowledge.

Fittest candidate selection: Node i selects β_i of its neighbors such that $Ml_i(t) < Ml_j(t)$, where j a neighboring node to i. In this way, the neighbors with the highest mobility level are selected, hoping to reduce latency. In the case where no neighbors with higher mobility level can be found, node i waits for a short period of time and repeats the neighbor discovery process in the hope that it either reaches a new neighborhood or new neighbors approach it.

Probabilistic candidate selection: To avoid long delays until finding suitable nodes with higher mobility level and also to reduce the strain imposed on these nodes, we compromise our selection criterion. Again node i selects β_i of its neighbors such that $Ml_i(t) < Ml_j(t)$; however, if no such neighbors are found, the rest of the nodes are examined probabilistically, in a way that favors nodes with high mobility. In detail, if node i has higher mobility level from all of its neighbors, it probabilistically favors the neighbors that have the highest mobility level using the probability $Ml_j(t)/Ml_i(t)$. Let p_j be the probability of sending a message to node j; p_j is calculated as follows:

$$p_j = \begin{cases} \dfrac{Ml_j(t)}{Ml_i(t)}, & Ml_j(t) \leq Ml_i(t) \\ 1, & Ml_j(t) > Ml_i(t) \end{cases}$$

Thus, the node examines the neighborhood information, and for each neighboring node, it performs a probabilistic choice using p_j until the message is sent to β_i neighbors.

12.2.4.7 Long Neighbor Selection

After calculating TR_i, node i needs to select a particular long neighbor to deliver the message to. Node i, queries nodes that are at distance between $TR_i - 1$ and TR_i from node i, and have smaller euclidean distance from the sink from node i. As mentioned earlier, this selection can influence the overall performance of the protocol, intuitively "fast" moving in "good" direction nodes at high mobility level should be preferred. However, always selecting the same "fast" nodes will result in uneven workload and strain their resources. We present two different strategies in the following for selecting the nodes to disseminate a message to.

Completely random selection: See 12.3.4.6 for details.

Fittest candidate selection: See 12.3.4.6 for details.

12.2.4.8 Inhibition of Obsolete Messages

Note that although selecting only β neighbors at a time will have the effect of reducing the rate, a message spreads throughout the network, the propagation of a message is arbitrary, and eventually it may be transmitted to every single node. Even when a node k delivers the message to the sink, the rest of the nodes that have a copy of the message will propagate the message to about β neighbors each. Nodes that already store a copy of the message will discard it; however, the message may still be flooded through the network at a slow pace. Here, we present a mechanism to reduce the spread of a message.

We introduce a hop counter h_c contained in each message transmitted; a node i before transmitting a message increases its h_c. Each node j that receives a message performs a deterministic check to decide whether to further propagate the message to its neighbors or to simply store the message in its delivery queue until a sink is located. The decision to whether to propagate or just store the message is done as follows:

$$Decision = \begin{cases} \text{Propagate message} & h_c < h_{opt} \\ \text{Store message} & h_c \geq h_{opt} \end{cases}$$

where h_{opt} is the optimal number of hops between node j and the sink as given by $h_{opt} = \lceil dist_{sink}(j)/R \rceil$, where $dist_{sink}(j)$ is the euclidean distance of node j from the sink and R is the wireless transmission range of nodes. Since the location of the sink (hence also $dist_{sink}(j)$) may not always be known, h_{opt} can be calculated by using another distance, for example, by setting $dist_{sink}(j) = D/2$. In this way, the inhibition decision depends on the distance the message has traveled (as given by h_c) with respect to the overall required distance (as given by h_{opt}). Thus, a message will not be propagated further than h_{opt} hops. On the other hand, messages that performed few hops are more likely to enter the forward queue. We note that this inhibition mechanism is executed every time a node tries to store a message to its forward queue in both direct and long transmissions.

12.2.5 Modeling Diverse Mobility

In most real-world scenarios, most nodes will move in many different and diverse ways. For example, a sensor attached on a vehicle will move fast on a trajectory that consists of a consecutive set of line segments. On the other hand, a pedestrian will tend to move slower over local trajectories with more curves. During these types of movement, small variations of speed are usual. Also, a node will most likely change the type of movement it follows after some time varying not only the average speed but also the type of trajectory it follows. Consider a person riding a bicycle to work, then spending several hours working (low mobility), then riding back home. For example, consider a person working in a university campus; for long periods of time, he/she moves slowly in a confined space (e.g., 10 by 10 m) as he/she goes about his/her work in the office. At some point, the person may start walking faster toward a specific direction as he/she goes to the next building where he/she continues his/her work reverting to the previous type of movement. These examples demonstrate the diversity and variability that may arise in networks of mobile sensors. Modeling real-life movement patterns is a subject of active research. Simplistic mobility patterns, such as random walk or random waypoint alone, cannot accurately capture the heterogeneous mobility characteristics we described. Here, we try to mimic several main types of movements inspired from the aforementioned observations. Using well-defined mobility models, we define a few characteristic mobility roles that are used to construct more complex mobility behaviors.

Working movement: We parameterize a version of random walk [8] to achieve slow movement with small variations away from the center of the movement. We define the function $\mathcal{M}_{work}$ with parameters [0.5, 1.5] m/s for choosing speed and by setting the movement distance toward a direction to be small, [1, 5] m. Nodes move slowly; the movement is mostly centered in the area around their initial position $\mathcal{M}_{work}$. Such movement can be approximated by a random walk [8] mobility function. We define the function to choose a direction and a distance to move toward that direction. Good parameters for obtaining a local movement are [0.5, 1.5] m/s for choosing speed and by setting the movement distance toward a direction to be small, for example, [1, 5] m. $M_{avg} \simeq 7.5$.

Walking movement: Nodes move more quickly than the working mobility and travel in smoother trajectories $\mathcal{M}_{walk}$. Such behavior can be obtained by using a variation of the boundless area [8] mobility model to define $\mathcal{M}_{walk}$, which is more rapid and less local than $\mathcal{M}_{work}$. When a node reaches the boundaries of the network area, we force it to reflect, that is, take a left turn of 45°. We bound the speed to vary between [1, 2] m/s, we set the time step $\Delta t = 2$ s; at each time step, we allow the speed to vary by $\Delta v = 0.25$ m/s and the direction to vary by $\Delta\alpha = 30°$.

Bicycle ride: This type of movement is similar to the walking movement except that the speed is usually greater and there are less direction changes $\mathcal{M}_{bic}$. Again we use our variation of the boundless area [8] mobility model; we bound the speed between [3, 6] m/s (10.8–21.6 km/h), we set $\Delta t = 3$ s, $\Delta v = 0.5$ m/s, and $\Delta\alpha = 30°$.

Vehicular movement: Vehicular movement $\mathcal{M}_{veh}$ is the faster of all, we use the probabilistic random walk [8]. In this mobility model, nodes move only toward predefined directions north, north east, east, etc. We vary the speed between [5.55, 10] m/s (20–36 km/h).

Mobility transitions: Assigning a mobility role is enough to diversify the mobility levels of the nodes. However, in realistic scenarios, nodes will change mobility roles. To model such dynamic mobility, we use a state transition diagram to change between mobility models. Each state of the diagram corresponds to a mobility role as defined earlier. From each state, a set of outgoing edges to one or more of the other states exist; each edge is associated with a probability of transition. Also, there is an outgoing edge that returns to the same state. The sum of all outgoing edges from a state is equal to 1. While on a state, the node follows the mobility defined by the corresponding mobility model. As soon as a new position needs to be selected, a probabilistic experiment is performed to choose a new state according to the state transition diagram, and then the corresponding mobility function is invoked to select the position and speed of the node. We also define a special state called the stop state in which the node remains still for a small period of time. The following diagrams define characteristic mobility transitions.

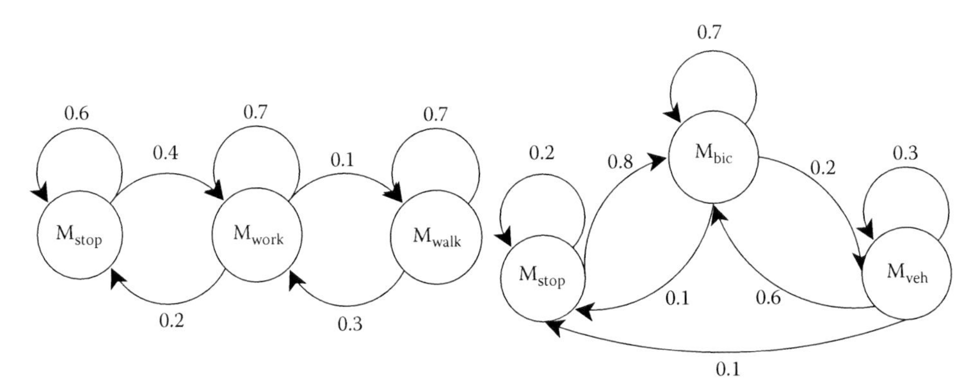

C1: Transitions between slow mobility roles

C3: Transitions for medium mobility with fast bursts

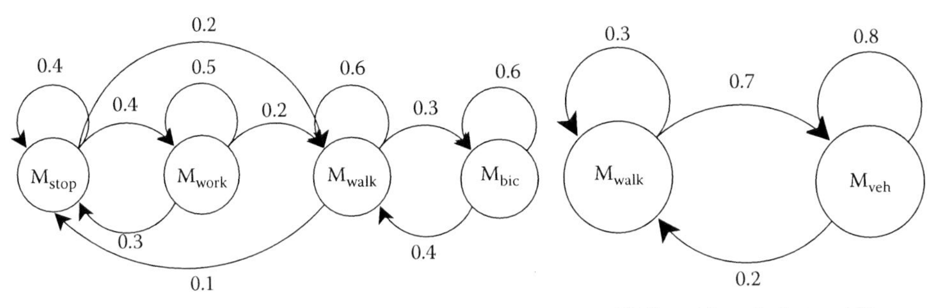

C2: Transitions for medium mobility level

C4: Transitions for fast mobility

12.2.6 Experimental Evaluation

We implement our protocols on the *ns-2* simulation platform version 2.33, using the TRAILS toolkit [9], which simplifies the implementation and evaluation of complex and dynamic mobility scenarios. Our configuration uses 802.11 for the MAC layer and our experimental results are affected by the collisions occurred in the wireless network, message retransmissions, etc. We set the network area to be $1000 \times 1000\,\text{m}^2$ and we always position $\mathcal{S}$ at (500, 500) the center of the network. We deploy in a random uniform manner, a number of 50, 100, 150, 200, 250, and 300 nodes in the network area, and repeat each experiment 70 times, with 95% confidence intervals calculated. The nodes and sink have significant energy resources (100 *J*) to prevent failures due to energy depletion. Also, we do not consider the possibility of other types of node failures. The sink $\mathcal{S}$ transmits beacon messages at a steady rate $\lambda_{Beacon} = 1$, that is, a beacon message per second. We assume that all sensor nodes record an instance of the environmental conditions producing 20 messages during the simulation. Messages are produced at random intervals, and on the average, new messages are produced at rate $\lambda_i = 0.025$ messages per second. Thus, the data generation phase lasts for about 800 s; we simulate the network for 3600 s in order to collect delayed data. The data are generated in packets of 36 bytes, while the size of a beacon message is 24 bytes. Each node uses fixed sized caches for the forward and delivery queues, and each cache can accommodate 64 messages; thus, there is a possibility of message drops due to caches exceeding their maximum size, that is, we avoid the ideal case of infinite buffers. The transmission range of both nodes and sink is set to $R = 70\,\text{m}$, which corresponds to the transmission power value of 0.00263 (−26 dBm) on ns-2. The characteristics of the radio module, that is, the values of ϵ_{trans}, ϵ_{recv}, and E_{idle}, were set to match as close as possible the specifications of commercially available sensors such as TelosB motes that uses the Chipcon CC2420 [4] transceiver.

Node movement: We assign different mobility roles to the nodes of the network. In a first experiment, we examined mixed mode scenarios where 25% of the nodes follow $\mathcal{M}_{work}$, 25% $\mathcal{M}_{walk}$, 25% $\mathcal{M}_{bic}$, $\mathcal{M}_{bic}$, and 25% $\mathcal{M}_{veh}$. The assigned mobility functions remain the same for a particular node during the simulation. In the second experimental setup, the mobility of the nodes changes during the simulation using the mobility transitions defined earlier (see Figure 12.2); we assign C1 (slow mobility) to 25% of the nodes, C2 (medium mobility) to another 25%, C3 (medium mobility with fast bursts) to another 25%, and C4 (fast mobility) to the remaining 25% of the nodes.

Protocol comparison: We implemented and evaluated in these settings three known protocols to use as a point of reference in our evaluation. We obtain the *simple flooding protocol* simply by setting $\beta = \infty$ (i.e., the node will send the message once to all its neighbors) without any adaptation or the message to inhibition mechanism. Also, this protocol does not execute the neighbor discovery protocol, instead it broadcasts a message simultaneously to all neighboring nodes. The second test case protocol is the *gossiping protocol* in which a message is sent randomly to one neighbor, and after the transmission, the message is discarded from the cache. So, only the initial message is traveling through the network and no copies are created. The third test case protocol is the *adaptive mobility protocol (AMP)* [29], which is an adaptive redundancy protocol for data propagation. We compare these three protocols with our *fixed* TR_i *protocol* for both hard-threshold and probabilistic decision criterion, which is a nonadaptive version of our main protocol; we set $TR_i \in \{2, 4, 6, 8\}$. Also, we comparatively study characteristic variations of our method corresponding to selected different design alternatives.

Metrics: Conducting these experiments, we measure several metrics that depict the behavior of our protocols. We call *success rate*, the percentage of data messages that were received by all sinks over the total number of generated messages. We measure the *energy* consumed at the sensor network due to communication as the average number of Joules consumed at each node. Note that we consider the motion of the nodes to be initiated by the objects/persons/vehicles they are attached onto, so the nodes themselves do not consume energy for movement. We also measure the *delivery delay*, which is defined as the average time interval between the creation of a message and the time when it is delivered to the sink.

12.2.6.1 Performance Findings

12.2.6.1.1 First Experiment

In the first set of experiments we present here, the sensor nodes are divided in four groups, where the nodes of each group follow a specific mobility role. First, we compare the best variation of our protocol with flooding, gossiping, and AMP. In Figure 12.23, we can see that the highest success rate is achieved by our adaptive *direction sensitive mobility protocol* (*DSMP*), in which the hard-threshold decision is selected and the fittest neighbor, either direct or long, is being selected (95%). The AMP is almost in a tie with the fixed DSMP. Flooding achieves a very low success rate due to the many packets dropped by the limited sized queues and the large number of collisions during transmissions. Gossiping achieves the lowest success rate due to the lack of replicas of the message being transmitted.

In Figure 12.24, we observe that the flooding protocol is the most energy consuming among the four compared protocols. The adaptive DSMP consumes about 25% more energy than the gossiping

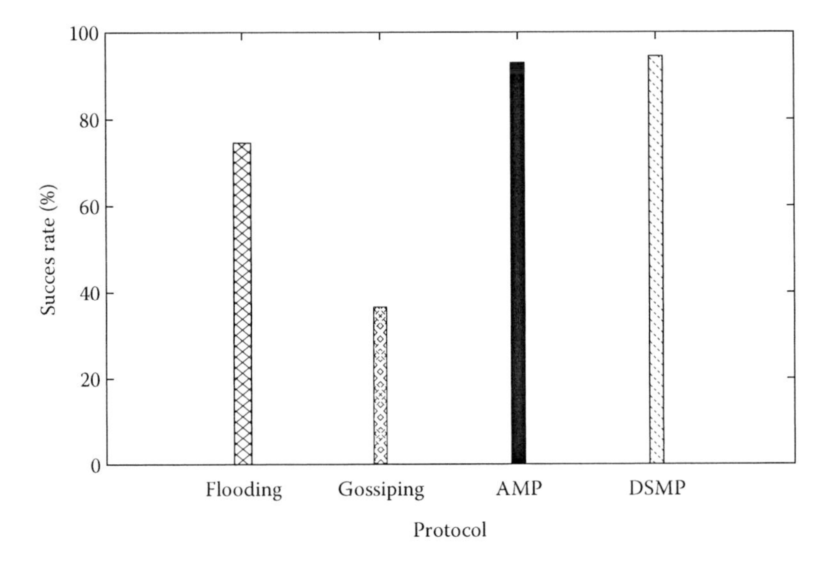

FIGURE 12.23 Success rate of the protocols when nodes are assigned a static mobility role.

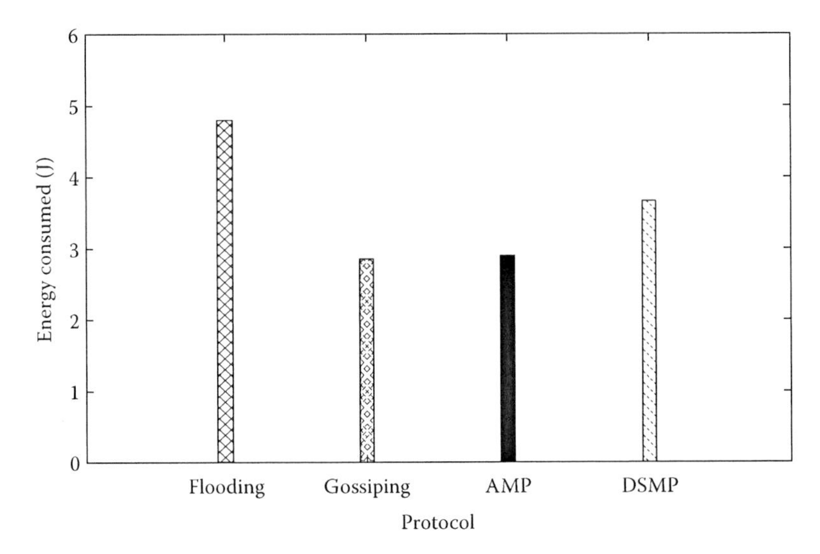

FIGURE 12.24 Energy dissipation of the protocols when nodes are assigned a static mobility role.

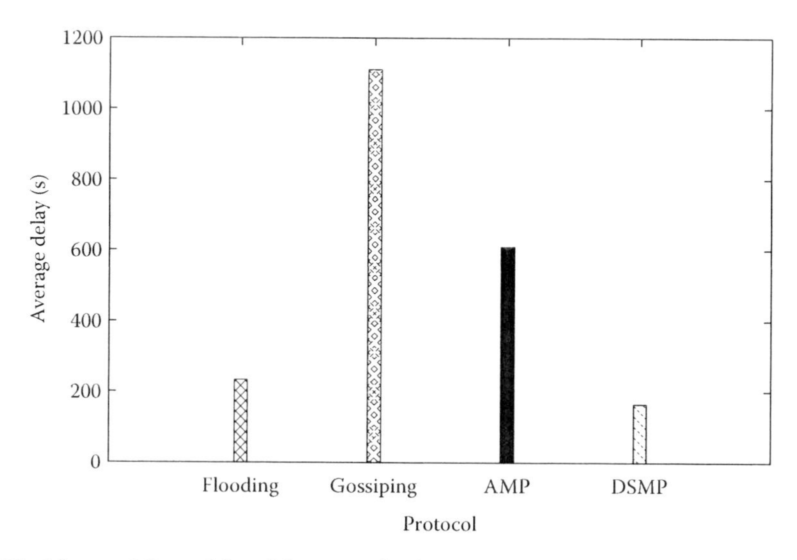

FIGURE 12.25 Message delivery delay of the protocols when nodes are assigned a static mobility role.

protocol and AMP. This is explained by the fact that the DSMP is the only one that uses the expensive, but fast, long transmissions.

The delivery delay is shown in Figure 12.25. Our adaptive DSMP is the fastest and improves about 360% the delay compared to the adaptive AMP. In DSMP, every message is transmitted/ferried exclusively toward sink, thus incurring the lowest possible delay. A very interesting result that can be explained from what has been provided earlier is that DSMP achieves better delivery delay than the fast one-hop flooding protocol. We can see that flooding has very low delay, but this is expected since messages farther away from the sink that exhibit long delays are most likely to be dropped and thus are not considered in the calculation of the delay. As expected, gossiping has the highest delay because of the absence of adaptation and replication in data dissemination.

12.2.6.1.2 Variations of Our Protocol

Due to lack of space, it was impossible to present experimental results for every possible variation of our protocol. As a result, we decide to calculate data redundancy β and length of jump TR_i using the neighbor discovery protocol. In addition, for neighbor selection, we decide to use the fittest candidate selection for both direct and long neighbors.

In Figure 12.26, we can see that the adaptive protocol with the hard-threshold decision criterion achieves the highest success rate. The hard-adaptive protocol is almost in a tie with the fixed $TR_i = 2$ protocol. The probabilistic adaptive protocol achieves a satisfactory success rate (89%). For both the hard-threshold and probabilistic case, as TR_i increases the delivery ratio decreases rapidly, because of the increasing number of unsuccessful transmissions due to the collisions in the MAC layer.

The energy dissipation of the protocols is shown in Figure 12.27. We observe for both cases of decision criterion (probabilistic and hard threshold) that, as the fixed TR_i increases, the number of collisions increases and the number of successful data transmissions decreases. As a result, the nodes cannot discover many neighbors because the control messages for neighbor discovery fail to reach their destination; and thus, our protocol is no longer operational as it relied on local information. Comparing the two different decision criteria, the hard-threshold criterion is the most energy conservative in any case.

The delivery delay is shown in Figure 12.28. The lowest delivery delay is achieved by the hard-adaptive protocol that is about 200% lower than the prob-adaptive protocol. For the fixed TR_i protocols, the best performance is achieved for $TR_i = 4$ in both hard-threshold and probabilistic decision criterion. As TR_i increases, the delivery delay deteriorates since the number of successful data transmissions decreases because of the vast number of collisions.

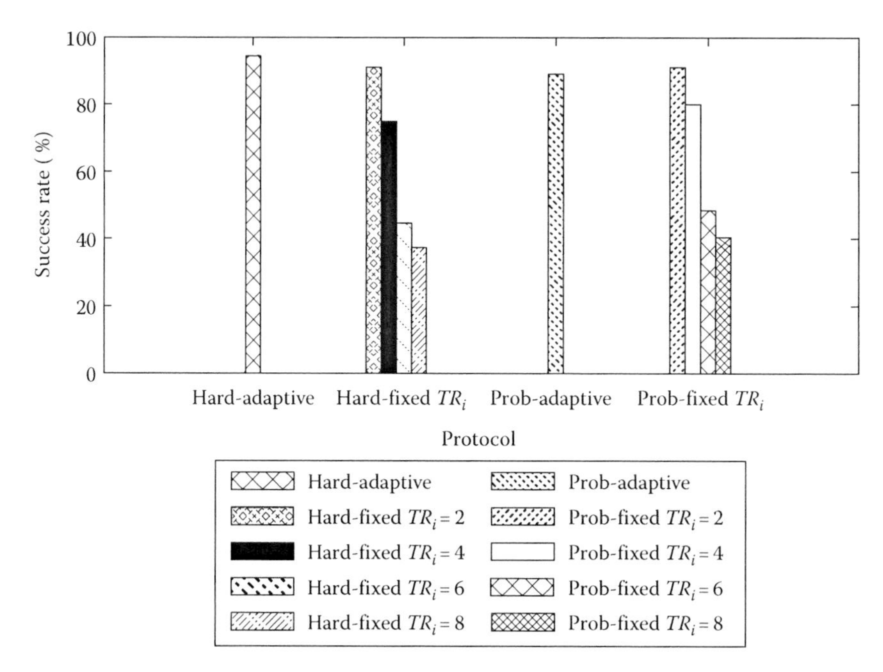

FIGURE 12.26 Success rate of our protocol's variations when nodes are assigned a static mobility role.

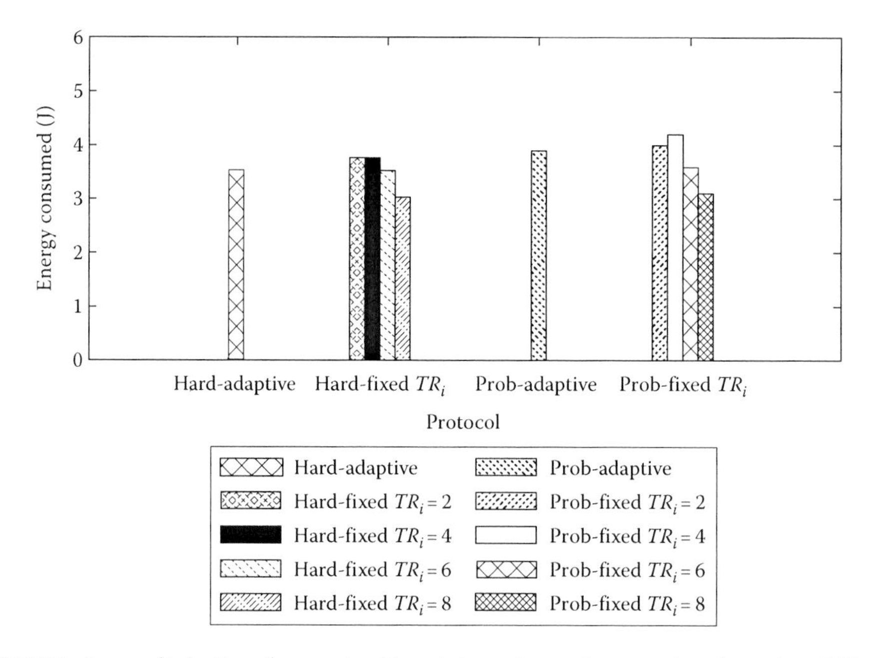

FIGURE 12.27 Energy dissipation of our protocol's variations when nodes are assigned a static mobility role.

12.2.6.1.3 Second Experiment

The results for this setup (varying mobility profiles) that are depicted in Figures 12.29 through 12.34 are similar in the sense that our protocol and its variations perform very good in this case as well.

To be more specific, Figure 12.32 shows that the AMP and DSMP adaptive protocols have the highest success rate (92%). All the four studied protocols have a little bit lower success rate than the case where the mobility roles are assigned statically.

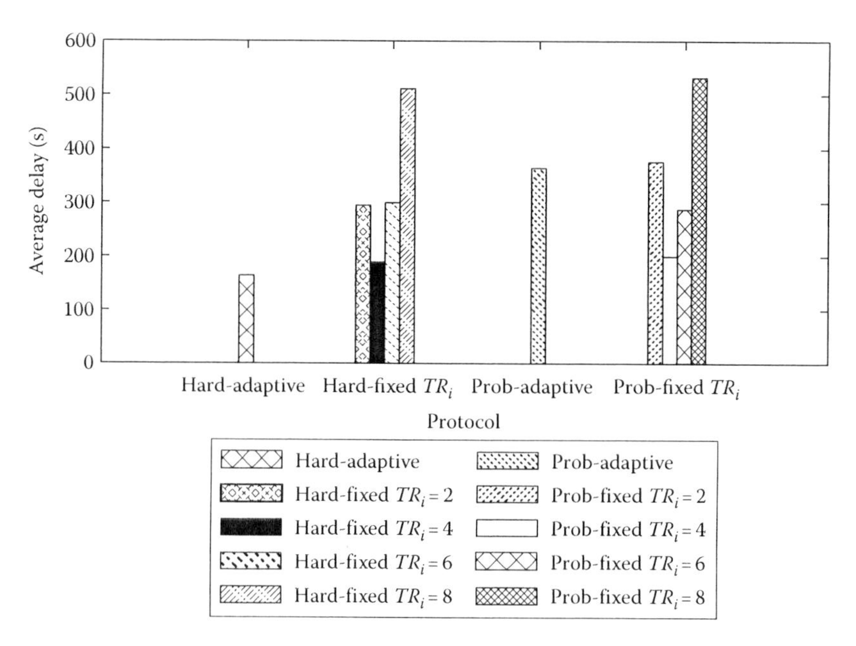

FIGURE 12.28 Message delivery delay of our protocol's variations when nodes are assigned a static mobility role.

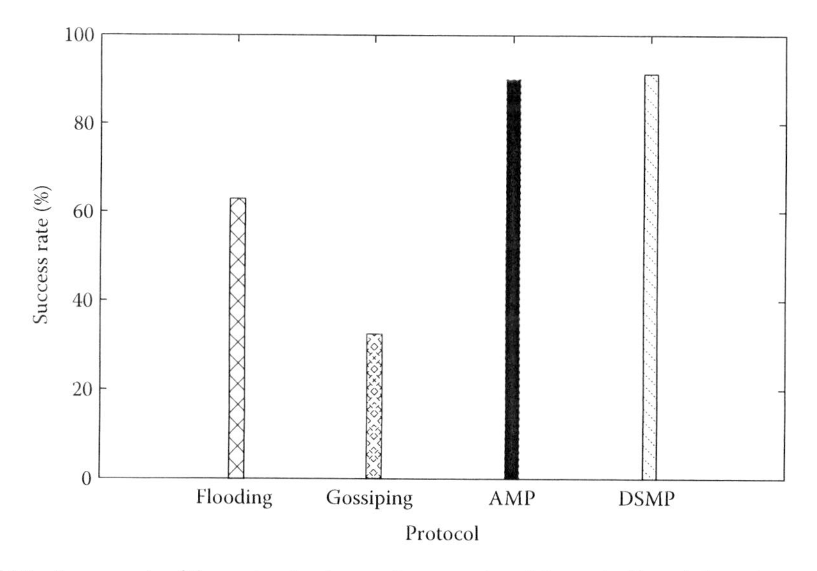

FIGURE 12.29 Success rate of the protocols when nodes are assigned dynamically mobility roles.

Figure 12.33 depicts the findings about energy consumption. We note that the AMP adaptive protocol has again the smallest energy consumption among all of the protocols. Interestingly, the DSMP adaptive protocol has better performance in this case than the static mobility role scenario, which proves that our protocol is efficient and adapts well in more complex and changing mobility scenarios.

Figure 12.34 depicts the delivery delay where all the studied protocols have the same performance as in the case of the statically assigned mobility roles.

12.2.6.1.4 Variations of Our Protocol

In Figure 12.32, we note that the adaptive protocol with the hard-threshold decision criterion achieves again the highest success rate (92%), while the other variations (<85%) for the varying mobility profiles

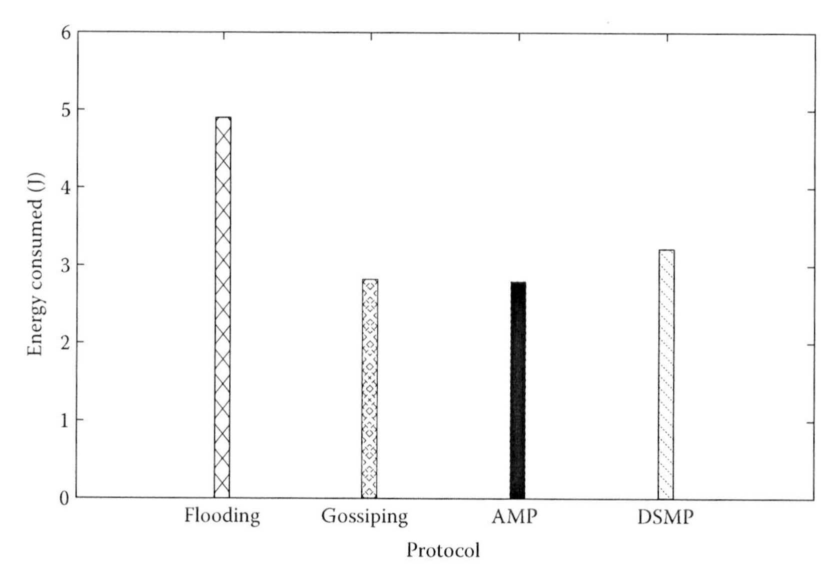

FIGURE 12.30 Energy dissipation of the protocols when nodes are assigned dynamically mobility roles.

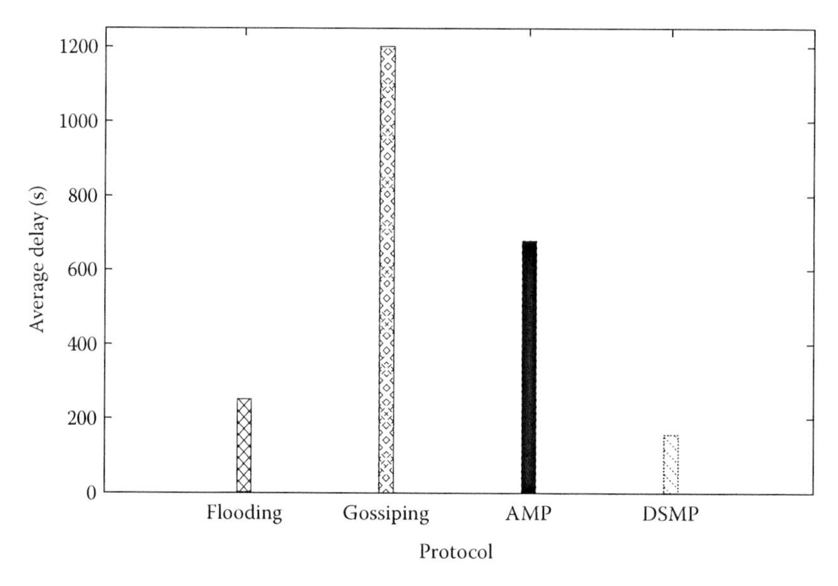

FIGURE 12.31 Message delivery delay of the protocols when nodes are assigned dynamically mobility roles.

scenario have a bit lower success rate than the case of statically assigned mobility profiles. The fixed $TR_i = 2$ variations for both the hard-threshold and probabilistic case and the probabilistic adaptive protocol achieve a satisfactory success rate (85%). As TR_i increases, the number of collisions increases and the success ratio decreases vastly.

The energy dissipation of the protocols is depicted in Figure 12.33. We note that the adaptive protocol with the hard-threshold decision criterion is the most energy efficient protocol by combining the information for the success rate metric and has better performance for the case of the dynamically changing mobility roles.

Figure 12.34 shows the data delivery latency. The lowest delivery delay is achieved by the hard-adaptive protocol that has better performance for the dynamic mobility profiles, and it has over 250% better performance than the prob-adaptive protocol. Again, for the fixed TR_i protocols, the best

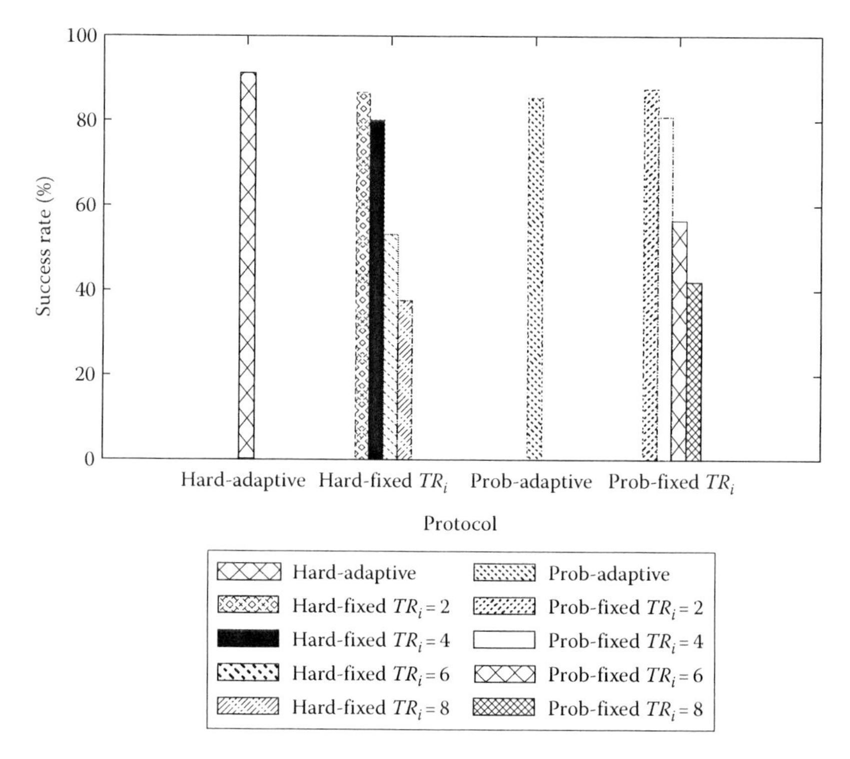

FIGURE 12.32 Success rate of our protocol's variations when nodes are assigned dynamically mobility roles.

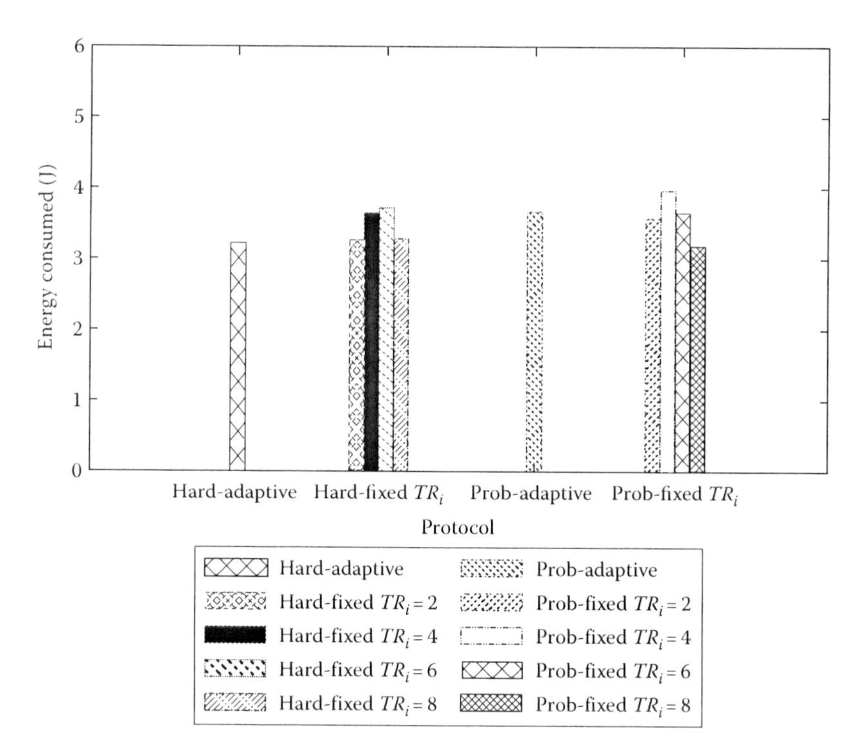

FIGURE 12.33 Energy dissipation of our protocol's variations when nodes are assigned dynamically mobility roles.

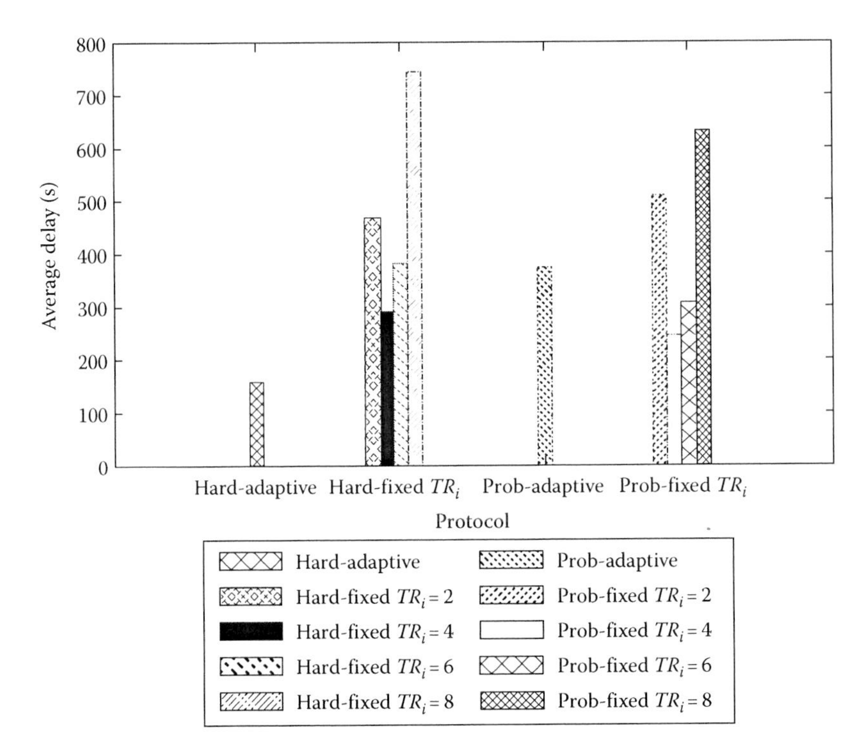

FIGURE 12.34 Message delivery delay of our protocol's variations when nodes are assigned dynamically mobility roles.

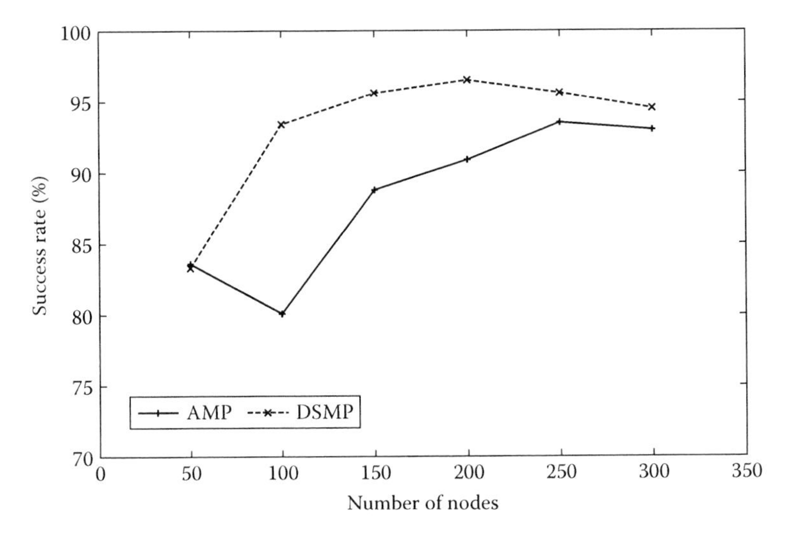

FIGURE 12.35 Success rate of the protocols for various densities.

performance is achieved for $TR_i = 4$ in both hard-threshold and probabilistic decision criterion. As TR_i increases, the impact of the collisions and the retransmissions to the average delivery delay is dominant.

The impact of density: We note that the figures mentioned earlier concern the case of 300 nodes. However, we have conducted experiments for various densities corresponding to 50, 100, 150, 200, 250, and 300 nodes. In Figures 12.35 through 12.37, we present the comparison of the two adaptive protocols (AMP and DSMP). We interestingly remark that our DSMP behaves better in sparse networks as well; this is due to its ability to jump over sparse or even disconnected areas.

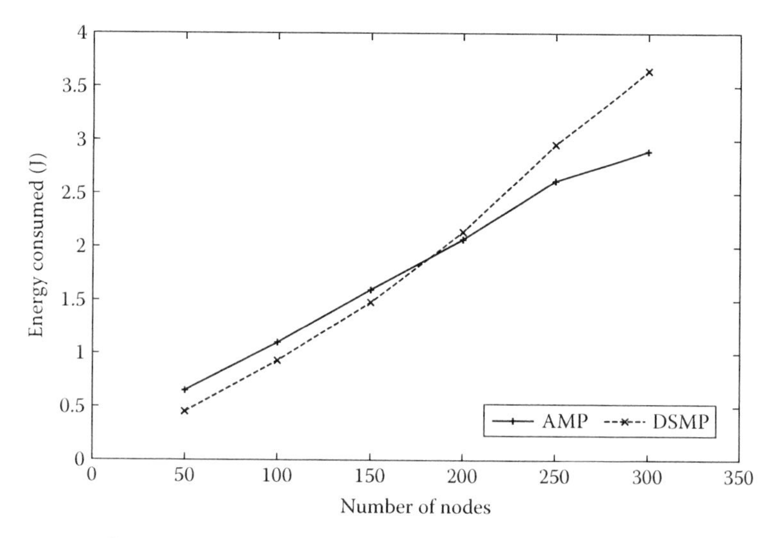

FIGURE 12.36 Energy dissipation of the protocols for various densities.

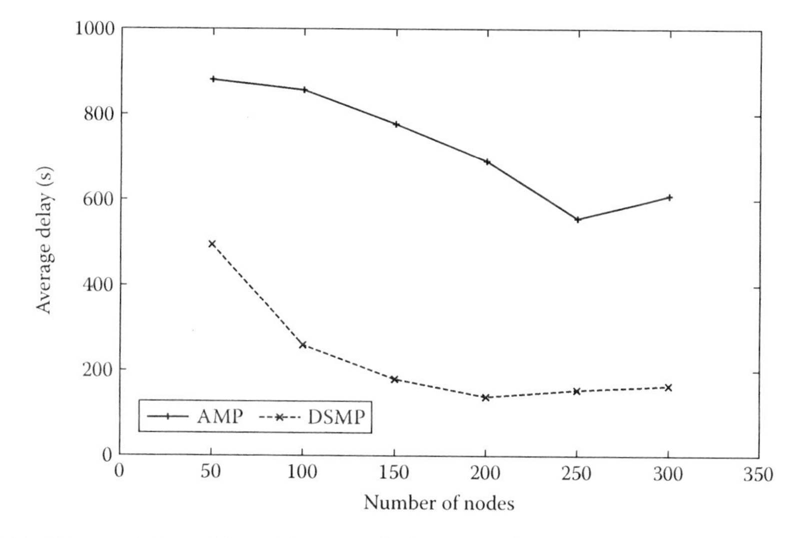

FIGURE 12.37 Message delivery delay of the protocols for various densities.

12.3 Conclusions and Future Work

In the first part of the chapter, we study the problem of greedy data propagation, aiming mainly to reduce the energy dissipation of the routing algorithm. Toward parameterized energy-latency trade-offs, we provide hybrid combinations of two greedy optimization criteria, as any single criterion does not simultaneously satisfy both energy efficiency and low latency.

We rigorously analyzed the DAR protocol, the LAR protocol, and the NFP protocol.

Also, we compared experimentally the aforementioned protocols and hybrid combinations of location-aware and direction-aware protocols toward a satisfactory performance trade-off. Although the DAR protocol increases latency, on the other hand, it achieves much better results regarding the energy consumption. On the contrary, LAR is the fastest way to transmit data toward the sink, while it is the most energy consuming. NFP is very cheap in terms of energy consumption, but it increases data delivery latency dramatically. The hybrid protocols' performance is in between the latency values of the DAR and the LAR and allow a fine-tuning of performance. Interestingly, the hybrid protocols beat both LAR and DAR as far as the energy balance is concerned.

In the second part of the chapter, we investigate networks of mobile sensors with diverse, highly changing mobility profiles. To abstract the sensor mobility dynamics, we propose a novel network parameter, the direction-aware mobility level, which captures how fast and close toward the sink each mobile node is moving. Exploiting the potential of mobility to serve as a low-cost replacement for connectivity and data propagation redundancy, we provide distributed, adaptive data dissemination protocols that significantly improve performance (especially latency) when compared to known methods. In particular, data ferrying is used when mobility is high, while in the case of low mobility, we either employ data redundancy or use long-distance wireless transmissions to accelerate data propagation.

In future work, we plan to extend our schemes in the case of multiple (static) sinks and also to come up with (possibly new) methods for sensor networks where the sinks are also mobile (in that case, lightweight distributed coordination techniques for sink mobility management may be needed as well). Actually, this research has a strong relation with geographic routing. We plan to indeed investigate the relation of our routing scheme that tries to maximize the speed toward the destination, whereas greedy geographic routing tries to maximize distance traveled to the destination as it forward data. Also, we plan to examine alternative probabilistic choices that could further improve performance.

References

1. L. Badia, E. Fasolo, A. Paganini, and M. Zorzi. Data aggregation algorithms for sensor networks with geographic information awareness. In *Proceedings of Wireless Personal Multimedia Communications (WPMC)*, San Diego, CA, 2006.
2. L. Bononi and M. Di Felice. A cross layered MAC and clustering scheme for efficient broadcast in VANETs. In *Proceedings of 1st IEEE International Workshop on Mobile Vehicular Networks (IEEE MoVeNet 2007)*, Pisa, Italy, October 8, 2007.
3. L. Bononi, M. Di Felice, and S. Pizzi. DBA-MAC: Dynamic backbone-assisted medium access control protocol for efficient broadcast in VANETs. *Journal of Interconnection Networks*, 10(4): 321–344, 2009, ISSN: 0219-2659, 1793-6713.
4. B. Bougard, F. Catthoor, D. C. Daly, A. Chandrakasan, and W. Dehaene. Energy efficiency of the IEEE 802.15.4 Standard in dense wireless microsensor networks: Modeling and improvement perspectives. In *Design, Automation and Test in Europe*, pp. 196–201, 2005.
5. A. Boukerche and K. Abrougui. An efficient leader election protocol for mobile networks. In *International Conference on Wireless Communications and Mobile Computing (IWCMC)*, Vancouver, British Columbia, Canada, pp. 1129–1134, ACM, New York, 2006.
6. A. Boukerche, D. Efstathiou, and S. Nikoletseas. Adaptive, direction-aware data dissemination for diverse sensor mobility. In *7th ACM/IEEE International Symposium on Mobility Management and Wireless Access Protocols (MobiWac 2009)*, Tenerife, Canary Islands, Spain, pp. 50–57, ACM Press, New York, 2009.
7. A. Boukerche, Y. Ren, and Z. Zhang. Performance evaluation of an anonymous routing protocol using mobile agents for wireless ad hoc networks. In *32nd IEEE LCN*, Dublin, Ireland, pp. 893–900, 2007.
8. T. Camp, J. Boleng, and V. Davies. A survey of mobility models for ad hoc network research. *Wireless Communications and Mobile Computing*, 2(5):483–502, 2002.
9. I. Chatzigiannakis, A. Kinalis, G. Mylonas, S. Nikoletseas, G. Prasinos, and C. Zaroliagis. TRAILS, a toolkit for efficient, realistic and evolving models of mobility, faults and obstacles in wireless networks. In *41st Annual Simulation Symposium (ACM)/IEEE ANSS*, pp. 23–32, 2008.
10. I. Chatzigiannakis, A. Kinalis, and S. Nikoletseas. Sink mobility protocols for data collection in wireless sensor networks. In *4th ACM International Workshop on Mobility Management and Wireless Access Protocols*, pp. 52–59, October 2006.

11. I. Chatzigiannakis, S. Nikoletseas, and P. Spirakis. Efficient and robust protocols for local detection and propagation in smart dust networks. *ACM Mobile Networks and Applications Journal*, 10(1): 133–149, 2005.
12. I. Chatzigiannakis, S. Nikoletseas, and P. Spirakis. Smart dust protocols for local detection and propagation. *ACM Mobile Networks and Applications Journal*, 10(1):9–16, 2005.
13. Z. Chen, H. Kung, and D. Vlah. Ad-hoc relay wireless networks over moving vehicles on highways. In *Proceedings of the 2001 ACM Symposium on Mobile Ad Hoc Networking and Computing (Mobihoc 2001)*, Long Beach, CA, October 2001.
14. A. Clementi, F. Pasquale, A. Monti, and R. Silvestri. Communication in dynamic radio networks. In *26th ACM Annual Symposium on the Principles of Distributed Computing (PODC)*, Portland, OR, pp. 205–214, 2007.
15. A. Clementi, F. Pasquale, and R. Silvestri. MANETS: High mobility can make up for low transmission power. In *36th International Colloquium on Automata, Languages and Programming (ICALP)*, Rhodes, Greece, 2009.
16. D. Efstathiou, A. Koutsopoulos, and S. Nikoletseas. Analysis and simulation for parameterizing the energy-latency trade-off for routing in sensor networks. In *13th ACM International Conference on Modeling, Analysis, and Simulation of Wireless and Mobile Systems (MSWIM)*, Bodrum, Turkey, pp. 205–209, 2010.
17. C. Efthymiou, S. Nikoletseas, and J. Rolim. Energy balanced data propagation in wireless sensor networks. *Wireless Networks Journal*, 12(6): 691–707, 2006.
18. M. Elhadef, A. Boukerche, and H. Elkadiki. Diagnosing mobile ad-hoc networks: Two distributed comparison-based self-diagnosis protocols. In *4th ACM International Workshop on Mobility Management and Wireless Access Protocols*, pp. 18–27, 2006.
19. E. Fasolo, C. Prehofer, M. Rossi, Q. Wei, J. Widmer, A. Zanella, and M. Zorzi. Challenges and new approaches for efficient data gathering and dissemination in pervasive wireless networks. In *Proceedings of IN-TERSENSE*, Nice, France, May 2006.
20. D. Gavalas, G. Tsekouras, and C. Anagnostopoulos. A mobile agent platform for distributed network and systems management. *Journal of Systems and Software*, 82(2): 355–371, 2009.
21. Y. Gu, D. Bozdag, R. W. Brewer, and E. Ekici. Data harvesting with mobile elements in wireless sensor networks. *Computer Networks*, 50(17): 3449–3465, 2006.
22. R. Haider and M. Y. Javed. Energy efficient greedy approach for sensor networks. In *Greedy Algorithms*, Chapter 8, 2008, ISBN 978-953-7619-27-5.
23. M. Ho and K. Fall. Poster: Delay tolerant networking for sensor networks. In *IEEE Conference on Sensor and Ad Hoc Communication and Networks (SECON)*, Santa Clara, CA, 2004.
24. T.-C. Hou and V. Li. Transmission range control in multihop packet radio networks. *IEEE Transactions on Communications*, 34(1):3844, 1986.
25. R. Jain, A. Puri, and R. Sengupta. Geographical routing using partial information for wireless ad hoc networks. IEEE Personal Communication, 2001.
26. P. Juang, H. Oki, Y. Wang, M. Martonosi, L. Peh, and D. Rubenstein. Energy-efficient computing for wildlife tracking: Design tradeoffs and early experiences with Zebranet. In *Proceedings of the 10th International Conference on Architectural Support for Programming Languages and Operating Systems (ASPLOS)*, San Jose, CA, 2002.
27. A. Kansal, A. Somasundara, D. Jea, M. Srivastava, and D. Estrin. Intelligent fluid infrastructure for embedded networks. In *2nd ACM/USENIX International Conference on Mobile Systems, Applications, and Services (MobiSys04)*, Norfolk, VA, 2004.
28. B. Karp and H. T. Kung. GPSR: Greedy perimeter stateless routing for wireless networks. In *International Conference on Mobile Computing and Networking (MobiCom)*, New York, 2000.
29. A. Kinalis and S. Nikoletseas. Adaptive redundancy for data propagation exploiting dynamic sensory mobility. In *Proceedings of the 11th ACM International Symposium on Modeling, Analysis and Simulation of Wireless and Mobile Systems (MSWiM)*, Vancouver, British Columbia, Canada, pp. 149–156, ACM Press, New York, 2008. Also, in *the Journal of Interconnection Networks*, 2009.

30. A. Kinalis and S. Nikoletseas. Scalable data collection protocols for wireless sensor networks with multiple mobile sinks. In *40th Annual Simulation Symposium (ACM)/IEEE ANSS*, Norfolk, VA, pp. 60–69, March 2007.
31. C. Konstantopoulos, A. Mpitziopoulos, D. Gavalas, and G. Pantziou. Effective determination of mobile agent itineraries for data aggregation on sensor networks. *IEEE Transactions on Knowledge and Data Engineering*, 99(PrePrints), 2009.
32. C. Konstantopoulos, A. Mpitziopoulos, D. Gavalas, and G. Pantziou. Exploiting the cloning capability of Mobile Agents for cost-effective data fusion in wireless sensor networks. In *IEEE Symposium on Computers and Communications (ISCC)*, pp. 963–968, 2008.
33. E. Kranakis, H. Singh, and J. Urrutia. Compass routing on geometric networks. In *Proceedings of 11th Canadian Conference on Computational Geometry*, Vancouver, British Columbia, Canada, p. 51, 1999.
34. K. Lee, U. Lee, and M. Gerla. *Survey of Routing Protocols in Vehicular Ad Hoc Networks: Developments and Challenges*, IGI Global, May 2010.
35. P. Leone, S. Nikoletseas, and J. Rolim. Stochastic models and adaptive algorithms for energy balance in sensor networks. *Theory of Computing Journal*, 47(2):433–453, 2010.
36. C. Liu and J. Wu. Scalable routing in delay tolerant networks. In *Symposium on Mobile Ad Hoc Networking and Computing (Mobihoc)*, Montreal, Quebec, Canada, 2007.
37. J. Luo and J.-P. Hubaux. Joint mobility and routing for lifetime elongation in wireless sensor networks. In *24th IEEE INFOCOM*, Miami, FL, 2005.
38. G. Pantziou, A. Mpitziopoulos, D. Gavalas, C. Konstantopoulos, and B. Mamalis. Mobile sinks for information retrieval from cluster-based WSN islands. In *International Conference on Ad Hoc Networks and Wireless (ADHOC-NOW)*, Murcia, Spain, pp. 213–226, 2009.
39. A. Papadopoulos and J. McCann. Towards the design of an energy-efficient, location-aware routing protocol for mobile, ad-hoc sensor networks. In *DEXA*, Zaragoza, Spain, 2004.
40. O. Powell, P. Leone, and J. Rolim. Energy optimal data propagation in wireless sensor networks. *Parallel Distributed Computing*, 67(3): 302–317, 2007.
41. S. Ross. *Stochastic Processes*. John Wiley & Sons Inc., New York, 1996.
42. T. Small and Z. Haas. The shared wireless infostation model: A new ad hoc networking paradigm. In *MobiHoc 03*, Annapolis, MD, 2003.
43. H. Takagi and L. Kleinrock. Optimal transmission ranges for randomly distributed packet radio terminals. *IEEE Transactions on Communications*, 32: 246–257, 1984.
44. A. Vahdat and D. Becker. Epidemic routing for partially-connected ad hoc networks. Duke Tech Report CS-2000-06, 2000.
45. W. Wang, V. Srinivasan, and K.-C. Chua. Trade-offs between mobility and density for coverage in wireless sensor networks. In *International Conference on Mobile Computing and Networking (MobiCom)*, Montréal, Québec, Canada, 2007.
46. Y. Wang and H. Wu. DFT-MSN: The delay/fault-tolerant mobile sensor network for pervasive information gathering. In *INFOCOM*, 2006.
47. W. Zhao, M. Ammar, and E. Zegura. A message ferrying approach for data delivery in sparse mobile ad hoc networks. In *MobiHoc '04*, Tokyo, Japan, 2004.

13

Traffic Analysis Attacks on Sensor Networks

Lu Yu
Clemson University

Richard R. Brooks
Clemson University

Ryan Craven
Clemson University

Harakrishnan Bhanu
Clemson University

13.1 Introduction

Among various evaluation criteria of wireless sensor networks (WSNs), the issue of security about sensor networks has attracted a great deal of attention over the previous decades. An adversary can compromise a sensor network by launching remote, software-based attacks, or physical attacks.

For sensor networks to be privacy preserving, data confidentiality is usually ensured through encryption of the message content, which is nevertheless not enough. An adversary can sniff out subtle information by monitoring the context of communications. In this chapter, we focus our discussion on a particular class of attacks—traffic analysis attacks. They are passive attacks based on analyzing the information gained in communication without interrupting it, making this type of attack harder to be detected by the victim network.

For example, the disclosure of both spatial and temporal data through traffic analysis may give the adversary enough knowledge to learn the relative or even actual location of a sensor through correlation with a prior knowledge about the physical location of the sensor. In traffic analysis attacks, the adversaries do not need to go into further details on the mechanism of the encryption. The information obtained by monitoring the sensor network can reveal distinct patterns, which allow an adversary to deduce heterogeneous types of information. Information that can be deduced includes the following:

- Who is talking (the source)?
- To whom they are talking (the destination)?
- What application-layer protocols are in use (SSH, HTTPS, etc.)?
- The amount of data contained within each packet.
- Packet timings.

Each of these pieces of information, especially when combined with similar information from other packets in a flow or stream, can be very revealing about the data within an encrypted message.

These attacks have made the conflicts between the area of traffic analysis and network security become prominent and attract more attention.

In WSNs, each sensor node sends the collected data to a base station via a multi-hop circuit through the network. The data flows are inevitably exposed on the network. The wireless communication adds mobility to sensor node, which, on the other hand, degrades the reliability and security of the system. Actually, it is quite unlikely to find an efficient prevention against eavesdropping due to the open nature of the wireless communication. This gives advantage to adversaries to learn the pattern involved in the network traffic. In addition, sensor data are typically routed along relatively fixed path from the sensor node to the base station, making WSNs even more vulnerable to traffic analysis attacks. Through analyzing traffic gleaned from the network, the adversary can infer all different kinds of information about the underlying network, such as the location of the sensor nodes and base stations, the identity of the user, the communication protocol used, etc.

13.2 Traffic Analysis Attack over the Internet

Two examples that illustrate the effectiveness of traffic analysis attacks on the Internet are presented in the next sections. In both illustrations, the Internet communications are using protocol tunneling tools. Thereby, successful implementations of traffic analysis attack over these "secure" networks further demonstrate the efficiency of this type of attack. By extracting the traffic pattern of the user from the timing information, the adversaries can infer the communication protocol used. This protocol could be the behavioral habit of users, which can be further used to identify the user or even deduce the communication relationship between nodes. It also could be the language used to communicate. These principles are directly applicable to distributed sensor networks as is discussed in Sections 13.2.1 and 13.2.2.

13.2.1 Protocol Detection through Tor Network

When anonymous communication networks were first proposed [1], they are defined by the mixes. The attractive features of Tor, for example, low latency, make it very suitable for common tasks, such as web browsing. In addition, with software-as-a-service becoming mainstream, more and more sensitive user data are delivered through the web. Accordingly, more and more people use Tor for the purpose of keeping their communications anonymously. However, the requirement of low latency makes it easier to confirm the timing signature of traffic (in both directions) over the circuit.

Traffic-monitoring attacks can be performed at different levels, one class of which treats the anonymous Tor network as a "black box" and only considers the times when the data packets step into the anonymous communication system and reaches the application or service outside the Tor network. The work presented in Ref. [2] stages a protocol detection experiment through Tor network based on recent advancements with pattern recognition tools [3,4].

13.2.1.1 Introduction of Tor

Since its introduction in 2002, Tor has made itself one of the most popular anonymous communication networks. As the third-generation onion routing network, Tor is a circuit-based low-latency overlay network, supporting transform of Transmission Control Protocol (TCP) streams. The research of launching effective attacks over Tor, as well as corresponding countermeasures, has become an active topic.

Tor is a low-latency anonymity system that operates as an overlay network on the Internet. An overlay network is a smaller subnetwork that has been built over a larger and usually preexisting network. Individual systems in the overlaid network communicate through virtual links that are physically transmitted by the underlying network but are encapsulated so that they stay logically separated from regular traffic. Examples of this type of network would include a virtual private network (VPN) or a peer-to-peer network, with the Internet being the underlying network.

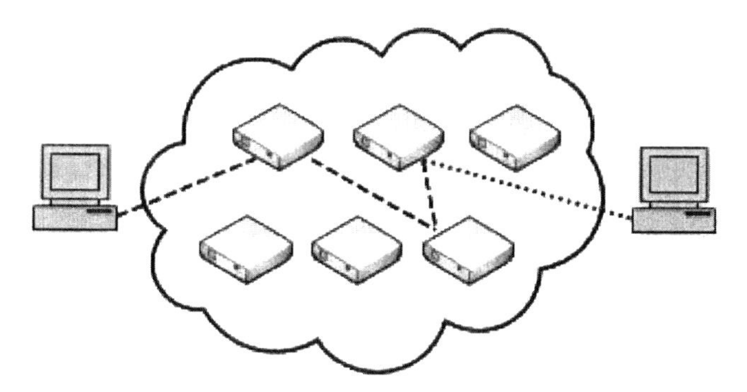

FIGURE 13.1 Example Tor connection.

Tor primarily consists of computers running any number of three types of services: relay, directory server, or client. Relays, also sometimes called nodes, routers, or onion routers, are the backbone of the network. Relays transfer data from clients to other relays, between two relays, or retrieve external resources for a client. By default, relays listen on TCP port 9001 for incoming requests. While they are active, they publish their status to a list of predefined directory servers. Directory servers catalog the information they have about various relays and vote among each other on which ones to list as running, valid, stable, etc. After the servers vote and all agree on a list, they come to a consensus. The consensus is published on a TCP port (9030 by default) where it can be downloaded by clients.

Figure 13.1 shows a typical Tor connection. The client service sets up a listener for TCP streams that the user wishes to route through the network. The listener is a SOCKS [5,6] proxy that listens on port 9050 by default. Before packets arrive to be proxied, the client initializes some circuits with the relays listed in the downloaded consensus. To create a circuit, the client chooses a relay to be the first node in the new circuit and sends it a creation request. Once created, the client uses the same circuit to extend out to a second and then third node. This way, the client chooses and knows each node in the path its traffic will take. Also, this gives the client the opportunity to set up symmetric key pairs with each node for performing encryption. Building the circuit is analogous to extending a spyglass from the source to the destination with the end of each ring being a node of the circuit.

The flowchart in Figure 13.2 summarizes the process a Tor client follows to prepare for a data transfer. Once the first piece of data for a new TCP stream arrives at the proxy, one of the previously created

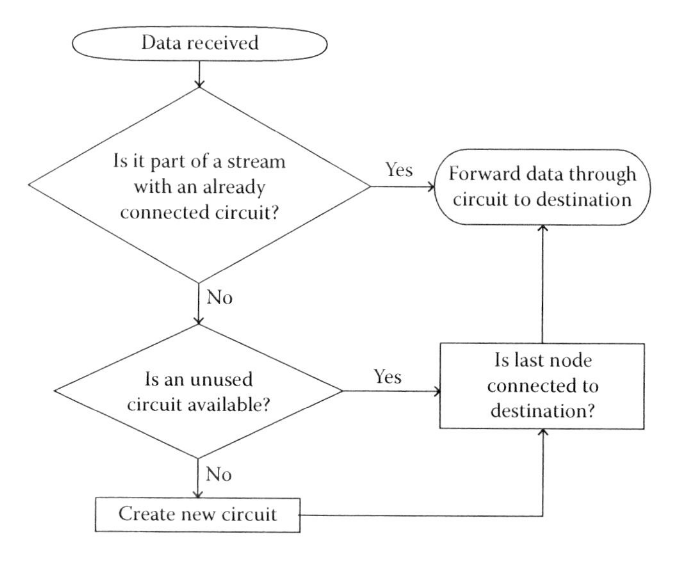

FIGURE 13.2 Preparation for data transfer.

circuits is selected for the transfer. If there are no circuits available, a new one is created. Before any data are transferred, the destination of the packet is sent through the circuit to the last node so that it can start the connection. This node is called the exit node since it is where all of the packets will exit the Tor overlay and continue on to their final destination. The exit node sends confirmation to the client once it has connected to the destination, allowing the data transfer to begin.

As each data packet is ready to be sent, it is split or padded into cells of 512 bytes, which are then iteratively encrypted using the key of each successive relay of the circuit. In other words, the original packet is encrypted with the key for the last node, which is wrapped in another packet and encrypted with the key for the second node, which is encapsulated again and encrypted with the key for the first node. The result is similar in nature to an onion. As each relay in the circuit receives the onion packet, it will decrypt and peel away a layer, forwarding on what remains.

When the destination replies to clients' request, the same process is followed in reverse order. The response packets are successively wrapped in relay cells at each node in the circuit. Once again, the cells are iteratively encrypted by each relay using the key set up for that circuit. When a client gets the response packet, his Tor client unwraps and decrypts each layer to return the original packet from server up through the SOCKS proxy and back to the application he is using. Obviously, there are many more details to the process, such as encryption schemes, integrity checking, congestion handling, and path selection. For more in-depth information on those topics, see the Tor specification [7].

13.2.1.2 Inferring Hidden Markov Models

A zero-knowledge hidden Markov model (HMM) inference algorithm [4] is applied to express the protocol a client uses when he talks to a server through the Tor network by collecting the inter-packet timings on the client. Unlike common HMM inference approaches [8], the approach directly associates state output symbols with state transitions and does not require a priori knowledge of the system.

The algorithm is straightforward. We are given a sequence $\mathbf{y} \in \mathcal{A}^*$ and an "a priori" value $L \in \mathbb{N}$, where $\mathbf{y}$ is a symbolic output of an unknown system and $\mathcal{A}^*$ is the set of all possible sequences. Reference [4] shows how the value of L can be established as part of the HMM learning process. In addition, if $\mathcal{A}^*$ is unknown, readers can refer to Ref. [9] for an alphabet inference algorithm. For values of i increasing from 0 to L, we identify the set of sequences W that are subsequences of $\mathbf{y}$ and have length i (when $i = 0$, the empty set is considered to be a subsequence of $\mathbf{y}$). Conditional distribution of the next symbol following each $x \in W$ is computed using the input data stream $\mathbf{y}$ and partition the subsequences according to these distributions. These distributions become states in the inferred HMM. If states already exist, we compare the conditional distribution of subsequence $\mathbf{x}$ to the conditional distribution of the existing states and add $\mathbf{x}$ to this state if the conditional distributions are found to be statistically equivalent. Distribution comparison is done using a χ^2 test with a specified level of confidence. The level of confidence chosen defines the type I error rate. Once state generation is complete, the states are further split to ensure that the inferred model's transition relation is deterministic. Reconstruction merges states when possible to avoid creating redundant states.

The time between sending each packet depends on the symbol associated with the transition. Each symbol is assigned a specified time delay in milliseconds and the server waits for that amount of time before sending the packet to the client. This technique relates the inter-packet delays to transitions of the HMM. In other words, the time delays between successive packets will be our observations of the underlying process. This is the behavior we expect in real protocols that the packet times will be related to processing required by a specific task in the process.

13.2.1.3 Experiment Configuration

This experiment is designed to be very simple but still required each of the steps to work properly to achieve the desired result. Using a private Tor network, two systems are configured with simple client and server processes. The client connects up to the server and just listens. The server, once the client has connected, sends data packets to the client based on a preloaded model. The model used by the server

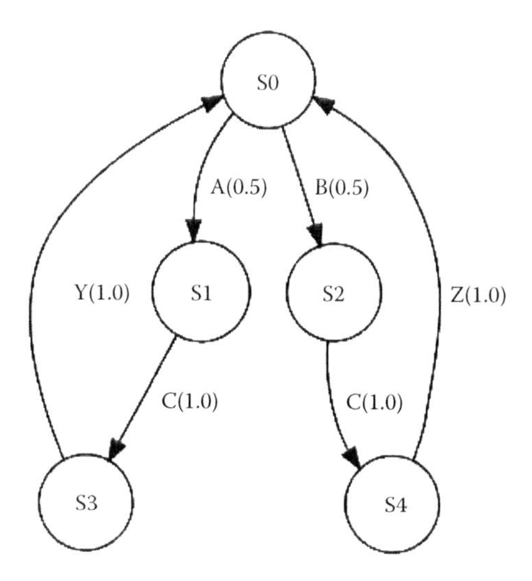

FIGURE 13.3 Original five-state model for the experiment.

in this experiment is shown in Figure 13.3. The server randomly selects a starting state in its model. To send each packet, an outgoing transition is taken with the associated probability. The server waits for that amount of time associated with the symbol of the transition and then sends a packet to the client.

13.2.1.4 Model Construction

To run the attack on Tor, training data are collected from a Tor connection where the protocol being used is unknown. The construction procedure of the HMM required by the attack is described by the flowchart in Figure 13.4. All data collection is done on processes sent through Tor. The TShark [10] program is used to capture packets within the network. Before any collected data can be used by the pattern recognition tools, it must be symbolized. To do this, the difference between each successive packet time ∇t is computed. We then symbolize the data by grouping them into ranges and assigning anything in that range a unique symbol such as *A* or *B*. After symbolizing the data captured from Tor, the zero-knowledge HMM inference algorithm introduced is used to create a model. The process starts with $L = 2$ and increases it as needed. To determine if an increase is needed, the proposed model confidence

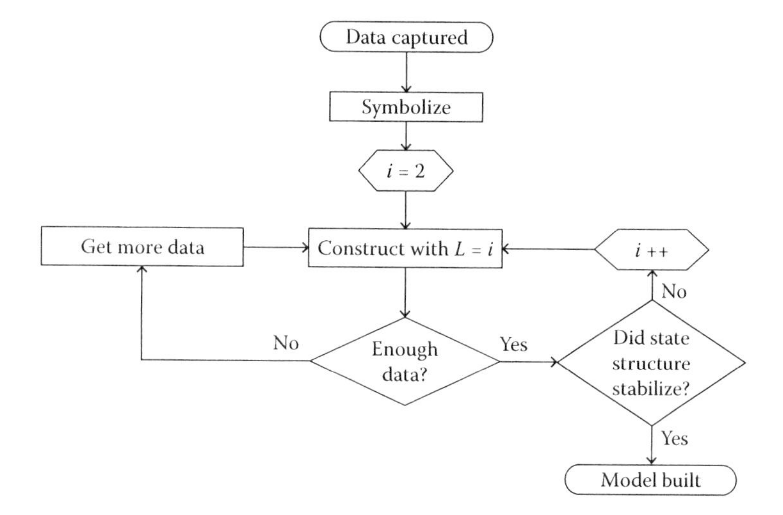

FIGURE 13.4 Flowchart summarizing the model construction process.

test algorithm is first applied, which is an application of the z-test [11] to make sure enough data are used to generate the model. Otherwise, the constructed model may not accurately represent the underlying process. Sometimes enough data (according to the model confidence test) cannot be captured. We will also discuss the reason for this and what can be done in order to still make use of the model.

13.2.1.5 Protocol Analysis

The process is first configured to generate 200,000 packets. After the process completed, differences of the captured packet times are calculated and symbolized. Figure 13.5 shows the model constructed with these first 200,000 packets. The model confidence test [11] is then run on the model, which allows us to determine the statistical significance of the inferred model, which, in turn, indicates whether or not the volume of training data is sufficient. The result shows that it requires 20,624,750 data samples. Because the amount of required data is so large, it has to be generated in lengths of 200,000 packets at a time. After a set of 200,000, we rebuild the model and run the model confidence test (z-test) again. The values computed should become more precise as the volume of observation data increases. Oddly enough, the required amount of data keeps increasing with each set. This is because different mis-symbolizations at different points in the data sequence cause even more states and transitions to be added to the model. All of these new events are of very low probability, which result in a lower minimum asymptotic state probability for each new set. This lower probability causes the confidence test to increase the amount of data required. The smallest asymptotic state probability and corresponding result from the confidence test are plotted against the number of packets captured in Figure 13.7.

The steady increase suggests it will not be easy to capture enough data to rebuild the model confidently. The generated model using all million packets, which contains 79 states and 274 transitions, is shown in Figure 13.6. The solution is to prune the states and transitions that are not statistically significant from the model. Analysis of the asymptotic state probabilities shows a large gap between 71 of the states and the other 8. These 71 states have probabilities below 0.06%, while the other 8 have probabilities above 8.2%. That is a break of over 2 orders of magnitude. This separation makes a good significance level for pruning. Following the pruning process, the model in Figure 13.8 shows the pruning results with a significance level of 0.01 (or 1%). By manually tracing the model, the resulted model of Figure 13.8 is essentially the same as the original model.

If we monitor two points on the network used by the same Tor session, we are able to identify that they are using the same protocol. If we further trace the protocol's path through the Markov model at the two points, it becomes clear that the two points are instances of the same protocol.

13.2.2 Detection of Protocol Tunneling Using Side-Channel Analysis

The traffic analysis attack on Tor presented in Section 13.2.1 takes advantage of the fact that Tor does not make any specific effort to reorder the packets in a flow or introduce extra latency. Luckily, for the "bad" guys, the similar vulnerability also resides in another widely deployed protocol tunneling tool—Secure Shell (SSH). As Song et al. demonstrated, the delays between keystrokes are preserved when using an SSH tunnel [12], so side-channel timing attacks based on traffic monitoring can be carried out to detect behaviors.

A practical application about how to detect tunneled protocols through SSH is presented in Ref. [13], in which HMMs are also used to analyze traffic information. As long as the protocol to be detected can be phrased as a transition process consisting of a finite number of states with stationary state transition probabilities, the approach is valid.

13.2.2.1 Observation Class Identification

The data used in the experiment are delays between specific keystroke pairs and are collected from native speakers [14] and Italians [15]. The HMM inference is based on the conditional probabilities inherent in English and Italian. For example, it is more likely to see a letter "u" showing next other than a letter "z" after the letter "q" is typed. The statistics of delays between keystrokes includes alphabets,

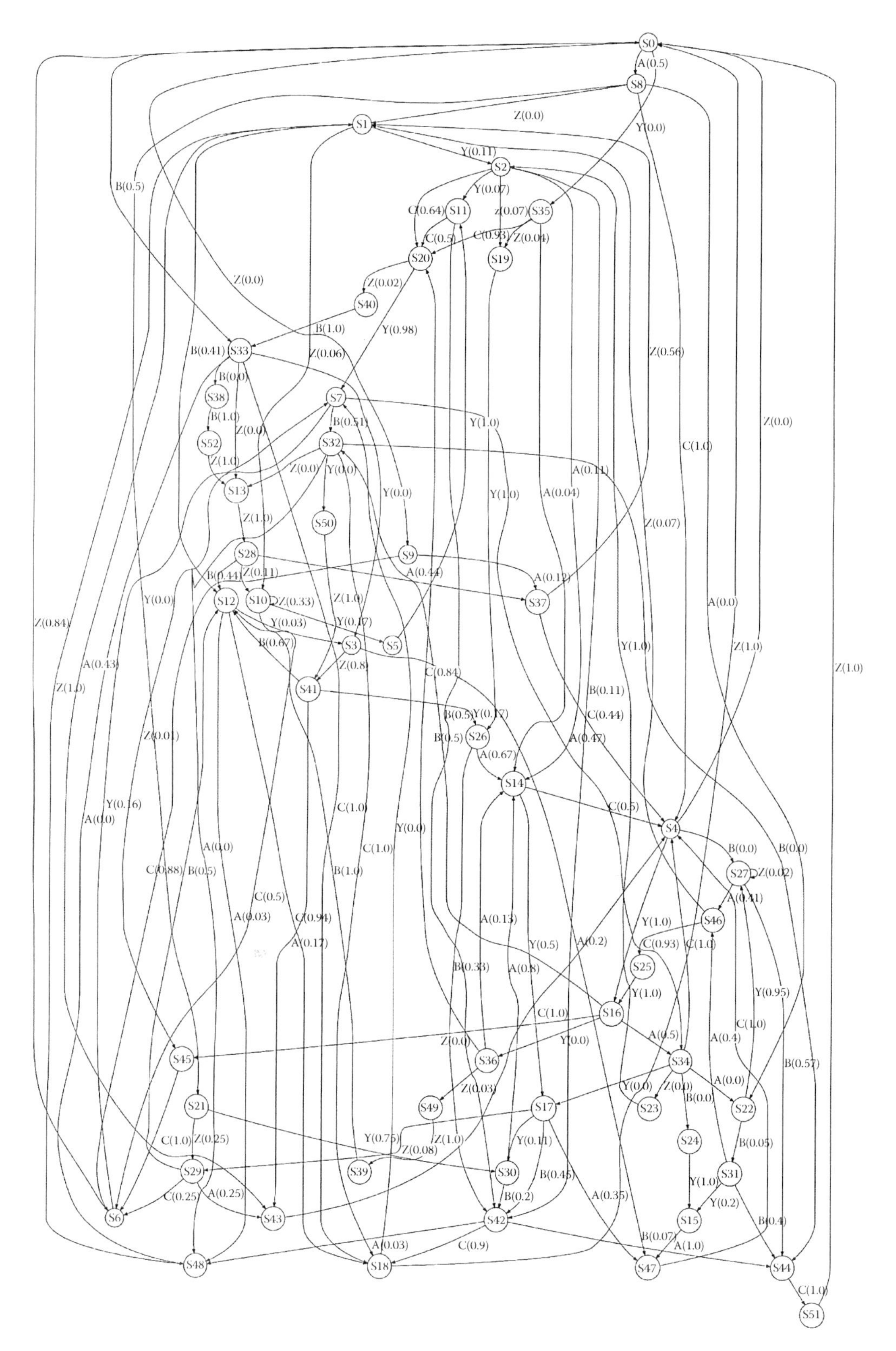

FIGURE 13.5 Model that is reconstructed from first 200,000 packets of captured data.

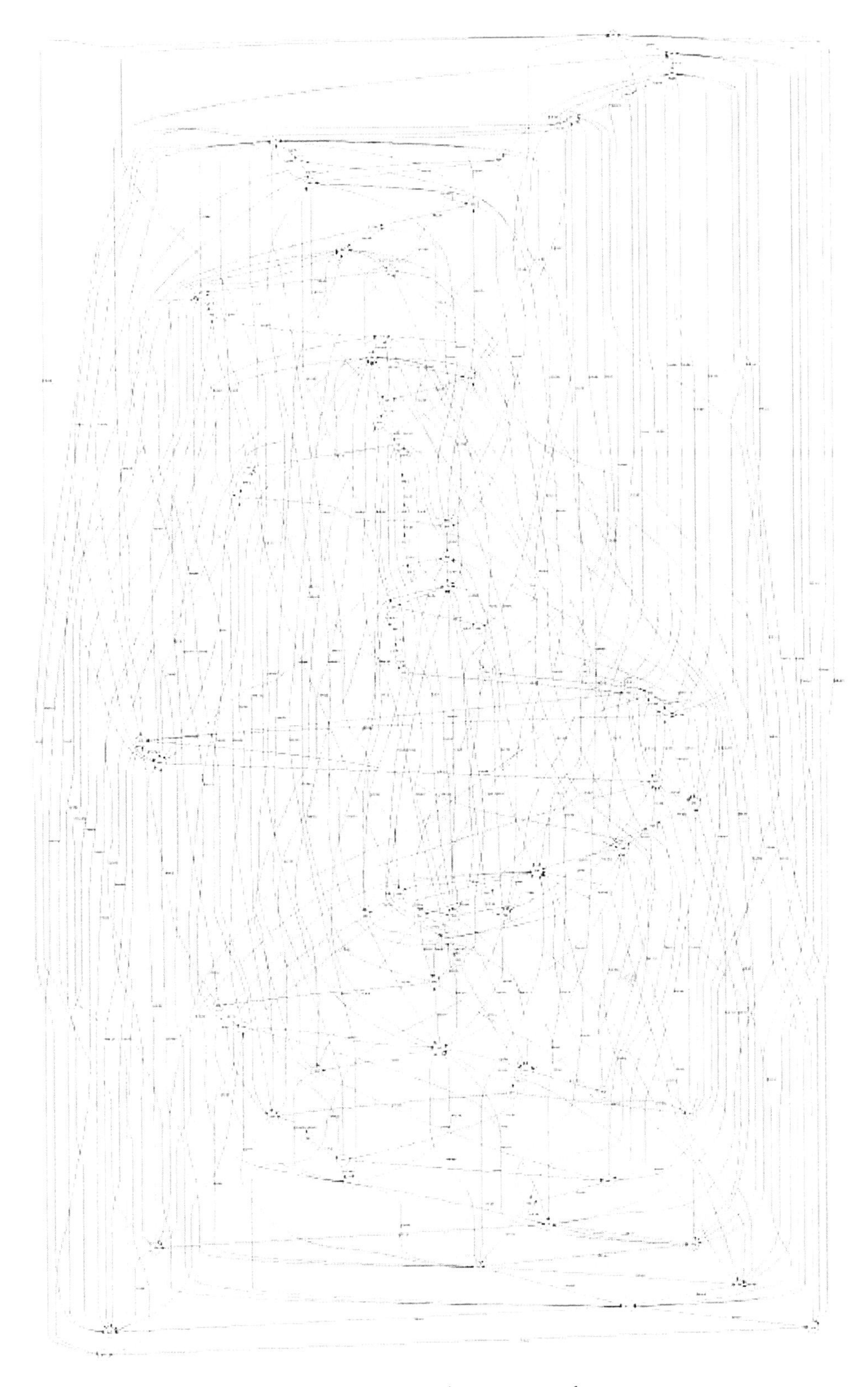

FIGURE 13.6 Plot of model confidence results as more data are captured.

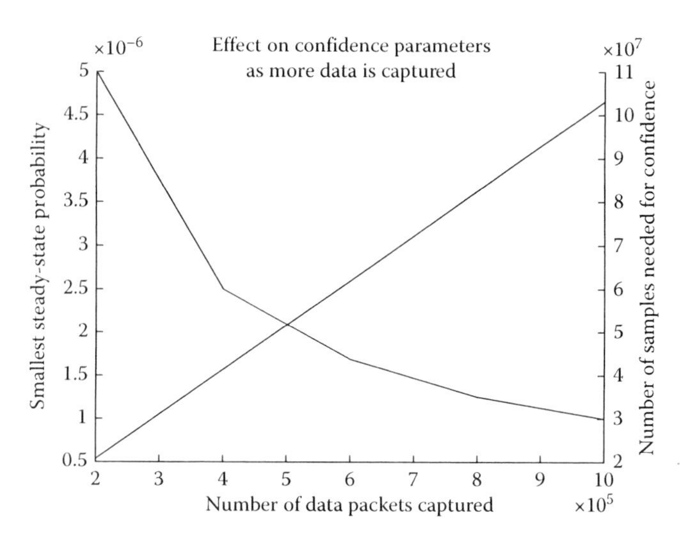

FIGURE 13.7 Model that is reconstructed from all million packets of captured data.

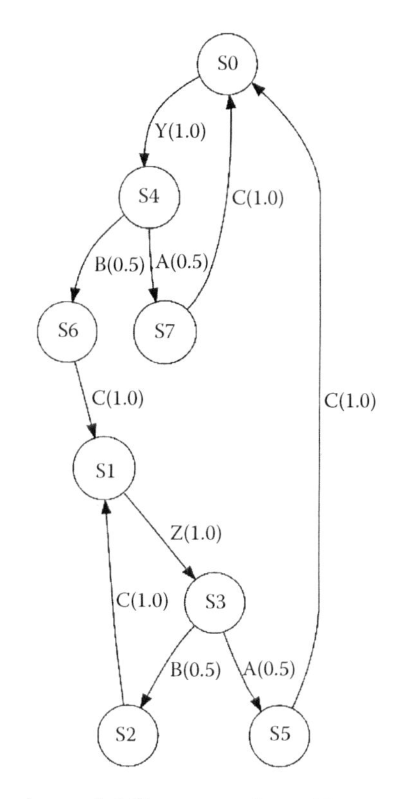

FIGURE 13.8 Result after pruning low-probability states and transitions.

numerals, enter, space, and backspace; a total of 39 characters as case are ignored. These values are used to populate a 39-by-39 delay matrix.

The training data set [14,15] does not include statistically significant samples of all key pairs. Interpolation was performed for such key pairs. For example, if the pair "AO" does not exist, the delays of its neighbors "AI," "AK," "AL," and "AP," if they exist, are averaged. This is similar to the process applied to images to remove spots, where neighboring pixel values are averaged to create the "missing" value.

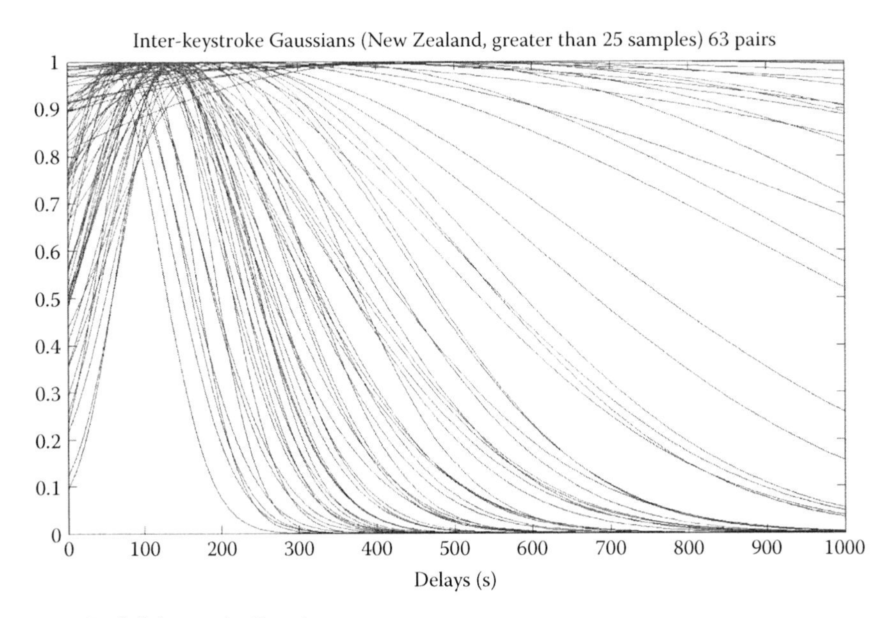

FIGURE 13.9 English keystroke Gaussian.

If sufficient data are present for the neighbor list of the destination key, the missing value is updated with the average of the neighboring key values. If insufficient data are present, however, the destination key is held constant and the neighbor list of source key is consulted. This process is repeated until the matrix remained constant across two passes.

After enough data are collected, we need to extract finite classes of observations from the continuous timing data because the HMM inference approach introduced earlier is based on symbolic observations [4]. The normal distribution of inter-keystroke delay of each key pair is plotted using the mean and variance extracted from the data. These plots are shown in Figures 13.9 and 13.10. A clustering approach named growing neural gas (GNG) [16] is adopted to determine distinct classes of key pair delays.

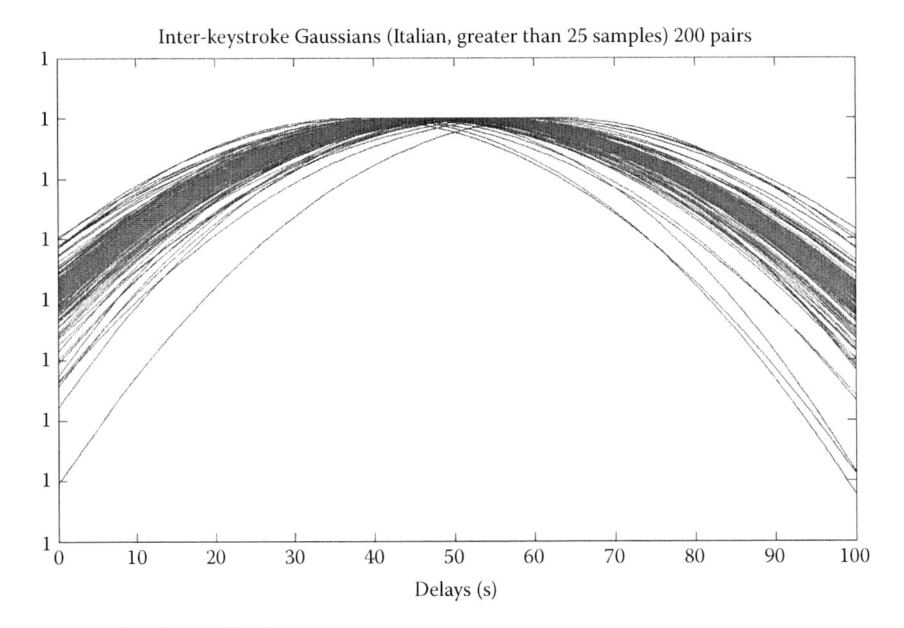

FIGURE 13.10 Italian keystroke Gaussian.

TABLE 13.1 Symbolization of English Keystroke Statistics

Lower Bound	Upper Bound	Mean	Symbol
0.00	125.00	95.14	A
126.00	182.00	153.17	B
183.00	236.00	209.04	C
237.00	287.00	261.29	D
288.00	329.00	311.21	E
330.00	364.00	345.07	F
365.00	414.00	382.01	G
415.00	494.00	445.05	H
495.00	625.00	541.29	I
626.00	10,000.00	707.73	J

TABLE 13.2 Symbolization of Italian Keystroke Statistics

Mean	Lower Bound	Upper Bound	Symbol
15.32	0.00	28.00	A
38.88	29.00	45.00	B
49.98	46.00	59.00	C
67.19	60.00	10000.00	D

The results are shown in Tables 13.1 and 13.2, respectively. Four clusters are identified for Italian and 10 for English. Bounds are determined by locating the midpoint between means. These bounds are then associated with symbols. All values in the table are in milliseconds. An upper bound of 10 s is used to prevent any symbols from being identified as null. The means are of similar inter-keystroke delays clustered by GNG. Boundaries are midpoints between neighbor means. Different capital letters are assigned to each range. If an inter-keystroke delay falls into a certain range determined by lower and upper bounds, it is symbolized by the assigned letter.

13.2.2.2 HMM Construction for English and Italian

The same method proposed in Ref. [4] is adopted to construct the HMM, representing the protocol tunneled through SSH. Training data for HMM construction were selected from texts available at Project Gutenberg [17]. Recent, 1900 or later, texts are taken and preprocessed to remove cases and special characters. Delays are assigned to each pair of successive remaining characters by using the previously constructed delay matrix as a lookup table. After all the delays are symbolized using Tables 13.1 and 13.2, they are divided into a testing set and training set. The constructed HMMs for English and Italian are shown in Figures 13.11 and 13.12, respectively.

The training data set used to construct English HMM contains approximately 1.1 million key pairs. However, the algorithm proposed in Ref. [11] indicates that creating a significant model for $L = 1$ would have required a training set of over 11 million key pairs. This is due to the existence of a number of low-probability transition events. In contrast, Italian only had four clusters, which makes it possible to achieve a statistically significant reconstruction with $L = 3$ using a similar volume of training data. There are a smaller number of possible transitions from each state, which results in fewer low-probability state transitions. The training process has a comparatively larger sample set available for determining probability distributions.

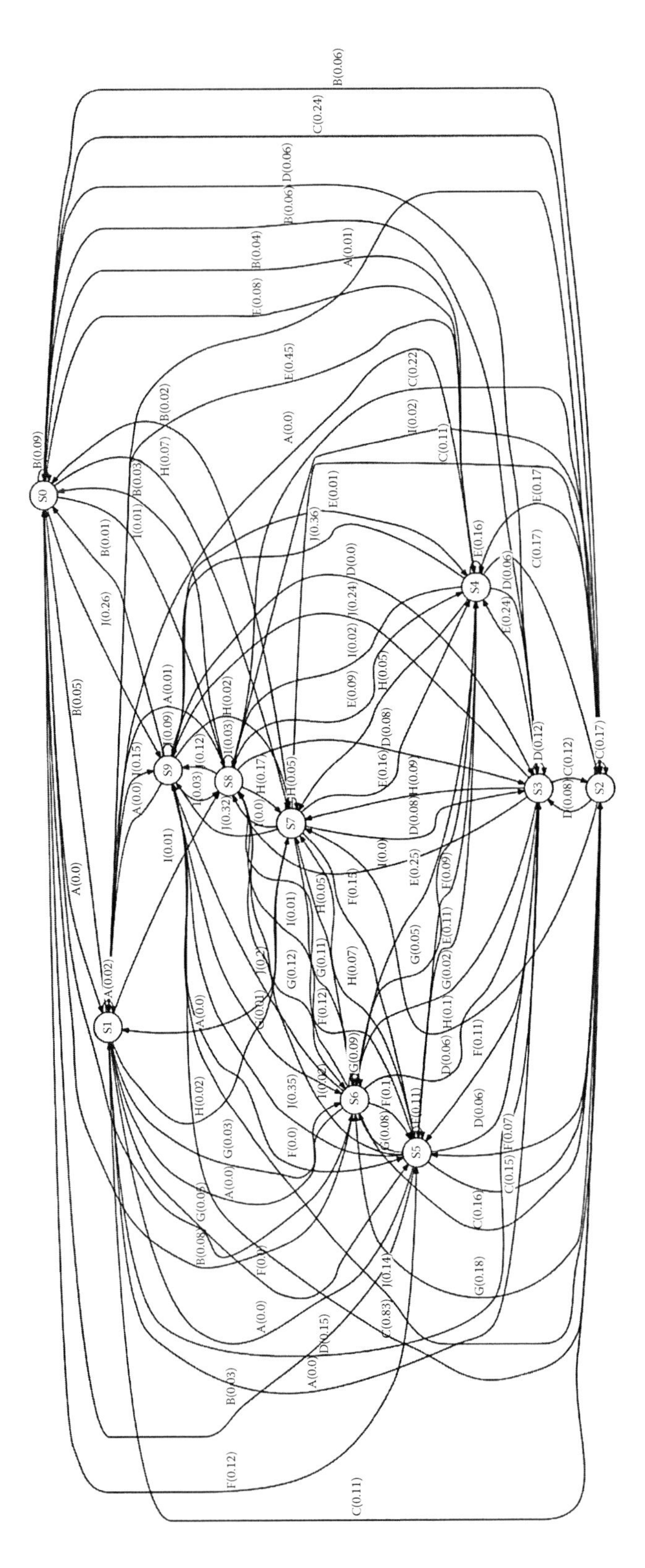

FIGURE 13.11 English HMM (10 states, 100 transitions).

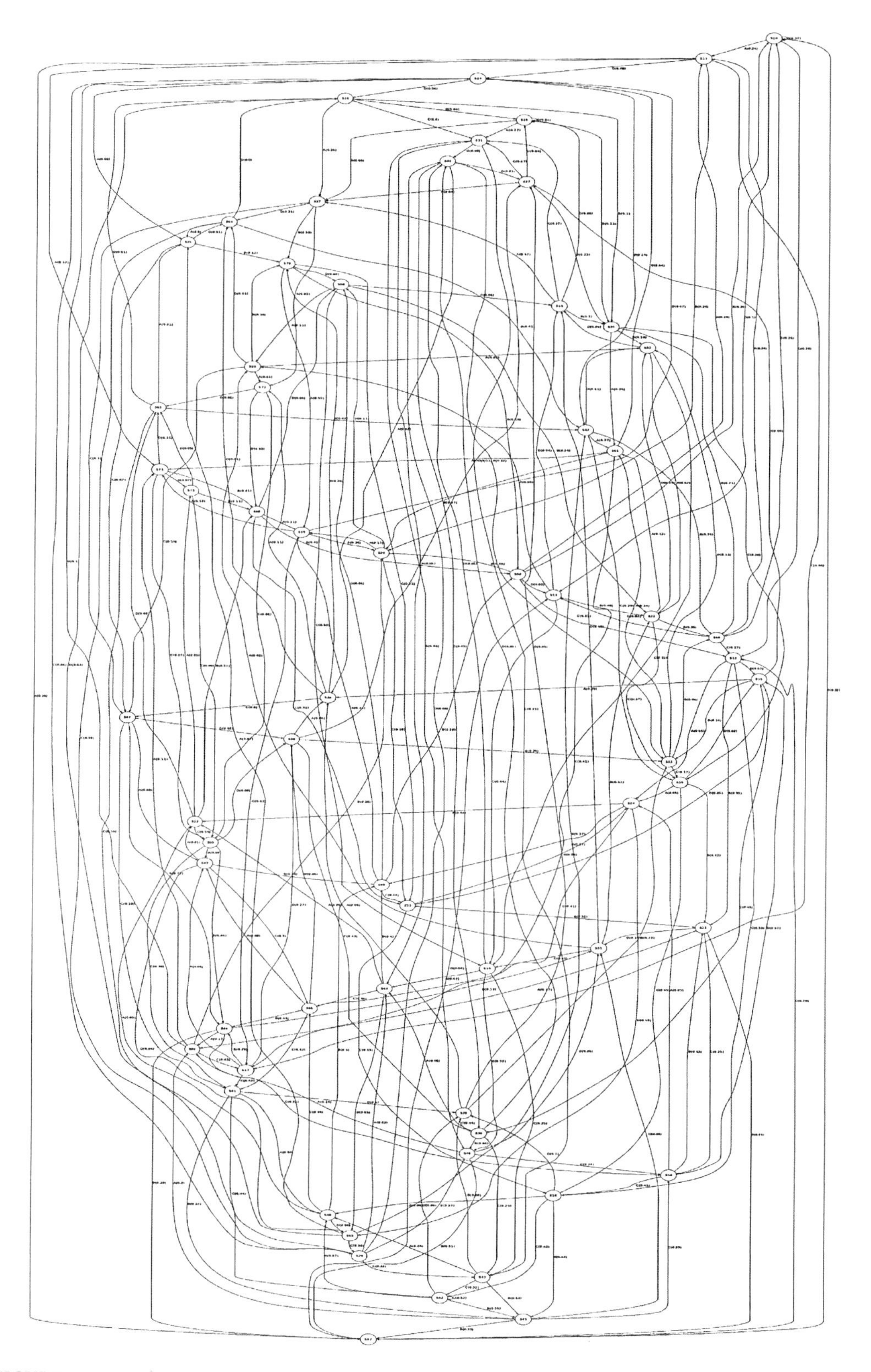

FIGURE 13.12 Italian HMM (64 states, 253 transitions).

13.2.2.3 Protocol Detection Using Model Confidence Interval

However, unlike traditional HMM inference approaches, the model inferred using Ref. [4] does specify a starting state. To test if a given observation sequence is generated by the inferred HMM, all possible starting states have to be taken into account. The confidence interval (CI) approach from Ref. [3] is used to solve this detection problem. It counts the number of times a particular state is entered and creates CI bounds for each exiting transition by dividing exiting transition counts by the entering transition count. The CI for the transition in question can then be found by solving Equation 13.1:

$$\left[p_{i,j} - Z_{\alpha/2}\sqrt{\frac{p_{i,j}(1-p_{i,j})}{c_i}},\ p_{i,j} + Z_{\alpha/2}\sqrt{\frac{p_{i,j}(1-p_{i,j})}{c_i}}\right] \tag{13.1}$$

where

$p_{i,j}$ is the transition probability from state i to state j
c_i is the entry counter for state i
Z is taken from the standard normal distribution

These probabilities are known to us, as we construct the models.

If the estimate falls within the CI defined by Equation 13.1, we accept the assumption that the given symbolic sequence is generated by the HMM with a false-positive rate of α. Otherwise, the assumption is rejected.

To map transition acceptance or failure across the model, the receiver operating characteristic (ROC) curves [3] are used to determine the ideal threshold for acceptance of false positives. This is done by identifying the point on the curve nearest to the point (0,1), corresponding to 0% false positives and 100% true positives. By allowing a false-positive rate equal to the threshold value, the true-positive rate is maximized. Consequently, if the rejection rate exceeds this threshold, the model is similarly rejected. This is because more false positives are encountered than ideal. However, if the acceptance rate passes this threshold, the model is accepted as a valid source for the presented symbol sequence.

The window-size calculation [18] shows that the minimum string length needed to differentiate between the two models, with a 95% true-positive rate, is 77 symbols. We therefore divide the testing data into sets of 77 symbols. A set of 800 English and Italian windows (400 for English and 400 for Italian) are chosen to test the ability of the constructed HHMs to detect the language being used in interactive SSH sessions.

13.2.2.4 Language Detection Results

The testing data are sent through interactive SSH connections following the keystroke pair delay distributions described in Section 13.2.2.1. The detection procedure redirects the output of a parsed TShark [10] capture to the constructed English and Italian HMMs with maximum likelihood [8] and CI detection criteria. The ROC curves for the tests are shown in Figures 13.13 through 13.16, where circular points compare English and Italian data streams; square points are for Malagasy (the national language of Madagascar) data streams; and diamond points are for English (Italian) data transmitted with Italian (English) timing.

From the plots denoted by circular markers in the ROC curves, it is clear that both the CI and maximum likelihood approaches are able to detect the language used. It is also found that with strings of 77 symbols, the optimal thresholds for acceptance are 89.0% for English detection and 0.0% for Italian detection when HMMs are used.

From the procedure of inferring observation classes, it is obvious that the range of Italian keystroke delays is a subset of English. All English inputs produce impossible transitions within 77 symbols while Italian text never produces probabilities outside the 95% CI. No impossible transition occurs when Italian text is parsed by the English Markov model. Also, the English conditional probabilities are less homogeneous. When English text is parsed, it would often produce observed transition probabilities outside the

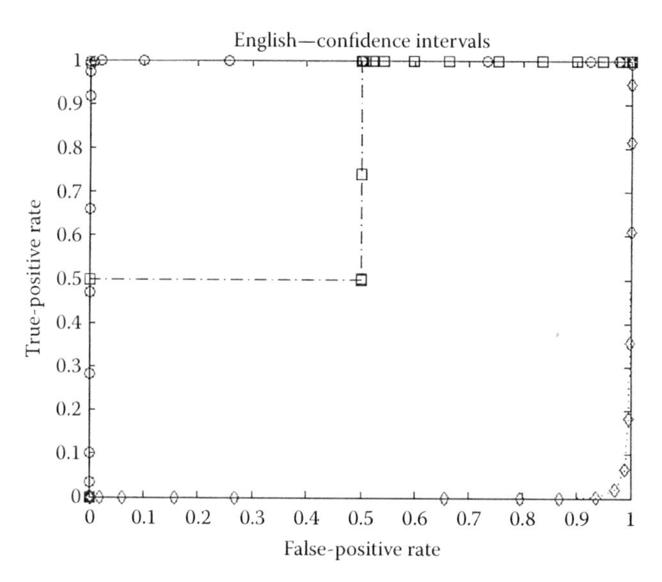

FIGURE 13.13 English CI results English versus Italian (circle), cross-symbolization (diamond), and Malagasy (square).

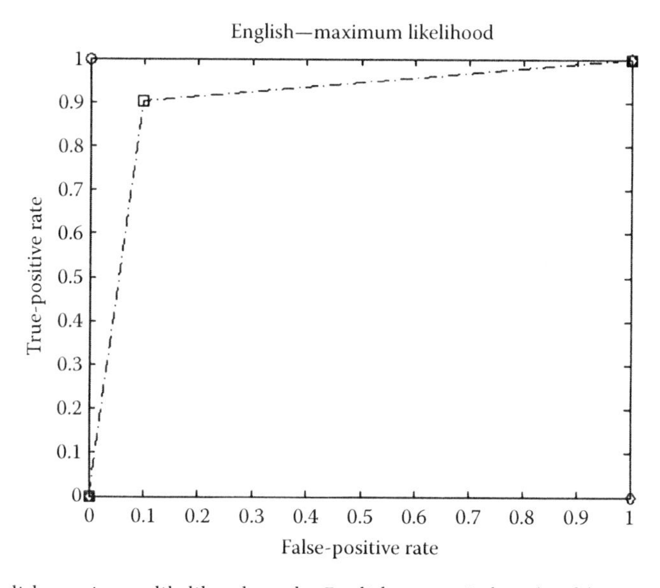

FIGURE 13.14 English maximum likelihood results English versus Italian (circle), cross-symbolization (diamond), and Malagasy (square).

95% CI. While this is to be expected approximately 5% of the time, our observations can be explained, in part, as an artifact of using the $L = 1$ approximation of the true process. However, the conditional probabilities in Italian text are quite different from English conditional probabilities, explaining the optimal 89% threshold, as shown in Tables 13.3 and 13.4. This threshold is able to reliably differentiate between the two languages as shown in Figure 13.13. The last column in each table, "Dis," is the distance from the curve at that point to the point (1,0) on the axes. The ideal threshold is reached when this distance is minimized. (shown in bold italics in Tables 13.3 and 13.4).

In cross-symbolization, English is symbolized with the Italian delay statistics and the Italian symbol space and vice versa. This is done to see which of the two phases of our process (symbolization or

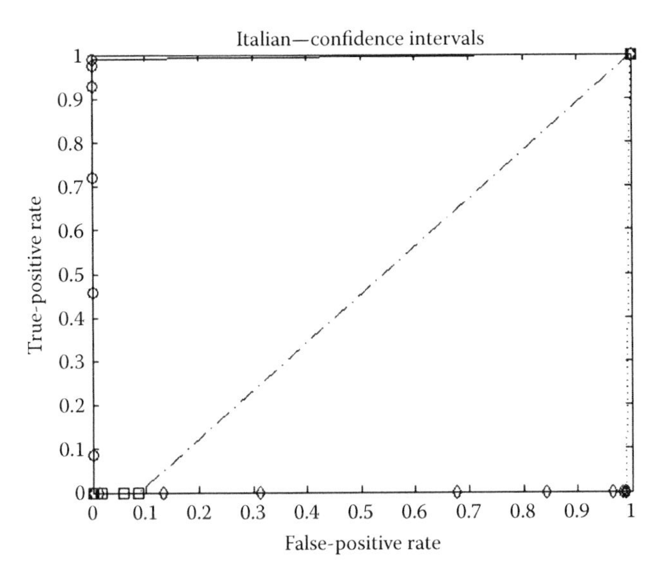

FIGURE 13.15 Italian CI results English versus Italian (circle), cross-symbolization (diamond), and Malagasy (square).

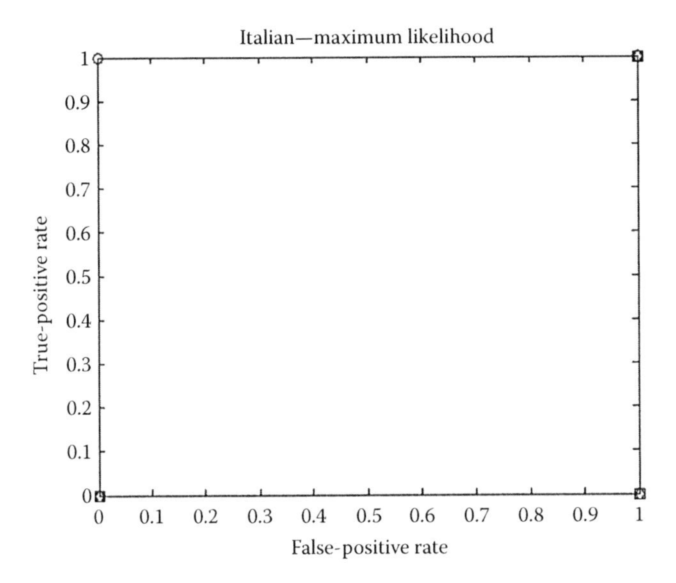

FIGURE 13.16 Italian maximum likelihood results English versus Italian (circle), cross-symbolization (diamond), and Malagasy (square).

HMM parsing) dominates the process. When the symbolizations are switched, the opposite language is identified. That is, Italian is identified in the case of English. From the results, shown by the plots with diamond markers in Figures 13.13 through 13.16, it is clear that the symbolization process dominates the approach. However, it is not clear if the behavior identified is a function solely of the symbolization or also due to language structure. Note that while English is a Germanic derivative language and Italian is a Latin derivative, both are in the Indo-European family and hence have substantial similarities.

To address this issue, the modern languages are compared with ancestor languages. This experiment shows that when Old English and Latin are symbolized in a similar way to their younger successors, they are detected as their younger counterparts. It appears that related languages can be identified using the proposed approach. This further supports the hypothesis that detection outcomes primarily

TABLE 13.3 ROC Statistics for English versus Italian

Th	TP	FP	TN	FN	Dis
0.00	401	401	0	0	1.000
		Repeated 79 times			
0.80	401	401	0	0	1.000
0.81	401	392	9	0	0.978
0.82	401	371	30	0	0.925
0.83	401	294	107	0	0.733
0.84	401	201	200	0	0.501
0.85	401	103	298	0	0.257
0.86	401	40	361	0	0.100
0.87	401	9	392	0	0.022
0.88	399	3	398	2	0.009
0.89	***399***	***0***	***401***	***2***	***0.005***
0.90	397	0	401	4	0.010

TABLE 13.4 Cross-Symbolization with CIs

Th	TP	FP	TN	FN	Dis
0.00	401	401	0	0	1.000
		Repeated 78 times			
0.80	401	400	1	0	0.998
0.81	401	400	1	0	0.998
0.82	401	371	30	0	0.925
0.83	401	297	104	0	0.741
0.84	401	195	206	0	0.486
0.85	401	105	296	0	0.262
0.86	401	37	364	0	0.092
0.87	400	7	394	1	0.018
0.88	***395***	***0***	***401***	***6***	***0.015***
0.89	389	0	401	12	0.030

depend on the procedure of symbolization. The final experiment uses Malagasy, the national language of Madagascar. Malagasy is selected as it uses no diacritical marks, can be represented with the Latin character set, and does not originate from Sanskrit (as English and Italian do); it is a member of the Austronesian language family. The only text available in Malagasy is a copy of The Bible [19]. This test produces curious results. For this experiment, the book of Genesis is symbolized with both the English and Italian statistics and symbol spaces. These symbolized strings are then analyzed using CIs and maximum likelihood. The results of these comparisons are presented in the plots with square markers in Figures 13.13 through 13.16. From these curves, it appears that detection is strongly influenced by language structure and not solely symbolization as previously thought.

At first glance, it appears that the ROC curves in Figures 13.13 through 13.16 favors the existing maximum likelihood measure over the CI approach. Since the strings used for are only 77 symbols long. Floating-point underflow might be if longer strings are implemented. Methods that can be used to avoid this, such as normalization at every step and the use of logarithms, would introduce more noise into an already noisy calculation.

It should also be noted that the CI approach is for detection, not for classification as maximum likelihood is. CI methods indicate the presence of a specific behavior in a given sample string. It can be used as a classifier, but that is not its intended use. As said earlier, to do so would require ROC curve inspection to determine a suitable threshold between behaviors.

In Ref. [3], Schwier et al. pointed out that with CIs there is a marginally higher false-positive rate. This is due to less noise being introduced than with maximum likelihood. Consequently, it becomes harder to make a decision. It should be noted that CI analysis can be performed online. This is not the case for maximum likelihood testing. Furthermore, presenting the data in windows is necessary for online use and for the differentiation between languages. While this windowing is not needed for maximum likelihood, it can be applied to it.

13.3 Transplantation of Traffic Analysis Attacks into WSNs

Figure 13.17 shows the general classification of network attacks and the classification of attacks on WSNs is given in Figure 13.18 [20]. It is not difficult to discern that these two dendrograms are very alike. The majority of general network attacks can find their counterparts in security attacks against WSNs. Traffic analysis is one of them.

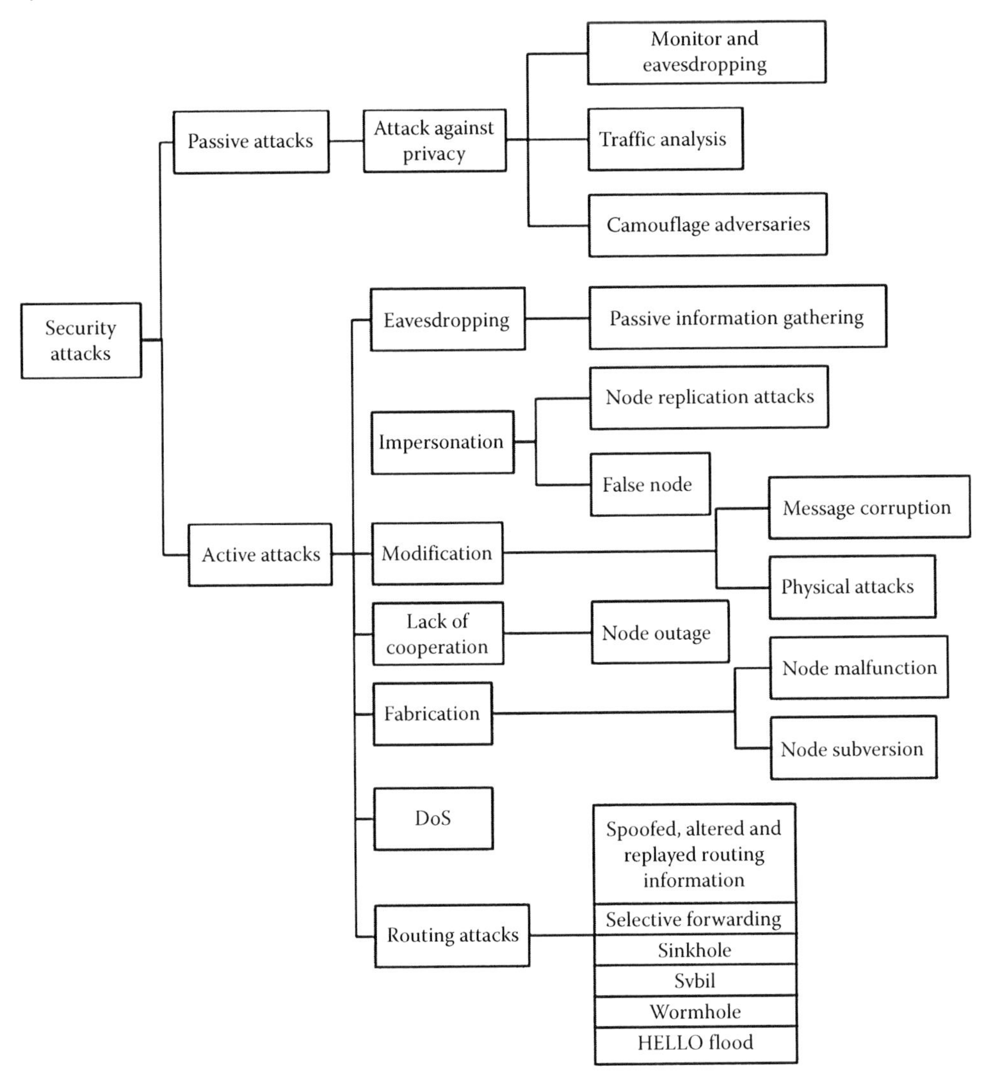

FIGURE 13.17 General classification of security attacks.

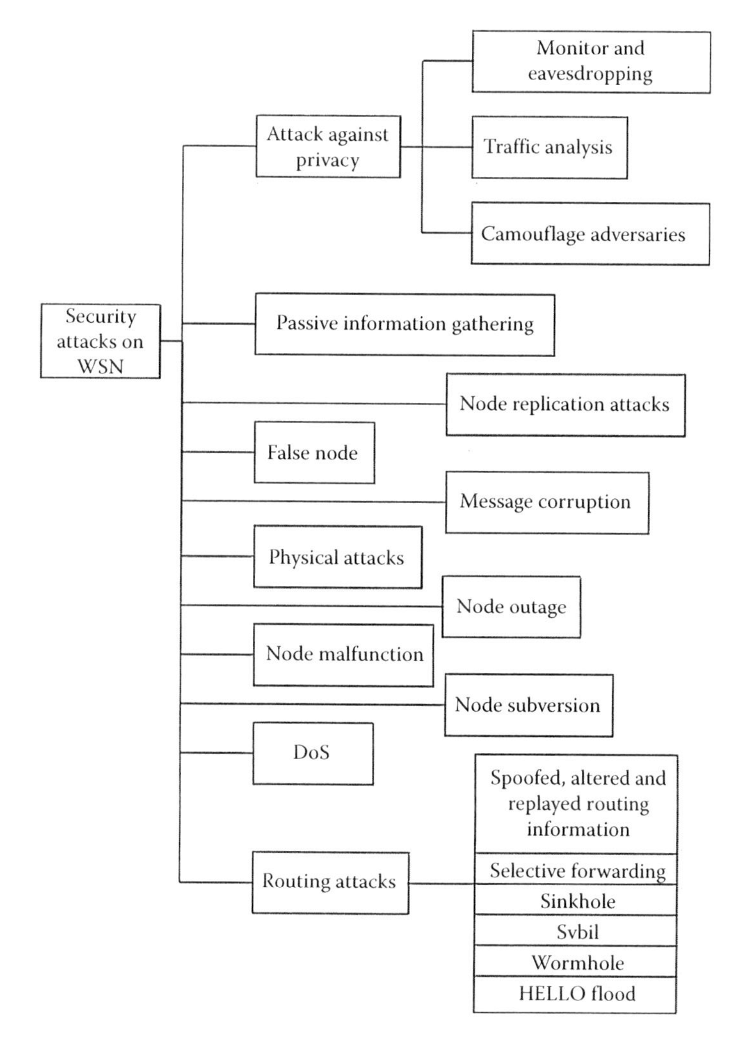

FIGURE 13.18 Classification of security attacks on WSN.

As shown in Figure 13.18, there are a wide assortment of attacks against WSNs. The security mechanisms that should be used to protect the network depend on the context and the requirements of a specific application. Generally, the objective of a security mechanism should fulfill one or multiple of the following functions: attack prevention, attack detection, and self-healing of infected portions. Considering the services offered by WSNs and their deployment fields, it is quite unlikely to come up with an effective means of protection against traffic analysis attacks.

In this section, we explain how to develop the attacks discussed in Section 13.2 over WSNs. The analysis would reveal that sensor networks are especially vulnerable to external and internal attacks due to their peculiar characteristics. We discuss about this in the following sections.

13.3.1 Eavesdropper Placement in WSN

In both attacks presented in Section 13.2, we assume that the attackers can access information exchange by eavesdropping on both end points of the session. This is an even more likely assumption in the case of WSNs because WSNs are usually deployed in critical situations. Considering their deployment fields (e.g., hostile environments as battle field) and the services offered by WSNs (e.g., continuous monitoring of their surroundings), adversaries can easily gain physical access to sensor nodes in many cases. Malicious users or machines can therefore reprogram the nodes or simply destroy them.

The majority of communications within sensor networks use wireless medium, such as RF spectrum, making the communications public to anyone. As a result, even if the sensor nodes are resistant to physical attacks, WSNs are still subject to various types of software-based remote attacks [21]. Any internal or external device can have access to the data transferred in the field. Compared with wired networks like Ethernet, signal broadcast over airwaves can be more easily intercepted with receivers tuned to the proper frequency. In this sense, traffic traveling through WSNs are more vulnerable to malicious or unauthorized eavesdroppers.

Furthermore, the traffic analysis attacks introduced in Section 13.2 can be classified as internal attacks according to the domain of the attacks. The attackers have to be "insiders" or have some nodes compromised in order to monitor the traffic flow tunneling through either Tor or SSH. However, attackers do not have to have access to the network elements and services to reproduce the same attacks over WSNs. Traffic monitoring can be carried out by nodes that do not belong to the domain of the network by taking advantage of the broadcast vulnerability.

Also, sensor nodes are usually not mobile. Once the adversarial monitors have gained access to the exchange traffic, the target network might be exposed to a long period of information leakage.

Overall, the broadcast nature of the transmission medium of WSNs intensifies the privacy problem because it makes large volumes of information easily available through remote access.

13.3.2 Sensitive Information of WSNs

The protocol detection through Tor described in Section 13.2.1 is based on the notion that Tor only minimally hides the end-to-end timing correlations since the timing signature of an anonymized stream is hardly distorted by low latency. The same applies to WSNs. Most sensor network applications rely on some form of time synchronization. Also, sensors may wish to compute the end-to-end delay of a packet as it travels between two pairwise sensors. A more collaborative sensor network may require group synchronization [22] for tracking applications. Therefore, timing information can also be used as a communication pattern for traffic analysis attack on WSNs.

In addition to timing information, there is other information that can be used or considered to be sensitive, such as the network identity of the communicating parties, the frequency of communications, the size of the messages, and the location and time at which the sensor's measurements are being sent. Techniques like time stamping, padding, using serial numbers, or, frequently, key redistribution can be used to prevent these sensitive patterns from being observed by third parties [23]. However, concealing these patterns is not trivial for sensor networks since the devices of the network, that is, sensor nodes, are constrained in terms of computational capabilities, memory, communication bandwidth, and battery power. As a result, great limitations are posed on the cryptographic algorithms and protocols. It is unrealistic for sensor networks to use techniques like onion routing used in Tor. Multiple layers of asymmetric encryption operations require severe computing overhead. Cryptographic schemes requiring excessive computational overhead are not allowed in WSNs.

Furthermore, an adversary can also monitor some other physical properties of the nodes, such as power consumption and electromagnetic emanation [24] caused by performing encryption. Given enough EM signal collected, if the recorded values are influenced by the secret key, the adversary can extract information about that key. These attacks are independently classified as side-channel attacks because the properties used for attack are side-channel information gained from the application of cryptography. This type of attacks are similar to traffic analysis attacks as they also utilize sensitive informations rather than seeking to decode cipher texts.

13.3.3 Further Attacks Based on Traffic Analysis

As we all know, a WSN is a typical ad hoc network. The link-layer protocols of WSNs ensure one-hop connectivity between neighbors, which is then extended by the network layer protocols to other nodes in the network. Based on this infrastructure, the server–client correlation through traffic analysis can be further used to identify the path a particular node used to send its data through. Due to the transmission range limit of sensor nodes, one can easily locate all the neighbors of a target node. Correlation based on traffic analysis can then be used to find the next hop in the path used to deliver its collected data to the base station.

As a passive attack, the acquired information from traffic analysis attacks can be used to perform active attacks. Once the adversaries acquire enough information about the underlying WSN, they can create fake events or hide problematic situations and can even introduce bogus control information [21]. Attacks of this type include message modification, message replay, message injection, etc. The effects of active attacks are destructive.

Acknowledgments

This material is based upon the work supported in part by the Air Force Office of Scientific Research contract/grant number FA9550-09-1-0173, NSF grant EAGER-GENI Experiments on Network Security and Traffic Analysis contract/grant number CNS-1049765, and NSF-OCI 1064230 EAGER: Collaborative Research: A Peer-to-Peer based Storage System for High-End Computing. Opinions expressed are those of the author and neither the National Science Foundation nor the U.S. Department of Defense.

References

1. D. Chaum. Untraceable electronic mail, return addresses, and digital pseudonyms. *Communications of the ACM*, 24(2):84–88, 1981.
2. R. Craven. MS thesis, *Traffic Analysis of Anonymity System*, Holcombe Department of Electrical and Computer Engineering, Clemson University, Clemson, SC, May 2010.
3. R. R. Brooks, J. M. Schwier, and C. Griffin. Behavior detection using confidence intervals of hidden Markov models. *IEEE Transactions on System Man and Cybernetics, Part B: Cybernetics*, 39(6):1484–1492, December 2009.
4. J. M. Schwier, R. R. Brooks, C. Griffin, and S. Bukkapatnam. Zero knowledge hidden Markov model inference. *Pattern Recognition Letters*, (30):1273–1280, 2009.
5. D. Koblas and M. R. Koblas. SOCKS. In *Proceedings of the 3rd USENIX Security Symposium*, Baltimore, MD, pp. 77–83, 1992.
6. M. Leech, M. Ganis, Y. Lee, R. Kuris, D. Koblas, and L. Jones. Socks protocol version 5. RFC number 1928, March 1996.
7. R. Dingledine, N. Mathewson, and P. Syverson. Tor: The second-generation onion router. In *Proceedings of the 13th USENIX Security Symposium*, San Diego, CA, pp. 303–320, August 2004.
8. L. R. Rabiner. A tutorial of hidden Markov model and selected application in speech recognition. In *Proceedings of the IEEE*, 77:257–285, February 1989.
9. C. Griffin, R. R. Brooks, and J. Schwier. A hybrid statistical technique for modeling recurrent tracks in a compact set. *IEEE Transactions on Automatic Control*, 56(8):1926–1931, 2010.
10. Gerald Combs. *TShark Manual*. http://www.wireshark.org/docs/man-pages/tshark.html
11. L. Yu, J. M. Schwier, R. M. Craven, R. R. Brooks, and C. Griffin. Inferring statistically significant hidden Markov models. *IEEE Transactions on Knowledge and Data Engineering, Accepted for Publication*, 2012.

12. D. X. Song, D. Wagner, and X. Tian. Timing analysis of keystrokes and timing attacks on SSH. In *Proceedings of the 10th Conference on USENIX Security Symposium*, Berkeley, CA, p. 25, 2001.
13. H. Bhanu, J. Schwier, R. Craven, K. Hempstalk, D. Gunetti, C. Griffin, and R. R. Brooks. Side-channel analysis for detecting protocol tunneling. *Advances in Internet of Things*. 1(2):13–26, 2011.
14. K. Hempstalk. Continuous typist verification using machine learning. PhD thesis, Department of Computer Science, University of Waikato, Hamilton, New Zealand, 2009.
15. D. Gunetti and C. Picardi. Keystroke analysis of free text. *ACM Transactions on Information and System Security*, 8(3):347–358, 2005.
16. B. Fritzke. Fast learning with incremental RBF networks. *Neural Processing Letters*, 1(1):2–5, 1994.
17. Gutenberg Project. Gutenberg. http://www.gutenberg.org/
18. J. Schwier, R. R. Brooks, and C. Griffin. Methods to window data to differentiate between Markov models. *IEEE Transactions on System Man and Cybernetics, Part B: Cybernetics*, 41(3):650–663, August 2010.
19. Adabibliq. http://www.madabibliq.org/
20. G. Padmavathi and D. Shanmugapriya. A survey of attacks, security mechanisms and challenges in wireless sensor networks. *International Journal of Computer Science and Information Security*, 4(1):313–320, 2009.
21. R. Roman, J. Lopez, and C. Alcaraz. Analysis of security threats, requirements, technologies and standards in wireless sensor network. In *Foundations of Security Analysis and Design V, LNCS 5705*, Bertinoro, Italy, pp. 289–388, 2009.
22. Y. Sankarasubramaniam, I. F. Akykildiz, W. Su, and E. Cayirci. A survey on sensor networks. *IEEE Communications Magazine*, 38(4):393–422, 2002.
23. E. Aivaloglou, S. Gritzalis, and S. Katsikas. Privacy protection mechanisms in sensor networks in C. Lambrinoudakis, L. Mitrou, S. Gritzalis, S. Katsikas (Eds). *Privacy Protection and Information and Communication Technologies*, Papasotiriou Publishers, Athens, Greece, pp. 221–245, 2010. (In Greek).
24. J. R. Rao, D. Agrawal, B. Archambeault, and P. Rohatgi. The EM side-channel(s): Attacks and assessment methodologies. In *Proceedings of the 4th International Workshop on Cryptographic Hardware and Embedded Systems, LNCS2523*, Redwood shores, CA, pp. 29–45, 2002.

14

Secure Key Distribution and Revocation for Advanced Metering Infrastructure

Nathaniel Karst
Babson College

Stephen B. Wicker
Cornell University

14.1 Introduction

The availability of accurate, real-time power consumption data has the potential to substantially increase both the efficiency and the quality of service offered by utility providers. Real-time demand statistics enable service providers to price electricity on an hourly or finer timescale. Customers who are provided with this pricing information can make better-informed choices as to when to perform energy intensive tasks. These local cost-saving behaviors, known as *economic demand response*, organically flatten global demand. A reduced peak-to-average demand ratio further benefits electricity providers and their customers by reducing the extent to which generators are taken on- and offline as power consumption fluctuates. The ensemble of technologies used to gather and analyze the electricity consumption data necessary to make demand response possible is known as *advanced metering infrastructure* (AMI). The Federal Energy Regulatory Commission (FERC) has estimated that as of 2010, AMI penetration in the United States reached 8.7% nationwide and over 13% in some areas, with over 500 groups offering demand response services [10]. This level of participation represents an 85% increase over a 2 year period. The advancement of AMI is also receiving significant fiscal support from the federal government. The American Recovery and Reinvestment Act has so far awarded over 790 million dollars for AMI development and deployment [10], and in addition, AMI projects have received funding through federal commitments to the modernization of the nation's electrical grid. We note that while the work presented here will focus on the applications of AMI to distribution of electricity, it is becoming more prevalent to use similar advanced metering technologies and techniques in water and natural gas utilities.

In the deployment of a distribution system featuring AMI, homes are first retrofitted with "smart" meters as replacements for traditional electromechanical meters. These smart meters monitor consumption in the usual way on an hourly or subhourly basis and in addition transmit usage data to a neighborhood-wide collection station at least once daily [10]. The collection station oversees the neighborhood-area network (NAN) by aggregating consumption statistics and sending a summary to the electricity provider. Transmissions from smart meters to collection stations are typically wireless to facilitate easy installation and upgrade of smart meters, while the backhaul from the collection station to the electricity provider is typically a wired connection. The 2010 FERC definition of AMI specifies that in return for the consumption data gathered from smart meters, electricity providers must supply customers with current utility price information at least once daily [10]; the definition does not specify the mode in which this information be conveyed, however. On top of this base level of service, one can imagine entrusting further functionality to the AMI, including providing fine grain (e.g., subdaily or more frequent) pricing data, sending pricing data directly to the home via the smart meter, emergency consumption reduction for outage avoidance, quality of service monitoring, and remote disconnect capabilities. A similar system is being developed for individual homes. Here, a home-wide base station collects electricity consumption from individual appliances and in return distributes utility pricing reports. Consumers can configure appliances to cycle on only if certain pricing conditions are met. General Electric's collection of Brillion-enabled appliances is an example of one such system. Privacy concerns in this context can be even more acute, as the power consumption signatures of individual appliances are sufficient to determine the make, model, and operating schedules of a home's monitored appliances.

The potential for AMI to be a transformative set of technologies is directly tied to its perceived security. Accurate prediction of system-wide demand requires large-scale participation, and universal adoption of a nonsecure system is not realistic. Customers are less likely to accept a system that does not closely guard their personal information or does not provide strong assurances that the pricing data and commands received at the home smart meter can be trusted. Both the customer-to-provider and provider-to-customer communication links in AMI raise security and privacy concerns. Privacy issues related to utility consumption data have been receiving attention at both the federal and local levels. The 2010 Guidelines for Smart Grid Cyber Security published by the National Institute of Standards and Technology (NIST) documents the need for the security of consumer data to remain a priority as smart grid technologies advance and are deployed [22]. The California Public Utilities Commission has recently dealt with privacy issues related to the availability of customer utility consumption data both to the customers themselves and to "other interested persons" [9]. More recently, President Obama has ordered a 60 day cyber security review, including the security of the nation's electric grid. These concerns are not unfounded. Lisovich et al. have shown that even the coarse-grain consumption information provided by AMI can be used to infer details as to what is taking place within a household [20]. Lerner and Mulligan have detailed potential types of abuse stemming from unencrypted AMI data and discussed Fourth Amendment implications to AMI data availability [19]. Without encryption on the provider-to-customer link, a system with AMI is vulnerable to attacks in which an adversary impersonates the utility provider. Depending on the level of control over individual homes given to the provider through a smart meter, such an impersonation could result in consequences ranging from incorrect pricing information to termination of services.

14.2 Rekeying Group Communication Systems

At its core, AMI is a collection of sensor nodes that collect data and wirelessly transmit encrypted reports to a base station that aggregates and processes the data. In the case where a sensor node cannot communicate directly to the base station, other sensor nodes act as intermediaries on a so-called *multi-hop path*. It is therefore desirable to have a high probability that any two users can securely communicate with one another. This type of communication system, called a *wireless sensor network* (WSN), has been studied in a variety of different contexts; for a survey of applications and technologies, see Ref. [1]. The wireless sensors themselves typically have tight constraints on computation, memory, and power. For data security, symmetric key cryptography protocols in which a single cryptographic key is used for both encryption and decryption have traditionally been favored over more powerful public key encryption techniques that demand greater memory and computational resources. The popular ZigBee wireless sensor platform, for instance, employs the 128 bit version of the Advanced Encryption Standard (AES). The use of symmetric key encryption does place its own requirements on a system's infrastructure: any two sensor nodes can communicate securely using symmetric cryptography if and only if both have access to one or more common cryptographic key. In a full-mesh Zigbee network operating in the commercial mode, every pair of users possesses a unique cryptographic key for this purpose; each user of n users then owns $O(n)$ link keys. For even moderately sized networks, this number of keys represents a considerable memory overhead for the individual sensor nodes.

In a more general framework, some subset of a network-wide collection of keys is distributed by the base station to each user in the network. How must these keys be distributed in order to ensure a given level of secure connectivity among the sensor nodes? To begin investigating this question, we formalize the notion of a key distribution by defining a *key pool* X and a collection of *key chains* $\mathcal{B}$ in which every set $B \in \mathcal{B}$ is a subset of X. In the mathematics literature, the pair $(X, \mathcal{B})$ is known as a *set system*.

Definition 14.1

A set system *is a finite set* X *together with a collection* $\mathcal{B}$ *of subsets of* X. *The elements of* X *are known as* points *and the sets in* $\mathcal{B}$ *are known as* blocks. *A set system is* uniform *if all blocks have the same cardinality. A set system is* regular *if every point occurs in the same number of blocks.*

We will use the terms keys and points, and key chains and blocks interchangeably, respectively.

Qualitatively speaking, large key chains relative to the size of the key pool result in higher probability that two users will share a common cryptographic key. In the extreme case, there is a single key in the key pool, and every sensor node has access to this key. (Such a key distribution can be found in the Zigbee residential security mode, for instance.) This solution is not preferred for critical applications, because if the single key is compromised, then no secure communication can take place between members of the network. The maximum key chain size supported by a sensor node is dictated by the node's memory constraints, and so for both mathematical and technical reasons, it is advantageous to focus on the case in which all sensor nodes have key chains of the same cardinality k, that is, uniform set systems. In one of the first WSN key distribution schemes, Eschenauer and Gligor proposed choosing k keys uniformly randomly from the network-wide key pool of size v in order to populate each node's key chain [14]. Then the relative sizes of k and v determine the probability that two arbitrary sensor nodes can communicate securely. While this key distribution is mathematically straightforward and easy to administrate, probabilistic claims about network connectivity are not sufficient in many communication scenarios; firm guarantees about worst-case performance are necessary. It is not surprising that the key distributions satisfying these additional requirements have much more structure than randomized distributions. We discuss one such class of key distributions in Section 14.4.

Advanced metering systems that provide feedback to consumers through smart meters have a level of infrastructure on top of the underlying WSN. In addition to communicating securely with the base station, each consumer must be able to decode the group-wide pricing information and commands being transmitted via the NAN collection station. Making the same assumptions about the necessity of data security and the limitations of the sensor nodes, the group communication stream is symmetrically encrypted by the collection station using a group-wide *session key*. Each user additionally owns a collection of *administrative keys* that are used, for instance, to distribute a new session key if the need arises. These administrative keys can also be used to facilitate multi-hop paths in the case where a sensor node cannot communicate directly to the base station. We typically use the term "key chain" in this context to refer to a user's collection of administrative keys only. Similarly, the term "key pool" in this context is used to refer to the collection of all administrative keys. A network in which sensor nodes unicast data to a base station and the base station in return broadcasts to all network members is known as a *group communication system*.

Fluid membership makes key distribution in group communication systems a difficult task. For instance, suppose a member leaves or must be ejected from the network. Clearly, the session key must be changed, because if not, then the ejected member can continue decoding the private group communication to which he/she is no longer privileged. For another case, suppose that a new member joins the network. It is possible that this new member has been recording the encrypted group communication stream before his/her addition to the network. Hence, if the session key is not changed after his/her arrival, the new user can decrypt old communications that he/she was not at the time privileged to hear. The problem of resecuring the network after a member leaves is more complicated than resecuring after a member joins. When a user leaves the network, the base station generates a fresh session key together with new administrative keys to replace those owned by the ejected user. The new session and administrative keys are collectively known as the *rekeying message*. The base station then securely distributes the rekeying message to all remaining privileged members. The technical difficulty arises in that this secure distribution must be accomplished without using any key owned by any ejected users. The exact manner in which secure rekeying is achieved can vary, and some methods are far more efficient than others.

For a concrete example of how the rekeying problem dictates the structure of the underlying administrative key distribution, first consider the case in which a single user with key ring E is ejected. In a default rekeying solution, the base station sequentially encrypts the rekey message with *every* administrative key not owned by the ejected user, that is, every key in $X \setminus E$, and broadcasts the encrypted messages. Then any remaining privileged user can gain access to the replacement keys if and only if he/she has access to at least one administrative key not owned by the ejected user. Hence, the system can be securely rekeyed after the ejection of one user if and only if no key chain is identical to or a proper subset of any other. In symbols, $B \not\subseteq C$ for every pair of distinct sets B and C in $\mathcal{B}$. This case was considered by Eltoweissy et al. [12].

In general, we are concerned with the system's ability to eject more than one user simultaneously. If a group of users is found to be colluding to compromise the network, they must be ejected together; if even one of the colluders receives the replacement session key, he/she can distribute it to his/her cohort and so neutralize the rekeying operation. Suppose again that the base station encrypts the rekey message sequentially with each administrative key not owned by any of the ejected members and broadcasts to the group. Then a remaining user can decrypt the replacement keys if and only if he/she has access to at least one key not owned by *any* of the ejected users. Mathematically speaking, the base station can successfully rekey the network after any r simultaneous ejections if and only if no user's key chain is included in the union of r other key chains. The formulation of this requirement is captured nicely in the definition of an r-cover-free family.

Definition 14.2

An r-cover-free family *is a set system* $(X, \mathcal{B})$ *in which any distinct blocks* $B_1, B_2, \ldots, B_r$ *and* A *in* $\mathcal{B}$ *satisfy*

$$A \not\subseteq \bigcup_{i=1}^{r} B_i \tag{14.1}$$

In words, no union of r blocks covers any other block.

Cover-free families have been rediscovered several times over the years, exactly because they are so useful in such a wide range of applications. Kautz and Singleton were the first to articulate the appropriate general axiom in their work on file retrieval using binary superimposed codes [16]. Since then, cover-free families have found numerous other applications in electrical engineering and computer science. Desmedt et al. proposed using cover-free families to protect spread spectrum communication systems from insider adversaries [11]. Cover-free families were put forward by Colbourn et al. for transmission scheduling in mobile ad hoc networks without knowledge of the network topology [7]. Wang and Pieprzyk used 2-cover-free families for anonymous membership broadcast schemes in which the base station can broadcast a message and only the intended recipient can deduce the message's destination [26]. Cover-free families were used as traceability codes by Staddon et al. [24]; here, illegally distributed content can be traced back to the users colluding to pirate the content. We note in passing that Stinson and Wei have formulated the most inclusive generalization of a cover-free family [25]; in our work here, the original, limited definition will suffice. Also note that 1-cover-free families and Sperner families are synonymous.

As early as the late 1980s, cover-free families were recognized for their ability to prevent collusion [21]. In the context of a WSN, if no user's key chain lies in the union of r other key chains, then no r users can collude to forge the key chain of any other user. More recently, however, Xu et al. have proposed using an r-cover-free family as the key distribution in a group communication system supporting up to r simultaneous user ejections [28]. In the rekeying phase, Xu et al. stipulate that the base station first determine a collection of keys for which every nonejected member owns at least one of these selected keys, and no ejected member owns any. We will simply call any such collection of keys a *rekeying set* where the exact key distribution and number of ejected users under consideration are clear. Then to rekey the network, the base station must encrypt the rekey message with each key in the rekeying set and broadcast the result to the network. It is obviously desirable to minimize the size of this collection because the no secure communication can take place until the network has been rekeyed. At the same time, the method for finding a minimal rekeying set should be as computationally inexpensive as possible to minimize load at the base station. Balancing rekeying latency and the computational complexity of determining a suitably small rekeying set is integral to the success of a rekeying solution.

Manufacturing a minimal rekeying set is related to hitting set, one of Karp's original NP-complete problems [15]. Hitting set can be phrased as an optimization problem in the following way: Given a set X and a collection $\mathcal{B}$ of subsets of X, what is the minimal cardinality of $H \subseteq X$ such that H "hits" every set in $\mathcal{B}$, that is, $H \cap B \neq \emptyset$, for every $B \in \mathcal{B}$? We can formalize the relationship between hitting set and secure rekeying in the following theorem.

Theorem 14.1

The problem of finding a minimal rekeying set after any number of user leaves in a group communication system with key distribution (X', $\mathcal{B}'$) is NP-hard.

Proof

We will reduce from hitting set, meaning we will show that every instance of hitting set corresponds to at least one instance of the rekeying problem; therefore, the rekeying problem is at least as difficult as hitting set. Let X be the support set and $\mathcal{B}$ the collection of subsets of X for which we want to find a minimal hitting set. Let E be a set disjoint from X, and suppose that E represents the union of the key chains of the ejected users. Let

$$B_i = B_i' \setminus E$$

$$\mathcal{B}' = \{B_i' : B_i \in \mathcal{B}\}$$

$$X' = X \cup E$$

so that $(X', \mathcal{B}')$ is a key distribution of the network before the user ejections. Then the hitting set problem on the set system $(X, \mathcal{B})$ is equivalent to the minimal rekeying problem on the key distribution $(X', \mathcal{B}')$ in the case where any users owning the key collection E is ejected from the network. Hence, hitting set is many-one reducible to the minimal rekeying set problem.

Wong et al. were the first to notice that the rekeying problem is NP-hard [27]. These authors referred to a reduction from set cover. Here, we have included a reduction from hitting set; this reduction is more natural and intuitive when considering key distributions as set systems.

Notice that the statement of the theorem makes no assumptions about the exact nature of the key distribution. And in general, the results of a NP-hardness theorem should be always taken with a grain of salt. The theorem does not claim that finding a minimal rekeying set is always hopeless endeavor, but rather that finding a minimal rekeying set for an *arbitrary* key distribution should be expected to be difficult. Therefore, the theorem naturally points toward only considering key distributions with a large amount of combinatorial and/or algebraic structure and hoping that this structure is sufficient to circumvent the natural difficulty of the problem.

The mathematical structure of a key distribution after the ejection of one or more members can be described by considering an associated *residual* set system.

Definition 14.3

The ε-residual of a set system $(X, \mathcal{B})$, $\mathcal{E} \subseteq \mathcal{B}$, is the set system $(X', \mathcal{B}')$ with

$$X' = X \setminus \left(\bigcup_{E \in \mathcal{E}} E \right)$$

$$\mathcal{B}' = \left\{ C \setminus \left(\bigcup_{E \in \mathcal{E}} E \right) : C \in \mathcal{B} \setminus \mathcal{E} \right\}$$

If $\mathcal{E}$ is a single set E, we will abuse notation and identify the ε-residual as the E-residual. If the choice of $\mathcal{E}$ is either irrelevant or understood, we will call the $\mathcal{E}$-residual simply the residual.

Using this formulation in a group communication context, the blocks in $\mathcal{E}$ represent the key chains of the ejected users. Then the residual set system can be thought of qualitatively as the original set system with all compromised keys having been removed from both the key pool and the key chain of every non-ejected member. In this way, determining a rekeying set for a distribution $(X, \mathcal{B})$ after ejecting users with key chains $\{E_1, E_2, \ldots, E_r\} = \mathcal{E} \subseteq \mathcal{B}$ is equivalent to manufacturing a hitting set of the $\mathcal{E}$-residual of $(X, \mathcal{B})$.

14.3 *t*-Designs

Combinatorial designs, uniform and regular set systems whose points satisfy incidence conditions, were first seen in the recreational mathematics of the mid-1800s. Before the early 1900s, formal mathematical investigation of combinatorial designs was limited to sporadic examples and limited nonexistence results. Design theory experienced a resurgence in the mid-1900s when statisticians developed designs to facilitate methodical experimentation and when geometers connected certain classes of designs with projective and affine geometries over finite fields. Since then, combinatorial designs have been used in numerous and diverse applications in science and technology; for some examples, see the review of computer science applications by Colbourn and Oorschot [8]. Still, a general understanding of even the existence of combinatorial designs has remained elusive. Significant progress has been made on classifying and documenting certain classes of designs, however. Before discussing these specialized objects, we present the general definition.

Definition 14.4

A set system $(X, \mathcal{B})$ *is a* t-(v, b, r, k, λ) design *if*

1. $|X| = v$
2. $|\mathcal{B}| = b$
3. *Every* $x \in X$ *occurs in exactly* r *sets in* $\mathcal{B}$
4. $|B| = k$ *for all* $B \in \mathcal{B}$
5. *Every* t*-subset of* X *appears in exactly* λ *blocks in* $\mathcal{B}$

We call r and k the *replication number* and *block size*, respectively. Note that even here the property that $\mathcal{B}$ is a set, that is, that there are no repeated blocks, is not required a priori. Designs containing no repeated blocks are often referred to as *simple*. For the balance of this work, we assume that every design that we introduce is simple. We will also omit from consideration the *complete design* that is composed of all k-subsets of X with $t \leq k$.

Conditions (1) through (4) impose uniformity and regularity on the set system. It is condition (5) that puts t-designs apart from all other set systems, and indeed it is not easily satisfied. Designs with $t = 1$ are redundantly defined, as condition (3) would imply that $r = \lambda$ in this case. We therefore always assume that $t > 1$. Designs with $t = 2$ have been relatively well studied and will be main mathematical tool used in the applications featured here. Some infinite classes of t-designs with small λ are known for $t > 2$. For instance, there is a 3-$(q^2 + 1, q + 1, 1)$ design, a so-called *Möbius* or *inversive plane*, for every prime power q [6]. However, no t-(v, b, r, k, λ) design is known to exist for any $t > 5$ and $\lambda < 4$ [6]. Determining how many if any t-$(v, b, r, k, 1)$ designs exist for large t is one of the largest open problems in design theory.

The parameters of a t-(v, b, r, k, λ) design are not independent of one another. Simple algebraic equations allow us to write any two of the parenthetical parameters in terms of the remaining three. The proofs of these well-known theorems are included here both for completeness and so that the reasoning and methods found in the original work presented later will have proper motivation.

Result 14.1

The parameters of a t-(v, b, r, k, λ) *design* $(X, \mathcal{B})$ *satisfy* $bk = vr$.

Proof

We will count the total number of points in all blocks in two ways. There are b blocks in $\mathcal{B}$, each containing exactly k points. Additionally, there are v points in X, each occurring in exactly r blocks. Hence, the total number of points in the design is $bk = vr$.

Result 14.2

The parameters of a t-(v, b, r, k, λ) *design* $(X, \mathcal{B})$ *satisfy* $\lambda(v - 1) = r(k - 1)$.

Proof

Fix $x \in X$. We will count the number of pairs (x, y), $y \in X$ and $y \neq x$, occurring in all blocks in $\mathcal{B}$. For all $v - 1$ choices of suitable y, there exist exactly λ blocks containing both x and y. For the right side of the equality, the point x occurs in exactly r blocks, and in each there are exactly $k - 1$ other distinct points.

Despite the fact that Results 14.1 and 14.2 imply that there exists a more concise notation that describes any t-(v, b, r, k, λ) design, we will continue to use this expanded version for clarity's sake, except in one particular class that will be introduced in the following.

Example 14.1

Let $X = \{0, 1, \ldots, 6\}$ *and define*

$$\mathcal{B} = \{\{0,1,3\}, \{1,2,4\}, \{2,3,5\}, \{3,4,6\}, \{4,5,0\}, \{5,6,1\}, \{6,0,2\}\}. \quad (14.2)$$

One can verify that $(X, \mathcal{B})$ *is* 2-$(7,7,3,3,1)$ *design. It is commonly known as the Fano plane.*

We note that in this particular case the number of points v equals the number of blocks b, and the replication number r equals the block size k. This additional structure is indicative of a larger class of 2-(v, v, k, k, λ) designs that have been by far the most extensively investigated in design theory literature.

14.3.1 Symmetric 2-Designs

Fischer's inequality states that for any 2-(v, b, r, k, λ) design, there are at least as many blocks as points, that is, $|\mathcal{B}| = b \geq v = |X|$. The class of 2-designs that meet this bound with equality are called symmetric 2-designs.

Definition 14.5

A symmetric 2-(v, k, λ) design *of order* $q = k - \lambda$ *is a* 2-(v, v, k, k, λ) *design.*

We sometimes refer to a symmetric 2-(v, k, λ) design simply as a symmetric 2-design if the parameters are either clear or irrelevant. Here, the term "symmetric" refers not to any geometrical property necessarily, but rather to the equivalence of the conditions on the numbers of blocks and points and on the block size and the replication number. For reasons that we will not delve into here, some authors prefer to use the term *square* to describe 2-(v, v, k, k, λ) designs. Symmetric designs have yet another interesting (and useful) equivalence between blocks and points.

Result 14.3

In a symmetric 2-(v, k, λ) *design, any two blocks share exactly* λ *points.*

Algebraic identities imply that if a symmetric 2-design satisfies $\lambda = 1$, then $v = q^2 + q + 1$ and $k = q + 1$ for some q. The symmetric 2-$(q^2 + q + 1, q + 1, 1)$ designs are known as *projective planes*. We can manufacture a projective plane of prime power order q using the following construction due to Singer [23]. Let $\mathbb{F}_q^3$ be the three-dimensional vector space over the finite field with q elements $\mathbb{F}_q$. Associate the one- and two-dimensional subspaces of $\mathbb{F}_q^3$ with the points and blocks of the design, respectively, with a block containing a point if and only if the associated vector space inclusion holds. One can verify that this construction satisfies the axioms from Definition 14.5. The details of the proof can be found in any elementary design theory textbook. We assume that any projective plane employed in the balance of the chapter will have been manufactured using Singer's construction. The existence of projective planes of non-prime-power order remains an open question, with the most recent advance having been made by Lam et al. in 1989 [17].

While we have so far concentrated on symmetric 2-designs with $\lambda = 1$, other symmetric 2-designs with $\lambda > 1$ do exist. These classes of symmetric designs are far less well understood that their projective plane counterparts. One class that will be featured in later applications are the symmetric designs with $\lambda = 2$.

Definition 14.6

A biplane *is symmetric* $2\text{-}(v, k, 2)$ *design.*

There are only finitely many biplanes known to exist, namely, for $k = 4, 5, 6, 9, 11, 13$ [6]. Moreover, it is widely conjectured that for any $\lambda > 1$, there are only finitely many symmetric designs. We see that when used as key distributions of group communication systems, biplanes offer interesting trade-offs when compared with projective planes.

Example 14.2

Let $X = \{0, 1, \ldots, 10\}$ *and define*

$$\begin{aligned} \mathcal{B} = \{&\{1,3,4,5,9\}, \{2,4,5,6,10\}, \{3,5,6,7,0\}, \{4,6,7,8,1\} \\ &\{5,7,8,9,2\}, \{6,8,9,10,3\}, \{7,9,10,0,4\}, \{8,10,0,1,5\} \\ &\{9,0,1,2,6\}, \{10,1,2,3,7\}, \{0,2,3,4,8\}\}. \end{aligned} \tag{14.3}$$

One can verify that $(X, \mathcal{B})$ *is a biplane of order 3, a symmetric* $2\text{-}(11, 5, 2)$ *design.*

14.4 Symmetric 2-Design Key Distributions

Çamtepe and Yener were the first to use combinatorial designs as key predistributions for secure WSNs [3]. Since then, other authors have explored the utility of various classes of designs to the key distribution problem (e.g., [4,18]). The projective plane scheme of Çamtepe and Yener in which the points and blocks of the projective plane are associated with keys and key chains, respectively, is in many senses optimal: all users have the same number of keys, providing regularity among memory requirements; each key is owned by a constant number of users, normalizing the potential cost of a key being compromised by an adversary; every pair of users share a single key and so possess the minimal amount of information required to communicate securely. These features do not come without cost, however; key distribution formed by a projective plane supports far fewer users than other key distributions. For use in group communication systems with nonstatic membership, we must also consider the ability of the key distribution to support secure rekeying. As discussed earlier, for a group communication system to support secure rekeying after any r simultaneous user leaves, the key distribution of the network must form an r-cover-free family.

Result 14.4

A symmetric 2-(v, k, λ) design $(X, \mathcal{B})$ forms a $\lfloor (k-1)/\lambda \rfloor$-cover-free family [13].

Proof

Fix a block $B \in \mathcal{B}$. Because the design is symmetric, any block not equal to B shares exactly λ points with B. Suppose there is a collection of blocks whose cardinality λ intersections with B are disjoint from one another. Then since B has k points, and each block in the selected collection covers exactly λ, the covering collection must have at least $\lceil k/\lambda \rceil$ members, and clearly no fewer will suffice to cover B. Then in particular, no collection of $\lceil k/\lambda \rceil - 1 = \lfloor (k-1)/\lambda \rfloor$ blocks will cover B. Since B was chosen arbitrarily, the pair $(X, \mathcal{B})$ forms a $\lfloor (k-1)/\lambda \rfloor$-cover-free family.

Results 14.3 and 14.4 shows that a symmetric 2-(v, k, λ) design is a natural candidate for a key distribution of a group communication system that must support full-mesh connectivity and up to $\lfloor (k-1)/\lambda \rfloor$ simultaneous user leaves. Projective planes are both the most plentiful known class of symmetric 2-designs and by Result 14.4 support the largest number of simultaneous user leaves relative to key chain size, and so we focus our attention here.

In 1978, Brouwer and Schrijver settled the problem of the hitting set cardinality of the residual of a projective plane in a pure mathematical context.

Result 14.5

Let $(X, \mathcal{B})$ be a projective plane of order q with q a prime power. Then the cardinality of a minimal hitting set of a E-residual of $(X, \mathcal{B})$ is $2q - 1$ for any $E \in \mathcal{B}$ [2].

There are several known constructions for the minimal hitting set of a projective plane after the removal of a single block. Here, we give one both for completeness and to motivate the strategy for proof of the two simultaneous ejection scenario. Let $E \in \mathcal{B}$ be the key chain of the ejected user, and let B be any other key chain. In a projective plane, any two blocks share exactly one point, so $B \cap E = x_e$ is a single key. For the same reason, the remaining q keys of B are sufficient to rekey any user not owning x_e. Each point in a projective plane occurs in exactly $q + 1$ blocks, so there are $q - 1$ key chains, excluding E and B that contain x_e. To complete the rekeying set, it suffices to include one key from each of these key chains. We have constructed a rekeying set with $q + (q - 1) = 2q - 1$ points, and the result mentioned earlier implies that this rekeying set is minimal.

The structure of projective planes is sufficiently strong to allow us to make claims about the two simultaneous user leaves scenario, as well. First, we must extend the minimality result to include this case.

Theorem 14.2

Let h_0 be the cardinality of the minimal hitting set of a set system $(X, \mathcal{B})$. For each set $\{E_1, E_2, \ldots, E_n\} = \mathcal{E} \subseteq \mathcal{B}$, the cardinality of the minimal hitting set of the $\mathcal{E}$-residual of $(X, \mathcal{B})$ is at least $h_0 - n$.

Proof

First, consider the case where $n = 1$, and choose an arbitrary $E \in \mathcal{B}$. Suppose for contradiction that the minimal hitting set of the E-residual of $(X, \mathcal{B})$ is less than or equal to $h_0 - 2$. In other words, only $h_0 - 2$ points suffice to hit all blocks except B. Hence, by adding only a single point from B to the hitting set of cardinality $h_0 - 2$ of the E-residual of $(X, \mathcal{B})$, we have formed a hitting set of cardinality $h_0 - 1$ of $(X, \mathcal{B})$, a contradiction of the minimality of h_0. So the minimal hitting set of the E-residual of $(X, \mathcal{B})$ is at least $h_0 - 1$ for any E.

The general result follows from n repeated applications of the finding mentioned earlier: any block removal from the set system reduces the minimal hitting cardinality by at most 1.

The result implies that for the case of two simultaneous leaves from a projective plane key distribution, the minimal rekeying set must have cardinality of at least $2q - 2$. This lower bound is achievable.

Theorem 14.3

Let $(X, \mathcal{B})$ be a key distribution formed by a projective plane of order $q > 1$. Then the cardinality of the minimal rekeying set after the simultaneous ejection of any pair of distinct users is $2q - 2$.

Proof

Let $E_1, E_2 \in \mathcal{B}$ be the key chains of the ejected users and define x_e to be their single shared key. Recall that every point in a projective plane of order q is contained in exactly $q + 1$ blocks. Since $q > 1$ by hypothesis, the key x_e is included in at least one key chain outside E_1 and E_2. Let B be one such key chain. Any key chain not containing x_e can be rekeyed by one of the q keys of $B \backslash \{x_e\}$ since every pair of blocks in a projective plane shares exactly one point. Each point in a projective plane of order q appears in exactly $q + 1$ blocks, so there are $q - 2$ key chains, excluding E_1, E_2, and B, which contain x_e, and it suffices to include one key from each of these key chains in the rekeying set. Hence, we have constructed a rekeying set with $q + (q - 2) = 2q - 2$ keys. By Result 14.5 and Theorem 14.2, this rekeying set is minimal.

Recall that the default rekeying solution takes $X \setminus (\cup_{E \in \mathcal{E}} E)$ as the rekeying set, where X is the key pool and $\mathcal{E}$ is the collection of key chains of the ejected users. Hence, for the default rekeying sets have cardinality $v - k = q^2$ and $v - (2q - 1) = q^2 - q + 2$ for a projective plane of order q key distribution with one and two simultaneous ejections, respectively. Using the minimal rekeying sets guaranteed by Result 14.5 and Theorem 14.3 instead of the default solution represents an improvement from rekeying sets of size $O(q^2)$ to $O(q)$ in both cases. For even moderately sized network, this represents significant savings.

One can imagine making similar arguments as the proof of Theorem 14.3 for the cases of more than two simultaneous ejections. However, due to the structure of symmetric 2-designs not all ejection scenarios are isomorphic, that is, the properties of the minimal rekeying set will be dependent on exactly which users were ejected together. Rather than enumerate a growing list of cases for increasing number of ejections, we will point out that a rekeying set is guaranteed to exist by Result 14.4 for any collection of q or fewer simultaneous user leaves and move on to more applicable but less rigorous methods.

Theorem 14.4

Given a key distribution $(X, \mathcal{B})$ based on a biplane of order $q \geq 2$, a minimal rekey set after the ejection of a single member contains at most q keys, and a minimal rekeying set after the simultaneous ejection of two members contains at most q keys.

Proof

We will upper bound the rekeying load in this scenario by explicit construction of suitable rekeying sets.

For the case of a single ejection, let $E \in \mathcal{B}$ be the key chain of the ejected user. Choose any $B \in \mathcal{B} \setminus \{E\}$, and define $X_e = B \cap E$. Every pair of blocks shares exactly two common points, and every pair of points occurs in exactly two blocks. Hence, the only blocks containing X_e are E and B themselves, and every other block contains at least one point in $B \setminus X_e$. Thus, the rekeying load for this scenario is at most $k - 2 = q$.

For the case of two simultaneous ejections, let $E_1, E_2 \in \mathcal{B}$ be the key chains of the ejected user. The replication number and the order of a biplane are related by $r = q + 2$. Since $q \geq 2$, we have $r \geq 4$. Hence, it is possible to choose $B \in \mathcal{B} \setminus \{E_1, E_2\}$ such that B contains a point x_e found in both E_1 and E_2. The collection $B \cap (E_1 \cup E_2)$ must contain exactly three points. If it contained two, then the two points $E_1 \cap E_2$ would occur in more than two blocks, a contradiction of the 2-design axiom. If it contained four, then the intersections $B \cap E_1$ and $B \cap E_2$, each having cardinality 2, would be disjoint, a contradiction on our choice of B.

Any block $B' \in \mathcal{B} \setminus \{E_1, E_2\}$ is hit by $B \setminus (E_1 \cup E_2)$ unless both of the points shared between B' and B occur in $B \cap (E_1 \cup E_2) = \{x_e, y, z\}$. Without loss of generality, suppose that the pair $\{x_e, y\}$ occurs in both B and E_1 and that the pair $\{x_e, z\}$ occurs in both B and E_2. The remaining pair $\{y, z\}$ appears in B and one other block U. Hence, this block U is the only block not hit by the set $B \setminus (E_1 \cup E_2)$. It suffices to take one point from U to complete the rekeying set. Hence, there exists a rekeying set of size $|B| - B \cap (E_1 \cup E_2) + 1 = k - 3 + 1 = k - 2 = q$.

Unfortunately, there is no known biplane equivalent of Result 14.5, and so far a proof has eluded the authors. We can, however, establish that in some cases the bounds from Theorem 14.4 are tight.

Take for instance the case of ejections from the biplane of order $q = 3$ featured in Example 14.2. To improve on the result mentioned earlier, a rekeying set with fewer than $q = 3$ keys must be constructed. Any key hits $k = 5$ key chains, any single pair of keys occurs in exactly two blocks. Hence, any pair of points hits $2k - 2 = 8$ key chains by inclusion–exclusion. There are $v = 11$ users. Clearly, no set of $q - 1 = 2$ keys is sufficient for either the one or two ejection case, and so the bounds from Theorem 14.4 are tight when $q = 3$.

Suppose now that the key distribution is based on the biplane of order $q = 4$, so that $k = 6$. A rekeying set improving on Theorem 14.4 has at most three keys. By inclusion–exclusion, any three keys hit at most $3k - 2$ choose (3, 2) + choose (3, 3) = 18 − 6 + 1 = 13 key chains. There are 16 blocks in the original biplane, and so a set of any three keys is insufficient to rekey the system after the ejection of one or two users, and so the bounds from Theorem 14.4 are tight when $q = 4$.

This is where this type of reasoning about minimal hitting sets in residuals of biplanes runs dry. The tactic used by Brouwer and Schrijver to determine cardinality of a minimal hitting set of the E-residual of a projective plane relies heavily on the vector space interpretation of the underlying set system. Unfortunately, there is no straightforward equivalent in the context of biplanes; these set systems are defined by axioms that are not in agreement with those of finite geometry, and to date, there has been no successful work in connecting biplanes of arbitrary order outside of their axiomatic definition.

14.5 Hitting Set Constructions

While, in general, finding a minimal hitting set is hard, there do exist approximation algorithms that guarantee that the hitting set produced will be at most a multiplicative factor times the minimal cardinality. Take for instance the following greedy algorithm for finding a hitting set of the set system $(X, \mathcal{B})$:

Greedy algorithm

(0) $H \leftarrow \varnothing; \mathcal{B}' \leftarrow \mathcal{B}; X' \leftarrow X$
(1) $H \leftarrow H \cup \{x\}$ with $\arg\max_{x \in X'} |\{B \in \mathcal{B}' : x \in B\}|$
(2) $X' \leftarrow X' \backslash \{x\}$
(3) $\mathcal{B}' \leftarrow \{B \setminus X' : B \in \mathcal{B}' \text{ and } B \cap H = \varnothing\}$
(4) If H is a hitting set of $\mathcal{B}$, return H; else, go to (1)

We note for clarity that if in Step (1) there are multiple valid choices for x, we select one uniformly at random. Chvatal showed that this algorithm produces hitting sets that are at most $\ln(k)$ times the minimal hitting set cardinality [5]. We see that in practice, the algorithm sometimes performs much better than this bound for residuals of symmetric designs.

The default rekeying solution is to take as the rekeying set all keys not owned by any ejected user. One can imagine that such a scheme does not provide competitive results. So instead we compare the greedy algorithm to a randomized construction in which keys chosen uniformly randomly without replacement from the key pool and until the accumulated collection forms a rekeying set.

Randomized algorithm

(0) $H \leftarrow \varnothing$
(1) $H \leftarrow H \cup \{x\}$ with $x \in X \backslash H$ chosen uniformly randomly
(2) If H is a hitting set of $\mathcal{B}$, return H; else, go to (1)

The data presented in Table 14.1 and Figure 14.1 detail the performance of the greedy algorithm when applied to the key distribution based on a projective plane of order 11. For each data point, users where

TABLE 14.1 Comparison of the Sample Mean μ (Sample Standard Deviation σ) Cardinalities with $N = 1000$ for Greedy, Random, and Default Rekeying Set Construction Algorithms for a Projective Plane of Order 11

No. of Ejections	Greedy	Random	Default
1	21.00 (0.00)	45.61 (7.59)	121.00 (0.00)
2	20.00 (0.00)	44.13 (6.76)	110.00 (0.00)
3	20.18 (0.43)	43.16 (6.36)	99.93 (0.25)
4	20.05 (0.53)	42.27 (6.43)	90.70 (0.50)
5	20.06 (0.62)	40.84 (5.73)	82.25 (0.70)
6	20.07 (0.65)	39.68 (5.40)	74.54 (0.92)
7	20.09 (0.72)	38.61 (4.94)	67.51 (1.11)
8	20.10 (0.76)	37.03 (4.91)	61.01 (1.39)
9	20.20 (0.79)	35.90 (4.35)	55.19 (1.53)
10	20.31 (0.86)	35.13 (4.04)	49.90 (1.65)
11	20.42 (0.92)	33.82 (3.68)	45.03 (1.80)

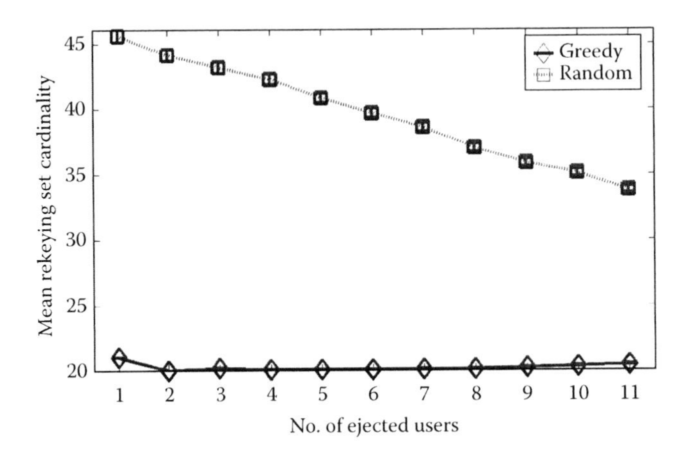

FIGURE 14.1 Plot of the data featured in Table 14.1 with error bar width representing four times the standard error $\sigma/\sqrt{1000}$.

chosen uniformly randomly without replacement and ejected from the network. The greedy and randomized algorithms were then performed on the residual set system and the cardinalities of these rekeying sets were recorded. (Note that since the projective plane of order 11 forms an 11-cover-free family according to Result 14.4, both algorithms are guaranteed to terminate.) The procedure was repeated $N = 1000$ times. Error bars in Figure 14.1 have width four times the standard error $\sigma/\sqrt{N}$, where σ is the sample standard deviation. Result 14.5 and Theorem 14.3 imply that the minimal cardinalities of a rekeying sets after one ejection and two simultaneous ejections for this key distribution are 21 and 20, respectively, and indeed the greedy algorithm produces such minimal rekeying sets for all N trials. For greater than two simultaneous ejections, the greedy algorithm exhibits near constant performance, requiring roughly 20 separate encryptions of the replacement keys in order to secure the network after the simultaneous ejection of 3–11 users. Note that the default rekeying scheme is competitive with neither the randomized nor greedy algorithm.

Table 14.2 features sample mean and standard deviation data from $N = 1000$ trials of the greedy and randomized algorithms applied to the key distribution based on a biplane of order 7. (Note that biplane of order 7 forms 4-cover-free family by Result 14.4.) Figure 14.2 displays these data; error bars in Figure 14.2 have width four times the standard error $\sigma/\sqrt{N}$, where σ is the sample standard deviation. The results are similar to those observed in the projective plane example with a few notable exceptions. For the cases of one ejection and two simultaneous ejections, the greedy algorithm sometimes produces hitting sets of cardinality 6. This performance beats the upper bounds guaranteed by Theorem 14.4. We have shown, however, that Theorem 14.4 is tight for $q \leq 6$. Together, these facts imply that the minimal hitting sets of biplanes have more interesting structure than those of projective planes. For more than two simultaneous ejections, the greedy algorithm has near constant performance. Note that the default rekeying scheme is not competitive with either of the other two algorithms.

TABLE 14.2 Comparison of the Sample Mean μ (Sample Standard Deviation σ) Cardinalities with $N = 1000$ for Greedy, Random, and Default Rekeying Set Construction Algorithms for a Biplane of Order 7

No. of Ejections	Greedy	Random	Default
1	6.97 (0.16)	11.66 (2.14)	28.00 (0.00)
2	6.99 (0.19)	10.66 (1.69)	21.00 (0.00)
3	6.91 (0.40)	9.75 (1.38)	15.57 (0.49)
4	6.85 (0.54)	8.91 (1.07)	11.48 (0.75)

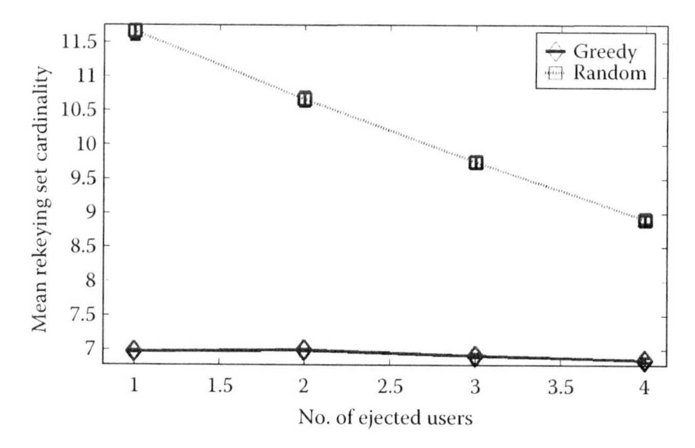

FIGURE 14.2 Plot of the data featured in Table 14.2 with error bar width representing four times the standard error $\sigma/\sqrt{1000}$.

14.6 Conclusion

We have presented in this work a scheme for a cryptographic key predistribution and secure rekeying based on symmetric 2-designs. These combinatorial objects are natural candidates due to their highly regular structure, rich literature, and well-documented construction algorithms. We provided results showing that a symmetric 2-(v, k, λ) design forms a $\lfloor (k-1)/\lambda \rfloor$-cover-free family. In a projective plane key distribution, this prevents up to $k-1$ users from pooling their key chains in order to impersonate another user. Moreover, this same structure allows secure rekeying of the network after up to $k-1$ simultaneous user ejections. We showed that, in general, the problem of finding a collection of keys suitable for rekeying a group communication system after even one user ejection is NP-hard. Fortunately, the combinatorial structure of symmetric 2-designs allowed us to circumvent the natural difficulty of this problem. We provided an algorithm for constructing a rekeying set after one or two simultaneous user ejections from a projective plane key distribution and produced a combination of known results and novel work to show that these rekeying sets are minimal. We provided a similar construction for biplane key distributions; we presented limited results showing that these rekeying sets are minimal for some orders and provided simulation results showing that the bound can certainly be improved for some orders. For more than two simultaneous ejections in either of these key distributions, we cited a well-known approximation algorithm for hitting set and documented simulation results detailing the good performance of this algorithm.

We presented a sample application of this technology in the form of AMI, on both the home and appliance levels. The information being passed from consumer to utility provider through the AMI presents a potential privacy hazard, and we concluded the WSNs being deployed to monitor end-user utility consumption as part of the AMI should employ symmetric cryptography.

References

1. I. Akyildiz, W. Su, Y. Sankarasubramaniam, and E. Cayirci. Wireless sensor networks: A survey. *Computer Networks*, 38(4):393–422, January 2002.
2. A.E. Brouwer and A. Schrijver. The blocking number of an affine space. *Journal of Combinatorial Theory, Series A*, 24(2):251–253, January 1978.
3. S. Çamtepe and B. Yener. Combinatorial design of key distribution mechanisms for wireless sensor networks. *IEEE/ACM Transactions on Networking*, 15(2):346–358, April 2007.
4. D. Chakrabarti, S. Maitra, and B. Roy. A key pre-distribution scheme for wireless sensor networks: Merging blocks in combinatorial design. *International Journal of Information Security*, 5(2):105–114, January 2006.

5. V. Chvatal. A greedy heuristic for the set-covering problem. *Mathematics of Operations Research*, 4(3):233–235, January 1979.
6. C. Colbourn and J. Dinitz. *The CRC Handbook of Combinatorial Designs*. Chapman & Hall, London, U.K., January 1996.
7. C. Colbourn, A. Ling, and V. Syrotiuk. Cover-free families and topology-transparent scheduling for MANETs. *Designs*, 32(1–3):65–95, January 2004.
8. C. Colbourn and P. Oorschot. Applications of combinatorial designs in computer science. *ACM Computing Surveys*, 21(2):223–250, 1989.
9. California Public Utilities Commission. Assigned commissioner and administrative law judge's joint ruling inviting comments on proposed policies and findings pertaining to the smart grid policies established by the Energy Information and Security Act of 2007, 2009.
10. Federal Energy Regulatory Commission. Assessment of demand response and advanced metering infrastructure: Staff report, February 2011. http://www.ferc.gov/legal/staff-reports/2010-dr-report.pdf.
11. Y. Desmedt, R. Safavi-Naini, H. Wang, L. Batten, and C. Charnes. Broadcast anti-jamming systems. *Computer Networks*, 35(2–3):223–236, January 2001.
12. M. Eltoweissy, M. Heydari, L. Morales, and I. Sudborough. Combinatorial optimization of group key management. *Journal of Network and Systems Management*, 12(1):33–50, January 2004.
13. P. Erdös, P. Frankl, and Z. Füredi. Families of finite sets in which no set is covered by the union of r others. *Israel Journal of Mathematics*, 51:79–89, January 1985.
14. L. Eschenauer and V. Gligor. A key-management scheme for distributed sensor networks. *Proceedings of the 9th ACM Conference on Computer and Communications Security*, Washington, DC, pp. 41–47, November 2002.
15. R.M. Karp. Reducibility among combinatorial problems in *Complexity of Computer Computations: Proc. of a Symp. on the Complexity of Computer Computations*, R.E. Miller and J.W. Thatcher (eds.), The IBM Research Symposia Series, New York, Plenum Press, pp. 85–103, 1972.
16. W. Kautz and R. Singleton. Nonrandom binary superimposed codes. *IEEE Transactions on Information Theory*, 10(4):363–377, 1964.
17. C. Lam, L. Thiel, and S. Swiercz. The non-existence of finite projective planes of order 10. *Canadian Journal of Mathematics*, 41:1117–1123, January 1989.
18. J. Lee and D. Stinson. A combinatorial approach to key predistribution for distributed sensor networks. *IEEE Wireless Communications and Networking Conference*, 2:1200–1205, January 2005.
19. J. Lerner and D. Mulligan. Taking the 'long view' on the fourth amendment: Stored records and the sanctity of the home. *Stanford Technology Law Review*, 3, January 2008. http://stlr.stanford.edu/pdf/lerner-multigan-long-view.pdf.
20. M. Lisovich, D. Mulligan, and S. Wicker. Inferring personal information from demand-response systems. *Security and Privacy, IEEE*, 8(1):11–20, 2010.
21. C. Mitchell and F. Piper. Key storage in secure networks. *Discrete Applied Mathematics*, 21(3), October 1988.
22. NIST. *Introduction to NISTIR 7628 Guidelines for Smart Grid Cyber Security*, The Smart Grid Interoperability Panel Cyber Security. Working Group, p. 7, September 2010.
23. J. Singer. A theorem in finite projective geometry and some applications to number theory. *Transactions of the American Mathematical Society*, 43(3):377–385, December 1938.
24. J. Staddon, D. Stinson, and R. Wei. Combinatorial properties of frame proof and traceability codes. *IEEE Transactions on Information Theory*, 47(3):1042–1049, 2001.
25. D. Stinson and R. Wei. Generalized cover-free families. *Discrete Mathematics*, 279:463–477, January 2004.
26. H. Wang and J. Pieprzyk. A combinatorial approach to anonymous membership broadcast. *Computing and Combinatorics: Lecture Notes in Computer Science*, 2387:387–413, 2002.
27. C. Wong, M. Gouda, and S. Lam. Secure group communications using key graphs. *IEEE/ACM Transactions on Networking*, 8(1):16–30, 2000.
28. L. Xu, J. Chen, and X. Wang. Cover-free family based efficient group key management strategy in wireless sensor network. *Journal of Communications*, 3(6):51–58, January 2008.

15

Measurement, Modeling, and Avoidance of IEEE 802.15.4 Radio Transmission Errors in Rotating Sensor Systems

Kuang-Ching Wang
Clemson University

Fan Yang
Clemson University

Yong Huang
Clemson University

Lei Tang
Clemson University

15.1 Introduction

Embedded wireless sensors complete with sensing, processing, and wireless communication capabilities have unleashed a wide range of sensing solutions for places hard to reach with wires due to hazardous conditions or object mobility. The removal of cables for traditional wired sensors also means significant cost reduction. For factory monitoring, wireless sensors are envisioned to be installed on machines, vehicles, and other facilities. A few pilot efforts have explored the feasibility of such factory sensing systems [1–4]. While some studies confirmed basic wireless sensor operations to be feasible in such settings, others found challenging levels of transmission errors in some machine environments (e.g., [5]). This chapter examines the nature of this problem and presents approaches for measurement, modeling, and avoidance of transmission errors in one such environment, specifically, in high-speed rotating machines.

IEEE 802.15.4 is one widely adopted physical (PHY) and medium access control (MAC) standard for today's wireless sensors [6]. It supports low data rate, low power consumption, and short-range transmission with low-complexity circuitry. Several industrial sensing standards have been based on IEEE 802.15.4, for example, ZigBee [7], ISA100.11a [8], and WirelessHART [9]. Numerous studies have reported IEEE 802.15.4 data transmission performance in offices, public places, forests, farmlands, and factories (e.g., [10–14]). However, very little work has examined conditions in moving machines. Reference [15] conducted transmission experiments with a sensor on a rotating spindle surface at tangential speeds from 1.26 to 6.28 m/s, sending a series of 2 byte probe packets and measured the number of received probe packets; the experiments showed that very few packets were lost and concluded that the rotation induced no significant communication errors. The rotation speed, however, was relatively low in industrial operation standards and the packets were sent at a relatively low rate. Rotating structures in machines often achieve several thousand revolutions per minute (rpm) rotation speed. For a rotor of 10 cm radius rotating at 2000 rpm, the tangential speed is about 21 m/s. Such conditions were studied in Ref. [5] with Zigbee radios on a computer numerical controlled (CNC) lathe, where significant transmission errors were observed. Specifically, with tangential speed around 13.6 m/s, the radios observed a packet error rate (PER) above 11%, high enough to be a concern for reliable operation. For example, vibration analysis for defect diagnosis of rolling bearings requires measurement at up to 3000 Hz sampling rates [16]. More accurate prediction of the radio's data transmission performance is certainly needed for such systems.

15.2 Potential Causes of Transmission Errors

Rotating mechanical structures are usually located within confined enclosures, with the structures themselves and their surroundings made of metallic materials. To monitor such structures, a wireless sensor system must consist of two parts: a measuring and transmitting node mounted on or embedded inside the rotating structure, and a receiving node mounted on a stationary part near the rotating structure. Figure 15.1 gives a simplified structure diagram of such a wireless sensing system. To be consistent, throughout the chapter, the rotating radio is always referred to as *the transmitter*, and the stationary radio is referred to as *the receiver*.

Willig et al. [17] examined fundamental problems for wireless industrial communication systems and attributed transmission errors to multipath propagation, Doppler shift, interference from other collocated wireless communication systems, and thermal and man-made noise. Multipath propagation affects the radio channel in two ways by causing, respectively, signal attenuation and intersymbol interference (ISI). Additionally, in a high-speed rotation scenario, the sensor radios' hardware reliability can also be a concern.

In Refs. [5,18], a series of experiments were conducted with the wireless sensing system described earlier to confirm the causes of transmission errors in a rotating machine. The studies measured the receiver's received signal strength indication (RSSI), PER, bit error rate (BER), and the frequency spectrum. The key findings were as follows:

- Multipath effects dominated received signal strength, causing large RSSI fluctuations when either the transmitter or the receiver moved very small distances.
- Doppler effects were insignificant based on analysis of symbol timing error (<1.44 ppm per symbol), carrier frequency shift (<112 Hz), and channel coherence time (>8.9 ms) at 2000 rpm. Note that IEEE 802.15.4 adopts a symbol duration of 16 μs.
- No significant machine electromagnetic noise was seen in measured frequency spectrum.
- Transmitter hardware (antenna) stability was not a concern based on the following: (1) errors were dependent on receiver location and some locations had very low errors and (2) same observed PER behavior with a radio with printed circuit board antennas.

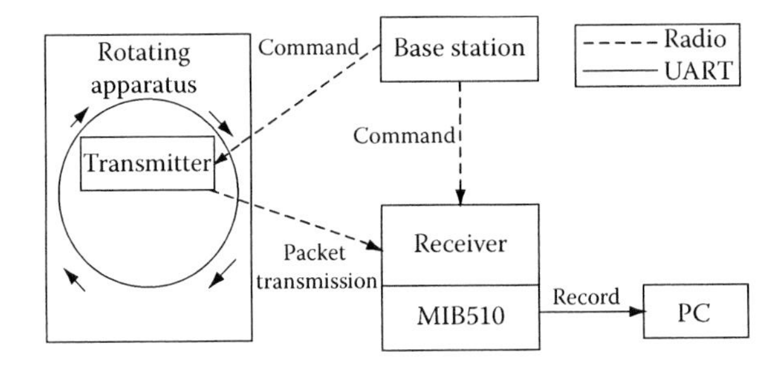

FIGURE 15.1 Wireless sensing system organization for monitoring a rotating shaft.

In conclusion, the received signal strength was the dominant factor for transmission correctness and multipath effects caused fast changes in the received signal strength. The conducted measurements are discussed in more details in the following section.

15.3 Radio Transmission Error Measurement

The measurement was conducted on two distinct rotation systems. One is a professional CNC lathe capable of controlled speeds up to 6000 rpm, and the other is a laboratory system we composed to conduct experiments with precise location measurement with 1° accuracy.

15.3.1 Experiment Setups: CNC Lathe and Laboratory Rotation System

The CNC lathe used in the experiments was a Talent 6/45 CNC lathe by Hardinge Inc. [19]. The lathe exterior is shown in Figure 15.2a. The lathe is equipped with a 60–6000 rpm speed-controllable spindle in a metallic enclosure and digital speed display. Figure 15.2b shows the lathe dimension and placement of the sensor radios. The transmitter was safely mounted on the spindle with duct tapes, and the receiver was attached to the lathe inner wall. During experiments, the machine window was closed and no people or other objects were allowed in the room to prevent any change to the signal propagation properties.

The laboratory rotation system consisted of a permanent magnet DC motor and a rotating plate. The motor was DAYTON 1F800 with speed up to 1750 rpm. A controller was equipped with the motor to tune rotation speed and switch rotation direction. The motor and the rotating plate were supported by brackets on a bench. Figure 15.3 shows the rotation system setup. The system can also work with a vector network analyzer (VNA) for spectrum measurement. To accommodate the VNA, the system used a rotary joint, MI-10-3 from MI Technologies, on the rotating plate to avoid cable twist. The rotary joint supported DC—8 GHz frequency and 500 rpm rotation speed.

The sensor radios are MICAz motes made by Crossbow Inc. [20]. Except for the transmitter and receiver, another mote served as a base station for control and synchronization. A Mote Interface Board (MIB510) was used as an interface between the receiver mote and a computer for data transfer over a universal asynchronous receiver–transmitter (UART) connection. Figure 15.1 shows the functional structure of this measuring system. Table 15.1 lists the key settings on the motes. Channel 26 was chosen because previous measurements showed the power of noise and interferences was consistently low at this frequency range in the CNC lathe. Two transmit power (0 and −25 dBm) were used, respectively. The experiments were repeated with two different receiver locations 3 cm apart, referred to as location 1 and location 2.

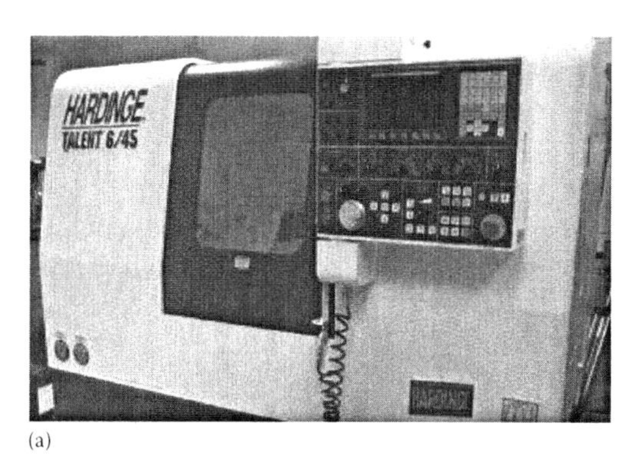

(a)

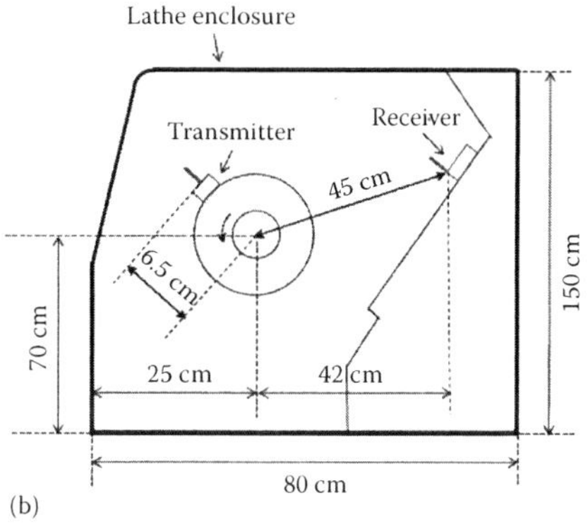

(b)

FIGURE 15.2 (a) CNC lathe exterior and (b) radio placement.

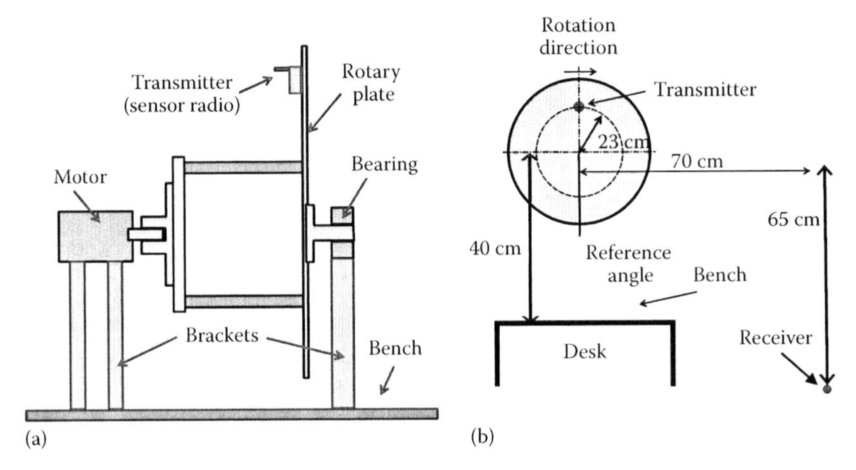

FIGURE 15.3 Laboratory rotation system (a) structure and (b) dimensions.

TABLE 15.1 Radio Configuration for Experiments

Configuration	Setup
Channel frequency (MHz)	2478.5–2481.5
Transmit power	0, −25 dBm
Packet interval	0.05 s
Packet size	30 bytes
Retransmission limit, N	5
Packets transmitted (training/operation)	5000/5000
PER threshold	0.2
Rotation speed (rpm)	242, 544, 845, 1027, 1324, 1510, 1812

15.3.2 Experimental Procedure

The base station began each experiment by sending a command to both the transmitter and the receiver to set transmission parameters including the transmit power, size and number of probe packets, interval between successive transmissions, and measurement delay, if needed. The acknowledgment (ACK) function on the transmitter and the receiver was disabled during the measurement experiments (for error avoidance experiments to be shown in Section 15.5, ACK would be enabled). After a preset delay for the CNC lathe to reach steady speed, the transmitter started transmitting probe packets to the receiver with incrementing identification (ID) numbers. For each received probe packet, we store in the flash memory its CRC check byte, RSSI, and the receiving time stamp. If the packet passed the CRC check, its ID number was stored as well. After each experiment, the stored information was relayed to the PC for analysis and permanent storage.

15.3.3 Measurement Results

15.3.3.1 PER versus Speed, BER versus Location

The PER versus speed curves measured at receiver location 1 is shown in Figure 15.4. The figure shows that at location 1, the PER proportionally increases with speed, while at location 2, the PER is between 0.001 and 0.01 with no clear dependency on speed. Further analysis of the received packet contents at bit granularity revealed that bit errors occurred in groups and the majority of error bursts occurred when the transmitter was passing the same location.

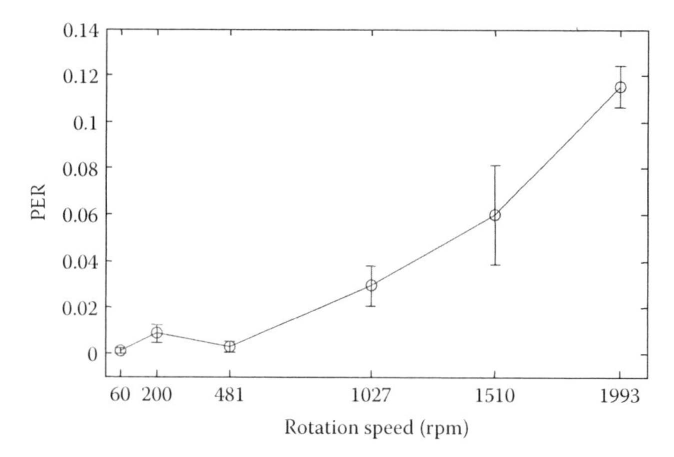

FIGURE 15.4 PER versus speed with receiver at location 1.

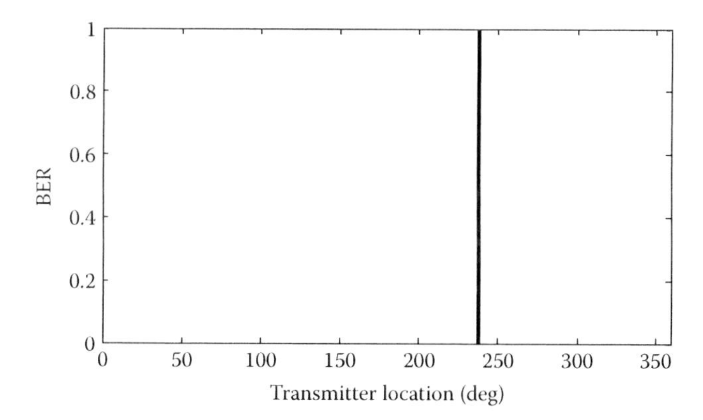

FIGURE 15.5 BER versus transmitter location with receiver at location 1.

For a stationary channel, the BER can be found from the measured PER from the equation

$$PER = 1 - (1 - BER)^n \tag{15.1}$$

based on the fact that bit errors occur randomly following a uniform random distribution. With the machine off, we measured PER with the transmitter at each location at 1° resolution to derive the BER. Figure 15.5 shows the BER measured with the receiver at location 1. As expected, the one spike in BER corresponded to the transmitter location that repeatedly caused error bursts in the rotating experiments. We refer to regions with high PERs like this as high-error regions. Identifying high-error regions is useful for designing error avoidance techniques as to be discussed in Section 15.5.

15.3.3.2 Path Loss versus Speed

As discussed in Section 15.2, path loss is the only significant factor affecting BER. Hence, it is essential to find out whether the path loss changes with speed. The measurement was done with the laboratory rotating system using an Agilent 8714 VNA [21]. Figure 15.6 shows the VNA-measured 2.48 GHz signal path loss for all transmitter locations under four different speeds (stationary, 100, 200, and 300 rpm). The good match confirms the consistent BER with different speeds.

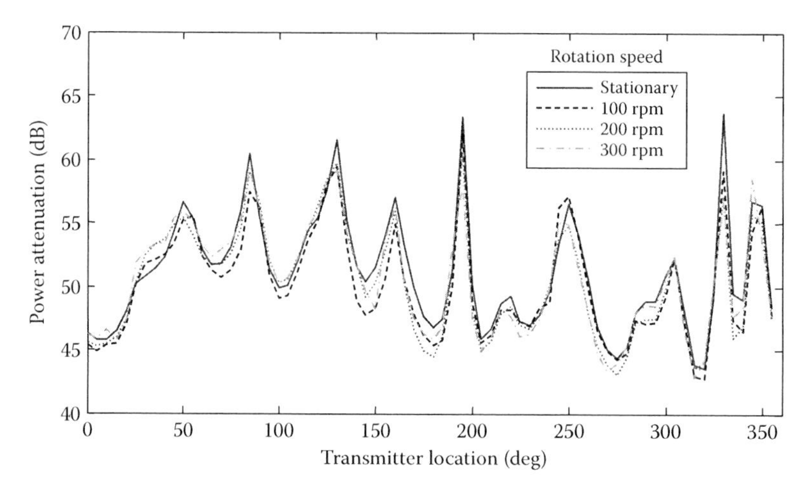

FIGURE 15.6 Path loss measured at different speeds.

15.3.3.3 Throughput versus Speed Using ARQ

Transmission errors result in packet losses. Standard practice adopts automatic retransmission methods like ARQ to tolerate such errors. With ARQ, the receiver always transmits an acknowledgement (ACK) to the transmitter when a packet is received correctly. Thus, if an ACK is not received within a specified time-out period, the transmitter knows the packet is lost and automatically initiates a retransmission. The process can repeat up to a specified number of trials before the transmitter discards the packet and stops retransmitting. There are other error handling approaches beyond ARQ as well. Evaluation of the actual data throughput achievable by a sensor system depends on the specific error handling method used.

Figure 15.7 shows the measured achievable throughput with ARQ and two error avoidance schemes at different speeds using the laboratory rotation system. The error avoidance schemes are introduced in the next section.

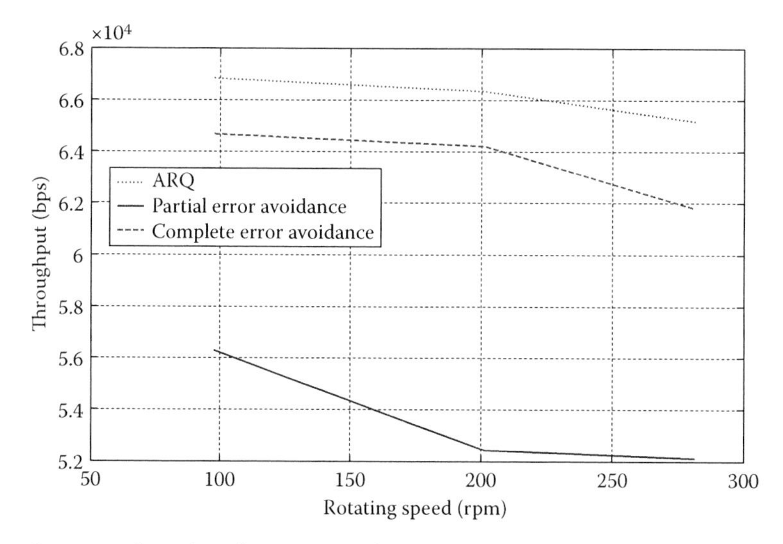

FIGURE 15.7 Saturation throughput for ARQ, complete, and partial error avoidance in the laboratory rotation system at 100–300 rpm.

15.4 Radio Performance Modeling

The rotating radio's performance can be modeled in different ways based on the assessment purpose. At a lower layer, the speed-dependent PER as observed in the experiments can be modeled. Interested readers may refer to Ref. [22] for such a modeling solution. Figure 15.8 shows the measured and modeled PER for the CNC lathe with the maximum error less than 0.007.

At a higher layer, the radio's data transmission throughput can be derived based on the PER model expression. As an example, the following derives the throughput expression achieved with the ARQ scheme based on the given PER expression. Throughput also depends on the transmitter's transmission timing pattern. Many industrial monitoring systems report sensor data at a constant rate. We consider constant packet rate transmission in our example.

In a constant packet rate transmission scheme, the transmitter transmits one packet in each constant interval. With ARQ, if the first transmission of a packet fails, the transmitter retransmits it until either it is correctly received or a limit of N transmission attempts has been made. For the system to be stable and not to overflow the transmitter's buffer, the average time needed to transmit a packet successfully (after needed retransmissions) must be much shorter than the packet interval.

The throughput is derived based on the following observation: with constant packet rate traffic, as long as all packets are eventually transmitted successfully with no more than $N - 1$ retransmission attempts, the throughput will be equal to S/I, where S is the packet size in bits and I is the packet interval. On the other hand, if any packet can fail for N transmission attempts, the throughput will degrade proportionally to the probability of such events. Let the probability of any packet successfully completing transmission with no more than N transmission attempts be defined as the average successful delivery ratio (SDR), the constant packet rate throughput is

$$\mathrm{TP}_{\mathrm{CPR}}(S,\omega,N) = \frac{S}{I} \times \mathrm{SDR}(\omega,N) \tag{15.2}$$

Note that SDR and, as a result, TP both depend on the rotation speed ω and N. As PER differs when the transmitter is at different locations, the average SDR must be found by accounting for the SDR for individual locations. Let $\mathrm{SDR}_L(\theta, \omega, N)$ be defined as the *local SDR* at location θ. This is the probability for a packet whose first transmission begins at location θ to, eventually, complete

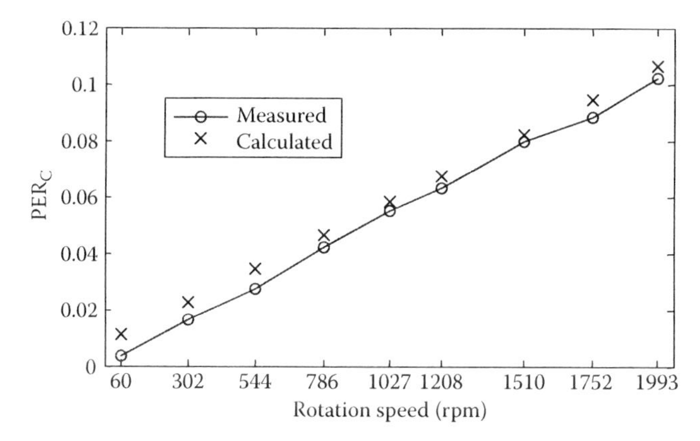

FIGURE 15.8 PER versus speed (calculated versus measured).

successfully after potentially a number of retransmissions. Assuming transmissions can be initiated at any random location with a uniform probability,

$$\mathrm{SDR}(\omega, N) = \int_{0^\circ}^{360^\circ} \mathrm{SDR}_L(\theta, \omega, N)\,d\theta \tag{15.3}$$

Consider an example with $N = 4$. The first transmission occurs at location θ; the probability for it to fail is the PER for location θ and let it be p_1. If the first transmission fails, the retransmission must take place at location $(\theta + d)$ mod 360°, where d is the transmitter's traveled distance during D_{cycle}, the time it takes to transmit one packet and wait for its ACK to return; hence, $D_{\text{cycle}} = S/R + t_{\text{ack}}$, where $R = 250$ kbps is the link rate and $t_{\text{ack}} = 54$ symbol durations is the expected ACK return time as specified by the IEEE 802.15.4 standard. Subsequent retransmissions, if needed, will take place at $(\theta + 2d)$ mod 360°, $(\theta + 3d)$ mod 360°, etc., and let their failure probabilities be p_i. Thus, the local SDR for a transmission starting at θ is

$$\begin{aligned}\mathrm{SDR}_L(\theta) &= (1 - p_1) + p_1(1 - p_2) + p_1 p_2(1 - p_3) + p_1 p_2 p_3(1 - p_4)\\ &= 1 - p_1 \cdot p_2 \cdot p_3 \cdot p_4\end{aligned} \tag{15.4}$$

where $p_i = \mathrm{PER}_L((\theta + (i - 1)d) \bmod 360°, \omega)$, $i = 1$–4. As defined in Ref. [22], $\mathrm{PER}_L(\theta, \omega)$ stands for the PER for a packet beginning transmission at location θ while the rotation speed is ω. Thus,

$$\mathrm{SDR}_L(\theta, \omega, N) = 1 - \prod_{i=1}^{N} \mathrm{PER}_L((\theta + (i-1)d) \bmod 360°, \omega) \tag{15.5}$$

15.5 Radio Transmission Error Avoidance

From the measurements and models, it was seen the error regions (i.e., transmitter locations that would lead to transmission errors) were highly predictable and high PER could result from even very small error regions. This has motivated us to develop error avoidance methods. By identifying error regions through a simple training method, the transmitter can avoid initiating transmissions in the error regions.

15.5.1 Assumed Radio Configurations

While the proposed method is easily applicable for different radios, for IEEE 802.15.4, a number of configurations can greatly improve the precise control of packet transmission time. Specifically, we consider the radio is operating in the *non-beacon transmission* mode with acknowledgement *enabled*, clear channel assessment (CCA) *disabled*, and random backoff *disabled*. CCA and random backoff are optional features of the IEEE 802.15.4 standard to avoid collision; when CCA and random backoff are enabled, a radio wishing to transmit a packet must listen to an idle channel for a fixed CCA period and a random backoff period before it can transmit. While they are important for avoiding collisions when multiple sensors operate in the same vicinity without coordination, we contend that in a fast rotating environment, the CCA and backoff wait can be a significant waste of the already scarce "good" periods for transmission, and such sensors should instead operate with well-coordinated transmission schedules instead of relying on CCA. Random backoff also makes it difficult for precise error avoidance.

15.5.2 Transmission Synchronization

When error avoidance is employed, the transmitter and receiver radios must have synchronized time. Various well-known synchronization methods can serve the purpose. Here, we adopt the reference broadcast synchronization (RBS) algorithm [23], a well-known and simple approach that uses a third radio, that is, the base station in our system, to synchronize the transmitter and receiver. In brief, with RBS, the base station initiates the time synchronization process with a broadcast transmission heard by both the transmitter and receiver. Upon hearing the broadcast, both nodes record the reception time stamp using their local clocks. Finally, the receiver sends its recorded time stamp to the transmitter such that the transmitter can calculate and correct their clock offset to complete the synchronization. The synchronization process can be initiated periodically to account for sensor nodes' clock drift.

15.5.3 Error Avoidance Algorithm

The error avoidance algorithm, as originally proposed in Ref. [24], has two phases: the training phase and the operation phase. In the training phase, the transmitter transmits probe packets to the receiver so that the receiver collects a sufficient history of both transmission successes and failures to establish the *error location profile* and sends it to the transmitter. In the operational phase, the transmitter uses the error location profile to compute the permissible time to transmit the next packet, avoiding the error regions.

Error avoidance and ARQ can be used together; that is, even though the transmitter already attempts to avoid a transmission failure, it may still occur and the transmitter retransmits a packet when it detects a failure. This can, however, be done in two ways. If the retransmission occurs immediately at the end of every ACK time-out, the method is referred to as *partial error avoidance*. On the other hand, if the retransmission is also scheduled with the error avoidance algorithm, the method is referred to as *complete error avoidance*.

15.5.3.1 Training Phase

The training phase essentially measures the PER profile for all transmitter locations around its rotation trajectory. It is assumed that the system has no specialized mechanism to determine the absolute position of the transmitter. Instead, it is expected the probes will be transmitted, opportunistically, with the transmitter at random locations around its trajectory. Moreover, a relative coordinate system is established with respect to the (unknown) location where the first probe was transmitted based on the probes' transmission time stamps; the location is referred to as the *chosen origin*. We assume the rotation speed is known since it is readily available in most modern rotational systems. Thus, knowing the time difference between two probes' transmission times, the distance between the transmitter locations can be calculated. ACK is turned off in the training phase.

The transmitter and receiver clocks are synchronized prior to entering the training phase. As the phase begins, the transmitter sends probe packets continuously at the specified interval. Each probe carries an incrementing sequence number. The receiver detects lost probes by detecting gaps in received sequence numbers. And since the probes are sent at constant intervals, the receiver is able to calculate all lost packets' time stamps by interpolation. Thus, a histogram of received and lost probes at all locations can be created with a specified spatial resolution, for example, 1°. The histogram can be used to calculate $PER_L(\theta, \omega)$ for all θ. Finally, the receiver identifies all regions with PER_L higher than a specified *PER threshold* as high-error regions and sends the information to the transmitter. Upon receiving this information, the transmitter enters the operation phase and starts sending packets based on the error avoidance algorithm.

15.5.3.2 Operation Phase

In the operation phase, the transmitter avoids transmitting packets in high-error regions. When the packet queue is not empty, the transmitter calculates its current location with respect to the chosen origin. A data transmission is permissible only if the entire expected transmission duration does not

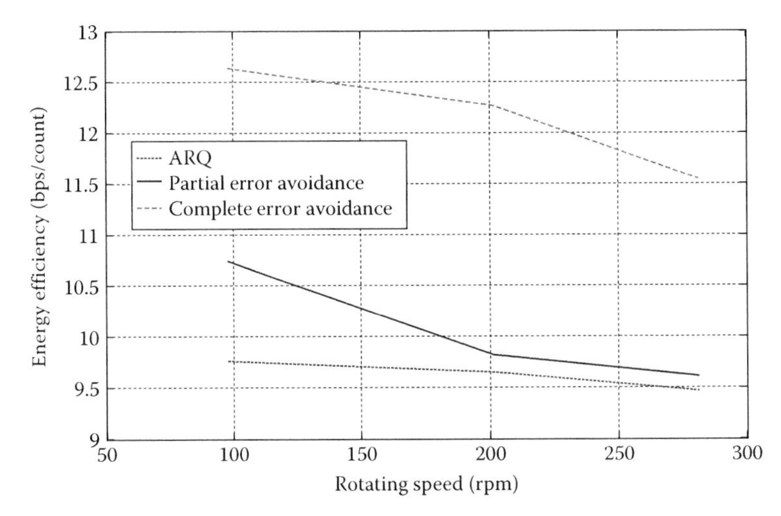

FIGURE 15.9 Energy efficiency for ARQ, complete, and partial error avoidance in the laboratory rotation system at 100–300 rpm.

overlap with any high-error regions; otherwise, the transmission will be delayed until the beginning of an earliest low-error region that is long enough for the transmission to complete. If error avoidance is combined with ARQ, either the partial error avoidance or complete error avoidance procedure will be incurred to handle retransmissions.

Figure 15.7 shows a comparison of the maximum achievable throughputs, referred to as the saturation throughput, obtained with ARQ, partial, and complete error avoidance. Figure 15.9 shows the energy efficiency (throughput/number of needed transmissions) for the same schemes. The extent of throughput improvement depends on the radio link's error region profile. The more high-error regions a link has, the larger throughput improvement will be achieved with error avoidance. The energy efficiency's improvement is even more pronounced to justify the importance of adopting an error avoidance scheme in such error-prone rotating environments.

15.6 Conclusions

The chapter describes the transmission error problem for IEEE 802.15.4 radios operating on fast rotating surfaces that are typical for machine monitoring systems. With experiments conducted on a commercial high-speed lathe and a laboratory rotating platform, the monotonic increase of transmission PER was demonstrated. The experiments also provided evidence that the signal strength variation caused by significant multipath effects in the machine was the primary cause of the transmission errors. While our prior work has derived mathematical models for predicting PER under rotation of different speeds, in this chapter, we showed a model for the throughput under given link path loss profiles and rotating speeds. In light of the cause of the transmission errors, an error avoidance approach was developed and shown to be effective in improving the achievable throughput of the monitoring system. The improved energy efficiency was even more significant than the throughput improvement, considering the substantial reduction of unproductive retransmissions with error avoidance.

Acknowledgment

The work reported here is in part supported by the National Science Foundation under grant number CMMI-0728035.

References

1. A. Tiwari, F.L. Lewis, and S.S. Ge, Wireless sensor network for machine condition based maintenance, in *Control, Automation, Robotics and Vision Conference*, Kunming, China, Vol. 1, pp. 461–467, 2004.
2. V. Sundararajan, A. Redfern, M. Schneider, and P. Wright, Wireless sensor networks for machinery monitoring, in *Proceedings of ASME International Mechanical Engineering Congress and Exposition (IMECE)*, Seattle, WA, pp. 1–9, 2005.
3. N. Ota and P. Wright, Trends in wireless sensor networks for manufacturing, *International Journal of Manufacturing Research*, 1(1), 3–17, 2006.
4. P. Wright, D. Dornfeld, and N. Ota, Condition monitoring in end-milling using wireless sensor networks (WSNs), *Transactions of NAMRI/SME*, 36, 177–183, 2008.
5. K.-C. Wang, L. Tang, and Y. Huang, Wireless sensors on rotating structures: Performance evaluation and radio link characterization, in *Proceedings of ACM Workshop on Wireless Network Testbeds, Experimental Evaluation and Characterization*, Chania, Greece, pp. 3–10, September 2007.
6. IEEE. IEEE Standard for Information Technology Part 15.4: Wireless medium access control (MAC) and physical layer (PHY) specifications for low-rate wireless personal area networks (WPAN), New York, IEEE Press, 2003.
7. ZigBee Alliance, ZigBee Specification, http://www.zigbee.org/Default.aspx, 2005
8. The International Society of Automation, ISA-100.11a-2009 Wireless systems for industrial automation: process control and related applications. http://www.isa.org/Template.cfm?Section=Standards2&template=/Ecommerce/ProductDisplay.cfm&ProductID =10766, accessed in May 2010.
9. HART Communication Foundation, Wireless HART technology, http://www.hartcomm.org/protocol/wihart/wireless_technology.html, accessed in May 2010.
10. D. Lal, A. Manjeshwar, F. Herrmann, E. Uysal-Biyikoglu, and A. Keshavarzian, Measurement and characterization of link quality metrics in energy constrained wireless sensor networks, in *Proceedings of the IEEE Global Telecommunications Conference (Globecom'03)*, San Francisco, CA, pp. 446–452, December 2003.
11. A. Cerpa, N. Busek, and D. Estrin, SCALE: A tool for simple connectivity assessment in lossy environments, Center for Embedded Networked Sensing, UCLA, Technical Report 0021, pp. 1–16, September 2003,.
12. N. Reijers, G. Halkes, and K. Langendoen, Link layer measurements in sensor networks, in *Proceedings of the 1st IEEE International Conference on Mobile Ad-hoc and Sensor Systems*, Fort Lauderdale, FL, pp. 224–234, 2004.
13. J. Werb, M. Newman, V. Berry, S. Lamb, D. Sexton, and M. Lapinski, Improved quality of service in IEEE 802.15.4 mesh networks, in *Proceedings of International Workshop on Wireless and Industrial Automation*, San Francisco, CA, pp. 1–6, March 2005.
14. L. Tang, K.C. Wang, Y. Huang, and F. Gu, Channel characterization and link quality assessment of IEEE 802.15.4-compliant radio for factory environment, *IEEE Transactions on Industrial Informatics*, 3(2), 99–110, 2007.
15. V. Sarkimaki, R. Tiaien, T. Lindh, and J. Ahola, Applicability of ZigBee technology to electric motor rotor measurements, In *Proceedings of the International Symposium on Power Electronics, Electrical Drives, Automation and Motion*, Taormina, Italy, pp. 137–141, 2006.
16. S. Orhan, N. Akturk, and V. Celik, Vibration monitoring for defect diagnosis of rolling element bearings as a predictive maintenance tool: Comprehensive case studies, *NDT & E International*, 39(4), 293–298, 2006.
17. A. Willig, K. Matheus, and A. Wolisz, Wireless technology in industrial networks, *Proceedings of the IEEE*, 93(6), 1130–1151, 2005.
18. L. Tang, K.C. Wang, and Y. Huang, Performance evaluation and reliable implementation of data transmission for wireless sensors on rotating mechanical structures, *Structural Health Monitoring*, 9, 113–124, 2009.

19. Hardinge, Inc. 2003. Talent 6/45 CNC Lathe. Available online at: http://www. hardinge.com, accessed in May 2012.
20. Crossbow Technology, Inc. 2007. MICAz Motes. Available online at: http://www.xbow.com, accessed in May 2012.
21. Agilent Technologies, 1999. 8714 ES RF Network Analyzer. Available online at http://www.home. agilent.com/agilent/, accessed in May 2012.
22. L. Tang, Packet error rate predictive model for sensor radios on fast rotating structures, PhD thesis, Clemson University, Clemson, SC, August 2010.
23. J. Elson, L. Girod, and D. Estrin, Fine-grained network time synchronization using reference broadcasts, In *Proceedings of the Fifth Symposium on Operating Systems Design and Implementation (OSDI 2002)*, Boston, MA, pp. 147–163, December 2002.
24. K.-C. Wang, J. Jacob, L. Tang, and Y. Huang, Transmission error analysis and avoidance for IEEE 802.15.4 wireless sensors on rotating structures, *International Journal of Sensor Networks*, 6(3/4), pp. 224–233, 2009.

16

Detecting Sybil Attacks in Vehicular Ad Hoc Networks*

Tong Zhou
Duke University

Romit Roy Choudhury
Duke University

Peng Ning
North Carolina State University

Krishnendu Chakrabarty
Duke University

16.1 Introduction

16.1.1 Objectives

Vehicular ad hoc networks (VANETs) are playing an increasingly important role in traffic control, accident avoidance, and management of parking lots and public areas. VANET is composed of vehicles equipped with short-range radios. Vehicles in VANET can connect to other vehicles and send/receive messages via the radio. Security and privacy are two major concerns in VANETs. Unfortunately, in VANETs, many privacy-preserving schemes are vulnerable to Sybil attacks, whereby a malicious user can pretend to be multiple (other) vehicles. In this chapter, we introduce a lightweight and scalable protocol to detect Sybil attacks. We go through the entire lifetime of securing vehicles' messages and identities, starting from key generation and distribution all the way to the Sybil attack detection. Simulation results are shown for a realistic test case to highlight the overhead for a centralized authority such as the Department of Motor Vehicle (DMV), the false-alarm rate, and the detection latency. The results also quantify the inherent trade-off between security, that is, the detection of Sybil attacks and detection latency, and the privacy provided to the vehicles in the network.

16.1.2 Background

VANETs are being advocated as a means to increase road safety and driving comfort, as well as to facilitate traffic control [5,7]. For example, cars can collectively sense information about traffic congestion and relay them to other cars, toll stations, or the DMV to facilitate traffic rerouting. Several other applications can become feasible if vehicles cooperate among themselves to achieve a common goal. When designing a cooperation-based system, it is important to address security and privacy concerns. The system needs to be robust to noncooperating entities and should ideally be able to detect/punish them quickly. To ensure the authenticity of messages propagated in VANET, a straight-forward method is to use public keys certified by a certification authority (CA) to sign the messages. The certified public keys are called "pseudonyms." On the other hand, in order to prevent vehicles from being tracked by identifying the keys that are used, each vehicle can switch among multiple pseudonyms, which are difficult to correlate to each other. With this approach, it is difficult for an attacker to identify vehicles by examining the used keys. The aforementioned scheme has been proposed by many researchers [1,2,9,11,15] and works efficiently.

Although the previous method protects the privacy of the vehicles, it leaves another security hole. Since it is difficult to tell whether two messages are from the same vehicle by examining their public keys, a malicious vehicle may pretend to be multiple vehicles (*a Sybil attack*) and then distribute false information. The deleterious effects of such attacks can cascade through the network.

In Refs [8,12], a lightweight solution is proposed to solve this issue. Vehicles only hold one valid pseudonym at a time and are expected to obtain a new pseudonym from a trusted road-side box (RSB) or from the online CA if the current pseudonym becomes invalid. In this scheme, it is critical that the vehicles have access to a CA when it needs to update its pseudonym. Without such an online infrastructure support, the vehicles are not able to obtain new pseudonyms and send signed messages. Also, if an attacker compromises an RSB, he/she can issue many certified pseudonyms to malicious vehicles, thus creating false messages in that area.

Yet another technique exploits directional antennas to identify the position/direction from which a message arrives [4]. A car launching a Sybil attack is expected to get caught because all the duplicate messages will come from the same position. However, in dense networks, localization errors can lead to frequent false positives. More importantly, a smart attacker may use directional antennas to mislead its neighbors about its directions.

In our proposed detection scheme, a vehicle is guaranteed to have a pseudonym to use at any time. The RSBs are only used for detecting pseudonym abuse; therefore, the compromise of RSBs does not have any significant impact on the security and privacy of the whole system. Moreover, our scheme does not require hardware support such as directional antennas, and the detection only relies on the messages.

16.2 System Model

In this section, we describe our assumptions regarding to the VANET system, capabilities of the attackers, and the Sybil attacks.

16.2.1 Assumptions on VANET Architecture

1. The *DMV* is the trusted party that maintains vehicle records and distributes certified pseudonyms to vehicles. The DMV has enough resources to generate pseudonyms quickly and store all the vehicle-related information and is referred to when any authoritative clarification is necessary. However, such DMV services are not designed for heavy network traffic—excessive communication can cause the DMV to become a bottleneck.
2. *Vehicles* are untrusted parties. They communicate with each other in a multi-hop manner. A message exchanged among vehicles is signed with a DMV-certified pseudonym.
3. *RSBs* are wireless access points. They are scattered along the road and connected to the DMV via a backhaul network, acting as intermediates to the DMV. The RSBs monitor vehicular activity, identify suspicious behavior, and report to the DMV for confirmation and punishment. The RSBs may be compromised, thus they cannot be used for critical functions—for example, the RSB cannot authenticate a message or distribute pseudonyms. However, they can be used to improve the scalability of a system.

The overall architecture of a VANET is shown in Figure 16.1.

16.2.2 Assumptions of Attackers' Actions

In this section, we discuss the actions of attackers that we are interested in.

1. *Announce false messages—false data injection*: A vehicle can sign a false message and then broadcast it. Such an attack cannot be detected from the message itself, since the message can be signed by a CA-certified pseudonym. This attack may be addressed by a majority voting scheme if there are more benign vehicles than attackers. However, the voting scheme fails if the attacker carries out Sybil attack by generating sufficient false identities to outvote benign vehicles, which we will discuss next.
2. *Pretend to be multiple vehicles by using multiple pseudonyms—Sybil attack*: In VANET, a vehicle can carry out a Sybil attack by using its multiple pseudonyms to sign messages. If a vehicle can be identified from a set of pseudonyms it uses, then the vehicle's privacy is compromised. As a result,

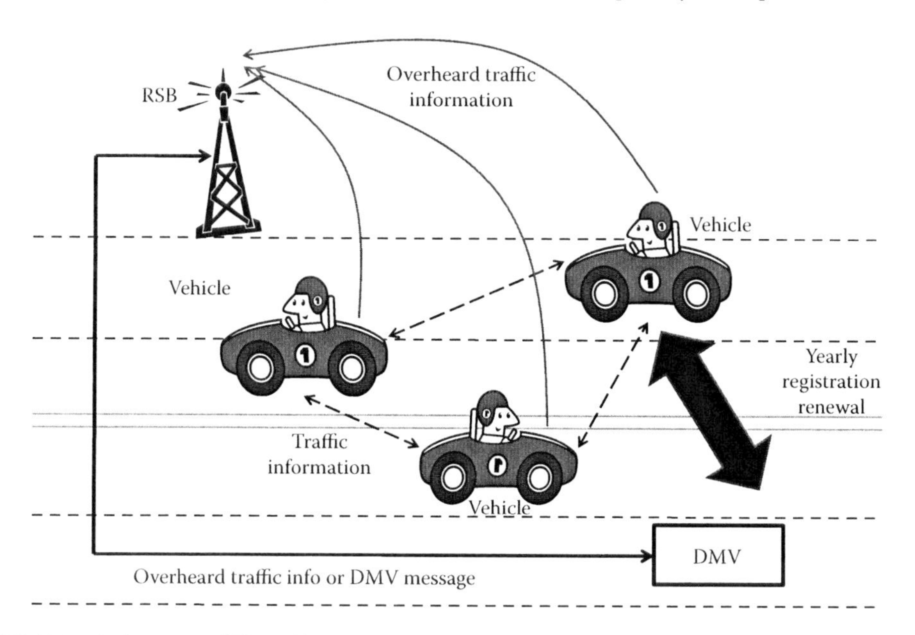

FIGURE 16.1 Architecture of VANET.

vehicles and RSBs that overhear multiple messages signed with an attacker's pseudonyms have no means of recognizing that these pseudonyms actually belong to the same vehicle. This chapter proposes a method to solve this problem.

3. *Compromised RSBs*: RSBs are semi-trusted parties, and some RSBs may be compromised by the attackers. We assume that RSB compromises can be easily detected by the DMV, thus being quickly revoked. However, attackers can still gain information stored in the RSBs. Therefore, a scheme's resilience to RSB compromise is determined by the amount of information released to and held by the RSBs.

In summary, it is necessary to have a framework allowing RSBs to detect a Sybil attack, without knowing the association between pseudonyms and unique vehicle IDs (i.e., without compromising privacy). The framework design needs to be scalable in terms of the workload it imposes on the DMV and needs to be robust against RSB compromise. Moreover, it is important to quickly detect the attack and perform subsequent punishments/revocations to minimize the impact of the attack. To address these concerns, we propose a new scheme, referred to as privacy-preserving detection of abuses of pseudonyms (P^2DAP).

16.2.3 Structure of Events and the Use of Pseudonyms

In vehicular network applications, vehicles are expected to broadcast specific events whenever they observe them. Counting the number of vehicles that send the same message is an important primitive that several VANET applications depend on. To achieve the notion of *same* or *different* events, we unambiguously define the format of "events."

An event is a tuple (t, l, e) *generated at a predefined time interval* $t \in \mathbf{T}$, *in a predefined region* $l \in \mathbf{L}$ *for an event type* $e \in \mathbf{E}$, *where* $\mathbf{T}, \mathbf{L}, \mathbf{E}$ *are distributed to vehicles.*

For example, we can start from 12:00 am. Each interval of 20 min is considered as an event time interval, and we consider from 1:00 pm to 1:20 pm as time interval t_0. Without loss of generality, we make an interstate highway segment from exit 279 to exit 280 as region l_0. We then take "car collision" as an event class e_0. Any vehicle reporting a car collision between exit 279 and exit 280 on the highway segment between 1:00 pm and 1:20 pm generates a report (t_0, l_0, e_0).

An attacker carries out the Sybil attack by abusing multiple pseudonyms. On the other hand, in order to avoid being tracked, benign vehicles can also use multiple pseudonyms to report events. In order to distinguish the benign use of pseudonyms from the abuse of pseudonyms, we now introduce the following restriction on the use of pseudonyms.

A benign vehicle can use only one pseudonym to sign one event.

If a vehicle uses multiple pseudonyms to sign an event such that others think there are multiple vehicles reporting the same event, the action is considered to be a *Sybil attack*, and the vehicle is deemed to be malicious.

16.3 Proposed P^2DAP Scheme

This section presents our scheme on handling Sybil attacks. The main purpose is to detect Sybil attacks and revoke malicious vehicles immediately after detection. A straightforward method is to forward all the reported events to the DMV and let the DMV examine the signatures of each message. On observing a single event (t_i, l_j, e_k) signed with two different pseudonyms of the same vehicle, the DMV considers that vehicle as an attacker. The drawback of this method is the heavy network traffic on the DMV. To reduce the communication overhead, we propose P^2DAP schemes in which RSBs perform most of the DMV's task. We also discuss how our schemes preserve privacy in case of RSB compromise.

16.3.1 Complete Two-Stage P^2DAP Scheme

We first propose the complete two-stage P^2DAP scheme, abbreviated as C-P^2DAP. Later, we propose a number of variances of C-P^2DAP to improve the performance of the scheme. In P^2DAP scheme, we delegate most of the detection to RSBs and involve the DMV only when suspected vehicles need to be confirmed as a Sybil attacker. However, since RSBs are not trusted entities, the vehicle information available to the DMV cannot be transferred to the RSBs. In view of these constraints, we divide the vehicles into groups and release the group information to RSBs. Such information allows RSBs to detect suspicious behavior but is not sufficient for RSBs to track vehicles because RSBs cannot distinguish a vehicle from a group of vehicles. To group the vehicles, we use the one-way hash function to hash the pseudonyms when generating them.

16.3.1.1 Initialization Step

Initially, the DMV knows the total number of vehicles and sequentially generates a sufficient number of yearly pseudonyms for all the vehicles. After generating a pseudonym p, the DMV first hashes $(p|\kappa_c)$ using a one-way hash function, where κ_c is a global key. It then selects a set of bits from the hashed result to create hash collisions. The selected bits are referred to as "coarse-grained hash value." After that, the pseudonym p is placed into a group, which stores the pseudonyms with the same coarse-grained hash values. In other words, for each pseudonym p_l in the mth coarse-grained group, we have $H(p_l|\kappa_c) = \Gamma_m$, where H is a one-way hash function, and Γ_m is the coarse-grained hash value for group m. We refer such groups as "coarse-grained groups." The key κ_c will be distributed to all the RSBs.

Next, the DMV calculates the hash value for the aforementioned p with a new key κ_f and selects a set of bits from the result. The bits selected from the new hash value are referred to as the "fine-grained hash value." The pseudonym p is then placed into a subgroup of the coarse-grained group, namely, fine-grained group, in which all the pseudonyms have the same fine-grained hash value. For each pseudonym p_l' in the nth fine-grained group under the mth coarse-grained group, we have $H(p_l'|\kappa_f) = \Theta_n$, where Θ_n is the fine-grained hash value for the subgroup n.

The aforementioned steps are referred to as "two-level hash" and are shown in Figure 16.2. The DMV keeps generating two-level hashing pseudonyms until all fine-grained groups contain enough pseudonyms for a vehicle's use. After that, the DMV loads a unique fine-grained group of pseudonyms to each

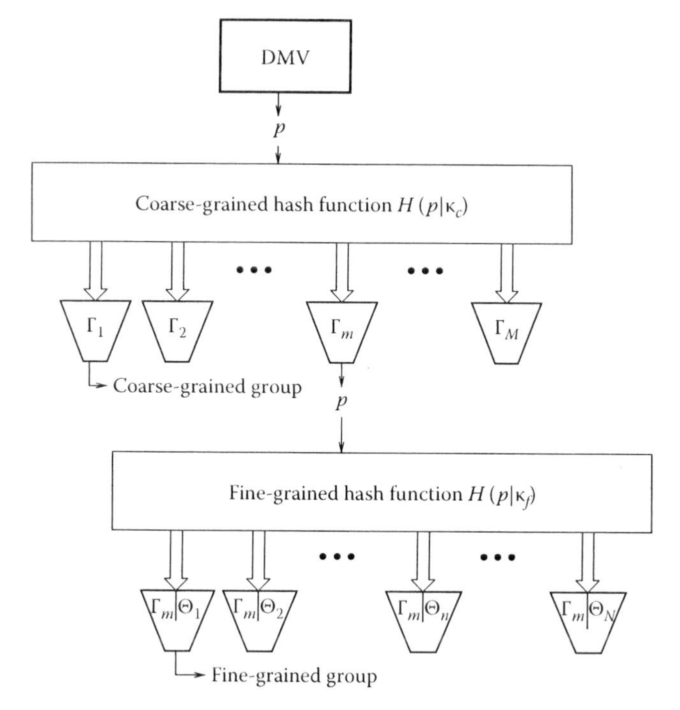

FIGURE 16.2 Generation and two-level hashing of a pseudonym p.

vehicle at the time of yearly vehicle registration and stores the corresponding $(\Gamma_m|\Theta_n)$ as the vehicle's secure plate number. From the aforementioned description, it is obvious that the mapping from secure plate numbers to vehicles is one to one. Thus, the DMV needs to carefully choose the length of Γ_m and Θ_n, such that the total number of available secure plate numbers is greater than or equal to the number of vehicles.

The two-level hashing saves storage for the DMV because the DMV can link a pseudonym to a vehicle by calculating its coarse-grained and fine-grained hash values and then comparing them with the secure plate number $(\Gamma_m|\Theta_n)$. This obviates the need of maintaining vehicle secure plate numbers and pseudonym association.

After the initialization stage, the DMV stores the secure plate number for each vehicle and secretly keeps the fine-grained hash key κ_f.

16.3.1.2 Generating Pseudonyms with Short-Period Keys

When generating the pseudonyms, we need to consider the lifetime of a coarse-grained key κ_c because an attacker gaining access to an RSB can partially learn the pseudonyms of all the vehicles for that lifetime. If the lifetime is too long, the privacy of the vehicles will be severely impaired. Therefore, κ_c should be given a short lifetime such as 1 or 2 days. We then modify the initialization stage as follows.

We divide a whole year's time into Ω time intervals. Each time interval can be 1 day. In the initial pseudonym generation, the DMV uses a set of coarse-grained keys K_c, instead of one key, κ_c, to hash the pseudonyms. Each key $\kappa_{c,\gamma} \in K_c$ is used to generate pseudonyms for the γth time interval. The pseudonyms that are hashed to Γ_m with the key $\kappa_{c,\gamma}$ will be put into the mth coarse-grained group, and they can only be used in the γth time interval. After that time interval, these pseudonyms will be discarded. To prevent malicious vehicles from using expired pseudonyms to sign events, the DMV uses different keys to generate certificates for pseudonyms in different time intervals. Thus, vehicles can recognize an expired pseudonym by examining its certificate. Initially, the DMV secretly holds all the coarse-grained hash keys. At the beginning of γth time interval, the DMV releases the key $\kappa_{c,\gamma}$ to the RSBs. By this approach, an RSB holds each valid coarse-grained key only for a short time. When an RSB is compromised, the attacker only obtains the coarse-grained hash key for the current time interval. We do not impose any restrictions on the fine-grained key κ_f because the DMV does not release it, and an attacker cannot obtain it by compromising an RSB.

Comparing to the long-period keys, this short-period key generation uses Ω coarse-grained hash keys instead of one, thus bringing an extra storage overhead to the DMV. At the same time, the computational overhead will increase, which we show in Section 16.5.2.

16.3.1.3 Sybil Attack Detection Step

When vehicles communicate, an RSB overhears all the vehicles within their communication range and puts the pseudonyms used to sign the event (t_i, l_j, e_k) in the list $L_{i,j,k}$. When all pseudonyms for the event (t_i, l_j, e_k) are collected, the RSB detects the Sybil attacks as follows: The RSB goes through each pseudonym p in the list $L_{i,j,k}$ and computes the coarse-grained hash value $H(p|\kappa_c)$. (Recall that κ_c is predistributed to all RSBs in the initialization step.) If $\exists p, p' \in L_{i,j,k}$ such that $H(p|\kappa_c) = H(p'|\kappa_c)$, then the RSB notices that two pseudonyms of the same coarse-grained hash value are used to sign the event (t_i, l_j, e_k). This can be either (1) a Sybil attack, where one vehicle is using multiple pseudonyms to report the same event, or (2) a *false alarm*, where an event is reported by two vehicles whose pseudonyms are in the same coarse-grained group. The RSB cannot discriminate between (1) and (2) and it sends the report to the DMV. The report contains the event (t_i, l_j, e_k), the hash value Γ, the pseudonyms whose coarse-grained hash value is Γ, the signatures of the event, and the certificates accompanying the pseudonyms.

On receiving an RSB report, the DMV first verifies the signatures and the coarse-grained hash value Γ to prevent a compromised RSB from implicating a benign vehicle. If the RSB proves to be bona fide, the DMV calculates the fine-grained hash value $H(p|k_f)$ for each pseudonym p in the RSB report.

If $\exists p, p'$ in the report such that $H(p|\kappa_j) = H(p'|\kappa_j)$, the DMV concludes that p and p' are from the same vehicle that has attempted a Sybil attack. The DMV then takes further action to punish or revoke the malicious vehicle.

In this scheme, a Sybil attack is guaranteed to be detected. However, when the vehicles are densely distributed, false alarms can happen often.

16.3.2 E-P^{2}DAP—Detecting Events Instead of Sybil Attack

In the C-P^{2}DAP scheme, an RSB reports to the DMV whenever it finds any set of pseudonyms that hash to the same coarse-grained values. Thus, when an event is reported by a large number of vehicles, C-P^{2}DAP can cause false alarms. Clearly, on a road with heavy traffic, such false alarms will create a heavy communication overhead on DMV. To address this problem, we observe that detecting each and every Sybil attack may not be necessary for practical VANET applications. We first make the following assumptions: (1) each false (faked) event is generated by only one malicious vehicle; (2) benign vehicles will not report false events. We then propose the event-P^{2}DAP scheme (abbreviated as E-P^{2}DAP), which does not detect all Sybil attacks, but only detects those creating false events.

For an event (t_i, l_j, e_k), the RSB collects a list of pseudonyms $L_{i,j,k}$ used to sign the event. If $\forall p,p' \in L_{i,j,k}$, $H(p|\kappa_c) = H(p'|\kappa_c)$, that is, all the pseudonyms used to sign (t_i, l_j, e_k) have the same coarse-grained hash value, then the event is probably sent from only one vehicle and is likely a faked event. In this case, the RSB generates a report with the same format as in C-P^{2}DAP and sends it to the DMV.

In this scheme, the DMV only needs to examine the pseudonyms in two cases: (1) an attacker reports a false event and carries out a Sybil attack; (2) a true event is reported by multiple benign vehicles whose pseudonyms have the same coarse-grained hash value, which is a false alarm. Obviously, the number of false alarms is likely to be small compared to the total number of the pseudonyms that RSBs process. Therefore, the RSBs are able to efficiently take over most of the pseudonym processing tasks, thus reducing the burden on the DMV.

16.3.3 T-P^{2}DAP—Detecting Collusion

One issue with the E-P^{2}DAP scheme is that it cannot detect colluding vehicles, that is, two or more malicious vehicles reporting the same faked event. In order to address the collusion, we propose threshold-P^{2}DAP (abbreviated as T-P^{2}DAP), described as follows. We assume each faked event is generated by a small number of colluding attackers instead of one attacker, but that number will not exceed a threshold τ. Then, for a pseudonym list $L_{i,j,m}$, the RSB calculates the coarse-grained hash value for each pseudonym $p \in L_{i,j,k}$ and obtains a set of coarse-grained hash values S_c. If $|S_c| \leq \tau$ and two or more pseudonyms in $L_{i,j,k}$ map to the same coarse-grained hash value, the RSB suspects the event to be fake and reports to the DMV. Similar to the aforementioned C-P^{2}DAP and E-P^{2}DAP schemes, the DMV in the T-P^{2}DAP scheme then examines the RSB report and finds out whether the event is from attackers.

Comparing to E-P^{2}DAP, T-P^{2}DAP is more resilient to collusion. Any false event reported by less than τ attackers can be detected by an RSB. Obviously, T-P^{2}DAP has a larger false-alarm rate than E-P^{2}DAP.

In this section, we introduced C-P^{2}DAP, E-P^{2}DAP, and T-P^{2}DAP. Note that the first two schemes are special cases of T-P^{2}DAP. In T-P^{2}DAP, if $\tau = 1$, an RSB detects a Sybil attack by verifying that the coarse-grained hash values of the pseudonyms used to sign a single event are the same. Thus, T-P^{2}DAP becomes E-P^{2}DAP. On the other hand, when $\tau \geq$ number of coarse-grained hash values, an RSB reports a Sybil attack once it finds the coarse-grained hash values of two pseudonyms used to sign an event are the same. In this case, T-P^{2}DAP becomes C-P^{2}DAP. Currently, the T-P^{2}DAP cannot detect the colluding vehicles if each malicious vehicle only reports a faked event with one pseudonym. However, such an attack is not a Sybil attack and is beyond the scope of this chapter. Section 16.5 shows more details of performance comparison between the three schemes.

16.4 Discussions: Improvements on P²DAP

In this section, we propose several improvements to the aforementioned P²DAP schemes from the perspectives of key revoking convenience and adaptivity.

16.4.1 Revoking the Pseudonyms of Malicious Vehicles

After a malicious vehicle is detected, the DMV should revoke all its pseudonyms. In this section, we discuss three possible approaches of revoking vehicles that can be combined with P²DAP.

The first two approaches are to use the revocation schemes proposed in Ref. [10], that is, revocation of the tamper-proof device (RTPD) and revocation using compressed certificate revocation lists (RC²RL). The RTPD requires the hardware support on the vehicle, in which a tamper-proof device (TPD) used to store pseudonyms and sign messages is installed on each vehicle. On observing a malicious vehicle, the DMV sends a revocation message to the TPD, and then the TPD will erase all the pseudonyms and stop signing messages. By this approach, the DMV can revoke a vehicle in a single message.

Different from RTPD, RC²RL does not assume any hardware support such as the TPD. Instead, it creates a bloom filter for all the pseudonyms to be revoked. The bloom filter is then broadcasted to all the vehicles. On receiving a message, a vehicle uses the bloom filter to verify its pseudonym and drop it if the pseudonym is found revoked. In Ref. [10], the size of each bloom filter is estimated as tens of kilobytes. Therefore, in order to revoke a vehicle, the DMV needs to flood tens of kilobytes throughout the network.

The third approach is to create a secret key for each vehicle that helps to identify its pseudonyms. Such a secret key is regarded as a "backdoor" in this chapter. The group signature (GS) schemes proposed in Ref. [1] can be used to generate such a backdoor. In the GS scheme, each vehicle is equipped with a group public key gpk_{CA} and a private signature key gsk_V and generates its own pseudonyms. On creating a pseudonym K_V^i, the vehicle generates a GS $\Sigma_{CA,V}(K_V^i)$ for K_V^i with the private key gsk_V and then calculates a certificate $Cert_{CA}^H(K_V^i)$ using the group public key gpk_{CA}. Other vehicles can verify K_V^i by validating $\Sigma_{CA,V}(K_V^i)$ using $Cert_{CA}^H(K_V^i)$. When the DMV needs to revoke a vehicle, it simply broadcasts the vehicle's private signature key gsk_V.

The idea of GS is adopted in the initialization stage of P²DAP, in which the DMV generates gsk_V and $\Sigma_{CA,V}(K_V^i)$ for the vehicle V when generating a pseudonym K_V^i. The DMV will distribute the pseudonyms, the GSs, and the group public key gpk_{CA} to the vehicles, and keep the private key gsk_V as the "backdoor" of V. The vehicles can then use the pseudonyms to sign the messages and use gpk_{CA} to verify certificates of the pseudonyms. When the DMV needs to revoke a vehicle V, it broadcasts gsk_V, like the DMV does in the GS scheme. On receiving gsk_V, a vehicle can verify whether a pseudonym is from vehicle V, thus being able to identify the messages from a revoked vehicle.

In this section, three different approaches of revoking pseudonyms are discussed. In these approaches, RTPD incurs small communication and computation overhead, but it requires hardware support. RC²RL and GS do not need hardware support, but they require large communication overhead and computation overhead, respectively. Each method has its pros and cons, and which one to choose depends on the system's available resources.

16.4.2 τ-P²DAP: Real-Time Adaptive P²DAP Scheme

This section discusses the P²DAP scheme adaptive to the real-time traffic. For a particular RSB, the number of nearby vehicles is a time-varying value. For example, the study in Ref. [13] shows that within 1 day, the traffic volume in a street near Incheon International Airport ranges from 10 to 3000 vehicles/h. Such fluctuations in traffic volume cause difficulty in using a single detection scheme to efficiently detect the attackers. When there are fewer vehicles, the E-P²DAP or the T-P²DAP with a small threshold is preferred, such that the malicious vehicles can be detected with a smaller cost. When the traffic volume is high, the C-P²DAP, or the T-P²DAP with a large threshold, is selected to better catch collusions. A method to solve this problem is to make each RSB adaptively choose detection scheme, the DMV, based on the traffic volume.

We then propose τ-P²DAP in which each RSB checks the total number of received packets for each reported event (written as N_{PE}) when it attempts to detect a Sybil attack. N_{PE} can then be a parameter to calculate the value of τ in the T-P²DAP, using the equation $\tau = \alpha N_{PE}$, where α is an estimated proportion of attackers among all vehicles. The value of α can be either predistributed to the RSBs or learned by the RSBs during detection of attackers. Using α, we estimate the number of nearby attackers and use this estimation as the value of τ.

With τ-P²DAP, the RSB is expected to adaptively report to the DMV based on the current traffic status. The algorithm may not always be the best one if the malicious vehicles intentionally mislead the RSB. For example, the malicious vehicles can use an extremely large number of pseudonyms to sign an event such that the RSB is forced to use C-P²DAP and incurs huge communication overhead to the DMV. Also, several colluding malicious vehicles can generate many events, each signed with a small number of pseudonyms, such that the RSB will use E-P²DAP algorithm and miss these events. However, in the first case, these malicious vehicles are guaranteed to be quickly detected, and the large communication overhead hence brought up will only last for a short period. In the second case, these faked events can be easily filtered out when there is a large enough number of benign vehicles around.

16.4.3 κ-P²DAP: Distribute Different Information to Different RSBs

This section discusses the adaptive P²DAP algorithm from the location perspective. Even in the same city or county, the traffic volume varies significantly on different roads. As shown in a survey of traffic volume in 2002 at Columbia, NY [6], on different streets in the Columbia county, the average traffic volume ranges from 100 to 25,000 vehicles/day. Thus, it is difficult to determine the number of coarse-grained groups when the P²DAP scheme is applied. A small number of coarse-grained groups will result in many false alarms on a highway, while a large number can harm the privacy of vehicles in a small community. To solve this issue, we propose κ-P²DAP, which extends the two-level hash keys to n-level hash keys to cope different traffic volumes. In this scheme, the DMV distributes different subsets of the hash keys to the RSBs based on the traffic volume near them.

When generating a pseudonym, instead of using two hash keys, the DMV uses q-hash keys $\{\kappa_1, \ldots, \kappa_q\}$ to calculate the hash values. With each hash key κ_i, the DMV calculates an ϵ bit hash value $\Gamma_{i,j} = H(p_j|\kappa_i)$ for the pseudonym p_j. The hash values for the pseudonym p_j is written as $\mathbf{\Gamma}_j = (\Gamma_{1,j}, \ldots, \Gamma_{q,j})$ and can be considered as an element of a q-dimensional space ν, in which each dimension has elements. The DMV keeps generating pseudonyms, until their hash values fill the space ν.

After all the pseudonyms are generated, the DMV distributes the pseudonyms that can hash to the same value with the keys $\{\kappa_1, \ldots, \kappa_\beta\}$ to the same vehicle, where β is chosen such that $2^{\beta_\epsilon} \geq N_V$. After that, DMV adaptively release the last δ hash keys $\{\kappa_{q-}\delta+1, \ldots, \kappa_q\}$ to the RSB, where δ is determined by the traffic volume around the RSB. If we want the RSB not being able to distinguish among vehicles passed by within an hour, we should have $\delta = \log_2 K_V/\epsilon$, where K_V is the number of vehicles passing by the RSB within an hour. Such an indistinguishability of one vehicle among N multiple vehicles is defined as *N-anonymity* [14], which is an important metric of privacy. More details about anonymity are discussed in Section 16.5.3.

One possible issue of κ-P²DAP is that an attacker can compromise an RSB that watches heavy traffic and use the knowledge to track vehicles on roads with less traffic. In this scenario, the vehicles will lose their privacy. Therefore, when releasing keys to an RSB, the DMV needs to consider both the local traffic and the cost of compromising the RSB. RSBs with more hash keys are expected to be more difficult to compromise. κ-P²DAP has increased computational overhead of the DMV and the RSBs. However, the communication overhead remains the same as the other P²DAP schemes discussed in Section 16.3.

16.5 Performance Evaluation

In this section, we evaluate the performance of P²DAP using the following metrics: computation/communication overhead of the DMV, privacy of the vehicles, and detection latency. We also examine the trade-off between detection latency and overhead. Results from the evaluation are expected to offer insights into the design of practical vehicular networks.

16.5.1 Simulation Setup

The P²DAP scheme is simulated in ns-2 version 2.29. We use the 802.11a MAC and PHY layer protocol, and the SHA-1 hashing as our hash function. In the initialization stage, we use SHA-1 hashing function to generate pseudonyms that are enough for our use. We also simulated two different scenarios—one comprised densely distributed vehicles (in Sections 16.5.4 and 16.5.5), and the other comprised sparsely distributed vehicles (in Section 16.5.6). The major simulation parameters are listed in Table 16.1.

In our simulation, we randomly generate events with different time intervals, locations, and types and then store them in a global array. When generating the events, we consider the length of time interval as 20 s, and the length of location segment as 250 m, with a total number of five different event types. Each vehicle periodically accesses the array, obtains the events with current time interval and the vehicle's current location, and broadcasts them. A vehicle also relays events heard from other vehicles.

We have defined four types of nodes: benign vehicle, malicious vehicle (attacker), RSB, and DMV. A benign vehicle frequently senses the events and signs and broadcasts them. Meanwhile, an attacker generates a random number of events, then signs each event with multiple pseudonyms and broadcasts them. Due to the small number of event types in our simulation, there is a high probability that two attackers report the same event, thus creating a colluding scenario. This behavior of attackers is called "semi-collusion" because there are times that they report events individually. We intuitively create such a behavior to test P²DAP's resilience to colluding attackers.

In the following sections, we present our results for the following schemes:

- C-P²DAP—detects all Sybil attacks.
- E-P²DAP—detects Sybil attacks that generate false events.
- T-P²DAP—detects collusions of a threshold number of attackers.
- τ-P²DAP—detects collusion with a traffic-volume-adaptive threshold number of attackers.

κ-P²DAP only differs from other algorithms in the initialization stage and does not have a different behavior when detecting Sybil attacks. Therefore, we do not evaluate its performance in our simulation.

TABLE 16.1 Parameters Used in Simulation

Parameters	Dense Vehicles	Sparse Vehicles
Street length (m)	2000	20,000
Communication radius (m)	200	50
Street width (lanes)	3	3
Lane width (m)	3	3
Vehicle speed (m/s)	25–35	25–35
Pseudonyms/vehicle per day	20	20
Vehicle packet rate (packets/s)	3	3
Simulation time (s)	400	800

16.5.2 Theoretical and Experimental Results: Computational Overhead of Generating Pseudonyms

Assume we have N_V vehicles in total, while each vehicle needs M pseudonyms. We also assume a hash function generating evenly distributed hash values. We first calculate an upper bound (defined as N_u) of the expected number of pseudonyms that the DMV needs to generate for all vehicles. We start from the case of $M = 1$ in which the problem is converted to a *coupon collector's problem* [3]. The expected number of generated pseudonym is $N_p \equiv N_V \log N_V + \mu N_V + \frac{1}{2} + o(1)$, where $\mu \approx 0.577$. Thus, for $M > 1$, $N_u = M \times (N_V \log N_V + \mu N_V + \frac{1}{2} + o(1))$. On the other hand, we can easily find from the definition of N_p that it has the lower bound of $O(MN_V)$. Therefore, we conclude that in order to generate a year's pseudonyms, the number of pseudonyms that the DMV needs to generate is between $O(MN_V)$ and $O(MN_V \log N_V)$.

We next calculate the cost of generating short-period pseudonyms. In this scenario, the pseudonyms of each vehicle are divided into d equal portions, and each portion is hashed with a unique key. Therefore, with each hash key, the DMV needs to generate M/d pseudonyms for each vehicle. In this case, we have

$$N_u = d\left(\frac{M}{d} \times \left(N_V \log N_V + \mu N_V + \frac{1}{2} + o(1)\right)\right) \tag{16.1}$$

$$= M \times \left(N_V \log N_V + \mu N_V + \frac{1}{2} + o(1)\right) \tag{16.2}$$

while the lower bound of the number of pseudonyms is still $O(MN_V)$. Therefore, the upper bound and the lower bound of the expected number of generated pseudonyms for short-period keys remain the same. Obviously, in the extreme case where each time interval only has one pseudonym, the expected number of generated pseudonyms will reach the upper-bound N_u.

We then use the simulator to generate the pseudonyms. Since we only need to obtain the number of pseudonyms generated by the DMV, we stopped the simulation right after the initialization stage. The comparison between the theoretical results and the experimental results of generating long-period pseudonyms are shown in Figure 16.3. It can be seen that the experimental results fall between the calculated upper bounds and lower bounds.

Also, in Figure 16.4, we show that short-period keys increase the expected number of generated pseudonyms. When dividing pseudonyms into 50 short periods, the number of generated pseudonyms almost doubles, yet it has not reached the upper bound N_u.

16.5.3 Experimental Results: Privacy

We first give the definition and metric of privacy in our scenario. If an RSB is compromised, the attacker can obtain the coarse-grained hash keys stored in the RSB, thus learning the coarse-grained hash values of all the pseudonyms. However, because the coarse-grained hash values are shared among multiple vehicles, the knowledge of a vehicle's coarse-grained hash value does not compromise its anonymity completely. Here, we are using the k-anonymity model in Ref. [14] to evaluate privacy; in order to avoid confusing k in the k-anonymity with our keys, we rename the model of privacy as N-anonymity and apply its definition to vehicular networks:

Given a set of vehicles $\{V_i\}_{1 \le i \le N_V}$, *a set of attribute values A and a one-way attribute function* $F: \{V_i\} \rightarrow A$, *the vehicle set is said to achieve N-anonymity if and only if for each attribute value* $a \in F(\{V_i\})$, *there are at least N occurrences of a in* $F(\{V_i\})$, *where* N_V *is the number of vehicles.*

From this definition, we see that the anonymity of the vehicles to an RSB equals to the number of fine-grained groups in each coarse-grained group. Therefore, we conclude that the anonymity of vehicles in

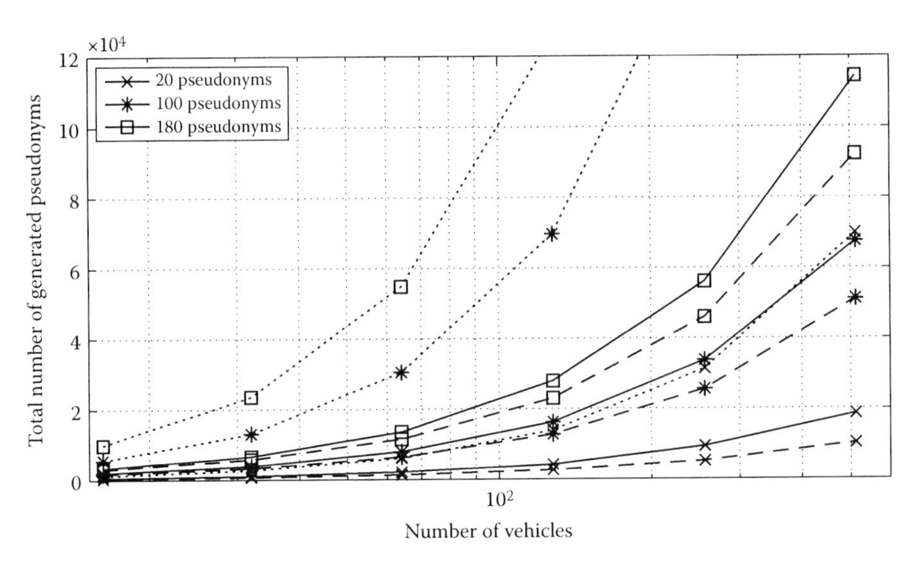

FIGURE 16.3 Computational overhead of the DMV when generating pseudonyms. Solid lines are the simulated results, dotted lines are the theoretically calculated upper bound values, and dashed lines are the theoretically calculated lower bound values.

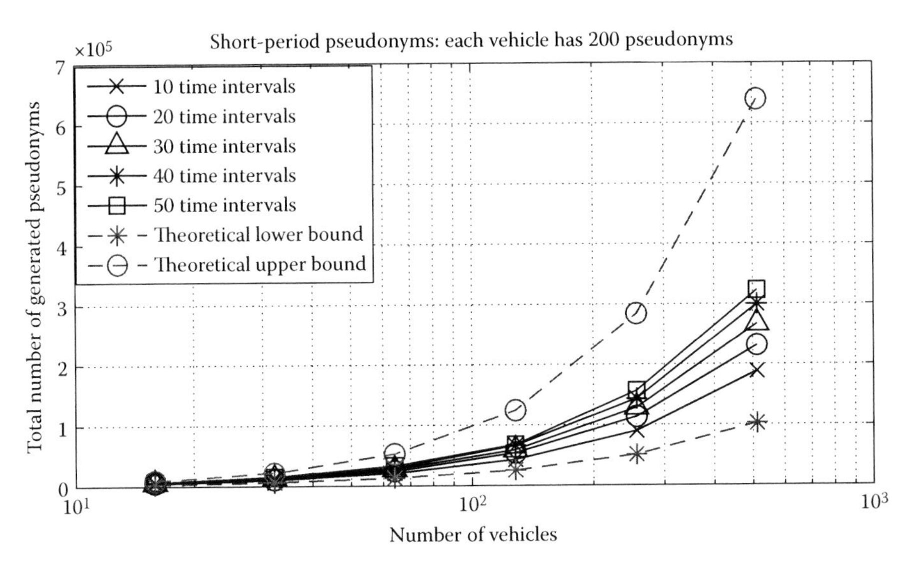

FIGURE 16.4 Computational overhead generating short-period pseudonyms. Each vehicle has a total of 200 pseudonyms.

case of RSB compromise is 2^{n_f}, where n_f is the number of bits in the fine-grained hash value. In other words, the anonymity is $M/\,2^{n_c}$, where n_c is the number of bits in the coarse-grained hash value. In order to study the privacy of vehicles in a subset of all the vehicles, we generate pseudonyms for 256 vehicles and randomly pick a subset of vehicles to examine their anonymity. The results are shown in Figure 16.5, from which we see the anonymity of the vehicles quickly converges to 0 when the number of bits of coarse-grained hash values goes to 5. For more vehicles, we expect a longer coarse-grained hash value is required to reduce the anonymity. For 2^{24} (more than a million) vehicles, we expect a 20 bit coarse-grained hash value can make the anonymity 0.

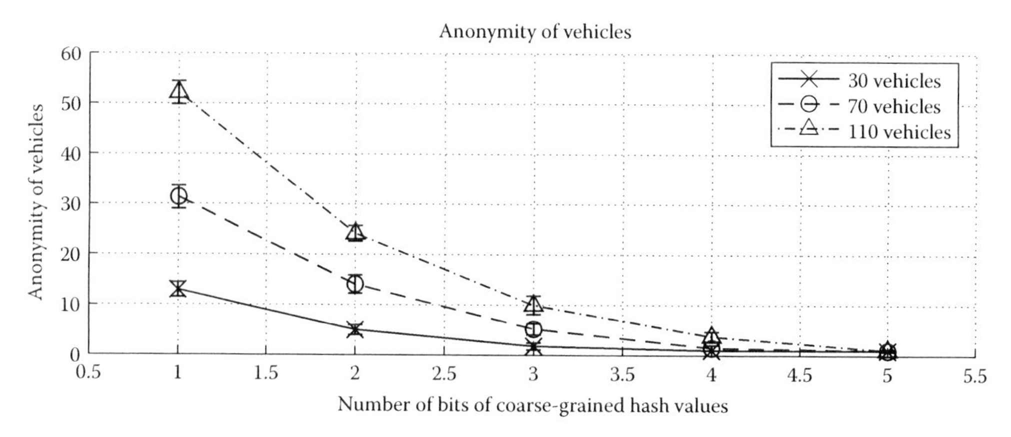

FIGURE 16.5 Anonymity of a subset of all the vehicles.

16.5.4 Experimental Results: Communication Overhead

16.5.4.1 Overhead on the RSBs

Figure 16.6 shows the number of packets processed by an RSB. From the figure, obviously the number of packets received by an RSB increases with the increase in the number of attackers or the number of benign vehicles. We include these results here for the later comparison of the overhead on the DMV and show the reduction of overhead with the introduction of P^{2}DAP.

16.5.4.2 Overhead on the DMV

We next examine the number of packets sent to the DMV when an RSB detects suspicious activities and reports to the DMV. This metric is indicative of the communication overhead over the backhaul network connecting the RSB and the DMV. Moreover, the number of packets forwarded by the RSB dictates the computation overhead of the DMV, since the latter must process each of these packets to detect/confirm a Sybil attack.

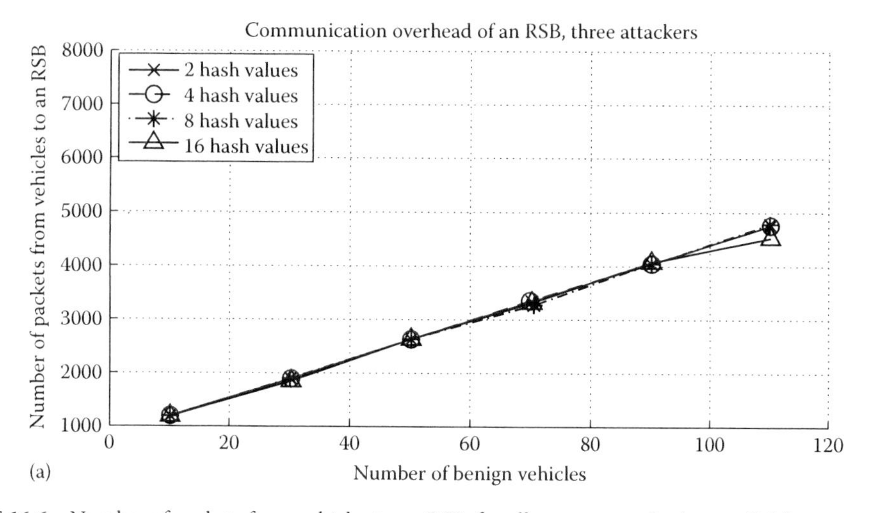

FIGURE 16.6 Number of packets from vehicles to an RSB: for all our proposed schemes. Subfigures a and b are displaying the results of experiments with different number of malicious users.

(continued)

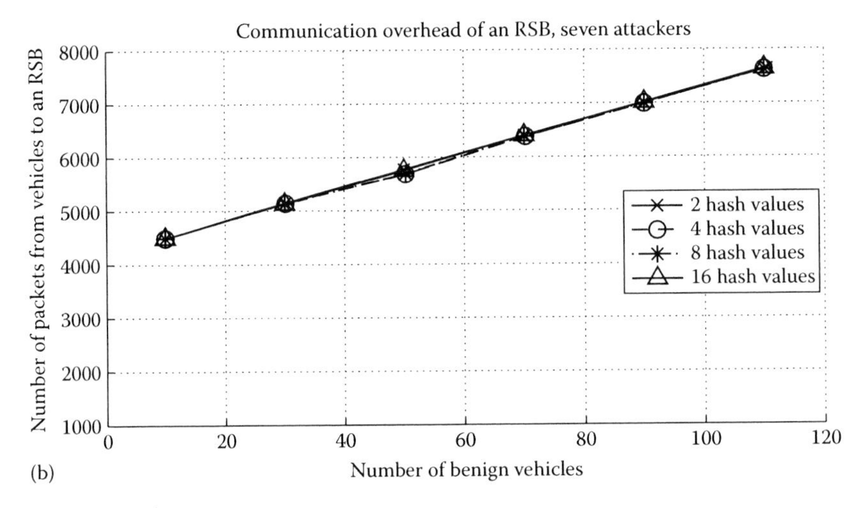

FIGURE 16.6 (continued)

- *Overhead on the DMV: C-P²DAP*

  We first show the results of the C-P²DAP scheme in Figures 16.7 and 16.8. We observe an increase in the number of transmitted packets when the number of coarse-grained hash values increases in Figure 16.7. This result is because for a suspected event with a given number of pseudonyms, when the number of coarse-grained hash values increases, the number of packets used to report the event increases. On the other hand, from Figure 16.8, we see the number of pseudonyms from an RSB slightly decreases for increasing number of coarse-grained hash values. While this result seems to contradict the results shown in Figure 16.7, it can be easily explained as follows: in C-P²DAP, a larger number of coarse-grained hash values result in a smaller number of false alarms.

  From both Figures 16.7 and 16.8, we see that the communication overhead of the DMV is very large, thus causing a huge computation overhead to the DMV as well. Therefore, we conclude that, though being able to detect all the malicious behaviors, the C-P²DAP is not scalable to large number of vehicles.

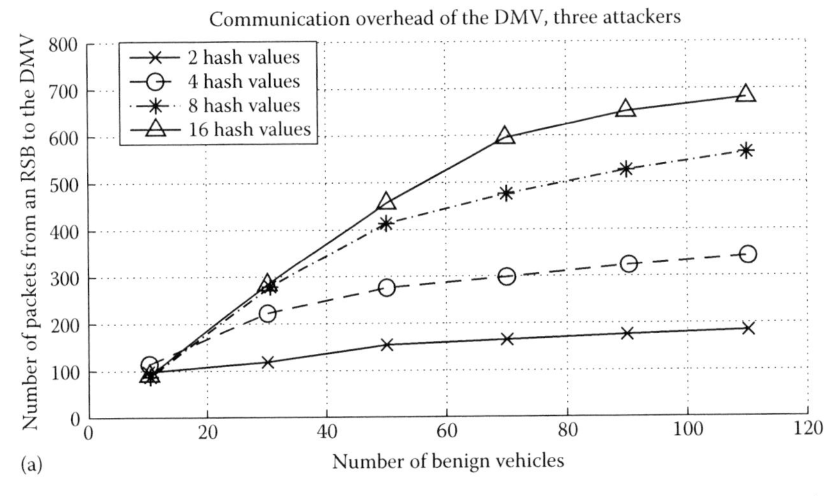

FIGURE 16.7 Number of packets sent from an RSB to the DMV: C-P²DAP. Subfigures a and b are displaying the results of experiments with different number of malicious users.

(continued)

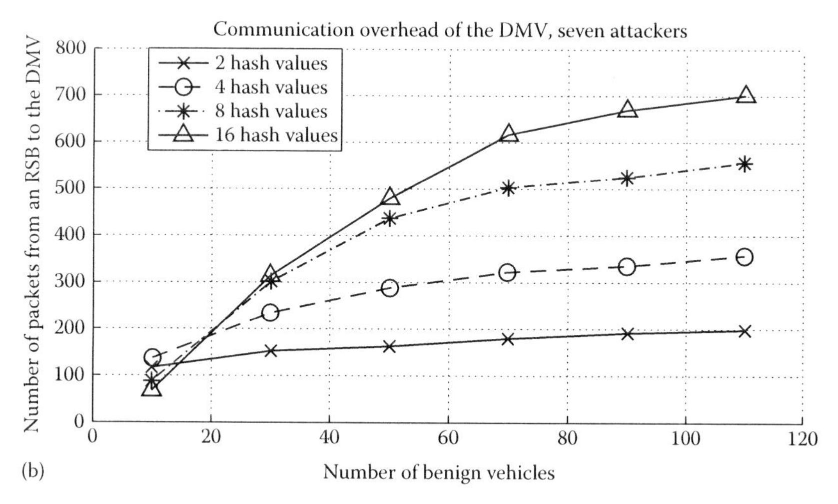

FIGURE 16.7 (continued)

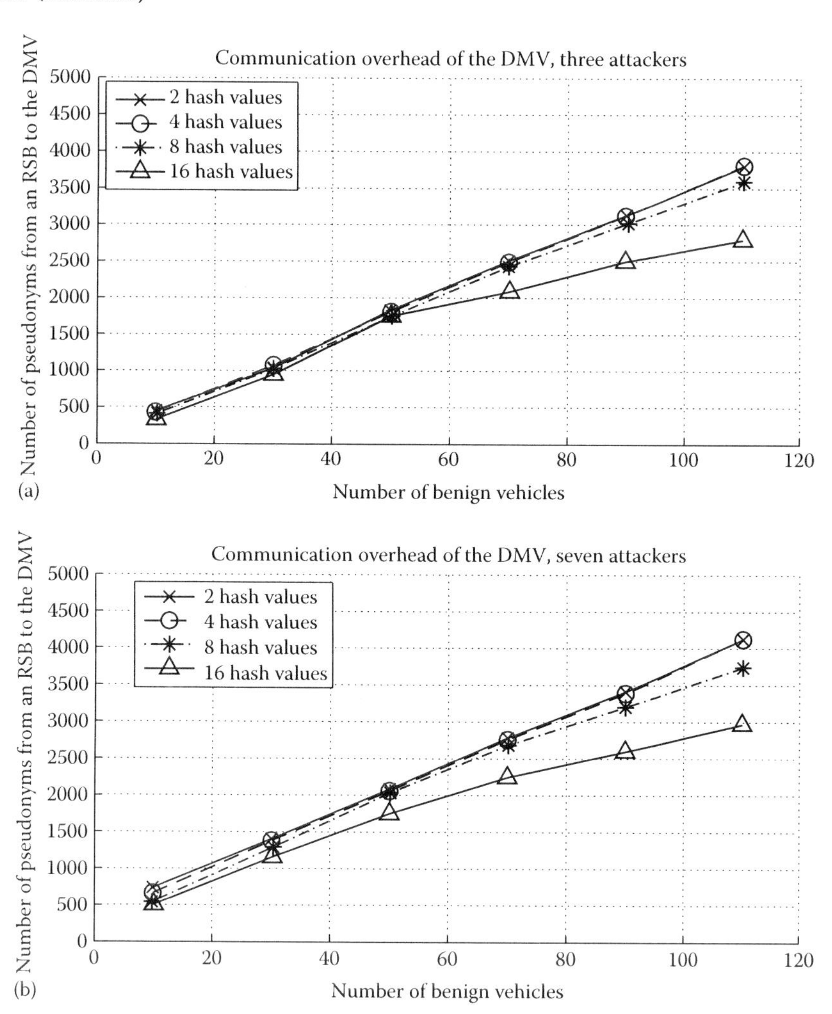

FIGURE 16.8 Number of pseudonyms from an RSB to the DMV: C-P[2]DAP. Subfigures a and b are displaying the results of experiments with different number of malicious users.

- *Overhead on the DMV: E-P²DAP*
  We next show the relationship between the number of packets from an RSB and the number of benign vehicles in E-P²DAP scheme in Figure 16.9. In the implementation of this scheme, we split each reported event to several packets such that each RSB report packet contains at most 20 pseudonyms. Therefore, it is not necessary to check the exact number of pseudonyms sent from an RSB because it can be easily estimated from the number of packets sent from the RSB. By comparing the results in Figures 16.9 and 16.7, we find that the packets received by the DMV is much less when using E-P²DAP, which means the E-P²DAP can efficiently distribute the job of detecting Sybil attack, the RSBs. Moreover, from Figure 16.9, we observe that the communication overhead of the DMV almost remains at the same level when the number of benign vehicles increases. We conclude from these observations that the E-P²DAP is scalable to large number of benign vehicles.
- *Overhead on the DMV: T-P²DAP*
  We then examine the DMV overhead in T-P²DAP. Similar to E-P²DAP, each packet from RSBs to the DMV contains an event and a maximum of 20 signing pseudonyms. If an event is signed with

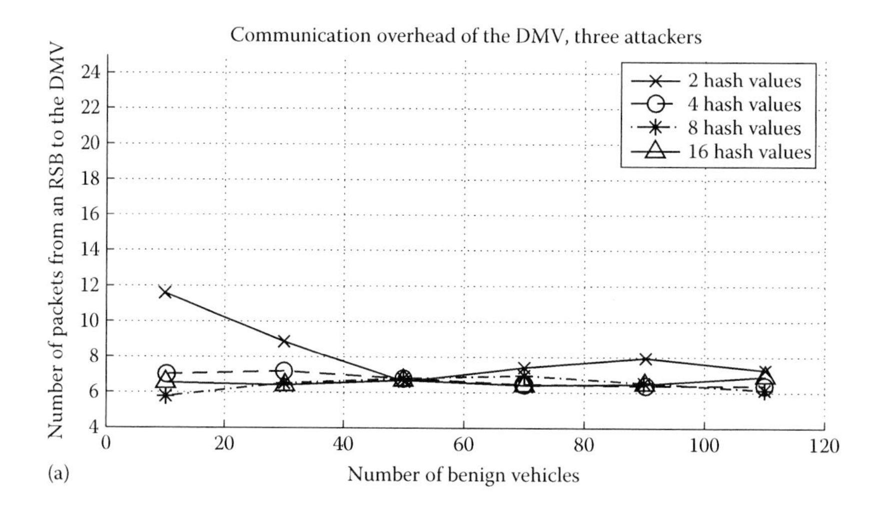

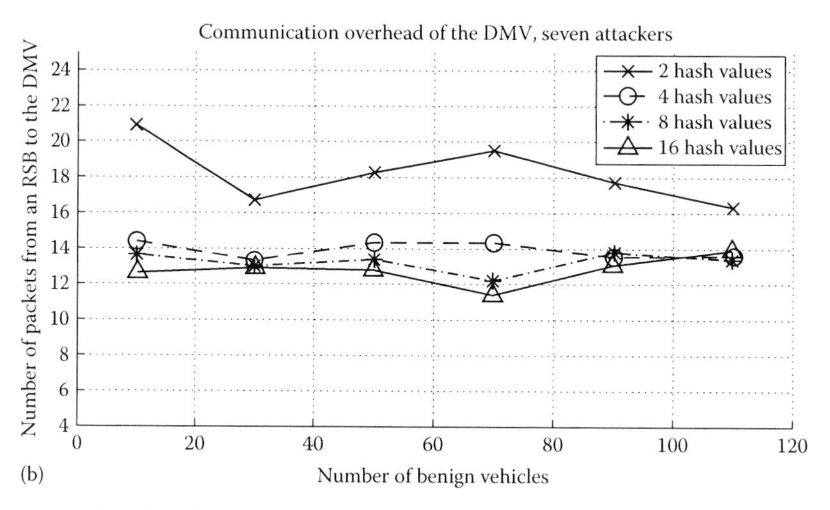

FIGURE 16.9 Number of packets from an RSB to the DMV: E-P²DAP. Subfigures a and b are displaying the results of experiments with different number of malicious users.

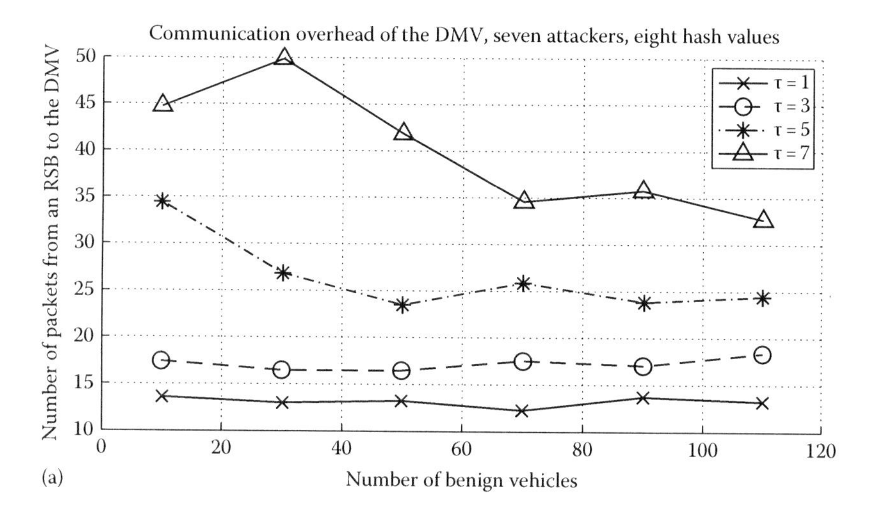

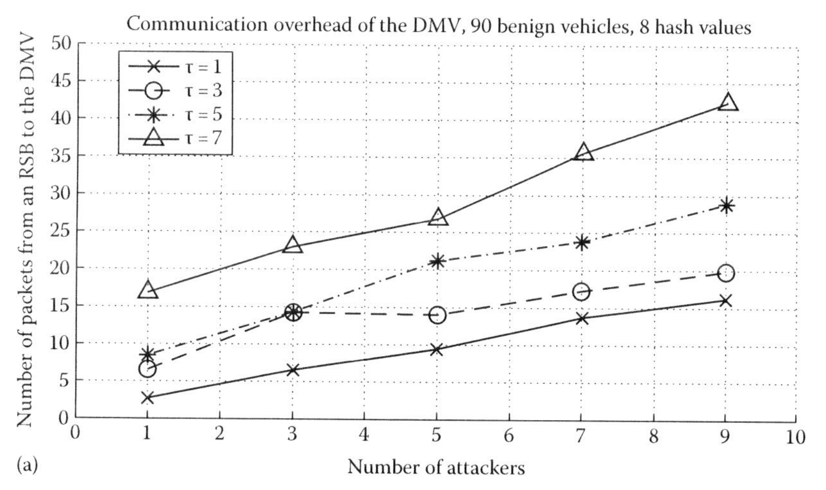

FIGURE 16.10 Number of packets from an RSB to the DMV: T-P²DAP. Subfigures a and b are displaying the results of experiments with different number of malicious users.

more than 20 pseudonyms, the RSBs will split the report into several packets. Figure 16.10 shows the communication overhead of the DMV in T-P²DAP scheme. We observe an increase in communication overhead when the value of τ increases or the number of attackers increases. However, the overhead is still much lower than C-P²DAP.

- *Discussion on the Three Schemes*

  As shown earlier, the C-P²DAP is costly. Comparing to the C-P²DAP, the E-P²DAP has a significant decrease of the communication overhead of the DMV. On the other hand, as a trade-off between C-P²DAP and E-P²DAP, the communication overhead of T-P²DAP is between them and is adaptive. One interesting observation by comparing Figures 16.7 and 16.10 is that even when τ is only one less than the number of coarse-grained hash values, the communication overhead of T-P²DAP is much less than the C-P²DAP.

16.5.5 Simulation Results: Latency for Detecting Malicious Vehicles

In our simulation, the latency Δt for detecting an attacker is defined as $t_{detect} - t_{attack}$, where t_{detect} is the time when the attacker is detected by the DMV, and t_{attack} is the time when the attacker first attacks.

16.5.5.1 C-P²DAP

The C-P²DAP guarantees that every Sybil attack can be detected; therefore, Δt is expected to be the shortest. As discussed in Section 16.3, an RSB makes one detection for suspected actions/events at each time interval. Therefore, the earliest time that an attack being caught is in the next time interval of that attack, and Δt is expected to be the length of the time interval.

In Figure 16.11, we show Δt for the C-P²DAP. We observe that Δt increases for an increasing number of vehicles or an increasing number of Γ (coarse-grained hash values), and Δt is obviously greater than the length of the time interval when the number of vehicles is greater than 90. All these differences are due to a same reason. (Recall that in C-P²DAP, a large number of benign vehicles will cause huge communication overhead on the DMV.) With a limited bandwidth between the RSB and the DMV, such overhead may cause delay for RSB report (note that in the simulation, the transmission rate between the RSB and the DMV is 3 packets/s), which explains the increase of Δt. Therefore, we conclude that the C-P²DAP scheme, although theoretically guarantees to detect every Sybil attack, may fail or have a large latency on a highly congested road. In real life, we can have much more

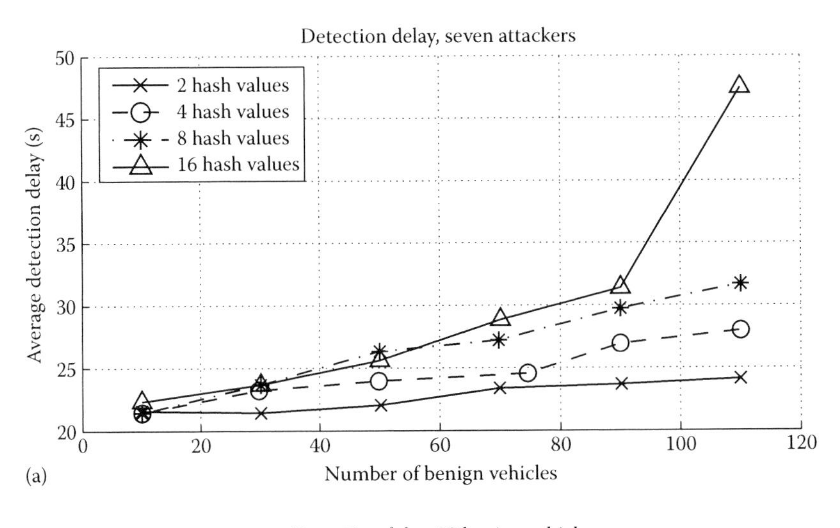

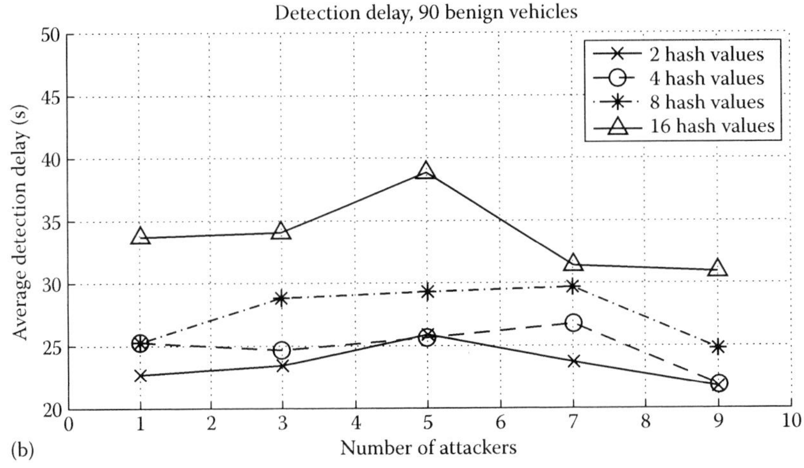

FIGURE 16.11 Detection latency: C-P²DAP. Subfigures a and b are displaying the results of experiments with different number of malicious users.

bandwidth between an RSB and the DMV to solve this issue. However, the result shows a constraint of C-P²DAP; it requires many computation and communication resources to guarantee the successful detections and short detection latency.

16.5.5.2 E-P²DAP

We next examine Δt of E-P²DAP scheme in Figure 16.12. We observe that the number of benign vehicles has little impact on Δt. This is because probability that all the vehicles reporting an event have the same value of Γ is small. On the other hand, we observe an increasing value of Δt when the number of attackers increases, which can be explained as follows.

Given semi-colluding attackers, the E-P²DAP scheme cannot detect colluding attackers. However, once an attacker reports an event by itself, it will be detected and then revoked. With more attackers being detected, the probability of attackers' collusion decreases, and the remaining attackers are more likely to

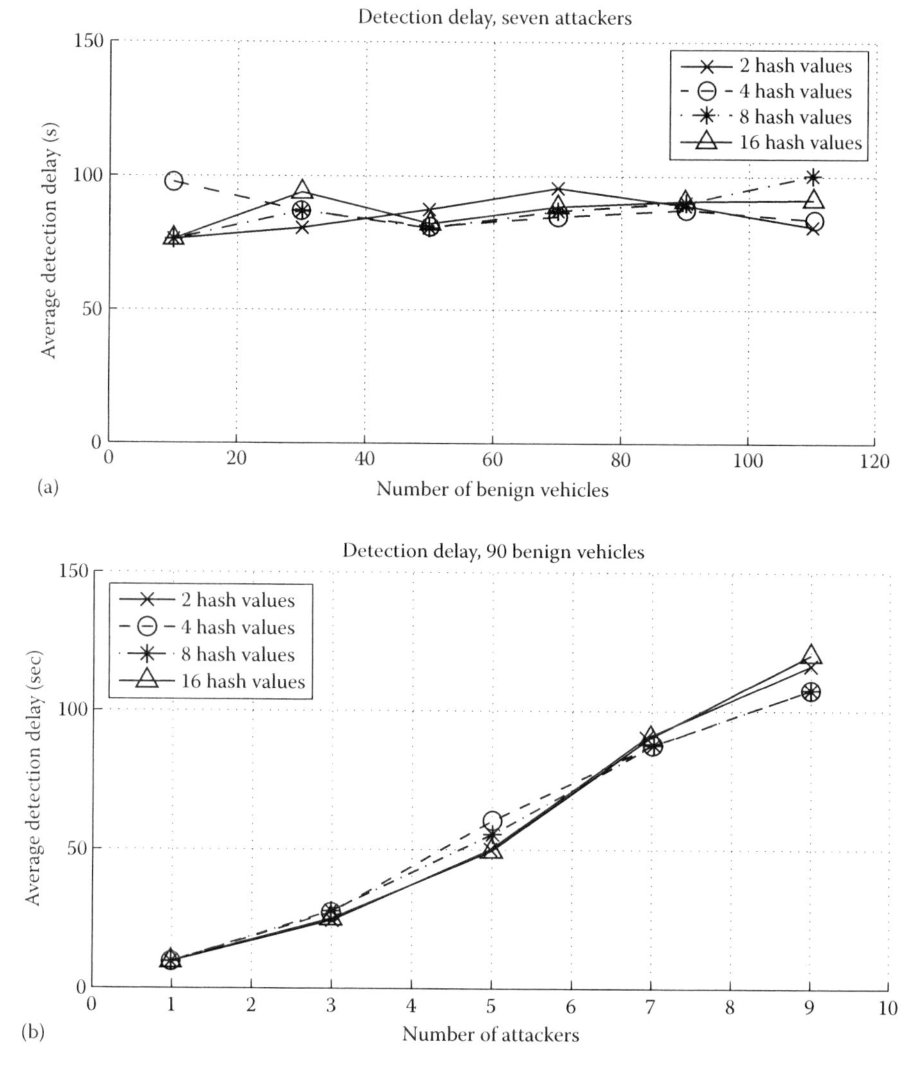

FIGURE 16.12 Detection latency: E-P²DAP. Subfigures a and b are displaying the results of experiments with different number of malicious users.

be detected. In this process, E-P²DAP detects attackers one after another. When there are more attackers, it takes more time for E-P²DAP to observe events signed by only one attacker, thus prolonging Δt.

16.5.5.3 T-P²DAP

The detection delay in T-P²DAP is shown in Figure 16.13. From the simulation results, we see that by increasing the value of τ, Δt is decreased to around 20 s, which is the length of time interval in the simulation. Such results mean that the semi-collusion is resolved by T-P²DAP. Besides, we observe that latency introduced by the communication overhead is not significant. Without this latency, the value of Δt of T-P²DAP is even better than that of C-P²DAP.

16.5.5.4 Discussion of the Three Schemes

When designing the three schemes provided earlier, according to the resilience to collusion for them, we would expect Δt to be

$$\Delta t_{C\text{-}P^2DAP} < \Delta t_{T\text{-}P^2DAP} < \Delta t_{E\text{-}P^2DAP}$$

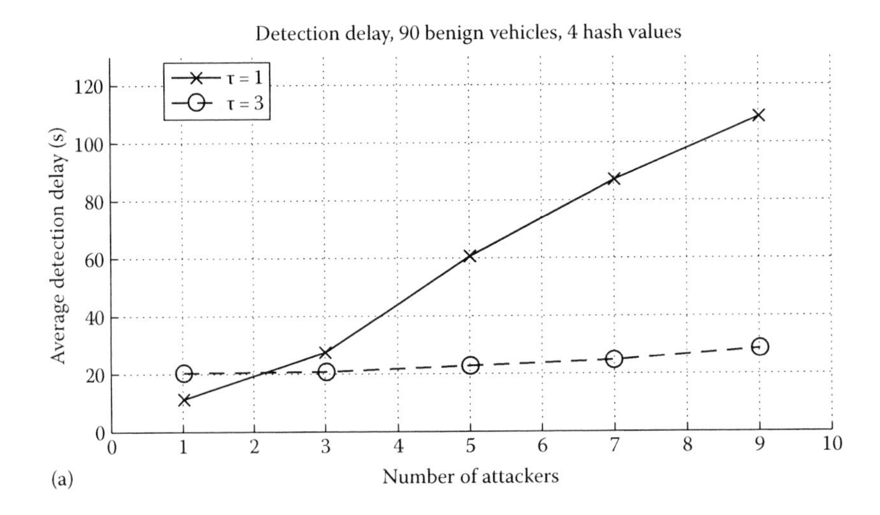

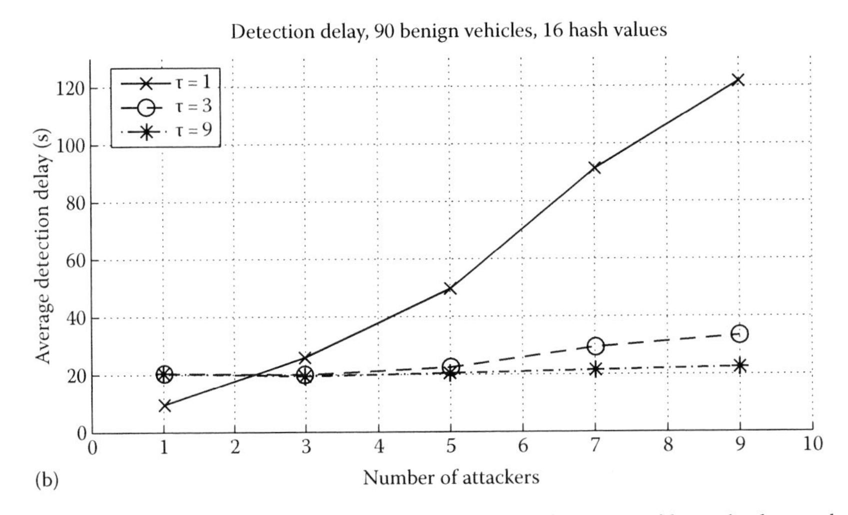

FIGURE 16.13 Detection latency, T-P²DAP, 90 benign vehicles. Subfigures a and b are displaying the results of experiments with different number of malicious users.

As shown in Figure 16.12, E-P²DAP has the highest Δt as expected. On the other hand, by comparing Figures 16.11 and 16.13, we observe that due to its light communication overhead, T-P²DAP could win over C-P²DAP in terms of detection latency in the scenario with constrained communication resources.

16.5.6 Scenarios with Sparser Vehicle Distribution

In Sections 16.5.4 and 16.5.5, we simulated a road of length 2000 m and vehicles with a radio range of 200 m. In such a scenario, each RSB is expected to hear almost all the event reports when the number of vehicles on the road is over 50. The aforementioned scenario can simulate the case where the vehicles are highly congested. While in some other cases, vehicles are more sparsely distributed, and not every packet from every vehicle can be captured by the RSB. To simulate such a case, we create a different scenario in which vehicles go back and forth on a road with a length of 20,000 m. Also, the communication radius is set to 50 m instead of 200 m. Thus, RSBs have less opportunity to overhear vehicles, and the traffic volume near an RSB fluctuates more. We compare the performance of τ-P²DAP and T-P²DAP under this scenario. Considering that both C-P²DAP and E-P²DAP are special cases of T-P²DAP, we do not individually analyze them in this section.

16.5.6.1 Communication Overhead of the RSB

We first show the communication overhead of the RSB such that we can compare them to the overhead of the DMV later. As Figure 16.14 shows, the overhead increases when the number of attackers or the number of benign vehicles grows. This result is similar to the scenario with densely distributed vehicles and is an expected behavior.

16.5.6.2 Communication Overhead of the DMV

In Figure 16.15, we show the communication overhead of the DMV for T-P²DAP and τ-P²DAP. In T-P²DAP, we use 4 bit coarse-grained hash values, whereas in τ-P²DAP, we use both 3 and 4 bit coarse-grained hash values. From the result of T-P²DAP, we observe slight fluctuations in the communication overhead when τ increases. This fluctuation happens because an increasing value of τ can have two opposite results: on one hand, when τ grows, the number of packets forwarded to the DMV grows as well, thus increasing the overhead; on the other hand, a larger value of τ also causes a shorter detection latency, which results in a smaller overall communication overhead. From Figure 16.15b, we see that

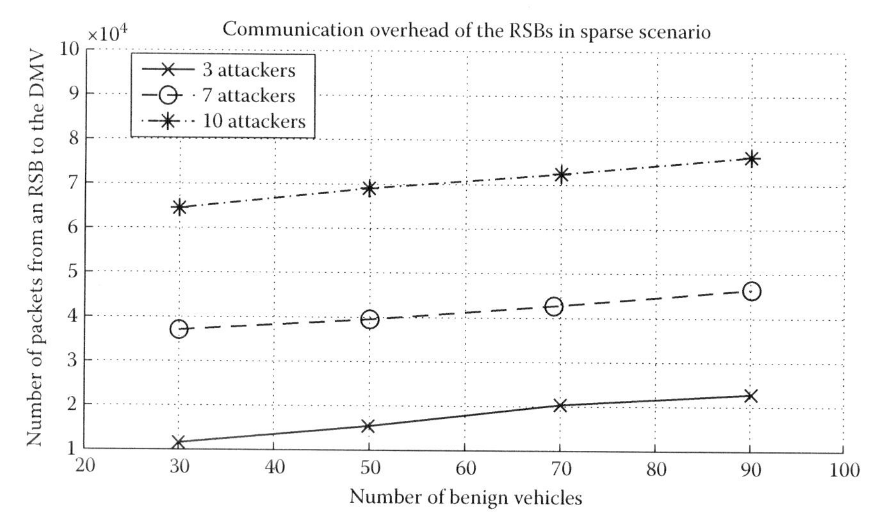

FIGURE 16.14 Communication overhead of an RSB in sparse vehicle scenario.

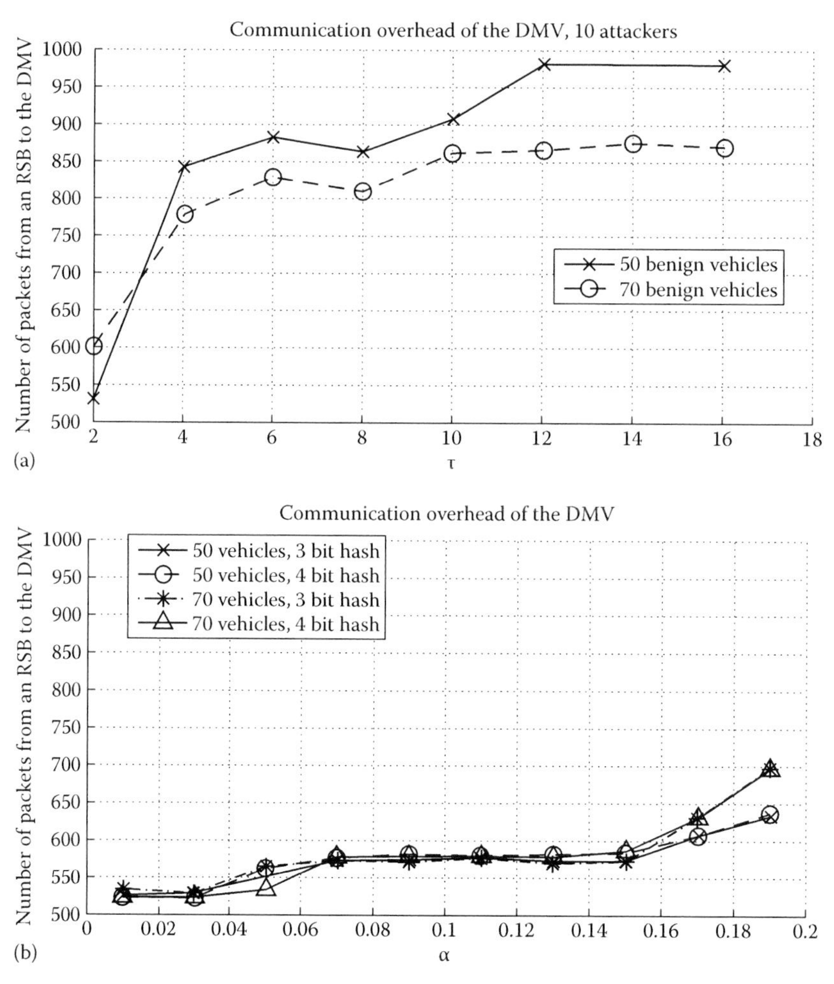

FIGURE 16.15 Communication overhead of the DMV, 10 malicious vehicles. Subfigures a and b are displaying the results of experiments with different number of malicious users.

when $\alpha < 0.15$, the communication overhead of τ-P²DAP remains at the same level of the T-P²DAP with $\tau = 2$. When $\alpha > 0.15$, the communication of τ-P²DAP increases dramatically. The results show that when $\alpha \leq 0.15$, we can obtain an acceptable communication overhead.

16.5.6.3 Detection Latency

We next check the value of Δt of the two schemes mentioned earlier. From Figure 16.16, we observe that Δt of τ-P²DAP quickly drops to 30 s when $\alpha > 0.05$ and then maintains at a constant level. On the other hand, T-P²DAP can achieve the similar Δt only when $\tau \geq 8$.

When combined the conclusion from the communication overhead, we conclude that when $0.05 < \alpha < 0.15$, τ-P²DAP achieves a good trade-off between overhead and latency, that is, Δt reaches 30 s at the cost of around 580 packets' communication overhead on the DMV. Such a trade-off cannot be achieved by using a fixed value of τ in T-P²DAP, with the following reason. In T-P²DAP, to achieve the same level of Δt, we require $\tau \geq 8$, whereas to achieve the same level of communication overhead, we require $\tau \leq 2$. The two requirements of the value of τ cannot be satisfied at the same time. Also, note the actual percentages of attackers are 12.5% (for 70 benign vehicles and 10 attackers) and 16.7% (for 50 benign vehicles and 10 attackers). From these observations, these proportions of attackers among all the vehicles are in the "best" range of α.

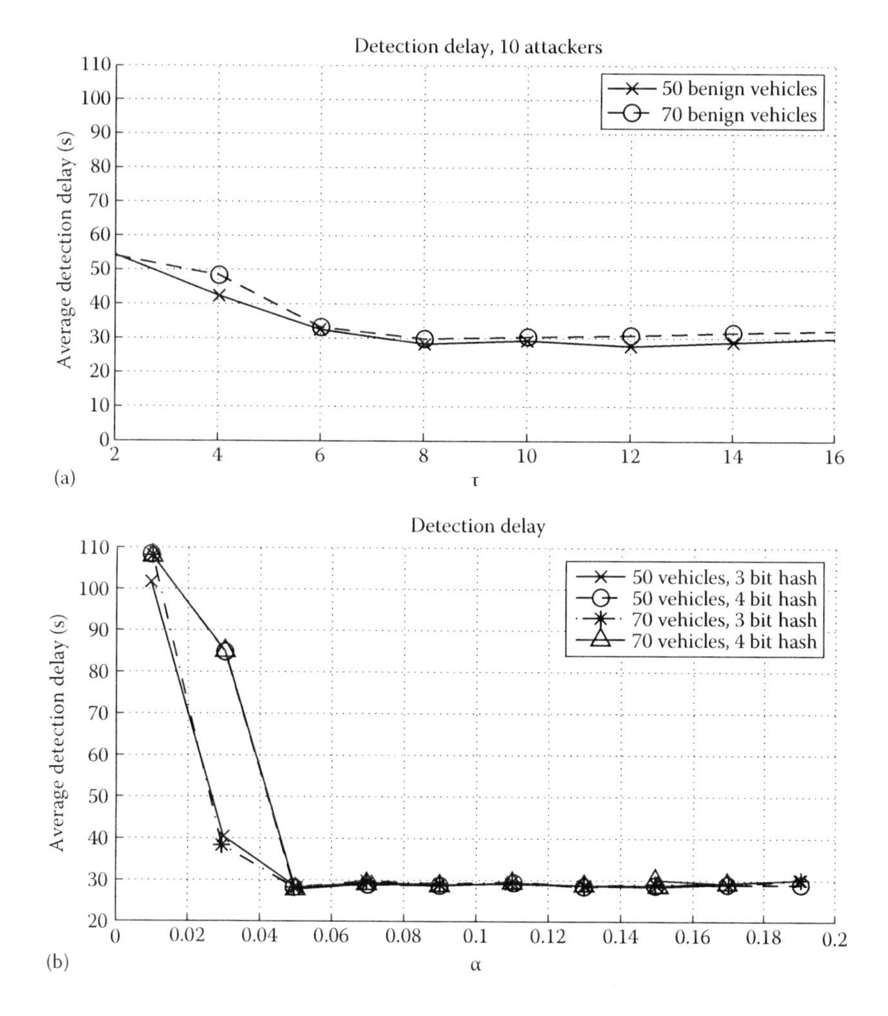

FIGURE 16.16 Average detection latency of 10 malicious vehicles. Subfigures a and b are displaying the results of experiments with different number of malicious users.

Therefore, we conclude that once a proper estimation of α can be made, τ-P²DAP can achieve a satisfying trade-off between communication overhead and detection latency.

16.6 Conclusion and Future Work

In this chapter, we propose a new method to detect Sybil attacks in VANET. The proposed method distributes the computation workload from the DMV to RSBs while releasing only a limited amount of information by using hash collusions. We also discussed some improvements on our scheme. Based on simulation results presented, we prove that the idea of distributing DMV workload to RSBs with limited information released is applicable in other VANET security and privacy applications.

One issue in our scheme is that it requires large storage overhead for both vehicles and the DMV. Therefore, one of our future works is to design an algorithm such that our algorithm can be applied on the pseudonym-generating algorithms in Refs [1,9]. By this approach, we will be able to further reduce the computational overhead of DMV and increase the privacy of vehicles. One interesting future work is to develop a machine-learning algorithm to predict the ratio and activities of malicious vehicles. With a good estimation of the ratio of attackers, P²DAP is expected to efficiently catch attackers with a small overhead and delay. Besides, the DMV can be involved for a centralized management of resources during the detection. Furthermore, the DMV can be distributed to different areas such as regional DMV, which matches the case in real life and forms a more powerful structure.

Other future work includes developing a more efficient method for partial pseudonym distribution. Moreover, we expect the ideas of distributing the DMV's duty to multiple RSBs for more applications than Sybil detection. Also, in our future experiments, real devices and 802.11p protocols are planned to be used.

References

1. G. Calandriello, P. Papadimitratos, A. Lioy, and J.-P. Hubaux. Efficient and robust pseudonymous authentication in VANET. In *ACM International Workshop on VehiculAr Inter-NETworking (VANET)*, Montreal, QC, Canada, pp. 19–27, 2007.
2. J. Y. Choi, M. Jakobsson, and S. Wetzel. Balancing auditability and privacy in vehicular networks. In *Proceedings of the 1st ACM International Workshop on Quality of Service & Security in Wireless and Mobile Networks*, Montreal, Canada, pp. 79–87, 2005.
3. P. Flajolet, D. Gardy, and L. Thimonier. Birthday paradox, coupon collectors, caching algorithms and self-organized search. *Discrete Applied Mathematics*, 39:207–229, 1992.
4. P. Golle, D. Greene, and J. Staddon. Detecting and correcting malicious data in VANETs. In *ACM International Workshop on VehiculAr Inter-NETworking (VANET)*, Philadelphia, PA, pp. 29–37, October 2004.
5. T. Kosch and M. Strassberger. The role of new wireless technologies in automotive telematics and active safety. In *8th Symposium Mobile Communications in Transportation*, Sophia Antipolis, France, 2004.
6. New York State Department of Transportation. List of state routes in Columbia county. http://www.dot.ny.gov/divisions/engineering/technical-services/hds-respository/columbiatubk.pdf, 2003.
7. W. Franz, Fleetnet: Communication platform for vehicular ad hoc networks. *First ACM International Workshop on Vehicular Ad Hoc Networks (VANET)*. Philadelphia, PA, October 2004.
8. B. Parno and A. Perrig. Challenges in securing vehicular networks. In *Fourth Workshop on Hot Topics in Networks (HotNets-IV)*, College Park, Maryland, pp. 11–21, 2005.
9. S. Rass, S. Fuchs, M. Schaffer, and K. Kyamakya. How to protect privacy in floating car data systems. In *ACM International Workshop on VehiculAr Inter-NETworking (VANET)*, San Francisco, CA, pp. 17–22, 2008.
10. M. Raya, D. Jungels, P. Papadimitratos, I. Aad, and J. P. Hubaux. Certificate revocation in vehicular networks. Technical Report LCA-Report-2006-006, Laboratory for Computer Communications and Applications (LCA), School of Computer and Communication Sciences, EPFL, Switzerland, 2006.
11. K. Sampigethaya, L. Huang, M. Li, R. Poovendran, K. Matsuura, and K. Sezaki. Caravan: Providing location privacy for VANET. In *Embedded Security in Cars (ESCAR) Workshop*, Cologne, Germany, pp. 29–37, 2005.
12. A. Studer, E. Shi, F. Bai, and A. Perrig. Tacking together efficient authentication, revocation, and privacy in VANETs. In *Conference on SEnsor, Mesh and Ad Hoc COmmunications and Networks (SECON)*, Rome, Italy, pp. 484–492, 2009.
13. W. Suh, H. Yun, K. S. Chon, and C. H. Park. Forecasting hourly traffic volume of airport access road: Case study of Incheon international airport. In *Annual Transportation Research Boards' (TRB) Meeting*, Washington, DC, 2005.
14. L. Sweeney. *k*-anonymity: A model for protecting privacy. *International Journal on Uncertainty, Fuzziness and Knowledge-Based Systems*, 10(5):557–570, 2002.
15. M. E. Zarki, S. Mehrotra, G. Tsudik, and N. Venkatasubramanian. Security issues in a future vehicular network. In *European Wireless (EuroWireless)*, Florence, Italy, pp. 270–274, 2002.

II

Adaptive Tasking

In the wake of failures or degradation of embedded sensors within a network, there is a need for architectures, models, and algorithms that create failure resistant networks. By adapting to different kinds of circumstances, the distributed sensor networks could negate the ill effects of various kinds of failures. Different issues such as software adaptation for networks, implementation of mobile code daemons for adaptive reconfiguration, mobile-agent-based computing for collaborative processing, and distributed services and mechanisms used to implement active querying in sensor networks are discussed in this part.

Madden and Gehrke describe database management aspects of sensor networks in Chapter 17. This includes query-processing issues. Queries can be defined at a higher level. The distributed database system is then responsible for assigning query components to proxy processes on the sensor nodes.

Brooks discusses software adaptation for networks of embedded processors in Chapter 18. A general framework, known as distributed dynamic linking, is described in detail. He motivates the use of mobile code to autonomously reconfigure software on individual nodes.

Brooks and Keiser concentrate on the implementation of mobile code daemons for adaptive reconfiguration in distributed systems in Chapter 19. They also provide a background on mobile code paradigm. Further, they discuss their design of a system based on lightweight mobile code daemons. They also provide an in-depth analysis of the indexing approach used for resource discovery in the mobile code, architecture, as well as the application-programming interface developed.

Qi et al. present the usage of mobile-agent-based computing paradigm for collaborative processing in sensor networks in Chapter 20. They discuss the principles of mobile-agent-based computing as well as its fundamental differences from the client- or server-based computing. Further, they also design and develop a mobile agent framework (MAF).

Lim has updated his chapter (Chapter 21) for the second edition, which discusses the purposes and benefits of distributed services, that are necessary for enabling sensor nodes to self-organize into impromptu networks that are incrementally extensible and dynamically adaptable to node failure and degradation, mobility and changes in task, and network requirements. The author further proposes an architecture for a self-organizing distributed sensor system, data-centric network protocols, distributed services for self-organizing sensor applications, application systems, etc.

In the updated chapter (Chapter 22), Krishnamachari focuses on a discussion of different mechanisms that can be used to implement active querying in sensor networks. He begins by discussing the simple idea of random walk and describes how it is used for different active querying techniques. He then discusses the possibility of sending active queries on predetermined trajectories, improvement of active query performance using reinforcement learning, and the use of geographic and senior information to direct the query.

Babadi et al. provide a new chapter (Chapter 23) that discusses in detail the target localization problems. The network self-organizes by dynamically allocating bandwidth while considering communication cost and bandwidth sharing.

Another new addition has been provided by Ho et al. (Chapter 24). They consider the problem posed by cloning, also known as node replication. Compromised nodes can be easily replicated and tampered with. They reduce the overhead of clone detection by proposing the structured deployment of sensor nodes.

Drs. Wang, Bagaria, and Chodwdhury provide a new chapter for the second edition (Chapter 25), which analyzes the use of sensor networks in intelligent transport systems. Specifically a topology-aware routing approach is implemented and tested. This approach leverages the structured nature of the application to provide a more efficient routing scheme.

Finally, Phoha et al. provide a new chapter (Chapter 26), which looks into the problem of fusing sensor inputs in urban environments. The problem is analyzed by simultaneously considering both the information space and the network space. This allows the two aspects of the sensor network to coevolve. Dynamic feedback loops are set up so that both aspects continually improve their ability to maintain system performance.

In summary, Part II discusses issues related to resistance to failures and adapt to different possibilities in distributed sensor networks.

17

Query Processing in Sensor Networks

Samuel Madden
University of California, Berkeley and Cornell University

Johannes Gehrke
University of California, Berkeley and Cornell University

17.1 Introduction

Recent advances in computing technology have led to the production of a new class of computing device: the wireless, battery powered, smart sensor. Traditional sensors deployed throughout buildings, labs, and equipment are passive devices that simply modulate a voltage based on some environmental parameter. In contrast, these new sensors are active, full fledged computers, capable not only of sampling real world phenomena but also filtering, sharing, and combining those sensor readings with each other and nearby Internet-equipped endpoints.

As an example of a specific instance of a sensor network platform, consider the small sensor devices called *motes* developed at UC Berkeley. Current generation motes are equipped with a 38.4 kbit radio, an 8 bit, a 7 MHz microprocessor, and a suite of sensors for measuring light, vibration, humidity, air pressure, magnetic field or gas or contaminate concentration. They are equipped with a small battery pack that provides sufficient energy for a few days of continuous operation but can be made to last for months or years if energy utilization is carefully managed. Motes run an operating system called TinyOS [20] that is especially suited to their capabilities. Networks of motes are usually deployed in an ad-hoc fashion. These ad-hoc networks differ from traditional networked environments in that motes in an ad-hoc network can locate each other and route data without any prior knowledge or assumptions about the network topology. Figure 17.1 shows a current generation mote in a small form factor with a weather sensor board.

FIGURE 17.1 A Mica2Dot mote next to a AA battery. The top board is a weather-sensor board with light, temperature, and humidity sensors. The middle board contains the processor, radio, and non-volatile Flash. The bottom board and silver cylinder are the battery and battery connector.

Smart-sensor technology has enabled a broad range of ubiquitous computing applications: the low cost, small size, and untethered nature of these devices makes it possible to sense information at previously unobtainable resolutions. Animal biologists can monitor the movements of hundreds of different animals simultaneously, receiving updates of location as well as ambient environmental conditions every few seconds. Vineyard owners can place sensors on every one of their plants, providing an exact picture of how various light and moisture levels vary in the *microclimates* around each vine. Supervisors of manufacturing plants, temperature controlled storage warehouses, and computer server rooms can monitor each piece of equipment, and automatically dispatch repair teams or shutdown problematic equipment in localized areas where temperature spikes or other faults occur.

Deployments such as those described above require months of design and engineering time, even for a skilled computer scientist. Some of this cost is hardware related and domain specific: for example, the appropriate choice of sensing hardware and device enclosure will vary dramatically if a network is designed for a forest canopy versus a sea floor. Aside from these domain specific considerations, however, there is a substantial collection of software functionality common to each of these deployments: they all collect and periodically transmit information from some set of sensors, and they all need to carefully manage limited power and radio bandwidth to insure that essential information is collected and reported in a timely fashion.

To that end, the primary goal of our research is to design and implement an architecture upon which such data collection applications can be built in dramatically less time. The key idea behind this architecture is that users specify the data they are interested in collecting through simple, *declarative* queries, just as in a database system, and that the infrastructure efficiently collects and process the data within the sensor network. In contrast to traditional, embedded-C based programming models where each is treated as a separate computational unit, these queries are high-level statements of logical interests over an entire network, such as "tell me the average temperature on the 4th floor of this building" or "tell me the location of the sensor with the least remaining battery capacity." The database system manages the details of data collection and processing (freeing the user from these concerns); in particular, it provides facilities for

- Dissemination of queries into the sensor network
- Identification of sensors which correspond to the logical names used in queries (e.g., "sensors on the fourth floor")
- Collection and processing of results from the network, over multiple radio hops and in a power efficient manner
- Energy conservation
- Acquisition of sensor readings from a variety of low-level hardware interfaces
- Storage and retrieval of collections of results in the network
- Adaptation of communication topology and data rates to optimize network performance

At Berkeley and Cornell, we have built several prototype *sensor network query processors* (SNQPs) which are instantiations of this architecture. The two systems, called Cougar [6] and TinyDB [31], respectively, run on a variety of different sensor platforms, including the Berkeley motes.

Aside from greatly simplifying the amount of work which users of sensor networks must do to prepare for a deployment of sensors, this query-processing based approach to sensor management has the potential to offer dramatic improvements in energy efficiency—the typical measure of performance in sensor networks—of these data collection applications. Again, this echos a well-known lesson from the relational database community: because declarative queries include no specification of *how* the required data is collected and processed, the system is free to explore many possible physical instantiations (plans) that have the same logical behavior as the user's query, and choose the one which is expected to offer the best performance. This process—*query optimization*—is central to the performance of our architecture.

In this chapter, we describe our experiences designing the TinyDB and Cougar query processors, discussing the unusual challenges and novel features required of a high-level data processing system in the world of sensor networks. We primarily discuss networks composed of homogeneous collections of Mica motes, though our work is general enough to be applicable outside of this regime—indeed, initial versions of Cougar were implemented on nodes from Sensoria Corporation—the first generation was running on Windows CE and the second generation was running on an embedded version of Linux.

17.2 Architecture for Query Processing in Sensor Networks

Sensor networks provide a surprisingly challenging programming and computing environment: the devices are small and crash-prone; the operating system that runs on them does not provide benefits like fault isolation to help mitigate such failures. Debugging is usually done via a few LEDs on the device. Programs are highly distributed, and must carefully manage energy and radio bandwidth while sharing information and processing with each other.

Because of limitations imposed by this impoverished computing environment, data collection systems in sensor networks are required to support an unusual set of software requirements, such as

- They must carefully manage resources, in particular power. Communication and sensing tend to dominate power consumption given the quantities of data and complexity of operations that are feasible on sensor networks. Furthermore, Moore's law suggests that the energy cost per CPU cycle will continue to fall [43] as transistors get smaller and lower voltage, whereas fundamental physical limits and trends in battery technology suggest that the energy to transmit data via radio will continue to be expensive relative to the energy density of batteries.
- This has led us to focus much of our work on minimizing and optimizing for the energy costs of query processing in this environment.

- They have to be aware of and manage the transient nature of sensor networks: nodes come and go, signal strengths between devices vary dramatically as batteries run low and interference patterns change, but data collection should be interrupted as little as possible.
- They must be able to reduce and summarize data on-line while providing storage, logging, and auditing facilities for off-line analysis. Transmitting all of the raw data out of the network in real time is often prohibitively expensive (in terms of energy) or impossible given data collection rates and limited radio bandwidth. Instead, small summaries, or *aggregates* (such as averages, moments, histograms, or statistical summaries) can be provided in real time. Many users, particularly scientists and the military, however, must be able to eventually collect and permanently store raw data, even if that data is not extracted from the network for several days or weeks.
- They must provide an interface that is substantially simpler than the based embedded-C based programming model of TinyOS. While being simple, this interface must also allow users to collect desired information and process it in useful ways.
- Users must be given the tools to manage and understand the status of a network of sensors that have been deployed, and it must be it easy to add new nodes with new types of sensors and capabilities.

Note that each of these points represents a dissertation worth of research each, much of which still remains undone. Our goal in this chapter is to survey the current state of the art, describing at a high level the software and languages we have developed to address these challenges.

17.2.1 Architectural Overview

Figure 17.2 shows a simple block diagram of a architecture for query processing in sensor networks. The main two pieces of this architecture are

- Server-side software that runs on the user's PC—the *basestation*—which, in its most basic form parses queries and delivers them into the network and collects results as they stream out of the network. In this article, we will not discuss many of the details of server-side query processing; see Refs [28,32] for more information.
- Sensor-side software that runs on the motes. As shown in the "Distributed In Network Query Processor" detail box on the left side of Figure 17.2, this software consists of a number of components built on top of TinyOS.

17.2.2 Introducing Queries and Query Optimization

In our architecture, queries are input at the server in a simple, SQL-like language which describes the data the user wishes to collect and ways in which he or she would like to combine, transform, and summarize it. The most significant way in which the variant of SQL we have developed differs from traditional SQL is that queries are *continuous* and *periodic*. That is, users register an interest in certain kinds of sensor readings (e.g., "temperatures from sensors on the 4th floor every 5 s") and the system streams these results out to the user. We call each period in which a result is produced an *epoch*. The *epoch duration*, or *sample period* of a query refers to the amount of time between successive samples; for this example, the sample period would be 5 s. As we discuss various aspects of our system, we will show some examples of our language syntax and discuss its other features (new and in common with traditional SQL) in more detail.

Just as in a traditional database system, queries describe a *logical* set of data that the user is interested in, such as "the average temperature on the 4th floor," but do not describe the actual algorithms and software modules, or *operators* which the system must use to collect the answer set. Typically, there are a number of alternative *plans*, or choices and orderings of operators for any given logical query. For example, to find the average temperature of the sensors on the fourth floor, the system might collect readings from every sensor, then filter that list for sensors on the fourth floor and compute their average *or* it might request that only sensors on the fourth floor provide their temperature, and then take the

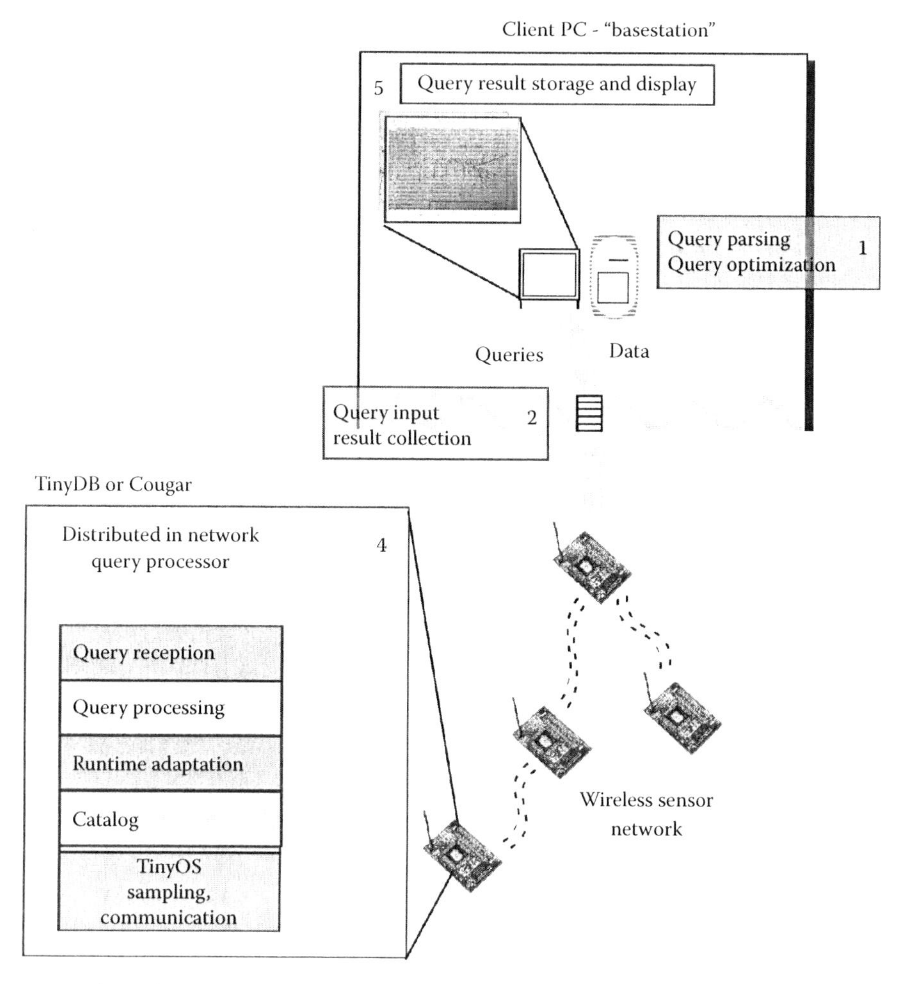

FIGURE 17.2 The architecture of an sensor network query processor. Numbers indicate the sequence of steps involved in processing a query.

average of all the values it collects. In a sensor network, we expect that the latter plan will always be a better choice, since it requires only sensors on the fourth floor to collect and report their temperature.

The process of selecting the best possible plan is called *query optimization*. At a very high level, query optimizers work by enumerating a set of possible plans, assigning a cost to each plan based on estimated costs of each of the operators, and choosing the plan of the lowest cost.

In a sensor network, this process of query optimization is done as much as possible on the server-side PC, since it can be quite computationally intensive. However, because the server may not have perfect state about the status of the sensor network, and because costs used to optimize a query initially may change over its lifetime, it is sometimes necessary to *adapt* running query plans once they have been sent into the network. We will see a few examples of this when we discuss query optimization in more detail.

17.2.3 Query Language

Queries in Cougar and TinyDB, as in SQL, consist of `SELECT-FROM-WHERE-GROUPBY-HAVING` blocks supporting selection, join, projection, aggregation, and grouping. The systems also include explicit support for windowing and subqueries (in TinyDB via *storage points*), and TinyDB also has explicit support for sampling. In queries, we view sensor data as a single virtual table with one column per sensor type. Tuples are appended to this table periodically, at well-defined intervals that are a parameter of the query. This period of time between each sample interval is the *epoch*, as described above. Epochs provide a convenient mechanism for structuring computation to minimize power consumption. Consider the following query:

```
SELECT nodeid, light, temp
FROM sensors
SAMPLE PERIOD 1s FOR 10s
```

This query specifies that each sensor should report its own id, light, and temperature readings once per second for 10 s. The virtual table sensors contains one column for every attribute available in the catalog and one row for every possible instant in time. The term *virtual* means that these rows and columns are not actually materialized—only the attributes and rows referenced in active queries are actually generated.

Results of this query stream to the root of the network in an online fashion, via the multi-hop topology, where they may be logged or output to the user. The output consists of an ever-growing sequence of tuples, clustered into 1 s time intervals. Each tuple includes a time stamp corresponding to the time it was produced.

Note that the sensors table is (conceptually) an unbounded, continuous *data stream* of values; as is the case in other streaming and online systems, certain blocking operations (such as sort and symmetric join) are not allowed over such streams unless a bounded subset of the stream, or *window*, is specified. Windows in TinyDB are defined as fixed-size materialization points over the sensor streams. Such materialization points accumulate a small buffer of data that may be used in other queries. Cougar has a similar feature called *view nodes* that can store intermediate query results similar to materialized views in relational database systems; data is pushed from sensors to view nodes and then either pulled through interactive queries or periodically pushed to other view nodes or a base station. We show the TinyDB syntax here for concreteness:

Consider, as an example the following query:

```
CREATE
STORAGE POINT recentlight SIZE 8 seconds
AS (SELECT nodeid, light FROM sensors
SAMPLE PERIOD 1s)
```

This statement provides a shared, local (i.e., single-node) location to store a streaming view of recent data similar to materialization points in other streaming systems like Aurora or STREAM [7,34], or materialized views in conventional databases.

Joins are allowed between two storage points on the same node, or between a storage point and the sensors relation, in which case sensors is used as the outer relation in a nested-loops join. That is, when a sensors tuple arrives, it is joined with tuples in the storage point at its time of arrival. This is effectively a *landmark query* [14] common in streaming systems. Consider, as an example, the following:

```
SELECT COUNT(*)
FROM sensors AS s, recentLight AS rl
WHERE rl.nodeid = s.nodeid
AND s.light < rl.light
SAMPLE PERIOD 10s
```

This query outputs a stream of counts indicating the number of recent light readings (from zero to eight samples in the past) that were brighter than the current reading.

TinyDB and Cougar also include support for grouped aggregation queries. Aggregation has the attractive property that it reduces the quantity of data that must be transmitted through the network, and thus can reduce energy consumption and reduce bandwidth usage by replacing more expensive communication operations with relatively cheaper computation operations, extending the lifetime of the sensor network significantly. TinyDB also includes a mechanism for user-defined aggregates and a metadata management system that supports optimizations over them.

Note that aggregation is a very powerful paradigm that has applicability that goes far beyond simple averaging. For example, the Cougar system has support for object tracking: nodes have a signal processing layer that generates signatures for objects that are in the vicinity of a sensor. Cougar implements a tracking operator as an aggregation over a region of sensor nodes, whose detections are aggregated into an estimation of a track containing the estimated speed and direction of an object. Overlap between regions ensures that an accurate track exists at all times.

In addition to aggregates over values produced during the same sample interval (e.g., as in the COUNT query above), users want to be able to perform temporal operations. For example, in a building monitoring system for conference rooms, users may detect occupancy by measuring maximum sound volume over time and reporting that volume periodically:

```
SELECT WINAVG(volume, 30s, 5s)
FROM sensors
SAMPLE PERIOD 1s
```

This query will report the average volume over the last 30 s once every 5 s, acquiring a sample once per second. This is an example of a *sliding-window* query common in many streaming systems [14,34].

When a query is issued in TinyDB or Cougar, it is assigned an identifier (id) that is returned to the issuer. This identifier can be used to explicitly stop a query via a "STOP QUERY id" command. Alternatively, queries can be limited to run for a specific time period via a FOR clause, or can include a stopping condition as a *triggering condition* or *event*; see our recent SIGMOD paper on *acquisitional query processing* [30] for more detail about these language constructs.

17.2.4 Query Dissemination and Result Collection

Once a query has been optimized, it is *disseminated* into the network. We discuss one basic communication primitive, a *routing tree*. A routing tree is rooted at either the base station or a storage point and it allows the root of the network to disseminate a query and to collect query results. This routing tree is formed by forwarding the query from every node in the network: the root initially transmits the query; all *child* nodes that hear it process it and forward it on to their children, and so on, until the entire network has heard about the query.

Each radio message contains a hop-count, or *level* indicating the distance from the broadcaster to the root. To determine their own level, nodes pick a *parent* node that is (by definition) one level closer to the root than they are. This parent will be responsible for forwarding the node's (and its children's) query results to the basestation. We note that it is possible to have several routing trees if nodes keep track of multiple parents. This can be used to support several simultaneous queries with different roots. This type of communication topology is common within the sensor network community and is known as *tree-based routing*.

Figure 17.3 shows an example sensor network topology and routing tree. Solid arrows indicate parent nodes, while dotted lines indicate nodes that can hear each other but do no use each other for routing. In general, a node may have several possible choices of parent; a simple approach is to chose the parent to be the ancestor node with the highest level. In practice, it turns out that making a proper choice of parent is quite important in terms of communication and data collection efficiency. Unfortunately, the details of the best known techniques for doing this are quite complicated and outside the scope of our discussion in this paper. For a more complete discussion of these and other issues, see Ref. [9].

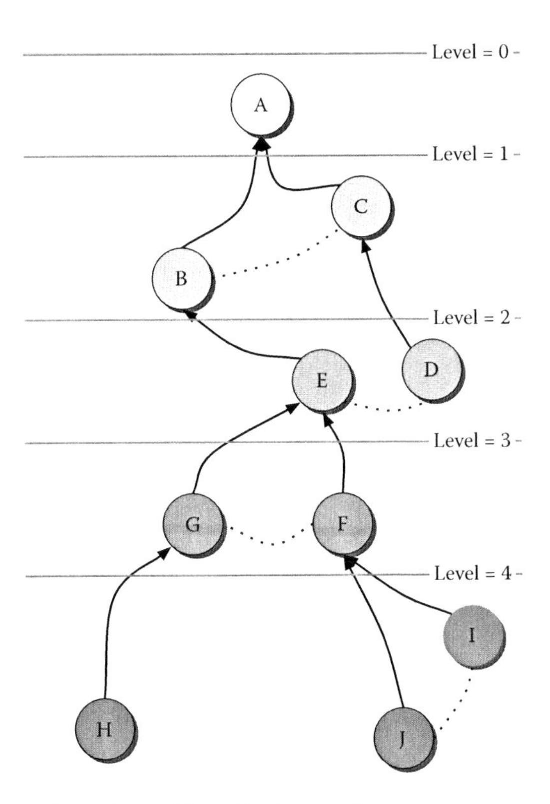

FIGURE 17.3 A sensor network topology, with routing tree overlay.

Once a routing tree has tree has been constructed, each node has a connection to the root of the tree which is just a few radio hops long. We can then use this same tree to collect data from sensors by having them forward query results up this path.

Note that simple routing structures such as routing trees are very suitable for our scenario: Sensor network query processors impose communication workloads on the multi-hop communication network that are very different from traditional ad-hoc networks with mobile nodes. Since the sensor network is programmed only through queries, there are very regular communication patterns, mainly consisting of the collection of sensor readings from a region at a single node or the base station.

Note that other types of routing structures beyond routing trees are necessary if the query workload has more than a few destinations since the overlay of several routing trees neglects any sharing between several trees and leads to performance decay. The discussion of such routing algorithms is beyond the scope of this chapter, but we have begun to explore such issues in our research.

17.2.5 Query Processing

Once a query has been disseminated, each node begins processing it. Processing is a simple loop: once per epoch, readings, or *samples* are acquired from sensors corresponding to the fields or *attributes* referenced in the query. This acquisition is done by a special *acquisition operator*. This set of readings, or *tuple*, is routed through the query plan built in the optimization phase. The plan consists of a number of operators that are applied in a fixed order; each operator may pass the tuple on to the next operator, reject it, or combine it with one or more other tuples. Any tuple that successfully passes the plan is transmitted up the routing tree to the node's parent, which may in turn forward the result on or may combine it with its own data or data collected from its other children. Table 17.1 describes some common query processing operators which are used in SNQPs.

TABLE 17.1 Common Sensor Network Query Processing Operators

Operator	Description
`data acquisition`	Acquire a reading (field) from a sensor or an internal device attribute. Example attributes are light sensor readings or free RAM in the dynamic heap.
`select`	Reject readings to do not satisfy a particular boolean predicate. For example, the predicate `temp > 80°F` rejects readings under 80°F.
`aggregate`	Combines readings together according to an *aggregation function*. For example, `AVG(light)` computes the average light value over each mote.
`join`	Concatenate two readings when some a *join predicate* is satisfied. An example predicate might be `mat-point.light > sensors.light` which joins (concatenates) all of the historical tuples in `mat-point` with current sensor readings for any pair of tuples where the current light value exceeds the historical value.

The acquisition operator uses a *catalog* of available attributes to map names referenced in queries into low-level operating system functions that can be invoked to provide their values. This catalog abstraction allows sophisticated users to extend the sensor network with new kinds of sensors, and also provides support for sensors that are accessed via different software interfaces. For example, in the TinyDB system, users can run queries over sensor attributes like light and temperature, but can also query attributes that reflect the state of the device or operating system, such as the free RAM in the dynamic memory allocator. Table 17.2 shows a list of some of the sensor and system attributes that are available on current generation sensors. This table includes energy per sample as an example of additional catalog metadata that can be used in query optimization.

Figure 17.4 illustrates query processing for the simple aggregate query "Tell me the average temperature on the 4th floor once every 5 s." Here, the query plan running on every node contains three operators: an acquisition operator, a select operator that checks to see if the value of the `floor` attribute is equal to 4, and an aggregate operator that computes the average of the `temperature` attribute from the local mote and the average temperature values of any that of mote's descendents (which happen to be on the fourth floor.) Each sensor applies this plan once per epoch, and the stream of data produced at the root node is the answer to the query. Note that we represent the partial computation of averages as {`sum, count`} pairs which are merged at each intermediate node in the query plan to compute a running average as data flows up the tree.

To make this scheme work, there are a number of implementation details that must be resolved: sensors must wait to hear from their children before reporting their own averages, and average records must be represented in such a way that they can be combined as they flow up the tree (in this case, as a sum and a count, rather than a simple average.)

TABLE 17.2 Some Sensors Available for Mica Motes and Their Power Requirements

Sensor	Notes	Energy per Sample (@3 V), mJ
Solar radiation [50]	Amount of radiation in 400–700 nm range that allows plants to photosynthesize	0.525
Barometric pressure [23]	Air pressure, in mb	0.003
Humidity [45]	Relative humidity	0.5
Passive infrared [33]	Temperature of an overhead surface	0.0056
Ambient temp [33]		0.0056
Accelerometer [1]	Measure movement and vibration	0.0048
(Passive) Thermistor [2]	Uncalibrated temperature, low cost	0.00009

"Tell me the average temperature of sensors on the 4th floor once every 5 s"

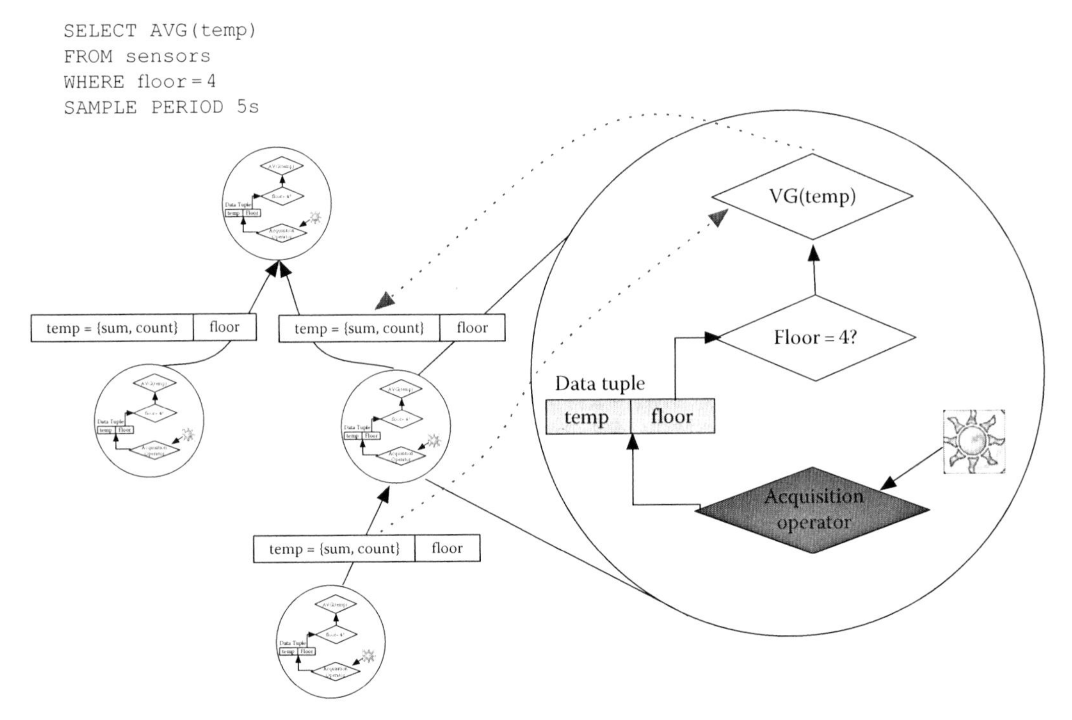

FIGURE 17.4 A sensor network executing a simple aggregate query.

17.3 Sensor Network Specific Techniques and Optimizations

Now that we have a basic outline of how query processing in sensor networks functions, we devote the rest of the chapter to the unusual kinds of optimizations and query processing techniques which arise in the context of SNQPs.

17.3.1 Lifetime

In lieu of a explicit `SAMPLE PERIOD` clause, we allow users to specify a specific query lifetime via a `QUERY LIFETIME` <x> clause, where $< x >$ is a duration in days, weeks, or months. Specifying lifetime is a much more intuitive way for users to reason about power consumption. Especially in environmental monitoring scenarios, scientific users may not be particularly concerned with small adjustments to the sample rate, nor do they understand how such adjustments influence power consumption. Such users, however, are very concerned with the lifetime of the network executing the queries.

Consider the following query:

```
SELECT nodeid, accel
FROM sensors
LIFETIME 30 days
```

This query specifies that the network should run for at least 30 days, sampling light and acceleration sensors at a rate that is as quick as possible and still satisfies this goal.

To satisfy a lifetime clause, the SNQP applies *lifetime estimation*. The goal of lifetime estimation is to compute a sampling and transmission rate given a number of Joules of energy remaining

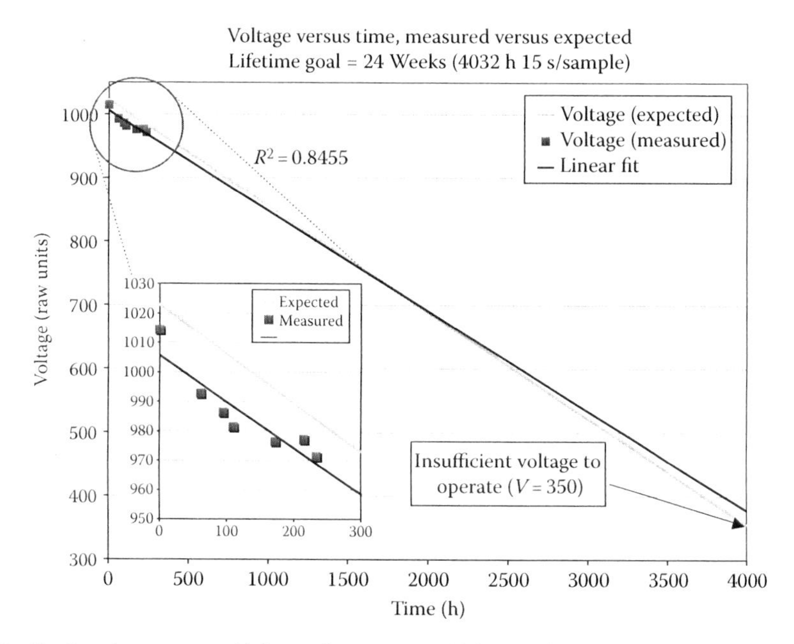

FIGURE 17.5 Predicted versus actual lifetime for a requested lifetime of 24 weeks (168 days).

(which can *usually* be estimated from the battery voltage on the mote) and a specific query or set of queries to run. Note that, just as with query optimization, lifetime estimation can be done when a query is initially issued at the PC, or may be applied periodically, within the network as the query runs. We have currently implemented the former approach in TinyDB, but the latter approach will be more effective, especially in a network with lots of nodes communicating in unpredictable ways.

To illustrate the effectiveness of this simple estimation, we inserted a lifetime-based query (`SELECT voltage, light FROM sensors LIFETIME x`) into a sensor (with a fresh pair of AA batteries) and asked it to run for 24 weeks, which resulted in a sample rate of 15.2 s/sample. We measured the remaining voltage on the device 9 times over 12 days. The first two readings were outside the range of the voltage detector on the mote (e.g., they read "1024"—the maximum value) so are not shown. Based on experiments with our test mote connected to a power supply, we expect it to stop functioning when its voltage reaches 350. Figure 17.5 shows the measured lifetime at each point in time, with a linear fit of the data, versus the "expected voltage" which was computed using a simple cost model. The resulting linear fit of voltage is quite close to the expected voltage. The linear fit reaches $V = 350$ about 5 days after the expected voltage line.

Lifetime estimation is a simple example of an optimization technique which the sensor network can apply to provide users with a more useful, expressive way of interacting with a network of sensors.

17.3.2 Pushing Computation

Among the most general techniques for query optimization in sensor network database systems is the notion of *pushing computation*, or moving processing into the network, toward the origin of the data being processed. We introduce two forms of this technique for aggregate queries.

A query plan for a simple aggregate query can be divided into two components. Since queries require data from spatially distributed sensors, we need to deliver records from a set of distributed nodes to a central destination node for aggregation by setting up suitable communication structures for delivery of sensor records within the network. We call this part of a query plan its *communication component*. In addition, the query plan has a *computation component* that computes the aggregate at the network root and potentially computes already partial aggregates at intermediate nodes. We describe two simple schemes for pushing computation; more sophisticated push-based approaches are possible [29,56].

Partial aggregation: For aggregates that can be incrementally maintained in constant space (or, using terminology from the database literature, for *distributive* and *algebraic* aggregate operators [17]), we push partial computation of the aggregate from the root node down to intermediate nodes. Each intermediate sensor node will compute partial results that contain sufficient statistics to compute the final result. AVERAGE is an example of a aggregate that has constant intermediate state and can be distributed in this way; the example given in Figure 17.3 illustrates the concept of pushing partial aggregation into the network.

Packet merging: Since it is much more expensive to send multiple smaller packets instead of one larger packet (considering the cost of reserving the channel and the payload of packet headers), we can *merge* several records into a larger packet, and only pay the packet overhead once per group of records. For exact query answers with aggregate operators that do not have a compact incremental state representation like the Median (these are called *holistic aggregates*), packet merging is the only way to reduce the number of bytes transmitted.

17.3.3 Cross-Layer Interactions

In the previous section, we saw that we can optimize aggregate operators through in-network aggregation, such as packet merging and partial aggregation at internal nodes. These techniques require internal nodes to *intercept* data packets passing through them to perform packet merging or partial aggregation. However, with the traditional "send and receive" interfaces of the network layer, only the root of the routing tree will receive the data packets. The network layer on an internal node will automatically forward the packets to the next hop toward the destination, and the upper layers will be unaware of data packets traveling through the node. Thus a node needs to the functionality to "intercept" packets that are routed through it, and the sensor network query processor needs a way to communicate to the network layer which and when it wants to intercept packets that are destined for another node.

One possible way to implement this interception is through network filters, which is the approach taken in Cougar. With *filters*, the network layer will first pass a packet through a set of registered functions that can modify (and possibly even delete) it. In the query layer, if a node n is scheduled to aggregate data from all children nodes, it can intercept all data packets received from it's children and cache the aggregated result. At a specific time, n will then generate a new data packet representing the incremental aggregation of it and its children's data and send it towards the root of the network. All this happens completely transparently to the network layer.

Another possibility to implement this interception is to collapse the network stack and to merge the routing layer with the application layer, the approach taken in TinyDB. In this case, the application has complete control over the routing layer, and each packet that is routed through a node is handled by the application-level routing layer.

Both of these approaches are instances of *cross-layer interactions*. In order to preserve resources, we believe that future generations of sensor networks will take an integrated approach toward the design of the system architecture, which crosscuts the data management and communication (routing and MAC) layers. We can distinguish two approaches: A *top-down* and a *bottom-up* approach. In the top-down paradigm, we design and adapt communication protocols and their interfaces to the particular communication needs of the sensor network query processor. In the bottom-up approach, we consider the task of adapting query processing techniques to a given routing policy. Cross-layer interactions are a very fertile area of sensor network research, and TinyDB and Cougar have only made preliminary steps in this direction.

17.3.4 Acquisitional Query Processing

At first blush, it may seem as though query processing in sensor networks is simply a power-constrained version of traditional query processing: given some set of data, the goal of sensor network query processing is to process that data as energy-efficiently as possible. Push-down strategies, such as those discussed above, are similar to push-down techniques from distributed query processing that emphasize moving queries to data.

There is, however, another fundamental difference between systems like sensor networks and traditional database systems, that has to do with the role of data acquisition in query processing. In this *acquisitional query processing* (ACQP), the focus is on the significant new query processing opportunity that arises in sensor networks: the fact that smart sensors have control over where, when, and how often data is physically acquired (i.e., *sampled*) and delivered to query processing operators. By focusing on the locations and costs of acquiring data, it is possible significantly reduce power consumption compared to traditional passive systems that assume the a priori existence of data. Acquisitional issues arise at all levels of query processing: in query optimization, due to the significant costs of sampling sensors; in query dissemination, due to the physical co-location of sampling and processing; and, in query execution, where choices of when to sample and which samples to process are made.

We have designed and implemented ACQP features in TinyDB. While TinyDB has many of the features of a traditional query processor (e.g., the ability to select, join, project, and aggregate data), it also incorporates a number of other features designed to minimize power consumption via acquisitional techniques. These techniques, taken in aggregate, can lead to orders of magnitude improvement in power consumption *and* increased accuracy of query results over systems that do not actively control when and where data is collected.

One of the fundamental techniques derived from ACQP has to do with when sensor readings are acquired during the processing of a query. Because the cost of sampling sensors can be quite high—a significant fraction of the total processing cost for expensive, calibrated, digital sensors like the Honeywell Magnetometer [21] used in current generation Mica Motes—it may be useful to postpone the acquisition of sensor readings until it is absolutely clear those readings will be needed.

As an example of a situation where postponing acquisition in this manner can be useful, consider the query below, noting that the cost to acquire a sample from the magnetometer [21] (the `mag attribute`) is several orders of magnitude greater than the cost to acquire a sample from a photo-resistor-based light sensor [2] on current generation Mica hardware –0.25 mW versus 0.09 μW.

```
SELECT WINMAX(light,8s,8s)
FROM sensors
WHERE mag > x
SAMPLE PERIOD 0.1s
```

In this query, the maximum of every 8 s worth of light readings will be computed, but only light readings from sensors whose magnetometers read greater than x will be considered. Interestingly, it turns out that, unless the x predicate is *very* selective, it will be cheaper to evaluate this query by checking to see if each new `light` reading is greater than the previous reading and then applying the selection predicate over `mag`, rather than first sampling `mag`. This sort of reordering, which we call *exemplary aggregate pushdown* can be applied to any exemplary aggregate (e.g., `MIN, MAX`), and is general (i.e., applies outside the context of sensor networks.) Furthermore, it can reduce the number of magnetometer samples that must be acquired by up to a factor of 80, which corresponds to a total power savings of about 20 mJ, or 2.5 mW (recall that 1 mW = 1 mJ/s). This is roughly half the power required to run the processor.

Thus, properly choosing the order in which data is acquired can dramatically reduce energy consumption and is an example of a novel kind of optimization that does not arise in other query processing environments.

17.4 Experiments with Data Collection

We have done a number of studies of the performance and behavior of our SNQP implementations, both in simulation to demonstrate the potential of our algorithms and approaches and in real-world environments to observe their overall effectiveness.

17.4.1 Berkeley Botanical Garden Deployment

During June and July of 2003, we began a deployment of the TinyDB software in the Berkeley Botanical Garden, located on Centennial Road in Berkeley, just East of the main UC Berkeley Campus. The purpose of this deployment was to monitor the environmental conditions in and around Coastal Redwood trees (the *microclimate*) in the Garden's redwood grove.

This grove consists of several hundred new-growth redwoods. Botanists at UC Berkeley [11] are actively studying these microclimates, with a particular interest in the role that the trees have in regulating and controlling their environment, especially the ways they affect the humidity and temperature of the forest floor on warm, sunny days.

The initial sensor deployment in the garden consists of 11 Mica2 sensors on a single 36 m redwood tree, each of which is equipped with a weather board that provides light, temperature, humidity, solar radiation, photosynthetically active radiation, and air pressure readings. The sensors are placed in clusters at different altitudes throughout the tree. The processor and battery are placed in a water-tight PVC enclosure, with the sensors exposed on the outside of the enclosure. A loose fitting hood covers the bottom of the sensor to protect humidity and light sensors from rain. The light and radiation sensors on the top of the assembly are sealed against moisture and thus remain exposed. Figure 17.1 shows an example of such a mote outside of it's PVC package.

Sensors on the tree run a simple selection query which retrieves a full set of sensor readings every 10 min and sends them toward the basestation, which is attached to an antenna on the roof of a nearby field station, about 150 ft from the tree. The field station is connected to the Internet, so, from there results are easily logged into a PostgreSQL database for analysis and observation. The sensors have been running continuously for about three weeks. We expect the sensors to function for about 40 days, and plan to eventually grow the deployment to approximately 100 nodes over 5 trees.

Figure 17.6 shows data from five of the sensors collected during the second week of July, 2003. Sensor 101 was at a height of 10 m, sensor 104 at an altitude of 20 m, 109 at 30 m, 110 at 33 m, and 111 at 34 m. 110 and 111 we fairly exposed, while the other sensors remained shaded in the forest canopy.

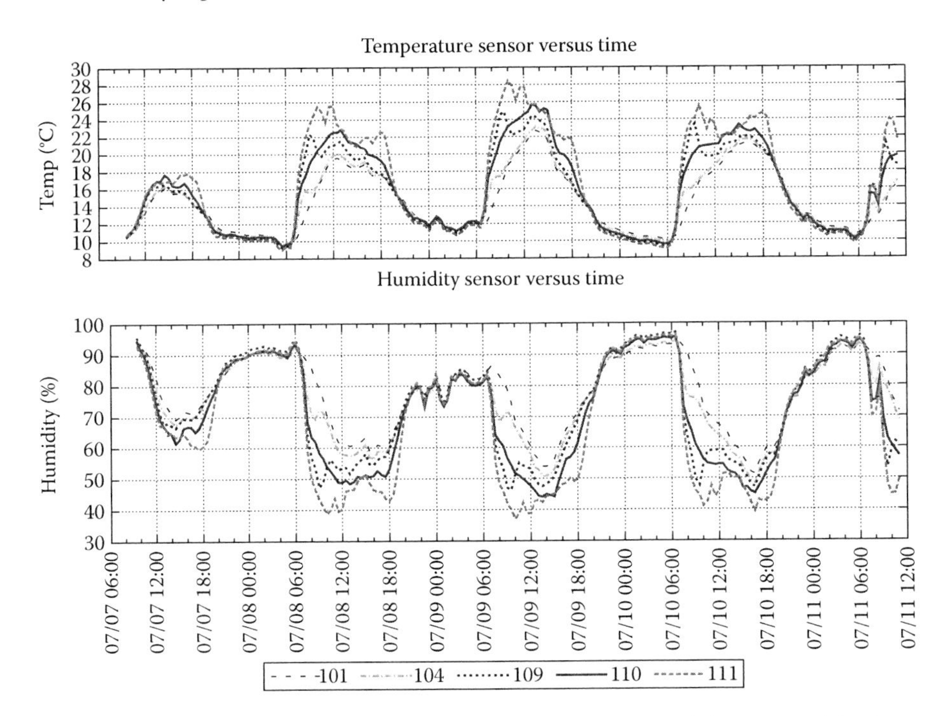

FIGURE 17.6 Humidity and temperature readings from five sensors in the Berkeley Botanical Garden.

The periodic bumps in the graph correspond to daytime readings; at night, the temperature drops significantly and humidity becomes very high as fog rolls in. Notice that 7/7 was a cool day, below 18°C and likely overcast as many summer days in Berkeley are. On days like this, all of the sensors record approximately the same temperature and humidity. Later in the week, however, it was much warmer, climbing as high as 28°C at the top of the tree. Note at these times, however, it can be as many as 10° cooler at the bottom of the tree, with a 30% higher humidity. This observation should be familiar to anyone who has ever been walking in a redwood forest and felt the cool dampness of the forest floor below the trees.

Although this is a fairly basic deployment, running only a simple query, we were able to program the sensors to begin data collection in just a few minutes. Far and away the most time consuming aspects of the deployment involved the packaging of the devices, obtaining access to the various spaces, and climbing the tree to place the sensors.

In future versions of this deployment, we hope to move to an approach where we log all results to the EEPROM of the devices and then just transmit summaries of that data out of the network, only dumping the EEPROM on demand or during periods where there is no other interesting activity (e.g., at night.) Once scientists believe that our hardware and software functions correctly, we believe they will be more likely to accept this style of approach.

Another interesting future direction related to this deployment has to do with tracking correlations between sensors and using those correlations to improve the efficiency of query processing. Note that, in Figure 17.6 temperature and humidity are highly correlated—knowing the humidity and sensor number allows one to predict the temperature to within a few degrees Celsius. This observation suggests an interesting query optimization possibility: for queries that contain predicates over temperature, one might could evaluate them by instead looking at humidity. If a humidity sample is needed for other purposes, or the energy costs of acquiring a humidity sample are low, this could be an energy-saving alternative.

17.4.2 Simulation Experiments

We also performed extensive simulation studies of our approach to show that it works well in a controlled environment—often, simulation is the only way to get repeatable results out of noisy, lossy sensor networks. For Cougar, we have a prototype of our query processing layer which runs in the ns-2 network simulator [4]. Ns-2 is a discrete event simulator targeted at simulating network protocols to highest fidelity. Due to the strong interaction between the network layer and our proposed query layer, we simulate the network layer to a high degree of precision, including collisions at the MAC layer, and detailed energy models developed by the networking community. In our experiments, we used IEEE 802.11 as the MAC layer [49], setting the communication range of each sensor to 50 m and assuming bi-directional links; this is the setup used in most other papers on wireless routing protocols and sensor networks in the networking community [22]. In our energy model the receive power dissipation is 395 mW, and the transmit power dissipation is 660 mW [22]; the radio was turned off outside of its "slot" in the routing tree. Sensors are randomly distributed in square region of increasing size, keeping the average sensor node density constant at 8 sensors per 100 m × 100 m. The root node is located in the upper left corner; we ran a simple query that computes the average over all sensors.

We compared three simple approaches:

- *Direct delivery*. This is the simplest scheme. Each source sensor node sends a data packet consisting of a record toward the leader, and the multi-hop ad-hoc routing protocol will deliver the packet to the leader. Computation will only happen at the leader after all the records have been received.
- *Packet merging*. In wireless communication, it is much more expensive to send multiple smaller packets instead of one larger packet, considering the cost of reserving the channel and the payload of packet headers. Since the size of a sensor record is usually small and many sensor nodes in a small region may send packets simultaneously to process the answer for a round of a query, we can *merge* several records into a larger packet, and only pay the packet overhead once per

group of records. For exact query answers with holistic aggregate operators like Median, packet merging is the only way to reduce the number of bytes transmitted [18].

- *Partial aggregation*. For distributive and algebraic aggregate operators [18], we can incrementally maintain the aggregate in constant space, and thus push partial computation of the aggregate from the leader node to intermediate nodes. Each intermediate sensor node will compute partial results that contain sufficient statistics to compute the final result.

The leftmost graph in Figure 17.7 illustrates the benefit of the in-network (push-down) aggregation approaches described above for a simple network topology. In the best case, every sensor only needs

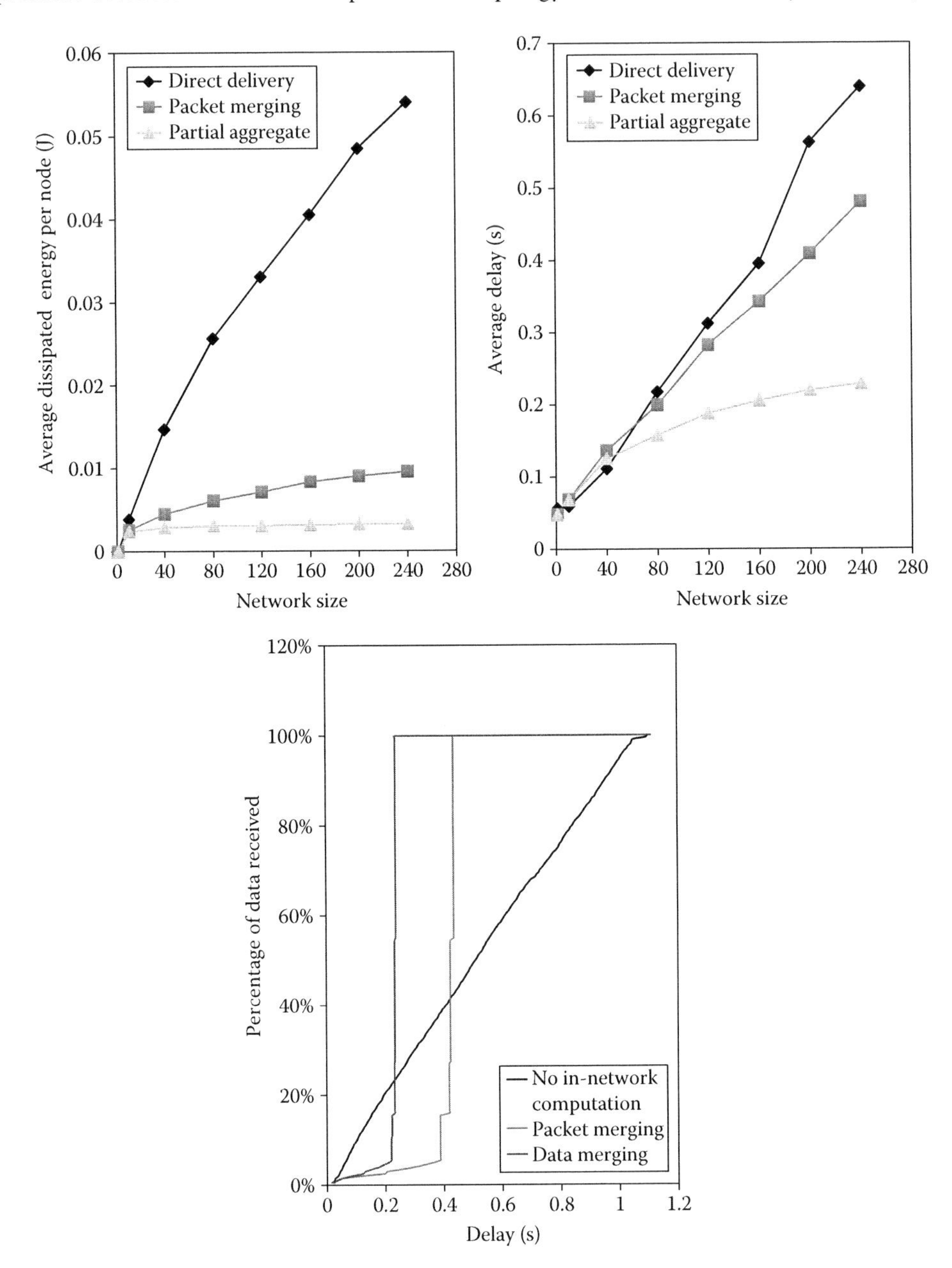

FIGURE 17.7 Simulation results comparing different approaches for answering an aggregate query.

to send one merged data packet to the next hop in each round, no matter how many sensors are in the network. The packet merge curve increases slightly as intermediate packets get larger as the number of nodes grows. Without in-network aggregation, a node *n* has to send a data packet for each node whose route goes through *n*, so energy consumption increases very fast.

We also investigated the effect of in-network aggregation on the delay of receiving the answer at the gateway node as shown in the middle graph in Figure 17.7. When the network size is very small, in-network aggregation introduces little extra delay due to synchronization, however as the network size increases, direct delivery induces much larger delay due to frequent conflicts of packets at the MAC layer. The rightmost graph in Figure 17.7 shows the cumulative distribution of the percentage of received sensor readings versus the delay. Without push-down aggregation, we continuously receive data, whereas with push-down aggregation we receive the final result in one packet.

17.5 Related Work

Related work on query processing for sensor networks can be classified into two broad areas: work on ad-hoc networks and distributed query processing.

17.5.1 Ad-Hoc Networks

Research on routing in ad-hoc wireless networks has a long history [26,44], and a plethora of papers has been published on routing protocols for ad-hoc mobile wireless networks [5,10,24,25,36,38,39]. All these routing protocols are general routing protocols and do not take specific application workloads into account. The SCADDS project at USC and ISI explores scalable coordination architectures for sensor networks [12], and their data-centric routing algorithm called directed diffusion [22] first introduced the notion of filters that we advocate in Section 17.3.3.

Ramanathan and Rosales-Hain developed protocols for adjusting transmit power in ad hoc networks [40] with the goal to improve connectivity in sparse networks and collisions in dense networks. PicoNet proposes an integrated design of radios, small, battery powered nodes, and MAC and application protocols that minimize power consumption [3]. Nodes are also scheduled to turn their radios on and off in order to conserve energy; however when a node needs to send a message to its neighbor it has to stay awake until it hears a broadcast message announcing its neighbor's reactivation. IEEE 802.11 supports ad hoc network configuration and provides power management controls [49]. Pearlman et al. [37] propose and energy dependent participation scheme, where a node periodically re-evaluates its participation in the network based on the residual energy in its battery. GEAR [62] uses energy-aware neighbor selection to route a packet toward a target region and restricted flooding to disseminate the packet inside the destination region; it addresses the problem of energy conservation from a routing perspective without considering the interplay of routing and node scheduling.

An energy-efficient MAC protocol called S-MAC has been proposed in Refs [58,59], where the nodes are locally synchronized to follow a periodic listen and sleep scheme. Each node broadcasts its schedule to its neighbors; the latter know exactly the time interval that the node is listening and (unlike PicoNet) only awake to send messages at that time. S-MAC does not explicitly avoid contention of the medium, but rather relies on an IEEE 802.11-style MAC for resolving collisions. This means that the receiving node must remain in the listening state for the (worst-case) time interval needed to resolve collisions, even its neighbors have no messages to send. GAF (Geographical Adaptive Fidelity) [54] is an algorithm that also conserves energy by identifying nodes that are equivalent from a routing perspective and then turning off unnecessary nodes.

17.5.2 Distributed Query Processing

An energy-efficient aggregation tree using data-centric reinforcement strategies is proposed in Refs [19,22]. In a recent study [16], an approximation algorithm has been designed for finding an aggregation tree that simultaneously applies to a large class of aggregation functions. A two-tier approach for data dissemination to multiple mobile sinks is discussed in Ref. [57].

There has been a lot of work on query processing in distributed database systems [8,27,35,60,61], but as discussed in Section 17.1, there are major differences between sensor networks and traditional distributed database systems. Most related is work on distributed aggregation, but existing approaches do not consider the physical limitations of sensor networks [48,55]. Aggregate operators are classified by their properties by Gray et al. [17], and an extended classification with properties relevant to sensor network aggregation has been proposed by Madden et al. [29]. Other relevant areas include work on sequence query processing [46,47], and temporal and spatial databases [63].

Due to space constraints, in this chapter we only introduced work on query processing, but there has been complementary recent work on in-network storage management. Ratnasamy et al. propose a distributed storage model based on hashing scheme [41,42]; this model is extended in Ref. [15]. Long-term storage in sensor networks is combined with multi-resolution data access and spatiotemporal data mining in Ref. [13].

17.6 Concluding Remarks and Future Challenges

We believe that a database approach to sensor network data management is very promising, and our initial experience with users of our technology has been has corroborated this belief: declarative queries offer the dual benefits of an easy-to-interface and an energy efficient execution substrate. Furthermore, our approach has unearthed a plethora of interesting additional research problems in this domain. Thus, we conclude this chapter by listing some future challenges that we believe are important unsolved problems in sensor network query processing.

17.6.1 Adaptivity

In the context of query processing, adaptivity usually refers to the modification or re-building of query plans based on runtime observations about the performance of the system. Adaptivity will be required to maximize the longevity and utility of sensor networks; examples of adaptation opportunities include

- *Query reoptimization*: Traditional query optimization, where queries are optimized before they are executed, is not likely to be a good strategy for long running continuous queries, as noted in Ref. [32]. This is because statistics used to order operators may change significantly over the life of a query, as we discuss below.
- *Operator migration*: Operators in a query plan may be placed at every node in the network, or only at a few select nodes. Choosing which nodes should run particular operators is tricky: for example, a computationally expensive user-defined selection function might significantly reduce the quantity of data output from a node. Conventional wisdom suggests that this filter should be propagated to every node in the network. However, if such a filter is expensive enough, running it may consume more energy than would be saved from the reduction in communication, so it may be preferable to run the filter at only a few select nodes (or at the powered basestation). Making the proper choice between these alternatives depends on factors such as the local selectivity of the filter, the number of hops which non-filtered data must travel before being filtered, processor speed and remaining energy capacity.
- *Topology adjustment*: Finally, the system needs to adapt the network topology. Existing sensor-network systems generally perform this adaptation based on low-level observations of network characteristics, such as the quality of links to various neighboring nodes [53],

but a declarative query processor can use query semantics to shape the network. For example, topology adaptation would be beneficial when computing grouped aggregates, since the maximum benefit from in-network aggregation will be obtained when nodes in the same group are also in the same routing sub-tree.

17.6.2 Nested Queries, Many to Many Communication, and Other Distributed Programming Primitives

TinyDB and Cougar support a limited form of nested queries through `STORAGE POINTS`—queries can be executed over the logged data in these buffers, thus providing a simple form of nesting. However, more complex, *correlated queries* that express fairly natural operations like "find all the sensors whose temperature is more than 2 standard deviations from the average" cannot be executed efficiently within the network. The reason for this is that this is really a nested query consisting of an inner query that computes the average temperature and an outer query that compares every node's temperature to this average. To execute this query in-network, some mechanism for propagating the average to the sensors is needed.

One important question is the location of storage points. Where should we place storage points in order to balance resource usage across the network while at the same time minimizing overall network resource usage?

The natural solution is a two round, gather-scatter protocol where some leader node (probably the root of the network) collects (gathers) the average and standard deviation and then disseminates (scatters) them to the entire network to answer the query. In-network processing can be a savings in this case because few sensors will have readings that are more than two standard deviations away from the average, and both average and standard deviation are algebraic aggregates that can be collected in an energy efficient manner.

For other queries, still more sophisticated communication primitives are needed; for example, consider the query "report all pairs of sensors within one meter of each other that have the same vibration reading." Answering this query without bringing all of the sensor data out of the network requires sensors to be able to exchange information with their 1 m neighborhood, which should be relatively communication efficient but is a different communication pattern than the tree-based protocols we considered in this dissertation.

Recently, in Cougar we proposed a simple technique for many-to-many communication in sensor networks based on the notion of a wave of communication flowing across a grid of sensors [51]. The main idea is to *schedule* sensors such that all collisions at the MAC layer will be avoided. Whitehouse et al. [52] propose a "neighborhood abstraction" that allows nodes to share information with their local neighbors. Incorporating these techniques into the implementations of TinyDB and Cougar and using them to support nested operations is an important area for future work.

17.6.3 Multi-Query Optimization

At any time, several long-running queries from multiple users might run over a sensor network. How can we share resources between these queries in order to balance and minimize overall resource usage? One possible approach is to apply standard multi-query optimization techniques from the literature such as exploiting common subexpressions. Another intriguing approach is to allow user-specified *approximations* to queries, and thus enable approximations of queries through answers of other queries.

17.6.4 Heterogeneous Networks

So far we only considered relatively homogeneous sensor networks where all nodes are equally powerful. Future networks will likely have several tiers of nodes with different performance characteristics. How can sensor network query processors take advantage of this heterogeneity? For example, if we had a set of more powerful and stable sensor nodes, they seem to be excellent candidates for storage points.

Acknowledgments

Cougar has been partially funded by the Defense Advanced Research Project Agency under contract F30602-99-2-0528, NSF CAREER Grant 0133481, the Cornell Information Assurance Institute, and by a gift from Intel. TinyDB has been has been supported in part by the National Science Foundation under ITR/IIS grant 0086057, ITR/IIS grant 0208588, ITR/IIS grant 0205647, ITR/SI grant 0122599, and by ITR/IM grant 1187-26172, as well as research funds from IBM, Microsoft, Intel, and the UC MICRO program.

References

1. Analog Devices, Inc. ADXL202E: Low-cost 2 g dual-axis accelerometer. http://products.analog.com/products/info.asp?product=ADXL202
2. Atmel Corporation. Atmel ATMega 128 microcontroller datasheet. http://www.atmel.com/atmel/acrobat/doc2467.pdf
3. F. Bennett, D. Clarke, J. Evans, A. Hopper, A. Jones, and D. Leask. Piconet: Embedded mobile networking. *IEEE Personal Communications*, 4(5):8–15, 1997.
4. L. Breslau, D. Estrin, K. Fall, S. Floyd, J. Heidemann, A. Helmy, P. Huang et al. Advances in network simulation. *IEEE Computer*, 33(5):59–67, 2000.
5. J. Broch, D. A. Maltz, D. B. Johnson, Y.-C. Hu, and J. Jetcheva. A performance comparison of multihop wireless ad hoc network routing protocols. In *Proceedings of the Fourth Annual ACM/IEEE International Conference on Mobile Computing and Networking (MOBICOM-98)*, pp. 85–97, Dallas, TX, October 25–30, 1998. ACM SIGMOBILE, ACM Press, New York.
6. M. Calimlim, J. Gehrke, and Y. Yao. Cougar project web page. http://cougar.cs.cornell.edu
7. D. Carney, U. Centiemel, M. Cherniak, C. Convey, S. Lee, G. Seidman, M. Stonebraker, N. Tatbul, and S. Zdonik. Monitoring streams—A new class of data management applications. In *VLDB*, 2002.
8. S. Ceri and G. Pelagatti. *Distributed Database Design: Principles and Systems*. McGraw-Hill, New York, 1984.
9. M. Chu, H. Haussecker, and F. Zhao. Scalable information-driven sensor querying and routing for ad hoc heterogeneous sensor networks. *International Journal of High Performance Computing Applications*, 16(3):293–313, 2002.
10. S. Das, C. Perkins, and E. Royer. Performance comparison of two on-demand routing protocols for ad hoc networks. In *Proceedings of the 2000 IEEE Computer and Communications Societies Conference on Computer Communications (INFOCOM-00)*, pp. 3–12, Los Alamitos, CA, March 26–30, 2000. IEEE, New York.
11. T. Dawson. Fog in the California redwood forest: Ecosystem inputs and use by plants. *Oecologia*, (117):476–485, 1998.
12. D. Estrin, R. Govindan, J. Heidemann, and S. Kumar. Next century challenges: Scalable coordination in sensor networks. In *Proceedings of the Fifth Annual ACM/IEEE International Conference on Mobile Computing and Networking (MobiCom-99)*, pp. 263–270, Seattle, WA, August 15–20, 1999. ACM SIGMOBILE, ACM Press, New York.
13. D. Ganesan, D. Estrin, and J. Heidemann. Dimensions: Why do we need a new data handling architecture for sensor networks? In *Proceedings of the First Workshop on Hot Topics in Networks (HotNets-I)*, Princeton, NJ, October 28–29, 2002.
14. J. Gehrke, F. Korn, and D. Srivastava. On computing correlated aggregates over continual data streams. In *Proceedings of the ACM SIGMOD Conference on Management of Data*, Santa Barbara, CA, May 2001.
15. A. Ghose, J. Grossklags, and J. Chuang. Resilient data-centric storage in wireless ad-hoc sensor networks. In *Proceedings of the Fourth International Conference on Mobile Data Management MDM 2003*, pp. 45–62, Melbourne, Australia, January 2003.

16. A. Goel and D. Estrin. Simultaneous optimization for concave costs: Single sink aggregation or single source buy-at-bulk. In *Proceedings of the 14th Annual ACM-SIAM Symposium on Discrete Algorithms*, Baltimore, MD, January 12–14, 2003.
17. J. Gray, A. Bosworth, A. Layman, and H. Pirahesh. Data cube: A relational aggregation operator generalizing group-by, cross-tab, and sub-total. In *ICDE*, New Orleans, LA, February 1996.
18. J. Gray, S. Chaudhuri, A. Bosworth, A. Layman, D. Reichart, M. Venkatrao, F. Pellow, and H. Pirahesh. Data cube: A relational aggregation operator generalizing group-by, cross-tab, and sub-totals. *Data Mining and Knowledge Discovery*, 1(1):29–53, 1997.
19. J. Heidemann, F. Silva, C. Intanagonwiwat, R. Govindan, D. Estrin, and D. Ganesan. Building efficient wireless sensor networks with low-level naming. In *SOSP*, Lake Louise, Alberta, Canada, October 2001.
20. J. Hill, R. Szewczyk, A. Woo, S. Hollar, and D. C. K. Pister. System architecture directions for networked sensors. In *ASPLOS*, Cambridge, MA, November 2000.
21. Honeywell, Inc. Magnetic sensor Specs HMC1002. http://www.ssec.honeywell.com/magnetic/spec_sheets/specs_1002.html
22. C. Intanagonwiwat, R. Govindan, and D. Estrin. Directed diffusion: A scalable and robust communication paradigm for sensor networks. In *Proceedings of MOBICOM 2000*, pp. 56–67, Boston, MA, 2000. ACM SIGMOBILE, ACM Press, New York.
23. Intersema. MS5534A barometer module. Technical Report, October 2002. http://www.intersema.com/pro/module/file/da5534.pdf
24. P. Johansson, T. Larsson, N. Hedman, B. Mielczarek, and M. Degermark. Scenario-based performance analysis of routing protocols for mobile ad-hoc networks. In *Proceedings of the Fifth Annual ACM/IEEE International Conference on Mobile Computing and Networking (MobiCom-99)*, pp. 195–206, 1999. ACM SIGMOBILE, ACM Press, New York.
25. D. B. Johnson and D. A. Maltz. Dynamic source routing in ad hoc wireless networks. In Imielinski and Korth, eds., *Mobile Computing*, Vol. 353 of The Kluwer International Series in Engineering and Computer Science. Kluwer Academic Publishers, Norwell, MA, 1996.
26. J. Jubin and J. D. Tornow. The DARPA packet radio network protocol. *Proceedings of the IEEE*, 75(1):21–32, 1987.
27. D. Kossmann. The state of the art in distributed query processing. *Computing Surveys*, 32(4): 422–469, 2000.
28. S. Madden and M. J. Franklin. Fjording the stream: An architecture for queries over streaming sensor data. In *ICDE*, 2002.
29. S. Madden, M. J. Franklin, J. M. Hellerstein, and W. Hong. TAG: A tiny aggregation service for ad-hoc sensor networks. In *OSDI*, 2002.
30. S. Madden, M. J. Franklin, J. M. Hellerstein, and W. Hong. The design of an acquisitional query processor for sensor networks. In *ACM SIGMOD*, San Diego, CA, 2003.
31. S. Madden, W. Hong, J. M. Hellerstein, and M. Franklin. TinyDB web page. http://telegraph.cs.berkeley.edu/tinydb
32. S. Madden, M. A. Shah, J. M. Hellerstein, and V. Raman. Continuously adaptive continuous queries over data streams. In *ACM SIGMOD*, Madison, WI, June 2002.
33. Melexis Microelectronic Integrated Systems. MLX90601 infrared thermopile module. Technical Report, August 2002. http://www.melexis.com/prodfiles/mlx90601.pdf
34. R. Motwani, J. Window, A. Arasu, B. Babcock, S. Babu, M. Data, C. Olston, J. Rosenstein, and R. Varma. Query processing, approximation and resource management in a data stream management system. In *CIDR*, 2003.
35. M. T. Özsy and P. Valduriez. *Principles of Distributed Database Systems*. Prentice Hall, Englewood Cliffs, NJ, 1991.
36. V. Park and S. Corson. Temporally-ordered routing algorithm (tora) version 1 functional speciation. Internet Draft, 1999. http://www.ietf.org/internet-drafts/draft-ietf-manet-tora-spec-02.txt

37. M. R. Pearlman, J. Deng, B. Liang, and Z. J. Haas. Elective participation in ad hoc networks based on energy consumption. In *Proceedings of the IEEE GLOBECOM*, Taipei, Taiwan, pp. 17–21, 2002.
38. C. Perkins and P. Bhagwat. Highly dynamic destination-sequenced distance-vector routing (DSDV) for mobile computers. In *ACM SIGCOMM'94 Conference on Communications Architectures, Protocols and Applications*, pp. 234–244, London, U.K., August 1994.
39. C. E. Perkins and E. M. Royer. Ad-hoc on-demand distance vector routing. In *Workshop on Mobile Computing and Systems Applications*, New Orleans, LA, February 25–26, 1999.
40. R. Ramanathan and R. Rosales-Hain. Topology control of multihop wireless networks using transmit power adjustment. In *Proceedings of the IEEE INFOCOM*, pp. 404–413, Tel Aviv, Israel, March 2000.
41. S. Ratnasamy, D. Estrin, R. Govindan, B. Karp, S. Shenker, L. Yin, and F. Yu. Data-centric storage in sensornets. In *First Workshop on Hot Topics in Networks (HotNets-I) 2002*, Princeton, NJ, October 2002.
42. S. Ratnasamy, B. Karp, L. Yin, F. Yu, D. Estrin, R. Govindan, and S. Shenker. GHT: A geographic hash table for data-centric storage. In *First ACM International Workshop on Wireless Sensor Networks and Applications (WSNA)*, Atlanta, GA, September 28, 2002.
43. T. Sakurai. Interconnection from design perspective. In *Proceedings of the Advanced Metallization Conference 2000 (AMC 2000)*, pp. 53–59, 2001.
44. N. Schacham and J. Westcott. Future directions in packet radio architectures and protocols. *Proceedings of the IEEE*, 75(1):83–99, 1987.
45. Sensirion. SHT11/15 relative humidity sensor. Technical Report, June 2002. http://www.sensirion.com/en/pdf/Datasheet_SHT1x_SHT7x_0206.pdf
46. P. Seshadri, M. Livny, and R. Ramakrishnan. Seq: A model for sequence databases. In P. S. Yu and A. L. P. Chen, eds., *Proceedings of the 11th International Conference on Data Engineering*, pp. 232–239, Taipei, Taiwan, March 6–10, 1995. IEEE Computer Society, Los Alamitos, CA.
47. P. Seshadri, M. Livny, and R. Ramakrishnan. The design and implementation of a sequence database system. In T. M. Vijayaraman, A. P. Buchmann, C. Mohan, and N. L. Sarda, eds., *VLDB'96, Proceedings of 22th International Conference on Very Large Data Bases*, pp. 99–110, Mumbai (Bombay), India, September 3–6, 1996. Morgan Kaufmann, San Francisco, CA.
48. A. Shatdal and J. F. Naughton. Adaptive parallel aggregation algorithms. In M. J. Carey and D. A. Schneider, eds., *Proceedings of the 1995 ACM SIGMOD International Conference on Management of Data*, pp. 104–114, San Jose, CA, May 22–25, 1995.
49. I. C. Society. Wireless LAN medium access control (mac) and physical layer specification. IEEE Std 802.11, 1999.
50. Texas Advanced Optoelectronic Solutions. TSL2550 ambient light sensor. Technical Report, September 2002. http://www.taosinc.com/pdf/tsl2550-E39.pdf
51. N. Trigoni, Y. Yao, A. Demers, J. Gehrke, and R. Rajaramany. Wavescheduling: Energy-efficient data dissemination for sensor networks. In submission.
52. K. Whitehouse and C. Sharp. A neighborhood abstraction for exploiting locality in sensor networks. In submission.
53. A. Woo and D. Culler. A transmission control scheme for media access in sensor networks. In *ACM Mobicom*, Rome, Italy, July 2001.
54. Y. Xu, S. Bien, Y. Mori, J. Heidemann, and D. Estrin. Topology control protocols to conserve energy in wireless ad hoc networks. Technical Report 6, University of California, Los Angeles, CA, Center for Embedded Networked Computing. Submitted for publication.
55. W. P. Yan and P.-Å. Larson. Eager aggregation and lazy aggregation. In U. Dayal, P. M. D. Gray, and S. Nishio, eds., *VLDB'95, Proceedings of 21st International Conference on Very Large Data Bases*, pp. 345–357, Zurich, Switzerland, September 11–15, 1995. Morgan Kaufmann, San Francisco, CA.
56. Y. Yao and J. Gehrke. Query processing in sensor networks. In *Proceedings of the First Biennial Conference on Innovative Data Systems Research (CIDR 2003)*, Asilomar, CA, January 2003.

57. F. Ye, H. Luo, J. Cheng, S. Lu, and L. Zhang. A two-tier data dissemination model for large-scale wireless sensor networks. In *Proceedings of the Eighth Annual International Conference on Mobile Computing and Networking (MobiCom)*, Atlanta, GA, September 23–28, 2002.
58. W. Ye, J. Heidemann, and D. Estrin. An energy-efficient MAC protocol for wireless sensor networks. In *Proceedings of the IEEE INFOCOM*, pp. 1567–1576, New York, June 2002.
59. W. Ye, J. Heidemann, and D. Estrin. Medium access control with coordinated, adaptive sleeping for wireless sensor networks. Technical Report ISI-TR-567, USC/Information Sciences Institute, Marina Del Rey, CA, January 2003.
60. C. Yu and W. Meng. *Principles of Database Query Processing for Advanced Applications*. Morgan Kaufmann, San Francisco, CA, 1998.
61. C. T. Yu and C. C. Chang. Distributed query processing. *ACM Computing Surveys*, 16(4): 399–433, 1984.
62. Y. Yu, R. Govindan, and D. Estrin. Geographical and energy aware routing: A recursive data dissemination protocol for wireless sensor networks. Technical Report UCLA/CSD-TR-01-0023, University of Southern California, Los Angeles, CA, May 2001.
63. C. Zaniolo, S. Ceri, C. Faloutsos, R. Snodgrass, V. S. Subrahmanian, and R. Zicari, eds. *Advanced Database Systems*. Morgan Kaufmann, San Francisco, CA, 1997.

18

Autonomous Software Reconfiguration

Richard R. Brooks
Clemson University

18.1 Problem Statement

Sensor nodes are cheap, small, resource constrained devices. Effective applications require decisions to be made, and actions to be taken, cooperatively by large groups of nodes. There are many good reasons:

- Nodes are placed in the immediate vicinity of a chaotic environment. Many nodes will be destroyed in the course of a mission.
- When nodes rely on battery power, they will die over time. Creating long sleep cycles for nodes can extend the lifetimes of each individual node.
- System robustness is aided by removing central points of failure.

In addition to these factors, node missions may change over time. There is an inherent contradiction. These requirements indicate the need for a large, complicated software infrastructure. On the other hand, this infrastructure would need to be implanted on nodes lacking the storage and cycles to support it. One way to overcome this would be to assign specific tasks to nodes, effectively spreading the software storage and execution burden over many nodes. The danger is that this added complexity would introduce brittleness to the system by creating multiple single points of failure.

This chapter motivates our use of mobile code to autonomously reconfigure software on individual nodes. Each node's role can change over time. Should a node performing a critical role be destroyed, another can take over that role. Should the system's environment change, for example new classes of entities for tracking are identified, the logic for identifying the entity can be painlessly added to the system. This comes at the cost of increased network communications. A hard constraint on embedded systems is thus replaced with a soft trade-off.

18.2 Resource Constraints

Applications of embedded systems, sensor networks and miniature robots are severely limited by constraints such as storage, power, computational resources, bandwidth, available sensors, and physical strength. Many of these constraints can be alleviated by using mobile code and data.

Currently, a hard constraint is imposed by the number of behaviors that can be stored locally. The ability to upload behaviors provides a potentially unlimited range of behaviors constrained by the time needed to upload the behavior. This resembles a computer's use of virtual memory, where memory storage on disk alleviates physical memory limitations. The number of accessible behaviors will go from tens (currently) to thousands.

Storage is in a hierarchy of memories each with varying costs and retrieval time. This provides the same benefits for robotic systems that virtual memory provides for computers. Figure 18.1 illustrates the different levels of memory in a traditional computer [Hwang 93] and the proposed system. In both, memory access speed and cost increase at each level of the hierarchy. Computers are currently designed to allow access to as large an amount of information as possible, as quickly as possible, within cost constraints. When information is not available at any level in the hierarchy, an interrupt propagates the request for information to the next level, until the request is satisfied.

Mobile code can organize and autonomously reconfigure software services in the same way. A sensor network software service broker is an off-board repository of mobile code defining behaviors and supporting tasks. If a node *N* attempts to execute a behavior *B* that is not stored locally, a network interrupt is generated. The interrupt is propagated to the service broker and *B* is uploaded to *N*. *B* may be cached locally to reduce network traffic.

Access to software is no longer constrained by local memory (Figure 18.1). This allows behaviors to be available to autonomous systems for rare events. Systems can also be updated in response to changing circumstances. For example, target recognition routines can be changed, as information becomes available about the types of systems the enemy has in the field.

Using mobile code also makes coordination simpler. Supervisory tasks are allocated to nodes at run-time, based on current conditions such as proximity to the task. Failure of one node can be overcome by transferring responsibilities and behaviors to another one in the vicinity. Multiple sensing modalities can be accommodated at run-time. We assume networked autonomous nodes with programmable microprocessors. They can download data and upload code. Mobile code and data support is layered onto existing miniature hardware.

Infrastructure for migrating code and data on demand has been implemented. It supports modeling arbitrary constraints, such as programs requiring specific processors, network connectivity, and hardware requirements. Figure 18.2 shows the roles in the current system.

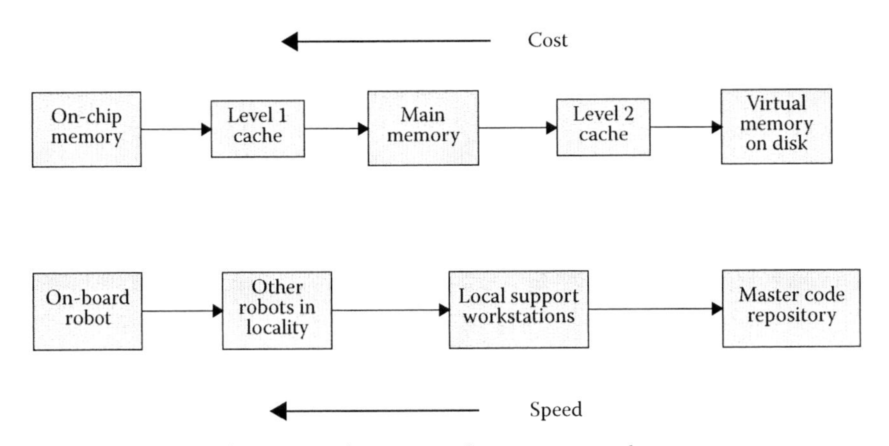

FIGURE 18.1 Memory hierarchies in virtual memory and in sensor networks.

Workstations	Worker	Supervisor
• Perform off-board processing • Contain library of mobile code implementing behaviors • Computational power ranging from Pentium to supercomputer • Used for computationally intensive tasks • Used when computational speed-up outweighs network delay 	• Performs one or more micro-behaviors • If the micro-behavior is not present it is uploaded • Inexpensive robots chosen for hazardous missions • Receives requests and combines them as appropriate • Processes data locally if more efficient than transmitting • May transmit data in compressed format to reduce latency 	• Chosen at run-time • Supervisor role may migrate from one robot to another • Coordinate and execute a macro-behavior • May also execute micro-behaviors • Particular hardware configurations may be needed • Located near workers executing micro-behaviors • Conserves latency, power, and bandwidth 

FIGURE 18.2 *Workstations* perform off-board processing. At run-time some nodes may execute macro-behaviors *supervising* other nodes. *Worker* nodes execute micro-behavior components of higher-level macro-behaviors. Multiple levels and specializations are possible.

18.3 Example Application Scenario

To support information dominance, unattended sensor network systems have been developed. New ones are being researched. They generally consist of ground deployed, battery powered, intelligent systems with sensors and wireless communications.

Communications is difficult for objects on or near the ground. Multiple propagation paths cause signal interference, limiting the communications range. To compensate, systems use directional communications or create multi-hop networks. In multi-hop networks, each sensor node communicates with neighbors located in the immediate vicinity. A message travels from node to node eventually reaching the information consumer.

For either approach, it is advantageous to place sensor nodes so that multiple communication paths are possible. The system is thus able to tolerate limited failures of single sensor nodes. This, and the use of multiple communications frequencies, can compensate somewhat for the presence of enemy jamming devices in the vicinity.

Since Unattended Ground Sensors (UGS) are often required in hazardous regions, it is reasonable to use unmanned vehicles to deploy them. Simultaneously deploying multiple sensor nodes requires coordination among multiple robots.

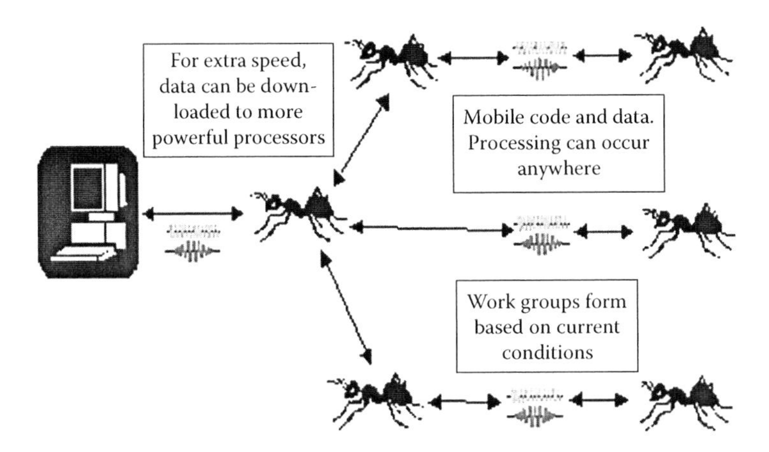

FIGURE 18.3 Mobile code allows the system to decide dynamically where processing should occur. Roles are assigned dynamically. Coordination and supervision are done locally. Sensor data travels up the network as needed. The entire network becomes an adaptive data processing device.

For this type of mission, it is reasonable to use networked autonomous vehicles with programmable microprocessors. They download data and upload code, as shown in Figure 18.3. Mobile code and data support is layered onto existing robots. It is necessary to model constraints, such as programs requiring specific processor types, and data requests requiring specific sensor types.

An operator defines top level behavior as a script of macro-behaviors. Behaviors are organized in a semiotic class structure reflecting the problem space. This structure is modeled by graphs of interacting automata [Phoha 97]. Under-defined aspects of the script are resolved at lower levels of the hierarchy. The system is controlled recursively from the top-down by macro-behaviors defining subsets of the network.

Macro-behaviors can be goals (e.g., *find_UGS*) or physics based task definitions (e.g., *lift_UGS*(X)). A macro-behavior M_a is a script of macro-behaviors and micro-behaviors. M_a is supervised by part of the subset of robots executing M_a. A micro-behavior M_i is the mobile code for a specific behavior on a specific robot platform. For example, the class of behaviors *go_to*(X) contains many micro-behaviors. The micro-behavior needed by specific hardware is chosen at run-time. Behaviors for robots mimicking crickets vary from behaviors for wheeled, or swimming robots. The semiotic class structure resolves these ambiguities. An example of the utility of this approach is shown in Figure 18.3.

The system is also controlled from the bottom-up by sensors providing information (*push*). Sensor interpretation is done by signal processing and sensor fusion [Brooks 98] routines distributed throughout an active network.

To decide between alternative actions use a *satisficing* approach, finding "good enough" alternatives in real-time by estimating performance as a function of resources, using experimental methods [Brooks 95]. This has been endorsed by a number of researchers [Brooks 95, Hooker 94]. Factors to consider include time, memory, uncertainty, bandwidth, and result quality. Decisions consider *meta-reasoning* to estimate decision overhead. Special attention should be given to hardware cost. For hazardous missions, low cost nodes can minimize loss.

18.4 Distributed Dynamic Linking

Assume a network with sensors attached to programmable microprocessors. Sensors can download data and upload code. Libraries of mobile code are available for use. Real-time networking control is built into the system; allowing adaptation to changing conditions. The sensor network should be generic to support multiple sensing modalities, by conforming to standard frameworks for intelligent sensing systems [Figueroa 94, IEEE 96]. Sensors integrate several phenomena. Physical phenomena are transmitted in a medium through an aperture to a detection device. Detected signals are processed and decisions made [Waltz 90].

Parts of our approach are appropriately implemented in low-level operating system or network software. Communications are based on an asynchronous model with time-outs. Synchronous communication can be emulated.

A service broker is a system for organizing and invoking software services. We implemented one for mobile code support. We discuss how this infrastructure works for Java, Forth, and C++. Java's support of mobile code makes it relevant to sensor networks. Other aspects are poorly suited for embedded applications:

- The full Java virtual machine required 1.5 MB of storage in 1998 [McDowell 98], excessive for some embedded applications. Since then, reduced virtual machines with partial functionality have been implemented.
- Embedded systems often require direct software interaction with hardware. Java's design intentionally omits this.
- Java's networking assumptions are not always appropriate.

Forth is commonly used for embedded programming. Like Java it is interpreted. It is not object oriented. Forth was designed for modular task construction and interaction with hardware. Many interpreters for Forth have small memory profiles [Tracy 89]. It has many favorable aspects of Java, without the unfavorable ones. C++ is widely available. It is a compiled object-oriented language with low-level constructs for hardware interaction. Java's syntax and structure are closely related to C++. C++ support provides completeness. Much of sensor network relevant code is written in C++ [Brooks 98, Press 94].

Java supports mobile code by compiling source code into a compressed byte code intermediate representation. Byte code is transmitted over the network to a virtual machine that executes it. In addition, a Remote Method Invocation (RMI) protocol allows Java programs to execute routines on remote machines [OMG 97]. RMI is similar to the Object Management Group (OMG) Common Object Request Broker Architecture (CORBA), and Microsoft Distributed Common Object Model (DCOM).

Java's mobile code model is inappropriate for C++ and Forth. The service broker approach is language agnostic, layering mobile code support onto existing languages. This approach can be easily adapted to other interpreted and compiled languages. Figure 18.4 shows our approach to mobile code support. It contains the following:

1. *Java*—Most necessary support is in the language definition. We require routines for
   a. Locating code on demand.
2. *Forth*—A Forth interpreter must be present on the machine for Forth programs to execute. Forth code is stored in clear text. Mobile code support requires routines that
   a. Locate programs on demand.
   b. Download routines as needed.
   c. Compress code for transfer.
   d. Decompress code for execution.
   e. Execute Forth programs remotely.
   f. Return results over network.
   g. Reclaim resources after processing.
3. *C++*—C++ is compiled before execution. Source code, object code or binary executables can be transmitted. Object and binary files are incompatible between machine types and operating systems. Storage limitations restrict the ability of programs to run on some nodes. Code transfer between machines may involve multiple steps: (1) Source code copied from machine *A* to machine *B*, (2) Code built on *B* with cross-compiler for machine *M*, (3) Binary executable copied from *B* to *M*, and (4) Program executes on *M*. Mobile code support requires routines that
   a. Locate programs on demand.
   b. Locate compilers (or cross-compilers) for classes of machines.

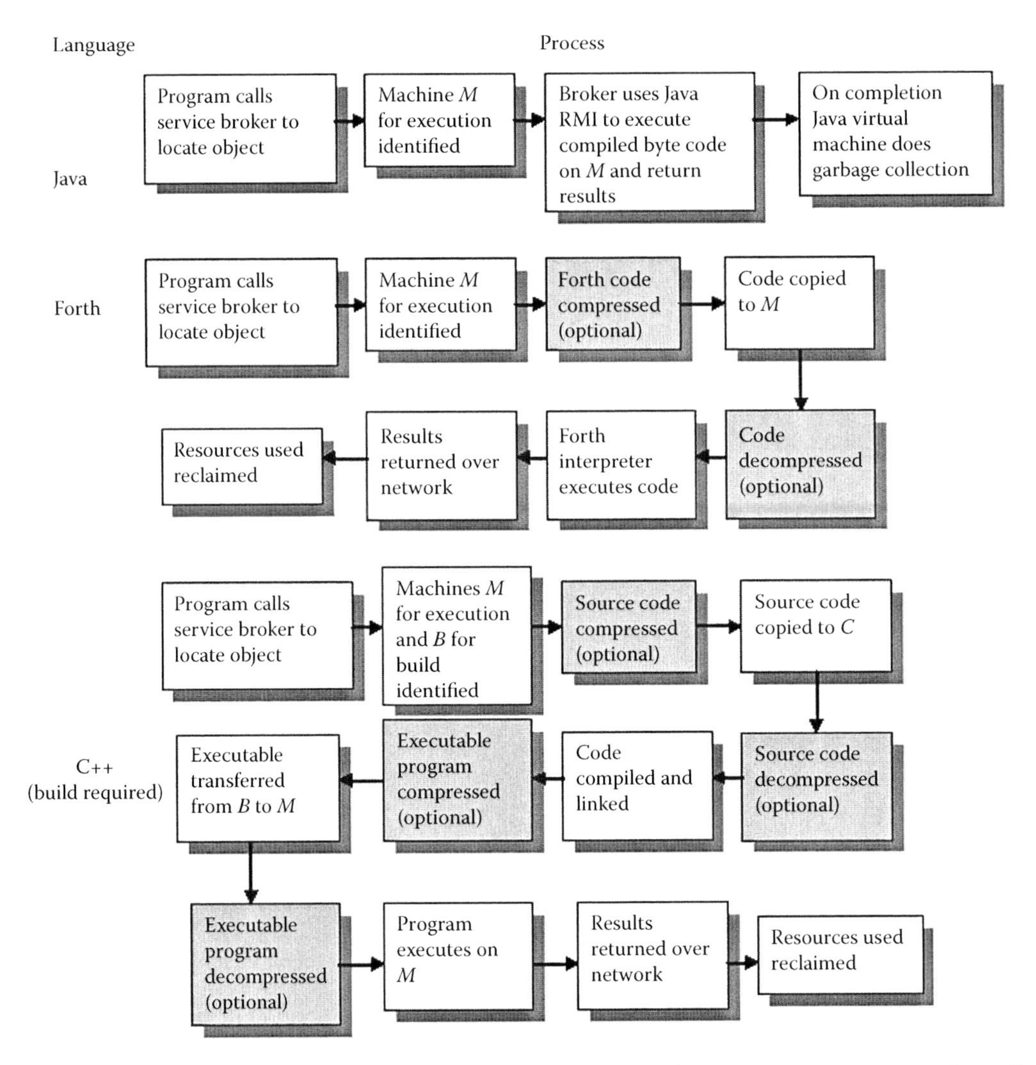

FIGURE 18.4 Typical processing performed by the service broker to invoke a service. More steps exist when calls are bound to service classes. Forth and Java portions are fairly standard. C++ libraries may consist of source, object, or executable files. C++ processing is shown for the case where the library stores source code.

c. Download routines as needed.
d. Compress code for transfer.
e. Decompress code for execution.
f. Execute C++ programs remotely.
g. Return results over network.
h. Reclaim resources after execution.

Libraries of mobile code can be developed and managed using this distributed service broker. When developing programs, designers bind calls to services or classes of services. If a call is bound to a class, the service broker chooses a service in the class at run time to efficiently use resources given current conditions. The service is chosen using profile functions derived using empirical methods. Resource-bounded optimization techniques make the choice. The amount of time spent choosing depends on the amount of time optimization can save. Time is unavailable for computing strictly optimal solutions on-line [Zilberstein 96]. *Satisficing* solutions provide "good enough" answers [Bender 96], the best found given current resources and constraints.

For calls bound to classes, hardware and software configurations are chosen at run time. It is undesirable to force all services bound to the same class to use identical parameter lists. In addition, programs and data may or may not be located on the same node. For transparency and efficiency, parameter formats may have to be modified at run time. For these reasons, interfaces must be malleable.

Run-time support finds adequate services, determines the likelihood of services fulfilling process requirements, and assigns resources to services. Process overhead is part of the calculations. It is a waste of time for the broker to spend 5 min calculating how to save 3 min. Our approach uses meta-reasoning. Optimization has a limited horizon, due to the dynamic nature of the system. Groups of components form flexible ad hoc confederations to deliver data in response to changing needs and resources. This is applicable to any sensing modality and can be used on any system containing networked processors, sensors, and embedded processors.

Services are registered interactively. Service information needs are part of the information registered. Interface requirements, similar to CORBA IDL interfaces [Seetharaman 98], are mapped to existing class parameters, or a new class is defined. There need not be a one-to-one correspondence between class and service parameters.

At design time, nodes where programs and data are located are undefined. Communication must be transparent to distributed processing. For parameter passing by value, this is not an issue. Parameter passing by reference can be difficult to implement. When data and program are on the same node, pass by reference should be used for efficiency. When they are on different nodes, pass by reference can usually be emulated with the ADA read-write parameter mechanism. The called program executes and results are written into the variable's storage. They are not always equivalent and work is needed to insure consistency in all cases.

Binding links to classes of services causes another problem. It is simpler to design a service broker by insisting all services in a class have an identical interface. This is restrictive for service providers. When registering methods, links are made between service and class parameters. This mapping must not always be one-to-one. Defaults may or may not be provided; their use is encouraged and supported. Similarly, consumers make links between program objects and class parameters. At run time, consumer and provider interfaces negotiate exact parameter passing methods. A protocol exchanges interface information. Inconsistencies may be found on both sides and alternatives proposed. This includes issues like data format (e.g. integer variables may need to be converted to floating point.) The process iterates until agreement is reached. Similar issues exist in inter-program communication for the program-program interface. The system emulates established methods using an asynchronous message passing protocol.

The final issue we consider is adaptation to system state. Information can be exchanged in a number of formats. It is reasonable to compress data for transmission over slow channels. Transmission over noisy channels requires redundant data. As noise (lack of noise) is detected in a channel, error checking increases (decreases). The meta-protocol starts with pessimistic assumptions of channel quality and modifies the protocol dynamically. Modifications are based on information from normal operations. Extra traffic for monitoring status is avoided.

18.5 Classifier Swapping

As a proof-of-concept demonstration of the mobile code daemon, we recently tested a system for swapping classification and tracking algorithms on the fly. This allows the sensor network to adapt to changes in its environment. We implemented mobile code support for sensor networks by creating a minimal infrastructure.

Target classification is a process in which sensor time-series data is used to assign target detections to one of a set of known classes. A vehicle driving by a sensor node could be a tank, a truck, a dragon wagon, a TEL, etc. Many classification techniques exist. In this test we used classifiers developed by

Hairong Qi of the University of Tennessee at Knoxville, Akhbar Sayeed of the University of Wisconsin [Brooks 03], and David Friedlander of the Penn State Applied Research Laboratory [Friedlander 02].

To choose between classifiers confusion matrices [Kohavi 98] were used. A confusion matrix is a matrix where the rows are the actual target classes and the columns are the predicted target classes. Each element of the matrix $e_{i,j}$ expresses the probability that the classifier returns codebook value j when target type i was actually present. For example, the matrix:

$$\begin{bmatrix} 0.90 & 0.10 \\ 0.25 & 0.75 \end{bmatrix}$$

could express the uncertainty in a classifier with two classes: tracked vehicle (class 1) and wheeled vehicle (class 2). In this case, the system correctly classifies class 1 90% of the time and class 2 75% of the time. In our case the code book values and target classes used were

Codebook	Target Name	Target Descriptor
0	Unknown	
10	sif_Buzzer	
11	sif_Motorcycle	Motorcycle
12	sif_TruckGas	Pickup Truck—Gas Engine
13	sif_TruckDiesel	Pickup Truck—Diesel Engine
14	sif_BuzzerRed	Red team
15	sif_BuzzerBlue	Blue team

Another part of the field test tested new classification techniques. Insufficient data was available before the tests to derive reliable confusion matrices for the three approaches used. Our classifier swapping tests used matrices fabricated to best illustrate the software functionality.

To support classifier swapping each node keeps a vector containing the set of target classes most recently detected by the sensor node.* The diagonal of the confusion matrix expresses the likelihood that a classifier is correct. The vector of target classes seen recently is multiplied against the diagonal of the confusion matrices for the three participating classifiers. This provides a measure of how well a given classifier should work given the current target mix.

All three participants used a unified classifier Application Programming Interface (API). This integrated the three different classifiers into the tracking process via a single call to the mobile code daemon. After each target was processed, the system determined which classifier was likely to work best with the current target mix. The daemon pre-fetches that classifier to be sure that it is present on the sensor node and uses it to classify the next target detected. This illustrates the distributed dynamic linking concept. A single classification call can trigger any of a number of implementations. The system automatically chooses the most appropriate one at run-time.

Implementation of this approach requires passing data to and from the classification program or library routine. The mobile code daemon can replace the routine used at will. We circumvent functionality traditionally given to the linker. We have used two different approaches to this problem: manufacturing call stacks and marshalling data to disk.

For the Windows 2000 and Windows CE versions of the daemon, we experimentally determined the layout of call stacks used by compilers. Calls to the daemon contain the data passed to the mobile code package. The daemon manufactures the appropriate call stack and passes it directly to the library routine. Passing data to standalone executables is done using a command line interface.

* We implicitly assume that the types of targets the node is likely to see in the near future resembles what it has seen lately (locality in time and space).

For the Linux port of the software, we took a different approach. A small wrapper process is integrated around calls to the daemon. Before calling the daemon, necessary data is marshaled (written to disk). The daemon executes the desired classification routine within another wrapper routine, which first reads the marshaled data into memory.

In the future, we plan on retaining the marshalling approach. It is more flexible and easier to integrate with multiple compilers. It also allows us to transparently pass data between nodes, as long as the program does not rely on the actual physical location in memory of the data.

As of this writing, the individual software components have all been written and tested. Some issues arose in porting parts of the system from the Intel X86 architecture to the SH4 architecture, which delayed the unified test. Plans exist to test the integrated system with field data in the immediate future.

Given a set of target tracks and a set of current detections, it is necessary to assign detections to tracks. This problem is known as *data association*. A number of methods have been developed for data association and none has been found to be clearly superior.

One widely accepted approach is multi-hypothesis data association [Bar-Shalom 93]. For each track multiple hypotheses are maintained regarding the target's codebook value and dynamics. Tracks are maintained using the most likely interpretation of a target's motion at any point in time.

The data association technique used is much simpler. For each detection, we extrapolate candidate tracks forward in time assuming there is no change in target heading or velocity. We compare detection position, heading, and velocity with the extrapolated data using a Euclidean metric. This essentially uses a nearest-neighbor approach. Each detection is mapped to the closest track.

[Brooks 03] shows how target dynamics can be built into the track extrapolation process using an Extended Kalman Filter (EKF). Any highly distributed data association approach will not be able to enforce global consistency as well as centralized approaches like in [Bar-Shalom 93].

The nearest-neighbor approach is straightforward and works well in many cases. In spite of this, it would be useful to use different techniques for track maintenance of different target classes. Different classes may have radically different dynamics. We have attempted to do this by integrating multiple trackers into our framework. Each class can be handled by a different tracking implementation. And the tracker implementation is responsible for the consistency of its own track.

As a demonstration of this concept, we integrated our tracking approach with a mobile agent tracking approach implemented by Hairong Qi at the University of Tennessee Knoxville. This application of distributed dynamic linking allows the sensor network to track a virtually unlimited number of target classes. This is possible in the same way that virtual memory systems provide modern workstations with a virtually unlimited amount of core memory. At any point in time, any node can recognize only a limited number of target classes. But the set of classes can be modified dynamically over the network. A hard constraint caused by local storage restrictions has been replaced by a soft constraint caused by network resource restrictions.

Note that when targets are sparse, multiple target tracking becomes a set of disjoint single target tracking problems. The ability of any system to track a dense cloud of targets will be limited by the sensor's abilities to differentiate between independent target detections.

18.6 Dependability

Our technology allows node roles to be chosen on the fly. This significantly increases system robustness by allowing the system to adapt to the failure of individual nodes. The nodes that remain exchange readings and find answers.

Consider a heading and velocity estimation approach using triangulation [Brooks 02a, Brooks 02b, Brooks 03a], at least three sensor readings are needed to get an answer. In the following, we assume all nodes have an equal probability of failure q. In a non-adaptive system when the "cluster head" fails, the system fails. The cluster has a probability of failure q no matter how many nodes are in the cluster. In the adaptive case, the system fails only when the number of nodes functioning is three or less.

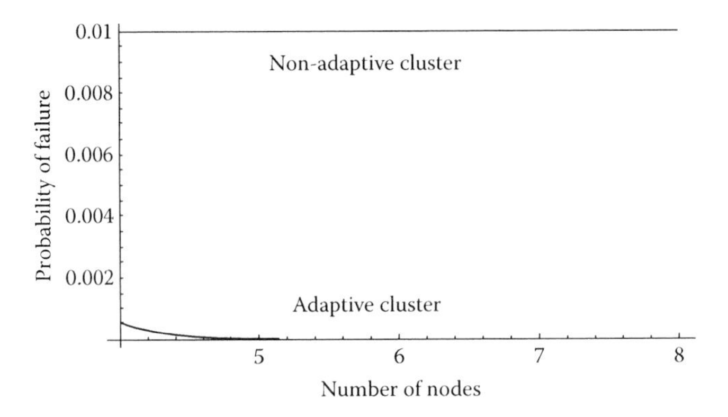

FIGURE 18.5 The top line shows probability of failure for a non-adaptive cluster. The bottom line shows probability of failure for an adaptive cluster. The probability of failure for a single node q is 0.01. The number of nodes in the cluster is varied from 4 to 8.

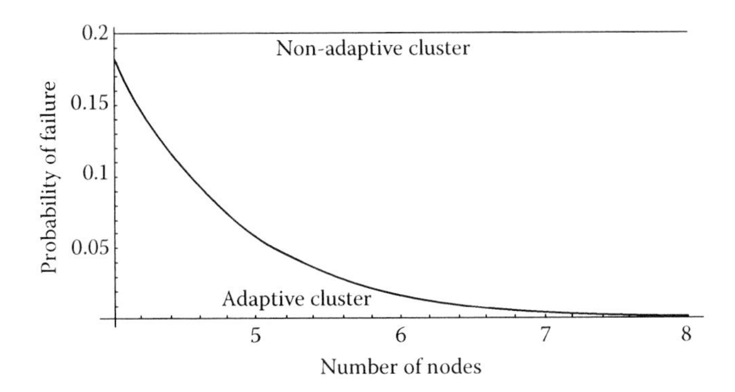

FIGURE 18.6 The top line shows probability of failure for a non-adaptive cluster. The bottom line shows probability of failure for an adaptive cluster. The probability of failure for a single node q is 0.2. The number of nodes in the cluster is varied from 4 to 8.

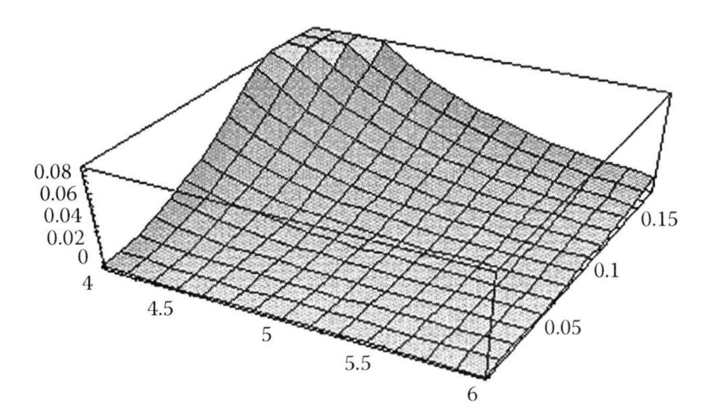

FIGURE 18.7 The surface shows probability of failure (z axis) for an adaptive cluster as the probability of failure for a single node q varies from 0.01 to 0.2 (side axis), and the number of nodes in the cluster varies from 4 to 6 (front axis).

Figures 18.5 through 18.7 illustrate the difference in dependability between adaptive and non-adaptive tasking. These figures assume an exponential distribution of independent failure events, which is standard in dependability literature. The probability of failure is constant across time. We assume that all participating nodes have the same probability of failure. This does not account for errors due to loss of power.

In Figures 18.5 and 18.6 the top line is the probability of failure for a non-adaptive cluster. Since one node is the designated cluster head, when it fails the cluster fails. By definition, this probability of failure is constant. The lower line is the probability of failure of an adaptive cluster as a function of the number of nodes. This is the probability that less than three nodes will be available at any point in time. All individual nodes have the same failure probability, which is the value shown by the top line. The probability of failure of the adaptive cluster drops off exponentially with the number of nodes. Figure 18.7 shows this same probability of failure as a function of both the number of nodes and the individual node's probability of failure.

18.7 Related Approaches

In this section, we compare the technology presented here to research in resource-limited optimization, and robotics.

Figure 18.8 illustrates important differences between our approach and existing interoperability frameworks. CORBA and RPC allow interoperability among programs on divers computing platforms, written in different languages. In CORBA an Interface Definition Language (IDL) program is written to make the conversion. It is assumed the developer knows the service that will be used a priori. The developer must have the IDL. A broker is called by the application to locate the service at run-time. Microsoft's DCOM has expanded this approach, placing a finite state machine between the broker and the service. Our service broker uses semantic networks to aid in locating services for use both at design time and run time.

Four main technologies exist for distributed program interoperability. The Distributed Computing Environment Remote Procedure Call (RPC), OMG Common Object Request Broker Architecture (CORBA), Microsoft's Distributed Component Object Model (DCOM), and Sun's Java Remote Method Invocation. All provide the infrastructure for a program on one machine to invoke a procedure that executes on another machine. This includes methods for parameter passing over computer networks.

RPC is a client server technology, which does not embrace the object model [Dogac 98]. CORBA is built on RPC concepts work and provides a standard for object definition and interaction. CORBA objects can be transient or persistent. They can be written in any of a number of languages, and invoked transparently irrespective of their location on the network. Objects are called using their interfaces,

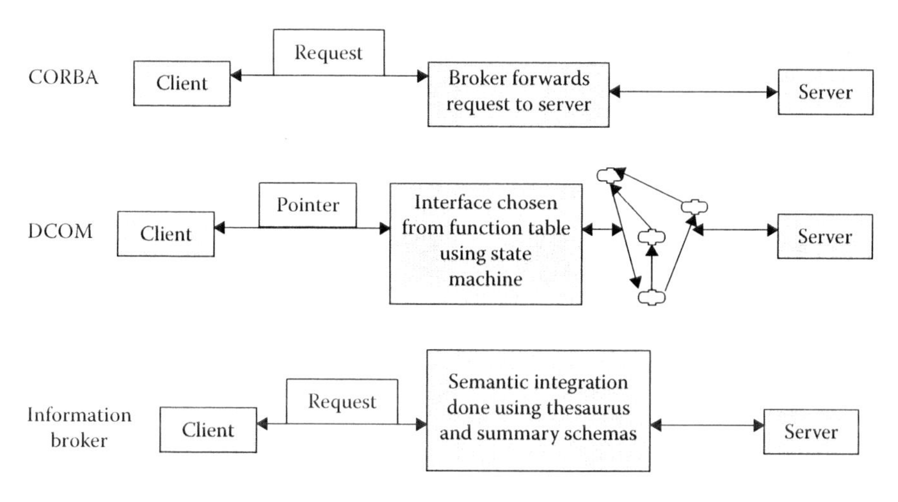

FIGURE 18.8 Contrast of the CORBA, DCOM, and info-broker integration models.

which are written in an Interface Description Language (IDL). They register their existence with an ORB. To invoke a service, the calling routine needs the IDL definition. The calling routine sends a message to the ORB, which completes the connection [Seetharaman 98].

DCOM is a binary level standard defined by Microsoft. It differs from CORBA in that it specifies multiple interfaces to a single data object; using a pointer to a function table. Object state is made part of the interface. This approach is more scalable than a pure object-oriented inheritance hierarchy [Wegner 96]. Details concerning DCOM implementation can be found in [Gray 98]. A detailed comparison of the use of CORBA and DCOM can be found in [Chung 98]. Interfaces exist for DCOM and CORBA interoperability.

Java is a distributed programming language developed by Sun Microsystems. Programs are compiled into a compressed byte-code. At run-time the byte-code is downloaded and interpreted by a virtual machine. Part of the language is a standard for Remote Method Invocation. RMI is roughly equivalent to a CORBA ORB, in that programs can invoke procedures to execute on remote machines. As with DCOM, interfaces between Java and CORBA exist [Curtis 97].

These interoperability implementations provide basic infrastructure for service brokering. OMG has also developed a standard for service brokering [OMG 97]. This standard allows a specific service to be chosen from a set of services of the same type at run-time. The type name must be known a priori. Choice between services of the same type is made by Boolean comparisons of a set of parameters. This is a very limited method for comparison. These parameters may be dynamic, but dynamic parameter values must be obtained from the service at run-time. This can induce significant delays. Support is also provided for ORB's passing service requests to other ORB's. No semantic interoperability support is available for service discovery.

The FINDIT prototype attempts to combine multi-database systems with the world-wide-web. Databases have associated co-databases containing meta-information. Coalitions are formed combining databases and meta-databases. The system is implemented on top of CORBA. Interfaces written in IDL are imported or exported. Only static meta-information, which is always true, is available about services. No mention is made of problems caused by use of differing terminology. Information is discovered by users browsing coalitions using the structured WebTassili query language [Benatallah 97, Bouguettaya 98].

Researchers at the TU Berlin have constructed a service broker for telecommunications applications, which resembles the run-time portion of our proposal. Their work uses intelligent agent technology to implement an active network; allowing agents to work within telecommunications switches [Breugst 98]. They divide agent technology research into the areas of mobile agents and communication among static agents. They concentrate on mobile agent technology. Another telecom broker implementation is discussed in [Sharma 97].

Xena is a service broker implemented at Carnegie Mellon University [Chandra 98], which also resembles the run-time portion of our proposal. This implementation concentrates on brokering for multiple services with different Quality of Service parameters. Services are chosen by user defined functions. Little support is provided for discovering services, since "general resource discovery is outside the scope of Xena." Global optimization is attempted using integer linear programming. Resource optimization is attempted by *semantic-preserving transformations*, such as converting MPEG to motion JPEG. For dynamic situations autonomously performing local optimizations will provide better results, since global information will generally not be available.

[Tanaka 97] describes a service broker prototype for telecommunications systems. Choices can be made among services of the same type. No support is provided for finding new types of services. Choices are made comparing cost. More complicated choices are not supported. [Yi 95] describes a real-time service broker. Real-time support is provided through the use of time outs. No support is provided for semantic ambiguity.

Another service brokering system is TSIMMIS, which supports the retrieval of heterogeneous objects. Wrappers are called mediators. Translators are procedures, which translate data formats [Chawathe 94]. Unfortunately, the assumption is made that attribute naming will be consistent across heterogeneous data sources. Research has shown this assumption to be questionable [Heiler 95]. Semantic ambiguity is not compensated for. A structured query language similar to SQL is used for query construction. The retrieval model is more closely aligned to Web browsing than information indexing.

18.8 Summary

This chapter discussed autonomous software configuration for sensor networks. Motivation was given for the use of mobile code in networks of embedded systems. The concept of distributed dynamic linking was described in detail. Examples where this approach has been fielded were provided, as well as a dependability analysis of a sensor network with node roles chosen at run-time. The analysis shows that system dependability can be greatly increased using this approach.

Acknowledgments and Disclaimer

This research is sponsored by the Defense Advance Research Projects Agency (DARPA), and administered by the Army Research Office under Emergent Surveillance Plexus MURI Award No. DAAD19-01-1-0504 and the Office of Naval Research under Award No. N00014-01-1-0859. Any opinions, findings, and conclusions or recommendations expressed in this publication are those of the authors and do not necessarily reflect the views of the sponsoring agencies.

References

[Bar-Shalom 93] Y. Bar-Shalom and X.-R. Li, *Estimation and Tracking: Principles, Techniques, and Software*. Artech House, Boston, MA, 1993.

[Benatallah 97] B. Benatallah and A. Bouguettaya, Data sharing on the web. *Proceedings of the First International Enterprise Distributed Object Computing Workshop*, Gold Coast, Queensland, Australia, pp. 258–269, October 24–26, 1997.

[Bender 96] E. A. Bender, *Mathematical Methods in Artificial Intelligence*. IEEE Computer Society Press, Piscataway, NJ, 1996.

[Bouguettaya 98] A. Bouguettaya, B. Benetallah, and D. Edmond, Reflective data sharing in managing internet databases. *Proceedings of the 18th International Conference on Distributed Computing Systems*, Amsterdam, the Netherlands, pp. 172–181, May 26–29, 1998.

[Breugst 98] M. Breugst and T. Magedanz, Mobile agents—Enabling technology for active intelligent network implementation. *IEEE Network*, 12(3), 53–60, 1998.

[Brooks 95] R. R. Brooks and S. S. Iyengar, Robot algorithm evaluation by simulating sensor faults. *Proceedings of SPIE*, 2484, 394–401, 1995.

[Brooks 98] R. R. Brooks and S. S. Iyengar, *Multi-Sensor Fusion: Fundamentals and Applications with Software*. Prentice Hall PTR, Upper Saddle River, NJ, 1998.

[Brooks 02a] R. Brooks, C. Griffin, and D. S. Friedlander, Self-organized distributed sensor network entity tracking. *International Journal of High Performance Computer Applications*, special issue on Sensor Networks, 16(3), 207–220, 2002.

[Brooks 02b] R. Brooks and C. Griffin, Traffic model evaluation of ad hoc target tracking algorithms. *International Journal of High Performance Computer Applications*, special issue on Sensor Networks, 16(3), 221–234, 2002.

[Brooks 03] R. R. Brooks, P. Ramanathan, and A. Sayeed, Distributed target tracking and classification in sensor networks. *Proceedings of the IEEE*, 91(8), 1163–1171, 2003.

[Brooks 03a] J. Moore, T. Keiser, R. R. Brooks, S. Phoha, D. Friedlander, J. Koch, A. Reggio, and N. Jacobson, Tracking targets with self-organizing distributed ground sensors. *2003 IEEE Aerospace Conference*, pp. 2113–2123, Invited paper, Big Sky, MT, March 2003.

[Chandra 98] P. Chandra, A. L. Fisher, C. Kosak, and P. Steenkiste, *Sixth International Workshop on Quality of Service (IWQoS 98)*, pp. 187–195, 1998.
[Chawathe 94] S. Chawathe, H. Garcia-Molina, J. Hammer, K. Ireland, Y. Papakonstantinou, J. Ullman, and J. Widom, The TSIMMIS project: Integration of heterogeneous information sources. *Proceedings of IPSJ Conference*, Tokyo, Japan, pp. 7–18, October 1994.
[Chung 98] P. E. Chung, Y. Huang, S. Yajnik, D. Liang, J. C. Shih, C.-Y. Wang, and Y.-M. Wang, DCOM and CORBA side by side, step by step, and layer by layer. http://www.bell-labs.com/~emerald/dcom_corba/Paper.html, 1998.
[Curtis 97] D. Curtis, Java, RMI and CORBA. http://www.omg.org/news/wpjava.html, 1997.
[Dogac 98] A. Dogac, C. Dengi, and M. T. Oszu, Distributed object computing platforms. *Communications of the ACM*, 41(9), 95–103, 1998.
[Figueroa 94] F. Figueroa and A. Mahajan, Generic model of an autonomous sensor. *Mechatronics*, 4(3), 295–315, 1994.
[Friedlander 02] D. Friedlander and S. Phoha, Semantic information fusion of coordinated signal processing in mobile sensor networks. *International Journal of High Performance Computing Applications*, 16(3), 235–241, 2002.
[Gray 98] D. N. Gray, J. Hotchkiss, S. La Forge, A. Shalit, and T. Weinberg, Modern languages and microsoft's component object model. *Communications of the ACM*, 41(5), 55–65, 1998.
[Heiler 95] S. Heiler, Semantic interoperability. *ACM Computing Surveys*, 27(2), 271–273, 1995.
[Hooker 94] J. N. Hooker, Needed an empirical science of algorithms. *Operations Research*, 42(2), 201–212, 1994.
[Hwang 93] K. Hwang, *Advanced Computer Architecture: Parallelism, Scalability, Programmability*. McGraw-Hill, New York, 1993.
[IEEE 96] IEEE draft standard for a smart transducer interface for sensors and actuators. IEEE P1451.2. Draft 2.01. August 1996.
[Kohavi 98] R. Kohavi and F. Provost, Glossary. *Machine Learning Journal*, 272, 1998.
[McDowell 98] C. E. McDowell et al., JAVACAM: Trimming Java down to size. *IEEE Internet Computing*, 2(3), 53–59, 1998.
[OMG 97] Object Management Group, Chapter 16 Trading object service specification, *CORBAServices*, March 1997. http://www.omg.org/library/csindex.htm
[Phoha 97] S. Phoha, E. Peluso, P. A. Stadter, J. Stover, and R. Gibson, A mobile distributed network of autonomous undersea vehicles. *Proceedings of the 24th Annual Symposium and Exhibition of the Association for Unmanned Vehicle Systems International*, Baltimore, MD, June 3–6, 1997.
[Press 94] W. H. Press et al., *Numeric Recipes in Fortran: The Art of Scientific Computing*. Cambridge University Press, Cambridge, U.K., 1994.
[Seetharaman 98] K. Seetharaman, The CORBA connection. *Communications of the ACM*, 41(10), 34–36, 1998.
[Sharma 97] R. Sharma, S. Keshav, M. Wu, and L. Wu, Environments for active networks. *Proceedings of the IEEE Seventh International Workshop on Network and Operating System Support for Digital Audio and Video*, pp. 77–84, May 19–21, 1997.
[Tanaka 97] H. Tanaka, Integrated environment for service-type repository management (IE-STREM). *Proceedings Global Convergence of Telecommunications and Distributed Object Computing TINA 97*, pp. 363–371, 1997.
[Tracy 89] M. Tracy, A. Anderson, and Advanced MicroMotion Inc., *Mastering Forth*. Brady, New York, 1989.
[Waltz 90] E. Waltz and J. Llinas, *Multisensor Data Fusion*. Artech House, Boston, MA, 1990.
[Wegner 96] P. Wegner, Interoperability. *ACM Computing Surveys*, 28(1), 285–287, 1996.
[Yi 95] S.-Y. Yi, J. Lee, and H. Shin, Operating system support for the trader in distributed real-time environments. *Proceedings of the Fifth IEEE Computer Society Workshop on Future Trends of Distributed Computing Systems*, pp. 194–199, 1995.
[Zilberstein 96] S. Zilberstein, Using anytime algorithms in intelligent systems. *AI Magazine*, 17(3), 73–83, 1996.

19 Mobile Code Support

Richard R. Brooks
Clemson University

T. Keiser
The Pennsylvania State University

19.1 Problem Statement

Chapter 18 provided background on why autonomous software reconfiguration is needed for sensor networks. It also discussed applications we have fielded, and showed that this increases system dependability. This chapter discusses our mobile code daemon implementations in detail. Along with design details useful for implementing mobile code systems, we provide background on mobile code in distributed systems. This information helps provide perspective on mobile code and how it can be integrated into distributed systems and ubiquitous computing implementations.

19.2 Mobile Code Models

Von Neumann, a father of computer science, invented flow charts, programming languages, and the serial "von Neumann" computer [von N 66]. His seminal concept of an automaton controlling another automaton [von N 66] can be viewed as a form of mobile code. The mobile code concept is evident in the 1960s remote job entry terminals that transfer programs to mainframe computers. Sapaty's wave system, created in the Ukraine [Sapaty 99] in the 1970s, provided full mobile code functionality. In the 1980s, packet radio enthusiasts in Scandinavia developed a forth-based approach to remotely transfer and execute programs through a wireless infrastructure [Nfor]. In the 1990s Java was developed by Sun Microsystems and became the first widely used mobile code implementation. Along the way, mobile code has been viewed from many different perspectives and paradigms.

Table 19.1 shows established mobile code paradigms [Fugetta 98, Milojicic 99, Tennenhouse 97, Wu 99]. There has been a clear progression of technology. Initially the client/server approach supported running procedures on a remote node. Remote evaluation augmented this by allowing the remote node to download code before execution. Code on demand, as implemented in Java, allows local clients to download programs when necessary. Process migration allows processes to be evenly distributed among workstations. Mobile agents permit software to move from node to node, following the agent's internal logic. Active networks allow the network infrastructure to be reprogrammed by

TABLE 19.1 Common Mobile Code Paradigms

Paradigm	Example	Description
Client/server	CORBA	Client invokes code resident on another node.
Remote evaluation	CORBA factory	Client invokes a program on remote node. Remote node downloads code.
Code on demand	Java, Active X	Client downloads code and executes it locally.
Process migration	Mosix, Sprite	Operating system transfers processes from one node to another for load balancing.
Mobile agents	Agent-TCL	Client launches a program that moves from site to site.
Active networks	Capsules	Packets moving through the network reprogram network infrastructure.

packets flowing through the network. Packets of data can be processed while being routed, through execution of encapsulating programs.

Paradigms differ primarily on where code executes and who determines when mobility occurs. Let's consider an example scenario. Given an input data file *f* on node n_f, a program *p* on node n_p to be executed, and a user *u* using node n_u, the paradigms in Table 19.1 would determine the following courses of action [Fugetta 98] (this problem is appropriate for neither process migration nor active networks):

- *Client-server*—Data file *f* is transferred from n_f to n_p. Program *p* executes on n_p and results are transferred to n_u.
- *Remote evaluation*—Program *p* is transferred to n_f, and executes there. Results are returned to n_u.
- *Code on demand*—Data file *f* and program *p* are transferred to n_u and execute there.
- *Mobile agents*—Program *p* is transferred to n_f and executes there. Program *p* carries the results to n_u.

Each approach will be efficient at times, depending on network configuration and the size of *p* and *f*. The model we propose accepts all of these alternatives. In this model, nodes, files, etc. can be specified at run time.

An important distinction exists between *strong* and *weak* code mobility [Fuggetta 98]. *Strong* mobility allows migration of both code and execution state. Programs can move from node to node while executing. This migration may even be transparent to the program itself (i.e., the program is not aware that it has migrated). *Weak* mobility transfers limited initialization data, but no state information, with the code.

The utility of strong migration is debatable, since it increases the volume of data transmitted as a process migrates [Zayas 87]. For load balancing, strong migration is worthwhile only for processes with long lifetimes [Harchol 97]. Mobile agents can be implemented using either weak or strong mobility. Differences of opinion exist in the literature as to whether distributed systems that handle migration transparently are [Milojicic 99] or are not [Fugetta 98] mobile code systems. We consider them mobile code systems.

In spite of the differences listed thus far, all mobile code systems have many common aspects. A network aware execution environment must be available. For Java applets, a web browser with a virtual machine downloads and executes the code [Vijaykrishnan 98]. A network operating system layer coupled with a computational environment provides this service for other implementations [Fugetta 98].

Some specific mobile code implementations are difficult to fit into the paradigms in Table 19.1 and warrant further discussion:

- *Postscript*—is one of the most successful mobile code applications, but is rarely recognized as mobile code. A postscript file is a program that is uploaded to a printer for execution. It produces graphic images as results.

- *Wave*—may be the earliest successful implementation of network aware mobile code. It was implemented in the Ukraine in the early 1970s [Sapaty 99]. Wave is a programming environment based on graph theory. Network nodes correspond to graph nodes. Network connections correspond to edges. Since distributed computing problems are often phrased in graph theoretical terms [Lynch 96], it is a very elegant approach.
- *Tube*—extends a LISP interpreter to distributed applications [Halls 97]. As an interpreted system, LISP is capable of meta-programming. Code can be generated and modified on the fly. The distributed interpreter is capable of robust computations and compensating for network errors.
- *Messenger*—uses mobile code as the backbone for implementing computer communications protocols [Tschudin 93]. Protocol Data Units (PDUs) are passed from transmitters to receivers, along with code defining the meaning of the PDUs. This concept is similar to Active Networks [Tennenhouse 97], but the approach is very different. Instead of concentrating on the mechanics of communication, this approach looks at the semiotics of message passing.
- *Jini*—adds distributed services, especially a name service, to the Java Remote Method Invocation (RMI) module [Jini 99]. Objects can be invoked on remote nodes. Jini is intended to extend Java technology to smart spaces, ubiquitous computing, and embedded systems. NISTs smart space researchers have found difficulties with this approach, since the Java Virtual Machine's 1.5 MB memory footprint [McDowell 98] is larger than the address space of most widely used embedded processors [Mills 99].

The paradigms discussed have been primarily oriented towards producing prototypes or commercial applications, rather than establishing the consequences of code mobility.

19.3 Distributed Dynamic Linking

In our work [Brooks 00a] we have developed methods to support transparent use of equivalent services with different implementations on heterogeneous machines. These concepts: *polymorphism* and *distributed dynamic linking* are described in this section. We allow daemons to choose nodes for executing tasks as a function of current network state. They are also enabled to choose between software implementations.

For code to migrate to a remote node, an active process must listen to the network on the remote node. This can be a user (Java applet invocation), or system (Java RMI) process. We use lightweight network daemons for this purpose [Brooks 00a]. The system currently runs on Windows NT, Windows CE, and Linux. Daemons accept remote requests for code transmission and execution. Data is transferred as needed. Daemons also monitor local system state (work load, disk space, congestion, etc.) Remote code can be identified explicitly by a URL, or implicitly. In the current implementation, mobile code is cached locally. When a call is made to execute a mobile code service, the daemon automatically retrieves it if it is not present. We have created this process, known as *distributed dynamic linking* [Brooks 00b]. The following calls allow software to directly manipulate the local software configuration:

- Verify local presence of mobile code modules.
- Pre-fetch mobile code for future use.
- Lock mobile code into local working set.

Data is transparently transferred between nodes by the daemon. Blocking and non-blocking calls are supported. A pipe abstraction in the API allows creation of distributed data-flow scripts on the fly. A program *p*1 runs on node *n*1 and its output *o*1 is sent over the network to node *n*2; triggering execution of program *p*2 on *n*2 using *o*1 as input.

Code is grouped in classes. Each class contains mobile code module implementations of a service [Brooks 00b]. Different implementations of the same service may exist in the repository. They may be executables or object libraries created for different operational environments, or they may be different algorithms for the same problem. When a call is made to execute a specific service on a specific node, the repository determines which implementation is appropriate for the target environment. This is equivalent to *polymorphism* in object-oriented languages, where different objects may be required to perform equivalent tasks. We have extended the concept to distributed environments and use it to tolerate heterogeneous hardware and operating systems. This project will extend the distributed dynamic linking and polymorphism concepts to create a single operational environment linking embedded processors, workstations, and supercomputers. The environment can choose appropriate software, hardware, and connecting for accomplishing a task.

Invocation information, and a profile are stored with each mobile code module. The profile expresses module capabilities as a function of resources used, and relevant constraints. The current implementation requires a target machine to be named by the calling program for execution of a mobile code module. It should be possible for daemons to choose the machine at run-time. Similarly, empirical methods characterize the resource needs of a given implementation in the profile.

Profiles model the relationship between time, system resources, and answer quality. Sun's law for speed-up:

$$S_n^* = \frac{\left(W_1 + G(n)W_n\right)}{\left(W_1 + G(n)W_n/n\right)}$$

(S_n^* is the speedup achieved with n PE's, W_j is the workload with degree of parallelism j, n is the number of PEs, and $G(n)$ is workload increase as memory increases) gives a theoretical basis for this relationship [Hwang 93]. Sun's law shows program speedup is a function of the number of PEs, available memory, and problem size. Choices between alternative implementations are made by finding the implementation whose profile uses the least resources, for the best results, given current conditions.

These methods allow us to start a task on any machine and have the machine construct a virtual enterprise to finish the task. The virtual enterprise recruits underutilized machines on the network. We create a satisficing response to the problem of how to allocate resources to the task [Bender 96]. The response is not optimal, but feasible and reasonable.

19.4 Daemon Implementation

The mobile code daemon we present is based upon a core network protocol called the Remote Execution and Action Protocol (REAP). This protocol is responsible for message passing between nodes within our network. On top of this packet protocol we have developed a framework to allow objects to serialize themselves and travel across the network. At a higher layer of abstraction we have written messages to handle remote process creation and monitoring, simple filesystem operations, and resource index operations. Use of the mobile code daemon in sensor network applications is documented in [Brooks 00a, Moore 03] (Figure 19.1).

The daemon is written in C++. The first version ran on the Windows NT and Windows CE operating systems. It has since been ported to the Linux operating system. The daemon structure is broken down into several core modules: foundation classes, the networking core, the random graph module, the messaging core, the packet router, the index server, the transaction manager, the resource manager, and the process manager. We will discuss each of these components in turn.

Before discussing the REAP daemon in detail, it is useful to discuss its underlying framework on which it is built. The framework abstracts many of the complexities of systems programming out of the core, into a set of libraries. Thus, we have written our own object-oriented threading and locking classes, whose current implementation calls into the threads library of the underlying operating system. We also rely heavily on a set of templated, multithreaded linked list, hash, and heap objects throughout the code. In addition, there are classes to handle singleton objects, the union-find problem,

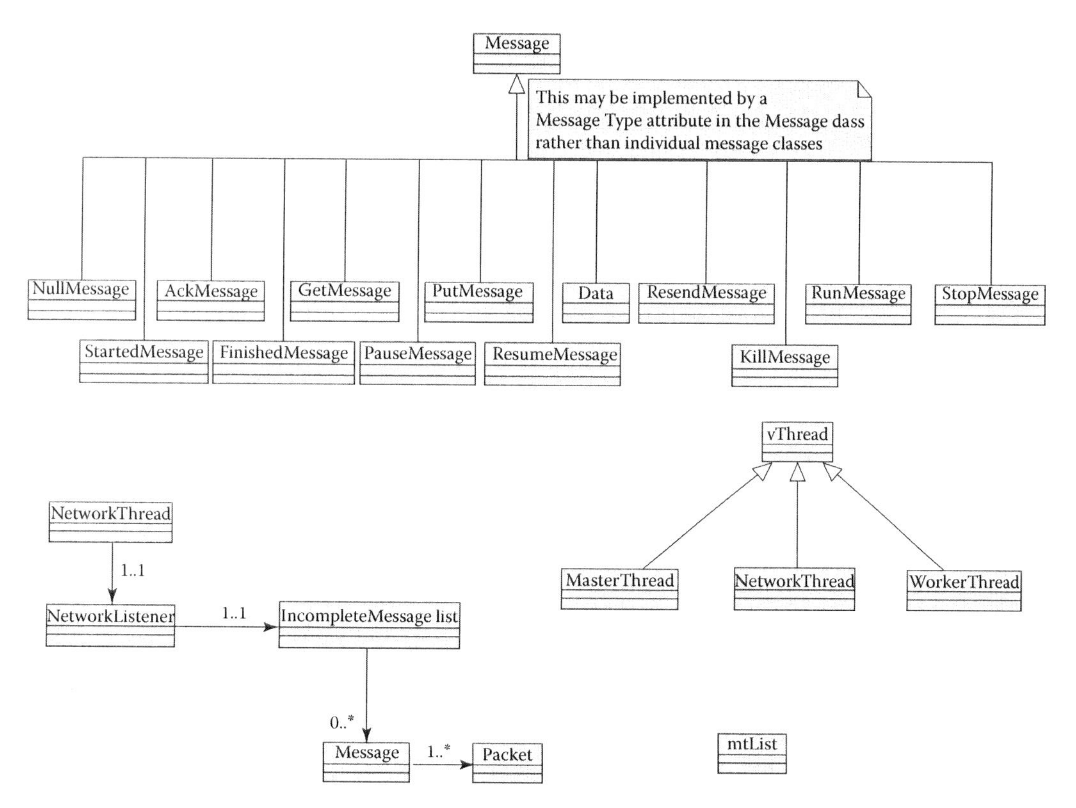

FIGURE 19.1 UML class structure of our current mobile code daemon.

and object serialization. Lastly, there is also a polymorphic socket library that allows different networking architectures to emulate unicast stream sockets, regardless of the underlying network protocol or topology. These socket libraries are explained in the discussion of the networking core.

The daemon is capable of communicating over several networking technologies. The major ones are: TCP/IP, Diffusion Routing [Intanagonwiwat 00], and UNIX domain sockets. The socket framework is designed so that new protocols are easily inserted into the daemon. To achieve this, an abstract base class\verb "Socket" includes all of the familiar calls to handle network I/O. Furthermore, all nodes are assigned a protocol-independent unique address. Opening a new socket involves looking up the network-layer address of a node in a local cache, and then opening the lower-level socket. When a cache miss occurs, a higher-level protocol is provided to find the network-layer address. The appropriate socket object is allocated based upon the network-layer address of the destination.

Diffusion provided some interesting challenges because it is not a stream-oriented unicast protocol. Rather, it provides a publish and subscribe interface, and is essentially a multicast datagram protocol. Thus, we had the choice of rewriting the REAP socket protocol as a datagram protocol, or building a reliable stream protocol on top of the Diffusion framework. It was deemed simpler to write a reliable stream protocol on top of Diffusion. In essence, we wrote a simplified userspace TCP stack. The current userspace stack employs the standard three-way handshake protocols for socket open and close, and it also employs a simple delayed-ACK algorithm. This system is implemented as an abstract child of the "Socket" base class. Our Diffusion driver then provides an implementation of our userspace TCP module. The Diffusion driver performs a role equivalent to the IP layer processing code in most kernels. It receives datagrams from the Diffusion daemon through callback functions, parses the headers to make sure the datagram has reached the correct destination, and then either discards the contents, or passes it up to the TCP layer. These steps were deemed necessary because Diffusion is a multicast protocol, and thus we could not rule out the possibility of datagrams reaching our socket object that were not actually destined for it.

Early on in the project, it became clear that persistent connections between the various nodes was essential. A single file transfer of a shared object could result in thousands of packets traversing the network, and session setup time was simply too long over TCP and Diffusion. To counteract this problem we implemented a system whereby sockets are kept open whenever possible. The first implementation of this system opened directly to a destination, and did not support multi-hop routing very well. Under this implementation, socket timeout counters were employed to close underutilized sockets. This method has inherent scalability problems, and we decided a better solution was required.

This better solution involves a multi-hop packet routing network built on top of a random graph of sensor nodes. Each node in the system has four graph parameters specified: minimum degree, maximum degree, cliquishness, and clique radius. The cliquishness parameter defines the probability of a new edge being formed to a node within the clique radius. The minimum degree and maximum degree parameters control how many neighboring nodes can exist at any point in time. The clique parameters allow us to control the size and connectedness of cliques within the graph. Cliques become more important when we investigate the index system.

To add a new edge, a random number is generated to decide whether or not to add a clique edge. Then a random node from the node cache is chosen based upon two filter criteria: the chosen node must have a minimum path length of two to this node, and its minimum path length must be less than or equal to the clique radius for a clique edge, or greater than the clique radius for a non-clique edge.

The messaging system implements the core of the REAP protocol. At its lowest levels, this consists of a packet protocol, on top of which serialized objects are built. The Packet'\class is nothing more than a variable-sized opaque data carrier that is capable of sending itself between nodes, and it also performs data and header checksumming. The header defines enough information to route packets, specify the upper-level protocol, and to handle multi-packet transmissions where the number of packets is known a priori. The options field consists of a 4-bit options vector, and a 4-bit header extension size parameter. The TTL field is used in the new multi-hop protocol to eventually destroy any packet routing loops that might form.

Higher level messaging functionality is handled by a set of classes that do object serialization, and by a base message class. The serialization class in REAP provides a fast method of changing common data types into network byte-ordered, opaque data. The key advantage to this serialization system is that it only handles common data types, and thus has much lower overhead than technologies such as XDR and ASN.1.

The base messaging class provides a simple interface to control destination address, source transaction information, possible system state dependencies for message delivery, and control over sending the message. In addition, it defines abstract serialization and reordering functions that are implemented by all message types.

The serialization class sits beneath the base message class and does the physical work of serializing data, packetizing the serialized buffer, and then injecting those packets into the router.

On the receiving end, packets are received by an object serialization class and inserted into the proper offset in the receive buffer. A union-find structure keeps track of packet sequence numbers, and once it detects that all packets have been received, the message is delivered to a message queue in the destination task structure.

Another interesting feature of the messaging system is the function called run. This function takes a task structure as an argument, and is generally intended to perform some action on the destination of the message. We will see an example of this later on when we discuss the index server.

The daemon packet router has several key responsibilities. The primary one is to use its internal routing tables to move packets from source to destination. The other primary function of the router is to coordinate the dissemination of multi-hop routing data.

The current method of determining multi-hop paths is through broadcast query messages. We gradually increase the broadcast TTL until a route is found, or a TTL upper limit is reached, at which point the node is assumed down. This methodology helps to reduce flooding, while making optimal paths likely to be found. A simple optimization allows a node to answer a multi-hop query if it has an answer in its routing table. Although this system is essentially a heuristic, it tends to work well because failed intermediate nodes are easily bypassed when their neighbors find that they cannot reach the next hop. Of course, this can lead to much longer paths through the graph, but support is integrated to warn of intermediate node failures, and multi-hop cache expire times help to reduce this problem by forcing refreshes occasionally. The multi-hop refreshes are carried out in unicast fashion, and a broadcast refresh is only used if a significant hop count increase is detected.

The actual routing of packets involves looking at the two destination fields in the packet header. First, a check is performed to determine whether the destination node identifier is equivalent to the current node's identifier, or the local loopback address, or one of several addresses that are defined for special purposes, such as broadcast to all members of a clique. The next check is to determine whether the destination process identifier is equivalent to that of the current process. If it is not, then the packet will need to be forwarded across a unix domain socket. If both of these tests pass, then the packet must be delivered to the appropriate task. Because packets do not contain sufficient routing data to deliver them to a specific task, we must recreate the high level message object in the router to determine the message's final destination.

Every task in a REAP process registers itself with the router during initialization. Once a task is registered, it can receive messages bound for any active ticket. Several special tickets are defined for every task that handle task status messages, and task-wide requests. Other tickets are ephemeral, and are allocated as needed.

An important component of the REAP daemon is the index system. This system implements a distributed database of resource available on the network. Each record in this database describes an object of one of the following types: index server, file, executable, library, pipe, memory map, host, or a task. Every record in the database has a canonical name, and resource locator associated with it. Both of these values are stored as human-readable strings. Besides this, metadata to allow for both data and metadata replication are present. The goal is to have a distributed cluster of index servers that transparently replicate each other's index records, and to have a resource control system that transparently replicates the actual data as well. At this point, the replication technology is only partially implemented.

The index system consists of the following modules: client, server, database, and the associated messaging protocol. The client is responsible for building a query message, sending the message, and either waiting for a response, or returning a response handle to the client in the case of an asynchronous call. The server consists of a pool of threads that poll for incoming messages on the server task structure. When a thread receives a message, it runs the query embedded in the message against the local database, and then sends the results back to the client in a query result message.

The query system is based upon a fairly extensible parse tree. The query language permits complex boolean filtering on most any variable defined in an index record. The index server is essentially a lightweight SQL server that is tailored to resource location.

The index infrastructure is mainly built upon two message types: a query message, and a result message. The query message consists of an operand tree, some query option flags, and possibly a list of index records. Once the query message reaches the server, it is received by a server thread, and the run function is called. This function performs a query against the index database object, and sends back a result message to the source node. Once these actions are complete, the run function returns, and then the index server deallocates the query object. The index server itself is nothing more than a pool of threads that accept a certain type of message, and then allow the messages to perform their actions. In this sense, the REAP messaging system implements the mobile agent paradigm.

The other major feature of the index system is a system to select code based upon destination system architecture and operating system. To handle this, system architecture and operating system are considered polymorphic class hierarchies. Every index record contains an enumeration defining its membership in each hierarchy. When a system requests object code or binary data, we must ensure that it is compatible with the destination system. Thus, every index query can filter based upon architecture, if desired. When a query indicates that architecture and/or operating system are a concern, then C++

dynamic_cast calls are made to ensure compatibility. Because we are using the C++ dynamic casting technology, supported architectures and operating systems are determined at compile time. It would not be a technically difficult modification to use human-readable strings, and runtime-defined polymorphic hierarchies. However, we chose the compile-time approach because it is faster, and the architectures and operating systems in our lab are relatively constant.

To give an example of how this technology would work, let's take an example of a sensor node having raw time series data that needs to be run through an FFT. Suppose a distributed process scheduler determines that it would be optimal to move the raw data to a wireless laptop that is deployed in the field. When the laptop goes to run the FFT, it queries the index database for a given FFT algorithm, and requests architecture polymorphic checking. Let's say this laptop has a processor with Intel's SSE and MMX extensions, but not the SSE2 extensions. When the index server processes the query, let's say it finds FFT algorithms that are compiled for 386, Pentium, SSE, Pentium 4, and Alpha EV5. When it filters these queries, it determines that it can cast the laptop into 386, Pentium, and SSE, but not Pentium 4 or Alpha EV5. The laptop will then attempt to download the optimal one, only dropping to slower implementations when it cannot download the fastest one.

All operations in REAP are addressed by their transaction address. This address consists of the four-tuple (node, process, task, ticket). These globally unique addresses permit flexible packet routing. A major goal of REAP is to permit network-wide interprocess communication through a simple high-level interface, without introducing high overhead. We will see how this goal is met when we discuss the resource management module of the REAP mobile code daemon.

In order to support the complex transaction routing system, a task control structure is required. All threads, and other major tasks have their own task structure. This structure is registered with the local packet router, and is where message structures get delivered. Its primary jobs are to handle message I/O, and to allocate tickets. Every active ticket has an associated incoming message queue, and thus it is possible in our framework to receive messages for specific tickets. As an added feature, message type filtering is supported at the task level. Any messages which fail to pass the filter are not delivered to the task, and are instead deallocated.

Another purpose of the transaction management system is task monitoring. We employ a publish/subscribe model for this purpose. Any task may request status information from another task by subscribing to its status information service, and then every status message published by that task will be sent to the subscribed task. At the moment, all status information is sent as unicast datagrams. The main purpose of this system is to notify the requester that its request has been received, and to notify it again when the request is completed. Other interesting applications of this technology could include distributed process schedulers that monitor the progress and system load on a cluster of nodes, and then schedule compute jobs to distribute the load to meet predefined criteria.

The resource management framework is tightly coupled with the index system. When a client program wants to access a resource, a query to the index system is made. The results returned can then be passed into the resource management object. The resource manager then attempts to open one of the resource from the result set. If possible, one resource from each canonical name in the result set will be opened. Thus, the resource manager is capable of overcoming node failures by looking for other copies of the same resource. The current implementation attempts to open one instance of every canonical name in parallel, and continues this iterative process as timeouts occur. Eventually, an instance of every canonical name will be opened, or the resource manager will run out of instances of a resource in the index result set.

The resource control system is built on top of a client–server framework. This framework was chosen because the types of resources we want to support are generally not concurrent objects. Thus, the resource management system consists of two REAP message types: a resource operation message, and a resource response message. Then, there are two types of resource objects: a client object, and a server object. For any given resource, there will exist exactly one server object, and one client object per task with an open handle to the resource. When a given client wants to perform an operation on the resource, it will send a resource operation message to the server object's transaction address.

The server will then call the run method of the message, and through a set of polymorphic calls described below, it will perform I/O operations on the server object. A response message will then be sent to the originating node.

The client and server resource objects are based upon an abstract interface that defines several common methods that can be used on UNIX file descriptors. The major base operations are: open, close, read lock, write lock, unlock, read, write, and stat. In all cases, blocking and non-blocking versions of these functions are provided, and the blocking functions are simply built on top of the non-blocking code.

As a simple performance improvement, client and server caching objects were constructed that perform both data and metadata caching. Since our distributed resource interface is essentially identical to the virtual filesystem interface that unix-like kernels give to applications, standard locking semantics can apply. Thus, our caching module simply looks at the numbers open read mode and write mode file descriptors to determine the acceptable caching strategy. For the multiple readers, and single writer cases, we allow client-side caching. For all other cases we must disable client-side caching. Thus, our caching semantics are identical to those used in the Sprite Network Filesystem [Nelson 88]. The REAP framework makes our implementation very simple because our mobile-agent based messages can easily turn on and off client caches with minimal overhead.

To demonstrate the power of this resource control model, we have built client and server objects to support a distributed shared memory architecture. Once again, we employ the abstract client–server caching model to increase performance.

The last major component of the REAP framework is process creation and management. This portion of the architecture consist almost entirely of message types. The primary message type is a process creation message. This message contains an index record pointing to the binary to execute. It also contains the argument and environment vectors to include, as well. A second message is process creation response message. This message simply contains the transaction address of the newly created process. Finally, task monitoring messages may be used to monitor the progress of a task using the publish/subscribe model discussed in the section on transaction management.

The REAP mobile code daemon permits us to experiment with many different mobile code paradigms over a fault-tolerant multi-platform framework. Because it provides a simple cross-platform, distributed interprocess communication framework, it is very useful for developing system of collaborating distributed processes. This approach is capable of mimicking all the major mobile code paradigms, as shown in [Orr 02]. Furthermore, its polymorphic code selection system permits us to use the optimal algorithm on a given system without significant user interaction. Finally, the distributed resource management system allows us to reduce bandwidth and permit concurrent use of resources without breaking normal concurrency rules.

19.5 Application Programming Interface

We provide here a summary of the application programming interface of the Reactive Sensor Networks Mobile Code System. It provides a high-level description of the system architecture. It describes the syntax and semantics of services, which may be of use to other projects in the Sensor IT program. These services are a subset of the full system. The initial implementation supported Windows machines using the Internet Protocol. The API presented can still be modified to fit the needs of other research projects. Please notify the author of this report of any requests and or comments. Nodes respond to the following API calls (semantics explained later):

status=**exec**(*program-class,input-data-vector,output-data-vector,resource-list,optional-command-line*)

status = **exec_noblock**(*program-class, input-data-vector, output-data-vector, resource-list, optional-command-line*)

status=**pipe**(*program-class,input-data-vector,output-data-vector,resource-list,optional-command-line*)

status = **kill_pipe**(*program-class, machine*)
status = **lock**(*program-class, machine*)
status = **unlock**(*program-class, machine*)
status = **load**(*program-class, machine*)
status = **register_program**(*class, URL, optional-command-line*)
status = **register_machine**(*machine, port*)
status = **list_machines**(*machine-info-vector*)
status = **list_classes**(*class-info-vector*)
status = **list_available_classes**(*machine, class-info-vector*)

Parameters

status—An integer value indicating call success or failure. ARL_MCN_SUCCESS indicates success. Other values indicate errors. A list of error returns will be provided with the final documentation.
program-class—In the initial delivery this will be a string name that uniquely identifies a program.
input-data-vector—A null terminated array of pointers to strings giving a list of URL's indicating files to be used as input.
output-data-vector—A null terminated array of pointers to strings giving a list of URL's indicating files to be used as output.
resource-list—A null terminated array of pointers to strings of resource identifiers. In the initial implementation, this will be a node name (e.g., strange.arl.psu.edu).
machine—A string containing a node name (e.g., strange.arl.psu.edu).
optional-command-line—A string defining the command line format for an executable or parameters for a DLL.
port—IP port number of the socket used by the ARL mobile code software to listen for mobile code requests.
machine-info-vector—A null terminated array of pointers to a data structure consisting of pointers to two fields: node_name, and port number.
class-info-vector—A null terminated array of pointers to a data structure consisting of pointers to three fields: class_name, URL, and default command line (possibly null).

API Call Semantics

exec—Executes in four phases: (1) Uploads all data in the vector *input-data-vector* and the program *program-class.* (If files are not currently located on the node.) (2) Executes the program. (3) Writes output to the URLs in *output-data-vector.* (4) Performs garbage collection. This call blocks until execution is complete. Returns the completion status of the program. The optional command line argument can be used to override the defaults given in *register_program.*
exec_no_block—Same as exec, but does not block. Returns a completion status indicating system acceptance (or non-acceptance) the call. The optional command line argument can be used to override the defaults given in *register_program.*
pipe—Executes in five phases: (1) Uploads the program *program-class.* (If the file is not currently located on the node.) *input-data-vector* identifies files on the node. (2) Executes the program. (3) Writes output to the URLs in *output-data-vector.* (4) Waits for modifications to files in *input-data-vector* and then goes back to step (2). (5) Performs garbage collection on receipt of the out of band signal from kill_pipe(). This call does not block. Returns a completion status indicating system acceptance (or non-acceptance) of the call. The optional command line argument can be used to override the defaults given in *register_program.*
kill-pipe—Sends an out of band message to *machine* indicating the pipeline should be terminated.
lock—Download a program to a node. Make this program unavailable for garbage collection.

unlock—Make a program previously unavailable for garbage collection, available for garbage collection.

load—If a local node attempts to execute a program not present locally, it can use this call to trigger a network interrupt. First nodes in the neighborhood will be signaled. If they have a copy of the program they will transfer the copy to the requesting node. Otherwise, the network interrupt will propagate through the multi-hop network up to the repository. If the program is found on a node on any of the hops, propagation of the request will stop and the program will be transmitted to the requestor. If the program is found on the repository, the program will be transmitted to the requestor. If the program is not found in the repository, an error condition is signaled.

register_program—Creates a link in the repository between the URL, the default command line and the unique class name. The default command line is either a string giving the command line arguments used when executing the class (including variables for input and output files), or a list of default parameters used when constructing a call to a function in a DLL. In the initial delivery, class name must be unique for each program. In later releases polymorphism and subclasses will be supported.

register-machine—Identifies a machine for use by the Mobile Code software.

list_machines—Returns a pointer to a machine-info-vector. One entry is given for every machine registered with the mobile code repository. This can be used to initialize a system list of nodes the machine can accept network connections from.

list_classes—Returns a pointer to a class info-vector. One entry is given for each class registered with the mobile code repository.

list_available_classes—Returns a pointer to a class info-vector. One entry is given for each class available for use on the node indicated by the *machine* parameter. If *machine* is null, the local node is used as a default.

19.6 Related Work

In this section we discuss related technologies. This discussion does not include mobile code paradigms. In this section we discuss ongoing research that has significant overlap with our approach.

Researchers from Berkeley and Oak Ridge National Laboratories have developed a linear algebra and scientific computing software library, PHiPAC, which adapts to its underlying hardware [Bilmes 98, Science 99]. The library probes the host's hardware to determine the memory and cache structure. Knowledge of the underlying substrate allows PHiPAC to modify computations and be more efficient. Performance increases of 300% have been reported. Researchers at Duke University have designed real-time task scheduling software that varies processor clock rates for tasks with different priorities and real-time deadlines. Energy consumption has been reduced by up to 70% [Swaminathan 00]. We consider these examples as evidence that adaptive software and hardware technologies have tremendous potentials for time and energy savings. Our work differs in that we look at applying similar concepts to mobile code implementations. We extend the ideas of these researchers by co-adapting hardware, software and network resources.

Some mobile code implementations may be considered adaptive. Java is probably the best-known mobile code implementation. Applets can be downloaded and executed locally. RMI allows applets registered with a service to be executed on a remote node. Use of a standardized language allows Virtual Machines running on different processors to execute the same intermediate code. Our approach differs greatly from Java's "write-once run-anywhere" implementation, which essentially restricts the system to a single standardized language like ANSI C. Any language that can link to our API can make mobile code calls. The mobile code modules can be written in any language with a compiler for the host machine, or interpreted language with an interpreter for the host machine. Our use of daemons to run remote invocations resembles RMI to a certain extent.

[Milojicic 99] describes a number of task migration approaches. They are divided into kernel or user space methods. Like our approach, Tui allows migration of processes among heterogeneous architectures.

Tui does this by compiling programs written in a standardized language once for each target machine. Our approach supports this, but also allows programs written in different languages to cooperate. Tui does not consider code morphing for efficiency improvement. Most implementations, including MOSIX, Sprite, etc. are designed for distributed computing environments with a high-speed dependable network infrastructure. Ours uses either IP or wireless and does not assume a dependable substrate.

A natural application of mobile code technology is network management. [Halls 97] describes using the tube system to efficiently manage ATM networks. [Breugst 98] uses mobile agents to provide advanced telecommunications services, such as call screening and call forwarding. These approaches differ from our work in that they use software solely to control the hardware substrate. We do not explicitly consider software control of the underlying network.

[Weiser 91] suggests the concept of ubiquitous computing. Computers as such should become invisible technology. Intelligence should be embedded directly into existing devices. General-purpose computers will no longer be necessary. At Xerox PARC a number of prototypes have been constructed following this general concept. Ubiquitous computing has inspired recent research in smart spaces and commercial products like Jini [Sun 99]. Ubiquitous computing can best be achieved through cooperating networks of embedded processors.

[Abelson 99] describes a novel research agenda for a new class of computational entities. Amorphous computing is computation as carried out by a network of cooperating automata. Each individual computational entity is fallible. It is hypothesized that nanotechnology will support low-cost construction of individual low power networked computational devices. Approaches embedding computation into living cells [Coore 99] have also been explored. Their approach to system design and evaluations uses CA abstractions and is relevant to the work proposed here. They assume that all nodes have identical software and do not consider mobile code. The standard CA model is modified to assume that the grid is not regular and communications are asynchronous. Synergy is possible between our work and amorphous computing. Their extensions to the CA model could be useful for our work. Some amorphous algorithms, such as the construction of global coordinate systems [Nagpal 99], are directly applicable to our work.

19.7 Summary

This chapter discusses mobile code support for sensor networks. It provides an overview of different mobile code implementations and research projects. We discussed in detail our concepts for distributed dynamic linking and the implementation of our mobile code daemons. We maintain that enabling software adaptation at this level greatly aids the ability of sensor networks to continue working in chaotic environments.

Acknowledgments and Disclaimer

This material is based on work supported by the Office of Naval Research under Award No. N00014-01-1-0859 and the Reactive Sensor Network Grant Award No. F30602-99-2-0520. Any opinions, findings, and conclusions or recommendations expressed in this presentation are those of the author and do not necessarily reflect the views of the Office of Naval Research. The authors would also like to thank the anonymous referees, whose inputs have greatly improved the paper.

References

[Abelson 99] H. Abelson et al., *Amorphous Computing*, AI Memo 1666, Massachusetts Institute of Technology, Cambridge, MA, August 1999. http://www.swiss.ai.mit.edu/projects/amorphous/paperlisting.html

[Bender 96] E. A. Bender, *Mathematical Methods in Artificial Intelligence*, IEEE Computer Society Press, Piscataway, NJ, 1996.

[Bilmes 98] J. Bilmes, K. Asanovic, C.-W. Chin, and J. Demmel, The PHiPAC v1.0 matrix-multiply distribution, Technical Report TR-98-35, International Computer Science Institute, Berkeley, CA, October 1998.

[Breugst 98] M. Breugst and T. Magedanz, Mobile agents—Enabling technology for active intelligent network implementation, *IEEE Network*, 12(3), 53–60, 1998.

[Brooks 00a] R. R. Brooks, E. Grele, W. Kliemkiwicz, J. Moore, C. Griffin, B. Kovak, and J. Koch, Reactive sensor networks: Mobile code support for autonomous sensor networks, in *Distributed Autonomous Robotic Systems DARS 2000*, pp. 471–472, Springer Verlag, Tokyo, Japan, October 2000.

[Brooks 00b] R. R. Brooks, Distributed dynamic linking, Penn State Invention Declaration, May 2000.

[Coore 99] D. Coore, Botanical computing: A developmental approach to generating interconnect topologies in an amorphous computer, PhD thesis, MIT Department of Electrical Engineering and Computer Science, Cambridge, MA, February 1999.

[Fugetta 98] A. Fuggetta, G. P. Picco, and G. Vigna, Understanding code mobility, *IEEE Transactions on Software Engineering*, 24(5), 342–361, 1998.

[Halls 97] D. A. Halls, Applying mobile code to distributed systems, PhD dissertation, University of Cambridge, Cambridge, U.K., 1997.

[Harchol 97] M. Harchol-Balten and A. B. Downey, Exploiting process lifetime distributions for dynamic load balancing, *ACM Transactions on Computer Systems*, 15(3), 253–285, 1997.

[Hwang 93] K. Hwang, *Advanced Computer Architecture*, McGraw-Hill, New York, 1993.

[Intanagonwiwat 00] C. Intanagonwiwat, R. Govindan, and D. Estrin, Directed diffusion: A scalable and robust communication paradigm for sensor networks, in *Proceedings of Mobicom '00*, Boston, MA, 2000.

[Jini 99] Jini Technology Helper Utilities and Services Specification, Sun Microsystems, Palo Alto, CA, 1999.

[Lynch 96] N. A. Lynch, *Distributed Algorithms*, Morgan Kaufmann Publishers, San Francisco, CA, 1996.

[McDowell 98] C. E. McDowell, B. R. Montague, M. R. Allen, E. A. Baldwin, and M. E. Montoreano, JAVACAM: Trimming Java down to size, *IEEE Internet Computing*, 2(3), 53–59, 1998. http://computer.org/internet/

[Mills 99] K. Mills, personal communication.

[Milojicic 99] D. Milojicic, F. Douglis, and R. Wheeler, eds., *Mobility: Processes Computers, and Agents*, Addison-Wesley, Reading, MA, 1999.

[Moore 03] J. Moore, T. Keiser, R. R. Brooks, S. Phoha, D. Friedlander, J. Koch, A. Reggio, and N. Jacobson, Tracking targets with self-organizing distributed ground sensors, *2003 IEEE Aerospace Conference*, IEEE, Big Sky, MN, March 2003.

[Nagpal 99] R. Nagpal, Organizing a global coordinate system from local information on an amorphous computer, AI Memo 1666, Massachusetts Institute of Technology, Cambridge, MA, August 1999. http://www.swiss.ai.mit.edu/projects/amorphous/paperlisting.html

[Nelson 88] M. Nelson, B. Welch, and J. Ousterhout, Caching in the Sprite network file system, *ACM Transactions on Computer Systems*, 6(1), 134–154, 1988.

[Nfor] Network Forth, http://www.sandelman.ottawa.on.ca/People/Michael_richardson/network-forth.html

[Orr 02] N. Orr, A message-based taxonomy of mobile code for quantifying network communication, MS thesis, Penn State Computer Science and Engineering, State College, PA, 2002.

[Sapaty 99] P. Sapaty, *Mobile Processing in Distributed and Open Environments*, Wiley, New York, 1999.

[Science 99] 'Self-tuning' software adapts to its environment, *Science*, 286, 35, 1999.

[Swaminathan 00] V. Swaminathan and K. Chakrabarty, Real-time task scheduling for energy-aware embedded systems, accepted for publication in *IEEE Real-Time Systems Symposium*, Orlando, FL, November 2000.

[Tennenhouse 97] D. L. Tennenhouse et al., A survey of active network research, *IEEE Communications Magazine*, 35(1), 80–86, 1997.

[Tschudin 93] C.-F. Tschudin de Bâle-ville, On the structuring of computer communications, PhD dissertation, Université de Genève, Genève, Switzerland, 1993.

[Vijaykrishnan 98] N. Vijaykrishnan, N. Ranganathan, and R. Gadekarla, Object-oriented architectural support for a Java processor, in *Lecture Notes in Computer Science*, Vol. 1445, pp. 330–354, Springer Verlag, Berlin, Germany, July 1998.

[von N 66] J. von Neumann, *Theory of Self-Reproducing Automata*, A. W. Burks, ed., University of Illinois Press, Urbana, IL, 1966.

[Weiser 91] M. Weiser, The computer for the 21st century, *Scientific American*, 94–100, 1991.

[Wu 99] D. Wu, D. Agrawal, and A. Abbadi, StratOSphere: Unification of code, data, location, scope and mobility, in *Proceedings of the International Symposium on Distributed Objects and Applications*, Edinburgh, Scotland, pp. 12–23, September 1999.

[Zayas 87] E. Zayas, Attacking the process migration bottleneck, in *Proceedings of the 11th ACM Symposium on Operating Systems Principles*, pp. 13–24, November 1987.

20

Mobile Agent Framework for Collaborative Processing in Sensor Networks

Hairong Qi
University of Tennessee

Yingyue Xu
University of Tennessee

Teja Phani Kuruganti
University of Tennessee

This chapter discusses the distributed computing paradigms used to support the collaborative processing in sensor networks. Sensor networks form a typical distributed environment, and the most popular computing paradigm deployed has been the client/server-based, where all the clients send the raw data to a processing center for data dissemination, as illustrated in Figure 20.1a. In some applications where the size of raw data is very large, the clients can perform some local processing and send a compressed version of the raw data or simply the local processing results to the processing center, as illustrated in Figure 20.1b. This scheme is widely used in distributed detection [33]. Sometimes, the client/server-based processing can be carried out hierarchically with multiple levels of processing centers as illustrated in Figure 20.1c to solve scalability problems [11].

Although popular, the client/server-based distributed computing is not suitable for applications developed in sensor networks as the sensor network possesses some unique characteristics that the client/server-based approach cannot accommodate. Here, we summarize these features as follows:

- Extremely constrained resources—limited communication bandwidth, power supply, and processing capability
- Sheer amount of sensor nodes—a sensor network can contain up to thousands of sensor nodes
- Fault-prone sensor nodes and communication links—due to harsh working environment and communication through unreliable wireless link
- Exceptionally dynamic nature—existing sensor nodes can stop functioning due to power depletion, new sensors can be deployed, and sensor nodes can be mobile

These properties request the distributed computing paradigm in sensor networks to be energy efficient, scalable, fault tolerant, and adaptive.

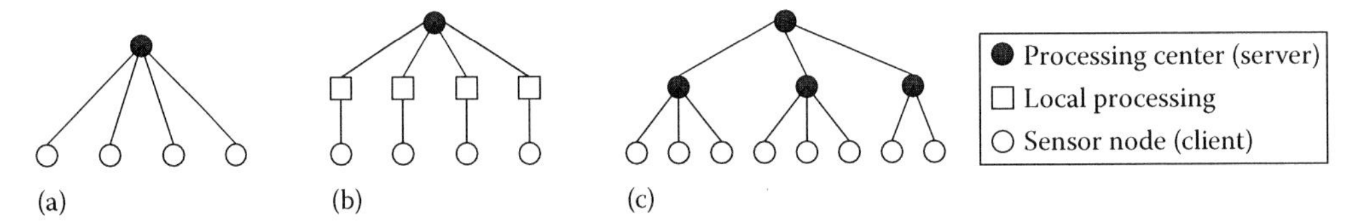

FIGURE 20.1 Illustration of the variants of client/server-based computing model. (a) Standard mode. (b) With pre-processing. (c) With hierarchy.

The client/server-based computing is able to achieve scalability with the employment of the hierarchical structure. However, it cannot respond to the load changing in real time. When more sensors are deployed, it cannot perform load balancing without changing the structure of the network. The client/server-based model can also achieve fault tolerance by using redundant information from multiple sensor nodes. However, it cannot provide energy efficiency at the same time. The processing centers behave like supernodes that need much higher energy and storage and computing capabilities. These will largely reduce the lifetime of the whole sensor network, especially if all sensor nodes in the network are of the same type (a homogeneous sensor network). It has been noted that the *lifetime* of a sensor network is defined from the time the sensor network is deployed to the time the first sensor node is out of power.

In this chapter, we present the usage of the mobile-agent-based computing paradigm for collaborative processing in sensor networks. We first discuss in Section 20.1 the principles of mobile-agent-based computing as well as its fundamental differences from the client/server-based computing. We then design and develop a mobile agent framework (MAF) in Section 20.2. In Section 20.3, we show how MAF is deployed in a collaborative target classification application. Finally, we summarize our discussion in Section 20.4.

20.1 Mobile-Agent-Based Distributed Computing

Mobile agent can be regarded as a special kind of software with its own attributes. Compared to traditional software, the unique features of a mobile agent are its *autonomy* and *mobility*. The mobile agent can execute autonomously [8]. Once dispatched, it can migrate from node to node performing data processing on its own, while software can typically only execute when being called upon by other routines. Mobile agents are preferred if an application requires to reduce network load, overcome network latency, and provide robust and fault-tolerant performance, etc. [14].

Although the role of mobile agents in distributed computing is still being debated mainly because of the security concern [9,18], several applications have shown clear evidence of benefiting from the use of mobile agents. For example, mobile agents are used in networked electronic trading [6], where they are dispatched by the buyer to the various suppliers to negotiate orders and deliveries and then return to the buyer with their best deals for approval. Instead of having the buyer contact the suppliers, the mobile agents behave like representatives, interacting with other representatives on the buyer's behalf, and alert the buyer when something happens in the network that is important to the buyer. Another successful example of using mobile agents is distributed information retrieval and information dissemination [10,12,20,35]. Agents are dispatched to heterogeneous and geographically distributed databases to retrieve information and return the query results to the end users. Mobile agents are also used to realize network awareness [3] and global awareness [29]. Network-robust applications are of great interest in military situations today. Mobile agents are used to be aware of and reactive to the continuously changing network conditions to guarantee successful performance of the application tasks.

The mobile-agent-based computing is first proposed to support the collaborative processing in sensor networks in Refs. [24,25]. Since then, a series of development has been underway to improve the design. Related papers include Refs. [36,37] for performance evaluation using simulation tools of distributed computing paradigms in sensor networks and Refs. [26,34] for successful applications of mobile agent in ground target classification.

20.1.1 Mobile Agent Attributes and Life Cycle

We consider the mobile agent as an entity of four attributes: *identification*, *itinerary*, *processing code*, and *storage*. The *identification* uniquely identifies the mobile agent. The *itinerary* specifies the route of agent migration. It can be predefined or derived adaptive to the change of network status. The *processing code* is the executable carried by the agent and performed at each local sensor node. This code is task adaptive that enables the network to perform different tasks according to the processing code. The *storage* is used to save partially integrated results from processing done at previously migrated nodes. Upon arriving at a local node, the mobile agent is able to start the processing code, terminate its execution, record the execution status, select the next stop, and resume processing at the next node, all autonomously.

To clarify the different states where the mobile agent might reside, Figure 20.2 uses a finite state machine (FSM) to describe the mobile agent life cycle. Upon *creation*, the mobile agent carries the processing code and *migrates* from node to node following the itinerary. Upon arriving at each sensor node, the mobile agent *executes* the processing code and the result is saved in the storage. If the node is the last node in the itinerary or if the accuracy of the result satisfies the requirement of a specific task, then the agent returns to its dispatcher and *terminates* itself; otherwise, the agent continues the migration and execution process.

From the previous discussion, we can identify two fundamental differences between the mobile-agent-based model and the client/server-based model: what is transferred over the network (agents with processing code vs. data) and where the processing takes place (local nodes vs. processing center). With these unique features, the mobile-agent-based processing provides a few important benefits in supporting collaborative processing in sensor networks. Besides scalability and fault tolerance that the client/server-based model also possesses, the mobile-agent-based model presents the following features:

- *Reliability*: Mobile agents can justify their itinerary such that they can always be sent when the network connection is alive and return results when the connection is reestablished. Therefore, the performance of the mobile-agent-based computing is not affected much by the reliability of the network.
- *Task adaptivity*: Mobile agents can be programmed to carry different task-specific processing codes that extend the functionality of the network.
- *Progressive accuracy*: A mobile agent always carries a partially integrated result, generated by nodes it already visited. As the mobile agent migrates from node to node, the accuracy of the integrated result is constantly improved, assuming the agent follows the path determined based on the information gain. Therefore, the agent can return results and terminate its itinerary at any time when the integration accuracy satisfies the requirement. This feature, on the other hand, also saves network bandwidth and computation time since unnecessary node visits and data transfers are avoided.
- *Balance between energy awareness and fault tolerance*: This is the most important feature of the mobile-agent-based computing. On the one hand, redundancy is employed to tolerate potential faults, that is, the mobile agent needs to integrate local results generated by multiple sensor nodes. On the other hand, the mobile agent can justify its itinerary on the fly, such that unnecessary node visits are avoided as mentioned earlier, and hence saving both transmission and receiving energy, as well as the computation energy.

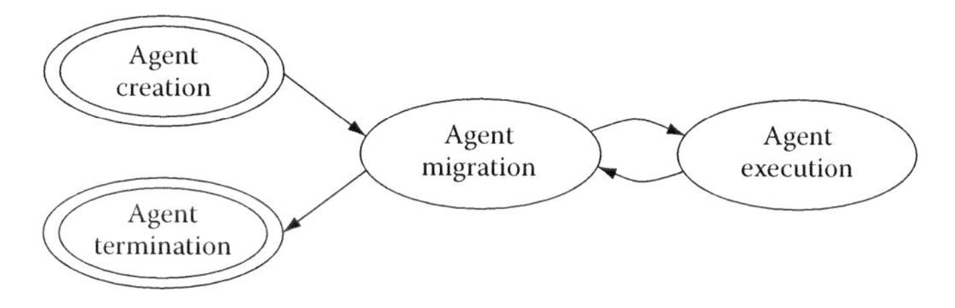

FIGURE 20.2 Mobile agent lifetime—an FSM illustration.

20.1.2 Performance Evaluation

Although the mobile-agent-based model possesses many advantages over the client/server-based model, it does not always perform better as the mobile agent creation and dispatch also bring overhead. The performance of different computing models also depends on many other parameters related to the network configuration.

We designed two metrics to help evaluate these two computing models. The *execution time* is defined from the time a task is initiated by the user until the time the user obtains the result. The *energy consumption* is the amount of energy consumed during the entire execution time.

We use a simulation software, called Network Simulator 2(*ns*-2) [15], to simulate the sensor networks. *ns*-2 is a discrete event simulator targeted at networking research. It is the most popular choice of simulator used in academy. To simplify the simulation, we assume that there are no events that simultaneously occur in the sensor field and that if an event does occur, all sensor nodes can detect it and collect the raw data. We choose a sensor field of $10 \times 10\,m^2$, a random node deployment model, a random waypoint model for node mobility (i.e., the node chooses a random destination and moves at a speed of 10 m/s), a network transmission rate of 2 Mbps, and a data processing rate of 100 Mbps. Interested readers are referred to Ref. [36] for detailed discussion about the simulation setup. Here, we present the simulation results when changing one of the following four parameters:

- The number of nodes in the network
- The number of mobile agents deployed
- The overhead ratio between the mobile agent computing and the client/server-based computing ($r_{oh} = o_f/o_a$), where the overhead of the mobile agent computing (o_a) comes from the agent creation, dispatch, and receiving, and the overhead of the client/server-based computing (o_f) comes from the time spent on large file access
- The ratio between the file size (s_f) and the mobile agent size (s_a) ($r_{size} = s_f/s_a$)

Figures 20.3 through 20.5 show the performance profiles of the two metrics when we change one parameter of the four listed earlier while keeping others unchanged.

In Figure 20.3, we change the number of nodes in the network from 2 to 30 and dispatch just 1 mobile agent. The overhead ratio is r_{oh} = 1:4 and the size ratio is r_{size} = 10:1. We observe that although both profiles in both subfigures grow as the number of nodes increases, the profile from the mobile-agent-based approach grows much slower than those from the client/server-based approach, indicating the better scalability of the mobile agent model, especially when the number of nodes is large. When the number of nodes is small (less than 15), the execution time of the client/server-based model is less than that of the mobile agent model. This happens when the mobile agent overhead is still a factor compared to

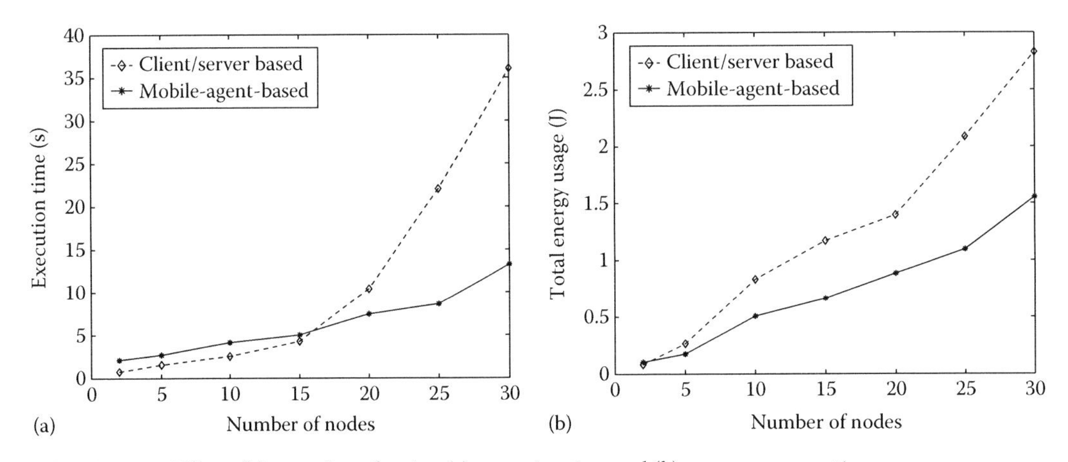

FIGURE 20.3 Effect of the number of nodes: (a) execution time and (b) energy consumption.

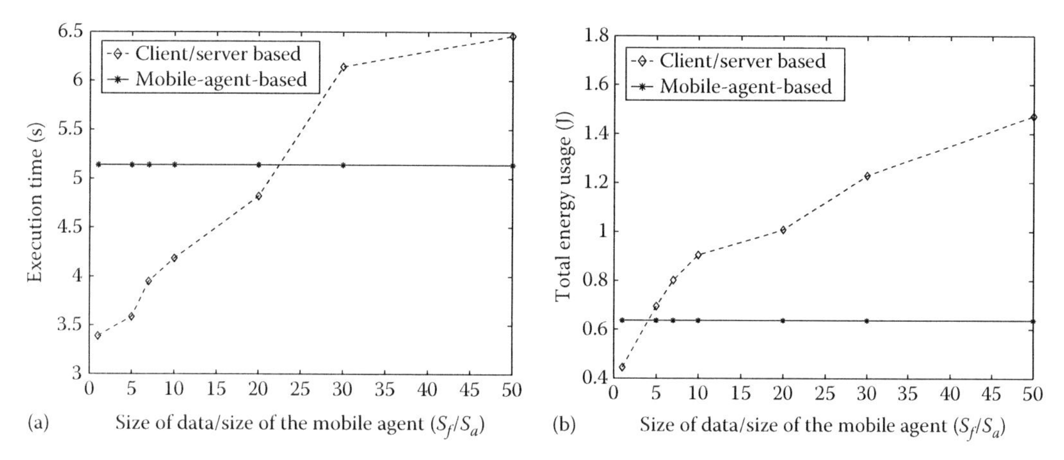

FIGURE 20.4 Effect of data size versus mobile agent size: (a) execution time and (b) energy consumption.

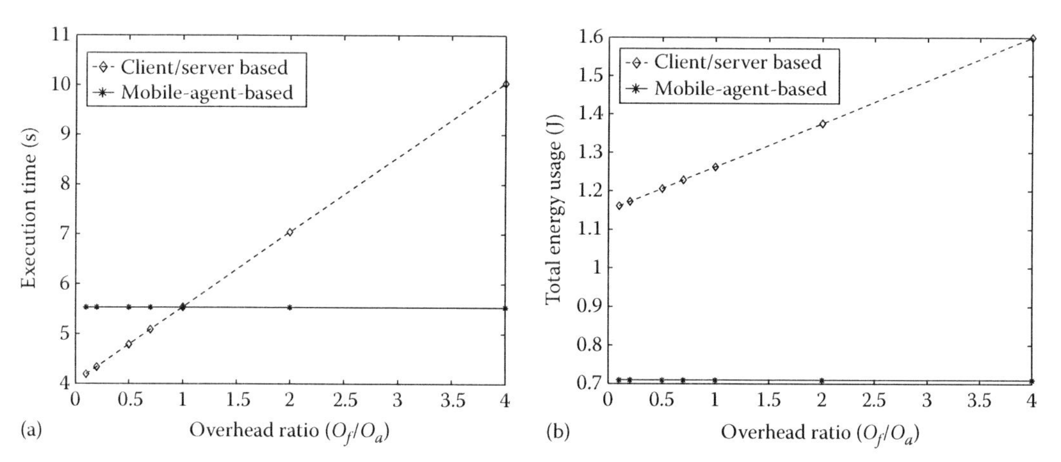

FIGURE 20.5 Effect of the overhead ratio: (a) execution time and (b) energy consumption.

the benefit it brings. However, when the node number goes beyond 15, the mobile agent model starts to show advantageous over the client/server model. Similar pattern can be observed from the energy consumption comparison except that the mobile agent approach shows advantages when the number of sensor nodes is still small.

Figure 20.4 shows the profiles of the two metrics when we fix the number of nodes at 15, number of mobile agents at 1, r_{oh} at 1:4, but change the r_{size} by keeping the mobile agent size at 1 kb and changing the file size from 1 to 50 kb, that is, changing r_{size} from 1 to 50. We expect to see a flat profile for both metrics for the mobile-agent-based model. From Figure 20.4a, we observe that when the transferred data size goes beyond 22 kb (i.e., r_{size} is greater than 22), the client/server-based approach starts to perform worse than the mobile agent approach in terms of the execution time. Similar pattern occurs to the energy consumption except that the turning point is r_{size} = 4:1. This comparison indicates how small the file size needs to be in order for the client/server-based model to perform better than the mobile agent model.

The third experiment evaluates the effect of the overhead ratio (r_{oh}). We again fix the number of node at 15 and the number of mobile agent at 1. We keep r_{size} at 10:1 but change r_{oh} from 0.1 to 4.0. We do this by keeping the overhead of mobile agent a constant and only changing the overhead of file access. As shown in Figure 20.5, when r_{oh} is greater than 1, the mobile agent model performs better, while the energy consumption always indicates a preference over the mobile agent paradigm even when r_{oh} is less than 1.

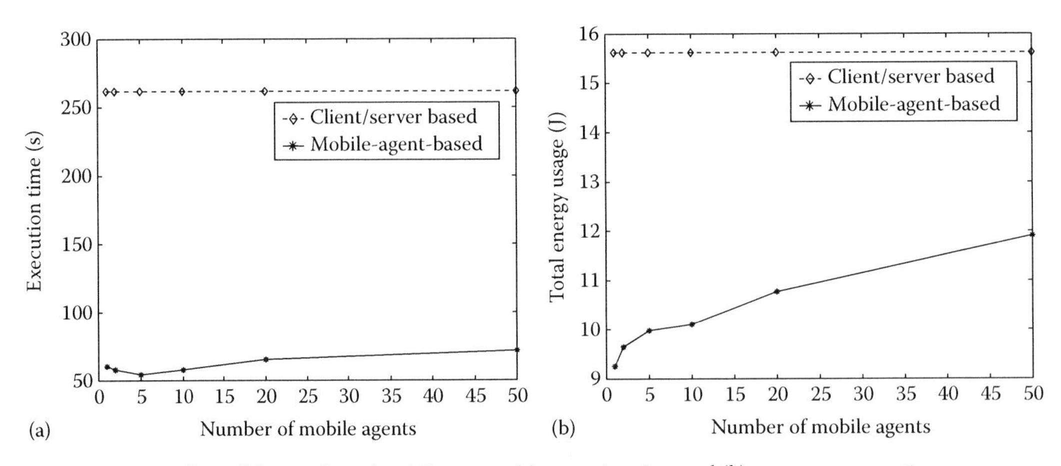

FIGURE 20.6 Effect of the number of mobile agents: (a) execution time and (b) energy consumption.

In the last experiment, we fix the number of sensor nodes at 100, r_{oh} at 1:4, and r_{size} at 10:1 but change the number of mobile agents dispatched from 1 to 50. The objective is to observe the effect of the number of agents in a densely deployed sensor field. Besides the obvious profile pattern we expect, that is, a constant profile from the client/server model and the always better performance of the agent model, we also observe an interesting phenomena in Figure 20.6a when the number of mobile agent equals 5. This is the place where the mobile agent spends the least amount of time for execution. The reason for this is that in order to handle the processing among 100 sensor nodes, more mobile agents should be dispatched to provide scalability. Therefore, the more agents, the less the execution time. However, while the number of agents increases, the overhead time spent on agent creation, dispatch, and receiving also increases. When the number of agent reaches 6, the overhead surpasses the benefits of the mobile agent and the execution time starts to increase. However, this growth pattern is not shown in the energy consumption profile.

The aforementioned simulations aim at helping the reader understand that the mobile-agent-based computing does not always perform better than the client/server-based computing and that different computing paradigms are most appropriate only for a certain network setup. While the client/server-based computing works better when there are only several nodes in the network or the raw data size is not large, the mobile-agent-based computing is more suitable for networks with a lot of sensor nodes. By carefully choosing the number of mobile agents dispatched, we can minimize the execution time of a task as well as reduce the energy consumption of the network.

20.2 Mobile Agent Framework

We design and develop the MAF to support collaborative processing in the context of sensor networks. The system architecture is shown in Figure 20.7. Different layers in this architecture perform different tasks and provide upward layer services. Compared to the traditional seven-layered

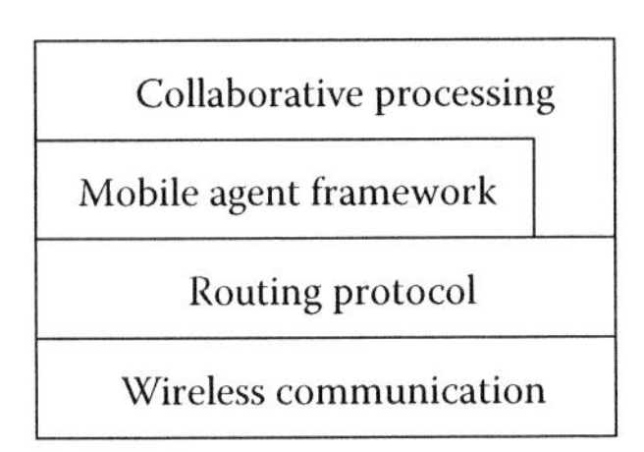

FIGURE 20.7 Architectural overview of the system.

network architecture, this is a unique integrated cross-layer design that is application oriented and data centric [7].

The collaborative processing layer hosts algorithms for the integration of the information derived from multiple sensors. The MAF layer provides the mobile-agent-based computing paradigm to achieve the collaborative processing task. MAF is realized using some routing protocols that facilitate communication over the wireless link. It has been noted that, as shown in Figure 20.7, the collaborative processing does not have to utilize the support from MAF, it can bypass MAF and use an application-oriented routing protocol like directed diffusion [7] directly.

There have been many mobile agent systems developed recently. Most of them use Java or the combination of C/C++ and a scripting language, such as IBM's Aglets [2], Dartmouth's Agent Tcl [1], General Magic's Telescript [32], etc. MAF is implemented purely in Python and has left flexible interface to other processing modules. Reference [17] provides a detailed list of benefits of Python. Raymond's article *Why Python* [27] is also a good resource. We summarize the benefits as follows:

- It is an object-oriented language from the ground up. Python is ideal as a scripting tool for OO languages like C++ and Java. It can easily *glue* components written in these languages.
- It is free and very well supported. Python comes with all popular Linux distributions, such as Debian, Caldera, Redhat, etc.
- It is portable. Python is written in portable ANSI C and compiles and runs on virtually every major platform in use today [17], including Linux, Unix, and Windows.
- It supports object serialization that is one way to save information between program executions. Object serialization is done by the Python pickle module, a standard function of the Python system. It converts Python in-memory objects to and from a single linear string format, suitable for shipping across network sockets, etc. [16]. The whole process is transparent to the end user. It needs to be clarified that object serialization only keeps the data space but not the execution status. No modification to the Python interpreter is needed in order to support this kind of *moderate mobility*.
- It is simple and suitable for rapid prototyping. Python is designed to optimize development speed. It is very efficient in proof-of-concept implementation.

In order to make use of the existing, mature integration modules, MAF also provides flexible interface where the integration modules developed in C/C++ can be dynamically linked and executed by the agents.

The two interfaces between the application routines (C/C++) and MAF services (Python) as well as the routing routines (C/C++) and MAF services (Python) are implemented by generating shared library modules accessible both by the Python and C/C++ code. These shared libraries are generated using a code development tool called simplified wrapper and interface generator (SWIG) [31]. SWIG is an open source software development tool that connects programs written in C and C++ with common scripting languages such as Python and Tcl/Tk.

The implementation interfaces is illustrated in Figure 20.8. We have also created a web site at Source Forge [25] with a complete implementation of the MAF in Python.

20.3 Application Examples

In this section, we take collaborative target classification as an example to show how the mobile-agent-based computing model can support collaborative processing. In order to use the mobile-agent-based computing, there are a few components need to be realized first:

- The format of local processing result
- An integration algorithm that can fuse local processing results from node to node
- An algorithm to determine the mobile agent itinerary

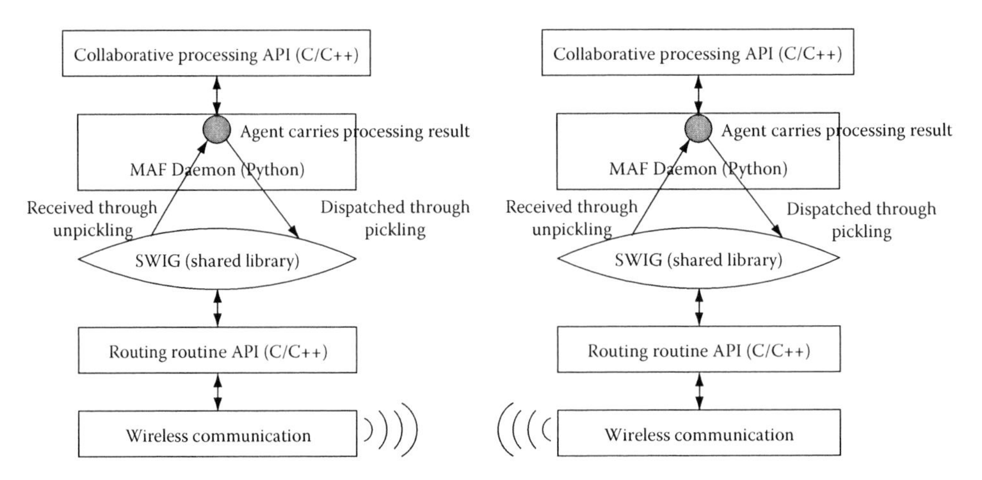

FIGURE 20.8 MAF implementation interfaces.

20.3.1 Format of the Local Processing Result

A reasonable choice of the format for local result is a *confidence range*. It indicates how confident the local node is about the local processing result based on data collected from the local sensors. For example, in the application example of target classification, the confidence range is the range of classification confidence over each possible target class. The sensor output might be expressed as "I am 40 to 70 percent sure that the target just went by me is a diesel truck, 20 to 30 percent sure that the target is an SUV." The confidence range can then be represented as a matrix using the lowest belief and the highest belief for each possible target class. In this case, the confidence matrix is

$$\begin{bmatrix} 0.4 & 0.7 \\ 0.2 & 0.3 \end{bmatrix}$$

where the first row in the matrix indicates the confidence range for target class "diesel truck" and the second row indicates the confidence range for "SUV."

The *confidence* itself can be modeled by different stochastic distributions, the simplest of which would be a uniform distribution, where equal weight has been put on each confidence value within the confidence range. Other appropriate distributions could be a Gaussian (more weight on the central confidence within the confidence range) or a Rayleigh (more weight on the low confidence within the confidence range) as shown in Figure 20.9.

A 1D array can serve as an appropriate data structure to represent the confidence range. Different resolution of processing determines the size of the array. We always assume the confidence ranges between 0 and 1. If the processing resolution is 0.05 and there are three possible targets, then the size of the 1D array is $(1/0.05 + 1) \times 3 = 63$ units of floating points.

20.3.2 Integration Algorithm

The integration algorithm needs to be both effective and simple for purposes of energy efficiency and real-time response. Interested readers are referred to Ref. [23] for a comparison of different integration algorithms. A good algorithm is the *overlap function*, which is first proposed in Ref. [21]. It is similar to a histogram function, where, according to the confidence range generated from multiple sensors, the overlap function *accumulates* the number of sensors with the same confidence value. Figure 20.10 illustrates the construction of overlap function for a set of six sensor nodes when the uniform and Gaussian distributions are used.

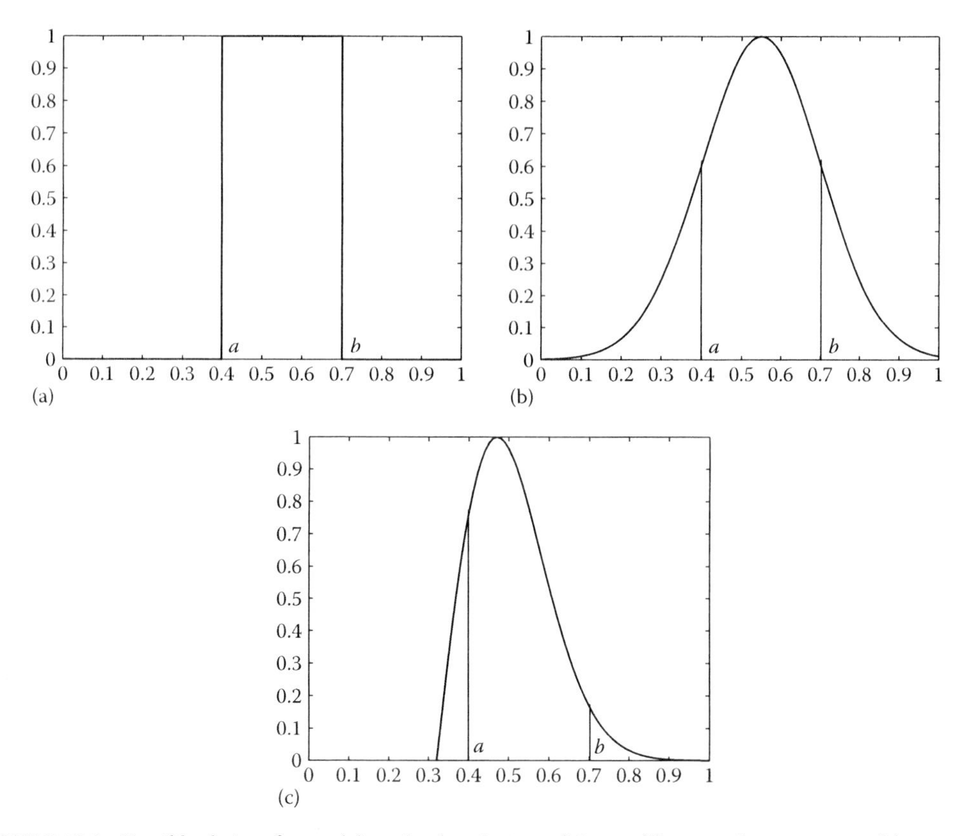

FIGURE 20.9 Possible choices for modeling the distribution of the confidence within certain confidence range $[a,b] = [0.4, 0.7]$: (a) uniform, (b) Gaussian, and (c) Rayleigh.

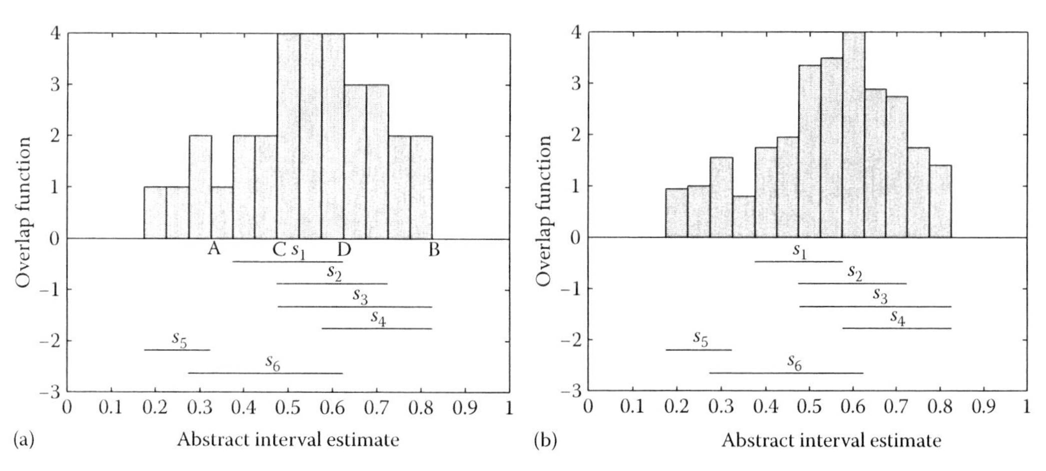

FIGURE 20.10 Overlap function for a set of six sensors using two distribution models (resolution is 0.05): (a) uniform and (b) Gaussian.

The integrated information lies within regions over which the maximal peaks of the overlap function occur with the largest spread. The original proposal for generating and analyzing the overlap function is centralized and can only take place at the processing center. A distributed integration of the overlap function is proposed in Ref. [30] for the mobile-agent-based computing such that at each stop of the itinerary, a *partially integrated* result can be generated from previously migrated sensor node outputs, and the accuracy of the information derived from this result can be used to justify if the mobile agent needs to continue with its migration. The distributed method would generate the same output as the centralized method if the mobile agents finish its itinerary.

A protocol is proposed in Ref. [26] in order to decide whether or not the mobile agent should stop its migration. The protocol includes the following three criteria:

1. The overlap function has its highest peaks ranging from $[n - f, n]$, where n is the number of nodes, $f = \lfloor (n-1)/3 \rfloor$ is the maximum number of faulty sensor node that the sensor network can tolerate. This equation comes from the Byzantine generals' problem [13] that specifies the number of faults (f) that certain number of sensor nodes (n) can tolerate.
2. The accuracy (multiplication between the height of the highest peak in the overlap function, the width or spread of the peak, and the confidence value at the center of the peak) calculated from the intermediate integration result at each sensor node has to be equal to or greater than 0.5.
3. Both 1 and 2 have to be satisfied in adjacent two migrations excluding the first sensor node.

The itinerary of mobile agent is critical in saving energy and achieving the required accuracy efficiently. Take the overlap function shown in Figure 20.10a as an example, different itinerary generates different partially integrated results. Figure 20.11 shows the step-by-step integration result as the mobile agent migrates among nodes s_1 to s_6, where s_4 and s_5 are faulty. In Figure 20.11a, the mobile agent itinerary is from s_1 to s_6 in sequence. We can see that the mobile agent can actually terminate the migration after the third stop (s_3) when all the criteria have been satisfied. On the other hand, in Figure 20.11b, we intentionally change the itinerary and hence the order of integration. We observe that the agent can stop only after the fourth integration.

20.3.3 Mobile Agent Itinerary

As shown in Figure 20.11, the mobile agent itinerary is critical in providing an energy-efficient and reliable solution for collaborative processing in sensor networks.

The problem of determining the mobile agent itinerary can be traced back to the famous traveling salesman problem (TSP) [28], where an *optimal* path (the shortest path in this context) is pursued for a salesman to travel through a set of cities. Although the problem is easy to solve when the number of cities is small, it becomes NP-complete when the number of cities is large, that is, the algorithm shows an exponential growth of run time with a growing number of cities [5]. The traveling agent problem (TAP) is discussed in Ref. [19] for determining the mobile agent itinerary in applications like information retrieval from the Internet. The ith site carries a probability of p_i of successfully providing the information needed by the mobile agent. In that paper, the authors show that if the latencies between any two nodes are constant, an optimal path can be obtained in polynomial time.

Both TSP and TAP are global optimization problems, where a centralized processing environment is anticipated. However, this is not the case in sensor networks, especially not suitable for the mobile-agent-based distributed computing. PARC developed an information-driven dynamic sensor collaboration for target-tracking application [4,38]. In order to achieve energy-efficient computing, it selects the next node that most likely improves the tracking accuracy based on both the information constraints and constraints on cost and resource consumption. Specifically, the approach formulates the tracking problem as a sequential Bayesian estimation problem. In order to estimate the track, each new sensor measurement is combined with the current estimate to improve the estimation accuracy. Therefore, the problem of selecting the next sensor can be formulated as an optimization problem as well. However, this optimization problem only requires local information. The design of mobile agent itinerary follows very similar idea.

We assume the sensor nodes exchange information among neighbors periodically or when there is a dramatic change over the information content. We design the content of the information exchange to include three pieces of information: the signal energy sensed, the remaining energy onboard the sensor

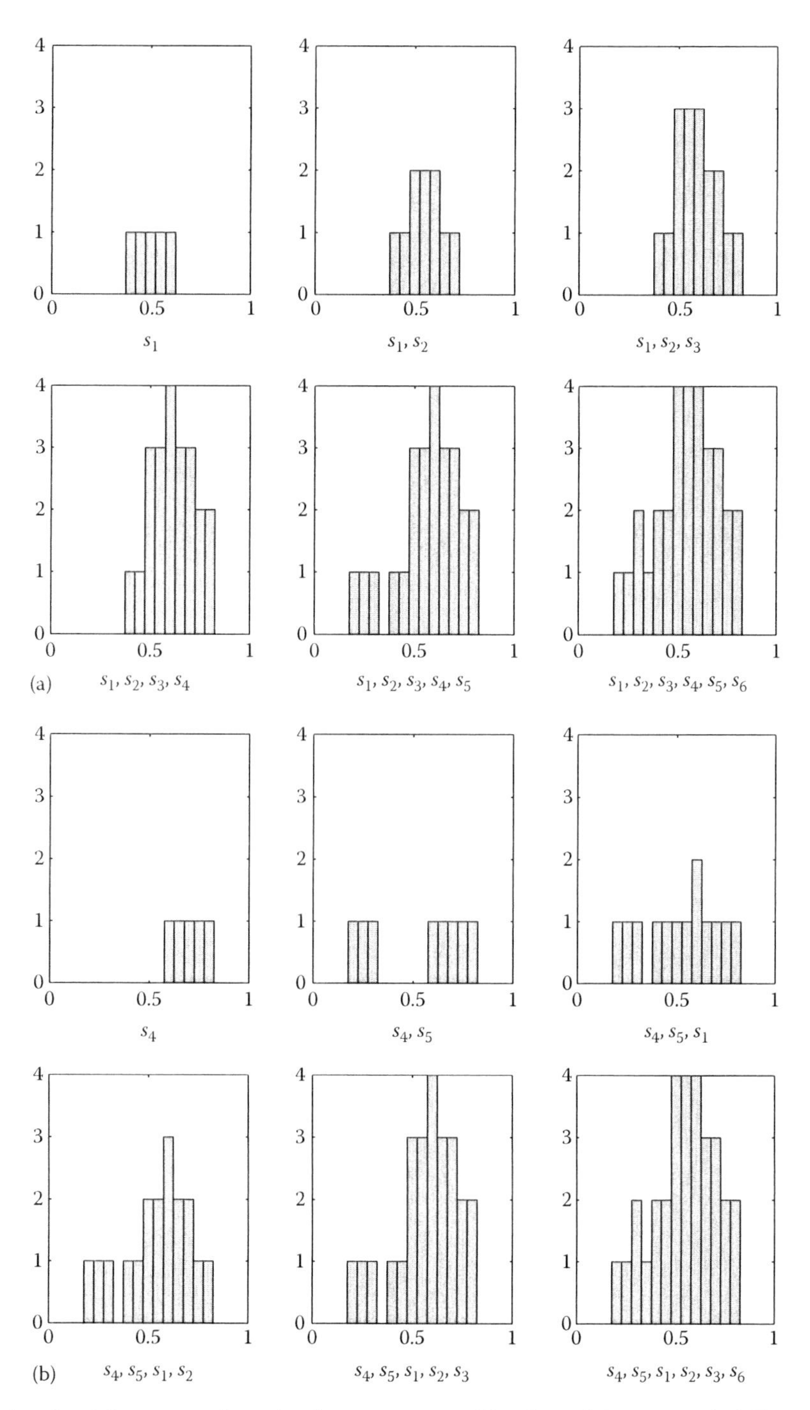

FIGURE 20.11 Partially integrated overlap function generated by the mobile agent with different integration order. (a) $s_1, s_2, s_3, s_4, s_5, s_6$ and (b) $s_4, s_5, s_1, s_2, s_3, s_6$.

node, and the location of the sensor node. For example, if there is a target passing by the sensor node, the signal energy sensed by the sensor could increase by a noticeable amount, or if the sensor node just moved from one site to another, the location of the node would change as well. When the agent arrives at a local node, it uses the information provided from the neighboring nodes to find out its next hop that should possess as high signal energy as possible, as much remaining power as possible, and as close distance to the current node as possible. This idea was partly implemented in Ref. [22].

An even simpler itinerary determination method is for the agent to choose its next hop randomly among neighboring nodes. The evaluation study is underway to compare the performance between random selection and optimal selection either using local information or global information. Since the sensor network is such a dynamic environment, we expect the random approach would present reasonably good result adaptive to the change of network topology.

20.4 Summary

This chapter discussed two distributed computing paradigms used to support the collaborative processing in sensor networks. We compared the performance between the client/server-based computing and the mobile-agent-based computing. We showed through simulation that if the network has a large number of nodes and the size of data file transferred is large, which is a typical setup in sensor networks, the mobile-agent-based model is advantageous to the client/server-based model. We described the implementation of the MAF, as well as its interfacing architecture. In the end, we addressed the three components in carrying out the mobile-agent-based processing, the format of local processing result, the integration algorithm, and the algorithm for determining the mobile agent itinerary. We showed that the agent itinerary is critical in providing both energy-efficient and fault-tolerant solutions. The mobile-agent-based model also helps realize adaptivity to both task and network topology changes.

Acknowledgment

This research was supported in part by DARPA under grant N66001-001-8946.

References

1. Agent TCL. http://agent.cs.dartmouth.edu/. Last accessed date: May 23, 2012.
2. Aglets. http://aglets.sourceforge.net/. Last accessed date: May 23, 2012.
3. W. Caripe, G. Cybenko, K. Moizumi, and R. Gray. Network awareness and mobile agent systems. *IEEE Communications Magazine*, 36(7):44–49, 1998.
4. M. Chu, H. Haussecker, and F. Zhao. Scalable information-driven sensor querying and routing for ad hoc heterogeneous sensor networks. *International Journal of High Performance Computing Applications*, 16(3):293–313, 2002.
5. G. Dantzig, R. Fulkerson, and S. Johnson. Solution of a large-scale traveling-salesman problem. *Operations Research*, 2:393–410, 1954.
6. P. Dasgupta, N. Narasimhan, L. E. Moser, and P. M. Melliar-Smith. Magnet: Mobile agents for networked electronic trading. *IEEE Transactions on Knowledge and Data Engineering*, 11(4):509–525, 1999.
7. D. Estrin, R. Govindan, J. Heidemann, and S. Kumar. Next century challenges: Scalable coordination in sensor networks. In *International Conference on Mobile Computing and Networking (MobiCom)*, pp. 263–270, Seattle, WA, August 15–20, 1999.
8. S. Franklin and A. Graesser. Is it an agent, or just a program?: A taxonomy for autonomous agents. In J. G. Carbonell and J. Siekmann, eds., *Third International Workshop on Agent Theories, Architectures, and Languages*, Vol. 1193. Springer-Verlag, New York, 1996. http://www.msci.memphis.edu/~franklin/AgentProg.html

9. C. G. Harrison, D. M. Chess, and A. Kershenbaum. Mobile agents: Are they a good idea? Technical Report RC 19887, IBM Thomas J. Watson Research Center, Yorktown Heights, NY, March 1995. http://www.research.ibm.com/massive/mobag.ps
10. M. Hattori, N. Kase, A. Ohsuga, and S. Honiden. Agent-based driver's information assistance system. *New Generation Computing*, 17(4):359–367, 1999.
11. Ivy. A sensor network infrastructure for the College of Engineering. http://www-bsac.eecs.berkeley.edu/projects/ivy
12. J. Kay, J. Etzl, G. Rao, and J. Thies. ATL postmaster: A system for agent collaboration and information dissemination. In *Proceedings of the Second International Conference on Autonomous Agents*, pp. 338–342, Minneapolis, MN, May 10–13, 1998.
13. L. Lamport, R. Shostuk, and M. Pease. The Byzantine generals problem. *ACM Transactions of Programming*, 4(3):382–401, 1982.
14. D. B. Lange and M. Oshima. Seven good reasons for mobile agents. *Communications of the ACM*, 42(3):88–89, 1999.
15. LBL, Xerox PARC, UCB, and USC/ISI. ns-2, The Network Simulator, http://www.isi.edu/nsnam/ns/. Last accessed date: May 23, 2012.
16. M. Lutz. *Programming Python*, 2nd edn. O'Reilly, Sebastopol, CA, March 2001.
17. M. Lutz and D. Ascher. *Learning Python*. O'Reilly, Sebastopol, CA, April 1999.
18. D. Milojicic. Trend wars—Mobile agent applications. *IEEE Concurrency*, 7(3):80–90, 1999.
19. K. Moizumi and G. Cybenko. The travelling agent problem. Technical Report, Dartmouth College, Hanover, NH, February 1998.
20. T. Oates, M. V. N. Prasad, and V. R. Lesser. Cooperative information-gathering: A distributed problem-solving approach. *IEE Proceedings—Software Engineering*, 144(1):72–88, 1997.
21. L. Prasad, S. S. Iyengar, and R. L. Rao. Fault-tolerant sensor integration using multiresolution decomposition. *Physical Review E*, 49(4):3452–3461, 1994.
22. H. Qi, S. S. Iyengar, and K. Chakrabarty. Distributed multi-resolution data integration using mobile agents. In *IEEE Aerospace Conference*, Vol. 3, pp. 1133–1141, Big Sky, MT, March 2001.
23. H. Qi, S. S. Iyengar, and K. Chakrabarty. Distributed sensor networks—A review of recent research. *Journal of the Franklin Institute*, 338:655–668, 2001.
24. H. Qi, S. S. Iyengar, and K. Chakrabarty. Multi-resolution data integration using mobile agents in distributed sensor networks. *IEEE Transactions on Systems, Man, and Cybernetics—Part C*, 31(3):383–391, 2001.
25. H. Qi and F. Wang. Mobile agent framework, 2000. http://maf.sourceforge.net
26. H. Qi, Y. Xu, and X. Wang. Mobile-agent-based collaborative processing in sensor networks. *Proceedings of the IEEE*, 91(8):1172–1183, 2003.
27. E. S. Raymond. Why Python? *Linux Journal*, 73, May 2000. http://www2.linuxjournal.com/lj-issues/issue73/3882.html
28. J. B. Robinson. On the Hamiltonian game (a traveling-salesman problem). *RAND Research Memorandum*, RM-303, 1949.
29. K. N. Ross, R. D. Chaney, G. V. Cybenko, D. J. Burroughs, and A. S. Willsky. Mobile agents in adaptive hierarchical Bayesian networks for global awareness. In *Proceedings of the IEEE International Conference on Systems, Man and Cybernetics*, pp. 2207–2212, San Diego, CA, October 11–14, 1998.
30. W. E. Snyder, H. Qi, J. Head, and C. X. Wang. Increasing the effective resolution of thermal infrared images. *IEEE Engineering in Medicine and Biology Magazine*, 19(3):63–70, 2000.
31. SWIG. http://www.swig.org. Last accessed date: May 23, 2012.
32. Telescript. General Magic. http://www.csc.liv.ac.uk/~mjw/pubs/ker95/subsubsectionstar3_4_1_1.html. Last accessed date: May 23, 2012.
33. R. Viswanathan and P. K. Varshney. Distributed detection with multiple sensors: Part I–Fundamentals. *Proceedings of IEEE*, 85(1):54–63, 1997.

34. X. Wang, H. Qi, and S. S. Iyengar. Collaborative multi-modality target classification in distributed sensor networks. In *Proceedings of the Fifth International Conference on Information Fusion*, Vol. 2, pp. 285–290, Annapolis, MA, July 2002.
35. J. S. Wong and A. R. Mikler. Intelligent mobile agents in large distributed autonomous cooperative systems. *Journal of Systems and Software*, 47(2):75–87, 1999.
36. Y. Xu and H. Qi. Performance evaluation of distributed computing paradigms in mobile ad hoc sensor networks. In *The Ninth International Conference on Parallel and Distributed Systems (ICPADS)*, pp. 451–456, Taiwan, Republic of China, December 15–17, 2002. IEEE, New York.
37. Y. Xu, H. Qi, and P. T. Kuruganti. Mobile-agent-based computing model for collaborative processing in sensor networks. *IEEE Global Telecommunications Conference (GLOBECOM)*, Vol. 6, pp. 3531–3535, Los Angeles, CA, December 2003.
38. F. Zhao, J. Shin, and J. Reich. Information-driven dynamic sensor collaboration for tracking applications. *IEEE Signal Processing Magazine*, 19(2):61–72, 2002.

21

Distributed Services for Self-Organizing Wireless Sensor Networks

Alvin S. Lim
Auburn University

21.1 Introduction

Distributed sensor systems are useful for gathering critical and real-time information from many dispersed integrated low-powered sensors and mobile devices [1–3] that could steer the control operations of dynamically changing enterprise systems, such as the battlefield, manufacturing, and commercial inventory and distribution systems. These mobile and miniaturized information devices are equipped with embedded processors, wireless communication circuitry, information storage capability, smart sensors, and actuators. These sensor nodes networked in an ad hoc way, with little or no fixed network support, to provide the surveillance, targeting, and feedback information for dynamic control of enterprises. Sensor devices are mobile, subject to failure, deployed spontaneously, and repositioned for more accurate surveillance. Despite these dynamic changes in configuration of the sensor network, critical real-time information must still be disseminated dynamically from mobile sensor data sources through a self-organizing network infrastructure to the components that control dynamic replanning and reoptimization of the theater of operation based on newly available information.

Since a large number of sensor devices may need to be quickly and flexibly deployed in impromptu networks, each sensor device must be autonomous and capable of organizing itself in the overall community of sensors to perform coordinated activities with global objectives. When spontaneously placed together in an environment, these sensor nodes should immediately know about the capabilities and functions of other sensor nodes and work together as a community system to perform cooperative tasks and networking functionalities. Sensor networks need to be self-organizing since they are often formed spontaneously from large number of mixed types of nodes and may undergo frequent configuration changes. Some sensor nodes may provide networking and system services and resources to other sensor nodes. Others may detect the presence of these nodes and request services from them.

21.2 Purposes and Benefits of Distributed Services

Distributed services are necessary for enabling sensor nodes to self-organize into impromptu networks that are incrementally extensible and dynamically adaptable to node failure and degradation, mobility of sensor nodes, and changes in task and network requirements. They enable sensor nodes to be agile, self-aware, self-configurable, and autonomous. Nodes are aware of their own capabilities and those of other nodes around them that may provide the networking and system services or resources that they need. Although nodes are autonomous, they may cooperate with one another to disseminate information or assist each other in adapting to changes in the network configuration. An impromptu community of these nodes may cooperate to provide continual coordinated services while some nodes may be newly deployed or removed from the spontaneous community.

Three fundamental mechanisms that support these self-organizing capabilities are service lookup, sensor node composition, and dynamic adaptation. Through a distributed implementation of these lookup servers, composition servers, and adaptation servers, other network and system services can be deployed and reconfigured spontaneously in the sensor network. They also dynamically adapt these services to device failure and degradation, movement of sensor nodes, and changes in task and network requirements. Application-specific network and systems services may be provided impromptu by sensor nodes and supporting nodes, including location services, naming and binding services, application-specific information dissemination and aggregation, caching and hoarding services, and security services. Critical sensor information can be disseminated through mobile transactions and dynamic query processing modules supported by the appropriate distributed services and network protocols to solve the problems of mobility, dispersion, weak and intermittent disconnection, dynamic reconfiguration, and limited power availability.

These distributed services enable sensor networks to overcome many of the following problems of very large and unstructured sensor networks that behave differently from traditional well-structured computer networks. First, many different types of sensors with a range of capabilities may be deployed with different specialized network protocols and application requirements. Data-centric network protocols are becoming common in sensor network [4,5]. With many mixed types of sensors and applications, sensor networks may need to support several data-centric network protocols simultaneously. Second, these mixed types of sensor nodes may be deployed incrementally and spontaneously with little or no preplanning. The networks must be extensible to new types of sensor node and services. They must be deployed spontaneously to form efficient ad hoc networks using sensors with limited computational, storage, and short-range wireless communication capabilities. They rapidly coordinate with each other to detect, track, and report activities. Furthermore, they disseminate the information efficiently through the impromptu network of sensors. Third, the sensor network must react rapidly to changes in the sensors composition, task or network requirements, device failure and degradation, and mobility of sensor nodes. Sensor devices may be deployed in very harsh environment and subject to destruction and dynamically changing conditions. The configuration of the network will frequently change due to constant changes in sensor position, reachability, power availability, and task requirements. The network protocols must be survivable in spite of device failure and frequent real-time changes. Sensor network must be secure in the face of this open and dynamic environment.

Since many services are application specific, different protocols for a certain service may be specified for different applications. However, they may interoperate through the three fundamental mechanisms provided in the self-organizing sensor network architecture. For instance, in some sensor network application, negotiation methods [6] may be preferred for information dissemination. Selected sensor nodes will register these negotiation services with the lookup server. Each of these sensor networks may establish their own services spontaneously and independently. Two different sensor networks may interoperate through filtering and translation services that must be defined to route information between the two sensor networks.

21.3 Related Distributed Services

Distributed services are useful in environments where sensor devices must be quickly and flexibly deployed in large number to coordinate through impromptu networks. Each sensor device must operate autonomously in determining the capabilities of the sensor nodes in the vicinity and participate with the entire community of sensors to achieve global objectives. Distributed lookup services allow remote sensor nodes to be located more efficiently. The end-to-end and group communication services between nodes over a wide area are more efficient when localized routing algorithms are restricted within clusters. Latency could be lowered for wide-area communication involving very large number of sensors. Discovery of services in mobile systems provides critical support for self-organizing sensor systems when sensors are being deployed and removed on the fly. In Jini [7], service discovery relies on mobile Java codes and is implemented based on TCP and UDP. It is not clear how these may be implemented using data-centric, ad hoc sensor networks with services based on more generic mobile codes. Service location protocol (SLP) [8] is an IETF protocol for service discovery that is designed solely for IP-based networks. Bluetooth [9] devices have a range of 10 m and can directly communicate with at most seven other Bluetooth devices in a piconet. Bluetooth Service Discovery Protocol (SDP) [9] allows devices to browse and retrieve services by matching service classes or device attributes. Only services within the range of the device are returned. Our lookup service may retrieve services that could be multiple hops from the requesting node.

Existing services in reconfigurable middleware[10], such as adaptive CORBA [11,12], Jini [7] and XML [13], require much larger memory and power than are available in most microsensor nodes. We instead develop lightweight distributed services that will execute with limited memory and computation power on microsensor nodes. Furthermore, existing adaptive middleware assumes a fixed TCP/IP network and will not work in ad hoc sensor networks.

Many self-organizing sensor network services have focused on ad hoc network routing and localized algorithms to route information in an energy efficient way through autonomous sensor devices. Directed diffusion routing protocol [14], based on the localized computation model, provides energy-efficient and robust communication for dynamic network with small incremental changes. For dynamic networks with large-scale changes and high level of mobility, directed diffusion may not adapt very well. Similar diffusion routing concept have also been presented in Hyphos [15]. Another localized protocol for information dissemination in sensor network uses meta-data negotiation to eliminate redundant transmission [6]. Other self-organizing network routing protocols are dynamic source routing [16] and destination-sequenced distance vector [17], although it is not clear if these algorithms are energy efficient enough for sensor networks. Our distributed services allow sensors to form high-level clusters and use directed diffusion within clusters. Clusters may be formed using localized algorithms [4] for coordinating among sensors to elect extremal sensors.

Distributed services may improve the efficiency of data-centric communication in sensor networks. Data-centric communication are useful since the identity of the numerous sensors are not as important as the data they contain [4,5]. The networking infrastructure may provide more efficient dissemination of data through replication, caching, and discovery protocols [5]. Communication protocols must be energy efficient since sensors [1,4] have very limited energy supply. In our architecture, sensors may cache and aggregate information. The discovery of these sensors can be made through the distributed lookup servers that are implemented using diffusion routing. Changes in the sensor network are propagated to other caching and aggregation services through the adaptation servers. Our framework facilitates consistent adaptation of networking and system services as well as distributed sensor applications.

Unlike other centralized control networks [18], our servers are associated only with sensors in a vicinity. Servers in different clusters will coordinate among themselves through information diffusion.

Recent advances in continual query and active database can be exploited for remote surveillance in sensor network through distributed services. The DIOM system [19] is an object-based database designed primarily for integrated access to heterogeneous data sources. We extend this system to support sensor data sources and mobile nodes. DIOM continual queries may repeatedly retrieve and update sensor data for target-tracking purposes. Cougar [20] is a distributed database designed specifically for network of sensors. Sensor devices are abstract data types (ADT) objects in an object-relational database. Sensor ADTs may contain asynchronous methods for retrieving readings from multiple sensors. Database operations, such as join, may be modified for these asynchronous methods.

21.4 Architecture of a Distributed Sensor System

We use an approach that integrates three mobility-aware system layers (Figure 21.1):

1. *Application systems*—for example, sensor information processing layer and collaborative signal processing
2. *Configurable distributed systems*—which provide distributed services to the application systems
3. *Sensor networking and physical devices' layer*—which routes messages through the ad hoc sensor network

The architecture avoids duplication of functionalities in the different layers and promotes efficient coordination between them. The sensor information layer contains collaborative signal processing applications, mobility-aware mediators and adaptive sensor query processing. The runtime reconfigurable distributed system contains distributed services for supporting mobile sensor applications. The network and physical layer contains data-centric network routing protocols, physical wireless transmission modules, and sensors that generate the raw data.

At the physical device layer, different physical sensor and mobile devices may be assembled impromptu and reconfigured dynamically in an ad hoc wireless network. Each sensor node contains battery power source, wireless communications, multiple sensing modality, computation unit, and limited memory.

Dual processors may be included for computation and real-time sensor processing. Three common sensing modalities are supported—acoustic sensing using commercial microphones, seismic vibration using geophones, and motion detection using 2-pixel infrared imagers. Wireless transceivers in the nodes provide communication between nodes, using time division multiplexing and frequency hopping spread spectrum. Neighboring nodes in each cluster communicate through a master node that establishes the frequencies used by the nodes. Each node contains a global positioning system (GPS) receiver that allows the node to determine its current location and time. GPS uses triangulation method with signals received from three satellites to calculate the location of the node with the accuracy of 1 m. However, without clear line of sight to the satellites as in urban environments, GPS cannot be used. As we describe later, message routing and query processing use these location information.

At the networking layer, ad hoc routing protocols allow messages to be forwarded through multiple physical clusters of sensor nodes. Directed diffusion routing is used because of its ability to dynamically adapt to changes in sensor network topology and its energy-efficient localized algorithms. To retrieve sensor information, a node will set up an interest gradient through all the intermediate nodes to the data source. Upon detecting an interest for its data, the source node will transmit its data at the requested rate.

The configurable distributed system uses the diffusion network protocol to route its messages in spite of dynamic changes in the sensor network. These distributed services will support applications systems, such as distributed query processing, collaborative signal processing, and other applications. The advantage of using these services is that application and system programs may use simpler communication interfaces and abstraction than the raw network communication interface and metaphor

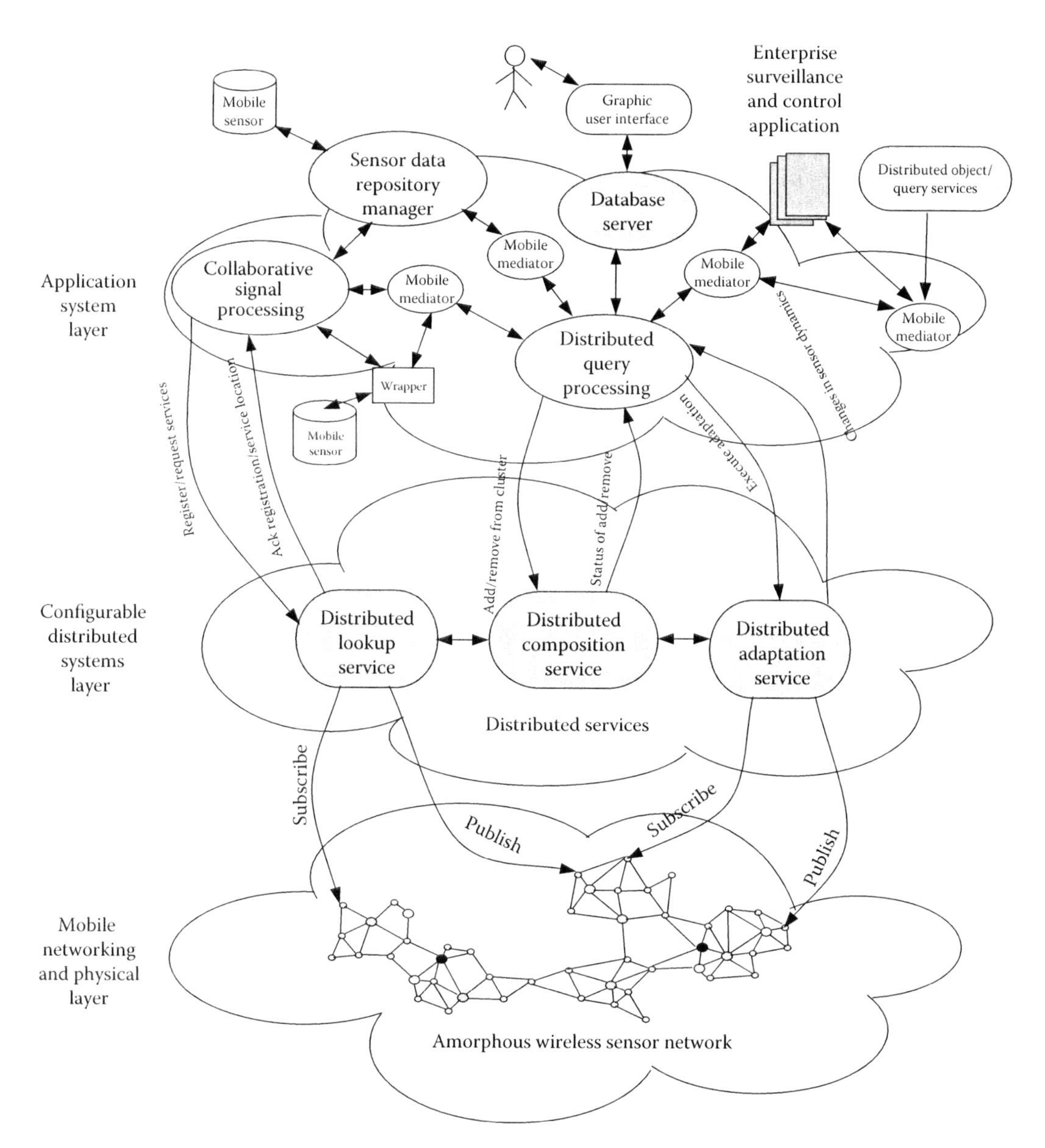

FIGURE 21.1 Architecture of a self-organizing distributed sensor system.

(e.g., subscribe/publish used in diffusion routing). Furthermore, these distributed services may enhance the overall performance, such as throughput and delay. These services will be implemented on top of the directed diffusion protocol that can still be used by applications concurrently with these distributed services. Directed diffusion can still be the preferred method for retrieving sensor data and will be used by some of the services. On the other hand, distributed services provide other forms of communication—such as interpersonal communication and impromptu establishment of community of services—required by other applications.

At the application system layer, distributed query processing and collaborative signal processing modules communicate with each other to support the surveillance and tracking functions of the enterprise. In sensor information systems, the cooperation between mobility-aware mediators, sensor agents, and collaborative signal processing modules provides efficient access to diverse heterogeneous sensor data, surveillance, and tracking information through the sensor network. The mobile sensor information layer is supported by three major components: interoperable mobile object, dynamic query processing, and mobile transactions. We will not discuss mobile transactions in this chapter. In the interoperable mobile object model, cooperative network of mobility-aware mediators and sensor agents will be configured to support interfaces to remote sensor data sources through multi-hop wireless network protocols.

21.5 Data-Centric Network Protocols

We use directed diffusion protocol [14] to implement all the distributed services and for retrieval of data through dynamically changing ad hoc sensor networks. Diffusion routing converges quickly to network topological changes, conserves mobile sensor energy, and reduces the network bandwidth overhead since routing information is not periodically advertised. Routing is based on the data contained in sensor nodes rather than unique identification. Directed diffusion is a type of reactive routing protocols that only update routing information on demand. In contrast, proactive routing protocols, such as link-state routing, frequently exchange routing information. For sensor networks that experience greater dynamic changes, reactive routing algorithms are more appropriate, whereas for those that are more static and experience infrequent topological change, proactive routing algorithms are more efficient.

Directed diffusion is a data-centric protocol, that is, nodes are not addressed by IP addresses but by the data they generate. A node names the data it generates by their *attribute–value* pairs. A sink node requests for a certain data by broadcasting an *interest* for the named data in the sensor network. The interest and gradient is established at intermediate nodes for this request throughout the sensor network. When a source node has a data that match the interest, the data will be "drawn" down toward that sink node using this interest gradient that was established. Intermediate nodes may cache, transform data, or direct interests based on previously cached data. The sink node can determine if a neighbor node is in the shortest path whenever it received a new data earliest from that node. The sink node will reinforce this shortest path by sending a reinforcement packet with a higher data rate to this neighbor node that forward it to all the nodes in the shortest path. Other nonoptimal paths may be negatively reinforced so that they do not forward data at all or at a lower rate.

Distributed services and applications use publish and subscribe APIs provided by directed diffusion. Through the subscribe() function, an application declares an interest that consists of a list of attribute–value pairs. The subscription is then diffused through the sensor network. A source node may indicate the type of data it offers through the publish() function. It then sends the actual data through the handle returned from the publish() function. The sink node then received the data that have propagated through the sensor network using a recv() function call with the handle returned from the subscribe() call. In the same way that data are disseminated through directed diffusion, the distributed services' information can similarly be propagated between the clients and servers, as described in the next section.

21.6 Distributed Services for Self-Organizing Sensor Applications

By augmenting sensor nodes as reconfigurable smart nodes through distributed services, we can simplify the development of self-organizing networks. These smart sensor nodes may be developed independently but may interact with other smart sensor nodes. Some smart sensor nodes may execute autonomously to provide networking and system services or control various information retrieval and dissemination in the dynamically changing sensor network [21]. To enhance the ability to reconfigure their networking, configuration, and adaptation functionalities, smart sensor nodes may make use of three main classes of distributed services: lookup service, composition service, and adaptation service (Figure 21.1).

The lookup service enables new system and network services to be registered and made available to other sensor nodes. Methods for calling the services remotely are also provided. The composition service allows clusters of sensor nodes to be formed and managed. The adaptation service allows sensor nodes and clusters to reconfigure dynamically as a result of sensor node mobility, failure, and spontaneous deployment. These servers enable sensor nodes to form spontaneous communities in ad hoc sensor networks that may be dynamically reconfigured and hierarchically composed to adapt to real-time information changes and events.

These distributed servers may be replicated for higher availability, efficiency, and robustness.

Distributed servers coordinate with each other to perform decentralized services, for example, distributed lookup servers may work together to discover the location of a particular remote service requested by a node.

21.6.1 Reconfigurable Smart Nodes

By exploiting these distributed services, sensor nodes can be enabled to be self-aware, self-reconfigurable, and autonomous. These sensor nodes, known as reconfigurable smart nodes, can be used to build scalable and self-organizing sensor networks. (In this chapter, we refer to reconfigurable smart sensor nodes as smart nodes or sensor nodes.) Smart nodes may represent sensor nodes, other types of mobile nodes, fixed nodes, or cluster of these nodes. They may simultaneously be service providers for other smart nodes and clients of services that other smart nodes provide. Smart nodes may be dynamically composed into impromptu networked clusters forming clustered smart nodes that work together to provide abstract services for the agile sensor network. They may also adapt rapidly to abrupt changes in the sensors' capabilities, events, and new real-time information. Very large networks with hundreds of thousands of sensors' nodes can be built by hierarchically composing reconfigurable smart nodes.

Smart sensor nodes may consist of hardware devices and software for interacting with the real-world systems. The hardware may contain computational, memory, wireless communication, and sensing devices. Smart nodes may contain control software for monitoring information from real-world devices such as simple sensors, engaging in distributed signal processing, and generating appropriate control signals to produce a desired result in the real-world system. The control software takes advantage of the functionalities provided by the networking and system software.

Smart nodes interact with other smart nodes through well-defined interfaces (for networking and systems operations) that also maintain interaction states to allow nodes to be dynamically reconfigured. These explicit interaction states and behavior information allow localized algorithms with the adaptation servers to maintain consistency when autonomous nodes and clusters are reconfigured dynamically, move around, or recover from failure. Smart node implementation and data (software and hardware) are encapsulated (hidden) from other nodes.

Different designers may independently develop smart nodes and their network and system services using different methods. For example, one designer may use a network protocol that is suited for a particular sensor application with its set of network requirements, such as low latency, power conservation, GPS capability, high error rate, and disconnection. In order to ensure consistency during dynamic reconfiguration and failure recovery of sensor nodes, adaptation servers may analyze the protocols using their specification model.

When new smart nodes are added to the sensor network, they register their services with a lookup server (Figure 21.2). Other nodes that require a service will discover the services available in a cluster through the lookup servers that return the location and interface of the service nodes. This is similar to Jini [7] that manages system-level services based on Java code executing in IP-based networks. On the other hand, reconfigurable smart nodes may provide lower-level networking services using generic mobile codes executing in data-centric sensor networks. Client nodes then interact directly with the service node. Smart nodes are self-aware of their own location, configuration, and services that they perform.

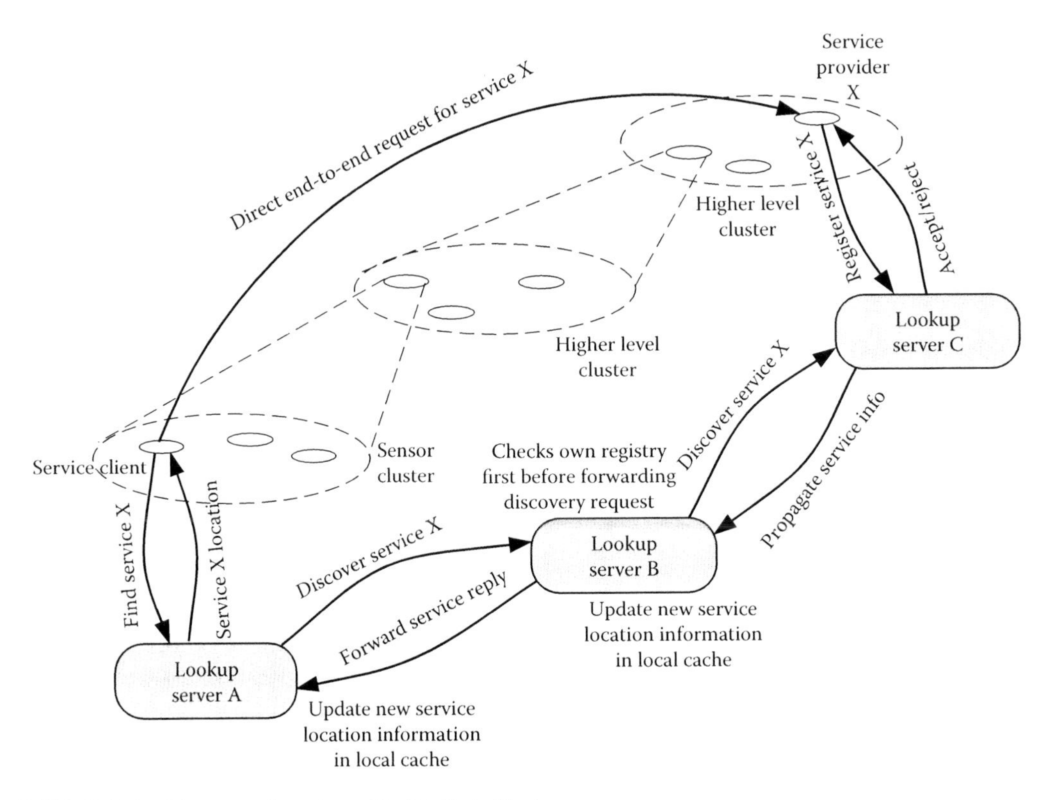

FIGURE 21.2 Discovery of services with distributed lookup server.

21.6.2 Lookup Services

A sensor node may deploy new network and system services for use by other nodes in a self-organizing sensor network. A sensor node that provides a service is called a service provider, and a node that uses the service is called a service client. Since service providers may be introduced or removed from the sensor network at any time, a lookup server is needed to keep track of the availability of these services. A sensor node may register a resource that it maintains or service that it can perform with a lookup server (Figure 21.2).

Each smart node has a home lookup server that keeps track of the location of the node when it moves.

A lookup server may contain information on services or resources at multiple clusters. Other nodes that require the service may request the service through a lookup server. If the service is recorded in the lookup server, it will return the location of that service to the requesting node. Otherwise, if the service is not recorded in the lookup server of the region, a discovery protocol is used to locate the service through other lookup servers (Figure 21.2). A request message is propagated to all the lookup servers, and the server that contains the service registration information will return the reply with the service location and service interface. It may also return the cluster name of that service. The lookup server that made the request will then cache that service location, interface, and cluster name information in its local registration cache. At regular frequency, service and resource registration information may be disseminated from one lookup server to other lookup servers in the agile sensor network.

Lookup servers support mobility of sensor nodes. When a sensor node moves to a different cluster at another location, it notifies, whenever possible, the previous lookup server that it is moving. When it arrives at another cluster in a new location, it will register with the new lookup server that will notify the previous lookup server. The new lookup server will propagate the change in the nodes' location to other lookup servers. The lookup server used by the sensor node that is interacting with the mobile node will notify the sensor node of the service location change. Existing interactions between the mobile node and other nodes will thus be handed over to the new location. Adaptation servers may be involved in the handover operation to preserve global consistency during the handoff, as discussed in Section 21.6.4, resulting in uninterrupted use of the service.

21.6.3 Composition Services

The compositional server manages various smart nodes that may be added to (or removed from) clusters in the agile sensor network. It also manages network abstractions (or group behavior) of clusters and hierarchical composition of clusters. The compositional server simplifies dynamic reconfiguration of services provided by each smart node or cluster. It also simplifies the development of large self-organizing sensor network by allowing individual node and cluster to be specified and designed independently while the compositional behavior and constraints on a cluster of components may be separately specified.

Compositional servers enable compositionality and clustering abstraction of sensor networks. To enhance adaptivity in sensor networks, each node is designed independently and the networking requirements with other nodes may be specified separately. This decoupling of autonomous smart nodes from their networking requirements enables smart nodes to be easily adapted, replaced, and reconfigured when triggered by dynamic events in the sensor network.

Clusters of smart sensor nodes may be formed under the management of a compositional server.

Hierarchical clusters are also possible for larger scale sensor networks. A cluster of sensors may also provide distributed services by coordinating the tasks among the sensors, such as aggregating summary information. Clustered smart nodes encapsulate the abstract capabilities provided cooperatively by the group of smart nodes. There will be a head smart node in the cluster that is responsible for the control of the cluster and intercluster communications and networking functions. Group communication to nodes in a cluster can be efficiently implemented by sending a message first to the cluster head which then multicasts it to the member nodes. Member nodes will elect a cluster head from the set of nodes with the most powerful networking and system capabilities. Smart nodes in a cluster may cooperate to perform the networking and system functions for the cluster. Synchronization constraints associated with the interactions among smart nodes may be specified in clustered smart nodes. The capability to specify hierarchical composite clusters enables designers to build large and complex sensor networks by clustering together smaller network-enabled sensor devices at each level.

The capability to reconfigure clustered smart nodes is made possible by abstracting out the interactions of the members of the cluster using a connector model. Connectors may be specified to map to physical communication links and guarded by a set of specified composition constraints. (We use the term physical communication link to denote links, routes, or connections established by the network.) The decoupling of the connectors from the physical links enables the clusters to be dynamically reconfigured as nodes fail, migrate away, or migrate into the area. In sensor networks, sensors can be modeled as a set of autonomous computational entities that can only interact via messages. We use this sensor object model, which is specified in a list of attributes and includes both functional and extra-functional properties of the sensor. These attributes help to define allowable patterns of interactions between sensor objects. The concepts of the connector, link, and composition constraints are described in the following sections.

21.6.3.1 Connector Model

A connector is a unit of adaptable communication. Data flow between sensor objects through communication channels called connectors. Connectors reflect data association between interacting sensor objects rather than just a physical communication channel between two sensor objects. In an unstable sensor network, physical communication channels may be broken. In contrast, sensor objects are expected to establish a stable and long-term relationship with the sensor objects of certain types (not just one specific

TABLE 21.1 Roles of an Endpoint

Role	Data Flow Direction	Applicable Connector Type
Server	Source + sink	One to one
Client	Source + sink	One to one
Producer	Source	All types
Consumer	Sink	All types
Peer	Source + sink	All types

sensor node). Besides transferring data, one distinctive feature of connectors is reconfigurability. Unlike physical communication channels, connectors can be reconfigured to achieve highly reliable data association between sensor objects. The connector model is characterized by the following:

1. Every connector has an identity. We called it Connector ID (CID) that differs from Link ID (LID). CID is generated by sensor node's connector layer and used by application as a handler of connector.
2. Associated with each connector is a connector type that indicates traffic direction and operations allowed for the connector. There are four types of connectors in terms of the number of source endpoints and sink endpoints of the connector. Endpoints of a connector are sensor objects attached to the connector. The four connector types are as follows:
   a. *One to one*: There are one source endpoint and one sink endpoint in this connector type. Damage to either endpoint (source or sink) will result in reconfiguration of the connector.
   b. *One to N*: There is one source endpoint and multiple sink endpoints. Data injected by the source are received by all sink endpoints. Failure of some sink endpoints may not necessarily result in reconfiguration of the connector.
   c. *N to one*: There are multiple source endpoints and one sink endpoint. Data injected by any of the *N* sources could only be received by the sink endpoint. Failure of some source endpoints may not necessarily result in reconfiguration of the connector.
   d. *N to M*: There are multiple source endpoints and multiple sink endpoints. Data injected by any source endpoint are received by all sink endpoints. Failure of some source or sink endpoints may not necessarily result in reconfiguration of the connector.
3. Endpoint of a connector actually comprises a service type and its role. The service type attribute defines only the sensor objects of that specific service type could be attached to this connector. When a sensor object is involved in forming a connector, it is required to be assigned a role. Connectors represent sensor objects' relationships or interactions in which each participant should perform a role. In Table 21.1, we list all the possible roles, data flow directions, and applicable connector types. "One to one" or "one to *N*" may not imply the data flow direction.
4. Composing operations over a connector are also guarded with state-based constraints.
5. For each connector, a constraint satisfaction function (CSF), if necessary, could be specified to make sure endpoints of the connector are compatible. A set of CSF could be provided by the composition service and/or applications. CSF index of a connector is used to search a specific CSF if there is a CSF table. The composition service may apply it to enforce constraint satisfaction among all endpoints involved in a connector. The composition server retrieves all composition requirements from each application-specific attribute of all connector's endpoint sensor objects. Then it runs the CSF to check their conformance to each other's requirements. CSF is actually performed by composition server.

21.6.3.2 Link Model

A link is an actual physical communication link between sensor nodes. It is a connector whose endpoints have been associated with actual sensor objects. It inherits all attributes of a connector except the

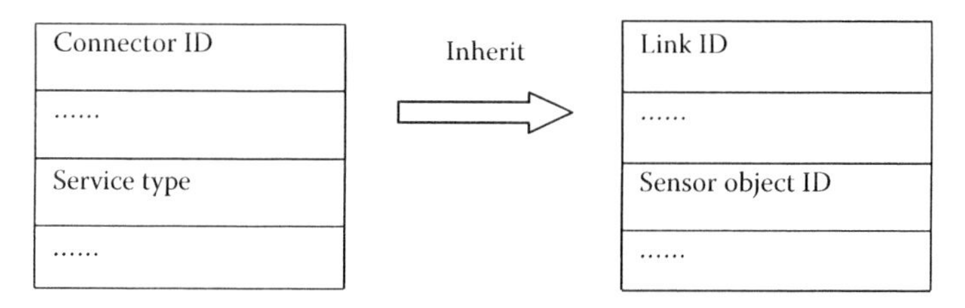

FIGURE 21.3 Attributes defined in link model.

connector ID and service type (Figure 21.3). In a link model, we use link ID to name a link. The endpoints of a link should contain the sensor objects that provide the required service type. From the sensor object ID, we can always retrieve the detailed sensor object information from the lookup server or local service information cache. So endpoint of a link contains two attributes: sensor object ID and its role.

21.6.3.3 Relationship between Links and Connectors

A connector describes interface dependency of different services. Any set of sensor objects with their service interfaces satisfying the dependency requirements of a connector can form a physical link through which the data are transmitted between them. For example, in Figure 21.4, *A*(*B*) stands for service *A*(*B*). a_1 and a_2 are sensor objects providing service *A*. b_1 and b_2 are sensor objects providing service *B*. A connector can be mapped to four links as illustrated in Figure 21.4. Therefore, a connector can be reconfigured by remapping it to a new link. A connector is always available to applications as long as there are some links to which the connector can be mapped.

21.6.3.4 Composition Constraints

Using the aforementioned sensor object model and connector model, applications or system services are able to compose sensors together by establishing connectors between sensor objects. However, a connector may not be established between any sensor object pair because the endpoints of a connector need to satisfy the specification for connector's interaction. Interaction specifications describe abstract patterns of interaction among roles of sensor objects. Connectors are built from sensor objects through a process of filling roles with sensor objects. Constraints are imposed over the composition process in that all parties of an interaction should satisfy the requirement of the specification.

Sensor objects contain extra-functional properties and codependence relations with interacting sensor objects. In order to make sure interfaces of all endpoint sensor objects attached to a connector are compatible, we have to check the following attributes of a sensor object before we use them in a role:

1. Check if the interface type of the endpoint sensor object matches the role. Matched role and interface type pairs are currently defined in Table 21.2.
2. Check if the extra-functional properties of sensor objects involved in a connector are consistent. For example, the average rate a producer can generate data should be less than the average rate the consumer can process incoming data.

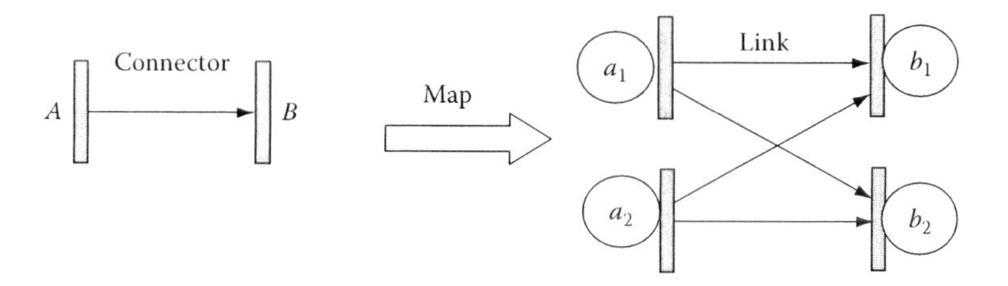

FIGURE 21.4 Map a connector to a link.

TABLE 21.2 Matched Interface Type and Role Pairs

Interface Type	Role
RPC	Server
	Client
Point-to-point STREAM	Producer
	Consumer
Point-to-multipoint STREAM	Peer
	Producer
	Consumer

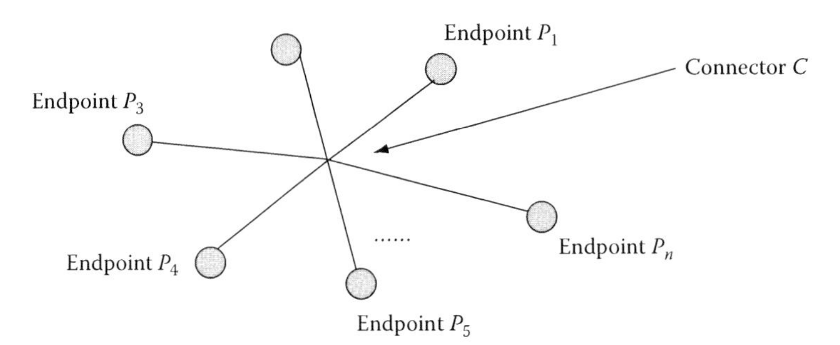

FIGURE 21.5 Codependency checking.

3. Codependence check is mostly application dependent. Since the sensor object model requires sensor objects to provide defined attributes as object information and the connector model allows applications to define their own CSF, composition server runs CSF with all sensor object information as input to check codependent consistency. We give a codependence checking example to make the codependencies concept clear.

Consider the example in Figure 21.5. There is an N–M type connector C that has n endpoints. The CSF for the connector C is that all endpoints of the connector should be located within a circle with radius equals 32 m. This constraint can be formulated as

$$\text{Max}\left[D=\left\{d \middle| \text{distance between } P_i \text{ and } P_j, i, j \in [1,n]\right\}\right] < 32 \times 2 \tag{21.1}$$

Any sensor object that wants to be integrated to this connector has to pass a codependency check performed by the composition service. The composition will enforce Equation 21.1 to make sure that the constraint of the connector is not violated.

21.6.4 Adaptation Services

Adaptation servers utilize information from the compositional server, lookup server, and analytical tools to control smart nodes during dynamic reconfiguration and failure recovery. Each smart node may execute autonomously to control different network operations in the sensor network and may interact and coordinate independently with other smart nodes to perform collaborative networking operations.

Adaptation servers monitor clusters of smart nodes during normal execution through the spontaneous signal from the sensors, probing of the smart nodes, or explicit network management directives for reconfiguration and failure recovery. When a runtime reconfiguration is requested or triggered, the adaptation server will generate the appropriate schedule of reconfiguration operations that will ensure the reconfigured and affected sensor nodes are globally consistent. To ensure correct adaptation and maintain consistency, the adaptation server makes use of analytical tools for dependency analysis and relevant information from compositional servers and lookup servers. When smart nodes are added or removed from the agile sensor network, a suite of analytical tools may be utilized to ensure that the sensor network still maintains its safety and liveness properties [22]. Smart nodes (or clusters of smart nodes) may be specified and analyzed independently.

The general model for adaptive systems can be applied to a wide range of situations that arise in the context of distributed systems. An adaptive system built on a distributed computing platform can be modeled as responding to changes with the following three phases:

1. *Change detection*: Monitoring for a possible change in the environment and verifying that the change has actually occurred. The change detection phase can take various forms depending on what type of system or change is being dealt with. For example, in a distributed system, it might involve monitoring message flow between sites or sending control/test messages to the change detection processes on other sites. The user could also cause the change when the requirement of the application or the service needed from the underlying system is changing.
2. *Agreement*: Reaching agreement among all sites that adaptation is required and assessing the change to decide if the action is necessary. The agreement phase often involves some kind of distributed agreement algorithm, which assesses the changes and decides whether the action is necessary based on the agreement policy. Depending on the situation, it may be a majority vote or agreement on maximum or minimum value.
3. *Action*: Changing the behavior of the system. The action phase can take various forms depending on the type of adaptation. The simplest case is simply changing some execution parameters, such as modifying the time-out period for a network protocol. Another possibility is reassigning roles, such as changing a centralized coordinator or reassigning roles in a primary/backup fault tolerance scheme. Slightly more involved would be taking some corrective action, such as recomputing routing tables, aborting deadlocked computations, or regenerating a lost token. In the extreme case, the action might consist of actually changing the algorithm used to provide the service.

21.6.5 API for Lookup Services

Applications use the lookup service through the following API. We focus primarily on the lookup service API in this chapter since it is the main service that is responsible for enabling sensor networks to be self-organizing. In the next section, we will describe how these API functions are used by the various application systems of a surveillance sensor network.

These API functions use the following parameters:

1. `service_type` is the generic type of service for which there may be several instances. For example, a type of service may be temperature monitoring, whereas specific instances of temperature sensors may be sensor *X* at location *Y*.
2. `service_name` is the specific name that identifies an instance of a service provider.
3. `input_list` is a list of attribute–value pairs containing the input parameters to the service invocation.
4. `output_list` is a list of attribute–value pairs containing the output values from the service invocation.
5. `lifetime` is the time period in which a service information will be stored in a lookup server.
6. `interface_type` is one of the following three types of interface that the callee of the service use:
   a. *Location or address*: This is used by the service client if the interface for interacting with the service provider is known. The service client only needs to retrieve the location or address of the service provider to be used for invoking the service request.
   b. *Interface definition*: This is used by the service client to retrieve the definition of the interface for interacting with the service provider. The service client must have the interpreter or compiler for the interface definition.

c. *Mobile code*: The service client retrieves the mobile code that implements the protocol for interacting with the service provider. The mobile code must then be dynamically linked to the service client.

Nodes that receive the service provider information can make service calls through a method provided in the following. The following is a description of the purposes and the side effects of the lookup service function calls:

1. `service _ call(service _ name, input _ list, output _ list, interface _ type)`
   This function allows a service client to find and make a call for a service when the service client does not know the location or address of the service provider and/or the interface for using the service. It is implemented as a combination of the `lookup _ service()` and `service _ exec()` calls described in the following. This function requires specific service name to be provided. Generic service type cannot be used since several service provider instances may match the service type.
2. `status = lookup _ service(service _ type, service _ name, input _ list, output _ list, interface _ type)`
   This function allows a service client to find the location or address of a service provider and/or the interface for using the service. If `service _ type` is defined and `service _ name` is NULL, then all service providers registered with the lookup server of that type are returned. Service lookup can also be based on cluster or predicate matching. The cluster information and predicate for matching service providers are contained in the `input _ list`. Depending on the `interface _ type` used, that is, location, interface definition, or mobile code, the respective results of the `lookup _ service()` call will be placed in the `output _ list`.
3. `status = service _ exec(service _ name, input _ list, output _ list, interface _ type)`
   This function allows a node to request for the service and gets the results back from the service provider. The input parameters are specified by the client in the `input _ list`. The service provider performs the requested service or remote procedure call and returns the results in the `output _ list`. The `interface _ type` defines the method used by the service client to communicate with the service provider.
4. `status = service _ register(service _ type, service _ name, lifetime, input _ list)`
   This function allows a service provider to register its service with a lookup server in the region. Services will remain in the lookup server for the lifetime specified by the service provider. In the `input _ list`, the service provider may supply one or more of the following service information: (1) location or address, (2) interface definition, or (3) mobile code. Lookup servers for different regions may coordinate with each other to update their list of service information.
5. `status = service _ deregister(service _ type, service _ name)`
   This function allows a service provider to remove its service from the lookup server registry.

21.6.6 API for Composition Services

Composition server is semi-decentralized and provides capabilities to facilitate cluster and group composition and management in a distributed environment. It allows sensors to form clusters and groups in line with their specific collaborative task requirements. The "group" concept is more generic than "cluster." Group denotes a collection of a number of sensors in any location. Cluster denotes an aggregation of a number of sensors located in a neighborhood. A cluster is a specific type of group. A group can be a logical set of sensors that contribute to a specific service, where group members may reside across several clusters.

The composition server maintains a database of all clusters/groups formed within a region. It is responsible for building groups, maintaining groups, and removing groups. The grouping service is provided to each node through a cluster service object. In each sensor node, sensor applications interact with a connector layer through a simpler composition service API that includes interfaces for creating (or removing) a group, joining (or leaving) a group, creating (or removing) a connector, and sending and receiving messages to and from a connector. The connector layer then communicates directly with the distributed composition server. The following is a description of the purposes and the side effects of the composition service API function calls:

1. `int createGroup(ObjectID * groupID, float x, float y, float radius)`
   The `createGroup()` function allows applications to create new groups. The API requires the application to provide the groupID and group region as parameters. The local connector layer will call the composition server to build a new group in its database. `groupID` is a pointer to an `ObjectID`. `groupID` consists of group type and group name pair. `x(y)` defines the `x(y)` coordinates of the center of the group region. radius specifies the radius of the group region.
2. `int deleteGroup(ObjectID * groupID)`
   An application deletes a group through calling the `deleteGroup()` interface. The composition server will not grant the request to erase the group unless there is no group member left within the group. `groupID` is a pointer to an `ObjectID` that consists of group type and group name pair.
3. `int joinGroup(ObjectID * groupID)`
   After a group is created, the application may join the group through the `joinGroup()` interface provided by the connector layer. The local connector layer will call the composition server to add the application into the group in its database. `groupID` is a pointer to an `ObjectID` that consists of group type and group name pair.
4. `int leaveGroup(ObjectID * groupID)`
   An application leaves a group through the `leaveGroup()` interface. The API also requires a group ID as the input parameter.
5. `int createConnector(ObjectID * remoteObj, int connectorType, int localRole, int remoteRole, int CSFIndex)`
   Applications can build an active connector to communicate to remote sensor objects through the `createConnector()` interface provided by the connector layer, which supplies the parameters to the composition server and notifies it to build a communication link between the sensor objects. The composition server configures all endpoints of a link by inserting an entry into each node's local connector table. The input/output parameters are as follows. `remoteObj` is a pointer to an `ObjectID`. `connectorType` defines the type of a connector. `localRole` specifies role attribute of local endpoint of the connector. `remoteRole` specifies role attribute of remote endpoint of the connector. `CSFIndex` (optional) specifies CSF index used by composition server to enforce composition constraints.
6. `int sendToConnector(int cid, NRAttrVec * input_list, NRAttrVec * output_list)`
   After building a connector, the applications may send data to the connector through the `sendToConnector()` interface provided by the connector layer. Depending on the communication modules, data will be transported either reliably or with best effort to the remote endpoints of the connector. The connector layer uses the communication modules and the local connector table to deliver application packets to the destination. If the supporting communication link is broken, the composition server will automatically repair it by dynamically changing the endpoints of the faulty physical connector to another functional sensor object that satisfies the constraints. All maintenance work is transparent to the applications. The `sendToConnector()` API is an overloaded function and may support either synchronous interface or asynchronous interface. The input/output parameters are as follows. cid specifies a connector handler. `input_list` is a pointer to a STL vector containing client's input data for the remote sensor object. Examples of communication style include RPC and Multicast. `output_list` is a pointer to a STL vector containing the feedback data coming from the remote sensor object.

7. `int recvFromConnector(int cid, NRAttrVec * output _ list)`
   Application may receive data from a connector through the `recvFromConnector()` interface provided by the connector layer. It is a synchronous interface, meaning the application will be blocked until data arrive. Based on the information of the local connector table, the connector layer receives packets from the other endpoints of the connector. The input/output parameters are as follows. cid specifies a connector handler. `output _ list` is a pointer to a STL vector containing incoming data.
8. `int deleteConnector(int cid)`
   The application deletes a connector using the `deleteConnector()` interface provided by the connector layer. The connector layer informs the composition server to update its local physical connector database. In addition, composition server will delete the corresponding connector entry from all participating endpoint objects' local connector tables. In other words, composition server reconfigures all endpoint sensor objects. cid specifies a virtual connector handler.

21.6.7 API for Adaptation Services

The client use the adaptation service APIs to send requests to the adaptation server. The client can register a new trigger, update an existing trigger, and remove a trigger using the following adaptation service APIs:

1. `Trigger _ ID Register _ Trigger(EventAttrVec *)`
   The `EventAttrVec` is an attribute vector, which contains several attributes that define the situation in which the trigger will be fired. For example, if we have this attribute in vector: "`FAULT _ RATE GE 30`", when the `FAULT _ RATE` reaches 30, this attribute may be matched by the attribute "`FAULT _ RATE IS 32`." The trigger thread will compare the events that are generated by the change detection function with the value of those attributes specified in the attribute vector and decide whether to activate the action or not. The vector also contains information for adaptation service, such as module name and group name. The return value of this API is `Trigger _ ID` if successful or –1 if failure occurs.

   When client calls `Register _ Trigger()`, it uploads the adaptation module that is specified by the user to the adaptation server. The adaptation server then starts an adaptation trigger thread, which will load the adaptation module that the client specified. The adaptation module contains the actual programs that define the operations for the change detection phase and the action phase. The adaptation server then executes the `change _ detection()` function of this module. If it detects some changes have occurred, the `change _ detection()` function issues an event using a message queue shared by the change detection thread and trigger thread. When the event that is received by trigger thread satisfies a trigger condition for an action, it will decide on the schedule for the adaptation based on the agreement policy and decide when to call the `action()` function to activate this adaptation transaction. The action may cause more changes that lead to other events, which may cause other triggers to be fired.
2. `Trigger _ ID Update _ trigger(Trigger _ ID, EventAttrVec *)`
   The attribute vector is similar to that used for `Register _ Trigger()`. This API updates the trigger with the new attributes set. It will return `Trigger _ ID` if successful or –1 if failure occurs. When the adaptation server receives this API call, it first stops the old trigger and kills the trigger thread. Then, it creates the new trigger based on the new `EventAttrVec` to replace the old trigger.
3. `int Remove _ trigger(Trigger _ ID)`
   This API removes the trigger from the adaptation server and kills the trigger thread. It returns 0, if successful and –1 if failure occurs.

21.7 Application Systems

The self-organizing sensor network architecture allows sensor information system designers to specify their own specialized protocols and services that are most appropriate and efficient for the specific application, although most sensor nodes may use generic protocols and services. Since these services may be application specific, different protocols for a certain type of service may be specified for different applications. Each of these sensor applications may establish their own services spontaneously and independently of each other. Various types of network services may be independently defined for different sensor nodes. In the following sections, we discuss some of these types of services.

21.7.1 Interoperable Mobile Object Model

The interoperable mobile object model is useful for retrieval of sensor information from sensor data sources to the surveillance application clients. The interoperable mobile object model extends the distributed interoperable object model (DIOM) [19] to sensor information networks. Application clients may access and update information sources in the sensor nodes through a group of mediators, object servers, sensor data repository managers, and sensor agents that communicate with each other using distributed services and diffusion routing (Figure 21.6).

An application end user poses queries to the database server that coordinates with the mediators to decompose, schedule, and route queries to the sensor information sources and collaborative signal processing modules. Mediators resolve the bindings of the sensor data sources through the object servers.

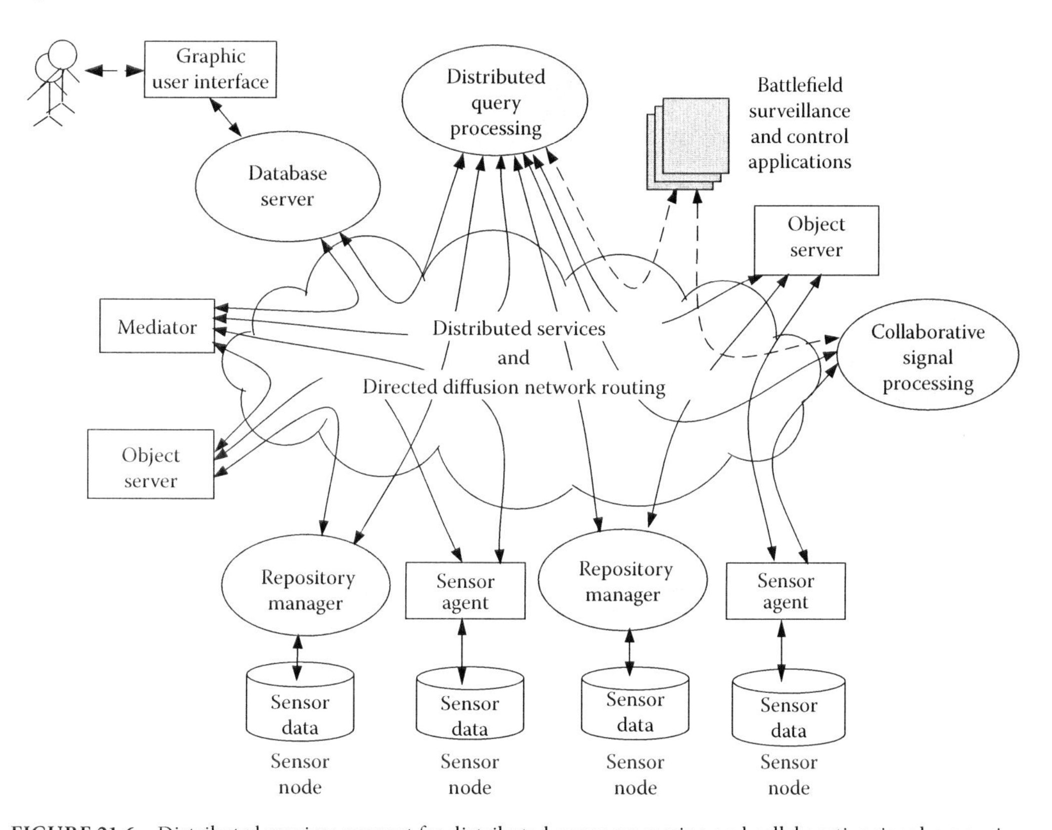

FIGURE 21.6 Distributed services support for distributed query processing and collaborative signal processing.

The binding maps the object unique identifier to sensor agents or repository manager at the sensor nodes specified by their location, data types, and application-specific information. Once the binding is resolved, mediators communicate with the sensor nodes directly. When sensor nodes move to new locations, routes may be changed through directed diffusion protocol if the change is incremental or through the distributed lookup service if the node moves quickly over a longer distance. Directed diffusion and the lookup services are responsible for solving the problem of mobility at the network routing and distributed service levels.

However, applications services such as mediators and object servers must also be aware of the mobility of sensor data sources and cache location binding information. Distributed lookup servers notify mediators and object servers of relocation of sensor nodes.

21.7.1.1 Mediators

Mediators cooperate with each other to implement dynamic query processing and mobile transaction in sensor networks. When a database server first receives a query, it may decompose it into multiple subqueries and forward them to selected mediators that may be associated with sensor object agents or collaborative signal processing agents for answering each subquery. Sensor agents (described in Section 21.7.1.3) provide accesses to their local sensor data sources. Collaborative signal processing agents (described in Section 21.7.2) provides more accurate tracking and surveillance data by combining readings from multiple sensors. Each mediator may dynamically determine and update its association with the sensor agents and signal processing agents in the region. Mediator may discover the sensor agents and signal processing agents in its region by calling `lookup _ service()` using a generic `service _ type` parameter to retrieve all agents that provide the type of service required to process the subquery.

The selection of the mediators for each subquery is based on database server's knowledge, stored in its cache, of the current location of the sensor data sources involved in the query. If the database server does not have location of the mediators, the database server may locate the selected mediators using the `lookup _ service()` call. It then calls `service _ exec()` to execute the subqueries in each selected mediator. Results of the subqueries are returned from these `service _ exec()` calls. The database server may alternatively use `service _ call()` that may automatically perform all the operations mentioned earlier by locating, executing the subqueries, and retrieving the results from each of the selected mediators. Mediators may be grouped into clusters using the composition service. Clusters may represent logical data partition and enable more efficient group communication to a cluster of mediators.

Dynamic query processing are performed within each mobile transaction. While the initial mediator is responsible for the overall transaction, the subsequent mediators are responsible for the subtransactions and their related locking and logging functions. Mediators are aware of mobility of sensor nodes through notification from the distributed lookup service. Sensor nodes that have moved to a new location may register, using `register _ service()` with another lookup server. The new lookup server will update the previous lookup server that sends a notification event to the mediator that is accessing the sensor node information. When sensor nodes have moved, mediators must reconfigure the routing and scheduling of its subqueries to current locations of the sensor nodes. In addition to mobility, mediators may reconfigure their subqueries when they detect disconnection and bandwidth variability.

21.7.1.2 Object Server

Object servers are primarily responsible for caching information on the location, availability, and attributes of sensor data sources. When a mediator is interested in retrieving particular sensor information, it can either inject an interest through the diffusion protocol or try to locate the sensor data source from an object server. The first option is preferred if the target sensors are within a certain range. The second option is preferred if the target sensors are located further away, in which case, the object server will return the location and application-specific attributes of the sensor data sources. Once the sensor data sources are located, the mediator communicates with them directly using directed diffusion.

Each mobile sensor node has a *home object server*. A large sensor network may contain many object servers, each responsible for a subset of sensor nodes. Each object server registers its service with a lookup server in its region using the `service _ register()` call. This enables other sensor nodes to locate and register its object information with the object server. The home object server caches the current location of sensor nodes whenever they move. If the sensor node moves to a new location in short distance, directed diffusion protocol will handle mobility and change in the network topology. Location updates at the home object server is unnecessary for mobility over a short distance. However, for mobility over a longer distance, sensors that have moved will try to update its location with its home object server. It first searches for the object server using the `lookup _ service()` call. Several lookup servers may be used to locate the object server. Then it updates its location with the object server using the `service _ exec()` call. Object servers keep track of location and application-specific information of objects, whereas lookup servers keep track of the location and system-level information of services.

In continual queries, data are continually sent from the sensor nodes to the mediators through the `service _ exec()` call. When sensor nodes move over a large distance, they will update its location with their home object servers through the distributed lookup services. They also update its information with the lookup server in the current region, which will propagate the information of all distributed lookup servers.

Changes in the sensor location will trigger the handover procedure in the `service _ exec()` call, which will redirect the distributed service call to retrieve data from the sensor new location. The update of the location cache in the object server will improve the performance of future queries.

If a sensor node's movement affects the performance of an ongoing continual query, the mediator may reevaluate the query decomposition. Alternate query route and schedule may be used for updates and queries to sensor nodes that have moved by tunneling subqueries directly to the foreign mediator. Early detection and notification of mobile sensor relocation by the lookup servers allows query routes and schedules to be updated to reflect current sensor node locations, thus improving the performance of distributed queries.

21.7.1.3 Sensor Agents

A sensor agent is a software module that serves one sensor data source. The sensor agent is built around the existing sensor data source to turn it into a local agent for the sensor object in order to make an existing sensor information source available to the network of mediators. Sensor agents enable different types of sensors to be deployed incrementally in the ad hoc sensor network. The local agent is responsible for accessing sensor information source and obtaining the required data for answering the query. Sensor agents are customized to integrate with applications methods used in the sensor nodes, such as autonomous identification and location management. Services provided by sensor agents may also include translating a subquery into a sensor information retrieval command or language expression, submitting the translated query to the target sensor information source, and packaging the subquery result into a mediator object.

The sensor agent is also responsible for local management for sensor data stored in the node. It contains the local locking mechanisms for the global concurrency control scheme and the local recovery mechanism based on write-ahead logs. Agents may convert data in raw format to interoperable sensor object format, turning each sensor into an agent for the interoperable mobile object.

Each sensor agent registers with a home object server and supply all its related application-specific object information, such as object identification and location. Other processes, such as mediators, may then locate the sensor objects through the object server. Sensor agent also register its service with the lookup server through the `service _ register()` call in order to handle sensor mobility and reconfiguration more efficiently at the distributed services' level.

When the sensor agent moves, it will inform the object server of its current location. However, it must first determine how to reach the object server through the `lookup _ service()` call and then updates the home object server through the `service _ exec()` call. The object servers may, in turn, notify the mediators of the change in sensor movement or connection variations. The mediators may determine whether to modify the subquery schedule.

21.7.2 Collaborative Signal Processing

Retrieval of sensor values from multiple sensors can increase the accuracy of the data in target recognition and tracking [23]. The multiple reading of the sensor values can be statistically combined to derive more accurate tracking data. Sensors in a region may be clustered together through the distributed composition server where their sensor reading may be combined using a weighted voting algorithm [24] to provide more accurate data. For each cluster, a sensor node may be elected to be the head of the cluster. Readings from multiple sensors in the cluster will be propagated to the cluster head through diffusion routing where the algorithm is applied. Results from each cluster head may be propagated to higher level cluster heads for data fusion.

In many sensor applications, data generated by the sensor node may be very large. The cost of propagation of these sensor data as in the previous scheme will be prohibitive. One solution to address this problem is to use mobile agents to migrate from node to node to perform data fusion using the local data [25]. Instead of transferring large among of data throughout the sensor network, this approach only transfers the mobile agent code that is smaller than the sensor data. The result is an improvement in the execution time of the collaborative signal processing algorithm.

A sensor node that needs to execute a mobile agent code must download the mobile code from a mobile code repository manager in its region. It may use the `lookup _ service()` call to first locate the repository manager that stores the relevant mobile code. It then calls `service _ exec()` to the repository manager to download the code. An alternative method is to store the mobile code with the lookup server.

The sensor node can then use `lookup _ service()` to retrieve the mobile code directly from the lookup server.

The entire surveillance area is split into several subareas. Each subarea is controlled by a signal processing agent, also known as processing element (PE), which may dispatch several mobile agents into that subarea. The signal processing agent may register itself with the lookup server using `service _ register()` to allow other nodes, such as the mediators, to access its tracking results. Each mobile agent will migrate from node to node to perform data fusion, for instance, using multiresolution data integration algorithm [25]. Each mobile agent, addressed by an identification number, contains an itinerary, data, method, and an interface. The identification is a two-tuple composed of the identification of the dispatcher and the serial number assigned by the dispatcher. The itinerary describes the migration route assigned by the dispatcher.

The data are the agent's private data that contains the integration results. The method describes the multiresolution data integration algorithm. The interface is the function by which the agent communicates with the PE and for the PE to access the agent's private data.

When mobile agents migrate from node to node, the results of the sensor integration algorithm from previous nodes are cached in the mobile agents. The state information must be transferred with the mobile agent as it migrates from node to node. Agent migration and state transfer are supported by the distributed adaptation service. As the mobile agents visit each node, it may register with the lookup server. The primary signal processing agent (PE) may find these mobile agents as they move around and retrieve results using the `service _ call()` function. A mobile agent may also retrieve intermediate results from other mobile agents through `service _ call()`.

For migration of mobile agents far beyond a region managed by a lookup server, the agent will register with another lookup server in the new region. A service client that try to contact the mobile agent through the first lookup server may need to use several intermediate lookup servers to get the information about the mobile agent current location.

Multiresolution signal processing algorithms may be implemented using hierarchical clustering of sensors provided by the distributed composition server. Results from a sensor cluster may be passed to agents responsible for signal processing for higher level clusters. Agents for higher level clusters may send messages to multiple sensor clusters using group communication method provided by the `service _ exec()` call by specifying the appropriate `service _ type` parameter.

21.7.3 Demonstration of the Distributed Services Support for Target Detection and Classification

Our distributed services support for target detection and classification was successfully demonstrated on the sensor network test bed at BBN (a Verizon company), Waltham, MA, at the SensIT PI meeting in November 6–8, 2002. The BBN test bed with 27 networked Sensoria WINS NG 2.0 sensor nodes distributed around a road (Figure 21.7). We used our distributed composition and lookup service to implement a system for collaborative signal and information processing (CSIP) that has been implemented over the BBN sensor network test bed. Our distributed services supported the execution of dynamic mobile agent and multisensor classifier (in collaboration with the University of Tennessee) and mobile code for classifier and tracker swapping (in collaboration with Penn State University). We also used Cornell University GUI for the live demonstration. Integration and distributed service support for Cougar distributed sensor database will be investigated in our future work.

Each Sensoria WINS NG 2.0 sensor node consists of two dipole antennas, a SH-4 main processor board, and a DSP coprocessor board with three sensor types (acoustic, seismic, and PIR sensors). The acoustic, seismic, and PIR time series data are analyzed at each node to produce target detection and classification information, which are then integrated with other sensors (using mobile agents) in a cluster for more accurate classification.

Retrieval of sensor values from multiple sensors can increase the accuracy of the data in target recognition and tracking. The multiple reading of the sensor values can be statistically combined to derive more accurate tracking data. Upon detection of a moving object in the vicinity, sensors in a region may be clustered dynamically together through the distributed composition server where their sensor reading may be combined using a weighted voting algorithm to provide more accurate data. For each cluster, a sensor node may be elected to be the head of the cluster. Readings from multiple sensors in the clusters are then integrated using mobile agents. In each cluster of five nodes, a mobile agent migrates to all the nodes to process the integration algorithm for classification. The itinerary of the mobile agent is computed using the `lookup _ service()` function, and the remote call to the remote mobile agent daemon to move a mobile agent is done through the `service _ exec()` function. An alternative is to propagate the multiple sensor data to the cluster head through directed diffusion routing where

FIGURE 21.7 BBN test bed with 27 Sensoria nodes.

the integration algorithm is applied. Results from each cluster head may be propagated to higher level cluster heads for data fusion and target tracking.

Figure 21.8 shows the results in which the detector and classifier correctly detected and identified moving walker and car targets. During the live demonstration, it was shown that the classifier results (car icon) popped up in the GUI about 2 or 3 s after the vehicles have passed each cluster. This was due to the time for the mobile agents to migrate to all the nodes in each cluster to aggregate more accurate results. Most of the migration time is due to the long delay of transmission through directed diffusion, although our distributed services help reduces the latency and network traffic. This delay time was an improvement over using the latest directed diffusion package because we used a new set of parameters from ISI/W that reduces the per-hop latency time from 2 s to about 100 ms.

In the stand-alone demonstration at the Westin Hotel Ballroom during the SensIT PI meeting, we demonstrated that classifier algorithms may be swapped dynamically using our distributed services when different objects are detected. Three types of classifier were used: UTK, PSU, and Wisconsin classifier. The appropriate classifier is chosen at runtime that will best classify the specific target correctly (classifiers performance are dependent on the type of target detected). We used dynamic linked libraries in our mechanism for swapping between different classifier codes at runtime in lieu of mobile code demon since the latter still does not execute reliably. We have been able to swap the classifier code using this method.

(a)

(b)

FIGURE 21.8 Results showing detection and identification of a walker and car in the BBN test bed: (a) detection of a *walker* by clusters of networked sensor nodes and (b) detection of a *car* by clusters of networked sensor nodes.

The aforementioned experiments show the utility of our distributed services in simplifying and improving the efficiency of executing distributed sensor fusion applications on top of diffusion networks. In particular, they show the utility of the lookup service, remote service execution, and composition service for running distributed sensor applications compared to running them directly over directed diffusion networks.

21.8 Conclusions

We have described how distributed sensor applications, such as sensor information retrieval and remote surveillance, can be supported by distributed services in self-organizing sensor networks. The three basic distributed servers are lookup servers, composition servers, and adaptation servers. Through these servers, sensor nodes may be placed together impromptu in spontaneous environments, and these sensor nodes will immediately know about the capabilities and functions of other sensor nodes and work together as a community system to perform cooperative tasks and networking functionalities. Newly deployed sensor nodes may provide new services. Other sensor nodes may locate and use these services spontaneously. These distributed services are implemented using directed diffusion network routing that provides energy-efficient and data-centric data communication. While diffusion routing can adapt dynamically to limited mobility and topological change, the distributed services supports large mobility and changes in sensor nodes. Diffusion uses data-centric communication model, whereas in distributed sensor applications, it may be more convenient to use end-to-end process-oriented communication. Sensor nodes that discover services provided by other sensor node may call these services through well-known interfaces, interpreted interface definition, or mobile codes downloaded from the lookup server. The benefits of using these distributed services is that application and system programs may use simpler communication interfaces and abstraction than the raw network communication interface and metaphor of the sensor network layer (e.g., subscribe/publish used in diffusion routing). Furthermore, these distributed services can be tuned to provide greater efficiency to improve the overall performance, such as throughput and delay.

References

1. Kahn, J.M. et al., Next century challenges: Mobile networking for smart dust, *ACM Mobicom*, New York, August 1999.
2. Pottie, G.J. and Kaiser, W.J., Wireless integrated network sensors, *Communications of the ACM*, 43(5), May 2000, pp. 51–58.
3. The ultra low power wireless sensors project. http://www-mtl.mit.edu/~jimg/project_top.html
4. Estrin, D. et al., Next century challenges: Scalable coordination in sensor networks, *ACM Mobicom*, Seattle, WA, August 15–19, 1999.
5. Esler, M. et al., Next century challenges: Data-centric networking for invisible Computing, *ACM Mobicom*, Seattle, WA, August 17–19, 1999.
6. Kulik, J. et al., Negotiation-based protocols for disseminating information in wireless sensor networks, *ACM Mobicom*, Seattle, WA, August 15–20, 1999.
7. Arnold, K. et al., *The Jini Specification*. Addison Wesley, Reading, MA, 1999.
8. Guttman, E., Service location protocol: Automatic discovery of IP network services, *IEEE Internet Computing*, 3(4), 71–80, 1999.
9. Specification of the bluetooth system. http://www.bluetooth.com/developer/specification/specification.asp
10. Agha, G., Adaptive middleware, *Communications of the ACM*, 45(6), 31–32, 2002.
11. Schmidt, D.C., Middleware for real-time and embedded system, *Communications of the ACM*, 43(5), May 2000, pp. 43–48.
12. Kon, F. et al., The case for reflective middleware, *Communications of the ACM*, 45(6), 33–38, 2002.

13. Quin, Liam, Extensible Markup Language (XML), http://www.w3.org/XML
14. Intanagonwiwat, C., Govindan, R., and Estrin, D., Directed diffusion: A scalable and robust communication paradigm for sensor networks, *ACM Mobicom*, Boston, MA, August 2000.
15. Poor, R., Hyphos: A self-organizing, wireless network, Master's thesis, MIT Media Laboratory, Cambridge, MA, June 1997.
16. Johnson, D.B. and Maltz, D.A., Dynamic source routing in ad-hoc wireless networks, In *Mobile Computing*, Imielinski, T. and Korth, H., Eds., Kluwer Academic Publishers, Dordrecht, the Netherlands, 1996, pp. 153–181.
17. Perkins, C.E. and Bhagwat, P., Routing over multi-hop wireless network of mobile computers, In *Mobile Computing*, Imielinski, T. and Korth, H., Eds., Kluwer Academic Publishers, Dordrecht, the Netherlands, 1996, pp. 183–206.
18. Echelon, The LonWorks Company. LonWorks Solutions. http://www.echelon.com/Solutions/
19. Liu L. and Pu, C., The distributed interoperable object model and its application to large-scale interoperable database systems, *ACM CIKM*, Baltimore, Maryland, November 28–December 2, 1995.
20. Bonnet, P., Gehrke, J., and Seshadri, P., Querying processing in a device database system, Technical Report Tr99-1775, Computer Science, Cornell University, Ithaca, NY, October 1999.
21. Lim, A., Architecture for autonomous decentralized control of large adaptive enterprises, *DARPA-JFACC Symposium on Advances in Enterprise Control*, San Diego, CA, November 1999.
22. Lim, A., Automatic analytical tools for reliability and dynamic adaptation of complex distributed systems, *Proceedings IEEE ICECCS*, Fort Lauderdale, FL, November 6–10, 1995.
23. Brooks, R.R., Chapter 26: Modern sensor networks, *CRC Handbook of Sensor Fusion*, CRC Press, Boca Raton, FL, 2002.
24. Saari, D.G., Geometry of voting: A unifying perspective, *Proceedings of Workshop on Foundations of Information/Decision Fusion with Applications to Engineering Problems*, DOE/ONR/NSF, Washington, DC, August 1996.
25. Qi, H., Iyengar S.S., and Chakrabarty, K., Distributed multi-resolution data integration using mobile agents, *Proceedings IEEE Aerospace Conference*, Big Sky, MT, March 10–17, 2001.

22

On-Demand Querying in Sensor Networks

Joon Ahn
IP Systems S&T
Ericsson, Inc.

Marco Zuniga
University of Duisburg-Essen

Bhaskar Krishnamachari
University of Southern California

22.1 Introduction

In the most abstract sense, a sensor network is a collection of nodes at which data are being produced (through sensing of the physical phenomena). For the network to be of practical use, at least some of these data or a processed version of it must be provided to at least one end user. The implementation of this information flow depends upon the application requirements as well as the resources available. In the simplest kinds of sensor networks, all the data generated may be sent continuously to a central data sink. This would naturally be resource intensive in terms of bandwidth as well as energy and may not be necessary if the end application requires only processed notification of events. Therefore, more sophisticated systems opt to route only selective information based on the in-network processing of one or more nodes' data, upon the issuance of a request (query) for this information.

In traditional networks, the primary task of the routing layer is to provide end-to-end connectivity so that arbitrary applications on two data terminals in the network may communicate with each other. Wireless sensor networks (WSNs) are different from traditional open networks in two crucial respects—the first is that they are often severely energy constrained, and there is a great need for optimizing protocols to minimize communication costs (which are often several orders of magnitude more expensive than computation). The second is that sensor networks tend to be significantly more application specific so that there is a much greater scope for optimizing protocols for a given application.

These features of WSNs argue for the implementation of cross-layer mechanisms in this domain that is very different from traditional networks. The end-to-end address-centric abstraction in which node IDs are the primary attributes for information flow can be replaced by data-centric techniques that allow data attributes and application information in a packet to be taken into account while deciding where to send it next, rather than relying on network addressing alone.

In these data-centric techniques, there are three general patterns of information flow, two of which are significant. The first is a well-formulated query for data emanating from a data consumer node (referred to as a sink). The second significant information flow pattern is the response information from one or more data-producing nodes (sources) potentially aggregated en route. The third (not always used) is an advertisement of available data from the sources, a sort of reverse-query looking to see if there are nodes interested in these data. When the second information flow pattern dominates, in terms of the amount of information being sent, the details of how the first (and possibly the third) pattern is implemented do not seriously affect network performance. This occurs primarily when the queries are for a continuous flow of information from the sources to the sink. In such a case, the query could be simply flooded through the network, with the cost of the flooded query being amortized over the much larger cost of the information flow it initiates. However, when the queries are for noncontinuous, instantaneous information, such a strategy could be very inefficient indeed. Such queries need to be handled more intelligently; or rather, they could be made intelligent. This is the basic premise behind the idea of on-demand querying that we describe in this chapter—query packets that move through the network, actively searching for the information, resolving themselves partially as they pick up information, and deciding, based on their contents and the state of the nodes they pass through, where they should head next and when they should terminate.

From a user perspective, it makes sense to treat the sensor network as essentially a distributed database (e.g., see Refs. [7,15] and the pertinent sections of this book). The user submits potentially complex queries for information from the network through a standardized query interface. The query is then disseminated and a response obtained, possibly after in-network aggregation/filtering such as those using operators (selects/joins). The distributed database perspective provides a unified way to handle different types of queries and, most importantly, allows for flexibility in how the information flow and routing patterns for different queries and their corresponding responses are implemented. On-demand queries fit into this vision because they will constitute a component of a portfolio of querying and routing techniques used to implement in-network processing of queries.

To clarify the design space for which on-demand queries are well suited, consider the following examples from different applications ranging from target tracking to habitat monitoring:

- "Where is target X currently located?"
- "Has there been rainfall in any part of the region in the past hour?"
- "Where is the chemical concentration the highest?"
- "Give k locations where species X's calls have been recorded"
- "Is the Boolean expression $(W \,\&\, X \,\&\, \overline{Y}) + (Z \,\&\, \overline{X})$ true?" (where W, X, Y, Z are binary conditions/subqueries such as temperature >50, humidity <0.5, etc.)

What is common to all these example queries is that they require a one-shot response based on current (or stored) data at the nodes. They may be complex (involving several subqueries), and they can be answered by replicated data within the network (there may be several nodes in the network that can provide a response to the same query).

This chapter focuses on a discussion of different mechanisms that can be used to implement on-demand querying in sensor networks. We begin with the simple ideas of flooding and random walk and describe how they are used for different on-demand querying techniques. We then discuss the possibility of sending on-demand queries on predetermined trajectories; improvement of query performance using reinforcement learning; and the use of geographic and sensor information to direct the query.

22.2 Flooding-Based Queries

Flooding is a simple unstructured querying scheme that needs no information at all about the existence and location (if it exists) of the target information. In this scheme, query packets are flooded to be delivered to every node in the network. Upon receiving the query packet, each node retransmits the packet to its neighboring nodes through broadcast, if the node does not have information that the query desires. If it does, the node sends the query answer to the querier.

Flooding is a basic operation and has many applications in resource discovery in wireless networks. Examples include route discovery in several routing protocols of wireless networks [20,30], sensor discovery in WSNs [19], and service discovery in wireless ad hoc networks [36].

With flooding it is guaranteed to find the target information (if it exists) and the shortest path to it if there is no loss of packets. It can also support very dynamic networks such as networks with high nodal mobility. On the other hand, it can waste a great deal of resources such as bandwidth and energy. When the network is dense, it can cause high contention and interference that, in turn, may cause excessive retransmissions eating up the precious bandwidth of wireless communication. Even if the target node is nearby, the query is flooded to the network because there is no easy way to stop the flood once it is initiated, which makes every node pay the transmission energy cost. This makes the communication cost always $O(n)$ regardless of the target location. However, the cost of flooding in querying can be justified when its response is a long-term continuous flow of packets, amortizing the cost over the response flow.

Ganesan et al. [13] presented a seminal experimental study of basic flooding on real testbed WSNs. Although it is simple from the perspective of protocol design, the flooding has been found in this study to exhibit surprising complexity in real settings that previous theoretical studies using circular connectivity models have failed to capture. The complexity is contributed by interactions between different parts of the system such as irregular changes of radio propagation at very low power levels, typical nonidealities in low-cost hardwares, radio processing, and protocol choices. The experiments have revealed that *long links* (over which a message directly reaches a node at an unusually long distance) are not uncommon and that the long links cause *backward links* (over which flooding packets go backward geographically toward the flood source). The authors have performed the link, MAC, and application layer analyses to identify long links and backward links as causes of two phenomena: (1) the wide variance in the distribution of node hop distance versus physical distance and (2) large coverage of some nodes in broadcasting. The large coverage enables query flooding to reach most nodes with fewer hops but may make the reverse paths fragile causing unreliable response delivery.

Flooding may be controlled by limiting the number of hops so that the flooding packet will not be transmitted beyond the maximum number of hop counts, or restricting the flooding distance (when the location information is available to nodes) so that the query is contained in a certain geographical area.

Johnson et al. [21] consider *expanding ring* mechanism for their famous DSR protocol for the target discovery while enhancing the resource inefficiency of flooding. Expanding ring search (ERS) mechanism issues a sequence of controlled flooding from a querier. The radius of flood (usually expressed as the maximum hop count from the issuer) is restricted to cover only a part of network in the controlled flooding. In ERS, the radius is increasing after each round of floods, if the previous controlled flood fails to resolve the query (in a desired time) as illustrated in Figure 22.1. This scheme is widely used in ad hoc routing protocols [11] as well as wired networks [5].

Any ERS can be characterized as a vector $u = \{u_1, u_2, \ldots, u_m\}$ that describes the sequence of successive time-to-live (TTL) values for controlled flooding in each step. For example, let $u = \{1, 5, 10\}$ for a network where the maximum hop count is 10. Then the ERS would proceed as follows: First the nodes within one hop are searched for the event through a controlled flood with TTL value of 1. If no copies of the event are located in this first step, then all nodes within five hops are searched for the event through a larger controlled flood. If still no copies of the event are located in the second step, then all nodes in the

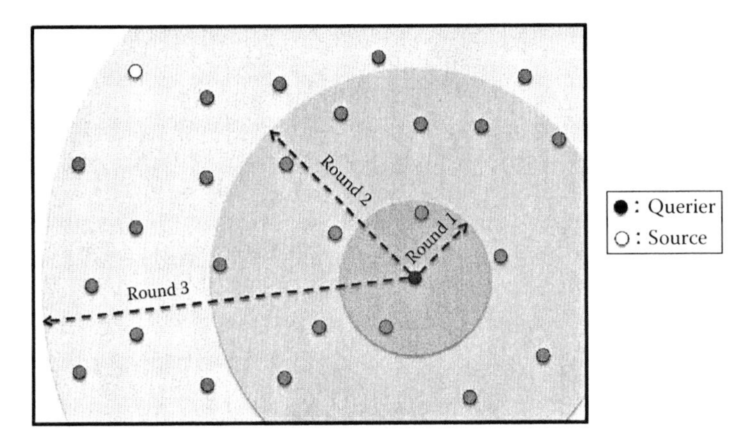

FIGURE 22.1 Illustration of ERS with three rounds.

network (within 10 hops) are searched. If at any step at least one copy of the event is located, the search terminates successfully at that step.

Because ERS can stop at a certain round without covering the entire network, it can save the network resources especially when the target is close to the querier. However, it may cause more waste if the sequence vector is poorly configured and the target is located far from the querier. It is because the nodes near the querier have to be covered multiple times as ERS progresses.

Cheng and Heinzelman [11] have shown that two-ring and three-ring schemes can reduce the search cost compared to a single attempt of pure flooding and have provided a general formula to determine good parameters for the two-ring and three-ring hop-based ERS schemes. They also have conducted a simulation study of ERS and claimed that it can obtain up to 10% energy saving without data replication, compared to the pure flooding, while the delay increases significantly. However, the optimum sequence vector is not identified and utilized in their study. Chang and Liu [10] have developed a dynamic programming solution to obtain the optimal TTL sequence vector that minimizes the expected search cost in terms of transmission number when the event spatial distribution is known a priori. They have found that the optimum sequence vector can be determined through the following dynamic programming that can be solved recursively for $0 \le k \le L-1$:

$$V(L) = 0 \tag{22.1a}$$

$$V(k) = \min_{k+1 \le l \le L} \{C_l + F(l \mid k)V(l)\} \tag{22.1b}$$

where

L is the radius (in terms of hop counts) of the network

C_l is the cost of controlled flooding up to l hops

$F(l|k)$ is the conditional tail distribution of the target given that the most recently used TTL value k has failed to locate the target

However, from their solution, it is not straightforward to obtain closed-form expressions for the optimum cost that may give intuition on the relationship between system parameters.

Intuitively, the performance of a TTL-based ERS improves with additional replicas. When there are more randomly placed replicas in the network, the likelihood that the event being searched for is located within a smaller number of steps, close to the sink, becomes higher. However, this reduction in the expected search cost comes at the expense of an increased communication cost for replication.

Krishnamachari and Ahn [24] have derived the closed-form approximation for the expected communication cost of ERS when there are multiple copies of the target information in the network. Using double bounds, they have found that the cost with r copies of the target information is proportional to $N/(r + 1)$ when the copies are spread uniformly in the network of N nodes. They have also considered the problem of optimizing the number of target copies minimizing the expected total communication cost of search and replication. Their finding is that the optimum cost can be obtained by the replication strategy where events are replicated with a frequency proportional to the square root of their query rates.

22.3 On-Demand Querying with Random Walks

Instead of flooding-based queries, we can build intelligent on-demand queries that refer to packets that contain the queries and proceed through the network moving from node to node in order to find responses to the queries. The simplest implementation of such a query is the execution of a random walk, in which the query packet is forwarded to a random neighboring node at each step.

This simple random walk requires no global IDs in the network, and little state at each node—at most, if no other routing mechanism is available in the network, each intermediate node may need to store the query ID and the ID of the neighbor from which it was received in order to route the query response back to the sink. If the query consists of several subqueries, random walks can also permit the partial resolution of the query as it moves through the network, picking up pieces of the information along the way.

The principal advantage of a random walk search as opposed to flooding the query is that—depending on the scenario—it may require significantly fewer transmissions. It should be noted that these savings in energy come at the expense of greater response latency, which must be acceptable for random walks to make sense. This advantage has motivated the use of random-walk-based search techniques in unstructured WSNs.

We shall discuss different variants of random-walk-based querying in detail, identifying some of the drawbacks of the basic random walk and how these limitations are solved through different techniques.

22.3.1 Search in Unstructured Peer-to-Peer Networks

Although there have been some proposals for structured data placement in sensor networks (e.g., using geographic hash tables [32] or hierarchical multiresolution structures [14]), most envisioned architectures for sensor networks are unstructured. This unstructured characteristic is also present on other peer-to-peer networks such as Gnutella, and thus, the search techniques studied in UP2P are pertinent to this discussion.

In Ref. [27], the authors consider the problem of searching for named resources in unstructured peer-to-peer networks. In these networks, there is no central directory where the information is stored, and there is no special topology or structured placement of data to make it easier to locate. This scenario is analogous to a one-shot query in sensor networks. Lv et al. [27] note that random walks greatly reduce messaging cost, but at the expense of greater delay. This delay depends on the degree of replication of the data being requested. For example, if there is only one node in the network containing the resource or file being searched for, the random walk may take unreasonably long to find the resource.

In order to reduce the delay of random walks in UP2P, the study evaluates two techniques. First, the study considers different replication strategies to decrease the delay (as we shall see, the Rumor Routing technique for sensor network also effectively performs replication in order to reduce the delay response of random walks). Second, the study examines the impact of increasing the number of simultaneous random walkers. As Lv et al. [27] point out "the expectation is that k walkers after T steps should reach roughly the same number of nodes as 1 walker in kT steps. Therefore by using k walkers we expect to cut the delay down by a factor of k." While this potentially solves the problem, it raises another—when should the multiple queries be terminated if they do not succeed? There are two solutions: one is to define a strict lifetime (TTL field) after which they terminate automatically, and the other is to check back periodically with the originating node to see if any of the other walks have terminated successfully.

While the latter solution is found to be well suited for Internet-based UP2P networks, this kind of explicit coordination will result in additional communication overhead that should be avoided as far as possible in sensor networks because of its energy costs.

22.3.2 Trade-Off between Flooding and Random Walks (ACQUIRE)

In Ref. [27], we have seen that random-walk-based on-demand queries can significantly reduce the communication overhead of query flooding. But it is certainly worth asking if this is always the case—are there conditions where something like flooding can be beneficial? Is there some strategy between the two extremes of flooding and carrying out a random walk that makes sense under different conditions?

Query flooding, as discussed earlier, makes sense when the querying cost can be amortized over the data that are being routed back to the sink, that is, for long-standing flows. However, even if we restrict ourselves to one-shot queries, flooding can be a reasonable solution if the queries are repeated, and data can be cached on the nodes. In other words, the first time a query is launched, it is flooded and the response for a query is cached. If there are additional queries for the same content, it can be answered directly from the sink's cache, instead of triggering a search within the network. This, of course, assumes that the cached data are still valid, highlighting the fact that the querying strategy might depend on the ratio of the validity time of cached data and the time between queries. These issues are explored in Ref. [33], which proposes a novel technique for on-demand querying known as ACQUIRE (which stands of ACtive QUerying In sensoR nEtworks).

The on-demand query in ACQUIRE proceeds as follows: the querying node first checks its cache and, if necessary, requests an update from all nodes that are within d hops. It uses the obtained information to partially resolve the query if possible and, if it still remains unsolved, forwards it to a random node that is another d hop away. The node receiving the forwarded query (referred to as the active node) then checks its own cache, does a d-hop look-ahead update, and forwards if necessary. The three-step process (examine cache, request update through local d-hop flood, and forward query) is repeated until the query is fully resolved. The most important point to note about ACQUIRE is that the d-hop look ahead allows it to be tuned flexibly. When d is 0, we have a pure random walk strategy; when d is comparable to the diameter of the network, it is the same as a query flood. There is a trade-off for different values of the look-ahead parameter; when the value of d is small, the query needs to be forwarded more often, but there are fewer update messages at each step. When d is large, fewer forwarding steps are involved, but there are more update messages at each step. Given this trade-off, we may inquire what determines the optimal choice of d in terms of minimizing the total number of transmissions in the network.

Reference [33] presents a mathematical model for the transmission cost of ACQUIRE as a function of d as well as the parameter c, which represents the expected number of updates per query. If $c = 0.01$, for example, an update is required at an active node only once in 100 queries (i.e., the requested information can be answered from the cache); if $c = 1$, caching is useless as updates are required every query.

Let σ be the expected minimum number of random nodes that must be queried in order to resolve a query. Then it is shown in Ref. [33] that E_{avg}, the expected number of transmission required to query for the information and obtain a response back to the sink using ACQUIRE with a d-hop look-ahead, is given as

$$E_{avg} = \begin{cases} \sigma\left[f(d)\right]^{-1}\left[c\big(f(d)+g(d)\big)+2d\right], & d \geq 1 \\ 2\sigma, & d = 0 \end{cases} \tag{22.2}$$

where

$f(d)$ is the expected number of nodes within d hops of an active node

$g(d)$ is the expected number of messages required for all nodes within d hops to respond to a query from an active node

It turns out that, depending on the value of c, E_{avg} either always increases with respect to d (in which case $d = 0$, random walk is the best strategy) or decreases first and then increases with respect to d (in which case some intermediate value of d represents the best strategy) or decreases with respect to d (in which case a large d corresponding to flooding is the best strategy). As expected, when the c is high and close to 1, that is, when data dynamics is high and caching is of no benefit, random walk is the best choice. When c is very low, close to 0, cached data remain valid for a longer time and flooding can be beneficial.

22.3.3 Several Walks Are Better than One (Rumor Routing)

Rumor Routing [8] is a technique for sensor networks that is similar in spirit to Ref. [27]. The queries are assumed to be of the type "Has event X occurred anywhere?" that is, for a specified event in the network at an unknown location. Rumor Routing includes two pieces—the creation of event paths (which point to the location where the event occurred) and on-demand queries (which search for event paths). Figure 22.2 shows a simple schematic illustration of this process. An event node (white node), which wishes to advertise an event, sends out event agents, which create a routing entry in the nodes they pass through leaving a pointer to the event. The entry may simply consist of the previous node that the agent visited on its way from the event. The event agents execute a random walk through the network. Later, when a sink (black node) wishes to query for that event, it sends out its own query, also as a random walk. When the query encounters an event path, it is routed along that path to the event. At this point, the requested information about the event could be routed back to the querying node (e.g., along the reverse path). Because of the wireless channel, event agents can be overheard by neighboring nodes, thus leaving a thicker trail. The event agents may be generated deterministically, or probabilistically, depending on whether one or more nodes detect an event.

Tunable parameters for Rumor Routing include the average number of agents generated for a given event. To minimize looping, both event agents and on-demand queries in the Rumor Routing scheme use TTL fields, but also recently visited nodes keep track of the agent and query IDs so that the agents and the queries can avoid visiting them. In simulation studies, Braginsky and Estrin [8] found that for a range of query-to-event ratios, Rumor Routing significantly outperforms query flooding (which is useful when there are many events and few queries) as well as event flooding (which is useful when there are many queries for few events)—the two basic schemes underlying the pull and push forms of directed diffusion.

For one-shot querying, the nodes along the event paths may be given summarized information about the event. Thus, one way to view Rumor Routing is that it improves the response time for random walk by replicating the sought information. The authors mention that Monte Carlo simulations show that the probability of two randomly placed lines intersecting in an arbitrary rectangular region is as high as 69%—this translates to a greater than 99% probability that five random lines

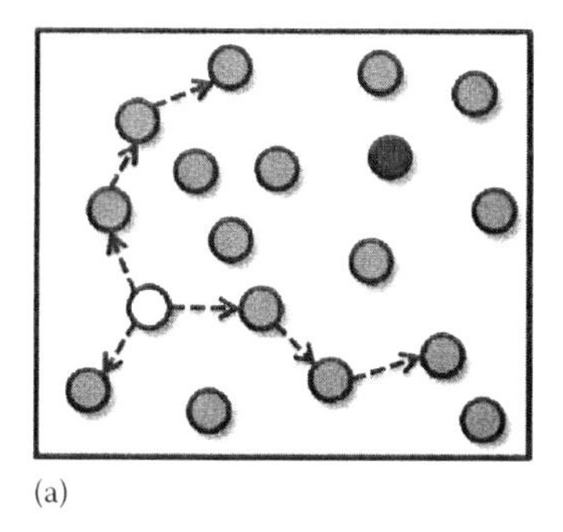
(a)

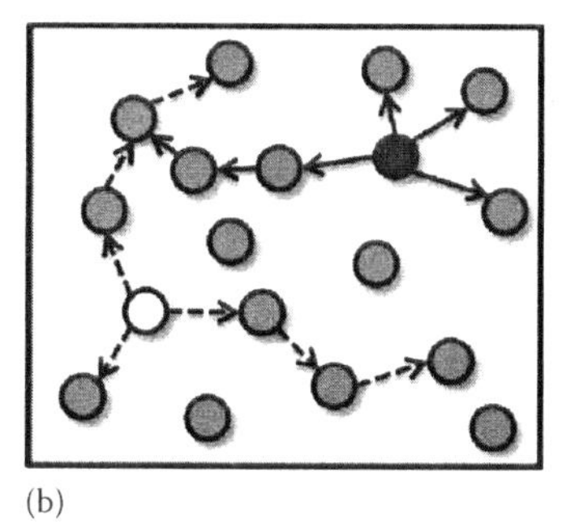
(b)

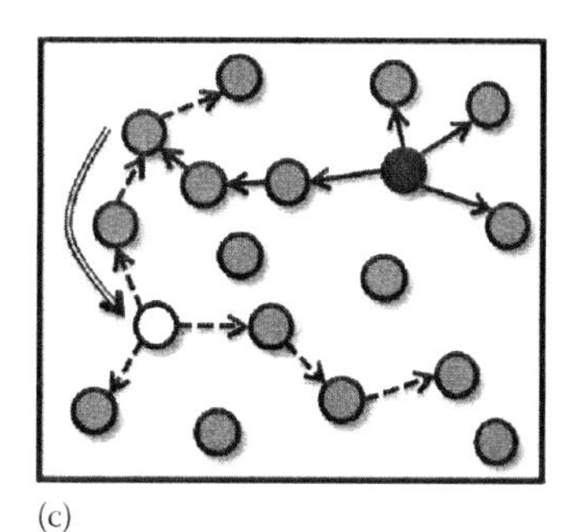
(c)

FIGURE 22.2 Schematic illustration of Rumor Routing: (a) agents creating event paths, (b) on-demand queries from sink, and (c) on-demand query routed to pertinent node after intersection with event path.

(corresponding to event paths) are intersected by a random line (corresponding to the on-demand query). This highlights the efficiency of using event paths to improve the performance of random-walk-based queries.

22.3.4 Importance of Data Replication

Rumor Routing conveys an important message. The performance of random-walk-based queries can be improved significantly when the sink and event nodes initiate random walks. But, how much can the network improve by following this approach? In Ref. [34], the author provides some insightful theoretical results that validate Rumor Routing and quantify the importance of using sticky searches together with data replication.

In Ref. [34], the author studies three different querying scenarios: (1) sink-only querying, where only the sink can initiate random walks that try to locate the source node, (2) sticky search, where both the sink and source initiate random walks, (same as Rumor Routing). The source node leaves a sticky trail with cached data along the path, and the query is successful when both walks intersect, and (3) spatially cached information, where the data of interest are uniformly replicated over the network and the query is successful if the random walk initiated by the sink locates any of the caches.

Clearly, scenarios (2) and (3) are expected to perform better because they use data caching, which is not used in scenario (1), but is the extra memory requirement justified? The main contribution of Ref. [34] is a remarkable answer to this question: "no matter how many finite number of source queries are used, we cannot match query techniques which utilize memory that is spatially distributed over the network." In other words, a sink-only approach can issue as many queries as it possibly can, but using data caching, as in scenarios (2) and (3), would always outperform the sink-only approach.

The analysis presented in Ref. [34] is framed within the following context. The sink node is positioned in the middle of the network area, and the event node is located at a normalized distance of 1 (i.e., close to the border of the network). The random walks have a (exponentially distributed) random lifetime with average t, if the information is not located within the lifetime of the walk, the query times out, and the search is unsuccessful. For sink-only search, the authors show that the probability that a query is unsuccessful decays as $\log(t)^{-1}$. When k independent queries are used, the probability decays as $\log(t)^{-k}$. When sticky search is used, the probability that a query is unsuccessful decays at least as $t^{-5k/8}$, where k is the number of walks sent by the sink and event nodes (k walks each). Finally, when a spatially periodic cache is used, the probability that a query is unsuccessful decays no faster than t^{-1}. Hence, as the authors point out, the use of memory on the network, on the sticky search and spatial caching scenarios, "enables the decay probability to change from logarithmic decay to polynomial decay."

Both caching strategies have polynomial time decay. However, the authors show that the memory requirement for periodic spatial caching is larger. Furthermore, spatial caching would require either a two-tier infrastructure or a high-degree cooperation (to disseminate data among the cache nodes), and either option would utilize more resources and add complexity to the operation of the network. Hence, the authors suggest that the appropriate strategy for querying large sensor networks would be a "sticky" search utilizing multiple walks.

22.3.5 Visiting a Fraction of Nodes May Be Sufficient

Random-walk-based queries can perform poorly if the data of interest are located on a single node and there is no data replication. In this scenario, the walk may need to visit all nodes (because the data are on the last unvisited node), and hence, in the worst-case scenario, random walks need to cover the entire graph. This metric is known as the cover time, and it is known to be of order $O(n\log^2(n))$ for a grid. However, the cost of finding unvisited nodes increases with time, that is, most of the time (and energy) of the query is spent on finding the last unvisited nodes. Considering this unbalanced effort, Avin and Brito [3] investigate the cost of covering a fraction of the network (partial cover time).

Starting from Matthew's theorem, which provides upper and lower bounds for the cover time, the authors prove that if instead of covering the entire network, the query needs to cover only a constant fraction of the graph, then Matthew's bound is reduced by an order of $\log(n)$. As stated by Avin and Brito [3], this result indicates that "on sufficiently large graphs, almost all the time used by a walk to cover the entire graph is spent trying to reach the last $\log(n)$ nodes." It is important to remark that the result is valid for any constant fraction c, for example, c could represent 50%, 80%, or 99% of the nodes in the network.

The authors show that visiting a constant fraction of a sensor network is sufficient for answering some interesting queries. For example, the study considers a scenario with four light sources and several thousands of nodes covering a square area. The aim is to generate a histogram of the light intensity. The authors show that covering 80% of the network leads to expected errors as low as 0.37% when compared with a complete coverage. The authors point out that this low error is due to two reasons. First, most physical events such as light, temperature, and humidity have a strong correlation in space, and hence, the information of neighboring nodes is similar. Second, in order to take advantage of spatial correlation, the query has to perform a uniform sampling of the network. The authors found that almost every unvisited node is within one or two hops from a visited node, which indicates an important balancing property of partial cover times; the query does not leave large parts of the network uncovered.

Another important contribution of this study is the concept of biased random walks. A simple random walk is likely to visit nodes that have already been visited, but in sensor network querying, there is no major incentive to do that, why not favor visiting unvisited nodes over visited nodes? The authors define a bias parameter p. Let d be the number of neighbors of the current node, and let d_u be the number of unvisited neighbors. Then instead of selecting any neighbor at random, an unvisited neighbor is selected with probability. A visited neighbor is selected with probability $(1 - p)/d$, and an unvisited neighbor is selected with probability $(1 - p)/d + p/d_u$. The authors show that this simple biased mechanism provides substantial improvements. In the next part of this chapter, we discuss in more detailed various random walk querying techniques with biased mechanisms.

22.3.6 It Is Important to Leverage Degree Heterogeneity

The literature presented in this section provides much insight into the scaling behavior of random walks on deterministic graphs (such as grids). However, regular graphs do not capture a major property of real-life networks: degree heterogeneity. Degree heterogeneity refers to the fact that in actual sensor network deployments, nodes do not have the same number of neighbors; some nodes have more neighbors than others. As Zuniga et al. [39] point out, degree heterogeneity is an important characteristic because simple random walks have a stationary distribution equal to $\pi(v) = d(v)/2m$, where $d(v)$ denotes the degree of node v and m the number of edges in the graph. In other words, not all nodes have the same probability of being visited by a random walk; nodes with higher degree are visited more frequently. There are two important reasons for the presence of degree heterogeneity in WSNs. First, random deployments are inherently nonregular graphs. Second, empirical studies [37] have revealed that hardware variance on the sensitivity and output power of radios lead to nodes with significantly higher (or lower) degree than the average.

In Ref. [39], the study leverages the presence of high-degree nodes. Considering that high-degree nodes have a higher likelihood of being visited, they are used as rendezvous points (local cluster heads). The main idea is intuitively simple, and it is based on the sticky search principle of Rumor Routing (where both sink and event nodes initiate random walks). Event nodes initiate pseudorandom walks, where at each step a node selects the neighbor with the highest degree. Upon reaching a local maximum, the event walk stops. The sink initiates a simple random walk. However, considering that the data have been copied to a node with a higher degree, there is a higher likelihood to visit the rendezvous node, and hence, solve the query earlier.

The main contribution of the authors in Ref. [39] is to show that "having even a small degree of heterogeneity can provide significant improvements in query performance." Using connections between random walks and electrical resistance [9], the authors derive an striking result for linear topologies: when cluster heads have a coverage k (that is, cluster heads cover k nodes to the right and k nodes to the left) and are uniformly distributed, a fraction of 4/5 k nodes being cluster heads can offer a reduction in query cost of $O(1 - 1/k^2)$. In intuitive terms, this translates to requiring less than 10% of the nodes being cluster heads to obtain 2 orders of magnitude improvement in query cost. For realistic two-dimensional (2D) topologies, the authors utilize absorption states techniques on Markov chains, and their numerical analysis indicates that "a small percentage of nodes being cluster heads (>10%) can lead to significant improvements in performance (between 30% and 70% depending on the coverage of the high-degree nodes)."

22.3.7 Querying Sensor Networks with Biased Random Walks (the Power of Choice)

An important limitation of most studies on random-walk querying is that some nodes are revisited all too frequently. In a practical setting, there is no incentive in revisiting a node that has no valid information. In Ref. [4], the authors propose a mechanism to alleviate the revisiting problem. The idea behind the proposed mechanism is simple: instead of selecting a node at random, poll d nodes ($d > 2$) and select the node with the minimum number of visits as the next hop.

The authors evaluate the performance of random walks with choice on networks with a square grid topology. The simulation results indicate some interesting results. First, even when two neighbors are selected instead of one, there is an unbounded improvement in the cover time (the time taken to visit all nodes in the network). Furthermore, in mesh networks with n nodes, which are known to have a cover time of $O(n\log^2(n))$ with simple random walks, the simulations indicate that the cover time is lower than $O(n\log(n))$. Second, the variance of the cover time is also reduced. This prevents rare cases when queries from taking a very long time. Third, random walk has positive effects on load balancing, which is a desired characteristic to increase the lifetime of sensor networks.

It is important to remark that random walk with choice consumes a little bit more memory and energy at each step (because it needs to keep track of visits at each node and requires some extra communication—depending on d). However, in practical settings, the inherent broadcast characteristic of wireless networks can provide this information by default. The practical evaluation of random walks is reviewed later on this chapter.

22.3.8 Testbed Evaluations

Most of the work on random-walk-based querying is either theoretical or based on simulations. While these studies provide important insights, they do not account faithfully for the nonidealities faced by the network in real deployments. In Ref. [1], the authors implement a three-way handshake protocol as a reliable implementation of a random walk and compare its performance with flooding in a 64-network test bed. The metrics considered are delay, reliability and transmission cost. The authors report a new insight into the behavior of practical random walks on sensor networks: flooding may be better suited for low-interference environments, while random walks might be a better option in networks with high interference. This result validates to some extent the claim made about the robustness of random walks (because it has to maintain the token alive and because it can adapt to temporal dynamics). The study also validates the high-degree heterogeneity of sensor nodes and their strong correlation with the number of visits that they are expected to receive (as mentioned before, this phenomena was leveraged in Ref. [37]).

In Ref. [38], the authors evaluate the querying performance of a strongly biased random walk. The walk is also based on a reliable three-way handshake mechanism to guarantee the survivability of

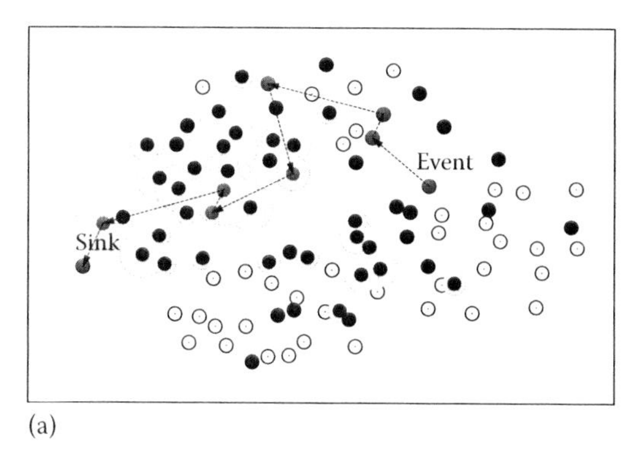

(a)

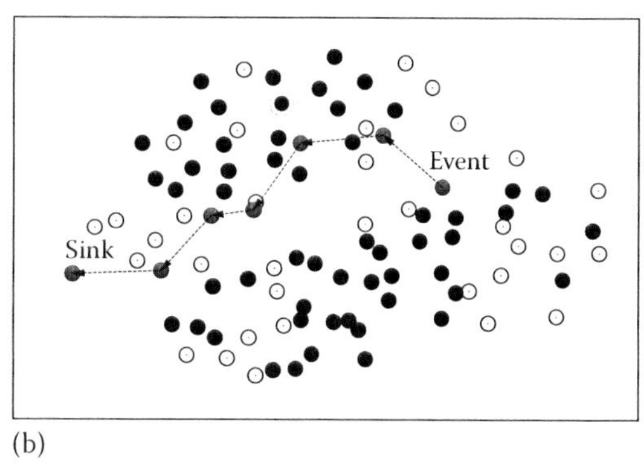

(b)

FIGURE 22.3 Testbed results for random walks: (a) simple random walks and (b) strongly biased walks. The black nodes represent visited nodes, the white nodes represent unvisited nodes, and the size of the gray circle around nodes captures the number of visits (the bigger the circle, the more revisits). In this example, the simple walk took 109 steps to reach the sink and covered 56 nodes, and the biased took 63 steps and covered 62 nodes. Not only biased random walks cover more with less energy and time (steps) but also have a more uniform load balance.

the walk. Nodes keep track of the number of visits, and at each step, the walk selects the nodes with the least number of visits (which could be zero); in case of tie, the node is selected at random. The authors evaluated this strongly biased walk on a test bed with 102 nodes (TWIST [16]) and through simulations utilizing realistic deployments. A comparison of the testbed results between simple and strongly biased random walks are presented in Figure 22.3a and b. The testbed and simulation results show an interesting result: the partial cover time (up to approximately 0.6 n, where n is the number of nodes) has order $O(n)$; this indicates a reduction of $O(\log(n))$ with respect to the bound found on theoretical studies. Furthermore, the results indicate that query utilizing a sticky search mechanism can be solved in $O(n)$ because both walks have a high probability of covering more than half of the network in linear time. Another important result of strongly biased walks is that the load balance is linear in time, that is, considering that a walk has performed s steps, each node has a bounded number of visits in the range $\left[\frac{s}{n}-c, \frac{s}{n}+c\right]$, where c is a constant.

22.4 On-Demand Queries with Direction

While random walks are a simple choice for the on-demand querying, they are essentially a form of blind search and can be improved upon by incorporating additional knowledge. In some cases, it may be desirable to send the queries on a predetermined trajectory, which can be done through source routing or using the routing on curves technique [29]. In some cases, it may be possible to route the queries to the locations where they are most likely to be answered based on the past history of similar queries, as exemplified by the LEQS technique [25]. In yet other scenarios, it would be helpful to exploit the query semantics and the spatial correlation of information available in network nodes to direct the query to its intended location, as exemplified by geographic forwarding techniques [28] and IDSQ/CADR [12]. There are also cases where it makes sense for sources of information to advertise their possessions following certain trajectories as well [26]. We shall now describe these techniques in turn.

22.4.1 Source Routing and Routing on Curves

In some situations, it will be desirable to send the on-demand queries through the network to sweep through a particular set of nodes, to undertake a prespecified trajectory. In traditional networks, one technique for doing this is source routing [35], in which the sequence of nodes to be visited is included in the packet header. In sensor networks, particularly those where fine-grained localization is available, a new technique has been proposed in Ref. [29] to send queries on geographically specified trajectories. With this technique, called routing on curves, an arbitrary spatial trajectory is described in the on-demand query packet, and at each step, the node that receives this packet forwards it to the neighbor whose location most closely matches the described trajectory. For example, this trajectory could be a straight line (useful, e.g., in Rumor Routing), and each node that receives the packet would forward it in the desired direction to the neighbor, which is closest to the described line. In order to effectively forward packets along the given trajectory, the sufficient nodal density and the capability by which nodes can position themselves relative to a coordinate system and estimate distances to their neighbors are required. Assuming such fine-grained geolocation information and density of nodes, this technique can approximate arbitrary trajectories very closely.

22.4.2 LEQS

Consider one-shot queries for the location of an identifiable object in the sensor field, an object that has a regular pattern of the location that can be described with a probability distribution. For example, this could be a target that is known to be always near one of three sensor locations with corresponding probabilities p_1, p_2, or p_3. If this distribution is stationary, there is the possibility of improving the efficiency of repeated queries for this object through learning. One mechanism that has been proposed to incorporating learning to improve the query efficiency is LEQS (learning-based efficient querying of sensor networks) [25].

LEQS is a simple localized and distributed learning algorithm for on-demand querying. In this algorithm, sensor nodes maintain weights indicating the probability with which a given query is forwarded to each neighbor. The query response is used to update these weights on the reverse path, effectively *training* the network to locate the object more and more efficiently over time.

Upon node deployment and setup, each sensor node identifies its immediate neighbors and sets up a vector of weights (one for each identifiable queried object) in a querying table. The weight represents the probability that a query for a particular object will be forwarded to a particular neighboring node. Initially, each neighbor of the node is assigned an equal weight. Each query starts from the sink and, with the probabilities denoted by the weights at each node, will be forwarded randomly from node to node until the target object is located. Thus, initially, when the node weights are all equal, this is equivalent to the unbiased random walk. A backtracking technique is incorporated to prevent looping. Once the object is located, the response of the query is then sent back directly to the sink on the reverse path of this query (using local state maintained at each node about where it received the query from) and this is when the weights are updated. The query response on the return path contains a counter that is incremented hop by hop so that all nodes on the reverse path get an indication of the number of hops that they were from where the query terminated successfully. This information is used in the weight update rule.

Over time (repeated queries), this learning policy in weight update trains the network to efficiently forward the queries toward locations where they are most likely to be resolved. While a low learning rate results in slow convergence, a learning rate that is too high may result in faster convergence but to a higher query cost solution.

22.4.3 Geographic Forwarding

The querying performance can also be improved by incorporating query semantics and exploiting spatial structure. A simple example of this is the use of geographic forwarding techniques, an excellent survey of which is provided in Ref. [28]. Such techniques are very useful for one-shot queries with geographic scoping (e.g., "what is the temperature at location (x,y)?"). The basic idea is to forward such a query at each step to a neighbor in the general direction of the destination. As pointed out in Ref. [28], there are several variations of greedy forwarding techniques. These include the (most forward within *R*) MFR technique [35] in which the packet is forwarded to the neighbor that is closest to the destination and the NF technique [17] in which the packet is forwarded to the nearest neighbor with forward progress (in order to minimize contention). Since greedy forwarding techniques are susceptible to local minima or voids where no forward progress can be made to the destination, it is possible to use the perimeter routing mode incorporated into the GPSR scheme [23].

While basic geographic forwarding schemes such as GPSR require each node in the network knowing its own geographical location, some advanced schemes can operate with relaxed requirements. Reference [31] proposes a geographic routing scheme, which requires only perimeter nodes to know their locations but no need for others, and then extends their algorithm for relaxed assumptions: (1) perimeter nodes know that they are on perimeter without knowing their positions; (2) all nodes have no information about either being on perimeter or location. To forgo the accurate geographic location information, this scheme maintains an overlaid virtual coordinate system that is constructed with the triangulation method.

22.4.4 Comb-Needle Technique

To expedite the query resolution, event information sources may advertise their possessions in addition to query forwarding. Reference [26] proposes comb-needle scheme where queries build a horizontal comb-like routing structure, while events follow a vertical needle-like trajectory to meet the "teeth" of the comb. Figure 22.4 illustrates the comb structure of query, the needle structure of event advertisement, and the query resolution process. Key tunable parameters are spacing between branches of the comb and the length of the needle, which turn out to consider on the query rate and event rate for better performance.

For better average search cost, the comb inter-spacing and the length of needle should be smaller if the event-to-query ratio is larger while they should be larger for the smaller ratio. The reverse-comb structure, consisting of the vertical comb for queries and horizontal needles for event, may be favorable when the query rate is larger than the rate of event generation, or when there are multiple queriers in the network. The query cost has shown to be $O(\sqrt{N})$, where N is the number of nodes in a grid network, and, due to the inherent feature of this scheme, the cost is hardly improved by replicating event unlike many other schemes.

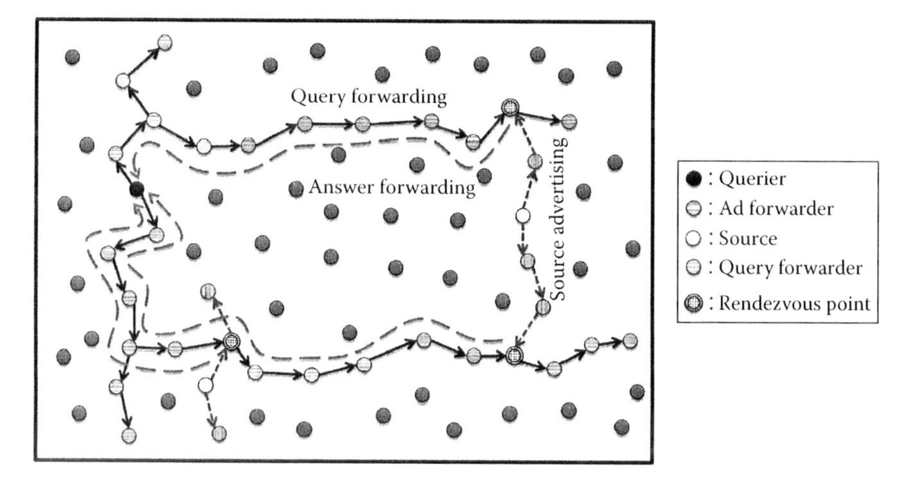

FIGURE 22.4 Illustration of comb-needle technique: query forwarders form the comb structure and ad forwarders form the needle structure. Rendezvous points send the query responses back to the querier.

An analytical comparison of the comb-needle approach and data centric storage is provided in Ref. [22]. Through the analysis based on a single-sink square-grid deployment, they have found that the structured hash-based data-centric storage generally performs better than the unstructured comb-needle mechanism when the query rates are high and event rates are low. In the case of high event rates, it is found that the comb-needle has better performance. They have also identified magic number thresholds ($\Theta \approx 39.78$) for event rates in the case of the aggregate queries that require information from all relevant nodes. The threshold is independent of network size or query probability.

22.4.5 Sensing-Driven Querying

On-demand queries in geographic forwarding can be conceived as a form of greedy search that is trying to minimize the objective function of the distance to the destination. A generalization of this is to consider arbitrary objective functions that depend on sensor readings. For example, queries such as "where is the target located?" may search to locate the node that maximizes the corresponding sensor reading (this could be a signal strength measurement). Because sensor readings are spatially correlated due to the underlying physical phenomena, a greedy search that follows gradients in the network can often provide efficient query performance. A scheme that utilizes such information is the IDSQ/CADR mechanism detailed in Ref. [12] (also described earlier in this book).

IDSQ provides a mechanism for sensor selection to maximize the information gain (balanced with its communication costs), while CADR (which stands for constrained anisotropic diffusion routing) is essentially an on-demand querying mechanism to determine how to route the queries dynamically through the network. It uses a composite objective function that incorporates both information gain and communication costs and routes the query at each step by choosing the neighbor that lies in the direction of the gradient of this objective function. Routing the query in this way is shown to maximize the information obtained while minimizing the communication costs. Techniques such as IDSQ/CADR demonstrate the ability to incorporate query semantics to decide the trajectory of queries on the fly, for even greater efficiency.

22.4.6 Biased Random Walks toward an Area or Trajectory

In the previous section, we have seen that random walks can be biased toward unvisited nodes, but these walks can also be biased to follow certain areas or less random trajectories. In Ref. [6], the authors study the effect of biasing a walk by-information and by-memory. A biased walk by-memory is similar in spirit to the study of power of choice, that is, at each step, a walk polls its neighborhood and selects unvisited nodes when possible. A bias by-information assumes that nodes can provide (with some probability) information about the location of the event node. For example, the authors evaluate the contact time on mobile networks. If node A has seen the event node 10 s ago and node B 20 s ago, then the random walk would have a higher probability of jumping to node A. Assuming a high density of nodes and a circular area where the event node is located in the middle and the sink node is located at the periphery of the circle, the authors show that the hitting time is upper bounded by $N^{1/p}$, where N is proportional to the radius of the circle, and p, $0.5 < p < 1$, indicates the biasing level of the walk. The authors validate through simulations the benefits of biasing the direction of random walks.

Another way to bias a random walk is by following a less random direction. In Ref. [18], the authors propose a guided random walk to reduce the hitting time when several replicas of the data are present on the network. Spiral [18], as the name suggests, tries to follow a spiral path from the sink (located at the center of the network) toward the edge of the network. The aim is to visit nodes that are near the sink before visiting more distant nodes. This visit is performed in a breadth-first search manner and stops when the first replica is found. In this way, the walk is prevented from querying further into the network, which could significantly increase the cost. The intuitive idea behind spiral is simple. Nodes establish their proximity to the sink based on hop distance. Then, the walk visits nodes based on a sliding window mechanism. Initially, the window is $[1, W]$, and when choosing the next hop, the node selects the node with the lowest level. Once no more nodes are available on the $[1, W]$, the window is increased to $[W + 1, 2W]$. The authors report that in a 500-node network simulation with an average degree of 20 and two copies of every data object, the total communication cost by Spiral is up to 72% of flooding, 81% of ERS, and 74% of a simple random walk.

22.5 Scaling Laws of Querying

Scaling laws of data-centric storage and querying have been investigated in Ref. [2]. The authors have used a constrained optimization framework to derive the scaling laws for both unstructured sensor networks, which use blind sequential search for querying (e.g., ERS), and structured sensor networks, which use efficient hash-based querying. One of the key findings in this work is that the scalability of a sensor networks' performance depends upon whether the increase in energy and storage resources with more nodes is outweighed by the concomitant application-specific increase in event and query loads.

With a constant spatial density of sensor nodes and fixed radio range, the authors have found that there exists a condition under which the querying mechanism in the sensor network operates in a scalable way in terms of per-node energy requirements. Particularly, the energy requirement per node is shown to be bounded if and only if

$$\left\{ m \cdot q^{1/2} = O(N^{(d-1)/2d}), \quad \text{for unstructured networks} \right. \tag{22.3a}$$

$$\left\{ m \cdot q^{d/d+1} = O(N^{(d-1)/d}), \quad \text{for structured networks} \right. \tag{22.3b}$$

where

m is the event occurrence rate
q is the query rate
N is the number of nodes
d is the spatial dimension of the deployment area

Satisfying this condition ensures that adding nodes to the network is beneficial in that the energy and storage resources they bring outweigh the additional event and query activity they induce. This can be seen from many perspectives. Satisfying this condition implies that (1) sensor networks require bounded energy and storage per node, (2) arbitrarily large networks can be operated successfully with a limited energy budget, and (3) the network lifetime increases with network size for a given energy budget. Another interesting finding of this work is that three-dimensional (3D) uniform deployments are inherently more scalable than 2D uniform deployments, which, in turn, are more scalable than one-dimensional (1D) uniform deployments.

In this study, they have not explicitly considered bandwidth capacity; it is implicitly assumed that the energy constraints will be more severe than bandwidth constraints in the system. However, if energy constraints are not significant (considered as an extreme case if all nodes could be wired for power), bandwidth issues could be the dominant consideration.

22.6 Conclusions

We have described a class of querying mechanisms that are suited for one-shot/instantaneous queries. Most of the on-demand querying mechanisms, which are essentially mobile-agent-based search techniques, are particularly useful because they require orders of magnitude less communication than simple flooding of queries. The transmission cost and the bandwidth inefficiency of flooding can be enhanced through techniques like ERS at the cost of extended delay. However, the range limit of the (controlled) flood has to be set up at the querier before its initiation, which makes it hard to build intelligent on-demand queries based on flooding schemes.

As discussed, the simplest intelligent query mechanisms involve the use of random walks. The tradeoff is that random walks result in high-latency for response, particularly when the queried information is sparsely located in the network. Solutions for this include (1) the use of multiple simultaneous walks and replication of the data, for example, through the setting up of event and sink paths, and (2) leveraging the graph properties of the network such as the heterogeneity on node degree. We also found that most of the querying cost is spent on visiting the last nodes of the network; this implies that applications requiring a partial coverage are better suited for random-walk-based querying.

We also examined several mechanisms to improve over random walks by introducing direction to the on-demand queries. This can be done explicitly through source routing or by prespecifying the trajectory using routing on curves. LEQS shows how the notion of query direction can be learned over time by exploiting history, if the queried object has an unchanging (or slowly changing) location pattern. Queries can also be directed to terminate quickly based on the query semantics by exploiting geographic and sensor information such as in GPSR and IDSQ/CADR. Comb-needle technique utilizes both push (source advertisement) and pull (query forwarding) mechanisms and attempts to obtain the best of both worlds. We also reviewed the scaling laws of data-centric networks that employ the on-demand queries. Operating a network in a scalable fashion essentially requires that the traffic load due to additional events and queries be outweighed by the improvement in energy and storage resources obtained as the network size increases.

While the existing literature has presented preliminary work and suggested possible design principles, there is still considerable room for future work on developing intelligent on-demand queries for different scenarios. On-demand query mechanisms will clearly form a useful part of the portfolio of querying techniques that will be deployed in practical sensor networks.

References

1. J. Ahn, S. Kapadia, S. Pattem, A. Sridharan, M. Zuniga, J.-H. Jun, C. Avin, and B. Krishnamachari. Empirical evaluation of querying mechanisms for unstructured wireless sensor networks. *ACM SIGCOMM Computer Communication Review*, 38:17–26, 2008. ACM ID: 1384612.
2. J. Ahn and B. Krishnamachari. Scaling laws for data-centric storage and querying in wireless sensor networks. *IEEE/ACM Transactions on Networking*, 17:1242–1255, 2009.
3. C. Avin and C. Brito. Efficient and robust query processing in dynamic environments using random walk techniques. In *Proceedings of the Third International Symposium on Information Processing in Sensor Networks, IPSN'04*, Berkeley, CA, pp. 277–286, 2004. ACM, New York. ACM ID: 984663.
4. C. Avin and B. Krishnamachari. The power of choice in random walks: An empirical study. In *MSWiM'06: Proceedings of the Ninth ACM International Symposium on Modeling Analysis and Simulation of Wireless and Mobile Systems*, pp. 219–228, 2006. ACM Press, New York.
5. Y. Baryshnikov, E. Coffman, P. Jelenkovic, P. Momcilovic, and D. Rubenstein. Flood search under the California split rule. *Operations Research Letters*, 32(3):199–206, 2004.
6. R. Beraldi. Biased random walks in uniform wireless networks. *IEEE Transactions on Mobile Computing*, 8(4):500–513, 2009.
7. P. Bonnet, J. Gehrke, and P. Seshadri. Querying the physical world. *IEEE Personal Communications*, 7(5):10–15, 2000.
8. D. Braginsky and D. Estrin. Rumor routing algorithm for sensor networks, In *Proceedings of the 1st ACM International Workshop on Wireless Sensor Networks and Applications (WSNA'02)*, ACM, New York, NY, pp. 22–31. doi = 10.1145/570738.570742. http://doi.acm.org110.1145/570738.570742.
9. A. K. Chandra, P. Raghavan, W. L. Ruzzo, and R. Smolensky. The electrical resistance of a graph captures its commute and cover times. In *Proceedings of the 21st Annual ACM Symposium on Theory of Computing, STOC'89*, pp. 574–586, 1989. ACM, New York.

10. N. Chang and M. Liu. Revisiting the TTL-based controlled flooding search: Optimality and randomization. In *Proceedings of the 10th Annual International Conference on Mobile Computing and Networking*, pp. 85–99, 2004.
11. Z. Cheng and W. B. Heinzelman. Flooding strategy for target discovery in wireless networks. *Wireless Networks*, 11:607–618, 2005, 10.1007/s11276-005-3516-7.
12. M. Chu, H. Haussecker, and F. Zhao. Scalable information-driven sensor querying and routing for ad hoc heterogeneous sensor networks. *International Journal of High Performance Computing Applications*, 16(3):293–313, 2002.
13. D. Ganesan, B. Krishnamachari, A. Woo, D. Culler, D. Estrin, and S. Wicker. Complex behavior at scale: An experimental study of low-power wireless sensor. Technical Report 02-0013, UCLA/CSD-TR, 2002.
14. D. Ganesan, D. Estrin, and J. Heidemann. Dimensions: Why do we need a new data handling architecture for sensor networks? *SIGCOMM Computer Communication Review*, 33:143–148, 2003.
15. R. Govindan, J. M. Hellerstein, W. Hong, S. Madden, M. Franklin, and S. Shenker. The sensor network as a database. Technical Report 02-771, USC CS, 2002.
16. V. Handziski, A. Köpke, A. Willig, and A. Wolisz. Twist: A scalable and reconfigurable testbed for wireless indoor experiments with sensor networks. In *Proceedings of the Second International Workshop on Multi-Hop Ad Hoc Networks: From Theory to Reality*, Florence, Italy, pp. 63–70, May 26, 2006. ACM, New York.
17. T.-C. Hou and V. Li. Transmission range control in multihop packet radio networks. *IEEE Transactions on Communications*, 34(1):38–44, 1986.
18. H. Huang, J. H. Hartman, and T. N. Hurst. Data-centric routing in sensor networks using biased walk. In *2006 Third Annual IEEE Communications Society on Sensor and Ad Hoc Communications and Networks, SECON'06*, Vol. 1, pp. 1–9, September 2006.
19. C. Intanagonwiwat, R. Govindan, and D. Estrin. Directed diffusion: A scalable and robust communication paradigm for sensor networks. In *Proceedings of the Sixth Annual International Conference on Mobile Computing and Networking, MobiCom'00*, Boston, MA, pp. 56–67, August 2000. ACM, New York.
20. D. B. Johnson and D. A. Maltz. Dynamic source routing in ad hoc wireless networks. In T. Imielinski and H. F. Korth, eds., *Mobile Computing*, Vol. 353 of The Kluwer International Series in Engineering and Computer Science, pp. 153–181. Springer, New York, 1996.
21. D. B. Johnson, D. A. Maltz, and J. Broch. DSR: The Dynamic Source Routing Protocol for Multihop Wireless Ad Hoc Networks. In *Ad Hoc Networking*. pp. 139–172, Addison-Wesley, Longman Publishing Co., Inc., Boston, MA, 2001.
22. S. Kapadia and B. Krishnamachari. Comparative analysis of push-pull query strategies for wireless sensor networks. In *International Conference on Distributed Computing in Sensor Systems (DCOSS)*, San Francisco, CA, June 2006.
23. B. Karp and H. T. Kung. GPSR: Greedy perimeter stateless routing for wireless networks. In *Proceedings of the Sixth Annual International Conference on Mobile Computing and Networking, MobiCom'00*, pp. 243–254, 2000. ACM, New York.
24. B. Krishnamachari and J. Ahn. Optimizing data replication for expanding ring-based queries in wireless sensor networks. In *WiOpt'06: 2006 Fourth International Symposium on Modeling and Optimization in Mobile, Ad Hoc and Wireless Networks*, Boston, MA, April 2006.
25. B. Krishnamachari, C. Zhou, and B. Shademan. LEQS: Learning-based efficient querying for sensor networks. Unpublished, 2003.
26. X. Liu, Q. Huang, and Y. Zhang. Combs, needles, haystacks: Balancing push and pull for discovery in large-scale sensor networks. In *Proceedings of the Second International Conference on Embedded Networked Sensor Systems, SenSys'04*, Baltimore, MD, pp. 122–133, November 3–5, 2004. ACM, New York.

27. Q. Lv, P. Cao, E. Cohen, K. Li, and S. Shenker. Search and replication in unstructured peer-to-peer networks. In *Proceedings of the 16th International Conference on Supercomputing, ICS'02*, New York, pp. 84–95, June 22–26, 2002. ACM, New York.
28. M. Mauve, A. Widmer, and H. Hartenstein. A Survey on position-based routing in mobile ad hoc networks. *IEEE Network*, 15(6):30–39, 2001.
29. B. Nath and D. Niculescu. Routing on a Curve. *SIGCOMM Computer Communications Review*, 33:155–160, 2003.
30. C. E. Perkins and E. M. Royer. Ad-hoc on-demand distance vector routing. *IEEE Workshop on Mobile Computing Systems and Applications*, 0:90, 1999.
31. A. Rao, S. Ratnasamy, C. Papadimitriou, S. Shenker, and I. Stoica. Geographic routing without location information. In *Proceedings of the Ninth Annual International Conference on Mobile Computing and Networking, MobiCom'03*, San Diego, CA, pp. 96–108, 2003. ACM, New York.
32. S. Ratnasamy, B. Karp, L. Yin, F. Yu, D. Estrin, R. Govindan, and S. Shenker. GHT: A geographic hash table for data-centric storage. In *Proceedings of the First ACM International Workshop on Wireless Sensor Networks and Applications, WSNA'02*, pp. 78–87, 2002. ACM, New York.
33. N. Sadagopan, B. Krishnamachari, and A. Helmy. Active query forwarding in sensor networks. *Ad Hoc Networks*, 3(1):91–113, 2005.
34. S. Shakkottai. Asymptotics of query strategies over a sensor network. In *INFOCOM 2004. 23rd Annual Joint Conference of the IEEE Computer and Communications Societies*, Hong Kong, China, Vol. 1, p. 557, March 7–11, 2004.
35. H. Takagi and L. Kleinrock. Optimal transmission ranges for randomly distributed packet radio terminals. *IEEE Transactions on Communications*, 32(3):246–257, 1984.
36. E. Woodrow and W. Heinzelman. SPIN-IT: A data centric routing protocol for image retrieval in wireless networks. In *Proceedings of 2002 International Conference on Image Processing*, Rochester, NY, Vol. 3, pp. 913–916, June 24–28, 2002.
37. M. Z. Zamalloa and B. Krishnamachari. An analysis of unreliability and asymmetry in low-power wireless links. *ACM Transactions on Sensor Network*, 3(2), 7, 2007.
38. M. Zuniga, C. Avin, and M. Hauswirth. Querying dynamic wireless sensor networks with non-revisiting random walks. In J. Silva, B. Krishnamachari, and F. Boavida, eds., *Wireless Sensor Networks*, Vol. 5970 of Lecture Notes in Computer Science, pp. 49–64. Springer, Berlin, Germany, 2010.
39. M. Zuniga, C. Avin, and B. Krishnamachari. Using heterogeneity to enhance random walk-based queries. *Journal of Signal Processing Systems*, 57:401–414, 2009. ACM ID: 1618411.

23

Improving Target Localization and Tracking by Dynamically Prioritized Frequency Band Allocation for Wireless Sensor Networks in Urban Environments

Behtash Babadi
Harvard University

Doina Bein
The Pennsylvania State University

Bharat B. Madan
The Pennsylvania State University

Shashi Phoha
The Pennsylvania State University

Vahid Tarokh
Harvard University

Yicheng Wen
The Pennsylvania State University

23.1 Introduction

In a multi-modal wireless sensor network, sensors of different modalities (or types) cluster together to accomplish the tasks of sensing, fusion, and management of cluster resources and to meet the application goals. A distributed sensor network consists of a large number of low-cost sensor nodes that are networked together for observing events that are localized in space and time and for fusing their data. Intuitively, it may be argued that the sensor fusion quality is proportional to the number of sensors that provide information-bearing sensor data. However, the desired fusion performance may require increased sensor density and, consequently, communication interference. To address this problem, we propose a spectrum-sharing algorithm that allows all active cluster members to transmit using a fixed number of shared frequency bands. The proposed spectrum-sharing technique is applicable to network topologies in which a data aggregator (or a cluster head) has to collect data from many senders or sensors. Once the channel allocation is done, some channel-sharing algorithm needs to be employed for communication between sensors in a cluster.

In this work, we make the following realistic assumptions: (1) the traffic generated by the sensors is bursty; when a traffic burst occurs, the data rate is much higher than the available link bandwidth; (2) sensor data compression can partially alleviate this problem; (3) sensors in the vicinity of an event generate more traffic and are assigned higher priority over sensors that are far away from the event and therefore do not generate traffic of interest; and (4) individual sensors are resource constrained and may not be very reliable. In our work, we have considered that the priority of a node is a combination of the weighting and the amount of traffic generated, and nodes of the same type or modality that are part of the same cluster have equal priorities.

The inherent scarcity of the spectrum makes it inevitable for the new generations of wireless networks (e.g., cognitive radios and smart sensor networks) to appeal to spectrum sharing. The problem of spectrum sharing has been well studied in various contexts such as digital subscriber lines (DSL) and wireless networks (see, e.g., Refs. [1–9]). It is well established in the literature that such techniques can significantly increase the efficiency of communications in resource-limited and low signal-to-noise ratio (SNR) regimes. Among the aforementioned strands of work, the greedy asynchronous distributed interference avoidance (GADIA) algorithm is appealing due to its low complexity, fast convergence rate, and near-optimal performance [1]. GADIA is a simple, low-complexity, and fast algorithm for dynamic spectrum allocation, tailored for the problem of sharing r available frequency bands among coexisting agents in the network in a distributed and asynchronous fashion. In GADIA, the agents, equipped with spectrum sensing devices, greedily choose the frequency band in which they experience the least amount of interference, in an asynchronous fashion. It has been shown that, given an interference symmetric model, GADIA converges to a fixed point that corresponds to a near-optimal solution by minimizing the aggregate interference of the network. Finally, the dynamics analysis of GADIA confirms its robustness with respect to time variations in the network activities [1].

In this chapter, we present a closely related algorithm, namely, intracluster greedy spectrum allocation (IGSA), which is adapted to the foregoing sensor network setting. IGSA is a generalization of GADIA to a multi-modal sensor network setting, where the cluster members have different priorities for transmitting their messages to the cluster head. Taking the priority-weighted aggregate interference of the cluster as the network utility, IGSA minimizes this value. Due to the specific priority weighting of the network utility, the convergence and most of the performance results of GADIA generalize to IGSA.

To validate the proposed technique of dynamic spectrum allocation in multi-modal wireless sensor networks, we consider the problem of detecting and tracking a moving object that is traveling through this network.

The chapter is organized as follows. Related work is presented in Section 23.2. In Section 23.3, we give some preliminary notions related to fusion-driven clustering of nodes and the prioritization of the data packets sent wirelessly by the sensor nodes. In Section 23.4, we present the IGSA algorithm for single-hop clusters and carry out its extension for multi-hop clusters. In Section 23.5, we present the detailed results of the experiments that show that IGSA performs better than GADIA when the sensory data are prioritized. We conclude in Section 23.6.

23.2 Related Work

Another existing class of protocols for channel sharing are the multichannel media access control (MAC) protocols. These protocols deal with channel sharing among multiple nodes in a peer-to-peer manner in order to minimize interference and achieve fairness. Also, most of them are based on "on-demand" requests for transmission. None of the proposed multichannel MAC protocols handle the priority of the data, that is, giving a less noisy channel to a higher priority node. Also, in case of a cluster, the messages need to be transmitted periodically only from the cluster members to the cluster head, so recomputing the channels for each transmission is a waste of time; thus, predefined channels for each cluster member saves time and computation cycles. Finally, most of these protocols perform in a synchronous fashion, which is not very appealing for ad hoc sensor networking. The multichannel MAC protocols can be separated into the following four groups [10]:

1. *Dedicated control channel*: One channel is used for controlling the access to other data channels [11–13].
2. *Common hopping*: Nodes that not exchanging data will cycle through all channels synchronously; two nodes stop hopping as soon as they make an agreement for transmission, and once the transmission ends, they restart the cycling process [14,15].
3. *Split phase*: One channel is used for controlling the access to itself and the other data channels. Time is divided into an alternating sequence of control and data exchange phases. During a control phase, all devices tune to the control channel and attempt to make agreements for channels to be used during the following data exchange phase [16–18].
4. *Pseudorandom hopping*: Each node picks a seed to generate an independent pseudorandom hopping sequence, and when idle, it follows its default hopping sequence. A node puts its seed in every packet it sends, so its neighbors eventually learn its hopping sequence. Nodes are assumed to hop synchronously [19].

The dynamic channel allocation (DCA) protocol [11] dynamically assigns channels to nodes in an "on-demand" fashion. The channel is released after the host completes its transmission. The control channel schedules the channels and the times for the transmission of data. A transmission schedule consists of a period when only control packets are transmitted (also called a *control window*) followed by a period when only data packets are transmitted (called a *data window*). The common channel used in the control window is called the *default channel*; the default channel is common to all nodes. The default channel is used as a control channel during the control window and as a data channel during the data window. Dynamic channel allocation with power control (DCA-PC) [12] builds on DCA protocol by adding the concept of power control that reduces the transmission power of the control packets so as to increase the reuse of the control channel. The dynamic private channel (DPC) protocol [13] is similar to DCA except that a negotiation protocol on which the data channel is to be used occurs between the transmitter and the receiver. A common hopping protocol uses all the channels for data exchange. It has two major disadvantages: when two nodes are exchanging data, they are unaware of the busy status of the other devices, and the hopping time penalty can be quite substantial. A split-phase protocol requires time synchronization among all devices, though the synchronization can be looser than for a common hopping protocol since devices hop less frequently. The duration of the data phase can be either fixed (protocol Multi-channel MAC (MMAC) of Ref. [17]) or variable (protocol multichannel accessing protocol (MAP) of Ref. [16], protocol LCM MAC of Ref. [18]), depending on the agreements made during the control phase.

To obtain reliable performance from individually less reliable sensors, multi-modal sensor nodes physically located near some event during a certain time period may be clustered dynamically [20–22]. Since the event is observed by a number of multi-modal sensors, due to the resource-constrained nature of the sensors (i.e., limited power, communication interference), it is imperative to fuse locally the data received from these sensors. The dynamic-space time clustering (DSTC) [20] algorithm groups homogeneous sensors in the space-time vicinity of an event to better observe and track it. To separate the application requirements from the autonomous organizing and managing requirements, Phoha and Ray [23] have proposed a layered architecture for a homogeneous sensor network in which the data fusion is handled by the *information space* module, while the communication is handled by the *network design space* module. In Ref. [24], the authors have expanded the network design space to include the heterogeneity and mobility of some sensors. To reduce interference in the intracluster communication, the proposed protocol, event-driven network controller (ENC) [24], asks selected sensors to either move outside the receiving range of the cluster head or stop transmitting. Additionally, to improve the quality of sensor fusion, sensors of certain modalities may be asked to move into regions where certain modalities do not have adequate presence. Moreover, to conserve the network bandwidth, each sensor may compress its sensed data before sending it for fusion. Consequently, the fusion algorithms should be able to aggregate the compressed sensor observations, decisions, or estimates to create a composite view of the situation.

23.3 Fusion-Driven Clustering and Spectrum Sharing

In Section 23.5, we consider the application of tracking a moving object in a multi-modal sensor network where each sensor compresses its data collected during a certain period of time in the form of probabilistic finite state automata (PFSA); the period of time is dependent only on the sensor modality/type. During the training phase, various PFSA are constructed. The ones of interest are stored as *subpatterns* into a *library of subpatterns*. A certain *semantic distance* between these subpatterns is enforced in order to ensure the determinism of choices; the semantic distance is a spatial signal quality measure. By definition, the function θ, called the *semantic distance*, measures the deterioration of the signal from its origination to the location of the sensor, and it has been shown in Ref. [25] that it increases as the target moves away from the sensor. For simplification, we will call it θ, but we have, in fact, a family of functions θ, one for each modality.

A composite pattern is the Cartesian product of subpatterns in each modality. If we have $M > 1$ different modalities, a composite pattern $G = \{G_1, G_2, \ldots, G_M\}$ is an M-tuple of subpatterns, one subpattern for each modality; a composite pattern has associated a composite metric [26]. The information space of each sensor node stores a set of composite patterns of interest, $\mathcal{G}$, in a so-called *databank*. The databank contains, for each stored composite pattern G^j, the minimum number of sensors of each modality that are required to observe the subpatterns of G^j, $N^j = (p_{j,1}, p_{j,2}, \ldots, p_{j,M})$, $p_{j,l} \geq 1, \forall 1 \leq l \leq \mathrm{M}$. A composite pattern has an associated *maximum lifetime* that represents the period during which the pattern is observable.

Let C_t be some cluster of multi-modal nodes that have decided on the structure of a composite pattern G at slow-time epoch $t \geq 0$. Each sensor i in the cluster, before deciding on the composite pattern, has decided at time t on a subpattern of G—let us call it $G(i)$. The physical distance between the sensor i and the target identified as a subpattern of interest $G(i)$ (and subsequently as a composite pattern of interest G) is a monotonically increasing function of the value $\theta(G(i), H_t(i))$, $\mathcal{D}:[0,\infty)\rightarrow[0,\infty)$, where $H_t(i)$ is the PFSA constructed at the slow-time epoch $t \geq 0$ from the data stream of the sensor s. If d is the distance of the sensor i from the target location, then $d = \mathcal{D}(\theta(G(i), H_t(i)))$.

As mentioned previously, the sensor nodes and the associated sensor network are resource constrained in terms of network bandwidth, node battery life, sensing range, and communication radius. Consequently, to deliver robust sensor fusion performance, it is important to dynamically control and optimize the available resources [23,27–29]. The heterogeneous (multi-modal) group of sensors looks for certain composite patterns of interests to decide whether a sensed event (or object) is of interest or not. Sensors that decide on the same composite pattern of interest form a *precluster* and select among themselves a *cluster head*, the one that has the best composite metric. The cluster head becomes responsible for forming, managing, and terminating the cluster, while performing data fusion as required by the application level. In forming the cluster, the sensory data from individual sensors that are part of the cluster needs to reach the cluster head and the other cluster members.

We assume a simple propagation model in which the power of the received signal is proportional to $d^{-\eta}$, where d is the distance between the sensor and the target and η is the path loss exponent, which is usually greater than 2. However, in real situations, the sensing propagation model may exhibit random attenuation (e.g., Rayleigh or Ricean distribution) in which case we will need to map the semantic distance to the physical distance via the Rayleigh or Ricean distribution. Furthermore, if the propagation model entails fading, we can use a parametric Ricean model to carry this mapping, where parameters are the moments of the Ricean distribution. However, these issues are beyond the scope of this chapter since our focus is on adaptive spectrum allocation.

Given a limited communication frequency band, the cluster head searches for r individual frequencies and gives one frequency to each cluster member based on its priority. We assume that the spectrum available is limited; thus, the number of cluster members is larger than r, and the cluster members need to compete in order to enjoy a less noisy communication environment.

We make the following assumptions:

1. Each sensor is part of at most one cluster at any moment.
2. In the network, there can be multiple clusters and multiple events to be observed at the same time.
3. When multiple clusters are present at the same time, we assume the use of cognitive radio-learning mechanisms to find the available radio spectrum [30].

A cluster head will search the channels and identify r available ones. Then it needs to make sure that no two channels are the same for two different clusters that have overlapping communication footprints. Assume that the cluster C_1 chooses the channels $\{f_{11}, \ldots, f_{1r}\}$ and the cluster C_2 chooses the channels $\{f_{21}, \ldots, f_{2r}\}$. Assume also that there are two channels f_{1i} and f_{2j} that are the same and there is a sensor X that is part of C_1 and another sensor Y that is part of C_2 such that they will both get assigned this channel to communicate data to the cluster head. If X is also close to C_2, then we have interference if both send their data at the same time. So C_1 and C_2 need to have an agreement to avoid the use of overlapping channels when they have spatial proximity. One solution is for C_1 and C_2 to run a "rendezvous" protocol and choose a common intercluster control channel. Another solution is to have C_1 broadcast a message on each of its selected channels. C_2 will then receive this message if its selected channels overlap with those of C_1. Then, one of these overlapping channels can be chosen as an intercluster control channel via an appropriate tie-breaking rule. The first solution is more economical in terms of communication costs [31–33].

23.4 Intracluster Greedy Spectrum Allocation Algorithm

The need for reliable communication, given the low transmission power virtue of the sensors, inspires the use of spectrum allocation techniques for maintaining high-capacity intracluster links. Let us assume that there are r frequency bands available and we have a single-hop cluster of $N + 1$ nodes ($N > r$) with a single cluster head c_0; we denote the cluster members by $c_1, c_2, \ldots, c_N$. We assume that the cluster members need to transmit their measured data to the cluster head for data fusion purposes; let P_i be the transmission power of the cluster member c_i. Each cluster member has a priority parameter, which reflects the urgency of its message; we denote by $W_i \in [0,1]$, $i = 1, 2, 3, \ldots, N$ the priority weight assigned to cluster member c_i. In order to increase the link capacities inside the cluster, the cluster head assigns different frequency bands to different cluster members to use for transmission. Let $s_i \in \{1, 2, \ldots, r\}$ denote the frequency band the cluster member c_i will use for communication.

23.4.1 Problem Definition

The priority weight inspires defining the weighted sum-rate of the cluster links, C_W, as the network utility:

$$C_W := \sum_{i=1}^{N} W_i C_i \tag{23.1}$$

where C_i is the capacity of the link between the cluster member c_i and the cluster head. More explicitly,

$$C_i := \log_2\left(1+\frac{P_i\alpha_{i0}}{\sum_{j\neq i} P_j\alpha_{j0}f(s_j,s_i)+N_0B}\right) \tag{23.2}$$

where

α_{ij} is the link gain between the cluster members c_i and c_j
$f(s_j,s_i)$ is the interference leakage between the frequency bands s_i and s_j
N_0B is the noise power

Maximization of the weighted sum-rate results in the class of iterative water-filling algorithms, which are known to converge to nonoptimal frequency assignment solutions [34]. However, it has been shown that maximizing the negated weighted aggregate interference results in near-optimal solutions [1]. Incorporating the priorities into the network utility gives rise to the *priority-weighted aggregate interference* given by

$$U_W := -\sum_{i=1}^{N} W_i P_i \alpha_{i0} \tilde{I}_{c_i} \tag{23.3}$$

where $\tilde{I}_{c_i}$ is the priority-weighted interference experienced by the message transmitted by the cluster member c_i on the channel s_i due to other cluster members transmitting on the same channel, given by

$$\tilde{I}_{c_i} := \sum_{j\neq i} W_j P_j \alpha_{j0} f(s_i,s_j) \tag{23.4}$$

The interference leakage $f(s_i, s_j)$ can be approximated by the Kronecker delta function $\delta(s_i, s_j)$

$$\delta(s_i,s_j) := \begin{cases} 1 & s_i = s_j \\ 0 & s_i \neq s_j \end{cases} \tag{23.5}$$

The objective is then to assign the frequency bands $s_i \in \{1, 2, \ldots, r\}$ to cluster member c_i, $i = 1, 2, \ldots, N$, so that the negated weighted aggregate interference is maximized. A schematic view of the intracluster frequency band allocation is depicted in Figure 23.1.

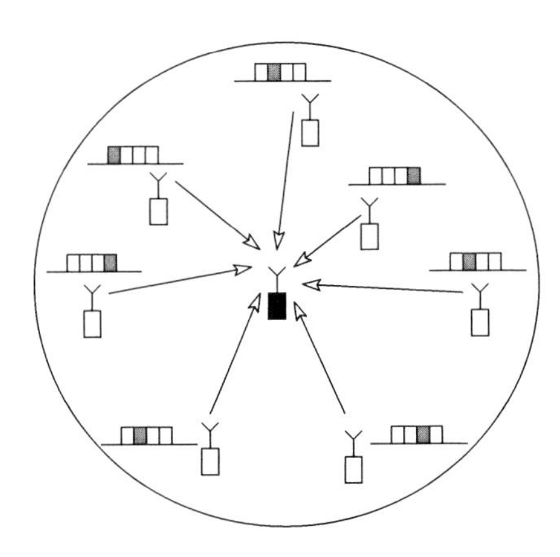

FIGURE 23.1 Intracluster frequency band allocation.

23.4.2 IGSA Algorithm for Single-Hop Clusters

The cluster head can compute all the link gains α_{i0} for $i = 1, 2, \ldots, N$, using the current locations of the cluster members. The ENC protocol [24] provides the cluster head with the locations of all the cluster members for estimation and tracking purposes. Given the locations, the cluster head can employ a channel model such as path loss or Hata model [35] in order to compute the interferences. For example, in the path loss model, the channel gain between the cluster head and cluster member c_i at distance d_{i0} is given by

$$\alpha_{i0} = \frac{1}{d_{i0}^{\eta}} \tag{23.6}$$

where η is called the path loss exponent. In the Hata model for urban areas,

$$10\log\alpha_{i0} = 69.55 + 26.16\log f_c - 13.82\log h_{te} - a(h_{re}) + (44.9 - 6.55\log h_{te}\log d_{i0}) \tag{23.7}$$

where

f_c is the frequency of transmission (MHz) (from 150 to 1500 MHz)
h_{te} is the height of the cluster head antenna (m)
h_{re} is the height of the cluster member antenna (m)
$a(h_{re})$ is the correction factor of the cluster member antenna (dB), which is a function of the size of the coverage area
d_{i0} is the distance between the cluster head and the cluster member (km)

This formula is able to capture the variations with respect to frequency, terrain modulations, and antenna height. In a sensor field, the distances between the sensors is short enough so that we can neglect large-scale fading phenomena, and a path loss model suffices to capture the nature of interference. However, the Hata model does not consider the random perturbation in the channel. This issue can be addressed by incorporating Rayleigh or Ricean fading models for environments with multipath fading. If the sensors are stationary, the channel model must also encompass frequency-dependent attenuation effects across the frequency bands. Doppler spread effects must also be taken into account for mobile sensors.

The cluster head then runs the IGSA algorithm that assigns frequency bands to the cluster members. To this end, the cluster head computes (not measures) the interference produced by some message sent by the cluster member c_i in each of the r frequency bands. Then it picks the frequency band with the least amount of interference and assigns it to c_i. It repeats this allocation process for the rest of the cluster members in a randomly chosen order. It is easy to show that the network utility following each step of the IGSA algorithm is nonincreasing [1]. Due to the lower boundedness of the network utility, the IGSA algorithm converges to a fixed point. Moreover, given the exponential convergence rate of such algorithms [1], by repeating this subroutine a total of three times, one can reach a small neighborhood of the fixed point with high probability. Finally, the cluster head reports the assigned frequency band to each cluster member using a dedicated so-called *common control channel*. (We note that the control channel is also used by clustering protocol—in our experiments we have used ENC protocol [24]—for nodes in the space-time vicinity of a target to select that cluster head.) Note that if the cluster head is able to estimate and separate the individual received powers from the cluster members, it can avoid utilizing an explicit channel model by replacing the expression $P_j\alpha_{j0}$ (on step 5 of the IGSA algorithm) with the received power from the jth cluster member.

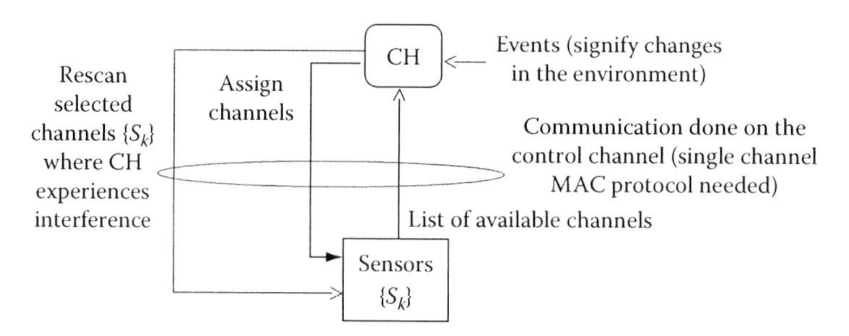

FIGURE 23.2 Cluster head rescans for available channels.

Algorithm 23.1 $IGSA\left(\{\alpha_{i0}\}_{i=1}^N, \{W_i\}_{i=1}^N, \{P_i\}_{i=1}^N, r\right)$

Inputs: $\{\alpha_{i0}\}_{i=1}^N, \{W_i\}_{i=1}^N, \{P_i\}_{i=1}^N, r$.

Output: $\{s_i\}_{i=1}^N$.

1: $s_i \leftarrow 1$ for all $i = 1, 2, \ldots, N$.
2: **for** $l = 1$ to 3 **do**
3: **for all** $i = 1, 2, \ldots, N$ **do**
4: **for all** $k = 1$ to r **do**
5: Compute $\tilde{I}_{c_i,k} := \sum_{j \neq i} P_j W_j \alpha_{j0} \delta(s_j, k)$.
6: **end for**
7: $s_i \leftarrow \arg\min_k \tilde{I}_{c_i,k}$.
8: **end for**
9: **end for**
10: Output $\{s_i\}_{i=1}^N$.

Next, we discuss some of the implementation issues of the IGSA algorithm. The IGSA algorithm can be implemented in a distributed fashion, if the cluster members are aware of the location information or are equipped with spectrum sensing devices. If the cluster members have access to location information, a distributed channel assignment algorithm based on graph multicoloring can be used (see, e.g., Ref. [36]). We expand next how the IGSA can be implemented in a distributed fashion. If the cluster members are equipped with spectrum sensing devices, they can participate in identifying available channels to reduce the load on the cluster head as follows. Each cluster member searches for available channels and communicates (using a so-called control channel) the list of available channels and the interference level of each channel to the cluster head over the shared control channel. Since all the clusters use the same control channel, single-channel MAC protocols can be used for this purpose (see Ref. [37] for a survey of MAC protocols for wireless sensor networks and additional references). The cluster head uses the received information to run the IGSA algorithm in order to allocate the channels to the cluster members based on their priorities. Scanning for available channels is triggered by the cluster head and can be either event driven (e.g., by monitoring the channel congestion or the interference level) or periodic (see Figure 23.2).

23.4.3 Priority Weighting

Let us consider the case when, among the cluster members $c_1, c_2, \ldots, c_N$, one of them needs to have assigned an exclusive frequency band while the others will share the rest of the available frequency bands; without loss of generality, let c_1 be that cluster member. In this case, the cluster head gives the highest priority to c_1 while all other cluster members are assigned lower but equal priority: $W_1 = 1$ and $W_i = 1/N$, for all $i \neq 1$, and the cluster head executes IGSA. The output of IGSA is that c_1 obtains an exclusive frequency band for itself, with a high SINR, say b_1, and all the other cluster members will share the rest of $r - 1$ frequency bands, $b_2, \ldots, b_r$.

We emphasize here that the fastest sensors do not necessarily imply a higher priority. For example, for the target-tracking problem described in Section 23.5, even though a video sensor may generate sensor data at the rate of 100 kB/s compared to a magnetic sensor that generates data at the rate of 1 kB/s, we assign a higher priority to the latter. This is justified on the basis that while a video sensor may be more important for long-distance surveillance, for target localization the utility of magnetic sensor data and pressure data is much higher. While analytics of assigning priorities to different sensors are important, we do not want to digress from the main problem of spectrum sharing being dealt with in this chapter.

In this chapter, we have chosen a simple prioritization scheme, based on our practical experience. As mentioned earlier, in tracking applications, video sensors are assigned higher priority, viz., magnetic and pressure sensors. Conversely, for localization applications, pressure sensors may have to be assigned higher priority than the video sensors. Assigning priorities is a nontrivial problem and requires application domain expertise. As mentioned in Section 23.6, in the future, we plan to investigate the use of a learning system for priority assignment. The learning system will deal with the problem of situation assessment using sensor system [38]. The learning system may reassign priorities depending on the degree of contribution of individual sensors to the common goal of situation assessment of the sensor network.

For estimating the location of an event, we give the highest priority to the type of sensors that offer the lowest error in estimating the target location (for the first experiment, the pressure sensors are the most important, followed by the magnetic and at last the video sensors). In another experiment, we give the highest priority to the type of sensors that have the largest sensing range (for the second experiment, the video sensors have the largest priority, followed by the pressure and magnetic with equal priorities).

23.4.4 Extending IGSA to Multi-Hop Clusters

The main benefits of multi-hop clustering are saving the energy of individual nodes and minimizing the interference at the cluster head. In multi-hop clusters, the data from a cluster member are indirectly communicated to the cluster head via one or more intermediate nodes (that are cluster members themselves). Let us assume that we have a k-hop cluster of $N + 1$ nodes with a single cluster head c_0; other cluster members are denoted by $c_1, c_2, \ldots, c_N$. It is assumed that the cluster members need to transmit their data to the cluster head for data fusion purposes. It is also assumed that there is a so-called *common control channel* that is used by cluster head or internal cluster members to communicate control data (such as channel assignments) to other cluster members. Each cluster member has an associated priority parameter that reflects the relative importance of its messages; let $W_i \in [0,1]$, $i = 1, 2, 3, \ldots, N$ denote the priority weight assigned to cluster member c_i. Let p_i be the parent of cluster member c_i in the spanning tree $T = (V,E_T)$ obtained from the cluster's communication graph $G = (V,E)$, which is either the cluster head itself or another cluster member that communicates to the cluster head through a smaller number of hops than c_i, $V = \{c_0, c_1, \ldots, c_N\}$ and $E_T \subseteq E$. For example, from the communication graph drawn in Figure 23.3a, the spanning tree could be the one drawn in Figure 23.3b.

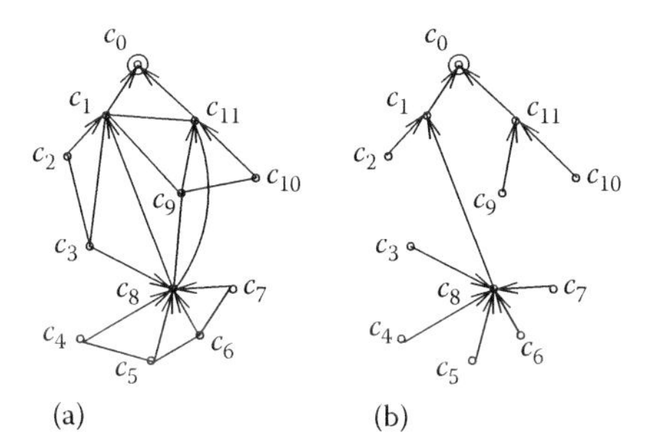

FIGURE 23.3 Intracluster frequency band allocation: (a) the communication graph and (b) the induced spanning tree.

When the cluster has r frequency bands allocated, a simple algorithm is to let each internal node in the tree run the IGSA algorithm in parallel with its siblings, in a top-down manner, one level at a time: first the root runs the IGSA algorithm to assign frequency bands to its children, then each of the children runs the IGSA algorithm to assign frequency bands to their own children using $r - 1$ frequency bands, excluding the frequency band each of them has been assigned, and so on and so forth. In the cluster example (Figure 23.3a), the IGSA algorithm will be executed by the cluster members c_0, c_1, c_{11}, c_8. This distributed method has a major drawback: one cluster member may end up sharing the channel allocated by its parent with many other cluster members that got the same channel from their parents. In our example, let us assume that there are four frequency bands available, f_1, f_2, f_3, and f_4, and the cluster members c_8 and c_9 have data of the highest priority. After running IGSA, c_0 can allocate f_1 to c_1 and f_2 to c_{11}. Then cluster member c_1 runs IGSA for its children and with $\{f_2, f_3, f_4\}$ frequency bands to be shared while cluster member c_{11} runs IGSA for its children and with $\{f_1, f_3, f_4\}$ frequency bands to be shared. Node c_1 can allocate f_3 to c_2 and f_4 to c_8, while c_{11} can allocate f_3 to c_{10} and f_4 to c_9. Thus, c_8 has to share the same channel with c_9.

To preclude situations in which cluster members with higher priority share the channels, whereas lower priority ones may end up with dedicated channels, there needs to be some coordination among the cluster members that run the IGSA algorithm and are located in the vicinity of each other, if they have to choose among overlapping sets of available frequency bands. We propose the distributed algorithm *merge priorities (MP)* that ensures up to a certain level that cluster members are given frequency bands based on their priorities. In a bottom-up manner, every internal node (including the root) collects the IDs of the descendants at no more than two hops away, namely, children and grandchildren. Then, in a top-down approach, starting from the root, nodes at the same distance to the root run a slightly modified IGSA algorithm with the two-hop descendants and $r - 1$ frequency bands, namely, the r frequency bands excluding the frequency band the node is using to communicate to its parent. The root runs IGSA with all the r frequency bands. In the slightly modified IGSA algorithm, called *mIGSA*, a frequency band is allocated to a node only *if* it has been assigned a frequency band previously. More precisely, step 7 is changed as follows:

"7: **if** $s_i = \emptyset$ **then** $s_i \leftarrow \arg\min_k \tilde{I}_{c_i,k}$"

Algorithm 23.2 $MP\left(\{\alpha_{i0}\}_{i=1}^N, \{W_i\}_{i=1}^N, \{P_i\}_{i=1}^N, r, T = (\{0, 1, \ldots, N\}, E_T)\right)$

Inputs: $\{\alpha_{i0}\}_{i=1}^N, \{W_i\}_{i=1}^N, \{P_i\}_{i=1}^N, r, T = (\{0, 1, \ldots, N\}, E_T)$.

Output: $\{s_i\}_{i=1}^N$.

1: **for all** $i = 1, 2, \ldots, N$ **do**
2: If c_i has at least one child at a 2-hop distance, then collect $CT_i = \{\forall c_j, c_j \text{ is a descendant of } c_i \wedge dist(c_i, c_j) \leq 2\}$.
3: **end for**
4: Root: collect $CT_0 = \{\forall c_j, c_j \text{ is a descendant of } c_0 \wedge dist(c_0, c_j) \leq 2\}$.
5: Run $IGSA(\{\alpha_{i0}, i \in CT_0\}, \{W_i, i \in CT_0\}, \{P_i, i \in CT_0\}, \{1, 2, \ldots, r\}$.
6: $d = 1$
7: **while** $\exists c_j, s_j = \emptyset$ **do**
8: Concurrently, for all nodes c_j at distance d to the root, execute $mIGSA(\{\alpha_{i0}, i \in CT_j\}, \{W_i, i \in CT_j\}, \{P_i, i \in CT_j\}, \{1, 2, \ldots, r\} \setminus \{\text{frequency allocated to } c_j\})$
9: $d = d + 1$
10: **end while**
11: Output $\{s_i\}_{i=1}^N$.

In our example, cluster head c_0 will run IGSA for $\{c_1, c_2, c_8, c_9, c_{10}, c_{11}\}$, the cluster member c_1 will run IGSA for $\{c_3, c_4, c_5, c_6, c_7\}$, and so on.

If one-hop interference model is assumed, the algorithm *MP* guarantees that the channel allocation is done in the order of priorities over the whole network. This way, a certain degree of fairness is ensured. If one-hop interference model is not assumed, to avoid the hidden terminal problem, a combination of some traversal of the tree and one-hop interference graph could be used instead.

23.5 Experimental Results

The purpose of the multi-modal sensor network is to track a particular vehicle moving in traffic by clustering nodes in its vicinity. We simulated a multi-modal sensor network for tracking mobile targets in an urban scenario in NS-2 as shown in Figure 23.4.

Pressure sensors, video sensors (cameras), and magnetic sensors were deployed in a Manhattan-like city; large blocks represent buildings that are separated by a two-way streets, 20 ft wide. In the ground of two-way streets, we embedded a total of 312 fixed pressure sensors. Adjacent pressure sensors are 2 m apart. Each pressure sensor generates a digital voltage ranging from 0 to 1023 due to the pressure applied when it senses an object above. Video sensors, mounted on the buildings (four per building), have adjustable view angles so that the sensing coverage can be easily changed. A video sensor takes snapshots of the scene within its view at some particular frequency. We positioned a total of 72 magnetic sensors on both sides of the street, as well as in the intersections. Magnetic sensors can detect the presence of large amounts of ferromagnetic material. Magnetic sensors were mounted on mobile platforms that can move freely along the streets; they can be quickly relocated if necessary.

Each sensor is equipped with three wireless channels: a so-called control channel (channel 0) used by the cluster head for transmitting data to the cluster members and two data channels (channels 1 and 2) used by the cluster members to transmit data to the cluster head and the other cluster members.

We consider two phases: (1) the *training phase* in which the target follows specific trajectories and data are collected from the actual sensor nodes in the laboratory for constructing PFSA of interest and (2) the *operational phase* in which the target moves freely, data are collected from the actual sensor nodes in the laboratory, and the elected cluster head either executes or it does not execute various spectrum-sharing algorithms. During the training phase, various PFSA are constructed. In the operational phase, once an event is observed by a sensor, the sensed subpattern is compared against the library of subpatterns, and semantic distances [26] to the stored subpatterns are computed. If the smallest semantic distance is less than a given ϵ, we conclude

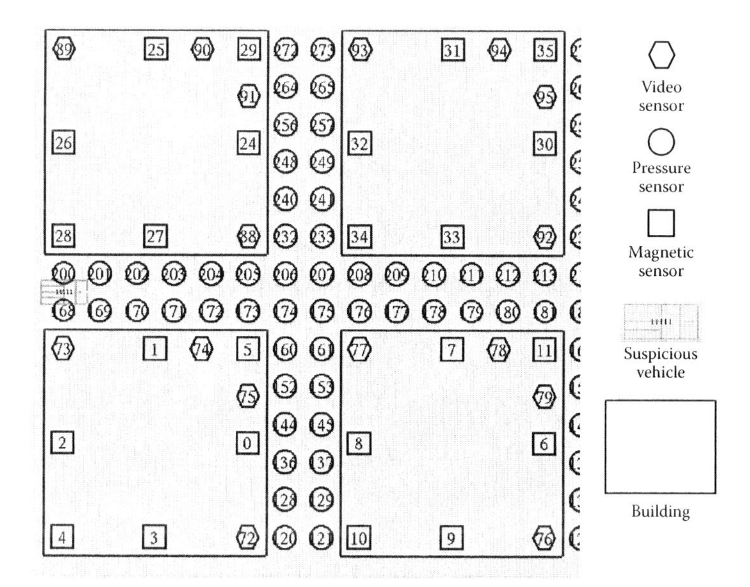

FIGURE 23.4 Multi-modal sensor network in the urban scenario.

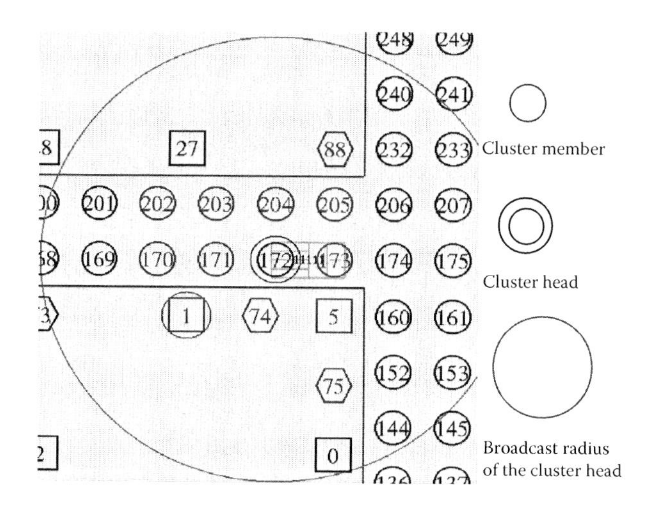

FIGURE 23.5 Multi-modal clustering in the sensor network.

that the sensor has detected that subpattern. The sensor then broadcasts the subpattern and waits for subpatterns from sensors of different modalities in order to construct a composite pattern of interest.

We consider that the vehicle of interest is moving at a constant speed in a rectangular profile that coincides with the physical location of the nodes 170 → 173 → 205 → 202 → 170 (see Figure 23.5). The ENC algorithm [24] yields a dynamic heterogeneous (multi-modal) cluster of 13 pressure, magnetic, and video sensors. The cluster formed is close to the physical location of the vehicle within a certain time frame and has the node 172 selected as the cluster head. We recall that the cluster head is responsible for collecting the sensing data streams from all the cluster members and estimating the position of the tracked target periodically. In our experiments, the data packets from the pressure and magnetic sensors are used for target localization (tracking) through a nonlinear least squares triangulation algorithm [39]; the data packets from the video sensors are used for tracking.

We run two experiments that involve the proposed IGSA algorithm when three channels are available. Channel 0 is used for the cluster head to transmit messages to the cluster members. The cluster head executes IGSA, and cluster members are assigned to use either channel 1 or channel 2 for sending their data to the cluster head based on their priority. The IGSA algorithm is executed on a time-varying number of cluster members. If a new sensor node joins or leaves a cluster according to ENC protocol, the cluster head will update the channel allocation by running the IGSA algorithm. In the first experiment (IGSA_1), we assign higher weights to the pressure sensors, followed by the magnetic sensors, and at last the video sensors. In the second experiment (IGSA_2), the video sensors have the highest weights, while the rest of sensors have equal, lower weights.

To show the effectiveness of the IGSA algorithm in improving the quality of data fusion within the cluster, we compare the results of these experiments among themselves and with other two experiments, one in which there is a simple communication between various members of the cluster using a single channel (channel 1) and another one in which the communication involves GADIA and the three channels (channels 0, 1, and 2). When the cluster head executes GADIA, the cluster members are assigned to use either channel 1 or channel 2 for sending their data to the cluster head based only on their physical location. We compare the results of these experiments in terms of the average packet received rate and the average localization error.

Table 23.1 shows the results of the experiments. When GADIA is executed at the cluster head, the packet received rate is larger for all sensor types, and also the position estimation error is smaller as compared to the case when the cluster head does not execute any channel-sharing algorithm. This is to be expected, since GADIA has been shown effective in improving the communication performance in the network. When IGSA_1 is executed at the cluster head instead of GADIA, by assigning higher priorities to the pressure and magnetic sensors, their packet received rates are improved and the localization is

TABLE 23.1 Comparison of Experimental Results

	Simple	GADIA	IGSA_1 (Priority)	IGSA_2 (Priority)
Pressure packets received rate	45.68%	86.05%	88.71%(10)	29.78%(1)
Magnetic packets received rate	49.66%	92.08%	81.63%(5)	54.95%(1)
Video packets received rate	22.54%	41.29%	35.37%(1)	54.20%(10)
Estimated position error (m)	0.3782	0.1161	0.0756	0.1857

TABLE 23.2 Comparison of Experimental Results in the Presence of Environmental Changes

	Simple	GADIA	IGSA_1 (Priority)	IGSA_2 (Priority)
Pressure packets received rate	44.68%	67.32%	92.22%	52.52%
Magnetic packets received rate	47.44%	79.76%	74.96%	27.37%
Video packets received rate	34.71%	32.08%	27.74%(1)	54.22%(10)
Estimated position error (m)	0.4304	0.1965	0.1354	0.2618

also better, although the video packet received rate (that has the lowest weight/priority) becomes worse. We recall that the video sensors do not contribute to the position estimation. In case of IGSA_2, since higher priority is assigned to the video sensors, we observe that the video packets have a better received rate than both the GADIA case and IGSA_1 case, and the localization becomes worse. Hence, IGSA_1 and IGSA_2 reflect the impact of the priority assignment on the packet received rate in each modality. Users are free to choose different weight assignments in order to achieve the performance required by different applications. For example, IGSA_1 is useful when a better localization of the target is needed, while IGSA_2 performs better when a better track continuation of the target is needed.

In an urban scenario, changes in the environment can affect the quality of the wireless channels that subsequently influence the fused estimation location and tracking of a target. We ran separate simulations that include the effects of the environmental changes on the wireless communication, namely, we consider a slowly varying wireless link quality. We assign a uniformly distributed random variable in the range [0.8,1] to each communication channel. This variable is refreshed every 50–100 s for each communication channel. The cluster head will rerun the IGSA algorithm for the two scenarios (*IGSA_1* and *IGSA_2*) every 40 s. The results are shown in Table 23.2. The estimation error is larger in each scenario due to changes in the environment that affect the packet received rate. The results are also consistent with the conclusions drawn in the case when the environment does not affect the channel quality. GADIA is effective in improving the communication performance in the network and reducing the estimation error versus the case when there is no channel sharing algorithm. IGSA_1 and IGSA_2 reflect the impact of the priority assignment on the packet received rate in each modality. The estimation error is reduced when the pressure sensors have a higher priority than the magnetic and video sensors.

23.6 Conclusion and Future Work

For a resource-constrained network in which a reduced number of frequency bands are available and need to be shared among the cluster members for transmitting their data to the cluster head, we propose an algorithm called IGSA that allocates dynamically the spectrum band to the cluster members based on their priorities. Our experiments run in NS-2 show the effectiveness of IGSA versus GADIA, a previously proposed spectrum algorithm based on the physical location of the cluster members, or no channel allocation whatsoever, in terms of packet received rate and estimation of the target location.

Another venue of research is to define the priority function as a learning process, where more complex patterns that involve a various number of sensors of certain modalities are observed over time, so the priority is time dependent. So, if at the beginning one has simple rules for spectrum allocation based on simple priority functions, later on the rules change as the node (the cluster head) learns about the environment and the variety of cluster members as they dynamically join or leave the cluster. As future work, we plan to explore two proposed prioritization schemes: (1) *directional prioritization* and (2) *contextual prioritization*.

In *directional prioritization*, the physical location of the sensor within the cluster dictates the priority value, irrespective of their type. For example, let us consider an urban scenario where a multi-modal sensor network composed of magnetic, video, and pressure sensors is responsible for locating and tracking people carrying lots of metals. In tracking people carrying lots of metals, the sensors physically located along the estimated direction of movement will need to have higher priority in transmitting their data compared with the rest of the sensors.

In *contextual prioritization*, the type of sensor dictates the priority value. Contextual prioritization emphasizes the importance of a certain subpattern within a composite pattern and sometimes having similar subpatterns observed by a number of sensors. For example, in locating a person in the same urban scenario considered earlier, the data coming from a magnetic sensor is more critical than the data coming from a video or pressure sensor. But later on, in tracking such a person, the data coming from the video sensors is more critical than the data coming from magnetic or pressure sensors.

Acknowledgments

This material is based upon work supported by, or in part by, the U.S. Army Research Laboratory and the U.S. Army Research Office under the eSensIF MURI Award No. W911NF-07-1-0376. Any opinions, findings, and conclusions or recommendations expressed in this publication are those of the authors and do not necessarily reflect the views of the sponsor.

References

1. B. Babadi and V. Tarokh. Gadia: A greedy asynchronous distributed interference avoidance algorithm. *IEEE Transactions on Information Theory*, 56(12), 6228–6252, 2010.
2. W. Yu, G. Ginis, and J.M. Cioffi. Distributed multiuser power control for digital subscriber lines. *IEEE Journal on Selected Areas in Communications*, 20(5), 1105–1115, 2002.
3. R. Cendrillon, J. Huang, M. Chiang, and M. Moonen. Autonomous spectrum balancing for digital subscriber lines. *IEEE Transactions on Signal Processing*, 55(8), 4241–4257, 2007.
4. J. Neel and J. Reed. Performance of distributed dynamic frequency selection schemes for interference reducing networks. In *Milcom*, Washington, DC, October 23–25, pp. 1–7, 2006.
5. N. Nie and C. Comaniciu. Adaptive channel allocation spectrum etiquette for cognitive radio networks. *ACM Mobile Networks and Applications (MONET), Special Issue on Reconfigurable Radio Technologies in Support of Ubiquitous Seamless Computing*, 11(6), 779–797, 2006.
6. D.C. Popescu and C. Rose. *Interference Avoidance Methods for Wireless Systems*. Kluwer Academic Publishing, Hingham, MA, 2004.
7. D.C. Popescu, O. Popescu, and C. Rose. Interference avoidance and multiaccess vector channels. *IEEE Transactions on Communications*, 55(8), 1466–1471, 2007.
8. C.W. Sung, K.W. Shum, and K.-K. Leung. Stability of distributed power and signature sequence control for CDMA systems—A game-theoretic framework. *IEEE Transactions on Information Theory*, 52(4), 1775–1780, 2006.
9. C. Peng, H. Zheng, and B.Y. Zhao. Utilization and fairness in spectrum assignment for opportunistic spectrum access. *ACM Mobile Networks and Applications (MONET)*, 11(4), pp. 555–576, 2006.

10. J. Mo, H.-S. W. So, and J. Walrand. Comparison of multichannel MAC protocols. *IEEE Transactions on Mobile Computing*, 7(1), 50–65, 2008.
11. S.-L. Wu, C.-Y. Lin, Y.-C. Tseng, and J.-P. Sheu. A new multi-channel MAC protocol with on-demand channel assignment for multi-hop mobile ad hoc networks. In *Proceedings of the 2000 International Symposium on Parallel Architectures, Algorithms and Networks (ISPAN)*, Dallas/Richardson, TX, December 7–10, pp. 232, 2000.
12. S.-L. Wu, C.-Y. Lin, Y.-C. Tseng, and J.-P. Sheu. A multi-channel MAC protocol with powercontrol for multi-hop mobile ad-hoc networks. *Computer Journal*, 45(1), 101–110, 2002.
13. W.-C. Hung, K.L.E. Law, and A. Leon-Garcia. A dynamic multi-channel MAC for ad-hoc LAN. In *Proceedings of the 21st Biennial Symposium on Communications*, Kingston, Ontario, Canada, pp. 31–35, June 2–5, 2002.
14. A. Tzamaloukas and J.J. Garcia-Luna-Aceves. Channel-hopping multiple access. In *Proceedings of IEEE ICC*, New Orleans, LA, pp. 415–419, June 18–22, 2000.
15. A. Tzamaloukas and J. Garcia-Luna-Aceves. Channel-hopping multiple access with packet trains for adhoc networks. In *Proceedings of the IEEE Device Multimedia Communications (MoMuC)*, Tokyo, Japan, pp. 1–6, October 23–26, 2000.
16. J. Chen, S. Sheu, and C. Yang. A new multichannel access protocol for IEEE 802.11 ad hoc wireless LANs. In *Proceedings of PIMRC*, Beijing, China, Vol. 3, pp. 2291–2296, September 7–10, 2003.
17. J. So and N. Vaidya. Multi-channel MAC for adhoc networks: Handling multi-channel hidden terminals using a single transceiver. In *ACM International Symposium on Mobile Ad Hoc Networking and Computing (MobiHoc)*, Tokyo, Japan, pp. 222–233, May 24–26, 2004.
18. R. Maheshwari, H. Gupta, and S.R. Das. Multichannel MAC protocols for wireless networks. In *Proceedings of IEEE SECON*, Reston, VA, pp. 1–10, September 25–28, 2006.
19. H.W. So and J. Walrand. McMAC: A multi-channel MAC proposal for ad-hoc wireless networks. In *Proceedings of the IEEE Wireless Communication and Networking Conference (WCNC)*, Hong Kong, pp. 1–5, March 11–17, 2007.
20. D. Friedlander, C. Griffin, N. Jacobson, S. Phoha, and R.R. Brooks. Dynamic agent classification and tracking using an ad hoc mobile acoustic sensor network. *EURASIP Journal on Applied Signal Processing*, 4, 371–377, 2003.
21. S. Phoha, N. Jacobson, D. Friedlander, and R.R. Brooks. Sensor network based localization and target tracking through hybridization in the operational domains of beamforming and dynamic space-time clustering. In *IEEE Global Telecommunications Conference*, San Francisco, CA, Vol. 5, pp. 2952–2956, 2003.
22. S. Phoha, J. Koch, E. Grele, C. Griffin, and B.B. Madan. Space-time coordinated distributed sensing algorithms for resource efficient narrowband target localization and tracking. *International Journal of Distributed Sensor Networks*, 1, 81–99, 2005.
23. S. Phoha and A. Ray. Dynamic information fusion driven design of urban sensor networks. In *IEEE International Conference on Networking, Sensing and Control*, London, U.K., pp. 1–6, April 15–17, 2007.
24. Y. Wen, D. Bein, and S. Phoha. Middleware for heterogeneous sensor networks in urban scenarios. In *Seventh International Conference on Information Technology: New Generations (ITNG 2010)*, Las Vegas, NV, April 12–14, 2010.
25. S. Phoha, A. Ray, I. Chattopadhyay, G. Mallapragada, and Y. Wen. Mathematical modeling of sensor network dynamics for control and stability. Technical Report TR eSensIF-08-01, Applied Research Laboratory, The Pennsylvania State University, University Park, PA, September 2008.
26. I. Chattopadhyay and A. Ray. Structural transformations of probabilistic finite state machines. *International Journal of Control*, 81(5), 820–835, 2008.
27. S. Ahn and K. Chong. A case study on message-oriented middleware for heterogeneous sensor networks, In *Embedded and Uniquitous Computing*, E. Sha et al. (eds) Vol. LNCS 4096, pp. 945–955. Springer Verlag, Berlin, Germany, 2006.

28. E.P. de Freitas, M.A. Wehrmeister, C.E. Pereira, and T. Larsson. Reflective middleware for heterogeneous sensor networks. In *Workshop on Reflective and Adaptive Middleware (ARM)*, Leuven, Belgium, pp. 49–50, December 1, 2008.
29. C. Muldoon, R. Tynan, M.J. O'Grady, and G.M.P. O'Hare. Realizing an agent-oriented middleware for heterogeneous sensor networks. In *ACM/IFIP/USENIX Middleware Conference Companion*, Leuven, Belgium, pp. 82–83, December 1–5, 2008.
30. V. K. Tumuluru, P. Wang, and D. Niyato. Channel status prediction for cognitive radio networks. *Wireless Communications and Mobile Computing*, 2010. doi 10.1002/WCM.1017.
31. A. Achtzehn, Z. Benenson, and C. Rohner. Implementing agreement protocols in sensor networks. In *IEEE International Conference on Mobile Adhoc and Sensor Systems (MASS)*, Vancouver, British Columbia, Canada, pp. 858–863, October 2006.
32. B. Horine and D. Turgut. Link rendezvous protocol for cognitive radio networks. In *IEEE International Symposium on New Frontiers in Dynamic Spectrum Access Networks (DySPAN)*, Dublin, Ireland, pp. 444–447, April 17–20, 2007.
33. M.D. Silvius, A.B. MacKenzie, and C.W. Bostian. Rendezvous MAC protocols for use in cognitive radio networks. In *IEEE Military Communications Conference (MILCOM)*, Boston, MA, pp. 1–7, October 18–21, 2009.
34. R. Etkin, A. Parekh, and D. Tse. Spectrum sharing for unlicensed bands. *IEEE Journal on Selected Areas in Communications*, 25(3), 517–528, 2007.
35. T.S. Rappaport. *Wireless Communications*, 1st edn. Prentice Hall, Upper Saddle River, NJ, 1996.
36. S. Gandham, M. Dawande, and R. Prakash. Link scheduling in sensor networks: Distributed edge coloring revisited. In *Proceedings of the 24th Annual Joint Conference of the IEEE Computer and Communications Societies (INFOCOM 2005)*, Miami, FL, Vol. 4, pp. 2492–2501, March 13–17, 2005.
37. I. Demirkol, C. Ersoy, and F. Alagoz. MAC protocols for wireless sensor networks: A survey. *IEEE Communications Magazine*, 44(4), 115121, 2006.
38. E. Blasch, I. Kadar, K. Hintz, J. Biermann, C. Chong, and S. Das. Resource management coordination with level 2/3 fusion issues and challenges. *IEEE A&E Systems Magazine*, 32–46, 2008.
39. C.T. Kelley. *Iterative Methods for Optimization*. SIAM Frontiers in Applied Mathematics, Philadelphia, PA, 1999.

24

Distributed Detection of Replica Node Attacks

Jun-Won Ho
Seoul Women's University

Donggang Liu
The University of Texas at Arlington

Matthew Wright
The University of Texas at Arlington

Sajal K. Das
The University of Texas at Arlington

24.1 Introduction

Wireless sensor networks are an important tool in security-sensitive scenarios, such as battle field surveillance, nuclear and chemical attack detection, intrusion detection, flood detection, weather forecasting, traffic surveillance, and patient monitoring [1]. In such scenarios, protecting the network from adversaries who seek to undermine the network's operations or steal its sensitive information is critical to the network's mission.

This task is made difficult, however, by the nature of the network's deployment. Generally, the network operator deploys a base station and a set of small sensor devices in the network field. The sensor devices form ad hoc networks, collaborate with each other to sense the phenomena associated with the assigned missions, and then send the sensory data to the base station. The sensors are deployed in potentially hostile environments where enemies may be present. Since wireless sensor networks (WSNs) usually need to be controlled remotely by the network operator, they are often deployed in an unattended manner. The unattended nature of wireless sensor networks can be exploited by adversaries. Specifically, an adversary can capture a sensor node and extract its keying materials. Using these keying materials, he/she can make *replica nodes*, which are sensor nodes with the ability to make and use keys that only legitimate sensor nodes should have [22]. He/she can then spread those replica nodes over the field and launch various attacks.

These *replica node attacks* are very dangerous to the operation of sensor networks. With a single captured sensor node, the adversary can create as many replica nodes as he/she wants. The time and effort needed to inject many replica nodes into the network is much less than the effort to compromise the equivalent number of original nodes. The replica nodes are controlled by the adversary but have keying materials that allow them to seem like authorized participants in the network. Protocols for secure sensor network communication would allow replica nodes to create pairwise shared keys with other nodes and the base station, thus enabling the nodes to encrypt, decrypt, and authenticate all of their communications as if they were the original captured node.

The adversary can then leverage this insider position in many ways. For example, he/she can simply monitor a significant fraction of the network traffic that would pass through these nodes. Alternatively, he/she could inject falsified data to corrupt the sensors' monitoring operation. A more aggressive attacker could undermine common sensor network protocols, including cluster formation and data aggregation, thereby causing continual disruption to the network's operations. Therefore, an adversary with a large number of replica nodes can easily defeat the deployed mission of sensor networks.

A straightforward solution to stop replica attacks is to prevent the adversary from extracting keying materials from sensor nodes by equipping them with tamper-resistant hardware. However, this solution may be too expensive to be practical for many applications. In addition, although tamper-resistant hardware can make it significantly harder and more time consuming to extract keying materials from captured nodes, it may still be possible to bypass tamper resistance for a small number of nodes in a reasonable amount of time. In this scenario, replica attacks are far more dangerous, as the adversary can leverage the compromise of a single node to make many malicious nodes. Compromising that many distinct nodes would take much more time and effort. We thus believe that it is necessary to develop distributed, software-based countermeasures to address the threat of node replicas.

24.1.1 Chapter Organization

In this chapter, we examine ways to defend against replica attacks. Most schemes in the literature are based on detecting these attacks, as prevention is only possible by preventing node compromise. We begin with a description of the work of Parno et al. [22] in Section 24.2, who first began work on this problem, as their methods provide a useful introduction to most of the other solutions we will discuss. The main idea of these schemes is to have nodes report *location claims* that identify their positions and attempt to detect conflicting reports that signal one node in multiple locations. While effective, their schemes require sensors to expend substantial energy due to computation and communication. The high overheads have motivated a number of researchers to seek alternative solutions with lower costs.

Before describing such alternatives, we first carefully lay out the assumptions and adversary model for the replica detection problem in Section 24.3, including assumptions regarding group deployment. In group deployment scenarios, sensor nodes are put together into groups and placed in the network such that all nodes know approximately where each group will be deployed. This should be realistic for many deployment scenarios, as sensor nodes can be dropped from airplanes or scattered over an area by hand. In such a case, the sensors can be preloaded with relevant knowledge about their own group's membership and all group locations. Then, the sensors in the same group should be deployed at the same time in the location given to that group.

Based on these assumptions, Section 24.4 describes a set of group-based replica detection schemes first presented in Ref. [14]. These schemes adapt the location claim idea from Ref. [22]. However, given that nodes are deployed in groups, these schemes allow most nodes to communicate without generating any location claims as long as they are able to directly send messages to at least one of their group members. This simple idea enables the schemes to significantly reduce the overhead of sending, receiving, and verifying location claims. Additionally, if we assume loose time synchronization for the sensors, the schemes can be adapted to allow nodes to accept messages from any node that has been deployed within a small window of its expected deployment time. This extension eliminates most of the overheads of the scheme, except when nodes are not deployed close to their expected deployment time nor close to their expected deployment location.

Section 24.5 compares these schemes to Ref. [22] in terms of the communication, computational, and storage overheads. In particular, we present the results of analysis and simulation that demonstrate that these group deployment schemes are very effective at detecting and blocking replica nodes while incurring substantially lower overheads than those in Ref. [22]. These detection schemes can be made arbitrarily efficient by increasing the accuracy of deployment knowledge. Networks with very accurate deployments, in both time and location, can reach practically zero overhead for benign nodes.

Section 24.6 describes several other proposed replica detection schemes. Two of these schemes are centralized, while two other schemes are decentralized and also draw on the location claim technique from Parno et al. [22], with variations designed to save on system overhead. We provide a table comparing the communication overhead of these schemes. We then conclude the chapter in Section 24.7.

24.2 Location-Based Replica Detection

In this section, we provide a brief overview of the replica detection schemes of Parno et al. [22], as these schemes were the first to address the problem and the first to use the location claim technique, which others have since applied. We begin with their centralized scheme, which will be helpful as we then describe the use of location claims in these and related approaches. Then we overview their schemes for distributed replica detection.

24.2.1 Centralized Scheme

First, Parno et al. discuss the drawbacks of a fully centralized technique [22]. They first outline how such a scheme would work: similar to a scheme proposed for Sybil detection by Newsome et al. [21], all nodes send a list of their neighbors and those neighbor's claimed locations to the base station. If any node reports itself as being in more than one location, the base station detects it as a replica node. While effective, this scheme suffers from a single point of failure in the form of the base station and is vulnerable to direct attacks on the base station, attacks on the nodes near the base station, and denial of service (DoS) or wireless jamming against the base station or its neighborhood. Furthermore, some networks may not have a powerful base station at all, in which case the scheme is not usable.

24.2.2 Location Claims

Note that the centralized scheme makes use of a *location claim*, in which a node reports its location. Since location claims are central to this and other schemes that we discuss in this chapter, we now explain the subtleties of how location claims work in replica detection. A location claim should include the claiming node's identity and location and a digital signature over these two pieces of information. The claim can then be forwarded to a detector. In the centralized scheme described earlier, the base station serves as the detector. An important point is that the claim must be checked by the neighboring node before it is forwarded. Specifically, the location of the claiming node must be close enough to be in radio range to the neighbor forwarding on the claim, based on the neighbor's own estimate of its location. If the claimed location is too far from the neighbor, based on a threshold distance, then the neighbor must not forward the claim. Otherwise, all of the replicas could simply claim they are in one location, despite being located all over the network. Furthermore, the neighbor should only communicate with a node that has sent a properly formed claim to the base station with an acceptable location. Any given node can be assured of this by forwarding the location claim itself, or the forwarding could be done by each neighbor probabilistically to provide a high chance of correct forwarding by at least one honest neighbor.

Note that if any neighbor is compromised, it will forward the replica's claims and other traffic anyway. Further, a node can have as many replica nodes as it wants in the radio range of a single honest node. We argue in Section 24.3 that either approach is wasteful for the attacker; simply put, the compromised node or the single replica, respectively, is just as effective in attacking the network without the (additional) replica.

From the detector's perspective, two location claims for the same identity can be verified using the signatures. If the claims are both valid, the detector has a simple proof that the identity and its keys are compromised, as either the identity is present in two locations or the signing key has been used to validate a lie. Thus, the use of signatures also prevents false positives. The two signed location claims can also be spread throughout the network to serve as an easily validated notice of node revocation. Because of these properties, signed location claims are a very valuable tool in protecting sensor networks from replica attacks, despite the high computational costs of signature generation and validation.

24.2.3 Distributed Detection

Based on the use of location claims, Parno et al. propose two distributed replica detection schemes: randomized multicast and line-selected multicast [22]. In the randomized multicast scheme, every node sends its location claim to its neighbors. Upon receiving the location claim, each neighbor forwards the location claim to randomly chosen witness nodes with a certain probability. A witness node that receives two conflicting location claims for a node decides that the node is a replica and initiates a process to revoke the node. In the line-selected multicast scheme, every claim-relaying node as well as every witness node participates in the replica detection and revocation process, leading to a reduction in the communication overhead of the randomized multicast scheme.

While effective, this requires every node to sign and send a location claim, as well as to verify and store the signed location claim of a number of other nodes. Thus, this requirement results in significant communication, computation, and storage overheads, which may cause the sensor nodes to run out of power. In the rest of this chapter, we focus on detection mechanisms that also provide high detection capability with lower overheads than these multicast schemes.

24.3 Assumptions and Model

In this section, we present the underlying assumptions for the replica detection problem, including the network model, attacker model, and design goals. We then describe a set of assumptions for group deployment that are central to the schemes we describe in Section 24.4.

24.3.1 Network Assumptions

We study replica detection in a two-dimensional *static* sensor network where the locations of sensor nodes do not change after deployment. We also assume that all direct communication links between sensor nodes are bidirectional, which is common in the current generation of sensor networks.

24.3.2 Attacker Model

We assume that an adversary may compromise and fully control a subset of the sensor nodes. However, we place some limits on the ability of the adversary to compromise nodes. If the adversary can compromise a major fraction of the network, he/she will not need or benefit much from the deployment of replicas. We assume that the attacker can identify and compromise a substantial fraction of the nodes in a small area. He/she will subsequently make replicas of one or more of these nodes and attempt to distribute them throughout the network. We do not specifically limit the attacker from compromising nodes throughout the network, but we note that there is little benefit to the attacker of having a replica node in the same area as another compromised node. The compromised node can just as easily report fake data, participate in local control protocols, and eavesdrop on messages sent through it. Furthermore, if the attacker needs one compromised node to accompany each replica node in the network, there will be a very high cost for launching replica node attacks. This assumption goes unstated but is implied by the use of signed location claims in other replica detection schemes [5,22]. It is also worth noting that the attacker does not benefit from having multiple replicas of a single node in the same small region. For example, in a false data injection attack, it would be easy to ensure that only one of the replica nodes' data values at a time is accepted by the data aggregators. In local control protocols, each node could only have input once, so multiple nodes with the same ID would not have more influence in a region than a single node.

24.3.3 Design Goals

Under the aforementioned system and attacker models, we identify three key design goals for replica detection schemes. First, replica nodes should be detected with minimal communication, computational, and storage overheads. Second, the detection schemes should be robust and highly resilient against an attacker's attempt to break them. More specifically, the schemes should detect replicas unless the attacker compromises a substantial number of nodes. Finally, there should be no false positives, meaning that only compromised and replica nodes would be detected and revoked. This is important to prevent the attacker from turning a replica detection scheme into a tool for DoS attacks.

24.3.4 Sensor Deployment

The key assumption in the schemes described in Section 24.4 is the use of a group deployment strategy. In this strategy, sensor nodes are grouped together by the network operator and programmed with the corresponding group information before deployment, with each group of nodes being deployed toward the same location, called the *group deployment point*. After deployment, the group members exhibit similar geographic relations. We argue that this is reasonable for any sensor network in which nodes are spread over a field, such as being dropped from an airplane or spread out by hand. A simple way to do this would be to keep the groups of nodes in bags marked with the group IDs and use a marked map with the group IDs on it. All that is needed is a map of the territory and a way to predetermine the deployment points, such as assigning a point on a grid to each group. This argument is further supported by the fact that the group deployment strategy has been used for various applications in sensor networks such as key distribution [8,20], detection of anomalies in localization [9], and public key authentication [10].

The deployment follows a particular probability density function (pdf), say f, which describes the likelihood of a node being a certain distance from its group deployment point. For simplicity, we use a two-dimensional Gaussian distribution to model f, as in Ref. [9]. Let (x_G, y_G) be the group deployment point for a group G. A sensor node in group G is placed in a location (x, y) in accordance with the following model:

$$f(x,y) = \frac{1}{2\pi\sigma^2} e^{-((x-x_G)^2+(y-y_G)^2/2\sigma^2)} \tag{24.1}$$

where

(x, y) is group deployment point

σ is the standard deviation of the two-dimensional Gaussian distribution

Given this predeployment knowledge, it is still nontrivial to perform replica detection in sensor networks due to the difficulty in *precisely* deploying sensor nodes at their expected group deployment points.

Hence, we may make wrong decisions when predeployment knowledge is used for replica detection, and there are errors during the actual deployment. How the system handles these deployment errors will be an important issue explored in the rest of this chapter.

24.4 Group-Based Replica Detection

This section presents the details of several group-based techniques to stop node replica attacks by using deployment knowledge. These were originally presented in Ref. [14]. We start by describing a basic approach and then show how to improve the security as well as the performance of this approach with additional knowledge about the deployment of sensor nodes. Table 24.1 lists notations used frequently in our discussion.

24.4.1 Scheme I: The Basic Approach

As mentioned earlier, these schemes assume that sensor nodes are deployed group by group and that each group is expected to be deployed toward a *deployment point* that can be predetermined. Prior to deployment, the network operator loads the predetermined deployment coordinates of every group onto every sensor node. Recall that sensor nodes in the same group are very likely to be close to each other after deployment. The schemes use this knowledge to stop node replica attacks. A simple way to do this would be to have each node only accept messages from members of its own group. However, this stops intergroup communication. We extend this idea by defining nodes that are close to their group deployment point as *trusted* and nodes that are far from their group deployment point as *untrusted*. Nodes will accept messages from trusted neighbors, while ignoring messages from untrusted nodes. Let us now describe the basic approach.

24.4.1.1 Protocol Description

Before deployment, the sensor nodes in the network are divided into g groups each with an equal number of nodes m and a unique group ID. The ID of every sensor node has two parts: the group ID and a unique ID within the group. Keying materials are also preloaded onto each sensor node for pairwise key establishment; we can use any key predistribution technique for sensor networks, such as the ones in Refs. [3,6,18,28]. Let us assume that every message in the system is protected by a pairwise key and that all messages include the ID of the sender in plaintext. This prevents simple spoofing attacks, as receivers will use the pairwise key matching the ID seen by all intermediate nodes.

Suppose that a sensor node u (a member of the group G_u) receives a request from its neighbor node v (a member of the group G_v) to forward a message. Node u first checks whether the distance between the deployment points of groups G_u and G_v is smaller than a predefined system-wide threshold distance, d. If so, u believes that node v is a trusted neighbor and accepts and forwards messages from node v. Otherwise, it will ignore all messages from node v. Intuitively, a node that is not near its own group members will be isolated and unable to send messages. The deployment points used in the aforementioned protocol are predetermined before the deployment and can be easily inferred from the group IDs. Neither u nor v can claim an incorrect deployment point even if they are compromised.

TABLE 24.1 Notations Used Frequently in This Chapter

N	Total number of sensor nodes
g	Total number of groups
m	Group size
b	Average neighbor size of a node
G_u	Group to which node u belongs
$\|\|$	Concatenation symbol

From the aforementioned description, we can clearly see that the basic scheme does not introduce any significant communication, computation, or storage overhead. Each node can immediately determine whether to forward the messages from other sensor nodes by only looking at their IDs. This approach, as the others proposed in this chapter, makes these decisions just once, whenever a node first contacts its neighbors. We now examine the security and effectiveness of the basic approach.

24.4.1.2 Security Analysis

Suppose that the adversary has already compromised node u and placed a number of replicas of u in the field. From the protocol description, we know that a replica node u' of u cannot change its group deployment point since it is predetermined before deployment. Hence, we see that only the nodes that are supposed to be deployed less than d meters away from the group deployment point of node u will accept replicas with ID u as their trusted neighbor nodes. In other words, node u's replicas will be accepted by the following set of nodes: $O = \{i|\text{dist}(i, u) \leq d\}$, where $\text{dist}(i, j)$ is the distance between the deployment points of groups G_i and G_j.

We now determine the maximum impact of a replica node attack. Let m be the number of nodes in each group, g be the total number of groups, and S be the size of the target deployment field. Assume that the predetermined group deployment points are randomly distributed in the field. Consider a particular group G_u. The average number of groups whose deployment points are no more than d meters away from the deployment point of group G_u can be estimated by $\frac{(g-1)\times\pi\times d^2}{S}$. We define the *affected* nodes as those nodes that will accept the replicas of a particular node u as their trusted neighbors. Thus, the average number of the affected nodes can be estimated by

$$\frac{m\times(g-1)\times\pi\times d^2}{S} \tag{24.2}$$

To concretely show the effect of replica nodes, we will examine a specific scenario. Suppose that $N = 1000$ sensor nodes are deployed in a square field of $S = 1000 \times 1000\,\text{m}^2$. Each sensor node can communicate with other sensor nodes in a radius of 50 m. These 1000 sensor nodes are divided into $g = 20$ groups with $m = 50$ nodes in each group. Figure 24.1 shows the effect of replica nodes. We can see that the number of affected sensor nodes increases with d. When $d = 81$ m, less than 2% of the nodes are affected. However, when $d = 185$ m, over 10% of the nodes become affected. Thus, we need to set d as small as possible to

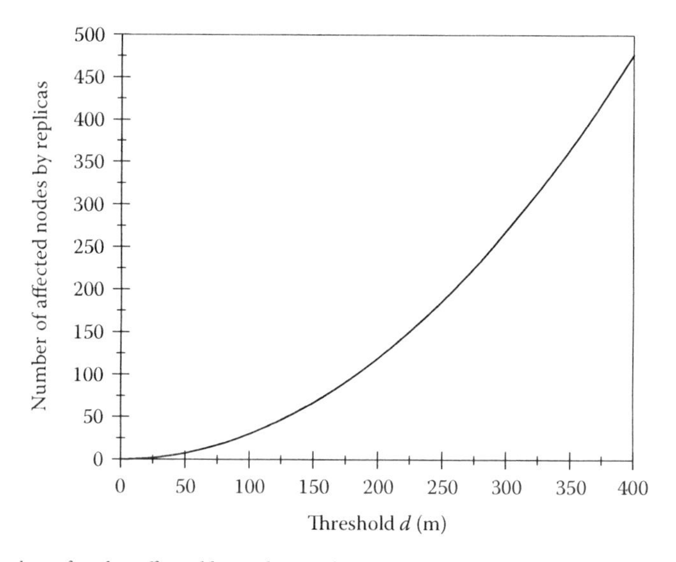

FIGURE 24.1 Number of nodes affected by replica nodes.

mitigate the impact of node replica attacks on the network. However, having a small value for *d* causes many nodes to be unable to communicate, as shown in the next part of our analysis.

24.4.1.3 Connectivity

In the basic approach, every sensor node will reject forwarding data packets for those untrusted neighbor nodes. However, according to the group deployment model, some benign nodes may be deployed at placement points that are far from their group deployment points. As a result, these benign nodes will be considered as untrusted and rejected by their neighbors. These are honest nodes, but they are unable to communicate and are effectively dead on arrival. In an extreme case, the network could be partitioned. We now examine *the number of nodes without connectivity* in benign situations.

To investigate the number of nodes disconnected from the rest of the network, we developed a simple simulation using the same network settings as earlier. Figure 24.2 shows the number of disconnected nodes for different values of threshold distance (*d*) and standard deviation (σ). We observe that the number of such nodes decreases rapidly with increasing *d*. However, increasing *d* will also allow the attacker to generate a larger impact (more replicas being accepted as trusted neighbors) on the network. When $\sigma = 50$, that is, the deployment is fairly accurate, almost all nodes have connectivity for $d > 100$ m. When $\sigma = 200$ and $d = 100$ m, nearly 30 nodes are affected by the attacker, but 104 nodes are without a connection. When $\sigma = 200$ and $d = 200$ m, the attacker impacts nearly 119 nodes while 26 benign nodes remain disconnected. Obviously, we have to make a trade-off for the choice of *d*, and that trade-off is significantly impacted by the accuracy of the sensor deployment.

24.4.2 Scheme II: Location Claim Approach

In the basic approach, sensor nodes only forward messages for their trusted neighbors. As discussed, when the deployment is not very accurate, many benign nodes may be rejected by their neighbors for message forwarding, and the sensor network may be poorly connected or even partitioned when the application requires high resilience against replica node attacks. To address this problem, the detection scheme can have sensor nodes also forward messages from untrusted neighbors as long as they provide provable *evidence* that they are not replicas. The evidence includes the location of the node requesting message forwarding. The evidence from an untrusted node will be sent to a predetermined location for replica detection; when two conflicting pieces of evidence reach this location, the replica will be detected.

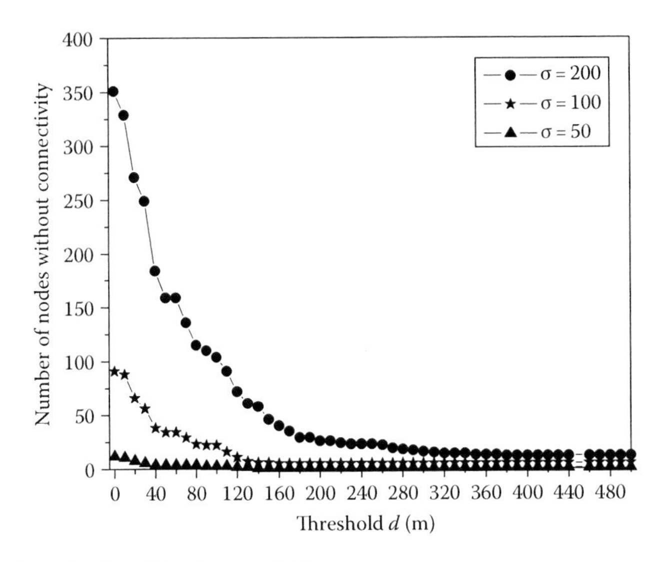

FIGURE 24.2 Number of nodes without connectivity.

24.4.2.1 Protocol Description

In Scheme II, in addition to the keying materials for pairwise key establishment, every sensor node also gets the keying materials for generating digital signatures. The scheme uses an identity-based public key scheme. It has been demonstrated that public key operations can be efficiently implemented in static sensor devices [13,17,25]. Moreover, most existing replica detection schemes for sensor networks [5,22,29] employ an identity-based public key scheme. Scheme I only requires the knowledge of group deployment points of sensors. The design of Scheme II also assumes that a secure software-based localization method has been employed in the network such as the ones in Refs. [2,16,19]. This allows a sensor node to discover its actual placement point after deployment. The scheme consists of four phases as described in the following.

Phase 1: Deployment. Let us define the *deployment zone* of a group as a circle centered at the group's deployment point with radius R_z. Here R_z is a configurable parameter of the system and the network operator should set it to balance security and performance. When discussing a given node u, we call the corresponding deployment zone of group G_u the *home zone*. After deployment, every node u discovers its real location, L_u. If u resides outside its home zone, it produces a *location claim* $C_u = \{u||L_u||Sig_u\}$, where Sig_u is the signature generated by u's private key.

Phase 2: Neighbor discovery. After deployment, every sensor node u discovers a set of neighbors $N(u)$ and asks for an authenticated location claim from every node $v \in N(u)$. Upon receiving a claim request, v sends u a location claim C_v if v is placed outside its home zone. Otherwise, it just sends u the message $M_v = \{v||L_v||MAC_{K_{uv}}(v||L_v)\}$, where MAC stands for message authentication code and K_{uv} is the shared secret key between u and v. Any node v will be removed from $N(u)$ if C_v or M_v fails to authenticate. Also, if v claims a location L_v such that the distance between L_u and L_v is larger than the communication range of u, then v will be removed from $N(u)$. After this, u checks every node $v \in N(u)$ to see if it is deployed in the right place (i.e., node v's home zone). This can be done by checking whether the distance between L_v and the deployment point of group G_v is less than R_z. If node v is indeed in its home zone, then u will mark node v as a *trusted* node; otherwise, it will mark v as *untrusted*.

Phase 3: Claim forwarding. A sensor node u will forward regular messages from its untrusted neighbor v, only if it has received and verified a location claim C_v. When u gets claim C_v, it forwards C_v to v's home zone with probability P_f to ensure that at least one claim reaches the home zone of G_u. If node u is going to forward the claim, it sends C_v to the deployment point of group G_v. Since most nodes in G_v will be deployed in v's home zone, delivering the claim to the deployment point of G_v will reach many sensor nodes in G_v.

Phase 4: Detection and revocation. Once an authenticated location claim reaches the home zone of node v, it will be flooded throughout the home zone of G_v. Such local flooding is inexpensive since it is confined to a small area. This ensures that every node in group G_v residing in v's home zone will know C_v. If the nodes in the home zone of v receive two conflicting claims, they conclude that node v has been replicated. The two conflicting claims can be used as evidence to revoke node v from the network. This can be most easily accomplished by broadcasting the conflicting claims throughout the network.

The scheme can employ either a coordinated or uncoordinated approach to broadcast the conflicting claims. In the coordinated approach, each node takes a role of broadcasting the conflicting claims in a round robin manner. In the uncoordinated approach, the node that broadcasts the conflicting claims is chosen in accordance with a pseudorandom number generator based on the ID value of the replica. Upon receiving the conflicting claims, every node can independently verify them and take revocation on node v. Note that revocation can be lazy, in that a node does not need to verify the claims until v attempts to be its neighbor. In this way, replicas are detected and revoked in a fully distributed manner.

24.4.2.2 Security Analysis

Suppose node u has been compromised. If all the replicas of u generate the same location claim C_u, they can only affect the nodes that are no more than r meters away from L_u, where r is the communication range of a node. This will not offer the adversary much benefit. In addition, if all the claims include locations that reside in the home zone of node u, they can only affect the nodes that are no more than $2r$ meters away from the deployment point of group G_u. This also does not give the attacker much benefit since it can only impact a small area. As a result, we focus on the probability $P_d(c)$ of detecting replica nodes when the attacker makes c different location claims outside of the home zone of node u.

Note that, for the calculation of $P_d(c)$, we ignore the possibility of compromised nodes failing to forward the location claim. If any significant fraction of the neighbors of a replica node are compromised, they already have substantial influence over that part of the network and hence adding replicas does not substantially help them, as discussed in Section 24.3.

For a particular location claim of node u, the probability P_z that a given node of group G_u resides in the home zone of u can be estimated by

$$P_z = \int_0^{2\pi}\int_0^{R_z} f(\rho\cos\theta, \rho\sin\theta)\rho d\rho d\theta = 1 - e^{-(R^{(2/z)}/2\sigma^2)} \quad (24.3)$$

where

σ is the standard deviation of the two-dimensional Gaussian distribution

R_z is the radius of the home zone of node u

Since these c claims are for locations that are outside of the home zone of node u, they will all be delivered to u's home zone. Assume that every claim can be reliably delivered to the home zone of u. Let m and f_c be the group size and the fraction of compromised nodes in the network. We estimate the probability P' that location claim can find at least one benign node in the home zone of node u. The probability P' can be estimated as

$$P' = 1 - [1 - P_z \times (1 - f_c)]^m$$

With P', we can estimate the probability $P_d(c)$ as

$$P_d(c) = 1 - (1 - P')^c - c \times P'(1 - P')^{c-1} \quad (24.4)$$

Using Equation 24.4, we study how $P_d(c)$ is affected by the different values of f_c and P_z. As shown in Figure 24.3, we see that Scheme II achieves a detection probability of $P_d(c) = 0.99$ as long as $P_z \geq 0.5$, even if 80% of the nodes in the home zone are compromised. If P_z reaches a very low value such as 0.1, and

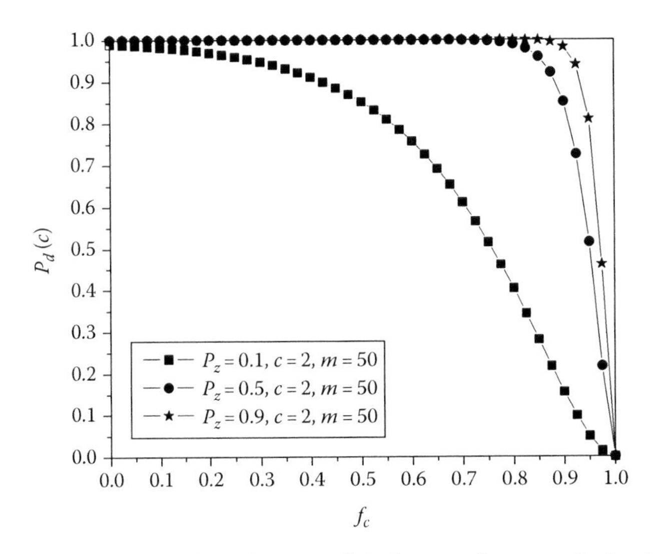

FIGURE 24.3 Detection probability $P_d(c)$ as a function of the fraction of compromised nodes f_c.

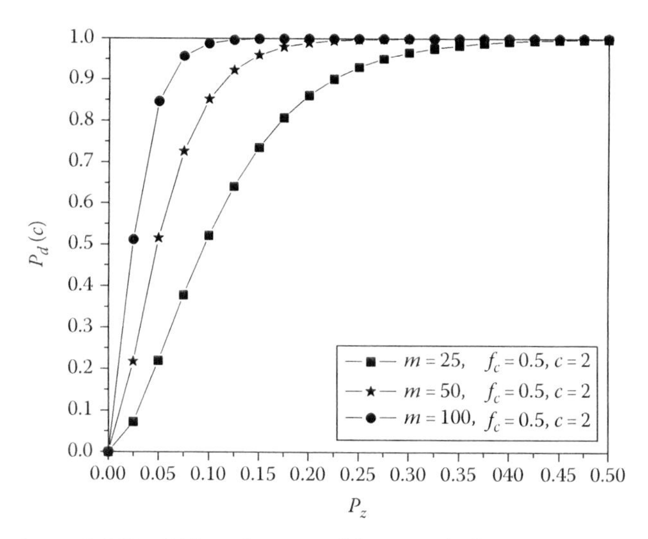

FIGURE 24.4 Detection probability $P_d(c)$ as a function of the group deployment accuracy P_z.

there is a high fraction of compromised nodes, the detection capability would be degraded. However, this scenario implies that the accuracy of group deployment was very poor.

We also study how m and P_z impact $P_d(c)$. From Figure 24.4, we observe that the replica detection probability tends to increase in proportion to the rise of group size (m) when $P_z \leq 0.4$. Moreover, Scheme II achieves a detection probability of $P_d(c) = 0.99$ as long as $m \geq 50$ for $P_z \leq 0.2$ and $f_c = 0.5$. This means that the group size of $m = 50$ is enough to achieve replica detection with very high probability of 0.99 even when the group deployment accuracy is low and there is a substantial fraction of compromised nodes.

Since every location claim is digitally signed, an adversary cannot make uncompromised nodes appear to be replicas by faking their location claims. Thus, the replica detection scheme works without any false positives.

24.4.2.3 Communication Overhead

We define communication overhead as the average number of location claims that are sent and received by nodes in the network. First, we emphasize that claims are only generated and sent once at the time when new untrusted nodes are deployed. Thus, Scheme II has lower overheads in the lifetime of the network than schemes with repeated claim generation and verification, such as in Ref. [5].

We begin by estimating the communication overhead in benign situations when there are no replica nodes in the field. In this case, the sensor nodes that reside in their home zone will not incur any additional overhead. Thus, on average, the fraction of sensor nodes whose location claims need to be forwarded to their home zones can be estimated by $1 - P_z$. This means that we often prefer a large probability P_z of being in the home zone for communication efficiency. This is also beneficial to replica detection, as shown earlier, so the accuracy of deployment is very important to the scheme. If deployment accuracy is poor, we can keep P_z high by increasing R_z. The downside is that it will increase the size of home zones within which replica nodes would be trusted.

In benign situations, the distance between the node and its home zone will be relatively small. Just as we defined home zones, we can also define *communication bands* around the home zone. Each communication band B_i is a circle around the group deployment point with radius $R_i = R_{i-1} + \alpha r$, where r is the communication radius of each node and α is a fraction set to ensure that every node in B_{i+1} can send a packet to at least one node in B_i. Band B_0 has radius R_z and is the same as the home zone. Inductively, this means that $R_i = R_z + i \times \alpha r$.

The probability that a node exists in a given band B_i is given by

$$P_i = \left(1 - e^{-(R^{2/i}/2\sigma^2)}\right) - \left(1 - e^{-(R^{2/(i-1)}/2\sigma^2)}\right)$$

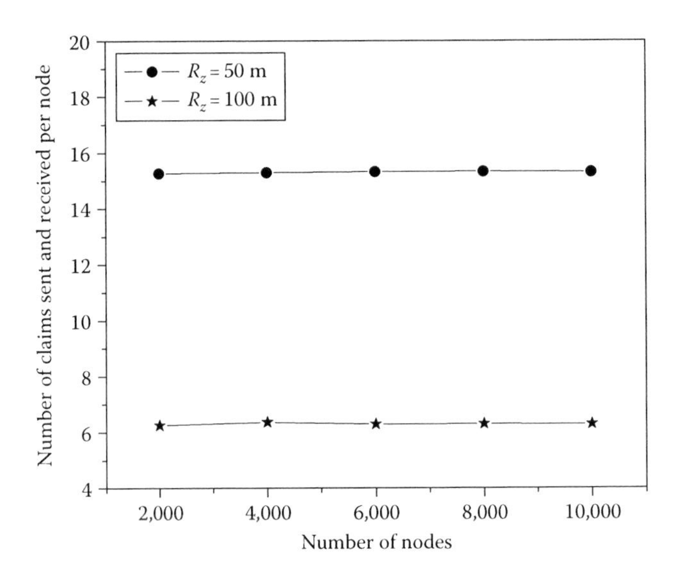

FIGURE 24.5 Communication overhead per node.

We can conservatively assume that if a node is in band B_i, then all of its neighbors further from the deployment point are in band B_{i+1}. Thus, the worst-case communication overhead per node can be estimated as

$$C_s = 2bP_f \sum_{i=1}^{\infty} P_i(i+2)$$

where b is the average neighbor size.

We developed a simple simulation experiment to investigate the average case of communication overhead per node. We simulated the scheme by varying the number of nodes from 2,000 to 10,000 in a field of 500 × 500 m^2. In this simulation, we set $\sigma = 50$, and let $m = 50$ be the number of nodes in a group and $r = 50$ m be the communication radius of each node. We also configured P_f in such a way that one claim per node is forwarded to the node's home zone. Figure 24.5 shows the communication overhead per node when $R_z = 50$, 100 m. We observe that the communication overhead is quite reasonable, less than 15.33 and 6.37 operations per node in all cases when $R_z = 50$ m and $R_z = 100$ m, respectively.

24.4.2.4 Computation and Claim Storage Overhead

We define computation and claim storage overhead as the average number of public key signing and verification operations per node and the average number of claims that need to be stored by a node, respectively. A sensor node residing outside its home zone only needs to perform a single signature generation operation on its location. Assume that every sensor node has b neighbors on an average and m is the group size. Each of these neighbors is outside of its home zone with probability $(1 - P_z)$. Thus, if a sensor node resides outside of its home zone, it needs to verify no more than $b(1 - P_z)$ signatures on average. When this node resides in its home zone, it also needs to verify and store those signatures forwarded to the home zone. Note that it only needs to verify and store up to two different signatures for the same node. The number of signatures to be stored and verified for the location claims forwarded to the home zone is bounded by $2 \times (1 - P_z) \times m$, even if there are replica nodes. Hence, the average number of signatures a sensor node needs to store and verify is at most $(1 - P_z)2m$ and $(1 - P_z)(b + 2m)$, respectively. We can see concretely how much overhead this incurs by studying a sample scenario. Using Equation 24.3 and setting $R_z = 2\sigma$, we get $P_z = 0.865$. Then for $b = 40$ and $m = 50$, each node needs to store 13.5 claims and perform 18.9 signature verifications.

24.4.2.5 Risk of Denial of Service

We note that the attacker could attempt a DoS attack by flooding fake claims throughout the system. This would cause increased computational and communication overhead and could also cause sensor nodes to run out of power. This is a risk in all replica detection schemes, including those in Refs. [5,22].

The following attributes of the design keep this from being an unacceptable risk:

- The communication costs of flooding could be generated via almost any protocol, including those for data gathering, routing, and control. The protocol introduces no new vulnerabilities in this regard.
- The computational costs of checking flooded signatures would be borne by only one node per signature: either the neighbor who verifies the claim or the receiver in the home zone if the claim is initiated through a compromised node.
- The number of claims initiated by any single node is limited. Each neighbor of a given node v will only verify and forward the first claim that v sends it, since that is all that is required for this scheme.

Thus, Scheme II introduces a limited additional risk of DoS, in line with the risks introduced by prior schemes. Preventing these risks completely would be challenging.

24.4.3 Scheme III: Multigroup Approach

Scheme II is very effective at detecting replicas with low communication and computational overheads. It also achieves high replica detection capability even if attacker compromises 80% of the nodes in a given home zone. However, the scheme could become ineffective for a more powerful attacker. In particular, if the attacker compromises or physically destroys all the nodes in group G_u's home zone, then none of the location claims for nodes in G_u will reach their destination. Alternatively, the attacker could populate the home zone of node v, G_v, with many replicas of v that act as black holes for location claims from other replicas of v.

We now describe Scheme III, which is designed to protect against this kind of aggressive adversary. In this scheme, every sensor node sends its neighbor's location claims to multiple groups rather than a single group. This greatly improves the scheme's robustness, as tremendous effort would be required by the attacker to undermine all the groups used in detection. While this scheme has higher communication overhead, it can provide a trade-off between the overhead and resilience to attack.

24.4.3.1 Protocol Description

The protocol is similar to Scheme II. The difference is how it selects the groups to which location claims are sent. Specifically, a neighbor u of an untrusted node v will select multiple groups, called *detector groups*, and forward v's location claims to each of the detector groups with probability P_f. Additionally, u will forward the claim to v's group, G_v. Any of these groups that receives conflicting claims will initiate revocation on the replica node, thus creating greater resiliance to node compromise.

The scheme aims to keep overhead low by selecting detector groups close to the home zone of the untrusted node. Specifically, node u selects the detector groups by using the communication bands of group G_v. Assume that v is placed in band B_i, where B_0 is v's home zone. For each band B_j between B_i and B_1, node u will select a detector group G_j from among the groups that have a deployment point in B_j. Node u can select the group by using a pseudorandom number generator based on the values of G_v and j, for example, taking a cryptographic hash of $(G_v||j)$. For some band B_j, there may be no group with a deployment point in B_j. In this case, u must identify the next closest communication band with at least one group deployment point, say, band B_l, and select the group pseudorandomly from that band.

For example, consider the communication bands of group G_v. Let us denote the detector groups for bands B_1 and B_2 by G_1 and G_2, respectively. As shown in Figure 24.6, node v's claims are forwarded to the home zones of groups G_1 and G_2 as well as its home zone.

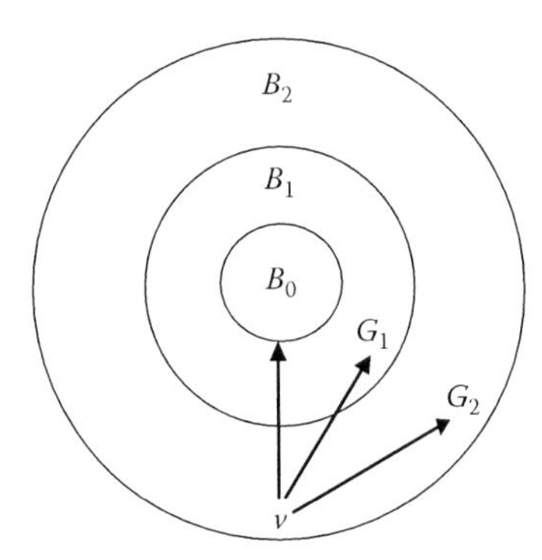

FIGURE 24.6 Multigroup claim forwarding.

24.4.3.2 Security Analysis

Similar to Scheme II, the network operator can configure the system parameters of Scheme III to ensure detection with very high probability. However, the main benefit of Scheme III is that the attacker needs to compromise multiple groups of nodes to prevent replicas from being detected. Let us assume that the attacker employs k replicas of node u in the network. Moreover, assume that k replicas are placed in bands $B_{\tau_1}, B_{\tau_2}, \ldots, B_{\tau_k}$, where $0 < \tau_1 \leq \tau_2 \leq \cdots \leq \tau_k$. Note that τ_k detector groups are selected for band B_{τ_k}. Also, u's group serves as a virtual detector group. Hence, at least two conflicting claims of u will be forwarded to the home zones of τ_{k-1} detector groups as well as u's home zone. Accordingly, the attacker needs to compromise $mP_z(\tau_{k-1} + 1)$ nodes in order to undermine the scheme. This means that the farther away the attacker places replicas from their home zone, the more nodes he/she needs to compromise to avoid replica detection.

24.4.3.3 Communication Overhead

Obviously, Scheme III requires more overheads than Scheme II, since a claim is forwarded to multiple home zones. However, if the group deployment accuracy is good, most nodes will be classified as trusted and even untrusted nodes will be placed closer to band B_0. Hence, Scheme III will work with relatively small additional overheads in a benign situation when compared with Scheme II. On the contrary, if the attacker places multiple replicas as far away from each other as possible, substantial overheads will be incurred by employing Scheme III. The scheme can further limit these overheads by extending the idea of Scheme I and blocking communications from nodes that are very far from their group deployment point. Note that, unlike in Scheme I, the network operator can choose a much larger value for the allowed distance d that a node can be from its deployment point. The value of d can be set to ensure that no legitimate nodes are blocked given reasonable deployment error, while still limiting overheads from untrusted nodes that are very far from their deployment points.

We now provide an analysis of the communication overhead. Assume that a node v is placed in the band B_i. Node v sends claims to the home zones of i detector groups as well as its home zone. Thus, the worst-case communication overhead per node C_m is given by

$$C_m = C_s + 2bP_f \sum_{i=1}^{\infty} P_i \sum_{j=1}^{i} (\beta_j + 1)$$

where

- C_s is the worst-case communication overhead per node in Scheme I
- β_j is the maximum number of bands between node v and the home zone of detector group selected for band B_j

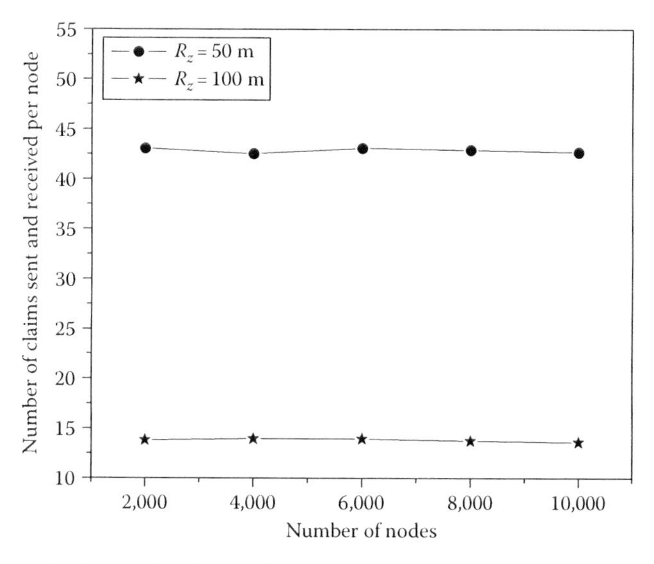

FIGURE 24.7 Communication overhead per node.

To investigate the average case of communication overhead per node, we extended the simulation environment used for studying Scheme II. Figure 24.7 shows the communication overhead per node when R_z = 50, 100 m. We observe that the communication overhead is less than 43.1 and 14.0 operations per node in all cases, when R_z = 50 m and R_z = 100 m, respectively. The communication overhead of Scheme III is approximately three and two times as much as that of Scheme II, when R_z = 50 m and R_z = 100 m, respectively. However, greater security resilience is achieved than Scheme II at the cost of this additional overhead. Thus, the communication overhead of Scheme III is still reasonable as a trade-off for greater security resilience.

24.4.3.4 Computation and Claim Storage Overhead

Similar to Scheme II, a sensor node residing outside its home zone only needs to perform a single signature generation operation on its location. Each node's claims are forwarded to multiple detector groups' home zones as well as its home zone. The average number of signatures that a sensor needs to store and verify is at most $2m(1 - P_z)$ in Scheme II. Thus, the average case of storage and signature verification overhead per node is bound by $2m(1 - P_z)(1 + D_{max})$ and $(1 - P_z)(b + 2m(1 + D_{max}))$, respectively, where D_{max} is the maximum number of times that a group serves as the detector group.

24.4.4 Deployment Time Checks

Thus far, the group-based schemes have used deployment locations to determine whether a node should be marked as trusted or not. As noted, it is possible that a substantial number of benign nodes are regarded as untrusted due to errors or inaccuracy in the placement of nodes. This increases the overheads of these group deployment schemes. To reduce these overheads, the schemes can additionally leverage the time of deployment as a measure of whether a node should be marked as trusted or not. The key idea is to preannounce the deployment time of each group and have nodes treat as trusted any node that initiates communications within a short time of its expected deployment.

This involves two important assumptions. First, it requires us to assume that it takes some time for an attacker to locate and compromise a sensor node. This need not be a long time, but let us assume that there is a minimum amount of time T_c that it takes to compromise a node once it has been deployed.*

* According to Ref. [15], it took approximately 1 min to compromise a node.

Second, we assume that the clocks of all nodes are loosely synchronized with a maximum error of ε. This can be achieved by the use of secure time synchronization protocols as proposed in Refs. [12,23,24]. We also note that if the network uses occasional broadcasts from the base station, authenticated time stamps could be included with little additional cost. Since time synchronization adds additional overheads, we do not recommend that this method be used unless time synchronization is already needed for some other reason, for example, to ensure correct time stamps on the data collected by sensors.

The deployment time checks can be used to enhance any of the schemes in this section, as they are just another way to decide whether a node can be trusted or not. If any node is trusted due to its location, then it remains trusted despite a late arrival to the network. While this means that replicas could be deployed in the home zone of the original compromised node, this is a limited attack of minimal value to the adversary. However, it would also be possible to require both a correct deployment time and a correct location to be trusted. This would increase security at the cost of additional overheads.

Due to redeployment or supplemental deployment, some nodes may be deployed later than others. If deployment time checks are used, newly joining nodes cannot determine the earlier deployment times of nodes already in the system. Replica nodes could take advantage of this by waiting until new nodes joined the network and asking them to become neighbors. Thus, new nodes should only use location information to determine whether to treat existing nodes as trusted. Note that this implies that deployment time checks cannot be used by themselves and should be used to enhance the performance of the location claim and multigroup approaches.

24.4.4.1 Protocol Description

We now describe the way the deployment time checks are used. When a group G_v of nodes are deployed, they will be preloaded with a time stamp T_v that is digitally signed by a trusted server. This time stamp indicates that the sensor nodes in G_v should finish neighbor discovery before time T_v. If they try to set up neighbor connections with other nodes after time T_v, they are considered to be untrusted nodes. The time stamp T_v should be a function of the deployment time T, the time T_c needed for compromising and replicating a node, and the maximum time synchronization error ε. Specifically, the network operator should set $T + T_n + \epsilon < T_v < T + T_n + T_c - \epsilon$, where T_n is the neighbor discovery time, such that no nodes should have clocks too fast to accept the new node, but no new node could be compromised and accepted in time. This means that $\epsilon < 0.5T_c$ determines the maximum amount of allowable error.

Consider a particular node u. Assume that its neighbors $N(u)$ has been marked as trusted or untrusted using the deployment location, according to one of the previously described group-based schemes. We attempt to distinguish more benign nodes from those marked as untrusted by checking the deployment times of those nodes. Specifically, for every node $v \in N(u)$ that is marked as untrusted, node u checks whether v was discovered before T_v, the deployment time of G_v. If yes, node v will be remarked as trusted.

24.4.4.2 Analysis

The timing-based approach clearly reduces the number of untrusted nodes even if there are large deployment location errors. Let P_t denote the probability that a node v is deployed within time T_v. The probability that v is marked as untrusted is $(1 - P_z) \times (1 - P_t)$. Thus, we expect $(1 - P_z) \times (1 - P_t) \times N$ claims instead of $(1 - P_z) \times N$, where N is the total number of sensor nodes. In calculating the communication overhead for Scheme II, we can multiply the probability of a claim from any band P_i by $(1 - P_t)$ to derive the likelihood of needing to send a claim from that band. Thus, the communication cost per node is $C'_s = (1 - P_t)C_s$. Similarly, the communication cost per node of Scheme III is $C'_m = (1 - P_t)C_m$.

The storage and computational overheads per node of Schemes II and III can similarly be estimated as $2m(1 - P_t)(1 - P_z)$ and $(1 - P_t)(1 - P_z)(b + 2m)$ and $2m(1 - P_t)(1 - P_z)(1 + D_{max})$ and $(1 - P_t)(1 - P_z)(b + 2m(1 + D_{max}))$, respectively.

This scheme thus provides substantial reductions in all overheads. For example, if $P_t = 0.9$, it reduces all communication, storage, and computational overheads by 90%. In addition, this approach can also enhances network connectivity, since more benign nodes can be correctly remarked as trusted.

24.5 Discussion

In this section, we compare the group-based detection schemes of Ho et al. [14] with those of Parno et al. [22] in terms of communication, computation, and storage (for location claims) overheads.

Scheme I is the most efficient solution in terms of overheads. In addition, this solution does not require any expensive public key operations or secure localization. We believe that this scheme is suitable to limit replica node attacks in applications that have strict requirements on energy consumption. The potential problem of Scheme I is that the system must trade off between the ability to defend against replica nodes near the group deployment point and the ability of some benign nodes to communicate. The trade off depends greatly on accuracy of sensor deployment to the group deployment points. However, the use of deployment time checks can reduce the sensitivity of the scheme to deployment accuracy such that the loss of communication only occurs when both time synchronization and sensor deployment are inaccurate.

Scheme II provides better security and availability than Scheme I at the cost of additional overheads. Compared with the previously proposed line-selected multicast scheme of Ref. [22], Scheme II has numerous advantages. First, Scheme II achieves near-perfect replica detection, while line-selected multicast achieves only about 80% detection in simulation [22]. Both schemes are robust to node compromise, except locally near the replica node, in which case the replica has little practical impact on the network. In terms of overheads, Scheme II is more efficient in communication, computation, and storage. line-selected multicast may use, for example, six *line segments* per node, each with an expected cost of $\sqrt{N}$ per line segment [22], where N is total number of sensor nodes. In the simulation, this was shown to cost approximately $\sqrt{N}$ packets sent and received per node [22].

As shown in Figure 24.5, Scheme II has at least 86% less communication overhead compared with Ref. [22], when $R_z = 100$ m. Even when $R_z = 50$ m, Scheme II has at least 66% less communication overhead than Ref. [22]. Although Scheme III requires more overhead than Scheme II, it still requires less overhead than Ref. [22]. In particular, Scheme III saves at least approximately 69% of communication overhead of Ref. [22], when $R_z = 100$ m. Even when $R_z = 50$ m, Scheme III saves approximately 44% as long as $N \geq 6000$.

In the line-selected multicast scheme, each node must verify and store $O(\sqrt{N})$ location claims. Specifically, the line-selected multicast scheme requires approximately 100 signatures per node when $N = 10{,}000$, whereas Scheme II can be expected to have less than 20 signatures per node in the sample scenario given in Section 24.4.2. Furthermore, the signature verification savings can be significant in the energy-limited lifetime of the sensor nodes. According to the experimental results of Ref. [17], iMote2 and TelosB mote sensors consume 3.51 and 21.82 mJ of energy for signature verification, respectively. When using an iMote2, we expect Scheme II to cost less than 70.2 mJ of energy per node for signature verification. Even considering the less power efficient TelosB mote, we would expect a cost per node of less than 436.4 mJ for signature verification. On the contrary, we estimate that line-selected multicast will cost approximately 351 mJ for the iMote2 and 2182 mJ for the TelosB when it requires 100 signature verifications per node.

Adding deployment time checks will significantly reduce the overheads of Scheme II by limiting location claims to replica nodes that are both placed outside their home zones and connect to their neighbors after the allowed time period. If $P_t = 0.9$, it reduces the overheads of Schemes II and III by 90%. In general, the more the effort by the network operator into making both the deployment time and deployment locations accurate, the lower the overhead in these schemes. It is not unreasonable to imagine networks in which, for example, the deployment times are sufficiently accurate to enable zero overhead for benign nodes.

24.6 Other Approaches

In this section, we describe other proposed approaches besides those of Parno et al. [22] and Ho et al. [14]. Two of these schemes [4,27] are centralized, while the other two [5,29] are decentralized. We also describe related issues and why addressing them does not protect against the replica attack.

We begin with the centralized schemes. Choi et al. [4] propose a replica detection scheme in which the network is considered to be a set of nonoverlapping subregions. An exclusive subset is formed in each subregion. If the intersection of subsets is not empty, it implies that replicas are included in those subsets. Each node generates subset information and forwards it to the base station. Each node, which participates in relaying subset information, inspects whether the intersections between subsets are not empty and detects replicas belonging to nonempty intersections. The base station performs this replica detection process to prevent replicas from not being detected by any relaying nodes.

Xing et al. [26] proposed a fingerprint-based replica node detection scheme. In this scheme, nodes report fingerprints, which identify a set of their neighbors, to the base station. The base station detects replicas when it receives two conflicting fingerprints from them.

We now briefly describe the two decentralized schemes. Conti et al. [5] proposed a randomized, efficient, and distributed (RED) protocol for replica detection. The main idea of Ref. [5] is similar to the line-selected multicast scheme except that witness nodes are deterministically chosen every time slot and claim-relaying nodes do not involve in replica detection process. RED improves the line-selected multicast scheme of Ref. [22] in terms of replica detection probability, storage, and computation overheads. However, RED still has the same communication overhead as the line-selected multicast scheme of Ref. [22]. More significantly, their protocol requires repeated location claims over time, meaning that the cost of the scheme needs to be multiplied by the number of runs during the total deployment time. Since the group-based schemes of Ref. [14], as described in this chapter, only require claims when new nodes are deployed, they are much more efficient than RED in terms of communication overhead over the lifetime of the network.

Zhu et al. [29] proposed localized multicast schemes based on the grid cell topology, where replicas are detected by letting location claims be multicasted to a single cell or to multiple cells. The main strength of Ref. [29] is that it achieves higher detection rates than the best scheme of Ref. [22]. However, Ref. [29] has similar communication overheads as [22] whereas the group-based schemes of Ref. [14] work as well with much lower overheads.

Table 24.2 shows the comparison results of replica detection schemes in sensor networks. Most of the replica detection works in Table 24.2 asymptotically produce the same communication overhead $O(N\sqrt{N})$ because each sensor node is required to send location claim or fingerprint information to randomly or deterministically selected detector nodes and the message delivery overhead between any two nodes is asymptotically calculated as $O(\sqrt{N})$, as shown in Ref. [22]. From Table 24.2, we see that the group-based schemes of Ho et al. [14] require less communication overhead than all other related work except Ref. [4]. This is because the group-based schemes utilize deployment knowledge to minimize communication overhead for replica detection. In Ref. [4], the base station is required for replica

TABLE 24.2 Comparison of Replica Detection Schemes

Approach	Schemes	Communication Overhead
Centralized	Choi et al. [4]	$O(N)$
	Xing et al. [26]	$O(N\sqrt{N})$
Decentralized	Conti et al. [5]	$O(N\sqrt{N})$
	Randomized multicast [22]	$O(N^2)$
	Line-selected multicast [22]	$O(N\sqrt{N})$
	Zhu et al. [29]	$O(N\sqrt{N})$
	Group-based schemes [14]	$O(N\sqrt{N} \times p)(0 \le p < 1)$

detection, and thus, replicas may not be detected if the base station malfunctions. The group-based schemes do not have this single point of failure problem because they fulfill replica detection without the help of the base station.

24.6.1 Related Issues

Replica attacks present a different problem from other node compromise issues in sensor networks [22]. For example, updating keys used in broadcast encryption can help protect data confidentiality when compromised nodes are discovered. It is also possible to do either centralized [11] or localized [3] key revocation upon discovery of compromised nodes. However, an attacker may seek to be stealthy to avoid detection of his/her replicas based on their behavior. He/she can still eavesdrop on messages passing through the replica nodes and carefully launch disruptive attacks that are difficult to attribute to the replicas, such as gray-hole attacks in which the replicas selectively drop packets. Key revocation only works when the attacking nodes are correctly detected.

Parno et al. [22] also point out the subtle difference with the Sybil attack [7]. In a Sybil attack, the attacker has each node use multiple identities. Contrast this with a replica attack, in which a *single identity* is used for *multiple nodes*. Nevertheless, we would expect identity reuse in both cases, as the Sybil attack would steal identities used elsewhere in the network—otherwise the attacker could simply move the compromised node around. Newsome et al. present techniques to overcome the Sybil attack in sensor networks [21], but these do not effectively address the replica attack [22].

24.7 Conclusions

In this chapter, we described the problem of detecting replica nodes in wireless sensor networks and several possible solutions. Replica attacks are a dangerous threat, as the attacker can leverage a few compromised nodes to perform network-wide attacks and substantially undermine the network's operations.

To overcome such attacks, a variety of schemes have been devised by researchers and present several trade-offs to consider for implementation and deployment. We first described the randomized multicast and line-selected multicast schemes of Parno et al. [22] as the originators of the popular location claim idea. This idea features heavily in the schemes of Ho et al. [14], which we spent much of the chapter discussing, and the other distributed schemes [5,29]. Two centralized schemes use other techniques [4,26], but these rely on a single point of failure.

The focus of the chapter has been on schemes of Ref. [14]. These schemes take advantage of nodes' knowledge of group deployment locations and deployment times to reduce the need to send and check location claims. This allows the schemes to reduce overheads while still ensuring high robustness to a powerful adversary. The schemes offer different trade-offs. Scheme I is simple and can be effective when the deployment location is accurate, and the system does not have strict security requirements. Scheme II achieves high replica detection capability with less communication, computational, and storage overheads than other approaches in the literature. Scheme III provides very strong resilience to node compromise at the cost of more overheads than Scheme II. Furthermore, the overhead of Schemes II and III can be significantly reduced by using deployment time checks.

One important advantage of the group-based approach of Ho et al. is the possibility of near-zero overheads [14]. By making the deployment times and locations more accurate, the network operator can arbitrarily reduce the overheads of these schemes without loss of security. At the same time, the approach is flexible and robust when there are errors, with costs rising only gradually when more errors are introduced. We showed that the overheads of these schemes are better than for the schemes of Ref. [22] for a reasonable range of deployment error estimates, even without assuming deployment time checks. Given these properties, network operators with a broad range of needs can employ group-based schemes to greatly reduce the threat posed by replica node attacks.

Acknowledgments

This work was partially supported by NSF grants IIS-0326505, DUE-0621280, CNS-0721951, and CNS-0916211. Any opinions, findings, and conclusions or recommendations expressed in this material are those of the authors and they do not necessarily reflect those of the National Science Foundation. The authors gratefully acknowledge the insightful comments of anonymous reviewers that helped improve the quality of the manuscript significantly.

References

1. I.F. Akyildiz, W. Su, Y. Sankarasubramaniam, and E. Cayirci. Wireless sensor networks: A survey. *Computer Networks*, 38(4):393–422, 2002.
2. S. Čapkun and J.P. Hubaux. Secure positioning in wireless networks. *IEEE Journal on Selected Areas in Communications*, 24(2):221–232, 2006.
3. H. Chan, A. Perrig, and D. Song. Random key predistribution schemes for sensor networks. In *IEEE Symposium on Security and Privacy*, Berkeley, CA, pp. 197–213, May 11–14, 2003.
4. H. Choi, S. Zhu, and T.F. La Porta. SET: Detecting node clones in sensor networks. In *IEEE/CreateNet Conference on Security and Privacy for Emerging Areas in Communication Networks (SecureComm)*, pp. 341–350, September 2007.
5. M. Conti, R.D. Pietro, L.V. Mancini, and A. Mei. A randomized, efficient, and distributed protocol for the detection of node replication attacks in wireless sensor networks. In *ACM Mobihoc*, Montréal, Quebec, Canada, pp. 80–89, September 9–14, 2007.
6. F. Delgosha and F. Fekri. Threshold key-establishment in distributed sensor networks using a multi-variate scheme. In *IEEE INFOCOM*, Barcelona, Spain, pp. 1–12, April 2006.
7. J.R. Douceur. The Sybil attack. In *Proceedings of the Workshop on Peer-to-Peer Systems (IPTPS)*, Cambridge, MA, March 7–8, 2002.
8. W. Du, J. Deng, Y.S. Han, S. Chen, and P. Varshney. A key management scheme for wireless sensor networks using deployment knowledge. In *IEEE INFOCOM*, Hong Kong, China, pp. 586–597, March 7–14, 2004.
9. W. Du, L. Fang, and P. Ning. LAD: Localization anomaly detection for wireless sensor networks. In *IEEE IPDPS*, Denver, CO, April 2005.
10. W. Du, R. Wang, and P. Ning. An efficient scheme for authenticating public keys in sensor networks. In *ACM MobiHoc*, Champaign, IL, pp. 58–67, May 25–27, 2005.
11. L. Eschenauer and V. Gligor. A key-management scheme for distributed sensor networks. In *Proceedings of ACM CCS*, Washington, DC, November 18–22, 2002.
12. S. Ganeriwal, S. Čapkun, C.C. Han, and M.B. Srivastava. Secure time synchronization service for sensor networks. In *ACM WiSe*, Cologne, Germany, pp. 97–106, September 2005.
13. V. Gupta, M. Millard, S. Fung, Y. Zhu, N. Gura, S. Eberle, and H. Chang. Sizzle: A standards-based end-to-end security architecture for the embedded Internet. In *IEEE International Conference on Pervasive Computing and Communications (PerCom)*, Koloa, Kauai, Hawaii, pp. 247–256, March 2005.
14. J. Ho, D. Liu, M. Wright, and S.K. Das. Distributed detection of replicas with deployment knowledge in wireless sensor networks. *Ad Hoc Networks*, 7(8):1476–1488, 2009.
15. C. Hartung, J. Balasalle, and R. Han. Node compromise in sensor networks: The need for secure systems. In Technical Report CU-CS-990-05, Department of Computer Science, University of Colorado at Boulder, Boulder, CO, January 2005.
16. Z. Li, W. Trappe, Y. Zhang, and B. Nath. Robust statistical methods for securing wireless localization in sensor networks. In *IEEE International Conference on Information Processing in Sensor Networks (IPSN)*, Los Angles, CA, pp. 91–98, April 25–27, 2005.

17. A. Liu and P. Ning. TinyECC: A configurable library for elliptic curve cryptography in wireless sensor networks. In *IEEE International Conference on Information Processing in Sensor Networks (IPSN), SPOTS Track*, St. Louis, MO, pp. 245–256, April 2008.
18. D. Liu and P. Ning. Establishing pariwise keys in distributed sensor networks. In *ACM Conference on Computer and Communications Security (CCS'03)*, Washington, DC, pp. 52–61, October 2003.
19. D. Liu, P. Ning, and W. Du. Attack-resistant location estimation in sensor networks. In *IEEE International Conference on Information Processing in Sensor Networks (IPSN)*, Berkeley, CA, pp. 99–106, April 26–27, 2005.
20. D. Liu, P. Ning, and W. Du. Group-based key pre-distribution in wireless sensor networks. In *ACM Workshop on Wireless Security (WiSe)*, Cologne, Germany, pp. 11–20, September 2005.
21. J. Newsome, E. Shi, D. Song, and A. Perrig, The Sybil attack in sensor networks: Analysis and defenses. In *IEEE Conference on Information Processing in Sensor Networks (IPSN)*, Berkeley, CA, April 2004.
22. B. Parno, A. Perrig, and V.D. Gligor. Distributed detection of node replication attacks in sensor networks. In *IEEE Symposium on Security and Privacy*, Oakland, CA, pp. 49–63, May 8–11, 2005.
23. H. Song, S. Zhu, and G. Cao. Attack-resilient time synchronization for wireless sensor networks. *Ad Hoc Networks*, 5(1):112–125, 2007.
24. K. Sun, P. Ning, C. Wang, A. Liu, and Y. Zhou. TinySeRSync: Secure and resilient time synchronization in wireless sensor networks. In *ACM CCS*, Alexandria, VA, pp. 264–277, October 2006.
25. H. Wang, B. Sheng, C.C. Tan, and Q. Li. Comparing symmetric-key and public-key based security schemes in sensor networks: A case study of user access control. In *IEEE ICDCS*, Beijing, China, pp. 11–18, June 2008.
26. K. Xing, F. Liu, X. Cheng, and H.C. Du. Real-time detection of clone attacks in wireless sensor networks. In *IEEE ICDCS*, Beijing, China, pp. 3–10, June 2008.
27. J. Yick, B. Mukherjee, and D. Ghosal. Wireless sensor network survey. *Computer Networks*, 52(12):2292–2330, 2008.
28. W. Zhang, M. Tran, S. Zhu, and G. Cao. A random perturbation-based scheme for pairwise key establishment in sensor networks. In *ACM Mobihoc*, Montréal, Québec, Canada, pp. 90–99, September 9–14, 2007.
29. B. Zhu, V.G.K. Addada, S. Setia, S. Jajodia, and S. Roy. Efficient distributed detection of node replication attacks in sensor networks. In *ACSAC*, Miami Beach, FL, pp. 257–267, December 10–14, 2007.

25

Topology-Aware Routing and Transmission Scheduling for Highway Sensor Networks

Kuang-Ching Wang
Clemson University

Devang Bagaria
Clemson University

Mashrur Chowdhury
Clemson University

25.1 Introduction

Countries around the globe are adopting intelligent transportation systems (ITSs) for monitoring and control of vehicle traffic on highways and major urban roads [1–3]. A typical state-of-the-art ITS collects and transports real-time traffic sensor data to one or a few traffic management centers (TMCs) for analysis and response. The communication network connecting sensors to TMCs thus determines the sustainable scope and response efficacy of the system. Most of today's ITSs rely on wired network links between sensors and TMCs, and the costs of deploying and maintaining the cables become one major limiting factor for ITS coverage. Looking into the future, connecting sensors with wireless networks is increasingly more favorable for their higher cost effectiveness and faster deployment [4,5].

Another factor limiting ITS growth is its centralized data collection practice that creates a network bottleneck at the TMCs. The apparent solution is to delegate selected tasks to a distributed set of controllers, such that each controller consumes local sensor data, makes local decisions, and controls local traffic control actions. These controllers can be embedded systems that intelligently discover and communicate with sensors in their vicinity as well as TMCs. The delegation approach is suitably served by a hierarchical system as shown in Figure 25.1. For example, sensors may convey real-time data to nearby controllers; controllers process the data to detect different traffic events to report to TMCs; TMCs

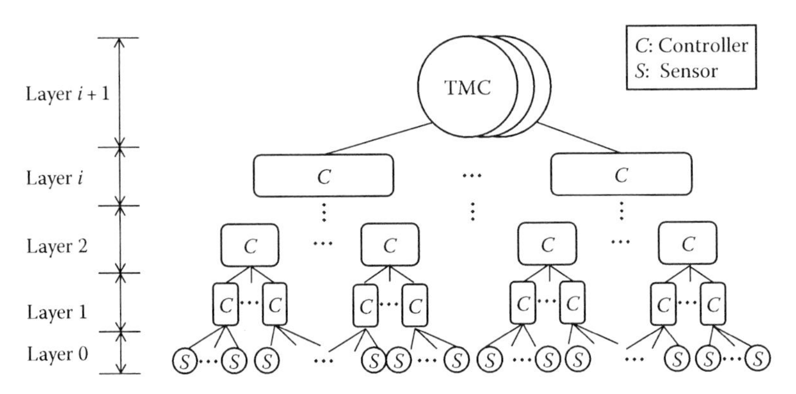

FIGURE 25.1 Hierarchical traffic sensor and controller system.

authorize the actions to be taken. TMCs can also set policies to authorize controllers to take certain automated actions without explicit approvals (e.g., announcing a reroute alternative on a highway message board). Nearby controllers may communicate with each other to aggregate data for improved detection accuracy. This chapter presents a closely coupled routing and scheduling scheme for such a hierarchical highway sensor and controller system based on a wireless ad hoc network. The solution utilizes topology awareness to achieve simultaneous high throughput, low latency, and energy efficient communication among sensors and multiple levels of controllers (including the TMCs).

We contend that the network must support communication over longer range and higher data rates to support the full range of existing traffic sensors (e.g., video-based sensors). The IEEE 802.15.4 standard has been widely publicized as "the" standard for wireless sensor networks (e.g., ZigBee and HART are two industrial sensor standards based on IEEE 802.15.4). Its 250 kbps data rate is, however, insufficient for our purpose. IEEE 802.15, also known as Bluetooth, is another wireless technology for low-power devices supporting up to 3 Mbps data rate [6]; supporting ad hoc networking with Bluetooth, however, is known to be complex and inflexible with its master–slave and time division multiple access (TDMA) architecture. Hence, we consider IEEE 802.11, also known as Wi-Fi, which has data rates up to 54 Mbps (802.11a/b/g) or beyond 100 Mbps (802.11n) and ranges from 100 m to beyond 30 km [7]. Nevertheless, the proposed scheduling algorithm adopts a slotted approach, making it easily adaptable for use with other carrier sense multiple access (CSMA)- and TDMA-based technologies.

The rest of the chapter is organized as follows. Section 25.2 reviews background and related work. Section 25.3 describes the network model. Section 25.4 describes the proposed routing method. Section 25.5 describes the proposed scheduling method. Section 25.6 presents a simulation study of the proposed schemes using ns-2. Section 25.7 concludes the chapter.

25.2 Background and Related Work

25.2.1 Traffic Control Practices (Thesis Citations)

The 2004 U.S. ITS survey conducted by the U.S. Department of Transportation summarizes the nation's latest traffic control operations and deployed infrastructure [8]. These systems adopt a centralized architecture with TMCs at the core and sensors and remote control devices on the roads. The functions of the different elements of this system are as follows:

- *Sensors* acquire real-time traffic data of monitored road segments. Widely adopted are loop detectors, closed circuit television (CCTV) cameras, video detection systems (VDS), infrared laser, acoustic, microwave radar, and piezoelectric strain gauge sensors. Typical metrics acquired are traffic speed, volume, road occupancy, and road conditions (e.g., pavement integrity; ice on pavements; fog; inclement weather conditions such as snow, rain, and tornadoes).

- *TMCs* collect, archive, and analyze sensor data. TMC operators detect traffic incidents, dispatch incident removal teams, and publish current traffic information to the public via Internet, highway advisory radio, TV, operator hotlines (511), in-car navigation systems, and call-outs to subscribed travelers.
- *Off-site control devices* are controlled by TMCs to regulate traffic flow. Mostly adopted are dynamic message signs (DMS), ramp meters, traffic signals, etc.

Traffic management systems typically perform the following traffic management operations:

- *Traffic regulation*: Controlling or reprogramming off-site control devices based on current and historical traffic data to regulate traffic flows against congestion.
- *Traveler information*: Providing travelers with current traffic condition, such as travel time estimates, driving advices, suggesting alternative routes, etc.
- *Incident prevention*: Disseminating real-time alerts, such as dangerous road conditions, misbehaving vehicles, and interfering entities, to prevent incidents.
- *Incident management*: Dispatching rescue teams to remove detected incidents.

The freeway performance measurement system (PeMS) in California [4] is a good example of a large centralized sensor system that links more than 25,000 loop detectors to the TMC. The system collects data from each detector every 30 s, accumulating more than 2 GB data per day in a central database. Sensors can be connected with different link media like fiber cables, general packet radio service (GPRS) wireless cellular links [5], and cellular digital packet data (CDPD) cellular wireless links [9]. Recently, the PEDAMACS project [10] unties the sensors from TMCs; instead, sensors in each cluster send data to a wireless gateway, which is then directly wired to TMCs. Multi-hop forwarding is used by sensors to send data to the gateway but not beyond the cluster. Data collection remains centralized and rooted at TMCs.

25.2.2 CSMA MAC Optimization

Energy-efficient medium access control (MAC) solutions have been widely studied for CSMA-based networks, mostly based on modifying IEEE 802.11 and 802.15.4 protocols [11–16]. These studies exploit the radios' ability to operate in four modes: idle listening, receiving, sending, and sleeping (turning off radio), and the fact that they consume nearly as much energy in idle listening as in the sending and receiving states. These studies explore different ways to put sensors in the sleeping state to trade off between energy efficiency and network performance. In Ref. [11], the S-MAC protocol establishes randomly generated common sleep schedules among neighboring sensors for them to sleep and wake up at the same time to transmit and receive messages; as a result, communication delay becomes longer due to multiple sleep cycles needed to forward a message across a multi-hop path. To reduce such delays, Ref. [13] proposed dynamic S-MAC (DS-MAC) to increase the wake-up duty cycle when observed delays exceed a set threshold. Delays can be further reduced by considering network topology and routing patterns. In DMAC [14], consecutive time slots were assigned to adjacent forwarding nodes in a data gathering tree to reduce the end-to-end latencies. Little work has been done on enhancing throughput, as most studies have considered sensor networks with low data rates. One exception is the TRaffic Adaptive MAC (TRAMA) [16], which was shown to achieve 20% higher throughput with IEEE 802.11 while saving 87% energy by putting nodes to sleep when they have no packets to handle. Independent of scheduling, time-out MAC (TMAC) [12] and WiseMAC [15] showed that opportunistic energy saving can be achieved by simply putting nodes to sleep in a time slot if no packets are detected for a small duration at the beginning of that slot. The distributed energy-aware MAC protocol (DE-MAC) [17] saves energy by increasing the sleep duration for the least energy node when the node's energy falls below a certain threshold. ZMAC [18] is a hybrid MAC protocol that incorporates the advantages of both CSMA and TDMA mechanisms by making nodes function in CSMA mode during low-traffic periods and switching them to TDMA mode during high-traffic periods. There has not been a study aiming at simultaneously increasing throughput, reducing latency, and increasing energy efficiency, which is the goal of the scheduling method introduced in Section 25.5.

25.3 Hierarchical Traffic Sensor and Controller Network

Hierarchical methods have been commonly used in traffic control since coordinated traffic signal control was introduced in early last century [19]. State-of-the-art real-time traffic control methods [20] have been based on hierarchical flow models at road segment, intersection, and road network granularities based on data collected by sensors deployed along the roads. Inheriting the segment–intersection–network hierarchy, we define a hierarchical traffic sensor and controller network model with the following properties:

- *The network consists of two types of devices: sensors and controllers.* Sensors perform sensing and report data to controllers. Controllers either process the data locally or relay them to higher-level controllers. TMCs are considered the highest level controllers.
- *A cluster can be formed by a group of controllers and sensors to perform a collaborative operation.* One controller and multiple sensors along a road segment form a lowest- level (level 1) cluster. Multiple level-i clusters can form a level-(i + 1) cluster with a designated level-(i + 1) controller.
- Routing among sensors and controllers is based on a hierarchical addressing scheme that inherits the U.S. highway reference system convention [21] in the format:

  [RID, Mileage, Direction, Cluster, Level, Class]

- Transmission scheduling is done by a level-1 controller in each level-1 cluster.
- Sensors on different sides of a road form two separate clusters due to the logical independence of their operations. The direction of a segment is defined with its road traffic's direction.

As shown in Figure 25.2, sensors on the same side of a road form a *segment*, all segments leading to the same controller form one *cluster* controlled by that controller, and the segments are said to be *associated*

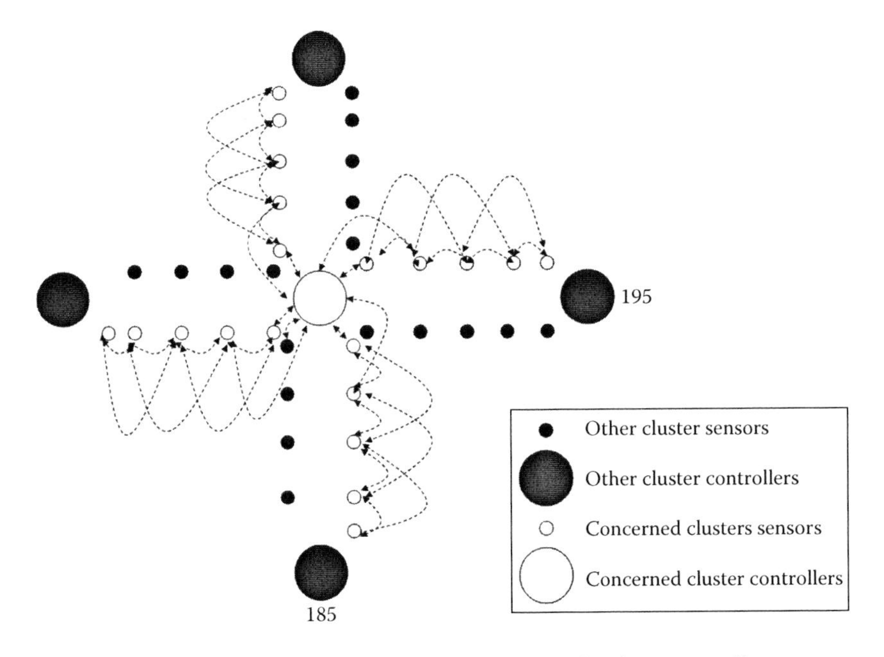

FIGURE 25.2 Single-intersection network with four clusters associated with one controller.

with that controller. The segment on the other side of a road is referred to as the controller's *neighbor-associated segment* since the segment is associated with another controller adjacent to the associated segments. Figure 25.2 illustrates associated and neighbor-associated segments for controller *C*2.

25.4 Routing

Routing in the network is done following the sensor–controller hierarchy based on the hierarchical addressing scheme. The address fields, specifically, include the following:

- *RID*—A character field for the highway ID on which the sensor is located; for example, I85.
- Milepost—A character field for the mileage on the highway where the sensor is located; for example, 0127.
- Level—A character field for the level at which the node is functioning; for example, 0 (sensor), 1 (level-1 controller), 2 (level-2 controller).
- Direction—A character field for the side of the road the node is located; for example, 1 (north bound), 2 (south bound), 3 (east bound), 4 (west bound).

While simple and intuitive, the address fully exposes the contextual information needed for message routing as well as most traffic control applications. Each sensor or controller can also have as many addresses as the roles it serves. For example, a sensor may be installed on a platform that also serves as a controller; they can share the same wireless network interface but use different addresses when communicating.

Routing takes place in three phases. Phase one is *topology discovery*, where a controller discovers all sensors that should be in its cluster, their hierarchical addresses, and GPS coordinates. Phase two is *hierarchical cluster formation*, where higher-level controllers discover lower-level controllers that should be associated to them. Phase three is *data routing*, where routing between any two nodes in the network is fully supported. Detailed protocol description and an ns-2 implementation of the routing method can be found in Ref. [22].

The topology discovery phase lets each level-1 controller discover in its vicinity all sensors that should be associated with it, that is, on road segments leading toward it. Through topology discovery, the controller collects all associated sensors' hierarchical address and geographical coordinates (which will be needed in for scheduling), while each sensor identifies the closest neighbor closer to the controller as the next hop to reach the controller. *Procedure*: Each level-1 controller can initiate this phase at any time by broadcasting a *topology request* message. Each sensor on the same road segment as the controller overhearing the message appends the message with its address information and rebroadcasts the message. Sensors on road segments not associated with the controller ignore the message. After a number of sensor broadcasts, the message will be heard by another controller at the other end of the road segment; this controller, sometimes referred to as an *adjacent controller*, would send all appended sensor information in a *topology response* message back to the initiating controller.

The hierarchical cluster formation phase lets each higher (>1)-level controller discovers in its vicinity all one-level-lower controllers that should be associated with it. Through this phase, the controllers discover their parents and children controllers in the hierarchical routing tree and retain the next hop node address toward them. *Procedure*: Each higher-level controller can initiate this phase after a topology discovery phase by broadcasting a *cluster-announcement* message, such that any controller of one level lower can identify the sender as its parent when receiving the message. When multiple announcements are heard, the closer sender is chosen to be the parent. A child controller then sends a *cluster-join* message to the chosen parent, who will record the child and the route to it. The routing tree is ultimately rooted at one or few TMCs.

The data routing phase is the productive phase where sensors and controllers forward data between any two nodes in the system using the routing tables created in the two prior phases. *Procedure*: A sensor always forwards all packets it generates or receives from other nodes to its parent controller. A controller forwards a packet based on the destination's RID. If the RID matches its own RID in its routing table, it sends the packet toward the adjacent controller with the same RID and milepost closer to the destination. If the RID is unknown to the controller, the message is forwarded to its parent controller. Note that most controllers are expected to be placed at an intersection of multiple crossing roads and hence possess multiple addresses with different RIDs.

25.5 Topology-Aware Scheduling

Based on the network properties defined in Section 25.3, the scheduling problem is to find a slot-based schedule for each device (sensor or controller) to transmit packets to its adjacent neighbors as directed by the routing algorithm, such that each device's end-to-end throughput, latency, and energy cost are simultaneously optimized.

When a sensor sends a packet, the destination is always its controller. When a controller sends a packet, its destination is always another controller on the other end of an adjacent road segment. Hence, each cluster always forwards traffic in only one direction and optimizes its schedule for that direction.

With IEEE 802.11 as the underlying MAC, the schedule need not guarantee 100% collision free. That is, the network, as any IEEE 802.11 network, is tolerant of collisions and resolves them with automatic retransmissions, while the schedule allows the network to minimize collisions to the minimum to achieve maximum throughput.

25.5.1 Heuristics

The scheduling algorithm is based on the following three heuristics:

- *Assign earliest and pipelined slots for adjacent nodes*: Minimum end-to-end latency results when packets are forwarded by adjacent nodes at the earliest opportunity; hence, for each slot scheduled for a node, its adjacent node should be assigned the earliest next nonconflicting slot. Maximum end-to-end throughput results when subsequent packets from the same source are sent with minimal time in between; hence, for each slot scheduled for the sensor closest to the controller (i.e., completing one round of forwarding across the road segment), a new round should be initiated by assigning the farthest node the earliest next nonconflicting slot in a pipelined fashion.
- *Rotate and weight assignment for road segments of a common intersection*: To minimize forwarding latency, the schedule must minimize backlog at any node by allocating sufficient slots to each segment to meet their demands. It is thus both efficient and fair to assign slots to each segment proportional to its traffic load. One way to achieve this is for road segments to rotate in receiving slot assignments, with the number of slots assigned to a segment per round weighted to be proportional to the ratio of their traffic loads.
- *Negotiate schedules of adjacent clusters*: The schedule is independently derived by each cluster's controller with only the cluster's topology information; hence, it is prone to conflicts with nodes in other clusters. Cluster schedules and topology should be exchanged among nearby controllers to resolve such conflicts.

25.5.2 Conflict Conditions and Capture Effect

Minimizing collision is the key goal for assigning time slots to nodes. As stated, our system does not require collision-free schedules. In the following, we review conditions for collision-free transmissions and explain the criteria used in our scheduling algorithm.

The following four rules are sufficient conditions for collision-free transmission:

1. The sender is the only node transmitting in the intended receiver's carrier-sensing range.
2. The sender is not in the carrier-sensing range of any active receiving node other than the intended receiver.
3. No two senders are in each other's carrier-sensing range.
4. The intended receiver is not in the carrier-sensing range of any other active receiving nodes.

Rules 1 through 3 assure conflict-free transmission of the data packet, while rule 4 assures the acknowledgment sent by the receiver does not interfere with transmissions of other scheduled receivers. Enforcing rule 4, however, turns out to be overly restrictive if the wireless capture effect is considered.

The capture effect [23] states if a conflicting signal I arriving during an ongoing desired transmission P, and I's power is less than κ times P's power, P's reception will proceed without error. κ is called the capture ratio and $0 < \kappa < 1$. After examining a wide range of scenarios, it was seen that collisions due to rule 4 violation have mostly occurred between a data transmission initiated (to a receiving node A) in the middle of an active acknowledgment (to a receiving node B) where the two receiving nodes are nearby and the two sending nodes are far from each other. Due to the short duration of an acknowledgment, such conditions seldom arise. In Sections 25.4, we show that by relaxing rule 4 and introducing different transmission time offsets to nodes transmitting in the same slot, throughput can be enhanced ~10% under high-load conditions.

25.5.3 Scheduling Protocol

The scheduling protocol begins right after the topology discovery phase of routing, and it consists of two consecutive phases: *per-cluster scheduling* and *intercluster negotiation.*

Per-cluster scheduling is based on a conflict graph derived by a cluster's controller based on collected sensor location following the definition in Ref. [24]. The conflict graph identifies all node pairs that violate any of the four conflict-free conditions (three if the last rule is relaxed). The pseudocode for the scheduling method is shown in Figure 25.3. Note that in order for an adjacent controller to send packets to this cluster's controller, the adjacent controller is considered the last node in a cluster to receive transmission slots within the cluster.

The method rotates through clusters in a fixed order assigning one slot to each cluster each time. A cluster is said to have completed one round when each sensor in the cluster has received one slot. Each cluster's round count can be scaled by a weighting factor to achieve nonuniform bandwidth allocation, for example, proportional to each cluster's traffic load. A new round of slot assignment begins only when all other clusters have completed equal rounds (or proportional many rounds when weighted) as the cluster.

Algorithm input:

1. *CG: conflict graph matrix*
2. *S: total slots allowed in one cycle, i.e., cycle length*

Algorithm variables

1. *T: current time slot to be scheduled*
   T.SendSet: set of nodes scheduled to send in T
   T.RecvSet: set of nodes scheduled to receive in T
2. *CS: current cluster to schedule*
   CS.NextCluster: next cluster to schedule after CS
   CS.Sender: current sender node in CS to schedule
   CS.FirstNode: first node in CS (farthest from controller)
   CS.StartSlot: slot assigned to CS.FirstNode this round
   CS.Sender.Parent: parent of CS.Sender, i.e. its receiver
   CS.rounds: rounds completed by CS so far
   CS.Weight: weighting factor for CS

START:

CS=1; CS.Sender=CS.FirstNode; T=0; T.SendSet=T.RecvSet=ϕ; CS.rounds=0;
while (any cluster can still find a valid slot in T=[0, S-1])

  //A cluster can proceed to next round only if all other clusters have
  //completed as many rounds.

FIGURE 25.3 Pseudocode for scheduling algorithm.

(*continued*)

```
        if CS.rounds > any other cluster's completed rounds
            CS = CS.NextCluster; //Proceed to next cluster
            T = CS.NextSlot
            continue;
        endif
    CONFLICT_TEST_LOOP:
        for each node n in T.SendSet,
            if CG(CS.Sender.Parent, n)>0,                    //check rule 1
                T=T+1; GOTO CONFLICT_TEST_LOOP; endif
            if CG(CS.Sender, n)>0,                           //check rule 3
                T=T+1; GOTO CONFLICT_TEST_LOOP; endif
        end for
        for each node n in T.RecvSet,
            if CG(CS.Sender, n)>0,                           //check rule 2
                T=T+1; GOTO CONFLICT_TEST_LOOP; endif
            if nonaggressive && CG(CS.Sender.Parent, n)>0,   //check rule 4
                T=T+1; GOTO CONFLICT_TEST_LOOP; endif
        end for
    // valid slot T for CS.Sender found
        if T<S,
            T.SendSet = T.SendSet + CS.Sender;
            T.RecvSet = T.RecvSet + CS.Sender.Parent;
        else
            GOTO NEXT_CLUSTER_TURN;
        endif
    //A new round starts at least one slot after first slot of prior round
        if CS.Sender == CS.FirstNode,
            CS.StartSlot = T;
        endif
    //CS completes one round
        if CS.Sender.Parent is a controller,
            CS.Sender = CS.FirstNode;
            CS.rounds = CS.rounds + 1/CS.Weight;
            CS.NextSlot = CS.StartSlot+1;
        else
            CS.Sender = CS.Sender.Parent;
            CS.NextSlot = T+1;
        endif
    NEXT_CLUSTER_TURN:
        CS = CS.NextCluster; T = CS.NextSlot;
    end while
```

FIGURE 25.3 (continued)

The slot assignment continues until one cycle (*S* slots, a configured parameter) has been fully scheduled and no more slots can be assigned to any cluster within the cycle. The controller disseminates the resulting schedule to each node as a sequence of assigned time slot indices.

To optimize relay latency in the direction toward the controller, each round of each cluster's slot assignment always begins from the node farthest from the controller and steps hop by hop until reaching the controller. To maximize throughput, slot assignment in a new round begins in the earliest "causal" time slot, that is, one slot after the previous round's first scheduled slot for the farthest node; this results in slot assignments of different rounds interleaved in a pipelined fashion. To minimize energy cost, a node only wakes up in their scheduled active (transmitting or receiving) slots. Moreover, similar to TMAC, a node will sleep even in a scheduled active slot if no transmissions are initiated within a specified time-out interval. The option to relax rule 4 leads to two different versions of scheduling. We refer to the relaxed method as the aggressive scheme and otherwise the nonaggressive scheme. Moreover, the aggressive scheme can suppress the already low collision probability due to the relaxation by having each node begins transmission in its time slot after a small offset that differs for each segment.

Multicluster negotiation begins right after per-cluster scheduling. The negotiation follows an iterative process, beginning with one controller designated as the seed. The seed controller retrieves the network topology and schedules from all neighboring controllers, compares them for any conflicts, and resolves such conflicts with a max-min heuristic to maximize the minimum throughput among all clusters. The seed controller does this in three steps. First, it compares and resolves its own schedule with all adjacent controllers' schedules. Second, it compares and resolves conflicts between schedules of every pair of adjacent controllers. Third, it concludes the suggested changes and distributes them to corresponding controllers. The three-step process then repeats by iteratively assigning each adjacent controller to act as seed controllers.

The distributed negotiation method can further enhance its throughput performance by sliding one schedule for a variable number of slots with respect to another schedule when comparing them for conflicts. Simulation shows that the sliding can achieve 4%–5% throughput enhancement dependent on the actual topology. The higher computation cost suggests that this enhancement should only be considered infrequently.

25.6 Simulation Study

The proposed scheduling methods were evaluated using the ns-2 network simulator [25]. In all experiments, the sensors and controllers have a transmit power corresponding to 125 m transmit range and 250 m carrier-sensing range; the radio capture ratio is $\kappa = 0.1$. Results were compared with IEEE 802.11 without our scheduling.

25.6.1 Single-Intersection Scenario

The single-intersection scenario simulates one controller connected to four road segments as shown in Figure 25.2. Each segment has five sensors on each side of the road, except the bottom segment with six sensors on each side of the road. Note that only sensors on the side with vehicle traffic moving toward the intersection are associated to that intersection's controller. Sensors on the other side of the road are associated to controllers on the adjacent intersections (shown black in Figure 25.2). Only the white sensors and the adjacent controller on each of the four segments are simulated in this scenario. All nodes are 100 m apart. Packets on each road segment are generated from the adjacent controller to the intersection controller with an exponential UDP flow of 50% duty cycle, 1000 bytes packet size, and selected generation rates.

The per-segment throughput, per-segment end-to-end delay, and per-node energy efficiency were studied. Figure 25.4 shows the per-segment throughput with respect to different per-segment traffic generation rates (all segments have the same generation rate). The 90% slope line is drawn to locate the saturated throughput, which is defined as the maximum achievable throughput with at least 90% arrival rate of the input traffic. Beyond the saturated throughput, the network is considered unstable and no longer acceptable for operation. As shown, IEEE 802.11 is able to achieve per-segment saturated throughput near 74 kbps, while the proposed nonaggressive method achieves 100 kbps, a 35% increase. The aggressive method further enhanced the throughput by 6% without offsets and 16% with offsets.

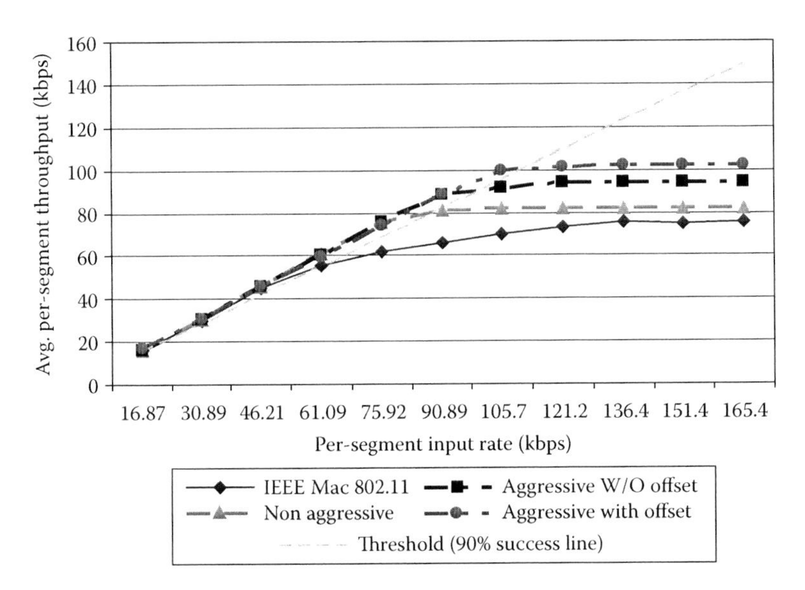

FIGURE 25.4 Per-segment throughput with different per-segment traffic generation rates.

Figure 25.5 shows the effects of dynamic scheduling weighted according to different average per-segment input load in different time intervals at a single intersection. The ratio of the traffic load on the four segments in the first, second, third, and fourth time interval is set as 1:1:1:1, 2:1:1:1, 1:2:2:2, and 1:2:3:4, respectively, and at all times, the total load of all segments is set equal to 368 kbps, the network capacity observed in the nonweighted scenario. As expected, the resulting per-segment throughputs reflect the configured weights. It was more interesting to note that the total network throughput achieved while scheduling with weights proportional to the actual traffic demands achieved the highest total network throughput as shown in Figure 25.5.

Figure 25.6 shows the average per-segment end-to-end delays measured with the aggressive method with offset and those measured with IEEE 802.11. The end-to-end delay consists of transmission time (initial random backoff + data transmission + interframe spacing + acknowledgment ≈ 0.025 s) and queuing time. The queuing time depends on the chosen slot interval; the longer the slot, the longer the average queuing time and end-to-end delay become. IEEE 802.11 always had delays lower than the

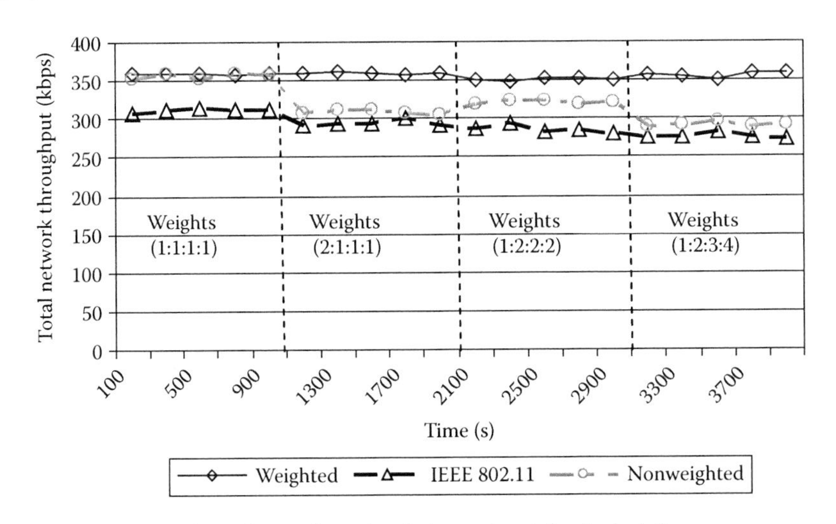

FIGURE 25.5 Total network throughput achieved with dynamic weighted scheduling.

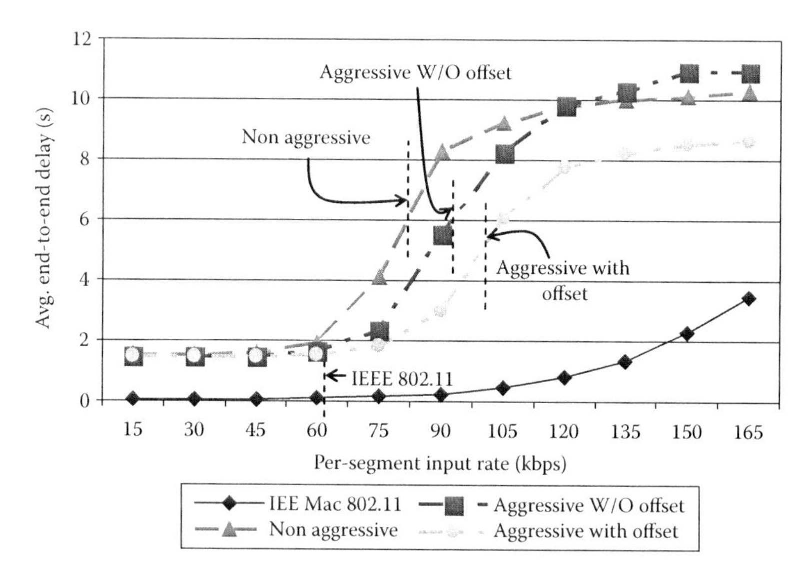

FIGURE 25.6 Average per-segment end-to-end delays.

scheduled methods; yet, as the slot size reduces, the scheduled methods' delays approached that of IEEE 802.11. Reducing the slot interval did not reduce the throughput, since the periodic cycle time was reduced proportionally. Also noticed was that the delay increased as the traffic load approached saturation, while the increase was consistently more gradual with the scheduled approach.

Figure 25.7 shows the average per-node energy efficiency in terms of the percentage of time a node's radio is turned on. The on-time percentage depends on the number of slots each node is scheduled and the input traffic rate since nodes are turned off if there is no traffic seen in the first 0.01 s of each slot. For the single-intersection network in Figure 25.2, the maximum possible on-time percentage is 38% (if a node is never turned off during its scheduled slot) since each node is scheduled an average 38% of all available time slots. It is seen for lower input rates a smaller slot time is more energy efficient while the differences due to slot time reduce with increasing input rates.

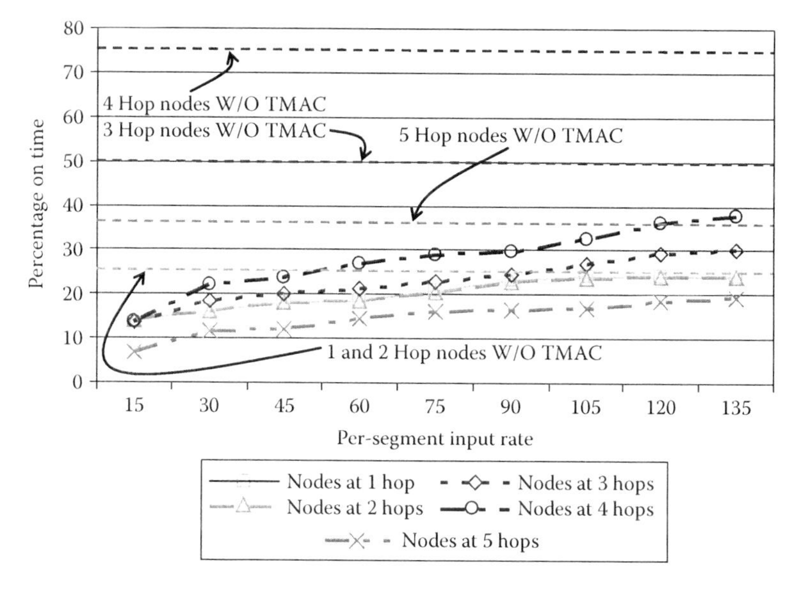

FIGURE 25.7 Average per-node energy efficiency.

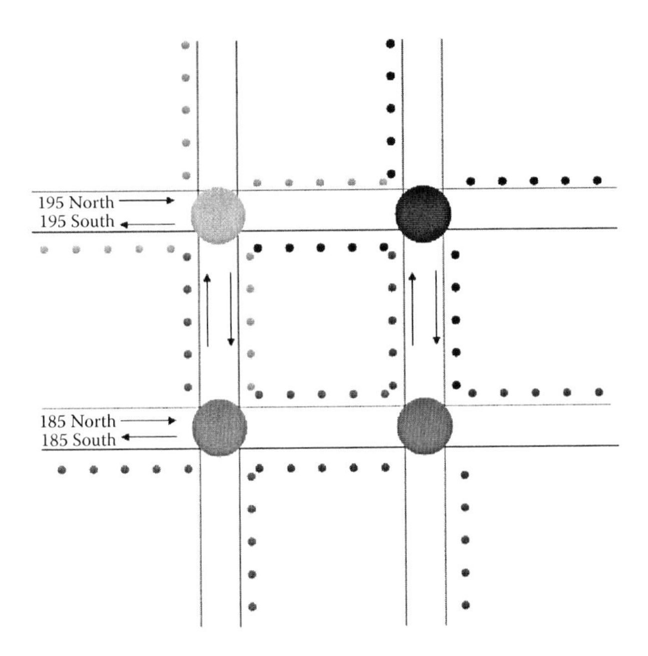

FIGURE 25.8 Four-intersection scenario.

25.6.2 Multiple-Intersection Scenario

The multiple-intersection scenario simulated four complete intersection sets as shown in Figure 25.8, including 4 intersection controllers and 16 associated clusters. Each cluster had 6 equally spaced sensors 100 m apart, and each cluster had one 50% ON/OFF exponential UDP flow sending packets from each adjacent controller to the intersection controller.

In addition to the proposed distributed schedule negotiation approach, a simple centralized alternative was simulated for comparison. The centralized approach was simply based on the single-cluster algorithm but considering all nodes in all four clusters. Controller 1 served as the highest level controller for the centralized method as well as the seed for the distributed method.

Figure 25.9 shows the resulting throughput for centralized and distributed methods (with and without schedule sliding) and IEEE 802.11. The centralized method achieved 60 kbps, the two distributed

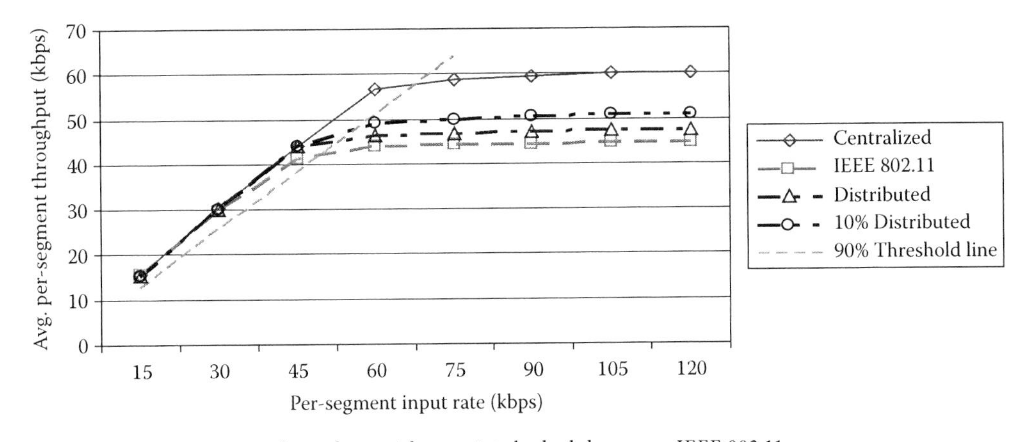

FIGURE 25.9 Per-segment throughput with negotiated schedules versus IEEE 802.11.

variants had 45 and 49 kbps, while IEEE 802.11 had only about 42 kbps saturated throughput. The throughput of the distributed method was expected to be lower than that of the centralized; nevertheless, it was shown that the differences were minimal.

25.6.3 Comparison with Other Work

In contexts close to our work, Refs. [16,26] have previously studied transmission scheduling methods for addressing inefficiencies in IEEE 802.11 and other CSMA-based MAC protocols. Specifically, Ref. [26] proposed to implement a thin layer above IEEE 802.11 (with RTS/CTS disabled) for a multi-hop IEEE 802.11 network to limit transmissions of nodes in loosely synchronized time slots to achieve access fairness in spite of asymmetric link and hidden terminal conditions. The solution was based on each node choosing a random slot to compete for transmission in each time slot such that everyone had a fair chance to grab the channel before others did. With testbed experiments, the method was shown to achieve the same throughput as IEEE 802.11 with improved fairness. Energy efficiency and delay were not considered in this work. Ref. [16] studied a conflict-free scheduling solution for wireless sensor networks to achieve highenergy efficiency. The method was also based on synchronized time slots; each sensor node periodically collected its two-hop neighbor topology and determined the contending priority of itself and others using a hash function consistent among all nodes. Since the priority was determined consistently at all nodes, each node knew exactly which slot it could transmit conflict free. With Qualnet simulation, the throughput was 20%–40% higher, delay was 1–2 orders of magnitude higher, and 25%–70% less energy compared to IEEE 802.11. Since the work considered a random ad hoc network and did not consider the traffic routing pattern, the delay was much higher than that achievable with our scheme.

25.7 Conclusion

In this chapter, a topology-aware routing and scheduling method was presented for IEEE 802.11-based highway traffic sensor and controller networks for high throughput, low latency, and energy-efficient communication. With a hierarchical controller–sensor network architecture, the methods utilized controllers to collect topology information of their associated sensors to assign collision-free time slots that maximized end-to-end throughput and minimized end-to-end latency for one-way traffic relayed across each sensor cluster. With the scheduling method, nodes could operate in sleep mode outside their scheduled slots for more than 60% of the time even when supporting the highest traffic load. With ns-2 simulation, both the centralized and distributed methods were shown to achieve 20%–40% higher throughput, slightly higher delay, and 65%–90% lesser energy compared to IEEE 802.11. With respect to other scheduling studies in close contexts, the proposed solution was shown to achieve comparable throughput and better delay and energy efficiency than the state-of-the-art solutions.

Acknowledgments

The authors are partially supported by South Carolina Department of Transportation under project SPR-676. Part of this chapter contents has been previously presented at the IEEE P2MNET 2009 workshop.

References

1. ITS Europe, Intelligent Transport Systems and Services for Europe, http://www.ertico.com/ [Accessed in May 2011].
2. ITS Australia, Intelligent Transport Systems Australia, http://www.its-australia.com.au/ [Accessed in May 2011].
3. U.S. Department of Transportation, Intelligent transport systems: Application overview, http://www.itsoverview.its.dot.gov/ [Accessed in May 2011].

4. University of California, Berkeley, Freeway Performance Measurement System (PeMS), http://pems.eecs.berkeley.edu/ [Accessed in May 2011].
5. Speedinfo Inc., SpeedInfo deploys real time traffic sensor network for SFO Bay area, http://www.speedinfo.com [Accessed in May 2011].
6. Bluetooth Special Interest Group, Bluetooth Technology, https://www.bluetooth.org/ [Accessed in May 2011].
7. Cisco Systems, Inc., Aironet 1400 Series Wireless Bridge, http://www.cisco.com/en/US/prod/collateral/wireless/ps5679/ps5279/ps5285/product_data_sheet09186a008018495c.html [Accessed in May 2011].
8. Research and Innovative Technology Administration, ITS Deployment Statistics Database, http://www.itsdeployment.its.dot.gov/ [Accessed in May 2011].
9. Federal Highway Administration, Intelligent transportation systems in work zones, http://ops.fhwa.dot.gov/wz/technologies/springfield/ [Accessed in May 2011].
10. S. Coleri-Ergen and P. Varaiya, Pedamacs: Power efficient and delay aware medium access protocol for sensor networks, *IEEE Transactions on Mobile Computing*, 5(7):920–930, 2006.
11. W. Ye, J. Heidemann, and D. Estrin, Medium access control with coordinated adaptive sleeping for wireless sensor networks, *IEEE Transactions on Networking*, 12(3):493–506, 2004.
12. T. V. Dam and K. Langendoen, An adaptive energy-efficient MAC protocol for wireless sensor networks, in *Proceedings of ACM Sensys*, Los Angeles, CA, pp. 171–180, November 5–7, 2003.
13. P. Lin, C. Qiao, and X. Wang, Medium access control with a dynamic duty cycle for sensor networks, in *Proceedings of IEEE WCNC*, Atlanta, GA, Vol. 3, pp. 1534–1539, March 21–25, 2004.
14. G. Lu, B. Krishnamachari, and C. S. Raghavendra, An adaptive energy-efficient and low-latency MAC for data gathering in wireless sensor networks, in *Proceedings of IPDPS*, Santa Fe, NM, pp. 224–231, April 26–30, 2004.
15. C. C. Enz, A. El-Hoiydi, J.-D. Decotignie, and V. Peiris, WiseNET: An ultralow-power wireless sensor network solution, *IEEE Computer*, 37(8):62–70, 2004.
16. V. Rajendran, K. Obraczka, and J. J. Garcia-Luna-Aceves, Energy-efficient, collision-free medium access control for wireless sensor networks, *Wireless Networks*, 12(1):63–78, 2006.
17. R. Kalidindi, L. Ray, R. Kannan, and S. Iyengar, Distributed energy aware MAC layer protocol for wireless sensor networks, in *Proceedings of the International Conference on Wireless Networks*, Las Vegas, NV, pp. 282–286, June 23–26, 2003.
18. I. Rhee, A. Warrier, M. Aia, J. Min, and M. L. Sichitiu, ZMAC: A hybrid MAC for wireless sensor networks, *IEEE/ACM Transactions on Networking*, 16(2):511–524, 2008.
19. M. Papageorgiou, C. Diakaki, V. Dinopoulou, A. Kotsialos, and Y. Wang, Review of road traffic control strategies, *Proceedings of the IEEE*, 91(12):2043–2067, 2003.
20. P. B. Mirchandani and K. L. Head, RHODES: A real-time traffic signal control system: Architecture, algorithms, and analysis, *Transportation Research: Part C*, 9(6):415–432, 2001.
21. Transportation Research Board, Highway location reference methods, NCHRP Synthesis Report #21, 1974.
22. D. Bagaria, Topology-aware transmission scheduling for distributed highway traffic monitoring wireless sensor networks, MS thesis, Clemson University, Clemson, SC, 2007.
23. R. R. Choudhury and N. Vaidya, MAC-Layer capture: A problem in wireless mesh networks using beamforming antennas, in *Proceedings of IEEE SECON*, San Diego, CA, pp. 401–410, June 18–21, 2007.
24. K. Jain, J. Padhye, V. N. Padmanabhan, and L. Qiu, Impact of interference on multi-hop wireless network performance, in *Proceedings of ACM MobiCom'03*, San Diego, CA, pp. 66–80, September 14–19, 2003.
25. The VINT Project, The network simulator—ns-2, http://www.isi.edu/nsnam/ns/ [Accessed in May 2011].
26. A. Rao and I. Stoica, An overlay MAC layer for 802.11 networks, in *Proceedings of ACM MobiSys*, Cologne, Germany, pp. 135–148, June 6–8, 2005.

26

Dynamically Adaptive Multi-Modal Sensor Fusion in Urban Environments

Shashi Phoha
The Pennsylvania State University

Doina Bein
The Pennsylvania State University

Yicheng Wen
The Pennsylvania State University

Bharat B. Madan
The Pennsylvania State University

Asok Ray
The Pennsylvania State University

26.1 Introduction

A sensor network operates on an infrastructure of sensing, computation, and communication, through which it perceives the evolution of physical dynamic processes in its environment. It is data centric [1] because data from sensor nodes are the most important to this application. A common characteristic of information fusion in sensor networks is that data streamed from different sources are brought together through a communication network. The following factors need to be taken into account:

- Typically, there are thousands of sensor nodes in a wireless sensor network. It is almost impossible to have a global centralized fusion center. In other words, the data fusion scheme has to be designed and implemented in a distributed fashion for wireless sensor networks.
- Each sensor node has limited battery power. A node dies quickly if the fusion algorithm consumes a lot of power, and the performance of the entire network becomes deficient with fewer functional nodes. Therefore, data fusion algorithms in sensor networks have to be computationally cheap and efficient so that the network has a decently long lifetime.

- Network resources are limited. Each communication link has its own maximum capacity. Sensors may collect data at a very high frequency rate but it is usually not possible to send large amounts of raw data through the overall network. There is a need to compress the raw data and extract only useful information from each source before the data is sent toward a place where fusion takes place.
- Communication delays are inevitable. It takes time for the data sent from the source to the destination. In general, the time delay is different for each transmission. For this reason, perfect fast time scale synchronization over multistream data from different sources is impossible for the wireless sensor networks.

Thus, data-level fusion is very costly to conduct in a sensor network due to the limited communication bandwidth and communication delay since it requires the sensors to stream the sensed data over the network. On the other hand, for decision-level fusion, a tiny sensor may not have enough data or computational power to make a decision for complex scenarios, and typically, decision-level fusion causes large information loss even before the final fusion step. Instead of involving all the nodes in the network in the clustering process, this chapter focuses on allowing only a subset of nodes that change over time and are geographically near to the mobile target to self-organize into one or more clusters, thus creating time-critical collaborative inference in the vicinity of the target. The sensor node would be able to distinguish between various types of data patterns in its knowledge database, which will allow it to selectively track a specific target among many mobile objects in an urban environment. A data pattern will be associated with a specific class of objects, and pattern-based tracking will allow us not only to pinpoint the object of a certain type among many other existing types but also to guarantee that only objects of that type are followed along a specific track. The tracking algorithm would be able to predict the target location and velocity even when some nodes fail and, in some cases, move sensors along the predicted trajectory to obtain better quality data [2].

Autonomous systems rely on various sensors to acquire information from the environment. A single sensor has its limitations in many aspects [3]. It may not provide enough information about the object and may not have enough sensing coverage. Sensor data is subject to a number of noise factors thus may not be reliable. It is also vulnerable to failure, and hence, it lacks robustness. There are many instances where it is not possible for a single sensor or source to derive inferences alone. Moreover, the sensed data are highly correlated in the vicinity of a stimulus. Over the decades, multisensor data fusion has been developed to solve a diverse set of problems with common characteristics. It seeks to combine the results of multiple sensors so that more accurate and robust information can be obtained and more reliable conclusions can be drawn.

Information fusion in a sensor network is deemed critical for both military and civilian applications and has found relevance in various applications such as tactical plan recognition [4,5], battlefield situation awareness problem [6], threat evaluation in air defense scenarios [7], and disaster response [8]. Motivated by challenging urban area applications that require adaptive sensor networks to dynamically cluster sensing, processing, and communication resources, a fusion-driven concept of the sensor network design is proposed in Ref. [9]. The design space of the data fusion algorithm is separated into two subspaces, information space and network design space; the network design space needs to adapt to the information space such that the statistical characteristics (predictability) of the ensemble of the original sensor data are preserved at the level of data fusion. An interface is added between the two spaces called *InfoNet Interface*, which seamlessly translates messages between the information space and network design space. The proposed architecture has the broader implication of providing insight and predictability in designing complex networks and evaluating their resilience to known or unknown operational perturbations. However, the technical details of each level as well as the interaction between the two were not presented. The information space is compatible with semantics of the fusion algorithms based on the concepts of symbol dynamics [10–12], finite state automata [13–15], and pattern recognition [16,17]; additionally, it generates the networking requirements to aid better fusion. The network design space is modeled as a network controller that takes inputs from the information space and adapts it to the requirements.

The chapter is organized as follows. Section 26.2 presents some related work to fusion-driven sensor networks. Section 26.3 presents the overall architecture of the sensor network. Subsequently, the modeling approach in the three interactive design subspaces is presented in the architecture, namely, information space in Section 26.4, network design space Section 26.5, and InfoNet interface Section 26.6, respectively. Some localization algorithms are presented in Section 26.7. Section 26.8 presents simulations in a multi-modal sensor network for tracking mobile targets. The chapter is summarized in Section 26.9.

26.2 Related Work

Data gathering is an important application of wireless sensor networks, in which the data collected by individual sensors is sent to a so-called base station (BS). Clustering schemes such as LEACH [18], PEGASIS [19], and HEED [20] for homogeneous networks (where nodes have the same initial energy), and LEACH-B [21], SEP [22], M-LEACH [M10], and EECS [24] for heterogeneous networks (where nodes start with various values for the initial energy) allow only the cluster heads, elected periodically, to aggregate the data from their cluster members and send it to the BS. Another important application of wireless sensor networks is tracking selective mobile targets crossing a sensor field [25–27], while ignoring others.

The sensors detect signal patterns in multiple, diverse, and spatially correlated sensor data streams that interact over the network. Based on these interactions, subsets of sensors may dynamically form *collaborative clusters* to estimate the position or the velocity of the target for tracking purposes.

In Ref. [28], the authors present an information-driven dynamic sensor collaboration that is a Bayesian approach to dynamically fuse the data from sensors to track mobile targets. This approach relies on reliable estimates of densities for belief states. Tracking mobile targets in sensor networks using a clustering approach is also discussed in Ref. [29]. In Ref. [30], a dynamic clustering algorithm based on Voronoi diagrams is presented for tracking targets in an acoustic sensor network. The authors of Ref. [31] assume that each sensor receives the same amount of information in the local region during a given time period. Their approach is too restrictive. Even within the region, sensors receive varied information corrupted with noise based on their physical location and the distance from the observed target. The cluster head is selected to be the node at the center of the cluster, which is too rigid. The authors of Ref. [32] use a centralized approach to select a subset of sensor nodes that cover a region and form a connected communication graph to answer a query within that region. The proposed approach is different, in that, cluster members are dynamically selected to be the ones that can sense the mobile target, in a dynamic space-time sliding window. Also, it is more general, in that, it can be applied to different sensing modalities and is flexible enough to incorporate various data fusion algorithms.

Dynamic clustering [33–35] generates flexible network topologies for sensors to cluster and collaborate on detecting targets moving through a sensor field. The dynamic space-time clustering (DSTC) protocol [35] is efficient and reliable in clustering nodes that observe the same pattern of interest such that in any cluster only one node, called the *cluster head*, is responsible for coordinating the communication among all other *cluster members*. Note that the nodes in a cluster only exchange messages with their cluster head, but the cluster head may send (or receive) messages to (or from) nodes outside the cluster. DSTC assumes that nodes have unique node IDs. Neighborhood information such as the list of the neighboring nodes and their physical distance to the current node is given by external protocols. We have expanded DTSC to include the heterogeneity of the sensors and the self-adaptation of the network topology due to nodes' mobility. In Ref. [36], we propose a fusion-driven distributed dynamic network controller, called multi-modal dynamic space-time clustering (MDSTC), for a multi-modal sensor network that incorporates distributed computation for in situ assessment, prognosis, and optimal reorganization of constrained resources to achieve high-quality multi-modal data fusion.

The Data Fusion Information Group (DFIG) [37] has proposed a seven-layer model for *information fusion* as shown in Figure 26.1. While *data assessment* involves signal conditioning, transformation, and signal state estimation, the objective of the *object assessment* layer is to estimate and predict *entity* states,

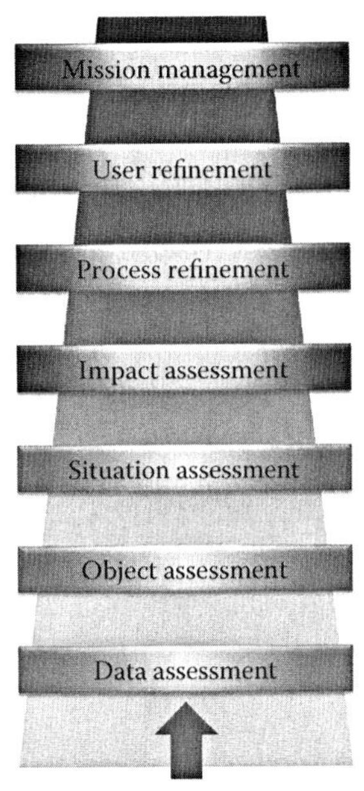

FIGURE 26.1 DFIG information fusion model.

such as entity type, position, and orientation, using data association. This is the layer where the fusion of information first occurs—signal features from different sensing modalities, observing the various footprints of an *entity*, are *fused* for estimation and prediction of the object states. This involves estimation of relationships between the footprints observed in various modalities for accurate and robust estimation of the observed *entity*. At the *situation assessment* layer, the objective is the estimation and prediction of the relations among entities identified at the object level for scene analysis and understanding.

The object assessment layer is the most developed layer. There are three basic well-studied problems involved in this layer [3]: data association, object attribute estimation, and object identification. For example, various approaches for data association have been developed in Ref. [38]. *Nearest neighbor* [39] is the simplest form of association algorithm. Other algorithms include *joint probabilistic data association* (JPDA) [40,41], *Lagrangian relaxation* [42], *artificial neural networks* [43], and *fuzzy logic* [44,45]. Object identification is nothing but a pattern classification problem [16].

Many techniques have been developed for object assessment while situation assessment is less well understood. In Ref. [46], the authors point out that situations should be modeled by some particular situation objects and some relations between these individual objects. The difficulty lies in how to properly model these relations.

The knowledge-based approach is mentioned in Ref. [47]. It starts with the modeling of the situation and then performs a pattern matching to identify the ongoing activity. Again it is observed that in either object assessment or situation assessment, pattern classification is an essential problem to solve in the information fusion architecture.

26.3 InfoNet Architecture

The proposed solution concept that addresses the design challenges mentioned earlier is presented. Different types of targets might require different configurations of the nodes to better track events. Thus, the sensor network needs to adapt to the tracking requirements during its lifetime.

The proposed approach is based on driving dynamic changes in the sensor network design space of an individual sensor in response to emerging changes in the statistical characteristics of the information

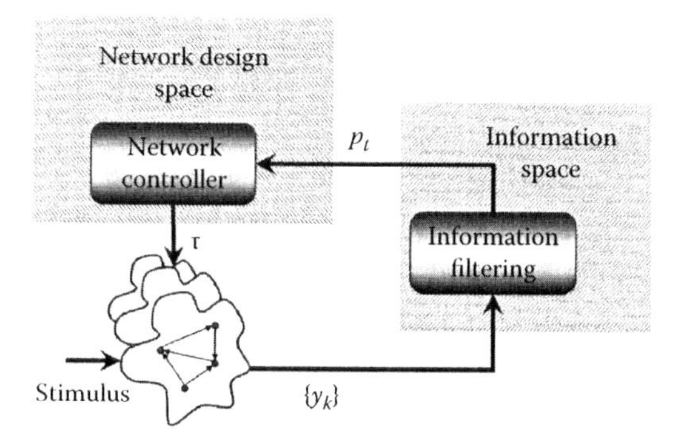

FIGURE 26.2 Fusion-driven network control.

fused from the sensor data. Rigorous mathematical modeling that captures the structural dynamics of the network is essential for engineering and verification of performance. Figure 26.2 illustrates the concept of the closed-loop network control that manipulates the network topology τ based on feedback information of evolving statistical patterns p_t derived from the data sequences $\{y_k\}$ from the sensors.

Effects of network structure on fusion processes are captured through symbolization and hierarchical abstraction in the information space. To enhance the quality and the resilience of data fusion, a distributed sensor network needs to be adaptively reconfigured, namely, the network topology must be able to change in real time based on the spatial-temporal information derived from the ensemble of the sensor data. To achieve this self-adaption, the sensor network is divided into two interactive subspaces, information space and network design space, with well-defined interactions between them (InfoNet interface). Figure 26.3 shows the architecture of a sensor network designed for mobile target tracking.

Information space of the sensor is used to derive the spatial-temporal statistics of the ensemble of the sensor data. Data fusion is posed as a multi-time-scale problem under the assumption of quasi-stationarity over the fast time scale (i.e., stationary over a sufficiently long time period) and possible nonstationarity caused by small deviations in the system behavior due to accumulating changes in the slow time scale. Each sensor addresses the issues of data compression and communication constraints by autonomously aggregating the data through symbolization and semantic construction of probabilistic finite-state automata (PFSA). The constructed PFSA are compared against a library of predetermined patterns of interest with an appropriate metric, thus allowing the sensor network to selectively track

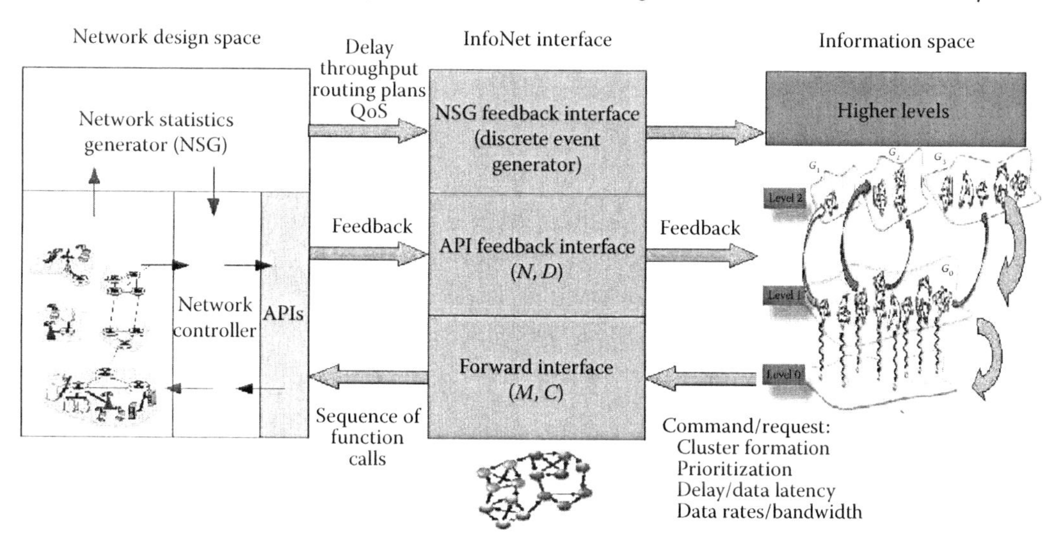

FIGURE 26.3 Detailed architecture of the design space.

interesting targets among many candidate targets [9]. Nodes observing and deciding on the same PFSA from the library are clustered dynamically, with additional network and physical requirements. This approach has the advantage of being robust to individual sensor failures.

In the *network design space*, the sensory data are modeled as a discrete-event dynamic system of interacting probabilistic automata, where sensor nodes may change their internal states through interactions with other nodes or the environment. To achieve this structural stability, the MDSTC protocol [36] is adapted to minimize communication between sensor nodes and to control the overhead while achieving a scalable operation of the sensor network for accurate mobile target tracking. The network design space is configured to adapt to the information space needs in a manner that preserves the statistical characteristics (predictability) of the ensemble of the original sensor data at each level of fusion. Reconfiguration of large sensor networks is achieved through adaptive sampling at individual sensors, sensor mobility, turning on/off of the existing sensors, bandwidth reallocation, protocol modification, or through resource redeployment. Structural stability in the face of such operational modifications must therefore be maintained along with performance. The adaptive parameters of the network design space include the sensor positions, the resource assignments, and the network connectivity.

InfoNet interface formalizes the interaction between information space and network design space. It consists of *forward* and *feedback* interfaces. The forward interface seamlessly translates information space requirements into commands and allows the network design space to act as an actuator to best meet the requirements. The feedback interface reports the result of the command back to the information space.

This architecture is used to build a sensor network that is highly scalable and robust, while providing the ability to track targets selectively and accurately.

26.4 Information Space Modeling

We present next the language-theoretic approach of modeling the information space of the sensor network. The quasi-stationarity of the physical process of interest is assumed over the fast time scale.

26.4.1 Object Identification for Single-Modality Sensors

Definition 26.1 ***(the sensor field)***

The sensor field $\mathcal{S}$ is defined to be the set of all sensors under consideration.

Typically, *Card*($\mathcal{S}$) is large ($>10^3$). Let $\mathcal{S}''$ denote all the sensors of identical sensing modality *m*.

Definition 26.2

The causal model construction map $\mathcal{H} : \mathcal{S}^m \rightarrow \mathcal{P}$ maps each sensor $s \in \mathcal{S}^m$ to a PFSA $P \in \mathcal{P}$, where $\mathcal{P}$ is the space of all possible PFSAs over the chosen symbol alphabet.

Each sensor in the sensor field collects the time-series data from the environment. For each modality, the alphabet is the same and also predefined. A PFSA is generated from the time-series data and the predefined alphabet. The map $\mathcal{H}$ describes this whole process. It is expected that the cardinality of the image Im($\mathcal{H}$) of the map $\mathcal{H}$ is significantly smaller compared number of sensors in the field. Since the sensors in $\mathcal{S}^m$ are identical, a similar partitioning is used on each sensor's data stream to yield a *PFSA* for each sensor $s \in \mathcal{S}^m$. A *PFSA* can be constructed using some existing algorithm (e.g., D-Markov [10] or CSSR [48]). While the map $\mathcal{H}$ is locally stationary in continuous time, it does evolve at a slower time scale, that is, the constructed *PFSA* evolves to reflect changes in the sensed environment.

Definition 26.3 *(atomic patterns of interest)*

An atomic pattern of interest G of a particular modality m is defined to be a PFSA over an a priori fixed alphabet.

We make the following assumptions:

1. From a library of atomic patterns of modality m that have been identified a priori, let the set of patterns of interest be denoted as

$$\mathbf{G}^m = \{G_i : i = 1, \ldots, K, \quad \text{where } K \in \mathbb{N}\} \tag{26.1}$$

   In most cases, $Card(\mathbf{G}^m) = K$ is small (~10).
2. The set of atomic patterns of interest $\mathbf{G}^m$ is totally ordered via the pattern characteristic function as follows:

$$\chi_{\mathbf{G}^m} : \mathbf{G}^m \to [0,1] \tag{26.2}$$

3. Each atomic pattern G encodes a particular dynamic process of interest. Further, if the same dynamic process occurs again, the observed PFSA from sensors possibly with noise effects from the environment are located within an open ball centered at G with some radius r in the metric space $(\mathcal{A},\theta)$.

Definition 26.4 *(metric on $\mathcal{P}$)*

The metric $\theta : \mathcal{P}\times\mathcal{P}\to[0,\infty)$ on the set of PFSA defined in Ref. [15] *is used in the sequel.*

Next define a parameterized equivalence relation $\simeq_\epsilon$ on $\mathrm{Im}(\mathcal{H})$, where $\epsilon > 0$.

Definition 26.5 *(ϵ-equivalence)*

For an arbitrary probabilistic machine $\mathcal{H}(s_i) \in \mathrm{Im}(\mathcal{H})$, a$G_j \in \mathbf{G}^m$ and a given $\epsilon > 0$,

$$\theta(\mathcal{H}(s_i), G_j) \leqq \epsilon \Rightarrow \mathcal{H}(s_i) \simeq_\epsilon G_j \tag{26.3}$$

Further let the set of equivalence classes of $\simeq_\epsilon$ be $\mathrm{E}_\epsilon = \{E^r : r = 1, \ldots, R\}$, that is, for $E_\ell, E_r \in \mathrm{E}_\epsilon$, the following conditions hold:

$$E^\ell \subseteqq \mathrm{Im}(\mathcal{H}) \tag{26.4a}$$

$$E^r \subseteqq \mathrm{Im}(\mathcal{H}) \tag{26.4b}$$

$$E^r \cap E^\ell = \varnothing \quad \text{for } r \neq \ell \tag{26.4c}$$

Definition 26.6 *(the precluster map)*

The precluster map $\Psi : \mathrm{E}_\epsilon \to \mathbf{G}^m \bigcup \{\varnothing\}$ *is defined as follows:*

$$\forall E^r \in \mathrm{E}_\epsilon, \Psi(E^r) = \begin{cases} G_i, & \text{if } \exists H \in E^r, \text{such that } \theta(H, G_i) \leq \epsilon \\ \varnothing, & \text{otherwise} \end{cases} \tag{26.5}$$

Definition 26.7 *(precluster)*

An equivalence class $E^\gamma \in \mathrm{E}_\epsilon$ *is defined to be a precluster if*

$$\Psi(E^r) \neq \varnothing \tag{26.6}$$

Lemma 26.1

The pre-cluster map is well defined if

$$\forall G_i, G_j \in \mathbf{G}^m, \quad \theta(G_i, G_j) > \epsilon \quad \text{for } i \neq j \tag{26.7}$$

Proof

It suffices to show that inequality (26.7) guarantees the following condition:

$$\forall E^r \in \mathrm{E}_\epsilon, (\exists G_i \in \mathbf{G}(\Psi(E^r) = G_i) \Rightarrow \forall H \in E^r, \forall j \neq i, \theta(H, G_j) > \epsilon) \tag{26.8}$$

The result follows immediately from the triangular inequality by noting that θ is a metric on the space of *PFSA* [15].

For all epochs in slow time scale, the following strict inequality has to be satisfied:

$$0 < \epsilon < \frac{1}{2} \epsilon^{\mathbf{G}^m} \tag{26.9}$$

where the critical upper bound $\epsilon^{\mathbf{G}^m}$ for possible choices of ε is given by

$$\epsilon^{\mathbf{G}^m} = \min_{G_i, G_j \in \mathbf{G}^m} \theta(G_i, G_j) \tag{26.10}$$

The aforementioned formalization has the following implications:

- Preclusters are 2ϵ spheres in the information space. They are mutually disjoint.
- The center of each precluster is occupied by a unique pattern of interest.
- To make clusters out of preclusters, further constraints arising from geometric and communication requirements must be imposed.

Decision stability in the information space is of critical importance for the algorithmic considerations of linguistic sensing. It follows from Definition 26.5 that the choice of ε is critical in ensuring that pattern identification can be carried out without ambiguity in the information space. We formalize the concept of decision stability as follows.

Definition 26.8 ***(margin of decision stability)***

For a given set of patterns of interest $\mathbf{G}^m$ *and a chosen* ϵ, *the margin of decision stability* λ *of a deployed sensor network is defined as follows:*

$$\lambda = \min_{G_i, G_j \in \mathbf{G}^m} (\theta(G_i, G_j) - 2\epsilon) \tag{26.11}$$

Definition 26.9 ***(stable, unstable and marginally stable networks)***

For a given set of patterns of interest $\mathbf{G}^m$ *and a chosen* ε, *decision stability status of a deployed sensor network is defined as follows:*

$$\left.\begin{array}{ll} \lambda < 0, & \text{Unstable} \\ \lambda = 0, & \text{Marginally stable} \\ \lambda > 0, & \text{Stable} \end{array}\right\} \tag{26.12}$$

26.4.2 Generalization to Multi-Modal Sensors

Sensors of a single modality have restrictions on what they can detect. For example, acoustic sensors cannot provide information of the weight of objects. Chemical sensors can only be aware of the chemical substance. A target of interest may require more than one type of sensor to be correctly identified.

Let $\mathbb{L} = \{\mathcal{L}_1, \mathcal{L}_2, \ldots, \mathcal{L}_N\}$ be the universal set of *atomic* patterns. The *atomic* pattern library $\mathbb{L}$ is a set of modal footprints identified from individual sensing modalities for various objects or events discovered via PFSA construction. Given the atomic pattern library, a naive way of generalizing the previous framework is called the *set-theoretic* approach. In this framework, higher level patterns, events and scenes, or contexts are modeled as subsets of $\mathbb{L}$. Thus, a composite pattern, representing an object or even an event, is a collection of elements from $\mathbb{L}$ and the composite pattern library is defined as $\mathbb{L}^* \subset 2^{\mathbb{L}}$. A set-theoretic approach to information fusion that is based on the DFIG information fusion model is shown in Figure 26.4. Objects (at level 1) are thought of as a collection of atomic patterns and scenes or situations as a collection of objects. We use this framework for selectively tracking mobile targets in a multi-modal sensor field in the urban scenario, which is described in detail in Section 26.8.

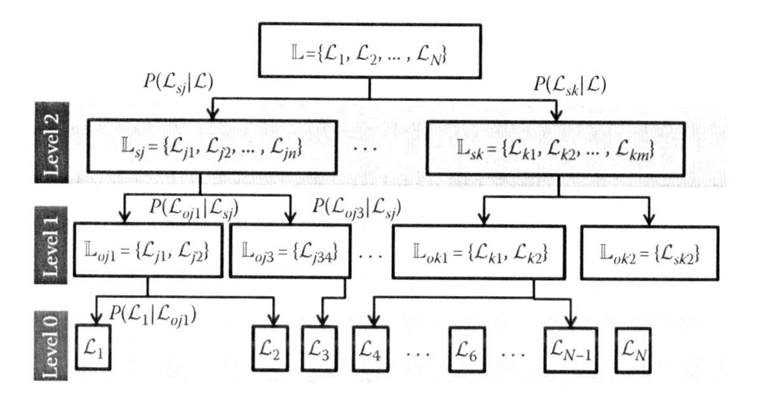

FIGURE 26.4 *Set-theoretic* approach to information fusion.

26.5 Dynamic Space-Time Clustering

In this section, we present the basic steps in clustering nodes in the space-time vicinity of an event; the clustered nodes will exchange information and the cluster head will perform data fusion. On the other hand, forming clusters increases the communication load of the network. During precluster formation, cluster heads are selected dynamically and are responsible for coordinating the local self-adaptation of the cluster members to the requirements of information space [9]; in this case, the cluster heads are responsible for event estimation and tracking. We assume that some sensors are mobile and have the capability to move under the guidance of their network controller. When requested by other sensors, some mobile sensor nodes could move within the area in which events are observed to proactively enrich both event-tracking and network resilience in order to ensure predictable performance for a given objective.

The protocol MDSTC [36] is implemented at the network space level and has seven basic operations for a cluster: precluster formation, creating a cluster, disbanding a cluster, adding a node to a cluster, removing a node from a cluster, node repositioning, and moving the data from the cluster head to another cluster member. The tasks of MDSTC are: (1) to keep track of the nodes in space-time vicinity of an event (precluster formation), (2) for a node selected as a cluster head, to keep track of cluster members (cluster formation and preservation), and (3) to request or act on requests to change the node's position when the application layer (information space) requires better quality data for data fusion (node repositioning). After the command carried by the message from the information space has been executed in the network design space, the network design space sends the feedback to the information space regarding the outcome of the execution. We present briefly these operations and how they relate to the InfoNet architecture.

As the target moves through the sensor field, a sensor generates a stream of data and then looks for patterns in it. The first node that recognizes a pattern requests the formation of a precluster. This node becomes the temporary precluster head, creates a unique cluster ID, and informs its immediate neighbors by sending a message that contains the pattern ID and the corresponding metric value that measures the node's suitability as a cluster head. On receiving such messages that arrive within a time frame, and depending upon whether they are free or not, a node may send back a message indicating its state to the precluster head and then possibly join the precluster. Alternatively, it may simply choose not to reply if the node does not see any pattern of interest. Eventually, the precluster head collects the state of each neighboring node that does reply and chooses the most suitable one as the cluster head.

The MDSTC algorithm checks periodically whether there is a cluster member with a better metric (closer to the node in terms of physical distance) for the observed pattern than the cluster head. In such case, that node becomes the new cluster head. This ensures the robustness of the cluster in case the cluster head fails or has moved significantly away from the target.

It is noted that a cluster is required to be of a minimum size for reliable sensing and communication robustness. When a cluster is deficient in size, the cluster head asks more nodes to join. A cluster head can also reduce the cluster size by notifying some cluster members that they are no longer part of the cluster. Additionally, cluster members may choose to remove themselves from the cluster and inform the cluster head. A cluster can exist for a decided period of time, called *max_lifetime*.

When the mobile target in the sensor network is in motion, a node other than a cluster head may observe the pattern better. Since the cluster head is selected as the node that currently observes the event the best, it is likely that a cluster is formed with the node that observes the event better than the cluster head. In some cases, the current cluster needs to be disbanded and another cluster is to be formed. The local network stability may suffer if clusters are created and disbanded as soon as some nodes observe better than the current cluster heads. A *lazy* approach is taken here to maintain stability: the cluster is allowed to live for at least some period of time, called *min_lifetime*. During the *min_lifetime*, the cluster head does not change due to a better composite metric for the same composite pattern. Once the *min_lifetime* expires and before *max_lifetime* expires, it is possible for a cluster member or a noncluster member to become a cluster head and force the current cluster head to disband the cluster.

In general, it is possible to have more than one current cluster in the space-time vicinity of an event. The reason for having multiple clusters (that never overlap) is that nodes that are not part of some cluster will try to form a precluster and may be successful.

26.6 InfoNet Interface

Data fusion is done in the information space using PFSA and nonlinear symbolic dynamics filtering. The network design space is a stack of standard protocols and function calls that communicate using application programming interface (API). The InfoNet interface provides the interface between the information space and the network design space.

Definition 26.10

A (network design space) API can be represented by a five-tuple $A = (\mathcal{F}, \mathcal{I}, \mathcal{O}, In, Out)$, where $\mathcal{F}$ is the set of all available function calls in the API, $\mathcal{I}$ is the set of input variables for all function calls, $\mathcal{O}$ is the set of output variables for all function calls; it is a four-tuple (k, Flag, f, N_p), where

- *k denotes the type of the API.*
- *Flag* = {0,1}. *Value 0 means that the network controller has successfully carried out the command and Value 1 means that the network controller could not execute the command.*
- *$f \in \mathcal{J}$, where $\mathcal{J}$ is a set of feedback function calls available in the API feedback interface. $\mathcal{J}$ and $\mathcal{F}$ are closely related in the sense that the interface calls some function in $\mathcal{J}$ to drive the network like a controller driving an actuator and the network calls some function in $\mathcal{F}$ to reply back to the API feedback interface about what the network controller has actually accomplished.*
- *N_p is a set of input arguments of a function $f \in \mathcal{J}$. In other words, N_p contains the parameters that the network controller will pass back to the API feedback interface.*

$In : \mathcal{F} \rightarrow 2$ represents the set of inputs for a function.

$Out : \mathcal{F} \rightarrow 2^{O}$ represents the set of outputs for a function.

The formal definition is necessary to provide a common framework in which all the network APIs can be defined. The formalism allows a common framework in which symbolic algorithms in the information space can interact with the function calls of the network APIs and vice versa.

In this approach, there is a need for an interface that converts the rigorous mathematical formalism of the information space to a set of heuristic rules that can be executed by the network design space and vice versa. In this section, two interfaces are defined, shown in Figure 26.3. A generic conversion algorithm will also be presented.

26.6.1 Forward Interface

The forward interface takes a command from the information space and translates it into a language understood by the network design space. APIs enable the network controller to exchange messages with the forward interface to achieve communication with the information space. According to Definition 26.10, the forward interface should provide a series of function calls and their associated input arguments to the APIs. The higher order API is provided to keep consistency when dealing with the two different design spaces (network and information spaces) and to provide a coherent framework for expandability in case future functions are to be added. Also, such a formal structure allows symbolic algorithms to monitor the statistics of the network at different levels (network packet, API function call, etc.) in order to reconfigure the network in response to stimuli.

Definition 26.11

The forward interface is modeled as $(\mathcal{M}, \mathcal{C})$, *where* $\mathcal{M}$ *is a collection of finite state automata and* $\mathcal{C}$ *is a library of input–output mappings. Each* $M^k \in \mathcal{M}$ *is a Mealy machine with* $M^k = \left(Q^k, \Sigma_I^k, \Sigma_O^k, \delta^k, F^k\right)$, *where*

- *k is the index of the machine because each machine is specifically designed for one API of the network design space.*
- Q^k *is the set of the states of* M^k.
- Σ_I^k *is the alphabet set of inputs of* M^k.
- Σ_O^k *is the alphabet set of outputs of* M^k *such that there is a bijective mapping* $\kappa : \Sigma_O^k \rightarrow \mathcal{F}$, *where* $\mathcal{F}$ *are the function calls in the kth API.*
- δ^k *is a mapping* $\delta : Q^k \times \Sigma_I^{k\text{TM}} Q^k$.
- F^k *is a mapping* $F^k : Q^k \times \Sigma_I^{k\text{TM}} \Sigma_O^k$.

Together with each M^k *there is a collection of mappings* $C^k \subseteq \mathcal{C}$ *defined as* $C^k = \{(I_p, \mathcal{I}, \varphi)\}$, *where*

- I_p *is the set of output parameters from the information space and also the input of the forward interface.*
- $\mathcal{I}$ *is the set of input parameters to network design space and also the output of the forward interface.*
- $\varphi \in C^k$ *is some mapping* $\varphi : I_p \rightarrow \mathcal{C}$.

Each node in the network can implement a subset of all the APIs available depending upon its role in the network. For instance, a sensor node in the network can implement cluster API, whereas a router can implement a routing API in addition to the cluster API. In this chapter, for each API, a finite-state Mealy machine (not to be confused with a machine as in network node) is designed to define the operation of the functions in the API. This finite state machine is clearly different for different APIs in the network.

It follows from Definition 26.11 that the inputs from the information space can be classified into two classes. One class $L_1 = \left\{s : s \in \Sigma_I^k\right\}$ is an input of a specific finite-state machine $M^k \in \mathcal{M}$ and becomes a sequence of function calls $\left\{w : w \in \Sigma_O^k\right\}$ in the network design space. The other class (L_2) will be handled by the mapping $\mathcal{C}$ and converted to the input arguments for the function calls. An illustrative example would be to assign the priority on a specific node. The priority value assigned to the node belongs to L_1, and the node ID in the information space is in L_2. $\mathcal{C}$ is actually composed of a library of mappings φ that can be shared among all the specific models of the forward interface. Typical examples of L_1 and L_2 class are the InfoNet to MDSTC messages.

Given the inputs from the information space, the following generic conversion algorithm for the forward interface decomposes them into two classes, class L_1 and class L_2, as follows:

1. Assume that $p \in L_1$ and $q \in L_2$.
2. Represent p as an input string $\sigma_1\sigma_2\sigma_1\sigma_3\sigma_2$. …
3. Feed the input string into the corresponding automata M_p.
4. Obtain the output string $s_3s_1s_1s_1s_2$… from M_p.
5. Obtain the input arguments from $\mathcal{C}(q)$.
6. According to the output string, execute the function calls with appropriate input arguments.

An example of the aforementioned automata formulation can be a simplified version of the state transition diagram of the DSTC protocol for an individual node in the sensor field as shown in Figure 26.5. A sensor node keeps collecting data and remains in the state *Alone* until the automaton built from the raw data matches a pattern in the pattern library. Then a node goes to the state *Initial Election*, waits for messages from neighbors, and attempts to elect a cluster head among the nodes that observe the same pattern. Based on the metric value, a precluster head is elected and a precluster thus is formed, which ultimately leads to a cluster after checking if some other constraints are satisfied. Finally, after a while,

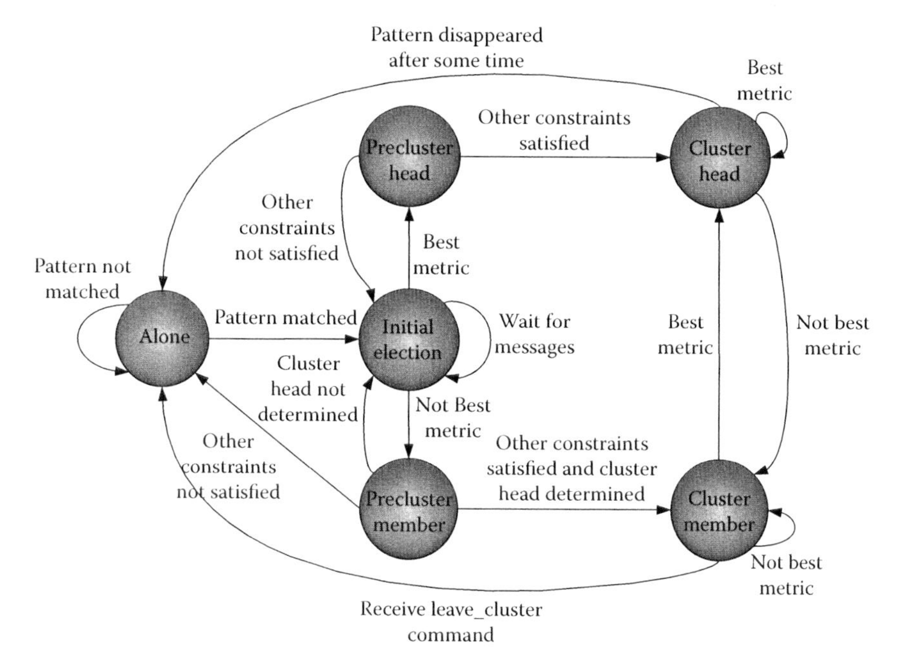

FIGURE 26.5 State transition diagram for a sensor node.

if the pattern is not observed anymore, the cluster head initiates the disbanding of the cluster. Then, the nodes go back to state *Alone* again. The same process repeats over and over for each node in the network.

26.6.2 Feedback Interface

The feedback interface passes the outputs from the network design space back to the information space. The outputs of the network design space are of two types. One type consists of API outputs such as whether the API function was executed successfully, etc. The other type consists of network statistics that describe the communication quality; based on the network statistics, the information space will be able to differentiate whether a bad data fusion is due to inappropriate cluster formation or to a bad communication.

Definition 26.12

The finite state automaton $\mathcal{N}$ of the API feedback interface for APIs outputs is the Cartesian product of two automata $N^k \times N^f$, where $N^k = (Q^k, \Sigma^k, \delta^k)$

- Q^k = {*Cluster Control API, Prioritization API, Delay API, Data Rate API*}.
- $\Sigma^k = Q^k$.
- $\delta^k\left(q_i^k, s\right) = q_i^k$ *for any* $s \in Q^k$ *and* $N^f = \left(Q^f, \Sigma_I^f, \Sigma_O^f, \delta^f, F^f\right)$.
- $Q^f = \mathcal{J}$ *is the set of all available feedback function calls in the API feedback interface.*
- $\Sigma_I^f = Q^f$.
- Σ_O^f *is the alphabet of updating actions.*
- $\delta^f\left(q_i^f, q_j^f\right) = q_j^f$ *for all* $q_i^f, q_j^f \in Q^f$.
- F^f is a mapping $F^f : Q^f \times \Sigma_O^f$™$\Sigma_O^f$.

Definition 26.13

The API feedback interface is a tuple ($\mathcal{N}$, $\mathcal{D}$), where $\mathcal{N}$ is the automaton from Definition 26.12, and $\mathcal{D}$ is a library of input-output mappings of the form (N_p, I_p, ψ), where ψ: $N_p \rightarrow I_p$.

$\mathcal{D}$ can be interpreted as the inverse mapping of $\mathcal{C}$, but $\mathcal{D}$ is not exactly as same as $\mathcal{C}^{-1}$.

The outputs of APIs have been presented in Definition 26.10 as a tuple $\mathcal{O} = (k, Flag, f, N_p)$. *Flag* is a single bit, informing the information space that the network is unable to fully complete the command that has been given. f and N_p carry all the information about what the network has already done. From the network design space, one will obtain $\mathcal{O} = (k, Flag, f, N_p)$. The algorithm is the same as the generic algorithm of the forward interface except that here k and f belong to the class L_1, denoted by p, and N_p belongs to the class L_2, denoted by q.

1. Represent p as an input string $\sigma_1\sigma_2\sigma_1\sigma_3\sigma_2$. ...
2. Feed the input string into the feedback automata $\mathcal{N}$.
3. Obtain the output string $s_3s_1s_1s_1s_2$... from $\mathcal{N}$.
4. Obtain the input arguments from $\mathcal{D}(q)$.
5. According to the output string, execute the updating actions with appropriate input arguments.

26.7 Localization Algorithms

There are many sophisticated tracking algorithms in the literature if some prior knowledge of the actual system dynamics is given. *Kalman filter* [50], also called minimum-variance filter, was first proposed in the 1960s, and it is the most commonly used technique in target tracking ever since. Kalman filter has been shown to be optimal for linear Gaussian systems. *Generalized pseudo-Bayesian model* [51,52] and *interacting multiple model* [53] use a bank of filters, each of which corresponds to one operational mode of the dynamic system. The concept of *particle filter* [54], also known as Monte Carlo Markov chain (MCMC), is developed to counter the situation where the noise is not Gaussian. Since our objective is to validate the designed fusion architecture not to design another sophisticated tracking algorithm, we have used in our simulations some simple localization algorithms that will be presented next.

26.7.1 CPA-Based Algorithm

Closest point of approach (CPA) tracking algorithm [55] was originally designed for an acoustic sensor field. However, the idea can be extended in this framework to other types of sensors or even to a multi-modal sensor field. When a target passes closest to the sensor node, the sensor's signal reaches a peak and the event is signaled. The captured event is called a *CPA* event. For example, in case of an acoustic sensor, when a noisy target approaches, the acoustic sensor can detect the loudness based on Doppler shift. The sensor detects the CPA event at the loudest point. The signal peak of the CPA event is directly related to the distance between the target and the sensor because the signal–noise ratio of sensing decreases with the distance. The closer the target is to the sensor, the higher signal peak of the CPA event is. The signal peak is used as weight on the sensor to triangulate the position of the target.

In the proposed formulation, it has been shown that the semantic distance detected by the sensor increases as the target moves away from the sensor [56]. The semantic distance can be used instead of the signal peak to calculate the weight. To apply the CPA-based algorithm for multi-modal sensors, one just needs to properly scale the weights for different modalities to the same level. The detailed algorithm can be found in Ref. [36].

26.7.2 Bayesian Approach

An event of interest is detected by a formation of a multi-modal cluster, and the cluster head is responsible for collecting the measurements from its cluster members of different modalities in general, each of which observed some aspects of the event and fusing those data. The objective for the cluster head is to estimate the position of the detected target.

Assume that there are M different sensing modalities in the sensor network. The linguistic sensing radius in modality m is denoted by ϵ^m, where $m = 1, 2, \ldots, M$. Each ϵ^m is a design parameter in the information space. Suppose there is a multi-modal cluster in the network $\mathcal{C}(\mathbf{G}) = \{s_i\}_{i=1}^N$ observing one composite pattern $\mathrm{G} = (G^1, G^2, \ldots, G^M)$ in the composite pattern library $\mathcal{G}$, where s_i denotes sensor i, a member of the cluster. Let m_i denote the modality of a particular sensor i and (x_i, y_i) denote the xy position of sensor i. Every cluster member s_i reports its semantic distance ϵ_i to the cluster head. Due to the clustering rule in the information space, it follows that $\epsilon_i \in [0, \epsilon^{m_i})$. The interval $[0, \epsilon^{m_i})$ is partitioned into finite subintervals $A_k^{m_i} = [a_k^{m_i}, a_{k+1}^{m_i})$, where $a_k^{m_i} < a_{k+1}^{m_i}$ and k are the index of the partition. Let S be the sensing region of the cluster $\mathcal{C}(\mathrm{G})$ beyond which a target cannot be detected. The physical space S is partitioned into some finite number of grids $\{l_j\}_{j=1}^K$, where K is the number of grids.

The fundamental assumptions are as follows:

1. The target can trigger a dynamic cluster formation. It implies that the target has a maximal moving speed v_{max}.
2. There is one and only one target in the footprint S of the cluster.
3. Each sensor is isotropic. The sensor measurement only depends on the distance between the target and the sensor itself and is not affected by the sensing angle.
4. The conditional probability $P\left[\{\epsilon_i \in A_k^{m_i}\} \mid T \in l_j\right]$ is the same for each atomic pattern G^m.
5. Given the position of the target $T \in l_j$, events $\left\{\epsilon_{i_1} \in A_k^{m_{i_1}}\right\}$ and $\left\{\epsilon_{i_2} \in A_k^{m_{i_2}}\right\}$ are independent for $i_1 \neq i_2$.

$$P\left[T \in l_j \mid \cap_i \{\epsilon_i \in A_k^{m_i}\}\right] = \frac{P\left[\cap_i \{\epsilon_i \in A_k^{m_i}\} \mid T \in l_j\right] P[T \in l_j]}{\text{Normalization factor}} \tag{26.13}$$

$$= \frac{\prod_i P\left[\{\epsilon_i \in A_k^{m_i}\} \mid T \in l_j\right] P[T \in l_j]}{\text{Normalization factor}} \tag{26.14}$$

$$= \frac{\prod_i P[\epsilon_i \in A_k^{m_i} \mid d_i \in B_i^n] P[T \in l_j]}{\text{Normalization factor}} \tag{26.15}$$

Equation 26.13 is the Bayes' formula and Equation 26.14 uses assumption 5. Equation 26.15 uses assumption 3 that states that the sensor measurement only depends on the d_i, the distance between the sensor (x_i, y_i) and the target (l_{x_j}, l_{y_j}).

Now it remains to find out the prior probability $P[T \in l_j]$ and the conditional probability $P[\epsilon_i \in A_k^{m_i} \mid d_i \in B_i^n]$. Without the measurements, nothing more can be drawn except the fact that the target is in the footprint of the cluster, that is, $T \in \mathrm{S}$. Therefore, it is natural to assume a uniform distribution over the physical region S. The probability of a target being in a grid is proportional to the area of that grid, namely, $P[T \in l_j] = \dfrac{Area(l_j)}{Area(S)}$. The conditional probability $P[\epsilon_i \in A_k^{m_i} \mid d_i \in B_i^n]$ needs to be trained from the experiments. Assumption 4 says the conditional probability only depends on the atomic patterns in the pattern library. Therefore, one only needs to train a limited number of such distributions. For each modality, several pairs of (ϵ, d) are generated from the training and the relative frequency is used to define the corresponding probability. This approach is implemented in NS-2 for a pressure sensor field (see Section 26.8).

26.8 Experimental and Simulation Results

We present several simulations in NS-2 [57] of the sensor network to validate the designed architecture of the fusion-driven sensor network. Figure 26.6 gives the schematics of the implemented modules on two nodes in NS-2. As one expects, each node has three subspaces, information space, network design space (network controller), and InfoNet interface. It also shows the sequences of function calls when node 1 initiates a cluster formation.

A pressure sensitive floor is used to collect data; it consists of an array of piezoelectric wires that serve as distributed pressure sensors. A coil of piezoelectric wire is placed under square floor tiles of size 0.65 m × 0.65 m such that each sensor generates an analog voltage due to pressure applied on it. The output of the voltage is in the range of 0–1023. A total of 144 sensors are placed in a 9 × 16 equidistant grid to cover the entire laboratory. Consider a Segway RMP robot moving in the sensor field.

The objective of the simulation is to detect and track spatiotemporal events of the behavior patterns of a Segway RMP. A Segway RMP moving in different types of motion trajectories are considered for illustration of detecting and tracking various behavior patterns.

In the *training* phase, a pattern library of the aforementioned behaviors is created by the following procedure:

- The raw time-series data are collected from each pressure sensor.
- A PFSA is built based on the data using D-Markov machine construction algorithm [10].
- The PFSA is added to the library by ensuring that it satisfies the fundamental equation of sensor network operation (see Ref. [56]).

In the *operational* phase, the spatiotemporal information of the Segway's movement is fused by clustering the sensors along the estimated path of the Segway. The sensed data is used to build PFSAs locally at each node using D-Markov machines, and the constructed PFSA is compared against the pattern library. In the *operational* phase, the spatio-temporal information of the Segway's movement is fused by clustering the sensors along the estimated path of the Segway.

The NS-2 simulator is used to simulate the network-related aspects of the experiments. The data collected by each sensor during the experiment is fed into a corresponding simulated node in the NS-2 environment to create a detailed simulation of the distributed sensor network. The algorithms associated with the information space, the network controller, and the InfoNet interface are implemented at each node. For the information space algorithms, a library of patterns (PFSAs) is created off-line and stored at every node. During the online operation of the sensor network, each node collects the raw data from its pressure sensor. After collecting a fair amount of data, a probabilistic finite state machine is built using the construction algorithm for a D-Markov machine [10]. The PFSA constructed in this manner is then matched with the existing set of patterns in the pattern library. The semantic sensing radius is chosen appropriately using the metric θ defined on the set of PFSAs [15]. Next, a precluster is formed among the neighboring sensors that observe the same pattern. This precluster eventually becomes a cluster, using the DSTC algorithm defined at the network controller level of each node of the pre-cluster, and a cluster head is elected.

The newly formed cluster has a lifetime, at the end of which the cluster is automatically disbanded. The timeout starts as soon as the cluster head stops observing the pattern for which the cluster was formed. While in the cluster, the cluster head fuses the data from its members to estimate the current position of the target. Figure 26.7 shows a snapshot of simulation setup after forming two clusters, represented by the large circles.

Figure 26.8 shows a run of the tracking algorithm. The actual trajectory of the Segway robot is shown as a line, and the estimated positions are shown by dots.

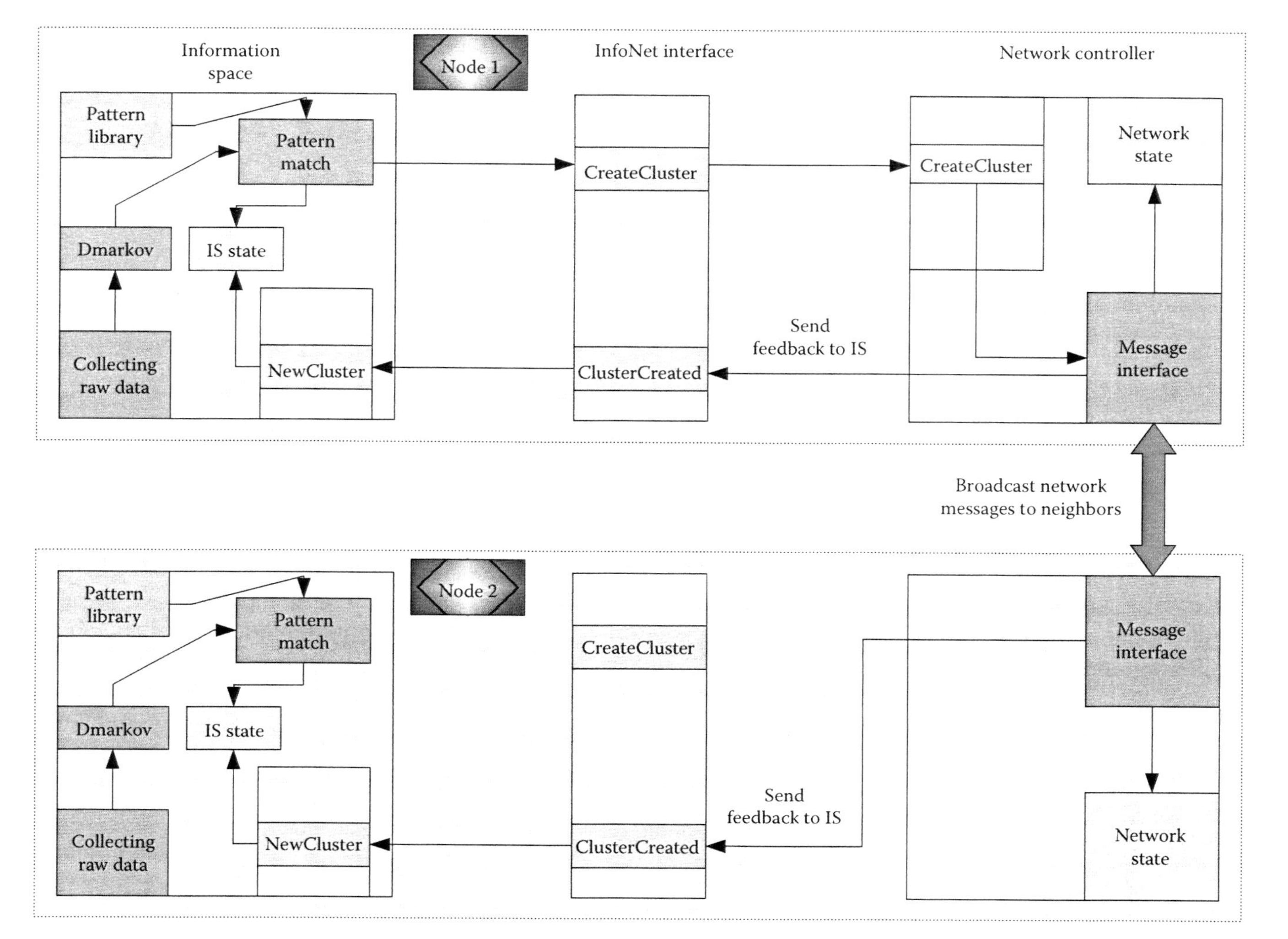

FIGURE 26.6 Implemented modules on two nodes in NS-2.

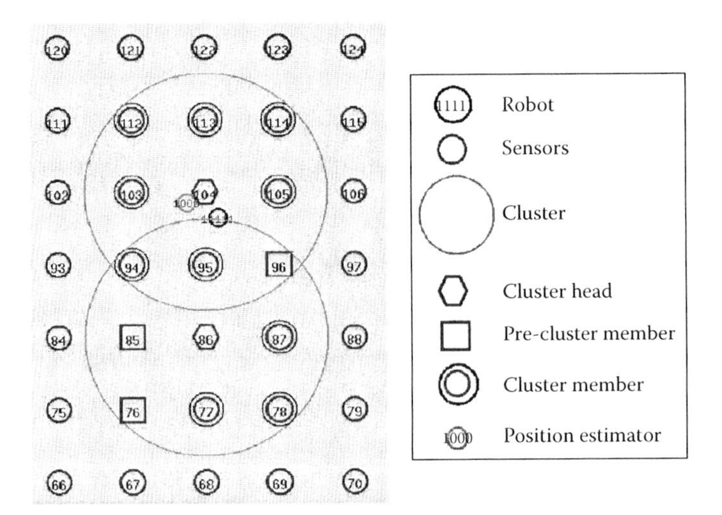

FIGURE 26.7 Snapshot of the robot tracking simulation.

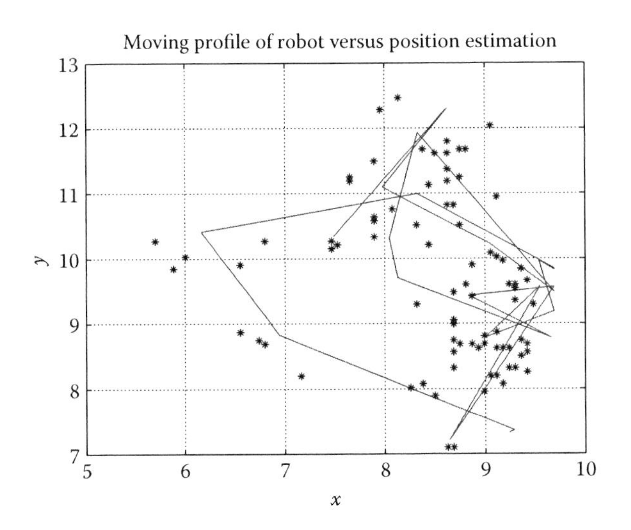

FIGURE 26.8 Robot trajectory versus estimated position.

The corresponding error in position estimation over time for every slow time epoch is shown in Figure 26.9.

Figure 26.10 shows a multi-modal sensor field in an urban scenario, where large blocks represent buildings separated by two way streets, 20 ft wide. There are three types of sensors deployed in this urban scenario, pressure sensors, video sensors (cameras), and magnetic sensors. Pressure sensors have the same characteristics as the ones implemented in the previous pressure sensor field, except that their sensing ranges are scaled up. These pressure sensors are embedded in the ground of two-way streets; they are immobile. Video sensors are mounted on the buildings, whose view angles are adjustable so that the coverage can be easily changed. Video sensors take snapshots of the scene within its view at some certain frequency. Thereafter, those static images are used to construct PFSA for cameras. Magnetic sensors can detect the variations and distortions in the Earth's magnetic field caused by ferromagnetic material. As suggested in Ref. [58], a moving ferromagnetic object can be modeled as a moving magnetic dipole

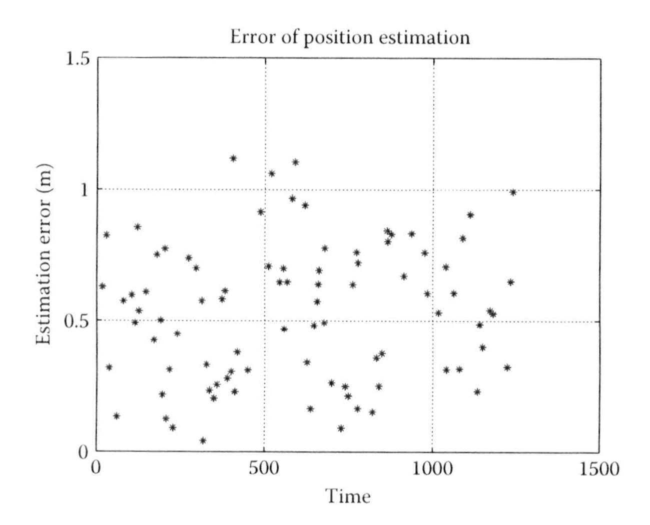

FIGURE 26.9 Error in position estimation.

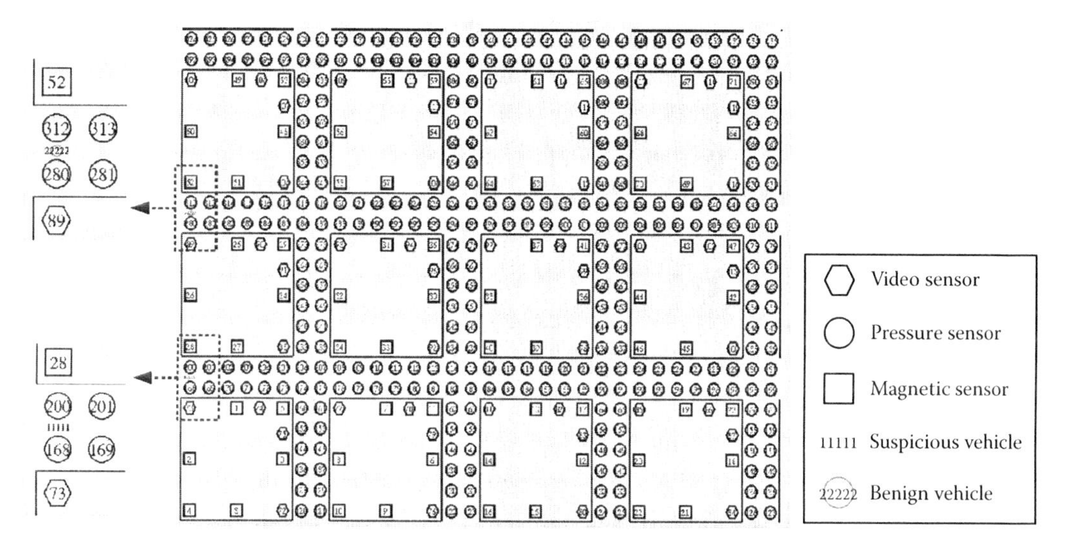

FIGURE 26.10 Multi-modal sensor field in an urban scenario.

centered at some (x,y,z) position. Magnetic sensors therefore are simulated to detect the magnetic field around a ferromagnetic object. Magnetic sensors are mounted on some mobile platforms that can move along the streets; they can be quickly relocated if necessary.

The objective of the multi-modal sensor network is to detect suspicious moving vehicles that carry large amounts of metallic material, localizing them, and keeping track of their trajectory. Other benign vehicles will be ignored, since they cannot trigger an atomic pattern match for magnetic sensor, thus a compound pattern cannot be identified; thereby, benign targets are intentionally ignored by the network. Both compound pattern detection and CPA-based tracking are implemented in this simulation.

Figure 26.11 shows a snapshot of the simulation in NS-2. We note that clusters are formed only around the suspicious vehicle when a composite pattern is identified. The arrows are the position and velocity estimates of the suspicious target, which indicates the direction of the target.

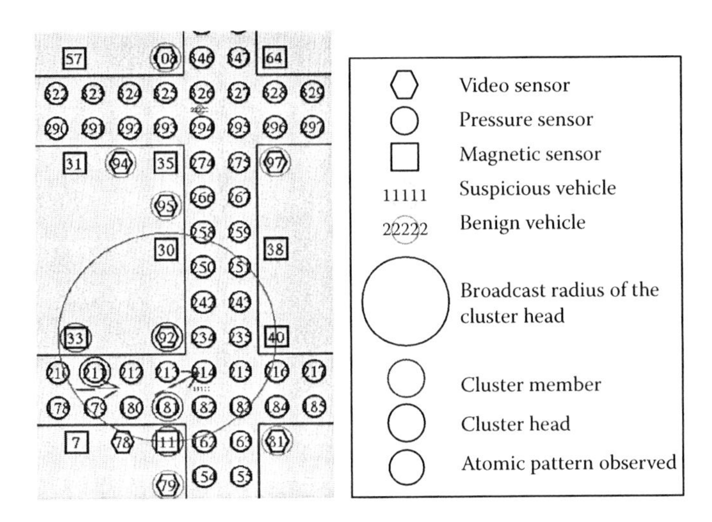

FIGURE 26.11 Tracking a suspicious vehicle in an urban scenario by composite patterns.

26.9 Conclusion and Future Work

We present the design and implementation of a sensor network for tracking mobile targets using multi-modal sensors. Patterns hidden in the time-series data observed by sensors are captured by PFSA after the data are symbolized into symbol sequences. The design space of the sensor network is decomposed into information and network design spaces with an InfoNet interface between the two. The DSTC protocol is introduced to allow the dynamic resource allocation of network to satisfy the need of information space for fusion. The overall architecture has been validated through simulation of selective target tracking using the NS-2 simulator. The designed network protocols run in a distributed manner that makes the whole design scalable.

Robustness of the information fusion algorithm needs to be investigated. Robustness means the fusion algorithm needs to reject small disturbances in the observed data. The robustness criterion can possibly be achieved by designing a method to choose the measure μ optimally for certain purposes. Once the optimal μ has been selected at the training stage, this robust fusion algorithm can be used at the operational phase. Furthermore, if the statistics of the data change significantly due to the change of the physical environment, the idea is to search for another measure μ to adapt the design algorithm rather than design a brand-new algorithm under a new environment. Potentially, this could be a very powerful technique for designing information fusion algorithms.

Acknowledgment

This material is based upon work supported by, or in part by, the U.S. Army Research Laboratory and the U.S. Army Research Office under the eSensIF MURI Award No. W911NF-07-1-0376. Any opinions, findings, and conclusions or recommendations expressed in this publication are those of the authors and they do not necessarily reflect the views of the sponsor.

References

1. M. Tubaishat and S. Madria. Sensor networks: An overview. *IEEE Potential*, 22:20–23, 2003.
2. Y. Zou and K. Chakrabarty. Distributed mobility management for target tracking in mobile sensor networks. *IEEE Transactions on Mobile Computing*, 6:872–887, 2007.
3. D. Hall and S.A.H. McMullen. *Mathematical Techniques in Multisensor Data Fusion*, 2nd edn., Artec House, London, U.K., 2004.

4. R. Suzic. A generic model of tactical plan recognition for threat assessment. In *The SPIE Defense and Security Symposium*, Orlando, FL, pp. 105–116, March 2005.
5. H.H. Bui. A general model for online probabilistic plan recognition. In *18th International Joint Conference on Artificial Intelligence (IJCAI-2003)*, Acapulco, Mexico, August 9–15, 2003.
6. S.M. Jameson. Architectures for distributed information fusion to support situation awareness on the digital battlefield. In *Fourth International Conference on Data Fusion*, Montreal, Canada, pp. 7–10, August 7–10, 2001.
7. F. Johansson and G. Falkman. A comparison between two approaches to threat evaluation in an air defense scenario. In *The Fifth International Conference on Modeling Decisions for Artificial Intelligence*, Sabadell (Barcelona), Spain, pp. 110–121, October 30–31, 2008.
8. H. Chen. *Terrorism Informatics: Knowledge Management and Data Mining for Homeland Security*, 1st edn., Springer, Berlin, Germany, 2008.
9. S. Phoha and A. Ray. Dynamic information fusion driven design of urban sensor networks. *IEEE International Conference on Networking, Sensing and Control*, London, U.K., pp. 1–6, April 15–17, 2007.
10. A. Ray. Symbolic dynamic analysis of complex systems for anomaly detection. *Signal Processing*, 84(7):1115–1130, 2004.
11. S. Gupta and A. Ray. Symbolic dynamic filtering for data-driven pattern recognition. *PATTERN RECOGNITION: Theory and Application*, Chapter 5, Nova Science Publishers, Hauppage, NY, ISBN 978-1-60021-717-3, pp. 17–71, 2007.
12. G. Mallapragada, I. Chattopadhyay, and A. Ray. Automated behavior recognition in mobile robots using symbolic dynamics filtering. *Proceedings of the Institute of Mechanical Engineers, Part I: Journal of Systems & Control Engineering*, 222(6):409–424, 2008.
13. J.E. Hopcrof, R. Motwani, and J.D. Ullman. *Introduction to Automata Theory, Languages, and Computation*, 2nd edn., Addison-Wesley, Boston, MA, 2001.
14. G. Mallapragada, I. Chattopadhyay, and A. Ray. Autonomous navigation of mobile robots using optimal control of finite state automata. In *IEEE Conference on Decision and Control*, San Diego, CA, pp. 2400–2405, December 13–15, 2006.
15. I. Chattopadhyay and A. Ray. Structural transformations of probabilistic finite state machines. *International Journal of Control*, 81(5):820–835, 2008.
16. R. Duda, P. Hart, and D. Stork. *Pattern Classification*, 2nd edn., Wiley-Interscience, New York, 2001.
17. M. Schmiedekamp, A. Subbu, and S. Phoha. The clustered causal state algorithm efficient pattern discovery for lossy data-compression applications. *Computing in Science and Engineering*, 8(5):59–67, 2006.
18. W.R. Heinzelman, A.P. Chandrakasan, and H. Balakrishnan. Energy efficient communication protocol for wireless microsensor networks. In *Proceedings of the 33rd Hawaii International Conference on System Sciences*, Washington, DC, pp. 3005–3014, January 4–7, 2000.
19. S. Lindsey and C.S. Raghavenda. Pegasis: Power efficient gathering in sensor information systems. In *Proceedings of the IEEE Aerospace Conference*, pp. 924–935, March 9–16, 2002.
20. O. Younis and S. Fahmy. Heed: A hybrid, energy-efficient, distributed clustering approach for ad hoc sensor networks. *IEEE Transactions on Mobile Computing*, 3(4):660–669, 2004.
21. A. Depedri, A. Zanella, and R. Verdone. An energy efficient protocol for wireless sensor networks. In *Proceedings of Autonomous Intelligent Networks and Systems (AINS)*, Menlo Park, CA, pp. 1–6, June 30–July 1, 2003.
22. G. Smaragdakis, I. Matta, and A. Bestavros. SEP: A stable election protocol for clustered heterogeneous wireless sensor networks. In *Proceedings of the Second International Workshop on Sensor and Actor Network Protocols and Applications (SANPA)*, Boston, MA, pp. 1–11, August 22, 2004.
23. V. Mhatre and C. Rosenberg. Design guidelines for wireless sensor networks: Communications, clustering and aggregation. *Ad Hoc Network Journal*, 2(1):45–63, 2004.

24. M. Ye, C. Li, G. Chen, and J. Wu. Eecs: An energy efficient cluster scheme in wireless sensor networks. In *Proceedings of IEEE International Workshop on Strategies for Energy Efficiency in Ad Hoc and Sensor Networks (IWSEEASN)*, Phoenix, AZ, pp. 535–540, April 7–9, 2005.
25. S. Phoha, T.F. La Porta, and C. Griffin. *Sensor Network Operations*. John Wiley & Sons, Inc., New York, 2006.
26. P. Biswas and S. Phoha. Self-organizing sensor networks for integrated target surveillance. *IEEE Transactions on Computers*, 55(8):1033–1047, 2006.
27. M. Lotfinezhad, B. Liang, and E.S. Sousa. Adaptive cluster-based data collection in sensor networks with direct sink access. *IEEE Transactions on Mobile Computing*, 7(7):884–897, 2008.
28. F. Zhao, J. Shin, and J. Reich. Information-driven dynamic sensor collaboration for tracking applications. *IEEE Signal Processing Magazine*, 19(2):61–72, 2002.
29. H. Yang and B. Sikdar. A protocol for tracking mobile targets using sensor networks. In *IEEE Workshop on Sensor Network Protocols and Applications*, Anchorage, AK, May 11, 2003.
30. W.-P. Chen, J.C. Hou, and L. Sha. Dynamic clustering for acoustic target tracking in wireless sensor networks. *IEEE Transactions on Mobile Computing*, 3(3):258–271, 2004.
31. H. Chen and S. Megerian. Cluster sizing and head selection for efficient data aggregation and routing in sensor networks. In *Proceedings of the IEEE Wireless Communications and Networking Conference (WCNC)*, Los Vegas, NV, Vol. 4, pp. 2318–2323, April 3–6, 2006.
32. H. Gupta, Z. Zhou, S.R. Das, and Q. Gu. Connected sensor cover: Self-organization of sensor networks for efficient query execution. *IEEE/ACM Transactions on Networking*, 14(1):55–67, 2006.
33. D. Friedlander, C. Griffin, N. Jacobson, S. Phoha, and R. Brooks. Dynamic agent classification and tracking using an ad hoc mobile acoustic sensor network. *EURASIP Journal on Applied Signal Processing*, 4:371–377, 2002.
34. S. Phoha, N. Jacobson, D. Friedlander, and R. Brooks. Sensor network based localization and target tracking through hybridization in the operational domains of beamforming and dynamic space-time clustering. *IEEE Global Telecommunications Conference*, 5:2952–2956, 2003.
35. S. Phoha, J. Koch, E. Grele, C. Griffin, and B.B. Madan. Space-time coordinated distributed sensing algorithms for resource efficient narrowband target localization and tracking. *International Journal of Distributed Sensor Networks*, 1:81–99, 2005.
36. D. Bein, Y. Wen, S. Phoha, B. Madan, and A. Ray. Distributed network control for mobile heterogeneous wireless sensor networks. *Journal of Parallel and Distributed Computing*, 71:460–470, 2011.
37. E. Blasch, I. Kadar, K. Hintz, J. Biermann, C. Chong, and S. Das. Resource management coordination with level 2/3 fusion issues and challenges. *IEEE A & E Systems Magazine*, 32–46, March 2008.
38. D. Smith and S. Singh. Approaches to multisensor data fusion in target tracking: A survey. *IEEE Transactions on Knowledge and Data Engineering*, 18(12), 1696–1710, 2006.
39. S. Blackman. *Multitarget Multsensor Tracking: Advanced Applications*, y. bar-shalom edn., Artech House, London, U.K., 1990.
40. Y. Bar-Shalom. Extension of the probabilistic data association filter to multitarget tracking. *Proceedings of Fifth Symposium on Nolinear Estimation and Its Applications*, San Diego, CA, September 23–25, 1974.
41. Y. Bar-Shalom and E. Tse. Tracking in a cluttered environment with probabilistic data association. *Automatica*, 11:451–460, 1975.
42. K. Pattipatti, S. Deb, Y. BarShalom, and R. Washburn. A new relaxation algorithm and passive sensor data association. *IEEE Transactions on Automatic Control*, 37:198–213, 1992.
43. M. Winter and G. Favier. A neural network for data association. In *Proceedings of 1999 IEEE International Conference Acoustic, Speech, and Signal Processing*, Phoenix, AZ, Vol. 2, March 15–19, 1999.
44. H. Hong, R. Chong-zhao, Z. Hong-yan, and R. Wen. Multi-target tracking based on multi-sensor information fusion with fuzzy inference. *Proceedings of the Sixth International Conference of Information Fusion 2003*, Cairns, Queensland, Australia, Vol. 2, pp. 1421–1425, July 8–11, 2003.

45. Y. Chen and H. Huang. Fuzzy logic approach to multi-sensor data association. *Mathematics and Computers in Simulation*, 52:399–412, 2000.
46. V. Gorodetsky, O. Karsaev, and V. Samoilov. On-line update of situation assessment: A generic approach. *International Journal of Knowledge-Based and Intelligent Engineering Systems*, 9:351–365, 2005.
47. M.L. Hinman. Some computational approaches for situation assessment and impact assessment. In *The Fifth International Conference on Information Fusion*, Annapolis, MD, pp. 687–693, Vol. 1, July 8–11, 2002.
48. C. Shalizi and K. Shalizi. Blind construction of optimal nonlinear recursive predictors for discrete sequences. In *AUAI'04: Proceedings of the 20th Conference on Uncertainty in Artificial Intelligence*, Arlington, VA, AUAI Press, Banff, Canada, pp. 504–511, July 7–11, 2004.
49. D. Friedlander, C. Griffin, N. Jacobson, S. Phoha, and R.R. Brooks. Dynamic agent classification and tracking using an ad hoc mobile acoustic sensor network. *EURASIP Journal on Applied Signal Processing*, 4:371–377, 2003.
50. R. Kalman. A new approach to linear filtering and prediction problems. *Transactions of the ASME, Journal of Basic Engineering*, 82:34–45, 1960.
51. G. Ackerson and K. Fu. On the state estimation in switching environments. *IEEE Transactions on Automatic Control*, 15(1):10–16, 1970.
52. H. Scott, J. Bruckner and G. Rea. Analysis of multi-modal systems. *IEEE Transactions on Aerospace and Electronic Systems*, 9(6):883–888, 1973.
53. Y. BarShalom and H. Blom. The interacting multiple model algorithm for systems with Markovian switching coefficients. *IEEE Transactions on Automatic Control*, 33(8):780–783, 1988.
54. N. Gordon, M. Arulampalam, S. Maskell, and T. Clapp. A tutorial on particle filters for online nolinear/non-Gaussian Bayesian tracking. *IEEE Transactions on Signal Processing*, 50(2):174–188, 2002.
55. R. Brooks, D. Friedlander, J. Koch, and S. Phoha. Tracking multiple targets with self-organizing distributed ground sensors. *Journal of Parallel and Distributed Computing*, 64:874–884, 2005.
56. S. Phoha, A. Ray, I. Chattopadhyay, G. Mallapragada, and Y. Wen. Mathematical modeling of sensor network dynamics for control and stability. Technical Report TR eSensIF-08-01, Applied Research Laboratory, The Pennsylvania State University, University Park, PA, September 2008.
57. The Network Simulator-ns-2. Information Sciences Institute at the University of Southern California. http://www.isi.edu/nsnam/ns/
58. A. Arora et al. A line in the sand: A wireless sensor network for target detection, classification, and tracking. *Computer Networks*, 46:605–634, 2004.

III

Self-Configuration

Part III concentrates on the organization of self-configuring distributed sensor networks. Autonomous sensor networks are distributed amorphous computing environments, consisting of a large number of unreliable nodes and communications links, subject to intermittent failure, likely destruction, and limited power resources. Adaptation to chaotic conditions, such as those faced by autonomous sensor networks, is best performed by self-organizing systems.

In a revised chapter (Chapter 27), Brooks proposes top-down control and bottom-up reconfiguration and self-organizing models to produce deliberative adaptation of the network to overcome the complexity inherent in real-time environments. Surveillance networks are subject to intermittent failure and require bottom-up emergent control push interacting with the users' pull and top-down control. This is illustrated using the concepts derived from the fields of biology and physics that give examples of dynamic self-organization.

Iyer et al. focus on collective intelligence for power-aware routing in mobile ad hoc sensor networks in Chapter 28. Emergent behavior gives rise to collective intelligence or the swarm intelligence, which has been used here to identify routes in mobile ad hoc sensor networks. In mobile ad hoc sensor networks, the problem manifolds in complexity due to mobility of the nodes. The authors have used specialized packets known as ants to establish routes. They also address the issue of scarcity of energy in mobile devices by making the algorithm power-aware.

In a revised chapter (Chapter 29), Brooks focuses on random network formalisms for designing, modeling, and analyzing survivable sensor networks. He emphasizes on how to use random graph formalisms for wired and wireless P2P systems. He also explains how to estimate network performance, network dependability, network redundancy, system dependability, system phase changes, vulnerability to intentional attack, whether or not the network is connected, and the expected number of hops between nodes. He uses concepts from graph theory, linear algebra, and probability.

Samar and Wicker describe the behavior of individual links in multi-hop systems in Chapter 30. These individual behaviors combine to create the statistical landscape, which the self-organization behaviors attempt to contain.

A new chapter for the second edition was provided by Li, and Iyengar (Chapter 31). This chapter aids self-organization by considering 3D localization problems, which lead to focusing on the subsequent geometric issues.

In summary, Part III focuses on designing, modeling, and analyzing of self-organization in distributed sensor networks.

27

Self-Configuration*

Richard R. Brooks
Clemson University

Matthew Pirretti
The Pennsylvania State University

Jacob Lamb
The Pennsylvania State University

Mengxia Zhu
Southern Illinois University

S. Sitharama Iyengar
Florida International University

27.1 Self-Organization

Surveillance networks can be viewed at many levels of abstraction. Effective systems must adapt at each level. These networks will be composed of a large number of autonomous, possibly heterogeneous, devices. At the highest level of abstraction, all devices must work in a coordinated manner as if they were a single entity. User interests vary dynamically. Top-down control is needed to ensure that tasks performed by the network fulfill operational requirements. Specific regions and/or target types will be of interest sometimes, but not at others. Connectivity maintenance requires high-level coordination.

Bottom-up self-organization is necessary to guarantee the system's ability to adapt to unforeseen events. The environment varies dynamically. Weather, wind direction, and ambient noise change effective sensing ranges [Swanson 2000]. All nodes have finite energy resources and eventually cease to function. Radio communications can be jammed. These factors require bottom-up reaction and reconfiguration in a fully autonomous mode.

Surveillance networks are distributed amorphous computing environments, consisting of a large number of nodes and communications links, subject to intermittent failure, likely destruction, and limited power resources [Brooks 1998]. Human configuration and control of the entire system is futile. Humans steer the network by declaring general interests. Translating higher-level tactical objectives into low-level activities is an ill-posed problem. Efficient resolution of this problem in a chaotic network of failure-prone nodes will require bottom-up emergent control *push* interacting with the user's *pull*.

* This chapter contains material from Chapters 44 through 47 of the first edition.

It is important to decentralize the system without sacrificing robustness and accuracy. Top-down control is need driven and deliberative. Bottom-up control is reactive and geared toward maintaining an orderly substrate on top of an underlying chaotic environment. The goal is to produce deliberative adaptation of the network to overcome the complexity inherent in real environments.

27.1.1 Top-Down Control

Scalable top-down control requires a hierarchical topology. Flat organizations need direct communication with all participating nodes. Once the number of components surpasses a critical value, the communications overhead becomes untenable. Similarly, the latency imposed by the volume of information that users must process in a large flat hierarchy makes decisions from higher levels obsolete by the time they can be formulated.

Within a hierarchy, each layer has its own responsibilities and information requirements. The top level provides general guidance. Each level in the hierarchy accepts commands from the higher level, refines them to fit local constraints, creates more detailed instructions, and distributes them to lower levels. Events occurring at lower levels are relayed to higher levels, which extract relevant information for their superiors, thereby reducing the volume of data. Higher-level instructions should not be overly restrictive to allow freedom for quick reaction as needed. This is an ill-posed problem [Phoha 1999], since conflicts and contradictions may arise as instructions percolate through the system. It can be expressed as a hierarchy of control languages, or interacting automata, and implemented as an adaptive hierarchy of discrete event dynamic system (DEDS) controllers [Rudie 1992, Phoha 1999].

Robust control methods extend continuous control enabling construction of plant controllers able to perform correctly and predictably within known noise bounds [Zhou 1996]. Given these known bounds, hybrid control methods can switch from robust controller C_1 to C_2 when the underlying dynamics of the system move into a mode of operation that is unstable for C_1 but not C_2 [DeCarlo 2000]. DEDS controllers exist either as the higher-level switching logic used by hybrid controllers or as methods for controlling plants whose underlying dynamics are unknown or unknowable [Rudie 1992, Phoha 1999].

DEDS control is amenable to the construction of hierarchical control structures, since usually only the dynamics of the lowest level in the control hierarchy could possibly be modeled using continuous variables and differential equations. Higher levels function by manipulating discrete variables indicating events occurring and behaviors active in the system. Two tools dominate for expressing and designing DEDS controllers: finite-state automata (FSA) [Rudie 1992] and Petri Nets [Brooks 1999].

FSA have the advantage of translating control strategies into formal languages. The language defines an abstract grammar for constructing strings using an alphabet defined by individual events. We have developed tools for analyzing system controllability and robustness by analyzing the formal language generated by the FSA. When this is extended to hierarchical control, each layer in the hierarchy works as a Mealy machine that does three things: (1) absorbs events generated by immediate superiors and subordinates, (2) modifies its internal state, and (3) generates new events to be absorbed by its immediate superiors and subordinates. We have used this approach in Defense Advance Research Projects Agency (DARPA)-sponsored research exploring air campaign planning and execution. Figure 27.1 shows a suppression of enemy air defenses air operations mission simulation using our hierarchy of discrete event controllers.

The approach shown in Figure 27.1 uses FSA controllers for each entity in the hierarchy. Adding a stack to the local controllers would allow context-free grammars, like modern programming languages to be recognized. Adding a tape would allow each participant to execute any computable function [Hopcroft 1979]. On the other hand, the FSA-based control hierarchy effectively defines a type of cellular automata (CA) variant. Since CAs of a given size are capable of performing arbitrary computations [Wolfram 1994], the use of FSA is not particularly constraining.

Petri Nets are advantageous for modeling resource contention and synchronization [David 1989]. We have used extensions of Petri Nets to model complex military operations [AC2C 1999]. A Petri Net is constructed using modular components to design an operation. In order to verify that the system is free of deadlock and livelock conditions, a Karp–Miller tree is constructed. The Karp–Miller tree is a

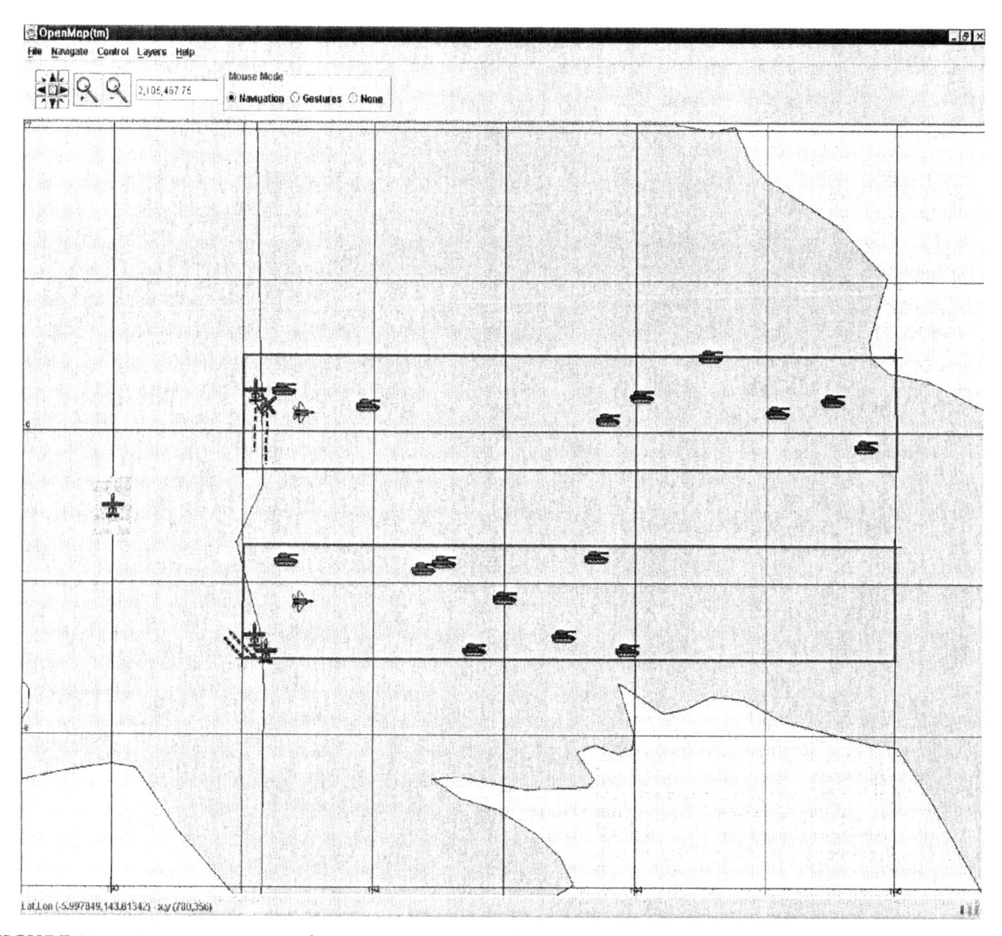

FIGURE 27.1 Air campaign simulation using C4iSim. Blue forces are controlled by a DEDS hierarchy.

FSA, where each state is defined by the number of tokens in each place of the Petri Net [Brooks 1999]. This state definition format is consistent with the construction of vector controllers [Li 1993, Li 1994]. Figure 27.2 shows a simple Petri Net and its associated Karp–Miller tree.

In this manner, DEDS FSA controllers can be automatically derived from Petri Nets. The Petri Net construct can guarantee the lack of deadlock and livelock. This design defines behaviors triggered in response to discrete events inferred from locally available information (including transmissions from other participating sensor nodes). The number of events considered is typically small, less than 100, and therefore, the bandwidth consumed by control messages is small. However, behaviors triggered by the DEDS control hierarchy can consist of sophisticated continuous control regimes or optimization methods [Phoha 2000a]. Making decisions with local information, when possible, saves power and bandwidth.

The role of an individual node may vary over time. For example, if a supervisory node fails, it should be possible to elect a replacement. We implemented this approach for networks of underwater oceanographic sampling robots [Phoha 2000b] by requiring that all nodes have identical software. We have implemented a more sophisticated approach for sensor network tasking in the DARPA SenseIT program [Brooks 2000a, Brooks 2000c]. Mobile code is used to overcome storage constraints that are ubiquitous in embedded systems. Nodes are initially configured with behaviors they are most likely to need. As needs change, programs are downloaded and storage reclaimed. In many ways, the local software configuration is managed like a local cache. We have implemented polymorphism and distributed dynamic linking to support this in networks of heterogeneous nodes. Control roles in the DEDS hierarchy may change to better suit environmental conditions.

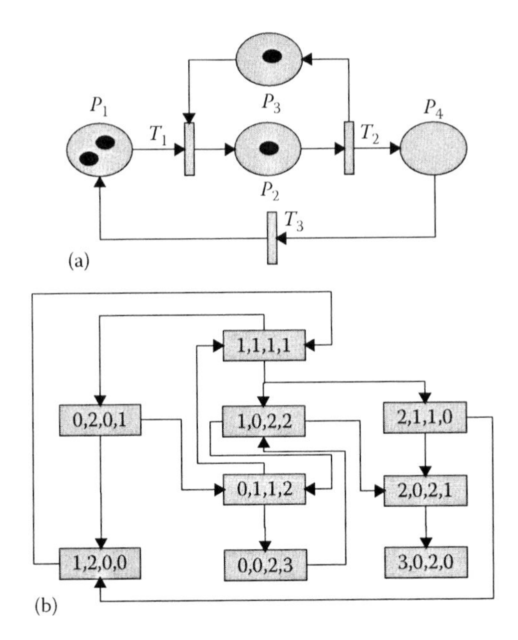

FIGURE 27.2 (a) Petri Net for a limited resource problem. (b) Its associated Karp–Miller tree.

The system must be robust. Its continued survival depends greatly on power conservation. Node longevity requires the following:

- Classification and detection decisions made as close to the sensor as possible.
- Network traffic routed in a manner that reduces redundancy.
- During periods when little of interest occurs, nodes enter a state of reduced power consumption.
- Node movement and beamforming occur rarely.

On the other hand, system robustness and operational considerations sometimes require the following:

- Critical information transmitted redundantly, taking independent paths through the network. This reduces single points of failure.
- Critical information made available to operational personnel with minimal latency.
- Information veracity verified through comparison with readings of multiple sensing modalities.
- Information sharing for beamforming and other expensive analysis techniques.
- Power expenditure for node movements to change fields of regard or communications paths.

27.1.2 Bottom-Up Reconfiguration

Equally important is the system's ability to respond autonomously to perceived changes in its operational environment. Self-configuration and adaptability are essential, since the nodes must adapt to maintain three separate hierarchies:

- *Control hierarchy*—The control hierarchy must be able to morph and reemerge, since connectivity may be broken at any point in the hierarchy.
- *Sensor coverage*—Changes in the environment (such as fog) or failures in the sensor network can result in the occlusion, or loss of coverage, of regions of interest. It may be necessary to reassign sensors or move nodes to reestablish coverage.
- *Network connectivity*—In addition to observing regions of interest, the network must report events to data consumers. To respond to jamming or low battery power, network configuration must be flexible. Multiple routes within the network should be maintained. It may be necessary for nodes to reposition themselves autonomously.

Since connectivity can be broken at any point in the hierarchy, self-organization must be supported to allow the hierarchy to reemerge. Candidate approaches to self-organization have been derived by researchers in artificial life studying insect colony coordination [Bonabeau 1997], in physics studying quantum optics [Haken 1983], and chemistry studying nonequilibrium thermodynamics [Nicolis 1977].

Each DEDS controller defines its local behaviors in response to an uncontrollable and frequently hostile environment. Tasks are performed in a distributed environment requiring coordination among multiple nodes. Communication may be corrupted with noise. Components are prone to failure. As network size increases, the probability of all components functioning at a given time decreases exponentially [Siewiorek 1982]. Nonlinear interactions exist between failure modes. Under these conditions, efficient operation cannot rely on static plans. Van Creveld has defined five characteristics of hierarchical systems need to adapt to this type of chaotic environment [Van Crefeld 1986, Czerwinski 1998]:

- Decision thresholds far down in the hierarchy
- Self-contained units at a low level
- Bottom-up and top-down information circulation
- Informal communications
- Commanders seek to supplement routine reports

Organizations using this approach have been successful in market economies, war, and law enforcement [Cebrowski 1998]. Our approach is influenced by and is consistent with these points. Each entity is self-contained and makes as many decisions as possible locally. Operational information and data travel in both directions in the hierarchy. The final two points imply informal activities that are difficult to implement in automated systems. In this section, we discuss local decisions made far down in the hierarchy to counteract the chaotic environment. Self-organization needs to occur to allow the system to adapt quickly to arbitrary unforeseen events.

The distributed control architecture can be modeled as a network of interacting automata. This is an abstraction of distributed systems that is a modification of CA [Adami 1998]. A CA is a synchronously interacting set of elements (network nodes) defined as abstract machines. A CA is defined by

- d, the dimension of the automata
- r, the radius of an element of the automata
- δ, the transition rule of the automata
- s, the set of states of an element of the automata

An element's (node's) behavior is a function of its internal state and those of neighboring nodes as defined by δ. Wolfram's study of CA [Wolfram 1994] has found four qualitative classes of CA:

- *Stable*—where all elements eventually evolve to the same state
- *Periodic*—where the global system evolves into regular structures.
- *Chaotic*—where no discernible structure emerges and the system remains in a constant state of flux
- *Interesting*—where no global stable regime occurs but areas of local stability emerge spontaneously

Chaotic regimes are clearly undesirable. Stable and periodic regimes are desirable but in some conditions may be inflexible and incapable of adapting to a changing environment. The interesting regime allows for spontaneous adaptation. It has been hypothesized that the ideal state for adaptation is stability on the edge of chaos [Kaufmann 1993], where chaos supports a quick search through new regimes when stability is no longer possible.

A DEDS model can produce a hierarchy of interacting automata that coordinate the overall system. This can be studied as a hierarchy of CAs of increasingly fine (coarse) resolution as traversed from the root (leaves) to the leaves (root). Inputs to each automaton at each time step are its own state and perceived states of its neighbors. For low levels of the hierarchy, the automata state includes the state of the environment in the region covered. Given the chaotic and uncontrollable nature of the underlying dynamics, it is likely that the regime of the lowest level is, and will remain, chaotic. Higher levels in the hierarchy should, however, retain a stable regime no matter what happens at the lowest levels of the control hierarchy. It is hypothesized that middle levels of the hierarchy should adapt, forming a bridge between the top-level stability and lower level chaos.

The communications and sensing coverage constructs are just as important as the control hierarchy. Since they depend both on system state and node position, they cannot be adequately modeled using a pure CA approach. Recent work [Portugali 2000] has adapted concepts developed as part of theoretical explanations of self-organization of laser beams [Haken 1983] to explain similar problems in urban planning. The model of free agents in a cellular space [Portugali 2000] allows agents to migrate between elements in a CA environment. The behaviors of agents and CA elements influence each other. In our approach, agents are sensing platforms that may or may not be mobile. CA elements are grid points defining surveillance regions. Surveillance targets can also be modeled as agents. In this way, we can study self-organization regimes for sensor coverage and maintenance of communication connectivity. We currently plan on only studying first-order effects. It may, in fact, be possible to study interactions between control hierarchy, sensor coverage, and communications connectivity maintenance, but this depends on the results of the modeling thrust.

27.1.3 Self-Organization Models

Many theories exist for self-organization and spontaneous emergence of order [Haken 1983, Kaufman 1993, Bonabeau 1997, Adami 1998, Brooks 2000b]. Those based on thermodynamic models [Nicolis 1977, Haken 1983] are likely to be untenable, since they are based on models defined by sets of differential equations. It is unlikely that we will be able to derive suitable sets of equations to adequately represent general circumstances. Some artificial life constructs [Bonabeau 1997, Adami 1998] are based on a few basic interactions among multiple participants:

- *Positive feedback*—It includes recruitment and reinforcement of behaviors.
- *Negative feedback*—It counterbalances positive feedback to stabilize the system.
- *Amplification of fluctuations*—Randomness and fluctuations are crucial to system adaptation.
- *Multiple interactions*—Simple behaviors at the microscopic level can provide intelligent adaptations at the macroscopic level.

These interactions form the basis of pheromone interactions used by insects for foraging and nest building as shown in Figure 27.3 [Bonabeau 1997, Adami 1998]. This approach can be seen as a way of constructing distributed feedback loops. It has been used in developing a number of transport and communications applications [Dorigo 1996].

Insect colonies use pheromones to provide positive and negative feedback signals [Bonabeau 1997]. This approach has been applied to distributed route planning [Bonabeau 1997] and military command and control [Parunak 1999]. We use pheromone mechanisms to encode information needed for coordination by the components in this mechanism. Other explanations of self-organization look at genetic constructs [Kaufmann 1993]. This concept is appealing, since DNA organization can be seen as a probabilistic grammar of an alphabet defined by a small number of amino acids. The use of abstract languages, to define order, parallels the FSA models of discrete event control. CAs model the distributed and synchronous nature of the problem. Synthetic pheromones provide primitives for communication and coordination. The network then formulates its own solution to the problem via local interactions.

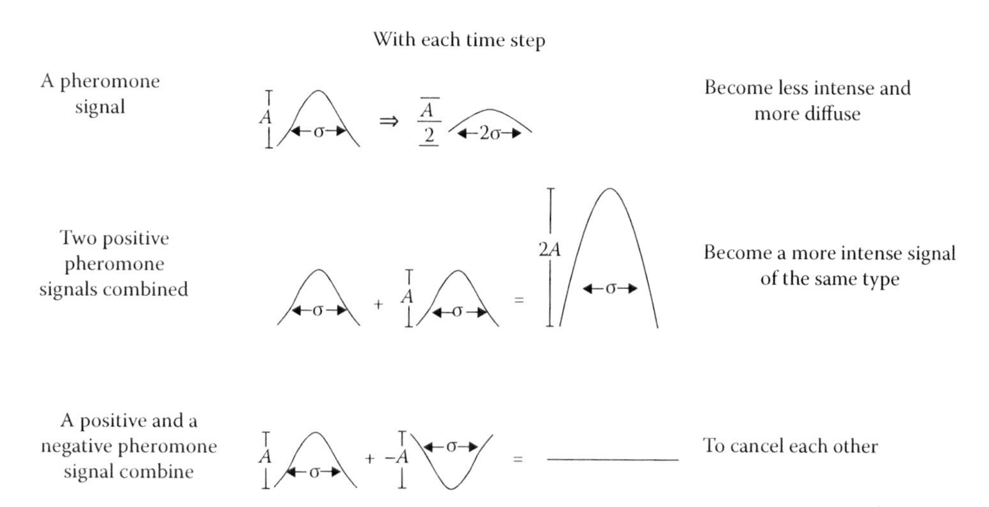

FIGURE 27.3 Pheromone abstraction provides a convenient way of expressing time-dependent positive and negative distributed feedback systems.

27.2 Emergence

Surveillance networks should be dependable entities composed of multiple, unreliable sensor nodes. This is a distributed amorphous computing environment consisting of a large number of nodes and communications links subject to intermittent failure, likely destruction, and limited power resources. Human configuration and control of this environment is futile. Humans steer the network by declaring general interests. Translating higher-level tactical interest into low-level activities is an ill-posed problem requiring a bottom-up emergent control *push* to interact with the user's *pull*. This system is a large complex entity whose behavior is defined by interactions among multiple participants.

The national infrastructure now consists of dynamic systems relying on interactions among multiple semiautonomous entities. The physics of these types of systems has been studied and understood in continuous [Haken 1978, Alligood 1996, Nicolis 1977 and discrete [Wolfram 1994, Delorme 1999, Brooks 2000a,b, Sarkar 2000] domains. Hybrid systems that integrate both are needed and have been neglected. Traditional dependability models rely on queuing theory and statistical methods [Siewiorek 1982, Rai 1990a]. Since many system pathologies are caused by dynamic nonlinear interactions between components [Kott 1999], it is essential to create these hybrid mathematical models and update control theory to provide a new generation of systems that adapt to chaotic environments.

Among the common characteristics of engineering designs are failure modes and pathologies. [Kott 1999] provides an initial taxonomy of pathologies for complex systems, which lists pathological behaviors and mechanisms that are often responsible. Pathologies are general descriptions of ways the system deviates from its goal states. The taxonomy includes ubiquitous problems such as race conditions, deadlock, livelock, oscillation, and thrashing. Reasons for pathology occurrence include lack of information, inappropriate thresholds, inconsistency, and excessive constraints [Kott 1999]. In this section, we describe relevant mathematical models of complex systems that express their underlying structure and pathologies.

Pathological behavior is endemic to complex systems, which are composed of multiple nontrivial components with nonlinear dynamic interactions. Components may not be homogeneous. Their internal composition may vary over time. Studies of interacting components indicate that extremely complex chaotic behavior occurs even in systems consisting of homogeneous, simple automata with static configurations [Wolfram 1994]. System behavior has both local and global aspects. Correct system behavior relies on synergetic effects emerging between local and global regimes.

Military logistics is an excellent example of a complex system. System inputs and outputs are material supply and demand, both of which are uncertain. This uncertainty cannot adequately be modeled as a stochastic variable, since many aspects are subject to influence by a hidden, pernicious, and intelligent adversary. Transport and storage is subject to failure and confusion. In addition, the overall system has uncertainties in sensing and actuation commonly referred to as *fog of war* and *friction* [Van Creveld 1980].

Advances in computing and mathematics make it possible to study and model interactions and dynamics of large complex systems, which were undecipherable until recently. [Haken 1978] and [Nicolis 1977] in physics and chemistry have determined the principles of self-organization in matter that hold for complex systems in biology [Eigen 1982] and sociology [Portugali 2000]. Applications of these principles have been found in diverse domains ranging from lasers [Haken 1978] to explaining behavior of insect colonies [Bonabeau 1997].

Systems are modeled as macroscopic entities made up of a large number of smaller elements [Haken 1978, Portugali 2000]. Self-organization is found when complex macroscopic behaviors occur as a nontrivial consequence of interactions between the individual elements. Self-organization occurs only in systems far from equilibrium [Nicolis 1977]. System behavior ranges from strict order to chaos.

We use the term emergent to characterize system behaviors with the following characteristics:

- They arise from interactions between multiple entities.
- All behaviors are based on local information.
- They contain a significant stochastic component.
- Positive feedback is used to encourage desirable behaviors.
- Negative feedback stabilizes the system.
- Global system behaviors contain spontaneous phase changes.
- The global behaviors are not obvious consequences of the local behaviors.

27.2.1 Continuous Models

The concept of emergence can be applied to many existing engineering systems and others currently being designed. The fact that recent studies of Internet traffic detect self-similarity in traffic flow [Leland 1994, Grossglauer 1999] supports our supposition that the basic tenets of synergetics [Haken 1978] will hold for critical infrastructure like sensor networks. In these engineered systems, a large number of complex individual entities merge to form a larger system. In many ways, the macroscopic behavior of the aggregate system emerges from interactions between individuals. These behaviors limit the ability to control the system from the top-down in ways that are not immediately evident. An example of this is fault propagation through complex coupled systems producing unforeseen macroscopic errors. These errors are difficult to foresee and/or correct, like the failure of the electric infrastructure in the western United States in the summer of 1996 [CNN 1996, PBS 1996].

For emergent behavior to arise, the systems must be complex and have a large number of degrees of freedom. Their behavior depends on a large number of system variables and has many modes of behavior. The effects of any given variable may be controllable or unstable, depending on the mode of operation [Haken 1978]. The change from one mode of system behavior to another occurs abruptly at a critical value of a single system variable [Jensen 1998].

Variables may be continuous or discrete. For continuous variables, a singular value decomposition (SVD) can be performed to decouple interactions and find those variables with the strongest influence on system behavior. The effects of some variables are controllable, while nonlinear interactions make the effects of variations of other variables inherently unstable and unpredictable. The set of controllable variables derived through the SVD thus provides a system with greatly reduced dimensionality.

At this point, the *slaving principle* can be used to provide structure in an otherwise chaotic system [Haken 1978]. It states that although the system as a whole is not entirely controllable, the controllable variables can be used to steer the system's evolution. They provide bounds for the chaotic actions forced by the unstable variables. The unstable variables are considered slaves of the controllable ones. This control system can be used until another critical value is reached, and the dominant mode of operation changes abruptly. It can even be used to force the system past another critical value. Methods for deriving and finding critical values, where the behavior of physical systems changes abruptly, are given in [Jensen 1998]. This is an extension of the basic concepts of robust control, where control strategies are shaped around instabilities and sources of error [Zhou 1996].

A matrix system representation can capture many essential system characteristics. If internal component structure is ignored, transportation models can be used. Material suppliers are sources y, and end users are sinks u. Transportation paths and storage depots are represented as limited capacity edges that form a network connecting sources to sinks. This problem can be formulated in terms of graph theory, linear programming, nonlinear differential equations [Smulders 1991, Haberman 1998], or control theory [Zhou 1996].

In terms of graph theory, many methods design systems that maximize flow through a network. In addition to algorithms specifically designed for flow maximization, dynamic programming is a common approach. In the case of stochastic systems, methods generally attempt to optimize the expected flow through the system. These models require knowledge of statistical distributions of the stochastic components, information consistency, and known functions that express supply and demand. These models are too static and deterministic. Pathologies will arise, but the model cannot express how or why.

One way of designing the system would be to try to have the supply match the demand, which could be phrased as a control system:

$$x' = Ax + Bu \quad y = Cx + Du \quad \text{or} \quad \begin{bmatrix} x' \\ y \end{bmatrix} = \begin{bmatrix} A & B \\ C & D \end{bmatrix} \begin{bmatrix} x \\ u \end{bmatrix} = G \begin{bmatrix} x \\ u \end{bmatrix}$$

where

x is the set of internal state variables
y is the supply
u is the demand

One then attempts to derive matrices A, B, C, and D to make y follow u. In the linear case, it is possible to solve the equations to optimize a set of known criteria. It is even possible to use stochastic noise variables and define a structured singular value, where the system remains stable and controllable as long as noise remains within limits [Zhou 1996]. Well-defined measures of stability, controllability, observability, and robustness exist in control theory [Zhou 1996].

The matrix formalism is still applicable when the system is nonlinear, even though its application becomes significantly more complex [Brogan 1991]. Note that in analyzing these control systems, the SVD is used to extract the principle components of the system. Singular values (eigenvalues) λ_i express the modes inherent in the system. Sorting the values in order of magnitude allows a threshold T to be set. Singular values larger than the threshold ($\lambda_i > T$) are considered significant and retained. Those smaller than the threshold ($\lambda_i < T$) are not significant and are discarded. This often allows the dimensionality of the problem to be reduced significantly.

Military and social systems are notoriously nonlinear in nature [Czerwinski 1998] with an unknown number of higher-order interactions. [Hagen 1990] expresses dynamic nonlinear systems with multiple operational modalities. These modalities are expressed by system singular values λ_i. Any operation mode is a function of interactions between these modes. The interactions can lead to instability, oscillations, chaotic dynamics, and most other pathologies. Any modality can be approximated by a linearization of the system using a matrix of the λ_i. Different control parameters are relevant for different system modalities. In particular, many system parameters become irrelevant, or subservient to another parameter, in specific behavior modes [Hagen 1990]. This can drastically reduce the number of degrees of freedom of a system, as long as underlying dynamics are known.

Although control theory is useful for expressing and discovering many pathologies, it still has significant limitations. Unfortunately in military logistics noise inputs are not adequately modeled as stochastic variables. They are at least partially controlled by the plans of an intelligent adversary. Inconsistencies in the system may be partially due to pilferage and/or sabotage. The confusion caused by chaotic conditions also leads to difficulties in collecting information and executing plans [Van Creveld 1980]. Internal system variables are subject to inconsistency and corruption. Many significant aspects of the plant are unknown and possibly even unknowable [Van Creveld 1980].

Alternatively, the network can be approached as a nonequilibrium system [Nicolis 1977]. Many biological and chemical systems exist where microscopic flux and chaos is offset by macroscopic order. In this model, stable macroscopic regimes emerge and provide a predictable system as long as system variables remain within certain limits. When critical values of system parameters are reached, the internal structure of the system changes radically. In many cases, these critical values are due to internal thresholds and effects of positive feedback. This is the basis of many fractal and chaotic interactions seen in nature [Jensen 1998]. Internal thresholds and positive feedback are also present in logistics systems. In the example of medical supplies, when more than a certain percentage of people contract an illness, the illness is more likely to spread. This type of positive feedback relation is hidden from a straightforward control model. Results from nonequilibrium systems are consistent with the pathology taxonomy in [Kott 1999]. Many pathologies are due to incorrect threshold values in a system. These threshold values are critical values where the system modifies its control regime. [Jensen 1998] discusses methods for deriving and verifying critical values where system behavior changes radically.

27.2.2 Discrete Models

[Kauffman 1993] explores the relation between higher-order interactions and system adaptation. He finds that in complex systems, ordered regimes are frequently separated by chaotic modes that adapt quickly to find new ordered regimes. His work applies differential equations, discrete alphabets, and CA to system analysis. His approach is consistent with the economics study in [Chang 1999], urban planning study in [Portugali 2000], and air operations study in [Brooks 2000]. These studies all explicitly consider differences in information availability to decision makers in distributed systems. Independent entities interact within a distributed space. It appears that discrete systems and CA are a reasonable approximation of the underlying continuous dynamics [Kauffman 1993]. It is also not necessary for all the underlying dynamics to be understood or modeled [Kauffman 1993, Portugali 2000]. For this reason, it appears that a model of agents in a cellular space is suited for expressing pathologies for systems with unknown underlying dynamics. It is also suited for expressing the interaction between global and local factors.

Some variables may be innately discrete, especially in engineered systems. In other cases, the dynamics may not be sufficiently known (or measurable) for models using differential equations to be practical. It is advisable to model these systems using the CA abstraction [Wolfram 1994, Delorme 1999]. CAs of sufficient size have been shown capable of performing general computation [Delorme 1999]. A CA is a synchronously interacting set of elements (network nodes) defined as a synchronous network of abstract machines [Sarkar 2000]. A CA is defined by

- d, the dimension of the automata
- r, the radius of an element of the automata
- δ, the transition rule of the automata
- s, the set of states of an element of the automata

An element's (node's) behavior is a function of its internal state and those of neighboring nodes as defined by δ. An element's (node's) behavior is a function of its internal state and those of neighboring nodes as defined by δ. The simplest instance of a CA is uniform that has a dimension of 1, a radius of 1, and a binary set of states. In this simplest case for each individual cell, there are a total of 2^3 possible configurations of a node's neighborhood at any time step. Each configuration can be expressed as an integer v:

$$v = \Sigma_{ji} * 2^{i+1}$$

where

i is the relative position of the cell in the neighborhood (left = −1, current position = 0, right = 1)

j_i is the binary value of the state of cell i

Each transition rule can therefore be expressed as a single integer r:

$$r = \sum_{v=1}^{8} j_v * 2^v$$

where j_v is the binary state value for the cell at the next time step if the current configuration is v. This is the most widely studied type of CA. It is a very simple many-to-one mapping for each individual cell. The aggregated behaviors can be quite complex [Delorme 1999]. Wolfram's work has found four qualitative equivalence classes for CAs [Wolfram 1994]:

- *Stable*—evolving into a homogeneous state
- *Repetitive*—evolving into a set of stable or periodic structures
- *Chaotic*—evolving into a chaotic pattern
- *Interesting*—evolving into complex localized structures

These states are useful for modeling complex systems. Under normal operating conditions, stable or repetitive behavior is desirable. When the system has been pushed out of equilibrium, chaotic and interesting interactions are desirable. They enable the system to quickly explore the set of possible adaptations, eventually finding a new configuration adapted to the new environment [Kauffman 1993].

The agents in a cellular space model [Portugali 2000] augment CA by adding agents that are defined as abstract automata. The agents migrate within the cellular space. Their behavior depends on their own state, the state of cells in their neighborhood, and possibly on a small number of global variables.

Cellular automaton are composed of an array of identically programmed automata, or "cells," which simultaneously interact with neighboring cells and advance in discrete time and space in any arbitrary number of dimensions. Each cell has its own states and rules to compute what its next new state shall be based on its own state and its neighbors' states. There is no centralized authority to control the global behavior; control is entirely through the local interaction among identical entities. This unique property of CAs makes them an invaluable tool to model self-organizing systems.

Although the local rules of the individual entities in most self-organized systems are well documented, there is still uncertainty as to the exact nature of the global behaviors, as knowledge of local behaviors does not imply that global behaviors will also be straightforward. Generalized probabilistic rules are proposed to specify the likelihood of each particular state, where each cell evolves into one particular state out of several possible states in a probabilistic fashion. This turns out to be a very powerful extension to traditional deterministic rules. More natural phenomena can be modeled with this extension included.

Although CAs are a natural way to model discrete dynamic and complex phenomena, this flexibility and generality comes with a cost, instead of several variables used in a partial differential equation to describe a continuous dynamical system, a large number of variables are needed for a CA system. This is because a large number of cells are required for a large-scale system with complex behaviors. Moreover, the number of time steps required for evolution is considerably large [Preston 84].

The Cantor tool is a CA simulator based on the generic automata with interacting agents (GAIA) model; the inclusion of agents allows for cells that can move throughout the simulation space interacting with other agents and cells alike. Cantor is capable of modeling dynamic, discrete time and discrete space event systems, which consist of large interacting individuals that display collective phenomena. Cantor was developed by the Applied Research Laboratory at the Pennsylvania State University in 2001 and has been successfully used to simulate many CA systems, including traffic engineering, network communication, sensor data routing, etc.

27.2.3 Characterization of Pathological Behavior

The question remains as to how to find the pathologies inherent in a given system, find concrete examples of pathologies in existing data, and recognize emerging pathologies online to correct problems as they arise. Starting from the pathology taxonomy in [Kott 1999], it is possible to analyze systems to find likely examples as long as internal dynamics, control signals, and stochastic distributions are known [Zhou 1996]. This is not the case. The problem is a difficult pattern recognition application of looking for amorphous patterns in a poorly defined space.

Where dynamics are known, continuous models of the known subproblems will be constructed. As with control and nonequilibrium systems, singular values will be computed for the continuous models. This extracts the most influential modes of operation and discard those with little impact [Hagen 1990, Brogan 1991]. Further, constructing hierarchies within the remaining control parameters can find those that are slaves to other parameters in given operational regimes [Hagen 1990]. This reduction of the number of degrees of freedom greatly simplifies pattern recognition and decreases sensitivity to noise [Pandya 1996].

Where dynamics are unknown, a model of agents in a cellular space [Portugali 2000] can be established. The model expresses the decision makers in the system, their interactions, and their environment. Continuous problems can be subsumed into the cellular space or agent as appropriate. Comparing the dynamics of this model to the dynamics recorded in databases of logistics scenarios allows for correction of modeling areas and derivation of statistical distributions for stochastic aspects of the problem. It may or may not be possible to satisfactorily model antagonistic aspects of the system.

Given historical data, an initial starting point is to create datasets that express scenarios in which a particular pathology occurs. Ideally, the singular values derived for the well-known continuous modeled subproblems would adequately describe the system as a whole. Using the singular values would then reduce the number of degrees of freedom in the pattern recognition problem and allow self-organizing memory to partition the system [Hagen 1990, Portugali 2000]. Adaptive resonance theory [Pandya 1996] and Kohonen maps [Hagen 1990] would be suited to creating systems that recognize emerging pathologies if this is the case.

If classification using the singular values of subsystems whose dynamics are well known is insufficient, data mining approaches may be needed to inferring additional diagnostic information [Tsaptsinos 1997, Tsumoto 1998, Tsymbal 1998]. This augments the original information with a number of rough set and fuzzy association rules. Interestingly, work with latent semantic analysis has shown that the SVD can be used to infer concepts and groupings from text data. It should therefore be possible to unify the semantic information in linguistic variables with the singular values from systems with known dynamics in a common vector space and use principal components analysis to derive a set of control variables of minimal size. Mined rules can influence the system globally, local system cells, or the agents in the cellular space.

27.3 Biological Primitives

Biological primitives are models of biological systems that exhibit the property of self-organization. These systems consist of numerous, usually homogeneous, biological entities interacting with each other with relatively simple behaviors in such a way as to exhibit global behaviors that are too complex to be exhibited by the individual entities. The individual entities interact utilizing only local information and lack any sort of "master plan" or centralized leadership. The literature from biology refers to the appearance of the global behavior from the interaction of numerous entities exhibiting simple behaviors as emergence.

Naturally occurring self-organizing systems are not unique to the realm of biology; there are also physical and chemical self-organizing systems where the individual entities are inanimate. An example of such a system would be the emergence of patterns in a sand dune due to the interactions among individual grains of sand [Camazine 01]. There are two important distinctions that should be made in regards to physical systems and biological systems with self-organizing properties. First, individual units in biological systems tend to be much more complicated than those from physical systems, as biological organisms represent more complexity than do inanimate objects. The second difference is that in a chemical or physical self-organizing system the individual units only follow the laws of physics, while biological systems additionally exhibit behaviors that are the result of an ever-changing genetic makeup. It is this second difference that makes biological primitives of particular interest to researchers; with time the system will fine-tune its behavior through the process of natural selection. With chemical and physical systems, there is no evolution, and the system will follow the same rules for all time [Camazine 01].

There are two different mechanisms by which self-organizing systems control themselves, positive and negative feedback. Positive feedback reinforces desirable behaviors by making it increasingly attractive to exhibit certain behaviors; negative feedback limits positive feedback so that positive feedback will not get out of control. This is done by making it increasingly unattractive to keep performing the desired behavior. The net result is that a global behavior emerges from the interplay of positive feedback, negative feedback, and the local behaviors of the entities in the system.

27.3.1 Why Apply Biological Primitives to Wireless Sensor Networks

Perhaps the best way to describe why biological primitives are applicable in the realm of wireless sensor networks (WSNs) is to first explain some of the difficulties that arise in WSNs. Our WSN application is for use in Military Operations in Urban Terrain (MOUT) scenarios. In these urban scenarios, radio communications are impeded by various obstructions which may cause the shortest path between two points to not be a straight line. The chaotic nature of MOUT missions causes unreliable and highly variable paths, causing only transient paths to exist. In spite of these difficulties, timely communications are required for the application to be useful. This is compounded with the fact that the global behavior of having an effective WSN routing methodology is extremely complicated. Further, traditional packet routing techniques have extremely poor performance, due to the assumptions that error rates at the link level are infrequent and that duplicate or missing packets are caused by congestion. Additionally, centralized control of this type of network is undesirable and unrealistic due to reliability, survivability, and bandwidth considerations, whereas distributed control does not have these issues [Chang 86]. Distributed control also has other advantages: (1) increased stability by avoiding single points of failure, (2) simple node behaviors replace complicated global behaviors, and (3) enhanced responsiveness to changing conditions since nodes react immediately to topological changes instead of waiting on central command.

The nature of these challenges makes it attractive to use a model based on self-organizing biological systems to solve the inherent difficulties associated with WSNs; a biological-based model would be intrinsically distributed, uses positive and negative feedback based on local information to attain the desired global behavior of routing, and could react very rapidly to transient activities that are common in a MOUT scenario by reacting immediately at a local level in lieu of waiting upon central command to determine the proper course of action to be carried out. The behaviors of the nodes in a biological system tend to be relatively simple, making design and verification straightforward once implemented in a WSN application. Further, with biological self-organizing systems, there is the ability of the system to evolve with time that could be potentially exploited to have a system that can tune itself to enhance performance over time.

27.3.2 Exemplar Models from Biology

For didactic reasons, several examples of self-organizing biological systems are provided. The first example, *Dictyostelium discoideum* a type of amoeba, explains the complex spiral pattern that these creatures form under specific circumstances. The second example, ant pheromone, explains how certain species of ants use pheromones to maintain short paths between their nest and food sites.

27.3.2.1 Dictyostelium discoideum

One interesting example of self-organizing systems is that of a certain species of cellular slime molds, amoebas, called *D. discoideum*. When there is a steady supply of food, each *D. discoideum* acts independently of its neighbors, eating and then splitting after it has eaten an ample amount of food. This process continues until the food supply is diminished, and the individual *D. discoideum* amoebas begin to aggregate together forming complex spiral-like patterns.

The mechanism by which the amoebas form these patterns is based on the secretion of an attractant known as cAMP. The amoebas secrete this substance during the starvation period every 5–10 min, while they move toward higher concentrations of cAMP. If the amoebas sense a cAMP pulse that is of a large enough magnitude, they will secrete an even larger quantity of cAMP, causing local hot spots where numerous amoebas have conglomerated. Eventually, as the concentration of cAMP increases, the amoebas start to become desensitized to this attractant, and consequently, they will curtail their cAMP production.

In this example, the global behavior of having the amoebas form complicated spiral patterns emerges from the individual behaviors of the amoebas, which are utilizing local information (the density of cAMP) as the only means to shape their behaviors. The cAMP attractant is acting as positive feedback for forming the global pattern, and the amoebas' eventual desensitization to cAMP acts as negative feedback by limiting the production of cAMP. For further information on the mechanisms of pattern formation in *D. discoideum*, see [Camazine 01].

27.3.2.2 Ant Pheromone

A second example from biology that exhibits self-organization is the method by which certain species of ants, for instance, *Lasius niger* the black garden ant, forage for food. Ants attempt to find paths between their nest and potential food sources. They release two different pheromones: (1) search pheromone when they look for food and (2) return pheromone when they have food and are returning to the nest. It is these pheromones that help the ants navigate where the food sources and the nest are located. Ants that are looking for food will typically follow the highest concentration of return pheromone; ants returning to the nest tend to follow the highest concentration of search pheromone. There are also scouting ants that move about randomly looking for new food sources.

The pheromones vary with time due to diffusion and evaporation. The initial pheromone laid down by the ants will begin to diffuse outward as time progresses. Additionally, pheromone will begin to evaporate, and eventually, the pheromone will have evaporated to the point where it is undetectable.

In this example, the global behavior of having the ants form short paths between their nest and various food sources emerges from the behavior of the ants and their pheromones. The positive feedback in this system is the ants laying down pheromones. Good paths, that is, those that go directly from the food to the nest, will be traveled by more ants and will consequently receive more pheromone. Negative feedback is exhibited from how the pheromone evaporates and diffuses, preventing the pheromone from attaining unbounded concentrations.

27.3.2.3 Tools Used

Most important to selecting a particular biological model to utilize in a WSN is to ensure that the global behavior of the biological model is consistent with the desired global behavior of the WSN; if the application and the base model are drastically different, it would be quite difficult to get the proper global behavior to emerge. Once a proper biological primitive has been selected, it can be tailored to better model the desired application, and with this done, the choice of tools to implement a solution will dictate the rest of the solution development. To develop our model, we have developed Cantor, which is a tool that aids in the creation of cellular automaton.

Cantor provides a generic infrastructure that is useful for designing CAs; specifically, it only requires users to provide the rules that the cells will follow and what information should be included in a cell's state. Some useful features of Cantor are that it supports any number of dimensions in the grid, it automatically creates data output files denoting the state of all of the cells for each generation, it allows the user to specify the creation of neighborhoods, and the user can specify a floorplan consistent with a MOUT application.

A sample trace from a model written using Cantor is provided in Figure 27.4. In this example, the cells are clearly defined by the regularly spaced horizontal and vertical lines. Further, there are elements of a MOUT application in this example, specifically the striped gray regions. These areas are walls, where no communication is possible, for instance, wireless communication is not possible through certain materials.

27.3.3 Ant Pheromone Model

The in-depth ant pheromone model discussion that follows is a case study of how it can be possible to use biological primitives to solve one particular problem in WSNs. The problem that was solved was that of information routing in a MOUT application. The model is a biological primitive based on ant foraging that forms short paths to and from a particular source and destination. The model is capable of adapting rapidly to changing system conditions; hence, the manner in which ants forage for food is a logical choice for use in a WSN application.

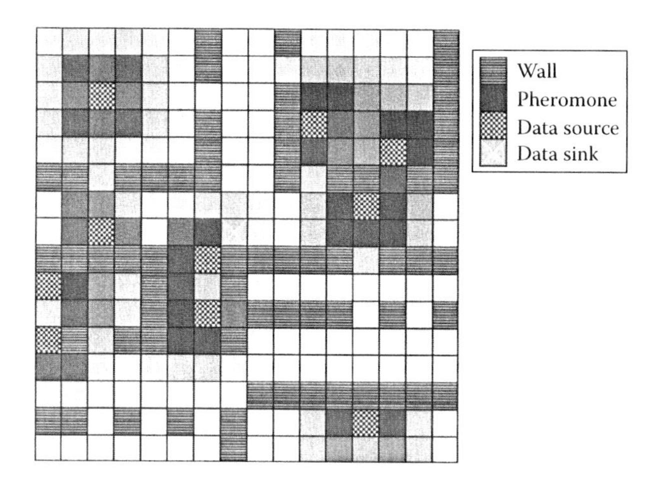

FIGURE 27.4 Sample trace from a model designed with Cantor.

27.3.3.1 How the Biological Model Was Modified

There are a few notable differences between the model that was created and the way in which ants forage for food. Most notably we have included certain aspects to our model to effectively study a MOUT scenario. Figure 27.5 shows an idealized MOUT terrain. Walls signify any obstruction that can block radio signals. Open cells are open regions allowing signal transmission. Open doors (closed doors) are choke points for signals that periodically allow (disallow) transmission. Finally, obstructions are intermittent disturbances that occur at random throughout the sensor field. Random factors are inserted to emulate common disruptions for this genre of network. Each square capable of transmission contains a sensor node. This amounts to having a sensor field with a uniform density. This provides an abstract example scenario approximating situations likely to exist in a real MOUT situation.

Another difference is that in our application and that of the biological system is the inclusion of a communication reduction technique known as gossip, which will be discussed later. Additionally, several principals reflect the fact that this model was specifically designed to implement a routing algorithm, which would not be particularly meaningful in the pure biological model (e.g., keeping track of path length).

27.3.3.2 Dorigo's Work

The work of Dorigo consisted of utilizing the foraging behavior of ants to solve various discrete optimization problems, such as the traveling salesman problem and the quadratic assignment problem [Dorigo 1996]. Di Caro and Dorigo went on to develop AntNet, an algorithm that solves the routing problem for packet-switched telecommunications networks [Di Caro 96].

27.3.3.3 How It Differs from Dorigo's

We have utilized a system similar to AntNet, except our system has been designed specifically for use in a WSN. This application requires the system to adapt to local transient behaviors much more rapidly than what was required from Dorigo's environment. Further, this application is based on wireless communications, which are by nature much less reliable than those found in wireline networks due to the much higher incidence of errors on the wireless link. To overcome these potential issues, the model had to be designed to be very adaptive to changing environmental factors and be very resilient to error conditions.

27.3.3.4 How It Applies to the Routing Problem

The manner in which the system was designed was by initially assuming that there are wireless sensor nodes distributed uniformly throughout the network of interest; specifically, the network is

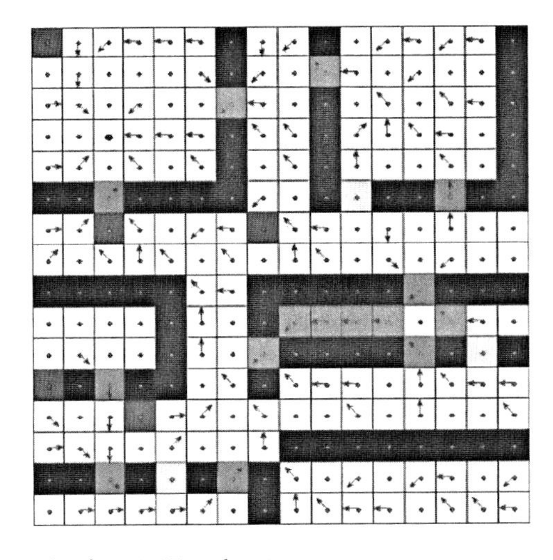

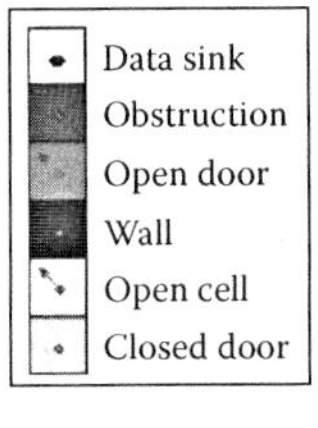

FIGURE 27.5 Example of a MOUT application.

initially connected. This does not mean to say that with time some of the nodes will not fail resulting in disconectivity. We have categorized the network into the following regions: (1) a region that is being covered by a sensor node that does not need to transmit or receive information, (2) a region being covered by a sensor node (a data source) that has information to send to another node in the network, (3) a region that is covered by a node that is the recipient of information from one or more nodes (a data sink), and (4) a region that is not being covered by a sensor node (e.g., the region may be unsecured or it may have properties that make radio transmission impossible).

Using this partitioning of the area under observation by the WSN allows the ant pheromone model to be used to set up routes running between the data sources and the data sinks, utilizing intermediate sensor nodes to pass along the information, all the while avoiding regions that for various reason do not permit communication.

Employing the ant pheromone model will allow the network to adapt rapidly to an ever-changing substrate, while maintaining good routes between the data sources and their respective data sink.

27.3.3.5 Tools Used

The Cantor tool has been utilized to design the routing algorithm. Using this tool requires the creation of rules that the cells and ants follow, and what information is contained in the state of an ant and the state of a cell. In what follows is a list of each of these rules and states for the cells and ants.

The following rules control the behavior of cells:

- *Diffusion rule*: Implements the diffusion behavior of the ant pheromone, enabling pheromone from one cell to diffuse into its neighboring cells taking into account neighboring cells that may obstruct communication (e.g., walls).
- *Disturbance rule*: Allows for a cell to change its state type at any given point in the execution of the algorithm. This is primarily used to model the transient nature of the application. For instance, one could convert a cell that initially allowed communication into one that does not allow communication.
- *Evaporate rule*: Every time step a portion of the ant pheromone gets dissipated due to the process of evaporation utilizing the following equation:

$$p(t) = p(t-1) * e^{-r} \tag{27.1}$$

where
 $p(t)$ is the pheromone level at time t
 e is Euler's constant
 r is the rate of evaporation

- *Gossip rule*: Implements the Gossip principle on the passing of pheromone information. Each generation a cell determines if it should take part in communicating its pheromone information. Gossip and how it relates to ant pheromone will be discussed later.
- *Haywire rule*: Allows cell malfunctions to be incorporated into the model. Cells that follow this rule select a random pheromone level instead of calculating it based on their neighbors state and their own state.
- *Preserve state rule*: The CA keeps track of two states for each cell and each ant called the previous state and the current state. The previous state, which is the current state from the prior generation, is used by the ant algorithm to formulate the current state. Using the previous state is done so that each cell in the grid will be utilizing the same information when it is updating its current state, thus avoiding race condition among the states. The function of this rule is to copy the current state from the last generation into the previous state of the current generation.
- *Spawn rule*: Causes each ant nest to periodically spawn an ant.

The following rules were developed to control the behavior of ants:

- *Lifetime rule*: Each ant has a lifetime associated with it, which is the number of generations a particular ant has been alive. This rule increments the lifetime of each living ant once per generation.
- *Movement rule*: This rule causes the ants to follow the movement behavior that has been described earlier. Additionally, this rule implements the two types of error conditions that the ants can exhibit, called haywire random and haywire weighted. Both error conditions cause the ants to follow some other behavior than the desired movement behavior. Haywire random is characterized by having the ants move randomly for an arbitrary amount of time; haywire weighted causes the ants to follow the same pheromone that they are releasing. Both are described further in the following.
- *Pheromone rule*: Each generation an ant will release a quantity of pheromone into the cell that it is currently occupying. The pheromone rule performs this function.

The following items were considered to be part of the state of a cell:

- *Communication ID*: Gives a unique ID to each data source, primarily used to keep track of where an ant was spawned from.
- *Communication State*: Defines what type of cell this is (e.g., data source, data sink, free cell).
- *Diffusion rate*: Determines how much pheromone will diffuse out of this particular cell and into neighboring cells each generation.
- *Evaporation rate*: Determines how much pheromone will evaporate out of this cell each generation.
- *Gossip probability*: The likelihood that this cell will take part in communicating its pheromone levels with its neighbors for a particular generation.
- *Will Gossip*: Indicates if the current cell will communicate with its neighbors during this generation.
- *Occupied*: Denotes if there is an ant present in the current cell.
- *Return levels*: A list of all return pheromone levels separated by the originating data source for each ant.
- *Search levels*: A list of all search pheromone levels separated by the originating data source for each ant.
- *Source count*: The maximum number of data sources in the CA.

The following items were considered to be part of the state of an ant:

- *Haywire random*: Denotes that the ant is malfunctioning and is moving randomly.
- *Haywire weighted*: Indicates that the ant is malfunctioning and is following its own type of pheromone.
- *Lifetime*: Indicates how many generations an ant has been in existence.
- *Source address*: Indicates the location of the data source that spawned the ant.
- *Next address*: Denotes the ant's location.
- *Random*: This is the probability that an ant will move randomly for one generation.
- *Status*: Indicates whether the ant is searching or returning.

In Figure 27.6, we provide a sample execution of the ant pheromone algorithm. Walls are areas that do not allow transmission. The data source is located in the upper right corner and is transmitting data to the data sink in the upper left corner. The search ants (return ants) are en route to the data sink (data source). The pheromone levels are represented by the intensity of the colored blocks that occupy most of the region; the blocks with more color represent a higher concentration of pheromone.

27.3.3.6 Derivation of Parameters

The state of the ants and the cells in the ant pheromone algorithm and many of the rules that control the CA depend on numerous parameter settings, so that the model can be modified with relative ease to fit

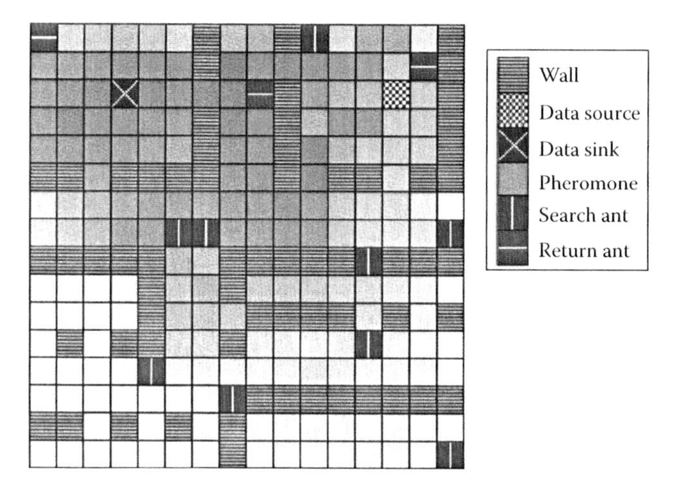

FIGURE 27.6 Example of ant pheromone algorithm designed with Cantor.

the application. Almost all of these parameters necessitate careful setup in order to get the algorithm to run at its peak performance. Determination of the proper parameter settings was calculated by varying one parameter at a time and executing the model for numerous runs.

To evaluate the parameters, two metrics have been utilized. Performance is the first metric and is defined as the mean number of hops needed for a round trip journey from the nest to food for all ants in the simulation. The second metric is power, which is defined as the percent of cells that change their state each generation. Communication overhead is the dominant consumer of power in WSN applications, which is why this metric gives a reasonable measurement of system power.

Figure 27.7 shows the floorplan that was used to derive all the system parameters. Notice that the flow of data would originate from the data source, go to the data sink, and then terminate at the data sink. Each possible parameter value was executed for several runs, until the performance results fell within a 95% confidence interval of 10 hops.

The following provides an explanation of what each of the critical parameters are, how they affect performance, and how they affect power.

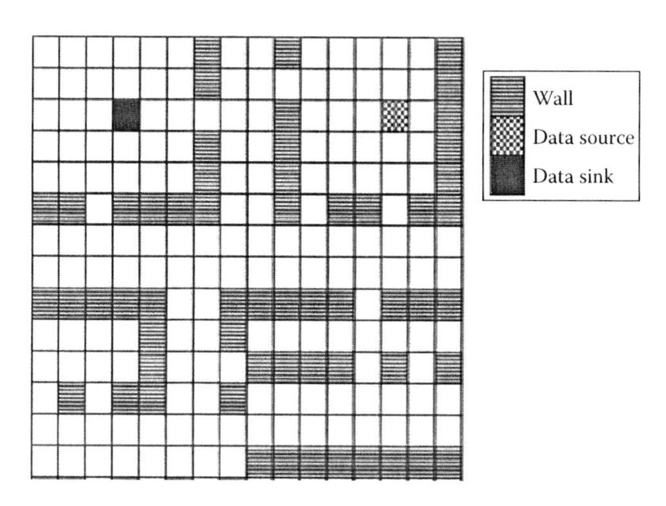

FIGURE 27.7 Floorplan used to derive system parameters.

27.3.3.6.1 Spawn Frequency

The rate at which a data source (ant nest) generates ants is controlled by a parameter setting that denotes the probability that a data source will spawn an ant in any particular generation. Figure 27.8a shows how varying this parameter affects performance; increasing spawn frequency improves performance. However, only a small performance gain is possible from increasing this parameter beyond 50%. We have also evaluated how spawn frequency affects power as is seen in Figure 27.8b. The power profile increases quite rapidly for spawn frequencies below 25%, which is not surprising, as if there are very few ants laying down pheromone, then the cells' states will not change as rapidly. Beyond 25% the rate that power increases becomes quite small.

27.3.3.6.2 Repulsion Ratio

In our initial implementation, a pathology was noticed where ants that were moving to and from the data sink would cluster together and never reach their destinations. The reason this would occur is that if two ants that were searching and returning were to come into contact with each other, it would be likely that their respective pheromones would act as local maxima, and as a result the ants would be trapped. To counteract this, we caused the ants to be repulsed by the pheromone they currently emit. As a result, ants are pulled toward the pheromone that they are not releasing and they are pushed away from the pheromone that they are releasing. A parameter called the repulsion ratio was created denoting the relative strength of repulsion compared to attraction. This compels ants not to stay in one area and solved the pathology.

Figure 27.9a plots performance versus the repulsion ratio. This parameter must be kept in the vicinity of 100% to achieve best performance as this graph indicates, meaning that a setting above 100% will keep the ants from grouping together. However, as the parameter is increased beyond 100%, performance begins to deteriorate since the ants begin to pay less attention to the ant trails. Figure 27.9b shows

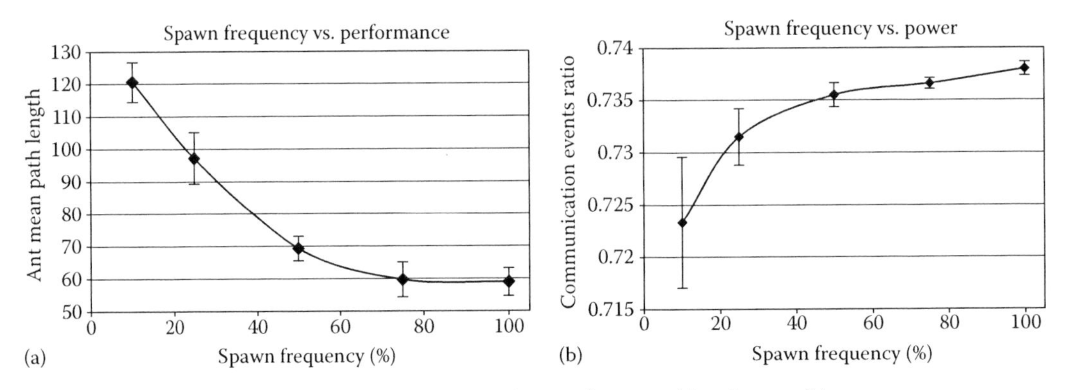

FIGURE 27.8 Effect of how rapidly ants are generated on performance (a) and power (b).

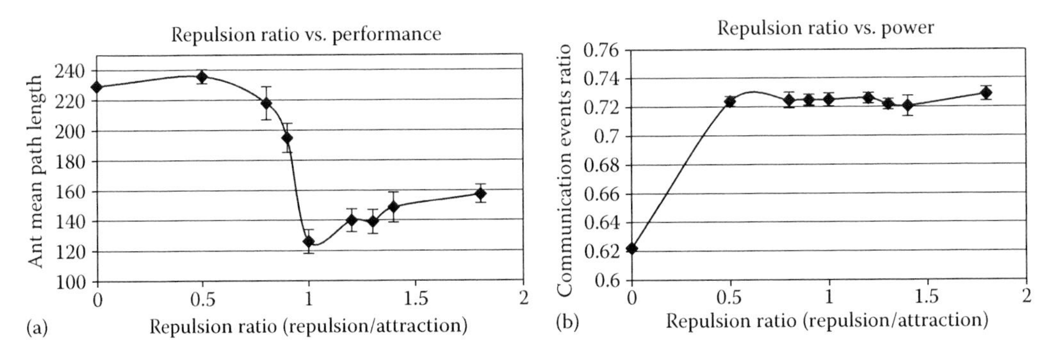

FIGURE 27.9 Effect of repulsion ratio on performance (a) and power (b).

how varying the repulsion ratio affects the power consumption of the algorithm. Analysis of this graph indicates that all parameter settings beyond 50% will have approximately the same power consumption. The low power exhibited by having a setting of 0% is a result of the ants clustering together and not moving, causing far fewer cell states to be affected by the ants, and as a result fewer messages are passed.

27.3.3.6.3 Evaporation

To keep the quantity of pheromone in the system from growing in an unbounded fashion, we have included a parameter denoted as evaporation. In each generation of the algorithm a certain percentage of the ant pheromone evaporates (i.e., it disappears). As Figure 27.10a indicates, any amount of evaporation tends to harm performance, as ant paths get erased before the ants can use them. Although it is not visible from this graph, parameter settings in the vicinity of 0.005 are optimal for performance. We have analyzed evaporation based on power as well, as Figure 27.10b indicates. Not surprisingly, increasing the value of this parameter helps to reduce the power of this algorithm. Unfortunately, any parameter setting that would result in significant power savings coincides with greatly decreased performance.

27.3.3.6.4 Random Ant Movement

Utilizing a random component in a computer algorithm is oftentimes beneficial toward increasing performance. Randomness was included into this algorithm by giving the ants a probability that they will move randomly for one generation rather than follow the pheromone gradient. The performance impact of this randomness is presented in Figure 27.11a. We have determined that a parameter value below 15% increases performance and beyond 15% starts to reduce performance. The optimal setting is around 5%. We have also evaluated random movement's effect upon power consumption. The results

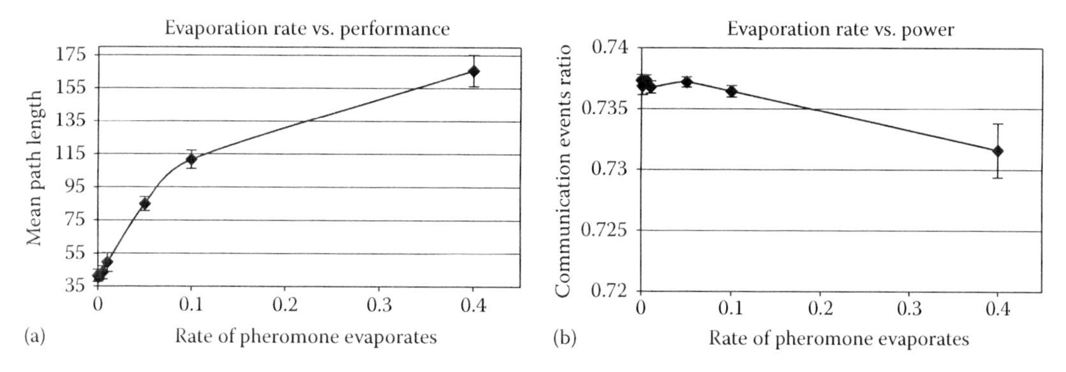

FIGURE 27.10 Effect of evaporation rate on performance (a) and power (b).

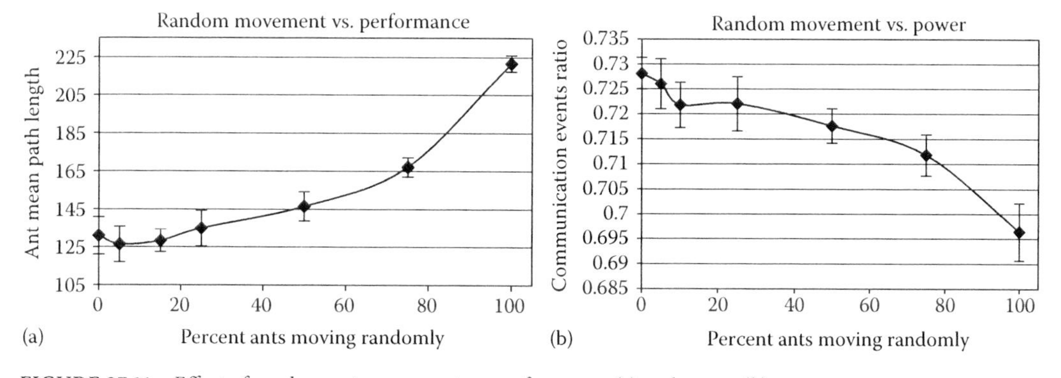

FIGURE 27.11 Effect of random ant movement on performance (a) and power (b).

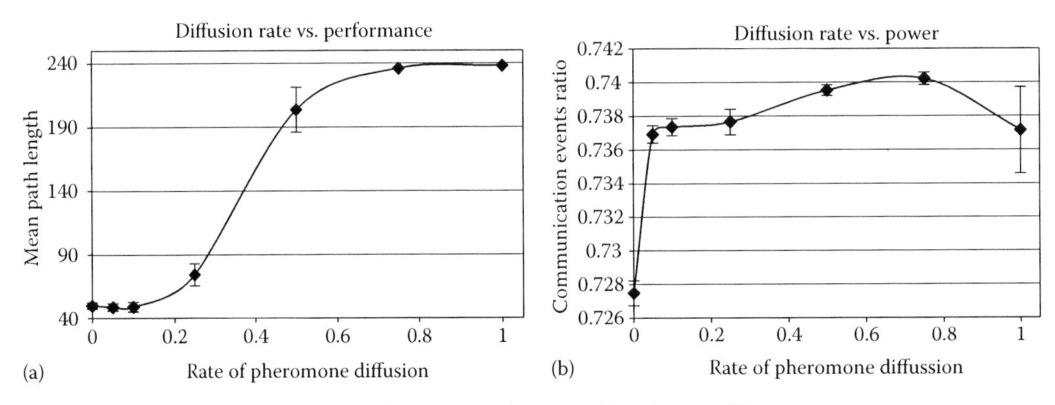

FIGURE 27.12 Effect of pheromone diffusion on performance (a) and power (b).

are provided in Figure 27.11b. This graph shows that having the ants move randomly is beneficial to reducing power consumption of the algorithm.

27.3.3.6.5 *Diffusion*

As stated previously, the ant pheromone spreads to other cells; the amount of pheromone that diffuses out of the cell and into its neighbors is controlled by the diffusion rate parameter. This parameter's effect upon performance is provided in Figure 27.12a. A setting of 0.05 appears to works best. A setting beyond 0.2 begins to have a large negative impact on performance. The effect of diffusion on power is illustrated with Figure 27.12b. This graph indicates that a low diffusion rate coincides with good performance and low power. The low diffusion setting exhibits low power since the pheromone does not move as quickly, while extremely high settings have low power since the pheromone moves so fast that a steady-state behavior occurs quite rapidly.

27.3.3.7 Exploration of Errors

To determine how robust ant pheromone is to errors, the algorithm was subjected to a number of different types of error conditions. These error conditions are as follows: (1) haywire random (a random selection of ants will move randomly for a random amount of time), (2) haywire weighted (a random selection of ants will follow the opposite pheromone), and (3) haywire cells (a random selection of cells will produce a random amount of pheromone). How these error conditions affect performance is illustrated in Figure 27.13a. Up to 50% of ants affected with haywire random increases performance. The effects of haywire cells and haywire weighted are similar, in that they reduce performance rather drastically up to about 25%, where further hits to performance begin to level off. In Figure 27.13b, we show how these error conditions affect the power consumption of the algorithm. The power consumption for haywire cells is quite dramatic; anything beyond 25% is near maximum power consumption. The decrease in power consumption attributed to haywire weighted can be attributed to how this behavior tends to cause the ants to conglomerate into groups and not move, which has the side effect of reducing communication.

27.3.3.7.1 *Conclusions on Errors*

From the preceding analysis, it has been concluded that ant pheromone allows for routing that is very robust to a number of errors, even when much of the network is being afflicted with various pathologies.

27.3.3.8 Why Pheromone Works

The following presents an informal explanation of why pheromone routing works. A CA that shows how pheromone evaporates and an informal proof have been provided.

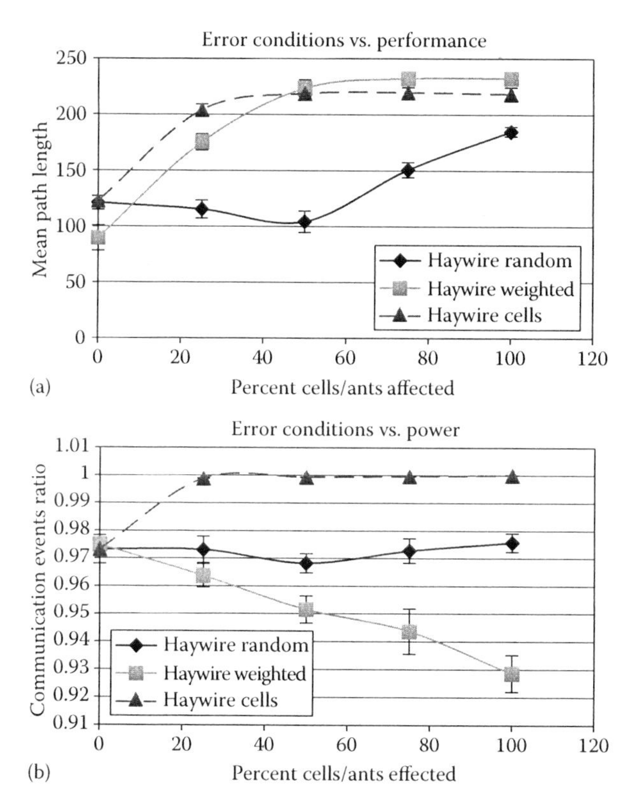

FIGURE 27.13 Effect of error conditions on performance (a) and power (b).

27.3.3.8.1 Pheromone Simulation

A simple CA was developed to show the behavior exhibited by the pheromone utilized in ant pheromone CA. This CA allows for an arbitrary collection of ant pheromone to be placed into a lattice, and with time it will evaporate and diffuse throughout the region.

Figure 27.6 shows one particular execution of this CA. In this example, a curve of ant pheromone is laid down at time 0, as is indicated by the black shape in Figure 27.14a. The intersection of the lines is provided to denote the center of the curve in each of the figures. The concentration of the pheromone is provided by intensity of the colors in the figure, where black is the highest concentration. Figure 27.14b shows the pheromone during generation 10; notice how the regions of highest pheromone concentration are below the original curve. As time progresses, the region of highest pheromone concentration begins to align itself with the straight line that connects the end points of the curve (i.e., the optimal path). As even more time passes, the path begins to fade into background noise.

Now consider what would happen if other ants were to approach this region. The following ants will mainly follow the highest concentration of pheromone. Shortly after the initial pheromone was laid down, the highest concentration is a slightly flatter curve, which is closer to the optimal path than the initial curve. As more time passes, the area of highest concentration will flatten further, until the pheromone gradient is essentially a circle with the pheromone level increasing as it gets toward the center. As a long time passes, the pheromone levels in a local region will tend to be the same, and the pheromone will act as background noise.

The intended point of this example is that ants following another ant's trail will tend to follow a more optimal path with time. If one considers the path of an ant to be composed of several consecutive curves,

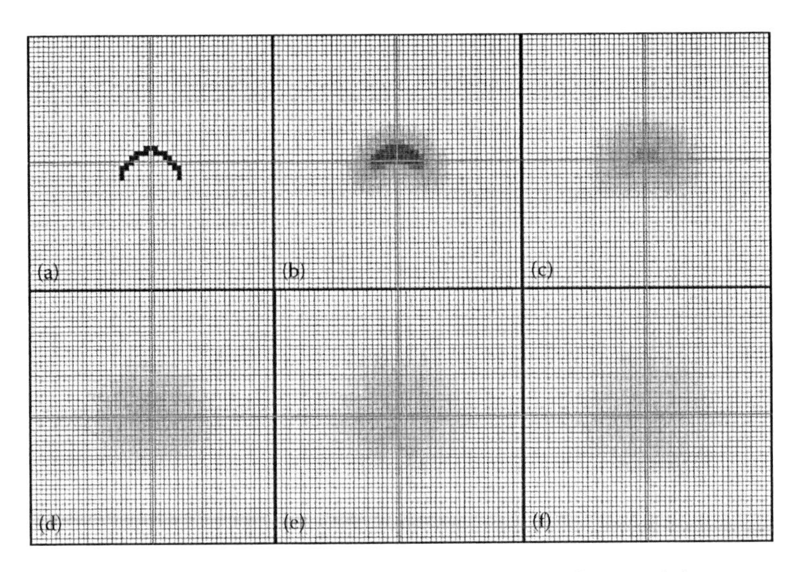

FIGURE 27.14 Pheromone diffusion at work. Part (a) shows that an initial curve of pheromone is laid down at time 0. As time progresses, the pheromone diffuses as shown by (b) at time 10, (c) at time 20, (d) at time 40, (e) at time 60, and (f) at time 100.

and that with time the pheromone will be at its highest concentration along a straight line connecting the beginning and end of the curve, the net result will be an increasingly optimal route that other ants can follow.

The second point to be taken is that transient paths that do not lead to the ant's destination will eventually be lost as background noise.

27.3.3.8.2 Pseudoproof

What follows is an informal proof of the concept illustrated in the previous simulation, namely, that ant pheromone will convert curved paths into optimal straight lines.

As stated in the last example, the path of an ant can be considered to be a sequence of consecutive curved paths, where a curve is defined as being two consecutive inflection points along an arbitrary continuous function, and the optimal route would be defined as the straight line connecting the two inflection points.

To help understand the behavior of ant pheromone, an iterative equation was developed. This equation is defined as

$$P_{x,y,t} = (1-E)P_{x,y,t-1}(1-\rho F) + L(x,y,t) + \sum_{i=1}^{\sqrt{N+1}} \sum_{j=1}^{\sqrt{N+1}} (1-E)P_{i,j,t-1}\frac{F}{N}\rho * s(i,j) \quad (27.2)$$

where

$P_{x,y,t}$ is the pheromone level at location (x, y) at time t
E is the evaporation rate in [0,1]
ρ is the Gossip probability in [0,1]
F is the diffusion rate in [0,1]
$L(x, y, t)$ is the path of ant for time t
N is the size of neighborhood

$$s(i,j)\{0, \quad \text{if} \quad i=j$$

To simplify matters, the following assumptions were made:

1. An instantaneously formed path of pheromone (e.g., the path in Figure 27.14) approximates the behavior of having an ant follow the same path laying down pheromone in each generation.
2. Perpendicular cross sections along the ant path are independent of each other. Specifically, it is assumed that the pheromones in one cross-sectional region do not affect into another cross-sectional area.
3. The evaporation rate should not be allowed to be exceedingly large.

Justification of assumptions:

1. This assumption can be explained best in two separate parts:
   a. Consider sections of the ant path that are close to each other. These sections will usually have been laid down within a short-time interval of each other. And since there is only a small disparity in time for these local sections of the ant path, the behavior of the entire ant path can be approximated by sections that are laid down instantaneously across local regions.
   b. Now consider sections of the ant path that are further apart from each other. These sections of the ant path were laid down in a relatively large time interval from each other. However, as the portions of the ant path are further apart, their respective pheromones will have negligible effect on each other, as is indicated by looking at Equation 27.2. This equation indicates that the further the pheromone travels, the more it gets diffused. Consequently, it should be clear that as two portions of the ant path get further apart, their pheromone interactions will get progressively weaker. As such it can be assumed that the entire ant path can be approximated as instantaneously laid down.
2. Consider the entire ant path as a conglomeration of sections that are perpendicular to the ant path. The true behavior of the ant pheromone is that each cross section will also affect its neighboring cross-sectional areas. Now consider the pheromone behavior of cross-sectional areas that are relatively close to each other. Since these sections are close to each other, they will be primarily affected by the same sections of the ant path, and hence their pheromone behaviors will be similar. Now consider the behavior of diffusion across neighboring cross-sectional areas. A particular cross section will diffuse equivalent quantities of pheromone into its neighbors. Likewise its neighbors pass pheromone back into it. If the cross-sectional area's pheromones are similar to that of its neighbors, then it can be assumed that the pheromone leaving the region is similar to the pheromone coming in. Thus, the cross-sectional areas can be considered to be independent of each other.
3. For the second assumption to be valid, the evaporation rate cannot be too high. Since this is entirely controlled by a parameter setting, this can be readily done. Fortunately, a high evaporation rate is not particularly interesting, since most ant paths are reduced to nothing quite rapidly making pheromone routing largely ineffective.

Given Equation 27.2, if one were to look at any point along the ant path during the instant that the ant laid down the pheromone, they would see the following behavior:

Initially the pheromone would all be located along the ant trail.

As time progresses, Equation 27.2 dictates that the pheromone will spread. Specifically, the pheromone will move one hop further from the initial ant path each generation. It can be seen from Equation 27.2 that in each cross-sectional area, the pheromone will resemble a curve. The peak of the curve shall always be located along the ant path. The further from the ant path a particular point is, the less its pheromone level shall be.

As time progresses, the curve of pheromone in a cross section will expand to new regions. The curve will also tend to flatten over time. This can be seen by examining Equation 27.2, and seeing that large local differences in pheromone will smooth out with time.

The ant's path divides the region into two portions. On one portion, the cross-sectional areas will point toward the center of the curve (call these sections type a), while in the other portion the cross-sectional areas will point away from the center of the curve (call these sections type b). Given the symmetry of the pheromone model, both of these regions will have the same amount of pheromone within them.

The type b cross-sectional region will diffuse into a larger area than the type a cross-sectional region. Therefore, the individual pheromone levels in the type a cross-sectional region will be higher than the pheromone levels in the type b cross-sectional region. This will cause the ants to want to follow the pheromone in the type a region, which contains the optimal path (i.e., the straight line connecting the curve's end points).

27.3.4 Gossip

Gossip is a communication method that can be used to replace message flooding. Consider the following example using flooding. A node in a sensor network wants to send a message to all other nodes in the network. So the node then broadcasts the message to all of its neighbors. These neighbors in turn broadcast the message to all of their neighbors. Each time a node receives such a message for the first time it will broadcast it. When a node receives a duplicate message, it will not broadcast it. Eventually, the entire network will have received the message, but at the expense of huge quantities of messages. With Gossip instead of always routing a received message to all of its neighbors, a node will probabilistically pass on the message, that is to say there is a probability that the node will broadcast the message. It has been shown that this technique greatly reduces the amount of messages required to get a message to all the nodes in a region, and with proper parameter settings, this technique can get a message sent to all nodes with a probability of nearly 1, with far fewer messages than what would have been required with flooding. An excellent example of a network that utilizes several different types of Gossip and shows the effect of varying key parameter values is provided in [Haas 02]. For more in-depth theory on Gossip, see [Chandra 01, Kempe 01, Wokoma 02].

The probabilistic message passing idea from Gossip has been applied to how pheromone behaves in the ant pheromone algorithm. Diffusion of pheromone in particular was modified. Now a node will have a certain probability that it will diffuse its pheromone upon its neighbors. By examining Equation 27.2, one should be able to see that including Gossip in this manner will slow down diffusion of pheromone. It will also make the overall pheromone gradient much more sporadic the less frequent pheromone information is shared.

To illustrate the effects of utilizing Gossip, two separate examples, similar to Figure 27.14, have been provided. Again an initial curve-shaped path of pheromone is laid down at time zero, as shown in Figures 27.15a and 27.16a, and as time progresses the pheromone diffuses throughout the region. In the example from Figure 27.14, every generation a cell would diffuse some of its pheromone to its neighbors. In Figure 27.15, there is a 0.25 probability that a cell will diffuse its pheromone, and in Figure 27.16, there is a 0.75 probability that a cell will diffuse its pheromone.

As these examples indicate utilizing Gossip in the pheromone tends to make the pheromone gradient much less smooth. Further, the less probabilistic a cell is to pass its pheromone information, the longer it takes to for the pheromone region to expand.

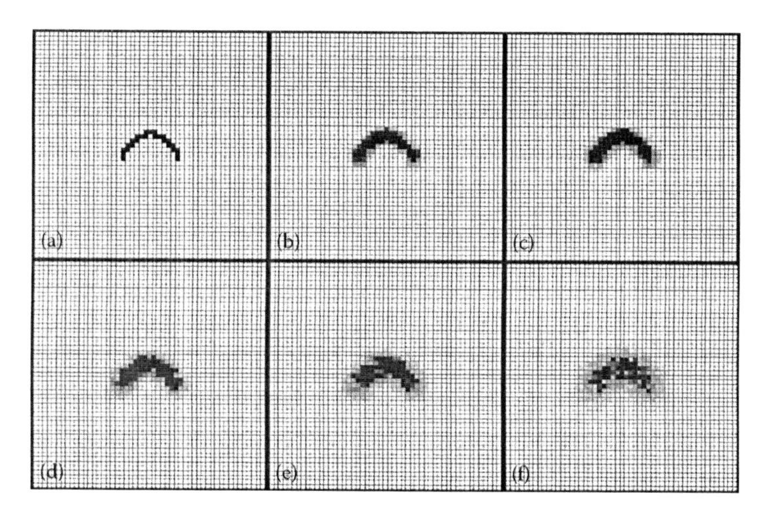

FIGURE 27.15 Effect of a Gossip probability of 0.25. Part (a) shows that an initial curve of pheromone is laid down at time 0. As time progresses, the pheromone diffuses as shown by (b) at time 10, (c) at time 20, (d) at time 40, (e) at time 60, and (f) at time 100.

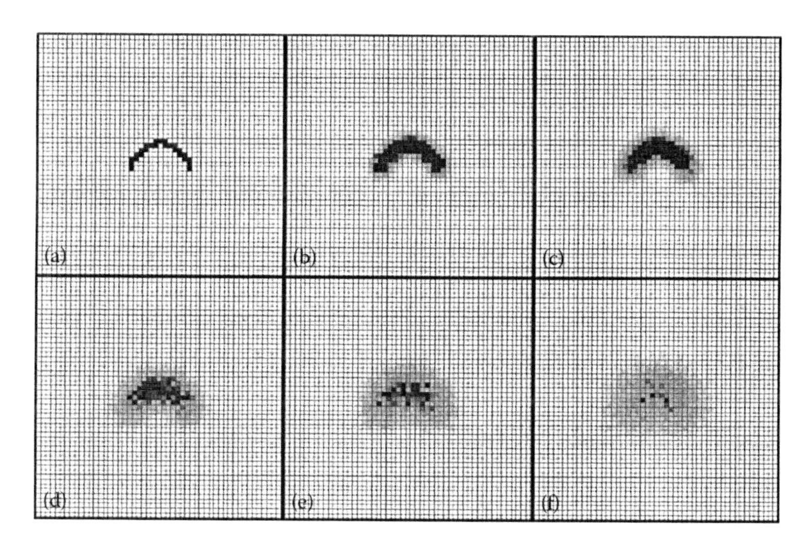

FIGURE 27.16 Effect of a Gossip probability of 0.75. Part (a) shows that an initial curve of pheromone is laid down at time 0. As time progresses, the pheromone diffuses as shown by (b) at time 10, (c) at time 20, (d) at time 40, (e) at time 60, and (f) at time 100.

27.4 Summary

Biological primitives can be a very important tool for studying and designing applications for WSNs. Much work has been done in the discipline of biology to model the behavior of naturally occurring self-organizing systems. A systematic way of modeling such complicated systems has been adapted to WSNs and is provided here. This chapter provides an in-depth case study of one particular biological primitive, ant pheromone, which was used to develop a routing algorithm in a WSN for a MOUT application. Explanation of the tools used, parameter settings, optimizations, performance, and power consumption are provided.

27.5 Physics and Chemistry Techniques in Wireless Sensor Networks

A WSN is a set of wireless sensor nodes that sense, process data, and communicate cooperatively. Sensor node location and network topology are typically not predetermined. Models for node deployment should thus be at least partially stochastic. WSNs have both military and civilian applications. WSN system designs need to simultaneously satisfy a number of constraints:

- WSN nodes typically rely on battery power, making eventual node failure inevitable. While wired network protocols are usually designed to provide a high quality of service (QoS), energy conservation is often a major issue for WSN implementations. There is a trade-off between QoS and system lifetime in these systems.
- WSNs use low-cost radios with a high error rate and limited bandwidth. Associated protocols need to have low communications overhead and high fault tolerance. Limited onboard processors and memory prohibits the use of overly complicated protocols.
- Sensor node failures and possible node mobility make network topology transient. The environment also has frequent and unpredictable perturbations. These topological disturbances demand self-organizing protocols, capable of adapting to these changes.
- WSNs must also have distributed control architectures to maintain a level of reliability, scalability, and flexibility that is not possible for centralized control systems.

27.5.1 Routing in WSN

Routing protocols determine the paths sensor nodes use to communicate. Ad hoc routing protocols have two main variants: (1) table driven and (2) demand driven.

Table-driven protocols maintain routing tables describing paths between all nodes in the network. Topological changes propagate throughout the network. Routing tables need to be modified on each node. The storage requirement of the routing tables and the transmission overhead of topological changes are the largest drawbacks of this approach. An example table-driven routing protocol is destination sequenced distance vector routing (DSDVR).

Demand-driven routing protocols run at the request of a source node. The control overhead is much less than for table-driven protocols. This approach has a route discovery phase, followed by a route maintenance phase, until either the destination becomes inaccessible or the route is no longer needed. An example protocol is ad hoc on-demand distance vector routing (AODV).

There is a vast range of complex physical, chemical, biological, and social systems in the real world. These systems consist of many interacting individuals following simple behaviors. Although individual behaviors appear simple, the collective behavior emerging from interaction among numerous individuals can be very complex. For instance, convection in fluid dynamics, particle diffusion in chemistry, and food foraging in social insects are all examples where individual components have simple behaviors, yet the group has a complex global behavior. These natural mechanisms compel computer scientists to design distributed algorithms, which can be emulated using simple computing devices. Cellular automaton, first conceived by John von Neumann in the early 1950s, serves as a tool for modeling collective behaviors. This chapter shows how the Ising model from physics and diffusion limited aggregate (DLA) from chemistry can help solve the routing problem in WSN.

27.5.2 Discussion of Two Examples from Physics and Chemistry

At first glance, there may appear to be little correlation between the Ising model and DLA and routing in a WSN. However, modifications to the two systems make them excellent models for solving the WSN routing problem.

27.5.2.1 Ising Model

The name of this model derives from the German physicist Ernst Ising who first introduced a simple mathematical model of magnetism in 1925. This model serves as one of the most important models of statistical physics. The Ising model is composed of atomic magnets that can be viewed as magnetic vectors pointing either to the north or the south pole. Suppose we have N such little magnetic spins (s_i, i = 1, 2, 3, ..., N) on a two-dimensional lattice with each pointing up ($s_i = +1$) or down ($s_i = -1$). Each spin interacts with its nearest neighbors in addition to an overall external magnetic field.

There are 2^N different configurations for N magnetic spins. The quantum Hamiltonian energy for a configuration of spins $\{s_i\}$ is given by

$$E[\{s_i\}] = -H\sum_{1}^{N} s_i - K\sum s_i s_j \tag{27.3}$$

Magnetization can be computed as

$$M = \sum_{1}^{N} s_i \tag{27.4}$$

The first term in Equation 27.3 represents the coupling energy between each spin and external field. H is the coupling constant of the external field. The second term explains the interaction energy between all neighboring spins. K defines the strength of spin–spin interaction; K is positive for a ferromagnetic bond (i.e., spins with parallel directions have lower energy) and negative for anti-ferromagnetic bond (i.e., spins with anti-parallel directions have lower energy). Here, we assume periodic boundaries (the right neighbor of a spin on the right edge is the corresponding spin on the left edge) and no external magnetic field. To find the ground state of the system, a brute force method would require considering all 2^N configurations, as is done for the traveling salesman problem by testing all possible routes among all N cities. Clearly, the Ising model is in the class of NP complete problems.

From Equation 27.4, we see that if most spins adopt the same direction, we can get a fairly large magnetization as shown by Figure 27.17b; otherwise, random spin arrangement with opposite directions counteract each other and results in an imperceptible magnetization as shown by Figure 27.17a.

The spin glass variation of the Ising model contains both ferromagnetic and anti-ferromagnetic bonds. In most cases, spins point in random directions, and no macroscopic magnetic field is formed because of cancellation among the individual magnets. However in some metals, like iron, numerous magnetic

(a)

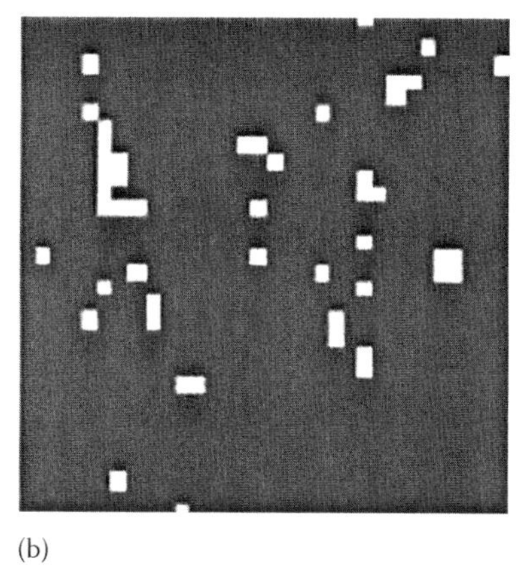

(b)

FIGURE 27.17 (a) random spin direction arrangement. (b) spin direction align. (Applet from http://bartok.ucsc.edu/peter/java/ising/keep/ising.html)

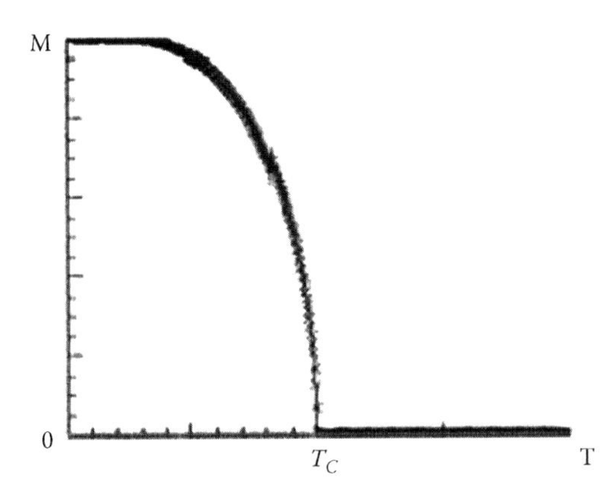

FIGURE 27.18 Magnetization versus temperature. (From Fitzpatrick, R., Lecture notes of Introduction to Computational Physics, 2002.)

vectors align to produce a perceptible macroscopic magnetic field. The phase transition between magnetization and nonmagnetization is tuned by a thermodynamic variable, temperature (similar to how water turns into ice when it is cooled). Energy minimization dominates under low temperature conditions, resulting in a macroscopic magnetization where most of the spins are aligned; entropy maximization dominates system behavior when the temperature is high. There is no net magnetization, since spins point in random directions. Figure 27.18 shows an approximation of net magnetization as a function of temperature. Magnetization is high at low temperatures and decreases as temperature increases. A sudden plunge is noticed when temperature exceeds critical temperature, T_c, and is a type of phase transition.

The probability for each possible microscopic spin configuration is defined by the Boltzmann distribution function:

$$P[\{s_i\}] = e^{-E(\{s_i\})/KT} \tag{27.5}$$

where

$E(\{s_i\})$ is the energy of system in state $\{s_i\}$
K is the Boltzmann constant
T is the temperature in degrees Kelvin

However, in order to make probabilities of all configurations add up to 1, a normalization factor is calculated in the following partition function:

$$Z = \Sigma_A e^{-E(A)/KT} \tag{27.6}$$

$E(A)$: Energy of system in all possible configurations.

$$P'[\{s_i\}] = \frac{e^{-E(\{s_i\})/KT}}{Z} \tag{27.7}$$

Equation 27.7 gives the actual probability of the system staying at configuration $\{s_i\}$.

27.5.2.2 Fractal

Highly regular geometries have been intensively studied by mathematicians throughout history, but it was not until the late 1970s that nested shapes with arbitrarily intricate pieces began to attract more attention [Wolfram 02]. Benoit Mandelbrot first proposed the notion of fractal, which is an irregular geometric object with an infinite nesting of structure at all scales. Fractals are commonly observed in nature such as coastlines, clouds, rivers, and trees. Fractals refer to those objects that have fractional dimensions as opposed to integral dimensions in Euclidean and Cartesian geometries. Fractional

FIGURE 27.19 Fern leaf.

dimension serves as a quantified index describing the complexity or texture of the object's surface. Another property of fractal is self-similarity; it looks similar at all level of magnification. For example, a fern leaf in Figure 27.19 looks similar either at a rough view or at a detailed view.

Euclidean space tells us that a line has one dimension, a plane has two dimensions, and a cube has three dimensions. Another dimension definition is based on the number of variables in a dynamic system. Hausdorff dimension, which was first introduced by Felix Hausdorff in 1919, defines an accurate way of measuring fractal dimension. Hausdorff dimension agrees with Euclidean dimension on regular shaped objects. Let us look at a simple example in Figure 27.20. We equally divide each spatial linear size by s. Equation 27.8 computes the dimension for each case:

$$D = \frac{\log(N)}{\log(s)} \tag{27.8}$$

where

N is the number of similar objects
S is the scaling factor

We show another example of how to calculate the numerical value for the fractal dimension of a coastline. Let us try to get the perimeter of the coastline in Figure 27.21. Suppose we first use a 10 unit ruler

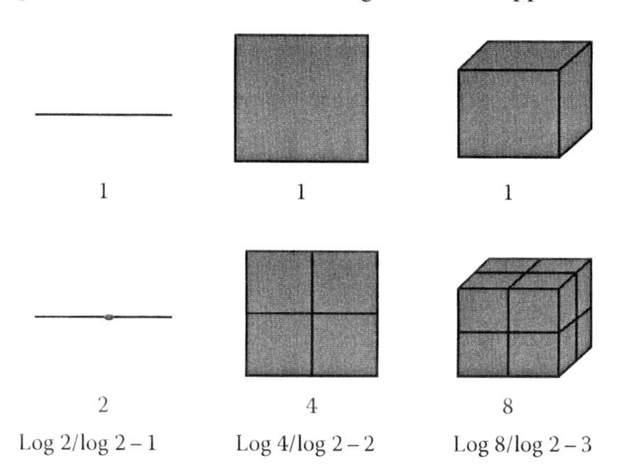

FIGURE 27.20 Example of Hausdorff dimension calculation when scaling factor s is 2.

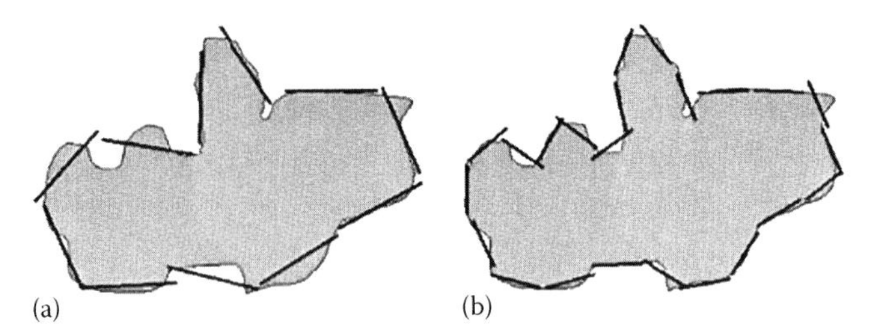

FIGURE 27.21 Coastline perimeter measurements. (a) 11 segments with 10 unit ruler. (b) 23 segments with 5 unit ruler.

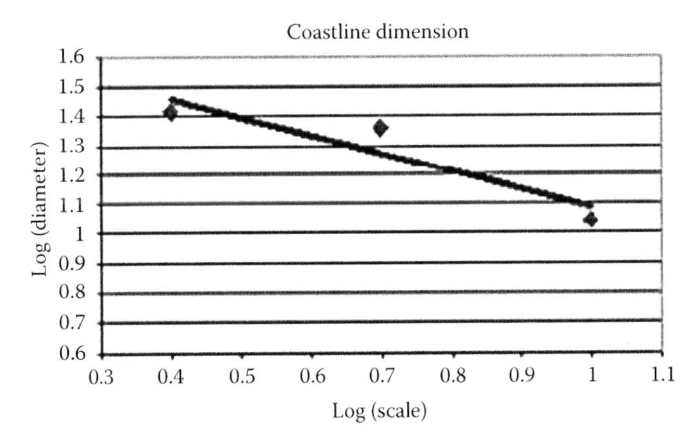

FIGURE 27.22 Log–log graph of coastline derived from an objects perimeter and the scale of the ruler that measured it.

to measure it, we get 11 segments (Figure 27.21a). If we use a smaller ruler with 5 unit ruler, we get 23 segments (Figure 27.21b), which means that we get a larger perimeter with smaller ruler. If the size of the ruler is progressively scaled down, we tend to get a perimeter that approaches an infinite value. The more irregular the coastline appears, the easier to observe this property.

The fractal dimension of an object is specified as being 1+ the absolute value of the ratio of the log of the increase in perimeter to the log of the increase in scale. Fractional dimension allows us to correlate these values with physical properties. Fractal dimensions of the following examples can be related to the following:

- Rocks: related to their erosion
- Fractures in oil bearing deposits: related to the most economical rate of extraction
- A topographic representation of disease spread: related to the agent's virulence
- The outline of a forest fire: related to its speed of spread and the difficulty of extinguishing it [Major 01]

Observe Figure 27.22 where the scale of the ruler versus the measured perimeter are plotted on a log–log graph, a best-fit line can be drawn through the points. The slope of this regression line is −0.623. Based on that, the fractal dimension is specified as being 1.623. If the coastline appears more irregular, the fractional dimension would be larger.

Studies show that there are two separate processes that create fractals: multiplicative iteration of random processes that create multifractal structure and additive processes that generate monofractals [Major 01]. A single exponent can define monofractal, while hierarchies of exponents are required to depict multifractal. The noninteger exponent indexes are called fractal dimension.

27.5.3 Idealized Simulation Scenario

Figure 27.23 shows an idealized MOUT terrain. Walls signify any obstruction that can block radio signals. Open cells are open regions allowing signal transmission. Open doors (closed doors) are choke

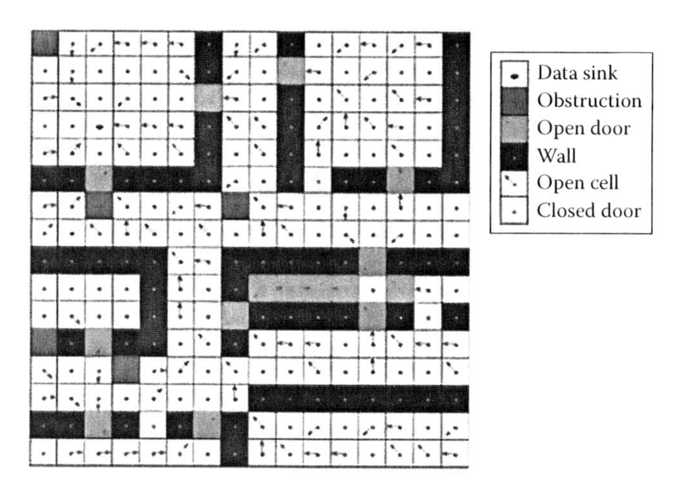

FIGURE 27.23 Example of a MOUT application.

points for signals that periodically allow (disallow) transmission. Finally, obstructions are intermittent disturbances that occur at random throughout the sensor field. Random factors are inserted to emulate common disruptions for this genre of network. Each square capable of transmission contains a sensor node. This amounts to having a sensor field with a uniform density. This provides an abstract example scenario approximating situations likely to exist in a real MOUT situation.

In Figure 27.23, denote the direction in which routing takes place; the neighboring cell that is being pointed to is the next hop on the way to the data sink. Roughly speaking, routes are chosen according to four different metrics:

- Maximum available power route: the route has the maximum amount of total available power.
- Minimum energy route: the route that consumes the minimal sum of transmission power.
- Minimum hop route: the route with the minimal number of hops to the data sink.
- Maximum minimum available power node route: the route along which the node with the minimum available power is larger than nodes with the minimum available power along other routes [Akyildiz 2001].

The minimum energy route scheme and minimum hop route scheme come up with the same results provided the same amount of transmission energy is required for every link.

As we discussed earlier, energy conservation is a crucial issue for WSNs. It would be very helpful if the end user could be aware of the remaining energy at each sensor node. Proper actions could be taken based on an energy map. For example, low activity sensor nodes could be put into sleep node to save energy. Supplemental deployment could be launched to replace sensor nodes with low battery. Cantor can show the energy map for each discrete time step in a particular simulation. Figure 27.24 shows an example of an energy map. Every node starts with full energy shown as green squares. Communication among the sensor nodes for route establishment and adaptation to topological disturbances are considered to be the only consumers of energy. The colors of the squares are used as an indicator for the remaining battery at each sensor node, namely, the darker the square is, the less remaining battery associated with that sensor node.

27.5.4 Apply Physics and Chemistry to WSN Routing

27.5.4.1 Spin Glass Model

A simplified spin glass model, which retains the essence of the Ising model, yet simple enough, to be solved and simulated by a computer, was derived from real-world iron magnetism. In fact, many breakthroughs in science were based on the study on some "toy models," emulating a much more sophisticated

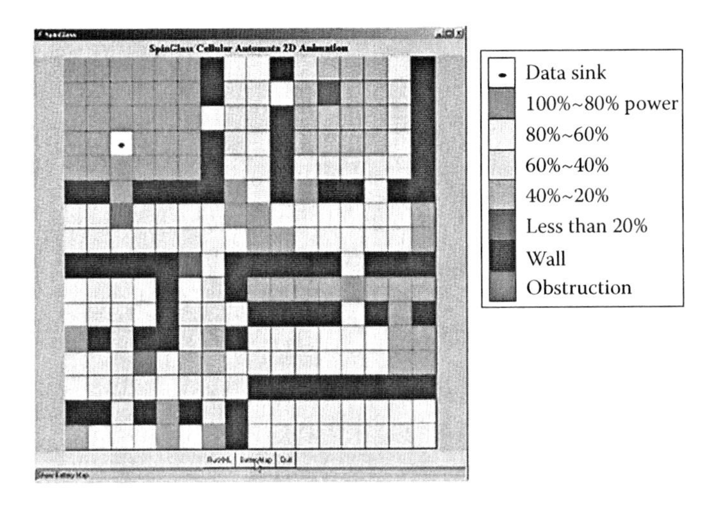

FIGURE 27.24 Energy map for WSN. The darker the color is, the less remaining battery.

real system. Although the correlation between spin glass model and routing in WSN may not be obvious, proper modifications to spin glass model make it very relevant to WSN routing.

In contrast to normal up and down spin directions, our sensor nodes can point to the eight different cardinal directions. This spin configuration belongs to Potts spins type, which allows each spin to have p different values 0, 1, ..., $p - 1$. It is different from classical physics in which a magnet can point to any directions it pleases. The nonzero energy difference associated with all directions is critical for the large-scale system correlation [Gottschalk 00].

In WSNs, hundreds or thousands of sensor nodes are deployed to monitor a particular region of interest. Data collected by the sensor nodes are typically sent to data sinks for postprocessing. Thus, traditional peer-to-peer routing philosophies, which establish a route between a source and destination address pair, are less pertinent. A routing mechanism capable of establishing data routes from sensor nodes to data sinks is needed.

Due to radio transmission limitations, data are relayed to the sink in a multi-hop manner. Multi-hop communication consumes less power than single-hop communications by keeping transmission power level low. In our spin glass routing model, a dynamic potential field defining the minimum transmission energy to reach a data sink is initially established through local interactions of the sensor nodes. Then, a potential field together with a kinetic factor define spin directions of each cell by following the Boltzmann distribution function in Equation 27.7. Nevertheless, the formidable number of all possible configurations required in the denominator of the function bans us from strictly following Boltzmann distribution function. A brute force method would require up to 8^N possible configurations of a grid with N cells. Instead, by only using local information, eight local configurations are calculated in the denominator in our model. This not only dramatically reduces computation load but also excludes global information. We use $T[n_i]$ to represent the potential energy value of node n_i. Node n_j, as one of eight neighbors of node n_i, has potential energy value denoted as $T[n_j]$. The probability that node n_i points to neighbor n_j, $P[n_{ij}]$ is given by Equation 27.9:

$$P[n_{ij}] = \frac{e^{-E(n_{ij})/KT}}{\sum_{k:\text{neighbor of node } i} e^{-E(n_{ik})/KT}} \tag{27.9}$$

where

$E(n_{ij})$ is the energy gap if node n_i points to neighbor n_j as $T[n_j] - T[n_i]$

$E(n_{ik})$ is the energy gap for node n_i to point to all of its eight neighbors

In our computer simulation, a random number is generated to see which direction a cell points to. We repeat this decision process for each sensor node in the lattice at each discrete time step. If we sweep the lattice for a sufficiently large number of times, the fraction of times for sensor nodes pointing to a specific direction will be close to the calculated true probability.

To investigate how the kinetic factor tunes the overall macroscopic phase in the spin glass model, we look at the relative probability of state A and state B in Equation 27.10:

$$\frac{P[A]}{P[B]} = e^{-D/KT} \tag{27.10}$$

If KT is much larger than the energy gap ($D = E(A) - E(B)$) between state A and state B, then the probability of taking either spin direction is approximately the same and the system is in high entropy state. If KT is much lower than D, the sensor node is far more likely to be in the lower energy state. Generally speaking, low-temperature systems exhibit better routing performance in terms of hop distance. T is important, because the shortest path is not the only important criteria. A large T may reduce the power drain on choke points by taking longer routes; an extremely low T can protect the system by reducing oscillations in the system. Moreover, T can be specified on a per-region basis, allowing flexible control over the terrain. A sequence of "snapshot" of the system will be captured and displayed as a time-dependent process in our simulation animation. Our approach works differently from the standard statistical physics simulation technique, Metropolis algorithm. However, the underlying idea is quite similar.

To quantify system adaptation, we measure the mean distance from each node to the data sink. Figure 27.25 shows the mean number of hops versus generation number (time step) for a low-temperature system (Low T), high-temperature system (High T), and a system with a topological disturbance (Disturb). Topological disturbances correspond to choke points in Figure 27.23 opening or closing. The system converges well when T is small, but not when T is large. The graph indicates that topological disturbances are accommodated after a number of fluctuating generations. It demonstrates that our system is capable of adapting to topological changes without the use of global information. Figure 27.26 further illustrates how mean distances are affected under various temperatures. We observe that there is an abrupt rise in mean hops if temperature is raised above 500 K, indicating that 500 K acts as a critical temperature in phase change.

The amount of messages sent during the route establishment phase is quantified to evaluate the scalability of our routing model. We also study how the spin glass model behaves under error conditions. Figure 27.27 demonstrates how system performance in terms of mean hops is affected by error conditions, which we consider to be when nodes randomly choose a spin direction instead of following Boltzmann distribution function or when nodes send incorrect potential value neighbors. It was observed that system is very sensitive to these errors. Performance drastically deteriorates as error percentage occurs even at a low level of 1%. Figure 27.28 illustrates the communication cost versus error conditions. As expected, the amount of messages exchanged increases largely due to error messages

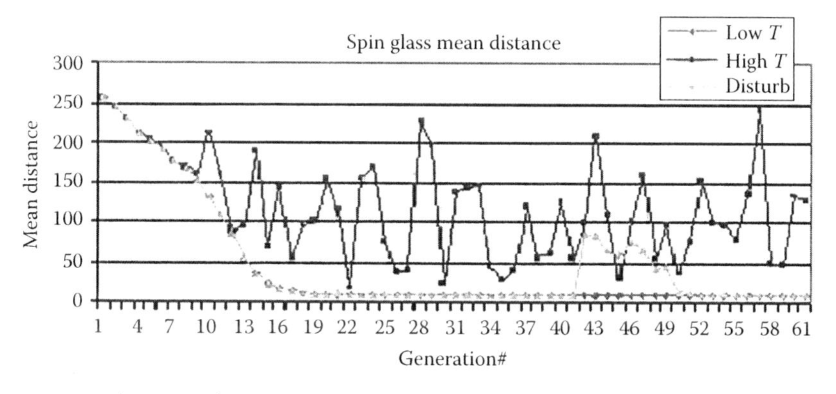

FIGURE 27.25 Spin glass mean distance convergence.

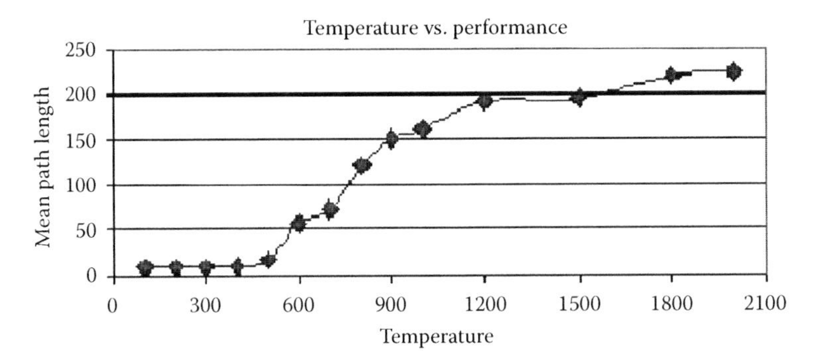

FIGURE 27.26 Effect of temperature on performance.

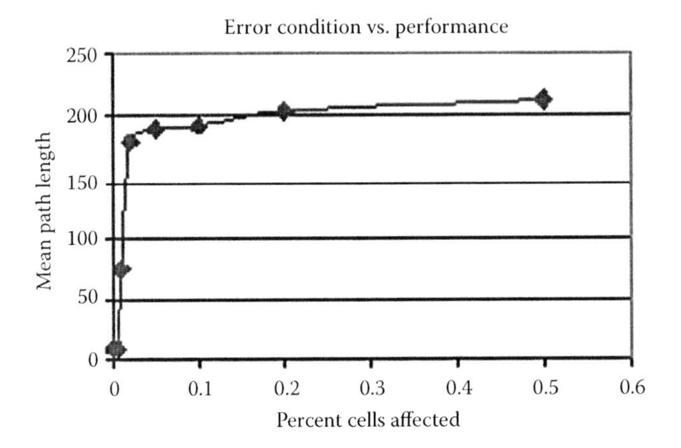

FIGURE 27.27 Error conditions versus performance.

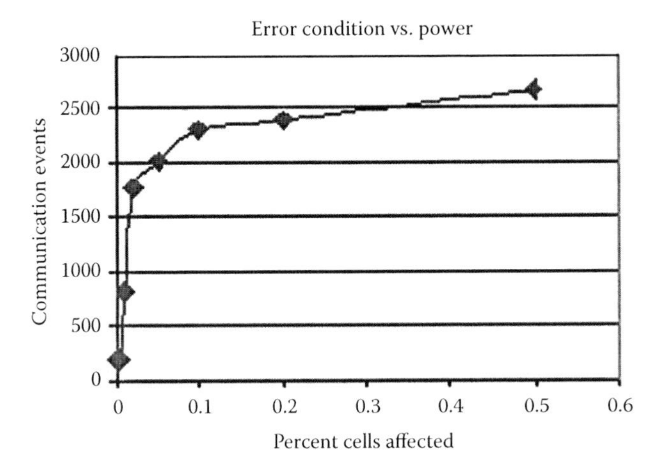

FIGURE 27.28 Error conditions versus power.

diffusing throughout the system. We conclude that although spin glass model achieves high performance, power consumption is high and error tolerance is very limited.

27.5.4.2 Multifractal

The classic irreversible fractal growth model for gas and fluid is called diffusion limited aggregation (DLA), first introduced by Witten and Sander (WS) in the early 1980s. Beginning with one foreign seed

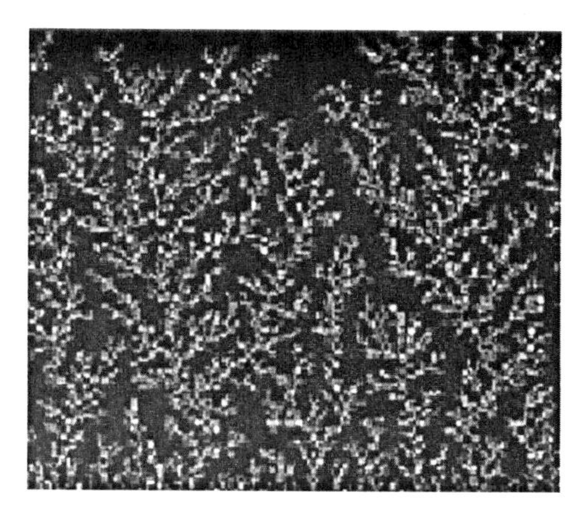

FIGURE 27.29 Diffusion limited aggregation.

or even a line segment, a random walk of gas or fluid particles becomes immobilized upon contact with the seed, if certain crystallization conditions are satisfied. Randomly diffusing particles keep sticking to each other forming an aggregate. The structure of this fractal is affected by many factors including crystallization growth inhibition exerted by the crystallization site to prohibit adherence by nearby particles. Interfacial surface tension and latent heat diffusion effects can physically explain this inhibition [Gaylord 96]. Such WS-like cluster examples can be found in metal electrodeposition experiments. Figure 27.29 shows an example of DLA.

DLA can be considered as a self-repelling random walk starting from the data sink. Let us look at a simple random walk in a Figure 27.30:

Construct a connectivity matrix P whose i, j entry represents the probability of going from node i to node j:

$$P = \begin{bmatrix} 0 & 1/2 & 1/2 & 0 \\ 1/2 & 0 & 1/2 & 0 \\ 1/3 & 0 & 1/3 & 1/3 \\ 0 & 0 & 0 & 1 \end{bmatrix} \quad (27.11)$$

If a mobile agent is initially at node B, the initial state vector is [0, 1, 0, 0]. After the first step, the probability distribution of this agent is P_1:

$$P_1 = P_0 \times P = \left[1/2, 0, 1/2, 0\right] \quad (27.12)$$

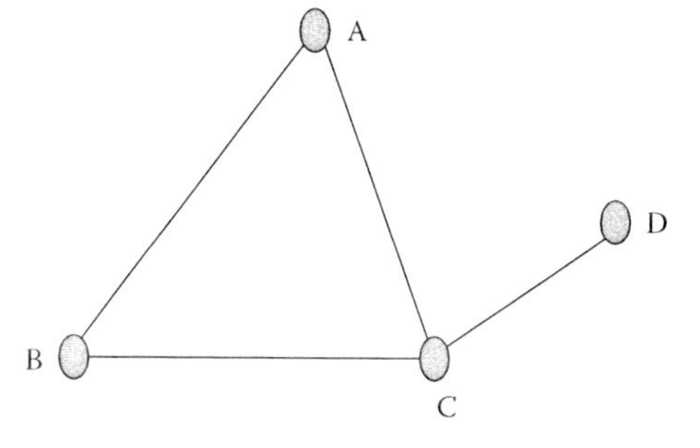

FIGURE 27.30 Random walk graph example.

We can easily get the state vector P_N after n discrete steps after P_0:

$$P_N = P_0 \times P^N \tag{27.13}$$

Equation 27.13 is actually a Markov Chain. In DLA model, a set of "stickiness" probabilities are specified based on the number of neighboring tree nodes. We can easily construct a connectivity matrix as shown in matrix P based on the specified probability set.

In our multifractal routing model, the data sink is set to be the single foreign seed. Each sensor node vibrates in a fixed coordinate of the lattice without randomly wandering around as gas or fluid particles do; however, we do not exclude the mobility of sensor nodes as our routing model can respond and accommodate topological change. A routing tree starts growing from the seed. A sensor node can possibly attach itself to the tree only if any tree node reaches its neighborhood. Based on the number of neighboring immobilized tree nodes, a set of probabilities of joining the routing tree is specified. Theoretically speaking, nodes are less likely to join in the routing tree as the number of neighboring tree nodes increases. In other words, node stickiness decreases as the number of neighboring tree nodes increases, also known as the self-repelling effect. Depending on different levels of self-repulsion we specified in the probabilities sets, the growth rate and the routing tree structure can be controlled.

Generally speaking, a sparse tree with high region coverage grown in reasonable amount of time steps is desired. As previously stated, probabilities are the only way for us to control our routing tree. In order to select an ideal probabilities set for a good routing tree, a fitness function is constructed in Equation 27.14 to evaluate the quality of the routing trees under different probability set. The fitness value is computed as

$$F_i = \frac{C_i}{T_i / b + N_i} \tag{27.14}$$

where

C_i represents region coverage in percentage
T_i represents discrete time steps
N_i represents number of tree nodes
b represents constant used to normalize time steps and number of tree nodes

The higher the fitness value is, the better the routing tree is. Constant b actually represents our trade-off between sparsity and routing time.

Figure 27.31 shows the mean number of hops per generation number (time step) with and without topological disturbances (as for the spin glass model in Figure 27.25). Communication cost as well as error tolerance are investigated as was done on the spin glass model. Malfunctioning nodes are not restrained by the desired multifractal behavior; two principle malfunctions have been modeled:

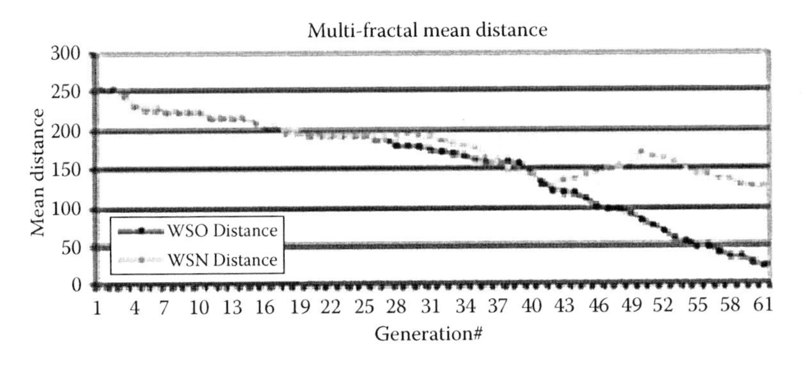

FIGURE 27.31 Multifractal mean distance convergence.

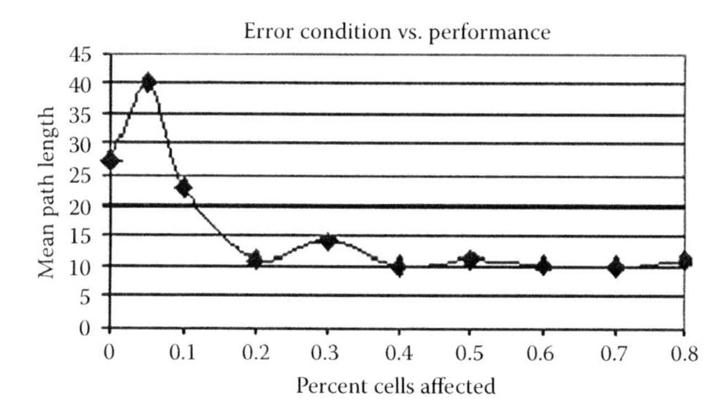

FIGURE 27.32 Multifractal effect of error condition on performance.

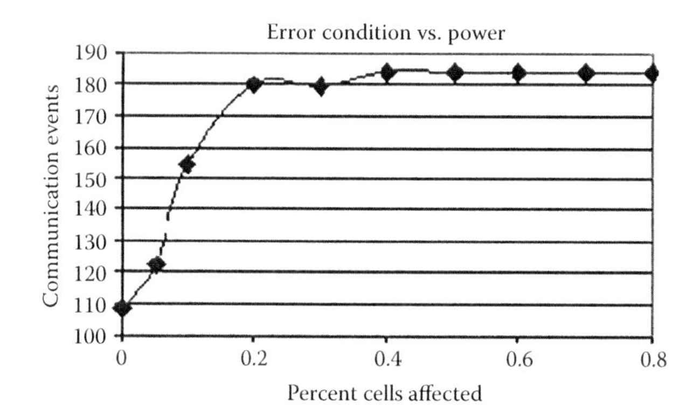

FIGURE 27.33 Multifractal effect of error condition on power.

(1) Faulty nodes have the same probability of joining the tree or not, and (2) faulty nodes randomly choose a neighbor tree node to attach to.

Figure 27.32 shows how the error condition affects the routing performance. This graph indicates that there is a slight increase in the mean hops when random errors come into play, but once more errors begin to occur, the mean hops begin to drop down to almost half and stay steadily while percent of cells affected goes beyond 20%. We may ask, "Does error condition actually improve routing performance?" The shorter mean hops come with the cost of denser trees. Recall that we want a sparse tree, which covers most of the region. However, the final tree consists of approximately 75% more tree nodes compared with the original zero-error tree. The fitness of the error-conditioned tree is actually not ameliorated.

We also investigate how error conditions affect the cost of communication. Figure 27.33 demonstrates that error conditions incur considerably more communication events; nevertheless, the extra communication cost is much lower than that of spin glass model under the same error conditions. It indicates that multifractal model is more error resilient in terms of performance and power.

27.5.5 Protocol Comparison and Discussion

Many routing protocols have been proposed for WSNs. The link-state routing algorithm requires global knowledge about the network. Global routing protocols suffer serious scalability problem as network size increases [Kurose 03]. Destination-sequenced distance vector algorithm (DSDV) is an iterative, table-driven, and distributed routing scheme that stores the next hop and number of hops for each reachable node. The routing table storage requirement and periodic broadcasting are the two main drawbacks to this protocol [Kurose 03]. In dynamic source routing (DSR) protocol, a complete record of traversed cells is required to be carried by each data packet. Although no up-to-date routing information is maintained in the intermediate nodes' routing table, the complete cell record carried by each packet imposes storage and bandwidth problems. AODV algorithm alleviates the overhead problem in DSR by dynamically establishing route table entries at intermediate nodes, but symmetric links are required by AODV. Cluster-head gateway switch routing (CGSR) uses DSDV as the underlying routing scheme to hierarchically address the network. Cluster head and gateway cells are subject to higher communication and computation burden, and their failure can greatly deteriorate our system [Royer 99]. Greedy Perimeter Stateless routing (GPSR) algorithm claims to be highly efficient in table storage and communication overhead. However, it heavily relies on the self-describing geographic position, which may not be available under most conditions. In addition, the greedy forwarding mechanism may prohibit a valid path to be discovered if some detouring is necessary [Karp 00].

The spin glass and multifractal models are related to the table-driven routing protocols by establishing routes from every cell to data sinks. These protocols ensure timely data transmission on demand without searching for the route each time. The ant pheromone model as discussed in Chapter 7.3 is related to the packet-driven protocols. Ants can be viewed as packets traversing from data sources to data sinks. All of the models we presented are decentralized, using only local knowledge at each node. They dynamically adapt to topological disturbances (path loss). Storage requirements for the routing table of spin glass and multifractal are low compared with most other protocols, while the ant pheromone's storage requirements are even lower than these two.

The temporally ordered routing algorithm (TORA) is a source-initiated and distributed routing scheme that shares some properties with the spin glass model. It establishes an acyclic graph using height metric relative to the data sink and also has local reaction to topological disturbances [Royer 99].

The kinetic factor in our spin glass model and the frequency of ant generation in the ant pheromone model provide the system with flexibility in controlling routing behaviors under various conditions. Route maintenance overhead is moderately high for the spin glass model.

The multifractal approach, as a probabilistic space-filling curve, has very light computation and communication load, and overhead is saved in route discovery and maintenance. This is at the cost of a higher distance to the data sinks. Route maintenance overhead for the pheromones is low due to the reduced number of nodes involved in each path. Since the multifractal model strives to cover the sensor field by using as few cells as possible, the sparse routing tree sparse conserves energy. The shortest routes to the data sink are not found using the multifractal model.

On the other hand, spin glass model is more sensitive to internal errors since any possible error may diffuse throughout the network. The multifractal and ant pheromone models are very resistant to internal errors. The time required for the ant pheromone algorithm to converge to a steady state is much longer than required by the other two adaptations. For applications requiring short data paths, the spin glass model is preferred. For overhead sensitive applications that require quick deployment, the multifractal model is a better candidate. If error resilience and low overhead are the principle requirements, then the ant pheromone model is appropriate. Hybrid methods or switching between methods at different phases may be useful.

Acknowledgment and Disclaimer

This research is sponsored by the Defense Advance Research Projects Agency (DARPA) and administered by the Army Research Office under Emergent Surveillance Plexus MURI Award No. DAAD19-01-1-0504. It is also sponsored by Air Force Research Laboratory, Air Force Materiel Command, USAF, under agreement number F30602-99-2-0520 (Reactive Sensor Network). This chapter is partially supported by the Office of Naval Research under Award No. N00014-01-1-0859. The U.S. Government is authorized to reproduce and distribute reprints for governmental purposes notwithstanding any copyright annotation thereon. Any opinions, findings, and conclusions or recommendations expressed in this publication are those of the authors and should not be interpreted as they necessarily represent the official policies or endorsements, either expressed or implied, of DARPA, Office of Naval Research (ONR), the Air Force Research Laboratory, or the U.S. Government.

References

[AC2C 1999] *Adaptive C2 Coalitions Air Campaign Model*, Penn State ARL Technical Report delivered to DARPA JFACC Program office and AFRL, December 1999.

[Adami 1998] C. Adami, *Introduction to Artificial Life*, Springer Verlag, New York, 1998.

[Alligood 1996] K. T. Alligood, T. D. Auer, and J. A. Yorke, *Chaos: An Introduction to Dynamical Systems*, Springer Verlag, New York, 1996.

[Akyildiz 2001] I. F. Akyildiz, W. Su, Y. Sankarasubramaniam, and E. Cayirci, Wireless sensor networks: A survey, *The International Journal of Computer and Telecommunications Networking*, 2001.

[Bonabeau 1997] E. Bonabeau, G. Theraulaz, J.-L. Deneubourg, S. Aron, and S. Camazine, Self-organization in social insects, *Working Papers of the Satna Fe Institute 1997*, http://www.santafe.edu/sfi/publications/Working-Papers/97-04-032.txt, 1997.

[Brogan 1991] W. L. Brogan, *Modern Control Theory*, Prentice Hall, Upper Saddle River, NJ, 1991.

[Brooks 1998] R. R. Brooks and S. S. Iyengar, *Multi-Sensor Fusion: Fundamentals and Applications with Software*, Prentice Hall, Upper Saddle River, NJ, 1998.

[Brooks 1999] R. R. Brooks, S. Phoha, and E. Peluso, Stability and controllability analysis of fuzzy Petri Net JFACC models, *DARPA-JFACC Symposium on Advances in Enterprise Control*, San Diego, CA, November 15–16, 1999.

[Brooks 2000a] R. R. Brooks et al., Reactive sensor networks: Mobile code support for autonomous sensor networks, *Distributed Autonomous Robotic Systems DARS 2000*, Springer-Verlag, New York, October 2000.

[Brooks 2000b] R. R. Brooks, Stigmergy—An intelligence metric for emergent distributed behaviors, *NIST Workshop on Performance Metrics for Intelligent Systems*, Gaithersburg, MD, August 2000.

[Brooks 2000c] R. R. Brooks, Distributed dynamic linking, Penn State Invention Declaration, May 2000.

[Camazine 01] S. Camazine, J. Deneubourg, N. R. Franks, J. Sneyd, G. Theraulaz, and E. Bonabeau, *Self-Organization in Biological Systems*, Princeton University Press, Princeton, NJ, 2001.

[Cebrowski 1998] A. K. Cebrowski and J. J. Garsta, Network-centric warfare: Its origin and future, *Proceedings of the Naval Institute*, 124(1), 28–35, 1998, http://www.usni.org/Proceedings/Articles98/PROcebrowski.htm

[Chandra 01] R. Chandra, V. Ramasubramanian, and K. Birman, Anonymous gossip: Improving multicast reliability in mobile ad-hoc networks, *Proceedings of 21st International Conference on Distributed Computing Systems*, Mesa, AZ, pp. 275–283, April 16–19, 2001.

[Chang 86] K. Chang, C. Chonge, and Y. Bar-Shalom, Joint probabilistic data association in distributed sensor networks, *IEEE Transactions on Automatic Control*, AC-31, 889–897, 1986.

[Chang 1999] M.-H. Chang and J. E. Harrington, Centralization vs. decentralization in a multi-unit organization: A computational model of a retail chain as a multi-agent adaptive system, *Working Papers of the Santa Fe Institute*, 1999.

[CNN 1996]CNN Online 'Domino Effect' Zapped Power in West, August 11, 1996, http://www.cnn.com/TECH/9608/11/power.outage/index.html

[Czerwinski 1998] T. Czerwinski, *Coping with the Bounds: Speculations on Nonlinearity in Military Affairs*, National Defense University, Washington, DC, 1998.

[David 1989] R. David and H. Alla, *Du Grafcet aux reseaux du Petri*, Hermes, Paris, France, 1989.

[DeCarlo 2000] R. A. DeCarlo, M. S. Brancky, S. Pettersson, and B. Lennartson, Perspectives and results on the stability and stabilizability of hybrid systems, *Proceedings of the IEEE*, 88(7), 1069–1082, 2000.

[Delorme 1999] M. Delorme, An introduction to cellular automata, in *Cellular Automata: A Parallel Model*, M. Delorme and J. Mazoyer (eds.), pp. 5–50. Kluwer Academic Publishers, Dordrecht, the Netherlands, 1999.

[Di Caro 96] G. Di Caro and M. Dorigo, AntNet: A mobile agents approach to adaptive routing, Technical Report IRIDIA/9712, Université Libre de Bruxelles, Brussels, Belgium, 1997.

[Dorigo 1996] M. Dorigo, V. Maniezzo, and A. Colorni. The ant system: Optimization by a colony of cooperating agents, *IEEE Transactions on Systems, Man, and Cybernetics-Part B*, 26(1), 29–41, 1996.

[Eigen 1982] M. Eigen, Ursprung und Evolution des Lebens auf molekularer Ebene, in *Evolution of Order and Chaos*, H. Haken (ed.), pp. 6–23. Springer-Verlag, Berlin, Germany, 1982.

[Fitzpatrick 2002] R. Fitzpatrick, Lecture notes of Introduction to Computational Physics, 2002.

[Gaylord 96] R. J. Gaylord and K. Nishidate, *Modeling Nature Cellular Automata Simulations with Mathematica*, Springer-Verlag Inc., New York, 1996.

[Gottschalk 00] T. Gottschalk and D. Davis, Hrothgar Project, Center for Advanced Computing Research at Caltech, Pasadena, CA, 2000.

[Grossglauser 1999] M. Grossglauser and J.-C. Bolot, On the relevance of long-range dependence in network traffic, *IEEE/ACM Transactions on Networking*, 7(5), 629–640, 1999.

[Haas 02] Z. Haas, J. Halpern, and L. Li, Gossip-based ad hoc routing, *Proceedings of the IEEE INFOCOM*, New York, 2002.

[Haberman 1998] R. Haberman, *Mathematical Models: Population Dynamics, and Traffic Flow*, SIAM, Philadelphia, PA, 1998.

[Hagen 1990] H. Hagen and A. Wunderlin, Application of synergetics to pattern formation and pattern recognition, *Self-Organization, Emerging Properties and Learning*, NATO ASI Series B: Vol. 260, pp. 21–30, Plenum Press, New York, 1991.

[Haken 1978] H. Haken, *Synergetics: An Introduction*, Springer-Verlag, Berlin, Germany, 1978.

[Haken 1983] H. Haken, *Synergetics*, Springer Verlag, Berlin, Germany, 1983.

[Harold 02] H. Brochmann, Introduction to fractal geometry, http://www.saltspring.com/brochmann/math/Fractals/fractal-0.00.html

[Hopcroft 1979] J. E. Hopcroft and J. D. Ullman, *Introduction to Automata Theory, Languages, and Computation*, Addison-Wesley, Reading, MA, 1979.

[Jensen 1998] H. J. Jensen, *Self-Organized Criticality*, Cambridge University Press, Cambridge, U.K., 1998.

[Karp 00] B. Karp and H. T. K. Ung, Greedy perimeter stateless routing for wireless networks, *Proceedings of the Sixth Annual ACM/IEEE International Conference on Mobile Computing and Networking*, Boston, MA, August 6–11, 2000.

[Kaufmann 1993] S. Kaufmann, *The Origins of Order*, Oxford University Press, New York, 1993.

[Kempe 01] D. Kempe, J. M. Kleinberg, and A. J. Demers, Spatial gossip and resource location protocols, *Proceedings of 33rd Annual ACM Symposium on Theory of Computing*, pp. 163–172, 2001.

[Kott 1999] A. Kott and B. Krogh, Toward a catalog of pathological behaviors in complex enterprise control systems, *Proceedings from November 1999 DARPA-JFACC Symposium on Advances in Enterprise Control*, San Diego, CA, pp. 1–6, November 15–16, 1999, http://www.darpa.mil/iso/jfacc/symposium/sess2-1.doc

[Kurose 03] J. F. Kurose and K. W. Ross, Computer networking a top-down approach featuring the Internet, AW Higher Education Group, a division of Pearson Education, 2003.

[Leland 1994] W. E. Leland, M. S. Taqqu, and D. V. Wilson, On the self-similar nature of Ethernet traffic (extended version), *IEEE/ACM Transactions on Networking*, 2(1), 1–15, 1994.

[Li 1993] Y. Li and W. M. Wonham, Control of vector discrete event systems I—The base model, *IEEE Transactions on Automatic Control*, 38(8), 1214–1227, 1993.

[Li 1994] Y. Li and W. M. Wonham, Control of vector discrete event systems I—Controller synthesis, *IEEE Transactions on Automatic Control*, 39(3), 512–531, 1994.

[Major 01] J. A. Major and Y. Lantsman. Part 1 Actuarial application of multifractal modelling, 2001.

[Nicolis 1977] G. Nicolis and I. Prigogine, *Self-Organization in Non-Equilibrium Systems*, Wiley, New York, 1977.

[Pandya 1996] A. S. Pandya and R. B. Macy, *Pattern Recognition with Neural Networks in C++*, CRC Press, Boca Raton, FL, 1996.

[Parunak 1999] H. Parunak and S. Brueckner, Synthetic pheromones for distributed motion control, *JFACC Symposium in Advances Enterprise Control*, San Diego, CA, November 1999.

[PBS 1996] Online Newshour, Blackout, July 3, 1996, http://www.pbs.org/newshour/bb/science/blackout_7-3.html

[Phoha 1999] S. Phoha, R. R. Brooks, and E. Peluso, A constructivist theory for distributed intelligent control of complex dynamic systems, *Symposium on Advances in Enterprise Control*, San Diego, CA, November 1999.

[Phoha 2000a] S. Phoha, N. Gautam, and A. Horn, Tactical intelligence tools for distributed agile control of air operations, *Second Symposium on Advances in Enterprise Control*, Minneapolis, MN, July 2000.

[Phoha 2000b] S. Phoha, E. Peluso, and R.L. Culver, A high fidelity AOSN simulator for intelligent control of networked ocean sampling assets, invited paper, special issue of the *IEEE Journal of Oceanic Engineering* on Autonomous Ocean Sampling Networks, 26(4), 646, 2000.

[Portugali 2000] J. Portugali, *Self-Organization and the City*, Springer-Verlag, Berlin, Germany, 2000.

[Preston 84] K. Preston, Jr. and M. J. B. Duff, *Modern Cellular Automata Theory and Applications*, Plenum Press, New York, 1984.

[Rai 1990a] S. Rai and D. P. Agrawal, eds. *Advances in Distributed System Reliability*, IEEE Computer Society Press, Los Alamitos, CA, 1990.

[Royer 99] E. M. Royer, A review of current routing protocols for ad hoc mobile wireless networks, *IEEE Personal Communication*, 6(2), 46–55, 1999.

[Rudie 1992] K. Rudie and W. M. Wonham, Think globally, act locally: Decentralized supervisory control, *IEEE Transactions on Automatic Control*, 37(11), 1692–1708, 1992.

[Sarkar 2000] P. Sarkar, A brief history of cellular automata, *ACM Computing Surveys*, 32(1), 80–107, 2000.

[Siewiorek 1982] D. P. Siewiorek and R. S. Swarz, *The Theory and Practice of Reliable System Design*, Digital Press, Bedford, MA, 1982.

[Smulders 1991] S. A. Smulders, *Control of Freeway Traffic Flow*, CWI Tract, Amsterdam, the Netherlands, 1991.

[Swanson 2000] D. C. Swanson, *Signal Processing for Intelligent Sensor Systems*, Dekker, San Francisco, CA, 2000.

[Tsaptsinos 1997] D. Tsaptsinos, Rough sets and ID3 rule learning: Tutorial and application to hepatitis data, *Journal of Intelligent Systems*, 8(1–2), 203–223, 1997.

[Tsumoto 1998] S. Tsumoto, Knowledge discovery in medical databases based on rough sets and attribute-oriented generalization, *The 1998 IEEE International Conference on Computational Intelligence*, Anchorage, AK, pp. 1296–1301, May 4–9, 1998.

[Tsymbal 1998] A. Tsymbal, S. Puuronen, and V. Terziyan, Advanced dynamic selection of diagnostic methods, *Eleventh IEEE Symposium on Computer-Based Medical Systems*, Lubbock, TX, pp. 50–54, 1998.

[Van Creveld 1980] M. L. Van Creveld, *Supplying War: Logistics from Wallenstein to Patton*, Cambridge University Press, Cambridge, U.K., 1980.

[Van Creveld 1986] M. L. Van Creveld, *Command in War*, Harvard University Press, Cambridge, MA, 1986.

[Wokoma 02] I. Wokoma, I. Liabotis, O. Prnjat, L. Sacks, and I. Marshall, A weakly coupled adaptive gossip protocol for application level active networks, *IEEE Third International Workshop on Policies for Distributed Systems and Networks*, 2002.

[Wolfram 1994] S. Wolfram, *Cellular Automata and Complexity*, Addison-Wesley, Reading, MA, 1994.

[Wolfram 02] S. Wolfram, *A New Kind of Science*, Wolfram Media, Inc., Champaign, IL, 2002.

[Zhou 1996] K. Zhou, J. C. Doyle, and K. Glover, *Robust and Optimal Control*, Prentice Hall PTR, Upper Saddle River, NJ, 1996.

28

Collective Intelligence for Power-Aware Routing in Mobile Ad Hoc Sensor Networks

Vijay S. Iyer
Indian Institute of Science

S.Sitharama Iyengar
Florida International University

N. Balakrishnan
Carnegie Mellon University and Indian Institute of Science

28.1 Introduction

Life can be defined as a state of functional activity and continual change peculiar to animals and plants before death. It can also be classified as something capable of reproducing itself, capable of adapting to an environment and also capable of independent actions not decided by some external agent.

28.1.1 Artificial Life

Artificial Life or Alife is the study of non-organic organisms, of life-like behavior beyond the creation of nature. In artificial life, the environment is originally is created by humans inside a computer. The rules of life are fairly universal and apply even out of natural setting. A set of rules is created for the creatures to follow. The creatures interact with each other and arrive at a solution (a state of global optimization).

One of the important characteristics of both life and Alife is emergence. Emergence is *something more than the sum of the parts.* In Alife simulations, global pattern emerge as a result of the behavior of the individual entities and interaction amongst the. Alife adopts a bottom-up approach, in which, only the behavior of the lower-level units are programmed, and the higher-level *collective behavior* is obtained from the interactions between these units [Kawata'94]. In nature many organisms exhibit collective behavior, which includes: flocking by birds, fishes in schools, nest building by social insects and foraging by ants.

28.1.2 Stigmergy

An important collective behavior mechanism that makes insects perform well in a large variety of tasks is called Stigmergy. Stigmergy is a form of indirect communication through the environment. The Stigmergy is of two forms:

- In the first form, the physical characteristics of environment may change as a result of carrying out some task-related action, such as a digging a hole by an ant, or adding a ball of mud to a growing structure by a termite. Subsequent perception of the changed environment may cause the next ant to enlarge the hole, or the termite to deposit its ball of mud on top of the previous structure. This type of influence has been called sematectonic [Bonabeau'94].
- In the second form, the environment is changed by depositing something, which makes no direct contribution to the task, but is used solely to influence subsequent behavior, which is task related. This is sign-based Stigmergy, which is highly developed in ants and other social insects. Ants use a special type of hormone, known as pheromone, to provide a sophisticated signaling system.

28.1.3 Trail Laying by Ants

As mentioned earlier, ants have a peculiar property of laying pheromone trails, while going in search of food or while returning back to their nests. Ants get attracted to the pheromone deposits depending on the trail strength. Pheromone is a hormone, which evaporates and diffuses into the atmosphere. At any given time the strength of trail encountered by another ant is a function of the original trail strength, and the time since the trail was laid.

Figure 28.1 shows more than one possible routes between the nest and the food source. Initially, an ant arriving at the junction makes a random decision with a probability of 0.5 of choosing one route over the other. Now assuming that, there are two ants in search of food and two more returning from the food source towards the nest. Lets assume that the ants in both the pairs chose to go in different routes.

After some time when those ants, which had chosen the shorter path, reach their destination first and scenario is depicted in Figure 28.2. We see that the pheromone concentration in the shorter path is more as compared to the longer path. The next coming at the junction would choose the shorter

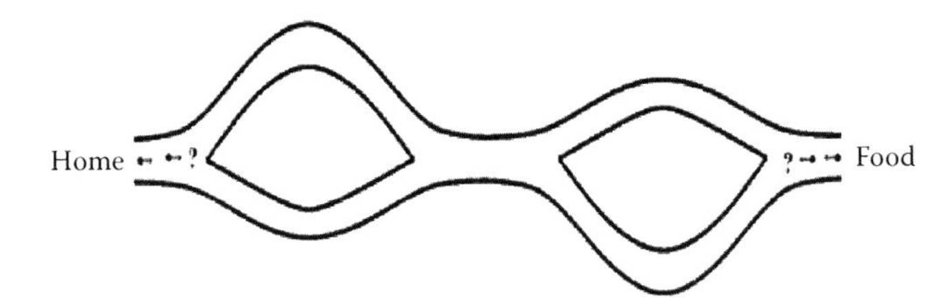

FIGURE 28.1 Ants have to make a decision.

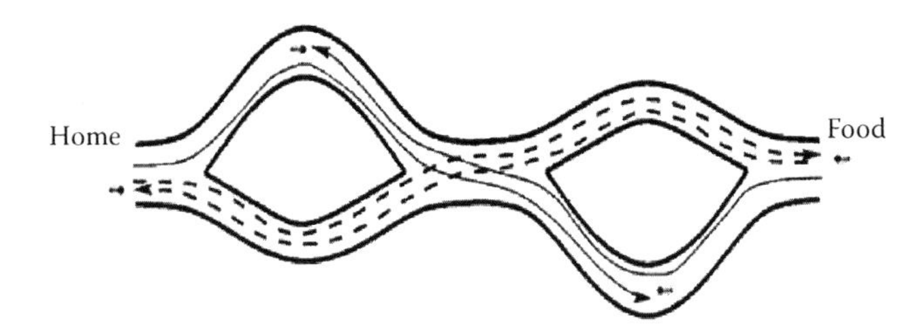

FIGURE 28.2 Ants laying pheromone trails.

route with a higher probability for reaching its destination. This in turn enhances the pheromone concentration in the shorter route, causing more and more ants to traverse the shorter route. As fewer ants travel through the shorter path, and the existing pheromone slowly evaporates, the trail in the longer path weakens and eventually disappears.

There is a kind of reinforcement learning that takes place in finding the optimal path. Over-training raises two problems:

- *Blocking problem*: This occurs when a route preciously found by ants is no longer available. It can take a long time for ants to find a new route, because the trails leading to the blocking route are sometimes very strong.
- *Shortcut problem*: This occurs when a new or shorter route suddenly becomes available. In this case the shorter route may not be easily found, because the existing trails are so strong that almost all the ants choose them.

28.2 Artificial Ants for Routing

We describe below a technique derived from the collective behavior of ants for power-aware routing in mobile ad hoc networks. We have used specialized packed (artificial ants) for finding shortest (at least shorter) routes in wireless a network. Such have already been made in telecommunication networks successfully for routing [Dicaro'97a] and congestion control [Schoon'96a].

28.2.1 Routing Table

The routing table used in this method is inspired from *Ant Based Control* [Dicaro'97a] and the *AntNet* [Schoon'96a] algorithms. The routing table assumes the structure of a probability table. The table gives the probability of choosing a neighbor for a given destination. For a network with n nodes the probability table in a particular node with m neighbors would contain $(n - 1) \times m$ entries.

A typical routing for a node would look like one showed in Table 28.1.

Table indicates the probabilities of choosing a route. The probabilities if the table sum up to 1.0. Updating the probabilities of the routing table represents the act of laying pheromones. When an ant reaches a node, it consults with the routing table, updates it and chooses the next hop based on the random decision but depending on the probabilities in the table. For example, consider an ant reaching node 1 which has the routing table exactly as given in the table above. The ants destination is node 3, say. The ant will choose the next hop as node 2 with a probability of 0.45 or node 4 with a probability of 0.55 after consulting the row corresponding to nodes 3 in the routing table.

28.2.2 Routing of Data Packets

Data packets move according to the entries in the routing table corresponding to their destination. Data packets move independent of ants. To determine the next hop for a data packet with a given destination, the row corresponding to the destination is looked up in the routing table. The neighbor with highest probability becomes the next hop.

TABLE 28.1

	Next Hop	
Dest.	Node 2	Node 4
2	0.95	0.05
3	0.45	0.55
4	0.02	0.98

28.2.3 Saturation Value

To overcome the blocking and shortcut problem (see Section ??), the freezing of routes should be avoided. Freezing of routes takes place when the probability of choosing the next hop becomes 1 for a particular neighbor and zero for others. This situation can be overcome by incorporation a saturation value above which none of the probabilities is allowed to rise. This allows finite number of ants to still traverse other routes even if the best route has been found out.

28.2.4 Modification of Probabilities

As described earlier in Section ?? δp depends on either the age of ants or the hop count and the energy left in the previous node. The prime requirement is to find methods for encouraging ants to establish shorter routes in the network. The measures for finding shorter routes is age of the ant or the number of hops the ant have covered. Since maintaining time in mobile ad hoc networks is a difficult, the hop count was used as a measure.

28.2.4.1 Hop Count

Since routing in MANETs is a multihop scenario, hop count can be used as a measure of distance between the source and the destination. Using hop count δp_h can be computed using Equation 28.4:

$$\delta p_h = \frac{h_f}{H} + h_c \tag{28.1}$$

where
H is the number of hops (nodes), including the source, an ant has traveled
h_f is the hop factor (typically 0.2–0.3)
h_c is a constant (typically 0.01)

28.2.4.2 Power-Aware Routing

Energy in a mobile device is a scarce resource, so the need for its conservation arises. Here a technique is presented to improve the energy utilization of the network. It is a well-known fact that if energy is tapped from a source like battery in chunks, the battery life extends [Balakrishna'02].

Life of a network depends on how many connections are up at any given time. Figure 28.3 shows a four-node wireless network. Lets assume the case of normal routing without power-awareness. The ants establish the route from node 1 to node 3 via node 4. After some time due to packet movement through node 4, energy in node 4 decreases eventually leading to its death. After the death of node 4, node 2 is selected as a hop for routing data packets from node 1 to node 4. At this point of time, only two routes are up.

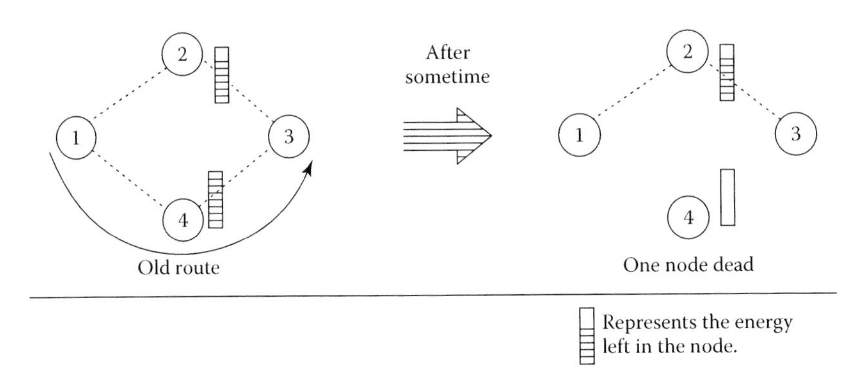

FIGURE 28.3 Normal routing without power-awareness.

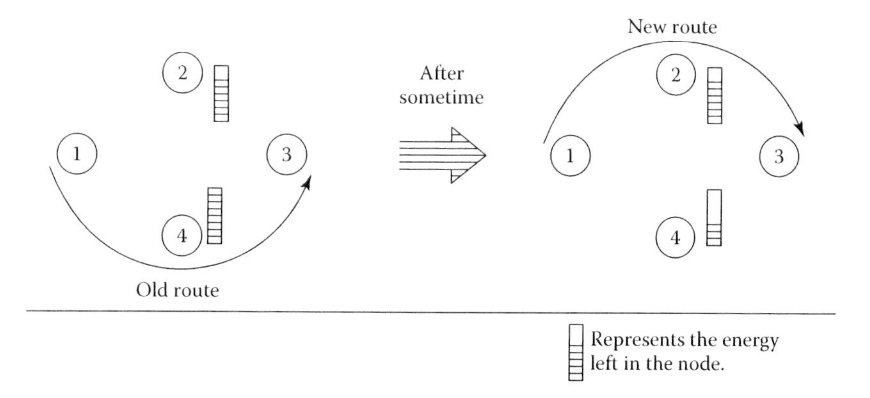

FIGURE 28.4 Change in route after sometime due to power-awareness in routing.

Figure 28.4 shows a way to improve the longevity of the network. Lets assume that the ants have established a route between node 1 and node 3 via node 4. Due to movement of packets the energy in the node 4 decreases. Ants after sensing a decrease in energy in node 4 change the route, and establish a new route from node 1 to node 3 via node 2. In this way the energy from the nodes are not continuously tapped, instead the node are given a time to charge themselves. And as we know if the energy is not drawn continuously from a source, then its life increases [Balakrishna'02].

Improving longevity of network by incorporating a power-awareness term in ∂p causes the links in the network to be up for a longer duration of time. δp_p is given by

$$\delta p_p = \frac{1}{a} \times \left(\frac{energy}{maxEnergy} \right)^2 \tag{28.2}$$

where

"*energy*" is the energy left in the previous hop at the time the ant was being transmitted
"*maxEnergy*" is the maximum energy a device has at the time of initialization
"*a*" is a constant (typically around 2–2.5)

28.2.5 Algorithm

1. Ants are launched from every node with randomly selected destinations.
2. The interval after which the ants are released is not fixed, but is a randomly selected between a fixed range.
3. Ants move randomly in the network but depending on the probabilities in the routing table, toward their destination (see Section 28.2.1).
4. Ants modify the probabilities in the routing table for the location they were launched from, by the increasing the probability of selection of their previous node by subsequent ants. The increase in probability is given by Equation 28.3:

$$p_{prev} = \frac{(p_{oldp} + \delta p)}{(1 + \delta p)} \tag{28.3}$$

where

p_{prev} is the new probability corresponding to the previous node
p_{oldp} is the original probability
δp is the increase in probability

The other entries in the table are decreased correspondingly as given by Equation 28.4.

$$p_{other} = \frac{p_{oldo}}{(1+\delta p)} \tag{28.4}$$

where

p_{other} is the new probability corresponding to other neighboring nodes
p_{oldo} is the original probability of other nodes

5. The increase in probabilities is a decreasing function of the hop count/age of ant, and energy left in the previous node and of the original probability (see Section 28.2.5).
6. To avoid freezing of trails, some saturation value (<1.0) is maintained, above which none of probabilities in the routing table is allowed to raise (see Section 28.2.4).

28.3 Results

The suggested algorithm was simulated for a mobile ad hoc network working on 802.11 protocol. The topology of the network was generated randomly and to add mobility to the network, nodes were moved at each time step of the simulation.

The following assumptions were made while designing the model:

1. There is no multi-path fading or distortion of signal
2. The coverage is perfectly circular
3. The transmission takes place only at 1 Mbps
4. The nodes were assumed to be situated in 2D plane
5. There in no bit-error occurring in the transmission, due to the medium in case there is no other simultaneous transmission

The energy consumption model of that of Lucent WaveLAN PC card was used [Feeney'01a]. Typical energy dissipation during sending and receiving data has been tabulated in Table 28.2.

28.3.1 Route Establishment

The simulation results have been produced here. Figures 28.5 and 28.6 shows the number of ants and time required to establish routes in a network with respect to the size of the network. The route

TABLE 28.2 Energy Dissipation by Lucent WaveLAN PC Card

	μW-s/byte	μW-s
Point-to-point send	1.9 × size	+454
Broadcast send	1.9 × size	+266
Point-to-point recv.	0.50 × size	+356
Broadcast recv.	0.50 × size	+56
	Non-destination $n \in S, D$	
Promiscuous recv.	0.39 × size	+140
Discard	−0.61 × size	+56
	Non-destination $n \in S, n \notin D$	
Promiscuous recv.	0.54 × size	+66
Discard	−0.56 × size	+24
Idle (ad hoc)	843 mW	
Idle (BSS)	66 mW	

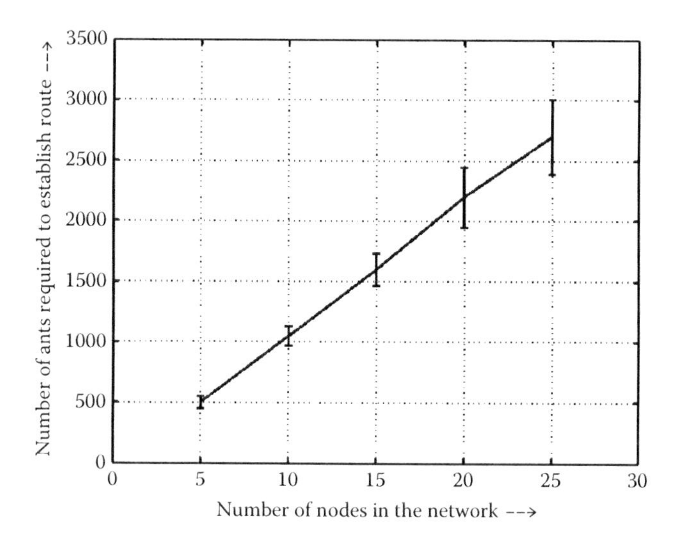

FIGURE 28.5 Number of ants vs. size of network.

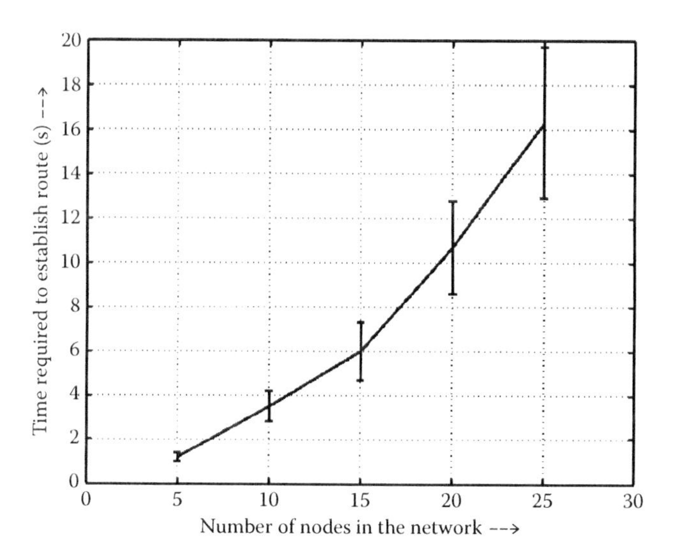

FIGURE 28.6 Time required vs. size of network.

establishment was considered complete, when there exists a unique path (when at least a path exists) between each pair of node in the network, which is at the maximum 10% longer than the shortest available path.

28.3.2 Energy Distribution

Longevity of a network depends on the number of critical nodes present in it. Critical nodes may be defined as those nodes, which have very less energy and are susceptible to die. Figures 28.7 and 28.8 show the graph between number of nodes vs. percentage decrease in energy. As is clear from the figure the use of power-awareness in routing has caused this number to decrease, thus enhancing the longevity of the network.

Figure 28.9 also proves the increase in energy awareness, due to the new routing methodology. Figure 28.9a shows the average energy of all the nodes in a 30-node network with and without power-awareness. Figure 28.9b shows the corresponding standard deviation.

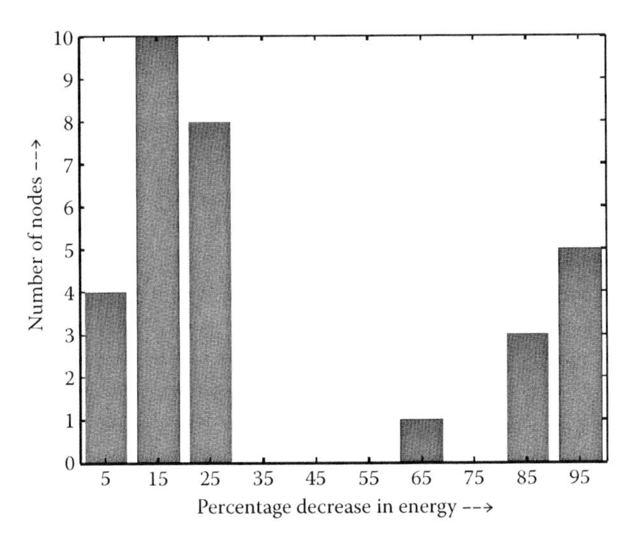

FIGURE 28.7 Energy reduction in nodes in normal routing.

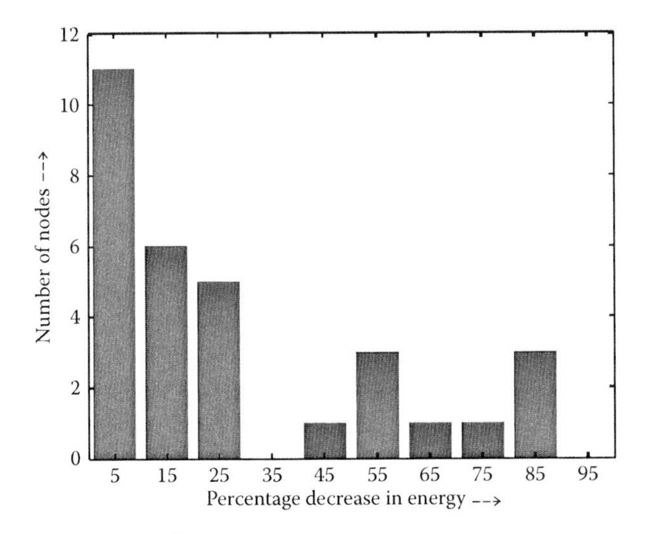

FIGURE 28.8 Energy reduction in nodes in power-aware routing.

28.3.3 Energy Access Pattern

Simulation performed over a network of 10 nodes shows the following access pattern for a particular node. From the energy access patter, it is clear that the energy consumption from that node has lowered over a period of time, which would improve the longevity of the network.

28.3.4 Effect of Noise

As mentioned in Section 28.2.4.2, we set a saturation value for the probabilities to avoid freezing of ant trails (routes). Figure 28.10 shows the average and standard deviation of energies of a network with and without saturation value. Clearly the saturation value has given a significant decrease in power consumption of the network (Figures 28.11 and 28.12).

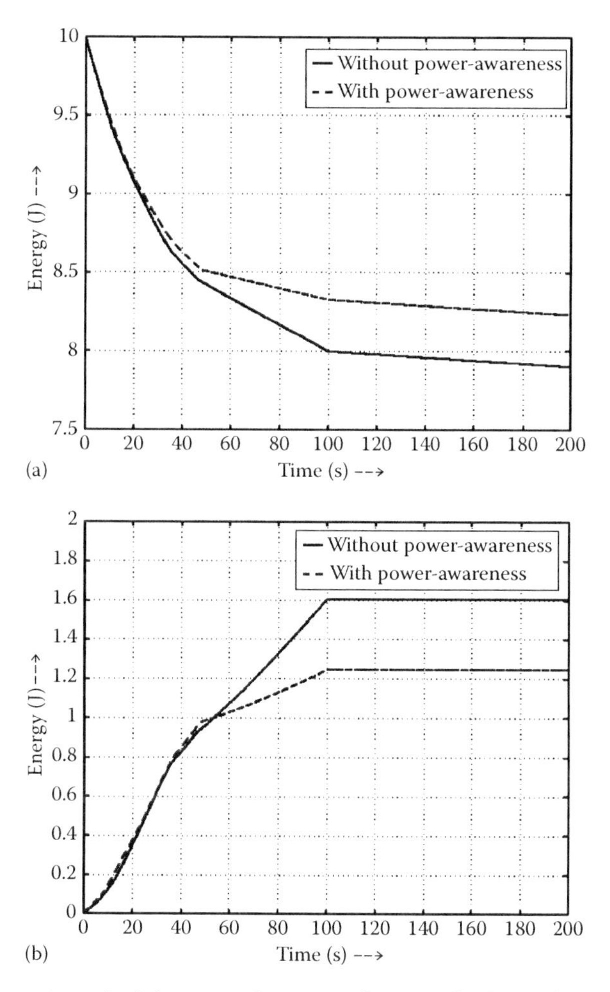

FIGURE 28.9 Average and standard deviation of energies of a network of 30 nodes with and without power-awareness factor. (a) Average energy of the network as a function of time. (b) Standard deviation of the energies as a function of time.

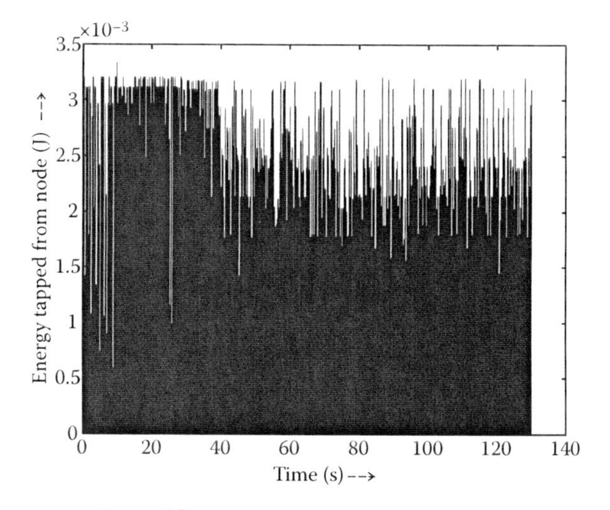

FIGURE 28.10 Energy access pattern without power-awareness.

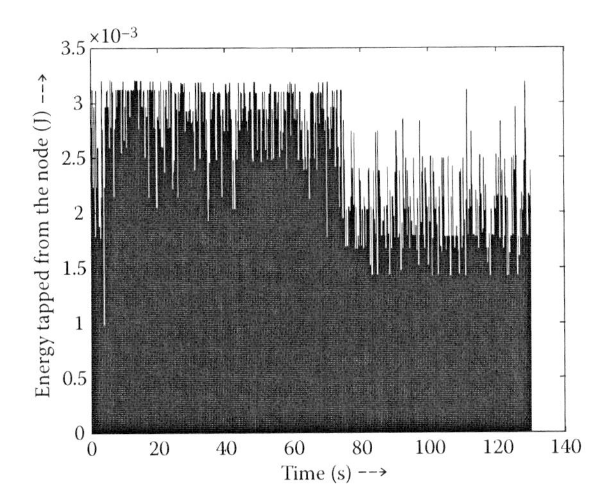

FIGURE 28.11 Energy access pattern with power-awareness.

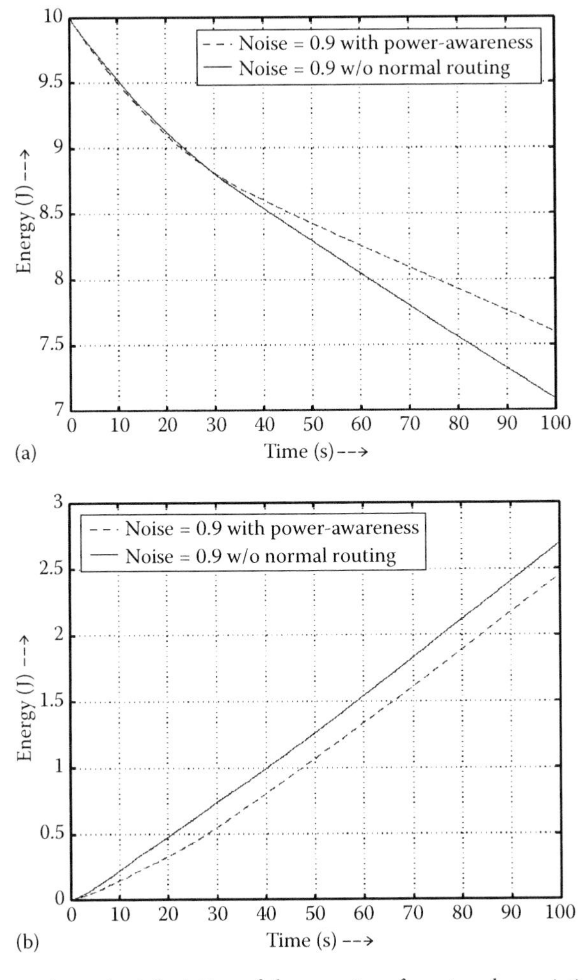

FIGURE 28.12 Average and standard deviation of the energies of a network consisting of 30 nodes with and without power-awareness factor when the saturation value is set to 0.9. (a) Average of energies of all the nodes in the network. (b) Standard deviation of the energies of all the nodes in the network.

28.4 Conclusion

Emergent behavior gives rise to collective intelligence or the swarm intelligence, which has been used here to identify routes in mobile ad hoc sensor networks. Mobile ad hoc sensor networks have tremendous application in surveillance, military and civil communications, personal communication and many other networks of prime importance. Routing being a problem of global optimization is highly complex. In mobile ad hoc sensor networks the problem manifolds in complexity due to mobility of the nodes. In this work we have used specialized packets known as ants to establish routes. We also the address the issue of scarcity of energy in mobile devices by making the algorithm power-aware. The algorithm is highly adaptive and scalable.

References

[Alcherio'01] Martinoli, A., Collective complexity out of individual simplicity, Invited book review on *Swarm Intelligence: From Natural to Artificial Systems*, E. Bonabeau, M. Dorigo, and G. Theraulaz. (eds.), MIT Press, Cambridge, MA, *Artificial Life*, 7(3), 315–319, 2001.

[Arabshahi'01] Arabshahi, P., Gray, A. et al., Adaptive routing in wireless communication networks using swarm intelligence, *Proceedings of 19th AIAA International Communications Satellite Systems Conference*, Toulouse, France, April 17–20, 2001.

[Balakrishna'02] Prabhu, B. J., Chockalingam, A., and Sharma, V., Performance analysis of battery power management schemes in wireless mobile devices, *IEEE WCNC'2002*, Orlando, FL, March 2002.

[Beckers'92] Beckers, R., Deneubourg, J. L., and Goss, S., Trails and U-turns in the selection of a path by the Ant Lasius Niger, *Journal of Theoretical Biology*, 159, 397–415, 1992.

[Beckers'94] Beckers, R., Holland, O. E., and Deneubourg, J. L., From local actions to global tasks: Stigmergy and collective robotics, *Artificial Life IV, Proceedings of the Fourth International Workshop on the Synthesis and Simulation of Living Systems*, Brooks, R. A. and Maes, P. (eds.), MIT Press, Cambridge, MA, pp. 181–189, 1994.

[Benjamin'01] Barán, B. and Sosa, R., AntNet: Routing algorithm for data networks based on mobile agents, *Inteligencia Artificial, Revista Iberoamericana de Inteligencia Artificial*, 3(12), 75–84, 2001.

[Bonabeau'99] Bonabeau, E., Dorigo, M., and Di Caro, G., Ant colony optimization: A new meta-heuristic, *Proceedings of 1999 Congress on Evolutionary Computation*, Washington, DC, pp. 1470–1477, July 1999.

[Bonabeau'00] Bonabeau, E. and Theraulaz, G., Swarm smarts, *Scientific American*, 282(3), 73–79, March 2000.

[Colorni'96] Colorni, A., Dorigo, M., and Maniezzo, V., The ant system: optimization by a colony of cooperating agents, *IEEE Transactions on Systems, Man, and Cybernetics, Part B*, 26(1), 1–13, 1996.

[DiCaro'97a] Di Caro, G. and Dorigo, M., AntNet: A mobile agents approach to adaptive routing, Technical Report IRIDIA/97-12, Université Libre de Bruxelles, Brussels, Belgium, 1997.

[DiCaro'97b] Di Caro, G. and Dorigo, M., AntNet: A mobile agents approach to adaptive routing in communication network, *Ninth Dutch Conference on Artificial Intelligence (NAIC '97)*, Antwerp, Belgium, November 12–13, 1997.

[DiCaro'97c] Di Caro, G. and Dorigo, M., Adaptive learning of routing tables in communication networks, *Proceedings of Italian Workshop on Machine Learning*, Torino, Italy, December 9–10, 1997.

[DiCaro'98a] Di Caro, G. and Dorigo, M., AntNet: Distributed stigmergetic control for communications networks, *Journal of Artificial Intelligence Research*, 9, 317–365, 1998.

[DiCaro'98b] Di Caro, G. and Dorigo, M., Ant colony routing, *PECTEL 2 Workshop on Parallel Evolutionary Computation in Telecommunications, Reading*, England, April 6–7, 1998.

[Dicaro'98c] Di Caro, G. and Dorigo, M., Mobile agents for adaptive routing, *Proceedings of 31st Hawaii International Conference on System Sciences*, Kohala Coast, Hawaii, IEEE Computer Society Press, Los Alamitos, CA, pp. 74–83, January 6–9, 1998.

[Feeney'01a] Feeney L. M., An energy consumption model for performance analysis of mobile ad hoc networks, *Journal of Mobile Networks and Applications*, 6(3), 239–249, 2001.

[Feeney'01b] Feeney L. M. and Nilsson, M., Investigating the energy consumption of an wireless network interface in an ad hoc networking environment, *IEEE Infocom 2001*, Anchorage, AK, April 2001.

[IEEE 802.11'99] IEEE 802 LAN/MAN Standards Committee, Wireless LAN medium access control (MAC) and physical layer (PHY) specifications, *IEEE Standard 802.11*, June 1999.

[Kassabalidis'01a] Kassabalidis, I. et al., Swarm intelligence for routing in satellite and sensor networks, *NASA Earth Science Technology Conference*, College Park, MD, August 28–30, 2001.

[Kassabalidis'01b] Kassabalidis, I. et al., Swarm intelligence for routing in communication networks, *IEEE GlobeComm*, San Antonio, TX, November 25–29, 2001.

[Kawata'94] Kawata, M. and Toquenaga, Y., From artificial individuals to global patterns, *TREE*, 9(11), 417–421, 1994.

[Langton'87] Langton, C. G., *Artificial Life, The Proceedings of an Interdisciplinary Workshop on the Synthesis and Simulation of Living Systems*, Los Alamos, NM, Langton, C. G., Taylor, C., Farmer, J. D., and Rasmussen, S. (eds.), Addison Wesley, Redwood City, CA, September 1987.

[Langton'90] Langton, C. G., Preface, *Artificial Life II, Proceedings of the Workshop on Artificial Life*, Santa Fe, NM, Langton, C. G., Taylor, C., Farmer, J. D., and Rasmussen, S. (eds.), Addison Wesley, Redwood City, CA, February 1990.

[Schoon'96a] Schoonderwoerd, R., Collective intelligence for network control, MS thesis, Delft University of Technology, Faculty of Technical Informatics, Delft, the Netherlands, May 1996.

[Schoon'96b] Schoonderwoerd, R., Holland, O.E., Bruten, J.L., and Rothkrantz, Ant-based load balancing in telecommunications networks, HP Labs Technical Report, HPL-96-76, May 21, 1996.

[Schoon'97] Schoonderwoerd, R., Holland, O.E., and Bruten, J.L., Ant-like agents for load balancing in telecommunications networks, *Proceedings of First ACM International Conference on Autonomous Agents*, Marina del Rey, CA, pp. 209–216, February 5–8, 1997.

29

Random Networks and Percolation Theory

Richard R. Brooks
Clemson University

29.1 Background

Random graph theory originated with seminal work by Erdös and Rényi in the 1950s. Until then, graph theory analyzed either specific graph instances or deterministically defined graph classes. Erdös and Rényi considered graph classes where the existence of edges between nodes was determined probabilistically. Their results were theoretically interesting and found applications in many practical domains [Barabási 02].

Erdös and Rényi used the same probability value to assign edges between any two nodes in the graph. As an extension to this in the 1990s, Strogatz and Watts studied "small world" graphs [Watts 99]. The term small world originates with Milgram's six degrees of separation model of social networks created in the 1960s. Strogatz and Watts' work considers networks where the probability of edges existing between nodes is not uniform. They were specifically interested in clustered graphs, where edges are more likely to exist between nodes with common neighbors. To study this phenomenon, they defined classes of pseudo-random graphs. These graphs combine a deterministic structure and a limited number of random edges. Their results have been used to analyze both social networks and technical infrastructures.

An alternative approach to studying similar systems has been proposed by Barabási [Barabási 02]. His group considered the probability distributions of graph node degree found in graph models of existing systems. This analysis shows that the probability of a node having degree *d* follows an inverse power law (i.e., is proportional to $d^{-\gamma}$ where γ is a constant). They also explain how this property can emerge from positive feedback in evolving systems. These models appear to be appropriate for studying a wide range of natural and large-scale technical systems. Important results from this model include quantification of the dependability of the Internet [Albert 00], and analysis of computer virus propagation [Storras 01].

Random graph concepts are also widely used in percolation theory [Stauffer 94]. Percolation theory studies flow through random media. The model of random media is usually built from a regular tessellation of an n-dimensional space. Edges may or may not exist between neighboring vertices of the tessellation with a uniform probability. Applications of percolation theory include oil extraction. We consider this model as an example of sensor networks with a planned wireless infrastructure.

Another random network model, given in [Krishnamachari 01], is used to study ad hoc wireless networks. A set of nodes is randomly distributed in a two dimensional region. Each node has a radio with a given range r. A uniform probability exists (in [Krishnamachari 01] the probability is 1) for edges being formed between nodes as long as they are within range of each other. This network model has obvious practical applications. Many of its properties resemble those of Erdös-Rényi graphs, yet it also has significant clustering like the small-world model. We use this model when analyzing ad hoc wireless systems.

Many network systems that would otherwise be intractable can be analyzed using random graph abstractions. This chapter is inspired by the design of adaptable peer-to-peer (P2P) computer networks, which are particularly suited to implementing sensor networks. Consider a P2P network with no centralized design, control, or plan. Nodes enter and leave the system of their own volition [Oram 01]. In P2P networks, all participants function simultaneously as both client and server, which is why they are sometimes called "servents." The existence or non-existence of an interaction between two nodes cannot be known in advance, making the random graph model an appropriate for some aspects of the system.

P2P networking gained recognition with the Napster and Gnutella implementations. Napster is a scalable approach to file dissemination.. On connecting to Napster, each user workstation uploaded to the Napster server a list of the local filenames. Filenames were on the order of tens of bytes. The files themselves were usually multi-megabyte MP3 music files. To retrieve a file, the user queried Napster's index and received a list of potential offerors. File exchanges occurred between individual nodes distributed at random across the Internet. This was very efficient, but with a single point of failure. A court order stopped the central index, effectively destroying the entire system.

Gnutella's distributed design is extremely robust. A Gnutella network of n nodes has n indexes. Each node keeps track of locally stored files. The network has no single point of failure. To stop the Gnutella service, it would be necessary to stop every node running Gnutella on the Internet. This is a desirable survivability property. On the other hand, a global search of a Gnutella network involves flooding the network with search packets. This is inefficient and scales poorly [Lv 02]. A relatively small number of concurrent requests can use all available network bandwidth creating an unintentional internal Denial of Service (DoS) attack.

We studied this tradeoff between efficiency and robustness, by generalizing the P2P design problem in [Kapur 02]. The paper shows how to determine the proper number of indexes (in the range $1 - n$) and packet time out values to support desired levels of robustness and Quality of Service (QoS). An essential part of the analysis was estimating the expected number of hops between nodes. To do so, we relied heavily on random graph analysis techniques described in this chapter.

We describe different classes of random graphs that P2P sensor networks can form using different connection strategies. Analysis of the random graph models provides insights for which strategy to use. The probabilistic connectivity matrices described here provide a uniform representation of the random graph classes considered and explain many of their more subtle properties. A method is given for identifying phase changes in system behavior and associated critical points. An algorithm is given for creating scale-free graphs with an arbitrary scaling factor. Methods are given for computing conditional probabilities in random graph systems.

29.2 Graph Theory

Traditionally, the tuple [V, E] defines a graph. V(E) is a set of vertices (edges). Each edge e is defined by (i, j) where i and j are the two vertices connected by e. Unless specified otherwise, we will consider only undirected graphs where $(i, j) = (j, i)$. Directed graphs (di-graphs) exist where $(i, j) \neq (j, i)$. An edge (i, j) is incident on vertices i and j. We do not consider multi-graphs where multiple edges connect the same end-points. The terms vertex and node will be used interchangeably. Similarly, edge and link are used synonymously.

The degree of a node is the number of edges incident on the node. For directed graphs, there are also the concepts of in-degree (out-degree) for the number of edges leaving (joining) the node. A graph is called regular, when all nodes in the graph have the same degree.

Many data structures can be used as practical graph representations. Common representations can be found in [Aho 74]. For example, a graph where each node has at least one incident edge can be fully represented by its list of edges. Another common representation of a graph, which we explore in depth, is the connectivity matrix. Connectivity matrix M is a square matrix where each element $m(i, j)$ is 1(0) if there is (not) an edge connecting vertices i and j. For undirected graphs this matrix is symmetric. Figure 29.1 shows a simple graph and its associated connectivity matrix.

As a matter of convention, the connectivity matrix diagonal consists of either zero's or one's. One's are frequently used based on the simple assertion that each vertex is connected to itself. We use the convention where the diagonal is filled with zeros. Justification is provided later.

A *walk* of length n is a set of edges expressed as an ordered list of n edges $((i_0, j_0),(i_1, j_1), \ldots, (i_n, j_n))$, where each vertex j_a is the same as vertex i_{a+1}. A *path* of length n is a walk where all i_a are unique. If j_n is the same as i_0, the path forms a *cycle*.

A connected component is a set of vertices where there is a path between every two vertices in the component. The graph in Figure 29.1 has two connected components. In the case of di-graphs, this would be called a fully connected component. A complete graph has an edge directly connecting any two vertices in the graph. A complete subgraph is a subset of vertices in the graph with edges directly connecting any two members of the set.

We use the following property of connectivity matrices: element $m^k(i, j)$ of the power k of graph G's connectivity matrix M (i.e., M^k) is the number of walks of length k from vertex i to vertex j on G [Cvetkovic 79]. This can be verified using the definition of matrix multiplication and the definition of the connectivity matrix. The computation of each element of $m^k(i, j)$ checks for the existence of a two hop path from i to j via any intermediate node.

You can compute successive powers k of M until $M^k = M^{k-1}$ to find the connected components of a graph. The resulting matrix will contain disjoint equivalence classes. Any zero element $m^k(i, j)$ in M^k signifies that elements i and j belong to different connected components (equivalence classes).

The diagonal of zero convention reduces the influence of cycles when computing powers of M. For example, we may look for the existence of a path of length 3 between nodes j and h. This should not include the walk (j, h), (h, j), (h, j). Maintaining a diagonal of zero in successive values of M^k does this.

We also present probabilistic connectivity matrices for random graph classes. These matrices replace the binary values for connectivity matrix elements $m(k, j)$ with the associated probabilities of an edge

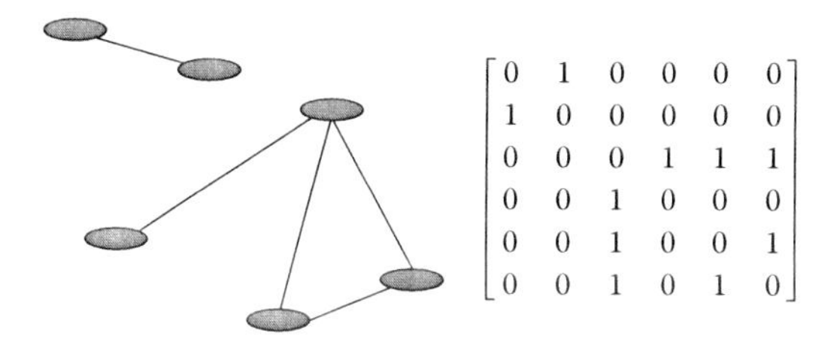

FIGURE 29.1 On the right is a graph of six nodes numbered from top to bottom. On the left is its associated connectivity matrix.

existing between nodes k and j. In which case, graph instances can be constructed by traversing the upper triangular half of the probabilistic connectivity matrix L and comparing the element value $l_{i,j}$ to a uniform random variable r in the range zero to one. If $l_{i,j} < r$, then an edge is inserted between nodes i and j. If not, none exists. A non-probabilistic connectivity matrix is made by setting $l_{i,j}$ to one in the first case, else it is zero. For the Erdös-Rényi and small-world graphs, this method is essentially identical to current graph construction methods. For scale-free graphs, this technique is new.

29.2.1 Erdös-Rényi Graphs

The first model we discuss in detail is the Erdös-Rényi random graph [Bollobás 01]. These graphs are defined by the number of nodes n and a uniform probability p of an edge existing between any two nodes. Let's use E for $|E|$ the number of edges in the graph. Since the degree of a node is the result of multiple Bernoulli trials with probability p, the degree of an Erdös-Rényi random graph follows a Bernoulli distribution. As the graph scales, the number of nodes n approaches infinity, and the degree distribution asymptotically converges to a Poisson distribution (Figure 29.2).

The expected number of hops between any two nodes in this graph grows proportionally to the log of the number of nodes [Albert 01]. Note that Erdös-Rényi graphs do not necessarily form a single connected component. When $E - n/2 << -n^{2/3}$ the graph is sub-critical and almost certainly not connected. A phase change occurs in the critical phase where $E = n/2 + O(n^{2/3})$ and in the supercritical phase where $E - n/2 >> -n^{2/3}$ a single giant component becomes almost certain. When $E = n \log n/2 + O_p(n)$ the graph is fully connected. [Janson 00] The expected number of edges for an Erdös-Rényi graph is $n(n - 1)\, p/2$).

To construct a probabilistic connectivity matrix for this graph, create an n-by-n matrix with all elements on the diagonal set to zero and all the other elements set to p. If n is 3 and p is 0.25, we get:

$$\begin{bmatrix} 0 & 0.25 & 0.25 \\ 0.25 & 0 & 0.25 \\ 0.25 & 0.25 & 0 \end{bmatrix} \tag{29.1}$$

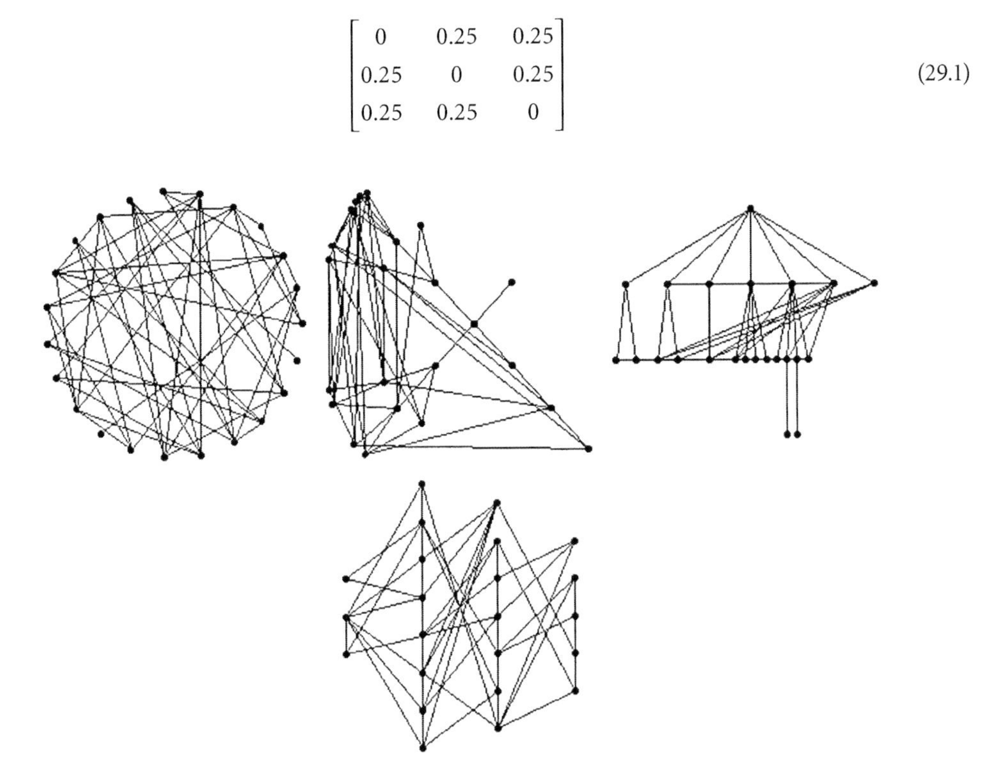

FIGURE 29.2 Example Erdös-Rényi graphs with n equal to 23 nodes and the probability p equal to 0.2. Clockwise from upper left: nodes in a circle, radial embedding, ranked embedding by geodesic distance from three nodes chosen at random, and rooted embedding from a random node.

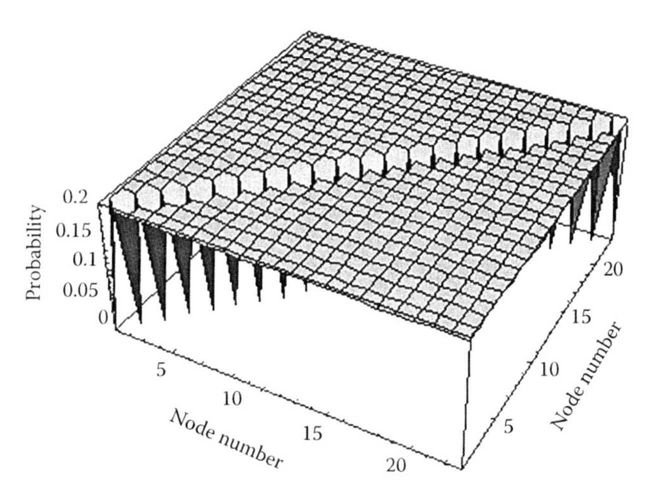

FIGURE 29.3 A three-dimensional plot of the probabilistic connectivity matrix for Erdös-Rényi graphs with $n = 23$ and $p = 0.2$. Diagonal values are zero. All other edges have the same probability.

These graphs are given for two reasons: (1) they are the best-known random graph class with many established properties and (2) they are simple and have tutorial value (Figure 29.3).

29.2.2 Small World

The second graph class considered is the family of "small world" graphs introduced in [Watts 99]. They have two characteristics: (1) a small expected value for the number of hops between any two nodes, and (2) significant clustering among the nodes. The rate of increase of the number of hops between nodes for these graphs is roughly equivalent to that of Erdös-Rényi graphs [Watts 99].

For example small-world (SW) graphs, we use the *connected caveman model* from [Watts 99]. To construct connected caveman graphs, use this procedure:

- Construct a set of complete sub-graphs.
- In each sub-graph, one edge is removed and replaced with an edge connecting the sub-graph to the next one; forming a cycle reminiscent of a "string of pearls."
- A fixed number of edges in the system are replaced with random edges.

[Watts 99] also explores other small world graph examples. The ideas presented here are also applicable to those examples. The connected caveman SW model has three parameters: (1) n the number of nodes in the graph, (2) c the number of nodes in each sub-graph, and (3) e the number of edges rewired.

The node degree distribution in this model is nearly constant. The mean node degree value is c (the number of nodes in the complete subgraph) minus one. Variance around the mean is caused by two sets of Bernoulli trials: (1) the probability an edge connected to the vertex is chosen to create the "string of pearls" raising (or reducing) the node degree by one, and (2) the likelihood of an edge attached to the vertex being re-wired.

This graph structure is discussed because: (1) the small expected number of hops between nodes is attractive for sensor network applications, (2) it is a structure that can be easily maintained as a distributed system, and (3) it has a partially deterministic structure. The small world connectivity matrix is constructed by essentially using the algorithm that creates a graph instance. For the connected caveman model example:

- Create an n by n matrix.
- Populate the diagonal of the matrix with c-by-c blocks of value one. If n is not a multiple of c, the last block will be n mod c by n mod c. This matrix is block diagonal.
- Set diagonal values to zero.

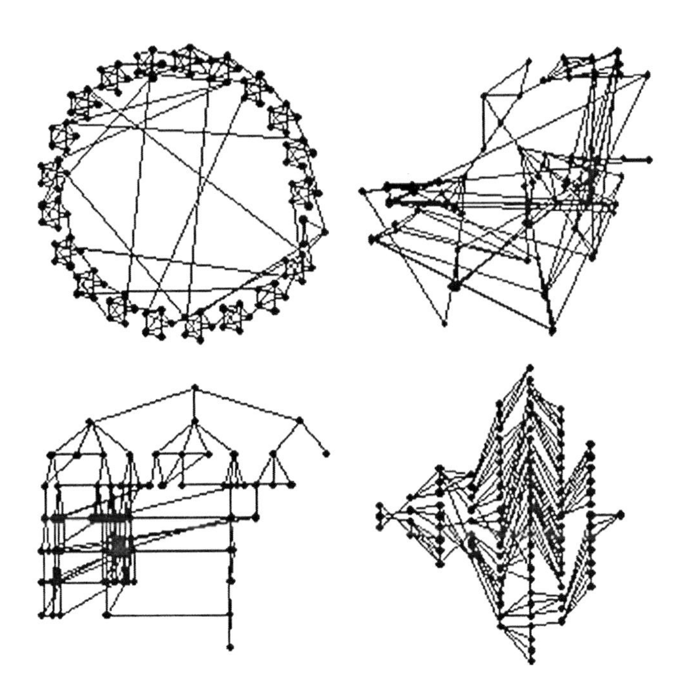

FIGURE 29.4 Example connected caveman graphs with 103 nodes (n) starting from connected subgraphs of 5 (c) members each. A total of 22 (e) edges were rewired at random. Clockwise from upper left: "string of pearls," ranked embedding by geodesic distance from nodes 101, 102, and 103, and radial embedding, rooted embedding from node 103.

- Connect the fully connected components. For all blocks: Block starting address j and last address $k = j + c$. Set elements $(k-1, k)$ and $(k, k-1)$ to zero. Set elements $(k+1, k)$ and $(k, k+1)$ to zero. Set element $(n-1, n)$ and $(n, n-1)$ to zero, and $(1, n)$ and $(n, 1)$ to one.
- Count the zeroes and ones in the matrix excluding the diagonal. The probability of gaining a connection by the rewiring step becomes $2e/$(number of zeroes). The probability of losing a connection by the rewiring step becomes $2e/$(number of ones).
- For all elements of the matrix except diagonals (which remain zero), if the element is one (zero) subtract (add) the probability of losing (gaining) a connection (Figure 29.4).

The resulting matrix expresses the probabilities of edges existing in the connected caveman model. For other examples, similar algorithms can easily be constructed. For our example, the matrix for ($n = 6$, $c = 3$, $e = 1$) is

$$\begin{bmatrix} 0 & 5/6 & 5/6 & 1/9 & 1/9 & 5/6 \\ 5/6 & 0 & 1/9 & 1/9 & 1/9 & 1/9 \\ 5/6 & 1/9 & 0 & 5/6 & 1/9 & 1/9 \\ 1/9 & 1/9 & 5/6 & 0 & 5/6 & 5/6 \\ 1/9 & 1/9 & 1/9 & 5/6 & 0 & 1/9 \\ 5/6 & 1/9 & 1/9 & 5/6 & 1/9 & 0 \end{bmatrix} \tag{29.2}$$

Figure 29.5 shows an example graph generated using the probabilities in matrices 2 and 3.

It is possible to modify step four of the probabilistic connectivity matrix generation procedure to choose block elements at random. Among other things, the matrix created in this manner is regular in

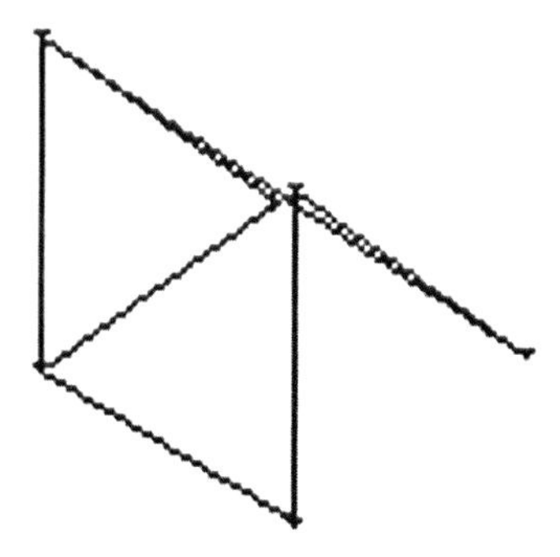

FIGURE 29.5 Connected caveman graph with $n = 6$, $c = 3$, and $e = 1$.

a certain sense. We call graphs with this property *rpm*-graphs (regular probability matrix). Perform the procedure above omitting step 4. At the end, modify the matrix as follows:

- All edges connecting components in the cluster have the same probability of being chosen $1/[c(c-1)]$.
- This value is subtracted from non-diagonal elements in the block representing the cluster on the diagonal.
- Each node in the current cluster has the same probability of being selected ($1/c$) for connection to the next cluster. Each node in the next cluster also is equally likely ($1/c$) to receive an edge.
- To represent this, each element potentially connecting the two clusters has a probability of $1/c^2$ added to it.

Note that when n is not zero modulo c, the block size of the final subgraph is not c but n modulo c. For the final block, the values above become $1/[c'(c'-1)]$ and $1/(c'c)$, where $c' = n$ modulo c. Notice that since each row is a permutation of the same probabilities, both this description and the Erdös-Rényi description are regular. The regular matrix description for ($n = 6$, $c = 3$, $e = 1$) is

$$\begin{bmatrix} 0 & 2/3 & 2/3 & 2/9 & 2/9 & 2/9 \\ 2/3 & 0 & 2/3 & 2/9 & 2/9 & 2/9 \\ 2/3 & 2/3 & 0 & 2/9 & 2/9 & 2/9 \\ 2/9 & 2/9 & 2/9 & 0 & 2/3 & 2/3 \\ 2/9 & 2/9 & 2/9 & 2/3 & 0 & 2/3 \\ 2/9 & 2/9 & 2/9 & 2/3 & 2/3 & 0 \end{bmatrix} \tag{29.3}$$

Note that the description of *rpm*-graphs is regular because each node has an equivalent probability density function and each row of the connectivity matrix is a permutation of every other row. The degree of each node thus has the same expected value. This does not mean that all nodes in a given instance of the *rpm*-graph will have the same degree (Figure 29.6).

29.2.3 Scale-Free

The third graph class is the scale-free model. It comes from empirical analysis of real-world systems, such as e-mail traffic, the World Wide Web, and disease propagation [Albert 01]. In scale-free graphs the node degree distribution varies as an inverse power law (i.e., $P[d] \propto d^{-\gamma}$). They are called scale-free because the power law structure implies nodes existing with non-zero probability at all possible scales. The expected number of hops for scale-free networks is smaller than the expected number of hops for Erdös-Rényi and small-world graphs [Albert 01]. Scale-free (SF) graphs are

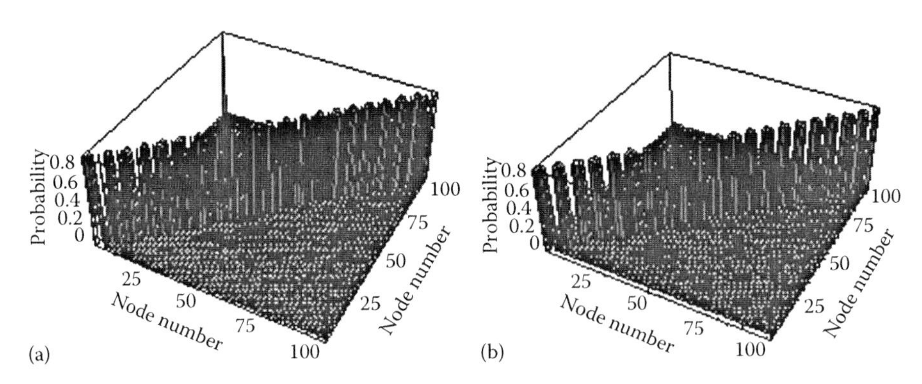

FIGURE 29.6 Three-dimensional plot of matrices for connected caveman model with $n = 103$, $c = 5$, and $e = 22$ (a) first method given (b) regular matrix. Note how clear the clustering is and how low the probability of other connections is.

defined by two parameters: number of nodes n, and scaling factor γ. Empirical analysis done by different research groups at different times find the Internet's γ parameter value ranging from 2.1 to 2.5 [Albert 01]). Of the random graph classes discussed, node degree variance in this class is the largest.

Scale-free networks are discussed for several reasons: (1) they appear in existing systems like the Internet and are of use for large-scale wired networks, (2) the small number of expected hops between nodes is attractive, (3) studies indicate that their structure has unique dependability properties [Albert 00], and (4) epidemiological studies using this model indicate that there may be parallels between biological pathogen propagation and computer virus propagation [Storras 01]. An algorithm constructing these graphs using positive feedback and producing graphs with $\gamma \approx 3$, can be found in [Barabási 99]. Barabási's use of positive feedback plausibly explains how SF systems emerge and why they are wide-spread.

Figure 29.7 illustrates how scale-free graphs differ from small world and Erdös-Rényi graphs. The majority of nodes have degree one or two, but there exists a small number of hub nodes with a very large degree. Erdös-Rényi and small world graphs have an almost flat architecture with node degree clustered about the mean. The hub nodes dominate the topology of the scale free graphs. The ranked embedding illustrates that it is extremely unlikely that a node would be many hops away from a major hub.

Creating a probabilistic connectivity matrix for scale-free graphs is challenging. Scale-free graphs are characterized by n the number of nodes and γ the scaling factor. The first step is to compute the probability distribution for node degree d. Remember $P[d] \propto d^{-\gamma}$. We compute the probability distribution, by finding a constant factor for all probabilities to sum to 1. Set

$$P[d] = bd^{-\gamma} \tag{29.4}$$

Since node degree ranges from 1 to $n - 1$,

$$1 = \sum_{d=1}^{n-1} bd^{-\gamma} \tag{29.5}$$

thus,

$$b = \frac{1}{\sum_{d=1}^{n-1} d^{-\gamma}} \tag{29.6}$$

We now have a closed form solution for the node degree probability distribution.

The next step is to determine how many edges are incident on each node. Construct a vector v of $n - 1$ elements, whose values range from 0 to 1. Each element k of the vector contains the value

$$v[k] = \sum_{d=1}^{k-1} bd^{-\gamma} \tag{29.7}$$

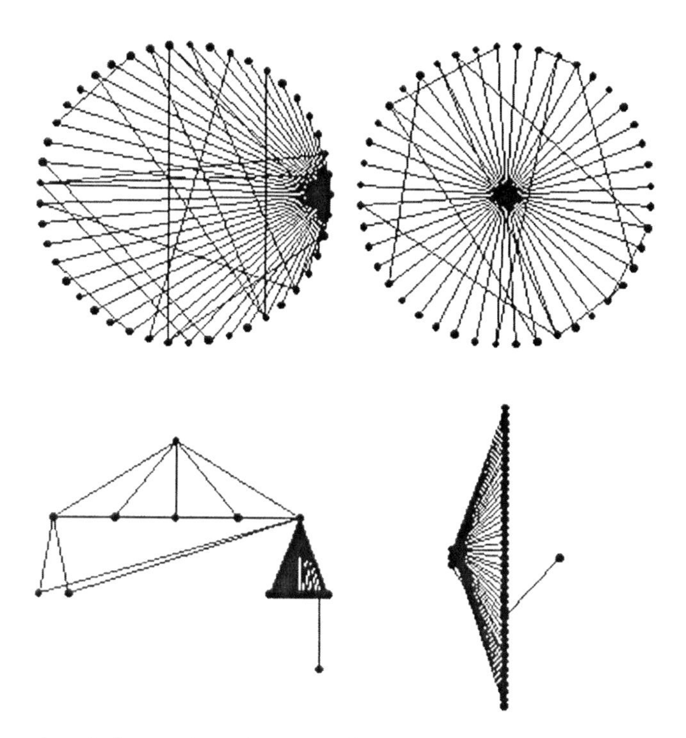

FIGURE 29.7 Example scale-free graphs with $n = 45$ and $\gamma = 3.0$. Clockwise from top left: nodes in a circle, radial embedding, ranked embedding in order of the geodesic distance from the three largest hubs, and rooted embedding with the root set as the second largest hub.

Vector element $v[0]$ has the value zero and element $v[n - 1]$ has the value one. Each element represents the probability of a node existing of degree less than or equal to k.

Each row of the probabilistic connectivity matrix represents the expected behavior of $1/n$th of the nodes of the class under consideration. We now construct a vector v' of n elements, the value of $v'[k]$ states how many edges are incident on node k. Set $v'[k]$ to the index of the largest element of v whose value is less than or equal to k/n.*

The elements of the connectivity matrix are probabilities of connections between individual nodes. These values are computed using the insight from [Barabási 99] that scale-free networks result from positive feedback. Nodes are more likely to connect to other nodes with many connections. The value of each matrix element (k, i) is therefore

$$P[k,i] = \frac{v''[i]v'[k]}{\sum_{m \neq k} v'[m]} \qquad (29.8)$$

The likelihood of choosing another node i to receive a given edge from the current node k is the degree of i divided by the sum of the degrees of all nodes except k. Summing these factors would give a total probability of one for the row. Since k has degree $v'[k]$ these probabilities are multiplied by $v'[k]$, so that the total of the probabilities for the row is k. This finishes the derivation of Equation 29.8.

We modify the values of Equation 29.8 in two ways. Since node degrees have an exponential distribution, the values of the bottom rows are often much larger than the other degrees. The result of (29.8) for

* It would also be possible to use the mean, or a weighted average of the index values that point to elements in the range $(k - 1)/n$ to k/n. Since Equation 29.8 can give values greater than 1, constraining matrix element values to the range $[0, \ldots, 1]$ flattens the degree distribution. Using the maximum index value counteracts this tendency.

values of k and l close to n can be greater than one. To avoid having elements of the matrix with values greater than one (i.e., probability greater than one), we compute the matrix elements in a double loop starting with k (outer loop) and i (inner loop) set to $n - 1$. The values of k and i are decremented from $n - 1$ to zero. If the value of Equation 29.8 is greater than one then the corresponding element is set to one and the value copied from $v'[k]$ for computing row k is decremented by one. This keeps all matrix elements in the range zero to one, so that they represent probabilities.

The other modification of element values that deviates from (29.8) forces the matrix to be symmetric. When computing a row k and $k < n - 1$, all elements for $i > k$ are set to be the same as the values computed for element (i, k). If the value of element (i, k) is one, the value copied from $v'[k]$ is again decremented. In some cases this may force the sum of row k to deviate from $v'[k]$. If the deviation is significant enough, the resulting connectivity matrix may only have a degree distribution that approximates the scaling factor γ. An example connectivity matrix for $n = 10$ and $\gamma = 2.0$ is (Figure 29.8):

$$\begin{bmatrix}
0 & 1/22 & 1/22 & 1/22 & 1/22 & 1/22 & 1/10 & 1/10 & 2/9 & 9/10 \\
1/22 & 0 & 1/22 & 1/22 & 1/22 & 1/22 & 1/10 & 1/10 & 2/9 & 9/10 \\
1/22 & 1/22 & 0 & 1/22 & 1/22 & 1/22 & 1/10 & 1/10 & 2/9 & 9/10 \\
1/22 & 1/22 & 1/22 & 0 & 1/22 & 1/22 & 1/10 & 1/10 & 2/9 & 9/10 \\
1/22 & 1/22 & 1/22 & 1/22 & 0 & 1/22 & 1/10 & 1/10 & 2/9 & 9/10 \\
1/22 & 1/22 & 1/22 & 1/22 & 1/22 & 0 & 1/10 & 1/10 & 2/9 & 9/10 \\
1/10 & 1/10 & 1/10 & 1/10 & 1/10 & 1/10 & 0 & 1/5 & 4/9 & 1 \\
1/10 & 1/10 & 1/10 & 1/10 & 1/10 & 1/10 & 1/5 & 0 & 4/9 & 1 \\
2/9 & 2/9 & 2/9 & 2/9 & 2/9 & 2/9 & 4/9 & 4/9 & 0 & 1 \\
9/10 & 9/10 & 9/10 & 9/10 & 9/10 & 9/10 & 1 & 1 & 1 & 0
\end{bmatrix} \tag{29.9}$$

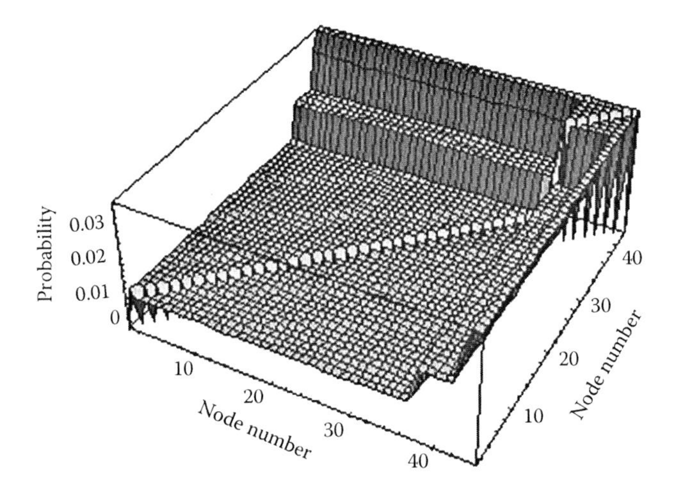

FIGURE 29.8 Three-dimensional plot of the connectivity matrix for a scale free graph with $n = 45$ and $\gamma = 3.0$. Note the zero diagonal and the high probability of connections to the hub nodes. Connections between hub nodes are virtually assured. Connections between non-hub nodes are very improbable.

29.2.4 Percolation Theory

Percolation theory studies flow through random media. Commonly, the random media is modeled as a regular tessellation of a d dimensional space where vertices are points connecting edges. It is also possible to consider arbitrary graphs. Two different models exist; site (bond) percolation is where vertices (edges) are either occupied or empty. We discuss only bond percolation, note that it is possible to create dual graphs to convert site (bond) percolation problems to bond (site) problems. These graphs can be considered models of wireless networks with a cellular infrastructure.

As an example, we construct probabilistic transition matrices for a rectangular tessellation of a two-dimensional space. This model requires three parameters: x the number of nodes in a row, y the number of nodes in a column, and p the probability of an edge being occupied. Note that n, the total number of nodes, is equal to xy. The matrix construction method is only valid for finite problems. Once the matrix has been constructed, however, scaling analysis can be performed to consider infinite ranges. Figure 29.9 shows an example graph.

Excluding edge effects in this tessellation, each node (i, j) has four immediate neighbors: $(i + 1, j)$, $(i, j + 1)$, $(i - 1, j)$, and $(i, j - 1)$. Each vertex is assigned the unique row position $(i + (j - 1)\, y)$ in the connectivity matrix. (This assumes that i (j) ranges from 0 to x (y) and makes the matrix row major. Readers that are dogmatic about C or FORTRAN can change these conventions at will [Press 1992].) Vertices outside the range $([1, \ldots, x], [1, \ldots, y])$ are ignored, since they are out of bounds. In the connectivity matrix all positions are set to 0, except that for each node (i, j) the positions in its row $(i + (j - 1)\, y)$ that correspond to its neighbors:

$$i-1+jy \quad i+1+jy \quad i+(j-1)y \quad i+(j+1)y \tag{29.10}$$

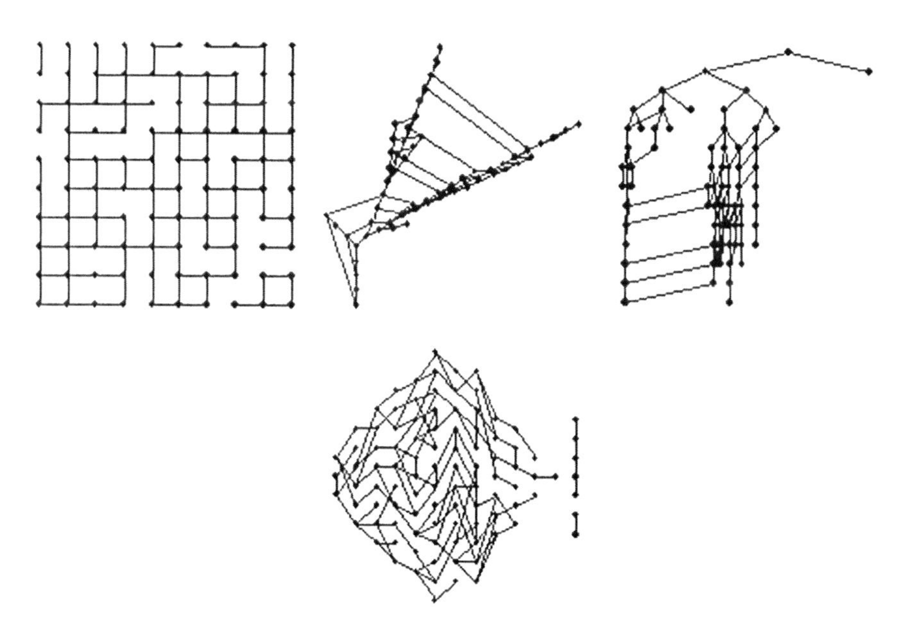

FIGURE 29.9 Different embeddings of a regular 10 by 10 matrix. Edge probability was set at 0.75. Top row from left to right: grid, radial embedding, and rooted embedding with node 50 as the root. Bottom row: ranked embedding from nodes 38, 39, and 40.

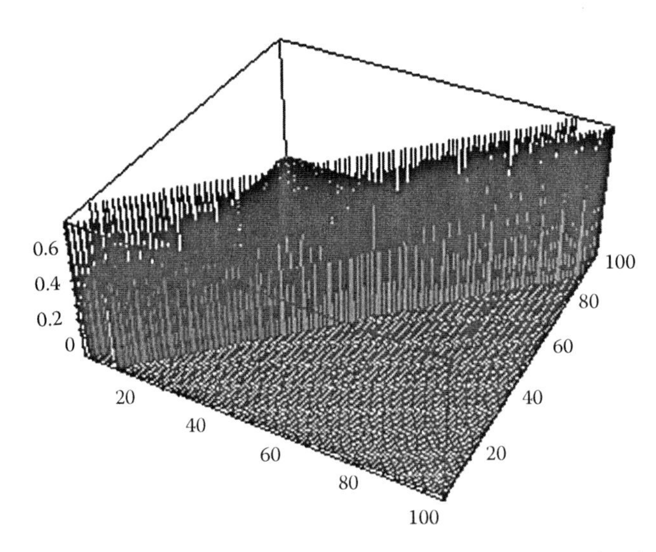

FIGURE 29.10 Three-dimensional plot of the connectivity matrix for a 10 by 10 grid with $n = 0.75$. It is a band-diagonal matrix.

are set to p. The matrix corresponding to a 3 by 3 grid with probability of 0.75 is (Figure 29.10)

$$\begin{bmatrix} 0 & 0.75 & 0 & 0.75 & 0 & 0 & 0 & 0 & 0 \\ 0.75 & 0 & 0.75 & 0 & 0.75 & 0 & 0 & 0 & 0 \\ 0 & 0.75 & 0 & 0 & 0 & 0.75 & 0 & 0 & 0 \\ 0.75 & 0 & 0 & 0 & 0.75 & 0 & 0.75 & 0 & 0 \\ 0 & 0.75 & 0 & 0.75 & 0 & 0.75 & 0 & 0.75 & 0 \\ 0 & 0 & 0.75 & 0 & 0.75 & 0 & 0 & 0 & 0.75 \\ 0 & 0 & 0 & 0.75 & 0 & 0 & 0 & 0.75 & 0 \\ 0 & 0 & 0 & 0 & 0.75 & 0 & 0.75 & 0 & 0.75 \\ 0 & 0 & 0 & 0 & 0 & 0.75 & 0 & 0.75 & 0 \end{bmatrix} \quad (29.11)$$

29.2.5 Ad Hoc Wireless

Scale-free networks provide good statistical descriptions of large, evolving, wired networks with no centralized control. Wireless networks are also of importance. In particular ad hoc wireless networks, which have no fixed infrastructure, are suited to analysis as a type of random graph. [Krishnamachari 01] explains a fixed radius model for random graphs used to analyze phase change problems in ad hoc network design.

The model in [Krishnamachari 01] places nodes at random in a limited two-dimensional region. Two uniform random variables provide a node's x and y coordinates. Two nodes in proximity to each other have a very high probability of being able to communicate. For this reason, they calculate the distance r between all pairs of nodes. If r is less than a given threshold, then an edge exists between the two nodes. In their work, many similarities are found between this graph class and the graphs studied by Erdös and Rényi. Their analysis looks at finding phase transitions for constraint satisfaction problems. These graphs differ from Erdös-Rényi graphs in that they have significant clustering, like the small world graph class. We will use the model from [Krishnamachari 01], except that they create an edge with probability one when the distance between two nodes is less than the threshold value. We will allow the probability to be set to any value in the range [0, …, 1]. Figure 29.11 shows an example range limited random graph.

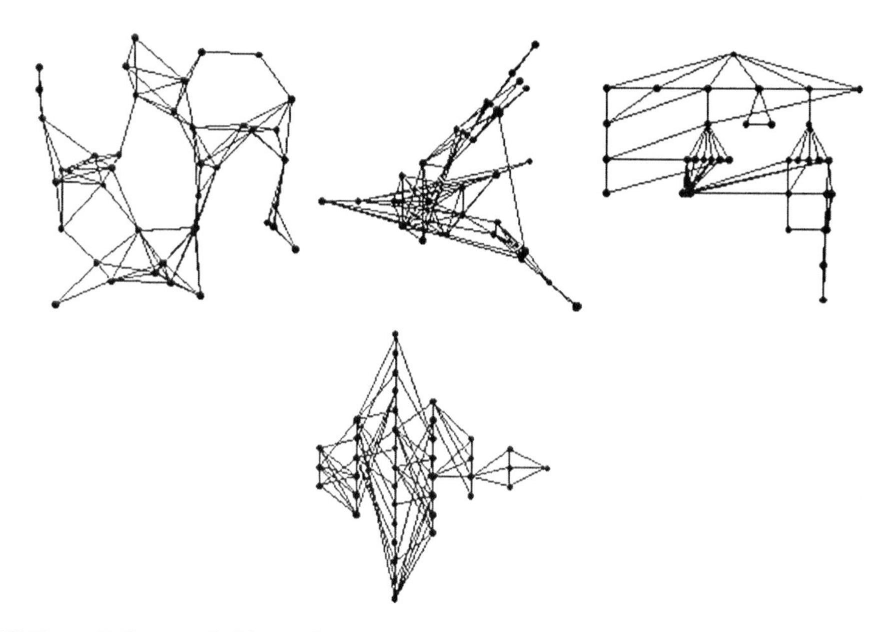

FIGURE 29.11 Different embeddings of a range limited random graph of 40 nodes positioned at random in a unit square region. The distance threshold was set as 0.25, and within that range edges exist with a probability of 1. Clockwise from upper left: geographic locations, radial embedding, ranked embedding from nodes 38, 39, and 40, and rooted embedding with node 40 as the root.

We construct range-limited graphs from the following parameters:

- *n*—the number of nodes
- *max_x (max_y)*—the size of the region in the *x* (*y*) direction
- *r*—the maximum distance between nodes where connections are possible
- *p*—probability that an edge exists connecting two nodes within the range

Construction of range-limited random graphs proceeds in two steps: (1) sort the nodes by either their *x* (or possibly *y*) coordinate and use order statistics to find the expected values of that coordinate, (2) determine probabilities for edges existing between nodes based on these expected values.

To construct the connectivity matrix for range-limited graphs, we consider the position of each node as a point defined by two random variables: the *x* and *y* location. Without loss of generality, we use normalized values for the *x*, *y*, and *r* variables limiting their range to [0, …, 1]. To calculate probabilities, we sort each point by its *x* variable. For the *n* nodes, rank statistics provide expected value $j/(n + 1)$ for the node in position *j* in the sorted list. Using Euclidean distance, an edge exists between two nodes *j* and *k* with probability *p* when

$$(x_j - x_k)^2 + (y_j - y_k)^2 \leq r^2 \tag{29.12}$$

By entering the expected values for nodes of rank *j* and *k* and re-ordering terms, this becomes

$$(y_j - y_k)^2 \leq r^2 - \left(\frac{j}{n+1} - \frac{k}{n+1}\right)^2 \tag{29.13}$$

We assume that the random variables giving the *x* and *y* positions are uniformly distributed and uncorrelated. The probability of this occurring is the probability that the square of the difference of two normalized uniform random variables is less than the constant value *c* provided by the right hand side of (29.13).

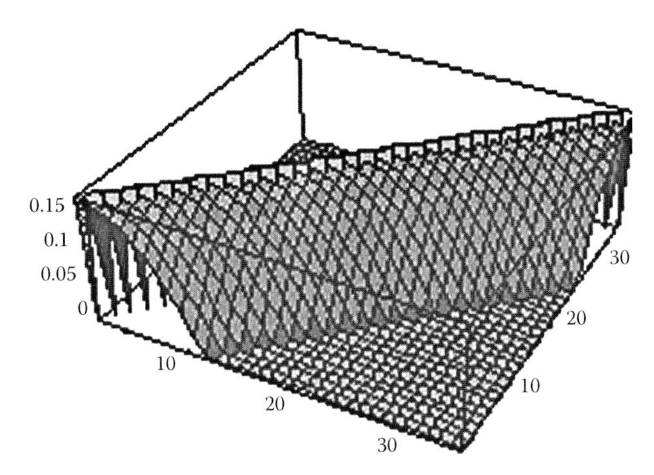

FIGURE 29.12 Three-dimensional plot of the connectivity matrix for a range limited graph of 35 nodes with range of 0.3.

Two uniform random variables describe a square region, where every point is equally likely. (29.13) is an inequality, so it defines a closed linear region. Because the right hand side is squared, two symmetric regions are excluded from the probability. The limiting points are when y_j or y_k are equal to the constant on the left hand side of (29.13). Algebraic manipulation provides the equation $2c - c^2$ for the probability of (29.13) occurring. An example matrix for six nodes in a unit square with $r = 0.3$ and $p = 1.0$ is

$$\begin{bmatrix} 0 & 0.134 & 0.0167 & 0 & 0 & 0 \\ 0.134 & 0 & 0.134 & 0.0167 & 0 & 0 \\ 0.0167 & 0.134 & 0 & 0.134 & 0.0167 & 0 \\ 0 & 0.0167 & 0.134 & 0 & 0.134 & 0.0167 \\ 0 & 0 & 0.0167 & 0.134 & 0 & 0.134 \\ 0 & 0 & 0 & 0.0167 & 0.134 & 0 \end{bmatrix} \tag{29.14}$$

Figure 29.12 shows a three-dimensional plot of an example matrix. Figure 29.13 compares the number of edges for range-limited graphs constructed directly versus those constructed using the probabilistic connectivity matrices as a function of n and r. The approximation is achieved by this approach is good, but far from perfect.

29.2.6 Cluster Coefficient

The clustering coefficient expresses the cliquishness of the network ($C = 1$ for a complete graph, $C = 0$ for a tree). It is the percentage of nodes two hops away that are also only one hop away. This can also be expressed as the percent of nodes reached in two hops that were already reached in one hop, or the likelihood that a friend's friend is also a friend of mine. It is defined as [Newman 01] (the two definitions are equivalent)

$$C = \frac{3(\text{Number of triangles in network})}{(\text{Number of connected triples of vertices})}$$

$$C = \frac{6(\text{Number of triangles in network})}{(\text{Number of paths of length two})} \tag{29.15}$$

We developed a new algorithm for computing C based on the fact that each element $a_{i,j}$ of a graph's connectivity matrix raised to the power k contains the number of paths of length k from i to j. We square

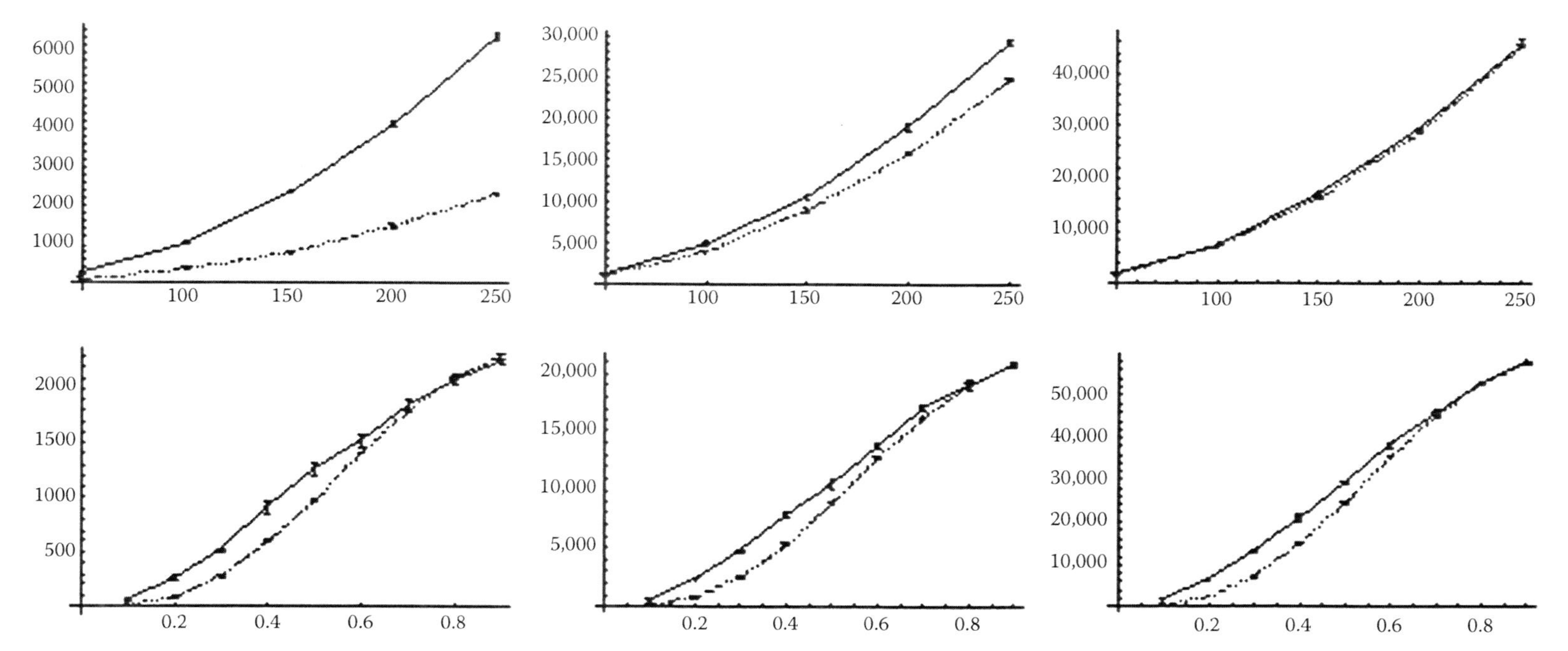

FIGURE 29.13 Number of edges (y axis) as a function of n and r (x axis). The top row varies n from 50 to 250 with r fixed. From left to right: $r = 0.2$, 0.5, and 0.7. Bottom row varies r from 0.1 to 0.9 while keeping n fixed. From left to right: 50, 150, and 250. For both rows of graphs the y-axis is the mean number of edges and the x-axis is the variable being varied. Five repetitions were made of each data point. 90% confidence intervals are shown.

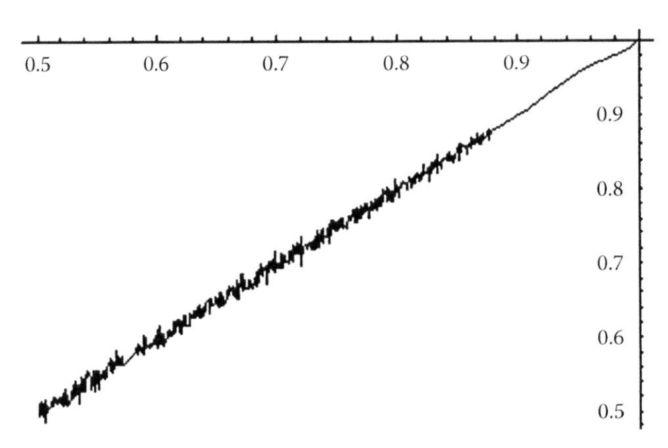

FIGURE 29.14 Cluster coefficient (y axis) versus edge probability (x axis) for Erdös-Rényi random graphs of 100 nodes. Results are similar for other graph sizes. As the graph size increases, the plot converges to a smooth curve.

the connectivity matrix M to compute M^2. We consider all non-diagonal elements of M and M^2 and compute two sums: the elements of M^2, and the elements of M^2 where the corresponding values of M are not zero. The second sum is the numerator and the first sum is the denominator of the cluster coefficient.

In [Watts 99] a different definition for cluster coefficient is given: the number of edges between nodes in the subgraph of nodes immediately adjacent to a given node are divided by the total number of edges possible given the degree of the node. The sum of these values is averaged. This definition has been deprecated [Watts 02] in favor of the definitions in [Newman 01]. The above definition is equivalent to computing this definition except that the ratio is over the entire graph instead of at each node. The approach above provides a more stable answer. The deprecated definition is

$$C = \frac{\text{(Number of edges between neighbors of the node)}}{\text{(Number of possible links in local neighborhood)}}$$

$$C = \text{Average}\left[\frac{\text{Number of edges between neighbors of the node}}{(d(d-1)/2)}\right] \tag{29.16}$$

Let's look at how cluster coefficient C relates to characteristics of random graph classes (Figures 29.14 and 29.15).

For small world graphs, we notice the same relationship between cluster coefficient and graph size growth independent of both parameter e and cluster size. Initially the graph is dominated by a single fully connected cluster, giving it a high cluster coefficient. Once multiple clusters exist in the graph, this value falls off rapidly. As the number of clusters increases the cluster coefficient asymptotically approaches a value close to the initial value. This is valid for the connected caveman model from [Watts 99]. We start our calculations with graph sizes larger than the cluster size. When graph size and the number if random rewirings are held fixed the cluster coefficient is sensitive to the cluster size. Initially, the cluster coefficient increases dramatically the value then asymptotically approaches one. As the number of randomly re-wired nodes increases the cluster coefficient decreases slowly (Figures 29.16 through 29.18).

For scale free graphs, increasing the graph size causes the cluster coefficient to decline asymptotically. The shape of the curve in Figure 29.19 leads us to believe that the Barabási algorithm for creating scale-free graphs is biased to create clusters in the initial phase of graph creation. We suspect this is an artifact of the algorithm, rather than a property of scale-free graphs. On the other hand, the average node degree appears to significantly influence the cluster coefficient. Figure 29.20 show cluster coefficient versus average node degree for scale-free graphs of size 100 (a) and 250 (b).

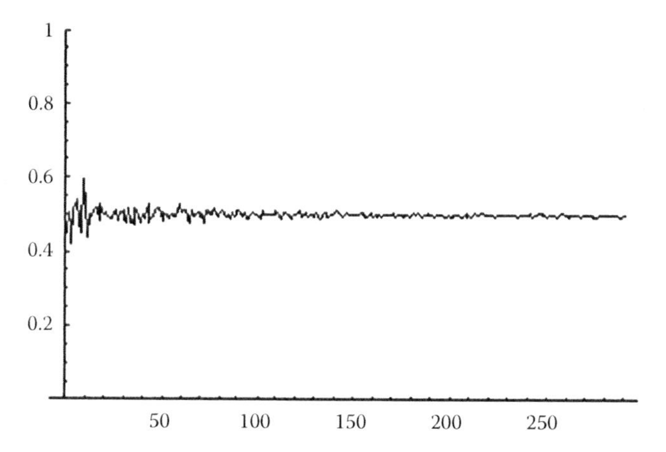

FIGURE 29.15 Graph of cluster coefficient (y axis) versus number of nodes in the graph (x axis) for Erdös-Rényi graphs, indicates that graph size does not significantly affect clustering coefficient mean value. The variance is greater at small node sizes. This effect is observable for most edge probabilities. As we have seen the edge probability strongly influences the cluster coefficient.

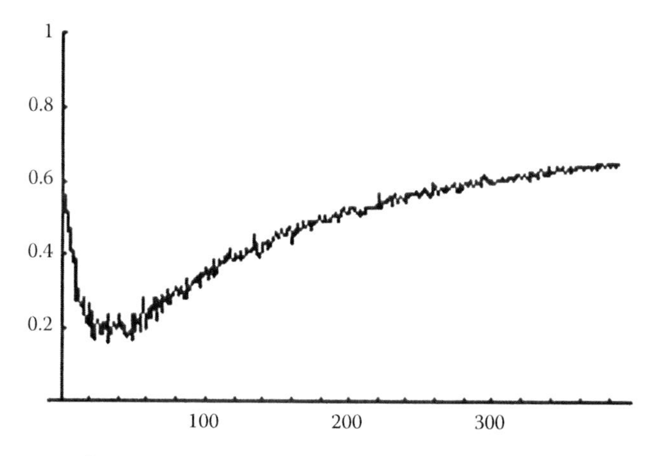

FIGURE 29.16 Cluster coefficient C (y axis) versus graph size (x axis) with fixed cluster size (7) and number of rewirings (100).

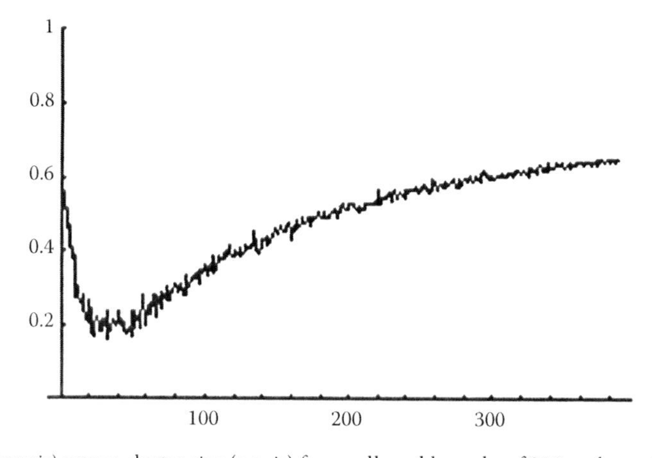

FIGURE 29.17 C (y axis) versus cluster size (x axis) for small world graphs of 200 nodes and 100 edges rewired.

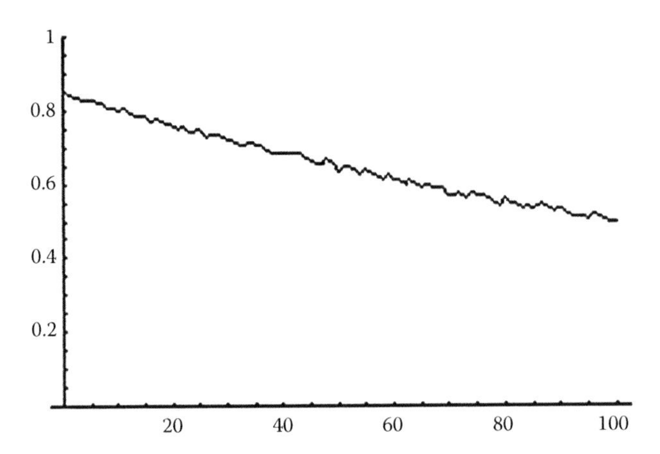

FIGURE 29.18 C (y axis) versus number of rewired nodes (x axis) for small world graphs of 200 nodes and cluster size of 7.

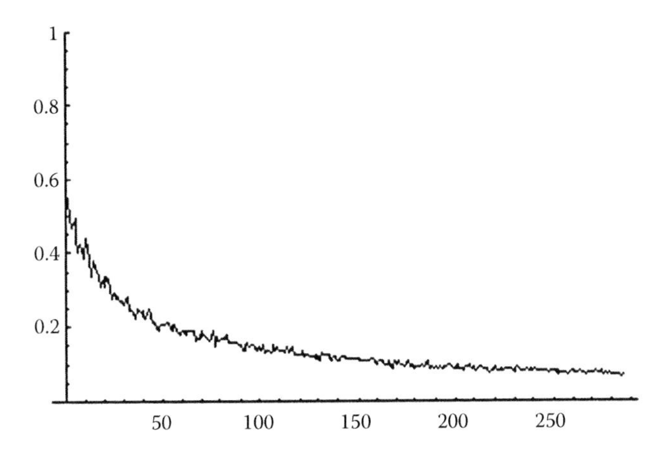

FIGURE 29.19 C (y axis) versus graph size (x axis) for scale-free graphs created using the Barabási algorithm with average node degree of 5.

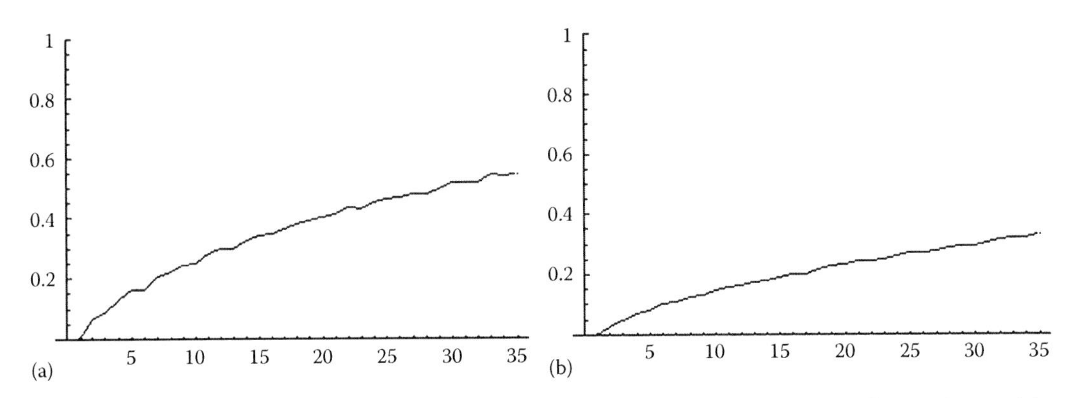

FIGURE 29.20 C (y axis) versus average node degree (x axis) for scale-free graphs created using the Barabási algorithm of 100 nodes (a) and 200 nodes (b).

29.2.7 Mutuality

[Newman 01] introduces a mutuality factor *Mu* (*M* in [Newman 01]) for the mean number of paths of length two leading to nodes two hops away:

$$Mu = \frac{mean[number\_of\_vertices\_2\_hops\_away]}{mean[paths\_of\_length\_2\_to\_those\_vertices]} = \frac{average\left[d/[1+C^2(d-1)]\right]}{d_{average}} \tag{29.17}$$

The estimator for z_2, the expected number of nodes first reachable in two hops, derived in [Newman 01] is

$$z_2 = Mu(1-C)(average[d^2] - average[d]) \tag{29.18}$$

Like *C* considering the number of triangles in the network, *Mu* looks at the number of rectangles. For a random graph

$$c = (average[d^2] - average[d])\frac{Mu}{n(n-1)/2} \tag{29.19}$$

To analyze these factors, 35 graphs in each random graph class were created. The Erdös-Rényi graphs had 100 nodes with a uniform probability of 0.04 for an edge existing between two nodes. The small world graphs had 100 nodes, connected components of 4 nodes, and 11 edges re-wired at random. The scale free graphs had 100 nodes and the average node degree set to 4. Note that all the graphs were the same size and with the same average node degree.

For Erdös-Rényi graphs, the mean value of *C* was 0.0423888 with variance 0.000164686. For the small-world graphs the mean value of *C* was 0.374865 and variance 0.000322814. For scale-free graphs the mean value of *C* was 0.127691 and variance 0.0000785896.

Mu accounts for the overcount of nodes two hops away when using *C* to estimate the number of nodes reachable by a given number of hops. We extend the mutuality concept to nodes 3 or more hops away and refer to this as $Mu[h]$. The question is whether or not values of $Mu[h]$ deviate significantly from 1. Our analysis indicates that it does.

For Erdös-Rényi graphs, average $Mu[h]$ factors are plotted in the top Figure 29.21. The error bars are $\pm 2\sigma/\sqrt{35}$. The bottom figure shows the inverse of the mean. We see that for between two and nine hops the *Mu* factor is very significant. For five hops five separate paths on the average existed to each node. The small error bars support the supposition that these values are likely to be consistent for the entire class of Erdös-Rényi graphs with these parameters.

It is interesting to note that the variance grows significantly after the value of $Mu[h]$ has peaked. This implies that the number of different paths is fairly uniform as the number of independent paths increases. When the number of paths starts to decrease the variance increases noticeably. The inference is that there are fewer paths, since most nodes have already been reached. We see later that the number of nodes reached by specific numbers of hops supports this inference. The number of paths to the remaining nodes becomes dependent on details of the graph structure of the instance and harder to predict.

For the set of small world graphs, the shape of the $Mu[h]$ function shown in Figure 29.22 is similar to the Erdös-Rényi $Mu[h]$ function. Note that the number of independent paths grows more slowly. This is tied to the relatively small number of edges that have been rewired.

As shown in Figure 29.23, $Mu[h]$ is most striking for scale-free graphs. The shape of the $Mu[h]$ function is the same as for Erdös-Rényi and small-world graphs. The number of independent paths grows more quickly than for either of the other random graph models. It is interesting to note that the variance is almost undetectable. These graphs were constructed using Barabási's algorithm where γ is

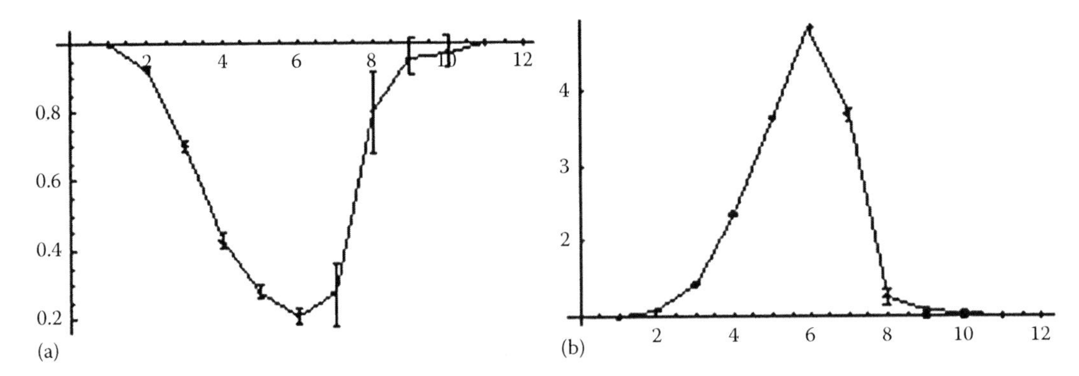

FIGURE 29.21 (a) Inverse of right graph. (b) Mean values of $Mu[h]$ (y axis) versus h (x axis) for Erdös-Rényi graphs.

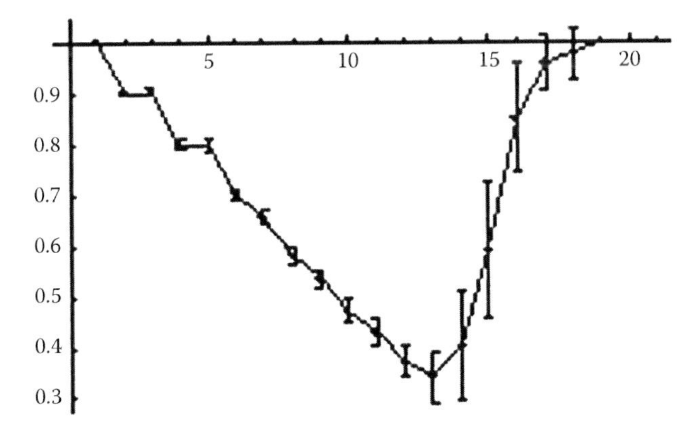

FIGURE 29.22 Mean values of $Mu[h]$ (y axis) versus h (x axis) for small world graphs.

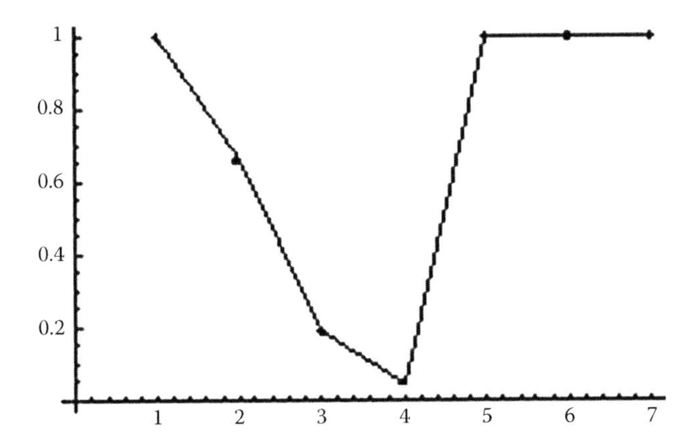

FIGURE 29.23 Mean values of $Mu[h]$ (y axis) versus h (x axis) for scale free graphs.

approximately 3. The results in Figure 29.23 are consistent with those in [Reittu 02] where the expected number of hops when $2 < \gamma < 3$ scale as log log n.

These results show that the $Mu[h]$ factor is even more significant when h is greater than 2. Aspects of this factor appear to be common to all three models. Figure 29.24 shows $Mu[h]$ for all three graph types.

We consider how $Mu[h]$ varies for the different graph classes by plotting $Mu[h]$ as parameters vary. For Erdös-Rényi graphs, the shape of the $Mu[h]$ curve is sensitive to average node degree, which is a

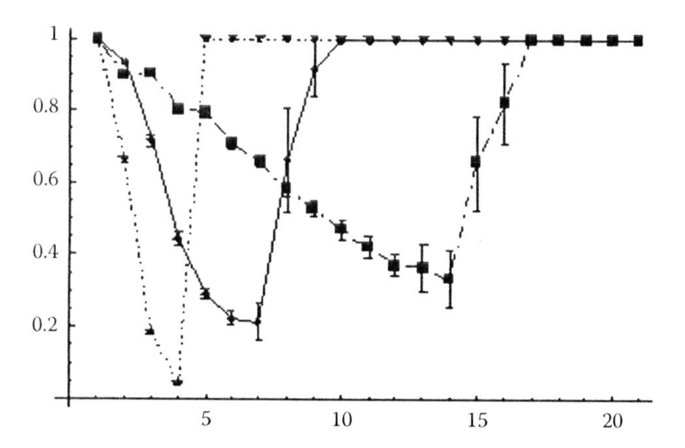

FIGURE 29.24 Mean values of *Mu*[*h*] (*y* axis) versus *h* (*x* axis) for the three graph types. From left: scale free, Erdös-Rényi, and small world.

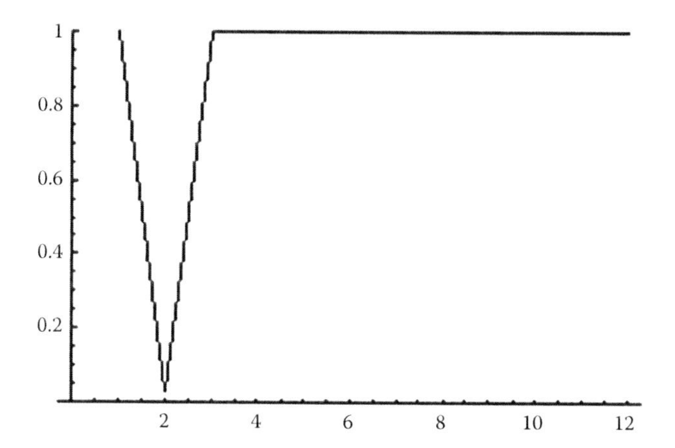

FIGURE 29.25 *Mu*[*h*] (*y* axis) versus *h* for an Erdös-Rényi graph of over 100 nodes and *p* over 20%.

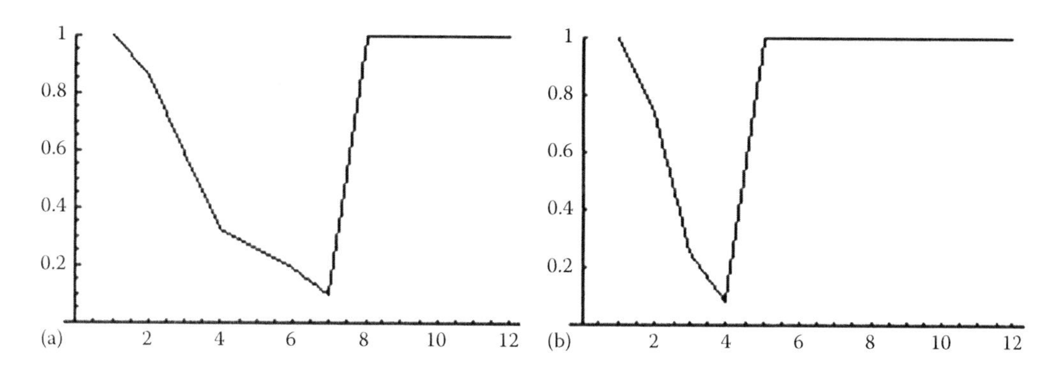

FIGURE 29.26 *Mu*[*h*] (*y* axis) versus *h* for Erdös-Rényi graphs of 100 nodes (b) and 130 nodes (a) with *p* of 10%.

product of the number of nodes in the graph times the probability of an edge between two random nodes. For graphs with more than 100 nodes and edge probability over 20%, the shape in Figure 29.25 with a large number of redundant paths of length two and no other redundant paths is typical. With graphs of 50 nodes, the same shape is typical with edge probability of 40% or higher. The shapes in Figure 29.26 occur for graphs of 100 and 130 nodes and probability of 10%. As the graph size increases, the shapes converge to the shape in Figure 29.25, with a spike at 2.

Figure 29.27 shows plots when 50 node graphs were generated. When the probability was less than 4% the plots tended to be flat lines, indicating that the graphs consisted primarily of isolated components. This was to be expected, since the expected value of node degree would be less than 2. Note how the shape of $Mu[h]$ tends toward a spike at two hops as the edge probability increases.

For small world graphs, there are three parameters to consider: size, cluster size, and number of rewired edges. In contrast to Erdös-Rényi graphs, increasing the graph size causes the shape of $Mu[h]$ to diverge. To illustrate this, look at the curves in Figure 29.28. Similar results were found when cluster size and number of re-wired edges were varied. Increasing the number of edges re-wired has the opposite effect, as is shown by Figure 29.29. Similar results were obtained for larger graph and cluster sizes.

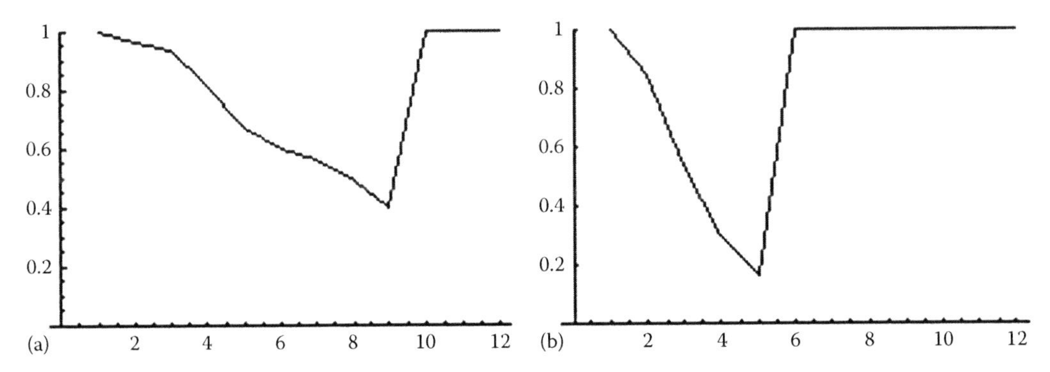

FIGURE 29.27 $Mu[h]$ (y axis) versus h for Erdös-Rényi graphs of 50 nodes with p of 4% (a) and 8% (b).

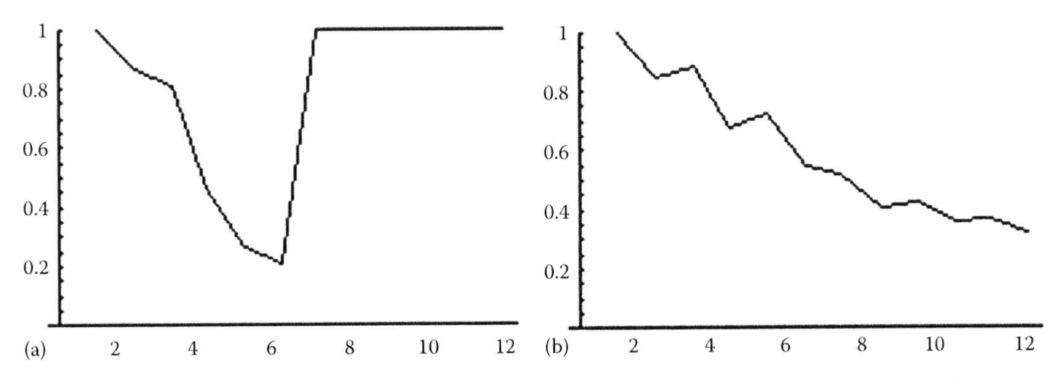

FIGURE 29.28 $Mu[h]$ (y axis) versus h for small world graphs of 35 nodes (a) and 135 nodes (b) with cluster size of 5 and 10 edges rewired.

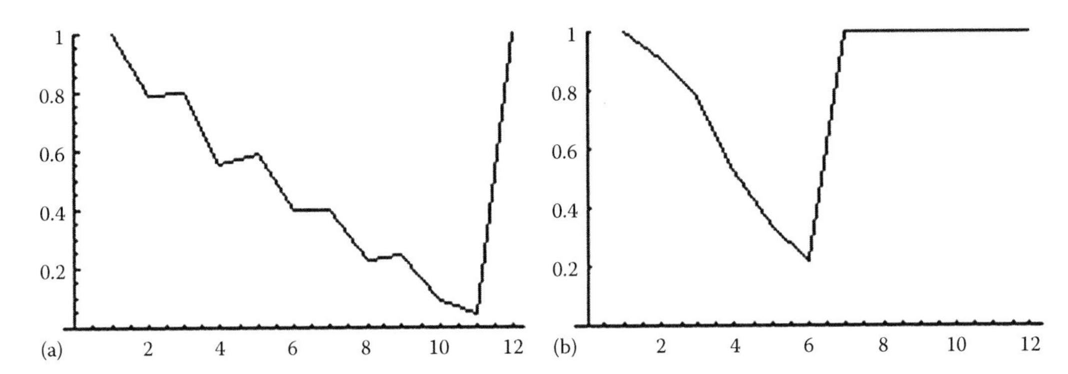

FIGURE 29.29 $Mu[h]$ (y axis) versus h for small world graphs of 50 nodes with cluster size of 5 when 4 edges (a) and 10 edges (b) were rewired.

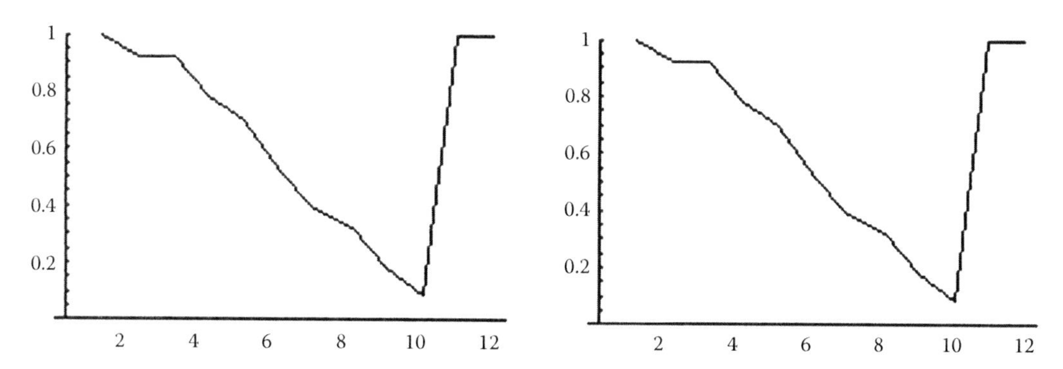

FIGURE 29.30 *Mu*[*h*] (*y* axis) versus *h* for small world graphs of 50 nodes with cluster sizes of 4 (a) and 10 (b) with 10 edges rewired.

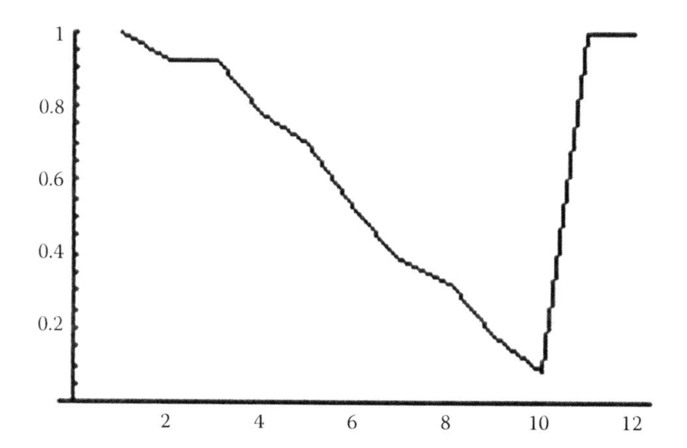

FIGURE 29.31 *Mu*[*h*] (*y* axis) versus *h* for scale free graphs of 50 nodes with average degree of 15.

Increasing graph cluster size shifts the *Mu*[*h*] values to the left, which eventually makes the spike narrower. This is illustrated by Figure 29.30. Similar results occur with other graph sizes and numbers of re-wired edges.

Scale-free graphs created using the Barabási algorithm have only two parameters, number of nodes and average node degree. The average node degree appears to cause *Mu*[*h*] to become narrower as it increases. The final result for large node degree looks similar to the Erdös-Rényi results. On the other hand, increasing graph size tends to have no major influence on *Mu*[*h*]. *Mu*[*h*] fpr scale free graphs seems to be narrower and steeper than roughly equivalent small world and Erdös-Rényi graphs (Figure 29.31).

29.2.8 Index Structure

Consider Napster and Gnutella two extremes of a range of possible P2P designs. How many indexes are desirable for these types of dynamic P2P networks? The application domain is a distributed servent sensor network using mobile code to reconfigure itself dynamically. This network is a prototype highly survivable distributed network service. Figure 29.32 gives a flowchart of the mobile code indexing system we implemented [Keiser 03].

Formally: Given a network of n nodes, how does the number of indexes affect global system QoS and dependability. QoS issues have been considered in [Kapur 02]. Here we analyze network dependability.

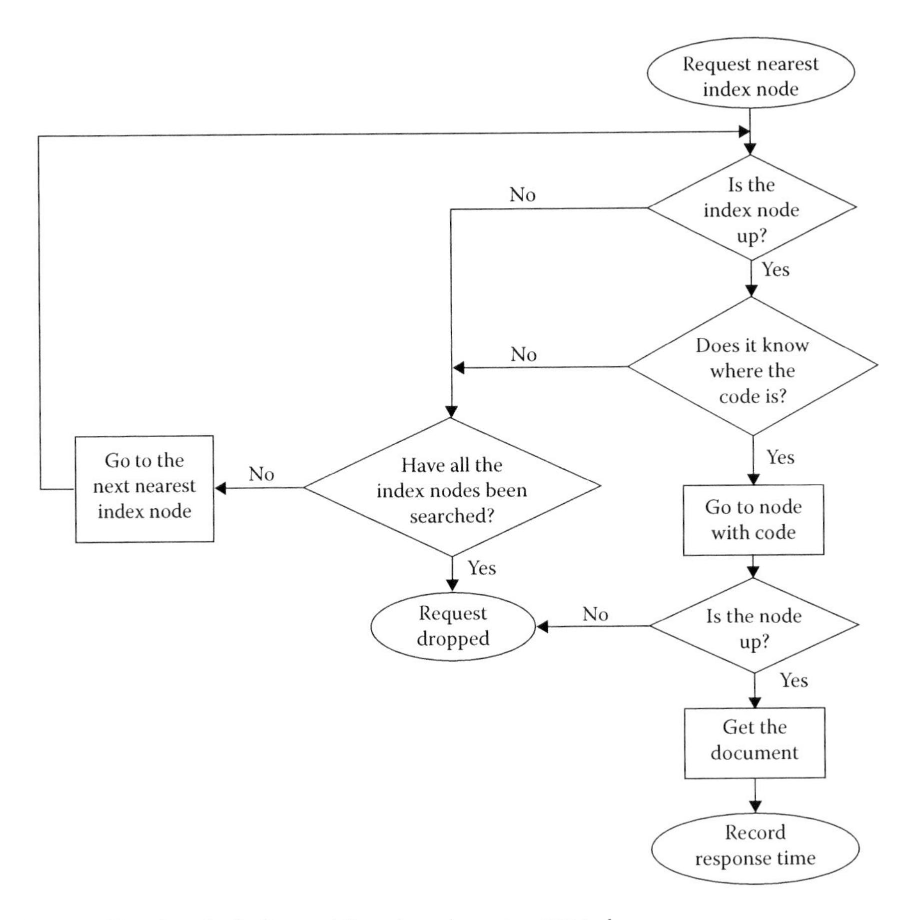

FIGURE 29.32 Flowchart for finding mobile code packages in a P2P index.

Specifically, we define dependability to be the probability that an arbitrary request can be completed. Indexes serve subgraphs of approximately equal size created from the original graph. Determining where to place indexes is equivalent to performing these two steps:

1. Perform the k-way partition problem in graph theory, where a graph is partitioned into k (in this case 1) partitions of equal size with a minimal number of connections between the partitions. The problem is NP-complete. Many heuristics have been developed to solve this problem for VLSI layout. Known approaches include self-organizing maps and the use of eigenvalues of the connectivity matrix.
2. Place the index at the centroid of each partition.

29.3 Probabilistic Matrix Characteristics

The applications we have found for the probabilistic connectivity matrix representation are mainly derived from the fact that each element l_{ij} of a connectivity matrix L raised to the power k expresses the number of walks from node i to node j. Since our matrix is probabilistic instead of computing the number of paths using matrix multiplication, we compute the probability that a path exists using a similar approach. A walk from i to j of two hops exists when there is an intermediate node m with edges from i to m and m to j. In this case, the probability of this path existing is the product of elements $l_{l,m}$ and $l_{m,j}$ of the probabilistic connection matrix. The total probability of the path existing is computed by calculating the inclusive or for all intermediate nodes. A convenient way of computing this is computing the complement of the probability that all paths do not exist:

$$l_{ij}^{h} = 1 - \prod_{k=1}^{n} (1 - l_{ik}^{h-1} l_{kj}) \quad (29.20)$$

This can be implemented by taking a standard matrix multiplication implementation and replacing the summation of products in the inner loop with Equation 29.20. Note that we constrain diagonal values to remain zero.

We illustrated how to construct these matrices for important graph classes. We now discuss the structure and meaning of the matrices. By definition, connectivity matrices are square with the numbers of rows and columns both equal to the number of vertices in the graph (n). Each element (j, k) is the probability of an edge existing between nodes j and k. Since we consider only non-directed graphs, (j, k) should equal (k, j). Care should be taken to guarantee that algorithms for constructing matrices provide symmetric results.

Theorem 29.1

The sum of each row (column) of the probabilistic connectivity matrix M provides the expected degree of every node in G.

Proof

The expected value of a random variable is defined as the sum of possible values of the random variable times the probability of the value. Each element (j, k) in row j is an independent Bernoulli random variable expressing the likelihood of an edge existing between nodes j and k. The expected number of edges between j and k is the value of (j, k) times one. The expected degree of a node is the expected number of edges with the node as an end point. The expected degree of node j is the sum of the number of edges connecting j with all other nodes. Since each element (j, k) of row j expresses the likelihood of an edge between j and k (k ranging from 1 to n), the expected degree of j is the expected value of the sum of k Bernoulli random variables with probabilities of value (j, k). Since the expected value of a sum of random variables is the sum of the expected values of the random variables, the theorem must be true.

Theorem 29.2

M is a probabilistic connectivity matrix of random graph class G. We define *rpm*-graphs (regular *p*robability *m*atrix) graphs as graph classes where each node has equivalent probabilities of being connected to other nodes. If a graph class defines rpm-graphs, all nodes in G are equivalent and every row (column) is a permutation of every other row (column).

Proof

A regular graph is one in which all nodes have the same degree. [Cvetkovic 79] provides a proof that each row of the connectivity matrix of a regular graph is a permutation of every other row, this proof also applies to our definition of *rpm*-graphs. With the exception that elements of the connectivity matrices for *rpm*-graphs do not have the value one or zero, but are in the range from one to zero. In an *rpm*-graph class because each row (column) is a permutation of every other row (column), the expected value of the degree of all nodes is equal. Note that instances of the *rpm*-graph class are almost certainly not regular graphs.

Theorem 29.3

For an *rpm*-graph class G, the value of the eigenvalue with the largest magnitude of its associated connectivity matrix is the expected degree of each node in G.

Proof

This proof follows directly from the proof of this assertion for non-random graphs given in [Cvetkovic 79].

29.3.1 Graph Partitioning

We create subgraphs of the P2P network, so that each subgraph is served by its own index. The following conditions for good partitions of the graph are straightforward. The partitions should be of equal (or near equal) size. The majority of the node connections should be within the region served by an index, to support efficient communication between each node and its local index as well as the subsequent exchanges of mobile code packages between nodes served by the index. This means that the number of edges between partitions should be minimized. This problem is known as the k-way partition problem, which has been shown to be NP-Complete [Bonabeau 98]. It is an important problem for many practical applications, such as VLSI design. Many heuristic methods have been formulated for finding approximate solutions to the problem, including self-organizing maps [Bonabeau 98] and eigenvalue decompositions of the graph connectivity matrix [Gu 01]. We do not consider how the k-way partition of the graph structure is performed. Any established heuristic could be used to derive an approximate answer.

We now consider relevant upper and lower bounds for the graph structures produced by the k-way partitioning of the original graph. These bounds describe the resulting subgraphs as structures similar to the original random graphs. This step is essential in determining the number of indexes (partitions) appropriate for the P2P infrastructure and analyzing the effect of partitioning on system performance.

For Erdös-Rényi random graphs, the index partition of the global graph forms a random graph of n/i nodes (discounting rounding errors). Two limiting cases exist for the nodes in the subgraphs:

Lower bound—same edge distribution probability as the global graph
Upper bound—same node degree distribution as the global graph

Proof

For the lower bound, it suffices to assign nodes to indexes at random. Edges are retained which connect nodes assigned to the same index. Other edges are deleted. Since a random graph is constructed by assigning edges between any two nodes with the same probability, removing any node constructs a random graph with the same probability but with $n - 1$ nodes. Creating the k-way partitions is equivalent to performing this process i-times each time removing n/i nodes. Each partition is a random graph with the same edge probability as the global graph, but consisting of only n/i nodes.

For the upper bound, we assume the successful output of the k-way graph partition algorithm. This means that we have k (in this case i) partitions of equal size with a minimal number of connections between partitions. An absolutely optimal partition occurs when there are no connections between partitions. In which case, no edges would be removed in creating the partitions. The node degrees are unchanged and the degree distribution for the global graph is unchanged. The assignment of nodes to subgraphs is unbiased. The nodes chosen should have the same distribution as the global graph. The change in sample size would significantly modify the variance. Because the global graph is connected, this case cannot occur in practice.

Since the number of connections between partitions is minimal, we expect the actual partitioning behavior to be closer to the upper bound as long as the number of indexes is small relative to the number of nodes. As the number of indexes approaches n, the lower bound is likely to be a better approximation.

For scale-free graphs, the upper bound for the index partitions is the same as for the random graph and the same proof is valid. Since the probability of an edge existing between two nodes is not defined for scale-free graphs, the lower bound suggested for random graphs is undefined and inappropriate. Instead, we use a lower bound defined by assigning nodes to indexes at random and assume that all edges are equally likely to cross index boundaries. This lower bound is a scale free network, but the node degree in the sub-network will on the average be a fraction ($1/i$) of the degree in the global network. This is a lower bound in part because it ignores clustering effects.

Proof of lower bound

For the lower bound, consider a total of n nodes in network. n/i nodes are in the subgraph served by an index. If an edge leaving a node is equally likely to lead to a node served by any given index, then the likelihood of it reaching any given index is $1/i$. The likelihood of the sub-graph not being the one the current node is in is $1 - 1/i$ or $(i - 1)/i$. A node with degree d would therefore have an expected degree of d/i.

For *small-world graphs*, two distinct cases exist for the structure inside the index partitions: (1) fewer nodes in the subgraph than in the initial fully connected components and (2) more nodes in the subgraph than in the fully connected components. For both cases, the subgraphs will be constructed by removing some of the small fraction of edges that have been re-wired. The subgraph will therefore keep the global structure of the small-world graph. It will be a graph that consists of a set of almost fully connected components connected by a small number of random vertices. Note that depending on the value of i, if $n \bmod i \neq 0$, some fully connected components will be split between indexes. This is not a significant modification of the structure, since within the split component almost always almost all nodes will be within a single hop of each other.

1. In almost all cases almost all nodes are within one hop of each other.
2. All nodes within a fully connected component are within one hop of each other. If the graph is made from fully connected components of size c. The subgraph would consist of approximately $n/(i^*c)$ fully connected components connected at random. How the random connections are handled depends on the number of edges re-wired at random.

29.3.2 Expected Number of Hops

For P2P infrastructure analysis, like the QoS study in [Kapur 02], an important issue is the expected number of hops between nodes in the network. In this section, we present two analyses of this problem. One analysis is based on empirical study. The other uses the connectivity matrices to directly compute the expected number of hops. This analysis is important for designing both wired and wireless sensor network systems.

To measure the rate of growth of the expected number of hops as a function of graph size, [Dorogovtsev 01] derives the following simple estimator for the average number of hops between any two nodes in an Erdös-Rényi random graph: $q_{ave} \approx \ln[n]/\ln[q_1]$. Showing that the average number of hops grows as the logarithm of graph size. Empirical results from [Watts 99] indicate that the relationship between graph size and average path length is similar for random and small world graphs. Scale-free graphs appear to grow even more slowly [Albert 01, Reittu 02]. Our empirical tests support this assumption.

[Newman 01a] provides a derivation using generating functions, which shows that the number of nodes h hops away can be approximated using only two factors: the number of nodes one hop away and the number of nodes two hops away. Results are approximate since detailed graph structures cause graph instances to deviate, and they assume a single fully connected component in the graph. According to them (using our notation),

$$q_h = \left(\frac{q_2}{q_1}\right)^{h-1} q_1 \tag{29.21}$$

The generating function approach ignores the structure of an individual graph instance. [Newman 01] provides a more accurate approach. A different derivation follows that produces a similar result. The result here is proposed for a slightly different problem and is slightly different. We use a recursion equation to find a simple estimate of the number of nodes reachable after exactly h hops.

The average number of nodes reachable after one hop is by definition d_{ave} (i.e., q_1). For the second hop, every node reached has on the average $d_{ave} - 1$ degrees free. By definition C percent of those $d_{ave}(d_{ave} - 1)$ degrees were already reached in one hop, leaving $(1 - C)$ percent connecting to new nodes. $Mu[2]$ compensates for the presence of quadrilaterals in the graph. $Mu[h]$ generalizes this. We apply the same logic to define q_3 and so on. The resulting equation is

$$\begin{aligned} q_h &= (d_{ave} - 1)q_{h-1}(1 - C)Mu[h] \\ q &= d_{ave} \end{aligned} \tag{29.22}$$

for $h > 1$ $q_h = (1 - C)^{h-1}(d_{ave} - 1)^{h-1} d_{ave} \prod_{i=2}^{h} Mu[i]$

A more accurate estimate of the number of nodes reachable in h hops is achieved by taking into account the greater likelihood of connecting to a node with a higher degree [Newman 01] (again $h > 1$, for $h = 1$ Equation 29.22 holds):

$$q_h = (1 - C)^{h-1} d_{ave} \prod_{i=2}^{h} Mu[i] \left(\sum_{d=1}^{n-1} d(d-1)^{h-1} p_d \right) \tag{29.23}$$

Figure 29.33 shows the relative quality of these estimators, using the same sets of graphs used to estimate the $Mu[h]$ factors. The solid lines have error bars and show the mean number of nodes reached after the given number of hops. The line connected by alternating dashes and dots is the generating function estimate. The dotted line shows the estimate function in Equation 29.23. To calculate the estimates for the class of graphs, we computed the C and $Mu[h]$ factors for each graph and used the mean value in the estimation functions. Note that all three graphs have the same general shape. We also plotted the values for individual graphs and the results were similar. The actual plot is an s-curve that approaches 100 fairly rapidly. The generating function estimate consistently overestimates the number of nodes reachable within h hops.

The top left graph of Figure 29.33 is the Erdös-Rényi data. Note that the actual plot does not reach 100 nodes, because some of the generated graphs were not fully connected. There was always one major component, but some nodes were not connected to it. The top right graph is for the small world graphs. Interestingly, underestimation is most pronounced in this case. The bottom graph shows results for the scale-free graphs. The estimate using C and $Mu[h]$ initially underestimates the number of nodes. It converges to the generating function estimate for Erdos-Renyi and scale-free graphs. For the small-world graphs the estimating function derived in this paper has an entirely different shape.

To de-couple the effects of the C and $Mu[h]$ factors, Figure 29.34 shows the same plots with all $Mu[h]$ values set to unity. Note that underestimation no longer occurs. The estimate is worse than that provided by the generating function.

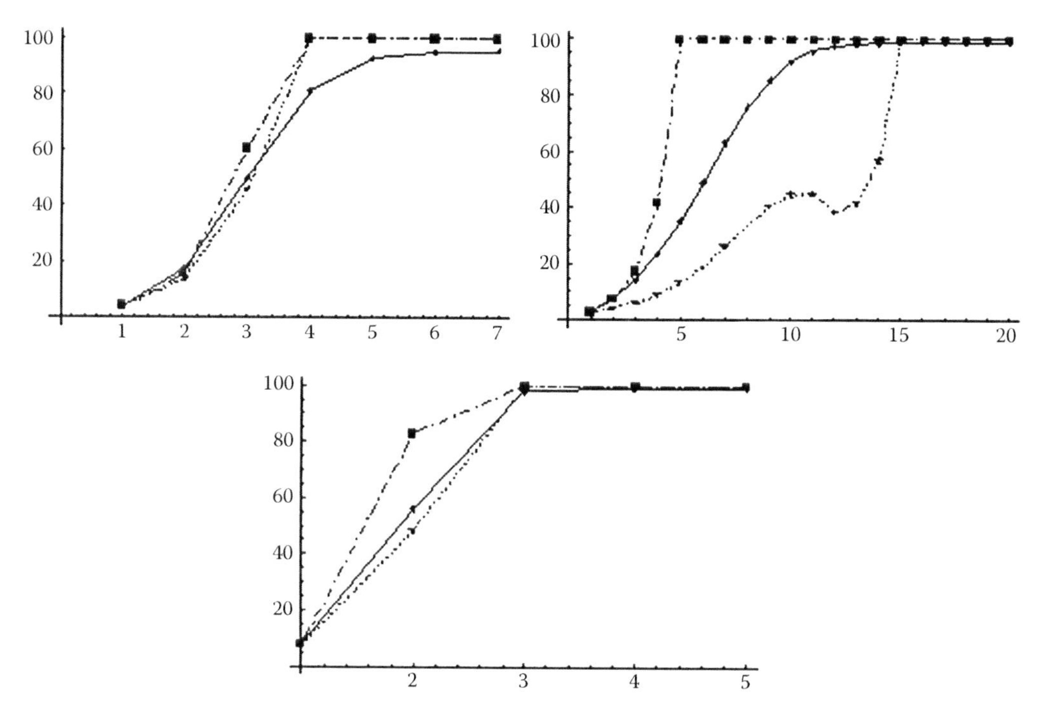

FIGURE 29.33 Average number of nodes reachable in h hops (y axis) versus h hops (x axis) for Erdös-Rényi (top left), small world (top right), and scale free graphs (bottom). Solid line is results from 35 appropriately generated random graphs with 95% confidence interval. Dashed line is from the [Newman 01] estimator. Dotted line is the estimate from Equation 29.23.

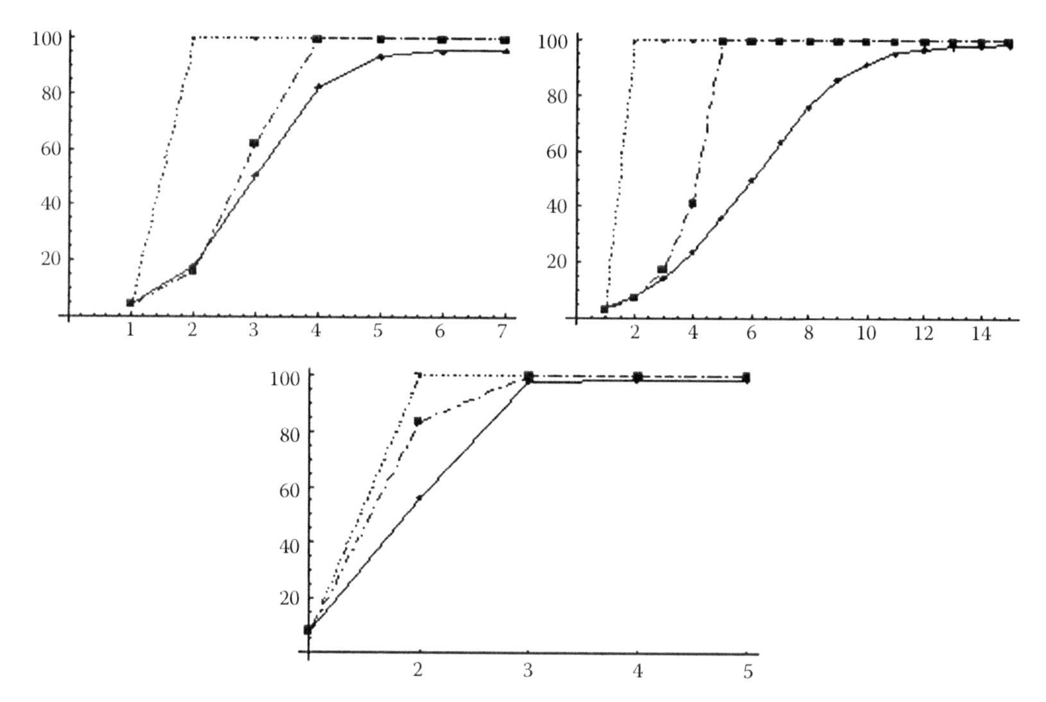

FIGURE 29.34 Average number of nodes reachable in h hops (y axis) versus h hops (x axis) for Erdös-Rényi (top left), small world (top right), and scale free graphs (bottom). Dashed line is from the [Newman 01] estimator with $Mu[1] = 1$. Dotted line estimate is Equation 29.23. For the two estimators all $Mu[h]$ values were set to one.

The expected number of hops between any two nodes chosen at random can now be calculated as

$$z_{exp} = \frac{(q_1 + 2(q_2 - q_1) + \cdots)}{n}$$

$$z_{exp} = \left(\frac{1}{n}\right)\left(d_{ave} + 2\left((1-C)Mu[2]\left(\sum_{d=1}^{n-1} d(d-1)p_d\right)\right)\right.$$

$$+ \sum_{h=3}^{\max hops} h\left(\left((1-C)^{h-1}\prod_{i=2}^{h} Mu[h]\left(\sum_{d=1}^{n-1} d(d-1)^{h-1} p_d\right)\right)\right.$$

$$\left.\left.- \left((1-C)^{h21}\prod_{i=2}^{h-1} Mu[h]\left(\sum_{d=1}^{n-1} d(d-1)^{h-2} p_d\right)\right)\right)\right) \tag{29.24}$$

Note that we estimate the number of hops reachable in exactly h hops z_h as $q_h - q_{h-1}$. The set of nodes reachable in h hops minus the set of nodes reachable in h – 1 hops. Since not all nodes reachable in h – 1 hops are reachable in h hops, this is not an exact measure. As we will see, it functions fairly well in practice although it causes a slight undercount.

As before, we plot the results from our graph test set versus the two estimators. The results are consistent with the cumulative graphs. As should be expected the discrepancies are more notable. For Erdös-Rényi graphs the estimation is reasonable. For scale-free the estimate is very good. For small world graphs the results are disappointing. The estimate is consistently low. A spike of nodes at the end compensates for this (Figure 29.35).

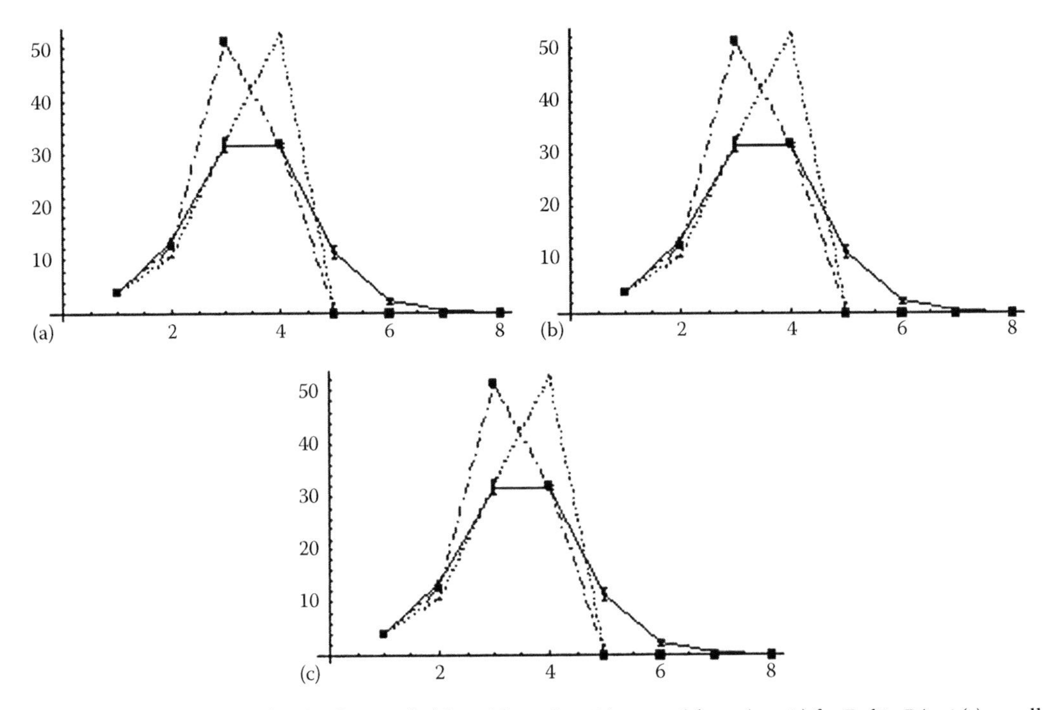

FIGURE 29.35 Number of nodes first reachable in h hops (y axis) versus h hops (x axis) for Erdös-Rényi (a), small world (b), and scale free graphs (c). Dashed line is from the [Newman 01] estimator. Dotted line is the estimate from Equation 29.24.

To verify the ability of these estimators to predict the expected number of hops in a random graph, we compare the actual expected number of hops between two nodes selected at random in our set of 35 test cases. The expected values are: 3.31463 (Erdös-Rényi), 4.39716 (small-world) and 2.36894 (scale-free). The generating function predictions are 3.24785 (Erdos-Renyi), 3.59476 (small-world) and 2.09639 (scale-free). The predictions from the derived estimation function are 3.52925 (Erdos-Renyi), 6.46731(small-world) and 2.47913 (scale-free). Both estimators function well. The generating function underestimates the expected number of hops. Equation 29.24 overestimates the number of hops. The only significant deviation appears to occur using the estimator function including clustering and mutuality parameters for small-world graphs.

We can also use the probabilistic connectivity matrices to estimate the expected number of hops between nodes. To do so, we use the following *theorem*: element (j, k) of M^z is the probability that a walk of length z existing between nodes j and k.

Proof

The proof is by induction. By definition, each element (j, k) is the probability of an edge existing between nodes j and k. M^2 is the result of multiplying matrix M with itself. Equation 29.16 is used to calculate each element (j, k) since all values are probabilities. As explained in Section 7, this calculates the probability of a path of length two existing between nodes j and k by exhaustively enumerating the likelihood of the path passing through each intermediate node in the graph. Using the same logic, M^z can be calculated from M^{z-1} using matrix multiplication to consider all possible intermediate nodes between nodes j and l. Where M^{z-1} has the probabilities of a walk of length $z - 1$ between j and k, and M has the values defined previously.

Example 29.1

Probabilities of walks of length three in an Erdös-Rényi graph of four nodes for $p = 0.6$ and 0.65:

$$M = \begin{bmatrix} 0 & 0.65 & 0.65 & 0.65 \\ 0.65 & 0 & 0.65 & 0.65 \\ 0.65 & 0.65 & 0 & 0.65 \\ 0.65 & 0.65 & 0.65 & 0 \end{bmatrix} \quad M^2 = \begin{bmatrix} 0 & 0.666 & 0.666 & 0.666 \\ 0.666 & 0 & 0.666 & 0.666 \\ 0.666 & 0.666 & 0 & 0.666 \\ 0.666 & 0.666 & 0.666 & 0 \end{bmatrix}$$
$$M^3 = \begin{bmatrix} 0 & 0.679 & 0.679 & 0.679 \\ 0.679 & 0 & 0.679 & 0.679 \\ 0.679 & 0.679 & 0 & 0.679 \\ 0.679 & 0.679 & 0.679 & 0 \end{bmatrix} \tag{29.25}$$

$$M = \begin{bmatrix} 0 & 0.6 & 0.6 & 0.6 \\ 0.6 & 0 & 0.6 & 0.6 \\ 0.6 & 0.6 & 0 & 0.6 \\ 0.6 & 0.6 & 0.6 & 0 \end{bmatrix} \quad M^2 = \begin{bmatrix} 0 & 0.59 & 0.59 & 0.59 \\ 0.59 & 0 & 0.59 & 0.59 \\ 0.59 & 0.59 & 0 & 0.59 \\ 0.59 & 0.59 & 0.59 & 0 \end{bmatrix}$$
$$M^3 = \begin{bmatrix} 0 & 0.583 & 0.583 & 0.583 \\ 0.583 & 0 & 0.583 & 0.583 \\ 0.583 & 0.583 & 0 & 0.583 \\ 0.583 & 0.583 & 0.583 & 0 \end{bmatrix} \tag{29.26}$$

Computing the expected number of nodes reachable in r hops is very straightforward in this case. If the graph description is regular (all rows are permutations of the same values), predicting this value is simple. First, compute M^r. The sum of the values of any row provides the expected number of nodes reachable in r hops. This value is the same for every node.

For scale-free graphs and small world graphs whose descriptions are not regular, compute the sum for each row. The maximum and minimum values of the sum provide the range of possible values. The average can easily be computed, and is a reasonable solution for small-world graphs. Since the degree of hub nodes in scale-free graphs is much larger than for most nodes (when γ is significantly greater than one), the mode value of the row sum is likely to be a better estimate for scale-free graphs.

29.3.3 Network Redundancy and Dependability

Given the expected number of hops in terms of the number of nodes and degree distribution, we consider the question of approximating dependability for paths of h hops for random graphs with arbitrary degree distributions using the same statistics. We start by assuming each link has the same probability of failure f. Note that, ignoring node failures, the dependability a of a serial path through the graph can be calculated using the fact that it can only working if all components work. This is given by the equation (a_h is the dependability of an h hop serial system) (see Figure 29.36):

$$a_h = (1-f)^h \tag{29.27}$$

This can be compared to the dependability of a parallel system of h components. In which case, the system fails only if all h components fail concurrently. Using the same variables a and f as previously, this gives (see Figure 29.37)

$$a_h = 1 - f^h \tag{29.28}$$

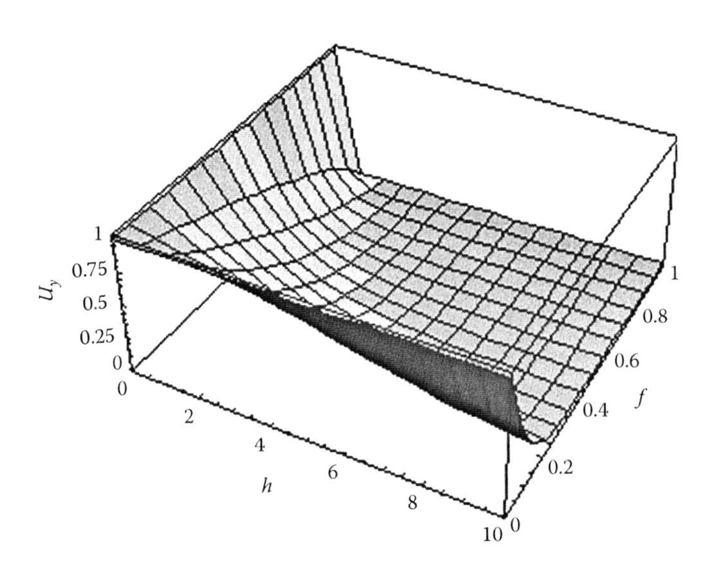

FIGURE 29.36 The availability a^h (vertical axis) of an h hop serial system. The number of hops h varies from 0 to 10 (front axis) and the probability of failure f varies from 0 to 1 (side axis).

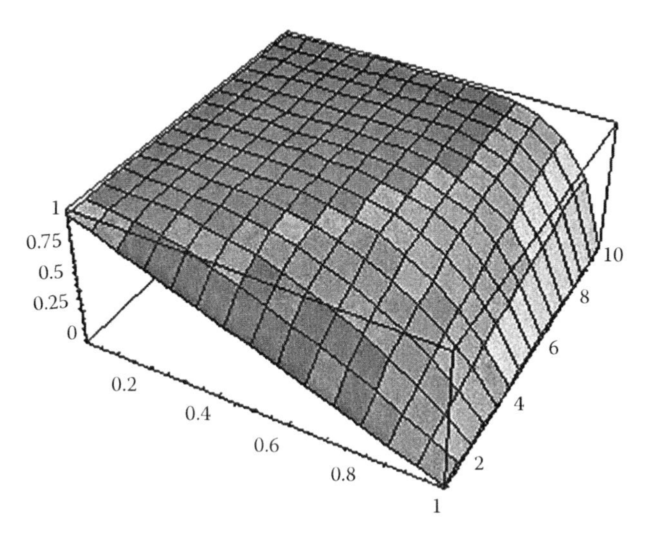

FIGURE 29.37 System dependability (z axis) for a parallel system versus number of components (y axis) and component dependability (x axis). This is a graphical representation of Equation 29.27.

Figures 29.36 and 29.37 illustrate well-established facts. For serial systems as the number of components increase, system dependability decreases rapidly and asymptotically approaches zero. The redundancy in parallel systems causes dependability to increase quickly and asymptotically approach one, as the number of components increases. Obviously, degradations in component dependability adversely affect both approaches.

All classes of random graph structures we consider can have non-negligible amounts of redundancy. We now analyze how this redundancy affects communications between nodes in the graph structures. Ideally, the statistics we have discussed should enable us to predict path dependability in these systems.

We have established ways of dividing random and pseudo-random graphs into partitions for indexing, authentication, etc. When this is done, we have parameters that describe the graph. Given these parameters, we can calculate statistics that describe clustering and mutuality in the graph. We are also able to estimate the expected number of hops required to service a request for a mobile code package [Kapur 02]. We now attempt to use this information to determine how partitioning the network affects dependability.

To determine system dependability as a function of the number of indexes (partitions), we need only calculate communications dependability as a function of the number of hops. We modify a to determine a new factor $\bar{a}$ which takes into account the network redundancy. Network redundancy provides multiple possible paths between two neighboring nodes. The single hop dependability adjusted to consider alternate paths of exactly r hops is $\bar{a}_r$, giving the following formula:

$$\bar{a} = 1 - \prod_{r=1}^{n-(h+1)} (1 - \bar{a}_r) \tag{29.29}$$

So that a better estimate of the dependability of an h hop path would be given by $\bar{a}^h$. Before proceeding, let's analyze what redundancy means in terms of availability. Figure 29.38 plots the dependability of a one-hop system with no redundancy and the dependability of a 50-node system where independent paths of all lengths are guaranteed to exist versus a probability of single link failure progressing from 1 to 0. Figure 29.39 shows the difference between the two. Note that the maximum dependability gain (about 25%) occurs at a probability of failure around 40%.

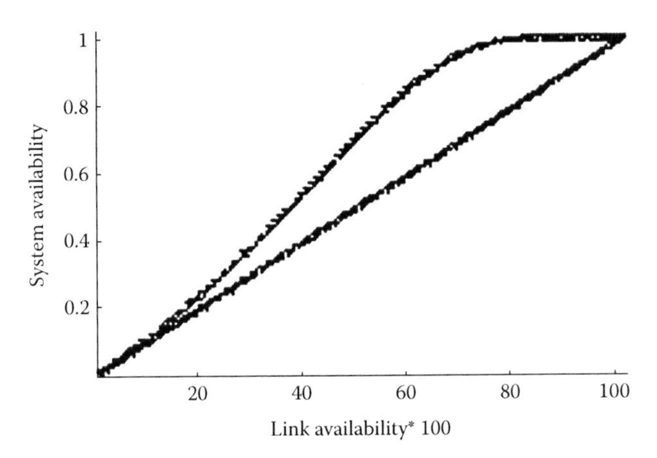

FIGURE 29.38 System dependability (y axis) versus component dependability times 100 (x axis). The bottom line is for a one-component system. The top line is for a system of 50 components with redundant links between all nodes.

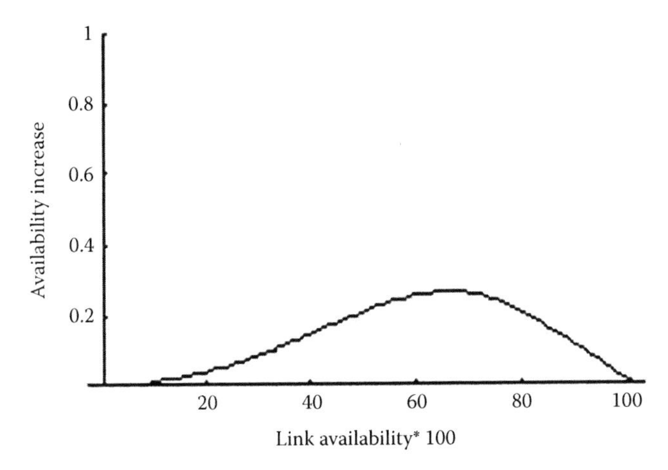

FIGURE 29.39 System dependability gain (y axis) versus component dependability times 100 (x axis) for a fully redundant system of 50 nodes.

We modify Equation 29.29 by considering redundant paths. In these graph classes multiple paths of varying lengths between two nodes frequently exist, and they can be used as alternatives to the shortest path of length h connecting the nodes. We attempt to quantify this effect by computing the expected amount of redundancy for individual edges (one hop paths) in the graph. Methods for computing the likelihood of redundant paths of varying lengths are described. These probabilities are then used to calculate a more realistic availability estimate for the h hop path using the modified per hop availability.

The variable $\hat{a}$ is used for the per hop availability of an edge, taking into account redundancy in the random graph. From Equation 29.1 a^r is the dependability of a path of r hops with no redundancy. Let's consider two adjacent nodes i and j. Equation 29.2 defines $\hat{a}_r$ the dependability of a path of r hops connecting nodes i and j parallel to the edge (path of length 1) connecting the nodes. Equation 29.30 takes into account the probability that this path exists:

$$\hat{a}_r = a^r P[q_r \mid q_1] \tag{29.30}$$

$P[q_r|q_1]$ is the probability that a path of r hops exists between two nodes given that a path of one hop (an edge) exists connecting the nodes. We now have paths of differing lengths connecting two nodes in parallel.

Remembering that h is the expected number of hops between two nodes in the network, our estimate of the dependability of the file transfer path becomes $\hat{a}^h$. This estimate has two weak points: (1) it is pessimistic in that it only considers redundancy for a single hop, and (2) it is optimistic in that it makes a tacit assumption that the paths of differing lengths share no edges.

For the sake of brevity, we introduce notation $P_{x,\ldots,y}$ for $P[\{q_x \wedge \cdots \wedge q_y\}]$. For example, $P[\{q_3 \wedge q_2 \wedge q_1\}]$ (the probability a node belongs to the set of nodes where paths of lengths 3, 2, and 1 connect it to another node) can be expressed as $P_{3,2,1}$ and $P_{2,\neg 1}$ means $P[\{q_2 \wedge \neg q_1\}]$ (the probability that a node belongs to the set of nodes with a path of length 2 and none of length 1 connecting it to another node).

Consider the effect of two hop paths:

$$\hat{a}_2 = a^2 P[q_2 \mid q_1] \tag{29.31}$$

$P[q_2|q_1]$ is the probability that a path of two hops exists between two nodes given that an edge connects the nodes. This differs subtly from the cluster coefficient. C is the percentage of two hop paths already reached in one hop. $P[q_2|q_1]$ is the percentage of one hop paths that also yield two hop paths. It can be computed by modifying procedure 2. It suffices to swap the roles of M^2 and M.

The variable $\hat{a}_3$ depends on $P[q_3|q_1]$, which we decompose into distinct cases depending on the existence (or absence) of paths of length 2:

$$\hat{a}_3 = a^3 P[q_3 \mid q_1] = a^3 \frac{P_{3,1}}{P[q_1]} = \frac{(1-f)^3}{P[q_1]}(P_{3,2,1}P[q_2] + P_{3,\neg 2,1}P[\neg q_2]) \tag{29.32}$$

$P[q_2]$ is q_2/n. $P[q_1]$ is q_1/n. Figure 29.40 shows the figures implied by the factors in (29.8). The right hand figure is two contiguous triangles, and occurs with probability $P[q_2|q_1]^2$.

The left hand object in Figure 29.40 is a rectangle. The probability of this object occurring can be calculated by the following procedure:

- First calculate M^2.
- Elements of M^2 that are greater than one, with corresponding element of M equal to zero, are rectangle vertices.
- $P_{3,\neg 2,1}$ is the number of columns of M^2 with elements that meet this criterion divided by n.

This procedure provides the final information needed to compute $\hat{a}_3$. Equation 29.32 can be completed giving

$$\hat{a}_3 = \frac{(1-f)^3}{q_1}\left(P_{3,2,1}q_2 + P_{3,\neg 2,1}\left(n - q_2\right)\right) \tag{29.33}$$

All $P_{i,\neg(i-1),\cdots,\neg 2,1}$ where the number of hops i is odd describe the probability of the node being a vertex of a polygon with an even number of sides $2(i-1)$. All sides are edges of the graph. Figure 29.6 illustrates this. The probability can be computed by extending procedure 4 to give procedure 5. The extensions in

FIGURE 29.40 On the left the figure implied by elements of the set $\{q_3 \wedge q_2 \wedge q_1\}$. On the right the figure implied by elements of $\{q_3 \wedge \neg q_2 \wedge q_1\}$.

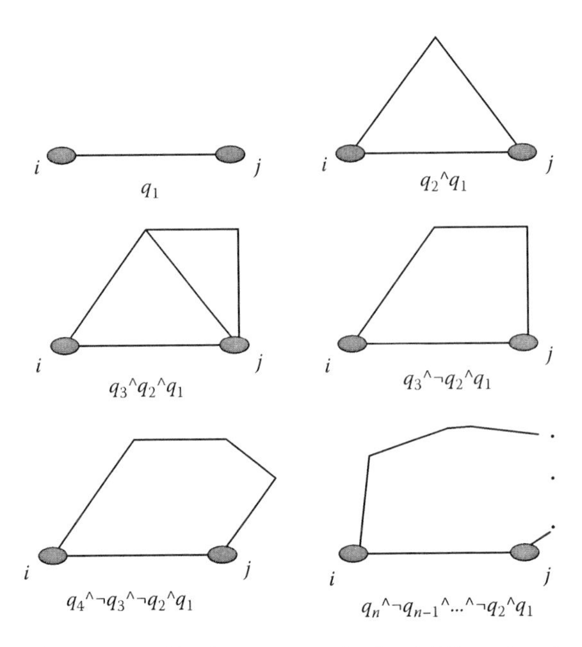

FIGURE 29.41 Geometric figures implied by the existence of paths of different lengths. If paths of length q_n and q_{n+1} exist, a triangle is implied. Paths of length q_i and q_j with $i > j$ and no paths of length k $i > k > j$. Implies a polygon with $(i - j) + 2$ sides.

step 2 of procedure 5 are needed to counteract the presence of loops in the calculation of powers of the connectivity matrix M (Figure 29.41).

Estimation of $P_{i,\neg(i-1),\ldots,\neg 2,1}$:

1. Initialize L' to the connectivity matrix L.
2. Perform the following for $i - 1$ iterations:
   a. Initialize L'' to equal L'
   b. Set diagonal elements of L' to zero
   c. Set all non-zero elements of L' to one
   d. Set all elements $l_{i,j}$ of L' to $l_{i,j} * l_{j,i}$
   e. Multiply L' by L giving L'
3. Count the columns of L' with elements greater than one and the corresponding elements of L^{i-1} to L equal to zero.
4. Divide this sum by n giving $P_{i,\neg(i-1),\ldots,\neg 2,1}$.

These values can be computed using an instance of the random graph class. As with $M[h]$, a more reliable value are found by computing the mean over several instances of the graph classes. If the number of hops is even, the procedure 5 will not work. In which case, estimate $P_{i,\neg(i-1),\ldots,\neg 2,1}$ by averaging $P_{i-1,\neg i,\ldots,\neg 2,1}$ and $P_{i-1,\neg(i-2),\ldots,\neg 2,1}$.

To compute $\hat{a}_n$ we follow a similar approach to computing $\hat{a}_3$. The value of $\hat{a}_n$ is $(1 - f)^n P[q_n|q_1]$, and

$$P[q_n \mid q_1] = \frac{P_{n,n-1,1}P[q_{n-1}] + P_{n,\neg n-1,1}P[\neg q_{n-1}]}{P[q_1]}$$

$$P_{n,n-1,1} = P_{n,n-1,n-2,1}P[q_{n-2}] + P_{n,n-1,\neg n-2,1}P[\neg q_{n-2}] \tag{29.34}$$

$$P_{n,\neg n-1,1} = P_{n,\neg n-1,n-2,1}P[q_{n-2}] + P_{n,\neg n-1,\neg n-2,1}P[\neg q_{n-2}]$$

Equation 29.34 is recursive, and continues until each element is a sequence exhaustively enumerating the existence or nonexistence of all paths of length 1 to *r*. The probability of each atomic element is the product of the probabilities of the polygons whose union defines the object. For example, $P_{4,\neg 3,2,1}$ describes the union of a triangle and a rectangle. It is equal to $P[q_2|q_1]$ times $P_{3,2,1}$.

When the recursion of Equation 29.34 terminates, each factor contains a variable of the form $P_{j,\ldots,1}$. Where *j* and 1 delimit a list of path lengths, some of which are negated. The limits of the list (*j*, 1) are never negated. This term needs to be translated into a probability.

Calculation of $P_{j,\ldots,1}$:

- Initialize probability $P_{j,\ldots,1}$ to 1.
- Start with 1 and count the number of negated paths until the next non-negated one. Call this number *k*.
- If *k* is 0, the polygon described is a triangle. $P_{j,\ldots,1}$ becomes $P_{j,\ldots,1} * P[q_2|q_1]$.
- If $k > 0$, the polygon described has $3 + k$ sides. The probability of it being a part of the figure described by $P_{j,\ldots,1}$ is described by $P_{k/2,\neg(k/2-1),\ldots,\neg 2,1}$. We have already described how to estimate this value for even and odd *k*. $P_{j,\ldots,1}$ becomes $P_{j,\ldots,1} * P_{k/2,\neg(k/2-1),\ldots,\neg 2,1}$.
- Replace one with the next non-negated path length. If that path length is less than *j*, start again at step two. Else, terminate the calculation.

We now compute an estimate for $\hat{a}$ the expected value of the dependability of an edge in a random graph. Considering the system as a set of parallel paths of differing lengths gives

$$\hat{a} = 1 - \prod_{j=1}^{diameter} (1 - \hat{a}_j) \tag{29.35}$$

The path dependability for the *h* hop path becomes $\hat{a}^h$.

This method has several shortcomings:

- It implicitly assumes independence for paths through the graph.
- Computation of Equation 29.34 has a combinatorial explosion. For paths of length *j*, 2^j factors need to be considered.
- Tables of $P_{i,\ldots,1}$ statistics are not readily available. The estimates we describe are computable, but computation requires many matrix multiplications. Stable statistics require computation using several graph instances. For large graphs, the amount of computation required is non-negligible.
- It ignores the existence of multiple redundant paths of the same length increasing per hop dependability. As shown in Figure 29.4, this factor is important.

Most of these can be overcome by realizing that the additional dependability afforded by a path drops off exponentially with the number of hops. It should be possible to stop the computation when a^r becomes negligible. Another factor to consider is that the diameter of the graph scales at worst logarithmically for these graph classes. The algorithm scales as the exponential of a logarithm making it linear. This approach reveals the redundancy inherent in these systems and how it can be effectively exploited.

Alternatively, $P[q_r]$ can be taken directly from the probabilistic connectivity matrices. For the Erdös-Rényi graph all non-diagonal elements have the same value, which is $P[q_h]$. For small-world and scale-free graphs, the average of all elements in M^h is a reasonable estimate of q_h. We suggest, however, using the minimum non-diagonal element value in the matrix. The average will be skewed by large values for connections in the clusters for small-world graphs and large values for hub nodes in scale-free graphs. The minimum value is the most common in both graph classes and provides a more typical estimate of the probability of connection between two nodes chosen at random.

29.3.4 Vulnerability to Attack

Empirical evidence that the Internet is a scale-free network with a scaling factor close to 2.5 is discussed in [Albert 01]. [Albert 00] analyzes the resiliency of the Internet to random failures and intentional attacks using a scale-free model. Simulations show that the Internet would remain connected even if over 90% of the nodes fail at random, but that the network would no longer be connected if only 15% of the best-connected hub nodes should fail. In this section, we show how this problem can be approached analytically. The techniques given here allow an analytical approach to the same problem:

- Construct a matrix that describes the network under consideration.
- The effect of losing a given percentage of hub nodes can be estimated by setting all elements in the bottom j rows and left j columns to zero, where j/n approximates the desired percentage.
- Compute C^2 and see whether the probabilities increase or decrease. If they decrease, the network will partition.
- Find the percentage where the network partitions.

Theorem 29.4

The critical point for scale-free network connectivity arises when the number of hub nodes failing is sufficient for every element of the square of the connectivity matrix to be less than or equal to the corresponding element in the connectivity matrix.

Proof

Using the algorithm given here, hub nodes correspond to the last rows and columns in the probabilistic connectivity matrix. When a hub node fails, all of its associated edges are removed from the graph. This is modeled by setting all values in the node's corresponding row and column to zero. Matrix multiplication is monotone decreasing. If all elements of matrix K are less than all elements in matrix K', then, for any matrix J, $JK < JK'$. When all two hop connections are less likely than one hop connections, then three hop connections are less likely than two hop connections, etc. Using the same logic as with Erdös-Rényi graphs, this will cause the network to tend to be disconnected. Therefore when the enough hub nodes fail so that all elements of M^2 are less than the corresponding elements in M the corresponding networks will be more likely to be disconnected.

29.3.5 Critical Values

Many graph properties follow a 0–1 law. The property either appears with probability 0 or probability 1 in a random graph class, depending on the parameters that define the class. Frequently an abrupt phase transition exists between these two phases [Bollobás 01, Krishnamachari 01]. The parameter value where

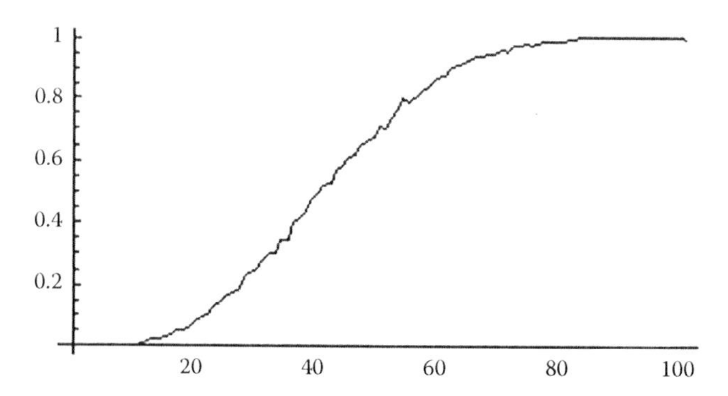

FIGURE 29.42 Shows empirical verification of theorem for Erdös-Rényi graph connectivity. Two thousand instances of Erdös-Rényi graphs of five nodes were generated as the edge connection probability varied from 0.01 to 1.00. The x-axis times 0.01 is the edge probability. The y-axis is the percent of graphs that were connected. The formula used in the theorem predicts the critical value around probability 0.4. When $p = 0.35$ (0.40) Equation 29.16 gives 0.357 (0.407).

the phase transition occurs is referred to as the critical point. The connectivity matrices defined in this paper can be useful for identifying critical points and phase transitions.

Theorem 29.5

For Erdös-Rényi graphs of n nodes and probability p of an edge existing between any two nodes, the critical point for the property of graph connectivity occurs when $P = 1 - (1 - P^2)^{n-1}$. When $P > 1 - (1 - P^2)^{n-1}$, the graph will tend to not to be connected. When $P < 1 - (1 - P^2)^{n-1}$, the graph will tend to be connected.

Proof

For Erdös-Rényi graphs, all non-diagonal elements of the matrix have the same value p. Diagonal elements have the value zero. The formula $1 - (1 - P^2)^{n-1}$ follows directly from these two facts and Equation 29.20. When the value of this equation is equal to p, two nodes are just as likely to have a two-hop walk between them as a single edge. This means that connections of any number of hops are all equally likely. When the value of the equation is less than p, a walk of two hops is less probable than a single hop connection. Since the equation is monotonically decreasing (increasing) as p decreases (increases). This means that longer walks are increasingly unlikely and the graph will tend not to be connected. By symmetry when the value of the equation is greater than p, the graph will tend to be connected (Figure 29.42).

29.4 Application to Network Life Extension

Consider a surveillance network charged with reporting when a member of a class of objects (*targets*) traverses a given surveillance domain (*terrain*). Reports are sent to a user community that we assume, for the sake of discussion, is external to the terrain. The network will be viable as long as it assures that: (1) an object traversing the terrain is detected (with acceptable error rates), and (2) the user community is alerted.

This criterion is a tautology: the network is viable as long as it performs its mission. To date, the implications of this tautology have been overlooked. For example, the following methods of determining network viability do not fit the criterion:

- Network connectivity—If full network connectivity is needed, the sensor network is a giant serial system. Network availability will fall exponentially with the number of sensor nodes (n), and thus large networks will have an unacceptable mean time to failure. For networks with any redundancy, some nodes can be isolated from the network without compromising its application.
- Sensing coverage—refers to placing nodes so that sensor detection regions have little overlap, but the system monitors the entire terrain. Since sensing ranges and coverage regions are unpredictable, problems with this "cookie cutter" approach are well known [Washburn 02], and often due to environmental influence [Swanson 2005]. Real-world approaches consider distributed surveillance as a tracking problem using sensors with finite space and time sampling rates [Brooks 03, Brooks 04]. Coverage approaches ignore sensor errors, background noise, and occlusion. In addition, coverage analysis creates a serial system, which fails when any component fails. Once again, we have a serial system where dependability falls exponentially with network size.

Network connectivity ignores sensing issues. Sensing coverage ignores wireless communications issues.

We propose a network viability criterion that is a direct consequence of the network model in [Krishnamachari 01], where nodes with a fixed communications range are placed at random in the terrain. Ad hoc networks with range limited communications exhibit phase change phenomena like those found in random graph [Bollobás 01] and percolation [Stauffer 94] theories. Random graph theory is a branch of graph theory that assigns probability distributions to the existence of edges between vertices. Percolation theory, a branch of physics, studies fluid flows in random media. Random media are modeled as tessellations of a terrain with probability distributions for the existence of edges between neighboring vertices.

In these models, network behavior has two phases. In the first phase, the probability of connection between nodes is small and the network has a large number of isolated components. As connection probability grows, the expected size of the largest component grows logarithmically. In the second phase, the network is dominated by a unique giant component that contains most of the system nodes. There are still isolated holes in the network. The size of the largest hole shrinks logarithmically as connection probability increases. The transition between these two phases is extremely steep. For random graphs, the curve of the maximum component size versus edge probability takes the form $e^{-e^{-c}}$. In percolation theory, the inflection point of this curve is referred to as the percolation threshold.

Percolation theory has established these properties for systems with a giant component:

- For systems above the percolation threshold, a path exists that connects the terrain's external boundaries.
- At the percolation threshold, property 1 is self-similar over scales.

Consider sensor networks with nodes either randomly placed [Krishnamachari 01], in a regular tessellation [Stauffer 94], or a weighted combination of the two. Sensor nodes are vertices in a random graph structure. Edges between vertices represent either an active communications link, or detection of a target passing between nodes. In practice, the edge probability distribution is the minimum of the two likelihoods.

Above the phase change (percolation threshold) a single giant component connects most of the sensor nodes ($O(n)$) [Stauffer 94]. It has at least one path connecting all the terrain's external boundaries (property 1). This property is true for subsets of the system across scales (property 2). Thus, for a sensor network with a giant component, targets traversing the network will be detected by at least one node that can report the detection to the user community. Therefore, the network fulfills our viability criterion.

This shows the network is viable while it has a giant component. In our simulation, we infer the loss of the giant component from the loss of property 1. When there is no path between the terrain's external boundaries, the giant component is fractured. Consider the worst-case scenarios for networks with initial configurations above the percolation threshold:

- A target entering the network cannot be detected and/or reported while in a hole. Since the largest hole above the percolation threshold is O(log n) [Bollabás 01, Stauffer 94], this is the upper limit of the target's ability to avoid detection in the initial network configuration.
- As nodes lose power: (1) maximum hole size grows logarithmically, (2) the network becomes sparse, and (3) the network approaches the percolation threshold. As long as we are above the percolation threshold property 1 holds and a target has to pass through a graph edge to traverse the terrain. Once it does so, a node connected to the giant component detects the target and notifies the user community. Property 2 says that property 1 holds for regions inside the terrain up to the percolation threshold.

29.4.1 Communications Range Management

Since transmission power is proportional to d^α, a natural optimization is to reduce communications range when possible. This results in an expected per node power savings rate of

$$E_c = \mu c d^\alpha \tag{29.36}$$

where,

E_c is the energy required for communication
μ is the rate at which the nodes generate packets
c is a constant
d is the distance between the source and destination nodes
α is the RF attenuation factor

$$\Delta E = \mu c (\Delta d)^\alpha \tag{29.37}$$

with the same variables as in Equation 29.36, except that ΔE is energy savings and Δd is reduction in transmission range.

Note that nodes can also increase the communications range as needed by increasing the energy used to transmit the signal. This increased energy drain is worthwhile when an intermediate node exhausts its battery power. The network can continue to function although individual nodes, in this example node B, will exhaust their power resources more quickly than would otherwise be the case.

29.4.2 Node Relocation

Since reducing communications range results in energy savings, we consider repositioning nodes to reduce the energy required to transmit data. On the other hand, moving nodes also requires energy.

Consider an arbitrary node n_0 in the network with n neighbors that it communicates with. Nodes n_1 to n_{n-1} transmit packets to n_0 and node n_n is the next hop on n_0's path to the user community.

From Equation 29.36, the energy consumed by communications between nodes n_0 and n_i is

$$E_i = \mu_i c d_i^\alpha \tag{29.38}$$

where

μ_i is as defined for Equation 29.36
d_i is the distance between nodes n_0 and n_i

We, therefore, need to minimize:

$$E = \sum_i E_i \tag{29.39}$$

Using the Pythagorean theorem, calculus, and the fact that x and y are orthogonal, this value is a minimum when the following two equations are satisfied:

$$\sum_i \mu_i (x - x_i)((x - x_i)^2 + (y - y_i)^2)^{\alpha/2-1} = 0 \tag{29.40}$$

$$\sum_i \mu_i (y - y_i)((x - x_i)^2 + (y - y_i)^2)^{\alpha/2-1} = 0$$

For $\alpha = 2$, this is a simple weighted average, for other values we have two equations with two unknowns. They can be solved using any number of root finding techniques, including Newton's method. Now, if we denote:

E_{0ci} as the energy required for i'th node to communicate with its next node with node "n_o" in its original position. The sum of these values equals the remaining node energy, ignoring sensing and computation energy.

E_{nci} as the energy required for i'th node to communicate with its next node considering node "n_o" in its expected new position, for the expected node lifetime in the original position.

$E_{conserved}$ as the energy saved by moving the node to the new position, then

$$E_{conserved} = \sum_{i=0}^{n-1} (E_{oci} - E_{nci}) \tag{29.41}$$

The energy required for a node to move distance d_1 is

$$E_m = m * d_1^{\alpha} \tag{29.42}$$

where

E_m is the energy required for movement
m is the energy required to move 1 unit
d_1 is the distance moved by the node in units
α is the RF attenuation factor

Nodes move only if

$$E_{conserved} > E_m \tag{29.43}$$

(i.e., Nodes move only when it leads to an expected net energy savings). Since intermediate nodes forward packets to the next hop on the path to the user community,

$$\mu_n = \sum_{i=0}^{n-1} \mu_i \tag{29.44}$$

There is a natural tendency for nodes to drift toward their data sink (closer to the edge of the terrain.) To slow this, we allow nodes to move only with probability p.

In addition to reducing the migration of nodes to the terrain edge, p also reduces jitter. Nodes moving simultaneously to lower energy positions would cause oscillations, with nodes searching for optimal positions in response to neighbor movement. This would reduce the network lifetime. In some ways, this resembles the use of simulated annealing. Empirical testing found that the value of 0.3 for p worked best in this scenario.

29.4.3 Local Data Agreement

The final energy optimization we consider reduces the volume of traffic in the network. Sensors have a non-negligible false positive rate. For every false positive, detection packets are forwarded to the user community. In addition, every node that detects a target sends packets to notify the users. This redundant information drains system resources. It is useful to aggregate information locally and reduce the number of packets traversing the network.

In addition, we consider the relationship between detection threshold and signal power. Just as the transmission power determines the communications range, detection thresholds determine effective sensing ranges and false positive rates.

A Receiver Operating Characteristic Curve (ROC) is defined by the ratio of the true positive and false positive likelihoods as the detection threshold varies. A normalized ROC is typically used to determine the optimal detection threshold, which is the point on the ROC closest to point (0, 1). This is where the detection ratio is the closest to the ideal.

Lowering the detection threshold allows the sensor node to detect targets with weaker signals at the cost of having higher false positive rates. Implicitly, detecting weaker signals allows sensors to detect targets at a greater distance, since signals emitted to the target also decay with d^{α}. Note that this attenuation factor is in general different from the communications attenuation factor. We compensate for the increased type 2 error rate by combining readings from multiple sensor nodes.

In this approach, neighboring nodes exchange detection packets. The final detection decision is taken by a majority vote. We use closest point of approach (CPA) events for detection and the space-time clustering approach described in detail. When a node receives a target signal, it monitors the signal as long as the signal strength is increasing. This occurs as long as the target is approaching the sensor. When the target passes the node, the signal power decreases and a CPA event is declared. When a node has a CPA event, it broadcasts a detection packet to its neighbors. After a predetermined time interval, nodes that have had CPA events look at the detection packets they have received. The node whose CPA event had the largest power then attempts to determine whether or not the detection event was a true positive.

To discriminate between true and false positives, we use these four probabilities:

p_t—likelihood of a true positive.
q_t—likelihood of a false negative (type 1 error = $(1 - p_t)$).
p_f—likelihood of a false positive (type 2 error).
q_f—likelihood of a true negative = $(1 - p_f)$.

In a neighborhood of n nodes, the decision-making node receives k detection packets.

We now assume that both type 1 and type 2 errors are statistically independent events. This may not always be the case, (e.g., the background noise signal may resemble the spectrum of the target). We assume as well that no detection packets are dropped. Given these assumptions, we use a binomial distribution to calculate the group likelihood P_t (P_f) that the event is a true (false) positive when exactly k detections are reported:

$$P_t = \binom{n}{k} p_t^k q_t^{n-k} \tag{29.45}$$

$$P_f = \binom{n}{k} p_f^k q_f^{n-k} \tag{29.46}$$

Equations 29.45 and 26.46 express the problem for cases when nodes have the same detection threshold and relatively close to each other. Should this not be the case, the right hand side may be modified to be the product of probabilities that are unique for each node.

The decision making node computes Equations 29.45 and 29.46 and accepts the hypothesis that a target is present when the value of Equation 29.45 is greater than or equal to the value of Equation 29.46.

This approach introduces some additional overhead, in the form of one hop for one packet for every node signaling a detection event. In exchange, it greatly reduces the network traffic caused by false positives. It also serves to aggregate detection events within a space-time window, which reduces the volume of information transmitted by the network without affecting network reliability.

When combined with the ability to dynamically vary the detection threshold, this approach allows the system to compensate for the loss of sensor nodes by increasing the effective sensing range of specific nodes as needed without compromising network performance.

29.5 Summary

Large-scale sensor networks will require the ability to organize themselves and adapt around unforeseen problems. Both of these requirements imply that their behavior will be at least partially non-deterministic. Our experience shows that mobile code and peer-to-peer (P2P) networking are appropriate tools for implementing these systems.

One difficulty with implementing adaptive infrastructures of this type is that it is difficult to estimate their performance. This chapter shows how random graph models can be used to estimate many important design parameters for these systems. They can also be used to quantify system performance.

Specifically we have shown how to use random graph formalisms for wired and wireless P2P systems, such as those needed for sensor networks. Specifically, we have shown how to estimate

- Network redundancy
- Expected number of hops
- System dependability
- QoS issues are handled in [Kapur 02]
- System phase changes
- Vulnerability to intentional attack

Notations

a	edge availability $(1 - f)$
$\hat{a}$	effective edge availability including graph redundancy effects
$\hat{a}_r$	effective edge availability including effects due to redundant paths of r or fewer hops
b	constant factor in a scale-free probability distribution
C	cluster coefficient
c	number of nodes in fully connected components of a connected caveman small-world graph
d	degree of a node
d_{ave}	average node degree
E	set of edges defining a graph
e	number of rewired edges in a small-world graph
f	probability of link failure $(1 - a)$
h	the expected number of hops between two nodes in a graph
L	graph connectivity matrix
$l_i j$	element i, j of L
$M[h]$	mutuality factor for r hops
n	number of nodes in the graph
p	probability of an edge between two nodes for Erdös-Rényi graph

$P_{x,\ldots,\neg y,\ldots,z}$	shorthand for $P[\{q_x \wedge \cdots \wedge \neg q_y \wedge \cdots \wedge q_z\}]$
q_r	expected number of nodes reachable in r hops
$\{q_r\}$	set of nodes reachable in r hops
V	set of vertices defining a graph
z_r	expected number of nodes first reachable in r hops
$\{z_r\}$	set of nodes first reachable in r hops
γ	scaling factor in a small world graph

Acknowledgment and Disclaimer

This chapter is partially supported by the Office of Naval Research under Award No. N00014-01-1-0859 and by the Defense Advanced Research Projects Agency (DARPA) under ESP MURI Award No. DAAD19-01-1-0504 administered by the Army Research Office. Any opinions, findings, and conclusions or recommendations expressed in this publication are those of the author and do not necessarily reflect the views of the Office of Naval Research (ONR), Defense Advanced Research Projects Agency (DARPA), and Army Research Office (ARO).

References

[Aho 74] A. V. Aho, J. E. Hopcroft, and J. D. Ullman, *The Design and Analysis of Computer Algorithms*, Addison-Wesley, Reading, MA, 1974.

[Albert 00] R. Albert, H. Jeong, and A.-L. Barabási, Error and attack tolerance of complex networks, *Nature*, 406, 378–382, 2000.

[Albert 01] R. Albert and A.-L. Barabási, Statistical mechanics of complex networks, arXiv:cond-mat/0106096v1, June 2001.

[Barabási 99] A.-L. Barabási and R. Albert, Emergence of scaling in random networks, *Science*, 286, 509–512, 1999.

[Barabási 02] A.-L. Barabsi, *Linked*, Perseus, Cambridge, MA, 2002.

[Bollobás 01] B. Bollobás, *Random Graphs*, Cambridge University Press, Cambridge, U.K., 2001.

[Bonabeau 98] E. Bonabeau and F. Henaux, Graph partitioning with self-organizing maps, http://www.santafe.edu/sfi/publications/Abstracts/98-07-062abs.html

[Brooks 03] R. R. Brooks, P. Ramanathan, and A. Sayeed, Distributed target tracking and classification in sensor networks, *Proceedings of the IEEE*, Invited Paper, 91(8), 1163–1171, 2003.

[Brooks 04] R. R. Brooks, D. Friedlander, J. Koch, and S. Phoha, Tracking multiple targets with self-organizing distributed ground sensors, *Journal of Parallel and Distributed Computing*, Special Issue on Sensor Networks, 64(7), 874–884, 2004.

[Cvetkovic 79] D. M. Cvetkovic, M. Doob, and H. Sachs, *Spectra of Graphs*, Academic Press, New York, 1979.

[Dorogovtsev 01] S. N. Dorogovtsev and J. F. F. Mendez, Evolution of networks, arXiv:cond-mat/0106096v1, June 2001.

[Gu 01] M. Gu, H. Zha, C. Ding, X. He, and H. Simon, Spectral relaxation methods and structure analysis for k-way graph clustering and bi-clustering, Technical Report, Department of Computer Science and Engineering, CSE-01-007, Pennsylvania State University, University Park, PA, 2001.

[Janson 00] S. Janson, T. Luczak, and A. Rucinski, *Random Graphs*, John Wiley & Sons, New York, 2000.

[Kapur 02] A. Kapur, N. Gautam, R. R. Brooks, and S. Rai, Design, performance and dependability of a peer-to-peer network supporting QoS for mobile code applications, *Proceedings of the Tenth International Conference on Telecommunications Systems*, pp. 395–419, September 2002.

[Keiser 03] T. Keiser and R. R. Brooks, Mobile code daemons for networks of embedded systems, *IEEE Internet Computing*, 8(4), 72–79, 2004.

[Krishnamachari 01] B. Krishnamachari, S. B. Wicker, and R. Bejar, Phase transition phenomena in wireless ad-hoc networks, *Symposium on Ad-Hoc Wireless Networks, GlobeCom2001*, San Antonio, TX, November 2001. http://www.krishnamachari.net/papers/phaseTransitionWirelessNetworks.pdf

[Lv 02] Q. Lv, P. Cao, E. Cohen, K. Li, and S. Shenker, Search and replication in unstructured peer-to-peer networks, *International Conference on Supercomputing*, http:/doi.acm.org/10.1145/514191.514206, 2002.

[Newman 01] M. E. J. Newman, Ego-centered networks and the ripple effect or why all your friends are weird, Working Papers, Santa Fe Institute, Santa Fe, NM, http://www.santafe.edu/sfi/publications/workingpapers/01-11-066.pdf

[Newman 01a] M. E. J. Newman, S. H. Strogatz, and D. J. Watts, Random graphs with arbitrary degree distributions and their applications, arXiv:cond-mat/007235, May 7, 2001.

[Oram 01] A. Oram, *Peer-to-Peer: Harnessing the Power of Disruptive Technologies*, O'Reilly, Beijing, China, 2001.

[Press 1992] W. H. Press, S. A. Teukolsky, W. T. Vetterling, and B. P. Flannery, *Numerical Recipes in FORTRAN*, Cambridge University Press, Cambridge, U.K., 1992.

[Reittu 02] H. Reittu and I. Norros, On the power law random graph model of Internet, submitted for review, 2002.

[Stauffer 94] D. Stauffer and A. Aharony, *Introduction to Percolation Theory*, Taylor & Francis, London, U.K., 1994.

[Storras 01] R. Pastor-Storras and A. Vespignani, Epidemic spreading in scale-free networks, *Physical Review Letters*, 86(14), 3200–3203, 2001.

[Swanson 2005] D. C. Swanson, Environmental effects, Chapter 11. *Distributed Sensor Networks*, eds. S. S. Iyengar and R. R. Brooks, pp. 201–212, Chapman & Hall/CRC Press, Boca Raton, FL, 2005.

[Washburn 02] A. R. Washburn, *Search and Detection*, 4th edn., *INFORMS*, Linthicum, MD, 2002.

[Watts 99] D. J. Watts, *Small Worlds*, Princeton University Press, Princeton, NJ, 1999.

[Watts 02] D. J. Watts, S. H. Strogatz, and M. E. J. Newman, Personal correspondence.

30

On the Behavior of Communication Links in a Multi-Hop Mobile Environment

Prince Samar
Cornell University

Stephen B. Wicker
Cornell University

30.1 Introduction

In ad hoc and sensor networks, the hardware for the network nodes is designed to be compact and lightweight to enable versatility as well as easy mobility. The transceivers of the nodes are thus constrained to run on limited power batteries. In order to conserve energy, nodes restrict their transmission power, allowing direct communication only with those nodes that are within their geographical proximity. To communicate with distant nodes in the network, a node relies on multi-hop communication, whereby the source's data packets get forwarded along communication links between multiple pairs of nodes forming the route from the source to the destination.

As ad hoc and sensor networks do not require any pre-existing infrastructure and are self-organizing and self-configuring, they are amenable to a multitude of applications in diverse environments. These include battlefield deployments, where the transceivers may be mounted on Unmanned Aerial Vehicles (UAVs) flying overhead, on moving armored vehicles or may be carried by soldiers on foot. They may be used for communication during disaster relief efforts or law enforcement operations in hostile environment. Such networks may be set up between students in a classroom or delegates at a convention center. Chemical, biological or weather-related sensors may be spread around on land or on flotation devices at sea to monitor the environment and convey related statistics. Sensors may even be mounted on animals (e.g., whales, migratory birds, and other endangered species) to collect biological and environmental data.

With such a varied range of applications envisioned for ad hoc and sensor networks, the nodes in the network are expected to be mobile. Due to limited transmission range, this implies that the set of communication links of a particular node may undergo frequent changes. These changes in the set of links of a node affect not only the node's ongoing communication, but may impede the communication of other nodes as well, due to the distributed, multi-hop nature of such networks.

As the capacity and communication ability of ad hoc and sensor networks are dependent on the communication links [5], it is important to understand how the links of a node behave in a mobile environment. In this chapter, we will analyze some of the important link properties of a node. The aim of the study is to gain an understanding of how the links behave and their properties vary depending on the network characteristics. The intuition developed can then be applied to design effective protocols for ad hoc and sensor networks.

The rest of the paper is organized as follows. In Section 30.2, we discuss related work on characterizing the link behavior in an ad hoc or sensor network. Various properties of the links of a node in a mobile environment are derived in Section 30.3. In Section 30.4, we validate the derived expressions with simulation results. Section 30.5 discusses some applications of the derived properties and Section 30.6 concludes the paper.

30.2 Related Work

Simulation has been the primary tool utilized in the literature to characterize and evaluate link properties in ad hoc and sensor networks. Some efforts have been directed at designing routing schemes that rely on identification of stable links in the network. Nodes make on-line measurements in order to categorize stable links, which are then preferentially used for routing. In Associativity-Based Routing (ABR) [16], nodes generate a beacon regularly to advertise their presence to their neighbors. A count of the number of beacons received from each neighbor is maintained in the form of associativity "ticks" which indicate the stability of a particular link. In Signal Strength-based Adaptive Routing (SSA) [3], received signal strength is also used in addition to location stability to quantify the reliability of a link. A routing metric is employed to select paths that consist of links with relatively strong signal strength and having an age above a certain threshold. Both of these approaches suffer from the fact that a link which is deemed stable based on past or current measurements may soon become unreliable as compared to those currently categorized as unstable, due to the dynamic nature of mobile environments.

The Route-Lifetime Assessment Based Routing (RABR) [1] uses an affinity parameter based on the measured rate of change of signal strength averaged over the last few samples in order to estimate the lifetime of a link. A metric combining the affinity parameter and the number of links in the route is then used to select routes for TCP traffic. However, shadow and multipath fading experienced by the received signal make the estimation of link lifetime very error-prone. [15] instead relies on information provided by a Global Positioning System (GPS) about the current positions and velocities of two neighboring nodes to predict the expiration time of a link.

Empirical distributions of link lifetime and residual link lifetime have been presented in Ref. [4] for different simulation parameters. Based on these results, two link stability metrics are also proposed to categorize stable links. The *edge effect* was identified in Ref. [9], which is the tendency of shortest routes in high density wireless networks to be unstable. This is because such routes are usually composed of nodes that lie at the edges of each others' transmission ranges, so that a relatively small movement of any node in the route is sufficient to break it. Estimated stability of links has been used as the basis of route caching strategies for reactive routing protocols [6].

Analytical study of link properties in a mobile network has been limited, partly due to the abstruse nature of the problem. Though a number of mobility models have been proposed and used in the literature [2], none of them is satisfactory for representing node mobility in general. The expected link lifetime of a node is examined for some simple mobility scenarios in Ref. [17]. It is shown that the expected link lifetime under Brownian motion is infinite, while under deterministic mobility it can be found explicitly, given the various parameters.

A random mobility model has been developed in Ref. [10], which is then used to quantify the probability that a link will be available between two nodes after an interval of duration t, given that the link exists between them at time t_0. This probability is then used to evaluate the availability of a path after a duration t, assuming independent link failures. This forms the basis of a dynamic clustering algorithm such that more reliable members get selected to form the cluster. However, selection of paths for routing using this criteria may not be practical, as the model considers a link to be available at time $t_0 + t$ even when it undergoes failure during one or more intervals between t_0 and $t_0 + t$. When a link of a route actively being used breaks, it may be necessary to find an alternate route immediately, instead of just waiting indefinitely for the link to become available again. Jiang et al. [7] tries to overcome this drawback by estimating the probability that a link between two nodes will be continuously available for a period T_p, where T_p is predicted based on the nodes' current movements.

A number of issues related to the behavior of links still remain unexplored. In this work, we develop an analytical framework in order to investigate some important characteristics of the links of a node in a mobile environment. The derived link properties can be instrumental in the design and analysis of networking algorithms, as illustrated by the discussion on the few example applications in Section 30.5.

30.3 Link Properties

Here, we derive analytical expressions for a number of link properties: (a) expected lifetime of a link, (b) probability distribution of link lifetime, (c) expected rate of new link arrivals, (d) probability distribution of new link interarrival time, (e) expected rate of link change, (f) probability distribution of link breakage interarrival time, (g) probability distribution of link change interarrival time, and (h) expected number of neighbors. These expressions will help us better understand the behavior of these properties and their dependence on various network parameters.

In order to model the network for the analyses, we make the following assumptions:

1. A node has a bidirectional communication link with any other node within a distance of R meters from it. The link breaks if the node moves to a distance greater than R.
2. A node in the network moves with a constant velocity which is uniformly distributed between a meters/second and b meters/second.
3. The direction of a node's velocity is uniformly distributed between 0 and 2π.
4. A node's speed, its direction of motion and its location are uncorrelated.
5. The locations of nodes in the network are modeled by a two-dimensional *Poisson Point Process* with intensity σ such that for a network region **D** with an area A, the probability that **D** contains k nodes is given by

$$\text{Prob}(k \text{ nodes in } \mathbf{D}) = \frac{(\sigma A)^k e^{-\sigma A}}{k!} \tag{30.1}$$

Assumption 1 implies that the Signal to Interference Ratio (SIR) remains high up to a certain distance R from the transmitter, enabling nearly perfect estimation of the transmitted signal. However, SIR drops beyond this distance, rapidly increasing the bit error rate (BER) to unacceptable levels. Though the shadow and multipath fading experienced by the received signal may make the actual transmission zone unsymmetrical, this is a fair approximation if all the nodes in the network use the same transmission power. This simplifying assumption is commonly used in the simulation and analysis of ad hoc and sensor networks.

Assumption 2 models a mobile environment where nodes are moving around with different velocities that are uniformly distributed between two limits. This high mobility model is chosen as it is challenging for network communication and can, thus, facilitate finding "worst-case" bounds on the link properties for a general scenario. It is to be noted that the intensity of mobility being modeled can be changed by appropriately choosing the two parameters, a and b.

Assumptions 2–5 characterize the aggregate behavior of nodes in a large network. Due to the large number of independent nodes operating in an ad hoc fashion, any correlation between nodes can be assumed to be insignificant. Although it is possible that some nodes may share similar objectives and may move together, a large enough population of autonomous nodes can be expected in the network so that the composite effect can be modeled by a random process.

Assumption 5 indicates the location distribution of nodes in the network at any time. Poisson processes model "total randomness," thus reflecting the randomness shown by the aggregate behavior of nodes in a large network. This assumption is frequently used to model the location of nodes in an ad hoc or cellular network. Using (30.1), it is easy to see that the expected number of nodes in **D** is equal to σA. Thus, σ represents the average density of nodes in the network.

30.3.1 Expected Link Lifetime

Figure 30.1 shows the transmission zone of a node (say node 1) which is a circle of radius R centered at the node. The figure shows the trajectory of another node (say node 2) entering the transmission zone of node 1 at A, traveling along AB, and exiting the transmission zone at B.

With respect to a stationary Cartesian coordinate system with orthogonal unit vectors $\hat{i}$ and $\hat{j}$ along the X and Y axes respectively, let the velocity of node 1 be $\overline{v}_1 = v_1\hat{i}$ and the velocity of node 2, which makes an angle θ with the positive X axis, be $\overline{v}_2 = v_2 \cos\theta\hat{i} + v_2 \sin\theta\hat{j}$. Hence, the relative velocity of node 2 with respect to node 1 is

$$\vec{v} \triangleq \vec{v}_{21} = \vec{v}_2 - \vec{v}_1 = (v_2 \cos\theta - v_1)\,\hat{i} + v_2 \sin\theta\hat{j} \tag{30.2}$$

Consider a Cartesian coordinate system $X'Y'$ fixed on node 1 such that the X' and Y' axes are parallel to $\hat{i}$ and $\hat{j}$ respectively, as shown in Figure 30.1. The magnitude of node 2's velocity in this coordinate system is

$$v \triangleq |\vec{v}| = \sqrt{v_1^2 + v_2^2 - 2v_1v_2\cos\theta} \tag{30.3}$$

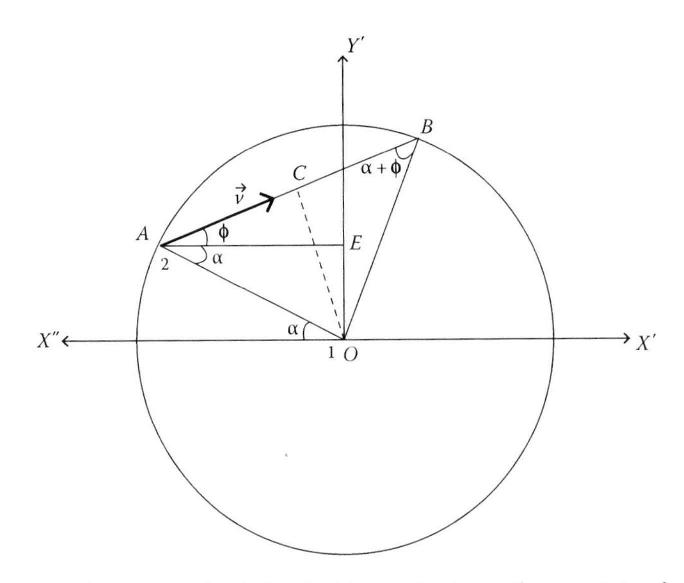

FIGURE 30.1 The transmission zone of node 1 at O with node 2 entering the zone at A and exiting at B.

and its direction of motion in this coordinate system, as indicated in Figure 30.1, is

$$\phi \triangleq \angle \vec{v} = \tan^{-1}\left(\frac{\sin\theta}{\cos\theta - v_1/v_2}\right) \tag{30.4}$$

Let the point of entry A of node 2 in node 1's transmission zone be defined by an angle α, measured clockwise from OX''. Thus, point A has coordinates $(-R\cos\alpha, R\sin\alpha)$ in the $X'Y'$ coordinate system. In Figure 30.1, $OA = OB = R$. AB makes an angle ϕ with the horizontal, which is the direction of the relative velocity of node 2. Line OC is perpendicular to AB. As OAB makes an isosceles triangle, $\angle OAB = \angle OBA = \alpha + \phi$. Therefore, $AC = BC = R\cos(\alpha + \phi)$. As θ and ϕ can have any value between 0 and 2π, the distance d_{link} that node 2 travels inside node 1's zone is

$$d_{link} = |2R\cos(\alpha + \phi)| = 2R|\cos(\alpha + \phi)| \tag{30.5}$$

Hence, the time that node 2 spends inside node 1's zone, which is equal to the time for which the link between node 1 and node 2 remains active, is

$$t_{link} = \frac{d_{link}}{|\vec{v}|} = \frac{2R|\cos(\alpha + \phi)|}{v} \tag{30.6}$$

The average link lifetime can be calculated as the expectation of t_{link} over v, ϕ, α:

$$\bar{T}_{link}(v_1) = E_{v\phi\alpha}[t_{link}(v, \phi, \alpha)] \tag{30.7}$$

Let the joint probability density function (PDF) of v, ϕ, α for nodes that enter the zone be $f_{v\phi\alpha}(v, \phi, \alpha)$. It can be expressed as

$$f_{v\phi\alpha}(v, \phi, \alpha) = f_{\alpha|v\phi}(\alpha | v, \phi) f_{v\phi}(v, \phi) \tag{30.8}$$

where

$f_{\alpha|v\phi}(\alpha|v,\phi)$ is the conditional probability density of α given the relative velocity $\bar{v}$
$f_{v\phi}(v,\phi)$ is the joint probability density of the magnitude v and phase ϕ of $\bar{v}$

Expressions for these PDFs are derived in Appendix 30.A.

Thus, the expected link lifetime can be calculated as

$$\begin{aligned}
\bar{T}_{link}(v_1) &= \int_{v=0}^{\infty} \int_{\phi=-\pi}^{\pi} \int_{\alpha=-\pi}^{\pi} t_{link} f_{v\phi\alpha}(v, \phi, \alpha) d\alpha d\phi dv \\
&= \int_{0}^{\infty} \int_{-\pi}^{\pi} f_{v\phi}(v, \phi) \left\{ \int_{-\pi}^{\pi} \frac{2R|\cos(\alpha + \phi)|}{v} f_{\alpha|v\phi}(\alpha | v, \phi) d\alpha \right\} d\phi dv \\
&= \frac{R}{2(b-a)} \int_{0}^{\infty} \int_{0}^{\pi} \frac{1}{\sqrt{v^2 + v_1^2 + 2vv_1\cos\phi}} \\
&\quad \times \left\{ u\left(\sqrt{v^2 + v_1^2 + 2vv_1\cos\phi} - a\right) - u\left(\sqrt{v^2 + v_1^2 + 2vv_1\cos\phi} - b\right) \right\} dv d\phi
\end{aligned} \tag{30.9}$$

In order to eliminate the unit step function $u(\cdot)$ from the integral in (30.9), one needs to identify the values of v which satisfy the following two inequalities:

$$\sqrt{v^2+v_1^2+2vv_1\cos\phi}-a\geq 0 \tag{30.10}$$

$$\sqrt{v^2+v_1^2+2vv_1\cos\phi}-b<0 \tag{30.11}$$

The range of $v \geq 0$ satisfying (30.10) and (30.11) are

- $v\in\left[0,-v_1\cos\phi+\sqrt{b^2-v_1^2\sin^2\phi}\right]$ if $0\leq\phi\leq\left(\pi-\sin^{-1}\left(\frac{a}{v_1}\right)\right)$
- $v\in\left[0,-v_1\cos\phi-\sqrt{a^2-v_1^2\sin^2\phi}\right]\bigcup\left[-v_1\cos\phi+\sqrt{a^2-v_1^2\sin^2\phi},-v_1\cos\phi+\sqrt{b^2-v_1^2\sin^2\phi}\right]$

  if $\left(\pi-\sin^{-1}\left(\frac{a}{v_1}\right)\right)\leq\phi\leq\pi$

Hence,

$$\begin{aligned}\bar{T}_{link}(v_1)=\frac{R}{2(b-a)}&\left(\int_0^{\pi-\sin^{-1}(a/v_1)}\int_0^{-v_1\cos\phi+\sqrt{b^2-v_1^2\sin^2\phi}}\frac{1}{\sqrt{v^2+v_1^2+2vv_1\cos\phi}}\,dv\,d\phi\right.\\&+\int_{\pi-\sin^{-1}(a/v_1)}^{\pi}\left\{\int_0^{-v_1\cos\phi-\sqrt{a^2-v_1^2\sin^2\phi}}\frac{1}{\sqrt{v^2+v_1^2+2vv_1\cos\phi}}\,dv\right.\\&\left.\left.+\int_{-v_1\cos\phi+\sqrt{a^2-v_1^2\sin^2\phi}}^{-v_1\cos\phi+\sqrt{b^2-v_1^2\sin^2\phi}}\frac{1}{\sqrt{v^2+v_1^2+2vv_1\cos\phi}}\,dv\right\}d\phi\right)\end{aligned} \tag{30.12}$$

(30.12) can be simplified to give

$$\bar{T}_{link}(v_1)=\frac{R}{2(b-a)}\left(\int_0^{\pi}\log\left|\frac{b+\sqrt{b^2-v_1^2\sin^2\phi}}{v_1+v_1\cos\phi}\right|d\phi-\int_{\pi-\sin^{-1}(a/v_1)}^{\pi}\log\left|\frac{a+\sqrt{a^2-v_1^2\sin^2\phi}}{a-\sqrt{a^2-v_1^2\sin^2\phi}}\right|d\phi\right) \tag{30.13}$$

In particular, if $a = 0$, the above expression reduces to

$$\bar{T}_{link}(v_1)=\frac{R}{2b}\left(\int_0^{\pi}\log\left|\frac{b+\sqrt{b^2-v_1^2\sin^2\phi}}{v_1+v_1\cos\phi}\right|d\phi\right) \tag{30.14}$$

(30.13) cannot be integrated into an explicit function. However, it can be numerically integrated to give the expected link lifetime for the chosen distribution of mobility in the network.

Figure 30.2 plots the expected link lifetime for a node as a function of its velocity. The velocity of the nodes in the network is assumed to be uniformly distributed between [0,40] m/s. As can be observed from the plot, the expected link lifetime for a node decreases rapidly as its velocity is increased. As an illustration, links last almost three times longer, on average, for a node moving with a velocity of 5 m/s as compared to a node moving with a velocity of 40 m/s. Also, as can be seen from (30.13), the expected link lifetime is directly proportional to the transmission radius R of a node.

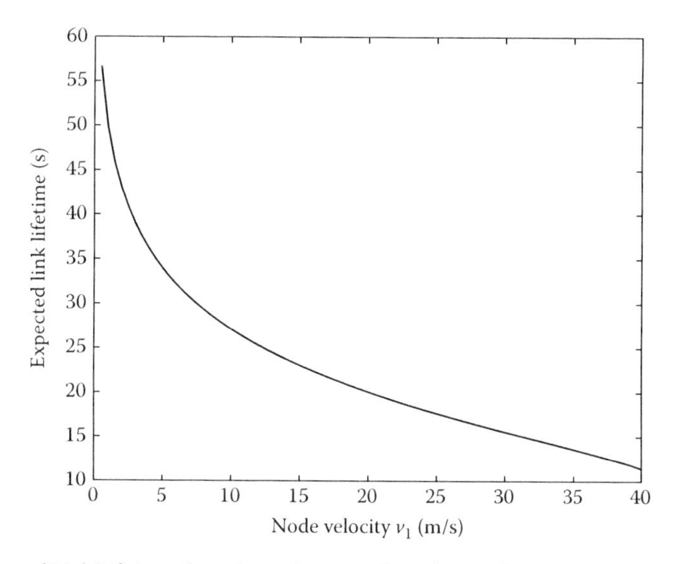

FIGURE 30.2 Expected Link Lifetime of a node as a function of its velocity, where $a = 0$ m/s, $b = 40$ m/s, and $R = 250$ m.

It is to be noted that Assumption 5 was not needed for determining the expected link lifetime and, thus, the derived expression is independent of the density of nodes in the network. This is because $\bar{T}_{link}(v_1)$ is averaged over link lifetimes corresponding to the range of velocities present in the network weighted by their probability density, without regard to how many or how often these links are formed.

30.3.2 Link Lifetime Distribution

For a particular node moving with a velocity v_1, the cumulative distribution function (CDF) of the link lifetime is given by

$$F_{link}^{v_1}(t) = \text{Prob}\{t_{link} \le t\} \tag{30.15}$$

Clearly, $F_{link}^{v_1}(t) = 0$ for $t < 0$. For $t \ge 0$, we have

$$\begin{aligned} F_{link}^{v_1}(t) &= \text{Prob}\left\{\frac{2R\,|\cos(\alpha+\phi)|}{v} \le t\right\} \\ &= 1 - \text{Prob}\left\{\frac{2R\,|\cos(\alpha+\phi)|}{v} > t\right\} \\ &= 1 - \text{Prob}\left\{|\cos(\alpha+\phi)| > \frac{vt}{2R}\right\} \end{aligned} \tag{30.16}$$

Now,

$$\begin{aligned} \text{Prob}\left\{|\cos(\alpha+\phi)| > \frac{vt}{2R}\right\} &= \text{Prob}\left\{-\cos^{-1}\left(\frac{vt}{2R}\right) - \phi \le \alpha \le \cos^{-1}\left(\frac{vt}{2R}\right) - \phi, v \le \frac{2R}{t}\right\} \\ &= \int_{\phi=-\pi}^{\pi} \int_{v=0}^{2R/t} \int_{\alpha=-\cos^{-1}(vt/2R)-\phi}^{\cos^{-1}(vt/2R)-\phi} f_{v\phi\alpha}(v,\phi,\alpha)\,d\alpha\,dv\,d\phi \\ &= \int_{\phi=-\pi}^{\pi} \int_{v=0}^{2R/t} f_{v\phi}(v,\phi)\left(\int_{-\cos^{-1}(vt/2R)-\phi}^{\cos^{-1}(vt/2R)-\phi} f_{\alpha|v\phi}(\alpha\,|\,v,\phi)\,d\alpha\right) dv\,d\phi \end{aligned} \tag{30.17}$$

Using the expression of $f_{\alpha|v\phi}(\alpha|v,\phi)$ from (30.46), (30.17) can be simplified to give

$$\text{Prob}\left\{|\cos(\alpha+\phi)|>\frac{vt}{2R}\right\}=\int_{\phi=-\pi}^{\pi}\int_{v=0}^{2R/t} f_{v\phi}(v,\phi)\cdot\sqrt{1-\left(\frac{vt}{2R}\right)^2}\,dvd\phi$$

$$=\frac{1}{\pi(b-a)}\int_0^{\pi}\int_0^{2R/t}\frac{v}{\sqrt{v^2+v_1^2+2vv_1\cos\phi}}\sqrt{1-\left(\frac{vt}{2R}\right)^2}\cdot\left\{u\left(\sqrt{v^2+v_1^2+2vv_1\cos\phi}-a\right)\right.$$

$$\left.-u\left(\sqrt{v^2+v_1^2+2vv_1\cos\phi}-b\right)\right\}dvd\phi \quad (30.18)$$

Substituting in (30.16), we get an expression for the cumulative distribution function of the link lifetime of a node moving with a velocity v_1:

$$F_{link}^{v_1}(t)=1-\frac{1}{\pi(b-a)}\int_0^{\pi}\int_0^{2R/t}\frac{v}{\sqrt{v^2+v_1^2+2vv_1\cos\phi}}\sqrt{1-\left(\frac{vt}{2R}\right)^2}$$

$$\cdot\left\{u\left(\sqrt{v^2+v_1^2+2vv_1\cos\phi}-a\right)-u\left(\sqrt{v^2+v_1^2+2vv_1\cos\phi}-b\right)\right\}dvd\phi \quad (30.19)$$

No closed-form solution for the integrals in (30.19) exists. However, (30.19) can be numerically integrated to give the cumulative distribution function of the link lifetime of a node moving with velocity v_1. Figure 30.3 plots the link lifetime CDF for different node velocities v_1, where a = 0 m/s, b = 40 m/s and R = 250 m.

The PDF $f_{link}^{v_1}(t)$ of link lifetime is found by differentiating (30.19) with respect to t. Figure 30.4 plots the PDF by numerically differentiating the curves in Figure 30.3. Note that for $v_1 > 0$, the point

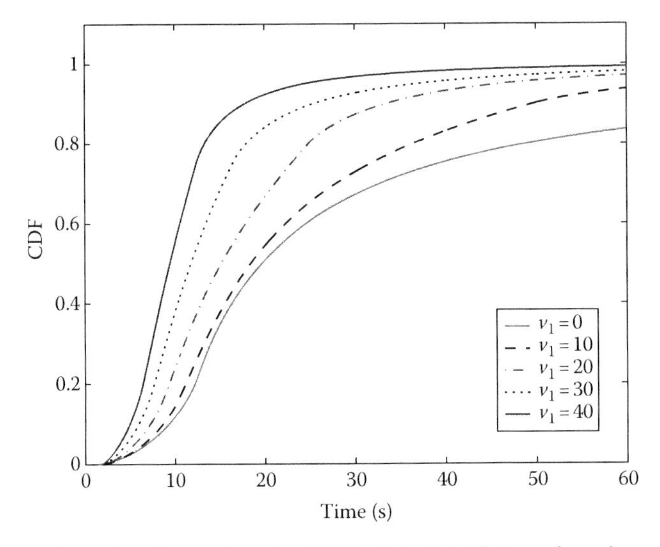

FIGURE 30.3 The cumulative distribution function (CDF) of the link lifetime of a node moving with velocity v_1, for a = 0 m/s, b = 40 m/s, and R = 250 m.

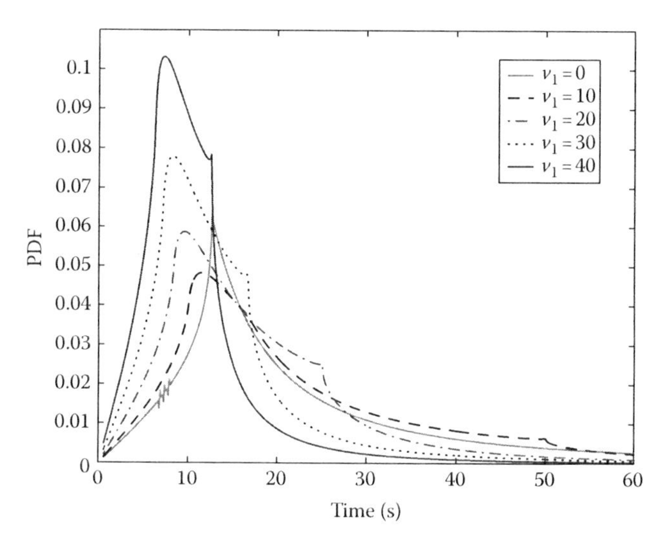

FIGURE 30.4 The PDF of the link lifetime of a node moving with velocity v_1, for $a = 0$ m/s, $b = 40$ m/s, and $R = 250$ m.

where the PDF curve is not differentiable corresponds to $t = 2R/v_1$. Also, it can be seen that the maxima of the PDF curve, which corresponds to the *mode* of the distribution, shifts towards the left as the node velocity increases.

As in Section 30.3.1, the derived expression does not depend on the density or location distribution of nodes in the network.

30.3.3 Expected New Link Arrival Rate

Consider Figure 30.5 which shows the transmission zone of node 1 moving with velocity $\bar{v}_1$ with respect to the stationary coordinate system XY, as defined before. For given values of v and ϕ, any node with relative velocity $\bar{v} = v\cos\phi\hat{i} + v\sin\phi\hat{j}$ with respect to node 1 can only enter node 1's transmission zone from a point on the semi-circle $\alpha \in [-((\pi/2) + \phi), (\pi/2) - \phi]$*, as seen in Appendix 30.A. Thus, a node with relative velocity $\bar{v}$ would enter the transmission zone within the next t seconds if it is currently located in the shaded region $\mathbf{D}_a$ of Figure 30.5, which is composed of all points at most vt meters away measured along angle ϕ from the semicircle $\alpha \in [-((\pi/2) + \phi), (\pi/2) - \phi]$.

The area of the shaded region $\mathbf{D}_a$ is $A = vt \cdot 2R$. Using Assumption 5, the average number of nodes in $\mathbf{D}_a$ is found to be equal to $2Rvt \cdot \sigma$. The average number of nodes in $\mathbf{D}_a$ with velocity $\bar{v}$ is equal to $2R\sigma vt \cdot f(v,\phi)dvd\phi$. This is just the average number of nodes with velocity $\bar{v}$ entering the zone within the next t seconds. The total expected number of nodes entering the zone within the next t seconds is found by integrating this quantity over all possible values of v and ϕ:

$$E\{\text{number of nodes entering the zone in } t \text{ seconds}\} = \eta(v_1)$$

$$= \int_{v=0}^{\infty}\int_{\phi=-\pi}^{\pi} 2R\sigma vtf(v,\phi)dvd\phi \qquad (30.20)$$

* α, as defined before, is the angle measured clockwise from the negative X axis of the coordinate system fixed on node 1.

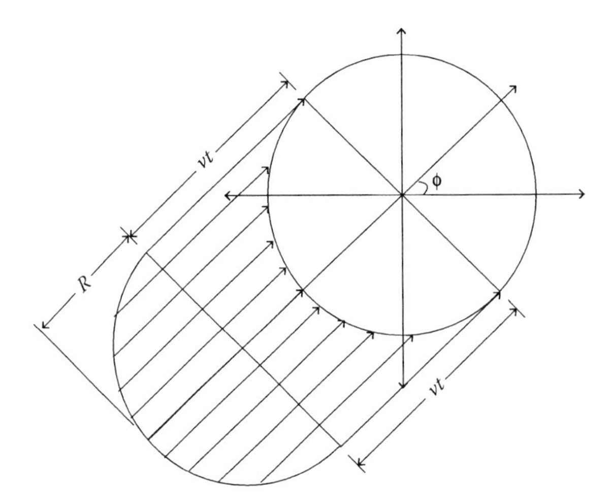

FIGURE 30.5 Calculation of expected new link arrival rate.

$f(v,\phi)$, the joint probability density of a node's relative velocity, has been derived in (30.56). Thus (30.20) can be expressed as

$$\eta(v_1) = \frac{R\sigma t}{\pi(b-a)} \int_{-\pi}^{\pi}\int_{0}^{\infty} \frac{v^2}{\sqrt{v^2+v_1^2+2vv_1\cos\phi}} \times \left\{ u\left(\sqrt{v^2+v_1^2+2vv_1\cos\phi}-a\right) - u\left(\sqrt{v^2+v_1^2+2vv_1\cos\phi}-b\right) \right\} dv d\phi$$

$$= \frac{2R\sigma t}{\pi(b-a)} \left(\int_{0}^{\pi} \int_{0}^{-v_1\cos\phi+\sqrt{b^2-v_1^2\sin^2\phi}} \frac{v^2}{\sqrt{v^2+v_1^2+2vv_1\cos\phi}} dv d\phi - \int_{\pi-\sin^{-1}(a/v_1)}^{\pi} \int_{-v_1\cos\phi-\sqrt{a^2-v_1^2\sin^2\phi}}^{-v_1\cos\phi+\sqrt{a^2-v_1^2\sin^2\phi}} \frac{v^2}{\sqrt{v^2+v_1^2+2vv_1\cos\phi}} dv d\phi \right) \quad (30.21)$$

The above can be simplified to give

$$\eta(v_1) = \frac{2R\sigma t}{\pi(b-a)} \left[b^2\left(\frac{v_1}{b}\right) - 2a^2\left(\frac{v_1}{a}\right) + a^2\left(\pi - \sin^{-1}\left(\frac{a}{v_1}\right), \frac{v_1}{a}\right) + \frac{v_1^2}{4} \times \int_{0}^{\pi} \{1+3\cos(2\phi)\}\log\left|\frac{b+\sqrt{b^2-v_1^2\sin^2\phi}}{v_1+v_1\cos\phi}\right| d\phi - \frac{v_1^2}{4}\int_{\pi-\sin^{-1}(a/v_1)}^{\pi} \{1+3\cos(2\phi)\}\log\left|\frac{a+\sqrt{a^2-v_1^2\sin^2\phi}}{a-\sqrt{a^2-v_1^2\sin^2\phi}}\right| d\phi \right] \quad (30.22)$$

where

(·) is the Complete Elliptic Integral of the Second Kind

(·,·) is the Incomplete Elliptic Integral of the Second Kind

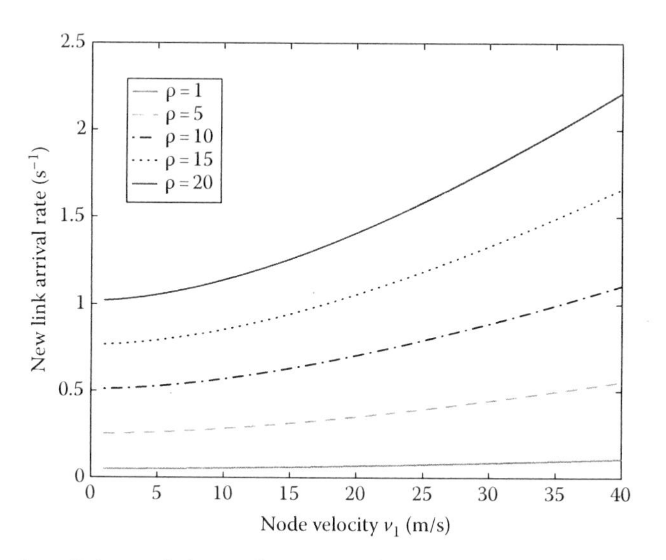

FIGURE 30.6 Rate of new link arrivals for a node moving with velocity v_1, where $a = 0$ m/s, $b = 40$ m/s, $R = 250$ m, and $\sigma = (\rho/\pi R^2)$ nodes/m².

Thus, the expected number of nodes entering the transmission zone per second, or equivalently, the rate of new link arrivals is given by

$$\dot{\eta}(v_1) = \frac{2R\sigma}{\pi(b-a)}\left[b^2\left(\frac{v_1}{b}\right) - 2a^2\left(\frac{v_1}{a}\right) + a^2\left(\pi - \sin^{-1}\left(\frac{a}{v_1}\right), \frac{v_1}{a}\right) + \frac{v_1^2}{4}\int_0^{\pi}\{1+3\cos(2\phi)\}\right.$$
$$\left.\times \log\left|\frac{b+\sqrt{b^2 - v_1^2\sin^2\phi}}{v_1 + v_1\cos\phi}\right| d\phi - \frac{v_1^2}{4}\int_{\pi-\sin^{-1}(a/v_1)}^{\pi}\{1+3\cos(2\phi)\}\log\left|\frac{a+\sqrt{a^2 - v_1^2\sin^2\phi}}{a-\sqrt{a^2 - v_1^2\sin^2\phi}}\right| d\phi\right] \quad (30.23)$$

When $a = 0$, $\dot{\eta}(v_1)$ reduces to

$$\dot{\eta}(v_1) = \frac{2R\sigma}{\pi b}\left[b^2\left(\frac{v_1}{b}\right) + \frac{v_1^2}{4}\int_0^{\pi}\{1+3\cos(2\phi)\}\log\left|\frac{b+\sqrt{b^2 - v_1^2\sin^2\phi}}{v_1 + v_1\cos\phi}\right| d\phi\right] \quad (30.24)$$

In Figure 30.6, we plot the expected rate of new link arrivals for a node moving with velocity v_1. While generating the curves, the values of the parameters are set to $a = 0$ m/s, $b = 40$ m/s, $R = 250$ m and $\sigma = (\rho/\pi R^2)$ nodes/m². Note that ρ represents the average number of nodes within a transmission zone.

An important point to observe from (30.23) is that the expected rate of new link arrivals for a node is directly proportional to the average density σ of nodes in the network. It is also directly proportional to the transmission radius R of a node.

30.3.4 New Link Interarrival Time Distribution

The cumulative distribution function of new link interarrival time is given by

$$F_{arrival}^{v_1}(t) = \text{Prob}\{\text{link interarrival time} \le t\} \quad (30.25)$$

$\mathbf{D_a}$, the shaded region of Figure 30.5, has an area $A = 2Rvt$. As seen in Section 30.3.3, a node with velocity $\bar{v} = v\cos\phi\hat{i} + v\sin\phi\hat{j}$ currently located in $\mathbf{D_a}$ will enter the transmission zone within the next t seconds. Thus, given $\bar{v}$, the probability that the link interarrival time is not more than t is equal to the probability that there exists at least one node in $\mathbf{D_a}$ with velocity $\bar{v}$. Therefore, using Assumption 5,

$$\begin{aligned}\text{Prob}\{\text{link interarrival time} \le t \mid v,\phi\} &= \text{Prob}\{\text{at least 1 node in } \mathbf{D_a} \mid v,\phi\} \\ &= 1 - \text{Prob}\{\text{no node in } \mathbf{D_a} \mid v,\phi\} \\ &= 1 - e^{-\sigma A} \\ &= 1 - e^{-2R\sigma tv} \end{aligned} \tag{30.26}$$

Hence the cumulative distribution function of new link interarrival time can be expressed as

$$F_{arrival}^{v_1}(t) = \iint_{v,\phi} (1 - e^{-2R\sigma tv}) f(v,\phi)\, dv\, d\phi \tag{30.27}$$

Substituting for $f(v,\phi)$ from (30.56),

$$\begin{aligned} F_{arrival}^{v_1}(t) = 1 - \frac{1}{\pi(b-a)} \int_0^{\pi}\int_0^{\infty} e^{-2R\sigma tv} \frac{v}{\sqrt{v^2 + v_1^2 + 2vv_1\cos\phi}} \\ \cdot \left\{ u\left(\sqrt{v^2 + v_1^2 + 2vv_1\cos\phi} - a\right) - u\left(\sqrt{v^2 + v_1^2 + 2vv_1\cos\phi} - b\right) \right\} dv\, d\phi \end{aligned} \tag{30.28}$$

The PDF $f_{arrival}^{v_1}(t)$ of new link interarrival time is given by

$$\begin{aligned} f_{arrival}^{v_1}(t) = \frac{d}{dt} F_{arrival}^{v_1}(t) = \frac{2R\sigma}{\pi(b-a)} \int_0^{\pi}\int_0^{\infty} e^{-2R\sigma tv} \frac{v^2}{\sqrt{v^2 + v_1^2 + 2vv_1\cos\phi}} \\ \cdot \left\{ u\left(\sqrt{v^2 + v_1^2 + 2vv_1\cos\phi} - a\right) - u\left(\sqrt{v^2 + v_1^2 + 2vv_1\cos\phi} - b\right) \right\} dv\, d\phi \end{aligned} \tag{30.29}$$

Figure 30.7 illustrates the new link interarrival time distribution for a node moving with velocity v_1, for $a = 0$ m/s, $b = 40$ m/s, $R = 250$ m, and $\sigma = 10/\pi R^2$ nodes/m^2. The corresponding new link interarrival time density for different node velocities v_1 is plotted in Figure 30.8. It can be observed that the new link interarrival time PDF curves drop rapidly as time t increases.

30.3.5 Expected Link Change Rate

Any change in the set of links of a node may be either due to the arrival of a new link or due to the breaking of a currently active link. Thus, the expected link change rate for a node is equal to the sum of the expected new link arrival rate and the expected link breakage rate. The expected new link arrival rate has been found earlier (see Equation 30.23).

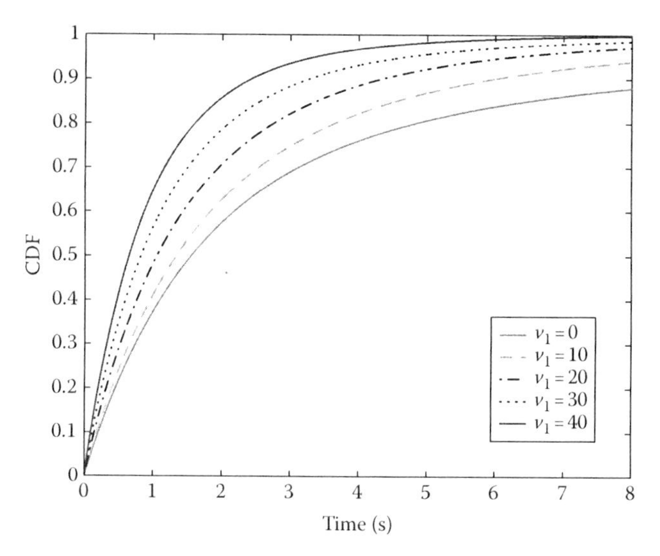

FIGURE 30.7 The cumulative distribution function (CDF) of new link interarrival time for a node moving with velocity v_1, for $a = 0$ m/s, $b = 40$ m/s, $R = 250$ m, and $\sigma = (10/\pi R^2)$ nodes/m^2.

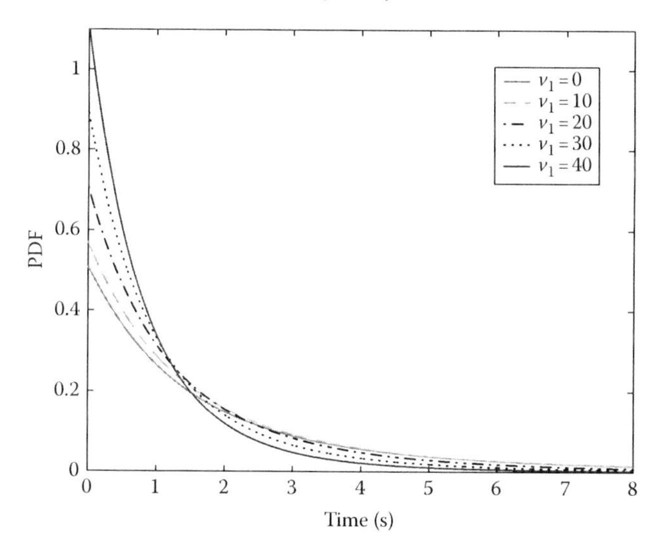

FIGURE 30.8 The PDF of new link interarrival time for a node moving with velocity v_1, for $a = 0$ m/s, $b = 40$ m/s, $R = 250$ m, and $\sigma = (10/\pi R^2)$ nodes/m^2.

In order to determine the expected link breakage rate, suppose that the network is formed at time $t = 0$. Let the total number of new link arrivals for a node between $t = 0$ and $t = t_0$ be $\eta(t_0)$ and the total number of link breakages for the node during the same interval be $\mu(t_0)$. Let the number of neighbors of the node at time $t = t_0$ be $N(t_0)$. Now,

$$\eta(t_0) - \mu(t_0) = N(t_0) \tag{30.30}$$

Dividing both the sides in (30.30) by t_0,

$$\frac{\eta(t_0)}{t_0} - \frac{\mu(t_0)}{t_0} = \frac{N(t_0)}{t_0} \tag{30.31}$$

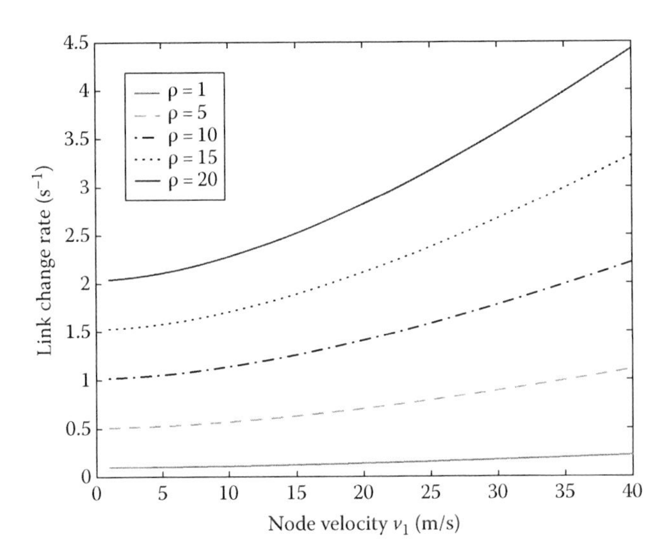

FIGURE 30.9 Expected link change arrival rate for a node moving with velocity v_1, where $a = 0\,\text{m/s}$, $b = 40\,\text{m/s}$, $R = 250\,\text{m}$, and $\sigma = (\rho/\pi R^2)$ nodes/m^2.

Taking the limit as $t \to \infty$ in (30.31), $\eta(t_0)/t_0$ equals the expected rate of new link arrivals $\dot{\eta}$ and $\mu(t_0)/t_0$ equals the expected rate of link breakages $\dot{\mu}$ (assuming ergodicity). If the number of neighbors of a node is bounded,* $(N(t_0)/t_0) \to 0$ as $t \to \infty$. This implies that $\dot{\mu} = \dot{\eta}$, that is, the expected rate of link breakages is equal to the expected rate of new link arrivals. Thus, the expected link change arrival rate $\dot{\gamma}(v_1)$ for a node moving with velocity v_1 is given by

$$\begin{aligned}\dot{\gamma}(v_1) &= \dot{\eta}(v_1) + \dot{\mu}(v_1)\\ &= 2\dot{\eta}(v_1)\end{aligned} \tag{30.32}$$

where $\dot{\eta}(v_1)$ is as expressed in (30.23).

The expected link change arrival rate as a function of the node velocity v_1 is plotted in Figure 30.9, where $a = 0\,\text{m/s}$, $b = 40\,\text{m/s}$, $R = 250\,\text{m}$, and $\sigma = (\rho/\pi R^2)$ nodes/m^2. Like $\dot{\eta}(v_1)$, $\dot{\gamma}(v_1)$ is also directly proportional to the average node density σ and the node transmission radius R.

30.3.6 Link Breakage Interarrival Time Distribution

In order to derive the link breakage interarrival time distribution, we proceed in a manner similar to Section 30.3.4. Consider Figure 30.10 showing the transmission zone of node 1. The shaded region $\mathbf{D}_b$ in the figure consists of all points not more than vt meters away along angle ϕ from the semicircle $\alpha \in [(\pi/2) - \phi, (3\pi/2) - \phi]$. It is easy to see that a node moving at an angle ϕ can break a link with node 1 only by moving out of its transmission zone from a point on this semicircle. Given its relative velocity $\bar{v} = v\cos\phi\hat{i} + v\sin\phi\hat{j}$, a node will leave the transmission zone of node 1 within the next t seconds—and thus breaking the link between the two—if it is currently located in $\mathbf{D}_b$. Note that $\mathbf{D}_b$ also includes the possibility of nodes that are currently outside the transmission zone of node 1 and have yet to form a link with it.

The area of the shaded region $\mathbf{D}_b$ is $A = 2Rvt$. For given v and ϕ, the probability that the link breakage interarrival time is not more than t is equal to the probability that there is at least one node in $\mathbf{D}_b$ with velocity $\bar{v}$.

$$\begin{aligned}\text{Prob}\{\text{link breakage interarrival time} \le t \mid v,\phi\} &= \text{Prob}\{\text{at least one node in } \mathbf{D}_b \mid v,\phi\}\\ &= 1 - e^{-\sigma 2Rvt}\end{aligned} \tag{30.33}$$

* Which is the case for any practical ad hoc or sensor network.

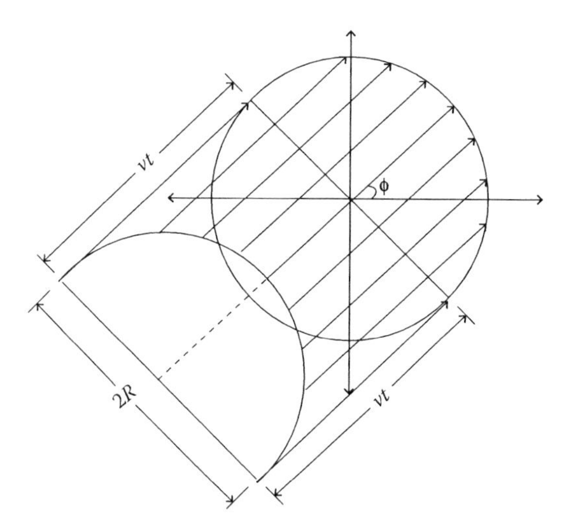

FIGURE 30.10 Calculation of link breakage interarrival time distribution.

Thus, the cumulative distribution function of link breakage interarrival time is given by

$$F_{break}^{v_1}(t) = \text{Prob}\{\text{link breakage interarrival time} \le t\}$$

$$= \iint_{v,\phi} \left(1 - e^{-2R\sigma t v}\right) f(v,\phi) dv d\phi \tag{30.34}$$

The right hand sides of (30.27) and (30.34) are the same, implying that the distributions of link breakage interarrival time and new link interarrival time are the same:

$$F_{break}^{v_1}(t) = F_{arrival}^{v_1}(t) = 1 - \frac{1}{\pi(b-a)} \int_0^{\pi}\int_0^{\infty} e^{-2R\sigma t v} \frac{v}{\sqrt{v^2 + v_1^2 + 2vv_1\cos\phi}}$$
$$\cdot\left\{u\left(\sqrt{v^2 + v_1^2 + 2vv_1\cos\phi} - a\right) - u\left(\sqrt{v^2 + v_1^2 + 2vv_1\cos\phi} - b\right)\right\} dv d\phi \tag{30.35}$$

Note that, using a different argument, it was already shown in Section 30.3.5 that the expected rate of link breakages is equal to the expected rate of new link arrivals.

30.3.7 Link Change Interarrival Time Distribution

Creation of a new link or expiry of an old link constitutes a change in the set of links of a node. Given its relative velocity $\bar{v} = v\cos\phi\hat{i} + v\sin\phi\hat{j}$, the existence of a node in the shaded region $\mathbf{D}_a$ of Figure 30.5 will cause the formation of a new link within the next t seconds. Likewise, a node with velocity $\bar{v}$ in the shaded region $\mathbf{D}_b$ of Figure 30.10 will cause the breaking of a link within the next t seconds. Figure 30.11 shows the union of these two shaded regions, $\mathbf{D}_c = \mathbf{D}_a \cup \mathbf{D}_b$. Given $\bar{v}$, a node currently located in the shaded region $\mathbf{D}_c$ of Figure 30.11 will cause a link change within the next t seconds.

The area A of $\mathbf{D}_c$ can be expressed as

$$A = \begin{cases} 2vtR + 2R^2\left(\sin^{-1}\left(\frac{vt}{2R}\right) + \frac{vt}{2R}\sqrt{1 - \left(\frac{vt}{2R}\right)^2}\right) & \text{if } vt \le 2R \\ 2vtR + \pi R^2 & \text{if } vt > 2R \end{cases} \tag{30.36}$$

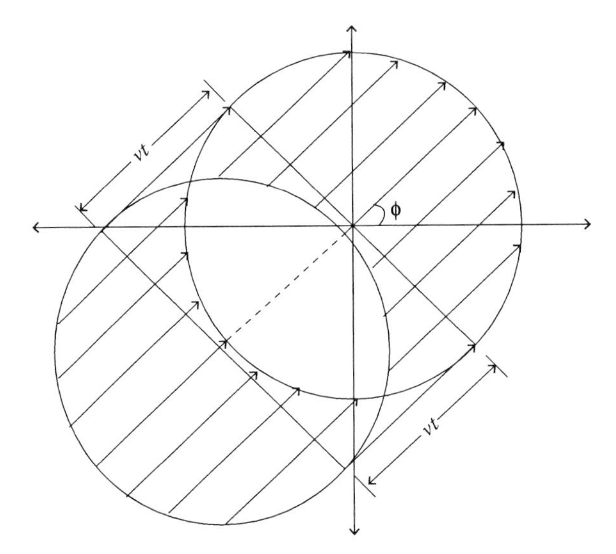

FIGURE 30.11 Calculation of link change interarrival time distribution.

From Assumption 5, as the nodes are assumed to be Poisson distributed,

$$\text{Prob}\{\text{no node in } \mathbf{D}_c \mid v,\phi\} = e^{-\sigma A} \tag{30.37}$$

Therefore, the link change interarrival time distribution is given by

$$\begin{aligned}
F_{change}^{v_1}(t) &= \text{Prob}\{\text{link change interarrival time} \le t\}\\
&= 1 - \text{Prob}\{\text{link change interarrival time} > t\}\\
&= 1 - \text{Prob}\{\text{new link interarrival time} > t, \text{link breakage interarrival time} > t\}\\
&= 1 - \iint_{v,\phi} \text{Prob}\{\text{no node in } \mathbf{D}_c \mid v,\phi\} f(v,\phi)\,dv\,d\phi\\
&= 1 - \iint_{v,\phi} e^{-\sigma A} f(v,\phi)\,dv\,d\phi\\
&= 1 - \frac{1}{\pi(b-a)}\left(\int_0^{\pi}\int_{2R/t}^{\infty} e^{-\sigma(2vtR+\pi R^2)} \frac{v}{\sqrt{v^2+v_1^2+2vv_1\cos\phi}} \cdot \left\{u\left(\sqrt{v^2+v_1^2+2vv_1\cos\phi}-a\right)\right.\right.\\
&\quad \left. - u\left(\sqrt{v^2+v_1^2+2vv_1\cos\phi}-b\right)\right\} dv\,d\phi\\
&\quad + \int_0^{\pi}\int_0^{2R/t} e^{-\sigma(2vtR+2R^2(\sin^{-1}(vt/2R)+vt/2R\sqrt{1-(vt/2R)^2}))} \frac{v}{\sqrt{v^2+v_1^2+2vv_1\cos\phi}}\\
&\quad \left.\cdot\left\{u\left(\sqrt{v^2+v_1^2+2vv_1\cos\phi}-a\right) - u\left(\sqrt{v^2+v_1^2+2vv_1\cos\phi}-b\right)\right\} dv\,d\phi\right)
\end{aligned} \tag{30.38}$$

It is not possible to explicitly evaluate the integrals in (30.38). In Figure 30.12, we plot the link change interarrival time distribution $F_{change}^{v_1}(t)$ for different node velocities v_1. $a = 0$ m/s, $b = 40$ m/s, $R = 250$ m

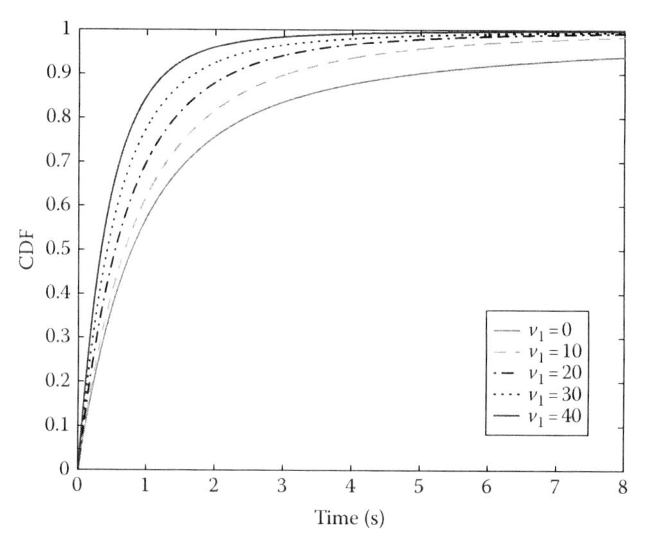

FIGURE 30.12 The cumulative distribution function (CDF) of link change interarrival time for a node moving with velocity v_1, for $a = 0$ m/s, $b = 40$ m/s, $R = 250$ m, and $\sigma = 10/\pi R^2$.

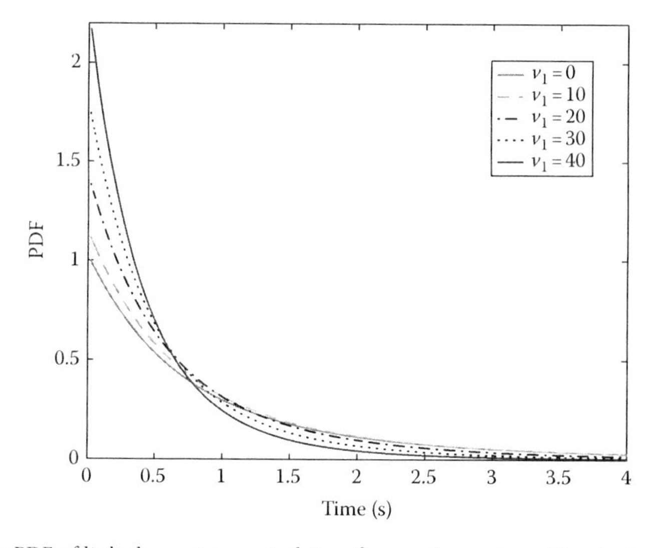

FIGURE 30.13 The PDF of link change interarrival time for a node moving with velocity v_1, for $a = 0$ m/s, $b = 40$ m/s, $R = 250$ m, and $\sigma = 10/\pi R^2$.

and $\sigma = (10/\pi R^2)$ nodes/m^2 have been used for the figure. In Figure 30.13, the corresponding link change interarrival time probability density $f_{change}^{v_1}(t)$ is plotted for the same parameter values.

It can be readily observed from the figure that the link change interarrival time density function decreases rapidly as time t increases. It is interesting to compare Figure 30.8 and Figure 30.13 which plot the PDFs of new link interarrival time (or link breakage interarrival time) and link change interarrival time respectively. The curves in Figure 30.13 appear to be scaled versions (by a factor of approximately 2, and then normalized) of the curves in Figure 30.8.

30.3.8 Expected Number of Neighbors

As the locations of nodes in the network are modeled as Poisson distributed random variables with intensity σ, the expected number of nodes located in an area A is equal to σA. This implies that the expected number of nodes in the transmission zone of a particular node is equal to $\sigma\pi R^2$. Therefore, the expected number of neighbors of a node is given by

$$\bar{N} = \sigma\pi R^2 - 1 \tag{30.39}$$

As expected, $\bar{N}$ increases with node density σ, but is independent of node mobility.

30.4 Simulations

In this section, we illustrate the validity of the analytically derived expressions of the link properties by comparing them to corresponding statistics collected from simulations.

The simulation set-up is as follows. The network consists of 200 nodes, where each node has a transmission radius R of 250 m. These nodes are initially spread randomly over a square region, whose sides are chosen to be equal to 1981.7 m each so that the node density σ turns out to be equal to $(10/\pi R^2)$ nodes/m^2 (or equivalently, $\rho = 10$). The velocity of the nodes is chosen to be uniformly distributed between $a = 0$ m/s and $b = 40$ m/s. A node's velocity is initially assigned a direction θ, which is uniformly distributed between 0 and 2π. When a node reaches an edge of the square simulation region, it is reflected back into the network area by setting its direction to $-\theta$ (horizontal edges) or $\pi - \theta$ (vertical edges). The magnitude of its velocity is not altered. The simulation duration is set to 240 min.

Statistics characterizing the link properties as a function of the node velocity v_1 are collected from the simulations. For the plots of Figures 30.14b, 30.15b, and 30.16b, the heights of the frequency bars have been normalized to make the total area covered by each of the histogram plot equal to unity.

From Figures 30.14 through 30.16, one can see that the theoretical curves are in fairly close agreement with the simulation results. The difference between the two is mainly attributed to the boundary effect present in the simulations. Nodes close to the boundary of the square simulation region experience fewer (or possibly, no) node arrivals from the direction of the boundary than otherwise expected. Also, when a node reaches the boundary of the simulation region, it gets reflected back into the network. Additional simulation studies suggest that the gap between the analytical and the experimental results decreases as the network size is increased while keeping the node density constant.

The results for the expected link breakage rate and the link breakage interarrival time density are similar to those in Figure 30.15 and are, therefore, omitted here.

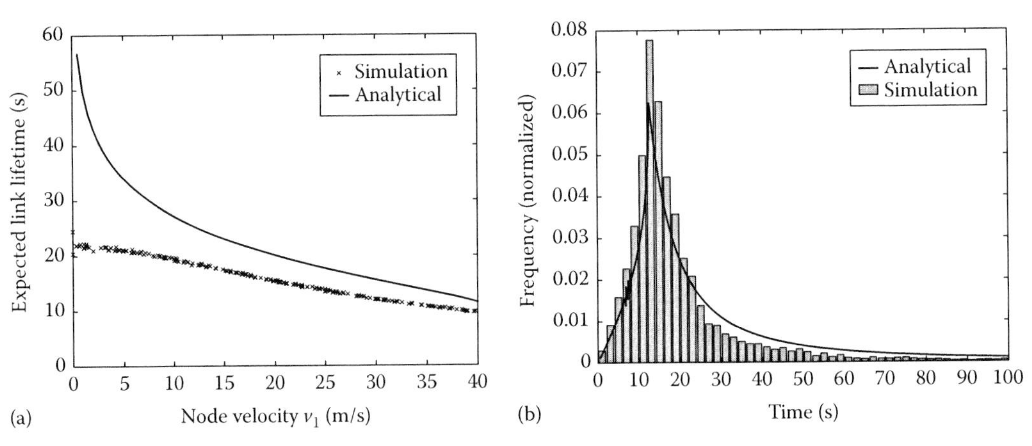

FIGURE 30.14 Comparison with simulation statistics: (a) Expected link lifetime (b) Link Lifetime PDF for a node with velocity $v_1 = 0$ m/s.

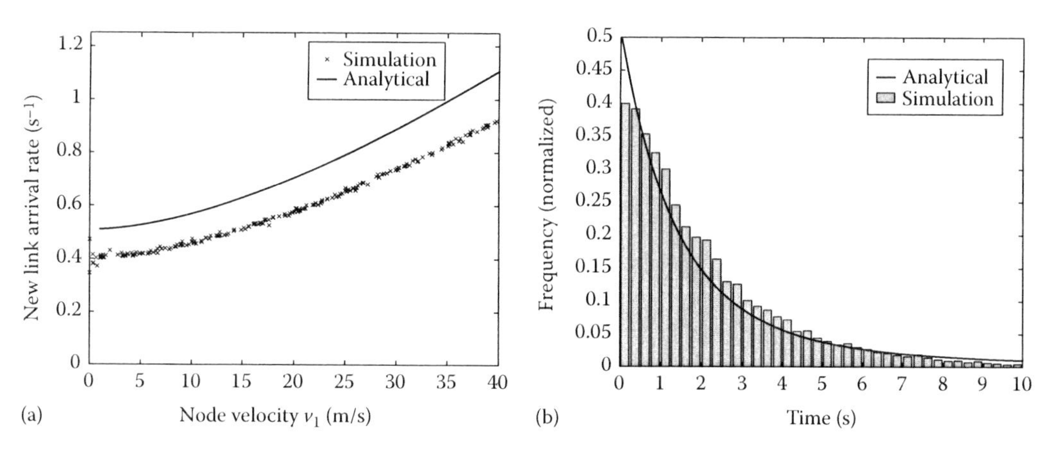

FIGURE 30.15 Comparison with simulation statistics: (a) Expected new link arrival rate (b) New link interarrival time PDF for a node with velocity $v_1 = 0$ m/s.

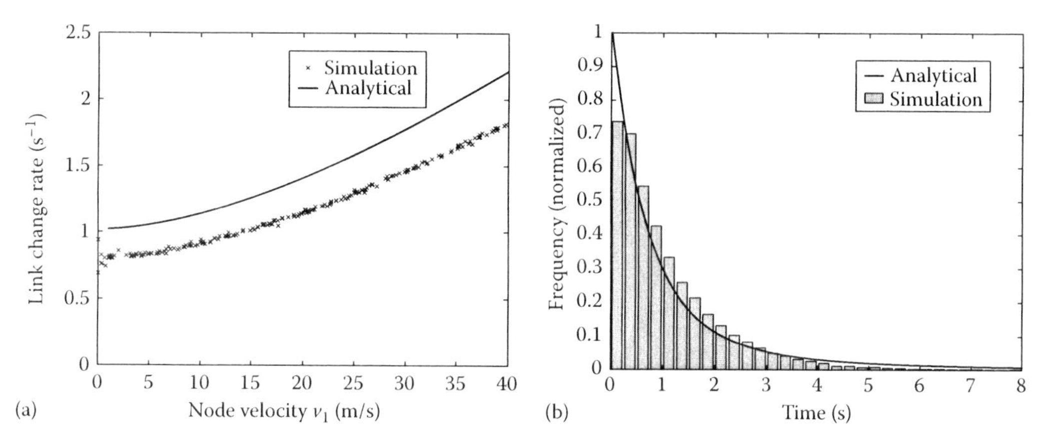

FIGURE 30.16 Comparison with simulation statistics: (a) Expected link change rate (b) Link change interarrival time PDF for a node with velocity $v_1 = 0$ m/s.

30.5 Applications of Link Properties

The various properties investigated in Section 30.3 characterize the behavior of the links of a node in a mobile environment. The derived properties can be used to design efficient algorithms for communication in ad hoc and sensor networks. They can also be used as a basis for analyzing the performance bounds of network protocols. In this section, we discuss some representative applications of the link properties studied in the previous section.

The link lifetime distribution can be used to examine the stability of links in the network. Once communication starts over a link, its residual lifetime distribution can be calculated as a function of the link lifetime distribution. Mathematically, the probability density $r_T^{v_1}(t)$ of residual link lifetime given that the link has been in existence for T seconds already can be expressed as

$$r_T^{v_1}(t) = \frac{f_{link}^{v_1}(t+T)}{1 - F_{link}^{v_1}(T)} \tag{30.40}$$

Here, $f_{link}^{v_1}(\cdot)$ and $F_{link}^{v_1}(\cdot)$ are the link lifetime PDF and CDF respectively, as derived in Section 30.3.2.

The residual link lifetime density can be used to evaluate the lifetime of a route in the network. For example, consider a route with K links and let $\mathbf{X}_1, \mathbf{X}_2, \ldots, \mathbf{X}_K$ be the random variables representing each of their residual lifetimes at the time when the route is formed, given that the links have already been in existence for $T_1, T_2, \ldots, T_K$ seconds respectively. Let $\mathbf{Y}$ be a random variable representing the lifetime of the route formed by the K links. As the route is deemed to have failed when any of the K links breaks, the route lifetime can be expressed as the minimum of the lifetimes of its constituent links:

$$\mathbf{Y} = \min(\mathbf{X}_1, \mathbf{X}_2, \ldots, \mathbf{X}_\mathrm{K}) \tag{30.41}$$

If we assume that the residual link lifetimes are independent and identically distributed (iid), then the distribution $F_Y(t)$ of $\mathbf{Y}$ can be calculated as

$$\begin{aligned} F_Y(t) &= \mathrm{Prob}\{\mathbf{Y} \le t\} \\ &= 1 - \mathrm{Prob}\{min(\mathbf{X}_1, \mathbf{X}_2, \ldots, \mathbf{X}_\mathrm{K}) > t\} \\ &= 1 - \mathrm{Prob}\{\mathbf{X}_1 > t, \mathbf{X}_2 > t, \ldots, \mathbf{X}_\mathrm{K} > t\} \\ &= 1 - \mathrm{Prob}\{\mathbf{X}_1 > t\} \cdot \mathrm{Prob}\{\mathbf{X}_2 > t\} \cdots \mathrm{Prob}\{\mathbf{X}_\mathrm{K} > t\} \\ &= 1 - (1 - R_{T_1}^{v_{1_1}}(t)) \cdot (1 - R_{T_2}^{v_{1_2}}(t)) \cdots (1 - R_{T_K}^{v_{1_K}}(t)) \end{aligned} \tag{30.42}$$

where $R_{T_i}^{v_{1_i}}(t)$ is the cumulative distribution function of the residual link lifetime of the ith link in the route, whose upstream node is moving with velocity v_{1_i}, given that the link was formed T_i seconds ago. $R_{T_i}^{v_{1_i}}(t)$ can be evaluated by integrating the corresponding density in (30.40).

The route lifetime distribution can be used to analyze the performance of routing protocols in ad hoc and sensor networks. It can also be used to provide Quality-of-Service (QoS) in the network. For example, the above framework can form the basis of schemes for selection of the best* set of routes for QoS techniques like Multi-path routing [11] and Alternate Path Routing [13].

Another application of the link properties is the optimal selection of the Time-to-Live (TTL) interval of route caches in on-demand routing protocols. For example, the work in Ref. [8] can be supplemented using the derived distributions in this paper to minimize the expected routing delay. It is also possible to develop alternate schemes to optimize other network performance metrics, if so desired.

Renewal theory [12] can be used to characterize the residual time $\mathbf{w}$ to arrival of the next link change after a given fixed instant t_0. Figure 30.17 shows the timeline where t_0 and $\mathbf{w}$ are indicated and the "x"s represent the arrival of link changes. The probability density $f_w^{v_1}(w)$ of $\mathbf{w}$ is given by

$$f_w^{v_1}(w) = \dot{\gamma}(v_1)\left[1 - F_{change}^{v_1}(w)\right] \tag{30.43}$$

where $\dot{\gamma}(v_1)$ and $F_{change}^{v_1}(w)$ are the expected link change arrival rate and the link change interarrival time distribution respectively, as found before. Similarly, given a fixed point t_0, the density of the residual time to arrival of the next new link or the next link breakage can be calculated by appropriately replacing the corresponding functions in (30.43).

* In terms of the particular QoS metric under consideration.

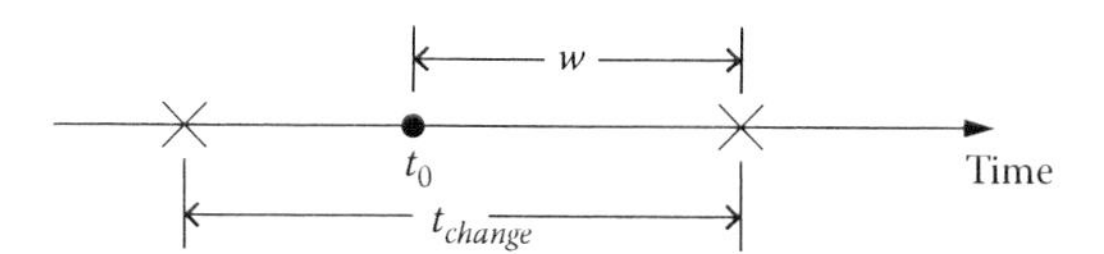

FIGURE 30.17 Timeline where the "x"s represent the arrival of link changes and t_0 is a fixed point.

Strategies for broadcasting of routing updates by proactive routing protocols have been proposed in Ref. [14]. These updating strategies are shown to lead to a considerable reduction in routing overhead while maintaining good performance in terms of other metrics. The design of these updating strategies are based on the assumption that link change interarrival times are exponentially distributed. However, the actual link change interarrival time distribution experienced by the nodes has been derived in Section 30.3.7. These updating strategies can be redesigned by utilizing the more realistic distributions as derived here. This would further improve the performance offered by these updating strategies.

30.6 Conclusions

Developing efficient algorithms for communication in multi-hop environments like ad hoc and sensor networks is challenging, particularly due to the mobility of nodes forming the network. An attempt has been made in this paper to develop an analytical framework that can provide a better understanding of network behavior under mobility. We derive expressions for a number of properties characterizing the creation, lifetime and expiration of communication links in the network.

This study can not only help analyze the performance of network protocols, but can also assist in developing efficient schemes for communication. This has been illustrated by the discussion on a few example applications of the derived link properties.

30.A Appendix: Joint Probability Density of v, ϕ, and α

Here we derive the joint PDF $f_{v\phi\alpha}(v,\phi,\alpha)$ for the nodes that enter the transmission zone of node 1:

$$f_{v\phi\alpha}(v,\phi,\alpha) = f_{\alpha|v\phi}(\alpha\,|\,v,\phi)f_{v\phi}(v,\phi) \tag{30.44}$$

$f_{\alpha|v\phi}(\alpha|v,\phi)$ is the conditional PDF of the angle α defining node 2's point of entry $(-R\cos\alpha, R\sin\alpha)$ into the transmission zone of node 1, given its relative velocity $\bar{v} = v\cos\phi\hat{i} + v\sin\phi\hat{j}$.* Now, given the direction ϕ of node 2's relative velocity, the node can only enter the zone from a point on the semi-circle $\alpha \in [-((\pi/2)+\phi), (\pi/2)-\phi]$. Consider the diameter of this semicircle, which is perpendicular to the direction of node 2's relative velocity. As nodes in the network are assumed to be randomly distributed, a node entering the zone with velocity $\bar{v}$ can intersect this diameter at any point on it with equal probability. This is illustrated in Figure 30.18, where the node's trajectory is equally likely to intersect the diameter QR at any point Q, $P1$, $P2$, ..., R on it, indicating that the probability of location of this point of intersection is uniformly distributed on the diameter.

In Figure 30.19, node 2 enters the transmission zone at T and travels along TV, which makes an angle ϕ with the horizontal. OT makes an angle α with OX''. QR is the diameter perpendicular to TV, defining the semicircle $\alpha \in [-((\pi/2)+\phi), (\pi/2)-\phi]$. Let OS = r, where S is the point of intersection of TV and QR. As OT = OV = R, it is easy to see that $r = R\sin(\alpha+\phi)$.

* Note that α is measured clockwise from the negative X axis.

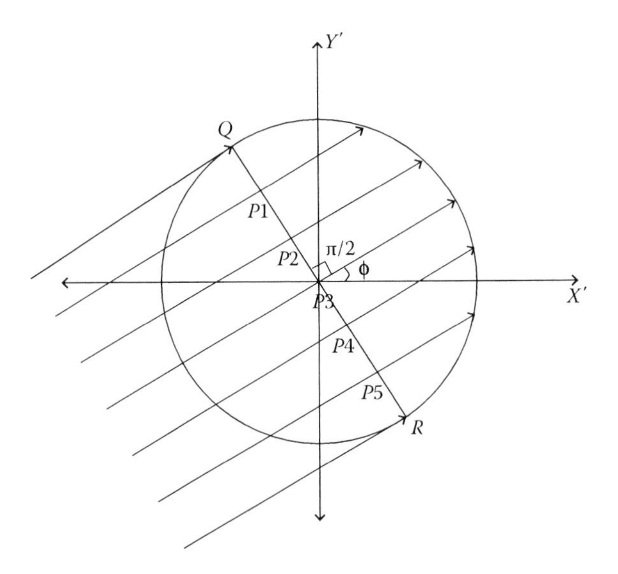

FIGURE 30.18 Given the direction of a node's relative velocity, it can intersect the diameter *QR* at any point on it with equal probability.

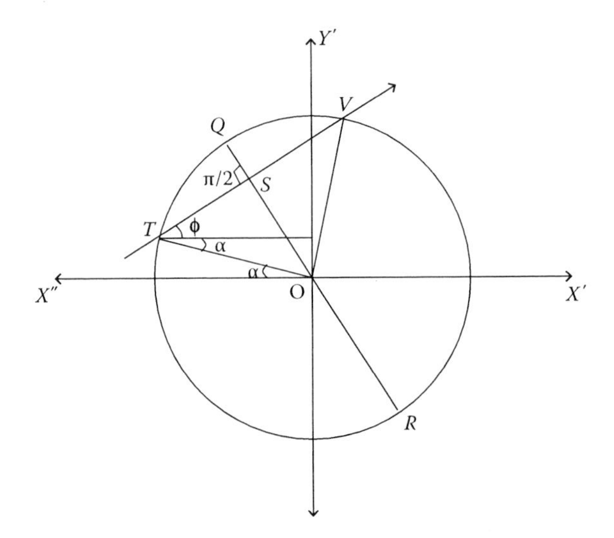

FIGURE 30.19 Calculation of $f_{\alpha|v\phi}(\alpha|v,\phi)$.

Let α be the random variable representing the angle defining the point of entry of node 2 in the zone. For $\alpha \in [-((\pi/2)+\phi), (\pi/2)-\phi]$,

$$F_{\alpha|v\phi}(\alpha \mid v,\phi) = \text{Probability}\{\alpha \le \alpha \mid v,\phi\}$$

$$= \int_{-R}^{r} \frac{1}{2R} dr$$

$$= \frac{r+R}{R}$$

$$= \frac{1}{2}(1+\sin(\alpha+\phi)) \tag{30.45}$$

Hence, by differentiating (30.45),

$$f_{\alpha|v\phi}(\alpha \mid v,\phi) = \begin{cases} \frac{1}{2}\cos(\alpha+\phi) & \alpha \in \left[-\left(\frac{\pi}{2}+\phi\right), \frac{\pi}{2}-\phi\right] \\ 0 & \text{otherwise} \end{cases}$$

$$= \frac{1}{2}\cos(\alpha+\phi)\left\{u\left(\alpha+\left(\frac{\pi}{2}+\phi\right)\right) - u\left(\alpha-\left(\frac{\pi}{2}-\phi\right)\right)\right\} \tag{30.46}$$

where, $u(\cdot)$ is the unit step function.

Note that for $\alpha \in [-((\pi/2)+\phi), (\pi/2)-\phi]$, $\cos(\alpha+\phi) \geq 0 \; \forall \phi \in [-\pi, \pi]$.

$f_{v\phi}(v,\phi)$ is the joint PDF of v and ϕ for the nodes that enter the zone. This is simply the density of the relative velocity $\bar{v}$ of the nodes in the network. It can be calculated by

$$f_{v\phi}(v,\phi) = \frac{f_{v_2\theta}(v_2^*, \theta^*)}{|J(v_2^*, \theta^*)|} \tag{30.47}$$

where

$f_{v_2\theta}(v_2^*, \theta^*)$ is the joint PDF of v_2 and θ

v_2^* and θ^* are the values of v_2 and θ that satisfy (30.3) and (30.4)

$$J(v_2,\theta) = \begin{vmatrix} \frac{\partial v}{\partial v_2} & \frac{\partial v}{\partial \theta} \\ \frac{\partial \phi}{\partial v_2} & \frac{\partial \phi}{\partial \theta} \end{vmatrix} \tag{30.48}$$

is the *jacobian* for the transformation.

Solving (30.3) and (30.4) for v_2^* and θ^* gives

$$\theta^* = \tan^{-1}\left(\frac{\sin\phi}{\cos\phi + v_1/v}\right) \tag{30.49}$$

$$v_2^* = \sqrt{v^2 + v_1^2 + 2vv_1\cos\phi} \tag{30.50}$$

Using (30.3) and (30.4) to get the derivatives for the *jacobian*,

$$J(v_2,\theta) = \begin{vmatrix} \frac{v_2 - v_1\cos\theta}{\sqrt{v_1^2 + v_2^2 - 2v_1v_2\cos\theta}} & \frac{v_1v_2\sin\theta}{\sqrt{v_1^2 + v_2^2 - 2v_1v_2\cos\theta}} \\ \frac{-v_1\sin\theta}{v_1^2 + v_2^2 - 2v_1v_2\cos\theta} & \frac{v_2^2 - v_1v_2\cos\theta}{v_1^2 + v_2^2 - 2v_1v_2\cos\theta} \end{vmatrix}$$

$$= \frac{v_2}{\sqrt{v_1^2 + v_2^2 - 2v_1v_2\cos\theta}} \tag{30.51}$$

Therefore,

$$J(v_2^*, \theta^*) = \frac{\sqrt{v^2 + v_1^2 + 2vv_1\cos\phi}}{v} \tag{30.52}$$

From Assumption 2, v_2 is uniformly distributed between a and b. Also, from Assumption 3, θ is uniformly distributed between 0 and 2π. Thus, their individual PDFs are given by

$$f_{v_2}(v_2) = \frac{1}{b-a}\{u(v_2 - a) - u(v_2 - b)\} \tag{30.53}$$

$$f_\theta(\theta) = \frac{1}{2\pi} \tag{30.54}$$

As v_2 and θ are assumed to be independent (Assumption 4), their joint PDF is simply the product of their individual density functions:

$$f_{v_2\theta}(v_2^\star, \theta^\star) = \frac{1}{2\pi(b-a)}\{u(v_2^\star - a) - u(v_2^\star - b)\} \tag{30.55}$$

Therefore, using (30.47), we get

$$f_{v\phi}(v,\phi) = \frac{1}{2\pi(b-a)} \frac{v}{\sqrt{v^2 + v_1^2 + 2vv_1\cos\phi}} \left\{u\left(\sqrt{v^2 + v_1^2 + 2vv_1\cos\phi} - a\right) - -u\left(\sqrt{v^2 + v_1^2 + 2vv_1\cos\phi} - b\right)\right\} \tag{30.56}$$

Hence, from (30.44), (30.46), and (30.56),

$$\begin{aligned} f_{v\phi\alpha}(v,\phi,\alpha) &= f_{\alpha|v\phi}(\alpha \mid v,\phi) f_{v\phi}(v,\phi) \\ &= \left(\frac{1}{2\pi(b-a)} \frac{v}{\sqrt{v^2 + v_1^2 + 2vv_1\cos\phi}} \left\{u\left(\sqrt{v^2 + v_1^2 + 2vv_1\cos\phi} - a\right)\right.\right. \\ &\quad \left.\left. -u\left(\sqrt{v^2 + v_1^2 + 2vv_1\cos\phi} - b\right)\right\}\right) - u\left(\sqrt{v^2 + v_1^2 + 2vv_1\cos\phi} - b\right)\Big\}\Big) \\ &\quad \cdot \left(\frac{1}{2}\cos(\alpha + \phi)\left\{u\left(\alpha + \left(\frac{\pi}{2} + \phi\right)\right) - u\left(\alpha - \left(\frac{\pi}{2} - \phi\right)\right)\right\}\right) \end{aligned} \tag{30.57}$$

Finally, the joint density of v, ϕ, and α is given by (30.57).

References

1. S. Agarwal, A. Ahuja, J.P. Singh, and R. Shorey, Route-lifetime assessment based routing (RABR) protocol for mobile ad-hoc networks, *IEEE International Conference on Communications 2000*, Vol. 3, pp. 1697–1701, New Orleans, LA, June 18–22, 2000.
2. T. Camp, J. Boleng, and V. Davies, A survey of mobility models for ad hoc network research, *Wireless Communication and Mobile Computing (WCMC)*: Special issue on *Mobile Ad Hoc Networking: Research, Trends and Applications*, 2(5), 483–502, 2002.
3. R. Dube, C. Rais, K.-Y. Wang, and S. Tripathi, Signal stability based adaptive routing (SSA) for ad hoc networks, *IEEE Personal Communications*, 4(1), 36–45, 1997.
4. M. Gerharz, C. de Waal, M. Frank, and P. Martini, Link stability in mobile wireless ad hoc networks, *IEEE Conference on Local Computer Networks (LCN'02)*, Tampa, FL, November 2002.

5. A.J. Goldsmith and S.B. Wicker, Design challenges for energy-constrained ad hoc wireless networks, *IEEE Wireless Communications*, 9(4), 8–27, 2002.
6. Y.-C. Hu and D.B. Johnson, Caching strategies in on-demand routing protocols for wireless ad hoc networks, *IEEE/ACM International Conference on Mobile Computing and Networking (MobiCom 2000)*, Boston, MA, August 6–11, 2000.
7. S. Jiang, D.J. He, and J.Q. Rao, A prediction-based link availability estimation for mobile ad hoc networks, *IEEE INFOCOM 2001*, Anchorage, AK, April 22–26, 2001.
8. B. Liang and Z.J. Haas, Optimizing route-cache lifetime in ad hoc networks, *IEEE INFOCOM 2003*, San Francisco, CA, March 30–April 3, 2003.
9. G. Lim, K. Shin, S. Lee, H. Yoon, and J.S. Ma, Link stability and route lifetime in ad-hoc wireless networks, *2002 International Conference on Parallel Processing Workshops (ICPPW'02)*, Vancouver, British Columbia, Canada, August 2002.
10. A.B. McDonald and T.F. Znati, A mobility-based framework for adaptive clustering in wireless ad hoc networks, *IEEE Journal in Selected Areas in Communications*, 17(8), 1466–1487, 1999.
11. P. Papadimitratos, Z.J. Haas, and E.G. Sirer, Path set selection in mobile ad hoc networks, *ACM Mobihoc 2002*, Lausanne, Switzerland, June 9–11, 2002.
12. A. Papoulis, *Probability, Random Variables, and Stochastic Processes*, 3rd edn., McGraw-Hill, New York, 1991.
13. M.R. Pearlman, Z.J. Haas, P. Scholander, and S.S. Tabrizi, On the impact of alternate path routing for load balancing in mobile ad hoc networks, *ACM MobiHOC'2000*, Boston, MA, August 11, 2000.
14. P. Samar and Z.J. Haas, Strategies for broadcasting updates by proactive routing protocols in mobile ad hoc networks, *IEEE MILCOM 2002*, Anaheim, CA, October 2002.
15. W. Su, S.-J. Lee, and M. Gerla, Mobility prediction and routing in ad hoc wireless networks, *International Journal of Network Management*, 11(1), 3–30, January/February 2001.
16. C.-K. Toh, Associativity-based routing for ad hoc networks, *Wireless Personal Communications*, 4(2), 103–139, 1997.
17. D. Turgut, S.K. Das, and M. Chatterjee, Longevity of routes in mobile ad hoc networks, *VTC Spring 2001*, Rhodes, Greece, May 6–9, 2001.

31

Localization in 3D Wireless Sensor Networks Based on Geometric Computation

Xin Shane Li
Louisiana State University

S. Sitharama Iyengar
Florida International University

31.1 Introduction

Geographic location information is critical in wireless sensor network. The distributed sensors should organize a coordinate system so that the physical coordinates of any node or any detected object can be determined. Integrating GPS receivers on all the sensors are too expensive for huge-sized networks, so localization using limited numbers of GPS resources is very important for many tasks in sensor networks. Localization must be accurate and reliable while the cost must be as little as possible. To solve this important problem, many algorithms, hardware, and applications have been explored recently. However, several key issues to support accurate localization for critical detections in complex and real environments have not been solved adequately (Figure 31.1).

To compute the localization, that is, to construct the global coordinate system from metrics measured from distributed sensors, most existing methods built the coordinates on a flattened metric space (e.g., on a 2D Euclidean plane), assuming sensors are distributed in a flat land. Such an assumption could often lead to significant distortions in reconnaissance tasks where very often sensors spread on curved terrains (2.5D) or underwater (3D, see the right figure, image from UCONN).

It is desirable to have a unified effective geometric-based localization paradigm for both centralized and distributed networks. It should work well for not only 2D sensors networks but also 2.5D (e.g., terrain scouting) and 3D (e.g., underwater reconnaissance and atmospheric monitoring) networks. It will also facilitate the robust localization in noisy and huge-sized military sensor networks and benefit many subsequent applications such as reliable routing, hole/boundary detection, and environment mapping and scouting.

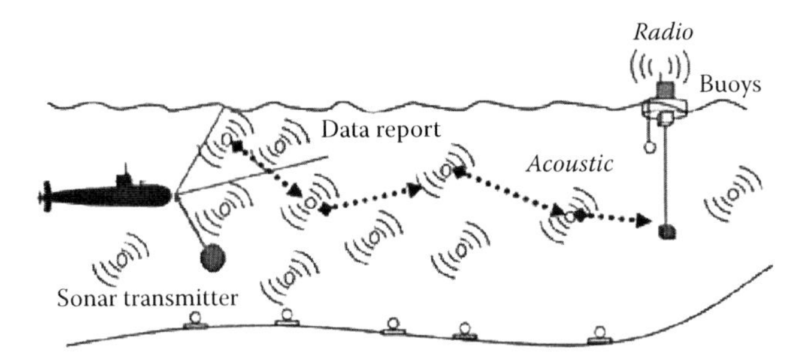

FIGURE 31.1 Accurate localization for critical detections in complex and real environments.

31.2 Research Problems and Ideas

We discuss a few challenging issues of localization in high-dimensional, topologically nontrivial, and huge-sized (therefore, maybe mere-connected) sensor networks and discuss potential solutions.

31.2.1 Optimizing Centralized Approaches

Global navigation satellite systems (such as GPS) have been widely employed for localization. However, integrating a GPS receiver on a sensor node is very expensive. Therefore, locating every node directly using GPS is prohibitive in large-scale sensor networks; sensor's coordinates are usually represented relatively, as rigid transformations away from some *beacon nodes* (also called anchor nodes). Beacon nodes (with GPS receivers, knowing their global coordinates) are expensive and ideally should be used as few as possible.

In a convex 2D (3D) region with no holes (voids), three noncollinear (four noncoplanar) beacons are required to unambiguously define a global coordinate system; while in more general environments with obstacles, topological holes, etc. (e.g., sensing in urban environments, underwater reconnaissance, and atmospheric monitoring), much more beacon points are necessary to ensure the accurate and robust localization within the sensor network.

Anchor placement has a significant impact on the accuracy of localization. Most existing research on this topic is based on experiments: usually existing literatures suggest that "localization accuracy improves if beacons are placed in a convex hull around the network," or "additional anchors in the center of the network are also helpful." Considerable evidence shows that planning beacon layout in the network is critical in improving the localization. Optimal anchor placement can be related to an *NP-hard problem* called *region guarding*. The problem is formulated as given a 3D region M-bounded by a discrete polygonal surface ∂M, where should a set of guarding camera $G = \{g_i\}$ be placed within M, so that any point in M can be visibly covered by at least one point in G. The optimal guarding, which has been shown to be NP-hard, seeks for the minimal guarding set G. Here, in the sensor localization problem, also due to the obstacles, walls, etc., a sensor within the interested network region may not be detectable from a given anchor. Therefore, we can consider the following discrete problem, given a set of spatially located sensors, upon which sensors we shall place the GPS, so that with fewest GPS we can build up the global coordinates that localize all the sensor nodes.

In Ref. [1], we propose an effective optimization framework using hierarchical integer linear programming to seek for the approximate optimal solution for 3D region guarding. This scheme converts the guarding problem into a set-covering problem and then computes its approximate solution. The conversion can be similarly applied here: sensor nodes are natural sampled points in the set and each of them can cover a subset of neighboring nodes with relatively high accuracy (such a reliable subset can then be associated with this node). The computation of intelligent placement of all the anchor sensors could certainly help the design of the localization sensor network; its hierarchical structure also suggests a way to design the centralized multitire networks.

In a sensor network with *mobile beacons*, beacons are moving around to assist other ordinary sensors in localizing themselves. Popular methods include radio-frequency-based mobile beacon placement [2], which is ubiquitous and experimentally effective. However, one wants to ask not only a fundamental but also a practical problem, how to move available beacons so that we can cover the entire network most thoroughly and efficiently? This reduces to a region covering problem with mobile guards. Generalizing the static guarding solvers [1] to mobile guarding problem can indicate optimal trajectories for moving beacons.

31.2.2 Optimizing Distributed Approaches Based on Geometric Mapping

Localization in a distributed sensor network can be considered as finding its least-distorted embedding in a lower dimensional space such as in 2D or 3D Euclidean space or on a 2D or 3D manifolds. The metric (distance between nodes) collected by sensors is usually distorted due to various noises, and our goal is to extract the metrics that respect the actual realization or the network in these spaces. This problem deeply roots in differential geometry and is often referred as surface and volumetric parameterization for 2D and 3D manifolds.

Popular localization approaches include multidimensional scaling (MDS) methods [3,4], neural network methods [5], and graph-rigidity-based methods [6], which all achieve high localization accuracy in 2D or 3D space. Accurate realization of the sensor network can also obtained using geometric curvature flow based on Riemannian geometry. Unlike the MDS-based methods that are time consuming and have low scalability, such an approach is efficient and directly applicable to 3D network with complicated topology; unlike neural-network-based methods that could be instable due to the nonconvexity of objective functions, such an approach could reveal the intrinsic geometry of the sensor network determined by its Riemannian metric and, in many cases, is convex whose unique global minimum can be effectively found in Ref. [7].

For example, in a 2D sensor network, the dense network can be treated as a discrete surface. The distance among points is determined by the Riemannian metric of the surface. Suppose the ground truth is a flattened surface. But when the metric is estimated by distances among sensor nodes, due to the measurement error, the estimated metric is curved; the localization problem now reduces to deforming the curved surface and embedding the detected metric to a region on the 2D plane with minimal stretching distortion. With effective surface parameterization algorithms [8] for 3D surfaces with arbitrary topologies or by controlling the boundary condition of the curvature flow, one can minimize the overall metric distortion. In surface with simple topology, this has been shown [9] to produce better localization compared with other popular approaches [3–5] mentioned earlier.

Also, the geometric realization and localization of sensor networks could be intrinsically curved, such that the localization of networks is deployed on a general 3D surface (e.g., terrains). While the target metric is curved, cross-surface parameterization techniques deforming the surface and reducing the curvature errors could be a natural tool to solve the localization. Similarly, for sensor networks that form 3D-manifolds (e.g., underwater networks that are intrinsically volumetric), volumetric parameterization algorithms [10] could provide effective solution for their localization.

31.2.3 Hierarchical Domain Partitioning for Localization in Huge-Sized Sensor Networks

One approach to increase the efficiency of algorithms in huge-sized sensor networks is through domain partitioning. Localization is firstly computed on each subnetwork; then subcoordinate systems are merged to form a global coordinate system. For example, a framework composed of cluster-based localization and a following subsequent intercluster transformation is discussed in Ref. [11] and tested in a simulated 2D floor plan with complicated walls (see the right figure, image from Ref. [11]). Given a geometric sensor network, one wants to study how to segment it into a set of subparts (whose localizations

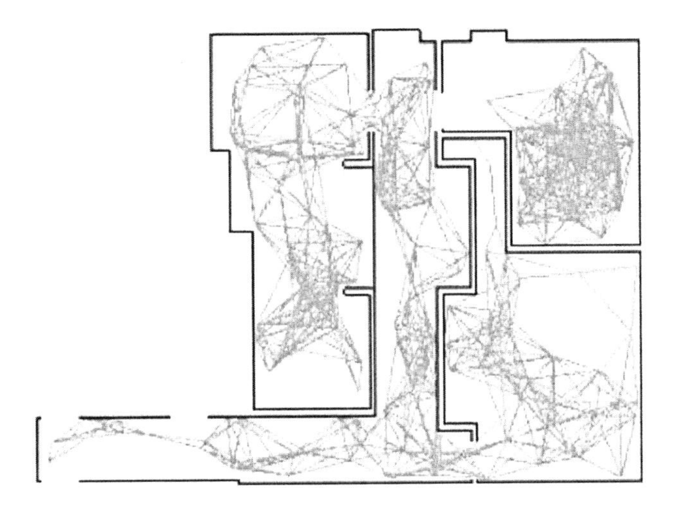

FIGURE 31.2 Shape partitioning methodologies.

are efficient). Segmenting complicated manifolds into topologically trivial subpatches is effective in solving the problem through divide-and-conquer schemes, especially when there are complex topological obstacles or geometric boundaries (e.g., walls or other sources that block the communication of sensors; so even they are close in Euclidean distance, they have long geodesic paths). Shape partitioning methodologies [12–14] deserve more exploration for their applications in this problem (Figure 31.2).

31.3 Further Discussions

The study of the localization problem gives rise to a unique cross-disciplinary area of sensor networking, computational geometry, and differential geometry. Abundant established theories and algorithms in computational geometry and Riemannian geometry can be applied to greatly facilitate the computation of the proper embedding of complicated network practically existed in engineering and scientific tasks. Efficient and accurate localizations are critical in order to facilitating subsequent tasks such as abnormal detection, reliable routing, etc. [15,16]. They can also be used in real-time autonomous robot environment exploration and mapping.

References

1. X. Li, W. Yu, X. Lin, and S. Iyengar, Optimizing autonomous pipeline inspection using 3D region guarding, *IEEE Transactions on Robotics*, to appear, 2011.
2. M. Sichitiu and V. Ramadurai, Localization of wireless sensor networks with a mobile beacon, In *Proceedings of IEEE International Conference on Mobile Ad-Hoc and Sensor Systems*, pp. 174–183, 2004.
3. V. Vivekanandan and V. Wong, Ordinal MDS-based localization for wireless sensor networks, *International Journal of Sensor Networks*, 1(3–4), 169–178, 2006.
4. Y. Shang and W. Ruml, Improved MDS-based localization, In *Proceedings of INFOCOM*, Piscataway, NJ, pp. 2640–2651, 2004.
5. G. Giorgetti, S. Gupta, and G. Manes, Wireless localization using self-organizing maps, In *Proceedings of IPSN*, New York, pp. 293–302, 2007.
6. Y. Wang, S. Lederer, and J. Gao, Connectivity-based sensor network localization with incremental Delaunay refinement method, In *Proceedings of INFOCOM*, Rio de Janeiro, Brazil, pp. 2401–2409, April 19–25, 2009.
7. X. Li, Y. Bao, X. Guo, M. Jin, X. Gu, and H. Qin, Globally optimal surface mapping for shapes of arbitrary topology, *IEEE Transactions on Visualization and Computer Graphics*, 14(4), 805–819, 2008.

8. M. Floater and K. Hormann, Surface parameterization: A tutorial and survey, In *Advances in Multiresolution for Geometric Modeling*, Spring-Verlag, New York, 2005.
9. M. Jin, S. Xia, H. Wu, and X. Gu, Scalable and fully distributed localization with mere connectivity, In *IEEE INFOCOM'11*, Shanghai, China, April 10–15, 2011.
10. X. Li, X. Guo, H. Wang, Y. He, X. Gu, and H. Qin, Meshless harmonic volumetric mapping using fundamental solution methods, *IEEE Transactions on Automation Science and Engineering*, 6(3), 409–422, 2009.
11. D. Moore, J. Leonard, D. Rus, and S. Teller, Robust distributed network localization with noisy range measurements, In *Proceedings of ACM Sensys-04*, 2004.
12. B. Chazelle and L. Palios, Decomposition algorithms in geometry, In *Algebraic Geometry and Its Applications*, Springer-Verlag, New York, pp. 419–447, 2005.
13. J. Keil, Polygon decomposition. In *Handbook of Computational Geometry*, J.-R. Sack and J. Urrutia, eds., Elsevier Science, Amsterdam, the Netherlands, 2000.
14. A. Shamir, A survey on mesh segmentation techniques, *Computer Graphics Forum*, 27(6), 1539–1556, 2008.
15. R. Sarkar, X. Zhu, J. Gao, L. Guibas, and J. S. B. Mitchell, Iso-contour queries and gradient descent with guaranteed delivery in sensor networks, In *Proceedings of 27th IEEE Conference on Computer Communications (INFOCOM'08)*, pp. 1175–1183, 2008.
16. R. Sarkar, X. Yin, J. Gao, F. Luo, and X. Gu, Greedy routing with guaranteed delivery using Ricci flows, In *Proceedings of Eighth International Symposium on Information Processing in Sensor Networks (IPSN'09)*, pp. 121–132, 2009.

IV

System Control

Wireless sensor networks are an important military technology with civil and scientific applications. Part IV emphasizes on deriving models and controllers for distributed sensor networks that consist of multiple cooperating nodes with each battery-powered node having wireless communications, local processing capabilities, sensor inputs, data storage, and limited mobility.

Zhu et al. (Chapter 32) focus on deriving a discrete event controller system for distributed surveillance networks that consist of three interacting hierarchies: sensing, communications, and command. The focus of their work is on deriving controllers using the following methods:

1. Petri Net
2. Finite state machine (FSM) using the Ramadge and Wonham approach
3. Vector addition control using the Wonham and Li approach

They compare the controllers in terms of expressiveness and performance and show that Petri Net model is concise and efficient, whereas the FSM model requires an offline state search, but its online implementation is less complex and the vector addition controller is essentially a Petri Net controller that enforces inequality constraints upon the system at runtime. They also present an innovative method for deriving finite state machine controller, which benefits from the use of Karp–Miller tree to represent all possible evolutions of the Petri Net plant model from an initial marking and Moore machine to generate control patterns in terms of current encoded state automatically.

In summary, Part IV elaborates on deriving a discrete event controller system for distributed surveillance networks.

32

Example DSN Control Hierarchy

Mengxia Zhu
Southern Illinois University

S. Sitharama Iyengar
Florida International University

Jacob Lamb
The Pennsylvania State University

Richard R. Brooks
Clemson University

Matthew Pirretti
The Pennsylvania State University

32.1 Introduction

In this chapter, we derive models of, and controllers for, distributed sensor networks consisting of multiple cooperating nodes. Each battery-powered node has wireless communications, local processing capabilities, sensor inputs, data storage, and limited mobility. An individual node would be capable of isolated operation, but practical deployment scenarios require coordination among multiple nodes. We are particularly interested in self-organization technologies for these systems. Network self-configuration is needed for the system to adapt to a changing environment [Bulusu 01]. In this chapter we derive hierarchical structures that support user control of the distributed system.

Our model uses discrete event dynamic systems (DEDS) formalisms. DEDS have discrete time and state spaces. They are usually asynchronous and non-deterministic. Many DEDS modeling and control methodologies exist and no dominant paradigm has emerged [Ramadge 87]. We use Petri Nets as will be described in Section 32.2, to model the plants to be controlled. Our sensor network model has three intertwined hierarchies, which evolve independently. We derive controllers to enforce system consistency constraints across the three hierarchies. Three equivalent controllers are derived using: (1) Petri Net, (2) vector addition and (3) Finite State Machine (FSM) techniques. We compare the controllers in terms of expressiveness and performance. Innovative use of Karp-Miller trees [David 92] allows us to derive FSM controllers for the Petri Net plant model. In addition, we show how FSM controllers can be derived automatically from control specifications in the proper format.

The remainder of the chapter is organized as follows. Section 32.2 gives a review of Petri Nets. Section 32.3 describes the structure of the network hierarchies. In Section 32.4, we provide control specifications. The controllers are derived in Section 32.5. Section 32.5 also provides brief tutorials on each approach. Section 32.6 provides experimental results from simulations run using the controllers.

32.2 Petri Nets

Carl Adam Petri defined Petri Nets as a graphic mathematical model for describing information flow in 1962. This model proved versatile in visualizing and analyzing the behavior of asynchronous, concurrent systems. Later research led to the direct application of Petri Nets in automata theory. Petri Nets are excellent for modeling the relationship between events, resources, and system states [Peterson 77].

A Petri Net is a bi-partite graph with two classes of nodes: *places* and *transitions*. The number of places and transitions are finite and non-zero. Directed arcs connect nodes. Arcs either connect a transition to a place or a place to a transition. Arcs can have an associated integer weight. DEDS state variables are represented by places. Events are represented by transitions. Places contain *tokens*.

The DEDS state space is defined by the *marking* of the Petri Net. A marking is a vector expressing the number of tokens in each place. A transition is enabled when the places with arcs incident to the transition all contain at least as many tokens as the weight of the associated arcs. The firing of a transition removes tokens from all places with arcs incident to the transition and deposits tokens in all places with arcs issuing from the transition. The number of tokens removed (added) is equal to the weight of the associated arc. The firing of a transition thus changes the marking of the Petri Net and the state of the DEDS system. Transitions fire one at a time even when more than one transition has been enabled. The system is non-deterministic in that any enabled transition can fire.

Mathematically, a Petri Net is represented as the tuple $S = (P, T, I, O, u)$ with P: Finite set of places, T: Finite set of transitions, I: Finite set of arcs from places to transitions, O: Finite set of arcs from transitions to places and u is an integer vector representing the current marking [David 92]. Figure 32.1 is a simple example of a Petri Net modeling the cycle of seasons. Safeness is a special issue to be considered. A Petri net is safe if all places contain no more than one token. However, a Petri net is called k-safe or k-bounded if no place contain no more than k tokens. An unbounded Petri net may contain infinite number of tokens in places and may have an infinite number of markings. Conversely, a bounded Petri net is essentially a finite state machine with each node corresponding to every reachable state. In deriving our controllers, we derive Karp-Miller trees from the Petri Nets [David 92]. Despite their name, Karp Miller trees are graph structures; they represent all possible markings a

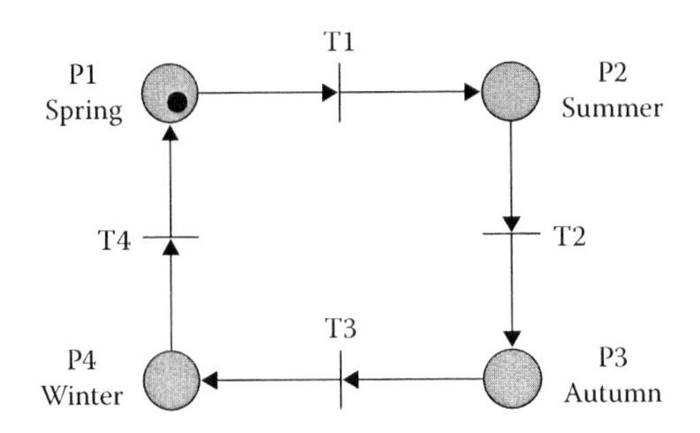

FIGURE 32.1 Petri Net model of the cycle of the seasons with four possible markings: {1000, 0100, 0010, 0001}.

Petri Net can reach from a given initial marking and ω is usually used to represent infinite number of tokens in a place if necessary. The algorithm for deriving the Karp-Miller tree is given in Section 32.5.

32.3 Hierarchy Models

32.3.1 Overview and Terminology

In an effort to thoroughly describe the functionality of a remote, multi-modal, mobile sensing network three issues must be addressed:

- Network communication—maintaining communications within the network.
- Collaborative sensing—coordinating sensor data interpretation.
- Operational command—assigning resources within the network and controlling internal system logistics.

Each hierarchy is composed of three separate levels:

- Root—is the top level of the hierarchy. It coordinates among cluster heads and provides top-level guidance.
- Cluster head—coordinates lower level controllers and propagates guidance from the root to lower layers.
- Leaf—performs low-level tasks and executes commands coming from the upper layers.

In this chapter, we provide a Petri Net plant model for each level of each hierarchy. The Petri Net models of the hierarchies can be found in Section 32.A.5 of the appendix.

We have identified numerous global consistency issues in the system that require a controller to constrain the actions taken by the hierarchies. These requirements were captured as control specifications and are used to derive the appropriate control structures.

Figure 32.2 shows the hierarchical relationship between the three nodes levels. To make the hierarchy adaptive, a cluster head can control any number of leaves. Similarly, a root node can coordinate an arbitrary number of cluster heads.

While there are three tiers within the network hierarchy design the design does not limit the physical network to only three levels. Networks which are intended to cover a large physical area, or to operate in a highly cluttered environment may require more nodes than can be effectively managed by three tiers. For this reason it is desirable to allow recursion within the hierarchy. Internal nodes can be inserted between the root node and cluster heads. Internal nodes are implemented by defining either root or

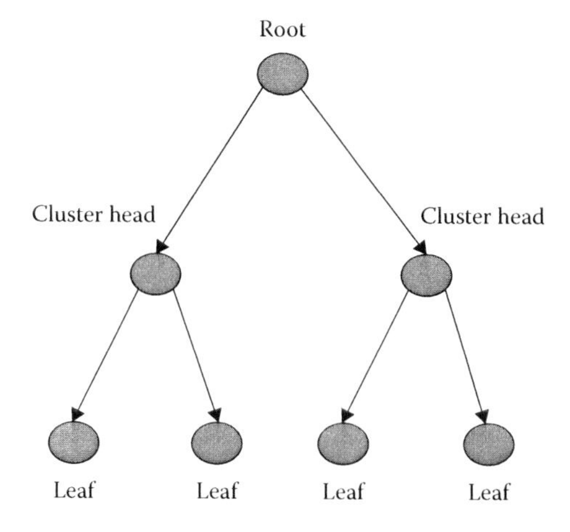

FIGURE 32.2 Relationships between three nodes levels.

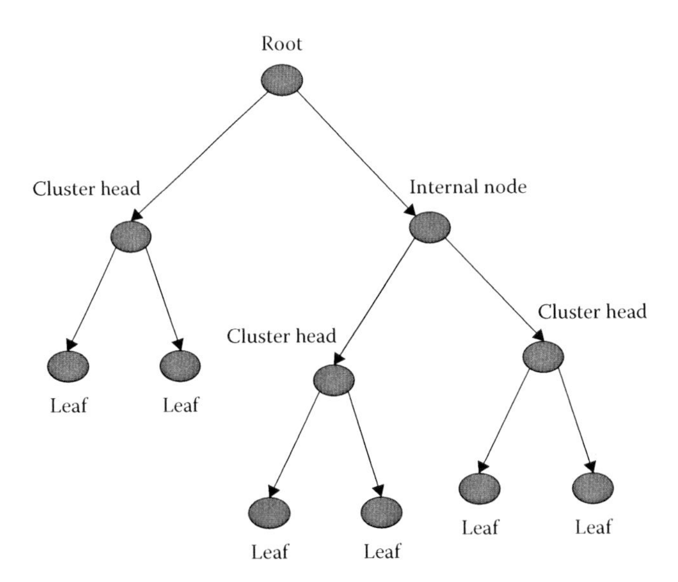

FIGURE 32.3 Example of a more complex structure.

cluster head nodes so that they can be connected recursively. This allows complex structures to arise as required by the mission. Figure 32.3 shows such a simple example.

In the network communication and collaborative sensing hierarchies, the root nodes are recursive. For example, in the network communication hierarchy, the root node's activities can be described in terms of interactions with a supervisor and data collection from a subnet. The root node expects from the subnet supervisor a set of data providing statistics on each covered area. Similarly, the root node reports to the subnet supervisor a set of data about the network or subnetwork that is being supervised by the root. In this manner, a communication network may in fact contain four or more levels. A network containing four levels would consist of a number of three-level subnets, each supervised by a root node. These root nodes at the third tier would each in turn report subnet statistics to an overseeing "master" root at the fourth tier. The master root would manage each of the three-level subnets according to subnet capacities. In other words, collections of cluster heads are subnets controlled by a root node. Combinations of cluster heads and root nodes can be controlled by another root node. In this manner the network may be expanded to manage an arbitrary level of complexity.

Recursion in the network communication and collaborative sensing hierarchies takes place at the root node; however, for the command and control hierarchy recursion takes place at the cluster head. As discussed previously the network communication and collaborative sensing network hierarchies are designed in a fashion in which supervising nodes at each level oversee the activities of subnets. This differs from the operational command hierarchy, where the top level of the hierarchy must be designed as a supervisor overseeing the network as opposed to a subnet. The mapping functions as well as topology maintenance require specific methods be implemented at the tier charged with overseeing the entire network. For this reason, the recursion in the operational command hierarchy is implemented at the cluster head level, the highest level in the hierarchy based on a supervisor-subnet philosophy. The root node controls a set of cluster heads. Cluster heads can coordinate leaf nodes and/or other cluster heads. The independent design and implementation allows recursion in different hierarchies to be designed at different tiers without complications.

A given physical node will have a "rank" or "level" in each of the three mentioned hierarchies. It is important to note that a nodes position in one hierarchy is completely independent of its ranking in the other two hierarchies (e.g., a node could be a root in the communication hierarchy, a cluster head in the command and control hierarchy, and a leaf in the collaborative sensing hierarchy. This allows for maximum flexibility in terms of network configuration as well as allowing the network the ability to configure the sensing clusters dynamically to best process information concerning an individual target event occurrence.

32.3.2 Operational Command

The combined operational command hierarchy controls allocation of nodes to surveillance regions, including mapping unknown territory and discovering obstacles. Further it also controls node deployment, and decisions to recall nodes. Figure 32.A.1, Section 32.A.5 demonstrates the interaction between the root, cluster heads, and leaf nodes.

The network reconfigures itself as priorities change. Initial node deployments are likely to concentrate nodes in regions: (1) where it is assumed enemy traffic will be heavy or (2) which are of strategic interest to friendly forces. Over time the network should find the areas where enemy traffic is actually flowing, which are likely to be different than initially anticipated. In a similar manner, the strategies of friendly forces are likely to change over time.

The root node manages network resources as well as oversees the following network functions: mapping the region of interest, node assignment, node reallocation, network topology and network recall. The root provides information about these functions to the end user and distributes user preferences and commands to appropriate subnets. A pictorial description of the root node is provided in the upper portion of Figure 32.A.1. Signals changes in strategy to the cluster heads.

Cluster heads (Figure 32.A.1 middle) manage the activities of subnets of leaf nodes and other cluster heads, generate topology reports, interpret commands from the root, calculate resource needs, and monitor resource availability.

Leaf node (Figure 32.A.1 down) responsibilities are limited to only a small portion of the total area being covered by the entire network. These nodes only consider the area they are currently monitoring and retain no global information. Each leaf node directly interacts with its environment, performing terrain mapping and providing position and status information as required by upper levels of the hierarchy.

32.3.3 Network Communications

The network communications hierarchy is implemented to maintain data flow in the presence of environmental interference, such as jamming and node loss. Actions the hierarchy controls include adjusting transmission power, frequency hopping schedules, ad-hoc routing, and movement to correct interference. The combined Petri net models in Figure 32.A.3 of the appendix describe how and when these actions are taken.

The Petri Net hierarchy describes a communications protocol between the nodes. Critical messages have associated acknowledgments. To ensure connectivity between nodes and their immediate superiors, all messages passing information up the hierarchy have matching acknowledgments. If an acknowledgment is not received, retransmission occurs according to parameters set by end users. When retransmissions are exhausted, a supervisor may have to be replaced. When communications with their supervisor is severed, leaf nodes (Figure 32.A.3 below) and cluster head nodes (Figure 32.A.3 middle) immediately enter a promotion cycle. The node waits for an indication that a replacement supervisor has been chosen. If none is received, the node promotes itself to the next level. It broadcasts that it has assumed control of the subnet and overtakes supervisory responsibility. If the previous supervisor rejoins the subnet, it may demote itself.

Lost contact between the root node (Figure 32.A.3 above) and the user is more difficult to address. Upon exhausting retransmissions, the root assumes contact has been lost and it is isolated from the network. The first action taken is to broadcast a message throughout the network indicating to the user that root contact has been lost. Each node tries to establish contact with the user and become the new root. If this fails, the network is put to sleep by a command propagated down the hierarchy. At this point it is left to the user to re-establish contact. While in this quiescent mode the network suspends operations, and responds only to a wake command transmitted by a member of the user community.

32.3.4 Collaborative Sensing

Coordination of sensor data interpretation is done using the collaborative sensing hierarchy shown in Figure 32.A.2. This hierarchy design is based partly on our existing sensor network implementation, which was tested at 29 Palms Marine Base in November 2001.

Initial processing of sensor information is done by the leaf node (Figure 32.A.2 below). Time series data is preprocessed. A median filter reduces white noise and a low pass filter removes high frequency noise. If the signal is still unusable, it is assumed either that the sensor is broken, or that environmental conditions make it impossible, and thus the node temporarily hibernates to save energy [Hall 97].

Each node has multiple sensors and may have multiple sensing modalities reducing the node's vulnerability to mechanical failure of the sensors and many types of environmental noise [Brooks 97].

After filtering, sensor time series are registered to a common coordinate system and given a time stamp. Subsequently, data association determines which detections refer to the same object. A state vector with inputs from multiple sensing modalities can be used for target classification [Luo 89]. Each leaf node can send either a target state vector or Closest Point of Approach (CPA) event to the cluster head. A cluster head is selected dynamically. Cluster heads (Figure 32.A.2 middle) take care of combining these statistics into meaningful track information.

Root nodes (Figure 32.A.2 above) coordinate activities among cluster heads and follow tracks traversing the area they survey. In this hierarchy, internal nodes are root nodes. They define the sensing topology, which organizes itself from the bottom up. This topology mimics the flow of targets through the system. It has been suggested that this information can guide future node deployment [Deb 02].

Sensing hierarchy topology can be calculated using computational geometry and graph theory. A root node can request topology data from all nodes beneath it. Voronoi diagrams are constructed given the locations of nodes. Maximal breach paths and covered paths can be calculated in this region. This data defines the system topology and the quality of service (surveillance) [Meguerdichian 00].

32.4 Control Specifications

Given the set of states G and the set of events Σ, the controller disables a subset of Σ as necessary at every state $g \in G$. Control specifications are defined by identifying state and event combinations that lead the system to an undesirable state. Each specification is a constraint on the system and the controller's behavior is defined by the set of constraints. Control of the DSN requires coordination of individual node activities within the constraints of mission goals. Each node has a set of responsibilities and must act according to its capabilities in response.

The controller is needed because the system has multiple command hierarchies. Each hierarchy has its own goals. When conflicts between hierarchies arise, the controller resolves them. We identified sequences of events that lead to undesirable states. Three primary issues were found that can cause undesirable system states: (1) movement of a node conflicting with the needs of another hierarchy; (2) nodes attempting to function in the presence of unrecoverable noise; and (3) retreat commands from the command hierarchy should have precedence over all other commands. Following is the set of constraints the controllers impose on the DSN:

CC—operational command
SC—collaborative sensing
WC—network communication

1. When a node is waiting for on-board data fusion, it should be prevented from moving by WC, CC, and SC. Also it should not be promoted by WC or by SC until sensing is complete.
2. Hibernation induced by unrecoverable noise or saturated signal in SC should also force the node to hibernate in WC and CC. (And vice versa, for leaf nodes only). Wakeup in SC needs to send wake-up to CC/WC.

3. While the cluster head is in the process of updating its statistics, its leaves should be prevented from moving by WC, CC, or SC.
4. While a cluster head node is receiving statistics from its leaf nodes, it should be prevented from moving by WC, CC, or SC.
5. When sensor nodes are in low power mode as determined by WC, or damaged mode as determined by CC, they should be prohibited from any moving for prioritized relocation or occlusion adjustments.
6. Retreat in CC should supercede all actions, except propagation of retreat command.
7. Nodes encountering a target signal in the SC should suspend mapping action in CC until sensing is complete.
8. Move commands in CC/WC should be delayed while node is receiving sensing statistics from lower levels in the hierarchy.

32.5 Controller Design

Each controller design method enforces constraints in its own way. Vector controllers use state vector comparison to determine the transitions that violate the control specifications. Petri net controllers use slack variables to disable the same transitions. Moore machines determine which strings of events lead to constraint violations.

Controller design is complicated by the existence of uncontrollable and unobservable transitions. Uncontrollable transitions cannot be disabled; unobservable transitions cannot be detected. When uncontrollable or unobservable transitions lead to undesirable states, the controller design process requires creating alternative constraints that use only controllable transitions. Ideally, the controller should not unnecessarily constrain the system. One particular methodology for creating non-restrictive controllers is described in [Moody 98].

Control specification is usually specified as: $l\mu \leq b$. l is a $N \times M$ matrix (number of control specifications by the number of places in the plant); μ is $M \times 1$ matrix representing number of tokens in each place of the plant. b is a $N \times 1$ integer matrix each element representing the total maximal allowed number of tokens in any combination of places.

32.5.1 Finite State Machine Controller

Verifying system properties, such as safeness, boundedness and liveness, is done using the Karp-Miller tree. It represents all possible states of the system. Figure 32.4 shows a Petri Net example and its associated Karp-Miller Tree [Peterson 77].

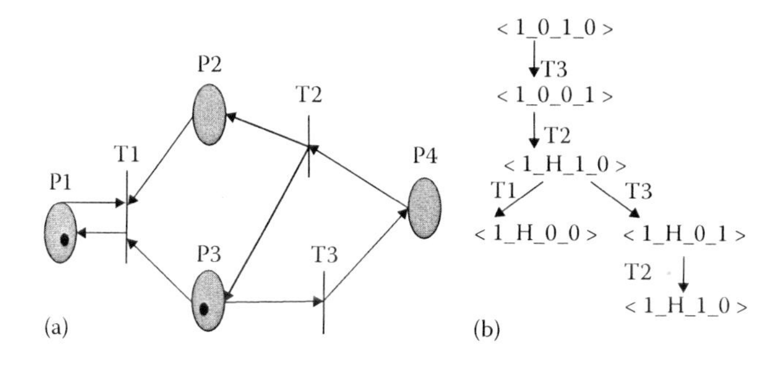

FIGURE 32.4 A sample Petri Net (a) and its associated Karp-Miller tree (b).

The following is the Karp-Miller algorithm [Murata 89]:

1. Label initial marking S_0 as the root of the tree and tag it as new.
2. While new markings exist do:
   a. Select a marking S.
   b. If S is identical to a marking on the path from the root to S, then tag S as old and go to another marking.
   c. If no transitions are enabled at S, tag S dead-end.
   d. While there exist enabled transitions at S do:
      i. Obtain the marking S' that results from firing T at S.
      ii. On the path from the root to S if there exists a marking S'' such that $S'(p) \geq S''(p)$ for each place p and S' is different from S'', then replace $S'(p)$ by ω for each p such that $S'(p) > S''(p)$.
      iii. Introduce S' as a node, draw an arc with label T from S to S' and tag S' as new.

Ramadge and Wonham described supervisory control of discrete event process using finite state automaton [Ramadge 87]. We generalized their contribution and proposed our own innovations.

All reachable state vectors could be infinite, but the Karp-Miller tree should be finite. Thus, we introduce the symbol omega (ω) in the Karp-Miller tree to indicate that the token number in the corresponding place is unbounded. A 5-tuple plant $\wp = (Q, \Sigma, \delta, q_0, Q_m)$ was obtained from the Karp-Miller tree, where Q: all legal and illegal states, Σ: all transitions, δ: next state function, q_0: initial state, Q_m: only legal states. Because the FSM generated without constraints contains illegal states, we enforce a state feedback map function on the plant to restrict its behavior. Let $\Gamma = (0, 1)^{\Sigma}$ be a set of control patterns. For each $\gamma \in \Gamma, \gamma : \Sigma_c \rightarrow \{0, 1\}$ is a control pattern of $|\Sigma_c|$ bits. An event σ is enabled if $\gamma(\sigma) = 1$. For uncontrollable transitions, $\gamma(\sigma)$ always equals 1. Then, we define an augmented transition function as

$$\delta_c : \Gamma \times \Sigma \times Q \rightarrow Q \tag{32.1}$$

According to

$$\delta_c(\gamma, \sigma, q) = \begin{cases} \delta(\sigma, q) & \text{if } \delta(\sigma, q) \text{ is defined and } \gamma(\sigma) = 1 \\ \text{undefined otherwise} \end{cases} \tag{32.2}$$

We interpret this controlled plant as $\wp_c = (Q, \Gamma \times \Sigma, \delta_c, q_0, Q_m)$, which admits external control [Ramadge 87].

The Moore machine is a 5-tuple, represented as $(S, I, O, \delta, \Gamma)$ where S: nonempty finite set of states, I: nonempty finite set of inputs, O: nonempty finite set of output, δ: next state function, which maps $S \times I \rightarrow S$, Γ: output function, which maps $S \rightarrow O$. The state feed back map can be realized by the output function of Moore machine, which defines a mapping between the current state and a control pattern for the current state.

Ramadge and Wonham acquire the state feedback map by enumerating all legal states in the FSM together with their binary control patterns. Introducing the Moore machine and state encoding automatically yields the control pattern from derived logical expressions in terms of their current state.

First, we trim the Karp Miller tree to reach a finite state automaton as a recognizer for the legal language of the plant. $[\log_2 N]$ bits are then used to encode N legal states. Since the choice of encoding affects the complexity of logic implementation, an optimal encoding strategy is preferred. The transition table is used to derive logical expressions in terms of binary encoded state for each controllable transition. State minimization is carried out to remove redundant states [Aho 86].

This approach to FSM modeled controller is unique in two respects. Instead of exploring the algebraic or structural property of a Petri Net as in the case of VDES and Petri Net controllers, it utilizes traditional finite automata to tackle the control problem of discrete event system. In addition, the introduction of the Moore machine to output controller variables guarantees real time response. The quick response is acquired at the cost of extensive searching and filtering of the entire reachable state space offline.

The FSM modeled controllers perform well for small and medium scale systems, but representation and computation costs would be prohibitive expensive for complex systems. One alternative is to model the system with Petri Nets. The current Petri Net state vector is converted to a binary encoded state and a binary control pattern is then calculated. Overhead is incurred while converting the state vector to binary encoded form, but the representative power of Petri Nets is greater than that of a FSM. Also, instead of the traditional brute force search of the entire state space, we examine only those transitions that have an elevated effect on the right hand side of our control specifications. All transitions are screened and only those that would result in an increase in the left hand side of the control specification ($l\mu \leq b$) as described in Section 32.5 are candidates for control. The binary control pattern bit for a particular transition is set to 1 when $l\mu \leq b$ continues to hold after the transition firing. For multiple control specifications, the binary control pattern for a particular transition is 1 if and only if the current state satisfies the conjunction of all the inequalities imposed by all constraints. In this case, the binary control pattern is software determined instead of hardware determined. The sample controller for our DSN can be found in Section 32.5.4.1.

32.5.2 Vector Addition Controller

The Vector Discrete Event System (VDES) approach represents system state as an integer vector. State transitions are represented by integer vector addition [Li 93]. The VDES is an automaton that generates a language over a finite alphabet Σ consisting of two subsets: Σ_c and Σ_{uc}. Σ_c is the set of controllable events that can be disabled by the external controller. Σ_{uc} is the set of uncontrollable events that can not be disabled by the controller. We use the following symbols:

$G_{uc} \in G$: The uncontrollable part of the plant G.
D: The incidence matrix of the plant constructed as in [David 92]. Places are rows. Transitions are columns. $x_{ij} = 1$ (-1) if an arc leads from place i to transition j, else $x_{ij} = 0$.
D_{uc}: The uncontrollable transition columns of the incidence matrix.
D_{uo}: The unobservable transition columns of the incidence matrix.
Σ: All transitions in the plant.
$\Sigma_{uc} \in \Sigma$: The subset of transitions that are uncontrollable.
$\Sigma_{uo} \in \Sigma$: The subset of transitions that are unobservable.
$L(G, \mu)$: The language of the plant starting with marking μ (i.e. the set of all possible sequences of transitions). The language can be directly inferred from the Karp-Miller tree, which we show how to compute in Section 32.5.1.
$\omega \in L(G, \mu)$: A valid sequence of transitions in the plant starting from the state μ.

Given a Petri Net with incidence matrix D and a control specification $l\mu \leq b$, a final state can be represented as a single vector equation as follows. Given a sequence of N events, $\omega \in L(G, \mu)$, the final state μ_N is given by

$$\mu_N = \mu_0 + Dq_1 + Dq_2 + \cdots + Dq_N = \mu_0 + D(q_1 + q_2 + \cdots + q_N) = \mu_0 + DQ_w \tag{32.3}$$

$Q_{\omega(i)} = |\omega|_{q_i}$ represents the number of occurrences of q_i in the event sequence. The number of event occurrences, independent of how the events are interleaved, thus defines the final state. We use the following Boolean equation:

$$f^*(\alpha, \mu) = \begin{cases} 1 & \text{if } \mu + Dq_\alpha \in [P] \\ 0 & \text{else} \end{cases} \tag{32.4}$$

where

$$[P] = \left\{\mu \middle| (\forall\omega \in L(G_{uc}, \mu)), l(\mu + D_{uc}Q_{uc,\omega}) \leq b\right\} = \left\{\mu \middle| \left(l\mu + \max_{\omega \in L(G_{uc}, \mu)} lD_{uc}Q_{uc,\omega}\right) \leq b\right\} \tag{32.5}$$

The transition associated with q_α is allowed to fire only if no subsequent firing of uncontrollable transitions would violate the control specification $l\mu \le b$.

In general, the maximization problem in Equation 32.5 is a non-linear program with an unstructured feasible set $L(G_{uc}, \mu)$. However, a theorem proven in [Li 94] shows that when G is loop free, for every state where

$$\mu \ge 0 \quad \text{and} \quad Q \ge 0, \quad \mu + DQ \ge 0 \Leftrightarrow (\exists \omega \in L(G, \mu))Q = Q_w \tag{32.6}$$

The computation of $[P]$ can be reduced to a linear integer program. The set of possible strings $\omega \in L(G, \mu)$ can then be simplified as

$$\left\{Q_{uc,\omega} \middle| \omega \in L(G_{uc}, \mu)\right\} = \left\{Q \in Z^K \middle| \mu + D_{uc}Q \ge 0, Q \ge 0\right\} \tag{32.7}$$

with this simplification of the feasible region, the set $[P]$ of allowed states becomes

$$[P] = \left\{\mu \middle| l\mu + lD_{uc}Q^*(\mu) \le b\right\} \tag{32.8}$$

where $Q^*(\mu)$ is the solution for

$$\max_Q lD_{uc}Q \qquad s.t. \begin{cases} D_{uc}Q \ge -\mu \\ Q \ge 0 \text{ (int)} \end{cases} \tag{32.9}$$

yielding Q^* as a function of μ [Wonham 99].

To confirm the controllability it suffices to test whether or not the initial marking of the system satisfies the equation

$$\mu_0 \in [P] \quad \text{or} \quad \left(l\mu_0 + \max_{\omega \in L(G_{uc}, \mu_0)} lD_{uc}Q_{uc,\omega}\right) \le b \tag{32.10}$$

If Equation 32.10 is not satisfied, no controller exists for this control specification [Li 94].

When illegal markings are reachable from the initial marking by passing through a sequence of uncontrollable events, it is an inadmissible specification. Inadmissible control specifications must take an admissible form before synthesizing a controller. Equation 32.5 is the transformed admissible control specification.

Essentially, a VDES controller is the same as a Petri Net modeled controller. A controller variable c is introduced into the system as a place with the initial value to be b minus the initial value of the transformed admissible control specification [Li 93]. A controllable event will be disabled if and only if its occurrence will make c negative. In our implementation, the controller examines all enabled controllable transitions. If the firing of a transition leads to an illegal state, system rolls back and continues looking for the next enabled transition.

32.5.3 Petri Net Based Control

[Li 93, Li 94] made significant contributions to the control of plants with uncontrollable events by specifying conditions under which control constraint transformations have a closed form expression. However, the loop free structure of the uncontrollable sub-plant is a sufficient but not necessary condition for control. [Moody 98] extended the scope of controller synthesis problems to include unobservable events, in addition to uncontrollable events already discussed in VDES. He also found a method for controller synthesis for plants with loops containing uncontrollable events.

In the Petri Net controller, a plant with n places and m transitions has incidence matrix $D_p \in Z^{n \times m}$. The controller is a Petri Net with incidence matrix $D_c \in Z^{n_c \times m}$. The controller Petri Net contains all the plant transitions and a set of control places. Control places are used to control the firing of transitions when control specifications will be violated. Control places cannot have arcs incident on unobservable or uncontrollable transitions. Arcs from uncontrollable transitions to control places are permitted. As with VDES, inadmissible control specifications must be converted to admissible control specifications before controller synthesis. An invariant-based control specification: $l\mu \leq b$ is admissible if $lD_{uc} \leq 0$ and $lD_{uo} = 0$.

If the original set of control specifications $L\mu \leq b$ contains inadmissible specifications, it is necessary to define an equivalent set of admissible specifications. Before proceeding with this step, we need to prove that the state space of the new control specifications lies within the state space of the original control specifications. Let $R_1 \in Z^{n_c \times n}$ satisfy $R_1\mu \geq 0\ \forall\mu$. Let $R_2 \in Z^{n_c \times n_c}$ be a positive definite diagonal matrix. If

$$L'\mu \leq b' \quad \text{where} \quad \begin{aligned} L' &= R_1 + R_2 L \\ b' &= R_2(b+1) - 1 \end{aligned} \tag{32.11}$$

1 is a n_c dimensional vector of 1's, then $L\mu \leq b$. The proof is given in [Moody 00].

To construct a controller that does not require inhibiting uncontrollable transition or detecting unobservable transitions, it is sufficient to calculate two matrices R_1 and R_2 which satisfy

$$[R_1\ R_2]\begin{bmatrix} D_{uc} & D_{uo} & -D_{uo} & \mu_0 \\ LD_{uc} & LD_{uo} & -LD_{uo} & L\mu_0 - b - 1 \end{bmatrix} \leq [0\ \ 0\ \ 0\ \ -1] \tag{32.12}$$

The first column in (32.12) indicates that $LD_{uc} \leq 0$; the second and the third column indicate that $LD_{uo} = 0$; and the fourth column indicates that the initial marking of the Petri Net satisfies the newly transformed admissible control specification. Using the admissible control specification, a slack variable μ_c is introduced to transform the inequality into an equality:

$$L'\mu + \mu_c = b' \tag{32.13}$$

thus,

$$\begin{aligned} D_c &= -(R_1 + R_2 L)D_p = -L'D_p \\ \mu_{c0} &= R_2(b+1) - 1 - (R_1 + R_2 L)\mu_0 = b' - L'\mu_0 \end{aligned} \tag{32.14}$$

Equation 32.14 provides the controller incidence matrix and initial marking of the control places.

In contrast with the VDES controller, a Petri Net controller explores the solution by inspecting the incidence matrix. Plant/controller Petri Nets provide a straightforward representation of the relationship between the controller and controlled components. The evolution of the Petri Net plant/controller is easy to compute, which facilitates usage in real time control problems. In our implementation, the plant/controller Petri Net incidence matrix is the output that results from the plant and control specification as input [Moody 00].

32.5.4 Performance and Comparison of Three Controllers

Figure 32.5 is an example of a Petri Net consisting of two independent parts with three uncontrollable transitions: T2, T3 and T5 with an initial marking of $[2\ 0\ 0\ 0\ 1\ 1\ 0]^T$. This net is a reduced form of our DSN. Its main purpose is to illustrate how the control issues are handled in our DSN. Results of the three approaches and comparison are given.

The behavior of the two independent Petri Nets should obey the control specifications. The first constraint requires that place P5 cannot contain more than two tokens. There cannot be more than two processes active at one time. The second constraint states that the sum of tokens in P2 and P6 must be less or equal to 1. This constraint implies that when a node is sensing in the scope of the collaborative sensing hierarchy, it is not allowed to move in the operational command hierarchy or vice versa. This mutual

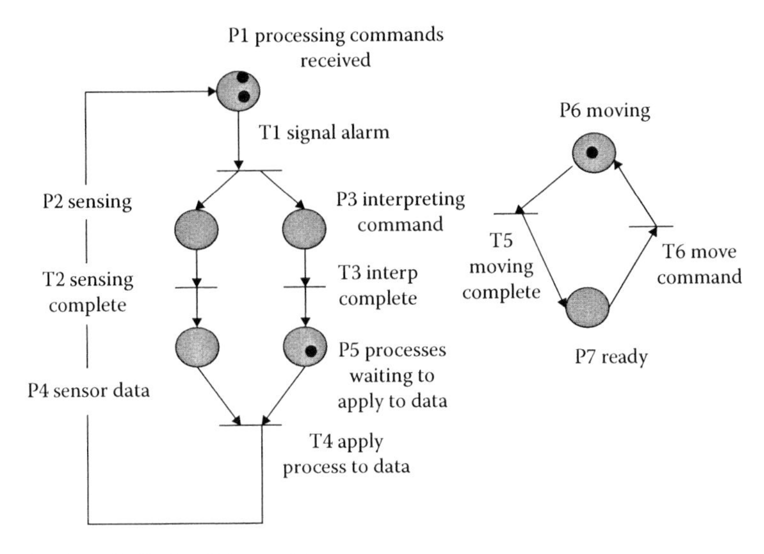

FIGURE 32.5 A reduced DSN Petri Net model.

exclusion constraint represents the major control task of enforcing consistency across independently evolving hierarchies in our DSN. Three uncontrollable transitions are sensing complete, interpreting complete and moving complete.

Two control specifications are

1. $\mu_5 \leq 2$
2. $\mu_2 + \mu_6 \leq 1$

32.5.4.1 Finite State Machine Controller

Detailed steps of how to construct a FSM controller for the reduced DSN model in Figure 32.5 are given. A reachability tree has thus been constructed from a Petri Net first. Some of the states that were generated from the Plant without constraints are

```
States{
s2_0_0_0_1_1_0,   s0_0_0_2_3_0_1,   s1_1_2_0_0_1_0,
s1_1_1_0_1_1_0,   s1_0_0_1_2_0_1,   s0_2_3_0_0_1_0,
s0_2_2_0_1_1_0,   s0_1_1_1_2_0_1,   s0_2_3_0_0_0_1,
s0_1_2_1_1_1_0,   s0_0_1_2_2_0_1,   s0_2_2_0_1_0_1,
s0_0_2_2_1_1_0,   s1_0_1_1_1_0_1,   s0_2_1_0_2_0_1,
s0_0_1_2_2_1_0,   s0_1_2_1_1_0_1,   s1_1_2_0_0_0_1,
s0_0_0_2_3_1_0,   s0_0_2_2_1_0_1,   s1_1_1_0_1_0_1,
s1_0_0_1_2_1_0,   s1_0_2_1_0_0_1,   s1_1_0_0_2_0_1,
s0_1_1_1_2_1_0,   s0_1_3_1_0_0_1,   s2_0_1_0_0_0_1,
s0_1_0_1_3_1_0,   s0_0_3_2_0_0_1,   s2_0_0_0_1_0_1,
s1_1_0_0_2_1_0,   s0_0_3_2_0_1_0,   }
s0_2_1_0_2_1_0,   s0_1_3_1_0_1_0,
s0_2_0_0_3_1_0,   s1_0_2_1_0_1_0,
s0_2_0_0_3_0_1,   s1_0_1_1_1_1_0,
s0_1_0_1_3_0_1,   s2_0_1_0_0_1_0,
```

We search the entire state space removing illegal states, which would either directly or indirectly violate the control specification. A new state space with 13 legal states as shown below is achieved. Four bits are needed to encode 13 states as encoded by A, B, C, and D. A Moore machine is constructed to output the binary control pattern based on the current encoded state:

States : Marking	Encode states
State S0:2000110	0000
State S1:2000101	0001
State S2:1110101	0010
State S3:1011101	0011
State S4:1001201	0100
State S5:1001210	0101
State S6:2010001	0110
State S7:1120001	0111
State S8:1021001	1000
State S9:1021010	1001
State S10:1011110	1010
State S11:2010010	1011
State S12:1100201	1100

Among six transitions, we cannot control T2, T3, or T5. These controllable transitions have firings which would lead to illegal states directly or indirectly. Based on knowledge of offline screening, T1 and T6 should be controlled. Thus, the binary control pattern has two bits. The transition table with encoded states for the Moore machine is

Present State	Next State						Output	
ABCD (Four Encoded Bits)	T1	T2	T3	T4	T5	T6	For T1	For T6
S0					S1		0	1
S1	S2					S0	1	1
S2		S3	S12				0	0
S3			S4	S6		S10	0	1
S4				S1		S5	0	1
S5				S0	S4		0	1
S6	S7		S1			S11	1	1
S7		S8	S2				0	0
S8			S3			S9	0	1
S9			S10		S8		0	1
S10			S5	S11	S3		0	1
S11			S0		S6		0	1
S12		S4					0	0

From the transition table, we can construct a Moore machine state diagram with 13 states, with 6 inputs and 2 outputs. The state feedback function, which outputs the binary control patterns based on current state, is used to regulate plant behavior by switching between control patterns.

The logical expression of a binary control pattern can be expressed as

$$\begin{cases} \text{T1} = \overline{A}\,\overline{B}\,\overline{C}D + \overline{A}BC\overline{D} \\ \overline{\text{T6}} = \overline{A}C(\overline{B}\,\overline{D} + BD) + AB\overline{C}\,\overline{D} \end{cases}$$

The logical implementation can be realized by hardware. The controller can immediately access the control pattern for each controllable transition based on its current encoded state without going through legal firing checking as in VDES or doing extra calculation involving added controller places and arcs as in Petri Net Modeled controller. The trade-off is offline legal state space searching. Following our search method used in the DSN, we simply check whether or not the current state vector satisfies the conjunction of: $\mu_3 + \mu_5 \leq 1$ and $\mu_2 + \mu_6 = 0$. If it does then the control pattern bit of T1 is 1, and if $\mu_2 + \mu_6 = 0$, then the control pattern bit of T6 is 1. This turns out to be efficient to compute and simple to express for a complex system.

32.5.4.2 Vector Discrete Event System Modeled Controller

A VDES controller is formed to control the same reduced DSN model shown in Figure 32.5. D is the incidence matrix of the plant as described in Section 32.5.2. D_{uc} is the uncontrollable portion of D as row to be places and column to be the uncontrollable transitions:

$$D = \begin{bmatrix} -1 & 0 & 0 & 1 & 0 & 0 \\ 1 & -1 & 0 & 0 & 0 & 0 \\ 1 & 0 & -1 & 0 & 0 & 0 \\ 0 & 1 & 0 & -1 & 0 & 0 \\ 0 & 0 & 1 & -1 & 0 & 0 \\ 0 & 0 & 0 & 0 & -1 & 1 \\ 0 & 0 & 0 & 0 & 1 & -1 \end{bmatrix} \qquad D_{uc} = \begin{bmatrix} 0 & 0 & 0 \\ -1 & 0 & 0 \\ 0 & -1 & 0 \\ 1 & 0 & 0 \\ 0 & 1 & 0 \\ 0 & 0 & -1 \\ 0 & 0 & 1 \end{bmatrix}$$

The goal of the controller is to enforce linear inequality on the state vector of G, usually in the form of $l\mu \leq b$. Our control specifications are $\mu_5 \leq 2$ and $\mu_2 + \mu_6 \leq 1$. Consider the first control specification

$$l_1 = [0\,0\,0\,0\,1\,0\,0] \qquad b_1 = 2$$

The initial marking satisfies the control specification, but the control specification is inadmissible because the uncontrollable firing of T3 would lead to a violation. Since the uncontrollable part of the system is loop free the inadmissible control specification can be transformed to an admissible control specification. Solve the $\max_Q lD_{uc}Q$ as discussed in Section 32.5.2:

$$s.t.\begin{cases} D_{uc}Q \geq -\mu \\ Q \geq 0\,(\text{int}) \end{cases}$$

By doing this, Effect of uncontrollable events firings on the control specification is taken into consideration:

$$\max_Q l_1 D_{uc} Q = [0\,0\,0\,0\,1\,0\,0] \begin{bmatrix} 0 & 0 & 0 \\ -1 & 0 & 0 \\ 0 & -1 & 0 \\ 1 & 0 & 0 \\ 0 & 1 & 0 \\ 0 & 0 & -1 \\ 0 & 0 & 1 \end{bmatrix} \begin{bmatrix} q_2 \\ q_3 \\ q_5 \end{bmatrix} = \max(q_3)$$

$$s.t.\begin{cases}\begin{bmatrix}0 & 0 & 0\\ -1 & 0 & 0\\ 0 & -1 & 0\\ 1 & 0 & 0\\ 0 & 1 & 0\\ 0 & 0 & -1\\ 0 & 0 & 1\end{bmatrix}\begin{bmatrix}q_2\\ q_3\\ q_5\end{bmatrix} \geq \begin{bmatrix}-\mu_1\\ -\mu_2\\ -\mu_3\\ -\mu_4\\ -\mu_5\\ -\mu_6\\ -\mu_7\end{bmatrix} \\ q_i \geq 0\end{cases} \Rightarrow \begin{cases}-\mu_1 \leq 0\\ \mu_2 \geq q_2\\ \mu_3 \geq q_3\\ -\mu_4 \leq q_2\\ -\mu_5 \leq q_3\\ \mu_6 \geq q_5\\ -\mu_7 \leq q_5\end{cases}$$

From above, it can be inferred that $\max(q_3)$ equals μ_3, as $\mu_3 \geq q_3$. The transformed admissible control specification is $l\mu + \max lD_{uc}Q^*(\mu) \leq b$, which is $\mu_5 + \mu_3 \leq 2$. The initial marking $[2\ \ 0\ \ 0\ \ 0\ \ 1\ \ 1\ \ 0]^T$ holds for $\mu_5 + \mu_3 \leq 2$, thus the controller exists for this control constraint [Wonham 99]. The second control specification is an admissible one, because no uncontrollable transition firing would lead to an illegal state.

In our controller implementation, the plant, together with two admissible control specifications is treated as input. The state space of the controlled system is

$$\begin{gathered}2\_0\_0\_0\_1\_1\_0, 1\_1\_2\_0\_0\_0\_1,\\ 2\_0\_0\_0\_1\_0\_1, 1\_0\_2\_1\_0\_0\_1,\\ 1\_1\_1\_0\_1\_0\_1, 1\_0\_2\_1\_0\_1\_0,\\ 1\_0\_1\_1\_1\_0\_1, 1\_0\_1\_1\_1\_1\_0,\\ 1\_0\_0\_1\_2\_0\_1, 2\_0\_1\_0\_0\_1\_0,\\ 1\_0\_0\_1\_2\_1\_0, 1\_1\_0\_0\_2\_0\_1,\\ 2\_0\_1\_0\_0\_0\_1,\end{gathered}$$

Each of these states satisfies the two control specifications mentioned previously.

32.5.4.3 Petri Net Modeled Controller

Finally, Petri net controller is built for the reduced DNS with the same control specifications. We have the same first control specification where $\mu_5 \leq 2$. Since the plant has no unobservable transitions, we only need to study uncontrollable transitions. The first step is to determine if the control specification is admissible.

The following indicates an inadmissible control specification as discussed in the second paragraph of Section 32.5.3:

$$[0\,0\,0\,0\,1\,0\,0]D_{uc} = [0\,1\,0] \geq [0\,0\,0]$$

It was observed that the third row of D_{uc}, if equal to $[0\ -1\ 0]$, could be used to eliminate the positive element 1 in the above equation. So,

$$R_1 = [0\,0\,1\,0\,0\,0\,0] \qquad R_2 = 1$$

$$L' = R_1 + R_2L = [0\,0\,1\,0\,0\,0\,0] + 1[0\,0\,0\,0\,1\,0\,0] = [0\,0\,1\,0\,1\,0\,0]$$

The initial marking satisfies the admissible control specification. The transformed admissible control specification is $\mu_5 + \mu_3 \leq 2$, which is the same as the admissible control specification from the above section. By introducing a new slack variable μ_c, the control specification becomes an equation:

$$\mu_5 + \mu_3 + \mu_c = 2$$

For the second admissible control specification: $\mu_2 + \mu_6 \leq 1$, we introduce another slack variable μ_c', and the second control specification becomes another equation: $\mu_2 + \mu_6 + \mu_c' = 1$

$$D = \begin{bmatrix} -1 & 0 & 0 & 1 & 0 & 0 \\ 1 & -1 & 0 & 0 & 0 & 0 \\ 1 & 0 & -1 & 0 & 0 & 0 \\ 0 & 1 & 0 & -1 & 0 & 0 \\ 0 & 0 & 1 & -1 & 0 & 0 \\ 0 & 0 & 0 & 0 & -1 & 1 \\ 0 & 0 & 0 & 0 & 1 & -1 \end{bmatrix} \quad \mu_0 = [2\ \ 0\ \ 0\ \ 0\ \ 1\ \ 1\ \ 0]^T$$

$$L = \begin{bmatrix} 0010100 \\ 0100010 \end{bmatrix} \quad b = \begin{bmatrix} 2 \\ 1 \end{bmatrix}$$

$$D_c = -LD = \begin{bmatrix} -1 & 0 & 0 & 1 & 0 & 0 \\ -1 & 1 & 0 & 0 & 1 & -1 \end{bmatrix} \text{ and } \mu_{c0} = b - L\mu_0 = \begin{bmatrix} 2 \\ 1 \end{bmatrix} - \begin{bmatrix} 1 \\ 1 \end{bmatrix} = \begin{bmatrix} 1 \\ 0 \end{bmatrix}$$

The resulting overall plant/controller incidence matrix is

$$D_{all} = \begin{bmatrix} -1 & 0 & 0 & 1 & 0 & 0 \\ 1 & -1 & 0 & 0 & 0 & 0 \\ 1 & 0 & -1 & 0 & 0 & 0 \\ 0 & 1 & 0 & -1 & 0 & 0 \\ 0 & 0 & 1 & -1 & 0 & 0 \\ 0 & 0 & 0 & 0 & -1 & 1 \\ 0 & 0 & 0 & 0 & 1 & -1 \\ -1 & 0 & 0 & 1 & 0 & 0 \\ -1 & 1 & 0 & 0 & 1 & -1 \end{bmatrix}$$

Two controller places can be added to the plant as P8 and P9 with initial marking of 1 and 0 respectively, as shown in Figure 32.6. In our implementation, the Petri Net modeled controller implementation computes a closed loop Petri Net incidence matrix based on the plant and the control constraints to be enforced without going through the above manual computation.

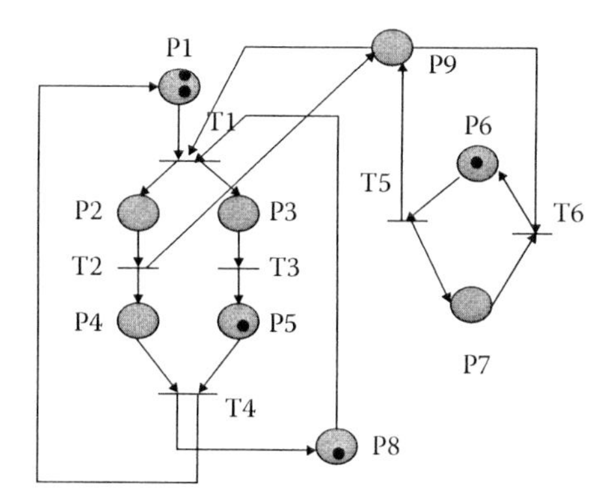

FIGURE 32.6 Plant/Controller Petri Net model.

Running our implementation program, we get a FSM as shown below:

```
states {
s2_0_0_0_1_1_0_1_0,
s2_0_0_0_1_0_1_1_1,
s1_1_1_0_1_0_1_0_0,
s1_0_1_1_1_0_1_0_1,
s1_0_0_1_2_0_1_0_1,
s1_0_0_1_2_1_0_0_0,
s2_0_1_0_0_0_1_1_1,
s1_1_2_0_0_0_1_0_0,
s1_0_2_1_0_0_1_0_1,
s1_0_2_1_0_1_0_0_0,
s1_0_1_1_1_1_0_0_0,
s2_0_1_0_0_1_0_1_0,
s1_1_0_0_2_0_1_0_0
}
transitions {
<s2_0_0_0_1_1_0_1_0,t5,s2_0_0_0_1_0_1_1_1,>,
<s2_0_0_0_1_0_1_1_1,t1,s1_1_1_0_1_0_1_0_0,>,
<s1_1_1_0_1_0_1_0_0,t2,s1_0_1_1_1_0_1_0_1,>,
<s1_0_1_1_1_0_1_0_1,t3,s1_0_0_1_2_0_1_0_1,>,
<s1_0_0_1_2_0_1_0_1,t4,s2_0_0_0_1_0_1_1_1,>,
<s1_0_0_1_2_0_1_0_1,t6,s1_0_0_1_2_1_0_0_0,>,
<s1_0_0_1_2_1_0_0_0,t4,s2_0_0_0_1_1_0_1_0,>,
<s1_0_0_1_2_1_0_0_0,t5,s1_0_0_1_2_0_1_0_1,>,
<s1_0_1_1_1_0_1_0_1,t4,s2_0_1_0_0_0_1_1_1,>,
<s2_0_1_0_0_0_1_1_1,t1,s1_1_2_0_0_0_1_0_0,>,
<s1_1_2_0_0_0_1_0_0,t2,s1_0_2_1_0_0_1_0_1,>,
<s1_0_2_1_0_0_1_0_1,t3,s1_0_1_1_1_0_1_0_1,>,
<s1_0_2_1_0_0_1_0_1,t6,s1_0_2_1_0_1_0_0_0,>,
<s1_0_2_1_0_1_0_0_0,t3,s1_0_1_1_1_1_0_0_0,>,
<s1_0_1_1_1_1_0_0_0,t3,s1_0_0_1_2_1_0_0_0,>,
<s1_0_1_1_1_1_0_0_0,t4,s2_0_1_0_0_1_0_1_0,>,
<s2_0_1_0_0_1_0_1_0,t3,s2_0_0_0_1_1_0_1_0,>,
<s2_0_1_0_0_1_0_1_0,t5,s2_0_1_0_0_0_1_1_1,>,
<s1_0_1_1_1_1_0_0_0,t5,s1_0_1_1_1_0_1_0_1,>,
<s1_0_2_1_0_1_0_0_0,t5,s1_0_2_1_0_0_1_0_1,>,
<s1_1_2_0_0_0_1_0_0,t3,s1_1_1_0_1_0_1_0_0,>,
<s2_0_1_0_0_0_1_1_1,t3,s2_0_0_0_1_0_1_1_1,>,
<s2_0_1_0_0_0_1_1_1,t6,s2_0_1_0_0_1_0_1_0,>,
<s1_0_1_1_1_0_1_0_1,t6,s1_0_1_1_1_1_0_0_0,>,
<s1_1_1_0_1_0_1_0_0,t3,s1_1_0_0_2_0_1_0_0,>,
<s1_1_0_0_2_0_1_0_0,t2,s1_0_0_1_2_0_1_0_1,>,
<s2_0_0_0_1_0_1_1_1,t6,s2_0_0_0_1_1_0_1_0,>
}
inputs {                          outputs {
+t1 +t2 +t3 +t4 +t5 +t6               }
}
```

All these states derived from the program are legal.

32.6 Case Study

32.6.1 Simulation Result

Software was developed to simulate the actions of a Distributing Sensing Network represented by a Petri Net plant model. The constraints listed in Section 32.4 were those to be monitored and enforced by each of the controllers. The Petri Net plant model of the DSN consisted of 133 places, 234 transitions, and roughly 1000 arcs. In order to enforce the plain language constraints, 44 inequalities of the form $l\mu \leq b$ were generated.

The Petri Net controller was implemented automatically by creating 44 control places that would act as the slack variables in a closed loop Petri Net. Arcs from these controller places influence controllable transitions in the plant net in an effort to enforce the constraints. Thus the controlled plant Petri net is simply a new Petri Net with additional places and arcs. Unlike the Petri Net controller, the VDES controller required no additional places or arcs to control the plant net. The VDES controller was implemented by examining every possible enabled firing given a plant state. The controller then examined the state of the system should each of these enabled firings take place, and disabled those transitions whose firings led to a forbidden state. This characteristic of VDES control illustrates a similarity with Moore machines; in Moore machines, the entire state space is explored offline and all forbidden strings are known a priori. In the case of VDES, exploration of reachable states is undertaken at each state dynamically and is limited to those states directly reachable from the current state.

The plant model was activated and the set of forbidden states were monitored at each transition firing. Without a controller of any kind in place, the plant model reached a forbidden state in less than 10,000 transition firings in each test. When the Petri Net or the VDES controllers were implemented, the plant model ran through 100,000 transition firings without violation. Thus, each controller was found to be effective in preventing the violation of system constraints, and the choice of which to use can be based upon issues such as execution speed. It was found that the relationship between the initial state and the controller specification was crucial. In complex systems such as the DSN, it is not difficult to specify an initial marking that will make the plant uncontrollable. Care must be taken to ensure that the system design and marking do not render the controller useless.

32.7 Discussion and Conclusion

Faced with the problem of synthesizing a controller for our large-scale surveillance network, we selected three methods as candidates and applied them to our system. Through comparison, we concluded that the approaches are roughly equivalent, each with pros and cons. Generally speaking, the three approaches can be classified into two categories: FSM belongs to traditional finite automata based controller category and Petri Net modeled and VDES belongs to the Petri Net based controller family.

The traditional Ramadge and Wonham control model is based on a classic finite automaton. Unfortunately, FSM based controllers involve exhaustive searches or simulation of system behavior and are especially impractical for large and complex systems. We eliminate illegal state spaces before synthesizing our finite automata, but the process is still a computationally expensive process for a system with a large number of states and events. Offline searching of the entire set of reachable states and the hardware implementation of the logical expression assures prompt controller response, which is crucial for those systems with strict real-time requirements. For a complex system such as a surveillance system, we use a modified version of an FSM modeled controller to avoid expensive computation and high representation cost. The controller is directly derived from the control specifications.

On the contrary, Petri Net based controllers take full advantage of the properties of the Petri Net. Their efficient mathematical computation employing linear matrix algebra makes real-time controlling and analysis possible, but they are still inferior to FSM in the performance of response time. Petri Nets offer a much more compact state space than finite automata and are better suited to model systems that exhibit a repetitive structure. Automatic handling of concurrent events is maintained as shown in [Moody 00, Wonham 99]. Vector discrete event system controllers explore the maximally permissive control constraint on the Petri Net with uncontrollable transitions by application of the integer linear programming problem, assuming that the uncontrollable portion of the Petri Net has no loops and the actual controller exists [Moody 00]. However, VDES does not consider unobservable events. The loop-free condition proves to be a sufficient, but not a necessary condition. Petri Net modeled controllers investigate the structural properties of a controlled Petri Net with unobservable events in addition to uncontrollable events. The integrated graphical structure of the Petri Net plant/controller makes system computation and representation straightforward.

The simulation results show that the system behavior is similarly and effectively constrained by any of the three approaches. Secondary concerns such as execution time and ease of representation can therefore guide decision on which approach to use.

32.A Appendix

32.A.1 Controllable Transitions

The following is a list of the controllable events shown in the control hierarchies. The transition number is the one shown in the relevant Petri net diagram. The hierarchies are denoted as: CC for operational command, SC for collaborative sensing, and WC for network communication. Event descriptions self-explanatory.

Trans.#	Hierarchy	Description
1	CC	Selecting donation region
2	CC	Altered coverage area
4	CC	Area unmapped/send map messages
5	CC	New node assigned to network
6	CC	Sending alterations
7	CC	New region priority
9	CC	Initial mapping complete
10	CC	Send deployment notices
11	CC	Resources found/sending coverage adjustments
12	CC	Wake command
13	CC	Receive cluster status message
14	CC	Sleep command
15	CC	Topology request from user/query clusters
16	CC	Resource request
17	CC	Recall command
18	CC	Receives cluster statistics
19	CC	Recall notices sent
20	CC	Network statistics to user
21	CC	Sending message
22	CC	Poll clusters

(*continued*)

(continued)

Trans.#	Hierarchy	Description
23	CC	Response TO, Respond to user
24	CC	Drain counter
25	CC	Response received
28	CC	Stop drain
30	CC	Altered coverage area
31	CC	Deployment command
32	CC	Coverage commands
33	CC	Wake command
35	CC	New node assigned to cluster
36	CC	Sending deployment notices
38	CC	Cluster topology request root
39	CC	Recall command
40	CC	Receive resource query
41	CC	Receive cluster status request
42	CC	Send recall notice
43	CC	Sending topology report
44	CC	Sending message
45	CC	Response TO
46	CC	Poll leaves
49	CC	Drain counter
50	CC	Response received
51	CC	Stop drain
55	CC	Coverage area adjusted
56	CC	New coordinates reached
58	CC	Send map update
61	CC	Update requested
63	CC	Statistics sent
64	CC	Wake command
65	CC	Recalled
66	CC	Receive resource query
67	CC	Retreat complete
69	CC	Response to CH
70	CC	Send message
71	SC	Receive message from user
72	SC	Receive message from CH
73	SC	Cluster boundaries and paths message
74	SC	Adjust the paths detectable probability threshold
75	SC	CH event summary
76	SC	Waiting TO, compute overall coverage
77	SC	Leaf movement for optimal cluster coverage
78	SC	Sensor movement for prioritized region
79	SC	Surveillance topology request
81	SC	Gap coverage not found
86	SC	Send fusion data message to user
88	SC	Receive event statistics
89	SC	Increase threshold
91	SC	Send to on board sensor fusion

(continued)

Trans.#	Hierarchy	Description
92	SC	Signal sensed
96	SC	Receiving message from root
97	SC	Move finished
100	SC	Send to onboard fusion
101	SC	Cluster optimal coverage movement
102	SC	Receive message from leaf
104	SC	Low noise
105	SC	Sleep wake
106	SC	Sensor movement for prioritized relocation
107	SC	Leaf node location and characteristics message
108	SC	Cluster self movement
109	SC	NonSelf movement
110	SC	Movement finished
111	SC	Adjust paths/detect probability
112	SC	Surveillance topology request
113	SC	Waiting TO
114	SC	Waiting TO
115	SC	Retain leaf node status
117	SC	Computing boundaries complete
118	SC	Latency TO
119	SC	Send to onboard sensor fusion
120	SC	Receive event statistics
121	SC	Threshold increased
123	SC	Send to onboard sensor fusion
126	SC	Movement complete
127	SC	Receive message from CH
130	SC	Prioritized location movement
131	SC	Surveillance topology request
135	SC	Sleep TO
136	SC	Leaf node move command
137	SC	Low noise
138	SC	On board fusion
139	SC	Movement complete
140	SC	Location and characteristics to CH
141	SC	Occlusion move complete
142	WC	Receive message from CH
147	WC	Message intact
149	WC	Self demotion
150	WC	Move complete
153	WC	Move complete
154	WC	Receive message from user
155	WC	Send message
156	WC	Request retransmit
157	WC	Move message
158	WC	Update user
160	WC	Signal power message
161	WC	Integrity adjusted

(*continued*)

(continued)

Trans.#	Hierarchy	Description
162	WC	General message
163	WC	Receive user ACK
165	WC	Frequency hopping message
166	WC	FH adjusted
167	WC	Processing complete
168	WC	Move complete
169	WC	SI failure
170	WC	Send retransmit request
174	WC	Move
179	WC	Receive message
182	WC	Demotion TO
184	WC	Send hello to root
185	WC	Message intact
190	WC	Request retransmit
191	WC	Move complete
192	WC	Receive message from root
193	WC	Move command
194	WC	Update root
195	WC	Receive root ACK
196	WC	Signal power message
197	WC	Integrity adjusted
198	WC	ACK TO
199	WC	Frequency hopping command
200	WC	FH complete
201	WC	General message
202	WC	Move complete
204	WC	Retain SH status
205	WC	Self demotion
206	WC	Processing complete
207	WC	SPI failure
208	WC	Send retransmit
210	WC	Wake message
211	WC	Send hello
213	WC	Move command
214	WC	Move complete
215	WC	Receive message from CH
216	WC	Event ACK not received
217	WC	Retain leaf status
218	WC	Signal power message
219	WC	Adjustment complete
220	WC	Frequency hopping message
221	WC	FH complete
222	WC	Interpreting signal integrity
223	WC	Send event summary
225	WC	General message
226	WC	Send message
229	WC	Wake message
230	WC	Processing complete
231	WC	Move complete

32.A.2 Uncontrollable Transitions

Following is a list of the uncontrollable events shown in the control hierarchies. The transition number is the one shown in the relevant Petri net diagram. The hierarchies are denoted as: CC for operational command, SC for collaborative sensing, and WC for network communication. Event descriptions are self-explanatory.

Trans.#	Hierarchy	Description
3	CC	Destroyed
8	CC	Insufficient resources for coverage alteration
26	CC	Demotion to cluster head
27	CC	Promotion to root
29	CC	Destroyed
34	CC	Sleep
37	CC	Attacked
45	CC	Timeout on response
47	CC	Promoted to cluster head
48	CC	Demoted to leaf
52	CC	Region mapping complete
53	CC	Path obstructed
54	CC	Destroyed
57	CC	Deployed
59	CC	Low power
60	CC	Damaged
62	CC	Attacked
68	CC	Sleep
80	SC	Waiting timeout
82	SC	Promotion
83	SC	Demotion
84	SC	Gap in coverage
90	SC	Background false alarm
93	SC	White noise
94	SC	Interference
95	SC	Spike noise
98	SC	Occlusion
99	SC	Excessive unrecoverable noise
103	SC	Saturation signal detected
113	SC	Waiting timeout
116	SC	Promoted
122	SC	Background false alarm
124	SC	Signal alarm
125	SC	White noise
128	SC	Interference/jamming
129	SC	Spike noise
132	SC	Occlusion
133	SC	Excessive unrecoverable noise
134	SC	Saturation signal detected
143	WC	Corrupt message
144	SC	Frequency hopping message
145	SC	Signal power message

(*continued*)

(continued)

Trans.#	Hierarchy	Description
146	WC	Position problem
148	WC	Demoted
159	WC	Re-contact user
171	WC	Corrupt message
172	WC	Dies
173	WC	Dies
177	WC	Promoted
178	WC	Overdue hello
181	WC	Dies
183	WC	Low power
187	WC	Signal power problem
188	WC	Frequency hopping problem
189	WC	Position problem
209	WC	Corrupt message
203	WC	Sleep command
212	WC	Promotion
224	WC	Target event sensed
227	WC	Dies
228	WC	Sleep
234	WC	Corrupt message

32.A.3 Petri Net Controller Implementation

The controller presented in this section adds new places and transitions to the Petri Net plant models to enforce the control specifications. It has been defined using the methodology described in the following list of constraints. Added controller places and arcs are given:

32.A.3.1 Define Controller Specifications

1. When a node is waiting for on-board data fusion, it should be prevented from moving by WC, CC, and SC. Also it should wait to be promoted by WC or by SC until sensing is complete.

$P57 + P120 < 2$	$P57 + P115 < 2$	$P57 + P107 < 2$	$P57 + P88 < 2$
$P57 + P96 < 2$	$P57 + P39 < 2$	$P57 + P130 < 2$	

2. Sleep state in SC caused by unrecoverable noise or a saturated signal should also force the sleep state in WC and CC, and vice versa for the case of a leaf node. Wakeup in SC needs to send wake-up to CC/WC.

   To enforce the above control specification, we added

   Inhibitor arc from P76 to all transitions in WC leaf hierarchy
   Inhibitor arc from P76 to all transitions in CC leaf hierarchy

3. Not a conflict issue, requires intra-plant transition to force all hierarchies into a reasonable state. Moving and self-location events cannot coexist.

$P79 + P120 < 2$	$P79 + P115 < 2$	$P79 + P107 < 2$	$P79 + P88 < 2$
$P79 + P96 < 2$	$P79 + P39 < 2$	$P79 + P130 < 2$	

4. While the node is in the process of dynamically updating the cluster head (receiving all statistics events), it should also be prevented from moving by WC, CC, or SC until a decision is made.

P47 + P120 < 2	P47 + P115 < 2	P47 + P107 < 2	P47 + P88 < 2
P47 + P96 < 2	P47 + P39 < 2	P47 + P130 < 2	

5. While the node is awaiting location and characteristics from a leaf (receiving all statistics events), it should also be prevented from moving by WC, CC, or SC until a decision is made.

P62 + P120 < 2	P62 + P115 < 2	P62 + P107 < 2	P62 + P88 < 2
P62 + P96 < 2	P62 + P39 < 2	P62 + P130 < 2	

6. Sensor in low power mode as determined by WC or damaged mode as determined by CC should be prohibited from any movements in SC as a result of prioritized relocation or occlusion adjustments.

   To enforce the above control specifications, we added

   Inhibitor arcs from P126 to transitions:

T130	T132	T136	T72	T53	T57	T58

   Inhibitor arcs from P31 to transitions:

T130	T132	T136	T232	T213	T193	T207	T157	T169

7. Retreat in CC should supercede all actions in WC/SC, except propagation of retreat command.

   To enforce the above control specifications, we added

   Inhibitor arc from P33 to all transitions leaving listening state
   Inhibitor arc from P22 to all transitions leaving listening state
   Arc from retreat signal to retreat place in all hierarchies

8. Entrance into the damaged state in CC should force entrance to the low power state in WC and vice versa.

   To enforce the above control specifications, we added

   Arc from T222 to P31
   Arc from T59 to P126
   Arc from T60 to P126

9. Nodes encountering a target signal in the SC should suspend mapping action in CC until sensing is complete

P70 + P28 < 2

10. Move commands in CC/WC should be delayed while node is receiving sensing statistics from below.

P62 + P120 < 2	P43 + P120 < 2	P62 + P130 < 2	P43 + P130 < 2
P62 + P115 < 2	P43 + P115 < 2	P62 + P107 < 2	P43 + P107 < 2
P62 + P96 < 2	P43 + P96 < 2	P62 + P88 < 2	P43 + P88 < 2
P62 + P39 < 2	P43 + P39 < 2	P62 + P28 < 2	P43 + P28 < 2

32.A.3.2 Controller Implementation for Unexplained Control Specifications

To Enforce Control Specification	Added Controller Places	Arc to Transitions	Arc from Transitions
P57 + P120 < 2	P137	T100, T213	T114, T214
P57 + P115 < 2	P138	T100, T207	T114, T202
P57 + P107 < 2	P139	T100, T193	T114, T191
P57 + P88 < 2	P140	T100, T158	T114, T153
P57 + P96 < 2	P141	T100, T169	T114, T168
P57 + P39 < 2	P142	T100, T72	T114, T75
P57 + P130 < 2	P143	T100, T232	T114, T231
P79 + P120 < 2	P147	T131, T213	T140, T214
P79 + P115 < 2	P148	T131, T207	T140, T202
P79 + P107 < 2	P149	T131, 193	T140, T191
P79 + P88 < 2	P150	T131, T157	T140, T153
P79 + P96 < 2	P151	T131, T169	T140, T168
P79 + P39 < 2	P152	T131, T72	T140, T75
P79 + P130 < 2	P153	T131, T232	T140, T231
P47 + P120 < 2	P157	T88, T213	T87, T214
P47 + P115 < 2	P158	T88, T207	T87, T202
P47 + P107 < 2	P159	T88, T193	T87, T191
P47 + P88 < 2	P160	T88, T157	T87, T150
P47 + P96 < 2	P161	T88, T169	T87, T168
P47 + P39 < 2	P162	T88, T72	T87, T75
P47 + P130 < 2	P163	T88, T232	T87, T231
P62 + P120 < 2	P167	T107, T213	T113, T214
P62 + P115 < 2	P168	T107, T207	T113, T202
P62 + P107 < 2	P169	T107, T193	T113, T191
P62 + P88 < 2	P170	T107, T157	T113, T150
P62 + P96 < 2	P171	T107, T169	T113, T168
P62 + P39 < 2	P172	T107, T72	T113, T75
P62 + P130 < 2	P173	T107, T232	T113, T231
P70 + P28 < 2	P190	T124, T58	T122, T52
P62 + P120 < 2	P200	T107, T213	T113, T214
P43 + P120 < 2	P201	T75, T213	T80, T214
P62 + P130 < 2	P202	T107, T193	T113, T191
P43 + P130 < 2	P203	T75, T232	T80, T231
P62 + P115 < 2	P204	T107, T207	T113, T202
P43 + P115 < 2	P205	T75, T207	T80, T202
P62 + P107 < 2	P206	T107, T193	T113, T191
P43 + P107 < 2	P207	T75, T193	T80, T191
P62 + P96 < 2	P208	T107, T169	T113, T168
P43 + P96 < 2	P209	T75, T169	T80, T168
P62 + P88 < 2	P210	T107, T157	T113, T150
P43 + P88 < 2	P211	T75, T157	T80, T150
P62 + P39 < 2	P212	T107, T72	T113, T75
P62 + P28 < 2	P214	T107, T57, T53, T58	T113, T52
P43 + P28 < 2	P215	T75, T57, T53, T58	T80, T52

32.A.4 Finite State Machine and Vector Controller Implementation

Boolean functions derived as the FSM controller are exerted on those controllable events to prevent violation of the control specifications. The controllable transitions are allowed to fire provided corresponding Boolean functions are satisfied.

The state vector of the system is the concatenation of state vector of a node in three different hierarchies. It is important to note here that the node roles in the hierarchies are independent. A node occupying the cluster head level in the sensor coverage hierarchy is allowed to occupy any of the three levels in the other two hierarchies and is not restricted in any fashion.

1. For transition 128, it can fire, i.f.f. it is enabled and the state status satisfies the conjunction of the following predicates:

P75 + P144 = 0	P75 + P137 = 0	P75 + P120 = 0	P75 + P130 = 0
P75 + P39 = 0	P75 + P145 = 0	P75 + P28 = 0	P75 + P115 = 0
P75 + P107 = 0	P75 + P88 = 0	P75 + P96 = 0	

2. For transition 241, it can fire, i.f.f. it is enabled and the state status satisfy the conjunction of the following predicates:

P75 + P144 = 0	P138 + P144 = 0	P80 + P144 = 0	P126 + P144 = 0

3. For transition 140, it can fire, i.f.f. it is enabled and the state status satisfy the conjunction of the following predicates:

P75 + P137 = 0	P138 + P137 = 0	P80 + P137 = 0	P126 + P137 = 0
P137 + P121 = 0	P137 + P120 = 0	P137 + P125 = 0	P137 + P133 = 0
P137 + P107 = 0	P137 + P111 = 0	P137 + P113 = 0	P137 + P112 = 0
P137 + P88 = 0	P137 + P91 = 0	P137 + P94 = 0	P137 + P98 = 0

4. For transition 213, it can fire, i.f.f. it is enabled and the state status satisfy the conjunction of the following predicates:

P75 + P120 = 0	P138 + P120 = 0	P137 + P120 = 0	P39 + P120 = 0
P54 + P120 = 0	P43 + P120 = 0	P59 + P120 = 0	P63 + P120 = 0
P49 + P120 = 0			

5. For transition 232, it can fire, i.f.f. it is enabled and the state status satisfy the conjunction of the following predicates:

P75 + P130 = 0	P138 + P130 = 0	P80 + P130 = 0	P59 + P130 = 0
P43 + P130 = 0	P63 + P130 = 0	P49 + P130 = 0	

6. For transition 55, it can fire, i.f.f. it is enabled and the state status satisfy the conjunction of the following predicates:

P75 + P39 = 0	P138 + P39 = 0	P80 + P39 = 0	P39 + P120 = 0
P39 + P121 = 0	P39 + P125 = 0	P39 + P133 = 0	P39 + P107 = 0
P39 + P111 = 0	P39 + P113 = 0	P39 + P112 = 0	P39 + P88 = 0
P39 + P91 = 0	P39 + P94 = 0	P39 + P98 = 0	P54 + P39 = 0
P43 + P39 = 0	P63 + P39 = 0	P49 + P39 = 0	P59 + P39 = 0

7. For transition 131, it can fire, i.f.f. it is enabled and the state status satisfy the conjunction of the following predicates:

P75 + P145 = 0

8. For transition 100, it can fire, i.f.f. it is enabled and the state status satisfy the conjunction of the following predicates:

P54 + P70 = 0	P54 + P115 = 0	P54 + P107 = 0	P54 + P146 = 0
P54 + P120 = 0	P54 + P88 = 0	P54 + P96 = 0	P54 + P39 = 0
P54 + P28 = 0			

9. For transition 117, it can fire, i.f.f. it is enabled and the state status satisfy the conjunction of the following predicates:

P54 + P70 = 0	P59 + P70 = 0	P63 + P70 = 0	P126 + P70 = 0

10. For transition 207, it can fire, i.f.f. it is enabled and the state status satisfy the conjunction of the following predicates:

P54 + P115 = 0	P59 + P115 = 0	P63 + P115 = 0	P75 + P115 = 0
P138 + P115 = 0	P80 + P115 = 0	P43 + P115 = 0	P49 + P115 = 0

11. For transition 193, it can fire, i.f.f. it is enabled and the state status satisfy the conjunction of the following predicates:

P54 + P107 = 0	P59 + P107 = 0	P63 + P107 = 0	P75 + P107 = 0
P138 + P107 = 0	P137 + P107 = 0	P80 + P107 = 0	P43 + P107 = 0
P49 + P107 = 0	P39 + P107 = 0		

12. For transition 91, it can fire, i.f.f. it is enabled and the state status satisfy the conjunction of the following predicates:

P54 + P146 = 0

13. For transition 235, it can fire, i.f.f. it is enabled and the state status satisfy the conjunction of the following predicates:

P138 + P144 = 0	P138 + P137 = 0	P138 + P120 = 0	P138 + P130 = 0
P138 + P39 = 0	P138 + P107 = 0	P138 + P115 = 0	

14. For transition 137, it can fire, i.f.f. it is enabled and the state status satisfy the conjunction of the following predicates:

P80 + P144 = 0	P80 + P137 = 0	P80 + P120 = 0	P80 + P130 = 0
P80 + P39 = 0	P80 + P115 = 0	P80 + P107 = 0	P80 + P96 = 0
P80 + P88 = 0			

15. For transition 96, it can fire, i.f.f. it is enabled and the state status satisfy the conjunction of the following predicates:

P59 + P70 = 0	P59 + P115 = 0	P59 + P107 = 0	P59 + P130 = 0
P59 + P120 = 0	P59 + P88 = 0	P59 + P96 = 0	P59 + P39 = 0

16. For transition 103, it can fire, i.f.f. it is enabled and the state status satisfy the conjunction of the following predicates:

P63 + P70 = 0	P63 + P115 = 0	P63 + P107 = 0	P63 + P130 = 0
P63 + P88 = 0	P63 + P120 = 0	P63 + P96 = 0	P63 + P39 = 0

17. For transition 222, it can fire, i.f.f. it is enabled and the state status satisfy the conjunction of the following predicates:

P126 + P144 = 0	P126 + P137 = 0	P126 + P70 = 0

18. For transition 53, it can fire, i.f.f. it is enabled and the state status satisfy the conjunction of the following predicates:

P75 + P28 = 0	P54 + P28 = 0

19. For transition 57, it can fire, i.f.f. it is enabled and the state status satisfy the conjunction of the following predicates:

P75 + P28 = 0	P54 + P28 = 0

20. For transition 218, it can fire, i.f.f. it is enabled and the state status satisfy the conjunction of the following predicates:

P137 + P121 = 0	P39 + P121 = 0

21. For transition 220, it can fire, i.f.f. it is enabled and the state status satisfy the conjunction of the following predicates:

P137 + P125 = 0	P39 + P125 = 0

22. For transition 234, it can fire, i.f.f. it is enabled and the state status satisfy the conjunction of the following predicates:

P137 + P133 = 0	P39 + P133 = 0

23. For transition 74, it can fire, i.f.f. it is enabled and the state status satisfy the conjunction of the following predicates:

P43 + P88 = 0	P43 + P96 = 0	P43 + P130 = 0	P43 + P120 = 0
P43 + P115 = 0	P43 + P107 = 0	P43 + P96 = 0	P43 + P88 = 0
P43 + P39 = 0			

24. For transition 157, it can fire, i.f.f. it is enabled and the state status satisfy the conjunction of the following predicates:

P43 + P88 = 0	P75 + P88 = 0	P54 + P88 = 0	P80 + P88 = 0
P59 + P88 = 0	P43 + P88 = 0	P63 + P88 = 0	P49 + P88 = 0
P137 + P88 = 0	P39 + P88 = 0		

25. For transition 169, it can fire, i.f.f. it is enabled and the state status satisfy the conjunction of the following predicates:

P43 + P96 = 0	P75 + P96 = 0	P54 + P96 = 0	P80 + P96 = 0
P59 + P96 = 0	P43 + P96 = 0	P63 + P96 = 0	P49 + P96 = 0

26. For transition 84, it can fire, i.f.f. it is enabled and the state status satisfy the conjunction of the following predicates:

P49 + P130 = 0	P49 + P120 = 0	P49 + P115 = 0	P49 + P107 = 0
P49 + P96 = 0	P49 + P88 = 0	P49 + P39 = 0	

27. For transition 196, it can fire, i.f.f. it is enabled and the state status satisfy the conjunction of the following predicates:

P137 + P111 = 0	P39 + P111 = 0

28. For transition 199, it can fire, i.f.f. it is enabled and the state status satisfy the conjunction of the following predicates:

P137 + P113 = 0	P39 + P113 = 0

29. For transition 209, it can fire, i.f.f. it is enabled and the state status satisfy the conjunction of the following predicates:

P137 + P112 = 0	P39 + P112 = 0

30. For transition 160, it can fire, i.f.f. it is enabled and the state status satisfy the conjunction of the following predicates:

P137 + P91 = 0	P39 + P91 = 0

31. For transition 165, it can fire, i.f.f. it is enabled and the state status satisfy the conjunction of the following predicates:

P137 + P94 = 0	P39 + P94 = 0

32. For transition 171, it can fire, i.f.f. it is enabled and the state status satisfy the conjunction of the following predicates:

P137 + P98 = 0	P39 + P98 = 0

32.A.5 Three Interwined Model Hierarchies for Sensor Network

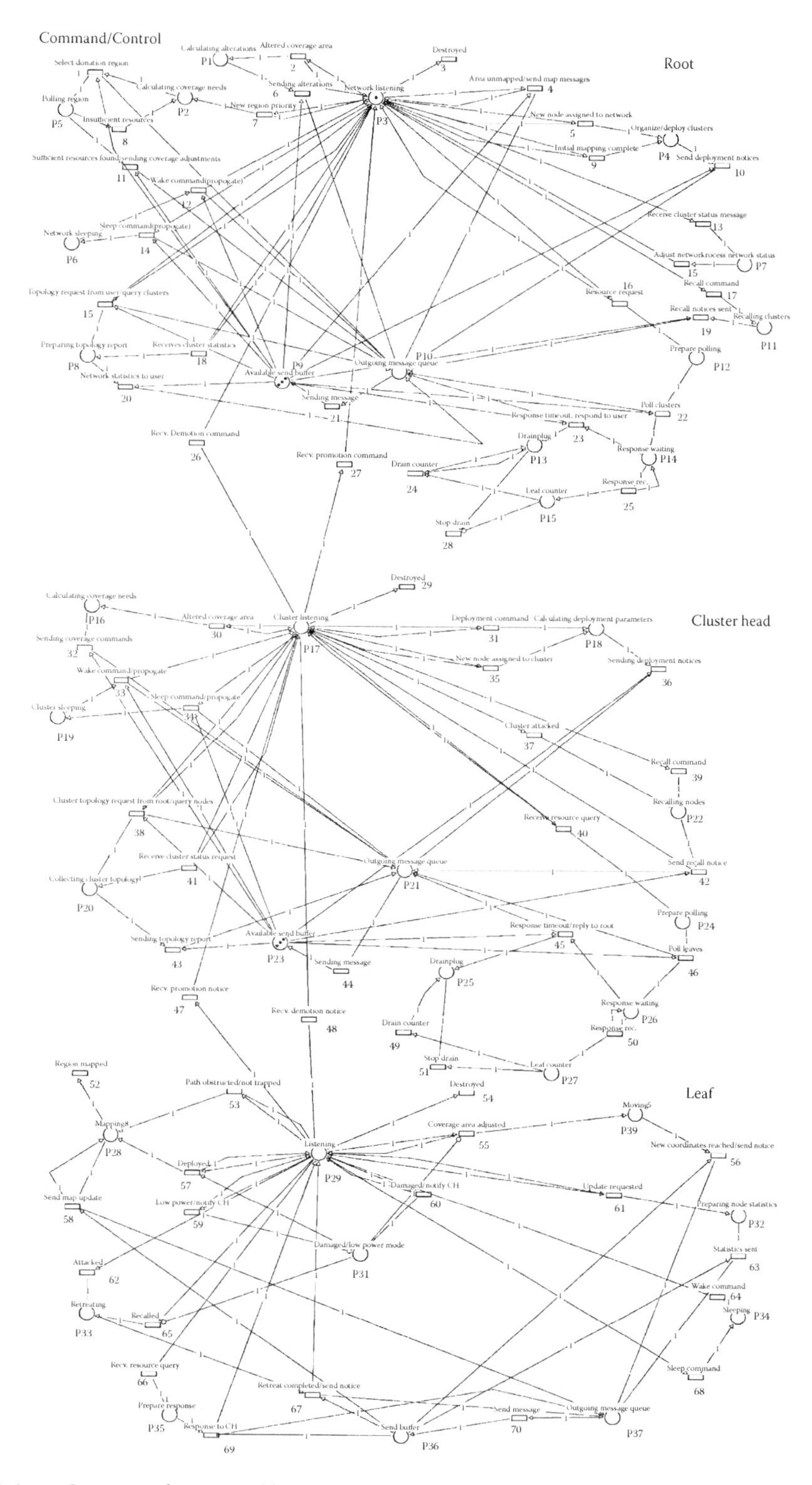

FIGURE 32.A.1 Operational command hierarchy.

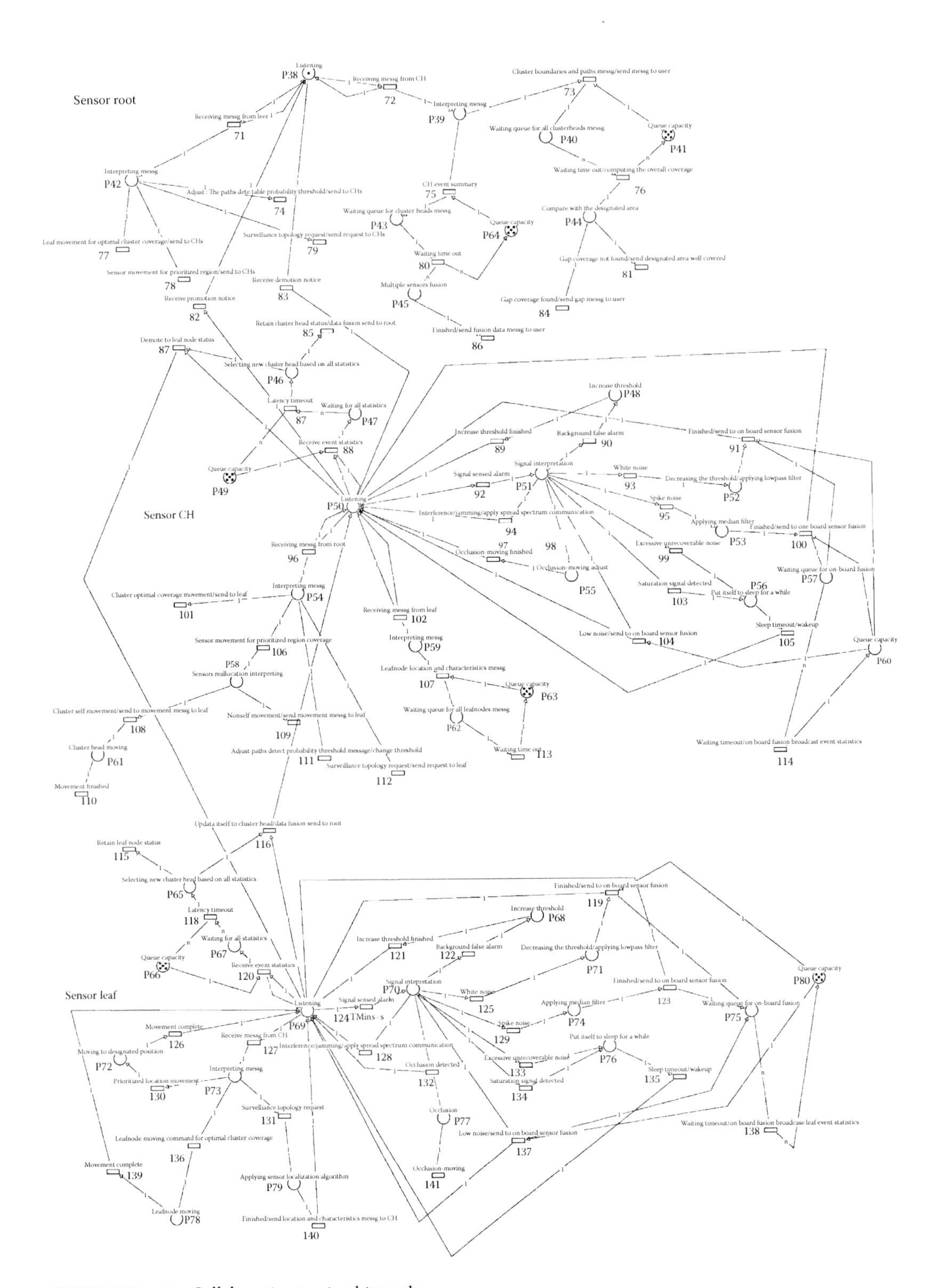

FIGURE 32.A.2 Collaborative sensing hierarchy.

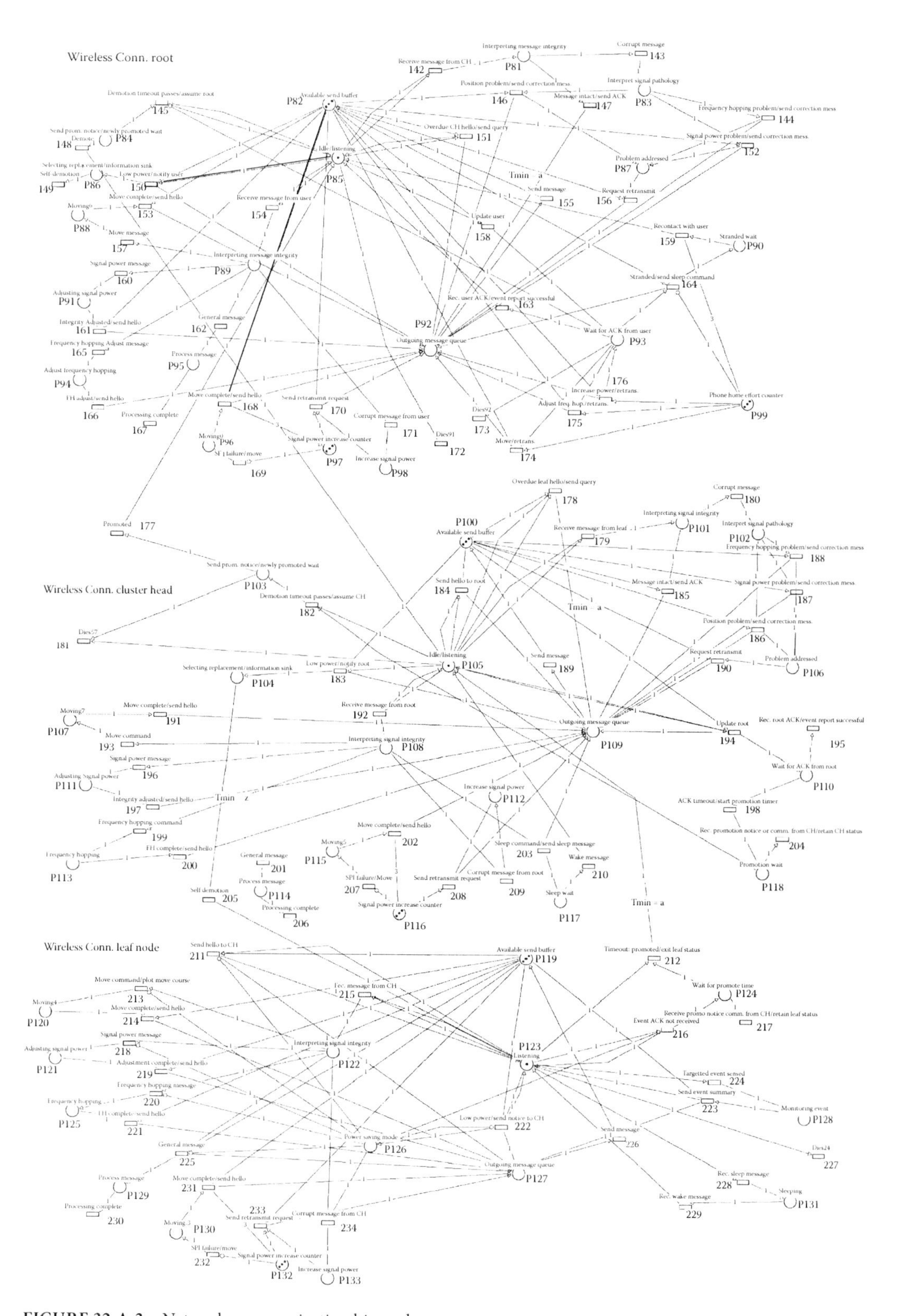

FIGURE 32.A.3 Network communication hierarchy.

Acknowledgments and Disclaimer

This material is based upon work supported by the U.S. Army Robert Morris Acquisition under Award No. DAAD19-01-1-0504. Any opinions, findings, and conclusions or recommendations expressed in this publication are those of the authors and do not necessarily reflect the views of the Army.

References

[Aho 86] A.V. Aho, R. Sethi, and J.D. Ullman, *Compilers: Principles, Techniques and Tools*, Addison-Wesley, Reading, MA, 1986.

[Brooks 97] R. Brooks and S.S. Iyengar, *Multi Sensor Fusion: Fundamentals and Applications with Software*, Prentice Hall Publication Co., Englewood Cliffs, NJ, 1997.

[Bulusu 01] N. Bulusu, D. Estrin, L. Girod, and J. Heidemann, Scalable coordination for wireless sensor networks: Self-configuring localization systems, USC/Information Sciences Institute, Marina Del Rey, CA, 2001.

[David 92] R. David and H. Alla, *Petri Nets and Grafcet Tools for Modeling Discrete Event Systems*, Prentice Hall, Upper Saddle River, NJ, ISBN: 0-13-327537-X, 1992.

[Deb 02] B. Deb, S. Bhatnagar, and B. Nath, A topology discovery algorithm for sensor networks with applications to network management, *Proceedings IEEE CAS Workshop on Wireless Communication and Networking*, Pasadena, September 2002.

[Hall 97] D.L. Hall and J. Llinas, An introduction to multisensor data fusion, *Proceedings of the IEEE*, 85(1), 6–23, January 1997.

[Li 93] Y. Li and W.M. Wonham, Control of vector discrete-event systems I—The base model, *IEEE Transactions on Automatic Control*, 38(8), 1214–1227, August 1993.

[Li 94] Y. Li, and W.M. Wonham, Control of vector discrete-event system Π—Controller synthesis, *IEEE Transactions on Automatic Control*, 39(3), 512–531, March 1994.

[Luo 89] R.C. Luo and M.G. Kay, Multisensor integration and fusion in intelligent systems, *IEEE Transactions on Systems, Man, and Cybernetics*, 19(5), 901–931, September/October 1989.

[Meguerdichian 01] S. Meguerdichian, F. Koushanfar, M. Potkonjak, and M.B. Srivastava, Coverage problems in wireless ad-hoc sensor networks, *Proceedings of IEEE INFOCOM*, 1380, 1387, 2001.

[Moody 98] J.O. Moody, Petri net supervisors for discrete event systems, PhD dissertation, Department of Electrical Engineering, Notre Dame University, South Bend, IN, April 1998.

[Moody 00] J.O. Moody, N.Y. Owego, and P.J. Antsaklis, Petri net supervisors for discrete event systems, *IEEE Transactions on Automatic Control*, 45(3), 462–476, 2000.

[Murata 89] T. Murata, Petrinets: Properties, analysis and applications, In *Proceedings of the IEEE*, 77(4), 541–580, April 1989.

[Peterson 77] J.L. Peterson, Petri nets, *ACM Computing Surveys*, 9(3), 223–252, September 1977.

[Ramadge 87] P.J. Ramadge and W.M. Wonham, Supervisory control of a class of discrete event progress, *SIAM Journal on Control and Optimization*, 25(1), 206–230, January 1987.

[Wonham 99] W.M. Wonham, Notes on discrete event system control, System Control Group, Electrical & Computer Engineering Department, University of Toronto, Toronto, Ontario, Canada, 1999.

V

Engineering Examples

Sensor networks have become an important source of information with numerous real-life applications. They are used for monitoring transportation and traffic control, contamination level in soil and water, climate, building structure, habitat, and the quality of perishable food items. In Part V, the emphasis is on monitoring of important components within the environment and cleaning of the data before decision making.

Mitchell et al. describe a filed test performed as a part of the Defence Advanced Research Projects Agency (DARPA) IXO initiative in Chapter 33. It took place at a military installation. The SITEX test was the first time that many of the technologies described in this book were used in a realistic military situation. The architecture, its components, and results of the test are described in detail.

Elnahrawy and Badrinath emphasize on online cleaning of the sensor data before any crucial decisions are taken in Chapter 34. Data collected from wireless sensor networks are subject to several problems and source of errors. These problems may seriously impact the actual usage of such networks and yield imprecise and inaccurate answers for any query on sensor data. The authors focus on probabilistic efficient and scalable approaches in reducing the effect of random errors, Bayesian estimation, traditional query evaluation, querying noisy sensors, online distributed learning, spatiotemporal dependencies and wireless sensors, detection of outliers and malicious sensors, and recovery of missing values.

Bridges and Biagioni focus on monitoring the phenology of endangered species of plants and their surrounding environment by sensor networks in Chapter 35. The weather information that is collected for phenological studies includes the air temperature, rainfall amount, relative humidity, solar radiation intensity, and wind speed and direction. They discuss monitoring devices such as digital sensor, thermocouple sensor, tipping bucket sensor, digital camera and networking and deployment of sensors, and data utilization.

Poor et al. have experience in the implementation and use of sensor networks for industrial applications. Their chapter (Chapter 36) provides application design rules. These rules are based on their experience. They show how these design principles can be used to implement systems using off-the-shelf components.

In a new chapter for the second edition (Chapter 37), Alvin Lim describes a sensor network test bed that is being used for target-tracking applications. Video sensors are connected using directed diffusion.

Ruj et al. provide another new chapter (Chapter 38) that looks into secure processing and key management. This chapter concentrates on lightweight key management tools and surveys the appropriate technologies.

Griffin et al. provide a new chapter that explores naval applications of these technologies and discusses maritime domain awareness applications (Chapter 39). It shows how target trajectories can be optimally grouped together to produce a grid that naturally groups similar target paths. The resulting continuous-to-discrete conversion aids in hidden Markov model inference. The final tools help in not only tracking targets but also predicting future positions.

Yu et al. provide a final chapter that describes an environmental monitoring application of sensor networks (Chapter 40). This network was implemented to gather water turbidity information in real time. It was implemented as part of an engineering capstone project. Multiple classes of sensor information related to water quality were gathered from a stream. They were then uploaded into a GIS database using GSM. Forestry researchers used the resulting datasets to monitor water quality.

In summary, Part V discusses specific sensor network implementations and lessons learned by fielding the systems. A number of new chapters in the second edition show how the field of sensor networks as well as the increasingly diverse set of applications of this technology have evolved.

33

SenSoft: Development of a Collaborative Sensor Network

Gail Mitchell
Bolt, Beranek and Newman Technologies

Jeff Mazurek
Bolt, Beranek and Newman Technologies

Ken Theriault
Bolt, Beranek and Newman Technologies

Prakash Manghwani
Bolt, Beranek and Newman Technologies

In 1999, the Defense Advanced Research Projects Agency (DARPA) established the Sensor Information Technology (SensIT) program to investigate the feasibility of employing thousands of autonomous, distributed, networked, multi-modal ground sensors to accomplish intelligence, surveillance, and reconnaissance (ISR) tasks. A large group of Principal Investigators was challenged to develop innovative hardware, algorithms and software to demonstrate the potential of distributed micro sensor networks. These investigators collaborated, with BBN Technologies as the system architect and integrator, to develop and demonstrate a system of distributed, networked sensor nodes with in-network data processing, target detection, classification, and tracking, and communication within and outside of the network. The SensIT architecture and SensIT Software system (SenSoft) described in this chapter are the results of this collaboration. This prototype distributed sensor network, and the research accomplished by the contributors in the process of achieving the prototype, represent a firm foundation for further development of and experimentation with information technology for distributed sensor systems.

33.1 Overview of SensIT System Architecture

A SensIT network is a system of *sensor nodes* communicating with each other within the network and also communicating with entities outside the SensIT network. The conceptual architecture for such a system is depicted in Figure 33.1. As shown here, a SensIT network is a robust, flexible collection of "smart" sensor nodes; nodes that include, in addition to sensing capabilities, processing and communications functionality that enable building ad hoc, flexible, and robust networks for distributed signal processing. For example, the network of nodes shown in Figure 33.1 might be deployed in hostile

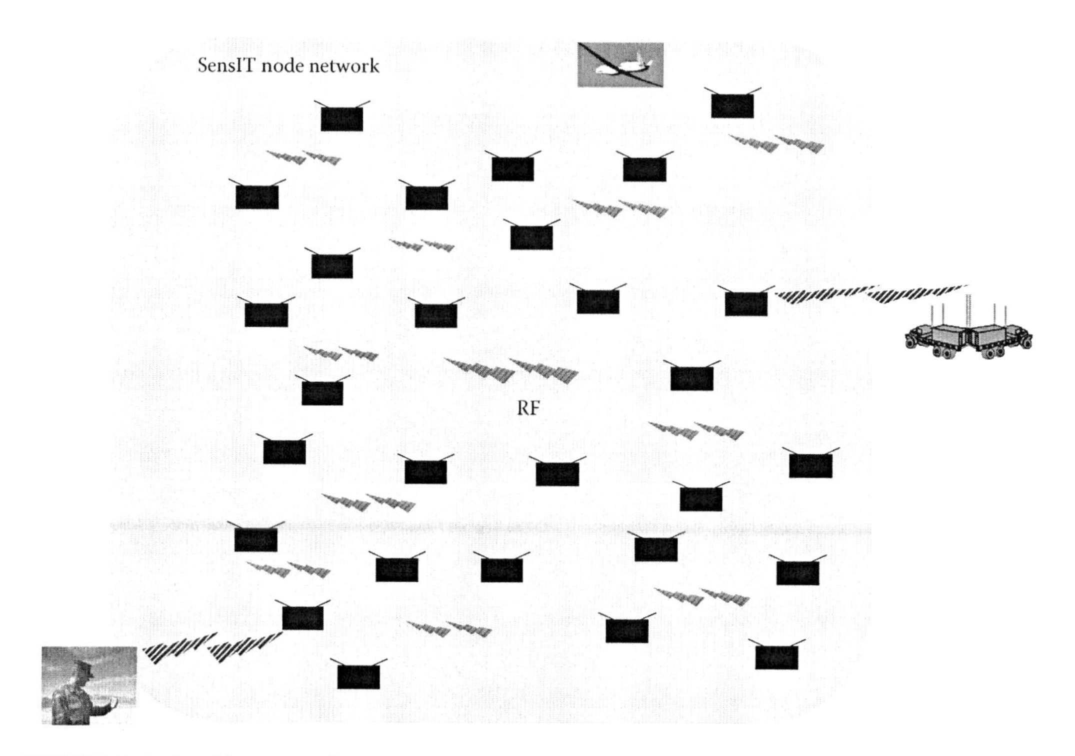

FIGURE 33.1 SensIT system architecture concept.

territory and receiving commands (tasks) from a mobile base station (the trucks). Similarly, a soldier in a remote platoon uses a handheld device to request information about activity sensed by the network, and an unmanned aerial vehicle (UAV) sends commands to request information that it relays to a ship waiting offshore. In each example a base station, that is, a user of the network, connects to a node in the network to give commands to, or receive information from, the network as a whole. The node with which a base station communicates is, at that point in time, the network's *gateway* node. The network gateway may or may not be the same node for all extra-network communications; the gateway node used by a base station site might even change if that site is mobile or if connectivity changes.

The key to making this happen is that data is obtained and commands are processed *within* the node network. Unlike distributed sensor systems that collect sensing information and move it to a centralized location for processing, the nodes in a SensIT network communicate with each other and cooperate to process the data themselves to accomplish tasks such as processing sensor signals and determining the presence and activity patterns of a target. As excellent example of centrally-processed distributed sensor systems consider the fields of aircraft-deployed sonobuoys used by naval forces to acoustically detect, classify, and track submarines. Although each individual sonobuoy may employ sophisticated in-buoy signal processing, there is no inter-buoy communication and all results are sent to a monitoring aircraft for integration and display. In SensIT individual sensor nodes employ sophisticated local processing but they also communicate local results with other nodes and then cooperate with those nodes to further process the results to detect, classify and track targets. Through node cooperation and interaction, the thesis is that the total amount of processing needed and the amount of data moved through the network or from the network to a display station may be reduced and, indeed, higher quality information may be obtained.

Commands to a SensIT network arise from within the network or from external sources, and are distributed and processed within the network. Commands from external sources are moved into the network through gateway nodes. The tasks needed to execute a command, and the nodes assigned to execute those tasks, are determined within the network. Similarly, requests from external agents for information obtained through in-network processing are processed within the network in such a way that the results are made available through the gateway. SenSoft (SensIT Software) is the name we give to the software applications and services that effect such in-network signal processing.

33.2 Prototype Hardware Platform

SenSoft was developed for the Sensoria Corporation's WINS NG 2.0 sensor nodes [WINS2]. These nodes prototype the processing power and other hardware functionalities that we expect to see in micro nodes anticipated for the future, and thus are a good platform for experimentation with the software and operational concepts for distributed micro sensor networks. Each WINS NG 2.0 node provides the flexibility to experiment with as many as four separate sensor inputs and to communicate with other nodes in the network. Each node also has embedded processing capability built on Linux operating system software, with a wide range of common utilities and applications (e.g., telnet, ftp, vi) available.

A WINS NG 2.0 node has an onboard digital signal processor (DSP), a Linux processor (Hitachi SH4), RAM and flash memory, a GPS receiver, and two embedded RF modems. A key feature supporting sensor signal experimentation is four analog sensor channels that can receive input from various kinds of sensors. These channels are read and processed independently by the DSP. These nodes provide the signal capture, network communications, and computer processing capabilities upon which a distributed sensor network can be built. As an experimentation platform, they also provide the flexibility to test various network solutions. For example, a sensor network can run as an RF communicating network, but the nodes can also use Ethernet for experimentation and debugging. The various communications modes are also valuable for experimentation with the capabilities and performance of Ethernet solutions vs. radio communications. And the generic sensor channels allow experimentation with a variety of types and combinations of sensing modalities.

The nodes also have a variety of external interfaces to support these capabilities, including

- Two PCMCIA slots for PCMCIA and CardBus expansion. We have used a slot, for example, to provide wireless Ethernet communications between nodes.
- A serial port for a console interface to the platform (useful for hardware debugging); an Ethernet port, allowing a node to connect to a LAN.
- Two antenna connectors for the embedded RF modems and an antenna connector for the GPS.

The two modems in each node are used to build network-wide RF connectivity using short-range local radio connections (see Figure 33.4). A Sensoria network is the union of many small local-area networks, each talking at a different RF frequency (the networks use frequency-hopping to prevent collision). Each node belongs to two different local networks and can pass messages between those networks; as a result, messages move within the larger network by "hopping" across the different local networks.

33.3 SenSoft Architectural Framework

The SenSoft framework describes how a network of nodes can be used to develop and experiment with algorithms and software for networked sensor processing. The architecture described in this chapter has two major types of component: *infrastructure* components and *signal processing* components. The *infrastructure* provides common functionality that enables the activities of collaborative *signal processing* applications in a network of SensIT nodes. Most of this functionality is present on all sensor node; some is only on nodes that can act as gateways; and some of the functionality is present on a command & control (C2) system that interacts with a SensIT network via a gateway.

The on-node software architecture is illustrated in Figure 33.2. In this figure the heavily outlined boxes indicate the functional components of the support infrastructure and the more lightly outlined boxes are signal processing components. Every node in a SensIT network has the infrastructure components indicated in the figure; the implementations of these components will be identical on all nodes. Similarly, each

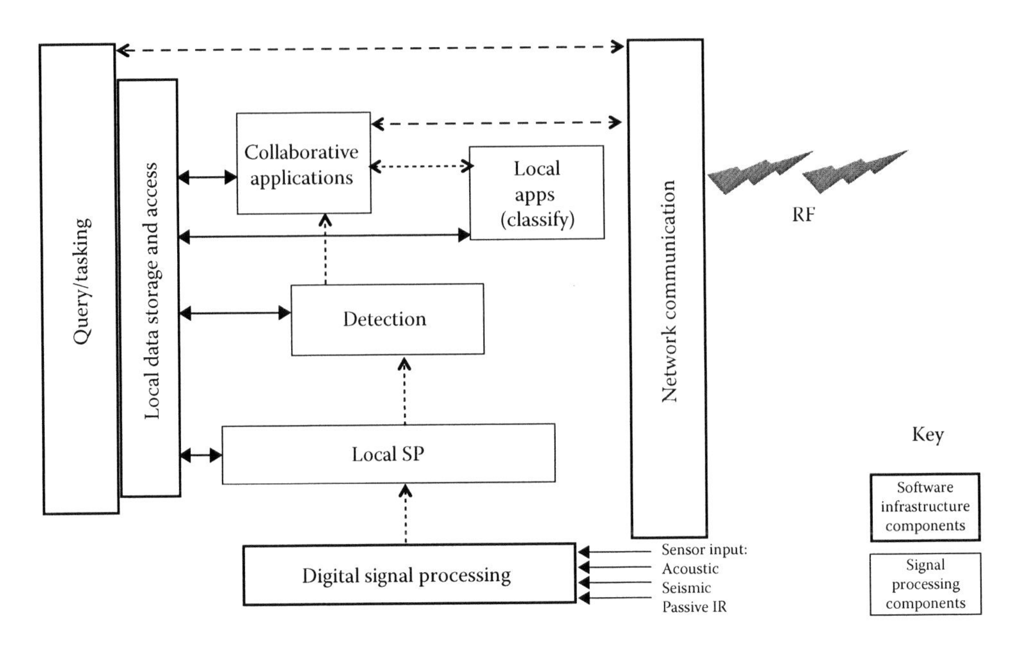

FIGURE 33.2 SenSoft on-node software architecture.

node may have signal processing (and possibly other) applications that use the infrastructure components' interfaces and data structures. Typically (or, at least in the implementations and experimentation we've seen thus far) the signal processing applications are also implemented in the same way on all nodes—placing the distributed aspects of the signal processing computation into the node software.

The architecture for command and control interaction with a SensIT network is illustrated in Figure 33.3. Note that this interaction combines on-node and off-node software (and hardware)

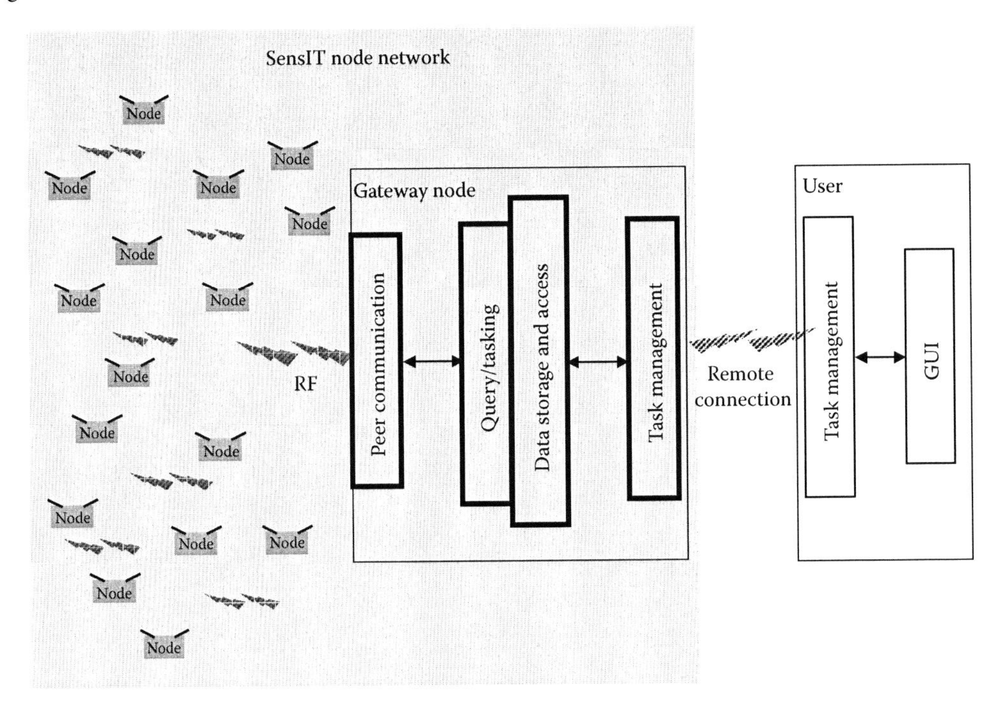

FIGURE 33.3 SenSoft command and control architecture.

components: one or more sensor nodes can be outfitted with task management software that allows them to interact with stations outside the node network. A user of the network (i.e., a C2 system) has a local version of the task management interface; and the C2 station has some sort of communication connectivity with a gateway node (we usually used IP over wireless Ethernet, although this is not a requirement).

33.4 Software Infrastructure

Let's examine the SenSoft infrastructure in a little more detail. On-node components of the infrastructure are illustrated in Figure 33.2; command and control components are needed both on- and off-node, and are illustrated in Figure 33.3. These infrastructure components perform common services for signal processing applications, thus reducing redundancy and facilitating communications within and between nodes. In addition, interfaces defined to each component are intended to facilitate replacement of any given component with a functionally equivalent component. The on-node components include

- *Hardware/firmware for digitizing the sensor signal inputs*: The Sensoria DSP on each sensor node can sample at a rate of up to 20 kHz per channel on each of the four separate analog input channels (in the WINS NG 2.0 nodes, all channels had to sample at the same rate). The DSP subsystem builds "sections" of 256 16-bit samples with identifiers and timestamps (and other useful information), and transparently moves the data in fixed-size blocks to circular, FIFO buffers in the SH-4 host system memory. Signal processing applications must access the time series data in these buffers in a timely fashion; timestamps and sequencing identifiers can be used to tell whether a block of samples has been overwritten in a buffer. A sampling API lets developers select different data rates, select a gain for each sampling channel, and start and stop sampling on each channel. Sensoria documentation [WINS2] provides more detailed information about the sampling API and its use.
- *Network routing for application-level communications between nodes*: Network communications software supports application-level data sharing between nodes and insulates the applications on a node from the mechanics of data transport. In SenSoft, data routing exposes a declarative interface to applications allowing them to specify what they want to send or receive; the network communications software then determines which messages need to be sent where (routing) and moves the messages over the particular transport mechanism. In SenSoft, data transport was typically over the Sensoria radios although the software can manage, and we also experimented with, IP-based transport (wired and wireless).
- *Local data storage for data persistence and communications within a node*: In addition to moving data between nodes, it is also necessary to store some data locally on each node for various lengths of time. This data is typically specific to the node—either describing the node, or obtained and analyzed at the node—and can have varying lifetimes that need to be managed by a data storage and access component. For example, the DSP buffers are a transient form of storage for time-stamped digitized signal data. Some of this data, and local analyses of this data, may need to be stored for longer periods than possible in the memory buffer (e.g., until no longer needed by the applications) and thus are more appropriately stored and managed by a local "database" system. (Note that a goal of the signal processing software is to reduce the amount of data that needs to be stored and transported.)

  Data requiring a longer lifetime includes data collected by a node that may be shared at a later point with other nodes, and data (typically, results) that will be moved and processed toward a gateway node for dissemination. For example, target detection event records may be computed and stored locally, and later sent to neighbors either to trigger actions on their part or in response to requests. A data storage and access component mitigates some of the complications of dealing with time delays in data acquisition across nodes.

Data about the state of a particular node or the state of the network itself can also be stored and managed at each node. Examples of this type of data include node configuration and status information (e.g., name, node location, sensor channel settings), code book values (assigning identifiers to targets, or names to values for display), or application-specific records (such as track updates). In these situations, the aggregate of local storage can be thought of as a database for the network; that is, the local data storage and access components are pieces of a distributed data management system for the SensIT network.

- *Query/tasking to control movement of queries or tasks into/from nodes, and to optionally perform higher-level query processing (e.g., aggregation)*: A query is a request for information; a task is a specification of action(s) that will generate data or information and thus can be thought of as defining how to answer a query. Given a request for information from outside the network, the query/tasking component is responsible for determining what processing needs to be done at which nodes in the network to provide the information. For example, a query for current activity within a network might involve all nodes reporting to their neighbors about target detections, and aggregation at certain nodes to provide an overview of amount of activity in different geographic areas of the network. Which nodes perform aggregation might be determined dynamically by the query/tasking component based on dynamic factors such as data availability. Similarly, requests for information that arise within the network and require information from multiple nodes would be managed by the query/tasking components of the nodes.
- *Task management at Gateway node(s)*: Each node that can be a gateway for the sensor network must have task management software that can interact with the related task management software at a C2 user station. A gateway node's software must be able to accept tasks (commands), translate those for distribution into the network (through query/tasking), collect results computed within the network, and format and transfer results to the correct C2 user.

A gateway node is one component of the command and control of any system that includes a user of sensor information and a collaborative sensor network (or multiple users, or multiple networks). Although we represent the user as a human, and thus assume need for a GUI for human/machine communications as illustrated in Figure 33.3, a collaborative sensor network will more likely be interacting within a larger operational system and communicating with other computer applications that command the network and use the sensor information obtained. Whether the user component is human or machine, it must include a task/query management component to mediate between the corresponding component at a gateway node and the user application. The user task management component works closely with task management on a gateway node, collectively forming a conduit for information and control flow between the sensor network and the user application (GUI or other).

In all SenSoft experiments, the user interface was a graphical interface for human use. A GUI application provides the ability to send tasks or queries to a sensor network and to display the results of those activities. Display of results creates some interesting issues. For example, should track updates be displayed as soon as they become available at the display station, or should they be displayed in the time sequence in which the events they describe occurred? How do I deal with a result (e.g., track update) that "appears" at the display minutes (or hours) after the event? Time sequencing of results across a large network assumes highly reliable and fast network communications.

33.5 SenSoft Signal Processing

The infrastructure architecture described above is intended to be very general and could be used to support a variety of network applications. The tactical goal (i.e., challenge problem) established for the SensIT community was to support surveillance types of operations with a system for target detection, localization, tracking, and classification. In SenSoft, processing to accomplish these signal processing tasks is done both locally, at individual nodes, and collaboratively within a network of nodes. The components that make up signal processing include

- *Local signal processing*: Local signal processing is performed on a single node, with multi-modal sensor signal data from only that node. This includes such processing as gain normalization (amplification), filtering (low/high-pass), downsampling, windowing/FFT, and often also includes threshold detection.
- *Target detection*: Target detection is accomplished through an analysis of sensor signals that determines some anomaly indicating presence of a target. Target detection will provide the location of the detection (i.e., the location of the node or sensor making the detection) and, depending on the sensing mode, may also locate the target.
- *Collaborative signal processing*: For most sensing modes, target localization and tracking are accomplished through higher level, collaborative signal processing. Nodes with knowledge of local events collaborate by sharing that information with other nodes and cooperating with those nodes to process the shared information to produce target locations and tracks.*
- *Target classification*: Classification is the identification of significant characteristics of a target from the signals received. These characteristics could include, for example, method of locomotion (tracked/wheeled) or relative weight (light/med/heavy) or could more specifically identify the target type (e.g., HMMWV) based on vehicle signature. Classification is typically accomplished locally, due to the large amount of data used. However, classification of a target could also be a cooperative effort (e.g., iterative processing through a sequence of nodes), and a classification component may also collaborate with other signal processing components. For example, classification could provide information to link a series of tracks as the movement of the same target (i.e., track continuation).

The signal processing components use the various infrastructure components to pass data and control among themselves on a single node and, for collaborative processing, between nodes. The components also interact with each other in semantically meaningful ways. For example, different algorithms for collaborative signal processing can require different types of target detection (e.g., CPA vs. signal threshold).

33.6 Component Interaction

The on-node infrastructure and signal processing components shown in Figure 4.2 and described above coordinate (approximately) as follows:

- The DSP moves time series data into on-node buffers where they can be retrieved by other components. This process is defined by the hardware, and described in the WINS 2.0 User's Guide [WINS2].
- A Local Signal Processing application reads and processes the time series sensor signal data from the DSP buffers. Note that each DSP buffer is circular, so an application must be sensitive to read timing to avoid losing data.
- Results of local signal processing are stored in program buffers or cache (Local Data Storage and Access) where they can be retrieved by or sent to other local or remote components. Local Signal Processing applications may include a Detection component, as noted earlier. Events detected are also stored in cache and shared with other nodes.
- As needed, Collaborative Applications obtain the results of digital signal processing, local signal processing, and/or detection processing from a group of nodes. The results are obtained as a consequence of queries posed by the application, or through tasks assigned to the application.

* A track is a location with a speed/direction vector attached.

These results can be processed locally, but the information and control can also be shared with collaborators on other nodes via the Communication component. Results of such collaborations can provide target localization, tracks and classification.

- Collaboration results are stored locally and also passed to other nodes and applications, via the communication mechanisms, for further collaborative processing. For example, a track computed on node A collaborating with nodes B, C and D could be (temporarily) stored at A and passed to other nodes as appropriate (e.g., in the direction of a target vehicle).

Command and control interactions between the sensor network and the user (as shown in Figure 4.3) can coordinate as follows:

- The task manager gathers user-level tasks and queries via GUI; the task manager may perform global optimization then instructs the query processor on the gateway node as to which tasks/queries are required of the network.
- The query processor optimizes where possible, and moves tasks into nodes for execution (Figure 4.2, described above).
- Task execution provides results via a local query processor; query processing may combine/aggregate the results at a higher level, and moves results to the gateway cache.
- Task manager reads results at the gateway and passes them on to the user.

33.7 An Example

The software infrastructure described in this chapter was designed and implemented in parallel with the research efforts of the SensIT participants; it has been implemented in a variety of ways to incorporate the (intermediate) results of many of those efforts, and at the same time provides a platform for research and experimentation by those participants. In this section we describe the implementation of one collection of infrastructure and signal processing software named *SenSoft v1*.

This system was demonstrated in the fall of 2002 on a network of more than 20 sensor nodes placed around a road as illustrated in Figure 33.4. In the figure, each sensor node is labeled with an integer name (1–27) and with a pair of radio assignments defining intra-network RF radio connectivity. The sensor network was connected from a gateway node (usually node 12) via a wireless Ethernet bridge to a command site (small network of PCs) located inside the building (X marks the approximate spot).

The two radio assignments given for each node are labeled as integer names for local area networks. Each local network is a star with a base radio (underlined label) and one or more remote radios; a base controls a time-division multiplexed cycle of messages with its remotes and each radio talks only within its network. For example, network 11 is based at node 11 (first radio) and consists of one radio at each of nodes 8, 11, 13, and 23. In Figure 33.4 there are 27 nodes and thus 54 radios collected into 17 local networks; each arrow points from a remote radio at a node to its corresponding base radio. As noted earlier, messages move through the network by hopping across the local networks. So, for example, a message might move from node 10 to node 16 by following network 12 to node 12, where it hops to network 15 and goes to node 15, where it then hops to network 16 and moves to node 16. Alternately, the same message might follow network 23 through node 23 to node 21, where it hops to network 17, etc.

The network communications component determines the routing that is needed for a message to move through the network. The specific network routing method used by most signal processing applications in SenSoft is ISI's subject-based routing using SCADDS data diffusion (Scalable Coordination Architectures for Deeply Distributed Systems [SCADDS]). This software provides a Publish/Subscribe API based on attributes describing data of interest; data is exchanged when a publisher sends data for which there are subscriptions and publications with matching attributes. Diffusion uses data attributes to periodically identify *gradients,* or best paths, from publishers to subscribers; gradient information is then used to determine how to route data messages. The efficiency of diffusion can be improved when geographic information is available—essentially, messages can be moved toward nodes that

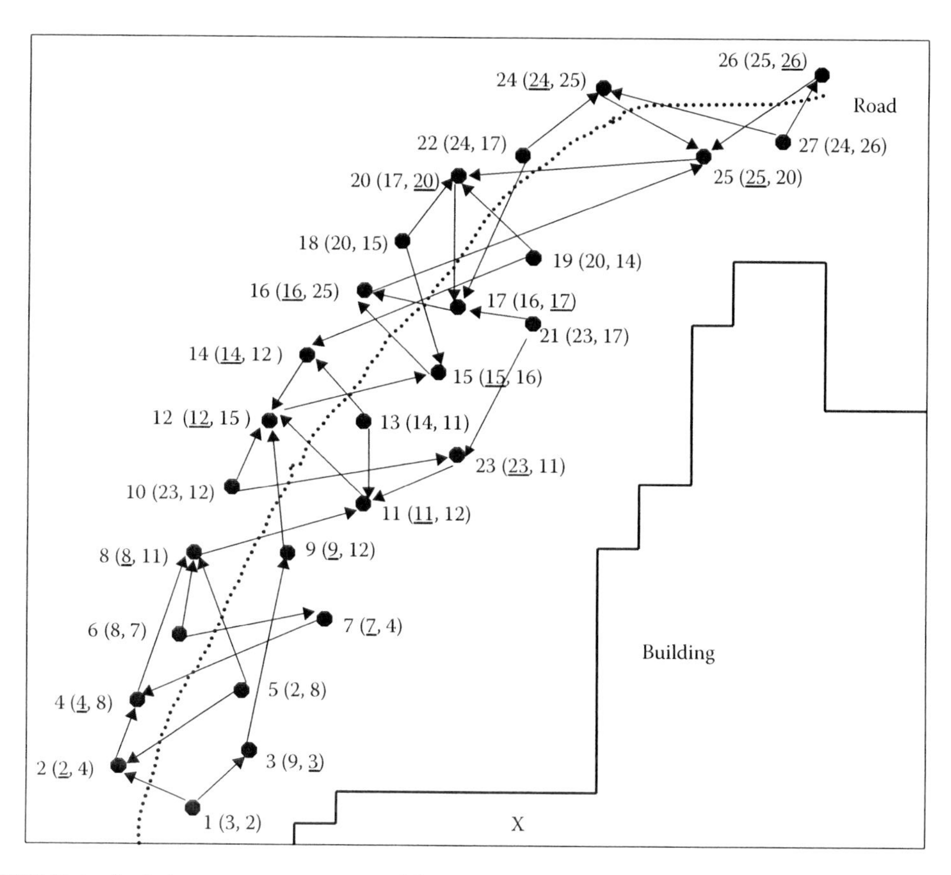

FIGURE 33.4 SenSoft experimentation network layout.

are geographically "in a line" with the target node. More detailed (and accurate) information about SCADDS can be obtained elsewhere in this book, and in ISI's SensIT documentation [SCADDS API].

Local data storage and access is provided in SenSoft v1 in two ways: BAE repositories and Fantastic Data Distributed Cache (FDDC). The BAE repositories are memory-mapped files primarily used for access to recently stored data. Repositories are defined and filled in processes that produce data; a subscription manager allows users in different processes to subscribe to the various repositories, maintains an event queue for each subscription, and notifies subscribers when new data (a repository event) is available. In SenSoft v1 repositories provide access to processed time series data (i.e., data is moved from buffers, processed, and shared via repository) and CPA events for classification and collaborative tracking.

In SenSoft v1, Fantastic cache provides longer-term storage of data, and storage of data that will be provided to a user through the gateway. FDDC is a distributed database spread over the node network—each node has a local cache; the collection of nodes forms a database system managed by the FDDC. In SenSoft v1 each local cache provides storage for and access to local node information (e.g., node name, location, sensor configurations, radio configurations), CPA event records (moved from BAE repository), and track update records.

FDDC is implemented as a server process that operates independently and asynchronously from the application programs. It presents an SQL-based interface to application programs, with limited versions of SQL statements such as CREATE TABLE, INSERT, UPDATE, DELETE, and SELECT, and additional statements such as WATCH, PUT, and UNDELETE that are particularly useful in the sensor network environment. It should also be noted that FDDC implicitly handles network routing of data (thus duplicating, to some extent, functionality provided by ISI). In particular, some queries posed against cache can result in the movement of data between nodes to respond to the query or to provide data redundancy for recoverability.

The demonstrated SenSoft v1 system did not include an *in-network query/tasking* application (although such an application was demonstrated in other implementations of SenSoft). As a result, signal processing application processes were always running within the sensor network, producing events and track updates which were all moved to a gateway for transmission to a user. In essence, the task "collect all information" is implicit in the execution of the SenSoft v1.

In SenSoft v1, the same implementations of local signal processing, target detection, tracking, and target classification are run at all nodes. Three sensor channels were used in most testing; one with a microphone, one with a seismometer, and the third with a passive infrared (PIR) detector.

BAE-Austin provides *local signal processing* of the acoustic, PIR and seismic signals. Time series data produced through the DSP is accessed from the buffer by BAE and processed according to the particular signal type. As noted earlier, sensor input on the WINS NG 2.0 nodes is restricted to be the same frequency for all channels, so BAE also downsamples as needed to achieve appropriate sampling rates for the sensor type. The processed signal data are stored in a BAE repository, along with sampling information. Time tags and gain settings for each data packet are stored in such a way that they are associated with their data block.

In SenSoft v1, BAE also provides target *detection* functionality. Local signal processing detects a target based on the strength of the received signature (e.g., its acoustic signal). Based on the time of occurrence of maximum signal strength (intensity), the detector also provides an estimate of the time of closest point of approach (CPA), when the target is closest to the sensor. The sequence of detection outputs is stored in a BAE Detection/Classification Repository; each output record contains detection levels, identification (codebook) values, and other information useful for tracking, data association, and collaborative signal processing.

As noted earlier, in SenSoft v1, event detections are also stored in Fantastic cache to provide database-style persistence and access to the detection information. To maintain backward-compatibility, cache storage of events replicates repository storage; a simple process runs on each node to read detection information from the BAE repository and store it in a cache table on the node.

In SenSoft v1, Penn State University Applied Research Lab (PSU/ARL) [PSU] provides *collaborative signal processing*, in which nodes collaborate to determine the track of a target. A PSU process subscribes to detection events from its local BAE repository and, via diffusion, can also subscribe to neighboring detections. In the presence of detection events a node will collaborate with its neighbors to determine whether there are enough events (within a pre-determined time period) to warrant track processing and then to elect a node to aggregate detection events among neighbor nodes. The elected node will calculate track update information, determine whether current update is a continuation of a previously recognized track (data fusion), and send track update notices to nodes geographically located in the general direction of target travel. Track update notices are published and sent via diffusion to the CSP applications running at neighbor nodes. Track update data is also moved into cache at a gateway node to be able to display it. In future implementations, we would want the CSP application to communicate directly with cache. However, for SenSoft v1, this integration could not be done and movement of track update data to a node gateway was accomplished by a special-purpose application at the gateway node that subscribed via diffusion to track updates and loaded all responses to gateway cache.

Target classification in SenSoft v1 is called as a function from the CSP process. The PSU SIF-based (Semantic Information Fusion) vehicle classification algorithm is called after data is received from a detection event. The routine requires (1) a pointer to a time series data block (in a BAE repository) and (2) start and stop times for the relevant data in the block. The routine then analyzes the data and returns, to the CSP process, a success/failure indicator and a features structure: success/failure tells whether the routine was successful at classifying the target; features include vehicle locomotion (wheeled or tracked) and vehicle weight (light or heavy), along with confidence values for both.

If classification is successful, SIF also supplies a codebook value for the target. For accurate display, this codebook must match the one used by GUI processes.

The University of Maryland provides the *task management* software for interaction between a node gateway and a user system. The Maryland software consists of two basic process types: a ForkServer and a Gateway. A ForkServer manages client (TCP/IP socket) connections with a gateway node, spawning a Gateway process for each connection (i.e., linking it to a client socket). In this way, a Gateway node can serve multiple simultaneous clients of varied types. Each Gateway process maps tasking commands sent by a client into executions in the network to produce the information indicated by the commands. These executions would typically be queries and tasks for the on-node query/tasking component, however in SenSoft v1 tasks were executed simply as queries to the gateway node's cache to extract the desired data.

The SenSoft v1 *GUI* consists of a tactical display linked to a user-tasking interface through which users specify the kinds of data to be displayed. The tactical display, built by Maryland using BBN's OpenMap software [OpenMap], shows an overhead picture (zoomable) of the geographic area of interest enhanced by graphical depictions of many aspects of the underlying sensor network (e.g., sensor node position and payload, detection events, and fused target track plots). For example, a target track displays as an icon whose shape depends on the target classification (if available) and a "leader" vector (logarithmically proportionate to the target speed and pointing in the direction of travel). The interface currently paints sensor nodes as small green circles with pop-up information that includes the node's geographic coordinates, node network name, and payload. Detection events (when requested) are displayed as purple squares placed at the reported detection position (which is typically the reporting node's location). Figure 33.5 shows a GUI view of the sensor network testbed used for the SenSoft v1 demonstration. This view displays track plots indicating two targets (in this example, people) moving towards each other along the road. The testbed web-cam view (ground truth) in the lower right corner of the display shows one of the targets. The detail pop-up boxes, highlighted along the left side, list the actual reported track values (classification, location, heading, velocity, and time) for each target.

The GUI is also used to specify queries and tasks to the network. Pop-up dialog boxes provided at the GUI allow the user to select the Gateway conduit and specify a task (e.g., Display Events) and constraints

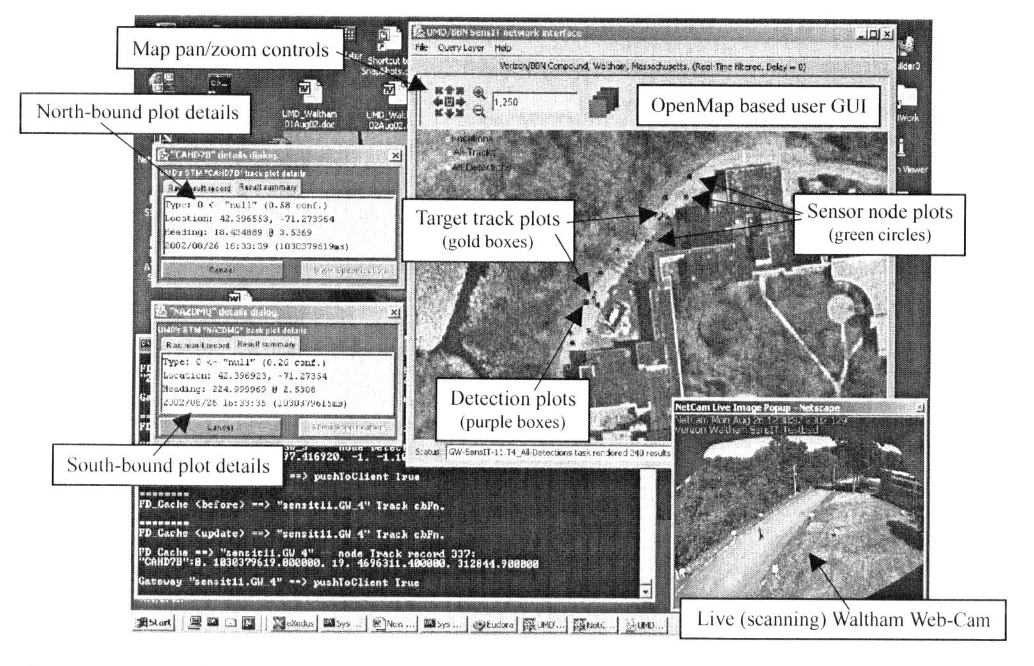

FIGURE 33.5 SenSoft GUI.

(e.g., only at node 17) by choosing from task entry menus. The conduit then translates user (and system) requests into Gateway commands, and moves (and deserializes) result data from the Gateway into result objects that can be displayed by the GUI.

33.8 Summary

The SensIT program encouraged a widely diverse research community to examine many issues pertaining to the information technology required to realize the vision of networked micro sensors in tactical and surveillance applications. The program established the feasibility of creating networks of collaborating sensors, defined an initial architecture and developed an initial software prototype, and laid the groundwork for continued research in areas such as network management, collaborative detection, classification, and tracking, data routing and communications, and dynamic tasking and query.

The SenSoft architecture and its instantiation as SenSoft v1 provide a useful experimentation platform in support of research and development in distributed sensor networks. For example, this system could be used to support development and experimentation in collaborative signal processing for multiple and arbitrarily moving targets, comparisons and validations of costs for various communication and routing approaches, development of languages and other approaches to tasking and executing tasks within a network, or simulation and test of new tactical concepts for sensor systems, to name a few areas.

A next important step for SenSoft would be the (re-)design and implementation of well-defined interfaces (both syntactic and semantic definitions) for the infrastructure components. Such an experimentation platform would be an even stronger foundation for research and development, and a great step toward realization of a fieldable, operational SensIT system of fully distributed, autonomous, networked ground sensors.

References

[OpenMap] http://www.openmap.org

[PSU] R. R. Brooks, C. Griffin, and D. S. Friedlander. Self-organized distributed sensor network entity tracking. *International Journal of High Performance Computing Applications*, 16(3), 207–219, 2002.

[SCADDS] C. Intanagonwiwat, R. Govindan, and D. Estrin. Directed diffusion: A scalable and robust communication paradigm for sensor networks, In *Proceedings of the Sixth Annual International Conference on Mobile Computing and Networks (MobiCOM 2000)*, Boston, MA, August 2000.

[SCADDS API] F. Silva, J. Heidemann, and R. Govindan. Network routing application programmer's interface (API) and walk through 9.0.1. December 9, 2002.

[WINS2] WINS NG 2.0 User's manual and API specification. Rev. A., Sensoria Corporation, San Diego, CA, May 30, 2002.

34 Statistical Approaches to Cleaning Sensor Data

Eiman Elnahrawy
Rutgers University

Badri Nath
Rutgers University

34.1 Introduction

Sensor networks have become an important source of information with numerous real life applications. Existing networks are used for monitoring several physical phenomena such as contamination level in soil and water, climate, building structure, habitat, and so on, potentially in remote harsh environments [17,24]. They also found several interesting applications in industrial engineering and inventory management such as monitoring the quality of perishable food items, as well as transportation and traffic control [1,23].

Detrimental actions are usually taken based upon the sensed information or sensor measurement, and therefore, the quality, reliability and timeliness are extremely important issues in such applications. Data collected from wireless sensor networks, however, is subject to several problems and sources of errors. The imprecision, loss and transience in wireless sensor networks, at least in their current form, as well as the current technology and the quality of the used, usually very cheap, wireless sensors contribute to the existence of these problems. This imply that these networks must operate with imperfect or incomplete information [8,24]. These problems however may seriously impact the actual usage of such networks. They may yield imprecise or even incorrect and misleading answers to any query on sensor data. Therefore, online cleaning of sensor data before any decision-making is crucial. In this chapter we will focus on probabilistic efficient and scalable approaches for this task. We shall discuss the following problems: reducing the effect of random errors, detection of outliers and malicious sensors, and recovery of missing values [5,6]. Before we proceed let us discuss some examples to show the significance of this problem.

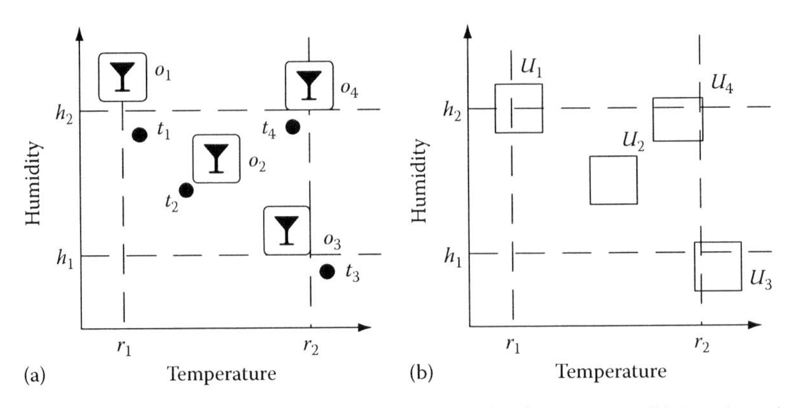

FIGURE 34.1 (a) Based on the observed readings items 1,4 will be thrown away. (b) Based on the uncertainty regions, only item 3 will be thrown away.

Example 34.1

Bacteria growth in perishable food items can be estimated by using either specialized sensors or by estimating it from temperature and humidity sensors attached to the items. Specialized sensors for are quite expensive. On the other hand, temperature and humidity sensors are much cheaper and more cost-effective. Therefore, the second alternative will usually be preferable. However, those cheap sensors are quite noisy since they are liable to several sources of errors and environmental effects.

Consider the scenario of Figure 34.1a, simplified for the sake of illustration. If the temperature and the humidity conditions of any item fall under or go over given thresholds, the item should be thrown away. Assume that the range of acceptable humidity and temperature are $[h_1, h_2]$, and $[r_1, r_2]$, respectively. t_i refers to the true temperature and humidity readings at item i, while o_i refers to the reported (observed) readings at item i. As shown in the figure and based on the reported noisy data, items 1, 4 should be thrown away while items 2,3 should remain. However, based on the true readings item 1 should remain while item 3 should be thrown away!

Example 34.2

Sensors becomes malicious and start reporting misleading unusual readings when they are about to run out of batteries. Serious events in the areas monitored by the sensors could also happen. In this case, the sensors also start reporting unusual readings. Detecting and reasoning about such outliers in sensor data in both cases is therefore an important task.

Consider the scenario shown in Figure 34.2. Frame (a) shows random observations that are not expected in dense sensor networks. On the other hand, frames (b) and (c) show two consecutive data samples obtained from a temperature-monitoring network. The reading of sensor i in frame (c) looks suspicious given the readings of its neighbors and its own last reading. Intuitively, it is very unusual that the reading of i will "jump" from 58 to 40, from one sample to another. This suspicion is further strengthened with knowledge of the readings in the neighborhood. In order for sensor i to decide whether this reading is an outlier, it has to know its most likely reading in this scenario. We shall show how we can achieve this later in this chapter.

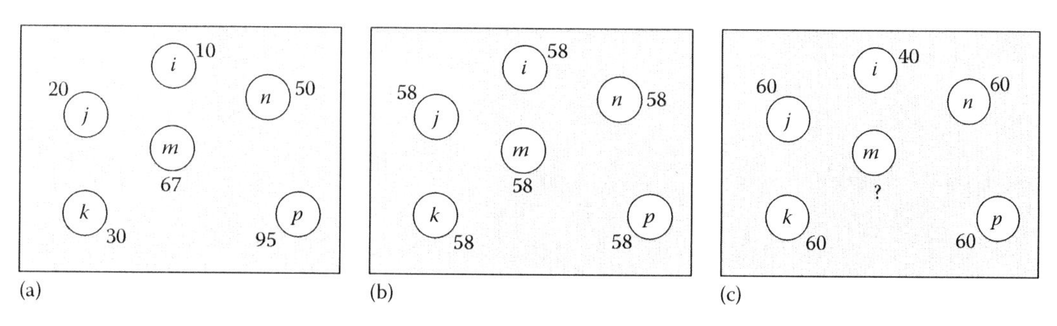

FIGURE 34.2 Frame (a) shows random observations that are not expected in dense sensor networks. Frames (b) and (c) show two consecutive data samples obtained from a dense network.

34.2 Bayesian Estimation and Noisy Sensors

Random errors and noise that affect sensors result uncertainty in determining the true reading or measurement of these sensors. Since these sensors are prone to errors they are uncertain about their true readings. Bayesian estimation can be utilized in order to reduce that effect, specifically by reducing the uncertainty associated with the noisy sensor data. Queries evaluated on the resultant clean and more accurate data are consequently far more accurate than those evaluated on the raw noisy data. Notice that the reading of each individual sensor is usually important, that is, fusion of readings from multiple sensors into one measurement to reduce the effect of noise is not usually an applicable solution. Therefore, we apply Bayesian estimation to every sensor. Even if multiple sensor fusion is possible, we can apply the approach discussed below to further enhance the accuracy of the result.

The overall technique for cleaning and querying such noisy sensors is shown in Figure 34.3. It consists of two major modules; a cleaning module and a query processing module. There are three inputs to the cleaning module: (1) the noisy observations reported from the sensors, (2) metadata about the noise characteristics of every sensor which we call the error model, and (3) information about the distribution of the true reading at each sensor which we call the prior knowledge. The output of the cleaning module is probabilistic uncertainty models of the reading of each sensor which we call the posterior, that is, a probability density function (pdf) of the true "unknown" sensor reading taking on different values.

The cleaning module is generally responsible for cleaning the noisy sensor data in an online fashion by computing accurate uncertainty models of the true "unknown" measurement. Specifically, it combines the prior knowledge of the true sensor reading, the error model of the sensor, and its observed

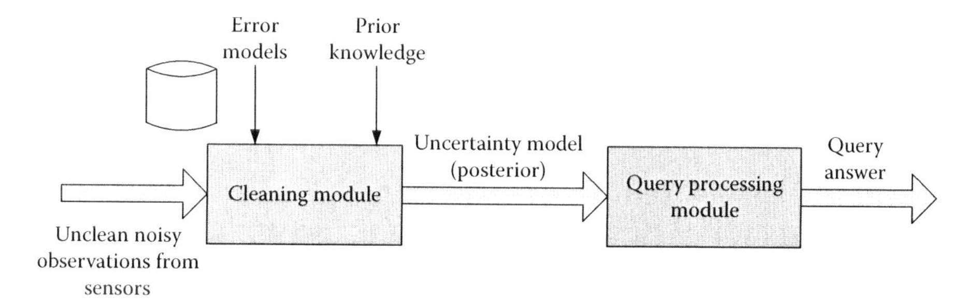

FIGURE 34.3 Overall framework.

noisy reading together, in one step and online using Bayes' theorem shown in Equation 34.1 (more information about Bayes' theorem can be found in [2,4,9]):

$$p(\theta \mid x) = \frac{\text{likelihood} \times \text{prior}}{\text{evidence}} = \frac{p(x \mid \theta)\, p(\theta)}{\int_{\Psi} p(x \mid \Psi)\, p(\Psi) d\Psi} \tag{34.1}$$

The likelihood is the probability that the data x would have arisen for a given value of the parameter θ and is denoted by $p(x|\theta)$. This leads to the posterior pdf of θ, $p(\theta|x)$.

The query processing module is responsible for evaluating any posed query to the system using the uncertainty models of the current readings. Since the uncertainty models are probabilistic (i.e., describes random variables), traditional query evaluation algorithms that assume a single value for each reading cannot be used. Hence, the query processing step performs algorithms that are based on statistical approaches for computing functions over random variables. A formal description of this overall technique is the topic of the nest three sections.

There are two places where we can perform cleaning and query processing, at the sensor level or at the database level (or the base-station). Each option has its advantages and limitations in terms of its communication and processing costs (which can be interpreted to energy consumption), and storage cost. It is usually difficult to come up with explicit and accurate cost models for each case since; there are many factors involved and some of them might be uncontrollable. The overall system capabilities, sensors' characteristics, application, etc., will help us decide which option to choose. Some experimentations can also guide our final decision. A discussion of these issues are beyond the scope of this chapter.

34.3 Error Models and Priors

There are numerous sources of random errors and noise in sensor data: (1) noise from external sources, (2) random hardware noise, (3) inaccuracies in the measurement technique (i.e., readings are not close enough to the actual value of the measured phenomenon), (4) various environmental effects and noise, and (5) imprecision in computing a derived value from the underlying measurements (i.e., sensors are not consistent in measuring the same phenomenon under the same conditions). The error model of each sensor is basically the distribution of the noise that affects it. We assume that it is Gaussian with zero mean. In order to fully define this Gaussian model we need to compute its variance. The variance is computed based on the specification of each sensor (i.e., accuracy, precision, etc.), and on testing calibrated sensors under normal deployment conditions. This testing can be performed either by the manufacturers or by the users after installation and before usage. Environmental factors or characteristics of the field should also be taken into consideration. The error models may change over the time and new modified models may replace the old ones. Notice that non-Gaussian models can also be used depending on the sensor's characteristics. The models in general are stored as a metadata at the cleaning module. Sensors are not homogeneous with respect to their noise characteristics, and therefore, each sensor type, or even each individual sensor may have its own error model.

Prior knowledge, on the other hand, represents a distribution of the true sensor reading taking on different values. There are several sources to obtain prior knowledge. It can be computed using facts about the sensed phenomenon, learning over time (i.e., history), using less noisy readings as priors for the more noisy ones, or even by expert knowledge or subjective conjectures. They can also be computed dynamically at each time instance if the sensed phenomena is known to follow a specific parametric model. For example, if the temperature of perishable items is known to drop by a factor of x% from time $t - 1$ to time t then the cleaned reading of the sensor at time $t - 1$ is used to obtain the prior distribution at time t. The resultant prior along with the error model and the observed noisy reading at time t are then input to the cleaning module in order to obtain the uncertainty model of the sensor at time t. This dynamic prior approach indeed resembles Kalman filters [14].

34.4 Reducing the Uncertainty

Let us assume that we have a set of n sensors in our network, $S = \{s_i\}$, $i = 1, \ldots, n$. These sensors are capable of providing their measurements at each time instance and reporting them to their base-station(s). Think of the reading of each sensor s_i at this instance as a tuple in the sensor database with attributes corresponding to its readings. Each sensor may have one or more attribute corresponding to each measurement. However, for simplicity of description, let us assume that each sensor measures a single phenomenon and that its measurement is real-valued. The following techniques can be fairly extended to accommodate multi-attribute and discrete-valued sensors.

Due to occurrence of random errors the observed value of the sensor o will be noisy, that is, it will be higher or lower than the true unknown value t. As we discussed in Section 34.3, the random error is Gaussian with zero mean and a known standard deviation $\sim N(0,\delta^2)$. Therefore, the true value t follows a Gaussian distribution centered around a mean $\mu = t$ and with variance δ^2, that is, $p(o|t) \sim N(t,\delta^2)$. We apply Bayes' theorem to reduce the uncertainty and obtain a more accurate model which we call the posterior pdf for t, $p(t|o)$. We combine the observed value o, error model $\sim N(0,\delta^2)$, and the prior knowledge of the true reading distribution $p(t)$ as follows:

$$p(t \mid o) = \frac{p(o \mid t)\, p(t)}{p(o)} \tag{34.2}$$

In fact, this procedure is generic. Although we explicitly assumed Gaussian errors, We do not have to restrict neither the error nor the prior distribution to a specific class of distributions (i.e., Gaussian). However, Gaussian distributions have certain attractive properties which makes them a good choice for modeling priors and errors. In particular, they yield another Gaussian posterior distribution with easily computed parameters as illustrated in the following example. This nice property enables performing the cleaning efficiently at the sensor level where we usually have restricted processing and storage. Moreover, Gaussian distributions are known to be analytically tractable, they are also useful for query processing and yield closed form solutions as we will show in the next section. Nevertheless, they have the maximum entropy among all distributions [4]. Therefore, approximating the actual distribution for the error and the prior by suitable Gaussian distributions is usually advantageous.

Example 34.3

In order to understand how Bayesian estimation works, let us assume that the reading of a specific sensor s is known to follow a Gaussian distribution with mean μ_s and standard deviation σ_s, that is, $t \sim N(\mu_s, \sigma_s^2)$ which is our prior. By applying Bayes' theorem and using some properties of the Gaussian distribution we can easily conclude that the posterior probability $p(t|o)$ also follows a Gaussian distribution $N(\mu_t, \sigma_t^2)$ [2,9]. Equations 34.3 and 34.4 show the parameters of this posterior:

$$\mu_t = \frac{\delta^2}{\sigma_s^2 + \delta^2}\mu_s + \frac{\sigma_s^2}{\sigma_s^2 + \delta^2}o \tag{34.3}$$

$$\sigma_t^2 = \frac{\sigma_s^2 . \delta^2}{(\sigma_s^2 + \delta^2)} \tag{34.4}$$

Why this Bayesian approach is superior? Suppose that we used a straightforward approach for modeling the uncertainty in sensor readings due to noise. That is, we will assume that the true unknown reading of each sensor follows a Gaussian pdf, centered around the observed noisy reading, with variance equals to the variance of the noise at this sensor, δ^2. Let us call this approach the no-prior approach. To prove the effectiveness of Bayesian estimation in reducing the uncertainty, let us consider the Bayesian mean squared error, $E[(t - \hat{t})^2]$ for the resultant posterior with parameters μ_t, σ_t shown in Equations 34.3 and 34.4, where t, $\hat{t}$ are the true unknown reading, and the posterior mean, respectively. We can then compare it with the no-prior approach.

The error or the uncertainty in the resultant posterior equals $\sigma_t^2 = \delta^2(\sigma_s^2/\sigma_s^2 + \delta^2)$ (refer to Ref. [12] for the proof). This amount is less than δ^2, the error (or uncertainty) in the no-prior approach. Therefore, Bayesian is always superior. Moreover, when the variance of the prior becomes very small as compared to the variance of the noise, or in other words, when the prior becomes very strong, the error of the posterior becomes smaller and the uncertainty is further reduced. Consequently, the Bayesian-based approach becomes far more accurate than the no-prior one. In general, if the prior knowledge is not strong enough, that is, if it has a very wide distribution compared to the noise distribution, then the Bayesian-based approach will still be superior, though not "very" advantageous in terms of estimation error. Fortunately, in many situations this is not the case. For example, consider situations where we have cheap and very noisy sensors scattered everywhere to collect measurements of a well-modeled phenomenon such as temperature, etc. A strong prior can be easily computed while the noise is expected to have a very wide variance.

Equation 34.3 also illustrates an interesting fact. It shows that the Bayesian-based approach in general compromises between the prior knowledge and the observed noisy data. When the sensor becomes less noisy, its observed reading becomes more important and the model depends more on it. At very high noise levels, the observed reading could be totally ignored.

34.5 Traditional Query Evaluation and Noisy Sensors

There are major differences between evaluation of queries over noisy sensor (uncertainty models) and exact data (single points). In uncertainty models, the reading of each noisy sensor at a specific time instance is considered a random variable (r.v.) described by the posterior pdf of that sensor and not necessarily by a single point with probability 1. Therefore, traditional query evaluation algorithms that assume single points cannot be used for noisy sensors. Another significant difference is illustrated in the following example.

Example 34.3

Consider that we have noisy temperature sensors in our network. We would like to know the maximum reading of those sensors that record a temperature ≥50°F at a specific time instance. However, we do not have a single estimate of the true reading of each sensor, but rather we have a pdf that represents the "possible" values of that reading. In order to determine whether or not a specific sensor satisfies this predicate (i.e., a temperature ≥50°F), we have to compute the probability that each sensor satisfies the predicate using its posterior pdf. When the probability is less than 1, which is highly expected, we will be "uncertain" whether the sensor satisfies the predicate or not. Even though there is a high chance that a specific sensor satisfies the predicate as its probability approach 1, for example, 0.8, neither the processing module nor any person can decide for sure. Therefore, there is no answer to this predicate and consequently we cannot decide which sensor is the maximum!

In order to overcome this difficulty without violating any statistical rules, we can modify our question by rephrasing it as "return the maximum value of those sensors that have at least a c% chance of recording at a temperature ≥50°F." We call c the "confidence level," and it is user-defined as part of the queries. Following this reasoning, we can now filter out all those sensors that have a probability less than $\frac{c}{100}$ of satisfying our query and return the maximum of the remaining sensors. This leave the problem of computing the maximum over a pdf which we will discuss shortly.

Definition 34.1

Confidence Level (c): The confidence level or the acceptance threshold c is a user-defined parameter that reflects the desired user's confidence. In particular, any sensor with probability $p < \frac{c}{100}$ of satisfying the given predicate should be excluded from the answer to the posed query.

34.6 Querying Noisy Sensors

Let us now discuss several algorithms for answering a wide range of traditional SQL-like database queries and aggregates over uncertain sensor readings. These queries do not form a complete set of all possible queries on sensors but they help illustrate the general approach to solve this problem. These algorithms are used in the processing module centrally at the database level, over the output of the cleaning module. They are generally based on statistical approaches for computing functions over one or more random variables. For simplicity of notation, we will use the term $p_{si}(t)$ to describe the uncertainty model $p(t|o)$, of sensor s_i.

34.6.1 Class I

The first class of queries returns the value of the attribute of the queried sensor (i.e., its reading). A typical query of this class is "What is the reading of sensor x?" There are two approaches for evaluating this class of queries. The first one is based on computing the expected value of the probability distribution and output it as follows:

$$E_{s_i}(t) = \int_{-\infty}^{\infty} t p_{s_i}(t)dt \tag{34.5}$$

where s_i is the queried sensor.

The second approach is based on computing the p% confidence interval of $p_{si}(t)$. The confidence factor p is user-defined with a default value equals to 95. The confidence interval is computed using Chebychev's inequality [3] as follows:

$$P(|t - \mu_{s_i}| < \epsilon) \geq 1 - \frac{\sigma_{s_i}^2}{\epsilon^2} \tag{34.6}$$

where μ_{si}, σ_{si} are the mean and the standard deviation of $p_{si}(t)$, $\epsilon > 0$. In order to compute ϵ we set $(1-(\sigma_{s_i}^2/\epsilon^2))$ to p and solve. The resultant p% confidence interval on the attribute will be $[\mu_{s_i} - \epsilon, \mu_{s_i} + \epsilon]$.

34.6.2 Class II

This class of queries returns the set of sensors that satisfy a predicate. A typical query of this class is "Which sensors have at least c% chance of satisfying a given range?" The range $R = [l, u]$ is specified by lower and upper bounds on the attribute value, l, u, respectively. The answer to this class is the set $S_R = \{s_i\}$ of those sensors with probability $\left(p_i > \frac{c}{100}\right)$ of being inside the specified range R, where $p_i = \int_l^u p_{s_i}(t)dt$ along with their "confidence," p_i. Although this is a simple range query the algorithm extends naturally to more complex conditions with mixes of AND and OR as well as to the multi-attribute case.

Example 34.4

Consider the scenario of Figure 34.1b where we have sensors of two attributes. Assume that the output of the cleaning module is that the reading of each sensor is uniformly distributed over the depicted squared uncertainty regions. The probabilities of the items being inside the given range are (*item* 1,0.6), (*item* 2,1), (*item* 3,0.05), (*item* 4,0.85). If the user-defined confidence level is $c = 50\%$, which is a reasonable confidence level, then only item 3 will be thrown away. This coincides with the correct answer over the true unknown readings, and is also more accurate than the answer on the noisy (uncleaned) readings.

34.6.3 Class III

The last class of queries that we consider is aggregate queries of the form "On those sensors which have at least c% chance of satisfying a given predicate, what is the value of a given aggregate?" Before evaluating the aggregate, we obtain the set S_R of those sensors that satisfy the given predicate using the algorithm of Class II. If the predicate is empty then all sensors in the network are considered in the aggregation, that is, $S_R = S$. In general, the aggregate can be a summary aggregate such as SUM, AVG, and COUNT aggregates or an exemplary aggregate such as MIN, MAX aggregates (this classification of aggregate queries into summary and exemplary has been extensively used among the database community).

To compute the SUM aggregate, we utilize a statistical approach for computing the sum of independent continuous random variables, also called convolution. To sum $|S_R|$ sensors, we perform the convolution on two sensors and then add one sensor to the resultant sum, which is also a r.v., repeatedly till the overall sum is obtained. Assume that the sum $Z = s_i + s_j$ of two uncertainty models of sensors s_i, s_j, is required. If the pdfs of these two sensors are $p_{si}(t)$, $p_{sj}(t)$, respectively, then the pdf of Z is computed using Equation 34.7 [3]. The expected value of the overall sum or a 95% confidence interval can then be computed and output as the answer similar to Class I:

$$p_Z(z) = \int_{-\infty}^{\infty} p_{s_i}(x) p_{s_j}(z - x) dx \tag{34.7}$$

Computing the COUNT query reduces to output $|S_R|$ over the given predicate. The answer of the AVG query however equals the answer of the SUM query divided by the answer of the COUNT query, over the given predicate. On the other hand, the MIN of m sensors in S_R is computed as follows. Notice that the MAX query is analogous. Nevertheless, other order statistics such as Top-K, Min-K, and median can be computed in a similar manner.

Let the sensors $s_1, s_2, \ldots, s_m$ be described by their pdfs $p_{s1}(t), \ldots, p_{sm}(t)$, respectively, and their cumulative distribution functions (cdfs) $P_{s1}(t), \ldots, P_{sm}(t)$, respectively. Let the random variable $Z = \min(s_1, s_2, \ldots, s_m)$ be the required minimum of these independent continuous r.vs. The cdf, pdf of Z, $P_Z(z)$, $p_Z(z)$ are computed using Equations 34.8 and 34.9, respectively [3]:

$$\begin{aligned} P_Z(z) &= \text{prob}(Z \le z) = 1 - \text{prob}(Z > z) \\ &= 1 - \text{prob}(s_1 > z, s_2 > z, \ldots, s_m > z) \\ &= 1 - (1 - P_{s_1}(z)) \cdots (1 - P_{s_m}(z)) \end{aligned} \tag{34.8}$$

$$p_Z(z) = -\frac{d}{dx}(1 - P_{s_1}(z))(1 - P_{s_2}(z)) \cdots (1 - P_{s_m}(z)) \tag{34.9}$$

34.6.4 Approximating the Integrals

The above algorithms involve several integrals that are not usually guaranteed to yield a closed form solution for all families of distributions. We recommended Gaussian priors and error models in Section 34.4. Here is another motivation for this recommendation. There are specific formulas for computing these integrals easily in the case of Gaussian distributions. For example, the marginal pdf of Gaussian is also a Gaussian, so is the sum of Gaussians (and consequently the AVG) [3]. Evaluation of Class I queries simply reduces to the mean parameter μ of the Gaussian uncertainty model in the single attribute case, and to the m -component mean vector in the multi-attributes case.

For other families of distributions where no known closed form solution exists, we can approximate the integrals by another suitable distribution. We then store these approximations in a repository at the query processing module. Therefore a large part of the computation is performed off-line and reused when needed, for example, by changing the parameters in pre-computed parametric formulas.

34.7 Spatio-Temporal Dependencies and Wireless Sensors

In the previous sections we showed how Bayesian estimation and statistics are used for reducing the effect of noise in noisy sensors and for querying them. In the rest of this chapter we discuss a statistical approach for detecting malicious sensors or serious anomalies in sensor data efficiently and online, as well as recovering missing readings. This approach is based on exploiting the spatio-temporal dependencies in sensor networks.

Sensor networks are usually dense for coverage and connectivity purposes, robustness against occlusion, and for tolerating network failures [7,15,18,20]. These dense networks are usually used for monitoring several well-defined real-life phenomena where redundant and correlated readings, specifically spatio-temporally, exist. In particular, there are spatial dependencies between spatially adjacent sensor nodes as well as temporal dependencies between history readings of the same node. Such dependencies, if defined appropriately, can enable the sensors to locally "predict" their current readings knowing both their own past readings and the current readings of their neighbors. This ability therefore provides a tool for detecting outliers and recovering missing readings.

Spatio-temporal dependencies can indeed be modeled and learned statistically using a suitable classifier, specifically by the use of a Bayesian classifier. The learning process reduces to learning the parameters of the classifier while the prediction process reduces to making inferences. The learning is performed online in a scalable and energy-efficient manner using in-network approaches that have been shown to be energy-efficient theoretically and experimentally [10,13,17]. The inference in Bayesian classifiers is also straightforward and can is afforded by the current generation of wireless sensors since it requires simple calculations. We shall show that in the next section. It is important however to notice that this solution is in general suitable to some classes of networks and applications where high spatio-temporal characteristics exist and can be learned, for example, networks used for monitoring temperature, humidity, etc., and for tracking.

34.8 Modeling the Dependencies

Spatio-temporal dependencies in wireless sensor networks are modeled using Makov assumption: the influence of all neighboring sensors and the entire history of a specific sensor on its current reading is assumed to be completely summarized by the readings of its immediate neighbors and its last reading. In other words, the features of classification (prediction) are (1) the current readings of the immediate neighbors (spatial), and (2) the last reading of the sensor (temporal) only.

Markov assumption is particularly used to make the problem feasible. Spatio-temporal data is a category of structured data. Precise statistical models of structured data that assume correlations between all observations are in general complicated and difficult to work with and therefore not used in practice.

This is specifically due to the fact that they are defined using too many parameters that cannot easily be learned or even explicitly quantified. For example, to predict the reading of a sensor at a specific time we need to know the readings of all its neighbors, the entire history of its reading, and the parameters or the probabilistic influence of all these readings on its current reading. Alternatively, Markov-based models that assume "short-range dependencies" have been used in literature to solve this difficulty [9,21]. The Markov assumption is very practical. It drastically simplifies the modeling and significantly reduces the number of parameters needed to define Markov-based models.

Figure 34.4 shows the structure of the Bayes-based model used for modeling the spatio-temporal dependencies. N is a feature that represents readings of neighboring nodes, H is a feature that represents the last reading of the sensor, and S is the current reading of the sensor. The different values of sensor readings represent the different classes. The parameters of the classifier is the dependencies, while the inference problem is to compute the most likely reading (class) of the sensor given the parameters and the observed correlated readings.

We model the spatial information using readings from "some" of the neighboring nodes, the exact number varies with the characteristics of the network and the application. Notice that the continuously changing topology and common node failures in wireless networks prohibit any assumption about a specific spatial neighborhood, that is, the immediate neighbors may change over the time. The model below is fairly generalizable to any number of neighbors as desired. However, a criteria for choosing the neighbors that yield the best prediction, if information about all of them are available at the sensor, is an interesting research problem. In our discussion let us assume a neighborhood that consists of two randomly-chosen indistinguishable neighbors.

Without loss of generality, assume that sensor readings represent a continuous variable that takes on values from the interval $[l, u]$. We divide this range, $(u - l)$, into a finite set of m non-overlapping subintervals, not necessarily of equal length, $R = \{r_1, r_2, \ldots, r_m\}$. Each subinterval is considered a class. These classes are *mutually exclusive* (i.e., non-overlapping) and *exhaustive* (i.e., covers the range of all possible readings). R can be quantized in different ways to achieve the best accuracy and for optimization purposes, for example, make frequent or important intervals shorter that infrequent ones. The classifier involves two features: the history H and the neighborhood N. The history represents the last reading of the sensor, while the neighborhood represents the reading of any two nearby sensors. H takes on values from the set R, while N takes on values from $\{(r_i, r_j) \in R \times R, i \leq j\}$, and the target function (class value) takes on values from R.

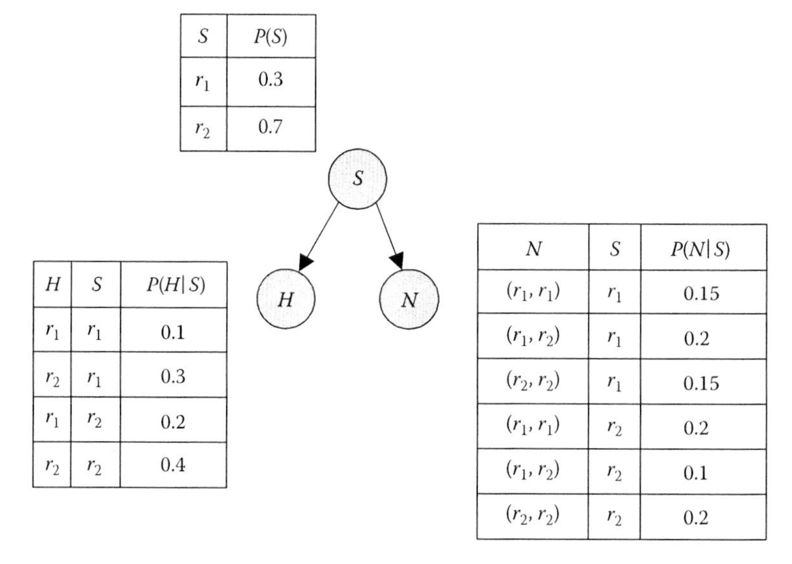

S	$P(S)$
r_1	0.3
r_2	0.7

H	S	$P(H\|S)$
r_1	r_1	0.1
r_2	r_1	0.3
r_1	r_2	0.2
r_2	r_2	0.4

N	S	$P(N\|S)$
(r_1, r_1)	r_1	0.15
(r_1, r_2)	r_1	0.2
(r_2, r_2)	r_1	0.15
(r_1, r_1)	r_2	0.2
(r_1, r_2)	r_2	0.1
(r_2, r_2)	r_2	0.2

FIGURE 34.4 The Bayes-based model for spatio-temporal dependencies.

To define the parameters of this model, let us first show how the inference is performed. Bayes classifier is generally a model for probabilistic inference; the target class, r_{NB}, output by the classifier is inferred probabilistically using maximum a posteriori (MAP) [9,19,22]:

$$r_{MAP} = \text{argmax}_{r_i \in R} P(r_i \mid h, n) \quad (34.10)$$

where h, n are the values of H, N, respectively. This can be rewritten using Bayes rule as follows:

$$r_{MAP} = \text{argmax}_{r_i \in R} \frac{P(h,n \mid r_i)P(r_i)}{P(h,n)} = \text{argmax}_{r_i \in R} P(h,n \mid r_i)P(r_i) \quad (34.11)$$

Since the denominator is constant for all the classes, it does not affect the maximization and can be omitted. From this formula we see that the terms $P(h, n|r_i)$, $P(r_i)$ should be computed for each h, n, r_i, that is, they constitute the parameters of the model. To cut down the number of training data needed for learning these parameters and consequently optimize the resources of the network we utilize the "Naive Bayes" assumption. It states that the feature values are conditionally independent given the target class. That is, we assume that the spatial and the temporal information are conditionally independent given the reading of the sensor. This assumption does not sacrifice the accuracy of the model. Although it is not true in general, "Naive" Bayes classifiers have been shown to be efficient in several domains where this assumption does not hold, even competing with other more sophisticated classifiers [9,19,22]. Based on this conditional-independence assumption we obtain the following:

$$r_{NB} = \text{argmax}_{r_i \in R} P(r_i)P(h \mid r_i)P(n \mid r_i) \quad (34.12)$$

The parameters now become (1) the two conditional probabilities tables (CPT) for $P(h|r_i)$, $P(n|r_i)$, and (2) the prior probability of each class $P(r_i)$. These parameters models the spatio-temporal dependencies at each sensor in the network and enables it to predict its reading.

34.9 Online Distributed Learning

The spatio-temporal dependencies or the parameters of the Bayesian classifier are learned from training sensor data in a distributed in-network fashion. The training data is the triples (h, n, r_t), where h represents the last reading of the sensor, n represents the current reading of two neighbors, and r_t represents the current reading of that sensor. This training information is available at each node when sampling the network since the shared channel enables "snooping" on neighbors broadcasting their readings. The snooping should be performed correctly in order to incorporate a very little cost and no communication complications. The neighbors can be, for example, the parent of the sensor and one of its children in case of a routing tree. To account for lack of synchronization, the node quantizes the time, caches the readings of the neighbors over each time slot, caches its own last reading, and use them for learning at the end of the slot. If at any slot the training instance is not complete, that is, some information is missing, this training instance is discarded and not used for learning.

If the sensed phenomenon is completely non-stationary in space, that is, if the dependencies are coupled with a specific location, the parameters are learned as follows. Each node estimates $P(r_i)$, $i = 1, \ldots, m$ simply by counting the frequency with which each class r_i appears in its training data, that is, its sensed value belongs to r_i. The node does not need to store any training data to perform this estimation; it just keeps a counter for each r_i, and an overall counter of the number of instances observed so far, all initialized to 0. It increments the appropriate counter whenever it observes a new instance. The CPT of H, N are estimated similarly. Notice that $P(H = h|r_i)$ is the number of times that ($H = h$ AND the sensor reading belongs to r_i), divided by the number of times the class is r_i. Since the node already keeps a counter for the latter, all it needs is a counter for each ($H = h$ AND the reading belongs to r_i), a total of m^2 counters. In order to obtain the CPT for $P(n|r_i)$, in the case of two indistinguishable neighbors, the node keeps $\frac{m^2(m+1)}{2}$ counters for each ($n = (r_i, r_j)$, $i = 1, \ldots, m, j = 1, \ldots, m, i \leq j$ AND the sensor reading belongs to r_i) since (r_i, r_j) is indistinguishable from (r_j, r_i). That is a total of $1+m+\frac{3}{2}m^2+\frac{m^3}{2}$ counters are needed. After a pre-defined time interval, a testing phase begins where the sensor starts testing the accuracy of the classifier. It computes its predicted reading using the learned parameters at each time slot and compares it with its sensed reading. It keeps two counters: the number of true predictions and the number of false predictions. At the end of the testing phase, the sensor judges the accuracy by computing the percentage of the correctly classified test data. If the accuracy is not acceptable according to a user-defined threshold, the learning resumes. The sensor repeats until the accuracy is reached or the procedure is terminated by the base-station.

On the other hand, if the phenomenon being sensed is not stationary over the time, the sensors re-learn the parameters dynamically at each change. They can be preprogrammed so that they re-learn at specific predefined time instances or they can detect the changes dynamically, for example, when the error rate of the previously learned parameters increases significantly. In both cases, the old learned correlations can be stored at the base-station and re-used if the changes are periodic. The sensors periodically send their parameters to the base-station to be recovered if they fail.

If the sensed phenomenon is stationary over the space, then the above learning procedure is modified to scale with the size of the network in an energy-efficient fashion using in-network aggregation. Specifically, the locally learned parameters at each node are combined together by "summing" the individually learned counters over all the nodes via in-network aggregation (SUM aggregate). So are the testing parameters (the two counters). The final overall parameters are then used by each sensor in the network. Notice that stationarity in space does not imply a static or a fixed topology. It rather implies that we can use training data from all the nodes to learn the parameters. It therefore enables collecting of a large number of training instances in a relatively short time. This approach also adapts to less perfect situations where the "stationarity in space" holds inside clusters of sensors over the geographic space. A separate model is learned and then used by the sensors in each of the regions. Finally, a hybrid approach for non-stationary phenomenon in time and space is formed in a fairly similar way. In all cases, Bayes-based models converge using a small training set [19]. The convergence also makes it insensitive to common problems such as outliers and noise given that they are usually random and infrequent, and duplicates since the final probabilities are the ratio of two counters.

A centralized approach for learning, where the parameters are learned centrally at the base-station, can be used in some situations. However, it is sometimes inferior to in-network learning with respect to its communication cost which is the major source of power consumption in sensor networks [7,17,20]. Moreover, in-network learning effectively reduces the number of forwarded packets which is a serious disadvantages of the centralized learning. In general, this decision is application-dependent and is driven by various factors such as the size of the training data.

To understand the tradeoffs between the distributed and the centralized approach, consider a completely non-stationary (in space) network, where learning is performed at each node, a centralized approach is inferior due to obvious communication cost. For stationary or imperfectly stationary networks the tradeoff is not that clear. We notice that in-network learning involves computing of a distributive summary aggregate while centralized learning can be viewed as computing of a centralized aggregate or as collecting of individual readings from each node [17]. Therefore, assuming a fairly regular routing tree, the communication cost of in-network learning is roughly $k \times O(m^3) \times O(n)$, where k is the number of epochs, m is the number of classes, and n is the number of nodes used for learning the parameters, which can be as large as the size of the network. This is equivalent to the cost of computing $O(m^3)$ summary aggregates k times. The cost of a centralized learning is roughly $p \times O(n^2)$, where p is the size of training data at each sensor which is "application-dependent." This is equivalent to the cost of computing p centralized aggregates (a detailed analysis of the cost of computing aggregates using in-network aggregation can be found in Ref. [17]). It has been shown that it yields an order of magnitude reduction in communication over centralized approaches. For a realistic situation where $p = 1000$, $k = 2$, $m = 5$, $n = 10$, the cost of a centralized learning is an order of magnitude higher. This difference further increases for perfectly stationary situations since n becomes very large. Even when m increases the difference remains significant. The above analysis fairly extends to non-stationarity in time.

34.10 Detecting Outliers and Recovery of Missing Values

Once the parameters of the Bayesian classifier are learned, it can be used for inference. In particular, the probability of sensor reading taking on different values, that is, being in different classes, r_i, $i = 1, \ldots, m$, is computed for every r_i from Equation 34.13, using the learned parameters, current readings of neighbors, n, and the last reading of this sensor, h. The class with highest probability is then output as the prediction:

$$P(r_i \mid h, n) \sim P(r_i)P(h \mid r_i)P(n \mid r_i) \tag{34.13}$$

Example 34.5

Consider the scenario shown in Figure 34.2. Assume that sensor readings can take values in the range [30, 60]. Assume that we divided this range into two classes, r_1 = [30, 45], r_2 = [45, 60]. Further assume that we have already learned the parameters, that is, the conditional probability tables (CPTs), shown in Figure 34.4. To infer the missing reading of sensor m in frame (c), we use the readings of sensors j, k in this frame, the history of m, $H = r_2$. We compute $P(r_1|h = r_2, n = (r_2, r_2)) \sim 0.3 \times 0.3 \times 0.15 = 0.0135$, while $P(r_2|h = r_2, n = (r_2, r_2)) \sim 0.7 \times 0.4 \times 0.2 = 0.056$. According to this, the second class is more likely. This indicates that the reading of sensor m is expected to be somewhere in the range [45, 60].

The ability of the sensor to predict its reading is a very powerful data-cleaning tool; it is used for detecting false outliers/serious anomalies, and approximating its reading when missing. To utilize the Bayesian model in outlier detection, the sensor "locally" computes the probability of its reading being in different classes using Equation 34.13. It then compares the probability of its most likely reading, that is, highest probability class, and the probability of its actual sensed reading. If the two differ significantly then the sensor may decide that its reading is indeed an outlier. For example, we follow the steps of the above example to compute the probability of sensor i taking on values in the ranges [30, 45], [45, 60]. We find that its reported reading, that is, 40, in Figure 34.2, is indeed an outlier since the probability of its reading being in [30, 45] ~ 0.0135 is small compared to [45, 60] ~ 0.056. Distinguishing anomalies from malicious sensors is somewhat tricky. One approach is to examine the neighborhood of the sensor at the base-station. In particular, if many correlated sensors in the same neighborhood reported alert messages then this is most likely a true serious anomaly.

The classifier is also used to recover missing values. The objective is to predict the missing reading of a specific sensor which is performed by inferring its class using Equation 34.12. The predicted class represents a set of readings (a subinterval) and not a single specific value. We can, for example, choose the median of this subinterval as the predicted single reading, and therefore the error margin in prediction becomes less than half the class width. Think of this approach as significantly reducing the uncertainty associated with the missing reading from $[l, u]$ to r_i, where $[l, u]$ is the interval of all possible readings, while r_i is a small subinterval of $[l, u]$. As the width of each class becomes smaller, the uncertainty further decreases. In general, there is a tradeoff between the complexity of the classifier and the uncertainty. Smaller subintervals translate to more classes, and consequently, to bigger CPTs that are hard to work with and to store locally at the sensor. Therefore, the width of "each" of the classes is chosen wisely; we assign small classes to important readings that require tight error margins.

Recovery of missing values is also generalized to in-network sampling where significant energy is saved. We "strategically" select a subset of the sensors to sense the environment at a specific time while predicting the missing readings within acceptable error margins, that is, we perform a sampling. The selection criteria is based on the geographical locations, remaining energy, etc. A complete re-tasking of the entire network can be performed; for example, when some sensors are about to run out of battery then their sampling rate is reduced, and so on. A basic algorithm is to control the nodes in a way such that they alternate sensing the environment by adjusting their sampling rate appropriately.

34.11 Future Research Directions

We discussed probabilistic approaches for online cleaning of sensor data. There are several challenges and open problems in this area that need further investigation. Wireless sensors are becoming very pervasive. New applications are emerging every day that rely on these sensors for decision-making. Therefore, quality and integrity of sensor data are very important problems. The future of wireless sensors lies in reasoning about and solving these problems "efficiently," in terms of the available resources, and "online." Existing research has always focused on providing low-level networking solution or customized solutions that work for specific applications [17,24]. In both cases, these problems persist though less severely. Hence, general purpose solutions are needed.

We discussed only simple traditional database queries in this chapter. In most of the algorithms the query-evaluation was centralized. A distributed version of these algorithms as well as addressing more complicated queries and optimization issues are an interesting research direction. It is important that the accuracy of the devised algorithms be suitable to the application at hand.

Generalizations to sampling and to heterogeneous sensors are challenging problems. Readings obtained from dense sensor network are sometimes highly redundant. In some cases, they may be complementary to each other. Therefore, queries can be evaluated on a sample of the sensors only. A large part of existing work on query processing in sensor networks has only focused on homogeneous clean data from all sensors [11,16,17]. However, sensors may not be homogeneous. They usually differ in their remaining energy, storage, processing, and noise effect. A repository is therefore needed at the database system to store metadata about the capabilities and the limitations of each sensor. The database system should be able to turn the sensors on/off or control their rate using proxies [16]. The underlying networking functionality should allow for such a scenario. Users may also define specific quality requirements on the answer to their queries as part of the query, for example, a confidence level, the number of false positives/negatives, etc. The challenge is how to minimize the number of redundant sensors used unnecessarily to answer a specific query while (1) meeting the given quality level (e.g., confidence) and (2) "best" utilizing the resources of the sensors. The sample size may need to be increased or specific more accurate sensors may have to be turned on in order to meet the given user's expectations. The sampling methods may have to be changed over the time (random, systematic, stratified, etc.). In general, this introduces another cost factor in decision making and actuation, query optimization and evaluation, and resource consumption. This problem is also related to the Bayesian classifiers. It is important to investigate the optimal number of neighbors needed and the effect of selecting the neighbors intelligently versus randomly on the accuracy of prediction. Several real deployment decisions in the approaches discussed in this chapter are application-dependent. Experimentations and characterization are needed for guiding such decisions.

Handling noise in multi-dimensional sensors is a fair extension of the single-dimension case. However, handling multi-dimensional outliers and missing values is far more complicated and is still an open problem. So is extending the quantized Bayesian classifier to the case of continuous classes. Finally, it is interesting to investigate non-Bayesian solutions to the discussed problems as well as cross-evaluation of these solutions with the Bayesian-based ones.

References

1. Bonnet, P., Gehrke, J., and Seshadri, P. Towards sensor database systems. In *Proceedings of the Second International Conference on Mobile Data Management* (January 2001).
2. Box, G. E. P. and Tiao, G. C. *Bayesian Inference in Statistical Analysis*. Addison-Wesley Publishing Company, Inc., New York, 1973.
3. Casella, G. and Berger, R. L. *Statistical Inference*. Duxbury Press, Belmont, CA, 1990.
4. Duda, R. O., Hart, P. E., and Stock, D. G. *Pattern Classification*, 2nd edn., John Wiley & Sons, Inc., New York, 2001.

5. Elnahrawy, E. and Nath, B. Cleaning and querying noisy sensors. In *Proceedings of the Second ACM International Workshop on Wireless Sensor Networks (WSNA'03)*, San Diego, CA (September 19, 2003).
6. Elnahrawy, E. and Nath, B. Context-aware sensors. Submitted for review.
7. Ganesan, D. and Estrin, D. Dimensions: Why do we need a new data handling architecture for sensor networks? In *Proceedings of First Workshop on Hot Topics in Networks (Hotnets-I)*, Princeton, NJ (October 2002).
8. Ganesan, D., Govindan, R., Shenker, S., and Estrin, D. Highly-resilient, energy-efficient multipath routing in wireless sensor networks. *Mobile Computing and Communications Review (MC2R)*, 1(2), 28–36 (2002).
9. Hand, D., Mannila, H., and Smyth, P. *Principles of Data Mining*. MIT Press, Cambridge, MA, 2001.
10. Heidemann, J., Silva, F., Intanagonwiwat, C., Govindan, R., Estrin, D., and Ganesan, D. Building efficient wireless sensor networks with low-level naming. In *Proceedings of the Eighteenth ACM Symposium on Operating Systems Principles*, Banff, Alberta, Canada (October 2001).
11. Hellerstein, J. M., Hong, W., Madden, S., and Stanek, K. Beyond average: Towards sophisticated sensing with queries. In *Proceedings of IPSN'03*, Palo Alto, CA (April 22–23, 2003).
12. Kay, S. *Fundamentals of Statistical Signal Processing, Volume I: Estimation Theory*. Prentice Hall, Upper Saddle River, NJ, 1993.
13. Krishanamachari, B., Estrin, D., and Wicker, S. The impact of data aggregation in wireless sensor networks. In *International Workshop of Distributed Event Based Systems (DEBS)*, Vienna, Austria (July 2002).
14. Lewis, F. L. *Optimal Estimation: With an Introduction to Stochastic Control Theory*. John Wiley & Sons, Inc., New York, 1986.
15. Liu, J., Cheung, P., Guibas, L., and Zhao, F. A dual-space approach to tracking and sensor management in wireless sensor networks. In *Proceedings of WSNA'02*, Atlanta, GA (September 28, 2002).
16. Madden, S., and Franklin, M. J. Fjording the stream: An architecture for queries over streaming sensor data. In *Proceedings of ICDE*, San Jose, CA (2002).
17. Madden, S., Franklin, M. J., and Hellerstein, J. M. TAG: A Tiny aggregation Service for ad-hoc sensor networks. In *Proceedings of Fifth Annual Symposium on Operating Systems Design and Implementation (OSDI)*, Boston, MA (December 2002).
18. Mainwaring, A., Polastre, J., Szewczyk, R., Culler, D., and Anderson, J. Wireless sensor networks for habitat monitoring. In *ACM International Workshop on Wireless Sensor Networks and Applications (WSNA'02)*, New York (2002).
19. Mitchell, T. *Machine Learning*. McGraw Hill, New York, 1997.
20. Pottie, G. and Kaiser, W. Embedding the internet: Wireless sensor networks. *Communications of the ACM*, 43(5), 51–58 (2000).
21. Shekhar, S. and Vatsavai, R. R. Spatial data mining research by the spatial database research group, University of Minnesota. *Specialist Meeting on Spatial Data Analysis Software Tools, CSISS, and NSF workshop on Spatio-Temporal Data Models for Biogeophysical Fields*, 2002.
22. Witten, I. H. and Frank, E. *Data Mining: Practical Machine Learning Tools and Techniques with JAVA Implementations*. Morgan Kaufmann, San Francisco, CA, 2000.
23. Wolfson, O., Sistla, P., Chamberlain, S., and Yesha, Y. The geometry of uncertainty in moving objects databases. In *Proceedings of International conference on EDBT* (2002).
24. Zhao, J., Govindan, R., and Estrin, D. Computing aggregates for monitoring wireless sensor networks. In *Proceedings of IEEE SNPA'03* (2003).

35

Plant Monitoring with Special Reference to Endangered Species

K.W. Bridges
University of Hawaii at Manoa

Edo Biagioni
University of Hawaii at Manoa

35.1 Introduction

The monitoring of populations of endangered plants is a model system that provides a focused challenge to our development of integrated sensor and remote network technologies, operations and interpretation. The concrete problems faced in the design, construction, and maintenance of such a system not only helps solve an urgent problem, but provides a general test bed that applies to many situations based on distributed sensor networks.

Many plant species are at risk of becoming extinct. These endangered populations are found through out the world and occur in a wide range of habitats. While some of these rare species are being monitored, most receive only cursory attention. Put simply, we know little about the biology of many of these species, particularly how they respond to environmental conditions.

The general objective of plant monitoring is to acquire a significant time series of data about individual plants, populations of the species or plant communities comprised of many species. In addition, a similar time sequence of environmental information is almost always gathered. Together, these data allow correlations between the plant life history events and the weather.

The plant life history events are called the "phenology" of the plant (Leith, 1974). There is generally a set of phenological stages through which a plant grows, including the seed, seedling, juvenile, subadult, and adult stages. Within these stages, other phenological events are recognized, such as periods of growth, flowering, leaf flushing and leaf fall. Different species, different habitats and different environmental conditions sometimes require adjustments to these general phenological stages and events. It is not just scientists who monitor plant phenology. The advance of the fall colors as deciduous trees prepare to drop their leaves is widely anticipated and closely monitored annual event by the entire population living in areas where this occurs. One of the remarkable properties of plant phenology, in general, is the close correlation to the local weather.

The weather information that is usually collected in phenological studies includes the air temperature, rainfall amount, solar radiation intensity and relative humidity. Wind speed and direction are sometimes included in the set of measurements.

The emphasis in this chapter is on making observations of rare and endangered plant species and their surrounding environment. While this represents the general requirements of plant monitoring, it adds some additional constraints that will be discussed later. The value of choosing this special group of plants is that such monitoring may be essential to saving and recovering these species. This is an urgent need and one that is, unfortunately, very poorly served with our current technology. This emphasizes that this problem is not only interesting from an engineering perspective, but it has great social value. Any progress in solving the monitoring problems will help in a large number of general situations and may also be critical to our properly maintaining part of our biological heritage.

In the United States, rare and endangered plant species are those that have come under federal protection with the Endangered Species Act (ESA) of 1973 (U.S. FWS, 2001). Scientists assess the population sizes, distributions and trends of the plants in a region. If any species has few individuals, is limited to a few sites and shows a trend of population decrease, then it is proposed as a candidate for "listing" (placing on the Endangered Species List). The candidates are carefully reviewed before they become officially listed species. Once on the list, the species is offered some federal protection. The ESA statue includes two key provisions: the species must be saved from extinction and there must be a plan for recovery it so that it is no longer in danger of extinction. This second provision, that of recovering the species, is aimed at the eventual removal of species from the list.

As of June 2003, there were 715 total U.S. flowering plant species on the ESA list with 144 in the threatened category and 517 endangered (U.S. FWS, 2003). These species occur throughout the United States, although approximately one third of the listed flowering plant species occur in Hawai‘i.

Understanding the habitat of endangered plant species is an obvious key to the maintenance of the existing populations. There are two parts to the habitat surveillance. Observations need to be made on the ESA-listed plants, and the characteristics of the environment in their immediate neighborhood need to be monitored. In addition, if we are to recover the species we must also know the environmental conditions in the surrounding region. Knowing the larger pattern of environmental conditions should give us some insight into why the current distribution of the species is limited. It may be, for example, that the rainfall is significantly different in the surrounding area and this limits the reproductive success of individuals in the drier areas or provides significant benefits to a competitive species.

The following section describes a system that meets the general requires of monitoring rare and endangered plants and their environments.

35.2 The Monitoring System

Any monitoring system that involves federally protected rare and endangered plants must not put the population in any further danger. While this clearly means that no destructive sampling can be done, it also prohibits changes to the local environment that might also harm the plants. This constraint sets some broad limitations on instrumentation.

Plant monitoring equipment must not have a physical effect on the plants. This includes shading the plants, modifying the soil conditions, intercepting rainfall or altering the wind pattern. In part, these are equipment size and proximity constraints. In addition, it is important that the monitoring equipment should not call attention to the plants. This implies that equipment should be as small as possible and, if possible, be able to be hidden.

To the extent possible, the plants and the environment should be monitored remotely. Visits to areas with ESA-listed plant species can negatively impact the environment (such as by soil compaction or transporting alien seeds into the area). As a result, the monitoring equipment should be designed to be highly reliable, able to survive field use for extended periods (at least several years) and require, at most, infrequent servicing (such as battery changes).

Traditional field weather stations are large, often with rainfall and temperature sensors standing about 2 m tall and wind sensors on a mast. Recordings of weather information are either periodically transferred in the field or the unit may be equipped with data transmission capabilities. Most of the weather stations that are used for long-term measurements are sufficiently close to habitation that they can be connected to telephone lines for data transfer. Some stations use cellular phone links. While these weather stations provide a key backbone of reliable, high-quality data, they are not well suited to the needs of rare plants. It is not just a size constraint. Endangered plant populations are generally not found conveniently located near communication facilities.

Installing equipment that will monitor both the endangered species and their environments obviously requires some physical proximity to the plants. At the same time, the equipment needs to be non-invasive. Two general and broadly complementary approaches have been used to meet these requirements. One strategy is to make all of the equipment as small as possible. The other is to hide the equipment. Both of these approaches have implications about the type of data that are collected. For example, standardized rainfall sensors (see below) have a 6 in. (15.4 cm) diameter collecting funnel. This is hard to disguise. While a smaller-diameter funnel would be possible, it may be better to consider a completely different design that doesn't attempt to directly measure rainfall amounts near the target plant population. Instead, it may be more appropriate to measure an aspect of rainfall that can be correlated with a standardized measure. This allows the larger equipment to be located at a great enough distance away that its presence does not draw attention to the endangered plants. An example of a surrogate measurement would be rainfall duration. This can be done with a sensor that is both small and quite unobtrusive, such as two parallel conductors that will be shorted together when wet.

The point that we would like to emphasize is that the environmental monitoring system does not have to be identical to traditional designs. A system that is built as a network of sensors provides many new opportunities for fundamentally different approaches.

Visual reconnaissance of the plants allows the collection of important data. Similar care must be taken in the design of the sensors to make sure that there is enough resolution to capture significant life-history events. For example, close-up images might be required to see the initiation of flowering. At the same time, these sensors should not be so close that variability within the plant is missed, or other important events are not seen.

Our experience with images that document a plant's life history events has emphasized the value of periodic high-resolution still images over video recordings. This isn't just a matter of data collection frequency. Still cameras generally have image sensors with a larger dynamic range and which possess better optical properties than video systems. This means that you are more likely to be able to see the needed details. Video is important for monitoring animals, but most plant phenological events can be satisfactorily captured by a time series of still images.

Images also have considerable value when trying to interpret the measurements of the other environmental conditions. Seeing the structure of the clouds in a picture helps improve the understanding of the solar radiation measurements, for example.

Having a near real time system is very important. There are some situations that will probably require on-site follow up to understand the full implications of a particular event. Remotely monitoring the field conditions, particularly during periods with critical weather, should provide enough information to decide when to make a trip to the study site. An example is heavy rain during a critical event such as seeding development. An on-site investigation, if it is timed right, will likely reveal the actual impact of the rainfall in ways that would be impractical to fully instrument. This example emphasizes that an important goal of the monitoring is to make sure that all field visits are timed for maximum effectiveness while minimizing routine activities around the plants.

The nature of most plant life histories calls for a monitoring system that will operate for several years. This means that renewable energy sources, such as solar panels, will probably be used. Alternatively, the system must operate on extremely limited power. This adds to the challenges in designing a system that will meet the constraints of use around endangered plant species.

35.3 Typical Studies Involving Plant Monitoring

There are many applications of wireless systems of sensors in plant monitoring. The description above has focused on the application to rare and endangered species. The generality of this system can be seen in other monitoring situations.

Crop monitoring, and the use of these data in models, is becoming a sophisticated agricultural management tool. There are many facets to such monitoring. These involve the use of different measurement scales (from satellite-based remote sensing to in-field sensor systems) and a range of sensors (from traditional weather instrumentation to multi-spectral systems). Even relatively simple systems, such as using NOAA air temperature data to calculate the day-degrees (the number of days with temperatures above some threshold temperature) has allowed predictions of when to harvest crops that have been used for many years. Crop models have become much more sophisticated, however, and can now be used to make a variety of predictions so that farmers can be more astute managers. Crop performance over large areas can be estimated (EARS, 2003). Most of these agriculturally-related systems monitor changes on a daily or weekly basis within a growing season.

At the other end of the temporal scale are those studies that monitor the occurrence of phenological events to help understand changes such as global warming. Plants (and animals) are often sensitive indicators of subtle environmental changes. The small temperature changes of the past century have already been seen in changes in more than 80% of the 143 studies of plants and animals analyzed by Root et al. (2003).

There are many situations where plants need to be monitored so that animal interaction events can be recorded. The types of events include pollination and herbivory. These may happen very infrequently and over a brief period. This contrasts with monitoring that involves measuring slow but relatively steady changes, such as the growth of an individual. If the animal can be detected, such as with sound or passive IR (PIR), the sensors can begin more intensive monitoring and image capture.

In summary, it is obvious that there are many types of systems that need to be monitored. The requirements differ based on the goal of the monitoring.

35.4 Sensor Networks

The emphasis of many systems of environmental measurement has focused on the temporal changes in the major climate factors, such as temperature and rainfall. While temporal patterns are obviously very important, it is likely that the spatial patterning of the environmental is equally important. The cost of placing many traditional sensors on a site, maintaining these sensors, and interpreting the data has been prohibitive except in a few well-funded studies. New designs of networked microsensors reporting on a near real time basis offer a promising alternative. The implementation of such a system involves a number of considerations that require careful planning.

The layout of a sensor network that investigates environments, especially those surrounding rare plants species, should be of a size and arrangement that will detect gradients, if they are present. For example, areas with strong topographic relief are very likely to have rainfall gradients, and if the elevation change is great enough, substantial temperature gradients as well. Discovering the gradient pattern and its magnitude is important since such microclimatic differences between the habitat in which a plant is growing and were it is absent may explain this distribution pattern. Therefore, the overall layout should be designed with careful attention to the hypothesized environmental patterns, as well as the general characteristics of the species being studied.

In some cases, the layout of the sensors may be needed to observe phenomena whose location is not known or is not easily predicted. An example, relative to rare plants, is the need to monitor herbivores that may be eating the plants. In many such cases, it is not clear ahead of time which species is a likely consumer or where they can be observed. It may take several modifications of the sensor layout before basic information is known. At that point, it may be possible to adopt a different sensor layout that examines the herbivory process in detail.

It has been mention before, but is worth repeating that the general location of the sensors, and the supporting ancillary equipment, should avoid changing the local environment especially in the vicinity of endangered species. The goal is to have a sensor system that improves access to what are otherwise remote (and perhaps fragile) areas. The overall system should have good long-term unattended operational capabilities. This should include appropriate redundancy in sensors, power, and networking components.

The connection of the sensor network to the Internet, or otherwise retrieving data to an attended base location, allows the near real time monitoring of the field site. Designing sensors and data analysis systems that function to alert researchers promotes the concept of limiting field visits to those times when critical events are occurring that will benefit from human observation. A variety of extreme events qualify as triggers for on-site follow-up visits, including intense rainfall, flooding, prolonged drought or intense winds. The system should also alert researchers when there has been a catastrophic failure of the system so that it can be repaired with minimal delay.

The sensor system does not need to consist of identical units. A system that has a variety of sensors, such as those that collect both rainfall amounts as well as wetness events (the periods with either precipitation or fog and clouds), is likely to improve the resolution of environmental information. A few rainfall-collecting sensors, which are large and hard to disguise, can be used in areas where their presence doesn't interfere with the plants. These can be enhanced, and to a certain extent correlated, with smaller moisture detectors that are located both near the collecting sensors as well as the plants. The combination of the two types of sensors is likely to give much more information about the amount and pattern of the moisture over the area being studied than if a single type of sensor is used. The important point is that some "non-traditional" sensors, especially when they are combined with traditional sensors in an appropriately designed network, are likely to provide a richer set of environmental information than has been available to researchers studying rare plant distributions.

35.5 Data Characteristics and Sensor Requirements

35.5.1 Weather Data

Air temperature, relative humidity, barometric (air) pressure, rainfall amount, wind speed and wind direction are standard measurements taken by weather monitoring stations. The air pressure measurements are generally not used in plant studies. In addition, solar radiation is a very useful measurement that should be included if possible.

Digital sensors are readily available for all basic weather parameters (e.g., Onset Corp., Dallas Semiconductor). Humidity measurements generally use a thin film sensor while temperature is measured with a thermocouple sensor. Some care is needed to make sure that these sensors are in proper enclosures, that is, shaded from direct sunlight and precipitation and with ample air circulation.

Rainfall amounts are accumulated using a tipping bucket sensor. These event-sensors generally record each 0.01 in. of rainfall. Rainfall is collected in a funnel, generally 6 in. (15.4 cm) in diameter. Hourly reporting is a standard measurement interval. Reporting is generally adjusted to report starting on the hour. If daily reporting is done, the accumulation is reset at midnight.

Wind speeds are measured in a variety of ways, all of which provide an instantaneous measurement value. These may be reported as an average, sometimes with a gust (1 min peak) value. Propeller devices have a minimal threshold below which they can't measure the wind speed, often around 1 m/s.

Wind direction, also determined as an instantaneous value, is generally reported as a compass direction. See Webmet (2003) for information on computing the vector mean wind speed and direction.

Solar radiation data are much less commonly reported. The radiation characteristics measured by the sensors vary considerably. Simple light measurements provide a very coarse value and may be adequate for general survey considerations. Critical measurements may require a photosynthetic light (PAR) sensor that closely matches the energy spectrum acquired by typical flowering plants.

35.5.2 Soil Data

Soil conditions, such as soil moisture, are often critical to the growth and survival of plants. Digital sensors for soil moisture are now available (Onset Corp). These are relatively temperature and salinity insensitive. They read the volume water content in the range from 0% to 40.5% with an accuracy of approximately ±3%. Soil temperature sensors are similar or identical to air temperature sensors.

Both soil temperature and soil moisture many vary substantially over short distances and depend on the soil composition, slope, type of vegetation and other factors. This suggests that sensors should be placed a several soil depths as well as in different locations.

35.5.3 Images

Periodic pictures of the site being monitored are very useful if they have sufficient resolution and dynamic range. Still images, such as those produced with 2 megapixel (or greater) digital cameras, meet these standards better than video images. In general, images should be timed to correspond to the collection of weather data (e.g., hourly). Color images, while not essential, may provide essential information such as differentiating between clear and cloudy sky conditions or helping to see the presence of flowers on a plant. Image collection has not been a standard part of the data collection protocol for plant monitoring. Our experience has shown that it can be particularly valuable if high-quality images are collected at consistent monitoring intervals over long periods of time.

35.5.4 Event Detection

There is a broad range of events that are of interest for plant monitoring. Many of these have a low probability of occurrence, but can have a dramatic (perhaps catastrophic) impact. Examples include fires, lightning, and floods. While lightning sensors are readily available, monitoring the occurrence of the other events requires the adaptation of other sensors. Additional important events that are not as closely associated with specialized sensors, such as grazing activity or pollination, may require analyses of images to determine their occurrence. Intrusion detection is a likely candidate to trigger image analysis, however the sensor requirements must be established relative to specific targets. Large grazing mammals present a qualitatively different problem than a pollinating insect.

35.6 Spatial and Temporal Scales: Different Monitoring Requirements

The precision of any particular sensor requires detailed analysis before it is selected for incorporation in any plant monitoring system. The basic issue is whether it is better to support fewer higher precision sensors or more that are lower precision. Researchers have traditionally used high-resolution sensors. The costs may be so great that monitoring is limited to a single set of sensors (e.g., one weather station). The benefit of such a system is that the accuracy allows its measurements to be compared with other similar systems in other areas. If there is a local gradient in the environment, however, a limited number of high precision sensors may not provide enough spatial coverage to measure the trend. As a result, the environmental factors limiting the plant distribution may not be detected.

The sensor accuracies generally used with the common environmental measurements are

- Temperature: ±1°F–2°F
- Humidity: ±3%
- Wind speed: ±1% to 3%
- Rainfall Amount: ±5%

There are a number of ways to measure solar radiation. Examples of differences in sensor costs can be seen comparing light measurement with a diode (at approximately $2 per sensor) and PAR sensors (at approximately $175 per sensor).

Medium precision systems, especially when they are widely deployed, appear to be well matched to the needs of monitoring heterogeneous environments. Design considerations should examine the use of low precision but very numerous sensors. It is likely that a network using such sensors holds some potential for efficiently uncovering some types of spatial patterns and temporal trends.

35.7 Network Characteristics

Sensors can be networked for a variety of purposes, the most common being for sending data from the field to a base station. Other goals might include coordination among sensors for event detection or computation within the network.

Some commercial weather stations transmit data from the weather station to a base-station receiver using 900 MHz spread spectrum modems. With appropriate antennas, line of sight distances of over 30 miles have been reported (Weathershop, 2003). Telephone lines are used if they are available. Other alternatives include radio data links or cellular telephone connections.

For weather measurements, little data is sent, so unless a very large number or high frequency of measurements must be transmitted, all these technologies are suitable. Other applications, including periodic high-resolution images, require higher data rates, though even a 700-kB jpeg image (typical from a 2-megapixel camera) once an hour only requires about 1600 bits/s on average. In comparison, a weather station transmitting 120 bytes once a minute only requires 2 bits/s, and 100 such weather stations still only require about 200 bits/s.

Traditional telephony and most cellular telephones are bandlimited to (before compression) about 64 and 9.6 kb/s, respectively. Newer cellular technologies allow data rates in the Megabit/second range. Satellite technology is capable of carrying large data rates, but the cost per bit may be high, as is currently the case for cellular technology.

Radio data links vary from lows of around 9.6 kb/s (many serial radios) to highs of 11 Mb/s (802.11b) and 56 Mb/s (the less common 802.11a). Radios provide low cost per bit, since the costs are related only to purchasing the hardware and providing electrical power.

All radio technologies have a distance that is variable depending on the antenna and the power level used. Power levels may be limited by the hardware, often to obey regulations. Antennas may also be a given for a given hardware (or limited by regulation), or may be selectable. In general, an antenna provides a gain by focusing data in a given direction. Omnidirectional antennas distribute the data 360° within a plane perpendicular to the axis of the antenna. The signal is strong within a number of degrees of the plane, for example 30° or 20°. The more the signal is focused near the plane, the greater the gain of the antenna, and the less power can be received away from the plane. Directional antennas instead focus the signal in a cone, with most of the signal strength within a certain angle from the axis of the cone. Again, the smaller the angle the greater the gain. Most directional antennas have higher gain than most omnidirectional antennas, but omnidirectional antennas used for communication on a plane (e.g., on the earth's surface) have no need to be aimed.

Antenna placement also affects range. When an antenna is near a conducting surface such as the surface of the earth, the power falls off very rapidly with distance, typically as r^4. This is due to the electromagnetic wave being reflected by the conductor, which typically leads to near cancellation of the wave. The signal for antennas that are far from any conductor, on the other hand, tends to only fall off as $1/r^2$. The actual attenuation of the signal with distance depends on a number of factors, including the directionality of the antenna and the overall geometry of the configuration of sender, receiver, and reflective surface.

With typical antennas and power levels, most radio modems will work over a distance of at most a few hundred meters, with 802.11 being similar but perhaps somewhat higher. Bluetooth, a relatively low power radio technology, is designed for a communication range of 10 m, though part of the standard (Class C) allows for communication up to 100 m. Cellular communication can in theory extend several miles from a cell phone tower, but cellular can support more bandwidth and transmissions if the cells are small, so current cellular systems tend to keep cells as small as possible. The range of satellite systems is limited to the area visible from the satellite itself, but this may be very large, since some satellites have a footprint covering most of the continental United States. In addition, many modern systems provide a number of satellites that can cover the entire planet.

With carefully aimed directional antennas and line-of-sight conditions, the same radio technologies can reach across many kilometers, though weather, including fog, clouds, and precipitations, can interfere with such transmissions.

The sensitivity to weather varies with the frequency of the radio signal. The 2.4 GHz microwave band, used in microwave ovens as well as 802.11, 802.11b, and Bluetooth, is among the most affected, though most radio frequencies in common use will be affected by weather and vegetation (which contains water). Other interference can also affect radio transmission. Technologies which use spread spectrum distribute the signal across different channels and are thus able to avoid more sources of interference than conventional single-channel technologies that don't use spread spectrum.

35.8 Deployment Issues

The deployment of sensors in a wireless sensor network used to monitor plants is affected by many factors, including accessibility (e.g., placing sensors or radios high in trees or in remote mountain locations), radio connectivity and coverage. Other factors may include the ability to conceal the sensor or to take pictures of specific plants. Some deployments may need to be made in dense vegetation. Visits to proposed deployment sites early in the planning process are essential.

The arrangement of the nodes is important. Researchers often want to control the precise location of the nodes. For example, it may appear to be conceptually useful to have a regular spacing among nodes, such as arranging them in a regular grid. The actual details of the site, however, will impose many constraints on locating nodes, especially if they need to be concealed. It is better to have a general plan and to make sure that there is some flexibility in its implementation. This calls for considerably more understanding of the communication properties of the nodes in the actual deployment conditions.

In a wireless ad-hoc sensor network, each unit relays the data collected by other units as well as generating its own data. In such a network, and if the units are not guaranteed to be 100% reliable, it may be desirable to place additional units to maintain radio connectivity should a few of the nodes fail.

The number of nodes used in the monitoring system is also a critical issue. In extreme cases, most of the power used by the network as a whole is used to transmit data. In such cases, and if the radio range falls off as $1/r^4$ as is usually the case, the power consumption is minimized by having the largest possible number of nodes with the shortest possible radio range. Doing this may also optimize the overall bandwidth (bits/second) of the network as a whole, though the benefits of this depend critically on the overall communication pattern. For example, if all the data are sent to a base station, then there is no benefit, as the bandwidth of the base station forms the bottleneck for the entire network.

Since minimizing the power may not minimize the cost, careful forethought is needed both at the radio selection stage and when planning the deployment. Cost is also usually divided into design cost, which is amortized when more of the units are built, and a per-unit (materials and assembly) cost, which is greater when more units are built.

One of the biggest challenges is to hide the instruments. Small items are relatively easy to conceal, but large components, such as solar panels, wind and rain sensors and cameras present some difficulties.

Long-term deployments require that all the instruments be adequately protected from the environment. Small openings seem to invite water, for example. Unprotected connectors may quickly corrode. The opposite type of protection is also important, as it is critical to keep the instrumentation from affecting the local environment. For example, corrosion or battery leaks could introduce toxins in the environment that could be detrimental to the organisms being studied.

35.9 Data Utilization

Making effective use of the data, once they have been collected, can be challenging. Typical situations range from those that are highly determined with a fixed set of questions to answer to those that are part of open-ended investigation.

If the goals of the data collection are known in advance, it is often possible to perform much of the necessary computation on the network nodes, decreasing the amount of data that must be transmitted or the distance over which the data are transmitted. In general, such goals may consist of collecting appropriate statistics and detecting specific events. Statistics (e.g., minimum, maximum, and average temperature) can often be computed in a distributed fashion so that data transmission is minimized. Event detection covers a broader range of computations, and may or may not be suitable for distributed implementation. Ultimately, nodes may only transmit once events are detected, potentially greatly reducing the power consumption of the network.

A completely different approach to data collection is to leave the data stored on the nodes (forming a distributed database) and allow queries to be performed on this distributed database. Some approaches combine queries with event detection, such as diffusion (Heidemann et al., 2001), or queries alone can be used. In such cases, the data are delivered on-demand, potentially substantially minimizing the amount of data transmission.

Once the data have been collected, they must be put to use. Typically the amounts of data are such that looking directly at the numbers has limited usefulness. Instead, most users prefer to use algorithms to visualize the data, for example by graphing temporal changes in the data or creating maps to display many data points at once. As for event detection, visualization is easiest when the goals of the data collection are known exactly. For example, a farmer wishing to know whether a sprinkler system has delivered the expected amount of water may study a map of current soil moisture, and perhaps trend different areas over different time periods. Scientists studying endangered plant species and trying to figure out why the species are threatened, on the other hand, may need to visualize the data in many different ways to identify cause-and-effect relationships that may be affecting the plants and their ecosystem. Even evaluating the health of an ecosystem may be somewhat challenging, if there is an absence of data defining what is normal and healthy.

References

EARS (Environmental Analysis and Remote Sensing). 2003. http://www.earlywarning.nl/earlywarning/index.htm

Heidemann, J., F. Silva, C. Intanagonwiwat, R. Govindan, D. Estrin, and D. Ganesan. 2001. Building efficient wireless sensor networks with low-level naming. In: *Proceedings of the Symposium on Operating Systems Principles*. Chateau Lake Louise, Banff, Alberta, Canada, pp. 146–159. http://www.isi.edu/~johnh/PAPERS/Heidemann01c.html

Leith, H. (Ed.), 1974. *Phenology and Seasonality Modeling*. Springer-Verlag, New York, 444pp.

Root, T.L., J.T. Price, K.R. Hall, S.H. Schneider, C. Rosenzweig, and J.A. Pounds. 2003. Fingerprints of global warming on animals and plants. *Nature* 421: 57–60.

U.S. Fish and Wildlife Service. 2001. http://endangered.fws.gov/esa.html

U.S. Fish and Wildlife Service. 2003. http://ecos.fws.gov/tess_public/html/boxscore.html

Weathershop. 2003. http://www.weathershop.com/WWN_rangetest.htm

Webmet. 2003. http://www.webmet.com/metmonitoring/table_of_contents.html

36

Designing Distributed Sensor Applications for Wireless Mesh Networks

Robert Poor
Ember Corporation

Cliff Bowman
Ember Corporation

More than 2000 years before Eckert and Mauchly conceived of the logic and electronics that would become ENIAC, Plato penned the words, *Necessity is the mother of invention.* This adage has particular relevance in the rapid growth of wireless mesh networks, where commercial, industrial, and military applications have spurred innovation and fostered technological advances. A broadening array of practical solutions to industry challenges now rely on distributed sensors linked wirelessly with networks based on mesh topologies.

Real-world applications based on wireless mesh networks and distributed sensors take a wide variety of forms. Imagine high-rise buildings in earthquake-prone southern California with embedded strain gauges contained in the structural members, delivering data wirelessly to monitor the integrity of the structure during seismic events. Freight shipments in ships or trucks are monitored for temperature, shock, or vibration using wireless sensors in the cargo area that store data en route and deliver it upon docking. Petroleum pumping stations in frigid regions maintain oil flow at a precise degree of viscosity using embedded sensors in the pipelines linked to a feedback mechanism that controls individual heaters. Environmental monitors in orchards and vineyards guide irrigation schedules and fertilization, and provide alerts if frost danger becomes evident. Water treatment facilities use wireless sensors to monitor turbidity levels at the final critical stages of treatment and issue warnings if the monitored values exceed limits.

The progression toward wireless sensor networks is inevitable. The decreasing costs and increasing sophistication of silicon-based integrated circuits, has lead to low-power processors, specialized chipsets, and inexpensive wireless components, encouraging broader acceptance of the technology. The evolution of the Internet offers an example of a clear progression, moving from one connection for many people (the mainframe mode) to a single connection for each person (the microcomputer/laptop model). The next stage in this progression, providing many connections per person, encompasses the sensor network model. Within this model, collections of sensors deliver data to a person through multiple connections.

As sensor networks become more ubiquitous, the sheer volume of deployed sensors makes it essential that these networks are designed to be self-maintaining and self-healing. Already the number of sensor devices exceeds the population of the planet and more than 7.5 billion new devices are manufactured every year. Developers who want to capitalize on the benefits of this technology must realize that sensor networks have distinct differences from conventional wired and wireless networks. Workable designs favor simplified deployment, low power consumption, and reliable, unattended operation.

Recent advances in wireless mesh network technologies open significant new opportunities for developers. The characteristic properties of mesh networks fit many types of embedded applications, where resources—such as power, memory, and processing capabilities—are constricted. With easy deployment and self-healing capabilities, mesh networks satisfy the primary requirements of well-designed sensor networks. Wireless mesh systems can be built using inexpensive, commonly available 8 bit processors.

This chapter describes the principles underlying application development for wireless mesh networks and provides several examples of real-world applications that benefit from this technology. Getting optimal performance from a wireless mesh network typically requires a fresh design approach. A straight translation of an existing wired network application to a wireless mesh implementation often yields disappointing results. Following a comparison of the popular network topologies, this chapter presents guidelines and design principles that lead to successful deployments of distributed sensor networks using wireless mesh systems.

36.1 Characteristics of Mesh Networking Technology

Mesh networking technology owes much of its increasing popularity to the inherent reliability of redundant message paths. This fail-safe approach to communication and control adapts well to implementations in manufacturing, public service utilities, industrial control,[1] and military applications.[2] Mesh networking offers a number of distinct benefits, including

- *Highly scalable network infrastructure*: Each node in a mesh serves as a relay point, resulting in a network infrastructure that grows along with the network. Because of this design framework, mesh networks support incremental installation paths. Initial investments in the technology are also minimized, since a very basic network can be quickly deployed and then extended as required.
- *Simplified deployment in distributed environments*: Deploying a mesh network is typically easier than deploying networks using other topologies, particularly when propagation varies widely over a geographic area or over time. Once deployed, mesh networks can automatically take advantage of "good" variations in propagation.[3]
- *Energy efficiency advantages*: Developers working on applications intended for embedded implementations can capitalize on certain characteristics of mesh networking. The success of many battery-powered embedded applications relies on achieving maximum energy efficiency, extending battery life as much as possible. Overall power drain attributable to r^n path loss in wireless mesh architectures tends to be lower because, on average, the value of r is less.[4] This lets developers significantly reduce transmitter power to reduce the corresponding power drain.
- *Minimal processing requirements*: Cost-effective embedded applications must often rely on low-power processors with limited memory. To overcome this challenge, software engineers and developers have constructed loop-free routing algorithms explicitly for mesh networks. These memory- and processor-efficient routing algorithms[5,6] make it possible for developers to implement large-scale networks using modest processors with very low power requirements.[7]

36.1.1 Design Considerations from a Developer's Perspective

Adapting existing applications to successful wireless mesh network implementations often requires re-evaluating fundamental design considerations. The data capacity and the capabilities of available wireless devices are typically more restrictive than the capabilities of an equivalent wired network. Developers evaluating wireless mesh projects should recognize that their existing messaging models don't translate seamlessly to available mesh devices. Throughput and the data capacity of the wireless mesh network become prime considerations, which can require rethinking the architecture of the network.

Upon further investigation, many developers discover a basic truth—the design of many embedded protocols is tightly linked to a wired medium. When developers simply translate an existing legacy application, the performance of the wireless mesh network can be disappointing. Often, the performance results do not reflect limitations of wireless mesh networking, but rather a misuse of the technology.

Although developers must sometimes integrate mesh applications with legacy systems, dropping irrelevant design practices that apply to outdated wired systems can frequently help improve efficiency and network performance. In real-world situations, designers often encounter challenges that fall somewhere between maintaining absolute interoperability with legacy systems and creating a standalone wireless mesh network with distributed sensors as a fresh design. Drawing on the guidelines presented in this chapter, design trade-offs can usually be managed in a reasonable way. Reliability, scalability, adaptability, and efficiency, the hallmarks of wireless mesh networks, represent achievable goals through intelligent engineering. Real-world examples of practical mesh networking implementations appear in Section 36.4.

36.2 Comparison of Popular Network Topologies

Two distinctive properties that help characterize communications networks are

- *Topology*: Topology refers to the pattern by which a network's nodes are organized. Popular network topologies are bus, star, and mesh. The network topology determines the kinds of connections that are possible between nodes. Essentially, the topology creates a framework that controls how individual network devices communicate.
- *Medium access*: Medium access defines the rules by which an individual node can transmit on the shared communication medium (bus, star, or mesh). These rules can dramatically affect network behavior and performance. A trend that is evident in recent designs is access responsibility distributed among the nodes.

Figure 36.1 illustrates the basic topological structures that differentiate bus, star, and mesh networks. Assume that a message must pass from the node A to the node F through each of these topologies. In all cases, the organization of the network topology determines the paths by which the message can travel. The mechanisms by which each node gains access to the shared communication path depend on the protocol applied to the selected topology and the medium access techniques in use.

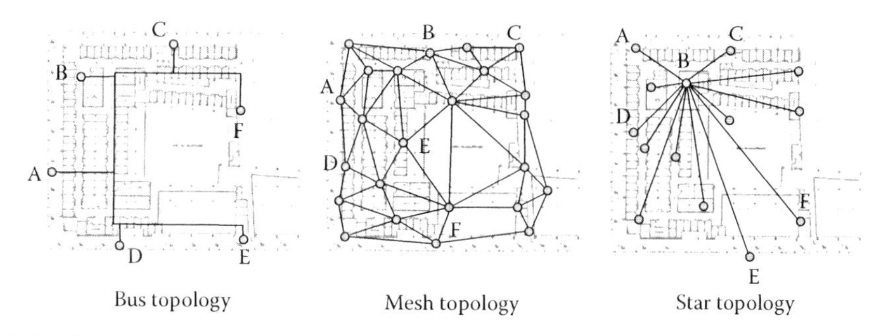

FIGURE 36.1 Bus, star, and mesh network topologies.

36.2.1 Transferring a Message within a Bus Topology

In the bus topology shown in Figure 36.1, every node can communicate with every other node—the message travels directly from node A to node F. Wired networks that operate in this manner include Ethernet LANs, Profibus, Modbus, and a number of proprietary systems that use the multi-drop RS-485 interface. Wireless networks, in some cases, also operate in a manner similar to a bus. An example of this is when a conference room of 802.11 devices is set to ad hoc mode.

Routing on a shared bus, however, is more complex than it appears. If two nodes attempt to transmit on the bus at the same time, their messages can collide, resulting in garbled information. Employing some form of medium access can minimize the chances of collisions. Some systems, such as Modbus, limit themselves to query/response messaging. For example, in a system employing Modbus, a *master* node owns the bus and a *slave* may transmit only when the *master* sends it a query. Other systems use scheduling schemes, such as the technique used with 802.11 ad hoc mode: each node can transmit only during a specified window of time assigned to it. A third strategy, implemented in Ethernet and known as carrier sense multiple access (CSMA), relies on carrier-sense hardware contained in each node. By sensing the state of a signal that indicates the bus is in use, each node can detect whether another node is already transmitting before it attempts to gain access to the bus.

Because messages travel directly from source to destination within a bus network, relay failure is not an issue. The vulnerability of bus systems, however, depends on the effectiveness of their medium access strategies, as well as in the integrity of the bus itself. In a Modbus network, where medium access is controlled exclusively by the master, communications are disrupted if the master node fails. Networks that rely on scheduling, where each node gets a specified window of time in which to transmit, also have a single point of vulnerability: the nodes typically depend on a synchronizing beacon to find their window. If the beacon station is lost, network recovery can take a significant amount of time. The technique used by Ethernet, based on detection of a signal that indicates the bus is in use, presents less vulnerability. Because access responsibility is distributed among the Ethernet nodes, the failure of a single node does not affect any of the other members of the network.

Bus systems by their nature all share a common vulnerability: the potential for losing access to an entire section of the network through bus failure. This can occur if a wired segment of the bus is cut or a wireless segment of the bus is jammed.

36.2.2 Transferring a Message within a Star Topology

Star networks employ a different method of organization. Within a star topology, each transmitted message travels a fixed path: node A can transmit only to the master node B, which then relays the message to node F. If separate cables are used to link each of the satellite nodes to node B, medium access does not present a problem. In the case of a shared medium, such as wireless, the common technique is to let node B determine which node can transmit. One example of this approach is the medium access scheme used within a Bluetooth piconet.

One inherent vulnerability of the star topology affects its reliability: if the master node fails, all communications on the network become disrupted. In shared-medium systems, such as Bluetooth, the member nodes can select another master and communications can be re-established after a delay. Recovery cannot be initiated, however, in some star topology configurations, such as when the single hub of a wireless LAN fails. In addition, if the path between the master and a node is blocked, that node can no longer participate in the network.

36.2.3 Transferring a Message within a Mesh Topology

Within a mesh network, messages can travel over multiple paths. A message transferred from node A to node F can be routed from A to B to F or from A to E to F. Many alternate paths can be used as well and this redundancy is a characteristic that increases the reliability of mesh networks. In a well-connected mesh network, the failure of a single node (node B, for example) only affects communications for that node. Messages previously directed through the failed node can be automatically rerouted.

Link failure, as occurs with the severing of a network cable or the blocking of an RF path, has much less effect on a mesh network than on other network topologies. The redundant routes available within a mesh network let traffic navigate around the broken link. This ensures that link failure cannot exclude a node from the network.

The nodes in a wireless mesh network typically use a shared RF channel, requiring some method to arbitrate medium access. The method commonly employed is CSMA. Since the hardware that supports CSMA is often built into each radio, implementing medium access can be fairly simple. As mentioned previously, the distributed strategy used by CSMA protects the network against the failure of a single node.

The medium access strategy that applies to mesh networks is similar to the strategy that applies to Ethernet LANs with one important difference: wired Ethernet LANs are usually bus networks,* so only one node can transmit at a time. In wireless mesh networks, nodes relay messages for each other, allowing the use of low-power transmitters. By reducing power to the point that transmissions reach only nearby nodes, the channel remains available for nodes that are beyond the range of the transmission. This phenomenon, known as *spatial multiplexing*, exists when multiple messages can travel simultaneously in different parts of the network. For example, as shown in Figure 36.1, traffic can pass between node A and node D at the same time that node C and node F exchange individual messages. The use of spatial multiplexing increases the effective data capacity of the network.

36.3 Basic Guidelines for Designing Practical Mesh Networks

Developers and engineers contemplating designs based on mesh networks can optimize their implementations by following these guidelines:

- *Distribute control tasks*: Mesh networks operate more effectively if tasks and messaging operations are distributed, rather than centralized. Centralizing tasks creates a network traffic pattern that focuses on the node controlling the process—messages either originate or terminate at an indicated node. Distributing control to several different points in a mesh network, particularly if these points are geographically separated, causes traffic around each point to flow independently. Multiple messages can be handled simultaneously (using the principle of spatial multiplexing), which effectively multiplies the capacity of the network. Distributing tasks has another benefit: messages don't need to travel as far across the network. By implementing multiple control points, the average distances from message source to destination can be shortened. This also enhances the overall reliability of the system. If a system relies on a single control point, the entire system gets shut down if that point fails. In a distributed system, however, even if individual components malfunction, the overall system can often continue to operate. Distributing tasks in this manner contributes to an increase in the long-term reliability of the system.
- *Use exception-based messaging to push the data*: To minimize network traffic and increase efficiency, rely on exception-based messaging to obtain data from nodes. Other techniques for exchanging messages, such as polling, generate a significant amount of superfluous network traffic. Exception-based messaging reduces network traffic in two ways: it eliminates the query initiating an exchange and it reduces the number of exchanges to those that indicate a noteworthy change in condition.

* LAN subnets are often wired as "Star Bus" architectures. Physically, they are star networks, but the hubs aren't really nodes—they merely repeat messages onto every arm of the star. Logically, the network functions as a bus.

TABLE 36.1 Typical Characteristics of a Wireless Mesh Network

Radio/MAC	IEEE 802.15.4 (CSMA)
Frequency band	2.4 GHz
Power output	+10 dBm
Routing	Multi-hop GRAd
Relaying strategy	Store and forward
Channel rate	250 kbps
Sustained network capacity[8]	40 kbps

- *Avoid query-response messages; let the network work*: Messaging techniques that depend on query/response methods or token passing reduce the efficiency of a mesh network. Traditional non-CSMA messaging models that perform well on earlier-generation network architectures may need to be adapted to mesh networks. Embedded protocols targeted for bus architectures typically rely on message-intensive models tied to query and response patterns or token passing to arbitrate access to the bus. The distributed nature of mesh networks favors the CSMA approach for efficient communication and the medium access strategies that apply to other technologies only add unnecessary overhead to a mesh network design.
- *Use local control and global monitoring*: Let the sensors and actuators communicate directly. The highest efficiency can be achieved in a mesh network by the distribution of tasks to lower-level devices in the network. For example, the control logic for operating an actuator can be embedded within the sensor and used to perform tasks as defined by the application. For binary or limited-state actuators, this can involve simply incorporating a table that specifies the threshold values. Decision-making that takes place through programmable logic controllers (PLCs) can be implemented through the individual sensors distributed throughout the network. Localized logic operations minimize reliance on a centralized processor. By reducing unnecessary processor communication, message transfers across the network can be minimized, improving overall efficiency.

36.3.1 Parameters for a Typical Mesh Network

Table 36.1 lists the typical characteristics of a commercially available mesh-networking suite based on the emerging IEEE 802.15.4 standard. Some of the other mesh networking technologies being developed for commercial use provide much higher network throughput; the data in the table, however, offers typical values for cost-sensitive embedded applications, such as condition monitoring and building automation.

As indicated in Table 36.1, the "Sustained Network Capacity" represents a fraction of the "Channel Rate." Several factors influence this situation. A half-duplex "store-and-forward" strategy reduces the rate by a minimum of 67%, CSMA introduces delays for "back-off" timing, and GRAd routing relies on a distance-based delay to select efficient pathways. The precise ratio between network capacity and channel rate in mesh networks will always depend on the nature of the implementation. However, factors similar to those mentioned are probably universal, suggesting that the ratio between capacity and channel rate will always be small. In the design of mesh-based solutions, these factors influence the strategy employed and make it necessary for the developer to be continuously aware of the available network capacity.

36.4 Examples of Practical Mesh Network Applications

The following examples illustrate a number of the principles discussed in the previous sections by highlighting design considerations in practical wireless mesh network applications. These examples are based on actual deployments, but the company names have been changed. In real-world deployments, mesh networks often require a hybrid approach that may include integrating components with existing legacy systems or combining wired and wireless network segments to achieve a design goal. Each example illustrates a particular type of challenge faced by developers in the field.

36.4.1 Equipping a Water Treatment Plant with Distributed Sensors

Indigo Waterworks provides water treatment services for a mid-sized community located in the Rocky Mountains. In an effort to simplify maintenance and reduce costs, senior management at the facility instituted a pilot study to determine whether wireless sensors could be used to transfer data to the central control room. The water treatment plant contained several potential RF interference sources, and staff members expressed concern that the environment would prove unsuitable for wireless applications. Sources of potential interference included the absorption effects of the water on the 2.4 GHz radio signals, the large amounts of iron piping running through the facility, the rebar contained in the thick concrete walls, and the electrical fields given off by the pump motors and switching gear. In this type of environment, there was an immediate concern that RF signals could not be transmitted reliably from sensors to the control room, and that the degree of difficulty in deploying the wireless network might make the entire project impractical.

The sensors used in this application measure the turbidity of the treated water at one of the final stages of the treatment process. The existing wired network in this facility spanned three floors and relayed data to a control room where specialists monitored the effectiveness of the treatment processes. Thick concrete walls, a winding stairwell, and several dozen feet separated the sensors from the control room. These factors complicated the situation for a wireless deployment.

The challenge in this example involved designing and implementing a parallel data delivery system to route information from the turbidity sensors to a mock control room situated beside the actual control room. The instruments were located three floors down, down a stairwell, and on either side of the stairwell. On one side, four sensors occupied a small pipe gallery built as part of the original facility. A later expansion added a larger pipe gallery on the opposite side; this gallery contained eight additional sensors. The deployment consisted of these 12 sensor nodes and additional relays to connect them with the mock control room.

36.4.1.1 Deployment Strategy and Implementation

The management at Indigo Waterworks wanted to answer a number of questions through this test deployment:

- How much time and effort would be required to deploy the wireless network?
- Would the installation and deployment require any special-purpose tools or additional equipment?
- Were specialized skillsets required for any personnel involved in the deployment?
- How reliable would the wireless network be, given the many possible sources of RF interference?
- Is a wireless network practical for the kinds of critical operations performed in a water treatment plant?
- Would the wireless system integrate effectively with the existing Modbus devices used at the facility?

For this environment, the sensor placement and wireless network communication links were organized as shown in Figure 36.2.

After a site evaluation that mapped the positioning of each of the functioning turbidity sensors, the wireless mesh design team placed simulated instruments next to each of the real instruments and deployed the network nodes. Then they installed the relay chain and linked it to the mock control room. Once the wireless network was up and running, information began coming in from each of the sensors, but the reliability from the original pipe gallery was not as high as expected.

A visualization tool that provides a complete evaluation of the network indicated that though most of the connections and links were functioning normally, other links were relatively weak. By employing a set of additional repeaters, the team managed to circumvent a significant barrier: a wall of reinforced

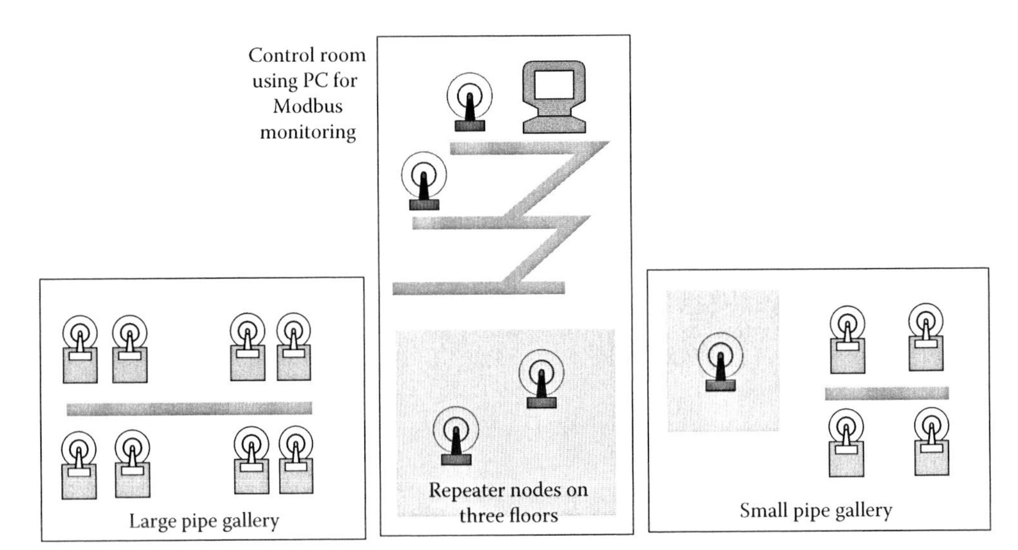

FIGURE 36.2 Deployment of wireless sensors in relation to the control room.

concrete that was 18 in. thick. Using a hole cut for an air duct, the team deployed wireless nodes on either side of the wall near the duct. This was all that was necessary to boost the signal strength sufficiently to bring up the reliability figures.

The connectivity at this point was significantly better. The design team began collecting data and performed a complete tabulation every 24 h to check the reliability of the connections.

36.4.1.2 The Results

During a 4 day interval, the wireless network and sensors functioned at a level of four nines reliability, which indicates that better than 99.99% of the reports were coming back and being successfully logged. The entire deployment, both the initial placing of the nodes and then the re-evaluation and placement of the additional repeater nodes, was completed within 3 h.

The Indigo Waterworks staff members were pleased by both the success rate of the message transfers and the fact that the RF interference proved to be less of an impediment than originally thought. Their instruments, which used the Modbus protocol, did not require any modification to work within the wireless environment. The design team had effectively encapsulated the Modbus packets. Devices throughout the network communicated without awareness that the connectivity involved wireless links. The deployment essentially provided a drop-in replacement for the wired network.

As encouraging as these results were, the nature of the deployment relied on polling techniques, a legacy requirement from the Modbus protocol, to acquire the sensor data. While this process worked very effectively for this particular implementation, the solution does not support the full scalability that can be achieved by a wireless mesh network that uses exception-based processing. Through exception-based processing, the approach could have been re-engineered so that the sensors only delivered data if the turbidity exceeded defined parameters. This type of re-engineering often requires balancing the efficiencies of pure wireless mesh design with the practicalities of a legacy protocol (in this case, Modbus). With intelligent design, engineers need not accommodate the requirements of an earlier protocol.

While this example illustrates the viability of wireless networking within a difficult RF environment, the design guidelines described earlier are not followed. The wireless networking essentially provides a drop-in replacement for existing equipment and helps reduce the costs associated with constructing cable conduits and pulling network cable throughout the facility. Since the nature of many water treatment plants dictates that they are built small and then extended as the needs of the surrounding community increase, the typical approach is to expand the existing plant rather than build a second facility.

If the scalability limits were not exceeded, the wireless network used in this example could be employed to effectively handle the sensor monitoring at a water treatment facility.

36.4.2 Designing a Process Control System Using a Wireless Mesh Network

BlackGold Inc. operates a petrochemical extraction facility in northern Alaska, pumping oil from the ground and heating it to a particular temperature to maintain desired viscosity. An existing system within one of their facilities used distributed sensors communicating by means of the Modbus protocol.

Temperature monitors are installed at several different points in the piping. The technique originally employed was to generate a series of Modbus queries to each instrument sensor in a round-robin fashion. New queries were generated as quickly as the instruments reported back and the results were fed to the controller, which turns heaters on and off for different sections of the pipe. Figure 36.3 illustrates the original system configuration.

36.4.2.1 Deployment Strategy and Implementation

A wireless mesh design team brought in to try to improve the process had to work out a solution that minimized the impact on the existing instrumentation. The team tackled the problem by starting at the data collection point and installing a wireless node onto it. The node simulated the entire network. It answered queries from the controller as quickly as the controller requested information. The link between the central controller and the wireless unit relied on the Modbus, but the query rate was too fast for the network to handle. To resolve the problem, the team engineered the solution so that the wireless node received information from each of the temperature sensors using an exception-based strategy. Any time that a temperature changed or a certain time window was exceeded, the temperature sensor generated a report. This information could then be cached and provided to the primary controller whenever the controller requested it. If the temperature changed, the node was set up to generate a report of that change immediately. This technique ensured that the wired controller would always have current information on the temperature status of every section of the pipe.

The 1 min time-out interval ensured that the system could detect a failure at one of temperature sensors. If the sensor failed to report, then the wireless node connected to the sensor would recognize a potential problem and attempt to contact that node. If the contact failed, the wireless node would report an error back to the primary controller. In this example, the solution was over-engineered in that the polling took place at more frequent intervals than required by the application. The time constant for heat loss in this piping system was on the order of 30 min. A sampling rate that delivered either changed

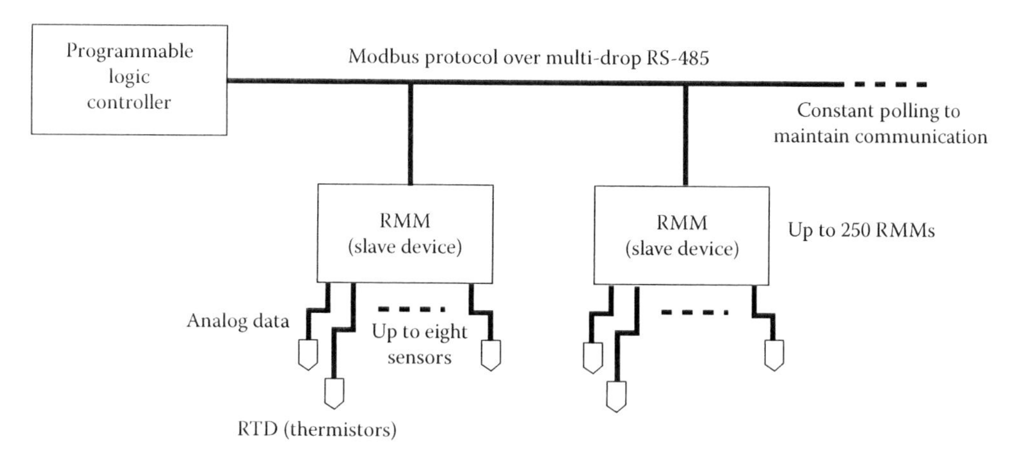

FIGURE 36.3 Original configuration using Modbus querying.

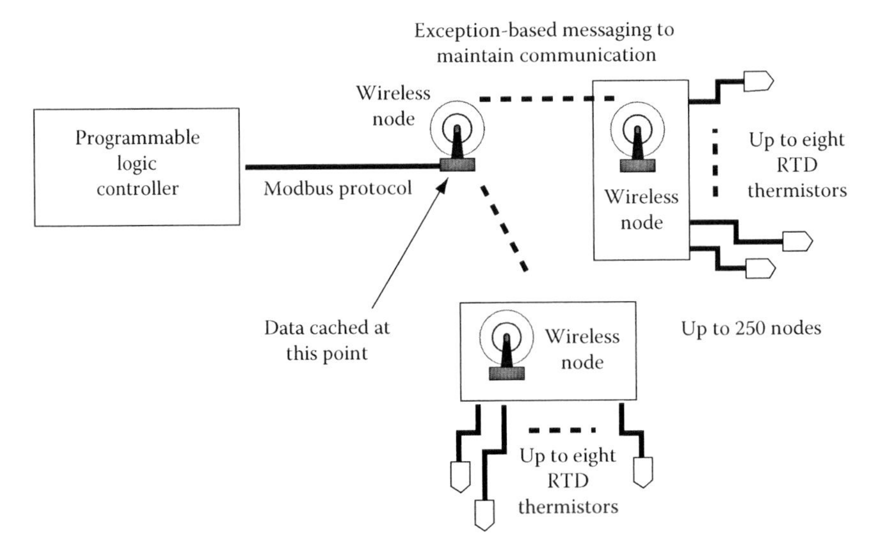

FIGURE 36.4 Wireless nodes and temperature sensors within heater feedback system.

data readings or a notification every few minutes would have provided adequate feedback to the heating system and ensured proper operation of the pumping units. The 1 min sampling rates were a conservative approach to this application.

Figure 36.4 illustrates the organization of the wireless nodes and temperature sensors in relation to the primary controller and Modbus.

The primary controller in this example consisted of a programmable logic controller set up to respond to predefined thresholds. The simple logic progression detects when the temperature drops below a certain value and turns on the heater in the corresponding section of pipe. When the temperature rises above a specified value, the PLC turns the heater off.

36.4.2.2 The Results

The use of exception-based monitoring demonstrated in this example reduced network load and improved reporting time without re-engineering any existing system component. In a polled system, designers typically schedule queries based on a worst-case analysis, generating traffic at regular intervals regardless of whether that traffic conveys useful information. Furthermore, the detection of state changes incurs a delay that averages one-half of the polling cycle because the state change is asynchronous. In exception-based messaging, the sensor generates a message immediately, relying on the MAC strategy to determine the earliest time the message can be transmitted. Through this approach, there is no inherent delay in reporting state changes.

In the BlackGold instance, instead of generating a new message every time that it is polled, the sensor generates a message based on the relevant criteria. In this case, either the change in temperature or the periodic timeout initiates the message transfer. In the overall system, the understanding is that unless a new message has been delivered, the current temperature data that is present is considered valid. That temperature value can be considered valid up to the point at which a sensor provides a different value.

The approach used by the design team supports a higher level of scalability than the previously described example, a benefit of exception-based monitoring. The limit in this particular situation applies to the PLC device and Modbus, which together can handle a maximum of 250 end points. But, the manner in which the wireless device communicated with the system was, in effect, spoofing the Modbus, which could permit a more extensive range of sensors to be deployed than the usual address limitations. This example demonstrated an effective technique for supporting multiple parallel buses using the Modbus protocol. The implementation could have been scaled to far exceed the conventional 250-node limit on Modbus activity.

36.4.2.3 Sensor Placement

The ability of sensor components in a wireless mesh network to contain a degree of intelligence and intercommunicate can help solve potential problems that occur in monitoring situations. As an example, at one BlackGold extraction plant a technician installing temperature sensors along the pipe every 5 ft failed to notice that he had placed a sensor very close to a steam vent. Instead of an ambient temperature of −40° (Fahrenheit), the sensor was indicating that the temperature was 40°. This caused the heating unit for that section of pipe to turn off and eventually the fluid in the pipe gelled, causing a major shutdown of the system.

Using a mesh network design where the nodes can communicate in a peer-based fashion, this type of problem could be eliminated through logical design. In such a design, the nodes would not only report temperature data back to the controller that turns the heater on and off, but each node would also check periodically with the neighboring nodes. Logically, in a group of nodes spaced along a 20 ft length of pipe, three of them would not be reporting −10° while one of them is reporting 40°. However, if the nodes are equipped with a type of voting logic, they can identify unexpected values, and those values can be flagged and used to generate alarms to service technicians. Because the mesh provides the flexibility to communicate from measuring point to measuring point, a group of nodes can function as a buddy system where the nodes are checking up on each other in addition to their normal tasks.

36.4.3 Monitoring Cargo Shipments Using Leaf Nodes

CoolTransport ships a variety of products, including produce and pharmaceutical items that require temperature control monitoring to prevent damage or degradation during transport. The cold chain management techniques employed by CoolTransport involve sensor placement within the cargo area to provide continuous monitoring of the ambient temperature. Through this monitoring process, the end customer can determine at the shipment destination whether the product was held at the appropriate temperature and whether or not to accept the cargo. Having used a variety of monitoring techniques, CoolTransport found an essential flaw in their approach. Monitored values from sensors implanted within the cargo could not be read without unpacking a significant amount of the shipment. If the customer decided at that point, because of sensor readings, to reject the shipment, a substantial amount of time and effort would be required to repack the cargo. For this reason and to satisfy additional requirements, CoolTransport set out to evaluate wireless techniques for relaying the monitored sensor values to an external network. Customers could then examine these values and decide whether to accept or reject the cargo before any items had been unpacked.

The particular requirements of this application suggested a hybrid approach that incorporated elements of both mesh and star networks. Because the temperature sensors used battery power and were out of network contact during shipping, they would not function as standard mesh nodes. On the other hand, these nodes could employ point-to-point style of messaging very naturally, thus avoiding the complexity of synchronizing their sleep cycles and expending battery power to relay for other nodes. Consequently, the network design employed a standard mesh within the loading facility and *leaf node* temperature sensors with more limited participation in the network.

In this model, the temperature sensors perform no network functions during transit and merely log data. At a predetermined docking point, these leaf nodes recognize the proximity of a wireless mesh access point and automatically convey the data collected during the transport period. This data can then be used to inform the customer of the temperatures maintained during shipment and consolidated at a central point, linked through a conventional wired network, for tracking and evaluation.

As an example of the problem faced by CoolTransport, one of their contracts involved large shipments of lettuce transported during the summer when ambient air temperatures along the trucking route often reached in excess of 100°. A shipment of lettuce represents a valuable commodity, but not an extremely valuable commodity, so the placement of one or two sensors and recording monitors within the truck's cargo area was considered sufficient to provide adequate temperature fluctuation readings to the customer. Reaching those sensors once the truck arrived at the loading dock, however, required that almost one third of the lettuce cartons be unpacked to gain access to the first sensor and its recorded data. If the customer made the decision to reject the shipment at that point, a very large number of cartons of lettuce would have to be repacked, compounding the losses of the trucking company. A solution that could provide a full accounting of the sensor's readings during transport could save time and reduce costs for both the customer and the shipper.

36.4.3.1 Deployment Strategy and Implementation

CoolTransport embarked on an approach whereby the sensors transported with the shipment would take measurements once a minute and record the monitored values in a log. By design, when the truck pulls up to the dock where the product is being unloaded, a standard mesh network is deployed at the docking facility. A node at each of the loading bays lets the shippers pull the truck up to the loading bay, back up, and open the door, and the temperature sensor that is inside completes its 1 min wait cycle. Once that measurement time elapses, the temperature sensor identifies the network, recognizes that it is at its destination, and proceeds to register itself on the network and offload the temperature info.

The fixed network at the loading facility gets that information back through the mesh network, which transfers it to a PC or another data display station. The displayed values of the temperature record indicate whether the shipment should be accepted or rejected, based on whether the temperature remained within acceptable values during the transport period. Figure 36.5 depicts the deployment configuration used in this example.

This collected information can be transferred using broadband channels to a central location. If it is a grocery store chain, for example, they can track it from central headquarters. From a communications standpoint, this network differs from a conventional network in that the nodes implanted in the truck

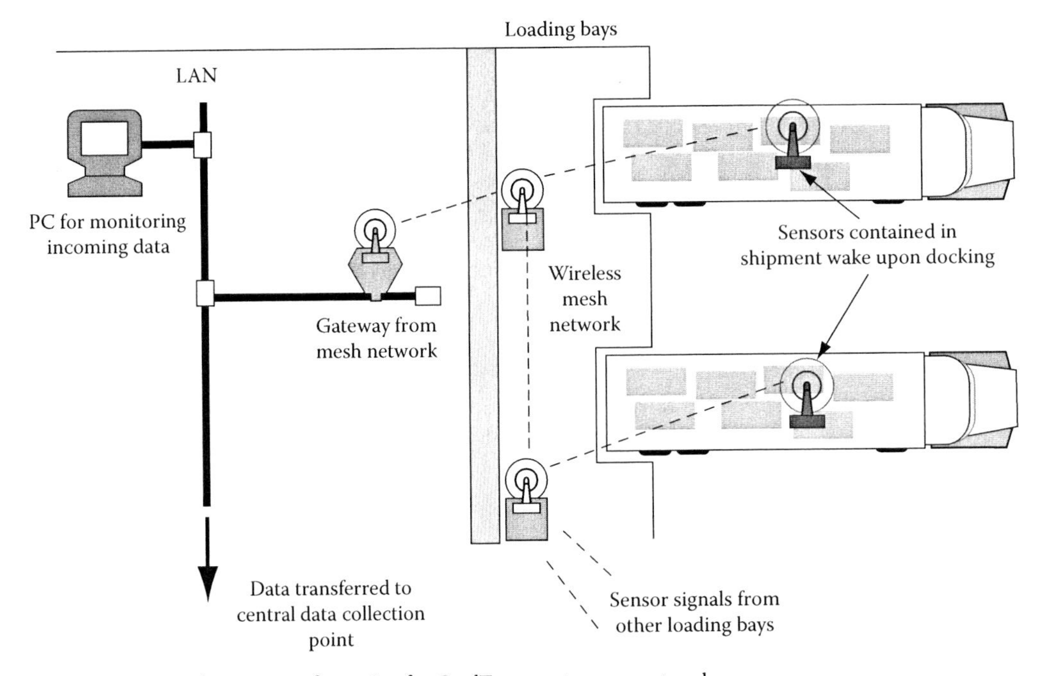

FIGURE 36.5 Deployment configuration for CoolTransport sensor network.

operate on extremely low power. These nodes, designed to run on watch batteries, have to operate for at least a year without replacement. Each of the sensors is reused—they have to be designed for long life.

The life expectancy can be maintained through a power cycle that is set up to maintain a very low duty cycle. Temperature sensors wake up once a minute, make a measurement, and listen, attempting to detect a wireless mesh network nearby. If they don't detect a network, they go back to sleep.

The other primary difference in this approach, compared to other distributed sensor architectures, is that the data flow does not take place from trailer-based unit to trailer-based unit. Data flow is always from the trailer-based unit through the wireless mesh network to a PC or other data collection point. The nodes residing in the trucks are different from typical member nodes of the mesh network. The term *leaf nodes* has been applied to distinguish their unique characteristics.

36.4.3.2 Wireless Mesh Configurations Employing Leaf Nodes

Leaf nodes do not function as full-fledged members of the wireless mesh network. On the outside perimeter of the wireless mesh the leaf nodes can talk point-to-point, communicating to the mesh node that resides at the docking point in the bay. The node at the bay, in essence, becomes a proxy in the mesh for the nodes that are being transported in the trucks.

This technique resolves a number of issues, including the following:

- *Reduces address data overhead*: Employing leaf nodes as handled in this example removes the need for creating a very large address space to accommodate all of the potential addresses in the network. For low-data-rate sensor networks, dedicating a substantial amount of the transmitted data to addressing schemes is counterproductive—the overhead is an unnecessary burden. By employing dynamic address allocation, the embedded sensor can wake up to join the network and get assigned a unique identification to communicate within the network. In this example, the node at the bay serves as a proxy so that the temperature-sensing node is never exposed to the network. Communication between the node at the bay and the temperature-sensing node can be mutually agreed and the proxy communicates the temperature values associated with the sensor to the wireless mesh network. The node does not need to be assigned an ID to transfer values.
- *Allows unsynchronized sleep cycles*: Nodes that relay on behalf of their neighbors must synchronize their sleep cycles. Because of clock drift, this is a significant problem for large networks with low duty-cycles. In the case of CoolTransport's application, synchronization is particularly difficult because the temperature sensors travel between networks that may not be synchronized at all. By eliminating the relaying requirement, leaf nodes may sleep in a completely unsynchronized manner, which greatly simplifies the implementation.
- *Eliminates cargo unloading*: Monitored values can be relayed to the mesh immediately following arrival without the need to unload any of the cargo within the shipment.

The temperature sensors store data collected during transport in a low-power SRAM, which can then be reset after the cargo is unloaded in preparation for the next monitoring operation. The logic driving the monitoring operations is contained in an ASIC, which also has very low power requirements.

36.4.3.3 The Results

The deployment of the hybrid wireless mesh network using leaf nodes proved successful, providing a valuable proof of concept for this technique. The leaf node technique can be effectively applied to many different varieties of sensors, depending on the nature of the cargo and the critical sensitivities. For example, the sensor might be equipped to measure humidity or maximum G forces or the presence of a particular chemical agent. The same principles can be used so that the monitored values are relayed to a proxy node at the dock upon arrival and then transferred through the wireless mesh network to a central data collection point.

This example differs from a classical sensor network structure, which usually trickles data through the network a few bytes at a time. In the CoolTransport example, the sensor stays out of communication with the network for a prolonged period, caching all data during that time. Upon docking and relinking with the wireless mesh network through the proxy, a substantial amount of data is transferred during a single hop, after which the sensor drops out of communication once again.

36.4.3.4 Scenarios That Favor a Leaf Node Approach

The leaf node approach provides favorable benefits in two distinct areas:

- *Extending battery-powered applications*: In distributed sensor applications that must extend battery life for lengthy periods, a wireless mesh network presents problems in that data transmissions to neighboring nodes can consume battery reserves. This conflicts with a key strategy for extending battery life, reducing the duty cycle periods so that the sensor is running as little as possible, ideally spending long periods in sleep mode. While in sleep mode, a node cannot relay for another node. In a full-fledged mesh network, some system of coordination must be used among the nodes to control wakeup and sleep cycles. The complexity of solving this problem can be a considerable challenge in many types of wireless mesh implementations. The leaf-node approach avoids the need for time synchronization by eliminating communication with the rest of the mesh network until the time at which the cached data is transferred.
- *Minimizing address space requirements*: For a low-end distributed sensor network, even the difference between a 2 B identifier and a 6 B identifier can be a crucial difference in the utilization of bandwidth. In this example, where nodes are moving between networks, even a 2 B address is limited to some 65,000 unique addresses. However, among a number of networks that may be visited by a node with one of these unique IDs, the likelihood of encountering an identical address is unacceptably high. In this example, the proxy node circumvents the need for the leaf nodes to maintain a large address space, acting as an intermediary in the communication with the rest of the wireless mesh network.

36.4.4 Devising a Wireless Mesh Security System

This example of a wireless mesh design illustrates a more well-rounded approach to the technology, taking better advantage of the design principles discussed in Section 36.3. A security company, IronMan Security, offers an access control and security system that consists of a central logging and control station and up to 500 devices. Typical supported devices include pass-card readers, keypads, electronic door locks, and sensors. The design specifications for this system required that transactions be completed within 1 s, including accepting input from a card reader or keypad, validating the entry, and activating the corresponding lock. The control station also has 1 s to process alarm conditions, such as intrusion detection. The system must also perform continuous self-monitoring and report any device failures that occur.

The original implementation for the IronMan system relied on a proprietary protocol operating over a multi-drop RS-485 bus. To manage medium access, the network was organized using a "master and slave" approach. Within this network, no slave can transmit except in response to a message from the master. To satisfy the operating criteria, the control station exchanges messages with each device at least once per second. Each exchange begins with a message from the master, which consists of either a default 5 B "status check" message or a command, such as "open the door." The default device response is a 5 B acknowledgment, but if the device has a condition to report (a user ID from a card swipe, for example, or a door-open alarm), it will send this data in a condition report. No response from the slave indicates a device failure and triggers a system failure report.

In Section 36.3.1, Table 36.1 provides the characteristics of a typical 802.15.4 wireless mesh product available today. As indicated in the following system parameters, the existing IronMan implementation requires an effective data rate of at least 282 kbps, which significantly exceeds the 40 kbps rate of a cost-effective mesh device.

The system parameters for the original implementation were

- Maximum number of devices: D = 500
- Control station time between queries: $T_q = 500\,\mu s$
- Processing time/condition Rpt (max): $T_p = 12.5\,ms$
- Maximum number of condition Rpts/second: $N_R = 5$
- Device response time (max): $T_d = 1\,ms$
- Total bytes exchanged/status check: $B_s = 10$

Comparing the existing IronMan system to the optimal design guidelines for wireless mesh networks, the wide disparity in the approach becomes evident. The existing design is based on a bus topology and offers no provision for the distributed control of medium access. Consequently, the system must centralize bus management in the control station and employ a query-and-response messaging model. Within this model, data cannot be pushed from the source, because the source does not know when it may transmit. Figure 36.6 shows the basic organization of the existing system.

By applying optimal mesh design principles, the handling of the condition reports can be managed more efficiently using exception-based messaging. Because each point in the wireless mesh has a built-in MAC, the query-and-response messaging to prevent bus contention can be eliminated. Rather than waiting for the next poll from the control station, a device can initiate a message as soon as it identifies a reportable condition. As a consequence, if a single condition report were the only traffic to the control station, the data rate required would be

$$\frac{R \times B_R}{1s - T_p - T_d} = 0.78\,\text{kbps} \tag{36.1}$$

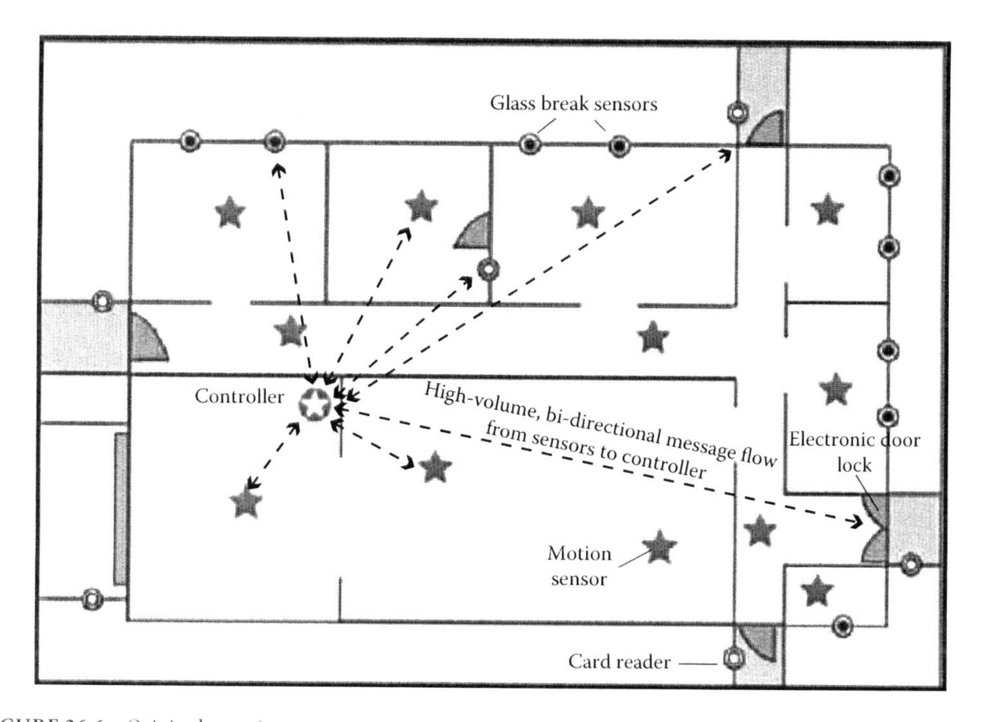

FIGURE 36.6 Original security system structure.

where R represents the maximum number of end-to-end retries, arbitrarily set here as R = 3. Because there could be as many as five such messages in the neighborhood of the control station simultaneously (which is probably a more conservative value than the actual requirement of five each second), the minimum throughput needed to support condition reports is approximately

$$0.78 \text{ kbps} \times N_R = 3.9 \text{ kbps} \tag{36.2}$$

Mesh devices can comfortably accommodate this data rate. Condition reports, however, are not the sole communications requirement. The self-diagnostic factor of the network is another consideration. The optimal mesh design principles state that polling from the control station is undesirable. Polling creates bottlenecks in the neighborhood of the station. To eliminate this problem, the individual devices can accomplish the necessary self-diagnostic operations.

System diagnostics can be implemented in a distributed manner by using a type of buddy system. Devices in these kinds of applications naturally tend to cluster. A card reader will typically be paired with a door lock. Sensors for motion and glass breakage will usually be deployed for each room. When these devices are commissioned, they can be placed into groups of *buddies* that monitor the transmissions from each other. If a node sends a message, the message counts as a transmission; if the node has not transmitted for a certain period of time, it will beacon. When one of nodes in the group fails to transmit for a certain period, a neighboring node polls it. If this polling gets no response, the neighbor can generate a condition report to alert the control station.

Distributing the self-diagnostic tasks among the nodes also distributes the associated messaging. Spatial multiplexing ensures that buddy groups that are geographically separated can perform self-diagnostic operations in parallel. By this technique, the network capacity can often be replicated at each group, so individual groups can be considered independently. For such groups, the available bandwidth for self-diagnostics is approximately 90% of the total or

$$40 \text{ kbps} - 3.9 \text{ kbps} = 36.1 \text{ kbps} \tag{36.3}$$

The largest number of nodes that might compose a group is not indicated, but assuming 10 nodes in the group (and assuming that each node must transmit at least once every 0.75 s) is reasonable. This assumes a fairly aggressive 1 s reporting time, reserving 0.25 s for a possible condition report. With the beacon message occupying 5 B, the traffic load per group would be a maximum of

$$\frac{10 \times 5\text{B}}{0.75\text{s}} = 0.67 \text{ kbps} \tag{36.4}$$

This value represents less than 1% of the total network capacity. This analysis becomes a more difficult if the groups are in proximity, creating a requirement that they share bandwidth. Even in a worst-case deployment, however, where all 50 possible groups completely overlap, the resulting traffic will not overload the network. Figure 36.7 shows the IronMan security system reorganized to benefit from an optimized wireless mesh design.

36.4.5 Successful Approaches to Application Design

As shown by both the design methodologies and examples in this chapter, wireless mesh networks and distributed sensors offer a number of advantages to developers who master the techniques of working within the framework of the technology. Benefits to be gained include fault-tolerance, ease of installation, incremental deployment, and greater processor efficiency. Achieving these benefits, however,

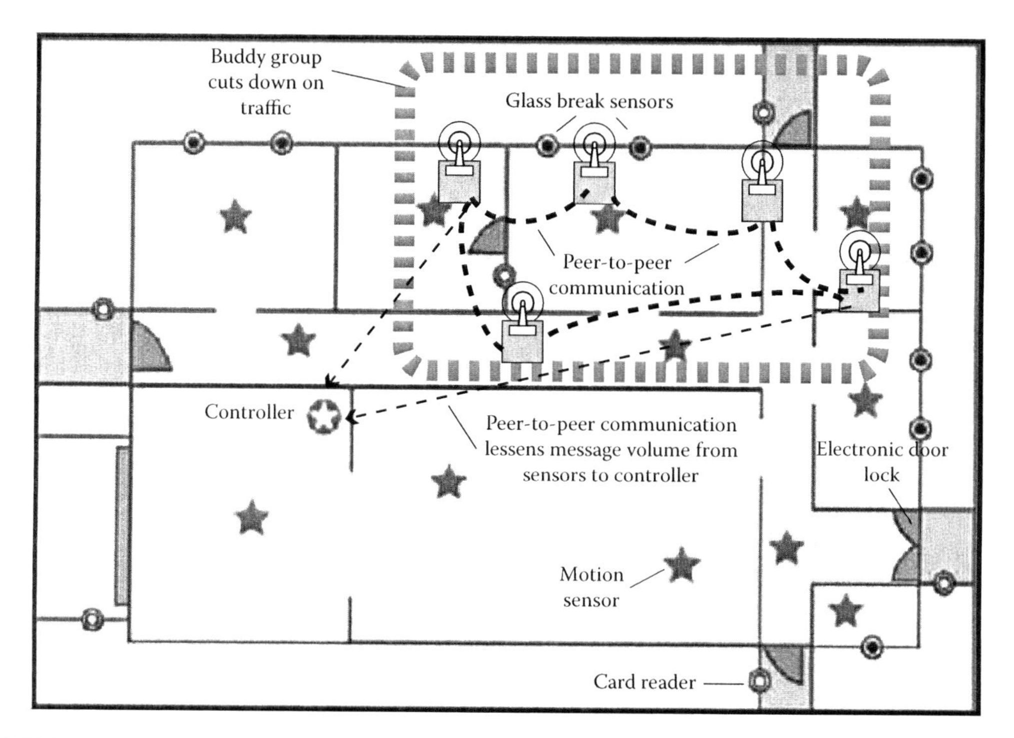

FIGURE 36.7 System employing wireless mesh design principles.

requires careful attention to the architectural model used; in particular, great care should be taken when adopting the familiar centralized organization and messaging models of wired systems. Very often, these systems make tacit assumptions about the communication medium. Often these assumptions do not apply to a practical wireless mesh application. By following the guidelines offered in this chapter, developers can construct efficient, practical applications and improve upon the design goals of wired systems.

References

1. Poor, R. and B. Hodges, Reliable wireless networks for industrial systems, Ember Corporation Technical White Paper, http://www.ember.com/products/whitepapers
2. Corson, S. and J. Macker, Architectural considerations for mobile mesh networking, IETF Network Working Group, May 1996.
3. Murphy, J., Mesh networks solve distribution dilemmas, *Wireless Europe*, November 2000.
4. Chandrakasan, A., R. Amirtharajah, S.H. Cho, J. Goodman, G. Konduri, J. Kulik, W. Rabiner, and A. Wang. Design considerations for distributed microsensor systems. *Custom Integrated Circuits Conference (CICC)*, San Diego, CA, May 1999.
5. Poor, R., Gradient routing in ad-hoc networks, MIT Media Laboratory, October 2001. http://www.media.mit/pia/Research/ESP/texts/poorieeepaper.pdf
6. Schurgers, C. and M.B. Srivastava, Energy efficient routing in wireless sensor networks, *MILCOM'01*, Vienna, VA, October 2001. http://www.janet.ucla.edu/~curts/papers/MILCOM01.pdf
7. Poor, R., Wireless embedded networking systems, *SensIT PI Meeting*, Waltham, MA, January 2002. http://dtsn.darpa.mil/ixo/sensit/PI_Briefs/Poor_Ember.ppt
8. Gupta, P. and P.R. Kumar, The capacity of wireless networks, *IEEE Transactions on Information Theory*, IT-46(2), 388–404, March 2000.

37

Wireless Sensor Network Test Bed for Target Tracking

Qing Yang
Montana State University

Alvin S. Lim
Auburn University

37.1 Introduction

The tremendous success in the development of technology in wireless communication and microelectromechanical systems (MEMS) has made it feasible to deploy large-scale wireless sensor networks (WSN) composed of relatively inexpensive sensors capable of collecting, processing, storing, and transferring information. Sensors can be widely deployed within a geographic area in order to gather physical data (e.g., sound and temperature) and to facilitate efficient and collaborative control of various natural and human-generated events. One interesting application of WSN is target tracking in a hostile environment, where physical access is accompanied by some form of danger, for example, battlefield. In this chapter, we focus on building the sensor network test bed for vehicle localization and tracking in WSN and use multi-hop transmission for communicating alerts generated by sensor nodes to special control nodes or base stations.

At the base station, there will be a set of cameras to capture the image/video of detected targets for further analysis through image processing software or by human beings in order to reduce false alarms.

We have designed and implemented a target tracking sensor network test bed system that integrates WSN with wired IP networks. It is useful for studies on advanced WSN applications including interoperating the enhanced closest point of approach (ECPA) algorithm for accurately computing target location, velocity, and direction with control software for controlling camera sensor nodes that capture image or video of the target at its predicted location. The software communicates over a data-centric sensor network software. Directed diffusion (DD), IEEE 802.11, and TCP/IP were integrated together to facilitate the requirements for both ad hoc acoustic sensor networks and the control center for video capture. We have developed the ECPA algorithm for target localization and tracking that is suitable for WSN composed of low-powered and inexpensive nodes. We have implemented the ECPA algorithm in the sensor network test bed and evaluated its performance in field experiments which show good accuracy and effectiveness in target localization and tracking.

Localizing and tracking vehicles with WSN is a challenging problem. First, the whole system must be energy-efficient. The size and number of messages must be minimized, since network communication consumes significantly more power than the local computation. Second, due to the limited capacity of each sensor, tracking mobile target requires multiple nodes to collaboratively exchange information between each other. Third, target position, velocity, and direction are required to control the camera. Since multi-hop communication from the active sensing region where target appears to the base station may incur considerable delay, the base station needs the target velocity information to predict the current target position at which the camera should point. Fourth, the localization algorithm must meet the real-time deadlines of the target tracking application. It must be as fast as (or faster than) the velocity of the vehicle.

To meet the design requirements, we designed and implemented an integrated system for target detection, tracking, and image/video capture of moving targets using collaborative mixed wireless sensor nodes connected by DD [10] and 802.11. The closest point of approach (CPA) algorithm [7,14] is adopted in our system because it can calculate target position, velocity, and direction without complex signal processing and can handle targets with high velocity. In fact, based on the analysis in [7], higher speeds will generate lower localization errors. The original CPA algorithm, however, cannot be directly applied to WSN, because specific deployment configurations of sensors are required [7,14]. For instance, in [7] the target trajectory must intersect the convex hull composed of sensors without evenly splitting the sensor field while in [14] the target trajectory must be located outside the convex hull of sensors. In a realistic system, the target will trigger a set of sensors that may not satisfy the configuration requirements. Therefore, neither of the CPA algorithms [7,14] may be directly used. To solve this problem, we developed an ECPA algorithm that first finds the estimated bearing of target trajectory and then computes the relative position information between sensors and the trajectory. In several field experiments, we successfully detected targets and predicted their location, velocity, and direction of travel with reasonable accuracy. We currently use only one cluster (five sensors) to demonstrate that the ECPA algorithm is practical or feasible and our goal is to eventually extend it to larger networks.

37.2 System Architecture

The target tracking system consists of two networks, 802.11 diffusion network and 802.3 IP network, that are interconnected by a gateway node (Figure 37.1). The sensor network is composed of sensors that use 802.11b as the MAC layer protocol and DD [19] as the routing protocol. Clusters are reactively formed around the moving target, where one sensor in the cluster will be elected as the cluster head based on the protocol in Ref. [6]. Other sensors collect the CPA times and send them to the corresponding cluster head. Then, the cluster head runs the ECPA algorithm and transmits the target tracking results to the gateway node through a multi-hop diffusion path. After that, the gateway node translates the diffusion packet into an IP packet that forwards the tracking results to the camera control node.

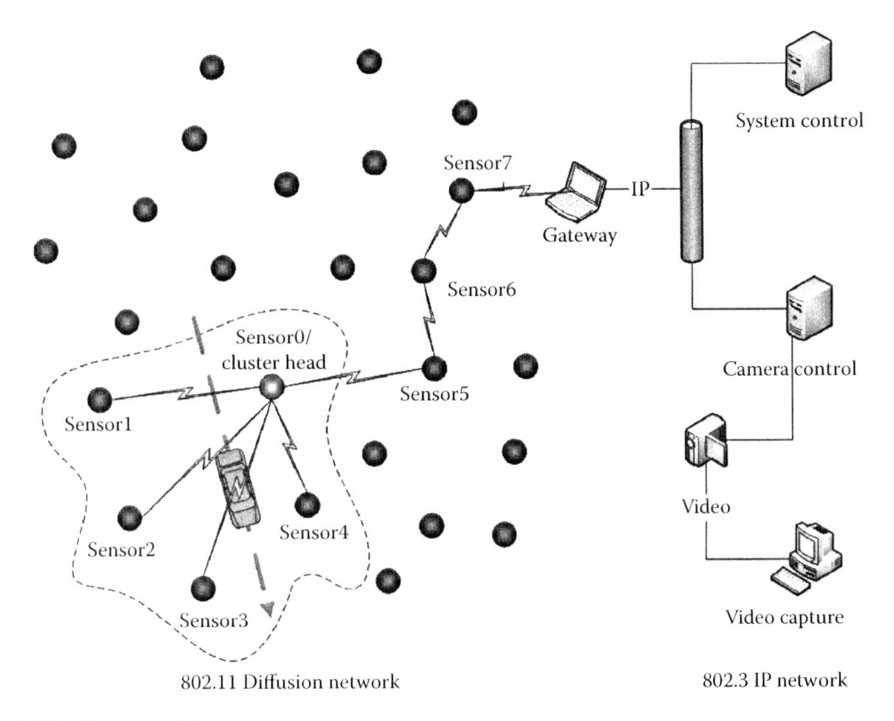

FIGURE 37.1 System architecture.

Finally, a command is sent to the camera through a serial cable so that it can pan and zoom in to the target at its predicted position. Because there is a delay for the tracking results to be transmitted from sensors to the camera, target may move further during this time period. Therefore, before sending the command to the camera, the camera control node need to recompute the (predicted) target position based on the delay and target velocity. We adopt this architecture to show that the ECPA algorithm works well in practice, although there are more efficient system implementations which are outside the scope of this chapter.

37.2.1 Key Components

There are six key components in our target detection and tracking system, each of them represents a role or a function. The six components are stated as follows.

Sensor: Sensor nodes monitor for targets, detect the measured CPA time, and report results to the cluster head through the DD network. Initially, sensors are in target monitoring mode where they continuously monitor for acoustic events. Once an acoustic threshold is exceeded, the sensor node switches into target detection mode. During the detection period, the CPA time is obtained as the instance when maximum amplitude of sound occurs.

Cluster head: The cluster head node receives the CPA time data from sensors over the DD sensor network as the input to determine the position, direction, and velocity of the target by running the ECPA algorithm. These data, along with the CPA time of the cluster head, are transmitted to the gateway node.

Gateway: The gateway node internetworks the diffusion network to the IP network. It forwards target motion information from the cluster head to the camera control node. The gateway represents the camera controller to the diffusion network and its primary task is to receive packets destined for the camera and convert them from diffusion packets to IP packets. This involves subscribing to target tracking data on the diffusion network and sending IP packets to the camera control node on the IP network.

Camera control: The camera control node receives packets from the gateway, and then predicts target position based on the target velocity and the time delay for package transmission from the sensors to the camera control node. It then pans and zooms the camera to point at the predicted location. The camera controller resides on a wired Ethernet network and listens for packets from the gateway on port 8899. When a packet is received, the camera controller issues movement commands to the camera over the serial port.

Video capture: The video capture node is responsible for interfacing with the video output of the camera. The output may be captured as a video or individual frame which may be saved as images.

System control: The system control node manages the execution of the remote nodes. This includes starting, monitoring, and stopping processes. We developed two methods for managing sensor networks: Interactive Sensor Network Execution Environment (ISEE) software and a direct method using SSH, which is an acronym for Secure SHell. It was designed and created to provide the best security when accessing another computer remotely. For simplicity, we used the direct method. The system control computer uses SSH to remotely login and issue commands on the sensor, cluster, and gateway nodes. We developed scripts to handle common tasks of the sensor system (e.g., starting and stopping processes). Since any command may be given over SSH, the system control node has complete control of all the nodes in the diffusion network and even the gateway and camera control node.

37.2.2 Internetworking Mixed Networks

The target tracking system is networked using several different network technologies. The sensors and cluster head communicate through DD running over an 802.11 network. The camera control and system control machines are connected with an 802.3 (Ethernet) network running IP. In order to internetwork the two networks, we developed an internetworking software which executes on the gateway node, which is configured with both an 802.11 interface and an Ethernet interface. The internetworking software converts diffusion packets into IP packets and vice versa. To accomplish this task, attribute-value pairs from the diffusion network must be mapped to an IP address corresponding to the machine on the IP network which is interested in the data from the diffusion network. The current application only requires one-way communication (from diffusion to IP) since the target tracking application does not need to communicate in the opposite direction. The camera software generates subscription (interest) for target detection data through the IP network which is translated by the internetworking software and flooded into the DD network. When target detection from a cluster head arrives at the internetworking software in the gateway node, it unpacks the data in a packet received from the cluster head and encapsulates it in an IP packet. This packet is addressed to the IP address of the camera control computer which has previously registered with the gateway to receive target tracking packets. The newly created IP packet is then sent to the camera controller on a prespecified port.

37.2.3 Collaborative Processing with Directed Diffusion

In our experiments (Figure 37.2), DD protocol was used in two ways. First, cluster heads broadcast interests (within the cluster area) for target CPA detection data; then, sensor nodes that detected the target will send back to the cluster head the CPA data that match the interest. Second, the camera node broadcasts the interest for target tracking data and the cluster head that computed these data will send the target location, velocity, and directions information to the camera node. The content of messages sent by the sensor to the cluster head includes CPA time, highest received signal strength, and sensor position. The messages sent by cluster head to the camera node contain CPA time, target position, speed, and moving direction. The diffusion attribute names and keys used in our system are summarized in Table 37.1. When target appears in the networks, sensors will send CPA data to the corresponding cluster head. Then, the ECPA algorithm running on the cluster head node follows three steps to compute

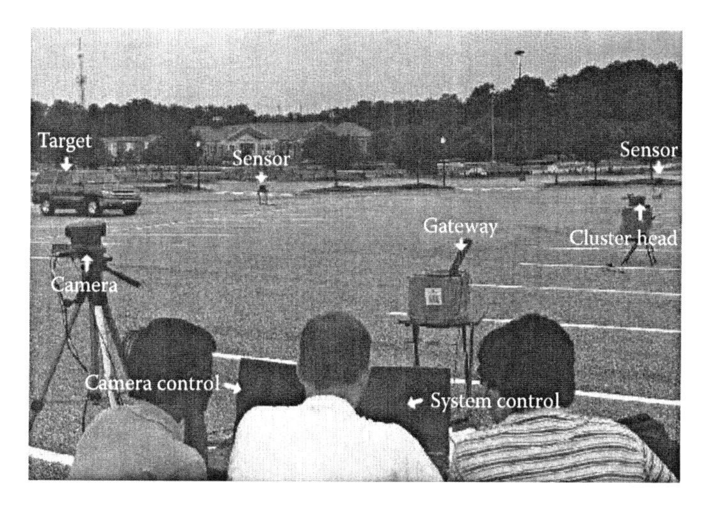

FIGURE 37.2 Experimental setup for target detection and tracking.

TABLE 37.1 Diffusion Attribute Names and Keys

Attributes	Key	Diffusion_Type	C++ Type
CPATime	6000	BLOB_TYPE	`struct timeval`
TimeStamp	6001	BLOB_TYPE	`struct timeval`
TaskName	6006	STRING_TYPE	`char[]`
SignalStrength	6007	INT32_TYPE	`int`
TargetLat	6008	BLOB_TYPE	`long`
TargetLong	6009	BLOB_TYPE	`long`
TargetSpeed	6010	BLOB_TYPE	`double`
TargetSlope	6011	BLOB_TYPE	`double`

the target motion parameters (position, speed, and moving direction). First, as shown in Figure 37.1, the cluster head (sensor0) determines the network configuration based on the CPA times sent by sensors (1–4). Then, sensor0 knows the estimated trajectory slope and is aware that it is the origin of the coordinate system (sensor1–4 are on the other side of the trajectory). Since this is the required network configuration by the original CPA algorithm (Figure 37.12), the target speed and four possible slopes can be computed. Second, the computed slope that is closest to the estimated value will be considered the right one. This matching process is important and necessary, because the original CPA algorithm usually fails after computing the four slopes as we mentioned previously. Third, the target location is calculated based on the received signal strength at the sensors (which will be described in the Section 37.5.3).

Then, the computed tracking results will be sent to the gateway node through a multi-hop route generated by the DD and forwarded finally to the camera control node. Since the message contains CPA time, target position, and velocity, the camera control node can predict the new target position if it knows the transmission delay which can be obtained by subtracting the target CPA time from current time. Till now, the camera knows where to locate the target and how fast to pan to capture the target based on the target speed.

37.2.4 Camera Control

The camera control node collects the tracking results generated from sensors (cluster heads) and issues command to the camera for capturing the target. The camera control node runs a simple UDP server listening for camera control packets. When packets are received from the bridge on the camera control port (8899 by default), the camera controller extracts the target location, velocity, and trajectory and moves the camera appropriately. The camera control application also receives the CPA time of the target. By calculating the delay between this time and the current time, the current position of the target can be computed. Camera movement occurs in two stages. First, the video camera moves to view the current position of the target. We call this target acquisition. Second, the video camera rotates to keep the target in view. We label this stage target tracking. Our current system used only one camera, but it is feasible to extend it to support multiple cameras which we will investigate in our future work.

37.3 Implementation of Wireless Sensor Network Test Bed

The test bed system architecture is implemented using commercial off-the-shelf hardware and specialized software based on PC104 for sensor nodes in the WSN. The specialized software includes dynamic services, a WSN middleware layer, and external agents (EAs) which permit access to WSN data from an external network.

37.3.1 Hardware

The PC104 CPU module (Figure 37.3), PFM-550S, is manufactured by Aaeon. It has a 533 MHz VIA Mark processor and has features such as 10/100 Base-TX Fast Ethernet port, one RS-232 port and one RS-232/485 port, four USB 1.1 ports, a SDRAM-SODIMM socket for up to 512 megabytes of RAM and supports type I compact flash cards. It supports 36-bit TL and 18/36-bit dual LVDS LCD panel, has a watchdog timer, and fully supports ISA. It is also fanless with an operating temperature of 0°C to +60°C. It requires +5 V for operation. More information can be found online at http://www.tri-m.com/products/aaeon/pfm550s.html.

The PCMCIA module (Figure 37.4), PCM-3115C, is manufactured by Aaeon. It is a two-slot PCMCIA module which supports two Type I/II cards or one Type III card. It complies with PCMCIA v2.1 and JEIDA v4.2. It has a 16-bit data bus and a busy status LED. It requires +5 V for operation. More information can be found online at http://www.tri-m.com/products/aaeon/pcm3115c.html.

The Power Supply module (Figure 37.5), PFM-P13DW2, is manufactured by Aaeon. It has an input range of +7 to +30 V and an output of +5 to +12 V. It is used to intake +12 V from the AC power supply or battery and convert it into the +5 V used by the CPU and PCMCIA modules.

FIGURE 37.3 PC104 CPU module from the top.

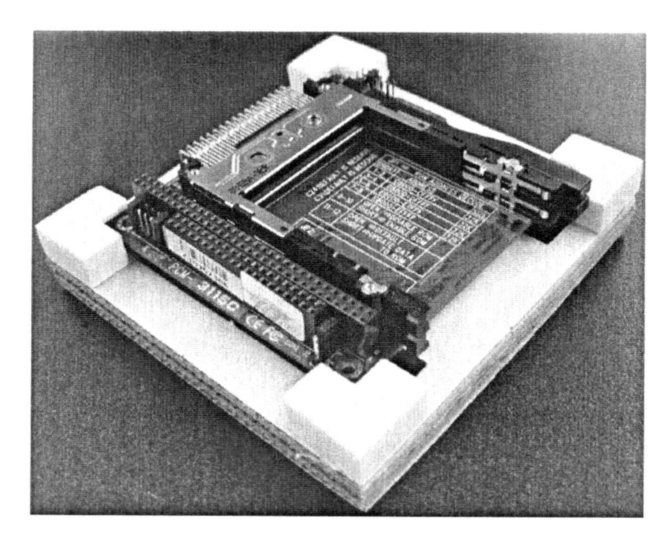

FIGURE 37.4 PC104 PCMCIA module from the top.

FIGURE 37.5 PC104 power supply module from the top.

The remaining hardware items are more commonplace items. For wireless communication, the Orinoco Gold wireless PCMCIA Local Area Network (LAN) card is used. The wireless LAN card is a 2.4 GHz radio which supports the four IEEE 802.11 High-Speed compliant speeds, 11, 5.5, 2, and 1 Mb/s, within the IEEE 802.11 standard for wireless LANs. A 2.4–2.5 GHz omnidirectional external antenna used to boost the wireless signal with a +5 dBi gain and a 60″ cable. A standard type I compact flash card is used for storing the program code and data. The RAM used is a 256 MB, 144 pin Synchronous Dynamic RAM high-density memory module at 133 MHz. Its technical specifications can be found at http://www.transcendusa.com/Support/DLCenter/Datasheet/TS32MSS64V6G 6755.pdf.

A plastic USB microphone, manufactured by Sound Professionals, is a mono, high sensitivity, omnidirectional microphone with headphone amplifier. Its dimensions measure 1.5″ × 1.0″ × 0.25″, and it can detect frequencies from 20 to 20,000 Hz. More information can be found online at http://www.soundprofessionals.com/cgi-bin/gold/item/SP-USB-MIC-1.

The AC Power Supply module (Figure 37.5), PFM-P13DW2, is manufactured by Sunny Computer Technology. It has an input range of 100–240 V and an output of +12 V. Accompanying the AC Power

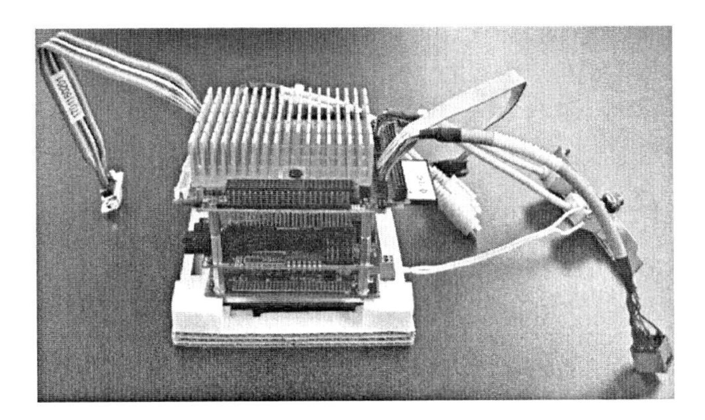

FIGURE 37.6 The completed PC104 sensor node without its protective casing.

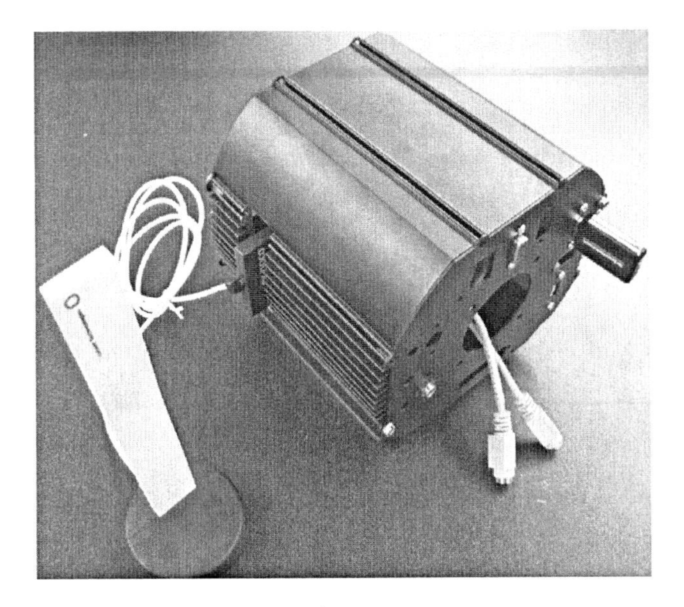

FIGURE 37.7 The completed PC104 sensor node with its protective casing.

Supply is the power input connector used to supply power to the PC104 Power Supply module. The +12 V AC Power Supply or the +12 V battery is connected to the power input connector. We solder wire onto the power input connector here in the lab to connect the wire to the power input connector. For more information, refer to http://www.sunny-euro.com/HTML/PRODUCTS/POWERSPL/SYS1183UP.html online.

The sealed lead acid battery is used to power the PC104 when away from the lab and outputs +12 V at 5.0 A·h. It is sealed and rechargeable. For more information, refer to http://www.power-sonic.com/site/doc/prod/86.pdf online.

A completed PC104 assembly can be seen in Figure 37.6 and in its protective casing in Figure 37.7. A close-up of the completed PC104 assembly can be seen in Figure 37.8.

37.3.2 Operating System

The PC104 sensor nodes use the Linux-based operating system (OS) Slax, v6.0.7, to run all sensor node application software. We chose Slax to use on the PC104 test bed, in part, because it is very small. We removed the graphical user interface (GUI) to further save space on the compact flash card. All software used in the test bed, including the OS, uses around 75 MB.

FIGURE 37.8 A close-up of the completed PC104 sensor node in its protective casing.

37.3.3 Dynamic Services

Dynamic Services (DS) is an integral part of this solution of information retrieval from the DD WSN, and there is a convenient for sensor node applications to be built on top of DS using DS as a service layer. DS is a middleware layer built on top of the DD protocol on wireless sensor nodes to provide the services necessary to facilitate IP-based information retrieval from sensor node applications built with this type of architecture. Note the difference between the normal DD network stack and the new network stack including the middleware layer DS shown in Figure 37.9. DS is placed between the DD network layer and the application layer. The services provided by the DS middleware layer to the sensor node application programmer allow the sensor node application programmer to ignore the details of the DD networking protocol. The sensor node application programmer need only worry about the neat and clean API to the DD protocol provided by DS. This flexible way to program these WSN nodes is one of the nicest features of the DS service layer.

DS also allows tasking of nodes, specifically for information or data production. Applications register the name of data they can provide with the DS service layer. Until an application receives an interest for this registered named data, it sleeps. Upon receipt of an interest for this registered named data, the

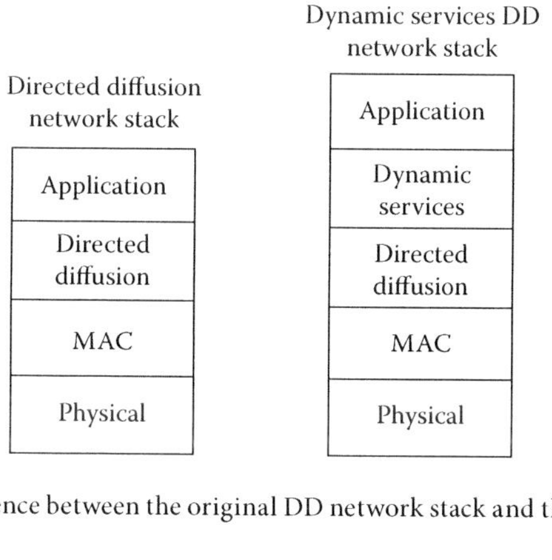

Directed diffusion network stack	Dynamic services DD network stack
	Application
Application	Dynamic services
Directed diffusion	Directed diffusion
MAC	MAC
Physical	Physical

FIGURE 37.9 The major difference between the original DD network stack and the modified network stack is the dynamic services service layer.

sensor node application is awakened by the DS service layer to begin producing the corresponding data. Sensor node applications then send out their produced data, again, through the DS service layer.

37.3.4 Device Roles

There are three main devices which act in this system. They are EAs, sensor nodes, and the gateway node. These will be introduced and summarized in the sections that follow.

37.3.4.1 External Agents

An IP-based host that is not a part of the WSN is termed an External Agent or EA. This is because, with respect to the sensor nodes in the WSN and the gateway node that connects the WSN to the Internet, the IP hosts are external to the WSN. This dotted line represents the boundary between EAs and the WSN. EAs use an API to query and task the sensor nodes in the WSN via the gateway node used to identify the WSN. It is assumed that EAs are full-edged computers with a full TCP/IP networking stack and can access the Internet or IP network on which the WSN resides.

The EA stack is shown in detail in Figure 37.10. Each EA application running on an EA node possesses its own copy of the EA API. EA applications use this API to submit requests to and receive data from remote WSNs.

37.3.4.2 Sensor Nodes

The wireless sensor nodes and the data they provide are one of the main components of the system. Sensor nodes must be programmed to recognize interests for which they can supply data. When a sensor node receives an interest for which it can supply data, routines are activated to actually produce the named data. All sensor nodes are assumed to have an antenna capable of wireless communication and sensors that are able to produce some data. Sensor nodes are also assumed to have at least a minimal multitasking OS as the middleware layer DS running on each node is its own unique process.

The DS-enabled DD networking stack is shown again with more detail in Figure 37.11. Each sensor node application possesses its own version of the DS API. Sensor applications use the API to communicate with the DS service layer, and DS communicates with sensor applications through the DS API.

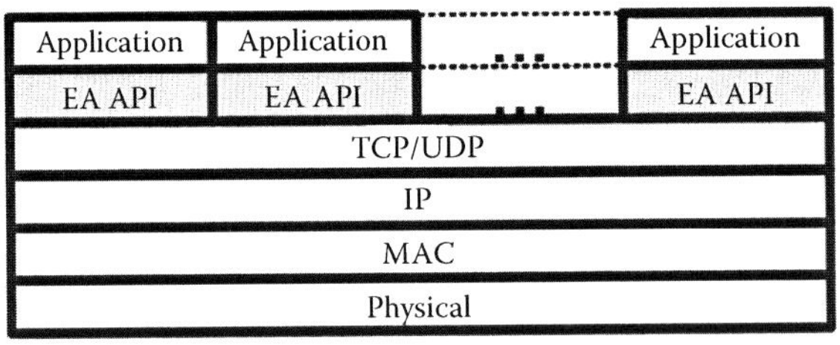

FIGURE 37.10 The network stack of EAs where the EA API provides services to the EA applications and utilizes the transport services of the lower layers.

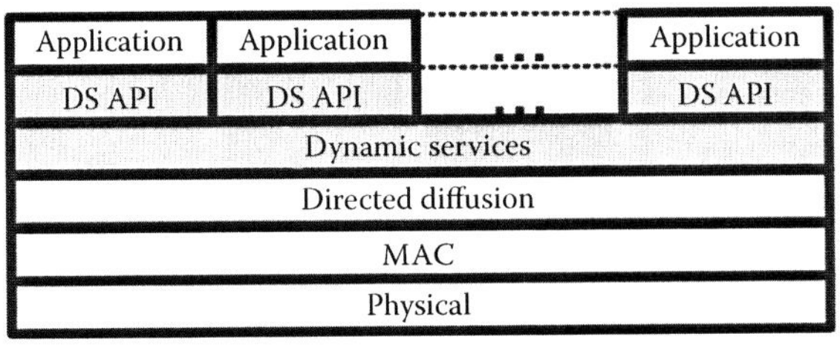

FIGURE 37.11 The dynamic services–enabled DD networking stack with the DS API and dynamic services layers.

37.4 Related Work on Target Tracking in Sensor Networks

The earlier-mentioned tracking sensor network test bed is used to study novel target tracking algorithms. There are many research efforts on target detection and tracking in WSN that describe several aspects of collaborative signal processing [5,13,23], target tracking with camera sensors [8], and real-time application for field biologists to discover the presence of individuals [2].

37.4.1 Target Localization Algorithms

Recently, a set of approaches [12,18,20] were proposed to solve the target localization and tracking problem with proximity binary sensors which report only 1 bit information to indicate if a target appears. Though the information transmitted in networks was reduced, the localization error was increased. As proven in Ref. [18], the achievable spatial resolution Δ in localizing a target trajectory is of the order of $1/(\rho R)$, where R is the sensing radius and ρ is the sensor density per unit area. Suppose there are 25 sensors, with sensing radius of 20 m, deployed in a $100 \times 100\,\text{m}^2$ area, then the lower bound of localization error will be 20 m.

In the area of the acoustic sensor networks, there are many solutions for target localization and tracking which can be divided into three basic categories: differential signal amplitude [17], direction of arrival (DOA) [2], and time difference of arrival (TDOA) [3,4]. The main problem with differential signal amplitude method is that distance calculation based on received signal strength is a very error-prone procedure. Because the accuracy of the received signal strength indication (RSSI) range measurements is highly sensitive to multipath, fading, non-line of sight scenarios, and other sources of interference, this method may result in large errors. These errors can propagate through all subsequent triangulation computations, leading to large localization error. The DOA approach requires a microphone subarray on each sensor, which will increase not only the cost of deployment but also the signal processing overhead on sensors. TDOA methods [3,4] make use of the relative time differences among sensors. To obtain the relative time difference, each sensor needs to first get the dominant frequency of acoustic spectrum, and then broadcast this information [3,4]. This may involve collaborative signal processing, such as FFT, which will increase the computational overhead. On the other hand, the audio data exchanged among sensors require more network bandwidth, a very precious resource in WSN.

37.4.2 Original CPA Algorithm

The CPA algorithm was originally designed for localizing low-flying aircraft by means of acoustic sensors [7,14]. One assumption of the CPA algorithm is that the target moves with a constant velocity (i.e., linear path) while passing through the set of nodes (either three [14] or four [7] sensors). This assumption is necessary because of the limited detection range of low-powered sensor nodes and the high speed of the target. For example, suppose there are 25 sensors uniformly distributed in a $100 \times 100\,\text{m}^2$ area, the sensing range is 20 m, and target speed is 30 miles per hour (MPH). Then within 5 s, the target will trigger on average 6 sensors around the trajectory, which meets the requirement of the CPA algorithm. We argue that within a short time interval, for example, 5 s, the target speed can be considered constant.

As shown in Figure 37.12, **v** and **r** denote target velocity and CPA to the origin of the coordinate system, respectively. Then the target path is given by

$$\mathbf{p}(t) = \mathbf{r} + \mathbf{v}(t - \tau), \quad \mathbf{r}^T\mathbf{v} = 0 \tag{37.1}$$

where

T denotes the transpose of a certain vector

τ is the instance at which the target moves through **r**

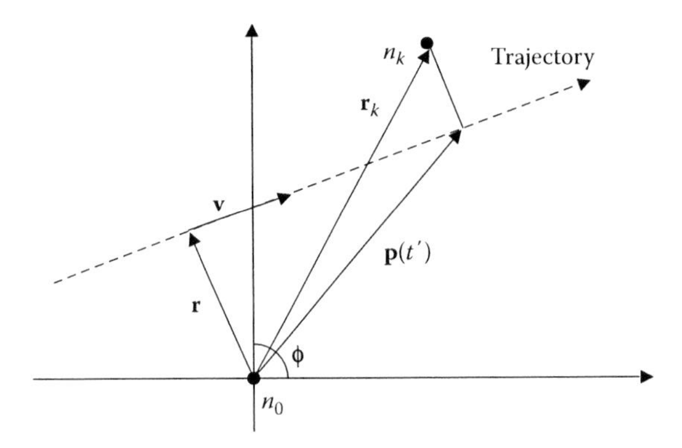

FIGURE 37.12 Illustration of CPA localization algorithm.

Let $\mathbf{r}_k$ denote the position vector of node k and t_k the measured CPA time. We define the CPA time as the time when target reaches the CPA, and measure CPA time as the instance at which the receiving amplitude is the highest at the sensor. Therefore, the difference between those two will be time of sound propagation from the CPA to the sensor. Then we can define t_k as

$$\left|\mathbf{p}(t')-\mathbf{r}_k\right| \to \min, \quad \left|\mathbf{p}(t')-\mathbf{r}_k\right| = c(t_k - t') \tag{37.2}$$

where
 $|\cdot|$ is the Euclidian norm
 c is the velocity of sound

Using Equation 37.1 to solve the minimal t and inserting the result into the Equation 37.2 yield

$$\left|\mathbf{r}-\mathbf{r}_k+(\mathbf{v}^T\mathbf{r}_k)\mathbf{v}/v^2\right| = c(t_k - \tau - \mathbf{v}^T\mathbf{r}_k/v^2) \tag{37.3}$$

where $v = |\mathbf{v}|$. The left side of Equation 37.3 can be rewritten as

$$\left|\mathbf{r} - \frac{\mathbf{r}_k\cdot\mathbf{r}\cdot\mathbf{r}}{\mathbf{r}\cdot\mathbf{r}} + \frac{(v^T\mathbf{r}_k)\mathbf{v}\cdot\mathbf{r}}{v^2\cdot\mathbf{r}}\right| \tag{37.4}$$

In the two-dimensional case, **r** and **v** are orthogonal, so Equation 37.4 can be simplified as

$$\left|\mathbf{r}-(\mathbf{r}_k\mathbf{r}^T)\mathbf{r}/|\mathbf{r}|^2\right| = |\mathbf{r}|\cdot\left|1-\mathbf{r}_k\mathbf{r}^T/r^2\right| \tag{37.5}$$

Thus, Equation 37.3 becomes

$$r\left|\mathbf{r}^T{}_k\mathbf{r}/r^2-1\right| = (t_k - \tau - \mathbf{v}^T\mathbf{r}_k/v^2) \tag{37.6}$$

For convenience, we place the origin of the coordinate system onto one of the sensors, for example, sensor0 ($\mathbf{r}_0 = 0$). Subtracting $r = c(t - \tau)$ from Equation 37.6 yields

$$r\left(\left|\mathbf{r}_k^T\mathbf{r}/r^2-1\right|-1\right)+(c/v^2)\mathbf{r}_k^T\mathbf{v} = c(t_k - t_0) = d_k \tag{37.7}$$

Let $\mathbf{r}_k = (x_k, y_k)^T$ and let ϕ and M be defined by

$$\mathbf{r} = r(\cos\phi, \sin\phi)^T, \quad (c/v^2)\mathbf{v} = \mathbf{M}^{-1}(-\sin\phi, \cos\phi)^T \tag{37.8}$$

Then, the Mach number M is positive or negative depending on whether the target crosses the CPA from right to left or left to right, as observed from the origin point. If the target trajectory intersects the line $\mathbf{r}_0\mathbf{r}_k$, which also means

$$\mathbf{r}_k^T \mathbf{r}/r^2 - 1 \geq 0$$

then Equation 37.7 becomes

$$x_k \cos\phi +_k \sin\phi + (-x_k \sin\phi + y_k \cos\phi)/M - 2r = d_k \tag{37.9}$$

Otherwise, while the trajectory does not intersect line $\mathbf{r}_0\mathbf{r}_k \left(\mathbf{r}_k^T \mathbf{r}/r^2 - 1 < 0\right)$, we have

$$-x_k \cos\phi - y_k \sin\phi + (-x_k \sin\phi + y_k \cos\phi) / M = d_k \tag{37.10}$$

Now, to solve for the target motion parameters **r**, **v**, and τ, three more sensors are needed. Thus, four sensors (including the one at origin of coordinate system) should be enough to solve the problem and compute the target motion parameters. However, the method in Ref. [7] will fail if the trajectory generates an even decomposition of the sensor field, in which two sensors are on one side of the trajectory and others are on the other side, because the resulting equation of M will be of fourth degree and must be solved numerically.

To solve the localization problem, the trajectory must unevenly divide the four sensors into two groups: three on one side and one the other side. To ensure this situation occurs, we need to collect information from five sensors.

The next section will discuss this process in detail. Now suppose the origin of the coordinate system is at n0 which is the lone node; then we have the following equation:

$$x_k \cos\phi + y_k \sin\phi + (-x_k \sin\phi + y_k \cos\phi) / M - 2r = d_k \tag{37.11}$$

where $k = 1, 2, 3$. After some algebra, which is described in detail in [22], we can obtain

$$1 + \frac{1}{M^2} = \frac{(\delta_1\xi_2 - \delta_2\xi_1)^2 + (\delta_1\eta_2 - \delta_2\eta_1)^2}{(\xi_1\eta_2 - \xi_2\eta_1)^2} \tag{37.12}$$

where

$\xi_k = (x_k - x_3)$
$\eta_k = (y_k - y_3)$
$\delta_k = (d_k - d_3)$

Now since all the variables δ_1, ξ_1, η_1, δ_2, ξ_2, and η_2 are known, then two possible M can be computed. As mentioned earlier, the Mach number M is positive or negative depending on whether the target crosses the origin from right to left, or vice versa. Therefore, by checking the position of nodes with the earliest and latest measured CPA times, the target direction can be determined easily.

From Equation 37.11, we can obtain the following expression [22]:

$$\left(-\xi_k + \frac{\eta_k}{M}\right)\cos\phi + \left(-\eta_k - \frac{\xi_k}{M}\right)\sin\phi - \delta_k = 0 \quad (k = 1, 2) \tag{37.13}$$

Given the computed M and $\cos\phi = \pm\sqrt{1-\sin^2\phi}$, four possible values of ϕ will be generated by solving the previous equation. According to the original CPA algorithm, by inserting the M and each ϕ into Equation 37.11, r can be obtained. If $r < 0$ or the computed trajectory does not match the assumed sensor field decomposition, the value of ϕ will be rejected; otherwise, it is correct.

We chose the CPA-based algorithm [7] to estimate target position, velocity, and moving direction. Unlike TDOA algorithms, the CPA algorithm only requires the CPA time be stored and exchanged among sensors. Since the CPA time may be stored as two bytes (one for the integer and the other for the decimal), message exchange overhead is low. As previously described, the original CPA algorithm [7,14] cannot be directly used for WSN because of the specific sensor deployment requirements. ECPA algorithm, however, does not have such network configuration requirements and it can overcome the localization error caused by the uncertainty of position and CPA time. Therefore, to the best of our knowledge, we are the first to apply an ECPA algorithm in a practical WSN, composed of acoustic sensors, to localize and track moving targets.

37.4.3 Directed Diffusion

We also adopt the DD [10] protocol for disseminating sensor data to the sink nodes because of its energy efficiency and scalability. Direct diffusion uses a publish/subscribe communication model whereby a sink node requests data by sending interests for a named data. As the interest is flooded through the network, each intermediate node establishes a gradient with its neighbors and enables data that match the interest to be "drawn" toward the sink. Sensor nodes with data that match the interest will forward an "exploratory data" that is propagated by intermediate nodes through established gradients to the sink. The sink sends a reinforcement message to the node that first forwarded the new data to it.

37.5 ECPA Algorithm

Due to the specific sensor deployment requirement and localization error caused by the uncertainty of position and CPA time, the original CPA algorithms cannot be applied in WSN. In this section, we will describe in detail how ECPA solves those problems.

37.5.1 Estimating Trajectory Slope

Since the CPA algorithm is applicable only if there are at least three nodes on one side of the target trajectory, we deploy five sensors. Among these five nodes, we denote the two with the earliest and latest measured CPA time as n_1 and n_3. And n_2 is the node which is located at the farthest position from the line n_1n_3, as shown in Figure 37.13. Assume the slope of target trajectory is $-1/k$, then the slope of line l (which is orthogonal to the trajectory) should be k. Thus, the formula of line l is

$$l : k \cdot x - y + (y_{n_2} - k \cdot x_{n_2}) \tag{37.14}$$

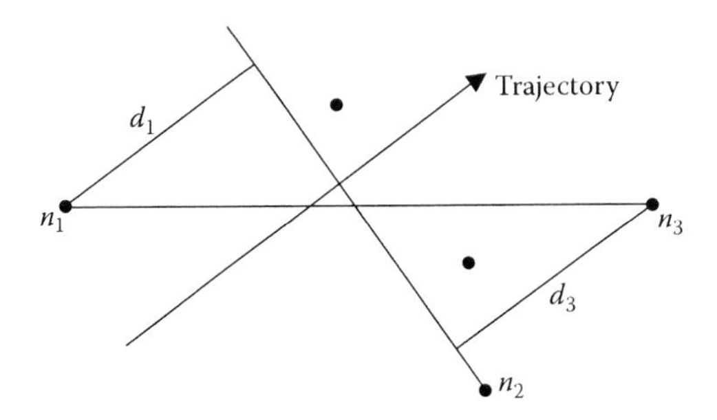

FIGURE 37.13 Estimating the slope of target trajectory.

Then the distance from n_1 and n_3 to this line l is

$$d_i = \frac{|k \cdot x_{n_i} - y_{n_i} + (y_{n_2} - k \cdot x_{n_2})|}{\sqrt{k^2 + 1}} \quad (i = 1, 3) \tag{37.15}$$

Since the target is moving at a constant speed, the ratio between d_1 and d_3 should be equal to $|(T_2 - \zeta_2) - (T_1 - \zeta_1)|/|(T_3 - \zeta_3) - (T_2 - \zeta_2)|$, where T_i is the measured CPA time at node n_i, and ζ_i is the time of sound's propagation from the CPA to node n_i. It can be further rewritten as

$$|(T_2 - T_1 + \zeta_1 - \zeta_2)|/|(T_3 - T_2 + \zeta_2 - \zeta_3)| \tag{37.16}$$

Compared to the speed of sound c, target velocity v in our system is very small, that is, $v \ll c$. Thus, we can obtain $d_1/d_3 \approx |(T_2 - T_1)|/|(T_3 - T_2)|$ because $\zeta_1 - \zeta_2 \ll T_2 - T_1$ and $\zeta_2 - \zeta_3 \ll T_3 - T_2$. By solving the earlier formula, we can get two possible results for k. Since line l must be located between line n_2n_1 and line n_2n_3, we can easily eliminate one and get the correct k. Therefore, the estimated slope of target trajectory is $-1/k$.

37.5.2 Estimating Network Configuration

We now know the slope of the trajectory (suppose it is k'). The next problem is to locate this trajectory since we need to know how it divides the hull composed of sensors. Let n_1 and n_2 be the two nodes with the smallest received signal amplitude, which also means they are at the farthest position from the trajectory. Then we have

$$d_i = \frac{|k' \cdot x_{n_i} - y_{n_i} + b|}{\sqrt{k'^2 + 1}} \quad (i = 1, 2) \tag{37.17}$$

Since the distance d_i is roughly proportional to the square root of the inversed receiving signal amplitude, we can solve Equation 37.17 and get two possible b values which give two possible locations of the trajectory. Since the two node n_1 and n_2 can be either on the same side of trajectory or on the different sides, we have two possible network configuration as shown in Figure 37.14a and b. In this figure, the two possible trajectories l (the correct one) and l' (the wrong one) are shown as solid and dashed lines, respectively. In Figure 37.14a where n_1 and n_2 are on the same side of the trajectory, if l' is considered the expected trajectory, it will contradict the assumption that both n_1 and n_2 are at the farthest position. In Figure 37.14b where n_1 and n_2 on the different sides, line l' denotes the trajectory being outside the convex hull of sensors that also contradicts the assumption that the trajectory intersects the convex hull. Eliminating the inadmissible solutions, we eventually can derive a formula for estimating target trajectory in the coordinate system. This formula may not be accurate enough to meet our target localization and tracking goals, but it is sufficient for determining the practical sensor deployment that meets the requirements of the CPA algorithm.

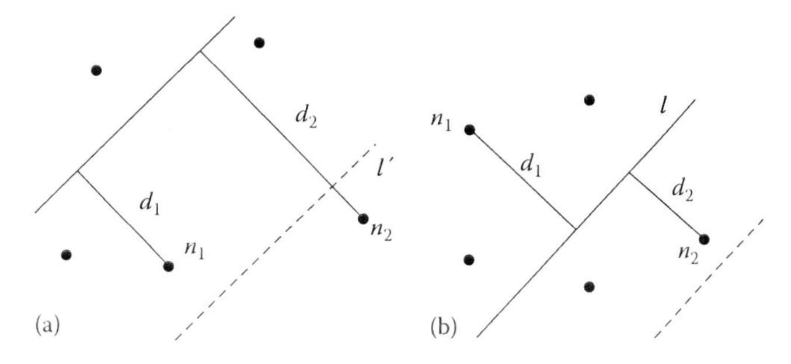

FIGURE 37.14 Estimating target trajectory location within a network. (a) Nodes n_1 and n_2 on the same side of trajectory. (b) Nodes n_1 and n_2 on different sides of trajectory.

37.5.3 Computing the Exact Trajectory

Based on the estimated trajectory, four sensors which are divided into two groups (one is on one side and the other three are on the other side) by the trajectory will be selected. Then from Equation 37.12, the Mach number M can be computed and inserted into Equation 37.13. Because $\cos\phi = \pm\sqrt{1-\sin^2\phi}$, four possible values of ϕ will be generated. However, because the uncertainty of nodes position and measured CPA time will cause r to be incorrectly computed, the original CPA algorithm will fail to compute the target location. As it will be shown in Section 37.7, the failure rate of the original CPA algorithm is very high (about 90%). Location of nodes obtained from GPS may be affected by the ionosphere error, satellite clock error, orbit error, troposphere error, and multipath error [9]. The measurement error of a C/A code receiver with either standard correlator or narrow correlator will range from 0.1 to 3 m [9]. In addition, the measured CPA time will be affected by the environment noise and the error of time synchronization between sensors.

Through mathematical analysis, we found that the error of the computed ϕ is highly dependent on the network configurations. For some node deployments, the error of the computed ϕ given by CPA will be very large; in this case, the estimated ϕ serves as a reference point and the computed ϕ that is closest to the estimated ϕ will be considered the result. Further analysis and simulations confirm that the selected ϕ value is very accurate. Then the final target trajectory location is computed based on the newly computed slope and the received signal strength at nodes, as previously stated. As this chapter focuses on the empirical evaluation, we omit the complete analysis of ECPA and refer the reader to our companion publications for detailed description.

37.6 Experimental Setup and Results

37.6.1 Experimental Setup

To evaluate the performance of our systems, we set up the target detection and tracking sensor networks at two field locations: a vacant parking lot at Auburn University, as shown in Figure 37.2, and at AU's National Center for Asphalt Technologies (NCAT) test tracks. We experimented with several wireless ad hoc sensor network configurations: (1) a basic sensor network and camera network configuration, (2) a network with three additional wireless network hops from the cluster head to the camera control node, and (3) a network supporting more directions (orientations) for the target to move.

In this chapter, we focus on the second network configuration whose logical topology is shown in Figure 37.15. To demonstrate that our system is practical, we conducted the experiments with one cluster (containing five sensors) although multi-cluster sensor networks could be deployed with more sensors and an appropriate clustering algorithm [6,16]. As shown in Figure 37.15, sensor0–4 forms the cluster in which sensor0 serves as the cluster head. Sensor5–7 are the relay nodes (in other clusters) that transmit target tracking results back to the gateway node (sensor8). All sensors locations are listed in Table 37.2, where sensors locations are measured through the universal transverse Mercator (UTM) global positioning system (GPS). The unit of latitude and longitude is meter. The target is a Chevrolet sport utility vehicle (SUV) running at the speed of 13 m/s, that is, 30 MPH. The two trajectories are those lines from (641359, 3608484) to (641366, 3608628) and (641402, 3608492) to (641308, 3608613).

Our results show that we can achieve our goals of detecting the target and predicting its location, velocity, and direction of travel with reasonable accuracy. The results from the algorithm that computes the target location and velocity are shown later. Our second set of results shows that the results of target detection algorithm can be used by a camera to take pictures or video of the target for identification purposes.

37.6.2 Hardware

Each of the sensor nodes is an x86-based laptop computer. The sensors including the cluster head are laptops with 800 MHz CPU and 82801CA/CAM AC'97 Audio Controllers. The gateway node, camera control, and system control are laptops with 1.2 GHz CPU and 512 M memory. The video capture

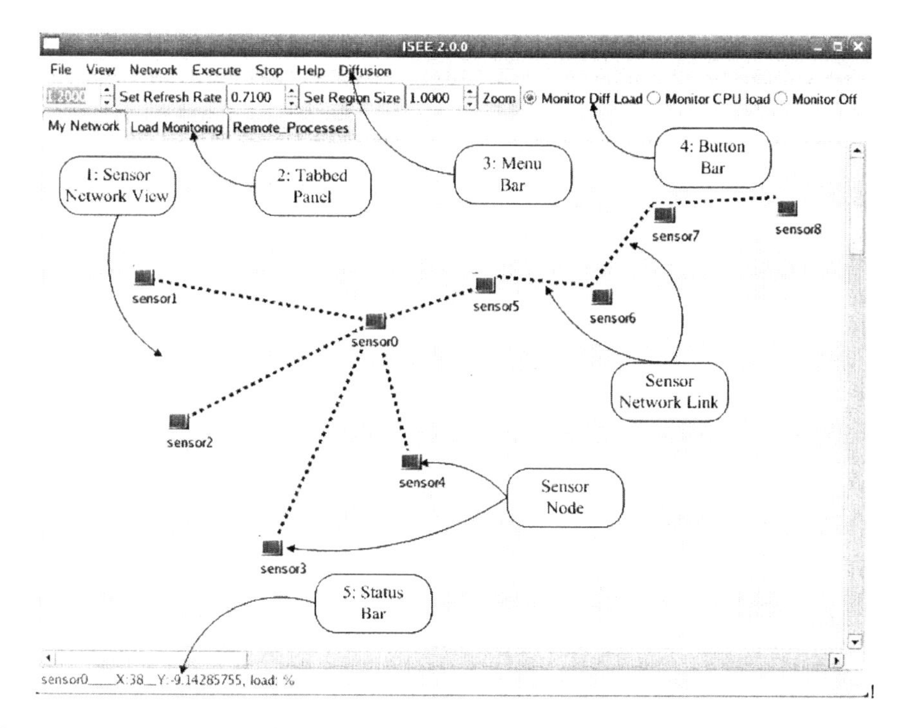

FIGURE 37.15 ISEE sensor network control GUI.

TABLE 37.2 Sensors Locations

	Latitude (UTM)	Longitude (UTM)
sensor0	641333	3608565
sensor1	641339	3608544
sensor2	641375	3608558
sensor3	641350	3608525
sensor4	641380	3608535
sensor5	641411	3608560
sensor6	641432	3608581
sensor7	641473	3608603
sensor8	641493	3608611

computer is a Pentium IV–based laptop (1.2 GHz CPU) with Windows XP OS. Each computer is also equipped with an IEEE 802.11b card for wireless connectivity. Video capture is performed using a Pinnacle 500 USB video converter which converts the camera's RCA output to digital video and interfaces with the computer using USB 2.0. The camera is a Sony EVID30, and the interface to the controller is RS232C, 9600 bps, serial port.

37.6.3 Sample Rate

Since the CPA time is only measured based on audio amplitude, it does not require sampling at high frequency. Though sampling at high frequency can improve the quality of sound, in practice, the recorded sounds in our system do not need to be of high quality as long as they are accurate enough for computing the CPA time. In the experiment, we used 4 kHz as the sampling rate since the quality of sound recorded with this frequency is clear enough for the post-experimental analysis. Informal experiments revealed that frequencies as low as 200 Hz were feasible. On the other hand, sampling rate cannot be reduced too much because it affects the accuracy of the CPA time's measurement. For example, if the sample rate is 200 Hz, that is, sampling every 5 μs, then the average accuracy of the CPA time will be around 1.25 μs.

37.6.4 Target Detection

A typical recorded sound (5 s long) of moving vehicle with sample rate 4 kHz is shown in Figure 37.16. The CPA time can be obtained as the instance when the highest amplitude occurs which is indicated by a vertical line in Figure 37.16. Obviously, longer recording times will cause higher detection delays, so the recording time should be as short as possible. To reduce the impact of environmental noise, we designed a simple yet effective algorithm to detect the measured CPA time. Through the digital signal processing (dsp) interface, we first recorded a period of data (e.g., 1 s) from the sound card, and then filtered out the spurious spike samples in it. As shown in Figure 37.16, the amplitude of recorded sound will keep increasing until it reaches the peak (shown as a vertical line), and then the amplitude decreases. Therefore, spurious spikes caused by noise should be ignored. After filtering, the sample will be divided into a number of time slices, such as 8 slices per second (0.125 s per slice). A target will be positively detected if and only if the following rules are satisfied: (1) the average amplitude level of the sound is larger than a predetermined threshold (e.g., −20 db) and (2) the zero crossing rate of the recorded data is above a certain level. This rule eliminates the sound of the wind, since the wind has a low frequency compared to the sound of moving vehicles.

37.6.5 Time Synchronization

Time synchronization among sensor nodes was achieved by Network Time Protocol (NTP), which synchronizes the clocks of sensors over packet-switched, variable-latency wireless networks. NTP typically provides accuracies of less than a few milliseconds on wireless networks [15], which are accurate enough for ECPA to successfully localize the target. Time synchronization of sensors is required because of the nature of the CPA localization algorithm, but this synchronization need only be done locally among the sensors. Therefore, the correctness of absolute time is not necessary in our system. In the experiment, we set the cluster head as the local time server and all sensors fetch the reference time from it. We noticed that after synchronization, the sensors did not need to be adjusted for 3–4 h in the experiment; therefore, the time synchronization process can be done at a relatively low frequency.

37.6.6 ISEE Sensor Network Control

As sensor networks become larger and more complex with sophisticated collaborative processing, developing these sensor networks requires a simple execution and monitoring environment

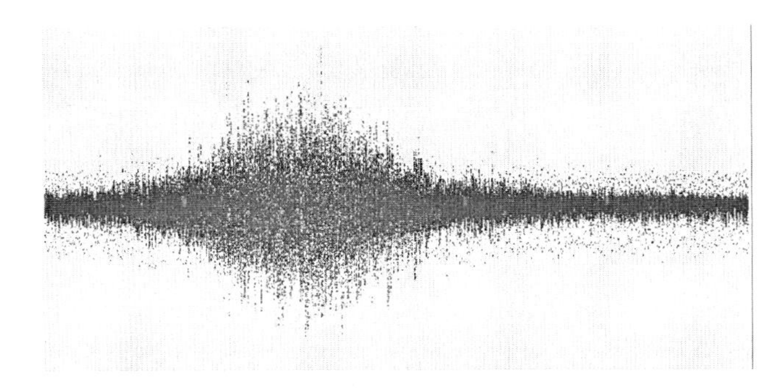

FIGURE 37.16 Audio recording of the moving vehicle.

for repeatable experimentation that allows for easy transfer to in situ real-world environments. We have developed ISEE [11] that allows for control and access to simulated, emulated, and real sensor networks. The control and access environment provides remote execution, logging, interaction, and analysis facilities independent of the implementation of the sensor network. ISEE uses a simple graphical interface for controlling and accessing sensor networks as shown in Figure 37.15. This framework allows for extensibility, scenario creation, and experiment repeatability. This provides visibility and repeatability to sensor network experimentation that may not be otherwise available. The combination of our framework with a distributed service framework allows for reactive and language-independent user and developer interaction to a sensor network. Such an environment is necessary as simulation has a tendency to warp reality when simulating sensor networks that are very sensitive to the environment and resource constraints. At the most, it requires fully immersive runtime and development environment. Large-scale sensor networks allow for the visualization of real-world environments. Such a runtime environment will transparently support any testing environment—simulated, emulated, or real.

37.6.7 Results of Computing Target Location and Velocity

ECPA computes the target location, velocity, and direction of travel based on the CPA time reported from the five sensors. Table 37.3 summarizes the results of the eight runs that we conducted with two target trajectories. The results show that the target location and velocity can be accurately computed. The average position error is only 1.78 m; the average velocity error is only 1.125 m/s; and the average trajectory error is 5.375°. For example, in the first run, the predicted target location is (3608558, 641360), only 1 m from the actual location (3608558, 641359). The computed speed is 13 m/s whereas the actual speed is 13 m/s (30 MPH), and the direction of travel is 90° compared to the actual direction of 87°.

Figure 37.17 shows the predicted position and actual path of the vehicle for the eight runs in our experiments. Note that the computed target speed and trajectory almost exactly match the actual values. The shorter dashed lines show the results from two runs in the AC direction and two runs in the CA directions (1st–4th runs in Table 37.3). These trajectories are very close to the actual one shown by the solid AC path. The longer dash lines show the results of two runs in the BD direction and two runs in the DB direction (5th–8th runs in Table 37.3). These trajectories (with one exception) are very close to the actual trajectory.

TABLE 37.3 Tests Results

	Target Position (m)		Speed (m/s)		Angle (°)	
Run	Computed	Error	Computed	Error	Computed	Error
1st	(3608558, 641360)	1	13	0	90	3
2nd	(3608558, 641360)	1	12	1	90	3
3rd	(3608559, 641357)	2.236	19	6	84	3
4th	(3608552, 641359)	6	12	1	111	24
5th	(3608545, 641358)	1	13	0	127	1
6th	(3608544, 641359)	1	13	0	130	2
7th	(3608544, 641359)	1	13	0	131	3
8th	(3608545, 641360)	1	14	1	132	4
Min		1		0		1
Max		6		6		24
Avg		1.78		1.125		5.375

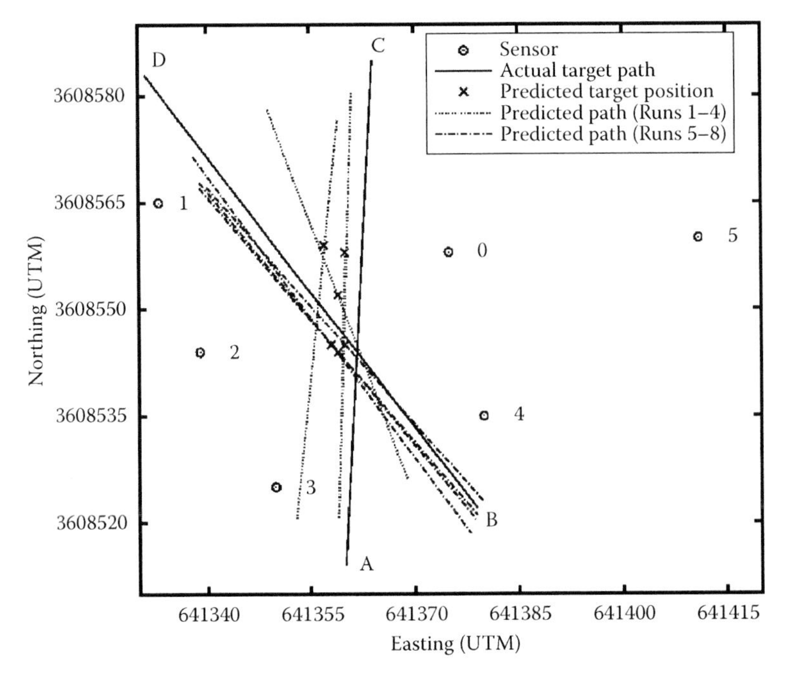

FIGURE 37.17 Plot of target tracking results showing computed target location and velocity match the actual value.

Compared to the results published in [5], ECPA gives more accurate results. For example, the average localization error of ECPA is 1.78 m, while the root mean square errors of Extended Kalman Filter (EKF), Lateral Inhibition (LAT), and EKF and LAT are 8.877, 9.362, and 11.306 m, respectively.

37.6.8 Video Capture of the Target

The camera first points in an initial position (in our case, west). The target then moves through the sensor field and quickly passes through the camera's field of view. The networked sensors then detect the target and calculate the target location and velocity. The predicted position of target is calculated based on the velocity and transmission delay from sensors to the camera. The camera then pans and zooms toward the predicted position to capture the target on video. The images from the video capture can be used to identify the target more accurately. Figure 37.18 shows the videos clips of this sequence of events where Figure 37.18a shows the target first appearing in the sensor network field and Figure 37.18c shows target at the predicted position.

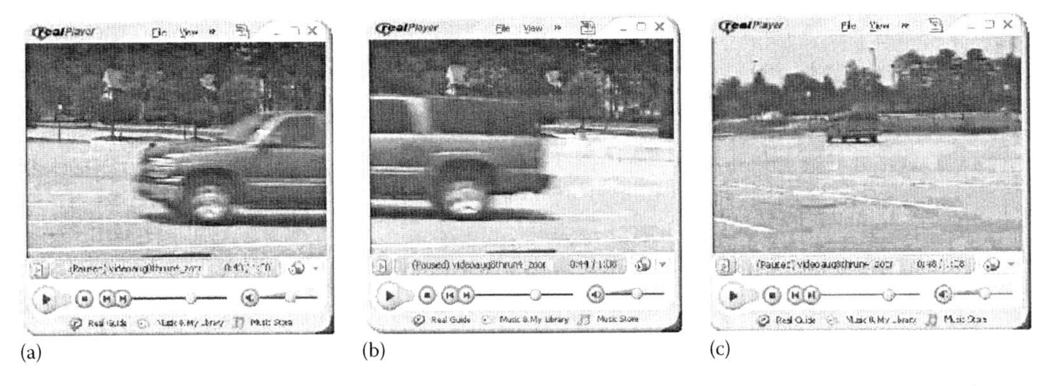

FIGURE 37.18 Video clips of the moving target. (a) Target entering the sensor network field. (b) Target leaving the sensor network field. (c) Target tracked at predicted position.

37.7 Simulation and Results

We make use of the extensions made to ns-2:27 by the Naval research laboratory [1], which provides the simulation with various physical phenomena such as acoustic, seismic, and chemical agents. The presence of physical phenomena in ns-2 is modeled with broadcast packets which are sent over a designated channel called the "phenom" channel. In the real world, detecting acoustic events is made more difficult by the environmental noise and sensitivity issues of the microphones. We assume that the acoustic packets experience a loss profile similar to 802.11 data packets, so the noise and packet losses are simulated by the extensions provided in [21]. In the simulation, we use the same node deployment as was used in the field experiment. As expected, the simulation results with target velocities of 13 m/s match the field test very well.

37.7.1 Successful Localization Rate

Due to the error of measured CPA time, the original CPA algorithm always failed while selecting the proper value. In most cases, the CPA algorithm can find $r > 0$, but after inserting this r into the CPA algorithm, the resulting trajectory is outside the convex hull composed of sensors. Thus, the localization process failed. As shown in Figure 37.19, the successful detection rate of CPA is at most 10%. Furthermore, when considering the error of sensors location, the situation becomes even worse. However, ECPA first chooses the ϕ which is closest to the estimated one, and then computes the trajectory location within the convex hull of sensors, thereby achieving a localization success rate of 100%.

37.7.2 Impact of Velocity

As stated in [7], the relative error on the target distance estimate is roughly proportional to the square root of the inverse Mach number. This means the localization error of ECPA should also decrease as target velocity increases.

This property is demonstrated in Figure 37.20, where the average velocity error decreased from 1.25 to 0.2 m/s, average direction error decreased from 5.5° to 0.5°, and the average location error decreased from 1.36 to 0.7 m. Note that when the target speed is 13 m/s (30 MPH), the error of estimated location, velocity, and direction will be 0.86 m, 0.78 m/s, and 3.1°, respectively. Indeed, these values are very similar to the errors experienced in the field (Table 37.3).

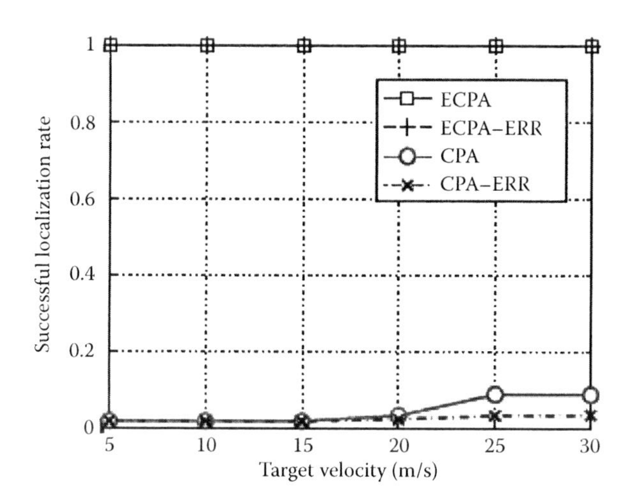

FIGURE 37.19 Successful localization rate.

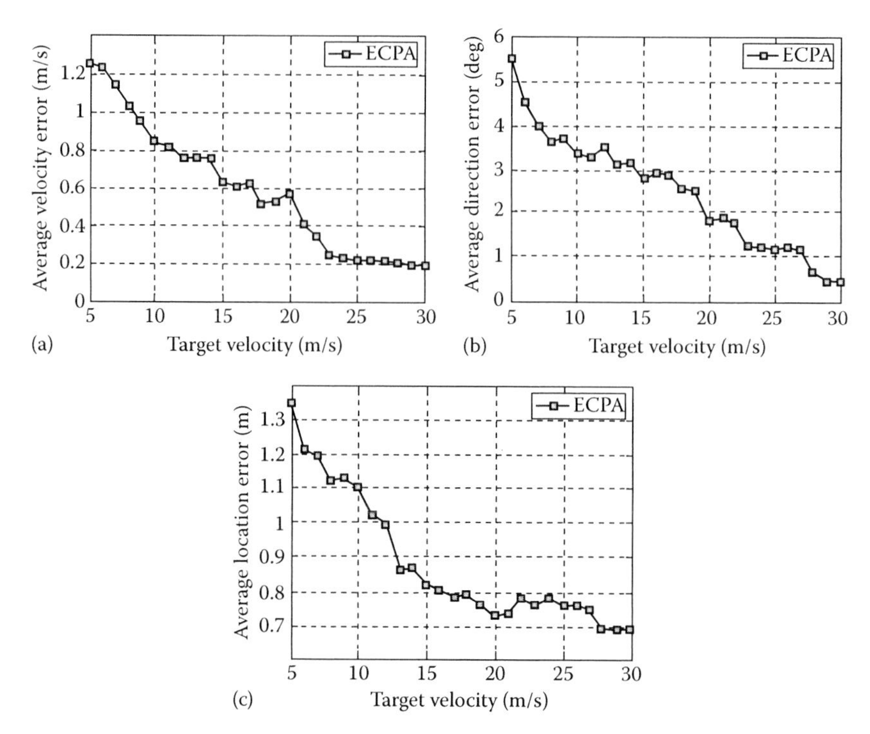

FIGURE 37.20 Impact of different target velocity. (a) Average velocity error versus target velocity. (b) Average direction error versus target velocity. (c) Average location error versus target velocity.

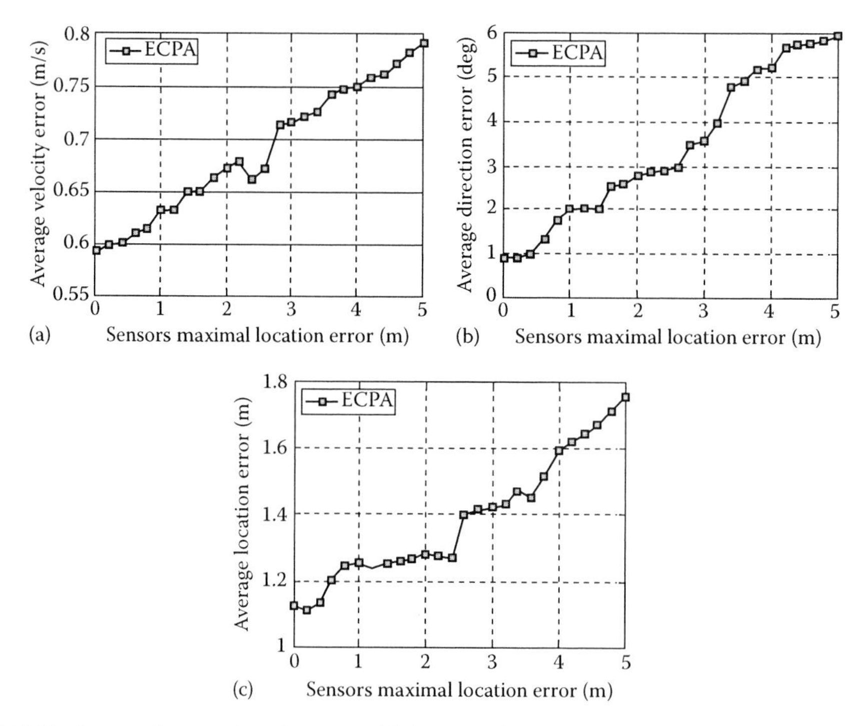

FIGURE 37.21 Impact of sensors location errors. (a) Average velocity error versus sensor maximal location error. (b) Average direction error versus sensor maximal location error. (c) Average location error versus sensor maximal location error.

37.7.3 Impact of Sensors Location Errors

Errors in sensor location range from less than 1 m to a few meters, depending on the localization technology. To understand the effect of this error, we simulated sensors location error as a uniform distributed function upon (0, Max], where Max increases from 0.1 to 5 m. As shown in Figure 37.21, when the location error of sensors increased, the target localization errors will also increase. Interestingly, ECPA is highly tolerant of the sensors location error because the estimated target information is quite close to the actual values even with a larger sensor location error. For example, the average error of target location increased only 0.7 m as sensors location error increased from 0 to 5 m. Moreover, the errors of estimated velocity and direction are still within the acceptable range.

37.8 Conclusion

For rapid deployment in the field, we have designed new compact sensor fusion nodes which are based on COTS components (such as PC104), that are small and inexpensive and can execute collaborative algorithms for reliably detecting and tracking targets. In addition, we will extend the network with multiple clusters to enable us to track targets moving along nonlinear paths. Finally, we will experiment with arbitrary deployment of sensors in random configurations and develop methods for automatically forming clusters in these topologies.

Our prototype target tracking system demonstrates the feasibility of detection, tracking, and image/video capture of moving targets using collaborative, wireless, acoustic sensor nodes. By using the ECPA localization algorithm, target position, velocity, and direction can be accurately computed and transmitted to the video camera. In addition, we successfully integrated the DD, 802.11, and TCP/IP network protocol together to facilitate the requirements for both ad hoc acoustic sensor networks and the control center for video capture.

The original CPA algorithm fails after calculating the velocity since the uncertainty of sensors' position makes it infeasible to obtain a positive value of r (the distance from a target to the reference nodes). Our ECPA scheme first estimates the trajectory and then obtains the sensors' deployment information to improve the estimate. Through the field experiments, ECPA has been shown to be an efficient and practical localization and tracking method for moving vehicles within WSN.

Acknowledgment

This research was supported in part by the U.S. Army Night Vision Electronic Sensors Directorate (NVESD) under Prime Contract Number DAAB07-03-D-C213-005, Subcontract No. SUB1170933RB.

References

1. Naval Research Laboratory, NRL's Sensor Network Extension to NS-2, 2012, http://cs.itd.nrl.navy.mil/work/sensorsim/index.php.
2. A. M. Ali, K. Yao, T. C. Collier, C. E. Taylor, D. T. Blumstein, and L. Girod. An empirical study of collaborative acoustic source localization. In *IPSN'07: Proceedings of the 6th International Conference on Information Processing in Sensor Networks*, Cambridge, MA, pp. 41–50, New York: ACM, 2007.
3. P. W. Boettcher and G. A. Shaw. Energy-constrained collaborative processing for target detection, tracking, and geolocation. In *IPSN'03: Proceedings of the 6th International Conference on Information Processing in Sensor Networks*, Palo Alto, CA, Volume 2634 of Lecture Notes in Computer Science, pp. 254–268, Berlin/Heidelberg, Germany: Springer, April 2003.
4. P. W. Boettcher, J. A. Sherman, and G. A. Shaw. Target localization using acoustic time difference of arrival in distributed sensor networks. *Battlespace Digitization and Network-Centric Warfare II*, 4741(1):180–191, 2002.

5. R. R. Brooks, P. Ramanathan, and A. M. Sayeed. Distributed target classification and tracking in sensor networks. *Proceedings of the IEEE*, 91(8):1163–1171, 2003.
6. W.-P. Chen, J. C. Hou, and L. Sha. Dynamic clustering for acoustic target tracking in wireless sensor networks. *IEEE Transactions on Mobile Computing*, 3(3):258–271, 2004.
7. F. M. Dommermuth. The estimation of target motion parameters from CPA time measurements in a field of acoustic sensors. *The Journal of the Acoustical Society of America*, 83(4):1476–1480, 1988.
8. A. O. Ercan, A. El Gamal, and L. J. Guibas. Object tracking in the presence of occlusions via a camera network. In *IPSN'07: Proceedings of the 6th International Conference on Information Processing in Sensor Networks*, Cambridge, MA, pp. 509–518, New York: ACM, 2007.
9. J. Rankin. An error model for sensor simulation GPS and differential GPS. In *Position Location and Navigation Symposium*, pp. 260–266, Cloud State University, MN: IEEE, April 1994.
10. C. Intanagonwiwat, R. Govindan, and D. Estrin. Directed diffusion: A scalable and robust communication paradigm for sensor networks. In *MobiCom'00: Proceedings of the 6th Annual International Conference on Mobile Computing and Networking*, Boston, MA, pp. 56–67, New York: ACM, 2000.
11. M. Ivester and A. Lim. Interactive and extensible framework for execution and monitoring of wireless sensor networks. In *Comsware 2006: First International Conference on Communication System Software and Middleware*, New Delhi, India, pp. 1–10, Piscataway, NJ: IEEE, 2006.
12. W. Kim, K. Mechitov, J.-Y. Choi, and S. Ham. On target tracking with binary proximity sensors. In *IPSN'05: Proceedings of the 4th International Symposium on Information Processing in Sensor Networks*, Los Angeles, CA, p. 40, Piscataway, NJ: IEEE Press, 2005.
13. D. Li, K. D. Wong, Y. H. Hu, and A. M. Sayeed. Detection, classification and tracking of targets in distributed sensor networks. *Signal Processing Magazine, IEEE*, 19(2):17–30, 2002.
14. E. E. Milios and S. H. Nawab. Acoustic tracking from closest point of approach time, amplitude, and frequency at spatially distributed sensors. *The Journal of the Acoustical Society of America*, 87(3):1026–1034, 1990.
15. D. L. Mills. Internet time synchronization: The Network Time Protocol. *IEEE Transactions on Communications*, 39(10):1482–1493, 1991.
16. R. Neelisetti, A. Lim, P. Agrawal, and Q. Yang. A robust clustering algorithm for target tracking in wireless acoustic sensor networks. In *IRADSN 2007: Innovations and Real-Time Applications of Distributed Sensor Networks (DSN) Symposium*, Shreveport, LA, pp. 17–25, 2007.
17. X. Sheng and Y. H. Hu. Energy based acoustic source localization. In *IPSN'03: Proceedings of the 6th International Conference on Information Processing in Sensor Networks*, Palo Alto, CA, Volume 2634 of Lecture Notes in Computer Science, pp. 285–300, Berlin/Heidelberg, Germany: Springer, April 2003.
18. N. Shrivastava, R. Mudumbai U. Madhow, and S. Suri. Target tracking with binary proximity sensors: Fundamental limits, minimal descriptions, and algorithms. In *SenSys'06: Proceedings of the 4th International Conference on Embedded Networked Sensor Systems*, New York, pp. 251–264, New York: ACM, 2006.
19. F. Silva, J. Heidemann, and R. Govindan. *Network Routing Application Programmers Interface (API) and Walk Through 9.0.1*. USC/Information Sciences Institute, Marina del Rev, CA, 2002.
20. J. Singh, U. Madhow, R. Kumar, S. Suri, and R. Cagley. Tracking multiple targets using binary proximity sensors. In *IPSN'07: Proceedings of the 6th International Conference on Information Processing in Sensor Networks*, Cambridge, MA, pp. 529–538, New York: ACM, 2007.
21. W. Xiuchao and A. L. Ananda. Link characteristics estimation for IEEE 802.11 DCF based WLAN. In *LCN'04: Proceedings of the 29th Annual IEEE International Conference on Local Computer Networks*, Tampa, FL, pp. 302–309, Washington, DC: IEEE Computer Society, 2004.
22. Q. Yang, A. Lim, K. Casey, and R. K. Neelisetti. Real-time target tracking with CPA algorithm in wireless sensor networks. In *SECON'08: 4th Annual IEEE Communications Society Conference on Sensor, Mesh and Ad Hoc Communications and Networks*, San Francisco, CA, pp. 305–313, 2008.
23. F. Zhao, J. Shin, and J. Reich. Information-driven dynamic sensor collaboration. *Signal Processing Magazine, IEEE*, 19(2):61–72, 2002.

38

Lightweight Key Management in Multi-Hop Wireless Sensor Networks

Sushmita Ruj
University of Ottawa

Amiya Nayak
University of Ottawa

Ivan Stojmenovic
University of Ottawa

38.1 Introduction

Sensor nodes are battery-powered resource-constrained devices, which sense temperature, pressure, mobility, seismic and acoustic waves, and other phenomena. They are deployed in wildlife monitoring, rescue operations, mining, healthcare, surveillance, tracking, military, and other applications.

Sensors communicate their measurements to one or more base stations (BS) which are scattered in the network. BS are powerful and more robust devices, compared to sensor nodes. Depending upon the organization of sensor nodes and BS, networks can be classified in several ways:

1. *Homogeneous and heterogeneous*: Networks in which all nodes are alike (homogeneous) or different (heterogeneous) in battery power and memory capacity.
2. *Distributed and hierarchical*: Networks in which there is no central server to coordinate the activities of the sensors (distributed) and networks with different levels in organization (hierarchical). In hierarchical networks, nodes are divided into clusters, with cluster heads (CH) coordinating the activities of the sensors and reporting to the BS.
3. *Randomly and deterministically deployed*: Nodes may be deployed randomly in the target region (random deployment, e.g., dropping from airships) or as networks in which the position of nodes is known with certain accuracy (deterministic deployment, e.g., placing at regular grid points).

4. *Static and mobile*: Nodes can be either static or mobile.
5. *Single and multi-hop networks*: In single-hop networks, all nodes are within communication range of each other. Key pre-distribution (KPD) in single-hop networks has been studied recently in details by Ruj et al. [53]. In a multi-hop network, some pairs are not within communication range and they communicate via intermediary nodes.

Whenever sensitive information needs to be sent, security issues need to be resolved. In this chapter, we discuss lightweight security issues in multi-hop networks. Since sensors are deployed in large numbers and operate unattended, the manufacturing cost has to be minimized. A MICA2 MPR400CB sensor has only 128 kB of programmable memory and 8 bit ATmega128L CPU [1]. This is powered by two AA batteries and has a data rate of 38.4 Kbaud in 500 ft. The constrained memory requirements make it difficult to carry out complex operations and provide security in presence of adversaries.

38.1.1 Attack Models

Sensor networks are prone to several types of threats, which can be classified as follows:

1. *Passive and active adversaries*: Passive adversaries eavesdrop on the network and learn the information being transmitted. They might later replay such messages (replay attacks) and cause inconsistencies in the network. Active adversaries may also alter messages, drop packets or selectively forward packets, modify routing information, inject false messages, replicate nodes (Sybil attack), exhaust resources, and cause denial of service (DoS) attacks.
2. *Internal and external adversaries*: Internal adversaries are nodes that belong to the network. These nodes are compromised (node compromise attack) and have malicious intent. External adversaries launch attacks from outside the network.

38.1.2 Security Requirements

To protect against attacks, the following security requirements have to be provided:

1. *Availability*: The services must be available as and when required. This is to ensure that the network is protected from DoS attacks [64], and receives all the packets as they are sent.
2. *Authenticity (entity and message authenticity)*: Entity authenticity ensures that the node which sends a message is a honest node, and sends correct information. Message authenticity ensures that the message is authentic, not tampered with by the sender or any intermediary node. Message authenticity is also referred to as *integrity* by several researchers [65]. Though the two definitions look the same, there are subtle differences. First, entity authenticity does not guarantee message authenticity, because the message might be tampered with, after transmission, either by intermediate nodes or by external adversary. Second, in certain networks (which are driven by rational choices), there is no strict classification of entity authenticity, because nodes can be selfish and send either good or malicious messages. In such cases, ensuring message authenticity is more important.
3. *Confidentiality*: The message sent by a node can be received only by the receiving node. No unauthorized node should be able to access the data. Confidentiality is ensured using cryptographic techniques.
4. *Scalability*: The network should be able to work properly when the size of the network increases.
5. *Tracing and nonrepudiation*: It might be possible to detect adversaries (tracing). Once adversaries are detected, they should not be able to deny their action (nonrepudiation).

38.1.3 Organization

We discuss key management issues in Section 38.2. In Section 38.3, we present several KPD schemes which have been studied in literature. We first discuss commonly used KPD techniques and then show how they are used in cluster-based multi-hop networks. Authentication issues are discussed in Section 38.5. A recent key distribution technique called triple key distribution is presented in Section 38.4. Secure routing and data aggregation are discussed in Sections 38.6 and 38.7, respectively.

38.2 Key Management

Key is a randomly generated large number (128 bits or more), which when fed into an encryption algorithm, along with the plaintext, generates a ciphertext. The ciphertext is sent across an insecure channel, so that even if an adversary intercepts the ciphertext, it is unable to decode the plaintext message.

There are several types of keys:

1. Unicast or pairwise key is a key shared between a pair of nodes. Ideally this key should be unique, so that a third node is unable to decrypt the message. However, since establishing pairwise key with each and every node in the network is space demanding, we look for ways which are computationally efficient and require less memory.
2. Multicast key is used in broadcast or multicast scenarios, where a message is to be sent to several nodes. Group key management has been extensively studied. We do not discuss them here. Surveys can be found in Ref. [49].

Session key establishment can be done using a key distribution center (KDC). A KDC is a resource-rich center and acts as a trusted arbiter for key establishment. Transport Layer Scheme (TSL) [15], SPINS [47], and Kerberos [60] are some examples of such schemes. Kerberos is an authentication protocol which is based on Needham and Schroeder's protocol [42]. The trusted server transmits session keys to nodes. They are calculated upon request from the long-term keys that nodes already share with the trusted party. This method is quite expensive for sensor networks.

Public key cryptography is widely used in several applications. Each user i has a pair of keys, public key, PK_i (which is known by other users in the network), and private key, SK_i, which is kept secret. If a user j wants to send a message to user i, then the message is encrypted using the public key of i, as $C = E_{PKi}(M)$. Only user i will be able to decrypt the message using its secret key SK_i. The encryption function $E_{PKi}(.)$ and decryption function $D_{SKi}(.)$ are such that $D_{SKi}(E_{SKi}(M)) = M$. The security of public key techniques relies on the hardness of problems like factoring (as in RSA [51]), or discrete logarithm problem (as in El Gamal cryptosystem [23]). The implementation of these algorithms is very expensive for sensor nodes. Moreover, there are instances of such schemes [24,26,38,39] using Elliptic curves, RSA, or PKI, which also require intervention of BS, which might not be always possible.

The main idea in symmetric cryptography is that communicating parties share some common keys. Transmitted message is encrypted with the common key to get the ciphertext. It can be decrypted using the same key to get back the message. If M is the message to be transmitted and K is the common key, then the ciphertext is obtained as $C = E_K(M)$. $E_K(.)$ is the encryption function. The other node can use K and decrypt C as $M = D_K(C)$. $D_K(.)$ is the decryption function. A simple example of an encryption (and decryption) function is XOR function, $C = M \oplus K$ and $M = C \oplus K$.

Symmetric key management in WSN involves distributing keys in WSN, before deploying them. It proceeds in three steps:

1. *KPD*: In this stage, keys are loaded into the sensor node. The efficiency of the next two stages depends upon this one. This stage is carried out before deploying the sensor nodes. The KPD algorithm is decided by the distribution center, taking into consideration the memory, battery power, application, and deployment mechanism. Key predistribution center (KDC) preloads the following information into each sensor:
   - Node identifier
   - Cryptographic keys, which are 128 bits
   - Identifiers of each key
   - Shared-key discovery algorithm (optionally it also may contain node-specific additional information) which decides a common key (if any) that a node shares with another node

Since the KDC is powerful and resourceful, we do not consider the cost of the algorithm. Surveys appear in Refs. [2,10,52,65].

2. *Shared-key discovery*: After nodes are deployed, two nodes wishing to communicate establish a common key. The shared-key discovery algorithm is embedded in the sensor and depends on the type of KPD scheme. In certain schemes, this algorithm takes as input only the identifier of the other node and in some it also needs the identifier of all keys. We will discuss the shared-key discovery algorithms in the next section.
3. *Path-key establishment*: If two nodes do not share a common key or they are not within communication range, then either there should be a path key (a common key between the two nodes, established through intermediary nodes) or there should be a secure routing mechanism to transmit packets from one node to the other. We will discuss secure routing protocols in Section 38.6. Suppose a path key needs to be established between source $S = n_0$ and destination $D = n_l$ through the intermediate nodes $n_1, n_2, \ldots, n_{l-1}$. Initially the source node encrypts the message M using the pairwise key it shares with node n_1 and transmits to n_1. n_1 decrypts it and re-encrypts it with the pairwise key between itself and n_2. This proceeds until the message reaches destination D. The problem with this approach is that all the intermediate nodes learn M. One way to overcome this is to use multiple paths [11]. The source node breaks down the message into parts $M_1, M_2, \ldots, M_p$, such that $M = \sum_{i=1}^{p} M_i$ Each message M_i is sent to the destination using a different path. An adversary needs to compromise at least one node in each path, to learn all parts M_i ($i = 1, 2, \ldots, p$), and reconstruct M. However, sending the message in p paths requires finding p hopefully disjoint paths, and communication overhead for using them. We will discuss other techniques in Sections 38.6 and 38.7.

While evaluating the performance of key management schemes, we mainly consider the following metrics [55]:

1. *Memory*: Memory requirement is generally measured by the amount of keying material stored in the node. This is generally the number of keys (of size 128 bits) that are stored.
2. *Computation*: Computation is measured by the number of CPU cycles needed to establish a common key between the nodes.
3. *Connectivity*: Connectivity is measured by the probability that two nodes can communicate with each other. One can consider two types of connectivity:
   a. *s-connected*: Strictly connected, in which communicating nodes are within their communication range and share common key.
   b. *p-connectivity*: Path-connected, where a path exists with two nodes at the beginning and end of it, and each link on the path consists of two intermediary *s*-connected nodes.
4. *Resilience to node compromise* [53]: The measure of the tolerance of the sensor network to node compromise. There are different measures of resilience:
   a. Number of keys compromised [11,17,34].
   b. Probability that a given link is compromised [33,56].
   c. Probability that a (uncompromised) node is disconnected from the rest of the network [56–58]. This means that the node is not connected to any other node in the network.
   d. Collusion resistance, minimum number of nodes to be compromised to compromise the entire network. Network is t-secure, if at least $(t+1)$ nodes are to be compromised, to compromise the whole WSN. This measure of resiliency has been used in Refs. [5,6,16,37].
5. *Scalability*: Scope of adding more sensor nodes in the network, without redistributing the keys in the existing nodes.

38.2.1 Shared-Key Discovery Algorithm

Details of shared-key algorithms can be found in Ref. [53]. Shared-key discovery can be done in many ways. The algorithm to be used depends on the KPD technique. We give three commonly used techniques as follows.

Suppose node n_i wants to find the keys that it shares with node n_j. Each of the algorithms runs on n_i and has input from n_j and outputs the common keys.

1. *Broadcasting key identifiers*: Each node transmits its k key identifiers. On receiving n_j's key identifiers, node n_i compares with its own set and finds out if there is a common key identifier. The corresponding key is used in communication. Keys can be sorted in $O(k \log k)$ time. Comparing the lists of k identifiers then takes additional $O(k)$ time. Each key identifier can be represented by $O(\log \upsilon)$ bits (since there are a total of υ keys). Thus, the communication overhead is $O(k \log \upsilon)$ bits per sensor.
2. An adversary who does not know the mapping of the identifiers to the keys cannot find the common key between two nodes, even though it can find the common key identifier. However, an adversary can launch a selective node capture attack. It compromises nodes which have the largest number of new keys. In this way, by compromising fewer nodes, it can compromise a large number of keys and disconnect several links.
3. *Using challenge-response protocol* [11,21]: To avoid transmitting the identifiers, node n_j transmits a list $\{\alpha, Ek, (\alpha), l = 1, 2, \ldots, k\}$, where α is a string of a chosen length (called the challenge). Algorithm proceeds as follows: n_i encrypts α with each of its keys and compares with $EK_l(\alpha)$, to check, if it has a common key.
4. *Using only the identifier of the sensor*: This class of algorithms is mainly used by deterministic schemes [33,56]. Node n_i takes as input the identifier of node n_j and returns one or more common keys. The additional information in the node enables this key derivation and prevents other nodes from doing so. There is no need to transmit the list of key identifiers. The shared-key discovery algorithm is embedded in the sensor and is specific to the KPD scheme, contrary to the previous two methods, which can be used for any predistribution schemes. However, the communication overhead is only $O(\log N)$ bits per node. The time complexity depends on the KPD algorithm. If a node is compromised, the adversary can only find the common keys between the node it has compromised and any other transmitting node. We will give an example in Section 38.3.2.

The problem with the first two approaches is that all the key identifiers are known and this fact can be exploited by an adversary to mount a selective node capture attack. Once an adversary compromises some nodes, it can learn the identifiers of other uncompromised nodes. In the last approach only the common key (if it exists) can be calculated by two nodes, without revealing information about other keys in the nodes. Thus, not only does it have lower communication overhead, but also it has more resiliency to selective node compromise.

38.3 KPD Schemes

In this section, we discuss KPD schemes for WSN. We first discuss some basic schemes and then show how they are used in multi-hop networks.

A simple way to distribute keys is to load each sensor with a master key. This is space optimal, but if a single node is compromised, then all the links in the network are compromised. At other extreme, each node can be loaded with unique pairwise key with each of the other nodes. This requires large $O(N)$ storage space, where N is the number of sensors in the network, but results in high resilience (none of the links between uncompromised nodes is affected even if a large number of nodes are compromised).

38.3.1 Key-Pool-Based Schemes

Eschenaur and Gligor [21] designed a probabilistic KPD scheme (the *basic scheme*), where keys are randomly chosen from a large key pool and placed in the sensors. Two sensors share keys with certain probability. Suppose the total number of nodes in the network is N, the size of the key pool is υ, and the number of keys in each node (called key chain) is k. Then two nodes do not share a common key when they have disjoint set of key chains. This occurs with probability

$$\frac{(\upsilon-k)}{\binom{\upsilon}{k}}=\frac{((\upsilon-k)!)^2}{(\upsilon-2k)!\upsilon!}$$

Therefore, the probability that two nodes share a common key is

$$1-\frac{((\upsilon-k)!)^2}{(\upsilon-2k)!\upsilon!}$$

However, here we assume single-hop network. Though this scheme is simple to implement, the shared-key discovery can be done by either transmitting all the key identifiers or using a challenge-response protocol, which have significant communication overheads.

Chan et al. [11] proposed several variations of the basic scheme. According to Q-composite scheme, two nodes are able to communicate, provided they share at least q keys. Key chains are chosen and preloaded in sensors as in Ref. [21]. Common keys are found in the same way. Let $K_1, K_2, \ldots, K_{q'}$ be the common keys (where $q' \geq q$). Then the link key is calculated as hash $(K_1 \,||\, K_2 \,||\ldots||\, K_{q'})$, where hash is a hash function, generally SHA–1 [61].

The choice of q determines the resilience and connectivity of the network. If q is small, the initial connectivity is higher. However, if nodes are compromised during the setup stage, the resilience decreases drastically. If q is large, the initial connectivity is low, but resilience is higher. This scheme provides better connectivity than the basic scheme.

The other scheme, proposed by Chan et al. [11], is a random pairwise scheme, where each node stores less than $(N–1)$ keys. For each node, k other node identifiers are chosen and a pairwise key is assigned for each of these nodes. This works if nodes are static, and k neighbors are deployed nearby to connect directly. However, if nodes are mobile, then on changing the position, the k nodes might not be in close proximity to communicate via the pairwise keys.

38.3.2 Polynomial-Based Schemes

Polynomial-based schemes are based on Blundo et al.'s scheme [6]. The scheme was first used for dynamic conferencing, and later in other applications. Blundo et al.'s scheme [6] for group key management works as follows: Let g be the size of the group. Let $GF(q)$ be a field. A symmetric c-degree polynomial in g variables $(P(x_1, x_2, \ldots, x_g))$ is constructed with coefficients in $GF(q)$. Each member i in the group is given a share, $P(x_1, x_2, \ldots, x_{i-1}, i, x_{i+1}, \ldots, x_g)$ of this polynomial. g members $n_1, n_2, \ldots, n_g$ can find a common key $P(n_1, n_2, \ldots, n_g)$, if they know all the g identifiers. For a pairwise scheme, group size $g = 2$. Thus, symmetric bivariate polynomial $P(x, y)$ of degree c is chosen with coefficients in the field $GF(q)$. q may be $2^8 + 1$ or $2^{16} + 1$. Each sensor n_i is assigned a univariate value $f_i(y) = P(i, y)$.

Two nodes n_i and n_j can calculate the pairwise key as $f_i(j) = f_j(i) = P(i, j)$. Node n_i takes as input the identifier of node n_j and evaluates $f_i(x)$ at j. The result is the common key. This algorithm takes $O(c)$ operations (if we apply Horner's rule).

If more than c nodes are compromised, then the polynomial $P(x, y)$ can be interpolated and the entire scheme is compromised. We say that the scheme is *c-secure*. Each node stores c elements from $GF(q)$ or equivalent to c keys. A node can calculate the common key, by knowing only the identifier of the other node. Thus, log N bits need to be transmitted.

Blom [5] had also used a similar concept, using matrices, which is also *c-secure*. During the predeployment phase, the KDC first constructs a $(c+1) \times N$ matrix M over a finite field $GF(q)$, q is a prime power, and N is the size of the network. M is public information. Then the BS creates a random $(c+1) \times (c+1)$ symmetric matrix D over $GF(q)$, and computes an $N \times (c+1)$ matrix $A = (D \cdot M)^T$, where V^T is the transpose of V. Matrix D needs to be kept secret (however, one row of $(D \cdot M)^T$ will be disclosed to each node). Since D is symmetric, $A \cdot M = (D \cdot M)^T \cdot M = M^T \cdot D^T \cdot M = M^T \cdot D \cdot M = (A \cdot M)^T$. Thus, $K = A \cdot M$ is a symmetric matrix. K_{ij} (or K_{ji}) is the pairwise key between node i and node j. Computing K_{ij} can be achieved using the following KPD scheme, for $w = 1, \ldots, N$. Store the wth row of matrix A and wth column of matrix M at node w. Thus, each node stores information about w elements of $GF(q)$, equivalent to $(c+1)$ keys. It has been proven that the previous scheme is c-secure if any $(c+1)$ columns of M are linearly independent. This property guarantees that no member other than i and j can compute K_{ij} or K_{ji} if no more than c members are compromised. Two communicating nodes can calculate the pairwise keys by exchanging the node identifiers. Thus, $\log N$ bits are transmitted. To evaluate the pairwise key, c multiplications are needed.

Du et al. [18] presented a pairwise scheme which uses multiple Blom's scheme [5]. Instead of one matrix, there are ω matrices and each node is randomly assigned shares from τ matrices. The storage requirement is thus $\tau(c+1) \log q$. Two nodes can generate pairwise key, provided they are assigned shares from the same matrix. Also since τ matrices are randomly assigned to each sensor, two communicating nodes have to broadcast the identifiers of the matrices, which requires $\tau(\log c)$ bits. The computation cost includes an additional cost of $\tau \log \tau$ of finding out if two nodes share the same key matrix. Each matrix can be calculated by compromising more than c nodes. The scheme will be fully compromised, if $(c+1)\tau$ nodes are compromised.

38.3.3 Combinatorial Schemes

Combinatorial schemes are deterministic schemes, in which keys are selected from the key pool according to some preselected pattern instead of selecting randomly. This construction is proposed by Mitchell and Piper [41]. They applied *combinatorial designs* to key distribution. This technique was revisited and applied to sensor networks by Çamtepe and Yener in Ref. [9], more than 25 years after the idea was first conceived.

Before proceeding to combinatorial KPD schemes, we give a brief idea about combinatorial designs. A *set system* or *design* [33] is a pair (X, A), where A is a set of subsets of X. Each subset of A is also called a *block*.

Depending upon the choice of construction of the subsets, designs can be of several types. The simplest and most widely studied combinatorial design is called a balanced incomplete block design (BIBD). A BIBD$(\upsilon, b, r, k, \lambda)$ is a *design* which satisfies the following conditions:

1. $|X| = \upsilon$, $|A| = b$.
2. Each subset in A contains exactly k elements.
3. Each element in X occurs in r blocks.
4. Each pair of elements in X is contained in exactly λ blocks in A.

Example 38.1

$X = \{0, 1, 2, 3, 4, 5\}$

$A = \{\{0, 1, 2\}, \{0, 1, 3\}, \{0, 2, 4\}, \{0, 3, 5\}, \{0, 4, 5\}, \{1, 2, 5\}, \{1, 3, 4\}, \{1, 4, 5\}, \{2, 3, 4\}, \{2, 3, 5\}\}$ is a BIBD (6, 10, 5, 3, 2).

A design can be readily mapped onto a KPD scheme. Following this design, a pool of υ keys is initially constructed. Each subset represents a key chain. For a key distribution scheme using BIBD designs, the key pool consists of υ keys and each sensor is preloaded with k keys according to the design. The maximum number of sensors that can be supported is b.

The KPD schemes which use combinatorial designs run in constant time and need to know only the identifier of the communicating node. Thus, combinatorial techniques have low computation and communications overhead. However, it is generally believed that probabilistic designs are more resilient to node compromise than deterministic designs. However, Xu et al. [66] showed that probabilistic techniques are only marginally more resilient.

Several combinatorial schemes have been proposed, since its inception by Çamtepe and Yener [9]. The schemes use different designs and trade-off between memory and resilience. Lee and Stinson [33] used transversal deigns [62]; Ruj and Roy used partially balanced incomplete block designs (PBIBD) [56]; and Blackburn et al. used Costas arrays [4].

The other class of combinatorial KPD uses codes. In Ref. [3], Al-Shurman and Yoo proposed a key management scheme based on maximum distance separable (MDS) codes to satisfy the properties of Cover Free Family (CFF) [20]. Ruj and Roy [58] used Reed–Solomon codes [50]. The main disadvantage of combinatorial KPD is that designs with the required parameters might not exist.

We have considered pairwise keys established between nodes which are within communication range. Multi-hop networks use these techniques as building blocks to communicate securely. We will describe polynomial-based KPD, combinatorial KPD in multi-hop networks in the following section. Secure communication can be ensured if message is sent securely between every pair of nodes in the multi-hop path, using basic schemes. However, there are some vulnerabilities, and we will discuss approaches to establish secure communication in multi-hop networks in Sections 38.6 and 38.7.

38.3.4 KPD in Cluster-Based Multi-Hop Networks

A number of KPD cluster-based networks have been proposed in which sensor nodes in each cluster send their data to a CH, which processes it and sends it to the BS. Jolly et al. [31] proposed a scheme in which each sensor node stores only two keys—one key is shared with a CH and the other is shared with the BS. Each CH stores a key with each of the sensor nodes in its cluster, exactly one key with one other CH and also a group key shared with all the CHs (sensor nodes able to communicate with two or more CHs). The BS is assumed to be secure and is assigned $(m+N)$ keys, where m and N are the number of CHs and sensor nodes, respectively. Each sensor belongs to an initial CH. Two sensor nodes in the same cluster can communicate via their common CH. Two nodes in different clusters can communicate via the respective CHs and gateways between CHs. If a node moves to a new cluster, then its old CH communicates with new CH and obtains the key for sensor to contact new CH. This scheme fails if one of CHs is dishonest, because it learns the keys of sensors in other clusters. All pairwise keys between CH and BS and also between CH and sensor are established using polynomials.

Scheme [31] was modified by Cheng and Agrawal [13]. They introduced the improved key distribution mechanism (IKDM). There are two symmetric polynomials used, one to establish pairwise keys between CHs and other to calculate pairwise keys between a sensor and its CH. The BS stores $(m+N)$ pairwise keys with each sensor node and CH. These keys are generated randomly and placed. Each CH shares a pairwise key with every other CH, calculated using a polynomial (similar to Blundo et al.'s scheme [6]). The pairwise key between a sensor node and the corresponding CH is calculated using a polynomial. Sensors communicate with each other via CHs.

Paterson and Stinson [45] presented two attacks on the previous scheme. The first attack is called the *interpolation attack* and takes place for certain parameters. The second attack though weaker can always take place and is called the *reconstruction attack*. Both attacks exploit the construction of the pairwise key between sensor nodes and nearest CH. During the formation of the pairwise keys, sensor nodes contain the list of l CHs. This list is transmitted to a CH, so it may be learned by the adversary. By compromising s CHs, the adversary learns sN/m such l subsets (blocks). The average number of occurrences of a CH $x \in \{1, \ldots, m\}$ in the sN/m blocks is sNl/m^2. If $sNl/m^2 = 1.25l$, then it can be shown that almost every CH will occur more than t times. Thus if an adversary compromises

$s = 1.25\ tm^2/(Nl)$ CHs after the IKDM process, then the expected number of CH polynomials that can be reconstructed using the interpolation attack is at least $m(1 - e^{-0.025t})$.

If s CHs are compromised, then sN/m sensors are exposed [13]. However, Paterson and Stinson [45] show that in fact $N/m\left(s+\left((m-s)\binom{s}{l}/\binom{m}{l}\right)\right)$ sensors are exposed. Suppose $\mathcal{J} = \{j_1, j_2, \ldots, j_s\}$ CHs are compromised, and then a sensor node which has a list B_i such that $B_i \subseteq \mathcal{J}$ is exposed, since all the keys K_i are compromised. The probability that $B_i \subseteq \mathcal{J}$ is $\binom{s}{l}/\binom{m}{l}$. There are $N - sN/m = N(m - s)/m$ sensor nodes whose nearest CH has not been compromised. Therefore, the expected number of reconstructed sensor node keys is $N(m-s)\binom{s}{l}/m\binom{m}{l}$.

The two schemes [13,31] also suffer from the disadvantage that sensor nodes cannot communicate with each other directly and can only communicate via the CH. If CHs become compromised, then all the sensor nodes in the cluster are compromised. To alleviate these problems, Das and Sengupta proposed a scheme [14] which also uses bivariate polynomials. One polynomial is used to establish pairwise keys between CHs and the BS. m polynomials (each of degree c) are used, one for each cluster, to establish pairwise keys between sensor nodes and CH in that cluster. The number of sensors in each cluster is less than c; thus, even if all the sensors are compromised, the bivariate polynomial will not be disclosed.

38.3.5 Combinatorial KPD in Multi-Hop Networks

In some applications where sensors are scattered over an adversarial area, we require that the complete disconnection of one region does not affect another region. For example, consider the scenario where sensors are deployed in a battlefield. Suppose the adversary captures all the sensors in one region. We must ensure that other regions are not affected by such a compromise. We also have to ensure secure communication not only between nodes in a particular region but also across regions. Ruj and Roy [57] described a scheme, where the target region of sensor deployed is partitioned into r^2 equally sized squares or grids as in [35,36]. There are two kinds of nodes having different battery power and storage capacities. The weaker nodes within each region can communicate with each other directly. The stronger specialized nodes, called agents, have higher battery power and storage. The weaker nodes can communicate across regions via agents. The keys in the small sensor nodes are predistributed using projective planes [9]. Each region consists of (p^2+p+1) nodes (p is a chosen parameter), each having $k=(p+1)$ keys. Sensors belonging to one region contain a set of keys that are disjoint from the sensors in some other region. This ensures that even if one region is totally disconnected, the other regions are not affected. Keys in the agents are predistributed using a modified form of transversal design [62], which is also a form of combinatorial design. For any square region $S_{i,j}$, a set of three agents $a^1_{i,j}$, $a^2_{i,j}$, and $a^3_{i,j}$ are deployed. The set of $(k + p + 1)$ keys assigned to the three agents are denoted by $B^1_{i,j}$, $B^2_{i,j}$, and $B^3_{i,j}$. Apart from the $(p+1)$ keys assigned from the set $P_{i,j}$, $B^1_{i,j}$ contains $\{(x, (xi + j) \bmod r): 0 \leq x < k\}$, $B^3_{i,j}$ contains $\{(x, r + ((j - xi) \bmod r)): 0 \leq x < k\}$, and $B^3_{i,j}$ contains $\{(x, 2r + ((xj + i) \bmod r)): 0 \leq x < k\}$.

Consider the keys of the form (x, y), where $0 \leq x < k$. If $y < r$ (r is a parameter), then the keys are called Type I keys; if $r < y < 2r$, then the keys are called Type II keys; and if $y > 2r$, then the keys are called Type III keys. We note that any agent contains keys of only one type. Depending on the type of keys an agent contains, agents may be of Type I, Type II, or Type III. The authors prove that three agents in each region are sufficient to ensure that two agents belonging to two different regions within communication range can communicate with each others directly. The weaker nodes belonging to two different regions communicate with the respective agents. To communicate with each other, the nodes need to exchange only their identifiers. The scheme may be extended to deployment schemes where the target region is divided irregularly.

Another hierarchical location aware KPD scheme was proposed by Younis et al. [68]. This scheme is called SHELL (Scalable, Hierarchical, Efficient, Location aware, and Lightweight). SHELL has three types of nodes: the sensor nodes, which sense and collect information from the deployment field, CHs which aggregate the information, process it, and send to the BS. The key distribution in sensor nodes is done in a deterministic way using Exclusion Basis System (EBS), developed by

Eltoweissy et al. [19]. It exploits the trade-off between the number of administrative keys, k and the number of rekeying messages m. A set of $(k + m)$ administrative keys is used to support a set of N nodes, where each node is assigned a distinct combination of k keys. The keys are chosen using a matrix, the details of which can be found in [19]. This is different from the polynomial schemes which we have discussed previously. The BS and CHs have preloaded pairwise keys. All nodes are assumed to be static.

38.4 Triple Key Distribution

So far we have discussed key distribution techniques, where two nodes communicate secretly. However, in many applications, three nodes need to share a common key. This common key shared between three nodes is called *triple key* and has been recently proposed by Ruj et al. [53]. Triple key is a special type of group key distribution, in which the group size is three. They help in passive monitoring [54], since one node may overhear if the second node has correctly forwarded an expected message to the third node. This idea will be exploited for secure routing and secure data aggregation (SDA), as elaborated in Sections 38.6 and 38.7, respectively.

Triple key distribution can be achieved using polynomials as follows. A symmetric polynomial $P(x, y, z)$ of degree c is chosen in three variables with coefficients in $GF(q)$, where q is a prime power. This means $P(x, y, z) = P(y, z, x) = P(z, x, y) = \cdots$. Each node n_i is given a share $P(i, y, z)$, which is the polynomial $P(x, y, z)$ evaluated at $x = i$. Three nodes n_i, n_j, and n_l can calculate the value of the common key as $K_{i,j,l} = P(i, j, l)$. This scheme is c-secure.

Triple key distribution can be applied to a hierarchical network, such that two sensor nodes share common keys with the CH and two CHs share common key with the BS. In this way, the communication can be monitored by the CH and the communication between two CHs can be monitored by the BS.

It is also possible to establish pairwise and triple keys simultaneously [54]. KPD takes place according to the following steps:

1. Set the node identifiers as $1, 2, 3, \ldots, N$.
2. Choose a prime q, for example, $2^{16} + 1$ or $2^{32} + 1$.
3. The key distributor chooses a symmetric trivariate polynomial $P(x, y, z)$ of degree c, with coefficients in Z_q.
4. For each node, n_i assign the share $P(x, y, i)$ $(z = i)$ to node n_i.
5. Pairwise key between node n_i and n_j is calculated as $P(0, i, j)$ by nodes n_i and n_j.
6. Triple key between nodes n_i, n_j, and n_l is calculated as $P(i, j, l)$ by nodes n_i, n_j, and n_l.

Node n_l does not know $P(x, y, z)$ or $P(0, y, z)$, so cannot calculate the pairwise key between nodes n_i and n_j.

Example 38.2

Let $q = 7$. Let $P(x, y, z) = x^2 + y^2 + z^2 + 4xy + 4yz + 4zx + x + y + z + 1$. Nodes are given the following shares:

Node n_1: $P(x, y, 1) = x^2 + y^2 + 4xy + 5x + 5y + 3$
Node n_2: $P(x, y, 2) = x^2 + y^2 + 4xy + 2x + 2y$
Node n_3: $P(x, y, 3) = x^2 + y^2 + 4xy + 6x + 6y + 6$
Node n_4: $P(x, y, 4) = x^2 + y^2 + 4xy + 3x + 3y$
Node n_5: $P(x, y, 5) = x^2 + y^2 + 4xy + 3$
Node n_6: $P(x, y, 6) = x^2 + y^2 + 4xy + 4x + 4y + 1$

Nodes n_1, n_2, and n_3 can calculate the triple key K_{123} as $P(3, 2, 1) = P(1, 3, 2) = P(1, 2, 3) = 2$. The pairwise key K_{23} of nodes n_2 and n_3 is $P(0, 3, 2) = P(0, 2, 3) = 5$. Other nodes do not know the polynomial $P(x, y, z)$ nor $P(0, y, z)$, and cannot calculate K_{23}.

38.5 Authentication Protocols

Message authentication can be achieved using digital signatures, which use public key techniques. Each node has a (public key, secret key) pair. Messages are signed with the secret key of a node and can be verified by the corresponding public key holders. This is the reverse of encryption process. These schemes involve expensive computation schemes, and we do not discuss them in details here.

The commonly used technique for message authentication is to apply *message authentication codes* (MAC). The (message, MAC value) pair is transmitted. The receiver calculates the MAC value for received message and compares with the received MAC value to verify its authenticity. A message authentication scheme should resolve the following two issues [7]:

- No node is able to calculate the secret key by observing the (message, MAC value) pairs.
- No node is able to compute another new correct (message, MAC value) pair from observing several such transmitted pairs.

MAC can be constructed in several ways, most popularly by using block ciphers [61] (Chapter 4). We will not discuss those techniques, but present some other commonly used authentication techniques for WSN. If the sender and receiver share a unique common key, then this key can be used in the MAC function. Here, the MAC function can be a simple hash function MAC = $h_K(M)$, where K is the common key. It is to be noted here that a hash function $h(m)$ is a function that takes a string m of variable length and converts it to a fixed length string. $h(m)$ is computationally easy to calculate (required polynomial time), however difficult to revert (takes exponential time). This implies that it is difficult to obtain m, given $h(m)$. It should also be difficult to find two messages m and m' which hash to the same value, $h(m) = h(m')$. The sender transmits (M, MAC). The receiver first calculates MAC′ = $h_K(M)$ and then checks with MAC. If there is a mismatch, then the message has been tampered with. This method requires that there should be a common key between sender and receiver. Thus, it is important to ensure that the common key is not known to adversary. The hash function generally used is SHA–1 [61].

A commonly used technique is to use chaining. Instead of using the same key K every time, it changes the key at intervals. Suppose the keys used are $K_1, K_2, \ldots$. In the ith session, a sender sends the following information to the receiver (M_i, MAC(M_i, K_i), K_{i-1}). The receiver buffers this information and when it later receives the key K_i (in the next session), it is able to verify the authenticity of the message M_i. The sender and receiver have to be loosely time-synchronized. An eavesdropper should not know K_i and thus cannot construct MAC(M_i, K_i), at session i.

This idea is the basis for Time Efficient Stream Loss-Tolerant Authentication (μTESLA) protocol [46]. The previous scheme is modified a bit, in that every message is not encrypted with a different session key. However, the same session key is used for t consecutive messages. The following packet P is sent by the sender:

$$P_{i+j} = (M_{i+j}, \text{MAC}(M_{i+j}, K_i), K_{i-1}) \quad \text{for } j = 1, 2, \ldots, t$$

All messages in the ith session are buffered at the receiver. In the $(i + 1)$th session, the receiver receives K_i and is able to verify the buffered messages. The size of t depends upon the memory capacity of the nodes.

μTESLA was used by Perrig et al. [47] to design an authentication protocol called *SPIN*. To construct a series of keys $K_0, K_1, \ldots, K_n$, the sender chooses K_n and constructs keys K_i = hash(K_{i+1}). In this way, user is not able to calculate K_i, given K_{i-1}. The rest of the steps are carried out as done earlier. A receiver can also authenticate the validity of a key K_i by checking if K_{i-1} = hash(K_i).

Zhu et al. [70] proposed another authentication protocol for sensor networks called Localized Encryption and Authentication Protocol (LEAP). LEAP is designed to support in-network processing, while at the same time restricting the security impact of a node compromise to immediate neighborhood of the compromised node.

The design of the protocol is motivated by the observation that different types of messages exchanged between sensor nodes have different security requirement and that a single keying mechanism is not suitable for meeting these different security requirements. LEAP supports the establishment of four types of keys for each sensor node: an individual key shared with the BS, a pairwise key shared with another sensor node, a cluster key shared with multiple neighboring nodes, and a group key that is shared by all the nodes in the network. The cluster key is used for in-network processing and passive participation. LEAP also includes an efficient protocol for internode traffic authentication based on the use of one-way key chains. An important feature of the authentication protocol is that it supports source authentication without precluding in-network processing and passive participation. It offers efficient protocols for supporting four types of key schemes for different types of messages, reduces battery usage and communication overhead through in-network processing, and uses a variant of μTESLA to provide local broadcast authentication. Disadvantages of this scheme are that it requires excessive storage with each node storing four types of keys and a one-way key chain, and computation and communication overhead dependent upon network density (the more dense a network, the more overhead it has).

Oliveira et al. [43,44] proposed a scheme SecLEACH, which uses node-to-node authentication (using MAC) in cluster-based hierarchical networks. It is a modification of the LEACH [27] and F-LEACH [22]. In a multi-hop network, authentication can be done either at each stage (at each intermediate node, node-to-node authentication) or only at the destination (source-destination authentication). Node-to-node authentication helps to detect malicious nodes early in the path. However, it is more expensive because every node has to verify messages that it receives.

38.5.1 Merkle Trees

The problem with earlier schemes is that the hash chains are recovered in a sequential manner. This is overcome by using *Merkle trees* [40]. Any value can be authenticated in any order. The concept of Merkle tree is as follows. Let $x_1, x_2, \ldots, x_n$ be the values to be authenticated. A binary tree is built with height d (such that $n \leq 2^d$) with the values as the leaves. The hashed values are then assigned to the intermediary nodes as shown in Figure 38.1. The root is also assigned a hash value calculated using its children. Without loss of generality, we have assumed that $n = 2^d$. In Figure 38.1, $i_j = \text{hash}(x_j)$, for $j = 1, 2, \ldots, 8$. $i_{12} = \text{hash}(i_1||i_2)$, $i_{34} = \text{hash}(i_3||i_4)$, $i_{56} = \text{hash}(i_5||i_6)$, $i_{78} = \text{hash}(i_7||i_8)$, $i_{1234} = \text{hash}(i_{12}||i_{34})$, $i_{5678} = \text{hash}(i_{56}||i_{78})$, and $i_0 = \text{hash}(i_{1234}||i_{5678})$.

During verification of a value, the value is sent along with the information at each of its siblings in the path from leaf to the root. The verifier can match the calculated value and check with that at the root. For example, if x_4 needs to be verified, then the values x_4, i_3, i_{12}, and i_{5678} and i_0 are sent. The value $h(h(i_{12}||h(h(x_4)||i_3))||i_{5678})$ is checked to see if it is equal to i_0. If the values match, then x_4 is authentic. Since hash functions are difficult to revert, knowing the value of $h(x)$ cannot help us recover x.

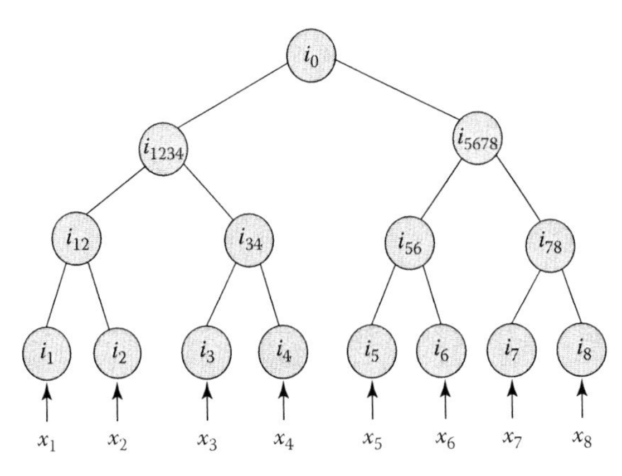

FIGURE 38.1 Merkle tree.

38.6 Secure Routing

To send some packets from a source to a destination, a route has to be established between them. In on-demand routing protocols, the route is established on the fly, only when needed. No prior routing information is maintained at each node. In dynamic source routing (DSR) [30], every data packet carries in its header the list of nodes to traverse to reach the destination. When a packet arrives at a node, the node checks if it is one of the intermediate nodes. It then forwards the packet to the next hop neighbor. If it does not belong to the list of intermediary nodes, then it drops the packet. Each of the intermediate nodes is aware of all other intermediate nodes, if the whole path is included in routing afterward; otherwise only the next and previous hops could be recorded and used. A compromised intermediate node can modify the header and change the destination, such that it is never found. It might change the intermediate nodes which might lead to loops and wastage of resources. To address this issue, we now discuss secure route discovery using triple keys.

38.6.1 Secure Route Discovery

At the route discovery stage, the source node S sends the address of the destination node encrypted with the pairwise keys of its neighbors. Each neighbor A, on receiving the encrypted message, decrypts it and re-encrypts it with the triple key K_{ABS} between another neighbor B, S, and itself. It sends separate message to every such neighbor B. B can decrypt and re-encrypt in a similar way as A. S also decrypts and checks if A sent the correct information to its neighbors. This process continues until the message reaches the destination D. D then notices the route of the message and sends back the reply, along established path. Any compromised intermediate node will be tracked, because the previous node always compares the received message with that overheard from forwarding nodes.

38.6.2 Onion Routing

We now assume that a path between source and destination is discovered, and a secret message between them is to be delivered. Most secure routing protocol, for example, [32,59,69] use *onion routing* [25], which is based on public key cryptographic techniques. In onion routing, the message is encrypted with the public keys of the intermediary nodes, in layers. At each node, a layer is removed, by decrypting with the private key of the node. This continues, until the message reaches the destination. Suppose S and D are the source and destinations, respectively. Let the intermediate nodes be $n_1, n_2, \ldots, n_l$. Each node n_i ($i = 1, 2, \ldots, l$) has a public and private key pair, p_i and s_i, respectively. The message M is encrypted in layers as $E_{p1}(E_{p2}(\ldots E_{pl}(E_{pD}(M)) \ldots))$. Node n_1 removes the first layer and finds out $E_{p2}(\ldots E_{pl}(E_{pD}(M)) \ldots)$ and sends the message to n_2 and so on. Finally, the destination node D is able to receive the message M. No external or internal node has any information about the intermediary nodes. A similar process takes place in the reverse path, D to S. Onion routing can thus be too expensive for the resource-constrained sensor nodes. However, it provides anonymity of connections.

Packets can be sent securely from the source to the destination using triple key, in either of the two ways:

1. Establish a common key K_{SD} between source S and destination D.
2. Send packets without establishing a prior common key.

The method is same and is described in the following section. The difference is whether or not the message itself is encrypted between any two hops. All nodes on selected path will learn the content of the message. If the content is the key between source and destination, then intermediate nodes learn that key. The key between source and destination may be decided outside routing process, and used during the process, with encrypted message being routed. In this case, intermediate nodes do not reveal the key and able to verify correct message forwarding without being able to interpret the message itself. The following section describes routing along selected path supported by triple keys verification.

38.6.3 Secure Routing between Source and Destination

Suppose nodes $S = n_0$ and $D = n_{l+1}$ want to send a message M between them accurately (without altering), using the intermediate nodes $n_1, n_2, \ldots, n_l$. S encrypts M using key $K_{0,1}$ and sends to n_1. Node n_1 decrypts M and encrypts it with using $K_{0,1,2}$ and sends to n_2. In this way, S can decrypt and check, if the message sent is correct. Any node n_i ($i = 2, \ldots, l - 1$) encrypts M using the triple key $K_{i-1,i,i+1}$. Node n_{i+1} upon receiving the encrypted M decrypts it using the triple key $K_{i-1,i,i+1}$. Then it re-encrypts M using the triple key $K_{i,i+1,i+2}$. Node n_i overhears the message broadcasted by n_{i+1} and checks if M is the same as that it had sent. If it does not match, then n_i sends a message back to the node S that n_{i+1} is compromised. Node n_i also sends a message to n_{i+2} using pairwise key $K_{i,i+2}$, constructed simultaneously with triple key (Section 38.4), even if n_i and n_{i+1} are not within communication range, and ignores the message M. So the use of triple key can also to identify a compromised node in the network.

There are however certain limitations of the earlier-given approach:

- Sometimes compromised nodes might not change M, but only overhear it, and be able to decrypt message later. Such compromised nodes cannot be detected.
- Suppose there are two or more consecutive compromised nodes. They can collude in the following way. Suppose the intermediate nodes are n_1, n_2, n_3, n_4. Suppose n_2 and n_3 are compromised. Then n_2 sends the correct message $E_{K1,2,3}(M)$ to n_3, and n_1 cannot detect n_2. However when n_3 changes M, then n_2 is able to detect but colludes with n_3 and does not report it to be a compromised node. In such circumstances, compromised nodes cannot be detected.
- If node n_1 is compromised and sends a wrong message that n_2 has changed the message, then it will be difficult to know if n_1 is lying or not.

For this reason, we might try to explore how to establish trust in the network.

The limitations of triple key in establishing a secure route for message delivery are that the route is not anonymous. The destination node knows the intermediate nodes in the route and even the intermediate nodes know other intermediate nodes. Privacy is also lost because the identities of all the intermediate nodes are known. An open problem will be to design a secure routing algorithm using triple key, to preserve the anonymity and privacy of the network. Exchanging identifiers for finding common key (as done in polynomial triple key establishment) compromises the privacy of the nodes. How to device a lightweight secure routing protocol without compromising privacy is left as a future work.

38.7 Secure Data Aggregation

Data aggregation refers to collecting data, processing the collected data, and sending toward the query node (generally the BS). An aggregator function might calculate average of the collected information (temperature/pressure), or sum the data collected by the sensor nodes. There are three types of nodes: normal sensor nodes which sense and collect information, aggregator nodes which process the information, using suitable functions, and the query node. This results in a tree structure, with the sensing nodes as leaf nodes and the BS as the root and all aggregator nodes as non-leaf nodes of the tree.

Adversaries can tamper with the collected data, modify the result thereby reporting a false result, or refuse to forward the collected or aggregated result. Most SDA techniques combine key management and authentication techniques. Both entity and data authentication are needed. Authentication can be carried out either hop-by-hop, meaning that at each step the authenticity of the data and the sender is verified, or it can be source to destination authentication, where the authenticity of the data is verified only at the destination. Though the first technique results in early discovery of adversaries, it results in higher computation costs.

SDA was first discussed by Hu and Evans [28]. The data are transmitted by the sensors along with the MAC. The intermediate nodes aggregate the results and send it to their parents. The BS authenticates the data using μTESLA protocol and then reveals the secret key to the intermediate nodes to validate the data. The drawback of this scheme is that the authentication is postponed, which wastes resources while calculating data. Przydatek et al. [48] proposed secure information aggregation (SIA), which resolved this problem and used node-to-node authentication using μTESLA instead of authentication at the BS.

Jadia and Mathuria [29] used pairwise keys instead of μTESLA. A node shares a pairwise key with its parent and the MAC is constructed using this pairwise key. The problem with this approach is that the pairwise keys need to be predistributed in the nodes.

Yang et al. [67] proposed SDAP, a Secure hop-by-hop Data Aggregation Protocol. It consists of two strategies: divide-and-conquer and commit-and-attest. The divide-and-conquer strategy divides the network tree into subtrees, which makes authentication and detection of compromised nodes easier. The commit-and-attest property means that the aggregator has to commit the aggregated result and cannot deny it later. Chan et al. [12] used Merkle tree for authentication.

Homomorphic encryption [8,63] is used to aggregate data securely. In this method, the encrypted data from the children do not need to be decrypted by an intermediary node in order to do the operations. Suppose two nodes x and y collect the messages m_1 and m_2. Let the aggregator function be multiplication, such that m_1m_2 is obtained. This is achieved using homomorphic encryption as follows: Nodes x and y calculate the ciphertext $c_1 = m_1^e$ and $c_2 = m_2^e$, respectively, and forward to their parent. The parent can calculate $c_1c_2 = m_1^e m_2^e = (m_1m_2)^e$ and forward it further. The parent will not know m_1 or m_2, but is able to forward the encrypted product m_1m_2. There are two problems with using homomorphic encryption in secure aggregation. First, it requires public key cryptographic techniques, which are demanding to resource-constrained sensor nodes. Second, to apply homomorphic encryption, the aggregator function should be simple like addition, multiplication, etc. To perform different operations, different homomorphic encryption schemes have to be chosen. This might not be feasible.

We may use triple key distribution to aggregate data securely. Each intermediate node shares a triple key with both its parent and child. We assume that the aggregator function is addition. Initially, leaf nodes say u, u', and u'' collect the information I_u, $I_{u'}$, and $I_{u''}$, respectively, and send to their common parent υ, encrypted with the pairwise keys $K_{u\upsilon}$, $K_{u'\upsilon}$, and $K_{u''\upsilon}$, respectively. Upon receiving the encrypted information from each of the children, υ performs different computations and sends the aggregated result I_υ to its parent w. This aggregated result $I_u + I_{u'} + I_{u''}$ is encrypted using the triple keys of $K_{u\upsilon w}$, $K_{u'\upsilon w}$, and $K_{u''\upsilon w}$, respectively. Node υ also sends to each child the aggregated result of all the other children, encrypted using the corresponding triple keys. So υ sends to node u the aggregated results $I_{u'} + I_{u''}$, of u' and u'', encrypted using K_{uvw}. u can then calculate the total aggregated result from its siblings and compare with the one already sent to w, which it can overhear and decrypt.

Suppose node υ tampers with any one or more of the information from its children (say u'). Instead of sending the aggregated message $I_u + I_{u'} + I_{u''}$, it sends $I_u + I_x + I_{u''}$ to w. Each of the children also receives the aggregated message. When υ sends the partial aggregate messages of the children, there is an inconsistency. When υ sends the result $I_x + I_{u''}$ to u, u does not notice any inconsistency. If node υ sends the result $I_u + I_{u''}$ to u', then u' notices that the aggregated result $I_u + I_{u'} + I_{u''}$ is not the same as it had overheard before. If node υ sends the result $I_u + I_{u''} + I_x - I_{u'}$, then node u' obtains consistent results. However, node w checks the new results and notices that it is inconsistent with the results obtained earlier. So, it can drop the message and report υ as a malicious node. This process continues throughout the path from leaves to the root node.

The limitation of triple key in data aggregation is that the aggregated results are sent not only once but also as many times as the number of children. Also, partial aggregated results are sent to all the nodes. So the communication overhead increases by a factor of the number of children of a node. The reduction of this communication overhead is left as a future work.

References

1. Crossbow Technologies. Available at http://www.xbow.com/. Last accessed date: April 27, 2012.
2. I. F. Akyildiz, W. Su, Y. Sankarasubramaniam, and E. Cayirci. Wireless sensor networks: A survey. *Computer Networks*, 38(4):393–422, 2002.
3. M. Al-Shurman and S.-M. Yoo. Key pre-distribution using MDS codes in mobile ad hoc networks. In *ITNG*, Las Vegas, NV, pp. 566–567. Washington, DC: IEEE Computer Society, 2006.
4. S. R. Blackburn, T. Etzion, K. M. Martin, and M. B. Paterson. Efficient key predistribution for grid-based wireless sensor networks. In *ICITS*, Calgary, Canada, pp. 54–69, 2008.
5. R. Blom. An optimal class of symmetric key generation systems. In *Proceedings of EUROCRYPT*, Paris, France, pp. 335–338, 1984.
6. C. Blundo, A. De Santis, A. Herzberg, S. Kutten, U. Vaccaro, and M. Yung. Perfectly-secure key distribution for dynamic conferences. In *Advances in Cryptology: Proceedings of CRYPTO'92*, vol. 740, pp. 471–486. Santa Barbara, CA: LNCS, 1993.
7. L. Butty'an and J.-P. Hubaux. *Security and Cooperation in Wireless Networks*. Cambridge, U.K.: Cambridge University Press, 2007.
8. C. Castelluccia, E. Mykletun, and G. Tsudik. Efficient aggregation of encrypted data in wireless sensor networks. In *MobiQuitous*, pp. 109–117. Washington, DC: IEEE Computer Society, Washington, DC, 2005.
9. S. A. Çamtepe and B. Yener. Combinatorial design of key distribution mechanisms for wireless sensor networks. In *ESORICS*, pp. 293–308, Sophia Antipolis, France, 2004.
10. S. A. Çamtepe and B. Yener. Key distribution mechanisms for wireless sensor networks: A survey, Technical Report TR-05-07. Rensselaer Polytechnic Institute, Computer Science Department, March 2005.
11. H. Chan, A. Perrig, and D. X. Song. Random key predistribution schemes for sensor networks. In *IEEE Symposium on Security and Privacy*, pp. 197–213. Washington, DC: IEEE Computer Society, Washington, DC, 2003.
12. H. Chan, A. Perrig, and D. X. Song. Secure hierarchical in-network aggregation in sensor networks. In A. Juels, R. N. Wright, and S. De Capitani di Vimercati, editors, *ACM Conference on Computer and Communications Security*, Alexandria, VA, pp. 278–287. New York: ACM, 2006.
13. Y. Cheng and D. P. Agrawal. An improved key distribution mechanism for large-scale hierarchical wireless sensor networks. *Ad Hoc Networks*, 5(1):35–48, 2007.
14. A. K. Das and I. Sengupta. An effective group-based key establishment scheme for large-scale wireless sensor networks using bivariate polynomials. In *COMSWARE*, pp. 9–16. Washington, DC: IEEE Computer Society, Washington, DC, 2008.
15. T. Dierks and C. Allen. The TLS Protocol Version 1.0. RFC 2246, January 1999.
16. W. Du, J. Deng, Y. S. Han, and P. K. Varshney. A pairwise key pre-distribution scheme for wireless sensor networks. In S. Jajodia, V. Atluri, and T. Jaeger, editors, *ACM Conference on Computer and Communications Security*, Washington, DC, pp. 42–51. New York: ACM, 2003.
17. W. Du, J. Deng, Y. S. Han, and P. K. Varshney. A key predistribution scheme for sensor networks using deployment knowledge. *IEEE Transactions on Dependable and Secure Computing*, 3(1):62–77, 2006.
18. W. Du, J. Deng, Y. S. Han, P. K. Varshney, J. Katz, and A. Khalili. A pairwise key predistribution scheme for wireless sensor networks. *ACM Transactions on Information and System Security*, 8(2):228–258, 2005.
19. M. Eltoweissy, M. H. Heydari, L. Morales, and I. H. Sudborough. Combinatorial optimization of group key management. *Journal of Network and Systems Management*, 12(1):33–50, 2004.
20. P. Erdös, P. Frankl, and Z. Füredi. Families of finite sets in which no set is covered by the union of two others. *Journal of Combinatorial Theory Series A*, 33:158–166, 1982.

21. L. Eschenauer and V. D. Gligor. A key-management scheme for distributed sensor networks. In *ACM Conference on Computer and Communications Security*, pp. 41–47, Washington, DC, 2002.
22. A. C. Ferreira, M. A. Vilaça, L. B. Oliveira, E. Habib, H. C. Wong, and A. A. F. Loureiro. On the security of cluster-based communication protocols for wireless sensor networks. In P. Lorenz and P. Dini, editors, *ICN (1)*, Volume 3420 of Lecture Notes in Computer Science, pp. 449–458. New York: Springer, 2005.
23. T. El Gamal. A public key cryptosystem and a signature scheme based on discrete logarithms. In *CRYPTO*, pp. 10–18, Santa Barbara, 1984.
24. G. Gaubatz, J.-P. Kaps, and B. Sunar. Public key cryptography in sensor networks—Revisited. In C. Castelluccia, H. Hartenstein, C. Paar, and D. Westhoff, editors, *ESAS*, Heidelberg, Germany, Volume 3313 of Lecture Notes in Computer Science, pp. 2–18. Springer, 2004.
25. D. M. Goldschlag, M. G. Reed, and P. F. Syverson. Hiding routing information. In *Information Hiding*, Ross J. Anderson (Ed.), pp. 137–150, Cambridge, UK, 1996.
26. N. Gura, A. Patel, A. Wander, H. Eberle, and S. C. Shantz. Comparing elliptic curve cryptography and RSA on 8-bit cpus. In *CHES*, Boston, pp. 119–132, 2004.
27. W. R. Heinzelman, A. Chandrakasan, and H. Balakrishnan. Energy-efficient communication protocol for wireless microsensor networks. In *IEEE Hawaii International Conference on System Sciences*, pp. 1–10, Hawaii, 2000.
28. L. Hu and D. Evans. Secure aggregation for wireless network. In *SAINT Workshops*, pp. 384–394. Washington, DC: IEEE Computer Society, Washington, DC, 2003.
29. P. Jadia and A. Mathuria. Efficient secure aggregation in sensor networks. In L. Bougé and V. K. Prasanna, editors, *HiPC*, Bangalore, India, Volume 3296 of Lecture Notes in Computer Science, pp. 40–49. New York: Springer, 2004.
30. D. Johnson and D. Maltz. *Dynamic Source Routing in Mobile Ad Hoc Networks in Mobile Computing*. Dordrecht, The Netherlands: Kluwer Academic Publishers, 1996.
31. G. Jolly, M. C. Kusçu, P. Kokate, and M. F. Younis. A low-energy key management protocol for wireless sensor networks. In *ISCC*, pp. 335–340. Washington, DC: IEEE Computer Society, 2003.
32. J. Kong and X. Hong. Anodr: Anonymous on demand routing with untraceable routes for mobile ad-hoc networks. In *MobiHoc*, pp. 291–302. New York, ACM, 2003.
33. J. Lee and D. R. Stinson. A combinatorial approach to key predistribution for distributed sensor networks. In *IEEE Wireless Communications and Networking Conference, WCNC 2005*, New Orleans, LA, pp. 1200–1205. 2005.
34. D. Liu and P. Ning. Establishing pairwise keys in distributed sensor networks. In S. Jajodia, V. Atluri, and T. Jaeger, editors, *ACM Conference on Computer and Communications Security*, Washington, DC, pp. 52–61. New York: ACM, 2003.
35. D. Liu and P. Ning. Location-based pairwise key establishments for static sensor networks. In S. Setia and V. Swarup, editors, *SASN*, Fairfax, VA, pp. 72–82. New York: ACM, 2003.
36. D. Liu and P. Ning. Improving key predistribution with deployment knowledge in static sensor networks. *TOSN*, 1(2):204–239, 2005.
37. D. Liu, P. Ning, and R. Li. Establishing pairwise keys in distributed sensor networks. *ACM Transactions on Information and System Security*, 8(1):41–77, 2005.
38. D. Malan, M. Welsh, and M. Smith. A public-key infrastructure for key distribution in TinyOS based on elliptic curve cryptography. In *First IEEE International Conference on Sensor and Ad Hoc Communications and Networks (SECON)*, pp. 119–132, Santa Clara, 2004.
39. D. J. Malan, M. Welsh, and M. D. Smith. Implementing public-key infrastructure for sensor networks. *TOSN*, 4(4), pp. 22:1–22:23, 2008.
40. R. C. Merkle. Secrecy, authentication and public key systems/a certified digital signature, PhD dissertation. Department of Electrical Engineering, Stanford University, Stanford, CA, 1979.
41. C. J. Mitchell and F. Piper. Key storage in secure networks. *Discrete Applied Mathematics*, 21:215–228, 1988.

42. R. M. Needham and M. D. Schroeder. Using encryption for authentication in large networks of computers. *Communications of the ACM*, 21(12):993–999, 1978.
43. L. B. Oliveira, A. C. Ferreira, M. A. Vilaça, H. C. Wong, M. W. Bern, R. Dahab, and A. A. F. Loureiro. SecLEACH—On the security of clustered sensor networks. *Signal Processing*, 87(12):2882–2895, 2007.
44. L. B. Oliveira, H. C. Wong, M. W. Bern, R. Dahab, and A. A. F. Loureiro. SecLEACH—A random key distribution solution for securing clustered sensor networks. In *NCA*, Cambridge, MA, pp. 145–154. Washington, DC: IEEE Computer Society, 2006.
45. M. B. Paterson and D. B. Stinson. Two attacks on a sensor network key distribution scheme of Cheng and Agrawal. *Journal of Mathematical Cryptology*, 2:393–403, 2008.
46. A. Perrig, R. Canetti, J. D. Tygar, and D. X. Song. Efficient authentication and signing of multicast streams over lossy channels. In *IEEE Symposium on Security and Privacy*, pp. 56–73, Oakland, 2000.
47. A. Perrig, R. Szewczyk, V. Wen, D. E. Culler, and J. D. Tygar. SPINS: Security protocols for sensor networks. In *MOBICOM*, pp. 189–199, Rome, Italy, 2001.
48. B. Przydatek, D. X. Song, and A. Perrig. Sia: Secure information aggregation in sensor networks. In I. F. Akyildiz, D. Estrin, D. E. Culler, and M. B. Srivastava, editors, *SenSys*, Los Angeles, CA, pp. 255–265. New York: ACM, 2003.
49. S. Rafaeli and D. Hutchison. A survey of key management for secure group communication. *ACM Computing Surveys*, 35(3):309–329, 2003.
50. I. S. Reed and G. Solomon. Polynomial codes over certain finite fields. *Journal of the Society for Industrial and Applied Mathematics(SIAM)*, 8:300–304, 1960.
51. R. L. Rivest, A. Shamir, and L. M. Adleman. A method for obtaining digital signatures and public-key cryptosystems. *Communications of the ACM*, 21(2):120–126, 1978.
52. S. Ruj. Application of combinatorial structures to key predistribution in sensor networks and traitor tracing, PhD thesis. Indian Statistical Institute, Kolkata, India, February 2009.
53. S. Ruj, A. Nayak, and I. Stojmenovic. *Fully Secure Pairwise and Triple Key Distribution in Wireless Sensor Networks using Combinatorial Designs*. Shanghai, China: IEEE INFOCOM, April 2011.
54. S. Ruj, A. Nayak, and I. Stojmenovic. Pairwise and triple key distribution in wireless sensor networks with applications. Draft submitted.
55. S. Ruj, A. Nayak, and I. Stojmenovic. Key Predistribution in wireless sensor networks when sensors are within communication range. In *Theoretical Aspects of Distributed Computing in Sensor Networks*. New York: Springer-Verlag, 2011.
56. S. Ruj and B. Roy. Key predistribution using partially balanced designs in wireless sensor networks. In *ISPA*, pp. 431–445, Niagara falls, Canada, 2007.
57. S. Ruj and B. Roy. Key predistribution using combinatorial designs for grid-group deployment scheme in wireless sensor networks. *ACM Transaction on Sensor Networks*, 6(1):14:1–14:28, 2009.
58. S. Ruj and B. K. Roy. Key predistribution schemes using codes in wireless sensor networks. In M. Yung, P. Liu, and D. Lin, editors, *Inscrypt*, Beijing, China, volume 5487 of Lecture Notes in Computer Science, pp. 275–288. Springer, 2008.
59. S. Seys and B. Preneel. Arm: Anonymous routing protocol for mobile ad hoc networks. In *AINA (2)*, Vienna, Austria, pp. 133–137. Washington, DC: IEEE Computer Society, 2006.
60. J. G. Steiner, B. C. Neuman, and J. I. Schiller. Kerberos: An authentication service for open network systems. In *USENIX Winter*, pp. 191–202, Dallas, USA, 1988.
61. D. R. Stinson. *Cryptography: Theory and Practice*, 3rd edn. Boca Raton, FL: CRC Press, 2006.
62. A. P. Street and D. J. Street. *Combinatorics of Experimental Design*. Oxford, U.K.: Clarendon Press, 1987.
63. D. Westhoff, J. Girão, and M. Acharya. Concealed data aggregation for reverse multicast traffic in sensor networks: Encryption, key distribution, and routing adaptation. *IEEE Transactions on Mobile Computing*, 5(10):1417–1431, 2006.

64. A. D. Wood and J. A. Stankovic. Denial of service in sensor networks. *IEEE Computer*, 35(10):54–62, 2002.
65. Y. Xiao, V. K. Rayi, B. Sun, X. Du, F. Hu, and M. Galloway. A survey of key management schemes in wireless sensor networks. *Computer Communications*, 30(11–12):2314–2341, 2007.
66. D. Xu, J. Huang, J. Dwoskin, M. Chiang, and R. Lee. Re-examining probabilistic versus deterministic key management. In *ISIT*, pp. 2586–2590, Nice, France, 2007.
67. Y. Yang, X. Wang, S. Zhu, and G. Cao. SDAP: A secure hop-by-hop data aggregation protocol for sensor networks. In S. Palazzo, M. Conti, and R. Sivakumar, editors, *MobiHoc*, Florence, Italy, pp. 356–367. New York: ACM, 2006.
68. M. F. Younis, K. Ghumman, and M. Eltoweissy. Location-aware combinatorial key management scheme for clustered sensor networks. *IEEE Transactions on Parallel and Distributed Systems*, 17(8):865–882, 2006.
69. Y. Zhang, W. Liu, and W. Lou. Anonymous communications in mobile ad hoc networks. In *INFOCOM*, pp. 1940–1951. IEEE, Florida, USA, 2005.
70. S. Zhu, S. Setia, and S. Jajodia. LEAP: Efficient security mechanisms for large-scale distributed sensor networks. In *ACM Conference on Computer and Communications Security*, pp. 62–72, Washington, DC, USA, 2003.

39

Maritime Domain Awareness

Christopher Griffin
The Pennsylvania State University

Richard R. Brooks
Clemson University

Jason Schwier
Clemson University

39.1 Introduction

In this chapter, we present an approach for solving the problem of modeling a vehicle's behavior as it traverses a closed n-dimensional region. A shorter version of this chapter is available in Ref. [1]. This problem is of fundamental importance to Maritime Domain Awareness, the problem of determining the position and behavior of ships at sea. Since it is impossible to track (in real time) every ship on the ocean, unless they are all using Automated Information System (AIS), we must periodically be able to predict the position of a ship based on its previous behaviors. The model we present is generalized and will operate with almost any vehicle that exhibits ship-like behavior. We discuss the requirements for the model subsequently.

We assume that the vehicle is recurrent, i.e., it will return to its point of origin. This is consistent with the notion that naval vessels have at least one common port-of-call. Behavior of this type can be described by collections of cycles or loops within a compact subset of $\mathbb{R}^n$. We use a hybrid statistical approach that incorporates a collection of linear statistical models combined with a discrete statistical model that governs transitions from one linear model to another. We allow the vehicle's position to be sampled with noise. We further allow the vehicle's behavior to contain a stochastic component; this stochastic component is limited only by our assumption that the probability distribution over the future behavior of the vehicle is a function of the previous positions and behaviors of the vehicle.

In Ref. [2], we developed symbolic transfer functions (STF), which are discrete stochastic input–output systems whose inputs and outputs are both purely symbolic. An STF is a doubly stochastic process. Each input is a random variable governed by a probability distribution over a finite set of input symbols (alphabet) that is prescribed by the previous n_1 input symbols. Each output is a random variable governed by a probability distribution over a finite set of output symbols that is prescribed by the previous n_2 inputs symbols and n_3 output symbols. We provide a formal definition for STF in Ref. [2] and in Section 39.3.

We apply STF to the problem of track modeling by assuming that the input symbols represent regions of space through which a track is passing while the output symbols represent specific linear functions that model the behavior of the vehicle more precisely.

More formally, let $\gamma(t) \subset D \subset \mathbb{R}^n$ be a continuous piecewise linear track in a compact subset D of Euclidean space. We suppose there is a partition $\mathcal{A}$ of D that defines regions of different behaviors for the vehicle under observation. For the remainder of this chapter, a behavior is a linear function of time drawn from a set $\mathfrak{A}$ that describes the motion of the vehicle under inspection for a finite period of time. Formal definitions are provided in Section 39.3.

The main result of this chapter is to provide an algorithm to construct a STF model of $\gamma(t)$, where the inputs to the STF are elements of a partition of D (denoted $\mathcal{A}$) and the outputs (denoted $\mathfrak{A}$) are the linear functions describing the piecewise motion of $\gamma(t)$.

39.1.1 Related Work

Modeling linear and nonlinear time series has been the subject of great attention. Techniques such as basic linear regression [3] and nonlinear regression [4] attempt to model observations γ_t as functions of a variable $t \in \mathbb{R}$. Autoregressive Integrated Moving Average Models (ARIMA) extend these notions by allowing γ_t to vary as a function of itself and past shocks [5]. Seasonal ARIMA models extend this notion by adding seasonal periodicity [6] while Fractional ARIMA [7] models add short- and long-range dependence, not expressible with classical ARIMA techniques. Finally (Generalized) Autoregressive Heteroskedastic Models (ARCH/GARCH) add heteroskedastic behavior to the error components of the time series allowing globally stationary and locally nonstationary error terms to be analyzed [8]. Perhaps the most general models are the stochastic differential and difference equations that use Weiner and Lévy processes to model stochasticity [9].

Approximate grid-based methods can be employed when the space of the time series is continuous but can be broken into a collection of discrete grid points trajectory modeled as a time series of grid points. The work in Refs [10–12] describes methods of using hidden Markov models (HMM; and in the case of Ref. [10] dynamic programming) to identify optimal estimators for the behavior of trajectories passing through the discretized state space. A critical element of this work is its lossy nature. Any discretization of a continuous space must naturally result in loss of information and thus a decrease in power of the resultant model. At the same time, simpler models with easier fitting parameters can occasionally be gained by reduction to a discrete model.

Two areas also related to the work presented here are nested partition optimization [13] and local regression methods [14]. The former uses a partitioned feasible region to iteratively search for a global optimum of an objective function. The region is repartitioned (in a way similar to the approach described in Section 39.4) and the optimal objective value refined within these partitions. In essence, a collection of restricted optimization problems are solved over a continuous space, much like a branch-and-bound method. In contrast, local regression uses several regressions to fit an individual nonlinear function. Each regression takes place over a subset of the function's support. In principle, this approach is similar to piecewise linear regression, except it employs nonlinear models when possible. Loader [14] describes a method for partitioning the support of the function to be fitted to find an appropriate local regression.

Unlike the models discussed earlier, which are purely continuous, hybrid statistical models of time series contain both discrete and continuous components. Ozonoff et al. [15] have applied hybrid models to the study of influenza epidemics. In this work, an upper-level model is given by a HMM [11] whose states correspond to a single output: a high-fidelity continuous periodic function describing the influenza-spread dynamics. Hidden transitions adjust the underlying dynamics of the system. Yin et al. [16] used hierarchical models for describing the behavior of vehicle tracks operating in a lab. In this case however, the authors employed a multilevel Markov model of behavior; this model was not hybrid in the sense of Ozonoff et al. Finally, there is an extremely rich literature in Hybrid Control [17] (Chapters 1 and 2), particularly Switched Linear Systems [18–21], which use multilayer mixed discrete and continuous models. This literature is vast and ever growing and is concerned primarily with the controllability and observability of such systems rather than in modeling them from data. Therefore, we believe that the work presented in this chapter may be complementary to this field.

39.1.2 Comparison of Previous Work to This Chapter

The work performed in this chapter attempts to fuse the simplicity of the grid-based and simple linear fitting approaches without incurring the intractability often found in approximate grid-based methods or poor fit that occurs when modeling a nonlinear process with linear models. Like Nested Partition Optimization [13] we use a partition refinement approach and like local regression techniques [14] we attempt to fit a collection of linear functions describing a nonlinear phenomena. Unlike both these approaches, our work attempts to

1. Work with nonlinear time series in track data
2. Include a discrete component, which can be used to control the linear model components and act as a predictor in their place when a linear fit is not possible
3. Provide continuous models in regions of the track that can be easily modeled using a linear estimator. This enhances the predictive power of the model over methods that use only discretized Markov models (see, e.g., Ref. [10] and other works already cited)

Our work is meant to be executed entirely offline. There are two reasons for this:

1. Our test problem, namely learning behaviors of ocean vessel tracks, allows us to learn vessel behaviors offline.
2. Translation of this approach to an online setting would involve the development of an online learning method for the discrete component of the model. Since we are using a nontraditional Markov learning algorithm, addition of this material would serve to cloud the underlying presentation.

Methods that generalize the algorithms presented in this chapter to an online setting can be dealt with in future work.

In short, the modeling method presented in this chapter can be thought of either as an end-in-and-of itself or as a starting-off point for the application of more complex online grid-based procedures or multiple applications of linear or nonlinear filters within different regions of the state space.

39.1.3 Chapter Organization

The remainder of this chapter is organized as follows. In Section 39.2, we provide a list of symbols used in this chapter for easy reference. In Section 39.3, we provide assumptions and define the regularity conditions that $\gamma(t)$ must obey. In Section 39.4, we provide our main problem and a theoretical existence result. In Section 39.5, we derive an algorithm to solve the main problem posed in Section 39.4. In Section 39.6, we provide an example on a robot tracking problem. We contrast the simulated example presented in Section 39.6 to the real-world tests results presented in Section 39.7. In this section, we present the results of our prediction algorithm on live ship tracks. Finally, we present conclusions and future directions in Section 39.8.

39.2 List of Symbols

$\mathbb{R}^n$	Euclidean n-space
$\gamma(t)$	A continuous, piecewise linear curve
D	A subset of $\mathbb{R}^n$ in which $\gamma(t)$ resides
$\mathcal{A}$	A partition of D
$\mathfrak{A}$	A collection of linear functions
$\mathbf{p}$	A time series of partition elements of $\mathcal{A}$
λ	A time series of linear functions from $\mathfrak{A}$
$\mathcal{F}_{\mathcal{A}}$	The set of discrete probability functions with support in $\mathcal{A}$
$\mathcal{F}_{\mathfrak{A}}$	The set of discrete probability functions with support in $\mathfrak{A}$
η	A function mapping sequence of length n_1 of elements from $\mathcal{A}$ into $\mathcal{F}_{\mathcal{A}}$
ξ	A function mapping sequence of length n_2 of elements from $\mathcal{A}$ and sequence of length n_3 of elements of $\mathfrak{A}$ into $\mathcal{F}_{\mathfrak{A}}$
$\mathcal{S}_{\mathcal{A}}$	A mapping from $\gamma(t)$ into sequences of elements of $\mathcal{A}$
$\mathcal{S}_{\mathfrak{A}}$	A mapping from $\gamma(t)$ into sequences of elements of $\mathfrak{A}$
ζ	The left shift sequence operator
$\mathcal{T}$	A symbolic transfer function
π	A partition scheme or rule for repartitioning D
Γ_p	A set of observations from curve γ corresponding to a particular element p in a sequence $\mathbf{p}$

39.3 Background

39.3.1 Assumptions

Let D be a compact subset of Euclidean space $\mathbb{R}^n$. A target (or vehicle) has a trajectory $\gamma(t)$ contained entirely within D. We assume that there is some partition of D (denoted $\mathcal{A}$) and that the linear components of $\gamma(t)$ are drawn from a set $\mathfrak{A}$. As $\gamma(t)$ moves through D, it specifies a sequence of partition elements (in $\mathcal{A}$) and sequence of linear functions (its piecewise components in $\mathfrak{A}$).

We make the following assumptions in this chapter:

1. The trajectory $\gamma(t)$ is piecewise linear and continuous.
2. The set $\Gamma = \{\boldsymbol{x} \in \mathbb{R}^n : \boldsymbol{x} = \gamma(t), t \geq 0\}$ is the finite union of a collection of simple closed curves contained in D (see Figure 39.1).
3. There is at least one point $\mathbf{x}_0 \in D$ shared by all curves making up $\gamma(t)$.
4. When $\gamma(t)$ reaches a point shared by at least two curves, a choice will be made as to which curve to follow. This choice may be probabilistic or deterministic. It may be a probabilistic or deterministic function of a finite portion of the sequence of partition elements it has defined, or a finite portion of the sequence of linear functions it has defined, or both.

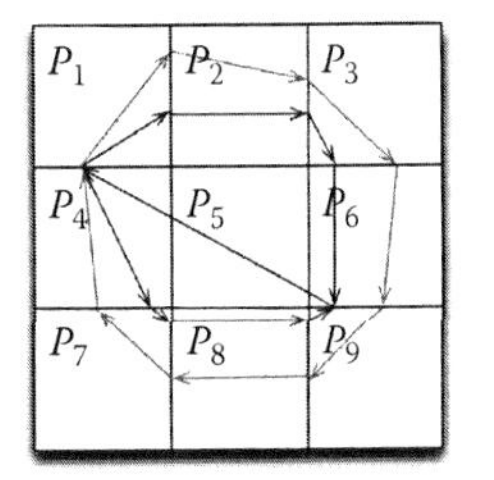

=
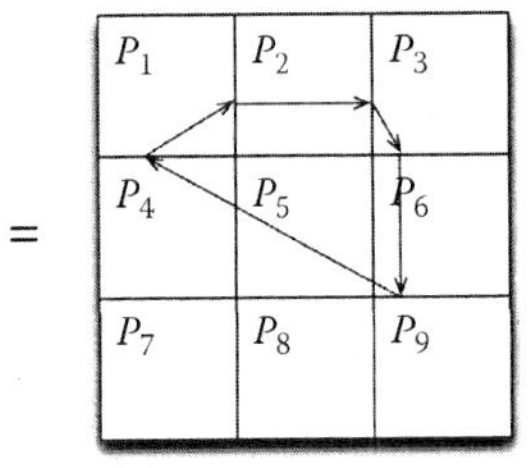

∪
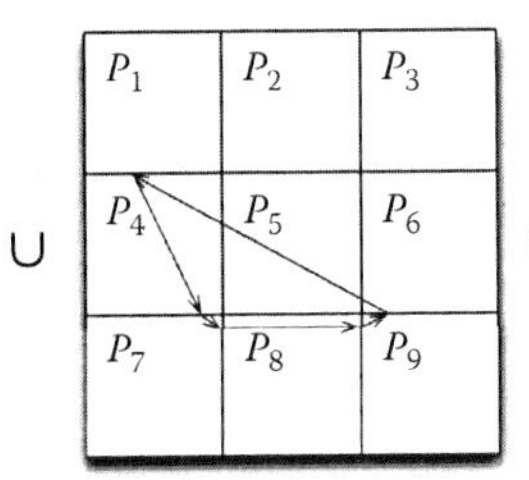

∪
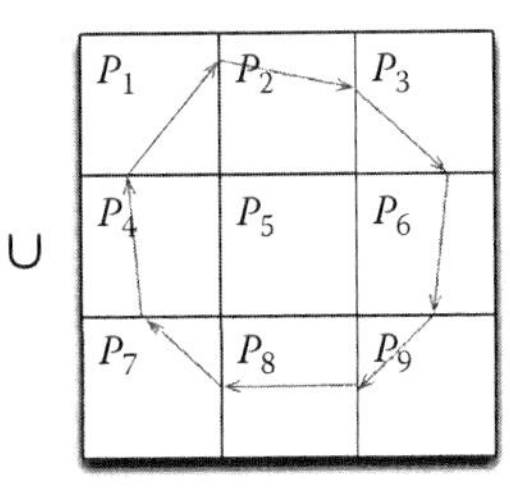

FIGURE 39.1 A vehicle trajectory that can be modeled using a symbolic transfer function with output alphabet a collection of linear functions. Note each time the vehicle moves from one region to another, a new linear behavior is executed. Each region may have multiple linear behaviors that are executed, one at a time.

In this chapter, we will call a track satisfying these three criteria a recurrent track in a compact set. Figure 39.1 illustrates such a trajectory $\gamma(t)$.

In practice, we will assume that noise is introduced into $\gamma(t)$ to produce

$$z(t) = \gamma(t) + \epsilon(t)$$

where $\epsilon(t)$ is a noise process with mean zero.

Our objective is to provide an algorithm to model any $\gamma(t)$ satisfying these three criteria with a hybrid statistical model defined by an STF. We will illustrate through experiment that the algorithm is tolerant to noise.

39.3.2 Probability Distribution Functions of Symbols and STF

Let $\mathcal{A}$ (the *input* alphabet) and $\mathfrak{A}$ (the *output* alphabet) be finite sets of symbols. A symbolic time series is a sequence: $\mathbf{x} = \ldots \mathbf{x}(-2)\mathbf{x}(-1)\mathbf{x}(0)\mathbf{x}(1)\mathbf{x}(2)\ldots$, where $\mathbf{x}(t)$ represents the symbol that occurred at discrete time t in $\mathbf{x}$. If $\mathbf{x}(t)$ is undefined, then we assume it is ε, the empty symbol. In this chapter, we use the left shift operator. Suppose $\mathbf{w} = \sigma_1\sigma_2\ldots\sigma_{n+1}\ldots$. Let $\zeta(\mathbf{w}) = \sigma_2\ldots\sigma_n\sigma_{n+1}\ldots$ be the left shift operator.

If $\mathbf{x}$ is defined for only a finite time interval $[0, n-1]$, we refer to $\mathbf{x}$ as a string having length n. By $\mathcal{A}^n$ we mean the set of all strings that have length n composed of symbols in $\mathcal{A}$. If $\mathbf{x}$ and $\mathbf{y}$ are two strings, then we will use the standard concatenation notation of $\mathbf{xy}$. Similarly if σ is a symbol in an appropriate alphabet, then $\mathbf{x}\sigma$ is simply the addition of the symbol σ to the string $\mathbf{x}$.

Consider a function $\eta : \mathcal{A}^{n_1} \to \mathcal{F}_{\mathcal{A}}$, where $\mathcal{F}_{\mathcal{A}}$ is the set of probability distributions with support $\mathcal{A}$. Then for any string $\mathbf{x}$ of length n_1 composed of symbols in $\mathcal{A}$, $\eta(\mathbf{x})$ is the conditional probability distribution given $\mathbf{x}$ with support $\mathcal{A}$. Specifically, $\eta(\mathbf{x})(\sigma_{\mathcal{A}})$ is the probability that symbol $\sigma_{\mathcal{A}}$ will occur after observing sequence $\mathbf{x}$.

If we have a function $\xi : \mathcal{A}^{n_2} \times \mathfrak{A}^{n_3} \to \mathcal{F}_{\mathfrak{A}}$ and given $\mathbf{x}$ a string of symbols drawn from alphabet $\mathcal{A}$ of length n_2 and $\mathbf{y}$ a string of symbols drawn from alphabet $\mathfrak{A}$ with length n_3, then $\xi(\mathbf{x}, \mathbf{y})(\sigma_{\mathfrak{A}})$ is the probability that the next output symbol is $\sigma_{\mathfrak{A}}$ after observing input $\mathbf{x}$ and previous output $\mathbf{y}$.

A STF [2] is a pair $\mathcal{T} = (\eta, \xi)$, where $\eta : \mathcal{A}^{n_1} \to \mathcal{F}_{\mathcal{A}}$ and $\xi : \mathcal{A}^{n_2} \times \mathfrak{A}^{n_3} \to \mathcal{F}_{\mathfrak{A}}$ and $\mathcal{A}$ and $\mathfrak{A}$ are appropriately defined finite alphabets. In Ref. [2], a procedure is defined for constructing an STF from sequences $\mathbf{x}$ and $\mathbf{y}$.

A symbolic input–output system governed by an STF will operate in the following way: at time t, the last n_1 symbols of the input string $\mathbf{x}$ are used to produce a conditional probability distribution $\eta(\mathbf{x}(t), \ldots, \mathbf{x}(t - n_1 - 1))$. A symbol $\mathbf{x}(t + 1) \in \mathcal{A}$ is generated according to this distribution. At the same time, the last n_2 symbols of $\mathbf{x}$ and n_3 symbols of $\mathbf{y}$ are used to produce a conditional probability distribution $\xi(\mathbf{x}(t), \ldots, \mathbf{x}(t - n_2 - 1), \mathbf{y}(t), \ldots, \mathbf{y}(t - n_3 - 1))$. A symbol $\mathbf{y}(t + 1) \in \mathfrak{A}$ is generated according to this distribution. Initial conditions must be specified if $\mathbf{x}$ and $\mathbf{y}$ are not long enough to contain the appropriate number of symbols or, in the formal case, an infinite left hand time frame can be assumed. A similar problem is observed in ARIMA processes (see Ref. [22]).

We note that at any time there is only one input string and one output string for an STF; no joint conditional probability distribution on multiple inputs or outputs is required. Finally, we refer to $\mathcal{T}$ as an STF because it is acting is a way that is similar to the classic transfer function, i.e., it is relating the inputs of a system to its outputs. The relationship is entirely formal and stochastic. It is not an extension of the classical transfer function, but a construct designed specifically for formal symbolic input–output systems satisfying the properties described earlier.

39.3.3 Hybrid Linear Models

Let $D \subset \mathbb{R}^n$ be a compact set and let $\mathcal{A}$ be a finite partition of D. Let $\mathfrak{A}$ be a set of linear functions mapping a single variable $t \in \mathbb{R}$ into $\mathbb{R}^n$. For the remainder of this chapter, $\mathcal{A}$ will be treated both as an input alphabet of symbols and as a partition of D. Likewise, $\mathfrak{A}$ will be treated as both an output alphabet and a collection of linear functions. That is, $\mathfrak{A}$ may be put into a one-to-one correspondence with a finite set of vectors $\lambda_i \in \mathbb{R}^n$ and we may think of the linear functions in $\mathfrak{A}$ as defined by $\lambda_i(t) = \lambda_i t$.

Let $\gamma: \mathbb{R} \to D$ be a continuous piecewise linear curve. If we are given the partition $\mathcal{A}$ and the set of linear functions $\mathfrak{A}$, then we may construct a sequence of partition elements **p** through which γ passes. Similarly we may construct a sequence of linear functions λ so that if at time t, γ is behaving according to a linear function λ, then $\gamma(t) = \lambda(t) + \gamma_0$, where $\gamma_0 \in \mathbb{R}^n$. Let us denote these two mappings by $\mathcal{S}_{\mathcal{A}}: \gamma(t) \mapsto \mathbf{p}$ and $\mathcal{S}_{\mathfrak{A}}: \gamma(t) \mapsto \lambda$.

39.3.4 Relation of STF to Assumptions

In this chapter, we study curves that produce behaviors $(\mathbf{p}, \lambda)$ that can be modeled by an STF $\mathcal{T} = (\eta, \xi)$. In this case, there are a sequence of times $t_1, t_2, \ldots, t_k, \ldots$ such that at time t_i, $\gamma(t_i)$ moves from partition element p_{i-1} to partition element p_i and changes from linear behavior λ_{i-1} to linear behavior λ_i. This is illustrated in Figure 39.1. Note that in each partition element, the linear function changes.

If trajectory $\gamma(t)$ can be modeled by hybrid linear models as we have defined it, then there is a predictor function mapping $\mathcal{A} \times \mathfrak{A} \times \mathbb{R}^n \to \mathcal{A}$. That is, suppose at position $\gamma_0 \in p_i$, $\gamma(t)$ behaves according to linear function λ. Using knowledge of the partition $\mathcal{A}$, the known linear function $\lambda(t)$, we estimate $p_{i+1} \in \mathcal{A}$ by finding the infimal t (denoted t^*) such that $\lambda(t) \notin p_i$. The partition element $p^* \in \mathcal{A}$ containing $\lambda(t^*)$ is p_{i+1}.

If we assume that $\mathcal{T} = (\eta, \xi)$ is known, then we assume that this prediction is consistent with the function η so that given $p_i, p_{i-1}, \ldots, p_{i-n_1}$ the prediction p_{i+1} is in the support of

$$\eta(p_i, p_{i-1}, \ldots, p_{i-n_1})$$

Likewise, the function λ_{i+1} is in the support of

$$\xi(p_i, p_{i-1}, \ldots, p_{i-n_2}, \lambda_i, \lambda_{i-1}, \ldots, \lambda_{i-n_3})$$

It should be clear at this point that the η function is used to predict the next partition element $\gamma(t)$ will enter, given a finite subset of the previous partition elements through which $\gamma(t)$ has passed. The ξ function is used to predict the next linear behavior $\gamma(t)$ will exhibit given a finite subset of the previous set of partition elements through which $\gamma(t)$ has passed and a finite subset of the previous set of linear behaviors $\gamma(t)$ has expressed.

The earlier statements suggest that η is not a necessary component of the model. We include the η function because the algorithm presented for constructing an STF $\mathcal{T}$ modeling the trajectory γ may at some point produce the *null linear function*. In this case, η can be used to predict the next partition element in which γ will be observed. This fact, combined with a partition refinement discussed in the next section, allows us to create a model in which trajectories with highly nonlinear behavioral changes can be described successfully with a minimal number of excess parameters.

39.4 Main Problem and Theoretical Constructions

We assume that we are given a track $\gamma(t)$ that obeys the four assumptions defined in Section 39.3. We assume that this track is sampled with noise as discussed in Section 39.3. Our main problems are

1. Determine some appropriate partition $\widehat{\mathcal{A}}$
2. Determine the set of linear behaviors $\widehat{\mathfrak{A}}$
3. Identify $(\hat{\eta}, \hat{\xi})$ that estimates the dynamics of $\gamma(t)$ using $\widehat{\mathcal{A}}$ and $\widehat{\mathfrak{A}}$

We assume that we are given an initial partition $\mathcal{A}_0$ of D (which may be D itself) and we will build $\widehat{\mathcal{A}}$ from this initial partition $\mathcal{A}_0$. We will also assume that we are given values of n_1, n_2, and n_3. The work in Ref. [23] discusses how to automatically obtain n_1. We can obtain values for n_2 and n_3 using an algorithm similar to the one provided in Ref. [23] because the algorithm provided in Ref. [2] is a variation of the CSSR algorithm [24,25], which is used in Ref. [23]. In Ref. [23], it is noted that a large amount of data may be required for estimating n_1 (and therefore n_2 and n_3). If this quantity of data is not available, the user must supply values for n_1, n_2, and n_3.

39.4.1 Some Theoretical Results

Theorem 39.1

If $\gamma(t)$ is a recurrent track in a compact set $D \subseteq \mathbb{R}^n$, then there is a STF modeling $\gamma(t)$.

Sketch of Proof

We first prove there is an appropriate partition of D. It is clear that we may construct a simple polytope covering each linear component of the curve $\gamma(t)$ because $\gamma(t)$ is composed only of a finite number of piecewise linear closed curves. It is further clear that this collection of polytopes can be designed so that they share common vertices at the change points of $\gamma(t)$. Examples of this are shown in Figure 39.2. The partition generated by the covering polytopes is the natural one consisting of these polytopes and the sets in the complement of these polytopes. Let $\mathcal{A}$ be this partition of D and let $\mathfrak{A}$ be the set of linear functions that compose the set of piecewise linear, continuous closed curves defining $\gamma(t)$.

Choose a linear segment of $\gamma(t)$. There is a linear function $\lambda(t)$ describing the behavior of $\gamma(t)$ on this segment and $\lambda(t) \in \mathfrak{A}$. That is, $\gamma(t) = \lambda(t) + \lambda_0$ as long as the target moves on this line segment. Furthermore, for this function $\lambda(t)$, there is a partition element p (a polytope) in which $\gamma(t)$ resides.

When $\gamma(t)$ transitions to its next behavior $\lambda'(t)$ (and hence next partition p'), there are two possibilities: (1) there is only a single next linear behavior or (2) there is an intersection of several closed curves and more than one choice of linear behavior is possible.

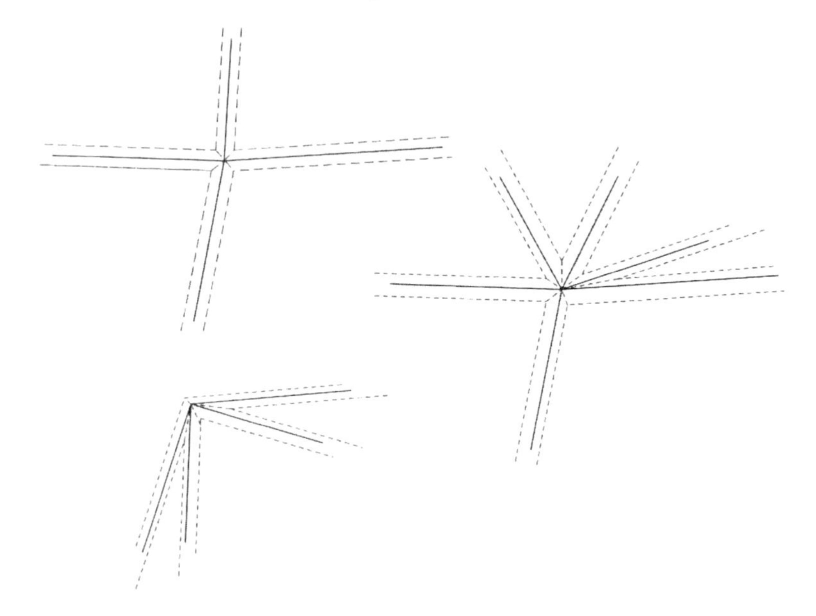

FIGURE 39.2 Example of polytope partitioning in two dimensions: A partition generated by polytopes of a piecewise linear and continuous curve $\gamma(t)$ can be generated by encasing each linear segment of $\gamma(t)$ in a polytope that shares a vertex with other polytopes encasing linear segments of $\gamma(t)$ that touch. The remainder of the partition is generated in the natural way.

In Case (1), it is clear that the next partition element is a function only of the previous partition element. Likewise, the next linear behavior is only a function of either the previous partition element or the product of the previous partition element and the previous linear behavior. In either case, we can define the functions η and ξ as needed.

In Case (2), Assumption 4 (Section 39.3) ensures that the next linear function will either be totally random, deterministic, or some probabilistic function that depends on previous behaviors of $\gamma(t)$ (that is, the previous partition elements through which $\gamma(t)$ has passed and linear functions that governed the behavior of $\gamma(t)$). This implies that there are n_2 and n_3 and a function $f \in \mathcal{F}_{\mathfrak{A}}$ that depends on $\mathcal{A}^{n_2}$ and $\mathfrak{A}^{n_3}$ so that

$$f = \xi(p_i, p_{i-1}, \ldots, p_{i-n_2}, \lambda_i, \lambda_{i-1}, \ldots, \lambda_{i-n_3})$$

where ξ is the second function required for an STF model and $f(\lambda') \neq 0$. As a result of this assumption, there is an n_1 and a probability distribution $g \in \mathcal{F}_{\mathcal{A}}$ so that

$$g = \eta(p_i, p_{i-1}, \ldots, p_{i-n_2})$$

and $g(p') \neq 0$. The function η is the first function required for the STF.

Thus we have shown that there is a partition of D and an STF describing $\gamma(t)$ under the assumptions made in Section 39.3.

Corollary 39.1

For any refinement of the polytope partition defined in the preceding proof, there is an STF modeling $\gamma(t)$ with this refined partition.

Proof

The assumptions in the proof of the theorem are not impacted by refinement of the partition. It follows at once that an STF exists, but with distinct values for n_1, n_2, and n_3.

39.5 Algorithm

Throughout this section, we will assume that we have a sample of the curve $\gamma(t)$ denoted by $\gamma_1, \ldots, \gamma_N$. Following the remarks in Section 39.3, we will assume that these samples may have been corrupted by noise.

39.5.1 Partitioning Scheme

Let $\mathcal{A}$ be a partition of the compact set $D \subseteq \mathbb{R}^n$. A *partitioning scheme* is a rule π that takes a set $p \in \mathcal{A}$ and returns a partition of p denoted by the sets $p^1, p^2, \ldots, p^L$. Formally

$$\pi : \mathcal{A} \rightarrow \bigcup_{p \in \mathcal{A}} 2^p$$

where 2^p is the power set of $p \in \mathcal{A}$.

These sets $p^1, p^2, \ldots, p^L$ can be used to define a new partition $\mathcal{A}_1$ of D as

$$\mathcal{A}_1 = \mathcal{A}_0 \setminus p \cup \pi(p) \tag{39.1}$$

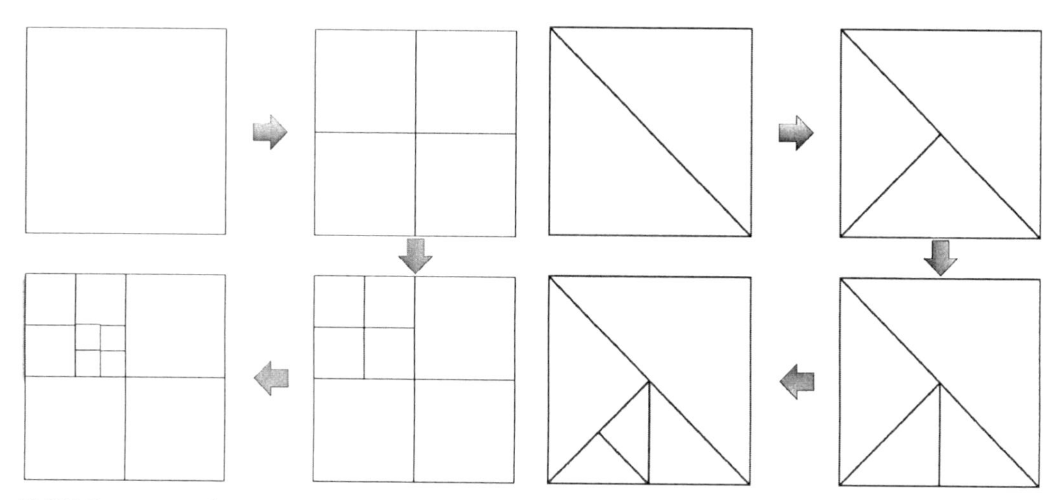

FIGURE 39.3 Simple partitioning schemes and a sequence of partitions: A partitioning scheme is a simple rule for replacing a partition of a set D with a new partition that is finer than the original.

If we assume that π can be extended to any set in its output, then we can use π to create an ever finer sequence of partitions by repeatedly applying π to elements of a partition $\mathcal{A}_k$, where $k \geq 0$. For the sake of brevity, we may write this as

$$\mathcal{A}_{k+1} = \pi(\mathcal{A}_k, p) = \mathcal{A}_k \setminus p \cup \pi(p)$$

to indicate that $p \in \mathcal{A}_k$ and we wish to refine set p into subsets $p^1, \ldots, p^L$ and create partition $\mathcal{A}_{k+1}$ by replacing p with these new subsets. Two examples for partition schemes are shown in Figure 39.3.

39.5.2 Null Linear Function

Suppose, by the nature of π, $\mathcal{A}_0$, and a sampling $\gamma_1, \ldots, \gamma_N$, we construct a partition element p with too few samples to fit a linear function. We call the linear function in this case *null* and write it as $\lambda_\emptyset$. In Section 39.6, we show a case in which the $\lambda_\emptyset$ function is used to cope with a highly nonlinear corner in $\gamma(t)$ in a region where several partition elements touch.

Note further, we will assume that linear fitting is possible whenever a sufficient quantity of data is available. We do not consider cases where techniques like Tikhonov regularization (see Ref. [26]) are needed to resolve ill-posed problems.

The occurrence of the null linear function (illustrated in the example in Section 39.6) is the reason we use two functions (η and ξ) instead of a single combined function. Depending upon the data available and partition used, we can estimate the next region a target will occupy even when no linear estimator is available.

39.5.3 Algorithm Description

Algorithm 1 will generate $\widehat{\mathcal{A}}$, $\widehat{\mathfrak{A}}$, and $\widehat{\mathcal{T}}$ given $\gamma(t)$. As it generates $\widehat{\mathcal{A}}$ and $\widehat{\mathfrak{A}}$, the functions $S_{\hat{\mathcal{A}}}$ and $S_{\hat{\mathfrak{A}}}$ will also be generated. These will change as the partition changes during algorithm execution. Knowing $S_{\hat{\mathcal{A}}}$ is easy once a partition is fixed. The function $S_{\hat{\mathfrak{A}}}$ is generated as the set $\mathfrak{A}$ is built and linear functions are mapped into partition elements (see Lines 10–16 of Algorithm 1). The input to the algorithm is an initial partition $\mathcal{A}_0$, a partitioning scheme π, and a finite sampling of $\gamma(t)$. We also assume that values for n_1, n_2, and n_3 can be provided, or computed using the work in [23]. An integer value M is also input to determine when insufficient data are available and the null linear function must be used. Using the sampling and the initial partition, the symbolic time series **p** is constructed from the function $\mathcal{S}_{\mathcal{A}_0}$. At the same time, the elements of the sample are partitioned into Γ_{p_i} for each partition element p_i in the **p**. The set of linear functions is initialized to the empty set.

Algorithm 1

Input: $\mathcal{A}_0, \pi, \gamma_1, \ldots, \gamma_N$, and n_1, n_2, n_3 (if not computed using [23]), $M \geq 3$

Output: $\widehat{\mathcal{A}}, \widehat{\mathfrak{A}}, \widehat{\mathcal{T}} = (\hat{\eta}, \hat{\xi})$

Procedure:

1: Set $j := 0$
2: Set $\mathbf{p} = \mathcal{S}_{\mathcal{A}_0}(\gamma_1, \ldots, \gamma_N)$
3: Divide $\gamma_1, \ldots, \gamma_N$ into sets $\{\Gamma_{p_i}\}_{p_i \in \mathbf{p}}$
4: Set $\widehat{\mathfrak{A}} = \emptyset$
5: **for** $p_i \in \mathbf{p}$ **do**
6: **if** $|\Gamma_{p_i}| \geq M$ {Γ_{p_i} has sufficient data to support a linear fit} **then**
7: Set $\lambda_{p_i} = \lambda_\emptyset$
8: $\widehat{\mathfrak{A}} = \widehat{\mathfrak{A}} \cup \{\lambda_{p_i}\}$
9: **continue**
10: Fit a linear function λ_{p_i} to Γ_{p_i} using techniques described in Ref. [27]
11: **if** goodness of fit for λ_{p_i} is below a threshold **then**
12: Set $\mathcal{A}_{j+1} := \pi(\mathcal{A}_j, p_i)$
13: $j := j + 1$
14: Go to Line 2
15: **else**
16: $\widehat{\mathfrak{A}} = \widehat{\mathfrak{A}} \cup \{\lambda_{p_i}\}$
17: Set $\widehat{\mathcal{A}} := \mathcal{A}_j$
18: Compute $\mathbf{p} = \mathcal{S}_{\widehat{\mathcal{A}}}(\gamma_1, \ldots, \gamma_N)$ and $\lambda = \mathcal{S}_{\widehat{\mathfrak{A}}}(\gamma_1, \ldots, \gamma_N)$
19: Compute $\widehat{\mathcal{T}} = (\hat{\eta}, \hat{\xi})$ using algorithms discussed in [2]
20: **return** $\widehat{\mathcal{A}}, \widehat{\mathfrak{A}}, \widehat{\mathcal{T}}$

At Line 5, we begin iterating over the sequence **p**. For each symbol p_i, we determine whether Γ_{p_i} has sufficient data for computing a linear estimator. If not, then the linear function corresponding to symbol p_i is $\lambda_\emptyset$ and we continue to the next symbol p_{i+1}. Otherwise, we compute a linear function modeling the data contained in Γ_{p_i} and denote it λ_{p_i}. Computing λ_{p_i} in Line 10 can be done using linear regression. The test for quality of fit in Line 11 can be accomplished using correlation coefficients, t-tests on coefficients, or any combination of these standard tests ([27]).

If the goodness-of-fit test fails, then the partition is further refined using π in Line 12. Otherwise, we continue on to Line 16 where we add the linear function to the set of linear functions $\widehat{\mathfrak{A}}$. This addition assumes that a statistical test is used to determine whether the function λ_{p_i} already exists within $\widehat{\mathfrak{A}}$. The simplest such test is a Student's t-test on the coefficients of the linear function estimator. This process is described in Ref. [27], as well.

The loop initialized at Line 5 is exited just after Line 16. If we reach Line 17, then for all p_i there is a linear function describing the behavior of the samples in Γ_{p_i}. At this point, we define $\widehat{\mathcal{A}}$ to be the partition constructed over the course of executing Lines 2–16. Using this partition, we compute the sequences **p** and λ. We then use algorithms described in Refs [2,23] to compute $\widehat{\mathcal{T}}$, which is returned.

39.5.4 Proof of Convergence

Preposition 39.1

Algorithm 1 will derive a partition $\widehat{\mathcal{A}}$ and set of functions $\widehat{\mathfrak{A}}$ that are used to produce an STF modeling $\gamma(t)$. The algorithm executes in a minimal number of steps, which may depend on $\mathcal{A}_0$ and the partition scheme π.

Proof

The fact that the algorithm terminates is ensured by the fact that, eventually, Step 12 will subdivide $\mathcal{A}_0$ to the point that λ_{p_i} will never fail the lack of fit test in Step 11 or too few points will be available in every Γ_{p_i} and Step 6 will be executed. The fact that $\gamma(t)$ is composed of linear functions ensures that one or the other of these two eventualities will be reached.

Once Step 17 is reached, there can be no more partitions in $\widehat{\mathcal{A}}$ that contain data and that cannot be modeled by a linear function. It is clear that this is the first time this occurs in the algorithm. The derived partition $\widehat{\mathcal{A}}$ and linear functions $\widehat{\mathfrak{A}}$ can be used to estimate an appropriate STF, provided that $\gamma(t)$ satisfies the assumptions we have set.

It is worth noting that the partition derived by Algorithm 1 most likely will not converge to the partition used in the Proof of Theorem 39.1. By changing Line 11 to require not only a goodness-of-fit test, but also a *uniqueness* test, so that each partition in $\widehat{\mathcal{A}}$ corresponds to at most one linear model, we could produce a partition that approximates the partition used in the proof of Theorem 39.1 for the partition elements that contain data from $\gamma(t)$. We did not do this because the predictive power of the STF should not be effected if two or more linear segments share a partition element since the ξ function in the STF takes into account both the inputs (partition element name) and the outputs (previous linear behavior). This is illustrated empirically in Section 39.6.

Preposition 39.2

If Step 12 is never executed, then computational complexity of Algorithm 1 is $O(N^4)$, where N is the number of samples of $\gamma(t)$.

Proof

The symbolic time series derived in Step 2 has length $O(N)$. The sets Γ_{p_i} computed in Step 3 has size at worst $O(N)$ (in the case when all the points are contained in a single partition element, then the size is precisely N). Computing the linear regression in Step 10 requires a matrix inversion, which is at worst $O(N^3)$ using Gaussian elimination. This is the most computationally complex operation in Steps 10–20. Step 19 can be made to run in $O(N)$ when the results of Ref. [28] are applied. Schmiedekamp et al. [28] shows a linear time algorithm to estimate η that can be extended to estimate ξ, by making appropriate changes to the algorithm presented in Ref. [2].

39.5.5 Discussion of Computational Complexity

Knowing the exact computational complexity of Algorithm 1 is difficult without knowing the specific partitioning scheme being used, the initial partition $\mathcal{A}_0$, and the extent of the nonlinearity of the original function $\gamma(t)$. These three components will determine the number of sub-partitions that need to be created. However, it is not difficult to construct an example where a large number of passes through Steps 5–19 are required in order to ensure the goodness of fit of the linear functions.

In the worst-case scenario, the algorithmic running time is exponential. Consider the case in which a rectangular region of initial area A_0 must be partitioned into rectangles of area less than A. Further, suppose the partitioning scheme being used is one in which each rectangular partition is divided into four new rectangles of equal area (see Figure 39.3). To obtain a single rectangle of area less than A, Steps 5–19 must be repeated at least $\log_4(A_0/A)$. If the end result is that we have all rectangles of area less than A, then Steps 5–19 must be repeated $\sum_{k=0}^{n} 4^k$ where $k = \log_4(A_0/A)$. Hence Steps 5–19 of the algorithm are repeated $O(4^{A_0/A})$. Thus we have

Preposition 39.3

The worst-case running time of Algorithm 1 is exponential in the quotient of the volume of the initial partition and the volume of the smallest partition created during algorithm execution.

A simple way to ameliorate this problem is to pass from linear functions to simple polynomial functions of time quadratics being the simplest example. The resulting model can still be computed using a linear regression scheme. However, if the function is highly nonlinear, then polynomial approximation may decrease the number of loops through Steps 5–19 by a marginal amount depending upon the nature of π and $\mathcal{A}_0$.

39.6 Behavioral Learning Example

To test our algorithm, we developed a simple piecewise linear trajectory simulation system. Data were generated in the form of (x, y, t) coordinates and a simple two-dimensional grid was used for the initial partition. Observations were perturbed by independent Gaussian noise. We employed a partitioning rule like the one shown in Figure 39.3a. The simulated vehicle (referred to as the target) was given a constant speed and a variable heading. As the target entered a new grid region, we recorded its initial point of entry (x_0, y_0, t_0) and computed a data set $\Delta = \{(x_i - x_0, y_i - y_0, t_i - t_0\}$ where i index all points occurring in a given partition.

To test for linear lack of fit in Step 11 of Algorithm 1, we computed a quadratic regression on both the x and y dimensions with respect to time, that is, we used Δ to fit the functions:

$$x_i = x_0 + v_x t_i + \frac{1}{2} a_x t_i^2 + w_i$$

$$y_i = x_0 + v_y t_i + \frac{1}{2} a_y t_i^2 + v_i$$

where w_i and v_i are statistical error terms used in linear modeling [27]. If the inferred values of a_x and a_y were both statistically insignificant, then a linear fit model was assumed. Statistical significance was tested using a standard t-test on the terms of the regression [3]. Matching of linear models needed Step 16 of Algorithm 1 was also done using t-tests on model terms. Computing $\mathcal{T}$ (Step 19) requires a distribution test [2,23]. As our distribution test, we chose the Kolmogorov–Smirnov test.

It is worth noting that Algorithm 1 does not prescribe any specific linear quality of fit test, linear model matching test, or specific tests for computing $\mathcal{T}$. The implementer is free to choose tests that

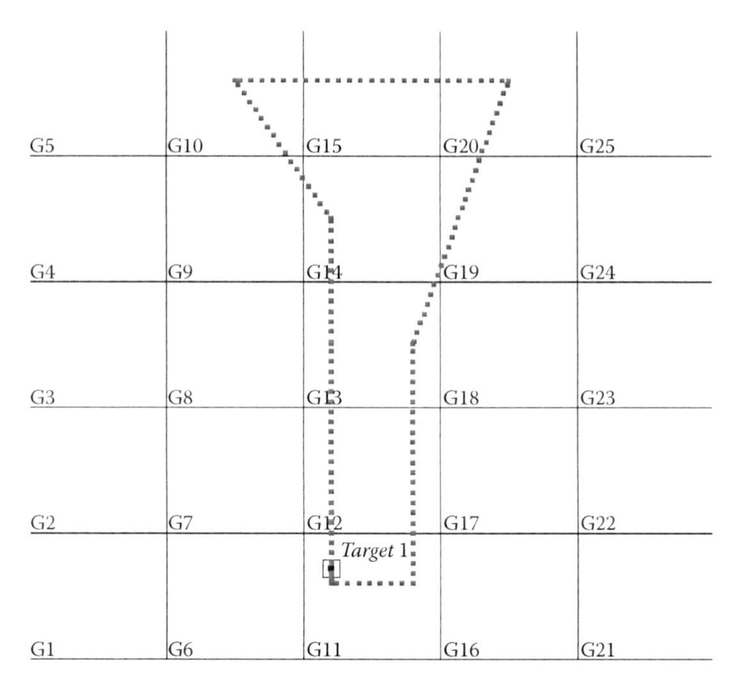

FIGURE 39.4 Deterministic sample path: A simple deterministic sample path was chosen to test the algorithm's ability to generate partition a region of space and fit linear functions.

work best for the given data set at hand. In our case, basic *t*-tests sufficed. More complex data sets or models may require more sophisticated testing, as would extensions to nonlinear models within partition regions.

We tested our algorithm using the sample path shown in Figure 39.4 perturbed by observation noise. Noise was introduced by adding a random number with distribution $\mathcal{N}(0, \sigma^2)$, where $\mathcal{N}$ is the normal distribution, to both the x and y values in the target's position.

In executing Algorithm 1, we assumed we were given

1. A sample of the track used by the simulated vehicle in the form of (x, y, t) points
2. The initial partition shown in Figure 39.4
3. The partition scheme (nested squares)
4. We set $n_1 = n_2 = n_3 = 1$ (though we later verified that the algorithm given in Ref. [23] could be used to estimate these parameters as well)

Using this information and Algorithm 1, we computed

1. A final partition $\widehat{\mathcal{A}}$
2. A set of linear function $\widehat{\mathfrak{A}}$
3. The STF model $\widehat{T}$ for the vehicle track provided
4. A set of estimators for future track behavior, which were compared for quality of fit (these are shown in Figures 39.5 through 39.7)

To test the efficacy of the linear fits, we compute residuals $\mathbf{r}$ with

$$r(t)^2 = (x(t) - \hat{x}(t))^2 + (y(t) - \hat{y}(t))^2 \tag{39.2}$$

where there is an exact linear estimator for the position (x, y) of the target at time t. That is, $r(t)$ is the distance between a pointwise estimator of the target's position and the ground truth.

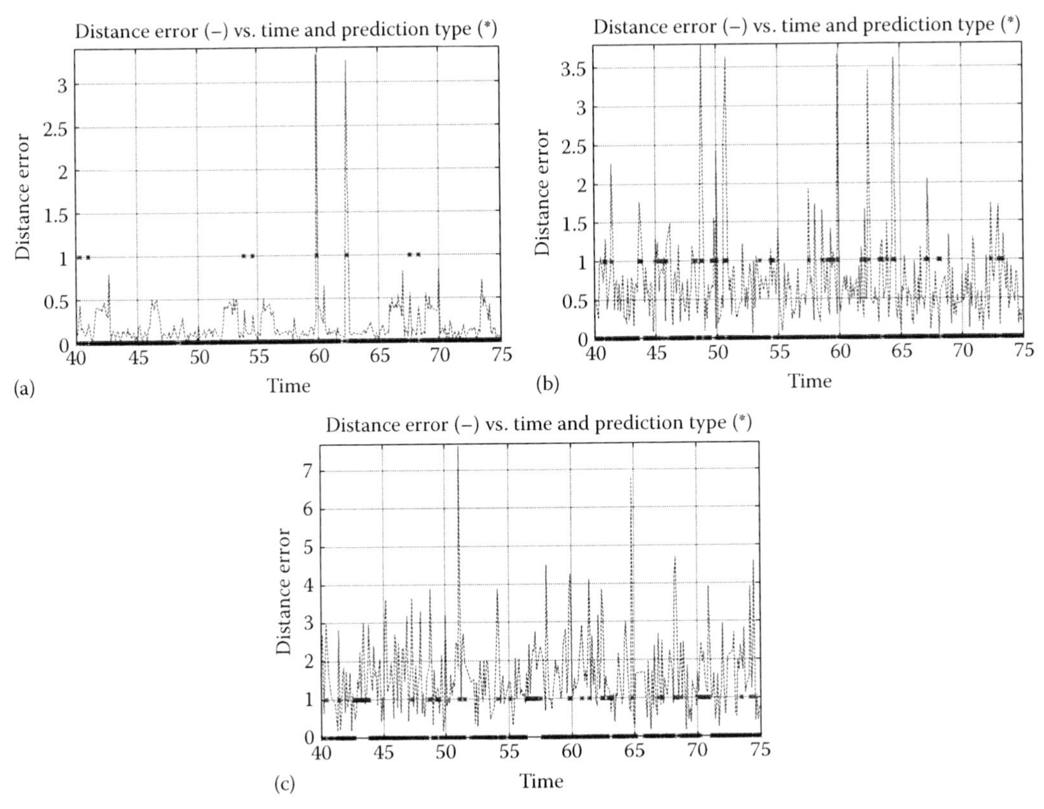

FIGURE 39.5 A plot of the function $r(t)$: Boundedness of the function suggests that most of the error associated with the predictions is due to the noise in the data. (a) $\sigma = 0.05$, (b) $\sigma = 0.3$, and (c) $\sigma = 0.8$.

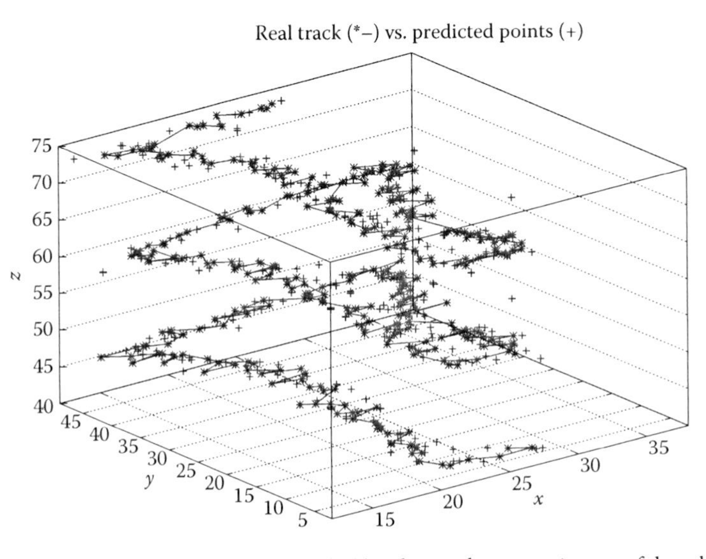

FIGURE 39.6 A track estimator and ground truth track: (*) indicates the true trajectory of the robot track modeled with noise. (+) indicates predicted positions of the track. We show only the plot when $\sigma = 0.8$.

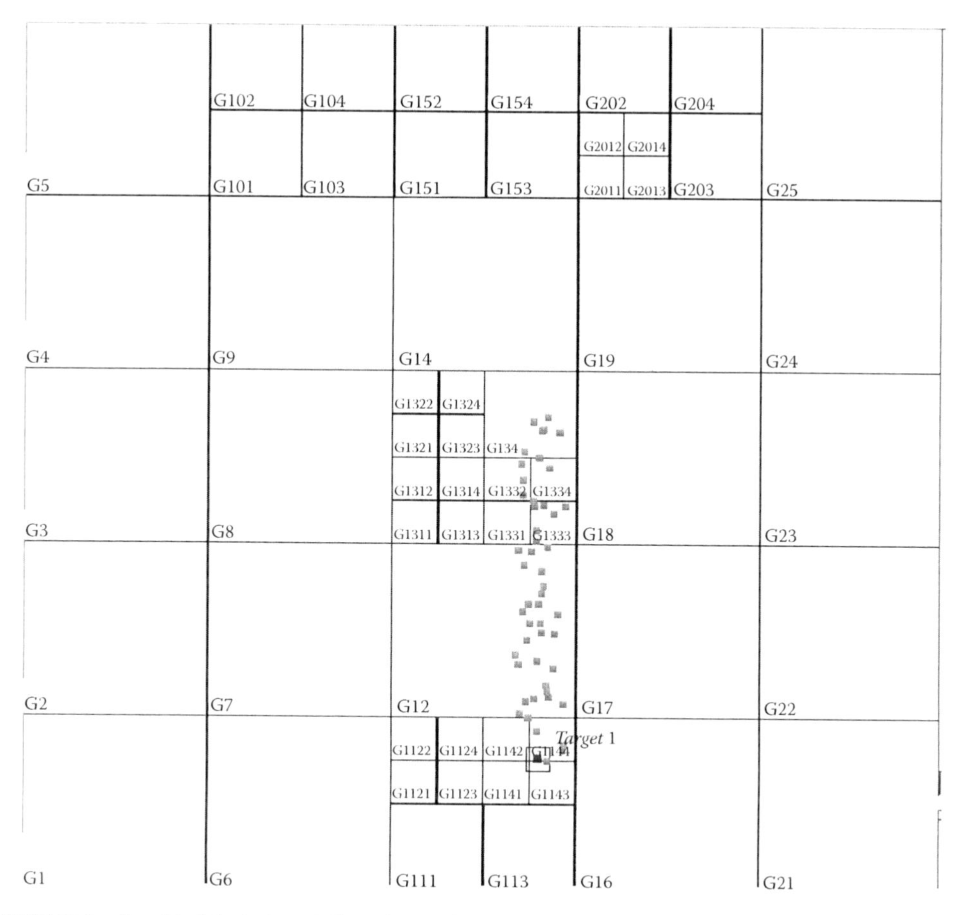

FIGURE 39.7 Graphical depiction of Algorithm 1: The original grid has been subdivided to deal with nonlinearities in the trajectory of the vehicle. We show only the plot when $\sigma = 0.8$.

When linear estimators were not available, we used the centroid of the estimated region as a position predictor. We recorded these occurrences and they are displayed along with the values of r. For all tests, the next grid region was predicted in the support of the η function 100% of the time. Thus in regions with dense partitioning, the partition element, estimated by η, can be used in place of the linear estimator (especially if one is not available). It is worth noting that the level of partition refinement and the number of times that the η function was used to predict track position (versus a linear estimator) are critical indicators of the ability of the STF to predict track positions, particularly when the track is switching from one linear behavior to another and exhibits a corner.

In Figures 39.5 through 39.7, we show the graphical results of our algorithm when used with $\sigma = 0.05$, $\sigma = 0.3$, and $\sigma = 0.8$. The quality of fit is displayed by the $r(t)$ plot and the 3D track plot. Spikes in the $r(t)$ plot correspond to regions where a partition centroid was used instead of a linear predictor. These were regions where insufficient data were available to make a linear predictor and the symbolic upper-level model was used to estimate the vehicle position. In Figure 39.5, the asterisks indicate whether or not the symbolic grid was used to fill in missing linear prediction information. A value of 1 indicates that it was, while 0 indicates that a linear model was used.

Mean square error for predictions is shown for the various cases in Tables 39.1 through 39.3.

Overall, the predictor performs well within reasonable noise levels. There is clearly an increase in error caused by the use of estimators from partition elements when the partitions are not sufficiently dense. This may be an indicator that a poor partitioning scheme is being used and an alternate one may be more appropriate.

TABLE 39.1 Mean Square Error for the Various Noise Levels

σ Value	Mean Error (Exact Predictions)	Mean Error (Region Predictions)	Total Mean Error
0.05	0.18136	0.61945	0.1915
0.3	0.57447	1.3927	0.68062
0.8	1.4508	2.456	1.6506

Note: This is illustrated graphically in Figure 39.5 for $\sigma = 0.8$.

TABLE 39.2 Standard Deviations of the Sample Errors

σ Value	Error StDev (Exact Predictions)	Error StDev (Region Predictions)	Total Error StDev
0.05	0.1616	0.7508	0.20667
0.3	0.29867	1.037	0.54079
0.8	0.76275	1.4913	1.0336

TABLE 39.3 Total Experimental Counts for the Various Noise Levels

σ Value	Sample Count (Exact Predictions)	Sample Count (Region Predictions)	Total Sample Count
0.05	2955	70	3025
0.3	2737	408	3145
0.8	2476	614	3090

Note: When used with other data, exact 95% confidence intervals on mean square error can be computed, if needed.

39.7 Real-World Data

We have had success in extending this approach to live ship tracks traversing the globe. Any curve that is continuous and recurrent but not necessarily piecewise linear can be approximated by a piecewise linear spline. It is clear we can apply this technique, in practice, to curves that are more general than the ones that satisfy Assumptions 1–4 in Section 39.3.

The available open source data are sparse and the detailed linear models of the type used in the simulated example cannot be used. Instead we collect data and compute simple linear models of speed and heading of the ships within the grid regions. For our grid structure, we use the Military Grid Reference System [29], which supports various levels of resolution from 100 km × 100 km to 1 m × 1 m. This method has been relatively successful at predicting ship behaviors and we continue to work on methods of improving it.

To test the ability of an STF model to predict the position of a vessel after a period of time, we collected initial training data from http://sailwx.info. Data were stored in a local database and this database is refreshed on a daily basis. We collected over 8 months worth of data for 13 distinct ships, representative of a variety of vessel classes (including cruise ships, Great Lakes trading vessels, and private craft).

We used the STF algorithm described earlier to produce a collection of vessel track models. In executing the algorithm, we provided values for n_1, n_2, and n_3. We then tested these models for their prediction power over the course of 3 days. Figure 39.8 shows an example of a ship predictor for the *Disney Magic*. In the figure, predictions were within 5 nautical miles (nm) of the true position of the craft after 24 h of observation.

The 13 ships that were monitored had various reporting characteristics. Some ships reported their position every hour, while other ships reported their position every few days. We used available reports to make predictions. This allowed us to verify our models at times ranging from 1 to 150 h. Figure 39.9 shows summary statistics for our experiment.

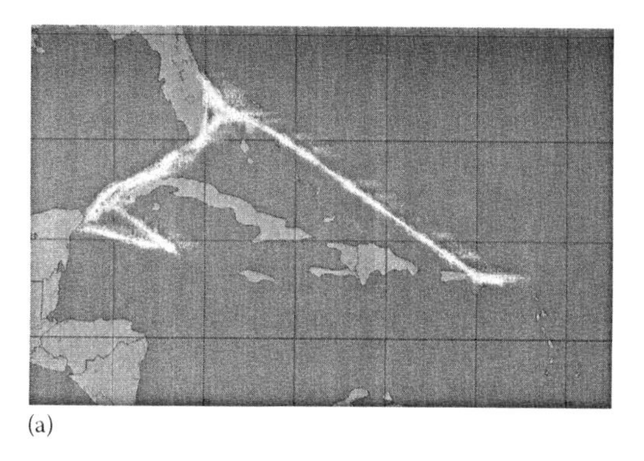

(a)

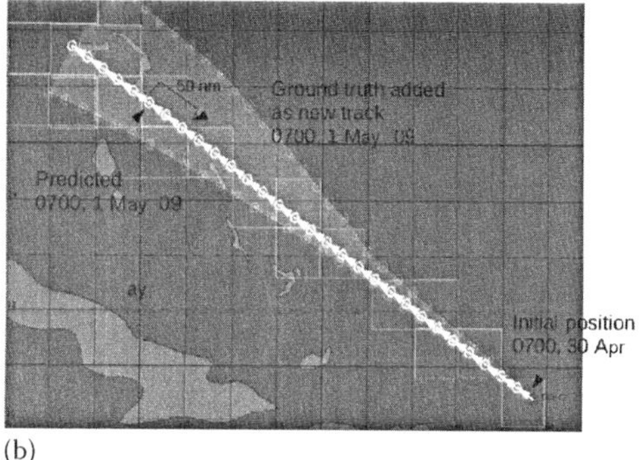

(b)

FIGURE 39.8 (a) The historic track of the *Disney Magic*. Note that it consists of a recurrent track composed of several closed curves. (b) Two day sample of the *Disney Magic* with predictions provided by an STF.

	Duration (h)	Error (nm)	Error/Time (nm/h)
Mean	20.8	32.95	2.83
Median	16.5	25	2.83
Standard error (95%)	14.06	14.73	1.38

FIGURE 39.9 Mean, median, and error of prediction in live ship tracks: This table shows the mean, median, and error for an experiment involving 13 ships tracked over 3 days. The ships self-reported their positions. Models built using the algorithms discussed in this chapter were used to predict vessels' positions. Ground truth was compared with the predicted results.

The table (shown in Figure 39.9) shows that the average error associated to prediction was 32.95 nm and that the average time for a prediction was 20.8 h. This means that our algorithm (on average) was predicting the position of a vessel 20.8 h in the future given an initial position. The mean values are clearly influenced by outliers, specifically the vessel that reported every 150 h. This is illustrated by the smaller median values for both duration of prediction and error.

39.8 Conclusions and Future Directions

In this chapter, we showed a hybrid hierarchical method for modeling a dynamical system whose future behavior is probabilistically determined by its previous behavior and its path through a collection of partition elements over the space through which the system evolves. This model assumes that the instantaneous behavior of the dynamical system is linear and chosen from a collection of available linear behaviors. The modeling approach is unique in that we require little a priori information on the system other than the fact that the dynamics are given by an STF.

This approach can be used to model the behavior of moving vehicles whose behaviors may be locally linear, but globally nonlinear. Since vehicles are often controlled by human agents, the behavior of the vehicle at any given time may be a function of its previous behavioral modes and its current position in space. This is precisely the type of behavior we capture in this model.

39.8.1 Future Directions

Many future directions of research are possible. One critical topic for future research is in determining the quantity of data required to obtain a specific level of accuracy in the model. The example in Section 39.6 used 3000 individual data points. This is significantly larger than the 100–200 points usually recommended in the literature [22] for using ARIMA models; in contrast, the number of posits in the ship tracking study usually did not exceed 1000. Throughout this work, we have simply assumed sufficient data to produce an accurate and statistically significant model. More work should be done to understand the number of data points needed to guarantee statistical significance in both the symbolic model and the statistical model.

Another interesting direction of work is in evaluating the trade-off between partition refinement and quality of fit, particularly in light of the fact that the computational complexity of the algorithm is entirely based on the level of refinement of the partition. It is clear that too much refinement will lead to a condition where all linear models are null and the system dynamics are described entirely by a Markov process on the partition elements. On the other hand, too little refinement can lead to poor model fit, particularly if this algorithm is used on a nonlinear track.

It is also possible to extend this work to cases where the local behavior is nonlinear and can be modeled in a regression framework. In this case, we appeal to the Weirstrass polynomial theorem. This will provide a more accurate model than a simple gross application of this technique to dynamical systems whose local behavior is nonlinear and probabilistically dependent on previous behaviors and its path through space.

Another interesting area of future work is in determining whether or not the use of a hybrid technique simplifies the complexity of data management for tracking. It is reasonable to suspect that the complexity of track management, data storage, and track matching may all be reduced since we are assigning track components to discrete regions and thus decreasing the number of data points that must be considered within each region. This conjecture could be evaluated in comparison to the standard Kalman filtering techniques already in use for ship tracking.

A final future direction for anomaly detection involves the online derivation of a model for $\mathcal{T}$. The Kalman filter can be used to determine whether the linear predictor produced by the hybrid model is still valid. Similarly, an online system for deriving η could be used to determine whether the hybrid model is still valid. We are currently working on a Bayesian technique to solve this problem that uses the fact that the conjugate prior of the multinomial distribution is known to be the Dirichlet distribution. This fact can be used to derive an optimal Bayesian estimator of the η function after any number of observations of the trajectory of $\gamma(t)$ as symbolized by $\mathcal{S}_{\mathcal{A}}$.

39.A Appendix: Algorithm for Deriving η of the STF

In this section, we provide a simple algorithm for deriving η the first component of the STF. This algorithm is essentially a minor modification of the Causal State Splitting and Reconstruction (CSSR) algorithm of Refs [24,30]. The CSSR algorithm of Shalizi et al. [24,30] can be used to derive a HMM from an observed sequence **y**, without a priori knowledge of the state structure of the underlying model. The resulting HMM is a Markov chain with labels on the edges. This is distinct from the more common form of HMM defined in Rabiner [11], in which each state corresponds to an output probability distribution and transitions through the model are unlabeled.

Details behind the CSSR algorithm are provided in Ref. [24]. Our approach is different. We evaluate substrings of **p** of length n_1. We use the empirical conditional distribution on the next symbol observed to estimate η. We simultaneously assign these substrings to states in the corresponding Markov chain representation of η. Because of the reduced complexity compared with the CSSR algorithm, our algorithm runs in $O(|\mathbf{p}|)$ time where $|\mathbf{p}|$ is the size of the input string.

Algorithm 2

Input:

Symbolic time series **p**

Alphabet $\mathcal{A}$

Length n_1

Output:

Stochastic Map η
Probabilistic Labeled Transition System G

Procedure:

1: Let W be the set of substrings of $\mathbf{p}$ with length n_1
2: **for all** $\mathbf{w} \in W$ **do**
3: Compute $\Pr(a \in \mathcal{A} \mid \mathbf{w}) := \dfrac{\#(\mathbf{w}a, \mathbf{p})}{\#(\mathbf{w}, \mathbf{p})}$
4: Define $f_{\mathbf{w}}(a) := \Pr(a \in \mathcal{A}|\mathbf{w})$
5: Define $\eta(w) := f_{\mathbf{w}}$
6: **return** η

The following formulas can be used to compute $f_{q_i|\mathbf{y}}$ and $f_{a\mathbf{x}|\mathbf{y}}$. Let $\#(\mathbf{x}, \mathbf{y})$ be the number of times the sequence $\mathbf{x}$ is observed as a subsequence of $\mathbf{y}$:

$$f_{\mathbf{x}|\mathbf{y}}(a) = \Pr(a \mid \mathbf{x}, \mathbf{y}) = \frac{\#(\mathbf{x}a, \mathbf{y})}{\#(\mathbf{x}, \mathbf{y})} \tag{39.3}$$

$$f_{q_i|\mathbf{y}}(a) = \Pr(a \mid q_i, \mathbf{y}) = \frac{\sum_{\mathbf{x} \in q_i} \#(\mathbf{x}a, \mathbf{y})}{\sum_{\mathbf{x} \in q_i} \#(\mathbf{x}, \mathbf{y})} \tag{39.4}$$

It is worth noting that the CSSR algorithm discussed in Ref. [24] and the algorithms discussed by the authors in Ref. [2] will return a probabilistic labeled transition system that is statistically equivalent to the one derived by Algorithm 2. The difference is the number of states contained in the probabilistic labeled transition system. The CSSR algorithm clusters states with similar conditional distributions into a single state; similarity is measured using the Kolmogorov–Smirnov algorithm. In our case, this will lead to a more efficient representation of the labeled transition system, but it is not necessary for our understanding of η or the dynamics of the system. Further, it leads to an algorithm with significantly higher computational complexity.

Preposition 39.A.1

The complexity of Algorithm 2 is $O(|\mathbf{p}|)$.

Proof

The number of substrings of $\mathbf{p}$ of length $k \le l$ is $|\mathbf{p}| - k + 1 = O(|\mathbf{p}|)$ as we parse strings of length $k + 1$ we can compute functions $f_{\mathbf{w}}$ for each $\mathbf{w}$ with length k. Hence Steps 1–4 may be computed in $O(|\mathbf{p}|)$. Steps 4–8 require each $\mathbf{w}$ to be evaluated with each element of $\mathcal{A}$. Using a hash table, the complexity of Step 13 can be made to be $O(1)$ and hence, Steps 9–12 require $O(|\mathcal{A}||\mathbf{p}|) = O(|\mathbf{p}|)$.

39.A.1 Deriving an Appropriate n_1

A key input element in Algorithm 2 is n_1, the length of substrings of symbolic time series $\mathbf{p}$ to use. If $\mathbf{p}$ is sufficiently long, we can derive the correct l to use from the data itself. Algorithm 3 will provide this value.

Algorithm 3

Input:

Symbolic time series $\mathbf{p}$
A p-value $p < 1$
A distribution equality test T {e.g., The Kolmogorov–Smirnov Test}
Integer $l_{\max}$, a search bound

Output:

The value of n_1 for Algorithm 2

Procedure:

1: Set $l := 0$
2: **while** $l \leq l_{\max}$ **do**
3: Let W_l be the set of substrings of $\mathbf{p}$ with length l
4: Let W_{l+1} be the set of substring of $\mathbf{p}$ with length $l+1$
5: Set ContinueLoop = **true**
6: **for all** $\mathbf{w} \in W_l$ **do**
7: **for all** $a \in \mathcal{A}$ **do**
8: **if** $a\mathbf{w} \in W_{l+1}$ **then**
9: Compute $f_{\mathbf{w}}$
10: Compute $f_{a\mathbf{w}}$
11: Set $p_{\text{test}} := T(f_{\mathbf{w}}, f_{a\mathbf{w}})$
12: **if** $p_{\text{test}} < p$ **then**
13: Set ContinueLoop = **false**
14: **break**
15: **if** ContinueLoop = **false then**
16: **break**
17: **if** ContinueLoop = **false then**
18: Set $l := l + 1$
19: **else**
20: **break**
21: **return** $n_1 := l$

Algorithm 3 works by comparing the distribution of each string of length l, $\mathbf{w}$ to histories $a\mathbf{w}$ of length $(l+1)$. If the distribution generated by $a\mathbf{w}$ is statistically different from that produced by $\mathbf{w}$, this implies that the embedding dimension of the system must be at least $(l+1)$; otherwise, the two distributions would be the same and no additional information would be acquired from the addition of the information a to string $\mathbf{w}$. Hence we have proved the following:

Preposition 39.A.2

If n_1 is known to be less than l_{max}, then Algorithm 3 will converge to the appropriate embedding dimension of the dynamical system $|\mathbf{p}| \to \infty$.

Clearly, Preposition 39.A.2 must be an asymptotic result because we are using a statistical testing method to determine whether or not the distributions generated in Lines 9 and 10 are equivalent.

Since we are performing many statistical tests, the probability of Type 1 error will become multiplicatively large as we repeat this procedure. We can apply the Bonferonni approach [27] to attempt to reduce this probability; however in general, this will only be successful with small alphabet sizes or dynamics in which the support on the next symbol is a small subset of the alphabet.

Preposition 39.A.3

The computational complexity of Algorithm 3 is $O(|\mathbf{x}|)$.

Proof

The loop from Steps 7–17 requires at most $|\mathcal{A}|$ steps. This loop is repeated for each string in $|W_l|$ which is $O(n)$. Hence, the complexity of Steps 7 to 14 and 6 to 16 is at worst $O(|\mathcal{A}||\mathbf{x}|)$. The loop would be executed at most $l_{\max}$ times, making the entire algorithm $O(l_{\max}|\mathcal{A}||\mathbf{x}|)$. Hence this algorithm is $O(|\mathbf{x}|)$.

39.A.2 Methods of Comparing $f_{\mathbf{w}}$ and $f_{a\mathbf{w}}$ in Algorithm 3

For any symbolic time series **w**, we can think of $f_{\mathbf{w}}$ as being a vector in the unit simplex of $\mathbb{R}^{|\mathcal{A}|}$. We can compare two such vectors using several techniques.

The most common technique is the Kolmogorov–Smirnov test [3] which compares the computed cumulative distribution function and relies on a fixed assignment of positions in the vector $f_{\mathbf{w}}$ to elements of $\mathcal{A}$. In this case, the distribution governing the Kolmogorov–Smirnov test is a random variable describing the supremum of a Brownian bridge [31]. As such, a p-value for each Kolmogorov–Smirnov test can be computed and a total p-value on all tests performed in Step 11 can be calculated.

The χ^2 goodness-of-fit test can also be used to compare $f_{\mathbf{w}}$ and $f_{a\mathbf{w}}$. In this case, we can replace the values of the distribution with $f_{\mathbf{w}}$ with their frequency counts [3]. It is also known that the χ^2 test is inappropriate if more than 10% of the elements of $f_{\mathbf{w}}$ fall below 0.05. As its name implies a p-value for the χ^2 test can be computed using the χ^2 test and the resulting p-values can be used as done earlier to compute a conjoined p-value for the entire process.

We can also consider the elements of $f_{\mathbf{w}}$ as being the probabilities associated with a multinomial distribution. When considered this way, the work by Gold [32] and Quesenberry and Hurst [33] are most germane. As the number of samples used to derive the multinomial distribution (the number of times **w** occurs as a subsequence of **x**) approaches infinity, the multinomial distribution tends to a multivariate normal distribution. As with the binomial distribution, we can use this fact to derive asymptotic confidence intervals on the values in $f_{\mathbf{w}}$. The simplest such confidence interval is given by

$$\Pr(a\,|\,\mathbf{w}) \in f_{\mathbf{w}}(a) \pm \chi\sqrt{\frac{f_{\mathbf{w}}(1-f_{\mathbf{w}})}{N}} \tag{39.5}$$

where $\chi^2 = \chi_1^2(\alpha/|\mathcal{A})$ although recent work suggests replacing χ by more accurate values [34]. Here $N = \#(\mathbf{w}, \mathbf{x})$. Using this result and classical hypothesis testing methods for proportions [3], we can construct a simple statistical test to determine whether $f_{\mathbf{w}}$ is equal to $f_{a\mathbf{w}}$ using multiple tests on the elements of the vectors.

By using the approach to testing the equality of $f_{\mathbf{w}}$ and $f_{a\mathbf{w}}$, we can use Duncan's multiple range test comparison method [27]. In this case, the use of the Bonferonni approach can be reduced.

Unfortunately, this method does not easily produce a p-value for determining the quality of this test, because the p-value in this case is computed directly using the formula for the multinomial distribution. Thus, the use of the Kolmogorov–Smirnov or χ^2 test is more effective at computing a p-value for the complete algorithm.

In the examples given in Section 39.6, we use the Kolmogorov–Smirnov test along with the Bonferonni method for computing an appropriate α value. Work by others suggests that in general this method is more powerful than the χ^2 test [24]. No other work has compared the Kolmogorov–Smirnov test with the use of Duncan's test applied to the proportions.

39.B Appendix: Algorithms for Deriving ξ of the STF and n_2 and n_3

In this section, we provide a simple algorithm for deriving η the first component of the STF. A variation on this algorithm was presented in Ref. [2]. We do not show how to retrieve a Mealey machine as it adds unnecessary length to the algorithm and is clear from the Algorithm 2 and the discussion in Section 39.3. Let $\#(\mathbf{w}_1, \mathbf{p}, \mathbf{w}_2, \lambda)$ be the number of times sequence $\mathbf{w}_1$ occurs in $\mathbf{p}$ and at the same time sequence $\mathbf{w}_2$ occurs in λ so that the ends of the two sequences coincide. (Recall, $\mathbf{w}_1$ may have different length $\mathbf{w}_2$). Also let $\#(\mathbf{w}_1, \mathbf{p}, \mathbf{w}_2, \lambda, a)$ be the number of times $\mathbf{w}_1$ occurs in $\mathbf{p}$ and at the same time sequence $\mathbf{w}_2$ occurs in λ so that the ends of the two sequences coincide and a immediately follows $\mathbf{w}_2$.

Algorithm 4

Input:

Symbolic time series $\mathbf{p}$ and λ
Alphabets $\mathcal{A}$ and $\mathfrak{A}$
Lengths n_2 and n_3

Output:

Stochastic Map ξ

Procedure:

1: Let W be the set of pairs of substrings of $(\mathbf{p}, \lambda)$ with length n_2 and n_3, respectively, synchronized in ending position
2: **for all** $(\mathbf{w}_1, \mathbf{w}_2) \in W$ **do**
3: Compute $\Pr(a \in \mathfrak{A} \mid \mathbf{w}_1, \mathbf{w}_2) := \dfrac{\#(\mathbf{w}_1, \mathbf{p}, \mathbf{w}_2, \lambda, a)}{\#(\mathbf{w}_1, \mathbf{p}, \mathbf{w}_2, \lambda)}$
4: Define $f_{\mathbf{w}_1,\mathbf{w}_2}(a) := \Pr(a \in \mathfrak{A} \mid \mathbf{w}_1, \mathbf{w}_2)$
5: Define $\xi(w_1, w_2) := f_{\mathbf{w}_1,\mathbf{w}_2}$
6: **return** ξ

Preposition 39.B.1

When we assume that $N = |\mathbf{p}| = |\lambda|$, then the complexity of Algorithm 4 is $O(N)$.

Proof

Since elements of the set W are synchronized in ending position in $\mathbf{p}$ and λ, it follows that the size of W is limited to $N - \max\{n_2, n_3\}$. Hence, the loop in Step 2 is executed at most $O(N)$ time. The sizes of $\mathcal{A}$ and $\mathfrak{A}$ are fixed. Hence, Steps 2–6 are computed in $O(N)$.

It is immediately clear that Algorithm 3 can be modified to compute estimators for n_2 and n_3 by simply comparing distributions over pairs of strings in an exhaustive way. Algorithm 5 formalizes this.

Algorithm 5

Input:

Symbolic time series pair $\langle \mathbf{p}, \lambda \rangle$
A p-value $p < 1$
A distribution equality test T {e.g., The Kolmogorov-Smirnov Test}
Integers $l^1_{\max}$ and $l^2_{\max}$, search bounds.

Output:

Values of n_2 and n_3 used in Algorithm 4.

Procedure:

1: $l^1 = 0$
2: **while** $l^1 \leq l^1_{\max}$ **do**
3: $l^2 = 0$
4: **while** $l^1 \leq l^1_{\max}$ **do**
5: Set ContinueLoop = **true**
6: Let W_{l^1,l^2} be the set of pairs of substrings of **p** and λ with lengths l^1 and l^2 respectively, synchronized in ending position.
7: Let $W_{l^{+}11,l^2}$ be the set of pairs of substrings of **p** and λ with lengths $l^1 + 1$ and l^2 respectively, synchronized in ending position.
8: Let W_{l^1,l^2+2} be the set of pairs of substrings of **p** and λ with lengths l^1 and $l^2 + 2$ respectively, synchronized in ending position.
9: **for all** $(u,v) \in W_{l^1,l^2}$ **do**
10: **for all** $a \in \mathcal{A}$ **do**
11: **if** $(au,v) \in W_{l^1+1,l^2}$ **then**
12: Compute $f_{u,v}$
13: Compute $f_{au,v}$
14: Set $p_{\text{test}} = T(f_{u,v}, f_{au,v})$
15: **if** $p_{\text{test}} < p$ **then**
16: Set ContinueLoop = **false**
17: **break**
18: **if** ContinueLoop = **false then**
19: **break**
20: **for all** $\alpha \in \mathfrak{A}$ **do**
21: **if** $(u,\alpha v) \in W_{l^1,l^2+1}$ **then**
22: Compute $f_{u,v}$
23: Compute $f_{u,\alpha v}$
24: Set $p_{\text{test}} = T(f_{u,v}, f_{u,\alpha v})$
25: **if** $p_{\text{test}} < p$ **then**
26: Set ContinueLoop = **false**
27: **break**
28: **if** ContinueLoop = **false then**
29: **break**

30: **if** ContinueLoop = **false then**
31: $l^2 := l^2 + 1$
32: **else**
33: **break**
34: **if** ContinueLoop = **false then**
35: $l^1 := l^1 + 1$
36: **else**
37: **break**
38: **return** $n_2 := l^1$, $n_3 := l^2$

Preposition 39.B.2

If n_2 and n_3 is known to be less than l^1_{max} and l^2_{max}, then Algorithm 5 will converge to the appropriate embedding dimension of the dynamical system as $|\mathbf{p}| \to \infty$.

Proof

The result follows by applying a similar argument to the one used to prove Preposition 39.A.2.

Preposition 39.B.3

When we assume that $N = |\mathbf{p}| = |\lambda|$, then the complexity of Algorithm 5 is $O(N)$.

Proof

Since elements of the sets W_{l^1+1,l^2}, W_{l^+11,l^2}, and W_{l^1,l^2+2} are synchronized in ending position in $\mathbf{p}$ and λ, it follows that the sizes are bounded by $O(N)$. Hence, the loop in Step 9 is executed at most $O(N)$ time since the sizes of $\mathcal{A}$ and $\mathfrak{A}$ are fixed and search in a known set can be performed in constant time using a hash table. The remainder of the proof follows with appropriate changes from the proof of Preposition 39.A.3.

Acknowledgments

This material is based upon work supported by, or in part by, the Office of Naval Research Code 311 contract/grant number N00014-06-C-0022. The authors gratefully acknowledge this support and take responsibility for the contents of this report.

Portions of Dr. Griffin's work were performed as a Eugene P. Wigner Fellow and staff member at the Oak Ridge National Laboratory, managed by UT-Battelle, LLC, for the U.S. Department of Energy under Contract DE-AC05-00OR22725.

References

1. C. Griffin, R. R. Brooks, and J. Schwier, A hybrid statistical technique for modeling recurrent tracks in a compact set, *IEEE Transactions on Automatic Control*, 56(8), 1926–1931, 2011.
2. C. Griffin, R. R. Brooks, and J. Schwier, Determining a purely symbolic transfer function from symbol streams: Theory and algorithms, in *Proceedings of American Control Conference*, Seattle, WA, June 11–13, 2008, pp. 4065–4067.

3. R. Hogg and E. Tanis, *Probability and Statistical Inference*, 7th edn. Pearson/Prentice-Hall, Upper Saddle River, NJ, 2006.
4. G. A. F. Serber and C. J. Wild, *Nonlinear Regression*. John Wiley & Sons, New York, 2003.
5. G. E. P. Box, G. M. Jenkins, and G. C. Reinsel, *Time Series Analysis: Forecasting and Control*. John Wiley & Sons, New York, 2008.
6. E. Ghysels and D. R. Osborn, *The Econometric Analysis of Seasonal Time Series*. Cambridge University Press, New York, 2001.
7. J. R. M. Hosking, Fractional differencing, *Biometrika*, 68(1), 163–176, 1981.
8. R. F. Engle, GARCH 101: The use of ARCH/GARCH models in applied econometrics, *Journal of Economic Perspectives*, 15(4), 157–168, 2001.
9. B. K. Oksendal, *Applied Stochastic Control of Jump Diffusions*. Springer, Berlin, Germany, 2007.
10. F. Martinerie and P. Forster, Data association and tracking using hidden Markov models and dynamic programming, in *Proceedings of Conference ICASSP*, 1992.
11. L. R. Rabiner, A tutorial on hidden Markov models and selected applications in speech recognition, *Proceedings of the IEEE*, 77(2), 257–286, 1989.
12. R. L. Streitt and R. F. Barrett, Frequency line tracking using hidden Markov models, *IEEE Transactions on Acoustics, Speech and Signal Processing*, 38, 586–598, 1990.
13. L. Shi and S. Ólafsson, Nested partitions method for global optimization, *Operations Research*, 48(3), 390–407, 2000.
14. C. Loader, *Local Regression and Likelihood*. Springer, Berlin, Germany, 1999.
15. A. Ozonoff, S. Sukpraprut, and P. Sebastiani, Modeling seasonality of influenza with hidden Markov models, in *Proceedings of the American Statistical Association*, 2006.
16. J. Yin, Q. Yang, D. Shen, and Z. Li, Activity recognition via user-trace segmentation, *ACM Transactions on Sensor Networks*, 2008.
17. A. Villa, *Hybrid Control Systems in Manufacturing*. Gordon and Breach Science Publishers, London, U.K., 1990.
18. G. Xie, D. Zheng, and L. Wang, Controllability of switched linear systems, *IEEE Transactions on Automatic Control*, 47(8), 1401–1405, 2002.
19. Z. Sun and S. S. Ge, *Switched Linear Systems: Control and Design*. Springer-Verlag, London, U.K., 2005.
20. M. Babaali and M. Egerstedt, Nonpathological sampling of switched linear systems, *IEEE Transactions on Automatic Control*, 50(12), 2102–2105, 2005.
21. Z. Ji, L. Wang, and X. Guo, On controllability of switched linear systems, *IEEE Transactions on Automatic Control*, 53(3), 796–801, 2008.
22. E. del Castillo, *Statistical Process Adjustment for Quality Control*. Wiley Interscience, New York, 2002.
23. R. R. Brooks, J. Schwier, C. Griffin, and S. Bukkapatnam, Zero knowledge hidden Markov model inference, *Pattern Recognition Letters*, 30, 1273–1280, 2009.
24. C. Shalizi, K. Shalizi, and J. Crutchfield, An algorithm for pattern discovery in time series, arXiv:cs.LG/0210025 v3, November 2002.
25. C. R. Shalizi and K. L. Shalizi, Blind construction of optimal nonlinear recursive predictors for discrete sequences, in *Proceedings of ACM International Conference on Uncertainty in Artificial Intelligence*, 2004.
26. P. C. Hansen, *Rank-Deficient and Discrete Ill Posed Problems*. Society for Industrial and Applied Mathematics, Philadelphia, PA, 1998.
27. D. C. Montgomery, *Design and Analysis of Experiments*. John Wiley & Sons, New York, 2001.
28. M. Schmiedekamp, A. Subbu, and S. Phoha, The clustered causal state algorithm: Efficient pattern discovery for lossy data-compression applications, *Computing in Science and Engineering*, 8(5), 2006.
29. National Geospatial Agency, Military map reading, http://earth-info.nga.mil (accessed November 18, 2009).

30. C. R. Shalizi, K. L. Shalizi, and J. P. Crutchfield, Pattern discovery in time series, Part I: Theory, algorithm, analysis, and convergence, Technical Report, Santa Fe Institute, Santa Fe, NM, 2002.
31. P. Glasserman, *Monte Carlo Methods in Financial Engineering*. Springer-Verlag, New York, 2004.
32. R. Z. Gold, Tests auxiliary to χ^2 tests in a Markov chain, *Annals of Mathematical Statistics*, 30, 56–74, 1963.
33. C. P. Quesenberry and D. C. Hurst, Large sample simultaneous confidence intervals for multinomial proportions, *Technometrics*, 6, 191–195, 1964.
34. K. Kwong and B. Iglewicz, On singular multivariate normal distribution and its applications, *Computational Statistics and Data Analysis*, 22, 271–285, 1996.

40

Environmental Monitoring Application

Lu Yu
Clemson University

Lianyu Zhao
Clemson University

Richard R. Brooks
Clemson University

40.1 Introduction

At Clemson University, Dr. Christopher Post was heading a Forestry research program to observe the effects of development on the rivers and streams of the local area. To accomplish this task, a need arose for a wireless sensor network to be deployed in rivers near ongoing development to monitor the depth and turbidity of the water. The network also needed to be capable of reliably transmitting sensor data back to a server at Clemson University. Due to the remote nature of the deployment, the analysis of mote data is used to detect and report malfunctions.

The Spring 2006 ECE 453 Software Practicum class, under Dr. Richard Brooks, was tasked with this project, assisted by two ECE MS students working under Dr. Post.

40.2 System Summary

A wireless sensor network with equipment to relay sensor readings back to campus was proposed as the solution to the assigned task. The solution consists of six primary systems: the sensor motes, relay system, root system, database system, graphical user interface (GUI) system, and system integrity system.

A mote is a low-power embedded device used to monitor sensors and uses a wireless radio to send sensor measurements. The sensor motes are composed of Moteiv TMote Sky motes and their associated turbidity and pressure sensors. The turbidity sensors determine the overall cloudiness of the water and the pressure sensors indirectly determine the current stream depth. The motes use their integrated radios to communicate with the relay systems.

Each relay system consists of a single-board computer, a cellular modem, and a base station mote. On a high level, the relay collects data from the motes and at a later time sends groups of data to the root system. The relay uses the mote to communicate with sensor motes and the modem to communicate with the portions of the project located on campus. The relay system is required because the low-power nature of the motes prevents them from communicating directly with the root system.

The root system component receives data from the relay systems deployed in the field and adds it to a database for long-term storage. It is a network service application that receives connections from the relay systems deployed in the field. The received data are then immediately inserted into the database. The root system is necessary to move the data from the Internet to the database in a secure fashion.

The database is then accessed by the system integrity and GUI systems. The use of a database provides for secure and reliable long-term storage of large volumes of sensor data as well as the innate ability to perform complex queries for desired information.

A web-based GUI allows a user to connect to the database from any computer on the Clemson campus. Through this GUI, the user can view and analyze the data collected from the site. The GUI simplifies the task of monitoring and managing the sensor network. In addition, the system integrity system monitors the database for errors and possible problems with the deployed motes and relay systems. The system integrity system automates the complex task of monitoring trends in the large volume of collected data for possible malfunctions.

Two project prototype deployments were conducted. The beta test occurred on Clemson University property and tested communication from the mote system, through the relay system, to the root system. This deployment was designed to test the important field-deployed components of the project. A full end-to-end test was performed at the deployment at the ICAR site in Greenville, SC. All components were utilized and tested during this week-long deployment.

40.3 System Architecture

40.3.1 Work Breakdown

See Figure 40.1.

40.3.2 Database System

The database provides long-term storage for the data coming in from the field through the root system until it is required for use through the user interface. It must be scalable and robust to be able to handle long-term use and potentially large quantities of data. It must be clean in design, secure, and provide intuitive input/output mechanisms so that both the root system and the user interface can interact with it with minimal effort. A single system platform must be selected to meet the computational requirements of the root system, database, and administration GUI.

40.3.3 GUI System

The administration GUI provides the user with the ability to set up and configure the system and monitor the system's current state. The GUI also provides for manipulating collected system data for analysis in exterior software. The GUI is clean, logical, easy-to-use, and documented (Figure 40.2).

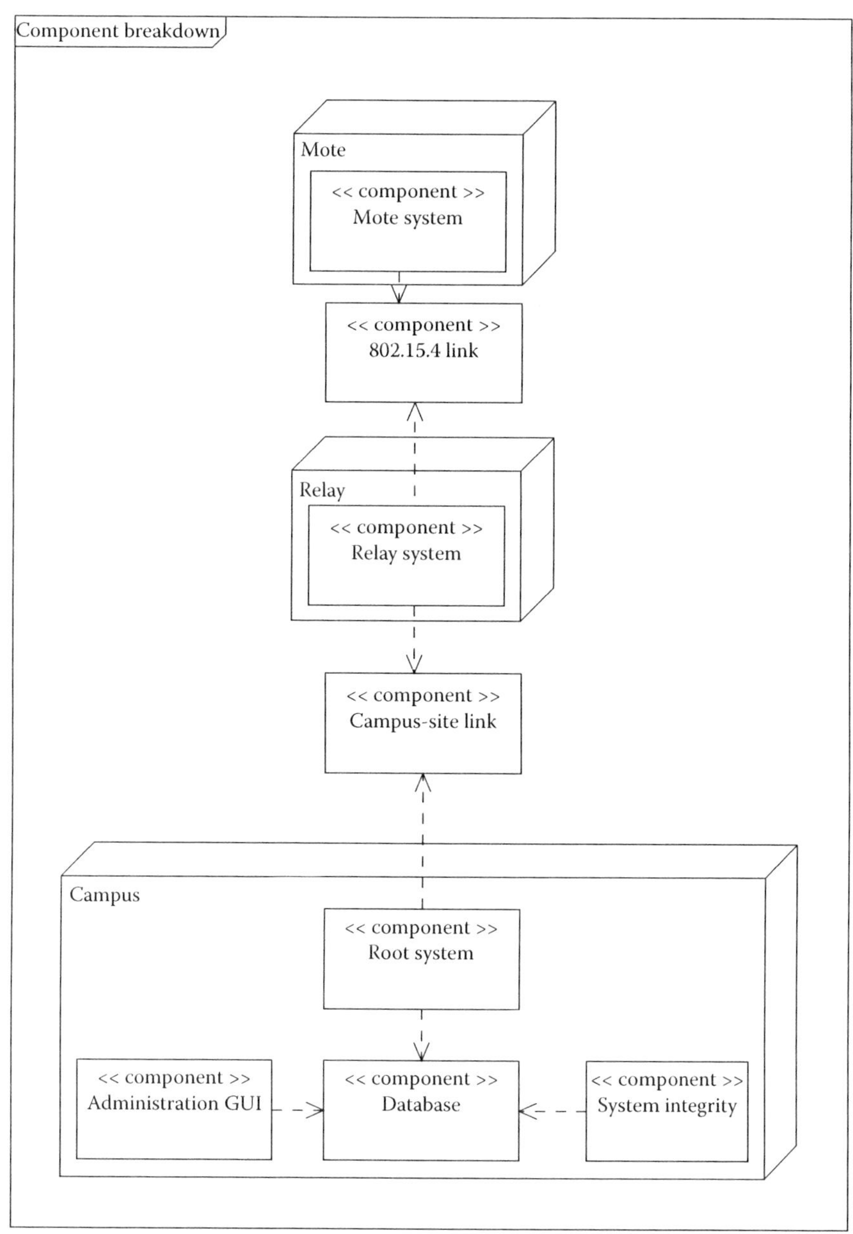

FIGURE 40.1 Component breakdown.

40.3.3.1 Components

1. *Network setup interface*: The network setup interface allows the user to add and delete motes and relays to the system. This includes functionality for displaying a geographical overview of the system and allows for the user to specify and change position data for each mote.
2. *Acquire and configure server*: A server for hosting the interface is acquired. The necessary software was installed and configured for use by the GUI component, the database component, and the root system component. This is joint effort with the appropriate groups.

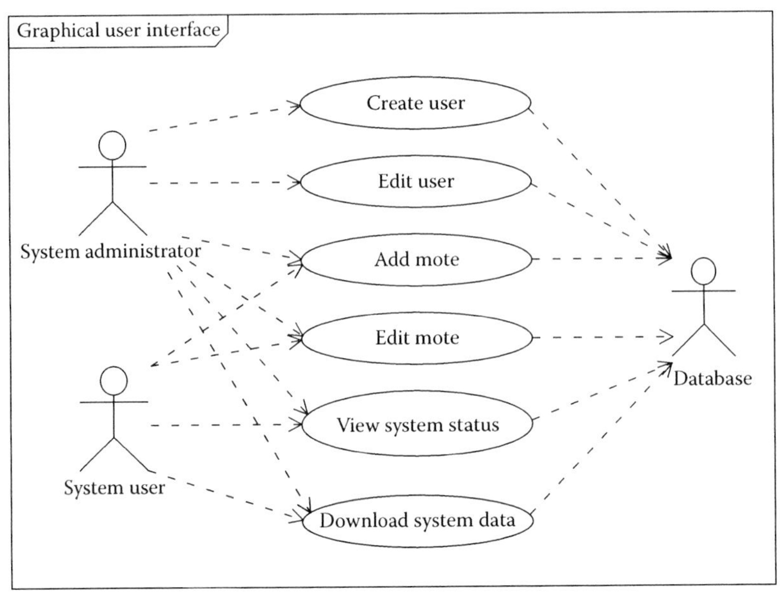

FIGURE 40.2 GUI use case diagram.

3. *User management*: The user interface has limited access based on an authorized user list. The GUI will have a login to enter the application and will have functionality for managing the authorized users from within the application.
4. *System status*: The GUI displays the current system status. The system status interface prominently displays any system alerts or messages. It also displays the geographic location of the motes and relays. The interface provides a log of all system events including configuration changes and network health warnings.
5. *Control message system*: The GUI allows the user to send control messages to the system. The GUI passes the control messages to the root system application for transfer to the on-site system. The control message system is coordinated with the root system group and the relay group.

40.3.4 Mote System

The mote system consists of a field network of mote sensors. The sensors periodically collect measurements on water turbidity and water depth (Table 40.1). The mote system communicates with the relay system. Each mote reports collected data and mote health.

40.3.5 Relay System and Root System

The relay and root systems work together to deliver sensor data from the network of mote nodes in the field to the campus where data analysis can occur without requiring direct physical access to the motes (Figure 40.3).

40.3.5.1 Component

1. *Relay system*: The relay system connects the mote systems in the field to the root system on campus. It must incorporate a ZigBee/802.15.4 radio to communicate with the nodes as well as a longer-range campus-site link. The relay collects data from the nodes and packages it for transmission as well as executing commands sent by the root system. A microcontroller was chosen to meet power and computational requirements.

TABLE 40.1 General Depth/Turbidity Cases

Depth	Turbidity	Result
Increasing	Increasing	OK
Increasing	Stable	Bad
Increasing	Decreasing	Bad
Stable	Increasing	Bad
Stable	Stable	OK
Stable	Decreasing	Bad
Decreasing	Increasing	Bad
Decreasing	Stable	Bad
Decreasing	Decreasing	OK

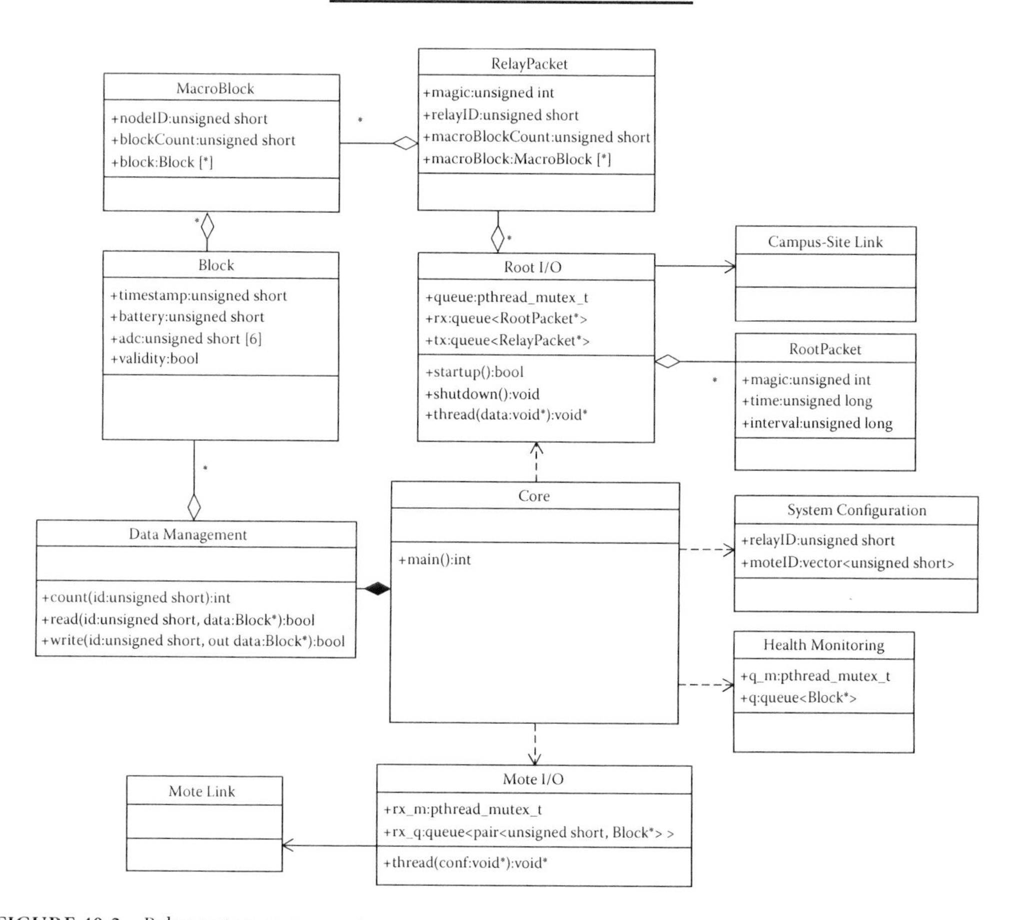

FIGURE 40.3 Relay system component.

2. *ZigBee/802.15.4 link relay system side*: For the relay system, the ZigBee/802.15.4 link is used to communicate with the mote systems. A radio was chosen to meet power requirements and be compatible with the mote hardware. Protocols were established to ensure correct communication.
3. *Campus-site link*: The campus-site link provides long-range communication between the root system and relay systems. On the relay system side, the link hardware was chosen to meet power requirements and computational overhead limitations. Overall, the link hardware must fulfill distance requirements and provide a useful data rate.

4. *Root system*: The root system connects the relay systems in the field to the administration GUI and database. It incorporates a longer-range campus-site link to communicate with the relay systems. It processes received data and stores it into the database. It executes commands initiated at the administration GUI. A system platform was selected to meet the computational requirements of the root system, database, and administration GUI.

40.3.5.2 System Integrity

System integrity deals with the quality of the entire system during its deployment in the field. It monitors the health of the nodes, ensures the validity of delivered data, analyzes the issue of securing data, and takes the appropriate steps to deal with known errors at all levels of the system.

40.4 Implementation

40.4.1 Database System

The Database Group uses a MySQL 4.x database to ensure security, encapsulation, and compatibility among the groups that require access to a database but are using many different types of programming languages. We decided to use MySQL instead of a proprietary file format in order to ease development and integration. Almost every major coding language has bindings or libraries for MySQL. With a quick, predefined, working interface into the data storage system, groups could spend the majority of their time working on more important aspects of their components. We decided to use MySQL instead of a different, commercial or otherwise, database is because MySQL is widely accepted and supported across many platforms and coding languages.

40.4.2 GUI

40.4.2.1 System Platform

The GUI is implemented on an x86 PC server running Ubuntu Linux 5.10 (Breezy Badger). Software versions used include Apache 1.3.33, MySQL 4.1, and PHP 5.0.5. The server has a reserved DCHP entry to ensure that it always has the same IP address. A static DNS entry for ece453.ece.clemson.edu has also been provided for ease of access.

40.4.2.2 Implementation Details

The GUI web pages use a combination of HTML, cascading style sheets (CSS), PHP, JavaScript, and Python. HTML is used for text markup, CSS for page layout, and PHP for server-side scripting. The interactive map is composed of two parts: a client-side JavaScript component and a server-side Python script. The server-side script combines data from a static map and the database using mapscript. The client-side script makes requests from the server as information is needed, allowing for an interactive map.

40.4.2.3 Use Cases

A web application is developed to provide a user interface to the environmental sensor network. This web application provides secure user management for access to the system. The interface provides functionality for setting up and configuring the system. System events such as warning and backup routines are prominently displayed for easy system monitoring. The interface provides the user with the data collected with by the system for each node and collectively as a system in an Excel spreadsheet format.

40.4.3 Mote System

The mote communications protocol was developed in conjunction with the relay system group. It is also the Mote group's responsibility to perform these tasks as well as trying to implement digital out on the motes, enable/disable the mote radio for power saving, and configure some hardware for the sensors.

40.4.4 Relay System

The relay system is composed of five major subsystems. All of which, except for the Data Management subsystem, run in separate threads to ease the timing constraints imposed by the associated hardware systems. The systems run on DT166 AMD Geode GX2 single-board computers using Moteiv Tmote Sky motes for communication with the mote systems and MultiTech MultiModem EDGE cell phone modems for communication with the root system.

40.4.4.1 Hardware

1. *System platform*: We use a DTRI DT166, AMD Geode GX2-based embedded system. This system has both low-power characteristics as well as supplying us with the USB ports necessary to connect our external devices. The systems have 128 MB of RAM and a 64 MB Flash disk used for storage. We run BusyBox/μC libC Linux out of a RAM disk that is loaded by the kernel at startup. Using a RAM disk, thereby disallowing the system to write to the Flash disk at runtime, helps to ensure longevity and reduces the chance of hardware failure.
2. *Mote link*: We use a Moteiv Tmote Sky, exactly the same model as that used as the mote systems deployed in the field, to create the Mote Link. They are powered and communicate through a USB interface (Figure 40.4).
   a. *Campus-site link*: We use MultiTech MultiModem EDGE USB cell phone modems. These modems provide a TCP/IP stack, reducing the implementation overhead in both the relay and root systems. It is USB bus-powered and can be powered down on command.
   b. *Power requirements*: Considering availability and cost issues, we choose an approximately 125 Ah deep-cycle marine battery. Due to the fact that the system does not draw the maximum current at any time, this battery is sufficient.
   c. *Enclosure*: We select a NEMA 4x enclosure, which is waterproof and lockable. This box also comes with a mounting plate that made it simple to mount the hardware in the enclosure.
   d. *Layout and assembly*: The system is first laid out on the mounting plate to achieve proper spacing. Holes are drilled in appropriate locations for the straps and screws used to secure components. Nylon zip ties are used to secure the geode and power regulator while the modem is bolted down using its supplied mounting bracket. The box also needs to be drilled to allow the power supply and antenna wires to enter. The hole for the power wire is placed near the power converter at the bottom. The antenna feed line has to be passed through a waterproof opening on the top of the box so a special waterproof cable pass-through is used.
   e. *Wiring*: Two conductor, 18 AWG wire is used to supply power, with the red wire used for +12 V and the black for ground. Upon entering the box, the power supply wire is connected to the input terminals of the regulator. The output from the regulator is connected to a size *N* (5.5 mm OD, 2.5 mm ID) barrel-type power connector to supply power to the geode. The cellular modem is connected to the Geode with a standard A/B USB cable. The included indoor cellular antenna is connected directly to the modem (inside the box). The base station mote connects directly to one of the Geode's USB ports. The antenna cable is connected to the mote's RP-SMA jack. The other end of the antenna cable ends in a male N-type connector, which connects to the female N-type on the antenna.
   f. *Antenna*: Any commercial 802.11b/g antenna is compatible since the mote's 802.15.4 radios operate on the same 2.4 GHz ISM band as wireless Ethernet. The HyperGain HG2412U 12 dBi omnidirectional (vertical) antenna is selected for our work.

40.4.4.2 Software

40.4.4.2.1 Core

The Core subsystem controls all of the different parts of the relay system. It is responsible for loading the system configuration from a file at startup and for launching the Root I/O, Health Monitoring,

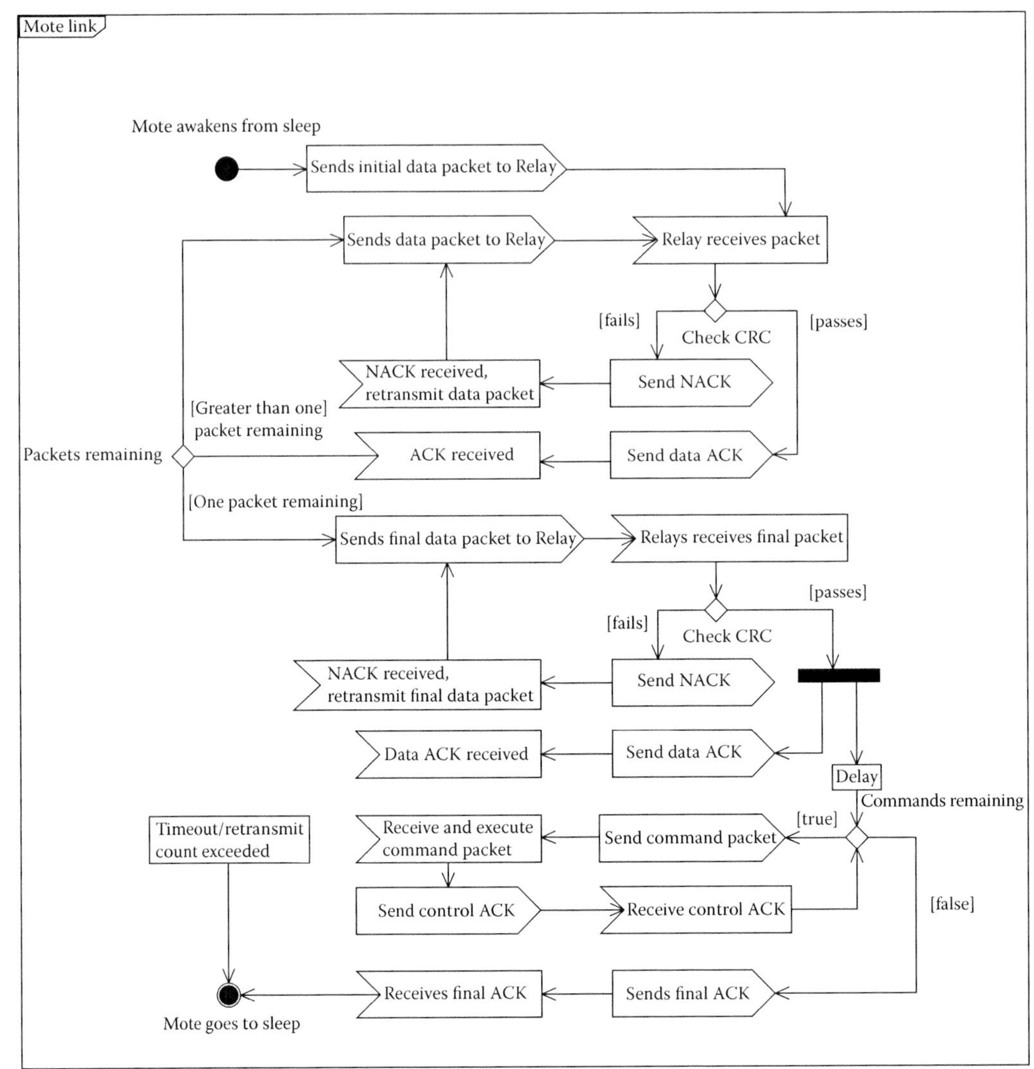

FIGURE 40.4 Mote Link protocol flow diagram.

and Mote I/O threads. It takes packets received by the Mote I/O subsystem and packets generated by the Health Monitoring subsystem and inserts them into the Data Management subsystem. At specified intervals, it removes a set of data from the Data Management subsystem and builds a relay packet to be sent to the root system by the Root I/O subsystem. The system configuration is stored in a simple text file in which the first line contains the relay system's ID and each following line contains a single mote ID that the relay system is supposed to communicate with.

It is necessary for the Core subsystem to operate as an independent thread due to the asynchronous, hardware-controlled behavior of the Mote I/O and Root I/O systems. The Core marshals access to the Data Management subsystem, preventing the possibility of deadlock. The Core eases the process of taking data from both the Mote I/O and Health Monitoring subsystems and sending it to the root system over the Root I/O subsystem.

40.4.4.2.2 Data Management

The Data Management subsystem stores the data from the motes and from the Health Monitoring subsystem until it is sent to the root system. Each node ID has a corresponding queue that holds the associated data. When the core receives a packet of data, it is added to the appropriate queue. When the Core is ready to send data to the root system, it pulls data from the queue, packages it, and then sends it to the root system using the Root I/O interface. The Data Management subsystem provides a simple interface to store and retrieve data being held. This abstraction improves the readability and reliability of the Core subsystem.

40.4.4.2.3 Health Monitoring

The Health Monitoring subsystem is responsible for monitoring the battery voltage and the signal strength reported by the cell phone modem and periodically placing a packet with this information into a queue which is then read from by the Core. The data are then sent to the root system along with the rest of the data held in the Data Management subsystem.

40.4.4.2.4 Mote I/O

The Mote I/O subsystem provides an interface between the USB-connected mote acting as a base station and the Core component. It implements the relay side of the mote–relay network protocol.

The Mote I/O thread installs a SIGALRM timer with 100 ms resolution used for timing in the network protocol. The thread also maintains a STL map of protocol state machines, one for each associated mote. Every time SIGALRM is raised, a counter is incremented to represent the number of 100 ms intervals that have elapsed since the start of the program, and the serial link is polled for a newly arrived packet. Each mote's state machine is serviced with either a NULL packet or the newly received packet depending on mote ID. The state machines themselves handle the actual network protocol, including the sending of the time synchronization packet to the motes. As data packets are received and validated, they are passed to the Core component using a STL queue.

The Mote I/O subsystem is needed to process the traffic between the motes and the rest of the relay system. It deals with the asynchronous nature of communicating with the motes and provides a layer of abstraction. This abstraction allows the other components of the relay to concern themselves with data received from the associated motes and not the underlying protocol and any errors that may occur in transmission.

40.4.4.2.5 Mote Communication Link

The API for this link is defined by the packet structure and its field descriptions as well as the flow diagram for the protocol.

40.4.4.2.6 Root I/O

The Root I/O subsystem is responsible for managing the cell phone modem which is used to communicate with the root system on campus. The Root I/O subsystem runs in an independent thread from the Core and provides a queue for the Core to write into to have data sent. Since it is necessary to synchronously initialize the modem and return this state to the Core, a function is provided which encapsulates the inter-thread communications necessary to signal the Root I/O thread to wake up, attempt to use the modem to connect to the root system, and then return the result of the attempt. This information allows the Core to make an informed decision about going ahead and constructing a packet to be sent or not. An additional function to signal the Root I/O subsystem to shut down the modem and go back to sleep once it has finished transmitting the pending data is also supplied.

Upon being woken up, the Root I/O subsystem attempts to initialize the modem and connect to the root system. Once this operation has completed, it indirectly returns the status back to the Core. If the modem is unable to establish a connection, then the Root I/O subsystem goes back to sleep, waiting for another signal from the Core. If the connection is successfully established, then the Root I/O attempts to send the first packet in the queue. If this succeeds, then the packet is removed from the queue and the memory is freed. The Root I/O then checks the modem for the arrival of data. If a Root Packet has arrived, and it appears to be valid, then it is enqueued in the receive queue so that it can be processed by the Core.

The Root I/O subsystem is needed to abstract the interface with the cell phone modem and manage the asynchronous nature of its operation. This abstraction simplifies the Core and controls access to the cell phone modem. The Root I/O presents an interface that combines synchronous and asynchronous behavior to make it easy for the Core to manage the Root I/O without having to synchronously operate it.

40.4.4.2.7 Campus-Site Communications Link

The Campus-Site Communications Link uses cell phone modem which incorporates a minimal TCP/IP stack. This reduces the protocol complexity that is necessary, but increases the amount of interfacing necessary with the modem. The protocol has only two types of packets, one for each direction of the link. The first type of packet, a Relay Packet, contains the blocks of data which have been sent to the relay system by the motes.

It is organized in a two-level format to reduce the amount of redundant data. Each node which has data in the packet has a corresponding MacroBlock, which contains the node's ID and the number of individual Blocks that are stored in the packet for that node. This helps to reduce the size of each packet, but it was later decided to also incorporate compression to obtain further reductions in the amount of data that must be sent.

The other type of packet, a Root Packet, contains the current time and the desired contact intervals for the relay system communicating with the root system and the mote systems communicating with the relay system. TCP/IP guarantees correctness of the delivery, so we do not incorporate any integrity checks. The code which sends the data waits until the modem verifies that the data were successfully delivering before returning back to the Root I/O.

Compression is adopted to reduce the cost of deployment. Our original Mote I/O dummy generates random values for each ADC field, restricted to the valid range of the field. Using this as input data, the standard TCP/IP Relay I/O component sent data to a test implementation of the root system with compression. The relay system is written to target the optimal package size: maximizing the amount of data, and thereby maximizing the compressibility, while preventing any problems with the modem's TCP/IP stack.

40.4.5 Root System

40.4.5.1 Overview

The root system is composed of three major subsystems. The Relay I/O is responsible for handling communications with the relay systems and interpreting the data that they have sent. The database interface is responsible for requesting the insertion of data into the database. The Core links the Relay I/O and database interface and preprocesses the sets of data which the Relay I/O receives into single element blocks which are then given to the database interface (Figure 40.5).

The root system is installed on the primary server located in the Clemson University networking lab. This is the same machine that the database is located on in order to simplify the interaction between the root system and the database. This also makes economical sense as the machines used for the root system and database would need to be dedicated servers, so installing both on the same machine prevents the potential waste of resources. Once compiled, the root system is executed and left running, constantly checking for connections on the proper port. Before actually receiving data from a true relay system node, the system is tested using a test module. This test connects to the root system and sends simulated data to it that can be examined for proper root system functionality.

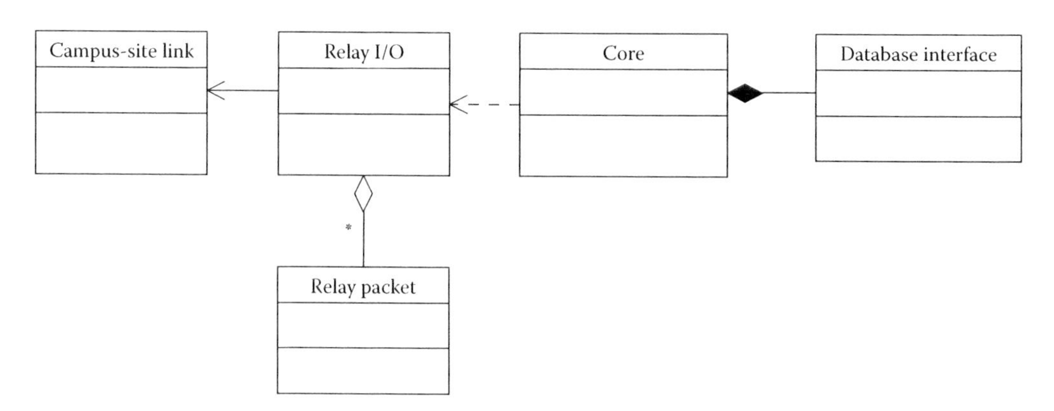

FIGURE 40.5 Root system components.

40.4.5.2 Core

After initialization, the core checks for a relay packet coming in. If there is data received from the Relay I/O, then the core accepts this data. The core deconstructs the relay packet which has a variable amount of blocks and converts it into the standard MoteBlock format that will be used by the database. It inserts these MoteBlocks into the queue where it will be stored until insertion into the database. By using a queue, we place a safeguard on our data in the small chance that the database is not functional while the core is still receiving data. Then the core inserts the queue data into the database. Until the queue is empty, or until the database sends back an error, data are read from the queue and inserted into the database. The data are not removed from the queue until that data are successfully entered into the database to prevent loss of data in the case of an error. This process is repeated indefinitely while the core is running and continues to receive data, convert it, store it, and then send it to the database.

The method used designates the core component as the primary program that controls access to the other components. This is found to be the simplest and most effective method. The core uses a queue as a buffer, entering data into the queue before sending it to the database. This ensures that data are not lost due to discrepancies in the link between the core and database. This along with the safeguards on the relay IO side of the core greatly ensures that no datum is lost while it is decoded and passed through the core.

40.4.5.3 Relay I/O

Relay I/O consists primarily of a method that is called from the core. The purpose of this method is to establish connections to all incoming relays, read and queue one complete packet from each connection, and send back the oldest packet in the queue. Connecting to all available relays has two steps. If there are no connections currently open, a blocking accept is performed to wait until a relay connects. If there are connections already open, the main socket is polled for activity to see if others are trying to connect. If there is activity, the new connections are made. If there is no activity, then this step is skipped and the reading phase starts. In order to keep track of the different connections, a class was made that holds the socket created for each connection as well as the methods that are used to control that connection. Once the reading phase starts, one packet is read from every connection in the connection vector and stored in the queue. The method `HandleTCPClient()` takes care of receiving this information. A passive server-side close is performed if a client is not active during this phase. It returns a 1 if the connection is still open and a 0 if it was closed and should be removed from the connection list. Once each connection has had its chance to read a packet, the oldest packet is returned and removed from the queue.

The design of the Relay I/O portion of this project is strongly influenced by our initial plan not to have a threaded program. Since it is possible for more than one relay to send data at the same time, it is important that the core is always ready to receive data. Normally, this would be done by forking a new process whenever a new client connected. Without using threads, other ways needed to be found to achieve this. That is where the connection class comes in. Each connection has its own socket that is maintained inside the connection class. This allows the core to always be receiving data on each of the connections even if they are not currently the ones being allowed to read. Reading the incoming information requires no extra thought. The information is read and the packets are reassembled and placed at the end of the queue. The one thing that initially posed a problem once this was developed was removing connections from the connection list. This is solved by a simple return value that tells the package to remove the connection from the connection list.

40.4.5.4 Database Interface

The database component is solved by implementing MySQL queries that stores the mote and relay information into the database. As the mote or relay data is sent to the database, a conversion is performed on the each mote's time stamp to create a standard Unix time. The mote time stamp conversion develops a bug in which as the clock of a mote drifts forward or backward in time, the time would not convert properly. The program might think that the time was from a different day. Thus, from observations of the current deployment, the time stamp is given a 10 min window ahead of the relay time stamp in which it is still considered part of the same day.

40.4.6 System Integrity

40.4.6.1 Functionality

The System integrity group's role is to ensure the integrity of data and to report via alerts and errors when data is abnormal. To implement this functionality, a library of C functions is written, compiled, and run on the lab computer wherein the database resides. The use case diagram is illustrated in Figure 40.6.

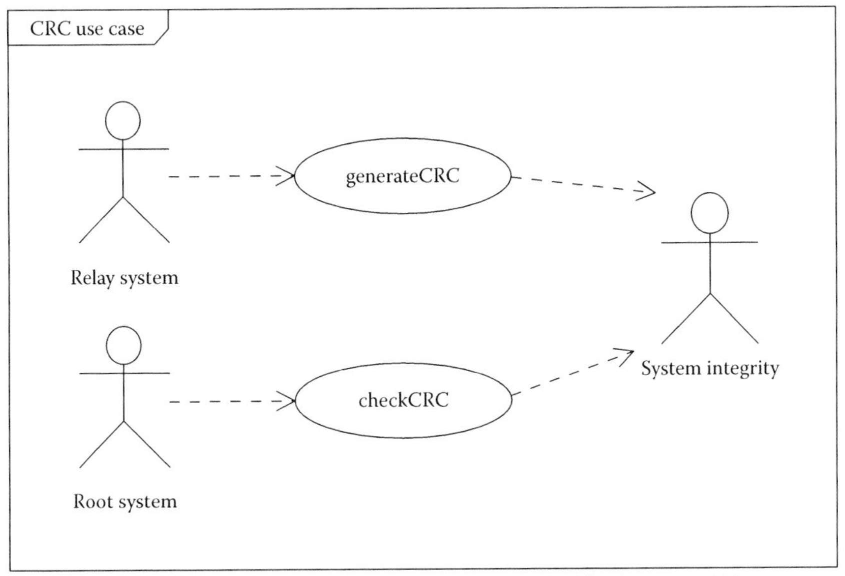

FIGURE 40.6 CRC use case diagram.

40.5 Component Testing

40.5.1 Database

The testing of the database is to prove that one can connect to the database, import data into it, leave it there, and retrieve it later as necessary. A test program which can test the functionality of every stored procedure is sufficient. Due to the fact that some stored procedures are inputting data, and some are outputting data, the full range of the database would be tested.

It is reasonable to use PHP to write the test code, since PHP is native to almost every Linux distribution, including the test server used to develop the database. Moreover, PHP is the code base used to develop the GUI. Testing it in a language that would be used by the GUI group will allow us to help troubleshoot problems encountered by the GUI group. PHP also has a well-developed MySQL library that comes native to the distribution.

40.5.2 GUI

Testing the GUI portion of the project is to use the system with limited knowledge. "An effective user interface is considered to be well designed when the program behaves exactly how the user thought it would"—Joel Splosky. This was one of the main goals when designing the interface. This test does not only determine the interfaces ability to handle correct inputs, but also tests for the ability to deal with the unintelligent user.

In testing the User Management portion of the GUI, both functionality, when it is used properly, and robustness, by executing incorrect commands and invalid values, are tested. Only the user administrator has the ability to use the User Management functions; thus we assume the user is indeed a system administrator for all tests involving adding/editing/or removal. It has also been tested using a regular user to ensure that they indeed do not have those capabilities.

Next step is the most important aspect of the interface, network management, on its ability to add and delete nodes. Attempts are made to add nodes with invalid values for x and y and add nodes on top of one another. The interface should display adequate functionality and robustness to handle all of these situations. Moving nodes from different points around the map is not necessarily a needed function, which is also conducted to accommodate a misbehaved user.

Diagnostics is a very critical portion of GUI. Without the ability to monitor the system, there is no way of knowing what happened in the field. The ability to test this function is only available after integration. To test this portion of the code, we need to rely on the information already sent back from the field to show its functionality.

The outline of component testing is listed as follows:

- Testing the Login System with Administrative Privileges
- Testing the User Management System with a System Administrator
- Testing the User Management System with a Guest Privilege
- Testing the User Management System with a User Privilege
- Testing the Network Management System as an Administrative User
- Testing the Network Management System as a Guest
- Testing the Network Management System as a User
- Testing the Diagnostic Page
- Testing the View Data and Graph Options

40.5.3 Mote System

Each of the following test cases ensures that the designated part of its component is running as developed in its protocol and running correctly.

- *Send nonfinal data*: Assuming that the mote has data to send. The base mote is listening and able to receive communication via zero hops. The network structure is fully operational between the motes themselves, the base mote, and the relay. When a mote wants to send sets, not including the final message, of sensor values to the relay, it constructs a packet with flags 1000000, data field including battery level as well as at least 3 but up to 6 ADC output values, and transmits that packet over the radio to the base mote. The relay picks that packet up off the base mote. The error condition of the packet not being received by the relay is handled by monitoring ACKs for sequence numbers as well as not deleting data or control messages from the buffer until they are ACKed. These messages are resent during the next "live session." There would also be a packet retransmit limit before the mote goes to sleep. A packet that fails the CRC check would not be ACKed. The steps are as follows:
  - Identify sets of sensor values on mote's RAM.
  - Form packets with appropriate info including the unchanging Tiny OS header as well as the CRC and the payload. The payload contains the data from at least three ADCs as well as the battery life with a flag for sent data and mote ID, sequence number, and time stamp.
  - Send() packet.
  - Wait for ACK from relay as per protocol flow.
  - If ACK is returned, then delete data from buffer. If NACK is returned, then resend data. If no ACK or NACK is sent, monitor timeout time and go to sleep or resend data depending on that time.
- *Send final data packet*: Assuming that this is the last packet of data, the mote would send during this "live" state and the network is fully operational including all motes and the relay. The mote sends the last packet of its data containing flag 10000001. The mote then waits for the ACK for that last packet of data. If an ACK is received, then the mote goes to sleep. If an ACK is not received, the mote would resend the data. If the timeout occurs, then the mote keeps the data in the buffer and goes to sleep. The steps are as follows:
  - Mote sends the final data packet.
  - Mote waits for ACK or NACK from relay.
  - Mote receives ACK or NACK.
  - Mote resends the packet on a NACK or goes to sleep if it receives an ACK.
  - Mote also goes into sleep state upon the expiration of the timeout.
- *Relay sends nonfinal and final data ACK*: Assuming that the network is operational. Data have been sent to the relay and that data are the final packet of data the mote needs to send. And the relay has received the data correctly. The relay then sends an ACK for a final data packet. The final data packet is marked with the flag 10000001. The ACK for this packet includes the sequence number of the sent packet. Steps for sending nonfinal ACK are as follows:
  - Mote sends data packet with flag 10000000.
  - Relay receives data correctly.
  - Relay sends data ACK to mote.

  Steps for sending final ACK are as follows:
  - Mote sends final data packet with flag 10000001.
  - Relay receives data correctly.
  - Relay sends final data ACK.

- *Relay sends NACK*: When the mote has sent data to the relay and the relay has received the data incorrectly. The relay will then send a NACK for said data in the following steps:
  - The mote sends a data packet to the relay.
  - The relay receives the data but it is in some way incorrect.
  - The relay sends the mote a NACK back.
- *Mote sends control ACK*: When the relay has sent the mote a control message and upon successful receipt of that control message, the mote will send an ACK that corresponds to the message in the following steps:
  - The relay sends the mote the control message.
  - If the mote receives the control message correctly, it will send an ACK for that control message back to the relay.
- *Relay sends sync clock message to mote*: If the relay sends the control message to the mote, then the relay and the mote have working clock implementations. The relay sends the mote a sync clock message. This message will contain a time stamp and is used to keep the mote clock in sync. The flag on this message is 00100001.
  - The relay sends a sync clock message to the mote with the flag 00100001.
  - The mote receives the message and syncs its clock with the time stamp given.
  - Mote sends ACK for control message.
- *Relay sends change sensing frequency control message to mote*: When the relay would like to change the sensing frequency, it sends a control message to the mote to change the sensing frequency of the mote. The control message contains the flag 00100010 as well as a time in seconds to set the frequency to. Steps are as follows:
  - The relay sends a packet containing the new frequency for the sensors as well as the flag 00100010.
  - If the mote receives the message correctly, it changes the sensing frequency.
  - The mote sends an ACK back to the relay for the control message.
- *Relay sends change transmission frequency control message to mote*: When the relay would like to change the transmission frequency, it sends a control message to the mote to change the frequency with which the mote sends data back to the relay. The packet sent from the relay contains the flag 00100100, as well as a time in seconds. The mote also ACKs this control message. Steps are as follows:
  - Relay sends control message with new transmission frequency and flag 00100100.
  - Mote receives control message and, if it is correct, sends an ACK back to the relay.
  - Mote changes transmission frequency for data back to the relay.

40.5.4 Relay System

A sophisticated test vector system is developed to facilitate the testing and verification of the relay system functionality. The test system is needed because generating traffic from multiple actual motes is not practical, nor possible prior to integration.

The test system reads a text file containing simulated mote packets and their times of arrival. When the abstracted mote interface is built in debug mode, alternate `readSerial()` and `writeSerial()` functions are provided. Upon being called for the first time, `readSerial()` initializes the test vector and parses the input file. Individual packets are stored in order of increasing arrival time. During each call to `readSerial()` (including the first), the arrival time of the next packet in the list is compared to current time elapsed since program starts. If the arrival time has been met or surpassed, `readSerial()` returns it; otherwise, it returns NULL. `writeSerial()` is currently an information sink that does nothing. However, the individual ACK, NACK, clock sync, and final ACK functions that call `writeSerial()` print basic information so that the status of the protocol stack can be monitored during testing.

The supplied test vector tests the following:

- Multiple motes sending interleaved data
- Receiving a packet with a bad CRC and recovery
- Multiple motes sending interleaved data, one of which consists of a single FINAL_SENSOR_DATA packet
- Multiple motes sending interleaved data, some of which the relay is not configured to service
- A timeout and recovery between packets of a single mote's data stream
- A timeout and recovery on the ACK of a CLOCK_SYNC command
- An unrecovered timeout during a single mote's data stream
- An unrecovered timeout on the ACK of a CLOCK_SYNC command
- A mote sending spurious non-SENSOR_DATA/FINAL_SENSOR_DATA packets in the middle of a data stream
- An extended period (1 min) with no motes transmitting
- An extended period (200 s) simulating normal activity (individual motes coming up, transmitting data, and going back to sleep)

The use of this test vector thoroughly tests the network protocol stack/state machine. It also exercises all other portions of the relay system. The queue interfaces between the various threads are all used since the full data flow is implemented (Mote I/O and Health Monitoring to the Core to Data Management to Core to Root I/O). Due to lack of hardware, the actual interface to the cellular modem is not yet tested, nor is the Health Monitoring component. The dummy modules are still used in these two cases. The data structures and communications mechanisms, however, are fully tested since the dummy modules use the same interfaces and structures as the final versions.

The result of running the relay system with the test vector is a long series of ACKs, NACKs, time syncs, and final ACKs. These indicate the actions taken by the network protocol. Interspaced within the protocol statements are the contents of packets sent via the Root I/O dummy module. In the test vector, packets with bad data carry zeros in the payload, which should not show up in the printed Root I/O packets.

40.5.5 Root System

40.5.5.1 Individual Testing

Each component of the root system has been tested at each individual's discretion primarily for troubleshooting. This includes using previously developed stub code to measure proper operation. The testing code is still included in the final version for demo purposes and future integration troubleshooting.

The original packet tester allows a user to simulate sending multiple relay packets containing multiple blocks worth of data from different motes over TCP/IP. The user would provide a port number and host name for the program to function correctly. With the correct input, the program would create multiple threads as desired for simulating multiple relay systems talking to the root system at the same time. Upon establishing a connection to the root system, each thread in the packet tester would send a predefined relay packet with various data in the appropriate fields for the root system to respond correctly. The packet tester would then wait for a response packet for a predefine time and then timeout upon receiving nothing. If it does receive a response packet containing a magic number and a time stamp, it will then print it out to the screen.

40.5.5.2 Whole System Testing

Once each individual component is functioning, the whole group meets several times to integrate each part. Then after integration, the whole system is tested with the packet tester. Later compression is added to the system and our group switched to using a tester written by Clint. This change does not affect the output as it is seamless to the end user. With this tester, we test the whole system again with the compression added which resulted in little change to the overall operation of the component.

Unlike the original packet tester, however, this one can only simulate a single relay system at a time. However, we have tested up to four independent instances of the test program connecting to the root system simultaneously.

40.5.6 System Integrity

40.5.6.1 CRC

The CRC that is implemented is a CRC-16 algorithm. The CRC algorithm generates a specific code based on the contents of the data and the divisor polynomial, also called a key. The idea behind the CRC-16 algorithm is that it generates a specific code that will be sent along with the data. Once the data are received, it can have the same algorithm performed on it and then have the two generated CRC codes compared; if they are the same, then the data are uncorrupted through transmission; otherwise the data are corrupted. The CRC error detection algorithm would fail if the error polynomial that corrupts the message is a multiple of the key that one was using. The CRC functions are written and tested, but not used in the implementation of the system.

40.5.6.2 Database

The database code carries two purposes. The first is to connect to the database, retrieve data, send messages, and disconnect. The second is to analyze the retrieved data and prepare the messages to be sent. Therefore, database connection test and database examination test are conducted. We omit the former test since its procedure is typical (Figure 40.7).

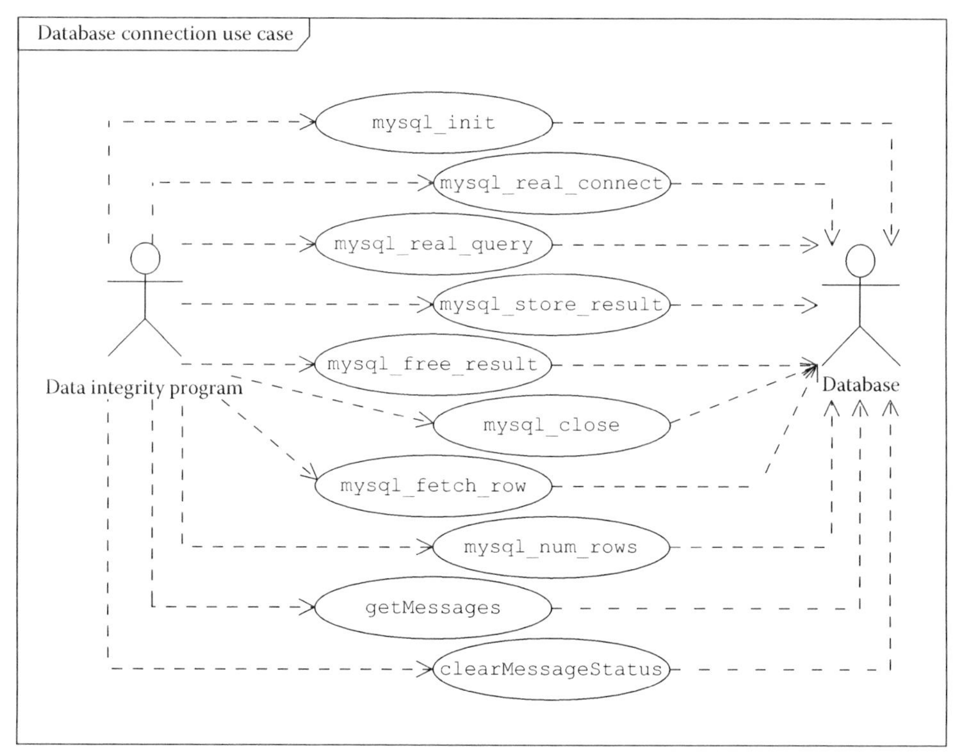

FIGURE 40.7 Database connection use case diagram.

The data examination code uses the data retrieved from the database. Each function serves to check the integrity of the system. The following descriptions will explain what checks are being performed and what output is expected.

- When the mote battery level falls below the threshold that is specified to be five, the following functions are utilized:
  - `moteBatteryCheck`

  The result of this battery level falling below the threshold is a "Low Mote Battery" alert sent to the database.
- When the mote battery level increases, the following functions are utilized:
  - `moteBatteryCheck`

  The result of the battery level increasing is an "Inconsistent Mote Battery" alert sent to the database.
- In the case in which the turbidity of the water increases when the depth does not, the following functions are utilized:
  - `moteTurbidityStability`
  - `moteDepthVariance`
  - `moteTurbidityVariance`
  - `moteVarianceValidity`
  - `moteTurbidityRange`

  The result of this case is an "Irregular Turbidity Reading" alert sent to the database.
- In the case in which the depth of the water increases when the turbidity does not, the following functions are utilized:
  - `moteTurbidityStability`
  - `moteDepthVariance`
  - `moteTurbidityVariance`
  - `moteVarianceValidity`
  - `moteTurbidityRange`

  The result of this case is an "Irregular Depth Reading" alert sent to the database.
- When the turbidity is greater than 600.0, less than 10.0, or outside of the acceptable range of 75% to 110% of the previous value, the following functions are utilized:
  - `moteTurbidityStability`
  - `moteVarianceValidity`
  - `moteTurbidityRange`

  The result of this case is an "Irregular Turbidity Reading" alert is sent to the database.
- When the turbidity readings become stuck at a single value, the following functions are utilized:
  - `dataStuckAt`

  The result of this case is an "Irregular Turbidity Reading" alert is sent to the database.
- When the depth readings become stuck at a single value, the following functions are utilized:
  - `dataStuckAt`

  The result of this case is an "Irregular Depth Reading" alert is sent to the database.
- When the depth variance of a particular mote does not correlate with the depth variable of the motes as a whole, for example, when one or more motes vary greatly with the overall trend of the whole system, the following functions are utilized:
  - `moteVarianceCheck (called previously: moteDepthVariance)`

  The result of this case is an "Irregular Depth Reading" alert being sent to the database.

40.6 Project Testing

Upon completion of the individual component implementation and testing phases of the project, the next task is integration between groups and components (Figure 40.8).

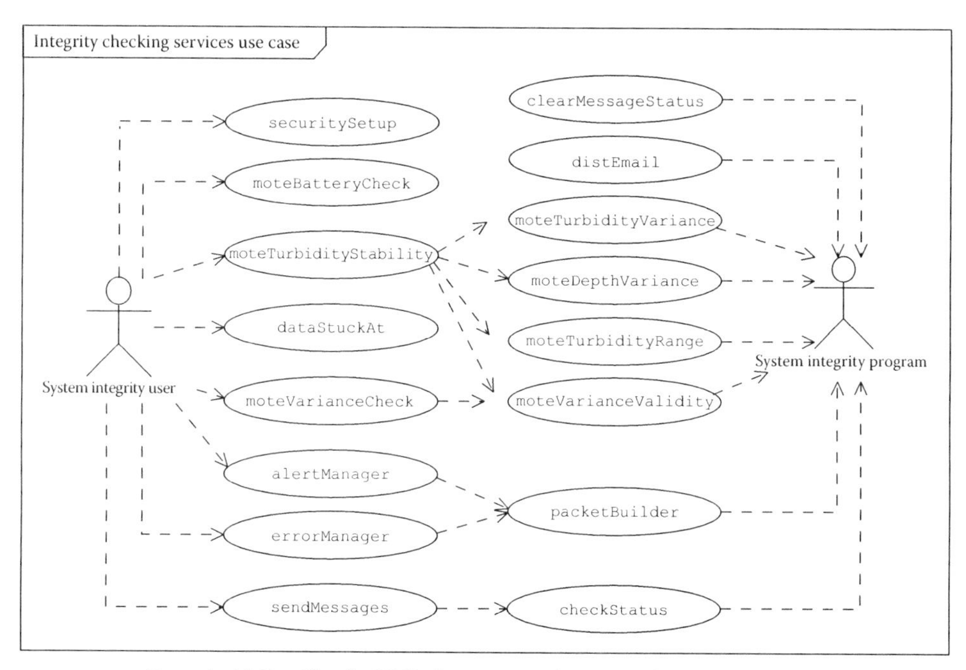

FIGURE 40.8 Integrity checking services use case diagram.

40.6.1 Integration Testing

40.6.1.1 Mote System and Relay System

A large amount of testing and integration occurred between the mote and relay systems. The two implementations of the network protocol designed for this link needed to be synchronized and thoroughly tested for bugs. Initially, several motes with the latest code revision were given to the relay system group to allow initial tests with the beta relay network protocol code. Once bugs were identified on both sides of the link, several group testing and debugging sessions were held during which both the mote and relay network protocol code were polished and prepared for deployment. Bugs were found involving sequence numbers and motes responding to packets addressed to other motes. Also, the complete network protocol has not been fully implemented on the motes due to time and system design constraints—proper response to NACK packets is currently missing. Also, the ability to modify sampling and transmission frequencies was developed but never enabled nor tested due to time constraints. Some limited field testing was also performed during the beta deployment (Figures 40.9 and 40.10).

40.6.1.2 Relay System and Root System

Integration testing between the relay and root systems was performed using an internally developed packet tester. This tester consisted of a special version of the relay code running dummy interfaces for the modem and mote. This allowed testing of the TCP/IP link between the two systems without the extra complication of using motes to generate data and the cell phone modem to transmit it. Some limited field testing was also performed during the beta deployment.

40.6.1.3 Root System and Database System

Due to time constraints and ongoing development, minimal offline testing was performed between the root system and the database system. During final mote/relay/root integration and protocol testing, the database was used to temporarily store the test packets generated.

FIGURE 40.9 Mote system laydown.

FIGURE 40.10 Unpacking motes.

40.6.1.4 Database System, GUI System, and System Integrity System

A majority of the integration testing between the database, GUI, and system integrity systems occurred concurrently. Actions were taken in response to problems that could only be discovered during a field deployment. Specifically, small structural and data content changes were made in the database to better cope with the nature and volume of data received.

40.6.1.5 End-to-End Testing

Two different end-to-end testing deployments were made. The first tested communication from the mote system, via the relay system, to the root system and took place in a stream owned by Clemson University. The second tested the fully system, from mote to GUI, and took place at the ICAR site in Greenville, SC.

40.6.1.6 Beta Deployment

The beta test deployment was performed in a stream owned by Clemson University in Pendleton, SC. Four motes with near-final code were deployed in the stream and a near-final version of the relay system was attached to a nearby fence. The deployed motes did not have fully functional sensor packages due to the (at that time) unresolved problems with the digital out and conditioning circuit. Several sets of data were received from the motes by the relay and transmitted to campus, but some bugs and general problems with the system were discovered. Most importantly, several issues with the cellular modem were discovered that caused the relay system to "get stuck" and not recover properly from problems in the cell phone network. Also, it was determined that the antenna used by the base station mote in the relay was not sufficient. It was too small and did not provide long enough range to reliably communicate with motes in the predicted deployment pattern. The problems discovered during the beta deployment were fixed prior to the ICAR deployment. Larger, much more effective base station antennas were acquired; the sensor package circuity was fixed; the digital out issues were bypassed; and the modem control code was made much more robust.

40.6.1.7 Deployment

The final prototype deployment occurred at ICAR site. This site has a small stream directly adjacent to heavy construction, thus providing the exact environment that Dr. Post had envisioned the system being used. A total of 10 mote systems were deployed in the stream, approximately every 200 ft. Five relay systems were deployed, one for every two motes to ensure that each mote was, without question, in range of its associated relay. Using the new antennas, it was determined that it would have been possible to use fewer relays, possibly as few as two.

The deployment can be considered an overall success since several mote–relay combinations were still operational after more than a week of deployment. As of this writing, almost 20,000 sensor readings have been automatically collected and entered into the database. End-to-end communication and functionality were verified with sensor readings taken by the mote systems appearing promptly in the GUI interface after each relay transmission sent them up to campus.

However, the ICAR deployment was not without its share of problems and newly uncovered bugs. From the very start, several motes were not successfully communicating with their associated relays. Several attempts were made to initiate communications, but it was determined that the silent motes were having communication problems due to poor antenna placement or, in one case, interference from the highway bridge between the mote and the relay. It was decided to leave these motes alone and concentrate on ensuring the remaining motes were properly operating and getting their data all the way to the GUI.

Approximately 12 h after the deployment was completed, the entire relay system went silent (around 4.00 AM the day after the deployment). A relay team member traveled back to the ICAR site to investigate the failure and determined that it was a combination of several issues. One of the relays had mechanical problems with a set of quick-disconnect terminals in the power line and was not receiving power. The remaining relays had various software problems, including inadvertent automatic shutdown of a modem, and various unknown problems with the cell phone network causing unrecoverable errors to occur in the modem interface. These errors are especially difficult to catch and bypass since it is not possible to simulate one in a controlled environment. Instead, one must wait until a cell phone error randomly occurs and then determine how to catch and bypass it. This issues stems from the complexity of interfacing the cell phone modem using its expanded AT command set and text written to and received from a serial port (Figures 40.11 through 40.16).

FIGURE 40.11 Assembly of motes and sensors in housings.

FIGURE 40.12 Fully assembled mote and sensor system.

40.7 Project Demonstration

The prototype deployment at Clemsons ICAR site was a week-long test of the full system. All developed hardware components were deployed to the field and data were successfully collected and transmitted back to campus. All software products were successfully used on campus to retrieve, store, display, and monitor collected data and system status. End-to-end testing methodology and results are discussed in Section 40.6. Following are photographs, screenshots, and data tables generated during the deployment and week-long testing period. They are a representative sample of the system deployment procedure and results.

FIGURE 40.13 Deployed mote system.

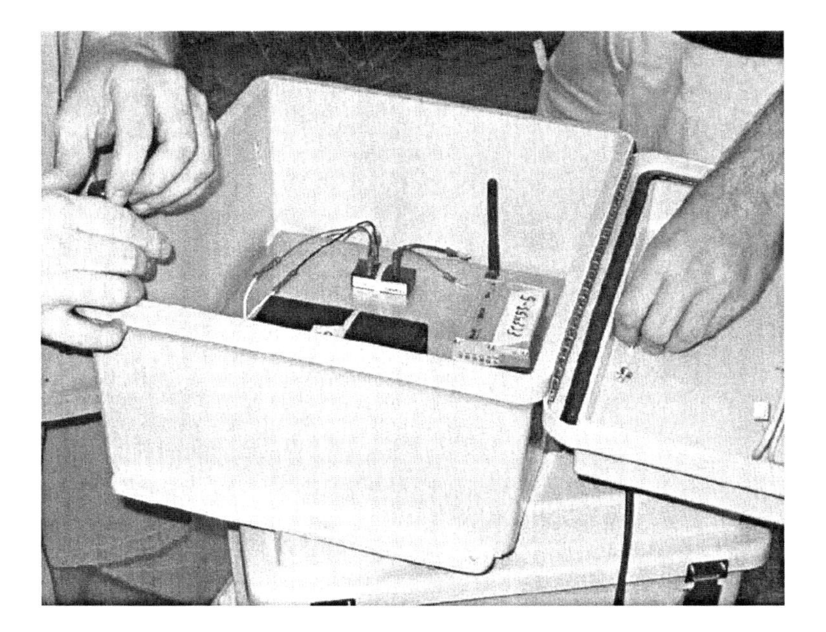

FIGURE 40.14 Assembly of a relay system.

40.7.1 Photographs of Deployment

See Figure 40.17.

40.7.2 Screenshots of GUI System

See Figure 40.18.

FIGURE 40.15 Beginning deployment of a relay system.

FIGURE 40.16 Deployed relay system.

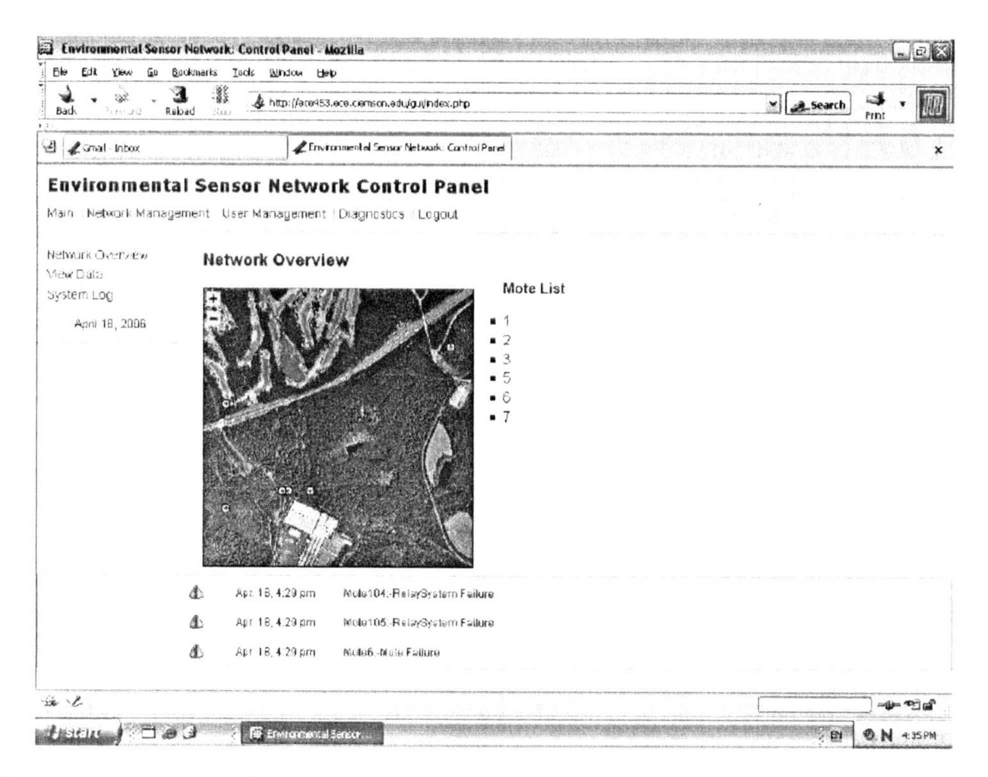

FIGURE 40.17 GUI screenshot showing overall network map and system integrity system messages.

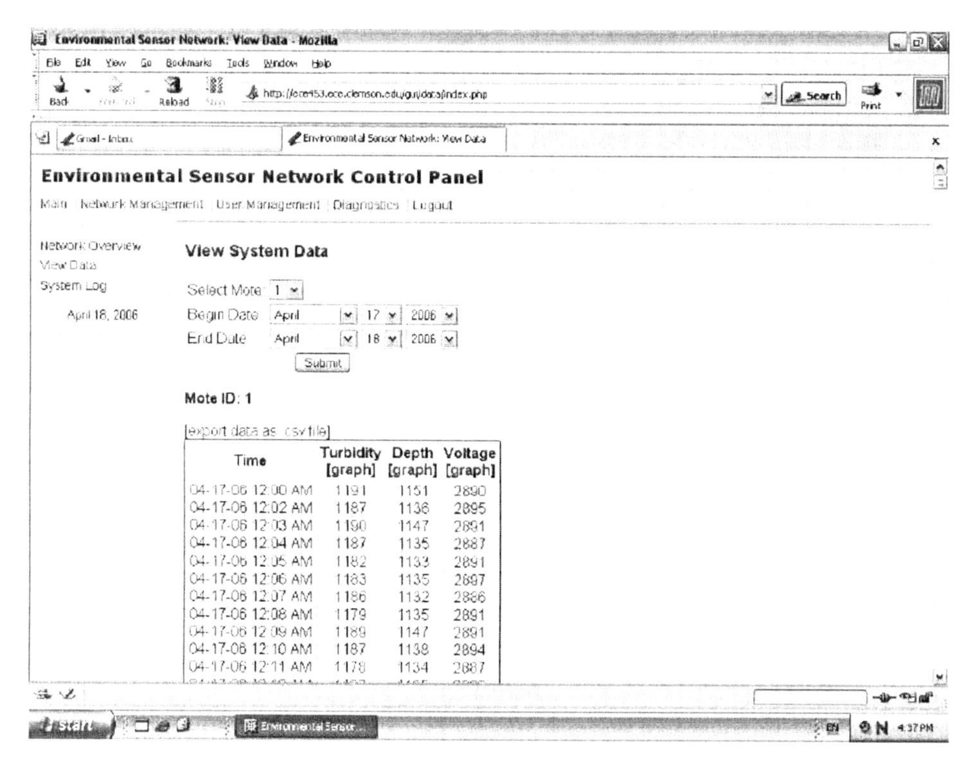

FIGURE 40.18 GUI screenshot showing a specific mote's data.

40.7.3 Sample Data

MoteID	Turbidity	Depth	Battery Voltage
1	850	955	3058
1	854	955	3055
1	850	957	3058
1	862	975	3056
1	850	958	3059
1	862	975	3055
1	858	959	3058
1	862	976	3058
1	855	955	3059
1	854	955	3059
1	863	977	3057
1	850	958	3058
1	862	976	3058
1	858	960	3055
1	863	978	3058
1	850	958	3058
1	850	958	3059
1	858	962	3058
1	851	958	3058

Acknowledgments

This material is based upon work supported in part by the Air Force Office of Scientific Research contract/grant number FA9550-09-1-0173, NSF grant EAGER-GENI Experiments on Network Security and Traffic Analysis contract/grant number CNS-1049765, and NSF-OCI 1064230 EAGER: Collaborative Research: A Peer-to-Peer based Storage System for High-End Computing. Opinions expressed are those of the author and neither the National Science Foundation nor the US Department of Defense.

Index

A

D

E

F

G

H

I

J

K

L

S

S

T

U

V

W

S